Insect Bioecology
and Nutrition for
Integrated Pest
Management

Insect Bioecology and Nutrition for Integrated Pest Management

Edited by
Antônio R. Panizzi
José R. P. Parra

CRC Press
Taylor & Francis Group
Boca Raton   London   New York

CRC Press is an imprint of the
Taylor & Francis Group, an **informa** business

CRC Press
Taylor & Francis Group
6000 Broken Sound Parkway NW, Suite 300
Boca Raton, FL 33487-2742

First issued in paperback 2016

© 2012 by Taylor & Francis Group, LLC
CRC Press is an imprint of Taylor & Francis Group, an Informa business

No claim to original U.S. Government works

Version Date: 20110825

ISBN 13: 978-1-138-19850-0 (pbk)
ISBN 13: 978-1-4398-3708-5 (hbk)

Library of Congress Cataloging-in-Publication Data

Insect bioecology and nutrition for integrated pest management / editors: Antonio Ricardo Panizzi, Jose R. P. Parra.
    p. cm.
    Based on the book Ecologia nutricional de insetos e suas implicações no manejo de pragas.
    Includes bibliographical references and index.
    ISBN 978-1-4398-3708-5
    1. Insects--Food. 2. Insects--Ecology. 3. Insects--Integrated control. 4. Insects--Nutrition. I. Panizzi, Antônio Ricardo. II. Parra, José Roberto P. (José Roberto Postali)

QL496.I3853 2012
595.7--dc23
                                           2011034453

Visit the Taylor & Francis Web site at
http://www.taylorandfrancis.com

and the CRC Press Web site at
http://www.crcpress.com

Contents

Foreword vii
Preface xi
Editors xiii
Contributors xv

Part I General Aspects

1. **Introduction to Insect Bioecology and Nutrition for Integrated Pest Management (IPM)** 3
   Antônio R. Panizzi and José R. P. Parra

2. **Nutritional Indices for Measuring Insect Food Intake and Utilization** 13
   José R. P. Parra, Antônio R. Panizzi, and Marinéia L. Haddad

3. **The Evolution of Artificial Diets and Their Interactions in Science and Technology** 51
   José R. P. Parra

4. **Molecular and Evolutionary Physiology of Insect Digestion** 93
   Walter R. Terra and Clélia Ferreira

5. **Insect–Plant Interactions** 121
   Marina A. Pizzamiglio-Gutierrez

6. **Symbionts and Nutrition of Insects** 145
   Edson Hirose, Antônio R. Panizzi, and Simone S. Prado

7. **Bioecology and Nutrition versus Chemical Ecology: The Multitrophic Interactions Mediated by Chemical Signals** 163
   José M. S. Bento and Cristiane Nardi

8. **Cannibalism in Insects** 177
   Alessandra F. K. Santana, Ana C. Roselino, Fabrício A. Cappelari, and Fernando S. Zucoloto

9. **Implications of Plant Hosts and Insect Nutrition on Entomopathogenic Diseases** 195
   Daniel R. Sosa-Gómez

Part II Specific Aspects

10. **Neotropical Ants (Hymenoptera) Functional Groups: Nutritional and Applied Implications** 213
    Carlos R. F. Brandão, Rogério R. Silva, and Jacques H. C. Delabie

11. **Social Bees (Bombini, Apini, Meliponini)** 237
    Astrid M. P. Kleinert, Mauro Ramalho, Marilda Cortopassi-Laurino, Márcia F. Ribeiro, and Vera L. Imperatriz-Fonseca

12. Defoliators (Lepidoptera) 273
    Alessandra F. K. Santana, Carla Cresoni-Pereira, and Fernando S. Zucoloto

13. Seed-Sucking Bugs (Heteroptera) 295
    Antônio R. Panizzi and Flávia A. C. Silva

14. Seed-Chewing Beetles (Coleoptera: Chrysomelidae, Bruchinae) 323
    Cibele S. Ribeiro-Costa and Lúcia M. Almeida

15. Rhizophagous Beetles (Coleoptera: Melolonthidae) 353
    Lenita J. Oliveira and José R. Salvadori

16. Gall-Inducing Insects: From Anatomy to Biodiversity 369
    G. Wilson Fernandes, Marco A. A. Carneiro, and Rosy M. S. Isaias

17. Detritivorous Insects 397
    Julio Louzada and Elizabeth S. Nichols

18. Insect Pests in Stored Grain 417
    Sonia M. N. Lazzari and Flávio A. Lazzari

19. Fruit Flies (Diptera) 451
    Carla Cresoni-Pereira and Fernando S. Zucoloto

20. Sap-Sucking Insects (Aphidoidea) 473
    Sonia M. N. Lazzari and Regina C. Zonta-de-Carvalho

21. Parasitoids (Hymenoptera) 515
    Fernando L. Cônsoli and S. Bradleigh Vinson

22. Predatory Bugs (Heteroptera) 539
    Vanda H. P. Bueno and Joop C. Van Lenteren

23. Predatory Beetles (Coccinellidae) 571
    Lúcia M. Almeida and Cibele S. Ribeiro-Costa

24. Green Lacewings (Neuroptera: Chrysopidae): Predatory Lifestyle 593
    Gilberto S. Albuquerque, Catherine A. Tauber, and Maurice J. Tauber

25. Hematophages (Diptera, Siphonaptera, Hemiptera, Phthiraptera) 633
    Mário A. Navarro-Silva and Ana C. D. Bona

Part III Applied Aspects

26. Plant Resistance and Insect Bioecology and Nutrition 657
    José D. Vendramim and Elio C. Guzzo

27. Insect Bioecology and Nutrition for Integrated Pest Management (IPM) 687
    Antônio R. Panizzi, José R. P. Parra, and Flávia A .C. Silva

Index 705

Foreword

After a deep freeze shriveled mulberry leaves near Avignon, France, silkworm growers needed alternative food plants to feed the newly hatched larvae. The growers sought out Jean Henri Fabre and begged for his help. The time was the late 1800s and Fabre was known in the region for his interest in plants and insects. "Taking botany as my guide," wrote Fabre, "suggested to me, as substitutes for the mulberry, the members of closely-related families: the elm, the nettle-tree, the nettle, the pellitory (*Anacyclus pyrethrum*, a composite). Their nascent leaves, chopped small, were offered to the silkworms. Other and far less logical attempts were made, in accordance with the inspiration of the individuals. Nothing came of them. To the last specimen, the new-born silkworms died of hunger. My renown as a quack must have suffered somewhat from this check. Was it really my fault? No, it was the fault of the silkworm, which remained faithful to its mulberry leaf." Fabre (1823–1915) pioneered the study of insect feeding behavior; he was puzzled by the host specificity of many insect species. This same puzzle continues to challenge scientists today.

It is estimated that of the more than one million species of insects that have been described thus far, about 45 percent feed on plants; the others are either saprophagous or predacious. Aside from the need to satisfy their pure scientific curiosity about the complexities of insects' feeding behavior and nutritional physiology, entomologists realize that what insects eat is, in great part, what makes them economically important for humans. In agriculture, insects compete with humans for the same food resources. In cities and villages they contaminate human habitats and, if not controlled, often destroy those habitats. In cities, villages, and farms, insects vector disease organisms to humans and their domestic animals. Understanding insect feeding and host selection behavior, their feeding ecology, and physiology provides indispensable tools in the continuous fight against insect pests and in the protection of those insects that are beneficial.

Following in Fabre's footsteps, generations of entomologists dedicated their careers to the study of insects and their relationship to food. An early extensive compilation on the subject was the 1946 book *Insect Dietary: An Account of the Food Habits of Insects* by Charles Thomas Brues. "A monophagous insect will deliberately starve to death in the absence of its proper food plant," wrote Brues, "and most oligophagous species with highly restricted dietary will do the same. We cannot appreciate such instinctive intolerance."

A potential key to solve the mystery of this complex behavior had already been found in studies by the Dutch botanist E. Vershaffelt, who in 1919 published a paper on interactions of the cabbage butterfly with its preferred crucifer host plants. For the first time, plant secondary compounds, in this case the glucosinolates, were implicated in the host selection process of an insect.

The role of secondary compounds was vigorously emphasized in the classic, albeit controversial 1958 paper by Godfried Fraenkel, published in *Science*. Fraenkel claimed that the *raison d'être* of plants' secondary compounds was defense against phytophagous insects. In time a few insect species evolve the capacity to overcome those defenses and eventually "learn" to use the same defensive compounds as cues to identify what becomes a preferred host plant. This hypothesis was later expanded and further substantiated in a paper published in 1964 by Paul Ehrlich and Peter Raven in which they advanced the concept of coevolution of insects and plants.

The role of secondary plant compounds in insect–plant interactions continued to gain support from research on a variety of insect–plant systems. Research proliferated on the effect on insect host selection and feeding behavior related to Cruciferae glucosinolates, Solanaceae alkaloids, Fabaceae anti-proteinase and phenolic compounds, and Umbelliferae furanocoumarins, to mention only a few.

Concurrent with the publication of Fraenkel's paper on plant defenses, however, another school of thought was rising among entomologists focused on aphid feeding behavior. In the early 1950s,

John S. Kennedy produced fundamental work in this area, claiming that nutrients were as important as secondary compounds in aphid host selection behavior. Kennedy proposed the "dual discrimination theory" later emphasized by A. J. Thorsteinson, in the 1960 *Annual Review of Entomology*. This theory postulated that secondary compounds, as token stimuli, as well as primary nutritional metabolites, were both implicated in the host selection process. Later it was demonstrated by detailed electrophysiological studies that insects possessed the sensorial capacity to detect both types of compounds. These findings gave rise to the "dual discrimination hypothesis."

While major strides were being made in the study of insect host selection behavior, equally important advances were made in insect nutritional physiology. By the mid-1900s insect nutritional requirements were well defined. Fraenkel's emphasis on secondary compounds related to his assumption that basic nutritional requirements of phytophagous insects were essentially similar and available in most green plants. This assumption was not supported by research on the European corn borer, *Pirausta nubilalis,* conducted by Stanley Beck at the University of Wisconsin. Beck's research and evidence from research from other laboratories showed that there were qualitative and quantitative differences in nutritional requirements that influenced an insect's range of host plants.

Knowledge of insect nutrition improved with development of chemically defined artificial media that satisfied all requirements for larval growth and development, and production of viable adults. Nutritional physiological studies required development of adequate measurement of food intake and utilization by insects. Landmark work in this area was done by Gilbert Waldbauer, a former student of Fraenkel's at the University of Illinois. Waldbauer adapted the nutritional indices developed for the study of domestic farm animals for application to insect alimentary physiology. These indices remain the backbone of research on insect nutritional physiology and ecology; our understanding of insect nutritional ecology after the 1970s is a direct result of these advances.

The field of insect nutritional ecology blossomed in the late 1970s through the early 1980s, mainly with the research of entomologists such as Mark Scriber and Frank Slansky, then at the University of Wisconsin. In the 1981 *Annual Review of Entomology* they stated, "nutritional ecology is central to proper interpretation of life history phenomena (e.g., manner of feeding, habitat selection, defense, and reproduction) both in ecological and evolutionary times." Thus nutritional ecology extends across many basic life sciences fields such as ecology, nutrition, behavior, morphology, physiology, life history, and evolutionary biology.

The editors of this volume developed professionally while the field of nutritional ecology was maturing into what it has become today, and were associated with institutions where most of the action was taking place. I had the pleasure of hosting José Roberto Postali Parra in my laboratory at the University of Illinois where he spent one sabbatical year as a visiting scholar. He conducted a series of elegant experiments comparing five basic methods used in the measurement of food intake and utilization by insects. That study consolidated the methodology and added powerful tools to the definition of the indices proposed by Waldbauer. Antônio Panizzi received his PhD at the University of Florida working with Frank Slansky. Panizzi had already established a reputation as a leading researcher on the nutritional ecology of Heteroptera: Pentatomidae associated with soybean. Upon returning to Brazil, both Parra, at the Escola Superior de Agricultura Luiz de Queiroz, in Piracicaba, São Paulo, and Panizzi, at the Centro Nacional de Pesquisa de Soja, in Londrina, Paraná, helped establish two centers of excellence in the research of insect nutrition and nutritional ecology of tropical insects.

The task of compiling the wealth of information that has accumulated since the publication of Brues's book would be overwhelming for a single author. For production of this volume, Parra and Panizzi assembled a cadre of Brazilian authors who represent the best in the field, along with several chapters in collaboration with international authorities who have spent time in Brazil. This volume offers the most authoritative compilation of up-to-date research on the ecology of insects with emphasis on nutrition and nutritional ecology, as well as the implications for the development of integrated pest management programs applied to the neotropics, arguably the most complex and diverse of the world's biogeographic zones. This volume is a landmark in a relatively young, multidimensional science, and will greatly contribute toward much-needed further research.

Were it possible for Fabre to witness the research developments of the past 120 years, he most certainly could now address the plight of the distressed silkworm growers. The industry today no longer depends on the health of the mulberry trees. Even though natural food still is the most efficient way to produce healthy silkworms, artificial diets have been developed that are suitable for maintenance of colonies should a crop of leaves fail. Had Fabre had this information, his reputation as a bona fide quack would have remained unblemished.

Marcos Kogan
Oregon State University

Preface

We initially conceived of a book on insect bioecology and nutrition as related to integrated pest management (IPM) back in 1985, at the Brazilian Congress of Entomology, held in Rio de Janeiro. The book (in Portuguese) was finally published in 1991, and was very well received by South American entomologists because it offered a much needed resource on the subject in a language accessible to both Portuguese and Spanish readers.

Eighteen years later the field has grown so much that we thought it was time to produce a second edition. Consequently, the book grew substantially in content. From the original nine it expanded to twenty-six chapters, including those on insect feeding guilds not covered in the first edition, plus chapters on insect feeding and nutrition covering subjects that have blossomed in those two decades. Examples are the role of symbionts on insect nutrition, chemical ecology versus food, and insect cannibalism. As result of this expansion of the subject matter, the book no longer could be considered a second edition of the 1991 title, but instead it was offered as an entirely new book. Therefore, in 2009 the book *Bioecologia e Nutrição de Insetos: Base para o Manejo Integrado de Pragas* (A.R. Panizzi and J.R.P. Parra, editors), Embrapa Informação Tecnológica, Brasília, 1,164 p., was published in Brazil.

Interest in the Portuguese version of the book prompted us to edit a further expanded version of the 2009 book, now to be published in English. The present publication by CRC Press basically is a translation of the 2009 book with updates and adaptations in most chapters and the inclusion of an entirely new one.

The book is organized in three parts. The first part (General Aspects) includes nine chapters, with an introductory chapter on insect bioecology and nutrition as basis for integrated pest management. The next two chapters cover nutritional indexes to measure food consumption and utilization, and the development of artificial diets and their interactions with science and technology. Chapters 4 through 9 cover molecular and evolutionary physiology of insect digestion, insect–plant interactions, symbionts and nutrition, multitrophic interactions mediated by chemical signals, insect cannibalism, and impact of entomopathogenic agents on insect behavior and nutrition.

The second part of the book is dedicated to specific feeding guilds, including ants, social bees, defoliators (Lepidoptera), seed-sucking bugs (Heteroptera), seed-chewing beetles, root-feeding beetles, gall makers, detritivores, pests of stored grains, fruit flies, sieve feeding aphids, parasitoids (Hymenoptera), predatory bugs (Heteroptera), predatory beetles (Coccinelidae), predatory lacewings, and hematophagous insects. Although not all feeding guilds were covered, we believe that the ones that were included should provide readers with a comprehensive view of the incredible diversity of the ways insects exploit feeding resources in nature, and how those resources affect insects' biology.

The final part of the book includes two chapters. The first is dedicated to the field of applied entomology known as host plant resistance. This chapter explores ways in which plant resistance influences insect bioecology and nutrition. The final chapter of the book presents a case study of heteropterans on soybean to illustrate how research on bioecology and nutrition may serve as a basis to design and deploy sophisticated and efficient integrated pest management systems.

We are aware that the field of insect nutritional ecology as defined by Frank Slansky, Jr., John Mark Scriber, and others in the 1980s (see references below) focuses on how insects deal with nutritional and non-nutritional compounds (allelochemicals), and how these compounds influence their biology and shape different lifestyles in evolutionary time. No attempt, in that original nutritional–ecological literature was made to fit the information within a framework applicable to the management of insect pests in agriculture. Therefore, we opted to avoid use of the expression "nutritional ecology," and adopted instead the more conservative terminology of insect bioecology and nutrition, with inclusion of chapters on applied aspects related to the main topic. Much of the research on which chapters were written was done in Brazil and based on its neotropical fauna. It is our hope that the complexity and diversity of the neotropics should afford readers from all zoogeographical regions to readily translate to their specific conditions the information provided herein.

A.R. Panizzi
Passo Fundo, Rio Grande do Sul, Brazil

J.R.P. Parra
Piracicaba, São Paulo, Brazil

REFERENCES

Scriber, J. M., and F. Slansky, Jr. 1981. The nutritional ecology of immature insects. *Annu. Rev. Entomol.* 26:183–211.

Slansky, Jr., F. 1982a. Toward the nutritional ecology of insects. In *Proc. 5th Inter. Symp. Insect–Plant Relationships*, 253–59. Wageningen, The Netherlands.

Slansky, Jr., F. 1982b. Insect nutrition: An adaptionist's perspective. *Fla. Entomol.* 65:45–71.

Slansky, Jr., F. and J. G. Rodriguez. 1987. Nutritional ecology of insects, mites, spiders, and related invertebrates: an overview. In *Nutritional Ecology of Insects, Mites, Spiders, and Related Invertebrates*, ed. F. Slansky, Jr., and J. G. Rodriguez, 1–69. New York: J. Wiley & Sons.

Editors

Antônio Ricardo Panizzi is a senior research entomologist at the Wheat Research Center at Embrapa (Brazilian Enterprise of Agricultural Research), in Passo Fundo, Rio Grande do Sul, Brazil. He earned his BS in agronomy in 1972 from the University of Passo Fundo, his MS in entomology in 1975 from the Federal University of Paraná, both in Brazil, and his PhD in entomology in 1985 from the University of Florida. He is the recipient of the Alexandre Rodrigues Ferreira Prize given by the Brazilian Society of Zoology and editor-in-chief of the *Annals of the Entomological Society of Brazil* (1993–1998). Currently, Dr. Panizzi is associate editor and president of the Entomological Society of Brazil. He served as a member of the advisor committee for agronomy (entomology) at the National Council for Scientific and Technological Development (CNPq) of Brazil from 1997 to 1999 and from 2009 to 2011. He has been an invited speaker at several congresses and symposia in different parts of the world and an invited scientist at the National Institute of Agro-Environmental Sciences, Tsukuba, Japan (1991). He coedited *Insect Nutritional Ecology and Its Implications on Pest Management* (Manole/CNPq, São Paulo, 1991) and *Heteroptera of Economic Importance* (CRC Press, Boca Raton, FL, 2000). Dr. Panizzi has published extensively on Hemiptera (Heteroptera), over 150 peer-reviewed publications, including an *Annual Review of Entomology* article on wild host plants of Pentatomidae. He teaches a course on insect nutritional ecology at the Federal University of Paraná, where he serves as advisor for MSc and PhD students. His current research focuses on the interactions of heteropterans (mostly Pentatomidae) with their wild and cultivated host plants, and the management of pest species on field crops.

José Roberto Postali Parra is a full professor of the Department of Entomology and Acarology at the College of Agriculture (ESALQ) of the University of São Paulo (USP). He is a member of the Brazilian Academy of Sciences and of the Academy of Sciences for the Developing World (TWAS). He was trained in insect rearing techniques and nutrition for biological control purposes at the University of Sao Paulo and the University of Illinois, Urbana–Champaign. Since the end of the 1960s, he has conducted research in biological control and is currently doing research on classical and applied biological control, with emphasis on *Trichogramma* and on the biological control of sugarcane and citrus pests. He has published more than 200 peer-reviewed papers and edited books, and he has supervised 55 MS and 40 PhD students. He was the president of the Entomological Society of Brazil and dean of ESALQ (USP), and he received awards from Embrapa, the Entomological Society of Brazil and the Zoological Society of Brazil. He also received honors from the state of São Paulo government and from the Brazilian government for his contributions in the biological control area. He is vice-president of the International Organization for Biological Control.

Contributors

Gilberto S. Albuquerque
Laboratório de Entomologia e Fitopatologia
Universidade Estadual do Norte Fluminense
    Darcy Ribeiro
Campos dos Goytacazes, Rio de Janeiro, Brazil

Lúcia M. Almeida
Departamento de Zoologia
Universidade Federal do Paraná
Curitiba, Paraná, Brazil

José M. S. Bento
Departamento de Entomologia e Acarologia
Universidade de São Paulo
Piracicaba, São Paulo, Brazil

Ana C. D. Bona
Departamento de Zoologia
Universidade Federal do Paraná
Curitiba, Paraná, Brazil

Carlos R. F. Brandão
Museu de Zoologia
Universidade de São Paulo
São Paulo, São Paulo, Brazil

Vanda H. P. Bueno
Departamento de Entomologia
Universidade Federal de Lavras
Lavras, Minas Gerais, Brazil

Fabrício A. Cappelari
Faculdade de Filosofia Ciências e Letras de
    Ribeirão Preto
Universidade de São Paulo
Ribeirão Preto, São Paulo, Brazil

Marco A. A. Carneiro
Laboratório de Padrões de Distribuição Animal
Universidade Federal de Ouro Preto
Ouro Preto, Minas Gerais, Brazil

Fernando L. Cônsoli
Departamento de Entomologia e Acarologia
Universidade de São Paulo
Piracicaba, São Paulo, Brazil

Marilda Cortopassi-Laurino
Laboratório de Abelhas
Universidade de São Paulo
São Paulo, Brazil

Carla Cresoni-Pereira
Faculdade de Filosofia Ciências e Letras de
    Ribeirão Preto
Universidade de São Paulo
Ribeirão Preto, São Paulo, Brazil

Jacques H. C. Delabie
Laboratório de Mirmecologia, Centro de
    Pesquisas do Cacau
Comissão Executiva do Plano de Lavoura
    Cacaueira
Itabuna, Bahia, Brazil

G. Wilson Fernandes
Instituto de Ciências Biológicas
Universidade Federal de Minas Gerais
Belo Horizonte, Minas Gerais, Brazil

Clélia Ferreira
Instituto de Química
Universidade de São Paulo
São Paulo, Brazil

Elio C. Guzzo
Embrapa Tabuleiros Costeiros
Maceió, Alagoas, Brazil

Marinéia L. Haddad
Departamento de Entomologia, Fitopatologia e
    Zoologia Agrícola
Universidade de São Paulo
Piracicaba, São Paulo, Brazil

Edson Hirose
Laboratório de Entomologia
Embrapa Arroz e Feijão
Santo Antônio de Goiás, Goiás, Brazil

Vera L. Imperatriz-Fonseca
Departamento de Ecologia
Universidade de São Paulo
São Paulo, São Paulo, Brazil

Rosy M. S. Isaias
Instituto de Ciências Biológicas
Universidade Federal de Minas Gerais
Belo Horizonte, Minas Gerais, Brazil

Astrid M. P. Kleinert
Departamento de Ecologia
Universidade de São Paulo
São Paulo, São Paulo, Brazil

Flávio A. Lazzari
Pesquisador Autônomo
Curitiba, Paraná, Brazil

Sonia M. N. Lazzari
Departamento de Zoologia
Universidade Federal do Paraná
Curitiba, Paraná, Brazil

Júlio Louzada
Departamento de Biologia-Setor de Ecologia
Universidade Federal de Lavras
Lavras, Minas Gerais, Brazil

Cristiane Nardi
Departamento de Agronomia
Universidade Estadual do Centro-Oeste do
   Paraná
Guarapuava, Paraná, Brazil

Mário A. Navarro-Silva
Departamento de Zoologia
Universidade Federal do Paraná
Curitiba, Paraná, Brazil

Elizabeth S. Nichols
Center for Biodiversity and Conservation
American Museum of Natural History
New York, New York

Lenita J. Oliveira † (deceased)
Laboratório de Insetos Rizófagos
Embrapa Soja
Londrina, Paraná, Brazil

Antônio R. Panizzi
Laboratório de Entomologia
Embrapa Trigo
Passo Fundo, Rio Grande do Sul, Brazil

José R. P. Parra
Departamento de Entomologia, Fitopatologia e
   Acarologia
Universidade de São Paulo
Piracicaba, São Paulo, Brazil

Marina A. Pizzamiglio-Gutierrez
Center for Analysis of Sustainable Agricultural
   Systems
Kensington, California

Simone S. Prado
Laboratório de Entomologia
Embrapa Meio Ambiente
Jaguariuna, São Paulo, Brazil

Mauro Ramalho
Departamento de Botânica
Universidade Federal da Bahia
Salvador, Bahia, Brazil

Márcia F. Ribeiro
Embrapa Semi-Árido
Petrolina, Pernambuco, Brazil

Cibele S. Ribeiro-Costa
Departamento de Zoologia
Universidade Federal do Paraná
Curitiba, Paraná, Brazil

Ana C. Roselino
Faculdade de Filosofia Ciências e Letras de
   Ribeirão Preto
Universidade de São Paulo
Ribeirão Preto, São Paulo, Brazil

José R. Salvadori
Faculdade de Agronomia e Medicina Veterinária
Universidade de Passo Fundo
Passo Fundo, Rio Grande do Sul, Brazil

Alessandra F. K. Santana
Faculdade de Filosofia Ciências e Letras de
   Ribeirão Preto
Universidade de São Paulo
Ribeirão Preto, São Paulo, Brazil

Flávia A. C. Silva
Laboratório de Bioecologia de Percevejos
Embrapa Soja
Londrina, Paraná, Brazil

Rogério R. Silva
Museu de Zoologia
Universidade de São Paulo
São Paulo, São Paulo, Brazil

Daniel R. Sosa-Gómez
Laboratório de Patologia de Insetos
Embrapa Soja
Londrina, Paraná, Brazil

Catherine A. Tauber
Department of Entomology
Cornell University
Ithaca, New York
and
Department of Entomology
University of California, Davis
Davis, California

Maurice J. Tauber
Department of Entomology
Cornell University
Ithaca, New York
and
Department of Entomology
University of California, Davis
Davis, California

Walter R. Terra
Instituto de Química
Universidade de São Paulo
São Paulo, São Paulo, Brazil

Joop C. Van Lenteren
Laboratory of Entomology
Wageningen University
Wageninen, The Netherlands

José D. Vendramim
Departamento de Entomologia, Fitopatologia e
 Acarologia
Universidade de São Paulo
Piracicaba, São Paulo, Brazil

S. Bradleigh Vinson
Department of Entomology
Texas A&M University
College Station, Texas

Regina C. Zonta-de-Carvalho
Centro de Diagnóstico Marcos Enrietti
Secretaria de Agricultura e do Abastecimento
 do Paraná
Curitiba, Paraná, Brazil

Fernando S. Zucoloto
Faculdade de Filosofia Ciências e Letras de
 Ribeirão Preto
Universidade de São Paulo
Ribeirão Preto, São Paulo, Brazil

Part I

General Aspects

1

Introduction to Insect Bioecology and Nutrition for Integrated Pest Management (IPM)

Antônio R. Panizzi and José R. P. Parra

CONTENTS

1.1 Introduction 3
1.2 Food 4
    1.2.1 Natural Food 4
    1.2.2 Artificial Diets 4
    1.2.3 Food Consumption, Digestion, and Utilization 4
    1.2.4 Multitrophic Interactions 5
        1.2.4.1 Symbionts 5
        1.2.4.2 Food and Chemical Ecology 5
1.3 Feeding Habits 6
    1.3.1 Feeding Habits of Social Insects 6
    1.3.2 Feeding Habits of Phytophagous Insects 6
    1.3.3 Feeding Habits of Carnivorous Insects 7
    1.3.4 Feeding Habits of Hematophogous Insects 7
    1.3.5 Other Feeding Habits 8
1.4 The Coverage and Implications of Studying Bioecology and Nutrition of Insects 8
1.5 Insect Bioecology, Nutrition, and Integrated Pest Management 8
1.6 Final Considerations 9
References 9

1.1 Introduction

All living organisms are, in general, a result of the food they consume. In the case of insects, many aspects of their biology, including behavior, physiology, and ecology are, in one way or another, inserted within the context of food. The quantity, quality, and proportion of nutrients present in the food, and of secondary or nonnutritional compounds (allelochemicals), cause a variable impact on the insect biology, shaping its potential reproductive contribution to the next generation (fitness).

Studies on bioecology and insect nutrition greatly evolved during the last 30 years, from the definition of the basic nutritional requirements for survivorship and reproduction (see Chapter 2) to the evaluation of their influence in insect behavior and physiology, with ecological and evolutionary consequences (see Chapter 5 on insect/plant interactions). This is called insect nutritional ecology and its concept and development occurred during the last 20-plus years (e.g., Scriber and Slansky 1981; Slansky 1982a, b; Slansky and Scriber 1985; Slansky and Rodriguez 1987). In this introductory chapter, we will touch in summary on the natural food and artificial diets, food consumption, digestion and utilization, multitrophic interactions including symbionts, and the interface between food and chemical ecology. Variable feeding guilds and the implications of bioecology and insect nutrition of pest species within the context of integrated pest management (IPM) will be covered as well.

1.2 Food

1.2.1 Natural Food

Natural food (i.e., food obtained in nature) has variable nutritional quality. From the time insects appeared on Earth (see Chapter 5), an evolutionary adaptation process started with the appearance of different lifestyles. This allowed insects to explore an array of foods in all of their diverse forms. If on one hand, insects adaptived to explore nutritional sources (e.g., vegetable and animal organisms), these, on the other hand, evolved to be less susceptible to being consumed, in an endless coevolutionary process. The fact that insects have a legendary ability to explore the more diverse habitats in search of food, with unique adaptive success, make them the only living creatures to challenge human beings in their total hegemony on Earth.

In addition to the variable quality of foods, their sazonality make exploration an even greater challenge. The abiotic environment, including temperature, humidity, and photoperiod makes natural food not constantly available, which "forces" the insects to adapt during less favorable periods. These adaptations include drastic changes in the physiology (e.g., diapause) and less pronounced changes (e.g., oligopause/quiescence). In both cases, energy storage such as lipids supports survivorship. Another strategy to face less favorable conditions is to search for suitable habitats through migration that demands storaged energy to sustain steady flight.

Natural foods vary in their quality, and often toxic secondary compounds or allelochemicals are present (see Chapters 5 and 26). Beyond chemical compounds, physical defenses (i.e., pilosity, thorns, and tough and thick tissues) make natural foods often out of reach and/or undigestible. Therefore, natural foods present a constant challenge to insects even to those specialized on certain foods (monophagous). Artificial diets may solve this problem for biological studies in the laboratory, but do not always yield favorable results (i.e., the case of artificial diets for pentatomids is still a challenge to be overcome).

1.2.2 Artificial Diets

The development of artificial diets for insects, mostly from the 1960s on, provided conditions for refinements on studies on their nutritional requirements. Today there are over 1,300 species of insects raised on these diets (see Chapter 3 and references therein). These advances in insect rearing using artificial diets allowed us to learn that some particular group of insects need nucleic acids and liposoluble vitamins in their diets. Sophisticated techniques were developed with artificial diets that allow raising parasitoids *in vitro* (i.e., excluding the natural host). Although artificial diets for parasitoids and predators have been developed (Cohen 2004), phytophagous insects of the orders Lepidoptera, Coleoptera, and Diptera concentrate 85% of the artificial diets. These diets allowed great advances in basic and applied studies in entomology, including insights in public education and in human and animal nourishment (see Chapter 3).

1.2.3 Food Consumption, Digestion, and Utilization

Insect nutrition can be focused on two aspects: qualitatively (i.e., the chemical nature of the nutrients) and quantitatively (i.e., the proportion of nutrients that encompass the food that is ingested, digested, assimilated, and converted into tissue for growth). The measurements of food consumption and utilization, including physiology and behavior in selecting host plants, leads to several applications not only on basic nutrition, but in the ecology of insect communities through host plant resistance and biological control (see Chapter 2 and Cohen 2004).

The basic concepts of food consumption and utilization were developed by nutritionists relating food quality and its effects on growth and development of animals. The interactions of nutrients and allelochemicals have been determined by nutritional indexes. These indexes allow understanding the impact of variable factors to the insect life, including temperature, humidity, photoperiod, parasitism, allelochemicals, cannibalism, and so forth (see Chapter 2 and references therein).

Coudron et al. (2006) proposed the term nutrigenomics or nutritional genomics that provide information on the impact of nutrition based on biochemical parameters through the investigation of how nutrition alters genetic expression. These studies with molecular markers of insects could be used to indicate initial responses to nutritional sources, providing cues to the biochemical, physiological, and genetic regulation of insect populations, with multiple implications.

1.2.4 Multitrophic Interactions

1.2.4.1 Symbionts

The success of insects as organisms able to colonize every habitat is due to their enormous ability to feed on an array of food sources. In addition, the exploitation of less suitable food resources is done through the association with microorganisms in a symbiotic process. This allows utilization of novel metabolic pathways with mutual benefits in the course of evolutionary time (see Chapter 6 and references therein).

A wide variety of microorganisms is involved in the feeding process of insects. These microorganisms include external symbionts that cultivate fungi, such as the Ambrosia beetles of the subfamilies Scolytinae and Platypodinae, ants of the subfamily Myrmicinae, tribe Attini, and termites of the subfamily Macrotermitinae, and internal symbionts such as protozoans and bacteria that can play a secondary role (bacteria in Heteroptera) or a primary role or be obligated (e.g., *Buchnera*, *Wigglesworthia*, and *Blochmannia*—see Chapter 6).

The study of insect symbionts has gained momentum due to the development of molecular techniques that allowed a better understanding of insect–symbiont interactions previously unknown. The development of complete genomes of endosymbionts with a wide ecological and phylogenetic diversity will open opportunities for better comparisons to test actual evolutionary models. The possibility to manipulate bacteria symbionts of insects' vectors of human diseases such as malaria, dengue, and Chagas open up potential strategies to reduce a bug's longevity or to mitigate the parasites that cause such diseases. With regard to crop pests, revealing the interactions of insects and their symbionts may yield sophisticated and efficient control measures. Once revealed, the role of symbionts, their manipulation through genomics, biochemical, or conventional means (e.g., elimination of symbionts using antibiotics) will create a real possibility to mitigate the impact of pests on crops (see Chapter 6, and Bourtzis and Miller 2003, 2006, 2009).

1.2.4.2 Food and Chemical Ecology

Trophic interactions of insects and their hosts include many chemical signs, the so-called infochemicals. These signs have great influence in finding hosts. There are constitutive volatiles: those normally produced and induced volatiles and those produced due to plant–herbivorous-natural enemy interactions, such as volatiles of plants eliciting pheromone production by insects (see Chapter 7 and references therein).

Chemical signs utilized by insects include allelochemicals that mediate interspecific interactions and in general aid in finding food, both for phytophagous and zoophagous, which act like allomones, kairomones, synomones, or apneumones, and pheromones, which act as intraspecific signaling. These latter include trail, aggregation, and sexual pheromones, which act in setting directions and sexual attraction, but also play a role in finding food. Some pheromones act in association with allelochemicals, increasing the success of finding the cospecifics (e.g., synergic action of aggregation pheromones and plant components—Reddy and Guerrero 2004).

Several pheromones are commercialized for management of pest species. Recently, these include the effects of plant volatiles on pests, predators, and parasites (see Chapter 7 and references therein). The discovery that plants attacked by herbivores react by activating indirect defenses by alerting predators and parasites of the presence of their specific hosts (De Moraes et al. 1998) had a great impact. This lead to investigations of biochemical mechanisms and ecological consequences of such interactions, and potential use of these compounds in agriculture (Turlings and Ton 2006). The myriad of trophic interactions among plants, herbivores, and their natural enemies (insect bioecology and nutrition and chemical ecology) open up a research area that is sophisticated and has great potential to be exploited in its most variable basic and applied aspects (see Chapter 7).

1.3 Feeding Habits

1.3.1 Feeding Habits of Social Insects

Feeding habits of social insects are among the most sophisticated of the Class Insecta. Ants (Hymenoptera) and social bees (Bombini, Apini, Meliponini (Hymenoptera)) included in this volume (Chapters 10 and 11, respectively) touch on this subject.

Ants function as important predators in trophic chains (Floren et al. 2002) and as main herbivores in tropical forests exploiting exsudates of phytophagous sucking insects (e.g., Homoptera) and flower nectaries (Davidson et al. 2003) beyond cultivated fungi for their nourishment. As predators and herbivores more important due to their abundance and wide distribution, in over 100 million years of evolution ants have had a major impact on other organisms and ecosystems (Holldöbler and Wilson 1990, see Chapter 10). Foraging strategies in ants are legendary and demonstrate a unique level of organization among living organisms (Fowler et al. 1991).

Social bees, similar to ants, are also highly specialized in their ability to explore nutritional resources and they also demonstrate sophisticated foraging behavior. Seeking pollen and nectar in flowers and honey production are two of the most complex biological systems among living organisms (see Chapter 11 and references therein).

1.3.2 Feeding Habits of Phytophagous Insects

The feeding habits of phytophagous insects are extremely variable, and include leaf chewers (Chapter 12), seed suckers (Chapter 13), seed chewers and borers (Chapters 14 and 18), root feeders (Chapter 15), gall makers (Chapter 16), frugivores (Chapter 19), and leaf, bud, and fruit suckers (Chapter 20) that are detailed in this volume.

Leaf chewers are species of the orders Coleoptera, Hymenoptera, and Lepidoptera, which in general are specialized in one or few plant families. Therefore, their evolutionary relations are narrow and the chemical defenses of plants to the leaf chewers are abundant (e.g., Bernays 1998). In general, caterpillars consume a relatively large amount of food, have big guts, and rapidly digest food. However, by being less selective, they often ingest plant parts that are not highly nutritional such as leaf veins or other metabolic poor tissues (see Chapter 12).

Seed suckers (true bugs) include heteropterans of several families that prefer to feed on immature seeds, although some feed on mature seeds. They insert their stylets (mandibles + maxillae) in the seed and inject salivary enzymes that make up slurry, which is sucked in, carrying the nutrients. Because of the feeding activity, total or partial damage occurs, creating seedlings with low viability. The impact of seed suckers on seed and fruit production is discussed at length in the literature of economic entomology due to its worldwide effect (see Chapter 13, and Schaefer and Panizzi 2000).

Seed chewers (borers) include species of Coleoptera and Lepidoptera, but only coleopterans have chewing mouthparts during the larval and adult stages. Among the coleopterans, seed weevils are a classical example (see Chapter 14). Their larvae develop exclusively from nutrients of seed contents, while adults feed on pollen and nectar. Although polyphagous, they prefer legumes of several species, most considered of economic importance (Vendramin et al. 1992).

Root feeding insects are represented mostly by coleopterans that feed on live root tissues. However, their feeding habits include boring roots, stems, and tubers, making galleries or cutting root tissues from outside (see Chapter 15 and references therein). Many larvae are able to feed on roots externally and adults feed on the foliage, not necessarily of the same species fed by larvae. Beyond feeding on roots, larvae may explore organic matter, decaying wood (xylophagy), feces (coprophagy), and dead animals (necrophagy) (Oliveira et al. 2003).

Gall makers are found on all orders of phytophagous insects (Hemiptera, Thysanoptera, Coleoptera, Hymenoptera, Lepidoptera, and Diptera) except for Orthoptera. Galls are characterized by being reactions of plants due to the damage caused by insects. They are classified as the organoid type, which show similar growth pattern as to the plant tissue and the plant structure colonized keeps its identity (e.g., intumescences and callosities), and the histioid type, which show a wide variety of abnormal growing tissues (see Chapter 16).

Frugivorous insects belong to several orders. In this book, fruit flies (Tephritidae) are covered in detail (Chapter 19). Tephritids are a fruit-feeding guild related to the feeding habits of immatures. Adults feed on fruits exsudates, bird feces, decaying organic matter, nectar, pollen, and so forth. Although larvae stay inside the fruits, they may feed on their own exoskeleton, on larvae of other insects and of their own, and on worms (Zucoloto 1993). Fruit fly females lay their eggs on the fruit skins, and larvae penetrate fruits as they hatch. Life cycle is completed on the ground where they pupate, originating a new adult (Christenson and Foote 1960).

Among insects that suck leaves, buds, and fruits (e.g., psilids, whiteflies, and other Stenorryncha specialized in phloem feeding) aphids are an interesting guild that will be covered in Chapter 20. Aphids (Hemiptera: Aphidoidea) penetrate the vegetable tissue to suck the sieve, affecting plant growth and causing localized or systemic lesions, aphids commonly transmit virus and this is a highly specialized insect/plant interaction. Several authors treat feeding and nutrition of aphids, with relevant aspects of the role of saliva, and adaptative mechanisms (see references in Chapter 20).

1.3.3 Feeding Habits of Carnivorous Insects

Some of the feeding habits of carnivorous insects include parasitoids (Hymenoptera) (Chapter 21), predatory hemipterans (Chapter 22), predatory beetles (Coccinelidae) (Chapter 23), and predatory lacewings (Neuroptera) (Chapter 24), which are presented in this volume.

Parasitoids (Hymenoptera) are insects that adapt to the parasitic way of life either utilizing the limited nutritional resources of the immatures or acquiring nutrients from adults. Larvae are adapted to maximize the utilization of nutritional sources in different ways (see Chapter 21). Their development is closely dependent of their hosts. Parasitoids can explore eggs, eggs and larvae, larvae, larvae and pupae, pupae or adults; they can be endo or ectoparasitoids, solitary or gregarious (Askew 1973).

Predatory hemipterans (Heteroptera) include several species of the genera Orius (Anthocoridae), *Geocoris* (Lygaeidae), *Nabis* (Nabidae), *Podisus, Brontocoris,* and *Supputius* (Pentatomidae), *Macrolophus* (Miridae), and *Zelus* and *Sinea* (Reduviidae). Many predators show phytophagy (see Chapter 22 and references therein). To reach the "perfect" nutrition, the ecological tritrophic interaction is involved; that is, the third level (the entomophagous), the second level (the host), and the first level (the plant that feed the host). Therefore, the coexistence of entomophagy and phytophagy is highly important for predatory heteropterans.

Predatory beetles (Coccinelidae) are among Coleoptera the most important predators. Feeding habits of larvae and adults are similar, and their mandibulae are similar. Many species feed on aphids, coccids, and mites; some species show phytophagy and their mandibulae are adapted to cut and chew plant tissue, mostly of plants that belong to the families Cucurbitaceae and Solanaceae (see Chapter 23). Coccinellids are efficient predators in finding and eating their prey in all environments, mostly preying on aphids (Hodek 1973).

Lacewings (Chrysopidae) are predators as larvae and as adults feed on nectar, pollen, and/or honeydew (Canard 2001). Prey are small arthropods, less mobile, and with soft tegument that allow being perforated by the mouth parts, such as mites, whiteflies, aphids, scales, eggs and small larvae of Lepidoptera, psocopters, trips, and eggs and small larvae of Coleoptera and Diptera (see Chapter 24 and Albuquerque et al. 2001).

1.3.4 Feeding Habits of Hematophogous Insects

Insects that feed on blood (hematophagous) are important in transmitting pathogenic agents. Species of Diptera, Hemiptera, Phthiraptera, and Siphonaptera, for example, are vectors of such agents, causing devastating diseases such as dengue, malaria, leishmaniasis, Chagas disease, and bubonic plague. Hematophagy is a feeding habit of immatures and adults of both genders, or exclusively by females that seek hosts for their oogenesis (Forattini 2002). Some species, although not hematophagous, cause allergic reactions due to the action of the saliva or by ingesting toxic compounds. Others develop inside their vertebrate hosts, feeding on tissues and blood, causing lesions and development of secondary infections associated with bacteria and fungi (see Chapter 25).

1.3.5 Other Feeding Habits

Other feeding habits less known include insects that feed on detritus. Detritus may contain relatively few nutrients as in the case of dead logs, feathers, and so forth, or a great amount of nutrients such as in carcasses and feces. Detritivory is a rather sophisticated feeding habit shown by insects belonging to several orders. However, this subject is little known (see Chapter 17).

1.4 The Coverage and Implications of Studying Bioecology and Nutrition of Insects

Studies in the area of bioecology and insect nutrition passed through a series of transformations. Initially, research efforts concentrated on determining feeding habits and qualitative nutritional needs (i.e., which basic nutrients such as aminoacids, vitamins, mineral salts, carbohydrates, steroids, lipids, nucleic acids, and water were needed for normal development and reproduction of insects). In this context, several classic studies were published a long time ago, such as the revision of insect nutrition and metabolism by Uvarov (1928), the feeding regime of insects by Brues (1946), and the dietary requirements of insects by Fraenkel (1959). These studies lead to the development of artificial diets later on (Singh 1977 and Singh and Moore 1985), which created conditions for mass rearing of insects in the laboratory with multiple purposes in integrated pest management programs. The quantitative approach, including concentrations and proportions of nutrients, followed. Several techniques to measure food consumption and utilization were developed and updated (Waldbauer 1968; Kogan 1986). Literature reviews on quantitative aspects of insect nutrition were published (Scriber and Slansky 1981; Slansky and Scriber 1982).

The so-called "insect dietetics" (Beck 1972; Beck and Reese 1976) or "quantitative nutrition" (Scriber and Slansky 1981) expanded to include insect physiology and behavior that vary according to the presence of different nutrients and secondary or nonnutritional compounds (allelochemicals). Beyond biotic, abiotic factors shape the behavioral and physiological patterns of insects, such as migration or diapause, with the decrease of temperature and photoperiod or increase in production of metabolic water when facing low humidity. These patterns cause, in the long run, ecological and evolutionary consequences with the appearance of new lifestyles (Slansky 1982). The development and evolution of this research is called nutritional ecology and its model was formed as follows: for a particular species and population, there is a set of states that result in the achievement of maximum fitness (i.e., the maximum reproductive contribution to the next generation). With the variability in the environment, biotic and abiotic, insects change their behavior and physiology in an attempt to compensate for less favorable conditions to achieve their maximum potential. Their responses implicate in ecological consequences for fitness (for details see Slansky 1982a, b; Slansky and Scriber 1985).

1.5 Insect Bioecology, Nutrition, and Integrated Pest Management

By definition, integrated pest management (IPM) includes the utilization of multiple control methods. For its implementation it is necessary to understand and plan the agroecosystem, to analyze the cost/benefit net result of its adoption, to understand the tolerance of the crops to insect damage, to know the right time for insect utilization, and finally, to educate people to understand and accept the IPM principles (Luckmann and Metcalf 1982; Kogan and Jepson 2007).

The concept of integration of several tactics for management of insect pests includes those related to the bioecology and feeding/nutrition of insects (see Chapters 26 and 27). Plants resistant to insects, with physical or/and chemical attributes that make them less suitable for the insect biology (antibiosis) or less preferred for feeding and/or oviposition (antixenosis) are good examples. In addition, ecological resistance by host escape such as noncoincidence of plant and insect phenology, and induced resistance by modification of the environment to negatively affect the insect biology are included here (e.g., Maxwell and Jennings 1980; Kogan 1982; Kogan and Jepson 2007).

The use of attractive plants (preferred food sources) to concentrate insects in order to manage them to mitigate their impact to crops is another tactic that is included in the context of insect feeding behavior. There are many examples of attractive plants that are used as trap crops; sometimes parts of these attractive plants are mixed with insecticides and used as bait, causing bugs to die (see Chapter 27). Furthermore, the cultivation of plants in a consortium creating agricultural landscape mosaics and/or growing crop plants in between uncultivated landscapes (e.g., Ferro and McNeil 1998; Elkbom et al. 2000) makes them less conspicuous and therefore less suitable for pests. Supplement of nutrients to attract natural enemies or to concentrate insect pests in a particular site to facilitate control and use of attractants, repellents, and agents that disrupt the feeding process are management tactics with a strong ecological–nutritional appeal. Most of those are yet to be fully exploited in IPM programs because they are poorly understood, seldom evaluated, and, therefore, little known.

1.6 Final Considerations

The study of insects under the scope of bioecology and nutrition (nutritional ecology) include the integration of several areas of research such as biochemistry, physiology, and behavior within the context of ecology and evolution (Slansky and Rodriguez 1987). A great amount of information is generated about the biology of insects that is accumulated over time; however, this data has not been analyzed in conjunction with the areas of knowledge mentioned above. The analysis of such data considering the holistic view of the bioecology and nutrition (nutritional ecology) certainly will generate questions whose answers are currently unknown. For instance, in considering an agroecosystem where we know the species of insects that inhabit it, questions can be raised such as the following: What are the effects of inter- and intraspecific competition of species to their biology and to the crop? How do insect pests and their associated natural enemies react to the fluctuation of temperature and change in photoperiod? How are the feeding behavior and physiology affected by a change in quality of food over time? How does a parasitized insect behave regarding feeding, reproduction, and dispersion? Which factors make secondary pests become primary pests? These and many other questions that are generated should be analyzed and answered considering the paradigm of bioecology and nutrition (nutritional ecology).

It is clear that many of the ecological, physiological, and behavioral processes are linked to the feeding and nutrition context. Therefore, it is important to develop acknowledgment on feeding preference, feeding habits, nutritional requirements, and their consequences to growth, survivorship, longevity, reproduction, dispersal, gregarism, and so forth. This will allow the design of control strategies that will include a myriad of tactics. For example, once aware of the feeding preference of an insect for a particular plant species, such a plant can be used as a trap to facilitate pest control; knowing the insect and plant phenologies, one can manipulate planting time to avoid insect damage to target plants. Furthermore, physical (e.g., pilosity, tissue hardness, thorns) and chemical (lack of nutrients, presence of toxic allelochemicals) create opportunities for their manipulation in order to mitigate the pests' impact. Studies on basic and applied aspects considering the bioecology and nutrition (nutritional ecology) will not only help to understand the different insect lifestyles but will also yield data to generate holistic integrated pest management programs.

REFERENCES

Albuquerque, G. S., C. A. Tauber, and M. J. Tauber. 2001. *Chrysoperla externa* and *Ceraeochrysa* spp.: Potential for biological control in the New World tropics and subtropics. In *Lacewings in the Crop Environment*, ed. P. McEwen, T. R. New, and A. E. Whittington, 408–23. Cambridge, U.K.: Cambridge University Press.

Askew, R. R. 1973. Parasitic Hymenoptera. In *Parasitic Insects*, ed. R. R. Askew, 113–84. New York: American Elsevier Publishing Co. Inc.

Beck, S. D. 1972. Nutrition, adaptation and environment. In *Insect and Mite Nutrition*, ed. J. G. Rodriguez, 1–6. Amsterdam, the Netherlands: North-Holland Publishing Co.

Beck, S. D., and J. C. Reese. 1976. Insect–plant interactions nutrition and metabolism. *Rec. Adv. Phytochem.* 10:41–92.

Bernays, E. A. 1998. Evolution of feeding behavior in insect herbivores. *Bioscience* 48:35–44.

Bourtzis, K., and T. A. Miller. 2003. *Insect Symbiosis*. Boca Raton, FL: CRC Press.

Bourtzis, K., and T. A. Miller. 2006. *Insect Symbiosis. Volume 2*. Boca Raton, FL: CRC Press.

Bourtzis, K., and T. A. Miller. 2006. *Insect Symbiosis. Volume 3*. Boca Raton, FL: CRC Press.

Brues, C. T. 1946. *Insect Dietary—An Account of the Food Habits of Insects*. Cambridge, MA: Harvard University Press.

Christenson, L. D., and R. H. Foote. 1960. Biology of fruit flies. *Annu. Rev. Entomol.* 5:171–92.

Cohen, A. C. 2004. *Insect Diets. Science and Technology*. Boca Raton, FL: CRC Press.

Coudron, T. A., G. D. Yocum, and S. L. Brandt. 2006. Nutrigenomics: A case study in the measurement of insect response to nutritional quality. *Entomol. Exp. Appl.* 121:1–14.

Davidson, D. W., S. C. Cook, and R. R. Snelling. 2004. Liquid-feeding performance of ants (Formicidae): Ecological and evolutionary implications. *Oecologia* 139:255–66.

De Moraes, C. M., W. J. Lewis, P. W. Paré, H. T. Alborn, and J. H. Tumlinson. 1998. Herbivore-infested plants selectively attract parasitoids. *Nature* 393:570–3.

Ekbom, B., M. E. Irwin, and Y. Robert. 2000. *Interchanges of Insects between Agricultural and Surrounding Landscapes*. Dordrecht, the Netherlands: Kluwer Academic Publishers.

Ferro, D. N., and J. N. McNeil. 1998. Habitat enhancement and conservation of natural enemies of insects. In *Conservation Biological Control*, ed. P. Barbosa, 123–32. San Diego, CA: Academic Press.

Floren, A., A. Biun, and K. E. Linsenmair. 2002. Arboreal ants as key predators in tropical lowland rainforest trees. *Oecologia* 131:137–44.

Fowler, H. G., L. C. Forti, C. R. F. Brandão, J. H. C. Delabie, and H. L. Vasconcelos. 1991. Ecologia nutricional de formigas. In *Ecologia nutricional de insetos e suas implicações no manjeo integrado de pragas*, ed. A. R. Panizzi, and J. R. P. Parra, 131–223. São Paulo, Brazil: Editora Manole.

Forattini, O. P. 2002. *Culicidologia Médica. Vol. 2*. São Paulo, Brazil: Editora da Universidade de São Paulo.

Fraenkel, G. 1959. A historical and comparative survey of the dietary requirements of insects. *N.Y. Acad. Sci.* 77:267–74.

Hodek, I. 1973. *Biology of Coccinellidae*. The Hague, the Netherlands: Academia, Prague & Dr. W. Junk.

Hölldobler, B., and E. O. Wilson. 1990. *The Ants*. Cambridge, MA: Harvard University.

Kogan, M. 1986. Bioassays for measuring quality of insect food. In *Insect–Plant Interactions*, ed. J. R. Miller, and T. A. Miller, 155–89. New York: Springer-Verlag.

Kogan, M., and P. Jepson. 2007. Perspectives in ecological theory and integrated pest management. Oxford, U.K.: Cambridge University.

Luckmann, W. H., and R. L. Metcalf. 1982. The pest-management concept. In *Introduction to Insect Pest Management*, ed. R. L. Metcalf, and W. H. Luckmann, 1–31. New York: John Wiley & Sons.

Maxwell, F. G., and P. R. Jenning. 1980. *Breeding Plants Resistant to Insects*. New York: John Wiley & Sons.

Oliveira, L. J., G. G. Brown, and J. R. Salvadori. 2003. Corós como pragas e engenheiros do solo em agroecossistemas. In *O uso da macrofauna edáfica na agricultura do século XXI: A importância dos engenheiros do solo*. 76–86. Londrina, Brazil: Embrapa Soja.

Reddy, G. V. P., and A. Guerrero. 2004. Interactions of insect pheromones and plant semiochemicals. *Trends Plant Sci.* 9:253–61.

Schaefer, C. W., and A. R. Panizzi. 2000. *Heteroptera of Economic Importance*. Boca Raton, FL: CRC Press.

Scriber, J. M., and F. Slansky, Jr. 1981. The nutritional ecology of immature insects. *Annu. Rev. Entomol.* 26:183–211.

Singh, P. 1977. *Artificial Diets of Insects, Mites, and Spiders*. New York: Plenum Press.

Singh, P., and R. F. Moore. 1985. *Handbook of Insect Rearing. Vol. 1*. Amsterdam, the Netherlands: Elsevier Science Publishers.

Slansky, Jr., F. 1982a. Toward the nutritional ecology of insects. In *Proc. 5th Int. Symp. Insect–Plant Relationships*, eds. J. H. Visser, and A. K. Minks, 253–59. Wageningen, the Netherlands.

Slansky, Jr., F. 1982b. Insect nutrition: An adaptionist's perspective. *Fla. Entomol.* 65:45–71.

Slansky, Jr., F., and J. G. Rodriguez. 1987. Nutritional ecology of insects, mites, spiders, and related inverte-brates: An overview. In *Nutritional Ecology of Insects, Mites, Spiders, and Related Invertebrates*, ed. F. Slansky, Jr., and J. G. Rodriguez, 1–69. New York: John Wiley & Sons.

Slansky, Jr., F., and J. M. Scriber. 1982. Selected bibliography and summary of quantitative food utilization by immature insects. *Bull. Entomol. Soc. Am.* 28:43–55.

Slansky, Jr., F., and J. M. Scriber. 1985. Food consumption and utilization. In *Comprehensive Insect Physiology, Biochemistry, and Pharmacology. Vol. 1*, ed. G. A. Kerkut, and L. I. Gilbert, 87–163. Oxford, U.K.: Pergamon Press.

Turlings, T. C. J., and J. Ton. 2006. Exploiting scents of distress: The prospect of manipulating herbivore-induced plant odours to enhance the control of agricultural pests. *Curr. Opin. Plant Biol.* 9:421–7.

Uvarov, B. P. 1928. Insect nutrition and metabolism: A summary of the literature. *Trans. Royal Entomol. Soc. London* 74:255–343.

Vendramim, J. D., O. Nakano, and J. R. P. Parra. 1992. Pragas dos produtos armazenados. In *Curso de Entomologia Aplicada a Agricultura*, ed. Fundação de Estudos Agrários Luiz de Queiroz–FEALQ, 673–704, Piracicaba: FEALQ.

Waldbauer, G. P. 1968. The consumption and utilization of food by insects. *Adv. Insect Physiol.* 5:229–88.

Zucoloto, F. S. 1993. Adaptation of a *Ceratitis capitata* population (Diptera, Tephritidae) to an animal protein base diet. *Entomol. Exp. Appl.* 67:119–27.

2

Nutritional Indices for Measuring Insect Food Intake and Utilization

José R. P. Parra, Antônio R. Panizzi, and Marinéia L. Haddad

CONTENTS

2.1 Introduction 14
2.2 Nutritional Indices for Measuring Food Intake and Utilization 16
    2.2.1 Experimental Techniques 17
    2.2.2 Quantity of Food Consumed 17
    2.2.3 Weight Gains by the Insect 18
    2.2.4 Measuring Feces 19
2.3 The Meaning of the Different Nutritional Indices 19
    2.3.1 Relative Consumption Rate 19
    2.3.2 Relative Metabolic Rate 19
    2.3.3 Relative Growth Rate 19
    2.3.4 Efficiency of Conversion of Ingested Food 19
    2.3.5 Efficiency of Conversion of Digested Food 19
    2.3.6 Approximate Digestibility (AD) 21
2.4 Methods Used to Measure Food Intake and Utilization 23
    2.4.1 Direct Method 23
        2.4.1.1 Gravimetric 23
    2.4.2 Indirect Methods 23
        2.4.2.1 Colorimetric Methods 23
        2.4.2.2 Isotope Method 24
        2.4.2.3 Uric Acid Method 24
        2.4.2.4 Trace Element Method 25
        2.4.2.5 Immunological Method 25
        2.4.2.6 Calorimetric Method 25
2.5 Comparison of Methods 28
2.6 Interpretation of Nutritional Indices Values 30
2.7 Food Consumption and Use for Growth in the Larval Phase 30
    2.7.1 Number of Instars 33
    2.7.2 The Cost of Ecdysis 35
    2.7.3 Food Intake and Utilization through Instars 35
2.8 Adult Food Consumption and Use for Reproduction and Dispersal 36
    2.8.1 Food Quality 36
    2.8.2 Food Selection and Acceptance 38
    2.8.3 The Role of Allelochemicals 39
2.9 Final Considerations 41
References 42

2.1 Introduction

Nutrition may be studied both qualitatively and quantitatively. Qualitative nutrition deals exclusively with nutrients needed from the chemical aspect. In this case, it is well known that, independently of the systematic position and the feeding habits of insects, qualitative nutritional needs are similar and that these needs, except for a general need for sterols, are close (with rare exceptions) to those of the higher animals. The basic nutritional needs of insects include amino acids, vitamins and mineral salts (essential nutrients) and carbohydrates, lipids, and sterols (nonesssential nutrients), which should be adequately balanced, especially in the ratio of proteins (amino acids): carbohydrates (see Chapter 3).

There have been many papers on nutrition since the beginning of the last century (Uvarov 1928), and after the revisions by Brues (1946) and Fraenkel (1953), especially since the 1970s, there have been a large number of publications on the subject (Rodriguez 1972; Dadd 1973, 1977, 1985; House 1972, 1977; Reinecke 1985; Parra 1991; Anderson and Leppla 1992; Thompson and Hagen 1999; Bellows and Fisher 1999; Cohen 2004). The development of artificial diets for insects, mainly since the 1960s, has refined research on nutritional needs (Singh 1977), and in 1985 there were artificial diets for more than 1,300 insect species (Singh 1985). This advance in rearing techniques resulted in the discovery that some restricted insect groups need nucleic acids and even liposoluble vitamins, such as A, E, and K_1. Sophisticated production techniques for parasitoids *in vitro* (excluding the host) have even been developed (Cônsoli and Parra 2002), with these authors referring to 73 parasitoid species reared *in vitro*, 16 Diptera and 57 Hymenoptera. The artificial diets used for phytophages today have the same composition as those developed in the 1960s and 1970s (see Chapter 3). A recent revision was published by Cônsoli and Grenier (2010).

It was some time before significant attention was paid to quantitative nutrition due to the technical difficulties in measuring food utilization. However, today, it is known that food intake and utilization is a basic condition for growth, development, and reproduction. Food quantity and quality consumed in the larval stage affects growth rate, development time, body weight, and survival, as well as influencing fecundity, longevity, movement, and the capacity of adults to compete. Larvae that are inadequately fed result in pupae and adults of "bad quality." For example, an artificial diet for *Pseudoplusia includens* (Walker) that does not contain wheat germ oil (a source of linoleic and linolenic acids) leads to deformed wings in all the adults (J. R. P. Parra, personal observation). Similar results were recorded by Bracken (1982) and Meneguim et al. (1997) for other lepidopterous species.

Quantitative (Scriber and Slansky 1981) or dietetic nutrition (Beck 1972) considers that not only are the basic nutritional requirements important for the insect but also the proportion (quantity) of food ingested, assimilated, and converted into growth tissue. This quantity varies not only in function of the nutrients but also the nonnutrient contents (such as the allelochemicals) in the food. Some researchers (e.g., Slansky and Rodriguez 1987a) considered quantitative nutrition more important. Thus, when the behavioral and physiological changes are examined in an ecological context (in constant change), by identifying the ecological consequences and the evolutionary aspects of such behavior, insect nutrition reaches a wider meaning, transforming it into the nutritional ecology. According to these authors, most, if not all these ecological, physiological, and behavioral processes in insects happen within a nutritional context, which includes feeding, growth, metabolism, enzyme synthesis, lipid accumulation, diapause, flight, and reproduction.

Since the measures of food consumption and use are the limit between feeding physiology and the selection behavior of the host plant, their study has a series of applications, not only in the basic area of nutrition, community ecology, and behavior, but also in applied areas of control through plant resistance and biological control (Kogan and Parra 1981; Cohen 2004; Jervis 2005).

The basic concepts of food consumption and use have been developed by nutritionists who related the quality of the food consumed with its effect on animal growth and development (Klein and Kogan 1974). The ecologists used this type of analysis as a basis for studies on community energy flows (Mukerji and Guppy 1970; Latheef and Harcourt 1972). Researchers in pest management can use consumption and growth rate measures to develop simulation models for determining pest economic injury levels (Stimac 1982) or even to evaluate which plant part is preferred by the insect (Gamundi and Parra, unpublished).

TABLE 2.1

Relative Consumption Rate, Relative Growth Rate, Relative Metabolic Rate, Approximate Digestibility, Efficiency of Convertion of Ingested Food, Efficiency of Convertion of Digested Food, Metabolic Cost, and Mortality for *Spodoptera frugiperda*, *Heliothis virescens*, and *Diatraea saccharalis* Larvae Reared on Artificial Diet at 25°C and 30°C, R-I of 60 ± 10%, and a 14 h Photophase

Nutritional Indices*	S. frugiperda		H. virescens		D. saccharalis	
	25°C	30°C	25°C	30°C	25°C	30°C
RCR (g/g/day)	0.5653 ± 0.1364a	0.5620 ± 0.0231a	0.8139 ± 0.0333b	1.2289 ± 0.0472b	0.8848 ± 0.0497b	1.0899 ± 0.0667a
RGR (g/g/day)	0.0835 ± 0.0060b	0.1407 ± 0.0040a	0.1500 ± 0.0036b	0.2315 ± 0.0041b	0.0771 ± 0.0016b	0.1142 ± 0.0026a
RMR (g/g/day)	0.2057 ± 0.1260a	0.0748 ± 0.0084a	0.3026 ± 0.0315b	0.5264 ± 0.0474b	0.1879 ± 0.0132b	0.4619 ± 0.0619a
AD (%)	42.04 ± 3.39a	38.90 ± 1.00a	54.55 ± 2.22b	60.64 ± 1.68a	31.43 ± 1.37b	50.91 ± 2.16a
ECI (%)	19.67 ± 1.44b	26.26 ± 1.40a	19.77 ± 0.95a	19.53 ± 0.56a	9.52 ± 0.39b	11.64 ± 0.60a
ECD (%)	53.30 ± 5.06b	67.06 ± 3.04a	40.07 ± 2.99a	33.73 ± 1.66a	31.87 ± 1.51a	24.23 ± 1.45b
MC (%)	46.70 ± 5.06a	32.94 ± 3.04b	59.93 ± 2.99a	66.27 ± 1.66a	68.13 ± 1.51b	75.77 ± 1.45a
Mortality (%)	58.0 ± 14.00b	52.0 ± 25.80a	18.0 ± 6.60a	20.0 ± 12.60a	12.0 ± 9.80a	20.0 ± 17.90a

Source: Souza, A. M. L., C. J. Ávila, and J. R. P. Parra, *Neotrop. Entomol.*, 30:11–17, 2001.

Note: RCR = relative consumption rate, RGR = relative growth rate, RMR = relative metabolic rate, AD = approximate digestibility, ECI = efficiency of conversion of ingested food, ECD = efficiency of conversion of digested food, MC = metabolic cost.

* Means followed by the same letter on the row do not differ significantly, based on the Tukey test (P ≥ 0.05), for each species in two temperatures.

The process that determines host plant selection by an insect, that is, the insect/plant relationship (see Chapter 5 for more details), is an application of the measures of insect intake and utilization. The interaction of allelochemicals and nutrients has been determined from nutritional indices that have helped in understanding the mechanisms of plant resistance to insects (Reese 1977). The study of nutritional indices may be done with natural or artificial diets and explains the phenomena that occur under variable conditions of temperature (Souza et al. 2001), relative humidity, photoperiod, parasitism, and even soil nutrients (Oliveira et al. 1990), allelochemicals, transgenic plants, enzymatic studies, or cannibalism (Nalim 1991). For artificial diets, a diet's nutritional suitability or even deciding which is the most suitable container can be done through measuring food intake and utilization. Thus, Souza et al. (2001), using food consumption and indices concluded that for *Spodoptera frugiperda* (J. E. Smith) the best rearing temperature is 30°C while for *Heliothis virescens* (F.) and *Diatraea saccharalis* (F.), there is no difference between 25°C or 30°C (Table 2.1).

Coudron et al. (2006) proposed the term "nutrigenomics" or "nutritional genomics," which has the aim of supplying information on the impact of nutrition on biochemical parameters by investigating how it alters the standards of global gene expression. Insect molecular markers would be identified that could be used as initial indicators of the response to different nutritional sources. Such molecular markers could be chosen based on the degree of expression and evaluated for their suitability as nutritional markers by examining development and expression by generation. Ideal markers would be those that are strongly expressed, that would appear in a development stage in the first generation, and that would be consistent over many generations. The authors demonstrated the first example with *Perillus bioculatus* (F.) (Heteroptera: Pentatomidae), by rearing it on optimum and suboptimum diets and analyzing the presence of expressed genes differentially in the two treatments. According to the authors, future research in this area can supply a better definition of biochemical, physiological, and genetic regulation of suitability, quality, and high performance in insect populations. It could be useful in evaluating the degree of risk of introduced natural enemies, since it is a faster method for identifying and evaluating potential alternative hosts; in a wider context, it could be important for the effective use of biological control and other control methods, as well as improving agricultural sustainability.

2.2 Nutritional Indices for Measuring Food Intake and Utilization

The first studies on insect food intake and utilization were made with natural foods, with no standardization and using methods with variable degree of precision, which resulted in much confusion. Waldbauer (1968) made a revision and standardized the indices for measuring consumption and use by herbivorous insects. Even today, this work is the basis for those researchers who study quantitative or diet nutrition, although Kogan and Cope (1974) and Scriber and Slansky (1981) have suggested some alterations that have been accepted by the scientific community. These indices are as follows:

a. Relative Consumption Rate (RCR)

$$RCR = \frac{I}{\overline{B} \times T}$$

b. Relative Metabolic Rate (RMR)

$$RMR = \frac{M}{\overline{B} \times T}$$

c. Relative Growth Rate (RGR)

$$RGR = \frac{B}{\overline{B} \times T}$$

d. Efficiency of Conversion of Ingested Food (ECI)

$$ECI = \frac{B}{I} \times 100$$

e. Efficiency of Conversion of Digested Food (ECD)

$$ECD = \frac{B}{I-F} \times 100$$

f. Approximate Digestibility (AD)

$$AD = \frac{I-F}{I} \times 100$$

The meaning of the variables of the different formulae is as follows:

T = Time of duration of feeding period

I = Food consumed during T

B = Food used during Ti

$$B - (I - F) - M$$

F = Undigested food + excretory products

M = (I – F) – B = food metabolized during T (part of assimilated food that was used in the form of energy for metabolism)

$\overline{I}$ – F = Food assimilated during T (represents the part of I which was used by the insect for conversion into biomass and for metabolism)

$\overline{B}$ = Mean weight of insects during T (some ways to determine this are described in Kogan 1986)

2.2.1 Experimental Techniques

The data needed to determine the indices include the quantity of food consumed in time T, the insect weight gain in the period T, and the total excretion (including exuviae, secretions, cocoons, and feces). Besides this data, the volume of CO_2 produced during respiration can be necessary in certain types of study. It is important that in the determination of nutritional indices a standard be adopted using the weights of fresh or dry materials for the parameters. It is preferable to use dry weights especially when the indices are determined in artificial diets since water loss from the medium is significant, making it difficult to make corrections for working with fresh weights. The indices calculated on the basis of fresh weight cannot be compared due to the difference in the percentage of water in the food, feces, and tissues of the insect. However, knowing the amount of water is fundamental for understanding adaptations to different lifestyles in which its use has important ecological consequences (Slansky and Scriber 1985). The selection of the period (e.g., the whole cycle, a stage, or one or two instars) to measure consumption and utilization is important. Periods defined physiologically offer the advantage of being able to be reproduced and be compatible with results from other experiments.

2.2.2 Quantity of Food Consumed

This parameter results from the difference between the amounts of food offered to the insect at the beginning of the experiment and what is left over at the end of the study period. The starting weight has to be determined as fresh weight with the dry weight obtained from the fresh and the dry weight of an aliquot, which should be as similar as possible to the food offered. When leaves were used, Waldbauer (1964) and Soo Hoo and Fraenkel (1966) found that great precision is possible by cutting leaves into two symmetrical parts along the midrib and using one part as the food and the other as the aliquot. These aliquots should be kept under the same conditions as the experimental batch. Food quality should be preserved, maintaining the humidity (especially with filter paper) in the recipient used and changing the food regularly (daily). In order that there is no alteration in the food to be offered the insect, ideally the determination should be done on the intact plant material without removing (e.g., leaf) the part to be consumed. Although this can be done by using cages that keep the insect together with the part to be consumed, this procedure is not always feasible. Thus, since the aliquot should be kept under the same conditions, the leaf could absorb water (if that is the case) and there might often be a "negative consumption" (the weight left over greater than the food offered). In order to avoid this problem, when the consumption of very small larvae is determined (first–second instars), then work should be done with groups of insects (Crócomo and Parra 1985). These errors are common, especially in early instars (Crócomo and Parra 1985, Figure 1; Schmidt and Reese 1986). In general, errors are reduced when studies are done measuring all the food consumed during the larval phase. Another aspect to be considered is that enough food should be supplied so that some is always left over. Therefore, previous knowledge of the insect's feeding habits is necessary so that enough food can be supplied that is sufficient for the study period. Waldbauer (1968), Crócomo and Parra (1979), and Crócomo and Parra (1985) proposed a series of formulae to determine the weight of food ingested, based on area or weight, but there are large differences between the methods (Kogan 1986).

2.2.3 Weight Gains by the Insect

Since the insect's dry weight cannot be measured at the start of the experiment, it is estimated from the percentage of dry weight of an aliquot of an identical larva, dried to constant weight (55°C–60°C). Very small larvae, such as some noctuids, should be weighed in batches of 100 because their very light weight does not register on normal balances (except for highly sensitive microbalances). Larvae should preferably be killed quickly thus avoiding the liberation of feces, or by freezing in a "freezer" or by immersing in liquid nitrogen, before drying the insect. The moment of weighing is fundamental to avoid errors. The food residue that remains in the gut can, at the beginning or at the end of the experiment, result in errors in determining weight gain. The gut may be empty before or soon after ecdysis, so that in general insects empty it before each molt. Keeping insects without feeding for a certain time does not always result in the elimination of all the feces (Waldbauer 1968), and some, when kept without food, retain more feces than the food they received due to the stress caused by interrupting feeding. Ecdysis can lead to errors of determination of weight gains. Thus, the insect reaches a maximum weight in each instar and loses

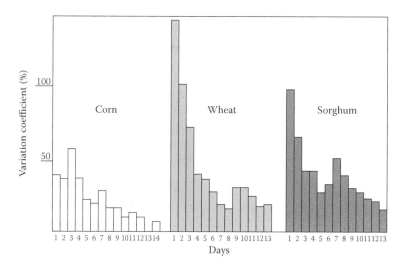

FIGURE 2.1 Coefficient of variation (CV) of the mean daily accumulated consumption of *Spodoptera frugiperda* larvae reared on corn, wheat, and sorghum. Temperature: 25°C ± 2°C, relative humidity: 60% ± 10%, and 14 h photophase. (From Crócomo, W. B., and J. R. P. Parra, *Rev. Bras. Entomol.*, 29, 225–60, 1985.)

weight during the molt because the molted cuticle and the energy used in ecdysis contributes to losses, which reach 45% (Waldbauer 1962). For studies with only one instar, the final instar should be preferred owing to the large amount of food consumed that will facilitate the weighings. The weight gain is calculated by subtracting the weight at the beginning of the experiment from that reached at the end.

2.2.4 Measuring Feces

Feces' dry weight can be measured directly, taking care to make frequent collections and dry the feces immediately to avoid decomposition and fungal growth. Feces fresh weight is difficult to measure due to water losses or gains. Feces of larval *Automeris* sp. (Lepidoptera) lose around 26% of their fresh weight in 24 hours (Waldbauer 1968). There are cases where it is difficult to separate out the feces, especially in artificial diets, because they are mixed in with the food. In these cases, the recommendation is to invert the rearing recipient so that the fecal pellets are collected in the lid. There are cases, such as in studies on stored grain pests, where it is impossible to separate the feces from the food. In these cases, indirect methods are used, such as the uric acid method (see uric acid method).

In the Department of Entomology and Acarology of ESALQ (Escola Superior de Agricultura Luiz de Queiroz), University of São Paulo (USP), in order to calculate these indices, special cards are used to collect the necessary data from the studies with artificial diets; today, information technology permits each person to elaborate a model for registering data.

2.3 The Meaning of the Different Nutritional Indices

2.3.1 Relative Consumption Rate

The relative consumption rate (RCR) represents the quantity of a food ingested per milligram of insect body weight per day, and is expressed as mg/mg/day. It can be altered in function of the amount of water in the food or the physical–chemical properties of the diet. Although insects consume a large percentage of food (more than 75%) in the last instar, in relation to the total amount of food consumed, the consumption is, proportionally to the size, greater in the first instars (Figure 2.2a and Table 2.2).

2.3.2 Relative Metabolic Rate

The relative metabolic rate (RMR) represents the quantity of food spent in metabolism per milligram of body weight (biomass of the insect per day) and is expressed in mg/mg/day (Figure 2.2b).

2.3.3 Relative Growth Rate

The relative growth rate (RGR) indicates the gain in biomass of the insect in relation to its weight and is expressed as mg/mg/day. It depends on host quality, the physiological state of the insect, and environmental factors (Figure 2.2c).

2.3.4 Efficiency of Conversion of Ingested Food

The efficiency of conversion of ingested food (ECI) represents the percentage of food ingested that is transformed into biomass. This index tends to increase up to the last instar. In the last instar, there are physiological changes and an extra energy expenditure in the stage before pupation, which provokes a proportionally lower weight gain in the insect in this instar (Figure 2.2d).

2.3.5 Efficiency of Conversion of Digested Food

The efficiency of conversion of digested food (ECD) is an estimate of the conversion of assimilated material into biomass by the biological system (represents the percentage of ingested food that is converted

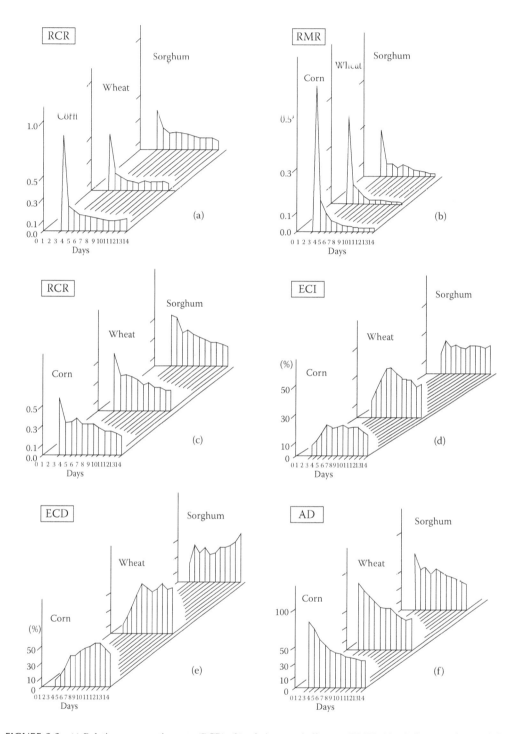

FIGURE 2.2 (a) Relative consumption rate (RCR), (b) relative metabolic rate (RMR), (c) relative growth rate (RGR), (d) efficiency of conversion of ingested food (ECI), (e) Efficiency of Conversion of Digested Food (ECD), and (f) approximate digestibility (AD) of *Spodoptera frugiperda* caterpillars feeding on leaves of corn, wheat, and sorghum. Temperature: 25°C ± 2°C, relative humidity: 60% ± 10%, and 14 h photophase. (From Crócomo, W. B., and J. R. P. Parra, *Rev. Bras. Entomol.*, 29, 225–60, 1985.)

TABLE 2.2

Food Consumed (%) per Instar of the Total Food Consumed during the Larval Stage

Insect and Food	% of Total Consumption per Instar						Reference
	I	II	III	IV	V	VI	
Agrotis orthogonia[*]							Waldbauer (1968)
Triticum aestivum	0.21	0.42	2.3	8.70	31.60	56.80	
T. durum	0.15	0.48	3.10	9.10	32.90	54.20	
Protoparce sexta[†]							
Tobacco leaves	0.08	0.53	1.90	10.50	86.40	–	Waldbauer (1968)
Agrotis ipsilon[‡]							
Corn leaves	0.06	0.18	0.77	2.60	10.40	86.00	Waldbauer (1968)
Pseudoplusia includens[a]							
Soybean leaves	0.60	0.35	2.33	6.53	14.96	75.08	Kogan and Cope (1974)
Eacles imperialis magnifica[‡]							Crócomo and Parra (1979)
Coffee leaves	0.37	1.43	3.78	15.13	84.87	–	
Lonomia circumstans[‡]							D'Antonio and Parra (1984)
Coffee leaves	0.18	0.46	1.30	4.14	13.90	80.02	
Alabama argillacea[*]							Carvalho and Parra (1983)
Cotton leaves	–	–	7.90§	11.26	81.00	–	
IAC–18							
Erinnyis ello ello[*]	0.37	0.93	3.49	15.38	79.83	–	Reis F° (1984)

* Measurement in dry weight.

† Measurement in fresh weight.

‡ Measurement in area.

§ Consumption from first to third instar.

into biomass). The ECD increases with insect development (Figure 2.2e). Variations can occur with age, as a variation of the RMR, lipid synthesis, and the rate of assimilation and activity by the organism (Slansky and Scriber 1985). The opposite of the EDC indicates the percentage of food metabolized into energy for maintaining life. Therefore, 100-ECD corresponds to the metabolic cost. Almeida and Parra (1988) demonstrated this cost to be greater at higher temperatures for *D. saccharalis* maintained on an artificial diet.

2.3.6 Approximate Digestibility (AD)

The approximate digestibility (AD) represents the percentage of food ingested that is effectively assimilated by the insect. This index is an approximation of the actual nutrient absorption through the intestinal walls, since the presence of urine in the feces makes accurate measurements of digestibility more difficult. In this case, fecal weight does not only represent the noningested food but added to this are the metabolic products discharged in the urine. The values obtained for approximate digestibility are, therefore, always less than the corresponding values of actual digestibility. This difference is minimal in phytophages. In general, digestibility diminishes from the first to the last instar (Figure 2.2f), with an inverse relationship between AD and ECD, since the smaller larvae digest the food better because they tend to select it, avoiding leaf veins that contain large quantities of fibers and feeding almost exclusively on parenchymatous tissue. Thus, most of the food consumed by the young larvae is spent in energy for maintenance and only a little is used for growth. In the older larvae, consumption is indiscriminate and includes leaf veins. In this way, less food is used for energy and a large amount is incorporated into body tissue, thus increasing the ECD. Digestibility is also affected by an unsuitable nutrient balance, water deficiency, or the presence of allelochemicals (Beck and Reese 1976). Nutritional indices have been discussed in great detail by Waldbauer (1968), Kogan and Cope (1974), Scriber and Slansky (1981), and Cohen (2004). According to Slansky and Scriber (1982) these nutritional indices vary considerably as follows: RGR = 0.03–0.39 mg/day.mg, RCR = 0.04–2.3 mg/day.mg, AD= 9%–88%, ECD = 18%–89%, and ECI = 0.6%–68%. A summary of these values for *S. frugiperda* fed on sorghum, corn, and wheat is shown in Figure 2.3.

Rates and efficiencies for the consumption of specific compounds can also be calculated. Waldbauer (1968), and Slansky and Feeny (1977) proposed the following terms to describe the use of nitrogen (N): rate of N; that is, milligrams of biomass of N gain/day (NAR); consumption rate of N, that is, milligrams of N ingested/day (NCR); and use efficiency of N (NUE,) which is calculated as follows:

$$NUE = \frac{\text{milligrams of biomass of N gained}}{\text{milligrams of N ingested}} \times 100$$

The rate of biomass of N gained is obtained by multiplying the dry weight gained by the mean percentage of N in a control larva. The conversion efficiency of N assimilated in biomass of N of the larva assumes that it is 100%. Since part of the N assimilated is feces, such as uric acid, allantoic acid, or other compounds, the NUE is underestimated from these calculations. Gamundi (1988) observed that the NUE is greater in soybean leaves (upper or lower) compared to Bragg soybean pods for *Anticarsia gemmatalis* Hübner, with less efficiency of nitrogen use in the larvae from pods compared to those from leaves. Lee

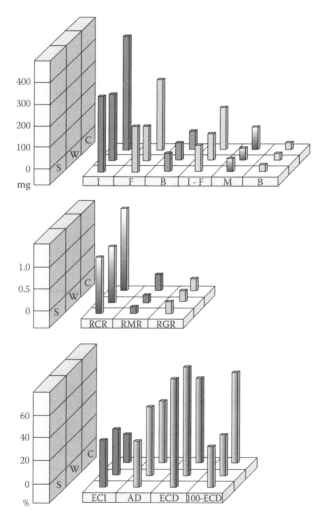

FIGURE 2.3 Means of parameters and nutritional indices obtained for *Spodoptera frugiperda* larvae fed with corn (C), wheat (W), and sorghum (S). Temperature: 25°C ± 2°C, relative humidity: 60% ± 10%, and 14 h photophase. (From Crócomo, W. B., and J. R. P. Parra, *Rev. Bras. Entomol.*, 29, 225–60, 1985.)

et al. (2004) studied the difference between the solitary and gregarious phases of *Spodoptera exempta* (Walker) and observed that the gregarious phase showed greater nitrogen conversion efficiency in a diet with a minimum of protein, and accumulated more lipids per quantity of carbohydrates consumed in a diet deficient in carbohydrate. Thompson and Redak (2005) studied the feeding behavior and nutrient selection in *Manduca sexta* (Cr.), and the alterations induced by parasitism of *Cotesia congregata* (Say). Unparasitized larvae regulate the absorption of proteins and carbohydrates in varying proportions. They consume equal amounts of nutrients independent of the protein:carbohydrate ratio and grow the same. If the level of the nutrient combination is reduced, the caterpillar abandons the regulation and feeds at random. Parasitized caterpillars do not regulate food absorption. Nutrient consumption varies considerably but growth is unaffected. If caterpillars are offered a choice of diet containing equal amounts of casein and sucrose but with variable fat (corn oil), they fail to regulate fat absorption although both the parasitized and the unparasitized caterpillars prefer the diet containing the high fat content.

2.4 Methods Used to Measure Food Intake and Utilization

2.4.1 Direct Method

2.4.1.1 Gravimetric

The gravimetric method is the most used method for measuring food intake and utilization. Although it demands a lot of time, it only needs a balance and a drying oven. It is difficult to measure food use in insects maintained on artificial diets or in situations where they live within the food substrate such as stored product pests, miners, borers, stem and fruit borers, and coprophages (Kogan 1986).

2.4.2 Indirect Methods

With indirect methods, products are added to diets, allowing the determination of consumption and use indirectly. The compound to be added should neither be toxic at the concentrations used nor be metabolized by the insects. Various compounds have been used, including lignin, amide, substances that occur in plant pigments ("chromogens"), coloring agents, iron oxide, barium sulfate, chromic oxide, and radioactive materials.

2.4.2.1 Colorimetric Methods

Dying agents are used to determine food consumption and use, such as chromic oxide (McGinnis and Kasting 1964), calco oil red N-1700 (Daum et al. 1969), solvent red 26 and soluble blue (Brewer 1982) and amaranth–acid red 27 (Hori and Endo 1977; Kuramochi and Nishijima 1980), quoted by Kogan (1986).

Among the many dying agents used, calco oil red N-1700 and solvent red 26 (Keystone Aniline and Chemical, Chicago, IL) give the best results. This method was developed by Daum et al. (1969) to measure ingestion by adult *Anthonomus grandis* Boh. The procedure is as follows: the dying agent is added to the diet at the rate of 100 to 1000 ppm and, in order to facilitate incorporation, it is dissolved in oil (e.g., corn, cotton, or wheat germ oil) (Hendricks and Graham 1970). The marking becomes more visible when the coloring agent is dissolved in corn or cotton oil compared to alcohol or acetone; larvae, prepupae, and pupae are washed with acetone to avoid external contamination by the dying agent. The calco oil red concentrated in larvae, pupae, and feces is extracted with tissue macerators and with acetone, the solution filtered, and the concentrations measured in a spectrophotometer, adjusting the wavelength scale to 510 um (Daum et al. 1969). If necessary, the coloring agent residue on the filter paper used should be extracted in a Soxhlet apparatus. Wilkinson et al. (1972) showed that this coloring agent can be added to diets of *Pieris rapae* (L.), *Helicoverpa zea* (Boddie), and *Trichoplusia ni* (Hübner) larvae, without harming development; this was also observed by Gast and Landin (1966), Lloyd et al. (1968) and Daum et al. (1969) for *A. grandis* and by Hendricks and Graham (1970) and Jones et al. (1975), to *H. virescens* and *H. zea*, respectively. Parra and Kogan (1981) observed that calco oil red, at the rate of 1 g/liter of artificial diet dissolved in wheat germ oil, affected food intake by *P. includens*, reducing it by approximately 50% compared to the control.

The approximate digestibility (AD) can be calculated without collecting the feces or measuring the food consumed, where:

$$AD = \frac{M_F - M_A}{M_F}$$

M_F = concentration of marker in the feces
M_A = concentration of marker in the food

If the weights of the feces or food consumed are known then the weight of food consumed can be calculated $= \frac{MF}{MA} \times$ wt. of feces or wt. of feces $= \frac{MA}{MF} \times$ wt. of food consumed.

The chromic oxide method consists of adding a known concentration of chromic oxide to the diet (4%) (dissolved in a basic medium) and determining its concentration in the feces by liquid oxidation of Cr_2O_3 to $Cr_2O^{-2}_7$, followed by a colorimetric measurement of the dichromate ion with diphenylcarbazide. Samples are digested with a mixture of perchloric acid–sulfuric acid–sodium molybdate for 30 minutes. The cold material digested is diluted with diphenylcarbazide and the mean absorbance of 540 um, compared to a control consisting of 9.5 ml of H_2SO_4 0.25 N and 0.5 ml of diphenylcarbazide. This method was described by McGinnis and Kasting (1964) for measuring the approximate digestibility of *Agrotis orthogonia* Morrison (Lepidoptera). According to these authors, this method was faster, more practical, and better than the gravimetric method. However, Daum et al. (1969) consider it an empirical chemical method that is easily influenced by the reaction time and temperature. Further disadvantages include the use of dangerous acids, such as perchloric and sulfuric acids, and the need for someone trained in analytical chemistry. McMillian et al. (1966) showed that chromic oxide inhibited feeding of *H. zea* and *S. frugiperda*. Instead of using colorimetric determinations with diphenylcarbazide, Parra and Kogan (1981) measured chromium directly with atomic absorption spectroscopy. With this method, the following formulae are used: food consumed (F) = (E × %Cr in the feces) + (B × %Cr in the insect), where F = food consumed, B = insect weight gain, and E = feces.

The food consumed is calculated indirectly and the other parameters are measured beforehand. From these values the ECI and the ECD can be calculated. Approximate digestibility (AD) = 1. (% of Cr in the medium/% of Cr in the feces.)

2.4.2.2 Isotope Method

Various isotopes have been used as markers in nutritional studies. Crossley (1966) used cesium 137 to measure the daily consumption of the third instar *Chrysomela knaki* Brow (Coleoptera) larvae feeding on *Salix nigra*. Marked sucrose or cellulose were added to the diet to estimate the food consumption of larval instars (first and fifth) of the lepidopteran, *A. orthogonia* (Kasting and McGinnis 1965), and at the end of the feeding period its feces and CO_2 were measured by radioactivity.

Food consumption by the migratory grasshopper was determined from Na through an abdominal injection (Buscarlet 1974). The CO_2 measured by radioactivity can be very high because in certain diets, mostly in those that are unsatisfactory, it can reach 75% of the total food ingested. Parra and Kogan (1981) observed for the *P. includens* larva, CO_2 equal to 32% of total consumption up to the sixth instar and 37% up to pupation. They used (^{14}C) glucose to measure the consumption and use by *P. includens* in an artificial diet. This glucose was dissolved in acetone and the solution, with an activity of 2.1 × 10^6 cpm/ml, was added to the artificial medium. Determinations were made during the complete larval development up to pupation with two larvae/rearing recipients. The CO_2 emitted was collected at sites containing 75 ml of carb-sorb. The equipment setup, measurement of the activity in a liquid scintillation counter, and the calculations have been described in detail (Parra and Kogan 1981; Kogan 1986).

2.4.2.3 Uric Acid Method

Bhattacharya and Waldbauer (1969a,b, 1970) used the uric acid method (spectrophotometer-enzymatic method) to measure food intake and utilization. This method is indicated when it is difficult to separate the feces from the medium, such as with stored products pests. The method is based on the fact that uric acid, which has an absorption peak of 292 μm is, in the presence of uricase, oxidized to alantoine, which absorbs less light of the same wavelength. Therefore, the concentration of uric acid can be calculated from the reduction in absorbance after treating with uricase. Extraction is made with an aqueous solution of lithium carbonate and the calculation of mg feces in the mixture is given by the formula:

$$\text{mg feces} = \frac{\text{mg uric acid in mixture}}{\text{mg uric acid per mg feces}} \downarrow + \text{feces wt. of sample}$$

Various authors have used this method to measure the food consumption of insects other than stored products pests. Chou et al. (1973) used it to measure food utilization by *Argyrotaenia velutinana* (Walker) and *H. virescens* and Cohen and Patana (1984) used it for *H. zea*.

2.4.2.4 Trace Element Method

The trace element method is a qualitative method that can, together with quantitative methods, be used in nutritional studies. Rubidium and cesium are used to mark insects in ecological studies (Berry et al. 1972; Stimman 1974; Shepard and Waddill 1976; van Steenwyk et al. 1978; Alverson et al. 1980; Moss and van Steenwyk 1982). These elements are rapidly absorbed by plant tissue and transferred to the insect through feeding. They can be detected by atomic absorption spectroscopy and knowing the concentration of the trace element in the food, insect, and feces, indices can be determined as in the chromic acid method (colorimetric method).

2.4.2.5 Immunological Method

The immunological method was used by Lund and Turpin (1977) to determine the consumption of *Agrotis ipsilon* larvae by carabids, Sousa-Silva (1980) to evaluate consumption of *D. saccharalis* larvae by predators, and Sousa-Silva (1985) in studies with *Deois flavopicta* (Stal) (Homoptera). Calver (1984) revised immunological techniques to identify diets.

2.4.2.6 Calorimetric Method

Food use can be determined based on the caloric equivalent instead of units of mass (Schroeder 1971, 1972, 1973, 1976; Stepien and Rodriguez 1972; Van Hook and Dodson 1974; Bailey and Mukerji 1977; Slansky 1978). Loon (1993), using the calorimetric method observed that *Pieris brassicae* L. (Lepidoptera: Pieridae) reared on artificial diet, developed with a greater metabolic efficiency than when reared on the host plant, *Brassica oleracea*. These differences were not detected when he used the gravimetric method. According to the author, metabolic efficiencies derived from calculations from gravimetric data are subject to random errors that distort the determination of the metabolic efficiency in plant studies. The heat of combustion of larvae, feces, and the medium is determined in a calorimeter using oxygen. This combustion heat is defined as being the energy liberated as heat when a substance is completely oxidized to CO_2 and H_2O. Waldbauer (1968) proposed the following indices:

Coefficient of metabolized energy (CME)

$$\text{CME} = \frac{\text{Gross energy in food consumed} - \text{gross energy in feces}}{\text{Gross energy in food consumed}}$$

Storage efficiency of energy ingested (ESI) (E)

$$\text{ESI (E)} = \frac{\text{Gross energy stored in body}}{\text{Gross energy in food consumed}} \times 100$$

TABLE 2.3

Comparison of the Use of Dry Material and Energy by *Bombyx mori*

Up to the End of the Fifth Instar				Up to the Recently Emerged Adult			
Dry Weight (mg)		**Energy (cal)**		**Dry Weight (mg)**		**Energy (cal.)**	
AD	37	CME	42	AD	37	CME	42
ECI	23	ESI (E)	28	ECI	8	ESI (E)	12
ECD	62	ESM (E)	67	ECD	22	ESM (E)	28

Instar	CME	ESI (E)	ESM (E)
I	52	32	61
II	49	29	90
III	42	28	66
IV	44	29	66
V	42	28	66

Source: Hiratsuka, E., *Bull. Seric. Exp. Sta.*, 1, 257–315, 1920; Waldbauer, G. P., *Adv. Insect Physiol.*, 5, 229–88, 1968.
Note: The large quantity of energy stored in the fifth instar is to supply the pupal and adult stages, which do not feed.

Storage efficiency of metabolized energy (ESM) (E)

$$\text{ESM (E)} = \frac{\text{Gross energy stored in body}}{\text{Gross energy in food consumed} - \text{gross energy in feces}} \times 100$$

Although energy use is greater than that of dry matter, both determinations are comparable (Table 2.3).

Slansky (1985) referred that more than 80% of the values for AD (CME), ECI (ESI), and ECD (ESM) calculated using energy values are greater than these indices calculated on the basis of dry weight, based on data from more than 65 species. The greatest values for AD based on energy are due to the large energy content of food and feces and those for ECI and ECD are due to the large energy content of insect biomass in relation to the food assimilated and ingested. The sources of error involved in the conversion of dry weight into energy are discussed (Slansky 1985).

Loon (1993) observed that the ECD calculated by the gravimetric method from feeding by *Pieris brassicae* (L.) was 58.34% and 57.10% in an artificial diet and a natural diet of *Brassica oleraceae*, respectively, with no difference between the values. On the other hand, when the respirometer method (calorimeter) was used, there were differences in the ECD values for the two substrates (9.19% and 11.72%, respectively), showing a limitation of the gravimetric method for studying phytophage food intake and utilization on plants.

Parra and Kogan (unpublished) observed a large quantity of residues of Si and Mn, probably originating from the wire used in this method's combustion process, as well as Mg, Al, and Ca from feces and the artificial medium (Table 2.4). Since the amount of residues was high (4.74% in artificial diet and 9.21% in feces), the variations in AD, ECI, and ECD between the gravimetric and calorimetric methods may be attributed to these residues in the feces and in the artificial diet. These elements were included in the gravimetric method and excluded from the calorimetric analysis. Besides this, there is usually a loss of

TABLE 2.4

Residues of the Combustion of Feces and Artificial Diet of *Pseudoplusia includens* Analyzed by Jarrel-Ash Plasma Atomcomp Model 975, Compared with the Wire Used to Measure the Combustion Heat

	Quantity (mg/g)				
	Si	**Mg**	**Al**	**Mn**	**Ca**
Feces	15.6	14.0	5845	535.7	154.7
Artificial diet	17.7	16.4	5583	618.8	145.2
Wire	23.4	62.6	924	1.7	424.0

Source: Parra and Kogan, unpublished.

TABLE 2.5

Gross Energy (Calories) in Food Consumed, in Feces, and Stored in the Body of the *Pseudoplusia includens* Caterpillar

Stage	Gross Energy in Food Consumed	Gross Energy in Feces	Gross Energy Stored in the Body
I–VI instar	1107.48 ± 119.47	429.80 ± 55.82	398.31 ± 25.65
I–Pupation	1251.98 ± 83.26	516.03 ± 43.59	365.75 ± 37.35

Source: Parra and Kogan, unpublished.

calories corresponding to the loss of lipids during the preparation of the pellets in the calorimetric process. In general, these lipids are not extracted before pelletizing (Schroeder 1972, 1973).

The values for AD, ECI, and ECD in Parra and Kogan's unpublished study based on gravimetric and calorimetric methods decreased when measured until pupation, compared to measurements made until the sixth instar (Table 2.5). The high quantity of stored energy up to the final instar is related to the gross energy stored for the pupal stage. The decrease in total energy was due to pupal metabolic activity that is not compensated by additional food "consumption." However, coinciding with results quoted in the literature, the nutritional indices obtained in the study were superior to those obtained gravimetrically (Table 2.6).

In studies of ecological energetics, some symbols of energy balance are used, based on Klekowski (1970) (cited by Stepien and Rodriguez (1972)). Thus, $C = P + R + U + F = D + F$, $D = P + R + U$, and $A = P + R$ [C = food consumption, P = production (body, exuvia, products from reproduction), R = respiration, U = urine and digestion residues, F = nonabsorbed part of consumption, D = digestion (part of the digested and absorbed food), FU = when difficult to separate F from U (considered together in these cases), A = assimilation (sum of production and respiration, food absorbed less feces)].

An organism's efficiency in using energy is evaluated by the following indices:

$$U^{-1} = \frac{P + R}{C} = \frac{A}{C} = \text{assimilation efficiency};$$

$$K_1 = \frac{P}{C} = \text{efficiency of use of energy consumed for growth (index of efficiency of gross production) and}$$

$$K_2 = \frac{P}{P + R} = \frac{P}{A} = \text{efficiency of use of energy for growth (efficiency index of net production)}.$$

Energy efficiency within and between trophic levels, including the determination of lipids, respiration, and the energy content of biological materials can be very useful for refining details in studies involving nutritional ecology (Slansky 1985). Details of these calorimetric measurements can be found in Petrusewicz and MacFadyen (1970) and Southwood (1978).

TABLE 2.6

Comparison of the Values for AD, ECI, and ECD Determined by the Gravimetric and Calorimetric Methods

Pseudoplusia includens		Indices		
Period	Method	AD	ECI	EDC
I–VI instar	G	56	25	44
	C	61	36	60
I–Pupation	G	52	22	43
	C	59	29	50

Source: Parra and Kogan, unpublished.
Note: G = Gravimetric, C = calorimetric.

According to Waldbauer (1972), the indices used in ecological energetics correspond to those used by insect nutritionists. Thus, U^{-1} is equivalent to AD, K_{-1} to ECI, and $K_{2\,to}$ ECD. The ecologists calculate the caloric values of R (respiration) (using respirometers) from the oxygen consumption of the study organism. This R includes the energy spent in metabolism and activity, and also the energy lost in the urine. R can be determined gravimetrically since it is equivalent to the caloric content of ingested food less the caloric content of the feces (Waldbauer 1972).

2.5 Comparison of Methods

The indirect methods are more sophisticated and are discussed by Waldbauer (1968), Parra (1980), and Kogan (1986). A comparative study between them was done by Parra and Kogan (1981) and Kogan and Parra (1981). A comparison of the precision of the results is shown in Table 2.7 and general characteristics of the different methods in Table 2.8. The time needed to process samples in the indirect methods varied from six (radioisotopes) to 18 times (Cr_2O_3) compared to the gravimetric method. All the indirect methods require a balance as well as other equipment for specific determinations. No greater precision was obtained from indirect methods, and with the calorimetric method where calco oil red was used, it was also observed that this dying agent affected insect development when added to the diet. Therefore, based on this study, the gravimetric method is the most suitable and cheapest of those studied. There are specific cases, such as for stored products pests, where indirect methods are preferable, since the separation of feces and food is impractical. In these cases, the uric acid method must be used (Bhattacharya and Waldbauer 1969a,b).

Kogan & Parra (1981) indicated the main sources of variation in these types of experiments: (1) individual insect variability in a population, (2) variations in diet humidities, (3) behavioral feeding

TABLE 2.7

Precision of Measurements of ECI and ECD for *Pseudoplusia includens* Reared on Artificial Diets by Five Methods[a]

Method	Precision (%)	
	ECI	ECD
Gravimetric	85.7	85.7
Colorimetric (Calco oil red) (COR)	33.3	34.7
Colorimetric[b] (Cr_2O_3)	80.0	82.1
Radioisotope	60.0	19.4
Calorimetric	88.9	80.0

Source: Parra, J. R. P., and M. Kogan, *Entomol. Exp. Appl.*, 30, 45–57, 1981.

[a] Precision = (1–standard deviation/mean) × 100.

[b] Done by atomic absorption.

TABLE 2.8

General Characteristics and Costs of Five Methods for Measuring the Food Intake and Utilization (of Artificial Diet) by *Pseudoplusia includens*

Method	Specimen	CO2	Diet	Cost	
				Equipment (U.S. $)	Time
Gravimetric	Live or dead	No	Natural or artificial	$1,300.00	5 min/individual
COR	Dead	No	Artificial	$1,500.00	1 hour/individual
Cr_2O_3	Dead	No	Artificial	$7,000.00	1.5 hour/individual
Radioisotope	Dead	Yes	Natural or artificial	$10,000.00	30 min/individual
Calorimetric	Dead	No	Natural or artificial	$5,000.00	1 hour/individual

Source: Parra, J. R. P., and M. Kogan, *Entomol. Exp. Appl.*, 30, 45–57, 1981.

differences resulting from adding components (coloring agents, chemical substances) to diets, (4) differences in diet utilization after ingestion, and (5) differences in sample handling.

Studies on food consumption and use based on nutritional indices conducted in Brazil are listed (Table 2.9). They include comparison of food substrates (natural and artificial), effect of pathogens or natural enemies on quantitative nutrition, and effect of different temperatures on nutrition and feeding behavior in a host.

TABLE 2.9

Some Studies Carried Out in Brazil Regarding Insect Food Intake and Utilization with the Indices That Were Determined

Insect	Host	Indices						Reference
		RCR	RMR	RGR	AD	ECI	ECD	
Eacles imperialis magnifica	Coffee	x	–	x	x	x	x	Crócomo and Parra (1979)
Spodoptera latifascia	Cotton, lettuce, soybeans	–	–	–	x	–	–	Habib et al. (1983)
Spodoptera frugiperda	Artificial diets	x	–	x	x	x	x	Susi et al. (1980)
Anticarsia gemmatalis	Artificial diets	x	–	x	x	x	x	Silva and Parra (1983)
Agrotis subterranea	Kale	x	–	x	x	x	x	Vendramim et al. (1983)
Alabama argillacea	Cotton	x	–	x	x	x	x	Carvalho and Parra (1983)
Lonomia circumstans	Coffee	x	–	x	x	x	x	D'Antonio and Parra (1984)
Heliothis virescens	Artificial diets	–	–	–	x	x	x	Mishfeldt et al. (1984)
H. virescens	Cotton	x	–	x	x	x	x	Precetti and Parra (1984)
S. frugiperda	Artificial diets	–	–	–	x	x	x	Parra and Carvalho (1984)
Erinnyis ello ello	Rubber	x	–	x	x	x	x	Reis F° (1984)
S. frugiperda	Corn, wheat, sorghum	x	x	x	x	x	x	Crócomo and Parra (1985)
Diatraea saccharalis	Artificial diets	–	–	–	x	x	x	Misfheldt and Parra (1986)
D. saccharalis	Artificial diets	–	–	–	x	x	x	Martins et al. (1986)
S. frugiperda	Artificial diets	x	–	x	x	x	x	Genthon et al. (1986)
Apanteles flavipes	*D. saccharalis*	–	–	–	x	x	x	Pádua (1986)
S. eridania	Sweet potato and *Mimosa scabrella*	x	x	x	x	x	x	Matana (1986)
D. saccharalis	Artificial diets	x	–	x	x	x	x	Almeida (1986)
Pseudaletia sequax	Artificial diets	x	–	x	x	x	x	Salvadori (1987)
A. gemmatalis	Soybeans	x	x	x	x	x	x	Zonta (1987)
S. frugiperda	Corn	x	x	x	x	x	x	Oliveira (1987)
A. gemmatalis	Soybeans	x	x	x	x	x	x	Gamundi (1986)
S. frugiperda	Corn	x	x	x	x	x	x	Nalim (1991)
D. saccharalis, H. virescens and S. frugiperda	Artificial diets				x	x	x	Souza et al. (2001)
S. frugiperda	Corn	–	–	–	x	x	x	Fernandes (2003)

Note: x = determined, – = not determined.

2.6 Interpretation of Nutritional Indices Values

The interpretation of results from quantitative nutritional research based simply on nutritional indices is not easy. In general, the highest indices indicate a greater nutritional suitability but the presence of allelochemicals, or even the interaction between nutrients and allelochemicals, can lead to erroneous results or interpretations. Sometimes, certain factors can cause lower digestibility that can result in the food being consumed in large quantities but with low growth rates. Besides this, the insect often shows a capacity to compensate a low consumption through greater use of the food. All these factors can alter the values of the nutritional indices cited and make their interpretation difficult. Thus, it is often necessary to associate the index values obtained with biological data from different food substrates or even based on data on insect behavior. In this case, other methods, such as cluster analysis, should be used (Kogan 1972; Parra and Carvalho 1984; Precetti and Parra 1984). Obviously, there are rare cases in which a simple analysis involving a test to compare means is sufficient to arrive at a satisfactory conclusion. The analyses done for the indices are based on the supposition that there is an isometric relationship between the variables in the numerator and the denominator, which does not always occur in biology. Raubenheimer and Simpson (1992) proved that when the relationship between the numerator and denominator of a nutritional index is not linear, the statistic F and its level of significance are altered, which can compromise the conclusion of a nutritional study. Another consequence of this fact is that the statistical power of the Tukey test to detect small treatment differences is much reduced using the indices. Since the interactive effect of the denominator and treatment are not measured in the analysis of the indices, the conclusions about the treatment effects are compromised. In spite of the advance of research on nutritional ecology (Slansky and Rodriguez 1987a), many conclusions are speculative today and need further studies to be properly supported.

Raubenheimer and Simpson (1992) presented a covariance analysis as an alternative for comparing treatments, considering one of the nutritional food intake and utilization indices (RCR, RGR, ECI, AD, and ECD) and indicated the RCR as the ideal index. The analysis of covariance (ANCOVA) better satisfies the statistical demands, supplying important information about data that is neglected by using a conventional form of analysis of variance that can lead to errors in evolutionary biology, morphometry, systematics, physiology, and plant ecology. Horton and Redak (1993) discuss the care that has to be exercised when using the ANCOVA. They further suggest that the ANCOVA could be used to evaluate the effect of the larval diet on adult fecundity after adjustments for larval consumption, the effects of the adult diet on fecundity after adjustments for food consumption, or the effects of the larval diet on adult size after adjustments for consumption in the larval stage. The ANCOVA has many advantages when using indices in biological data, including increased power of tests of hypotheses, more information on data groups, greater reductions in the error of the dependent variable, and greater reduction in the incidence of untrue treatment effects. Besides this, it analyzes the interaction of the dependent variable and treatments.

An example of the values of RGR, RCR, AD, ECI, and ECD of a hypothetical consumption of a noctuid lepidopteran on two artificial diets is shown in Table 2.10. Considering the RCR index in the tables referred to from the conventional analysis of variance and from the ANCOVA. From these results, diets A and B differ statistically (P = 0.0096) when the ANCOVA is used and do not differ (P = 0.1280) when analyzed using Tukey's *t*-test. The SAS program for covariance analysis is presented following Table 2.10.

2.7 Food Consumption and Use for Growth in the Larval Phase

Immatures tend to choose an appropriate food to consume it in balanced proportions to promote optimum growth and development, making originating adults reproductively competitive. This choice involves adaptations and strategies for each species including capacity to compensate for unsuitable conditions.

TABLE 2.10

Values of AD, ECI and ECD, RCR, RGR, Independent Variable Y and Covariance (X) of a Hypothetical Noctuid Lepidopteran on Two Artificial Diets (A and B), with the Respective Analyses of Covariance and Variance for RCR

			Diet A			
AD%	ECI%	ECD%	RCR	RGR	Y	COV (X)
42.59	16.91	39.71	0.5914	0.1	0.7333	1.2400
40.03	16.72	41.77	0.5981	0.1	0.8822	1.4750
39.63	16.14	40.73	0.6196	0.1	0.8365	1.3500
43.19	20.09	46.52	0.4978	0.1	0.8536	1.7150
43.78	18.99	43.39	0.5265	0.1	0.7897	1.5000
53.03	16.67	31.44	0.5998	0.1	0.6898	1.1500
50.54	15.08	29.84	0.6631	0.1	0.9483	1.4300
47.53	28.97	60.96	0.3451	0.1	0.6385	1.8500
43.86	18.42	42.00	0.5428	0.1	0.6947	1.2800
42.30	17.14	40.52	0.5835	0.1	0.8752	1.5000
42.42	18.45	43.49	0.5421	0.1	0.9378	1.7300
49.40	17.05	34.51	0.5866	0.1	0.7332	1.2500
63.34	28.00	44.22	0.3571	0.1	0.5428	1.5200
41.28	19.52	47.29	0.5123	0.1	0.5943	1.1600
41.17	21.32	51.79	0.4691	0.1	0.6051	1.2900

Y = independent variable = portion of food ingested by insect = numerator of RCR.
Covariable (X) = (dry weight of insect/2) * experimental time = denominator of RCR.

			Diet B			
AD%	ECI%	ECD%	RCR	RGR	Y	COV (X)
99.03	37.65	38.02	0.2656	0.1	0.1859	0.7000
41.76	46.84	112.16	0.2135	0.1	0.7899	3.7000
41.16	16.00	38.87	0.6250	0.1	0.7563	1.2100
34.52	14.50	41.99	0.6898	0.1	0.8140	1.1800
43.94	16.23	36.94	0.6160	0.1	0.7885	1.2800
44.21	16.21	36.66	0.6170	0.1	0.7528	1.2200
40.66	16.46	40.48	0.6075	0.1	0.6926	1.1400
81.70	14.23	17.42	0.7026	0.1	0.6393	0.9100
60.33	11.97	19.83	0.8357	0.1	0.3050	0.3650
51.01	17.25	33.81	0.5798	0.1	0.7654	1.3200
85.35	16.79	19.67	0.5955	0.1	0.6074	1.0200
51.15	10.67	20.86	0.9372	0.1	0.2109	0.2250
96.22	11.20	11.64	0.8931	0.1	0.2590	0.2900

Y = independent variable = portion of food ingested by insect = numerator of RCR.
Covariable (X) = (dry weight of insect/2) * experimental time = denominator of RCR.

Analysis of variance and Tukey test for data on nutritional index RCR for diets A and B.

Causes of Variation	GL	Sum of Squares	Mean Squares	F-Test	Pr > F
Diet	1	0.06080541	0.06080541	2.47	0.1280
Residual	26	0.63968088	0.02460311		
Total	27	0.70048629			

(continued)

TABLE 2.10 (Continued)

Values of AD, ECI and ECD, RCR, RGR, Independent Variable Y and Covariance (X) of a Hypothetical Noctuid Lepidopteran on Two Artificial Diets (A and B), with the Respective Analysis of Covariances and Variance for RCR

Groups	Mean	N	Diet
A	0.62910	13	B
A	0.53566	15	A

Means followed by the same letter do not differ statistically (P ≤ 0.05)
Analysis of Covariance (ANCOVA) for the data of the nutritional index RCR for the diets A and B.

Causes of Variation	GL	Type I Error	Mean Squares	F-Test	Pr > F
Treatments (Diets)	1	0.21309866	0.21309866	7.92	0.0096
COV (X)	1	0.31338411	0.31338411	11.65	0.0023
COV*DIET	1	0.00017631	0.00017631	0.01	0.9362

SAS Program (9.1) for Analysis of Covariance (ANCOVA) of nutritional indices

```
data ENTO;
input TREAT $ NU CO;
datalines;
DA 0.73333 1.24
DA 0.8822 1.4750
DA 0.8365 1.35
DA 0.8536 1.715
DA 0.7897 1.5
DA 0.6898 1.15
DA 0.9483 1.43
DA 0.6385 1.85
DA 0.6947 1.28
DA 0.8752 1.5
DA 0.9378 1.73
DA 0.7332 1.25
DA 0.5428 1.52
DA 0.5943 1.16
DA 0.6051 1.29
DB 0.1859 0.7
DB 0.7899 3.7
DB 0.7563 1.21
DB 0.8140 1.18
DB 0.7885 1.28
DB 0.7528 1.22
DB 0.6926 1.14
DB 0.6393 0.91
DB 0.3050 0.365
DB 0.7654 1.32
DB 0.6074 1.02
DB 0.2109 0.225
DB 0.2590 0.29
;proc print;
run;
ods html;
   ods graphics on;
proc glm data=ENTO;
```

(continued)

TABLE 2.10 (Continued)

Values of AD, ECI and ECD, RCR, RGR, Independent Variable Y and Covariance (X) of a Hypothetical Noctuid Lepidopteran on Two Artificial Diets (A and B), with the Respective Analyses of Covariance and Variance for RCR

```
class TRAT;
model NU=TRAT CO TRAT*CO;
lsmeans TRAT/adjust=tukey pdiff;
run;
    ods graphics off;
    ods html close;
```

Where:
  NUN = numerator of RCR or independent variable Y
  CO = denominator of RCR or covariable (X)
  TREAT = treatment, in this case diet A and diet B

2.7.1 Number of Instars

The number of instars is constant and varies in most insects from four to eight. However, there are some Odonata that have 10 to 12 ecdyses and some Ephemeroptera that have 20 or more (Table 2.11). There are various rules that try to forecast the degree of insect growth, such as from Dyar's rule (Dyar 1890; the cephalic capsule of Lepidoptera larvae grows in geometric progression, increasing in width at each ecdysis, at a constant rate for a certain species and on average 1.4), which is valid for many Lepidoptera, Archaeognata, Hymenoptera, Coleoptera, and Hemiptera. Other rules, such as from Przibram (Batista 1972), are postulated originating from the supposition that insect growth is harmonic. Since this growth is generally nonharmonic, heterogenic, or allometric, this rule is not applicable because according to it, "at each ecdysis there should be an increase of each body part in the same proportion as the whole body." In studies done with 105 insect species, Cole (1980) showed that at each ecdysis all the linear dimensions are increased by 1.52 and 1.27 times, respectively, for holometabolic and hemimetabolic insects. There are various factors besides those intrinsic to the species, which cause a variation in the number of instars, such as hereditary factors (Albrecht 1955; Moreti and Parra 1983) (Table 2.12), rearing method (crowded or isolated) (Long 1953; Peters and Barbosa 1977), temperature (Ferraz et al. 1983; Kasten and Parra 1984) (Table 2.13), nutrition (Parra et al. 1977; Reis 1984; Matana 1986) (Table 2.14), sex (Roe et al. 1982), and parasitism (Reynolds et al. 1984; Orr and Boethel 1985).

There is no direct correlation between the duration of the life cycle and instar number (Slansky and Scriber 1985), and depending on the insect's habits a change in instar may be necessary. Thus, an insect that wears down its mandibles when feeding may need a more constant change (Slansky and Rodriguez 1987b) compared to another one that feeds on more tender food. An insect that needs to maintain its agility in each instar cannot increase its weight very much. Thus, in order not to follow the normal sequence of weight gain during the instar (Figure 2.4), the insect tends to have ecdyses at shorter intervals (Daly 1985). In unfavorable conditions, an insect tends to have more instars (Roe et al. 1982; Nealis 1987; Parra et al. 1988).

Females, due to their reproductive activity, are generally bigger, with a longer development time, and therefore may have an additional instar (Slansky and Scriber 1985). Besides this, males tend to be born beforehand in order to facilitate mating (protandry). Larger size differences are observed between the sexes in those species whose adults do not feed. Weight is at least doubled at each instar and those larvae that are not mobile (Lepidoptera larvae) have larger increments than those that have to move around to find food (e.g., certain beetles, cockroaches) (Capinera 1978; Vendramim et al. 1983).

TABLE 2.11

Number of Larval Instars for Different Insect Orders

Insect Order	Number of Instars
Archaeognata	10–14
Zygentoma (or Thysanura)	9–14
Ephemeroptera	10–40
Odonata	10–12
Blattaria	(3) 6–10 (8)
Mantodea	5–9
Grylloblattodea	8
Orthoptera	5–11
Phasmida	8–12
Isoptera	5–11
Dermaptera	4–6
Embioptera	4–7
Plecoptera	22–33
Zoraptera	–
Heteroptera	(4) 5 (9)
Homoptera	3–5
Thysanoptera	5–6
Psocoptera	6
Phthiraptera	3–4
Strepsiptera	7
Coleoptera	3–5 (10)
Raphidioptera	3–4
Megaloptera	10
Neuroptera	3–5
Mecoptera	4
Siphonaptera	3
Diptera	3–6
Trichoptera	5–7
Lepidoptera	(3) 5–6 (11)
Hymenoptera	3–6

Source: Sehnal, F., In *Comprehensive Insect Physiology Biochemistry and Pharmacology*, ed. G. A. Kerkut, and L. I. Gilbert, 1 86, v. 2, Pergamon Press, Oxford, UK, 1985.

TABLE 2.12

Percentage of Larvae of *Heliothis virescens* That Reached the Sixth Larval Instar, Reared on Cotton Leaves (IAC-17) for Four Successive Generations

Generation	Male (%)	Female (%)
F_1	63.0	33.3
F_2	95.0	95.0
F_3	100.0	88.9
F_4	100.0	100.0

Source: Moreti, A. C. C., and J. R. P. Parra, *Arq. Inst. Biol.*, 50, 7–15, 1983.

Note: Temperature = 24°C ± 2°C, relative humidity = 65% ± 5%, 14 h photophase.

TABLE 2.13

Effect of Temperature on the Number of Insect Instars

Species	Temperature				Reference
	20°C	25°C	30°C	35°C	
Spodoptera frugiperda	7	6	6	6	Ferraz et al. (1983)
Alabama argillacea	6	6	5	5	Kasten and Parra (1984)

TABLE 2.14

Effect of Nutrition on the Number of Instars in Two Lepidopteran Species

Species	Host		Reference
Spodoptera eridania	Cotton	Soybeans	Parra et al. (1977)
Number of instars	6	7	
Spodoptera eridania	Sweet potato	*Mimosa scabrella*	Matana (1986)
Number of instars	6	7	

2.7.2 The Cost of Ecdysis

The molting process has a high energy cost and the caloric and nutritional content of a molted cuticle may represent >20% of total larval biomass production. The insect often compensates for this loss by reabsorbing the internal layers of the old cuticle before ecdysis and consuming (and even digesting) parts of the cuticle. About 33% of the lipids accumulated by the penultimate nymphal instar of *Acheta domesticus* (L.) (Orthoptera) are metabolized at ecdysis to the last instar and these lipids are only reconstructed on the second day after molting. From 19% to 34% of existing lipids in the "premolts" are used in the four ecdyses of *B. mori*, as well as 65% to 73% of carbohydrates existing in the premolts are used during ecdyses (Hiratsuka 1920). Therefore, to grow, increase weight, and accumulate energy reserves, insects need to alter body composition or make better use of food.

2.7.3 Food Intake and Utilization through Instars

The rates of consumption, metabolism, and growth tend to reach a peak at the beginning or near the middle of the instar, and efficiencies tend to decrease (Waldbauer 1968; Scriber and Slansky 1981; Crócomo and Parra 1985). There is a tendency to accumulate lipids from the first to the last instars, especially in the Holometabola that use energy to produce the cocoons. About 30% of the energetic content of the last larval instar of *B. mori* is used to make the cocoon (Hiratsuka 1920). Lipid accumulation also occurs in the Hemimetabola but in this case, instead of the accumulation happening via cocoon as in the Holometabola, it occurs in the last larval instar. There are cases of lipid accumulation that are not governed by this general rule when the insect enters diapause.

For the adult insect to be reproductively competitive, there are two larval characteristics that must be satisfied: the size, which can influence the choice for mating and its success as well as the capacity to disperse, and the weight, which is indicative of the nutrients and energy stored, and that influence the search for mating, dispersal flights, and fecundity. The minimum size and weight depend on the species lifestyle, environmental conditions, food availability, and neurohormonal control. A list of minimum weights that permit the pupation of different species includes values for Lepidoptera, from 13% to 26% (in dry weight matter) and from 25% to 60% (in fresh weight matter), in relation to the normal species weight (Slansky and Scriber 1985).

Food consumption of the last two instars is at least 75% of total (Waldbauer 1968) (Table 2.1). In this way, given the difficulties of separating the feces and even detecting the weight gain or food consumed (depending on the insect size) in the first instars, it can be said that the determination of nutritional indices only in these instars is sufficient for many types of study. In general, the relative indices tend to decrease from the first to the last instars due to the lipid reserves (less metabolic activity). Since females

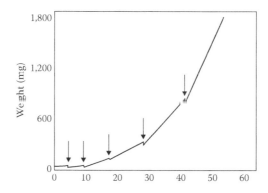

FIGURE 2.4   Standard weight increases in grasshoppers of the *Locusta* genus. (From Chapman, R. F., *The Insects: Structure and Function*, Harvard University Press, Cambridge, MA, 1982.)

are generally bigger, they consume more food to accumulate eggs and also because they have a longer development time, and in many cases, an extra instar. Despite this, the differences between the sexes in the efficiency of food use are small (Slansky and Scriber 1985).

2.8 Adult Food Consumption and Use for Reproduction and Dispersal

The main function of the adult is reproduction, and in many cases, dispersal. These functions depend on the interaction and integration of physiological processes and behaviors that are intimately correlated with food consumption and utilization. The production of eggs or progeny involves energy and nutrient accumulation by the female, which makes her consume more and gain more weight than the males. Egg production is affected by biotic and abiotic factors acting directly on adult performance and indirectly on larval development. Some components of the reproductive process and its relationship with food consumption and utilization are discussed by Slansky and Scriber (1985). Mating attraction and acceptance can depend on pheromone production that can be influenced by the absorption of pheromone precursors. Mating access and acceptance, which depend on body size, can be influenced by food in the larval stage and food quality can also affect this acceptance. In mating, the male can contribute nutritionally through secretions of accessory glands and spermatophores. For ovogenesis and oviposition, nutrient accumulation by larvae, food quality and quantity in the adults, the amount of nutrient deposited by the female in each egg, and the presence of suitable larval food as a stimulant for oviposition, can all be important.

2.8.1 Food Quality

Food quality depends on physical attributes (e.g., hardness, surface hairiness, shape) that influence insect capacity to consume and digest the food as well as allelochemicals and nutritional components. Allelochemicals, such as alkaloids, cyanogenic glicosides, glucosinolates, lignins, protein inhibitors, tannins, terpenoides, lipids and toxic amino acids, and hormones and antihormones, can act as food attractants and stimulants or as deterrents and repellents (Kogan 1977; Norris and Kogan 1980; Berenbaum 1985; Ishaaya 1986). The nutrients have already been fully discussed and in order to obtain nutrients in balanced proportions for optimal growth and development, the insect makes interconversions or syntheses, excretions, and selective concentrations or often counts on the fundamental help of microorganisms. The impact of the different aspects of food quality varies within and between different food categories (guilds). The amounts of water and nitrogen are fundamental for evaluating this behavior. Chewing insects show the best performance for food with high values of nitrogen (N) and water. Mattson (1980) (cited by Hagen et al. 1984), on correlating the ECI with N concentration for many herbivores found that this efficiency varied from 0.3% to 58%. The lowest values (1%) are associated with aquatic insects or terrestrial insects that feed on wood poor in N, litter, and detritus. The highest values (40% to 50%) were for those insects that feed on seeds, phloem sap, pollen, and nectar. The biggest conversion (more than 50%) was registered for

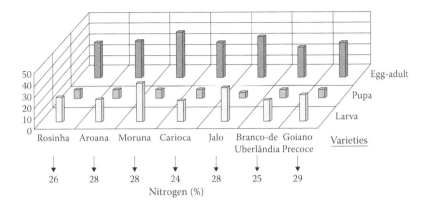

FIGURE 2.5 Duration of the larval, pupal, and total life cycle phases (egg, adult) of the lepidopteran, *Spodoptera frugiperda*, in seven artificial media and the respective percentages of nitrogen per variety used, temperature: 25°C ± 1°C, relative humidity: 70% ± 10%, 14h photophase. (From Parra, J. R. P., and S. M. Carvalho, *An. Soc. Entomol. Brasil*, 13, 305–19, 1984.)

parasitoids and predators. This author concludes that organisms feeding on nitrogen-poor diets consume more food than those who feed on nitrogen-rich diets. It is the amount of available N that limits insect growth, development, and fecundity. Thus, Parra and Carvalho (1984) observed that there was no correlation between *S. frugiperda* development and the total existing protein in dry bean varieties used in its artificial diet. The insect developed better on the diet in which N was supposedly more available (Figure 2.5). A list of the N content of different insect foods was compiled by Slansky and Scriber (1985). This N is variable in quantity and quality depending on the nutritional source (e.g., leaf, fruit, nectar, pollen, wood, and detritus). The content varies from 0.08% to 7% (in dry weight) depending on the plant part and the phase of the plant cycle. The highest concentrations are registered in new growing tissues and in propagules such as seeds and bulbs. N concentration tends to diminish according to leaf age, down to 0.5% at abscission. There is more N in the phloem than the xylem although the sap has low N levels (0.0002%–0.6%). The nitrogen fixers (legumes and nonlegumes) show variations of 2% to 5% (in dry weight). Gymnosperms have half the N of angiosperms, that is, 1% to 2% and 2% to 4%, respectively. Pollen and nectar are rich sources of N.

The limits between food quality categories are not so distinct. Thus, the amino acid L-canavanine is toxic to some insects, and therefore functions as an allelochemical. However, for other insects it is a source of N (Rosenthal et al. 1982). This explanation is valid for many phenols (Bernays and Woodhead 1982), and it is evident that these relationships depend on the quantities of these allelochemicals present in the food and on their persistence during use.

Insect performance can be affected by biotic and abiotic factors. For herbivores, plant quality varies with leaf age, the plant's growth conditions (temperature, soil fertility), disease and parasitoid infection, previous damage by other insects, and even chemical action. The influence of pathogens on food consumption and use is discussed by Mohamed et al. (1982) and Sareen et al. (1983), and for parasitoids by Slansky (1978),

TABLE 2.15

Effect of Parasitism by *Cotesia flavipes* on the Consumption and Use by Larvae of *Diatraea saccharalis* Reared on Artificial Diet

	Nutritional Indices		
Caterpillars	**AD**	**ECI**	**ECD**
Not parasitized	82.42a	13.47a	16.38b
Parasitized	70.28b	14.10a	20.74a

Source: Pádua, L. E. M., Ph.D. Thesis, University of São Paulo, Piracicaba, Brazil, 1986.

Note: Temperature = 25°C, relative humidity = 70% ± 10%, photophase = 14 h. Means followed by the same letter on the same column do not differ significantly, based on the Tukey test (P ≥ 0.05), for parasitized and not parasitized D. saccharalis by C. flavipes.

Brewer and King (1980, 1982), Slansky (1986), and Pádua (1986) (Table 2.15). The effect of temperature on nutritional indices has also been reported (Bhat and Bhattacharya 1978; Almeida and Parra 1988), the influence of fertility on consumption and use (Al-Zubaidi and Capinera 1984; Oliveira 1987), the effect of physiological stress on the quantitative nutrition of *Agrotis ipsilon* (Hufnagel) (Schmidt and Reese 1988), and the effect of diflubenzuron and its analogue, trifluron, on *Spodoptera littoralis* (Boisduval) (Radwan et al. 1986).

2.8.2 Food Selection and Acceptance

The semiochemicals (intra- or interspecific) are involved in the physiological or behavioral interactions between organisms. Among the numerous semiochemicals that insects respond to, many are associated with plants, and Fraenkel (1953) called attention to the secondary substances (allelochemicals). Thus, certain allelochemicals (alomones) protect plants from herbivores or pathogens, avoiding oviposition, reducing feeding and digestive processes and modifying food assimilation. Kairomones, on the other hand, favor insects, attracting them, stimulating them to oviposit and feed, and to use such compounds as precursors for hormones, pheromones, and alomones. An allelochemical may be a deterrent for one species and phagostimulant for another species. Thus, a substance that is a deterrent for an insect generalist can be a stimulant for a specialist. The main chemoreceptors responsible for food rejection or acceptance are located in the maxillary palps. Food characteristics are taken into consideration at feeding (e.g., color, shape, size, sound, temperature, texture, hardness) and chemical aspects (e.g., smell and taste) (Maxwell and Jennings 1980). The insect feeds on variable quantities of food to obtain different nutrients and also digests and assimilates this food with variable efficiencies. According to Slansky and Scriber (1985) the rate of relative consumption is variable [0.002–6.90 mg (day × mg)] with higher values for the Lepidoptera. It is difficult to remove water from some foods, and in this case, as in stored products pests, the insect drinks free water or absorbs water from water vapor or may produce metabolic water. In other cases, the insect avoids water loss by building a cocoon or wrapping itself in leaves or even lowering cuticle permeability. Often, essential nutrients are unavailable and the insect adapts and obtains them through various processes (Slansky and Rodriguez 1987b).

The synchrony of the life cycle stage with periods when nutrients are more available is one of these processes. Thus, chewing insects feed on new leaves that are nutrient-rich, sucking insects are in synchrony with the emission of plant buds or fruiting, parasitoids are synchronous with their hosts, the activities of bees are in synchrony with flower phenology, and so on. There is a harmony between these synchronisms with photoperiod, temperature, and host hormones. Another process is the modification of food quality. Gall-forming insects (see Chapter 16) alter plant tissue content by forming galls and this often results in an increase, for example, of lipids, influencing plant hormone production, and there are cerambycids that kill off the branches to interrupt nutrient flow and parasitoids that increase the available nutrients of the host hemolymph that stimulates consumption by the host.

There are special conditions of the digestive tract that permit the separation of usually indigestible complexes. Thus, insects that consume tissues with tannins can have an alkaline mesenterum that reduces the formation of indigestible protein/tannin complexes. The gut pH can be important for symbiotic microorganism growth. The cerambycids degrade all classes of structural polysaccharides and for this reason constitute the largest insect family that feeds on wood. On the other hand, the Lyctidae, which do not have the capacity to degrade cell walls, have a much smaller number of species. The synthesis of cellulose (by insects that feed on wood) and keratinase (by insects that feed on keratin-rich compounds) permit the use of unavailable substances. In *Sitophilus oryzae* (L.) (Coleoptera: Curculionidae), amylase activity is greater than in *S. granarius* (L.); for this reason, the first species uses more amide and grows more. Many bruchids have potent amylases and some species use trypsin inhibitors in the seeds as a nitrogen source. Other species leave the proteases of the gut and use free amino acids, thus avoiding the effects of trypsin inhibitors. Another specialization shown by insects is the time food takes to pass through the gut. A longer stay can facilitate nutrient extraction. Wood-feeding termites maintain the food in the gut for 13 to 15 h compared to the 4 to 5 h for a fungus-feeding termite (Slansky and Scriber 1985).

Alternating food is another adaptation shown by insects to obtain a balanced nutrition. There are a small number of species that alternate between unsuitable and suitable foods (Chang et al. 1987). This is the case with aphids, which alternate between herbaceous hosts and trees. This phenomenon also occurs in seed-sucking insects, termites, and soil arthropods. Also, the microorganisms can play an important role acting externally with their own chemical action or supplying more easily digested nutrients concentrated in their own biomass.

Nutrient conservation may also occur. The food assimilated is conserved by the insect digesting and absorbing the internal part of the cuticle, before ecdysis, or even consuming the chorion and the exoskeleton left behind in this process. From 3% to 27% of larval biomass (energy and N) can be lost with cuticle separation. However, digestive enzymes can be absorbed by the digestive tract during metamorphosis. Other types of nutrient conservation involve uric acid metabolism and the use of nutrients that have an allelochemical function. Coprophagy can allow a more complete nutrient use as well as facilitate the consumption of nutrient-rich bacteria.

Another adaptation is the transfer of nutrients between stages of the life cycle. The performance of each life stage basically depends on the success of the previous stage in obtaining, synthesizing, and accumulating nutritional substances in appropriate quantities. This is more evident for stages that do not feed (egg, pupae, and some adults), but the influence of previous stages is also significant in insects that feed. Females of *Aedes aegypti* L. can complete the previtellogenic phase of the ovarian cycle without feeding if they have not been reared in superpopulated conditions as larvae. However, if reared under very crowded conditions, they need blood or sugar to complete this phase (Slansky and Scriber 1985). In some species in which the adult normally feeds during the stage before egg production, a limited number of eggs can be laid if the females do not feed, depending on the micronutrients transferred during metamorphosis. Often, the micronutrients transferred by the egg phase are sufficient to satisfy the needs of the subsequent phases of the life cycle, at least for one generation. Therefore, nutritional studies should be carried out for successive generations. Finally, nutrient transfer between individuals may occur. Included in this category are cannibals (including autoparasitism), the production and consumption of nonfertilized eggs, specialized glandular secretions (e.g., a female consuming, digesting, and absorbing internally the spermatophore and seminal fluid). Trophallaxy in social insects and coprophagy can be considered. These exchanges allow not only exchanges of nutrients but also of symbionts and chemical products associated with caste regulation in social insects.

2.8.3 The Role of Allelochemicals

Allelochemicals play an important role in host selection and are very important in the tropics because this number tends to be higher than in temperate regions (Edwards and Wratten 1981) due to insect pressure throughout the year. However, insects have developed mechanisms to avoid them. Thus, seed-sucking insects avoid the toxins in the seed coat, perforating them with the stylets and feeding only on the cotyledons. Insects that suck the xylem and phloem can avoid allelochemicals in the same way. The coccinellid, *Epilachna tredecimnotata* (Latreille), makes circular holes in the plant in order to avoid deterrent substances produced when the leaf is damaged (Slansky and Scriber 1985).

Enzymatic desintoxication is also used by many arthropods to metabolize allelochemicals and thus avoid their toxicity. Many plant-sucking insects inject detoxifying phenolases with the saliva. Besides this, some insects avoid allelochemicals by producing surfactants or making the digestive tract alkaline or even through rapid excretion. Insects avoid photoactive compounds by feeding at night or even inside the leaves that they roll up (Slansky and Scriber 1985). Fungus-eating insects have certain species of microorganisms whose function is to detoxify allelochemicals. The interaction of nutrients and allelochemicals can affect food suitability. The tannins can block protein availability, forming complexes. Fox and Macauley (1977) found high levels of condensed tannin in some *Eucalyptus* species and low levels in others. The ECI values for *Paropsis atomaria* Olivier (Coleoptera) were similar when the insect fed on different plant species and the authors concluded that tannins and other phenols did not affect nutritional physiology. In some grasshoppers, hydrolyzed tannin is damaging when it passes through the peritrophic membrane but there are no damaging effects in other species (Bernays 1978). To pass through the gut, nutrients must be in a suitable form. Thus, the proteins are broken up into amino acids, due to the proteases that are produced which reduce protein availability. The plants can produce higher levels of these inhibiting enzymes after being attacked by insects and then transfer them to other parts. Green and Ryan (1972) describe this for the coleopteran, *Leptinotarsa decemlineata* (Say).

The allelochemicals, with the dehidroxy-ortho group in the aromatic ring, can quelate essential minerals. Gossypol reduces assimilation of *H. zea* larvae although it has no effect on *H. virescens.* Sinigrine reduces assimilation of *Papilio polyxenes* Fabr. (Lepidoptera) and many other allelochemicals reduce insect growth (Beck and Reese 1976; Berenbaum 1978). Studies on antihormones extracted from plants such as *Ageratum houstonianum* (Bowers et al. 1976), which act in the early stages or in adults, and of juvenile hormone analogues, which act in the last stages of metamorphosis, show an ecological function for these components (Harborne 1982).

Insects can compensate the low nutritional quality by consuming more food (Crócomo and Parra 1985; Simpson and Abisgold 1985) or altering the efficiency of use. Slansky and Wheeler (1992) observed that this attempt to compensate can lead to ingestion of higher doses of allelochemicals for example. Thus, *Anticarsia gemmatalis* Hub. tends to eat more on poor diets and eat more on diets that contain an allelochemical such as caffeine, and there will be changes in food use, growth, and survival. A similar effect was found by Lee et al. (2004) in an unbalanced diet with regard to the ingestion of cellulose by a generalist lepidopteran. Pompermayer et al. (2001) also observed that diet composition is important for the effect of proteinase inhibitors and for *D. saccharalis,* supporting what Broadway and Duffey (1986) and Jongsma and Bolter (1997) had observed previously. Warbrick-Smith et al. (2006) observed that *Plutella xylostella* L. reared for successive generations on a carbohydrate-rich diet progressively developed the ability to eat excess carbohydrate without converting it into fat, showing that the excess storage of fat has an adaptive cost. These studies always need considerable care because the insect can adapt physiologically to the proteinase inhibitors, as seen in *S. frugiperda,* which alters the complement of proteolitic enzymes of the mesenteron (Paulillo et al. 2000); the same happens with *H. virescens* (Brito et al. 2001) and with *S. frugiperda* and *D. saccharalis* (Ferreira et al. 1996).

Chang et al. (2000) demonstrated that a cistein proteinase from corn plants caused a reduction in the efficiency of digestion and absorption of *S. frugiperda* feeding on a natural diet and also on an artificial diet containing this substance; this lower efficiency is due to damage to the peritrophic membrane caused by the enzyme in corn (Pechan et al. 2002). At other times, when the food supply is limited, the insect must search more, increasing its dispersal capacity and even increasing the range of food used. After fasting, enzymatic action is reduced due to a lower metabolic activity. With total fasting, oviposition can cease. When food is limited, the adult can reduce the rate of oviposition (reducing embryogenesis or resorbing eggs). When starving is total, there is an early transformation into pupae, which weigh less. In many cases, the consumption rate can return to normal and a normal insect can result if food is offered after starving. Often the insect promotes changes associated with feeding, such as morphological adaptations (e.g., mouth parts, legs, spine formation), changes in the number of sensors (e.g., number of coeloconic sensilla on the palps), changes in the size of the gut or even internal structures, as in cockroaches, which facilitates the establishment of microorganism colonies. However, the insect shows adaptive strategies that include specializations, suitable development and size, defense, and responses to environmental variations (abiotic factors, starving, food quality, endoparasitism, density and competition, migration) (see Scriber 1984 and Slansky and Scriber 1985 for more details).

Many nutritional studies are done with artificial diets, which show comparable results to those obtained for natural diets. Diets are suitable for certain types of research, such as the evaluation of specific nutrients, their concentrations, and the determination of the concentrations of allelochemicals that affect the insect/ plant relationship (Giustolin et al. 1995). However, some care is necessary in these studies because many diets require cellulose (Vendramin et al. 1982), which can change the approximate digestibility. Besides this, the allelochemicals added to diets can interact in a different way with the nutrients in relation to natural food. Thus, artificial diets should not have phytoalexin induction in damaged tissue as happens in nature. The anticontaminants added to the diets (Greenberg 1970; Sikorowski et al. 1980; Funke 1983; King and Leppla 1984; Reinecke 1985) can affect the existing symbionts and even interfere in enzyme detoxification. In other types of studies of insect/plant relationships, plant extracts added to artificial diets are used. In these cases, there is a possibility that chemical resistance may be destroyed in the preparation of the extract and even undue dilution of the responsible chemical substance for the insect response. However, these studies with plant extracts can give good results (Martins et al. 1986). These authors, studying resistance in rice varieties to *D. saccharalis,* used plant extracts in artificial diets, eliminated the physical factor of the resistance, and through chemical analysis detected possible sources of resistance to this insect.

Whenever artificial diets are used as vehicles to test antinutrients and toxins, there must be some characteristics present for the results to be reliable, including (1) the substance to be tested, whether it be an allelochemical, a protein crystal, or a virus, should not be chemically affected (changed) by the diet; (2) the substance should not affect the palatability and/or attractivity of the diet; (3) it should not be avoided by specialized feeding mechanisms (such as extra oral digestion); (4) the formulation ingredients of the basic diet cannot mask or change the effects of the substance that is being tested; and (5) the diet should be totally suitable for offering and maintain healthy characteristics (no microorganisms) (Cohen 2004).

2.9 Final Considerations

Nutritional quality requirements in insects are similar, independent of their systematic position and feeding habits (identity rule). However, the proportions of the nutrients needed vary significantly between insect species (principle of nutritional proportionality), which results in a very diverse set of feeding habits. Also, food choice is not only determined by the nutritional components but also by the physical characteristics and by allelochemicals in the diets. Thus, the way foodstuff is ingested, digested, assimilated, and converted into growth tissues depends on these components within an ecological and evolutionary context (nutritional ecology). These nutritional analyses involving interactions at the different trophic levels require mostly the determination of nutritional indices discussed in this chapter, which demand meticulous studies and which, depending on insect type and size, can lead to mistakes, especially for determinations done with the first instars.

These types of studies evolved considerably during the 1970s and 1980s, as supported by the 347 citations on food consumption and use mentioned by Slansky and Scriber (1982, 1985). This evolution did not stop the continuation of many problems in their determination (sources of variation) and in the interpretation of the values obtained (see Section 2.6). At some trophic levels, more research is needed on physiology, nutrition, genetics, and behavior, especially for host/parasitoid relationships. There are innumerable aspects of mating, adult feeding, oviposition, development of immature forms and diapause (in colder regions) that need to be researched and discovered (Thompson 1986). In these cases, the attempt to rear the natural enemy *in vitro* is still a challenge. It is expected that with new techniques in molecular biology, these challenges will be overcome. In Brazil, where research with these nutritional indices began at the end of the 1970s (Crócomo and Parra 1979) and where there are few research groups directly involved, the problems are still greater. In general, the research is limited to the determination of nutritional indices and an unbiased analysis of the results obtained, which does not reach the expected level of detail. It is suggested that there be more interaction, principally of entomologists and biochemists, since through this association many of the complicated mechanisms that involve insect/plant relationships can be evaluated. In the last few years, new analytical methods of data (use of covariant analysis; see Table 2.10), studies with transgenic plants (Fernandes 2003) (Table 2.16), and even nutrigenomics development will be able to elucidate the intricate mechanisms of insect nutrition.

TABLE 2.16

Consumption of Leaf Area by *Spodoptera frugiperda* in Conventional and Transgenic (MON810) Corn during Three Laboratory Generations

	Generations		
	F_1	F_2	F_3
Treatment	Leaf Consumption (cm²)		
Conventional corn	201.44 ± 5.12 a A*	215.58 ± 6.10 a A	214.98 ± 6.24 a A
MON810	164.67 ± 4.44 b A	171.50 ± 7.15 b A	177.48 ± 5.38 b A

Source: Fernandes, O. D., Ph.D. Thesis, University of São Paulo, Piracicaba, Brazil, 2003.

Note: Temperature = 28°C ± 1°C, relative humidity = 60% ± 10%, 14h photophase.

* Means followed by the same small letter in the columns and large letter in the rows are not different between themselves by the Tukey test ($P < 0.05$).

REFERENCES

Albrecht, F. O. 1955. La densité des populations et al croissance chez *Schistocerca gregaria* (Forsk.) et *Nomadacris septemfasciata* (Serv.); la meu d'ajustement. *J. Agric. Trop. Bot. Appl.* 11:109–92.

Almeida, R. M. S. 1986. Nutrição quantitativa e influência da densidade populacional no desenvolvimento e fecundidade de *Diatraea saccharalis* (Fabr., 1974) (Lepidoptera: Pyralidae). Piracicaba, Ms. Sc. Dissertation, University of São Paulo, São Paulo, Brazil.

Almeida, R. M. S., and J. R. P. Parra. 1988. Nutrição quantitativa de *Diatraea saccharalis* (Fabr., 1974) em dieta artificial e em diferentes temperaturas. *XV Congr. Bras. Zool.*, p. 182.

Alverson, D. R., J. N. All, and P. D. Bush. 1980. Rubidium as marker and simulated inoculum for the black-faced leafhopper *Graminella nigrifrons*, the primary vector of maize chlorotic virus of corn. *Environ. Entomol.* 9:29–31.

Al-Zubaidi, F., and J. L. Capinera.1984. Utilization of food and nitrogen by the beet armyworm *Spodoptera exigua* (Hübner) (Lepidoptera: Noctuidae) in relation to food type and dietary nitrogen levels. *Environ. Entomol.* 13:1604–8.

Anderson, T. E., and N. C. Leppla. 1992. *Advances in Insect Rearing for Research and Pest Management.* Boulder, CO: Westview Press.

Bailey, C. G., and M. K. Mukerji. 1977. Energy dynamics of *Melanoplus bivittatus* and *M. femurrubrum* (Orthoptera: Acrididae) in a grassland ecosystem. *Can. Entomol.* 109:605–14.

Batista, G. C. 1972. *Fisiologia dos insetos.* Piracicaba: ESALQ.

Beck, S. D. 1972. Nutrition, adaptation and environment. In *Insect and Mite Nutrition: Significance and Implications in Ecology and Pest Management*, ed. J. G. Rodriguez, 1–3. Amsterdam, the Netherlands: North-Holland Publishing.

Beck, S. D., and J. C. Reese. 1976. Insect–plant interactions: Nutrition and metabolism. In *Biochemical Interaction between Plants and Insects*, ed. J. W. Wallace, and R. L. Mansell, 41–92. v. 10, New York: Plenum Press.

Bellows, T. S., and T. W. Fisher. 1999. *Handbook of Biological Control.* New York: Academic Press.

Berenbaum, M. 1978. Toxicity of a furanocoumarin to armyworms: A case of biosynthetic escape from insect herbivores. *Science* 201:532–4.

Berenbaum, M. 1985. Brementown revisited: Interactions among allelochemicals in plants. *Rec. Adv. Phytochem.* 19:139–69.

Bernays, E. A. 1978. Tannins: an alternative viewpoint. *Entomol. Exp. Appl.* 24:44–53.

Bernays, E. A., and S. Woodhead. 1982. Plant phenols utilized as nutrients by a phytophagous insect. *Science* 216:201–3.

Berry, W. L., M. W. Stimman, and W. W. Wolf. 1972. Marking of native phytophagous insects with rubidium: A proposed technique. *Ann. Entomol. Soc. Am.* 64:236–8.

Bhat, N. S., and A. K. Bhattacharya. 1978. Consumption and utilization of soybean by *Spodoptera litura* (Fabricius) at different temperatures. *Indian J. Entomol.* 40:16–25.

Bhattacharya, A. K., and G. P. Waldbauer. 1969a. Quantitative determination of uric acid in insect feces by lithium extraction and the enzymatic spectrophotometric method. *Ann. Entomol. Soc. Am.* 62:925–7.

Bhattacharya, A. K., and G. P. Waldbauer. 1969b. Faecal acid uric as an indicator in the determination of food utilization. *J. Insect Physiol.* 15:1129–35.

Bhattacharya, A. K., and G. P. Waldbauer. 1970. Use of fecal uric acid method in measuring the utilization of food by *Tribolium confusum. J. Insect Physiol.* 16:1983–90.

Bowers, W. S., T. Ohta, J. S. Cleere, and P. A. Marsella. 1976. Discovery of anti-juvenile hormones in plants. *Science* 193:542–7.

Bracken, G. K. 1982. The bertha armyworm, *Mamestra configurata* (Lepidoptera: Noctuidae). Effects of dietary linolenic acid on pupal syndrome, wing syndrome, survival and pupal fat composition. *Can. Entomol.* 114:567–73.

Brewer, F. D. 1982. Development and food utilization of tobacco budworm hybrids fed artificial diet containing oil soluble dyes. *J. Ga. Entomol. Soc.* 17:248–54.

Brewer, F. D., and E. G. King. 1980. Consumption and utilization of a soyflour–wheat germ diet by larvae of the tobacco budworm parasitized by the tachnid *Eucelatoria* sp. *Entomophaga* 25:95–101.

Brewer, F. D., and E. G. King. 1982. Food consumption and utilization by sugarcane borers parasitized by *Apanteles flavipes. J. Ga. Entomol. Soc.* 16:185–92.

Brito, L. O., A. R. Lopes, J. R. P. Parra, W. R. Terra, and M. C. Silva-Filho. 2001. Adaptation of tobacco bud-worm *Heliothis virescens* to proteinase inhibitors may be mediated by the synthesis of new proteinases. *Comp. Biochem. Physiol.* 128:365–75.

Broadway, R. M., and S. S. Duffey. 1986. Plant proteinase inhibitor: mechanism of action and effect on the growth and digestive physiology of larval *Heliothis zea* and *Spodoptera exigua. J. Insect Physiol.* 32:827–33.

Brues, C. T. 1946. *Insect Dietary: An Account of the Food Habits of Insects.* Cambridge: Harvard University Press.

Buscarlet, L. A. 1974. The use of ^{22}Na for determining the food intake of the migratory locust. *Oikos* 25:204–8.

Calver, M. C. 1984. A review of ecological applications of immunological techniques for diet analysis. *Aust. J. Ecol.* 9:19–25.

Capinera, J. L. 1978. Variegated cutworm: consumption of sugarbeet foliage and development on sugarbeet. *J. Econ. Entomol.* 71:978–80.

Carvalho, S. M., and J. R. P. Parra. 1983. Biologia e nutrição quantitativa de *Alabama argillacea* (Hübner, 1818) (Lepidoptera, Noctuidae) em três cultivares, de algodoeiro. In *VIII Congr. Bras. Entomol.*, p. 78.

Chang, N. T., R. E. Lynch, F. Slansky, B. R. Wiseman, and D. H. Habeck. 1987. Quantitative utilization of selected grasses by fall armyworm larvae. *Entomol. Exp. Appl.* 45:29–35.

Chang, Y. M., D. S. Luthe, F. M. Davis, and W. P. Williams. 2000. Influence of whorl region from resistant and susceptible corn genotypes on fall armyworm (Lepidoptera: Noctuidae) growth and development. *J. Econ. Entomol.* 93:477–83.

Chapman, R. F. 1982. *The Insects: Structure and Function*, Cambridge, MA: Harvard University Press.

Chou, Y. M., G. C. Rock, and E. Hodgson. 1973. Consumption and utilization of chemicals defined diets by *Argyrotaenia velutinana* and *Heliothis virescens. Ann. Entomol. Soc. Am.* 66:627–32.

Cohen, A. C. 2004. *Insect Diets. Science and Technology.* Boca Raton: CRC Press.

Cohen, A. C., and R. Patana. 1984. Efficiency of food utilization by *Heliothis zea* (Lepidoptera: Noctuidae) fed artificial diets or green beans. *Can. Entomol.* 116:139–46.

Cole, J. B. 1980. Growth ratios in holometabolous and hemimetabolous insects. *Ann. Entomol. Soc. Am.* 64:540–4.

Cônsoli, F. L., and J. R. P. Parra. 2002. Criação "in vitro" de parasitóides e predadores. In *Controle Biológico no Brasil: Parasitóides e Predadores*, Ed. J. R. P. Parra, P. S. M. Botelho, B. S. Corrêa-Ferreira, and J. M. S. Bento, 239–275. São Paulo: Manole.

Cônsoli, F. L., and S. Grenier. 2010. In vitro rearing of egg parasitoids. In *Egg Parasitoids in Agroecosystems with Emphasis on Trichogramma*, ed. F. L. Cônsoli, J. R. P. Parra, and R. A. Zucchi, 293–314. Dordrecht: Springer.

Coudron, T. A., G. D. Yocum, and S. L. Brandt. 2006. Nutrigenomics: a case study in the measurement of insect response to nutritional quality. *Entomol. Exp. Appl.* 121:1–14.

Crócomo, W. B., and J. R. P. Parra. 1979. Biologia e nutrição de *Eacles imperialis magnifica* Walker, 1856 (Lepidoptera, Attacidae) em cafeeiro. *Rev. Bras. Entomol.* 23:51–76.

Crócomo, W. B., and J. R. P. Parra. 1985. Consumo e utilização de milho, trigo e sorgo por *Spodoptera frugiperda* (J. E. Smith, 1797) (Lepidoptera, Noctuidae). *Rev. Bras. Entomol.* 29:225–60.

Crossley, Jr., D. A. 1966. Radioisotope measurement of food consumption by a leaf beetle species, *Chrysomela knabi* Brown. *Ecology* 47:1–8.

D'Antonio, A. M., and J. R. P. Parra. 1984. Biologia e nutrição quantitativa de *Lonomia circumstans* (Walker, 1855) (Lepidoptera, Attacidae) em cafeeiro. *IX Congr. Bras. Entomol.*, p. 19.

Dadd, R. H. 1973. Insect nutrition: Current development and metabolic implications. *Annu. Rev. Entomol.* 18:381–420.

Dadd, R. H. 1977. Qualitative requirements and utilization of nutrients: Insects. In *CRC Handbook Series in Nutrition and Food, Section D. Nutritional Requirements*, ed. M. Rechcigl Jr., 305–46. v. 1, Boca Raton, FL: CRC Press.

Dadd, R. H. 1985. Nutrition organisms. In *Comprehensive Insect Physiology Biochemistry and Pharmacology*, ed. G. A. Kerkut, and G. I. Gilbert, 319–90. v. 8, Oxford, UK: Pergamon Press.

Daly, H. V. 1985. Insect morphometrics. *Annu. Rev. Entomol.* 30:415–38.

Daum, R. J., G. H. McKibeen, T. B. Davich, and R. McLaughlin. 1969. Development of the bait principle for boll weevil control. Calco Oil Red N-1700 dye for measuring ingestion. *J. Econ. Entomol.* 62:370–5.

Dyar, H. G. 1890. The number of molts of lepidopterous larvae. *Psyche* 5:420–2.

Edwards, P. J., and S. D. Wrattens. 1981. *Ecologia das Interações Entre Insetos e Plantas.* São Paulo, Brazil: EDUSP.

Fernandes, O. D. 2003. Efeito do milho geneticamente modificado (MON810) em *Spodoptera frugiperda* (J. E. Smith, 1797) e no parasitóide de ovos *Trichogramma* spp. Ph.D. Thesis, University of São Paulo, Piracicaba, Brazil.

Ferraz, M. C. V. D., J. R. P. Parra, and J. D. Vendramim. 1983. Determinação das exigências térmicas de *Spodoptera frugiperda* (J. E. Smith, 1797) (Lepidoptera, Noctuidae) em condições de laboratório. *VIII Congr. Bras. Entomol.,* p. 17.

Ferreira, C., J. R. P. Parra, and W. R. Terra. 1996. The effect of dietary glycosides on larval midgut β gluco-sidases from *Spodoptera frugiperda* and *Diatraea saccharalis. Insect Biochem. Molec. Biol.* 27:55–9.

Fox, L. R., and B. J. Macauley. 1977. Insect grazing on *Eucalyptus* in response to variation in leaf tannins and nitrogen. *Oecologia* 29:145–62.

Fraenkel, G. 1953. The nutritional value of green plants for insects. *Symposia IXth Int. Congr. Entomol.,* pp. 90–100.

Funke, B. R. 1983. Mold control for insect rearing media. *Bull. Entomol. Soc. Am.* 29:41–4.

Gamundi, J. C. 1988. Biologia comparada e nutrição quantitativa de *Anticarsia gemmatalis.* Hübner, 1818 (Lepidoptera, Noctuidae) em folhas e vagens de soja. Ms. Sc. Thesis, University of São Paulo, Piracicaba, Brazil.

Gast, R. T., and M. Landin. 1966. Adult boll weevils and eggs marked with dye fed in larval diet. *J. Econ. Entomol.* 59:474–5.

Genthon, M., R. P. Almeida, and J. R. P. Parra. 1986. Nutrição, metabolismo respiratório e taxa de conversão protéica de *Spodoptera frugiperda* (J. E. Smith, 1797) em dieta artificial. *X Congr. Bras. Entomol.,* p. 45.

Giustolin, T. A., J. D. Vendramim, and J. R. P. Parra. 1995. Desenvolvimento de uma dieta artificial para estu-dos do efeito de aleloquímicos sobre *Scrobipalpuloides absoluta* (Meyrick). *An. Soc. Entomol. Brasil* 24:265–72.

Green, T. R., and C. A. Ryan. 1972. Wounds-induced proteinase inhibitor in plant leaves: A possible defense mechanism against insects. *Science* 175:776–7.

Greenberg, B. 1970. Sterilizing procedures and agents, antibiotics and inhibitors in mass rearing of insects. *Bull. Entomol. Soc. Am.* 16:31–6.

Habib, M. E. M., L. M. Paleari, and M. E. C. Amaral. 1983. Effect of three larval diets on the development of armyworm, *Spodoptera latisfascia* Walker, 1856 (Noctuidae, Lepidoptera). *Rev. Bras. Zool.* 1:177–82.

Hagen, K. S., R. H. Dadd, and J. Reese.1984. The food of insects. In *Ecological Entomology,* ed. C. B. Huffaker, and R. L. Rabb, 80–112. New York: John Wiley & Sons.

Harborne, J. B. 1982. *Biochemical Aspects of Plant and Animal Coevolution.* New York: Academic Press.

Hendricks, D. E., and H. M. Graham. 1970. Oil soluble dye in larval diet for tagging moths, eggs, and sper-matophores of tobacco budworms. *J. Econ. Entomol.* 63:1019–20.

Hiratsuka, E. 1920. Researches on the nutrition of the silk worm. *Bull. Seric. Exp. Sta.* 1:257–315.

Hori, K., and M. Endo. 1977. Metabolism of ingested auxins in the bug *Lygus disponsi*: conversion of indole-3-acetic acid and gibberillin. *J. Insect Physiol.* 23:1075–80.

Horton, D. R., and R. A. Redak. 1993. Further comments on analysis of covariance in insect dietary studies. *Entomol. Exp. Appl.* 69:263–75.

House, H. L. 1972. Insect nutrition. In *Biology of Nutrition,* ed. R. N. Fiennes, 513–73., v. 18, Oxford, UK: Pergamon Press.

House, H. L. 1977. Nutrition of natural enemies. In *Biological Control by Augmentation of Natural Enemies,* ed. R. L. Ridgway, and S. B. Vinson, 151–82, New York: Plenum Press.

Ishaaya, I. 1986. Nutritional and allelochemic insect–plant interactions relating to digestion and food intake: Some examples. In *Insect–Plant Interactions,* ed. J. R. Miller, and T. A. Miller, 191–223, New York: Springer-Verlag.

Jervis, M. A. 2005. *Insects as Natural Enemies: A Practical Perspective.* New York: Springer-Verlag.

Jones, R. L., W. D. Perkins, and A. N. Sparkes. 1975. *Heliothis zea:* Effects of population density and a marker dye in the laboratory. *J. Econ. Entomol.* 68:349–50.

Jongsma, M. A., and C. B. Bolter. 1997. The adaptation of insects to plant protease inhibitors. *J. Insect Physiol.* 43:885–95.

Kasten, Jr., P., and J. R. P. Parra. 1984. Biologia de *Alabama argillacea* (Hübner, 1818) I. Biologia em diferen-tes temperaturas na cultivar de algodoeiro IAC-17. *Pesq. Agropec. Bras.* 19:269–80.

Kasting, R., and A. J. McGinnis. 1965. Measuring consumption of food by an insect with carbon-14 labelled compound. *J. Insect Physiol.* 11:1253–60.

King, E. G., and N. C. Leppla. 1984. *Advances and Challenges in Insect Rearing.* New Orleans: USDA.

Klein, I., and M. Kogan. 1974. Analysis of food intake, utilization and growth in phytophagous insects—A computer program. *Ann. Entomol. Soc. Am.* 67:295–7.

Kogan, M. 1972. Intake and utilization of natural diets by the Mexican bean beetle *Epilachna varivestis* a multi-variate analysis. In *Insect and Mite Nutrition*, ed. J. G. Rodriguez, 107–26. Amsterdam, the Netherlands: North Holland.

Kogan, M. 1977. The role of chemical factors in insect/plant relationships. *Proc. XV Int. Congr. Entomol.* Washington, DC, 211–27.

Kogan, M. 1986. Bioassays for measuring quality of insect food. In *Insect–Plant Interactions*, ed. J. R. Miller and T. A. Miller, New York: Springer-Verlag.

Kogan, M., and D. Cope. 1974. Feeding and nutrition associated with soybeans. 3. Food intake, utilization and growth in the soybean looper *Pseudoplusia includens*. *Ann. Entomol. Soc. Am.* 67:66–72.

Kogan, M., and J. R. P. Parra. 1981. Techniques and applications of measurements of consumption and utiliza-tion of food by phytophagous insects. In *Current Topics in Insect Endocrinology and Nutrition*, ed. G. Bhaskaran, S. Friedman, and J. G. Rodriguez, 337–62. New York: Plenum Press.

Kuramochi, K., and Y. Nishijima. 1980. Measurement of the meal size of the horn fly, *Haematobia irritans* (L.) (Diptera: Muscidae) by the use of amaranth. *Appl. Entomol. Zool.* 15:262–9.

Latheef, M. A., and D. G. Harcourt. 1972. A quantitative study of food consumption, assimilation and growth in *Leptinotarsa decemlineata* (Coleoptera: Chrysomelidae) on two host plants. *Can. Entomol.* 104:1271–6.

Lee, K. P., S. J. Simpson, and D. Raubenheimer. 2004. A comparison of nutrient regulation between solitari-ous and gregarious phases of the specialist caterpillar, *Spodoptera exempta* (Walker). *J. Insect Physiol.* 50:1171–80.

Lee, K. P., D. Raubenheimer, and S. J. Simpson. 2004. The effects of nutritional imbalance on compensatory feeding for cellulose-mediated dietary dilution in a generalist caterpillar. *Physiol. Entomol.* 29:108–17.

Lloyd, E. P., R. J. Daum, R. E. McLaughlin, F. C. Tingle, G. H. McKibeen, E. C. Burt, J. R. McCoy, M. R. Bell, and T. C. Cleveland. 1968. A red dye to evaluate bait formulations and to mass mark field populations of boll weevils. *J. Econ. Entomol.* 61:1440–4.

Long, D. B. 1953. Effects of population density on larvae of Lepidoptera. *Trans. R. Entomol. Soc. Lond.* 104:543–84.

Loon, J. J. A. 1993. Gravimetric vs respirometric determination of metabolic efficiency in caterpillars of *Pieris brassicae*. *Entomol. Exp. Appl.* 67:135–42.

Lund, R. P., and F. T. Turpin. 1977. Serological investigation of black cutworm larval consumption by ground beetles. *Ann. Entomol. Soc. Am.* 70:322–24.

Martins, J. F. S., J. R. P. Parra, and L. H. Mihsfeldt. 1986. Resistência de variedades de arroz à *Diatraea saccharalis* e sua associação com características biofísicas e bioquímicas das plantas. *X. Congr. Bras. Entomol.*, p. 155.

Matana, A. L. 1986. Efeito do alimento no ciclo de vida e na nutrição, e exigências térmicas de *Spodoptera eridania* (Stal, 1871) (Lepidoptera: Noctuidae). Ms. Sc. Dissertation, Federal University of Paraná, Brazil.

Maxwell, F. G., and P. R. Jennings. 1980. *Breeding Plants Resistant to Insects.* New York: John Wiley & Sons.

McGinnis, A. J., and R. Kasting. 1964. Calorimetric analysis of chromic oxide used to study food utilization by phytophagous insects. *J. Agric. Food. Chem.* 12:259–62.

McMillian, W. W., K. J. Starks, and M. C. Bowman. 1966. Use of plant parts as food by larvae of the corn earworm and fall armyworm. *Ann. Entomol. Soc. Am.* 59:863–4.

Meneguim, A. M., J. R. P. Parra, and M. L. Haddad. 1997. Comparação de dietas artificiais, contendo diferentes fontes de ácido graxos, para criação de *Elasmopalpus lignosellus* (Zeller) (Lepidoptera: Pyralidae). *An. Soc. Entomol. Brasil* 26:35–43.

Mihsfeldt, L. H., and J. R. P. Parra. 1986. Comparação de dietas artificiais para criação de *Diatraea saccharalis* (Fabricius, 1974) (Lepidoptera: Pyralidae). In *X Congr. Bras. Entomol.*, p. 67.

Mihsfeldt, L. H., J. R. P. Parra, and H. J. P. Serra. 1984. Comparação de duas dietas artificiais para *Heliothis virescens* (F., 1781). In *IX Congr. Bras. Entomol.*, p. 70.

Mohamed, A. K. A., F. W. Brewer, J. V. Bell, and R. J. Hamalle. 1982. Effect of *Nomuraea rileyi* on consump-tion and utilization of food by *Heliothis zea* larvae. *J. Ga. Entomol. Soc.* 17:256–63.

Moreti, A. C. C., and J. R. P. Parra. 1983. Biologia comparada e controle de qualidade de *Heliothis virescens* (Fabr., 1781) (Lepidoptera-Noctuidae) em dietas natural e artificial. *Arq. Inst. Biol.* 50:7–15.

Moss, J. I., and R. A. Van Steenwyk. 1982. Marking pink bollworm (Lepidoptera: Gelechiidae) with cesium. *Environ. Entomol.*11:1264–8.

Mukerji, M. K., and J. C. Guppy. 1970. A quantitative study of food consumption and growth in *Pseudaletia unipuncta* (Lepidoptera: Noctuidae). *Can. Entomol.* 102:1179–88.

Nalim, D. M. 1991. Biologia, nutrição quantitativa e controle de qualidade de populações de *Spodoptera frugiperda* (J. E. Smith, 1797) (Lepidoptera: Noctuidae) em duas dietas artificiais. Ph.D. Thesis, University of São Paulo, Piracicaba, Brazil.

Nealis, V. 1987. The number of instars in jack pine budworm, *Choristoneura pinus pinus* Free. (Lepidoptera: Tortricidae) and the effect of parasitism on head capsule width and development time. *Can. Entomol.* 119:773–8.

Norris, D. M., and M. Kogan. 1980. Biochemical and morphological bases of resistance. In *Breeding Plants Resistant to Insects*, ed. F. G. Maxwell, and P. R. Jennings, 23–61. New York: John Wiley & Sons.

Oliveira, L. J. 1987. Biologia, nutrição quantitativa e danos causados por *Spodoptera frugiperda* (J. E. Smith, 1797) (Lepidoptera: Noctuidae) em milho cultivado em solo corrigido para três níveis de alumínio. Ms. Sc. Dissertation, University of São Paulo, Piracicaba, Brazil.

Oliveira, L. J., J. R. P. Parra, and I. Cruz. 1990. Nutrição quantitativa da lagarta-do-cartucho em milho cultivado para três níveis de alumínio. *Pesq. Agropec. Bras.* 25:235–41.

Orr, D. B., and D. T. Boethel. 1985. Comparative development of *Copidosoma truncatellum* (Hymenoptera: Encyrtidae) and its host, *Pseudoplusia includens* (Lepidoptera: Noctuidae) on resistant and susceptible soybean genotypes. *Environ. Entomol.* 14:612–6.

Pádua, L. E. M. 1986. Influência de nutrição, temperatura e umidade relativa do ar na relação *Apanteles flavipes* (Cameron, 1891): *Diatraea saccharalis* (Fabricius, 1792). Ph.D. Thesis, University of São Paulo, Piracicaba, Brazil.

Parra, J. R. P. 1980. Métodos para medir consumo e utilização de alimentos por insetos. In *Anais VI Congr. Bras. Entomol.*, ed. Z. Ramiro, J. Grazia, and F. M. Lara, 77–102. Campinas: Sociedade Entomológica do Brasil.

Parra, J. R. P. 1987. Nutrição quantitativa de Lepidoptera. *IX Congr. Bras. Entomol.* Campinas, SP.

Parra, J. R. P. 1991. Consumo e utilização de alimentos por insetos. In *Ecologia nutricional de insetos e suas implicações no manejo de pragas*, ed. A. R. Panizzi, and J. R. P. Parra, 9–65, São Paulo, Brazil: Manole.

Parra, J. R. P., A. A. C. M. Precetti, and P. Kasten Jr. 1977. Aspectos biológicos de *Spodoptera eridania* (Cramer, 1782) (Lepidoptera: Noctuidae) em soja e algodoeiro. *An. Soc. Entomol. Brasil* 6:147–55.

Parra, J. R. P., and M. Kogan. 1981. Comparative analysis of method for measurements of food intake and utilization using the soybean looper, *Pseudoplusia includens* and artificial media. *Entomol. Exp. Appl.* 30:45–57.

Parra, J. R. P., R. C. Estevam, P. S. M. Botelho, and J. A. D. Aguiar. 1988. Respiratory metabolism of *Diatrea saccharalis*. *Int. J. Cane. Agric.*, Sugar Cane, Spring Supplement, pp. 19–23.

Parra, J. R. P., and S. M. Carvalho. 1984. Biologia e nutrição quantitativa de *Spodoptera frugiperda* (J. E. Smith, 1797) em meios artificiais compostos de diferentes variedades de feijão. *An. Soc. Entomol. Brasil* 13:305–19.

Paulillo, L. C. M. S., A. R. Lopes, P. T. Cristofoletti, J. R. P. Parra, W. R. Terra, and M. C. Silva-Filho. 2000. Changes in midgut endopeptidase activity of *Spodoptera frugiperda* (Lepidoptera: Noctuidae) are responsible for adaptation to soybean proteinase inhibitors. *J. Econ. Entomol.* 93:892–6.

Pechann, T., A. C. Cohen, W. P. Williams, and D. S. Luthe. 2002. Insect feeding mobilizes a unique plant defense protease that disrupts the peritrophic matrix of caterpillars. *Proc. Natl. Acad. Sci. USA* 99:13319–23.

Peters, T. M., and P. Barbosa. 1977. Influence of population density on size, fecundity, and development rate of insects in culture. *Annu. Rev. Entomol.* 22:431–54.

Petrusewicz, K., and A. MacFadyen. 1970. *Productivity of Terrestrial Animals—Principles and Methods*. Int. Biol. Proj. Handbook 13, Oxford, UK: Blackwell.

Pompermayer, P., A. R. Lopes, W. R. Terra, J. R. P. Parra, M. C. Falco, and M. C. Silva-Filho. 2001. Effects of soybean proteinase inhibitor on development, survival and reproductive potential of the sugarcane borer, *Diatraea saccharalis*. *Entomol. Exp. Appl.* 99:79–85.

Precetti, A. A. C. M., and J. R. P. Parra. 1984 Nutrição quantitativa de *Heliothis virescens* (Fabr., 1781) em três cultivares de algodoeiro. *III Reunion ANC. Algodoeiro*, Recife, p. 148.

Radwan, H. S. A., O. M. Assal, G. E. Aboelghar, M. R. Riskallah, and M. T. Ahmed. 1986. Some aspects of the action of diflubenzuron and trifluron on food consumption, growth rate and food utilization by *Spodoptera littoralis* larvae. *J. Insect Physiol.* 32:103–7.

Raubenheimer, D., and S. J. Simpson. 1992. Analysis of covariance: An alternative to nutritional indices. *Entomol. Exp. Appl.* 62:221–31.

Reese, J. C. 1977. The effects of plant biochemical on insect growth and nutritional physiology. In *Host Plant Resistance to Pests*, ed. P. A. Hedin, 129–52. Washington, DC: American Chemical Society.

Reese, J. C. 1979. Interaction of allelochemicals with nutrients in herbivore food. In *Herbivores: Their Interaction with Secondary Plant Metabolites,* ed. G. A. Rosenthal, and D. H. Janzen, 309–29. New York: Academic Press.

Reinecke, J. P. 1985 Nutrition: Artificial diets. In *Comprehensive Insect Physiology Biochemistry and Pharmacology*, ed. G. A. Kerkut, and L. I. Gilbert, 319–419, v. 4. Oxford, UK: Pergamon Press.

Reis, F. W. 1984. Influência de clones de seringueira na biologia e nutrição de *Erynnyis ello* (L., 1758) (Lepidoptera: Sphingidae). Ms. Sc. Dissertation, University of São Paulo, Piracicaba, Brazil.

Reynolds, G. W., C. M. Smith, and K. M. Kester. 1984. Reductions in consumption, utilization and growth rate of soybean lopper (Lepidoptera: Noctuidae) larvae fed foliage of soybean genotype PI 227687. *J. Econ. Entomol.* 77:1371–5.

Rodriguez, J. G. 1972. *Insect and Mite Nutrition: Significance and Implications in Ecology and Pest Management.* Amsterdam, the Netherlands: North-Holland Publishing.

Roe, R. M., A. M. Hammond, Jr., and T. C. Sparks. 1982. Growth of larval *Diatraea saccharalis* (Lepidoptera: Pyralidae) on an artificial diet and synchronization of the last larval stadium. *Ann. Entomol. Soc. Am.* 75:421–9.

Rosenthal, G. A., G. C. Hughes, and D. H. Janzen. 1982. L-canavanine, a dietary nitrogen source for the seed predator *Caryedes brasiliensis* (Bruchidae). *Science* 217:353–5.

Salvadori, J. R. 1987. Biologia, nutrição e exigências térmicas de *Pseudaletia sequax* Franclemont, 1951 (Lepidoptera: Noctuidae) em dieta artificial. Ms. Sc. Dissertation, University of São Paulo, Piracicaba, Brazil.

Sarren, V., Y. S. Rathore, and A. K. Bhattachary. 1983. Influence of *Bacillus thuringiensis* on the utilization of *Spodoptera litura* (Fabricius). *Z. Ang. Ent.* 95:253–8.

Schmidt, D. J., and J. C. Reese. 1986. Sources of error in nutritional index studies of insects on artificial diet. *J. Insect Physiol.* 32:193–8.

Schmidt, D. J., and J. C. Reese. 1988. The effects of physiological stress on black cutworm (*Agrotis ipsilon*) larval growth and food utilization. *J. Insect Physiol.* 34:5–10.

Schroeder, L. A. 1971. Energy budget of larvae of *Hyalophora cecropia* (Lepidoptera) fed *Acer negnudo*. *Oikos* 22:256–9.

Schroeder, L. A. 1972. Energy budget of cecropia moth, *Platysamia cecropia* (Lepidoptera: Saturniidae) fed lilac leaves. *Ann. Entomol. Soc. Am.* 62:367–72.

Schroeder, L. A. 1973. Energy budget of the larvae of the moth *Pachysphinx modesta. Oikos* 24:278–81.

Schroeder, L. A. 1976. Effect of food deprivation on the efficiency of utilization of dry matter, energy and nitrogen by larvae of the cherry scallop moth *Calocalpe undulata. Ann. Entomol. Soc. Am.* 69:55–8.

Scriber, J. M. 1984. Host-plant suitability. In *The Chemical Ecology of Insects*, ed. W. J. Bell, and R. T. Cardé, pp. 159–202. London: Chapman and Hall Ltd.

Scriber, J. M., and F. Slansky, Jr. 1981. The nutritional ecology of immature insects. *Annu. Rev. Entomol.* 26: 183–211.

Sehnal, F. 1985. Growth and life cycles. In *Comprehensive Insect Physiology Biochemistry and Pharmacology*, ed. G. A. Kerkut, and L. I. Gilbert, 1–86, v. 2. Oxford, UK: Pergamon Press.

Shepard, M., and V. W. Waddil. 1976. Rubidium as a marker for Mexican bean beetle *Epilachna varivestis* (Coleoptera Coccinellidae). *Can. Entomol.* 108:337–9.

Sikorowski, P. P., A. D. Kent, O. H. Linging, G. Wiygul, and J. Roberson. 1980. Laboratory and insectary studies on the use of antibiotics and antimicrobial agents in mass rearing of bollweevils. *Anthonomus grandis. J. Econ. Entomol.* 73:106–10.

Silva, R. F. P., and J. R. P. Parra. 1983. Consumo e utilização de alimento artificial por *Anticarsia gemmatalis* Hübner. *VII Congr. Bras. Entomol.* p. 18.

Simpson, S. J., and J. D. Abisgold. 1985. Compensation by locust for changes in dietary nutrients: behavioral mechanisms. *Physiol. Entomol.* 10:443–52.

Singh, P. 1977. *Artificial Diets for Insects, Mites, and Spiders.* London: Plenum Press.

Singh, P. 1985. Multiple-species rearing diets. In *Handbook of Insect Rearing*, ed. P. Singh, and R. F. Moore, 19–24, v. 1. Amsterdam, the Netherlands: Elsevier Sci. Publ.

Slansky, Jr., F. 1978. Utilization of energy and nitrogen by larvae of the imported cabbage worm, *Pieris rapae* as affected by parasitism by *Apanteles glomeratus. Environ. Entomol.* 7:179–85.

Slansky, Jr., F. 1985. Food utilization by insects: Interpretation of observed differences between dry weight and energy efficiencies. *Entomol. Exp. Appl.* 39.47–60.

Slanky, Jr., F., and G. S. Wheeler. 1992. Caterpillars compensatory feeding response to diluted nutrients leads to toxic allelochemical dose. *Entomol. Exp. Appl.* 65:171–86.

Slansky, Jr., F. 1986. Nutritional ecology of endoparasitic insects and their host: An overview. *J. Insect. Physiol.* 32:255–61.

Slansky, Jr., F., and J. G. Rodriguez. 1987a. *Nutritional Ecology of Insect Mites, Spiders and Related Invertebrates.* New York: John Wiley & Sons.

Slansky, Jr., F., and J. G. Rodriguez. 1987b. Nutritional ecology of insects, mites, spiders, and related invertebrates: an overview. In *Nutritional Ecology of Insects, Mites, Spiders, and Related Invertebrates*, ed. F. Slansky Jr., and J. G. Rodriguez, 1–69. New York: John Wiley & Sons.

Slansky, Jr., F., and J. M. Scriber. 1982. Selected bibliography and summary of quantitative food utilization by immature insects. *Bull. Entomol. Soc. Am.* 28:43–55.

Slansky, Jr., F., and J. M. Scriber. 1985. Food consumption and utilization. In *Comprehensive Insect Physiology Biochemistry and Pharmacology*, ed. G. A. Kerkut, and L. I. Gilbert, 87–163, v. 4. Oxford, UK: Pergamon Press.

Slansky, Jr., F., and P. Feeny. 1977. Stabilization of the rate of nitrogen accumulation by larvae of the cabbage butterfly on wild and cultivate food plants. *Ecol. Monogr.* 47:209–28.

Soo Hoo, C. F., and G. Fraenkel. 1966. The consumption, digestion, and utilization of food plants by a polyphagous insect, *Prodenia eridania* (Cramer). *J. Insect Physiol.* 12:711–30.

Sousa-Silva, C. R. 1980. Uso de radiotraçador e serologia no estudo das relações alimentares entre a broca da cana-de-açúcar *Diatraea saccharalis* (Fabr., 1974) e artrópodes predadores. Ms. Sc. Dissertation, University of São Paulo, Piracicaba, Brazil.

Sousa-Silva, C. R. 1985. Serologia aplicada ao estudo de *Deois flavopicta* (Stal, 1854) (Homoptera: Cercopidae). Ph.D. Thesis, University of São Paulo, Piracicaba, Brazil.

Southwood, T. R. E. 1978. *Ecological Methods.* New York: Holsted Press.

Souza, A. M. L., C. J. Ávila, and J. R. P. Parra. 2001. Consumo e utilização de alimento por *Diatraea saccharalis* (Fabr.) (Lepidoptera: Pyralidae), *Heliothis virescens* (Fabr.) e *Spodoptera frugiperda* (J.E. Smith) (Lepidoptera: Noctuidae) em duas temperaturas. *Neotrop. Entomol.* 30:11–7.

Stepien, Z. A., and J. G. Rodriguez. 1972. Food utilization by acarid mites. In *Insect and Mite Nutrition: Significance and Implication in Ecology and Pest Management*, ed. J. G. Rodriguez, 127–51. Amsterdam, the Netherlands: North-Holland Publishing.

Stimac, J. L. 1982. History and relevance of behavioral ecology in models of insect population dynamics. *Fla. Entomol.* 65:9–16.

Stimann, M. W. 1974. Marking insects with rubidium: Imported cabbage worm marked in the field. *Environ. Entomol.* 3:327–8.

Susi, R. M., J. R. P. Parra, and W. B. Crócomo. 1980. Comparação de corantes para medir consumo e utilização de alimento por *Spodoptera frugiperda* (J. E. Smith, 1797) em dieta artificial. *VI Congr. Bras. Entomol.*, pp. 5–6.

Thompson, S. N. 1986. Nutrition and in vitro culture of insect parasitoids. *Annu. Rev. Entomol.* 31:197–219.

Thompson, S. N., and K. S. Hagen. 1999. Nutrition of entomophagous insects and other arthropods. In *Handbook of Biological Control*, ed. T. S. Bellows, and T. W. Fisher, 594–652. New York: Academic Press.

Thompson, S. N., and R. A. Redak. 2005. Feeding behavior and nutrient selection in an insect *Manduca sexta* L. and alterations induced by parasitism. *J. Comp. Physiol.* 191:909–23.

Uvarov, B. P. 1928. Insect nutrition metabolism. A summary of the literature. *Trans. R. Entomol. Soc. Lond.* 74:255–343.

Van Hook, R. J., and G. I. Dodson. 1974. Food energy budget for the yellow-poplar weevil, *Odontopus calceatus* (Say). *Ecology* 55:205–7.

Van Steenwyk, R. A., G. T. Ballmer, A. L. Page, and H. T. Reynolds. 1978. Marking pink bollworm with rubidium. *Ann. Entomol. Soc. Am.* 71:81–4.

Vendramin, J. D., A. R. R. Souza, and J. R. P. Parra. 1982. Ciclo biológico de *Heliothis virescens* (Fabricius, 1781) (Lepidoptera, Noctiudae) em dietas com diferentes tipos de celulose. *An. Soc. Entomol. Brasil* 11:3–11.

Vendramim, J. D., F. M. Lara, and J. R. P. Parra. 1983. Consumo e utilização de folhas de cultivares de couve (*Brassica oleracea* L. var. *acephala*) por *Agrotis subterranea* (Fabricius, 1974) (Lepidoptera-Noctuidae). *An. Soc. Entomol. Brasil* 12:129–44.

Waldbauer, G. P. 1962. The growth and reproduction of maxillectomizeds tobacco horworms feeding on normally rejected non-solanaceous plats. *Entomol. Expl. Appl.* 5:147–58.

Waldbauer, G. P. 1964. The consumption, digestion and utilization of solanaceous and non-solanaceous plants by larvae of the tobacco horworm *Protoparce sexta* (Lepidoptera Sphingidae). *Entomol. Exp. Appl.* 7:252–69.

Waldbauer, G. P. 1968. The consumption and utilization of food by insects. *Adv. Insect Physiol.* 5:229–88.

Waldbauer, G. P. 1972. Food utilization. In *Insect and Mite Nutrition: Significance and Implications in Ecology and Pest Management*, ed. J. G. Rodriguez, 53–5. Amsterdam, the Netherlands: North-Holland Publishing.

Warbrick-Smith, J., S. T. Behmer, K. P. Lee, D. Raubenheimer, and S. T. Simpson. 2006. Evolving resistance to obesity in an insect. *Proc. Natl. Acad. Sci.* 103:14045–9.

Wilkinson, J. D., R. K. Morrison, and P. K. Peters. 1972. Effects of calco oil red N-1700 dye incorporated into a semi-artificial diet of the imported cabbage worm, corn earworm, and cabbage looper. *J. Econ. Entomol.* 65:264–8.

Zonta, N. C. C. 1987. Consumo e utilização de alimento por larvas de *Anticarsia gemmatalis* Hübner, 1818 (Lepidoptera, Noctuidae), infectadas com *Nomurea rileyi* (Farlow) Samson. Ms. Sc. Dissertation, Curitiba, Federal University of Paraná.

3

The Evolution of Artificial Diets and Their Interactions in Science and Technology

José R. P. Parra

CONTENTS

3.1   The Importance of Rearing Insects in the Laboratory 52
3.2   History of Artificial Diets 55
3.3   Ways of Obtaining Insects and Types of Rearing 59
      3.3.1   Field Collecting 59
      3.3.2   Maintaining Populations on Natural Hosts 59
      3.3.3   Maintaining Populations on Artificial Diets 59
      3.3.4   Small-Scale Rearing 60
      3.3.5   Medium-Sized Rearing 60
      3.3.6   Mass Rearing 60
3.4   Terminology Used in Artificial Diets 61
3.5   General Principles of Nutrition 62
3.6   Types of Artificial Diets 63
3.7   Feeding Habits and Different Insect Mouthparts 64
      3.7.1   Feeding Habits 64
      3.7.2   Types of Mouthparts 65
            3.7.2.1   Mouthparts of Adult and Immature Insects 66
3.8   Physical, Chemical, and Biological Needs for Feeding 67
      3.8.1   Physical Stimuli 67
      3.8.2   Chemical Stimuli 67
      3.8.3   Biological Stimuli 68
3.9   Nutritional Needs for Growth 68
      3.9.1   Specific Nutritional Needs 68
            3.9.1.1   Amino Acids 68
            3.9.1.2   Vitamins 69
            3.9.1.3   Mineral Salts 70
            3.9.1.4   Carbohydrates 70
            3.9.1.5   Sterols 71
            3.9.1.6   Lipids 71
            3.9.1.7   Nucleic Acids 71
            3.9.1.8   Water 71
      3.9.2   Nutrient Storage 72
      3.9.3   Symbionts 72
3.10 Diet Composition 73
      3.10.1 General Components 73
      3.10.2 Adult Requirements 74
3.11 Rearing Techniques 76
3.12 Sequence for Preparing an Artificial Diet 77
3.13 Examples of Artificial Diets 79

3.14 Minimum Sanitary Precautions for Insect Rearing in Artificial Media 79
      3.14.1 Room for Diet Preparation 79
      3.14.2 Room for Adults 80
      3.14.3 Room for Larval Development 80
3.15 How to Begin an Artificial Diet 81
3.16 Evaluation of Artificial Diets 82
3.17 Causes of Failures and the Advantages of an Artificial Diet for Insects 85
3.18 The Future of Artificial Diets 85
References 86

3.1 The Importance of Rearing Insects in the Laboratory

The rearing of insects in the laboratory is of fundamental importance for solving problems of pure or applied entomology (Kogan 1980). The advance of research in modern entomology depends on insect availability in the laboratory so that studies do not suffer from lack of continuity or depend on the natural occurrence of the study insect, especially agricultural pests.

There are insects that can be easily reared in the laboratory and maintained at high populations. This is the case of *Drosophila*, which has been easily reared for many years and has been the main organism used in genetic research. In the same way, the silkworm, *Bombyx mori* L., has given rise to one of the largest industries in the world since 2000 B.C. in Asia (Cohen 2004), and the rearing of *Apis mellifera* L. since ancient Egypt (Cohen 2004) has increased agricultural yield by pollinating various crops important in human consumption. However, many insects, especially phytophages, require detailed study for mass rearing.

There have been revisions of insect nutrition and feeding habits since the beginning of the last century, such as those of Uvarov (1928) and Brues (1946), but the big advance occurred with the studies of G. Fraenkel of the University of Illinois, after the 1940s, in his research on the nutritional needs of stored products pests.

The big advances in rearing techniques on artificial media occurred in the 1960s, 1970s, and 1980s, especially in the developed countries. In the bibliographic compilation done by Singh (1985), artificial diets for more than 1,300 insect species belonging to most of the orders of agricultural importance were described (Table 3.1). Dickerson et al. (1980) listed around 1,000 insect colonies corresponding to 480 species (representing 109 families) maintained in 200 laboratories in the United States and other countries. Edwards et al. (1987) brought this list up to date and included a further 693 species reared in 263 facilities.

TABLE 3.1

Artificial Diets for Different Orders of Agricultural Importance

Order	Number of Species
Lepidoptera	556
Coleoptera	284
Diptera	279
Hemiptera	93
Hymenoptera	67
Orthoptera	24
Isoptera	5

Source: Singh, P., In *Handbook on Insect Rearing*, Vol. I, ed. P. Singh,
           and R. F. Moore, 19–44, Elsevier Amsterdam, the Netherlands,
           1985.

The first phytophagous insect to be reared in Brazil on an artificial diet was *Diatraea saccharalis* (Fabr.), the sugarcane borer, in 1969, using the artificial diet developed by Hensley and Hammond (1968), at Piracicaba, São Paulo, in research on biological control developed by Domingos Gallo, then head of the Department of Entomology (ESALQ/USP).

The validity and/or importance of rearing insects in the laboratory is clear. Knipling (1979) declared that one of the most important advances in entomology was the progress made by scientists in managing to rear an almost unlimited number of insects at a reasonable cost. The theme "insect rearing" is one of the most relevant since this topic has become a major part of the modern science of entomology (Leppla and Adams 1987; Cohen 2004; Schneider 2009).

The publication of Smith's (1966) classic book encouraged innumerable studies, such as those by Rodrigues (1972), Singh (1977), Leppla and Ashley (1978), Dickerson et al. (1980), Dadd (1985), Edwards et al. (1987), King and Leppla (1984), Singh and Moore (1985), Reinecke (1985), Ashby and Singh (1987), Parra (1991), Anderson and Leppla (1992), Thompson and Hagen (1999), Parra (2007, 2008), and Parra et al. (2002), which developed this area of entomology. After this, specific books on the subject became less common, perhaps due to the apparent failure to rear parasitoids and predators *in vitro*. In 2004, Cohen published the book *Insect Diets—Science and Technology*, covering basic aspects but also the technology involving knowledge of food science, including chemistry, physics, and microbiology, and the effect of the components on the manufacturing of the artificial diet during the processing. Although the advances in diets for parasitoids (Cônsoli and Parra 2002; Cônsoli and Grenier 2010) and predators (Cohen 2004), there is no doubt that 85% of artificial diets are concentrated on phytophages of the orders Lepidoptera, Coleoptera, and Diptera.

Depending on the study area, it is evident that insect field populations can be manipulated or insects can be kept in natural hosts in laboratories, insectaries (mesh screens), or in incubators. However, owing to the development of artificial media, it became possible to rear large number of insects necessary for studies in integrated pest management (IPM) programs, with total control over populations. Thus, it was possible to obtain significant advances in basic areas such as nutrition, toxicology, production of recombinant proteins and drugs, transgenic plants, biochemistry (enzyme studies), biotechnology, endocrinology, genetics, behavior, ecology, and taxonomy. With the development of insect mass rearing, applied research on biological control, plant resistance, insect pathology, genetic control (male sterilization and the use of lethal genes), disease vectors, the production of pheromones and kairomones (behavioral control methods), and chemical control, has been developed. This advance happened over the last 30 years, since Singh (1977) listed 154 publications on artificial diets from 1908 to 1950, and 1,807 from 1951 to 1976. From 1976 to 2010, there were few publications containing new knowledge on artificial diets, especially for phytophages.

In 1974, an international newsletter titled *Frass* (*Insect Rearing Group Newsletter*) was started in the United States to improve communication and cooperation as well as to solve the problems of scientists concerned with insect rearing. It has information on new diets, ingredient prices, the addresses of researchers working with different species, health and laboratory models, quarantine, and quality control (Dickerson and Leppla 1992).

A layout of the relationships between insect rearing and the diverse areas of entomology, focusing on pest management and sustainable agriculture, was drawn up by Parra (2008) (Figure 3.1). The maintenance of insect colonies in the laboratory is indispensable to modern strategies of pest management since in both basic and applied research programs, a continuous supply of insects is needed.

When the objective is research, insect colonies can be kept on natural media because large populations are not always necessary. However, whether on natural or artificial media, basic biological and behavioral information are fundamental for developing research that will support the different control methods (Parra 2000). In other cases, when the colonies are used for actual control, the number of insects to be used should be large, and in this case, artificial diets are mostly used. This gives rise to mass rearing, which has a series of peculiarities with not only entomological problems but also those related to a real factory involving the production of millions of insects.

In general, insect rearing is necessary for studies on insect plant resistance, insecticide trials (biological products, pathogens, growth regulators, new agrochemical groups), small-scale production of natural enemies (parasitoids, predators, and pathogens), studies on nutritional needs, mass production of natural

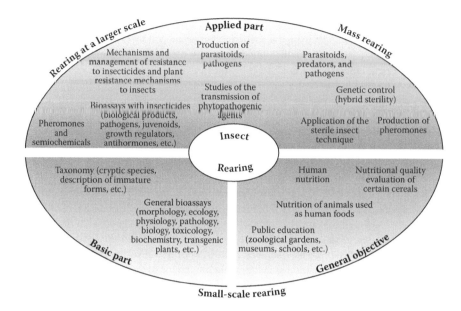

FIGURE 3.1   Relationships between insect rearing and various areas of entomology. (Modified from Parra, J. R. P., in *Encyclopedia of Entomology*, 2nd Edition, Vol. 3, ed. J. L. Capinera, 2301–05, Springer, 2008.)

enemies (parasitoids, predators, and pathogens), mass production for sterility programs, mass production for genetic control (hybrid sterility), evaluation of the nutritional quality of cereals (more economically than with other animal tests), nutrition of animals used in human feeding (fish, birds, and frogs), bioassays (morphology, biochemistry, physiology, pathology, biology, toxicology, ecology), studies on pheromones and semiochemicals, studies on taxonomy (cryptic species, description of immature forms), studies on insect and insecticide resistance (resistance mechanisms and management), studies on insect transmission of phytopathogenic agents, bioassays in biotechnology and molecular biology (especially for evaluation of transgenic plants), drugs, symbionts, enzymes, and other biochemical aspects. Insect colonies are also used in public education (zoos, museums, and schools) and as human food, since they represent important protein sources (Table 3.2) (Parra 2007, 2008).

TABLE 3.2

Relative Nutritional Value of Some Types of Insects

Insect	Protein %	Fat
Isoptera (termites)		
A live sample	23.2	28.3
A fried sample	36.0	44.4
Orthoptera (grasshoppers)		
12 dry samples	60.0	6.0
Diptera		
Three domestic fly pupae	63.1	15.5
Hymenoptera (ants)		
Adults		
Females	7.4	23.8
Males	25.2	3.3

3.2 History of Artificial Diets

The first insect to be reared axenically from egg to adult in an artificial diet (composed of peptone, meat extract, amide, and minerals) was *Calliphora vomitoria* (L.), by Bogdanov in 1908. In 1915, Loeb reared *Drosophila* sp. for five generations on a diet composed of grape sugar, sugar from sugarcane, ammonium tartarate, citric acid, potassium monoacid phosphate, magnesium sulfate, and water. Guyénot (1917) maintained *Drosophila ampelophila* Loew colonies with good results on an exclusively artificial diet. The cockroach species, *Periplaneta orientalis* (L.) and *Blatella germanica* (L.), were successfully reared by Zabinski (1926–1928) in a medium composed of egg albumin, amide, sucrose, agar, and a mixture of salts. In the 1940s, Fraenkel and his collaborators reared a large number of insects and stored products pests on a casein-based diet (Singh 1977).

The first attempt to rear a phytophagous insect on an artificial medium was made by Bottger (1942), who used a diet for *Ostrinia nubilalis* (Hübner) consisting of casein, sugars, fats, salts, vitamins, cellulose, agar, and water. Later, Beck et al. (1949) developed a diet for *O. nubilalis* composed of highly purified natural pure chemical products, also including an extract from corn leaves for supplying an unidentified growth factor (later identified as ascorbic acid by Chippendale and Beck (1964)). In 1949, House started a series of classic studies on applied aspects of insect nutrition. In 1950, K. Hagen, in Berkeley, California, launched the basis of parasitoid and predator nutrition and diets. Ishii (1952) and Matsumoto (1954) used diets that contained extracts of the host plants for *Chilo supressalis* (Walker) and *Grapholita molesta* (Busck). Vanderzant and Reiser (1956) reared the pink bollworm, *Pectinophora gossypiella* (Saunders), on a diet that did not contain plant extracts. From these initial experiments, a large number of insects have been reared on diets that consist entirely of pure chemical products and nutritional substances that are completely strange to the insect's natural food. In 1959, Fraenkel includes the concept of "secondary substances" to understand insects' feeding mechanisms. Ito (1960) in Japan began classic nutritional studies with *Bombyx mori* (L.).

The rearing of Hemiptera on artificial substrates was done by Schell et al. (1957), with the species *Oncopeltus fasciatus* (Dallas) and *Euschistus variolorius* (Palisot de Beauvois). The rearing techniques for aphids in the laboratory were developed in parallel in the United States and Canada by two research groups. Thus, Mittler and Dadd (1962) succeeded with *Myzus persicae* (Sulzer) in the United States and Auclair and Cartier (1963) with *Acyrtosiphon pisum* (Harris) in Canada. Mittler (1967) developed studies on the biochemistry, biophysics, and behavior of aphid nutrition. Gordon (1968) stated the principle of quantitative nutrition in insects, and Waldbauer (1968) standardized the indices for studying insect quantitative nutrition. The first references to rearing parasitoids in the laboratory on artificial media were made by Yazgan and House (1970) and Yasgan (1972) with the species *Itoplectis conquisitor* (Say) (Hymenoptera, Ichneumonidae).

One of the biggest advances in rearing techniques for Lepidoptera and other phytophages in the laboratory was due to the introduction of wheat germ in the diet formulations for *P. gossypiella* (Adkisson et al. 1960) and for *Heliothis virescens* (Fabr.) (Berger 1963). With some modifications, the formulations of these two authors constitute the basis for many insect diets. Such revisions of the history of artificial diets are based on Singh (1977), Singh and Moore (1985), and Cohen (2004).

A list of insect species reared is shown in Table 3.3. Revisions of the rearing of natural enemies were also done by Waage et al. (1985). In the last 30 years, research on artificial diets for parasitoids and predators *in vitro* has intensified and revisions on this subject have been made by Thompson (1986), Thompson and Hagen (1999), Cônsoli and Parra (2002), and Cônsoli and Grenier (2010). One of the few cases of successful rearing of a parasitoid *in vitro* is the production of *Trichogramma* by the Chinese in artificial eggs with a polyethylene chorion (Li Li Ying et al. 1986). The artificial medium is composed of the pupal hemolymph of *Antheraea pernyi* (Guérin-Méneville) [or *Philosamia cynthia ricini* (Boisd.)], chicken egg yolk, malt, and Neisenheimer salts (these are oviposition attractants). This medium can be used to rear various *Trichogramma* species if necessary, and using various thicknesses of plastic owing to the ovipositor size of the species being reared. The parasitoids are produced on plastic rings or cards that contain a large number of parasitoids. The Chinese now have computerized machines to produce thousands of *Trichogramma in vitro* per day for field liberation

TABLE 3.3

Taxonomic Distribution of Species Reared on Artificial Diets

Order/Family	Number of Species Reared
Coleoptera	284
Anobiidae	4
Bostrychidae	3
Buprestidae	1
Bruchidae	1
Cerambycidae	69
Chrysomelidae	11
Coccinellidae	69
Cucujidae	3
Curculionidae	28
Dermestidae	18
Elateridae	5
Lyctidae	2
Meloidae	5
Nitidulidae	5
Pythidae	1
Ptinidae	1
Scarabaeidae	17
Scolytidae	33
Tenebrionidae	7
Trogositidae	1
Dermaptera	1
Labiduridae	1
Dictyoptera	5
Blattellidae	4
Blattidae	1
Diptera	279
Anthomyiidae	10
Calliphoridae	19
Ceratopogonidae	19
Chironomidae	14
Chloropidae	4
Culicidae	61
Cuterebridae	1
Dolichopodidae	1
Drosophilidae	34
Glossinidae	1
Muscidae	16
Mycetophilidae	1
Mystacinobiidae	1
Oestridae	1
Phoridae	3
Piophilidae	1
Psilidae	1
Psychodidae	18
Sarcophagidae	8
Scatopsidae	1
Sciaridae	22

(continued)

TABLE 3.3 (Continued)

Taxonomic Distribution of Species Reared on Artificial Diets

Order/Family	Number of Species Reared
Sciomyzidae	1
Simuliidae	8
Sphaeroceridae	2
Syrphidae	3
Tabanidae	5
Tachinidae	4
Tephritidae	18
Tipulidae	1
Hemiptera	93
A. *Heteroptera*	22
Alydidae	1
Anthocoridae	6
Lygaeidae	1
Miridae	4
Nabidae	1
Pentatomidae	3
Reduviidae	5
Scutelleridae	1
B. *Homoptera*	71
Aphididae	50
Cercopidae	1
Cicadellidae	8
Coccidae	3
Delphacidae	7
Pseudococcidae	2
Hymenoptera	67
Aphelinidae	1
Apidae	4
Bethylidae	1
Braconidae	4
Cephidae	1
Chalcididae	2
Encyrtidae	1
Lepidoptera	556
Arctiidac	15
Bombycidae	2
Carposinidae	1
Cochylidae	1
Cossidae	2
Gelechiidae	10
Geometridae	32
Heliconiidae	1
Hepialidae	2
Hesperiidae	7
Lasiocampidae	6
Limacodidae	1
Liparidae	3
Lycaenidae	12

(*continued*)

TABLE 3.3 (Continued)

Taxonomic Distribution of Species Reared on Artificial Diets

Order/Family	Number of Species Reared
Lymantriidae	6
Lyonetiidae	1
Megalopygidae	1
Megathymidae	6
Noctuidae	217
Notodontidae	4
Nymphalidae	15
Oecophoridae	3
Olethreutidae	17
Papilionidae	3
Pieridae	13
Pyralidae	65
Riodinidae	1
Saturniidae	10
Satyridae	9
Eulophidae	4
Formicidae	35
Ichneumonidae	8
Megachilidae	1
Pteromalidae	1
Trichogrammatidae	4
Isoptera	5
Kalotermitidae	1
Rhinotermitidae	3
Termitidae	1
Mallophaga	3
Trichodectidae	3
Neuroptera	8
Berothidae	1
Chrysopidae	7
Orthoptera	24
Acrididae	10
Gryllidae	14
Phasmida	1
Phasmatidae	1
Siphonaptera	3
Pulicidae	3
Sesiidae	4
Sphingidae	6
Tineidae	1
Tortricidae	75
Yponomeutidae	4
TOTAL	1,329

Source: Singh, P., In *Handbook on Insect Rearing*, Vol. I, ed. P. Singh, and R. F.
            Moore, 19–44, Elsevier Amsterdam, the Netherlands, 1985.

(Dai et al. 1992; Liu et al. 1992). Cônsoli and Parra (2002) listed 73 parasitoid species reared *in vitro*, with 16 belonging to the Diptera and 57 to the Hymenoptera, with most representatives from the families Tachinidae (12 species) and Trichogrammatidae (18 species), respectively. In 1985, an oligidic diet was developed by Cohen, opening up the perspective of a series of diets for predators and parasitoids.

In Brazil, the first rearings *in vitro* were done by Parra and Cônsoli (1992) for *Trichogramma pretiosum* Riley. Studies with *T. pretiosum* (diet improvements) and *T. galloi* Zucchi (Cônsoli and Parra 1996a,b; 1997a,b; 1999a,b; Gomes et al. 2002) followed. Promising results were also obtained with the ectoparasitoid, *Bracon hebetor* Say (Magro and Parra 2004; Magro et al. 2006), and more recently with *Trichogramma atopovirilia* Oatman and Platner (Dias et al. 2010).

3.3 Ways of Obtaining Insects and Types of Rearing

3.3.1 Field Collecting

This is the oldest and most acceptable method for conservative entomologists since it deals with wild populations. However, such populations have the disadvantages of not occurring regularly and having unknown origins, nutrition, and ages, which can limit various types of study. In some cases, such as for studies of insect plant resistance, which are lengthy, such regularity of occurrence can delay these programs even more.

3.3.2 Maintaining Populations on Natural Hosts

This method requires considerable labor but is fundamental for some insect groups such as Hemiptera and Thysanoptera; in this case, plants easy to grow and handle are preferred (not always the natural hosts). Depending on the region, a greenhouse with temperature, relative humidity (RH), and photoperiod controls may be necessary. Precautions should be taken with small species (thrips, whiteflies), because if they are not kept in cages with fine mesh netting, a mixture of species may result. However, such colonies can be maintained since the rearing of whitefly species on host weeds are the most cited examples (Costa 1980).

3.3.3 Maintaining Populations on Artificial Diets

The diet should contain all the nutrients needed by the insect (proteins, vitamins, mineral salts, carbohydrates, lipids, and sterols), and some groups even need nucleic acid. However, this is not sufficient because the absence of certain physical properties and phagostimulants (physical and chemical), as well as a nutrient equilibrium, can result in unsuitable insect development. Insects dependent on symbionts may present obstacles to formulating an artificial diet (Parra 2007). Thus, a correctly formulated artificial diet has physical properties and contains chemical products that stimulate and maintain feeding, contains nutrients (essential and nonessential) in balanced proportions for producing optimal growth and development, and should be free of contaminating microorganisms. A diet has to be liquid for aphids, semiliquid (i.e., with a lot of water) for chewing insects, with a consistency that offers resistance to the insect's mouthparts, and a powder or in fragments for cockroaches or stored products pests. The more water there is in a diet, the greater the contamination problems.

A suitable artificial diet is one that results in high larval viability, produces insects whose larval stage is the same as in the wild, produces adults with a high reproductive capacity, may be used for more than one species and if possible for more than one insect order, whose components are cheap (easily available in the market), has a total viability greater than 75%, and maintains insect quality through the generations.

There are basically three types of insect rearing applicable to natural enemies: (1) small-scale rearing, (2) medium-sized rearing, and (3) mass rearing.

3.3.4 Small-Scale Rearing

These are the so-called rearings for research that can be expanded for applied research, especially in cases of biological control that need inoculative releasings. In July 1998, the parasitoid *Ageniaspis citricola* Logvinovskaya was imported into Brazil to control *Phyllocnistis citrella* Stainton. In this case, the parasitoid was reared by three full-time laboratory technicians and released over the whole of São Paulo state (Parra et al. 2004). During the releasings (from 1998 to 2002), around 1 million individuals were produced using a simple technique (Figure 3.2).

3.3.5 Medium-Sized Rearing

These are larger rearings for developing control methods.

3.3.6 Mass Rearing

Generally involve operations similar to those in a factory that serve as a support for a biological control program or other control method. The definitions of mass rearing vary considerably. According to Finney and Fisher (1964), mass rearing is "economic production of millions of beneficial insects,

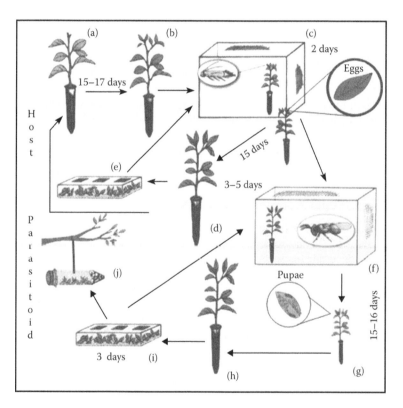

FIGURE 3.2 Production system of *Ageniaspis citricola* on the citrus leafminer, *Phyllocnistis citrella*. (a) recently pruned citrus plant in container, (b) plant giving out shoots, (c) shoots infested with *P. citrella* eggs, (d) shoots with *P. citrella* pupae, (e) pupae for maintaining pest colony in plastic containers, (f) shoots with eggs/larva of the first instar of *P. citrella* offered to the parasitoid, *A. citricola*, (g) shoots with parasitoid pupae, (h) pruning of shoots with parasitoid pupae, (i) *A. citricola* pupae in plastic containers for maintaining laboratory populations, and (j) parasitoid pupae in plastic tube for field release. (Modified from Chagas, M. C. M., et al., In *Controle Biológico no Brasil: Parasitóides e Predadores*, ed. J. R. P. Parra, P. S. M. Botelho, B. S. Corrêa-Ferreira, and J. M. S. Bento, 377–94, Manole, São Paulo, Brazil, 2002.)

in an assembly line, with the object of producing, with the minimum of man-hours and space, the maximum number of fertile females in the shortest possible time at a low cost." Mackauer (1972) and Chambers (1977) combined the economic aspect with the biological. According to the latter author, mass rearing is the "production of insects capable of reaching objectives at an acceptable cost/benefit ratio and in numbers exceeding from 10,000 to 1 million times the mean productivity of the population of native females." Leppla and Adams (1987) defined mass rearing as "a systematic activity, automated, in integrated facilities, with the objective of producing a relative large supply of insects for distribution."

Mass rearings of natural enemies are commercialized as biological insecticides. One of the insects most produced in the world today is *Trichogramma* spp. and extensive areas are "treated" with this parasitoid. In Russia, there is a large number of biofactories producing millions of insects per day (Parra and Zucchi 1986), reaching an annual production of 50 billion insects; in Mexico, 28 billion are produced per year and in some South American countries, such as Colombia, a high number of insects is produced (Parra and Zucchi 2004). These mass rearings involve the daily production of millions of insects, and in fact are like a production line for any product. Thus, for control of *Cochliomyia hominivorax* (Coquerel) through sterile insect techniques in the United States, from 50 to 200 million sterile flies were produced and released per week. At the end of the program, it was found that for each female in nature, 49 sterile females had been released, a much bigger proportion than what had been theoretically forecasted, which was 1 : 9. In such a "factory," more than 300 people were employed. In this case, apart from biological problems with the rearing, there are others, such as the inventory, purchase, and storage of materials and the maintenance of facilities and equipment. The greater the increase in insects produced, the more problems there are with the facilities, costs, microorganisms (contaminants), and insect quality control, and automation has to be considered. This automation should be encouraged for a production greater than 3,000 to 5,000 adults per week. Details of mass rearings can be found, among others, in Smith (1966), Ridgway and Vinson (1977), Starler and Ridgway (1977), Leppla and Ashley (1978), King and Leppla (1984), Singh and Moore (1985), van Lenteren and Woetz (1988), Parra (1990, 1992a,b, 1993, 1997, 2002), Parrela et al. (1992), van Driesche and Bellows Jr. (1996), Ridgway and Inscoe (1998), Bellows and Fisher (1999), Etzel and Legner (1999), Bueno (2000), and van Lenteren (2000, 2003).

Many companies in Europe and in the United States commercialize natural enemies commonly used in greenhouses. This started in the United States and Europe in the 1970s, and today companies from Brazil are interested in commercializing natural enemies. However, still there is lack of legislation to stop "adventurers" who could reduce the credibility of biological control (the so-called production ethic of Hoy et al. 1991).

Mass rearing of insects tends to increase due to the pressure of society against the use of pesticides. Therefore, the market for natural enemies is growing. More than 125 species are now available throughout the world to control 70 pests, including those of protected crops (whiteflies, two spotted spider mite, aphids, leafminers, thrips) (van Lenteren 2003). The number of natural enemies in Brazil is increasing and around 20 companies commercialize species of *Trichogramma, Cotesia flavipes* (Cam.), *A. citricola*, among others, besides predatory mites. The unavailability of natural enemies to the farmer is one of the reasons for the small use of biological control in Brazil (Parra 2006).

3.4 Terminology Used in Artificial Diets

In the beginning, the terminology of insect diets was very confusing. The terms artificial, synthetic, purified, and chemically defined were used by different researchers to describe diets that contained substances whose purity varied.

Diet means everything the insect eats to satisfy its physiological needs. Today, there still are confusions or ambiguities between natural and artificial diets. *Natural diet* is a group of hosts with which insects are normally associated; that is, they are the foods the insect ingests in the wild. *Artificial diet* is food supplied by humans in an attempt to substitute natural food with another more accessible or more convenient one from the technical or economic points of view. Therefore, artificial diets can be plants normally not used by the insect in the wild. For example, to rear the coffee root mealybug *Dysmicoccus cryptus* (Hempel), recently sprouted potatoes are used instead of coffee, or in the case of scales of orange trees (to produce coccinellids and nitidulids), citrus plants are substituted by squash or other cucurbits. For this reason, many researchers prefer to use the term artificial media instead of artificial diets to avoid ambiguity.

Dougherty (1959) classified artificial diets (or artificial media) into holidic, meridic, and oligidic, according to the purity of the components and this terminology is now accepted internationally. *Holidic diets* are diets (media) whose components all have a chemically defined or known composition. *Meridic diets* contain one or more ingredients whose composition is unknown or inadequately defined. *Oligidic diets* contain unpurified organic components, principally raw organic materials. According to the degree of purity of the diets, they may be classified as *axenic* diets in which only one species exists in the culture medium (symbionts are excluded). *Sinxenic* diets are those in which two or more species (monoxenic, dixenic, polixenic) are reared together in the culture medium; if all the species are known, the culture is called gnotobiotic (Ashby and Singh 1987). *Xenic* diets are those in which one species is reared without excluding unknown symbionts.

Other concepts related to insect bioecology and nutrition include *nutrition*, which is the study of insect food needs. *Qualitative nutrition* deals exclusively with nutrients needed from the chemical point of view. *Quantitative nutrition* considers not only the basic nutritional needs of the insect as important but also the proportion (quantity) of food ingested, digested, assimilated, and converted into growing tissues. *Essential nutrients* are compounds that have to be included in the diet because they cannot be synthesized either by the insect's metabolic system or by symbionts. They include vitamins, amino acids, and certain mineral salts. *Nonessential nutrients* are elements that have to be consumed to produce energy and are converted in such a way that the insects can use them through a metabolic process. They include carbohydrates, lipids, and sterols. At least one species from two orders is reared on a *diet for multiple orders* (e.g., diets for Lepidoptera and Diptera). *"Specific" diet for an order* is when species of at least two families of the same order are reared (e.g., species from the families Noctuidae and Tortricidae–Lepidoptera). *"Specific" diet for a family* is when species of two or more genera of the same family are reared (e.g., *Heliothis* sp. and *Spodoptera* sp. — Noctuidae). *"Specific" diet for a genus* is when two or more species of the same genus are reared on the same diet [e.g., *H. virescens* and *Heliothis subflexa* (Guenée)]. *Specific diet* is one used for monophagous insects.

3.5 General Principles of Nutrition

Based on House (1966), Singh (1977) defined three general principles of insect nutrition:

1. Identity rule. *The quality nutrition of insects is similar irrespective of their feeding habits and systematic position.* Thus, a chewing insect, a sucking insect, or a parasitoid have the same qualitative needs although the form in which this diet is offered varies, whether it be microencapsulated for a parasitoid (Thompson 1986), on a parafilm membrane for an aphid (Kunkel 1977), or in agar for phytophages. Reinecke (1985) discussed the types of diets according to insects' mouthparts and Grenier (1994) and Cônsoli and Parra (1999b, 2002; Cônsoli and Grenier 2010) discussed the problems in rearing parasitoids *in vitro*.

2. Principle of nutritional proportionality. *Metabolically suitable proportions of nutrients are required for normal nutrition.* Thus, the proportion of nutrients is of fundamental importance, mostly proteins : carbohydrates; the proportion varies from insect to insect (Dadd 1985, Table 3.4). Nutrient equilibrium varies according to insect age. *Schistocerca* spp. needs more carbohydrate in the last nymphal instar. *O. nubilalis* does not need carbohydrates in the first three larval instars. In these cases, reserves for the first instars come from the egg. Also, there are nutrient and nonnutrient (allelochemicals) interactions.

TABLE 3.4

Proportion (in Weight) of Proteins in Relation to Carbohydrates for Different Insect Species

Species	Proteins (Amino Acids)	Carbohydrates (Sugars)	Proportion
Cochliomyia hominivorax	90	0	∞
Itoplectis conquisitor	61	20	3.1
Musca domestica	55	21	2.6
Agria housei	49	20	2.5
Chrysopa carnea	41	56	0.7
Helicoverpa zea	40	32	1.3
Pectinophora gossypiella	37	45	0.8
Anthonomus grandis	35	32	1.1
Schistocerca gregaria	43	43	1.0
Bombyx mori	40	42	1.0
Blatella germanica	30	65	0.5
Tribolium confusum	20	74	0.3
Myzus persicae	23	72	0.3

Source: Dadd, R. H., In *Comprehensive Insect Physiology, Biochemistry and Pharmacology*, Vol. 4, ed. G. A. Kerkut, and L. I. Gilbert, 313–90, Pergamon Press, Oxford, UK, 1985.

3. Principle of cooperating supplements. *Supplementary sources of nutrients, as provided by nutrient resources and sometimes symbionts, may play an important role in the nutrition of almost any insect.* Symbionts (e.g., bacteria, fungi, yeasts, protozoans) can be the main food source (e.g., fungi for ants, beetles, mosquitoes, and *Drosophila*), can help in the digestion by secreting enzymes into the intestine, can convert food internally from an unusable into a usable form (termites and cockroaches), can supply auxiliary growth factors (vitamins), or execute biochemical functions that allow the insect to survive and grow on an unsuitable diet.

Research on symbionts has been intensified in the last few years, and today, one of the most studied symbionts is *Wolbachia*, a α-proteobacteria that is present in many insects (see Chapter 6). This type of bacteria is transmitted via the egg cytoplasm and can use various mechanisms to manipulate the reproduction of its hosts, including the creation of reproductive incompatibility, parthenogenesis, and feminization. They are also transmitted horizontally between arthropod species and have been studied in the egg parasitoid, *Trichogramma* spp. (Werren 1997). The effect of mutualistic endosymbionts, such as *Buchnera*, has been exhaustively studied in aphid nutrition. These symbionts and their evolutionary role have been discussed (Bourtzis and Miller 2003, see also Chapter 6).

3.6 Types of Artificial Diets

A correctly formulated artificial diet has physical properties and contains chemical products that stimulate and maintain feeding, contains nutrients (essential and nonessential) in balanced proportions to produce optimum growth and development, and should be free of contaminating microorganisms. There is a direct correlation between the problems associated with the formulation of artificial diets and their water content. The following classes of diets with their respective formulation include the following. (1) *Diets as powder or fragments.* Homogeneous mixtures are easy to produce and there is no microorganism contamination (e.g., diets for stored products pests, grasshoppers, and cockroaches). (2) *Semiliquid diets for chewing phytophages.* The mixtures are more difficult to obtain and there is immediate microorganism contamination if suitable measures are not taken. These diets are the ones most commonly used in biological control and alternative pest control programs. (3) *Liquid diets for sucking phytophages.* These diets are time consuming to prepare and are quickly contaminated by microorganisms. Other common problems are related to the distribution of the diets within the artificial parafilm membranes and the maintenance of nutrients in solution. This is the case with diets for aphids and certain Heteroptera (stinkbugs). (4) *Liquid diets for endoparasitoids.* They show the same problems as the previous diets as well as problems associated with the washing of the insect in the liquid medium. This involves micro-encapsulation and many problems arise as discussed by Grenier (1994, 1997), Cônsoli and Parra (2002), and Cônsoli and Grenier (2010).

3.7 Feeding Habits and Different Insect Mouthparts

The consistency and structure of a suitable diet are governed by the feeding habits and the type of an insect's mouthparts. Thus, knowledge of the feeding habits is fundamental for rearing insects on natural or artificial diets. The type of metamorphosis is also important for insect rearing and biological studies.

3.7.1 Feeding Habits

Insect food is very diversified and includes material of animal, vegetable, and organic origin. This is one of the characteristics of insects that make them specialize in different ways of eating these foodstuffs. According to Brues (1946) and Frost (1959), all living and dead organisms are consumed, including leaves, nectar, pollen, seeds, wood, sap (fluid from the xylem and phloem), fungi, animal meat, blood, hair, feces, and wax. There are other less common cases, like the tsetse fly, in which the developing larva feeds on the mother's internal uterine glands; there are various flies, mantids, and mites where the adult female can eat the male whereas flies from the Cecidomyiidae family produce galls, which are pedogenic, and have the larva developing internally in the mother's body, and in some cases, eating it. Some insects cultivate their own food while others have symbionts to help their nutrition (see Chapter 6). There are adult insects that consume the same food as the larval stage and some that do not feed. Feeding diversity includes chewing, piercing–sucking, siphoning, and sponging.

Brues (1946) believes that any classification of feeding habits is arbitrary; however, animals have traditionally been separated into four main categories according to the trophic level of their food. Thus, they are classified into herbivores, carnivores, saprovores, or detritivores and omnivores. These large categories can be divided into feeding classes (guilds) based on the specific type of food consumed [phyllophages that consume leaves; carpophages that eat fruits; nectivores (nectar), fungivores or mycetophages (fungi), and so on] and in the way in which they are consumed (within the phyllophages, there are chewers, dilacerators, raspers, miners, gall producers, and sap suckers).

Herbivores are those that consume plant tissues, and according to Chapman (1982), correspond to about half of the insect species. The phytophages and mycetophages are in this category. The insect orders that are mainly phytophagous include Orthoptera, Lepidoptera, Hemiptera, Thysanoptera, Phasmatodea, Isoptera, Coleoptera (Cerambycidae, Chrysomelidae, and Curculionidae), Hymenopetera (Symphyta), and some Diptera. Most insects feed on the higher plants while aquatic larvae of the Ephemeroptera, Plecoptera, and Trichoptera feed on algae. Larvae, which feed on fungi, are frequent in the Diptera and Coleoptera. Among the coprophages, the fungi make up part of their diet and the termites cultivate their own fungus. *Carnivores* include parasitoids and predators. Within the parasitoids, there are ecto-parasitoids and endoparasitoids. In the first group are representatives of the Phthiraptera (Siphonaptera, Anoplura) and some Dermaptera, Heteroptera, such as *Cimex*, and some Reduviidae and various Diptera (mosquitoes, Simuliidae, Ceratopogonidae, Tabanidae, and Pupipara). There are many blood suckers, including some on vertebrates. Sometimes, both sexes suck blood, such as in the Siphonaptera and the tsetse fly, or only females, such as in the Nematocera and Brachycera. In the latter case, the females also feed regularly on nectar, which is the only food of the males. Most of the endoparasitoids are only parasitoids as larvae and include all the Strepsiptera, Ichneumonoidea, Chalcidoidea, and Proctotrupoidea, among the Hymenoptera, and Bombyliidae, Cyrtidae, Tachinidae, and some Sarcophagidae, among the Diptera. Predominantly predatory groups include Odonata, Dictyoptera (Mantodea), Heteroptera

(Reduviidae and others), larvae of Neuroptera, Mecoptera, Diptera (Asilidae and Empididae), Coleoptera (Adephaga, larval stage of the Lampyridae, and most Coccinellidae), and Hymenoptera (Sphecidae and Pompilidae). *Saprovores or detritivores* occur mainly in higher insects whose larvae are different from the adults. Dead organic matter is the source of the most common food for many larvae of Diptera and Coleoptera. In these habitats, the microorganisms perform an important role in the diets. *Omnivores* consume more than one of these foods mentioned previously.

The feeding categories place insects into functional groups, and as such, there is a taxonomic mix. Thus, certain larvae of Coleoptera, Hymenoptera, and Lepidoptera can fall into the same category as the leaf chewers. Similarly, functional categories of parasitoids and predators can include herbivores and carnivores (e.g., a black cutworm caterpillar is a predator of small plants and an aphid is a "parasite" of its host plant).

For bioecology and nutrition, this classification of feeding categories is based on the fact that the composition of different foods varies according to the proportion of nutrients and allelochemicals. Besides this, some foods are more easily found and consumed and thus more abundant than others, so obviously there must be a variation in the selection pressure. Thus, according to the food, insects will show adaptations in the consumption and use of these foods, and therefore, within each feeding class there will be similarities governed by each species evolution. This will result in a diversity of adaptations within each class, including differences in size, dispersal ability, feeding specialization, and defense mechanisms against predators and parasitoids (see Chapter 2).

3.7.2 Types of Mouthparts

Based on Gallo et al. (2002) mouthparts can be classified as follows:

1. *Shredder or chewing.* It is considered primitive and has all the mouth parts: a pair of mandibles, a pair of maxillae, upper labrum, lower labium, an epipharynx, and a hypopharynx. Some parts can be slightly modified but this does not affect their functions; that is, the shredding or chewing of food. Insects have mouthparts that freely move and project into the mouth cavity (ectognathous). This type of mouthpart is present in most orders.

2. *Labial sucking.* Also called piercing–sucking. Shows mouthparts modified into stylets or atrophied with the exception of the upper labrum, which is normal and little developed. The lower labium becomes a tube, called haustellum, which houses the other stylets. The lower labium does not have a piercing function and the sucking up of food is the function of the mandibles, epipharynx, and hypopharynx. The maxillae, which have serrated edges, have a perforating function. According to the number of stylets enclosed by the lower labium, this type of mouthpart can show the following subtypes. (a) *Hexachaetous,* when there are six stylets, as discussed previously (two mandibles, two maxillae, one epipharynx, and a hypopharynx). This occurs in the Diptera (mosquitoes, *Culicoides,* tabanids). (b) *Tetrachaetous,* where four stylets are present (two mandibles and two maxillae). The epipharynx and the hypopharynx are atrophied. This occurs in Hemiptera. (c) *Trichaetous,* with three stylets. In Thysanoptera, there is a mandible (the left one because the right one is atrophied) and two maxillae; in Phthiraptera–Anoplura, the maxillae joined, lower labium and hypopharynx; in Siphonaptera there are two maxillae and an epipharynx. (d) *Dichaetous,* only two stylets (Diptera). In stable flies, the stylets are represented by the upper labrum plus the epipharynx (labroepipharynx) and hypopharynx, with a piercing function. In house flies these two stylets are rudimentary. Mouthparts are transformed into a proboscis adapted for licking.

3. *Sucking maxillary.* In this type, the modification only occurs to the maxillae with the rest of the parts atrophied. Thus, the maxillary galeae are transformed into two long pieces that have internal channels, so that when joined together they form a channel where the food is ingested by suction. The different part form a long tube that is wound up (at rest) and is found in Lepidoptera.

4. *Licking.* The upper labrum and the mandibles are normal. The mandibles are adapted for piercing, cutting, transporting, or molding wax. The maxillae and the lower labium are elongated and joined together, forming a licking organ. The glossae are transformed into a type of tongue with which the insects remove nectar from flowers and have a dilated end (flabellum). Occurs in Hymenoptera.

3.7.2.1 Mouthparts of Adult and Immature Insects

Depending on the type of mouthparts in the immature or adult stages, insects can be divided into three groups. (1) *Menorhynchous*, which have a sucking labial mouthpart in both larval and adult stages (e.g., Thysanoptera and Hemiptera). (2) *Menognathous*, which have a chewing mouthpart in the larvae and adults (e.g., Coleoptera, Orthoptera, Blattodea, Mantodea, and Isoptera). (3) *Metagnathous*, which have a chewing mouthpart in the larvae and in the adult a sucking maxillary type (Lepidoptera), licking type (Hymenoptera), or labial sucking type (Diptera).

According to Cohen (2004), although there are different arrangements of insect mouthpart structures, they are adapted to three types of diets: only liquid diets, only solid diets, and a mixture of liquid and solid diets. Among those that eat only liquid diets are the Homoptera, including aphids, cicadas, spittlebugs, and scales. Many Homoptera feed on xylem or phloem sap, with some feeding on cell liquids. Most adult Lepidoptera (butterflies and moths) feed on nectar. The various groups that feed on vertebrate blood (Siphonaptera, Heteroptera, and flies) are the "true liquid feeders" although their food is a mixture of blood cells suspended in a plasma matrix. Few Heteroptera (true bugs) feed on plant sap that are originally liquids (family Blissidae). However, most true bugs, as well as many flies, beetles (Coleoptera), Neuroptera, and Hymenoptera, which feed on materials that were originally solids and are converted into liquid mixtures before ingestion, were considered by Cohen (1995, 1998) as if they were feeding on a solid food turned to liquid or having extraoral digestion. For most of the remaining insects, consuming solid food using shredding or chewing mouthparts is the rule. Table 3.5 lists some insect orders, with the type of mouthparts, typical food, and whether there are artificial diets available for them. The digestive process and all its variations resulting from the evolution of the different orders are discussed in Chapter 4.

TABLE 3.5

The Mouthparts of Different Insect Orders, Their Typical Food, and whether an Artificial Diet Is Available

Order	Mouthparts	Typical Food	Artificial Diet
Protura	Chewer	Detritus	No
Collembola	Chewer	Detritus	No
Thysanura	Chewer	Detritus	No
Thysanoptera	Sucker	Plants (+ insects)	No
Dictyoptera	Chewer	Detritus, etc	Yes
Orthoptera	Chewer	Plants	Yes
Homoptera	Sucker	Plant sap	Yes
Heteroptera	Sucker	Mixtures	Yes
Siphonaptera	Sucker	Blood	Yes (limited success)
Mallophaga	Chewer	Vertebrate detritus	No
Ephemeroptera	Chewer	Detritus	No
Plecoptera	Chewer	Detritus in fresh water	No
Neuroptera	Chewer (+ sucker)	Insects	Yes
Coleoptera	Chewer (+ sucker)	Mixtures	Yes
Lepidoptera	Chewer	Plants	Yes
Diptera	Sucker	Mixtures	Yes
Hymenoptera	Chewer	Mixtures	Yes

Source: Cohen, A. C., *Insect Diets: Science and Technology*, CRC Press, Boca Raton, FL, 2004.

3.8 Physical, Chemical, and Biological Needs for Feeding

A diet that contains all the nutrients can fail to produce the development of a certain insect if there is no stimulus to start feeding. The principal stimuli are described below.

3.8.1 Physical Stimuli

The consistency and structure of a suitable diet are governed by the insect's feeding habits and the type of mouthparts. Thus, for stored products pests, grasshoppers, and cockroaches, granulated or powder media are preferable; phytophagous insects and borers need solid diets with a high water content; fly larvae develop best in gelatinous diets; for mosquito larvae, the food needs to be suspended or dissolved in water; and for suckers it is common to supply the food through a parafilm membrane. The physical properties of the diet, such as hardness, texture, homogeneity, and water content, play an important role.

Diets can be modified physically by adding cellulose, which is not digested by insects. It functions as a stimulant, acting as a diluent so that more food is ingested. Many insects show improved feeding and growth if cellulose is added to their diets, including *B. mori*, *Schistocerca gregaria* (Forskal), and *Locusta migratoria* L. However, the function of the cellulose is to alter the diet texture making it "rougher" and facilitating the passage of the food through the gut.

There are other inert substances, such as clays and insoluble materials, which can be used as diluents in diets to supply suitable texture and "roughness." Slices of whitened *Eucalyptus* sulfate have been the most used materials in Brazil (Vendramim et al. 1982). These slices are obtained from factories making cellulose and after being ground up in a blender are added to the diet and have a similar effect as the alfacel (α-celulose) added to artificial diets used in other countries, at a higher cost.

A diet's consistency can be difficult to adjust because most phytophages need a high water content but also a firm surface against which they can pressure their mouthparts. The polysaccharide agar is the preferred substance for controlling the consistency because it is compatible with the diet's ingredients. On the other hand, it has the disadvantage of containing traces of minerals that makes determinations of mineral needs more difficult. It forms a very firm gel at 1.5% concentration or above. Most diets are prepared with 3% agar although there may be variations in an insect's development depending on its concentration in the diet.

Various studies are being conducted in order to substitute agar that represents around 60% to 70% of the total diet cost. The possible substitutes for agar were discussed by Leppla (1985). The product that has had most success in this substitution is *carrageenan* (CIAGAR), which, although it is also extracted from marine algae, is cheaper. It is a polysaccharide composed of galactose, dextrose, and levulose. There are two forms of *carrageenan*: one extracted from cold water, forming a viscous solution; the other extracted from hot water, forming a gel and which is used in insect diets. Various other gelling agents have been tested, such as alginates, gelatins, gums, glutine, soybean lecithin, and CMC (carboxymethylcellulose), but not always successfully. In some diets, agar can be substituted with sugarcane bagasse and wood shavings. However, problems can occur with the water content in the diet, and in the former case, mite infestations are common.

For the cotton boll weevil, *Anthonomus grandis* (Boheman), the form of the diet is important for stimulating oviposition. Double the quantity of eggs is obtained when the diet is offered in cylindrical pieces with a curved surface compared to a flat-surfaced diet. The first instar larvae of *Lymantria dispar* (L.) only feed if the diet is placed on the internal wall of the rearing chamber in the form of a plate imitating a leaf (Singh 1985).

3.8.2 Chemical Stimuli

Phagostimulants liberate feeding behavior and stimulate insects to feed. To continue feeding insects depend on stimulants although these are not necessarily the same for each developmental stage. There are nutrients that stimulate feeding (sugars, amino acids, salts, sterols, vitamins, organic compounds, and organic acids) and the compounds do not have any nutritive value (secondary metabolic compounds) (allelochemicals), including flavonoids, quinones, tannins, phenylpropanes, isoprenoids, triterpenes (acetogenins and phenylpropanes), and isothioacyanates, phaseolunatin, and catalposids (cyanogenetics and other glucosides). A list of phagostimulants for insects was prepared by Reinecke (1985).

Nutrients are important in diets, of which sugars are the most important phagostimulants, followed by proteins (amino acids) and sterols. Sometimes, besides chemical compounds, physical aspects must also be considered. Thus, sucking insects, for example, only feed through a parafilm membrane. In other cases, plant extracts can be phagostimulants.

3.8.3 Biological Stimuli

Many biological needs of insects are species-specific and not directly related to the nutrition. However, physiological conditions, such as age and diapause, influence the evaluation of the results of nutritional experiments (Singh and Moore 1985).

3.9 Nutritional Needs for Growth

For most insects, a nutritionally complete diet in an axenic culture should contain all or most of the following elements: proteins or amino acids (10 are essential), carbohydrates, fatty acids, cholesterol, choline, inositol, pantotenic acid, nicotinamide, thiamin, riboflavin, folic acid, pyridoxine, vitamin B_{12}, carotene or vitamin A, tocoferol, ascorbic acid, minerals, and water (Vanderzant 1974).

It is known that, independently of the systematic position and feeding habits of insects, their qualitative nutritional needs are similar and these needs, except for a general need for sterols, are, with rare exceptions, close to those of the higher animals. Thus, insects have as basic nutritional needs, amino acids, vitamins and mineral salts (essential nutrients) and carbohydrates, lipids and sterols (nonessential nutrients), which should be suitably balanced, especially regarding the ratio proteins (amino acids) : carbohydrates.

There have been innumerable studies on insect nutrition since the beginning of the century (Uvarov 1928) and after the revisions by Brues (1946) and Fraenkel (1953), there were a large number of publications on the subject, especially after the 1970s (Rodriguez 1972; House 1977 quoted by Dadd 1977; House 1977; Dadd 1985). Mainly from the 1960s, research into the development of insect artificial diets was refined regarding the nutritional needs (Singh 1977), and artificial media for more than 1,300 insect species were described (Singh 1985). This advance in rearing techniques lead to the discovery that some restricted insect groups need nucleic acids and even liposoluble vitamins, such as A, E, and K_1.

3.9.1 Specific Nutritional Needs

3.9.1.1 Amino Acids

Amino acids are necessary for the production of structural proteins and enzymes. They are normally present in the diet as proteins since these are made up of amino acid links (peptide links). Therefore, the value of any protein ingested by an insect depends on its amino acid content and the insect's ability to digest it. Consequently, proteins or amino acids are always essential for a developing insect's diet and are needed in high concentrations for optimal growth. It is known that more than 20 amino acids are present in animal and vegetable proteins. However, in general, insects need at least 10 essential amino acids for growth and development (arginine, histidine, isoleucine, leucine, lysine, metionine, phenylalanine, treonine, tryptophan, and valine), with the others being synthesized from these. The 10 essential amino acids are also needed by adult insects for egg production. Many species, however, can obtain them from larval food and the adults do not need to ingest them. But for optimum egg production, many species should ingest them as adults (e.g., anautogenous mosquitoes, Cyclorrhapha Diptera, predators, Parasitica Hymenoptera whose hosts feed when adults, and some butterflies). Although there are exceptions, the amino acids are used in the laevorotatory form. For the aphid, *A. pisum,* metionine is used in both the laevorotatory (L) and the dextrorotatory forms (D) (Dadd 1977). The D isomers can be found in antibiotics and bacteria cell walls.

In many insects, the nonessential amino acids are needed for growth. Thus, glutamic and aspartic acids are essential for the growth of the silkworm, *B. mori*, which can be further favored with the addition of alanine, glycine, or serine. Good development of the aphid *M. persicae* depends on cysteine, glutamic acid, alanine, or serine (Dadd 1977). The transformation of an amino acid into another one depends, up to a point, on their structural similarity. For example, phenylalanine is essential for insects, and as a rule, tyrosine is not. However, the latter can be synthesized from phenylalanine since their structures are close. Similarly, cystine and cysteine can be synthesized by adding metionine to insects' diets.

The free amino acids constitute the largest or only nutrient of sucking insects specialized in feeding on xylem and phloem because the sap contains little or no protein. Therefore, aphids do not need proteases. Besides the Homoptera, other insects such as termites, wood-feeding beetles, and cockroaches, although they feed on nitrogen-rich sources will obtain them through symbiotic associations.

Nitrogen (N) has a very important role in metabolic processes and in genetic coding, and generally limits insect growth and fecundity (Scriber and Slansky 1981). Hematophagous insects obtain N from their host's blood and the carnivores (in the final larval stages) obtain N, either by feeding on the whole animal tissue or from the host's hemolymph. On the other hand, mycophagous and saprophagous insects derive their N from microorganisms.

3.9.1.2 Vitamins

Vitamins are organic substances, not necessarily interrelated, and which are needed in small quantities in the diet since they cannot be synthesized. They act in the metabolic processes, supplying the structural components of enzymes. The water-soluble vitamins (vitamins of the B complex) are essential for practically all insects. Thus, thiamin, riboflavin, nicotinic acid, pyridoxine, and pantothenic acid are essential for most insects, while biotine and folic acid are essential for some. Other vitamins are necessary for only some restricted insect groups. For example, *Tenebrio molitor* L. needs carnitine, although other insects such as *Dermestes* and *Phormia* are capable of synthesizing it. Other vitamins, such as cyanocobalamin (vitamin B_{12}), although not synthesized by higher plants, can affect cockroach (*B. germanica*) growth and increase silk production in *B. mori* (Dadd 1977). The presence of vitamin B in some insects is due to their association with microorganisms (symbionts).

Choline, although it has a distinct function regarding the complex B vitamins, is needed in much higher doses than the typical vitamins and is essential for all insects. This and mesoinositol are often called lipogenic factors because as subcomponents of phosphatidylcholine (lecithin) and phosphatidylinositol (types of phospholipids), they are involved in the structure of the lipidic membrane and in lipoprotein transport. In insects, choline is synthesized as in mammals by the transmethylation of ethanolamine. Besides this, its need in insects is linked to the fact that choline is a precursor of acetylcholine. Carnitine (vitamin Bt) is chemically related to choline and although nonobligatory for many insects it is indispensable for the Tenebrionidae (Coleoptera). In the oxidation of fatty acids, carnitine has an important biological function in the transport of acetyl coenzyme A from cytosol to the mitochondria in insects. Inositol is a component of phospholipids and essential for most phytophages. Ascorbic acid, or the vitamin C of human nutrition, is present in green plant tissues and has been found essential for insects of the orders Coleoptera, Lepidoptera, Hemiptera, and Orthoptera. Ascorbic acid has been shown to be nonobligatory for insects which do not eat green plants, such as stored products pests, cockroaches, grasshoppers, parasitoid Hymenoptera, and wood borers, demonstrating that these insects have the capacity to synthesize it (Dadd 1977). Although little is known of the biochemical functions of vitamin C, in insects it has both a phagostimulant (Ave 1995) and antioxidant function (Gregory 1996).

Another group of vitamins are liposoluble and include vitamin D (a steroid), which influences calcium absorption and metabolism in vertebrates but is unnecessary in insects. However, vitamin A (retinol) or the provitamin beta-carotene is essential for the formation of visual pigments in insects. Japanese researchers have shown the importance of vitamin A for the silkworm (*B. mori*).

Vitamin E (alpha-tocoferol) is an antisterility factor in vertebrate nutrition. It is used in artificial diets with an antioxidant function (to avoid breaking up polyunsaturated fatty acids) (Gregory 1996), a function also performed by ascorbic acid in many diets. However, there is evidence that it can affect reproductive performance.

Vitamin K, needed for normal vertebrate blood coagulation, was for a long time considered unnecessary for insects. However, McFarlane (1978) observed that vitamin K_1 has, like alpha-tocoferol, the phytil side of the molecular chain that would affect spermatozoid viability in insects, but this is the only evidence of the role of vitamin K in insects.

3.9.1.3 Mineral Salts

Very little is known of inorganic nutrition in insects because it is difficult to manipulate simple radicals in diets. It is known that insects need considerable amounts of potassium, phosphate, and magnesium but little calcium, sodium, and chlorine for growth and development. It is difficult to determine the amounts of these latter salts that insects need because since the needs are small (traces) and they are often supplied through the impurities in other diet components. Mineral salts are important for the ionic equilibrium and insect membrane permeability, often acting as enzyme activators or part of unidentified respiratory pigments (in this case, Cu). Recent studies over successive generations, in which the symbionts have been eliminated, have demonstrated how essential Cu, Fe, Zn, and Mn are for insects. Iron is very important in various biological processes, including enzymatic reactions, production of the ecdysis hormone, cuticle formation, and various metabolic processes. Selenium is an antioxidant. Mn and Zn are enzymatic cofactors. Due to the lack of knowledge in this area, salt mixtures for vertebrates (e.g., Wesson salts, mixture of salts n 2 USP XIII, mixture of salts M-D n 185, mixture of salts USP XIV) are used in insect diets but they are probably overestimated and contain many more minerals than the insects really need (Cohen 2004).

3.9.1.4 Carbohydrates

Carbohydrates are the main energy source for insects. They can be converted into fats for storage and contribute to the production of amino acids. Thus, the carbohydrates, fats, and proteins are involved in cycles of reactions for energy production. Probably most insects can use the common sugars and the omission of a sugar or digestible polysaccharide harms their development. Thus, in general, insects need large quantities of carbohydrates in their diets. Grasshoppers of the genus *Schistocerca* need at least 20% sugar in their artificial diets to obtain good growth. For *Tribolium* sp. (Coleoptera), maximum growth is reached with 70% sugar in the diet. Most larvae of phytophagous insects need some type of carbohydrate (various sugars or polysaccharides, depending on the digestive enzymes present). Insects that feed on seeds and cereals need around 20% to 70% carbohydrate of the solid nutrients in the diet while aphids need 80%. There are insects whose immature stages do not need sugars but which are required by the adult (e.g., mosquitoes). Therefore, carbohydrate needs vary between species and often between the immature stages and adults of the same species. For example, the larva of *Aedes* sp. (Culicidae) can use amide and glycogen while the adults do not. Therefore, the carbohydrates are used mainly as energy sources, as phagostimulants, and in many cases (as in *M. persicae*), are necessary for growth, a longer life, and fecundity. However, carbohydrates can be substituted with proteins (amino acids) or fats. This substitution will depend on the insect's ability to convert the proteins and fats into products that can be used in the transformation cycles as well as the speed with which these reactions occur. In *Galleria mellonella* L. (Lepidoptera), the carbohydrates are totally substituted with wax. The mosquitoes *Aedes aegypti* Rockefeller and *Culex pipiens* L. complete their development without carbohydrates (Dadd 1977).

Some insects use a large number of carbohydrates that depend on the ability to hydrolize polysaccharides, the speed with which different substances are absorbed, and the presence of enzymes that can introduce these substances into metabolic processes. *Locusta, Schistocerca,* and stored products pests use a large number of carbohydrates. For example, Tribolium needs amide, manitol (alcohol), rafinose (trissacharide), sucrose, maltose, and celobiose (dissacharides), as well as the monosaccharides, mannose and glucose, among others. Some insects do not use polysaccharides while others only use a small number of carbohydrates. *C. supressalis* (Lepidoptera) only uses sucrose, maltose, fructose, and glucose. Llpke and Fraenkel (1956) found that the pentoses do not promote growth and can be toxic, perhaps by interfering with the absorption and oxidation of other sugars that are normally used.

3.9.1.5 Sterols

Sterols are essential for almost all insects since they are incapable of synthesizing them from acetates like vertebrates. There are cases, as in aphids, in which the symbionts synthesize sterols, permitting the aphids to develop in a diet without this nutrient. However, in most insects, a source of sterol is necessary for growth and reproduction and the class used is that similar to cholesterol with a hydroxyl group in the third position. Sterols have various functions such as promoting ovigenesis and larval growth, being responsible for the cuticle sclerotization, having a metabolic and anti-infection role, and being steroid hormone precursors, as discovered in *B. mori* (Dadd 1977). The sterols also have an important structural function in the cell membrane and in lipoprotein transport. Many phytophages ingest phytosterols and can transform them into cholesterol. There are other cases in which the dehydrocholesterol, or ergosterol, can be included in the diet but they do not supply all the sterol needs. These are called sparing agents and are only capable of substituting the structural role without being able to perform cholesterol's metabolic functions. Although there are exceptions, the total needs of the various insects studied are satisfied with cholesterol or estigmasterol.

3.9.1.6 Lipids

Lipids are esters of one or more fatty acids and glycerol, which are formed from enzymatic hydrolysis in the insect's gut. Fats are the main form in which energy is stored, but except in specific cases and in small quantities, they do not normally form essential components of the diet. Since only small quantities of fats exist in the leaves, they could not be an important energy source in phytophages, and even in *G. mellonella* (the greater wax moth), bee wax is not essential for the diet. Insects synthesize lipids from proteins and carbohydrates. However, some fatty acids, such as linoleic and linolenic acids, are not synthesized by insects. Linoleic acid is essential for *Anagasta kuehniella* (Zeller) (Lepidoptera) and for grasshoppers of the genus *Schistocerca*, because related to the formation of lipidic phosphatides, when absent it affects ecdysis in these insects. This acid also interferes in wing formation in *A. kuehniella* and *Pseudoplusia includens* (Walker) (J. R. P. Parra, personal observation); in *A. kuehniella* it also affects insect emergence. In some cases, linolenic acid can replace linoleic acid.

3.9.1.7 Nucleic Acids

Nucleic acids or their components (nucleotides, nucleosides, and bases) form another category of growth factors soluble in water and which are necessary for building ribonucleic acid (RNA) and deoxyribonucleic acid (DNA). The higher animals can biosynthesize all the nucleotides they need. Among the insects, except for the Diptera, they are not required exogenously. Even the Diptera can synthesize them but so poorly that they can limit their growth. These insects complete larval development in diets without any nucleic acid but very slowly. *Drosophila melanogaster* Meigen and *Musca domestica* L. can complete their development without this component but if RNA (or more specifically adenine) is added, their development improves (Dadd 1977). House (1961) also refers to this need for the Coleoptera, *Sclublus granosus* Fahrer.

3.9.1.8 Water

Insects, like all organisms, need water and most terrestrial insects contain at least 70% of water, which varies from 46% to 92%. The ingestion of water can be direct or by removing it from the environment. Stored products pests, for example, *A. kuehniella* (Lepidoptera) can survive with only 1% water in the food. The water can be produced metabolically by fat oxidation to maintain water equilibrium, as seen for certain desert Tenebrionidae, which produce metabolic water at the cost of converting lipids (Dadd 1977).

After studying some nutritional indices, Scriber and Slansky (1981) verified that final instar larvae (which are leaf feeders) perform better when they feed on leaves with 75% to 95% water content. Insects that feed on tree leaves compared to those that feed on forbs (broad-leaf herbaceous plants) make better use of the first type of food owing to the lower water content, which coincides with a faster decline in nitrogen. The adults of many holometabolous insects "drink" water with some exceptions for larvae and nymphs. The eggs and larvae of many insects absorb water. *T. molitor* produces one generation per year when fed on dry food but with access to water it can produce up to six generations per year. Larvae of *Syrphus ribessi* (L.) (Diptera) aestivate when desiccated but in contact with water absorb it through the anal papillae and return to normal activity (Dadd 1977). Many insects require high moisture content in their food; the water dilutes the nutrients, and more are consumed. The conversion efficiency can be increased by dilution. Feeny (1975) found there was a reduction in this efficiency with less water in the diet for the Lepidoptera, *Agrotis ipsilon* (Hufnagel), and *Hyalophora cecropia* (L.). However, an optimum amount of water does not correspond to an optimum food conversion because there is an interaction between the efficiency and the amount of dry material ingested. There are cases in which plants poorer in nitrogen were consumed more efficiently by *Pieris rapae* (L.) (Lepidoptera) than nitrogen-rich plants (Feeny 1975).

3.9.2 Nutrient Storage

An essential nutrient is often not required in the diet because reserves were accumulated before feeding. There are two important nutrient sources: the vitellus of the egg and the fatty bodies of larvae and adults. Since eggs are small, they cannot store macronutrients such as glucose, but vitamins, for example, can be stored in such a way as to satisfy larval nutritional requirements. There are differences in storage capacity even among the micronutrients. Thus, cockroaches from the genus *Blattella* contain large quantities of linoleic acid but no thiamine. *Blattella* eggs contain sufficient amounts of inositol for development up to the third nymphal instar. However, in *Schistocerca* sp., the amount of beta carotene stored in the egg is sufficient for complete larval development. When eggs are obtained from adults with a beta carotene deficiency this component should be added to the diet for normal insect development (Singh and Moore 1985).

Nutrients can be stored in large quantities (even macronutrients) in larval and adult fatty bodies. This can be seen in adult Lepidoptera that do not feed and where the adult metabolic processes depend on reserves from the immature stages. Some grasshoppers can store certain nutrients in fatty tissue and if they are allowed to eat grasses during the first two larval instars and are then fed with carbohydrate-poor food, they can live without this component due to previously accumulated reserves. *A. grandis* (Coleoptera) larvae store enough choline and inositol for egg development even if these micronutrients are excluded from the adult diet (Chapman 1982). The food can often originate from tissue degradation; that is, result from the autolysis of flight muscles (e.g., egg development in some mosquitoes and the immature stages of aphids).

3.9.3 Symbionts

Insects can have symbiotic relationships with bacteria (Blattodea, Isoptera, Homoptera, Heteroptera, Anoplura, Phthiraptera, Coleoptera, Hymenoptera, and Diptera), flagellate protozoans (wood cockroaches and termites), yeasts (Homoptera and Coleoptera), and fungi (Hemiptera, *Rhodnius*). Those species that have no associations with microorganisms are called asymbiotic. The symbionts can live freely in the gut, as in the case with flagellate microorganisms that inhabit the hind portion of the gut of cockroaches and termites and feed on wood. Bacteria of plant suckers live in the cecum of the last segment of the midgut. Most of the microorganisms are intracellular and can occur in various parts of the body. The cells that house these symbionts are known as mycetocytes and they can join together into organs known as mycetomas. The mycetomas are large polyploids and occur in different tissues. Generally, they are distributed irregularly in fatty tissue but may be irregular cells in the midgut epithelium, or in the ovarioles, or free in the hemolymph. In holometabolous insects, the mycetomas are only found in immature forms. There are some cases of insect–symbiont associations that are casual and others in which this association is constant.

The role of symbionts in insect nutrition is very large (see Chapter 6). Thus, they can be the main food source (e.g., fungi for ants, beetles, mosquitoes, and flies), help in digestion by secreting enzymes in the intestine, convert internally unusable food into a usable form (e.g., in termites and cockroaches), supply auxiliary growth factors (e.g., vitamins and sterols), or even carry out biochemical functions to enable an insect to survive and grow on an inadequate diet (fixation of atmospheric nitrogen, desintoxication of metabolic residues, and allelochemicals). In general, arthropods have symbiont organisms only if they feed on inadequate diets during their life. Inadequate foods include wood and stored grains (rich in cellulose but protein deficient), wool, hair, feathers (rich in queratin but vitamin-deficient), plant juices (nitrogen-deficient), and blood or serum (deficient in water-soluble vitamins). Although symbionts can be eliminated by superficial sterilization of the eggs, centrifugation, heat treatment, microsurgery, or chemotherapy, the study of nutrition is complicated when there is any microorganism–insect association. These symbionts are transferred from one generation to the next through contaminated food, eggs, or by specialized processes during ovigenesis. The importance of these symbionts, such as *Wolbachia* and *Buchnera*, is being studied by research groups throughout the world. Today, it is known that a large percentage of insects have such symbionts, many with still unknown functions. For more details see Bourtzis and Miller (2003) and Chapter 6.

3.10 Diet Composition

3.10.1 General Components

The general components of an artificial diet are listed in Table 3.6. Making up a diet with each component by itself would be inviable as a routine laboratory activity or for producing insects in a mass rearing. Therefore, ingredients are used that supply each nutrient (proteins, vitamins, mineral salts, carbohydrates, lipids, and sterols) at a lower cost. Since diet preparation involves heating, there may be degradation of proteins, vitamins, and even anticontaminants during the preparation. At other times, there may be problems in obtaining a product in the market making adaptations necessary. Thus, results may not be repeated since a product may not always be elaborated in the same way: for example, toasted or fresh wheat germ, and water-soluble or insoluble anticontaminants. Cohen (2004) demonstrated the nutritional differences between toasted and fresh wheat germ. Also, since the products are not pure and may be badly stored, they can introduce contaminants (e.g., agar) and interfere in insect rearing.

The discovery that wheat germ could be a protein source for diets (Singh 1977) was a significant advance because it contains all the nutrients with the possible exception of ascorbic acid. It has 18

TABLE 3.6

General Components of an Artificial Diet

Protein Sources	Lipid and Sterol Sources	Mineral Salts Sources
Casein	Vegetable oils	Diverse mixtures
Wheat germ	Cholesterol	(e.g., Wesson salts)
Soybeans	Linolenic acid	
Dry beans	Linoleic acid etc	Carbohydrate sources
Brewer's yeast		Sucrose
Corn	Anticontaminants	Glucose
	Fungistatic agents	Frutose
	Antibacterial agents	
Gelling agents	Antioxidants	Vitamin sources
Agar		(Fortifying mixtures)
Alginates and similar		(e.g., Vanderzant)

common amino acids, sugars, triglycerides, phospholipids (choline and inositol), complex B vitamins, tocoferol, carotene, 21 mineral elements, and more than 50 enzymes, as well as having a phagostimulatory action. At the beginning of dietary studies, egg albumin was an important protein source but since it coagulates on heating, its use was discontinued.

Casein, soybean, dry bean, brewer's yeast, and corn are other much-used protein sources. Since dry beans are poor in certain amino acids, such as metionine and cistein, they are normally accompanied in diets with another protein source to complement these deficiencies. The same occurs with soybean, which is poor in aminosulfurs. Corn, with lysine and tryptophan added, can be an important protein source. Insect behavior differs depending on the varieties of soybeans corn and dry bean used in diets. Dry bean with different colored seed coats and varied tannin contents result in variable development of *Spodoptera frugiperda* (J. E. Smith) due to insect digestibility. Thus the Carioca variety has shown itself more suitable to Brazilian conditions (Parra and Carvalho 1984), whereas white corn has shown itself suitable for many species, such as *D. saccharalis* (Parra and Mihsfeldt 1992). This is also valid for natural diets, since Costa (1991) observed that *Nezara viridula* (L.) showed a varietal preference when reared on soybean seeds. In the United States, the dry bean variety Pinto Bean is easily found as a constituent of various artificial diets (Singh 1977).

Vegetable oils are the source of lipids and sterols. Among them, wheat germ oil and corn, flax, and sunflower oils are the most used. When linolenic and linoleic acid requirements are very high, such as for *Elasmopalpus lignosellus* (Zeller), and since these pure acids are expensive, an option is made for sunflower oil, which is richer in the acids (Meneguim et al. 1997). Sucrose, fructose, and glucose are the carbohydrates most used.

The diet pH may play an important role in insect development; thus, an insect that lives in the wild in an acid environment "would prefer" a lower pH while one that lives in an alkaline environment "would prefer" a diet with a pH greater than 7. Funke (1983) demonstrated that in acid media, fungi develop less, and therefore less anticontaminant is needed in the diet. Kasten Jr. et al. (1998) and Parra et al. (2001) found that the citrus fruit borer, *Gymnandrosoma aurantianum* (Lima), has a prolonged larval stage and greater mortality in more acid diets, which explains this insect's preference in developing in mature fruits that have a less acid pH. Thus, the pH is important for food palatability, influences the texture and smell, microorganism contamination, affects enzymatic and nonenzymatic reactions, and can be important in food preservation (Cohen 2004) (see Chapter 4).

In general, the antioxidants used are ascorbic acid (vitamin C), α-tocoferol (vitamin E), and vitamin A. The beneficial and harmful effects of these antioxidants have been described (Cohen and Crittenden 2004; Cohen (2004).

Cellulose is purchased as α-cellulose (known as alfacel). In Brazil, slices of whitened *Eucalyptus* sulfate are used and ground up in a blender (Vendramin et al. 1982), giving excellent results for *H. virescens* and other species.

A list of vitamin mixtures and mineral salts used in diets is presented in Cohen (2004). This is one of the big problems for those starting studies with artificial diets because the mixtures cannot be found already formulated for sale. Therefore, their composition has to be known so that they can be prepared in laboratories or drug stores. When various components are used, besides the macronutrients supplied, trace substances are added, and these are sometimes important but not always easy to detect (Table 3.7). For this reason, it is said that studies with artificial diets are part of practical nutrition since they are formulated by trial and error, considering the characteristics mentioned.

3.10.2 Adult Requirements

In hemimetabolic insects, nymphal feeding and adult feeding are generally similar. In holometabolic insects, in which both stages feed in the same way, the situation is similar; that is, nutrition in the adult is an extension of larval nutrition; in some Coleoptera studied, adult needs are like those of the larvae and the accumulation of certain nutrients in the larval stage allows significant egg production if the nutrients are absent from the adult diet. The reproduction of adult holometabolic insects that do not feed (e.g., Lepidoptera) totally depends on larval reserves; the intermediate situation in which the adult food is completely different from that of the larva (Lepidoptera, Hymenoptera, Diptera) has attracted more

TABLE 3.7

Substances Used in Insect Diets and the Macro- and Micronutrients Supplied

Ingredient	Macro Nutrient Supplied	Substances—Trace[a]
Proteins		
Casein, soybeans, dry beans, wheat germ, corn, etc.	Amino acids	Fatty acids, cholesterol, sugars, vitamins and minerals
Albumin	Amino acids	Vitamins and minerals
Amino acids	Amino acids	Other amino acids, isomers
Carbohydrates		
Sugars	Simple sugar	—
Amide	Simple sugars	Amino acids, vitamins
Lipids		
Vegetable oils	Fatty acids	Sterols, carotene, tocoferol
Phospholipids	Fatty acids, choline, inositol	Sterols, carotene, biotin, tocoferol
Fatty acids	Fatty acid	Isomers
Sterols	Esterol	Others sterols
Mixtures of salts	Cations, anions	Others minerals
Vitamin mixtures	Vitamins	—
Cellulose	None	Minerals
Ágar	None	Minerals

Source: Smith, C. N., *Insect Colonization and Mass Production*, Academic Press, New York, 1966.

[a] Traces of important minerals can be present in all diet components.

attention. It was first thought that in such cases, males and females only needed water and carbohydrates; however, it was found that females needed protein foods to continue egg laying; recent research shows that adult females require varied salts, lipids, and vitamins, as well as amino acids or proteins, for optimal longevity and fecundity (Cohen 2004).

The carbohydrate concentration required can vary between insects. For example, while maximum oviposition for *Trichoplusia ni* (Hübner) is obtained with 8% sucrose (Shorey 1962), for *Leucoptera coffeella* (Guérin-Méneville) the best results are obtained with 10% sucrose and 5% glucose (Nantes and Parra 1978). Maximum oviposition for *Anticarsia gemmatalis* (Hübner) is obtained with the following formula (Campo et al. 1985): sorbic acid (1 g), honey (10 g), methyl parahydroxibenzoate (nipagin) (1 g), sucrose (60 g), and distilled water (1 liter). Maintain this mixture in a refrigerator and when used, mix 75% of this solution with 25% of beer. For *Helicoverpa zea* (Boddie), beer also gives good results due to the presence of yeast. For Syrphidae (predators), Hymenoptera and Diptera (parasitoids), oviposition is obtained mixing honey dew, nectar, and pollen or honey, sugar, raisins, pollen, protein extracts, and minerals. Sometimes there are big differences in related groups. Tachinids are not very demanding since they only need water and sugar (Singh and Moore 1985).

There are more demanding insects, such as fruit flies, whose diet should contain enzymatic hydrolyzed. Some Lepidoptera, such as *D. saccharalis*, do not need carbohydrates when adults because water is more important and enough for oviposition (Figure 3.3) (Parra et al. 1999).

Although some entomophages require special foods, most hymenopterous parasitoids can produce eggs with a carbohydrate source such as honey (Waage et al. 1985). Even for those that do not feed, such as *Aphitis melinus* De Bach, Heimpel et al. (1997) showed that the availability of honey is important and interacts with the host, affecting the longevity and fecundity and directly affecting egg resorption. For *Trichogramma* spp., the supply of honey increases longevity and fecundity, especially when there is an opportunity to parasitize (Bleicher and Parra 1991). For others, such as the mymarid, *Anaphes iole* Girault, lack of food increases its efficiency (Jones and Jackson 1990). The ichneumonid *Phygadeuon trichops* Thomson requires honey, soybean flour, and yeast, with a supplement of sugar, powdered milk, soybean flour, and yeast (10:10:1:1). Some Chalcididae, Pteromalidae, Braconidae, and Ichneumonidae

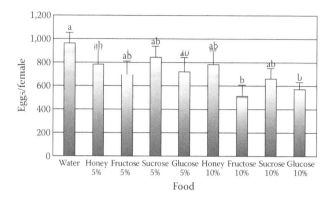

FIGURE 3.3   Oviposition (number of eggs/female) of *Diatraea saccharalis* fed with various sources of carbohydrates (25°C ± 10°C, 60% ± 10% R. H. and 14 h photophase). (Modified from Parra, J. R. P., et al., *An. Soc. Entomol. Brasil*, 28, 49–57, 1999.)

need the host's fluid. For parasitoids, a mixture of honey and pollen (1 : 1) applied on a rough surface (wood chips) and offered inside cages can function (Singh and Moore 1985).

Adult predators may require complex diets. For Chrysopidae, Morrison (1985) used a diet composed of equal parts of sugar and yeast flakes, mixed with water to form a thick paste. This mixture contains 65% protein and is sufficient for a high fecundity. Sometimes, chemical substances (potassium chloride and magnesium sulfate) can have a synergistic effect on parasitoid oviposition (Singh and Moore 1985), as well as amino acid peptides and proteins (Nettles 1986; Kainoh and Brown 1994).

3.11 Rearing Techniques

The choice of rearing containers can affect insect health and nutrition. If the insects are reared individually, the possibility of disease spread and contamination is reduced. Rearing insects individually eliminates cannibalism although there are cases in which even though species are gregarious they can turn cannibalistic if the insects are grouped together or if diets are inadequate.

In the United States, insects are reared in plastic cups with an approximate volume of 30 ml, having cardboard lids (or plastic). There are also ice cream cups and cups made of cardboard and transparent plastic. Ignoffo and Boening (1970) described plastic trays with 50 individual compartments (each compartment measuring 2.8 × 4.1 × 1.6 cm and with a volume around 15 ml). After placing the diet and "inoculating" with insects, these trays are closed with a fine layer of transparent plastic, aluminum foil, or paper. Various other types of containers have been described, even paraffin paper bags for multiplying larvae and obtaining pathogens. In the United States, there are specialized firms (e.g., BIO-SERV Inc.) that produce cheap and disposable containers commercially. The big problem with containers is the evaporation of the diet, which causes alterations in its texture and palatability. In Brazil, rearing in plastic containers (coffee cups) is being tried but these are often perforated, especially by borers, due to their thin walls. Glass containers (2.5 cm diameter by 8.5 cm height) have given good results, including for laboratory mass rearing of *C. flavipes*, parasitoid of the sugarcane borer, *D. saccharalis*.

There are some types of containers used in rearing insects on artificial diets, including plastic containers with a transparent plastic, self-sealing lid, which allow the insect to be observed and also allows oxygen exchange. To be considered suitable, a container should have the following characteristics: cheap, transparent, easily available on the market, made of a material nontoxic to the insect, and able to maintain humidity.

It is necessary to have an idea of the amount of diet consumed by each larva in order to know how many individuals can be reared in each container (Table 3.8). Although phytophagous insects eat foods with high water contents, much rearing has been successfully done with foods that have low water

TABLE 3.8

Effect of the Number of *Spodoptera Exígua* Larvae per Rearing Container*

Number of Larvae per Container	% Pupae	Pupal Weight (g)
50	94	0.126 a †
100	96	0.125 a
200	90	0.112 b
300	80	0.097 c

Source: Stimmann, M. W., R. Pangaldan, and B. S. Schureman, *J. Econ. Entomol.*, 65, 596–7, 1972.

* Container of 3.78 L.

† Means followed by at least one letter in common are not different at the 5% probability level.

contents since the lower water content reduces interference by contaminating agents. On the other hand, when insects are young, if there is water condensation on the container walls, this can cause death. Water can also interfere in the concentration of nutrients in the diet.

3.12 Sequence for Preparing an Artificial Diet

To explain the sequence for preparing an artificial diet, *D. saccharalis*, the sugarcane borer, will be used as an example since it was the first phytophagous insect to be reared on an artificial diet in Brazil. Rearing begins with the collection of eggs or pupae from other laboratories that already have good quality colonies or by field-collecting larvae. Often, rearing can be started with adults collected in light traps.

The diet is prepared by mixing the ingredients (except the agar) in water and then in a blender. The agar is dissolved separately in boiling water. The two preparations are then mixed and homogenized using an electric stirrer and the still-hot diet is transferred to rearing containers (in this case, glass containers). To avoid degradation, the anticontaminants and vitamins should be added to the diet when its temperature is about 60°C to 65°C.

Fifty adults (20 males and 30 females) of *D. saccharalis* are placed in PVC cages 10 cm in diameter and 22 cm high, lined internally with moist sulfite paper to receive the eggs and closed with voile cloth and elastic or with one of the parts of a Petri dish. Although many workers supply a 10% honey solution for adults, which is replaced every two days, Parra et al. (1999) have shown that they do not need carbohydrates since water is more important and sufficient for normal egg laying. Besides being required for egg laying, water for wetting the oviposition site is fundamental because otherwise the eggs will dry out due to the many aeropyles they have (Figure 3.4).

The sulfite paper with the eggs will be treated with 5% formaldehyde for 5 minutes and then washed with distilled water for a further 5 minutes. The eggs may also be treated in the following sequence: 0.2% formaldehyde (2 minutes), distilled water (2 minutes), and copper sulfate (2 minutes). Between 10 and 15 treated eggs will be transferred to the diet tubes inside aseptic chambers.

After inoculating the diet tubes with the treated eggs, the tubes are placed in wooden or wire frames. Part of the *D. saccharalis* population is for parasitoid production (95%) and part (5%) for further laboratory rearing. In this case, the pupae are taken from inside the glass tubes, sexed, and removed to adult cages.

To obtain eggs, a temperature of 20°C to 22°C, atmospheric relative humidity of 70% ± 10%, and a 14-hour photophase are suitable and for the development of eggs, larvae, and pupae, a temperature of 30°C, relative humidity of 70% ± 10%, and a 14-hour photophase. Based on this example, it can be seen that for more insect production, rooms with different temperature conditions may often be necessary since the needs can vary for each developmental stage.

It is possible to plan production by simply basing it on the temperature requirements of the insects. Thus, the insects' temperature requirements are evaluated by the thermal constant (K), expressed in

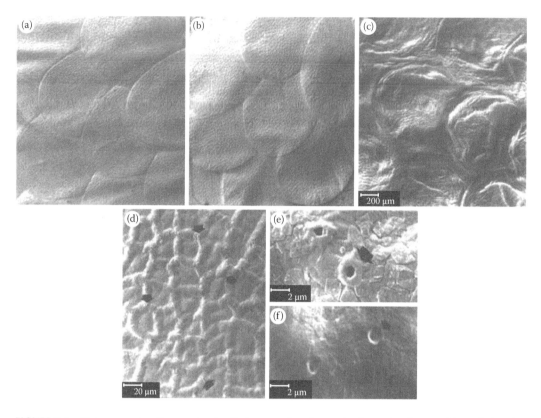

FIGURE 3.4 *Diatraea saccharalis* eggs, showing the large number of aeropyles that makes them sensitive to desiccation when the adults are offered with (a) water and moistened paper, (b) only moistened paper, (c) dry paper, (d) details of the aeropyles on the chorion, (e) detail of open aeropyle, and (f) detail of closed aeropyle. (Modified from Parra, J. R. P., et al., *An. Soc. Entomol. Brasil*, 28, 49–57, 1999.)

degree days, which has been used for many years to forecast plant growth. This constant originated from the hypothesis that the duration of development, considering the temperature, is a constant and is the sum of the temperatures calculated from a lower temperature limit called the threshold temperature (Tt). Since insects are poikilothermic; that is, follow the environmental temperature, and this thermal constant can also be applied to their development. Thus, K = D (T – Tt), where K = thermal constant (degree days), D = duration of development (days), T = atmospheric temperature (°C), and Tt = threshold temperature.

The Tt can be calculated by various methods (Haddad and Parra 1984; Haddad et al. 1999). Once the lower temperature is determined, the insect cycle can be estimated in a room whose temperature is registered or controlled. For example, in a rearing room kept at 25°C, the duration of the egg, larval, and pupal stages can be estimated for an insect whose thermal requirements are given (Table 3.9). Thus, simply put,

TABLE 3.9

Forecast of Insect Development Based on Thermal Requirements of the Different Development Stages in an Incubator Kept at 25°C

Stage	Tt (°C)	K (GD)	Estimate
Egg	11.5	79.48	5.9 ≈ 6 days
Larva	12.2	156.53	12.2 ≈ 12 days
Pupa	15.1	67.81	6.8 ≈ 7 days

Note: Tt = threshold temperature; K = thermal constant.

if the threshold temperature of the egg stage is 11.5°C, it will use 13.5°C per day (in a room maintained at 25°C) as energy until the 79.48 degree-days required for embryonic development are reached. This logic is valid for the other development stages and also for natural enemies.

3.13 Examples of Artificial Diets

Singh and Moore (1985) have referred more than 1,300 insect or mite species reared on artificial diets in different parts of the world. Dozens of diets for Lepidoptera, Coleoptera, and Diptera have been developed in Brazil, including diets for Heteroptera (Panizzi et al. 2000; Fortes et al. 2006) and Dermaptera (Pasini et al. 2007). Some examples are shown in Table 3.10.

3.14 Minimum Sanitary Precautions for Insect Rearing in Artificial Media

As a consequence of the increased use of artificial media for insect rearing, it has become necessary to use anticontaminants to control yeasts, fungi, bacteria, viruses, and protozoans. If these microorganisms are not eliminated they can eliminate laboratory populations because their dispersion is facilitated in mass rearing. Some general anticontaminants will be cited and for this reason it is suggested that before sterilizing the outer surfaces of eggs, artificial media, or pupae, preliminary tests on concentrations and exposure times be done for each species, since an incorrect application can affect insect development.

3.14.1 Room for Diet Preparation

There should be a suitable room for weighing out the ingredients and preparing the diets. In this room, the placing of eggs or larvae in containers with artificial media will also be done as well as the preparation of the adult food.

TABLE 3.10

Some Artificial Diets Developed or Adapted in Brazil

Species	Order	Family	Reference
Tuta absoluta	Lepidoptera	Gelechiidae	Mihsfeldt and Parra (1999)
Elasmopalpus lignosellus	Lepidoptera	Pyralidae	Meneguim et al. (1997)
Ceratitis capitata	Diptera	Tephritidae	Pedroso (1972)
Anastrepha fraterculus	Diptera	Tephritidae	Sales et al. (1992)
Sphenophorus levis	Coleoptera	Curculionidae	Degaspari et al. (1983)
Stenoma catenifer	Lepidoptera	Elaschistidae	Nava and Parra (2005)
Helicoverpa zea	Lepdoptera	Noctuidae	Justi Jr. (1993)
Anthonomus grandis	Coleoptera	Curculionidae	Monnerat (2002)
Psudaletia sequax	Lepidoptera	Noctuidae	Salvadori and Parra (1990)
Cerconota anonella	Lepidoptera	Occophoridae	Pereira et al. (2004)
Agrotis ipsilon	Lepidoptera	Noctuidae	Bento et al. (2007)
Agrotis subterranea	Lepidoptera	Noctuidae	Vendramim et al. (1982)
Gymnandrosoma aurantianum	Lepidoptera	Tortricidae	Garcia and Parra (1999)
Platynota rostrana	Lepidoptera	Tortricidae	Nava et al. (2006)
Phidotricha erigens	Lepidoptera	Pyralidae	Nava et al. (2006)
Argyrotaenia sphaleropa	Lepidoptera	Tortricidae	Nava et al. (2006)
Cryptoblabes gnidiella	Lepidoptera	Pyralidae	Nava et al. (2006)
Utetheisa ornatrix	Lepidoptera	Arctiidae	Signoretti et al. (2008)

3.14.2 Room for Adults

The adults should be kept in a separate location since the wing scales of Lepidoptera contain many microorganisms that will be easily transported by air currents. Eggs collected from adult cages should be transferred to the diet preparation room where they will be disinfected.

3.14.3 Room for Larval Development

The larvae will also be kept in a separate location until pupation, when the pupae will be transferred to the adult location.

Besides these three rooms, isolated locations are necessary for eliminating the residues since the latter are foci of contamination. The walls and the counters of these locations, as well as the shelves, should be lined with tiles, Formica, or similar material so that they can be cleaned daily with disinfectants, quaternary ammonium compounds, 5% sodium hypochlorite, 37% to 40% formaldehyde. Sodium hypochlorite is often used since apart from being cheap it has good stability and solubility and has a low mammalian toxicity. Besides these advantages, it inactivates proteins and eliminates viruses, bacteria, fungi, algae, and protozoans because it is a strong oxidant. Floors should also be cleaned daily with these products. Epoxy paints or equivalent materials can be substituted for tiles and Formica in rearing laboratories. Leppla and Ashley (1978) should be consulted for greater details on installations for insect rearing (including mass rearing).

Sterilization of equipment and containers should be carried using an *autoclave*. Heating is done under pressure (15 pounds at 121°C). The sterilization time varies from 10 to 15 minutes, depending on the type of material and its volume. It is suitable for solutions, rubber tubes, instruments, sand, and boxes (not plastic).

Irradiation: Ultraviolet (UV) (2,650 Å) and gamma rays are sufficient against bacteria, fungi, and virus. The sterilization lamps are mercury and transmit at 2,567 Å, and the UV is also used for plastic and paraffins.

Dry heat produced by ovens gives good sterilization of rearing containers, glassware, and cotton.

Sterilization of equipment and containers using chemical agents include *heavy metals.* Soluble salts of Hg, Ag, and Cu are bactericides. These salts were components of solutions that were used to sterilize insect eggs. A well-known solution is White's, which has $HgCl_2$ 0.25 g, NaCl 6.50 g, HCl 1.25 ml, ethyl alcohol 250.00 ml, and distilled water 750.00 ml. In this mixture, the alcohol increases the toxicity of the $HgCl_2$. There are cases in which adding alcohol lowers the toxicity, such as for the phenols and formaldehyde. Heavy metals can be used individually, such as $HgCl_2$, at 0.1%, for an exposure period of up to 4 minutes. Although these heavy metals are still efficient they are not currently recommended due to human health risks.

Ammonium quaternary compounds such as sodium hypochlorite. This component is used at concentrations of 0.01% to 5% for variable time periods (e.g., NaOCl 0.05% at 30 minutes exposure; NaOCl 0.2% at 7 minutes; NaOCl 2.5% at 5 minutes; NaOCl 5.0% at 3 minutes).

Formaldehyde. Examples include formaldehyde 10% at 20 minutes exposure, formaldehyde 20% at 10 minutes exposure. In both cases, after treatment, wash with water for 10 minutes.

Sodium hydroxide. Examples include NaOH 1% at 20 minutes exposure, NaOH 2% at 10 minutes (add formaldehyde at 2% and wash in 70% alcohol).

Copper sulfate ($CuSO_4$), glacial acetic acid (CH_3COOH), and trichloroacetic acid (CCl_3COOH) should have their concentrations tested for each insect species.

Sterilization of the medium (artificial diet) is done with chemicals used in sterilizing media, such as formaldehyde 0.03% to 0.3%, methyl parahydroxibenzoate (nipagin) 0.04% to 2.00%, butyl and propyl parahydroxibenzoate, sodium hypochlorite 0.01:0.2:1.0%, propionic acid, sodium benzoate, benzoic acid, potassium sorbate, sorbic acid 0.05% to 0.15%; streptomycin, penicillin, and aureomycin. Physical agents such as irradiation or autoclavation can be used to sterilize media. Adjusting the pH can be used as an alternative method for avoiding contaminations. Alverson and Cohen (2002) determined the ideal concentrations of various anticontaminants for *Lygus hesperus* (Knight) diets based on the number of emerging adults. Thus, some aspects have to be considered when choosing anticontaminants: dosage (test for each product), diet humidity, pH, and product formulation, insect stage exposed, effect on growth in the generation under study and subsequent ones, effect on symbionts, and insect order. Although the Diptera and Coleoptera are more sensitive to anticontaminants, there are lepidopterous families significantly affected by sorbic acid (Dunkel and Read 1991). The sterilization of pupae can be done by washing in sodium hypochlorite (e.g., at 0.2% for 15 minutes) followed by washing with distilled water.

Soares (1992) characterized the responses of insects attacked by different pathogens as follows:

- *Fungi*: slow movements and decrease in growth rate, general color change from rose-colored to red in *Beauveria* species and from yellow to brown in other cases, presence of cells with yeast or hyphae in the hemolymph, tendency for mummification, and sporulation on the tegument surface.
- *Bacteria*: reduction or cessation of feeding, slow movements and reduced levels of activity and feeding, septicemia or presence of cells with bacteria in the hemolymph, color change in hemolymph with a progression to septicemia from milky to dark brown, infected larvae turn darker, dysentery, and flaccid body.
- *Virus*: lepidopterous larvae cease feeding, affected tissues change color (e.g., white milky to granular and blue to iridovirus), retarded growth, slow movements, dysentery, tissue disintegration and liquefaction of dead, infected larva, and paralysis.
- *Microsporidia*: white milky aspect in transparent insects, whitish aspect of gut, Malpighian tubes, fat bodies, or other tissues, black melanized blemishes on the tegument, slow movements, locomotory loss, abnormal development, reduced feeding, white fecal exudate, presence of typical elliptical spores in the infected tissues.
- *Rickettsia and Chlamydia*: slow movements, dysentery, swelling of abdomen due to infection of fat body by *Rickttsiella*, and discoloration of infected larva, often turning to chalky white.

Soares (1992) developed some measures for avoiding contaminations with pathogens in insect rearings, including use of initial colony with no disease; quarantine for field-collected material; use of insectary with suitable planning; sterilization of eggs, pupae, and diets; use of suitable sanitary measures in the insectary; and monitoring of the contamination and quality of the insects reared.

3.15 How to Begin an Artificial Diet

If an insect is collected from any host, ideally a chemical analysis of the insect and the site attacked (e.g., leaves, fruits, roots) should be done so that the chemical compounds can be identified and a diet formulated based on the results. This was done for *B. mori,* with a detailed chemical analysis of the mulberry leaves and the feeding insects, although this is not always possible.

The first step is to use a diet that serves for a species or closely related genera, or a diet that is efficient for rearing many species (see Section 3.4). If satisfactory results are not obtained, formulate a general diet such as that shown in Table 3.11. If success is still lacking, try to formulate a diet from the natural food. In this case, there are various options, including macerating the vegetable part attacked in a blender, placing the vegetable part in an oven at 120°C for 1 to 2 h, lyofilizing the material, or placing the vegetable part in liquid nitrogen. The first two processes are cruder and generally give unsatisfactory results. The integrity of the material is maintained especially in the last two cases. After obtaining the material by any of the processes, add a phagostimulant (sucrose), anticontaminants, agar, and an antioxidant (ascorbic acid or α-tocoferol are the most used).

TABLE 3.11

Composition of a General Insect Diet

Ingredient	Quantity (g/100 g)
Protein	
Casein	3.500
Wheat germ, etc.	3.000
Carbohydrates	
Sucrose	3.000
Glucose	0.500
Lipids	
Linoleic acid	0.250
Sterols	
Cholesterol	0.050
Minerals	
Wesson salts	1.000
Vitamins	
Vanderzant mixture	2.000
Gelling and volume agents	
Agar	2.500
Cellulose	10.000
Microorganism inhibitors	
Streptomycin	0.015
Nipagin (methyl parahydroxibenzoate)	0.112
Sorbic acid, etc.	0.300
Water	72.170
KOH 4M	0.500

Source: Singh, P., *Insect Sci. Appl.*, 4, 357–62, 1983.

3.16 Evaluation of Artificial Diets

There are various ways of evaluating a diet's suitability based on morphological, biometric, nutritional, and life-table criteria. Morphological abnormalities can appear when an unfavorable diet is used. House (1963) described some abnormalities and their relationship with nutritional deficiencies. Rodrigues Fo. (1985) found abnormalities in larvae and pupae of *H. virescens* when reared on artificial diets. The main deformities were as follows: in larvae (Figure 3.5), expansion of the frons and swelling of the mandibles (A), uncharacterization of the vertex (B), and fusion of the cephalic capsules at ecdysis with superposition of exuviae (C). In pupae (Figure 3.5), a normal pupa (D) compared to pupae with retention of morphological larval characters (E,F), atrophy of wings characterized by exposition of the third and fourth uromeres which appear without pigments in the area that should be covered by normally sized wings (G), atrophy and deformation of antennae (H), deformation of uromeres with uncharacterization of the terminalia (I), specific or general displacement of organs, principally antennae, and mouthparts (J,K), tumors affecting wings and surrounding regions, and formation of "water bag" (L). Adults show deformities especially on the wings that may be due to fatty acid (linoleic or linolenic) deficiencies or even to interaction of these fatty acids with high temperatures. Often, the wing deformities can be associated with nonnutritional problems, such as lack of space for the adults to open their wings or even low-atmospheric RHs. Studies on this subject are rare and most only refer to deformities in a general fashion.

For a holometabolic insect, the following parameters can be used: egg stage (incubation period or embryonic development, viability of stage or mortality (%), coloration that can be variable in function of the diet and even serve as an indication of whether there was fertilization or not). The eggs are

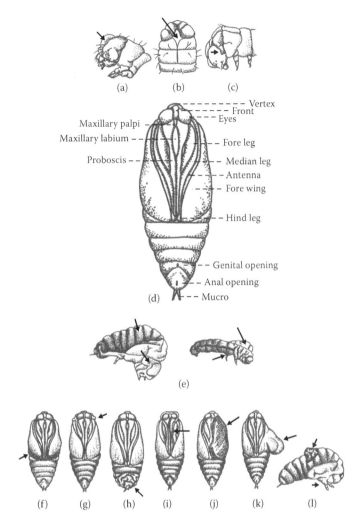

FIGURE 3.5 Deformities in larvae and pupae caused by nutritional deficiencies in artificial diets. (a) Expansion of frons and swelling of the mandibles, (b) deformation of the vertex, (c) fusion of cephalic capsules at ecdysis with superposition of exuviae, (d) morphologically normal pupa, (e) retention of morphological larval characters, (f) atrophy of wings, (g) atrophy and deformation of antennae, (h) deformation of uromeres with deformation of terminalia, (i) specific or generalized (principally antennae and mouthparts) displacement of organs, (j) tumors affecting wings, (k) formation of "water bag" on the wing, and (l) tumors in regions surrounding the wings. (Modified from Rodrigues Fo., I. L., Ms. Sc. Dissertation, University of São Paulo, Piracicaba, SP, Brazil, 1985.)

very sensitive to drying out and should be kept in places (e.g., moistened filter paper) with a relative humidity greater than 60%. In the larval stage (number of instars determined by measuring the width of the cephalic capsule in Lepidoptera (Parra and Haddad 1989; Haddad et al. 1999), duration of each instar, duration of larval period, larval viability, or mortality (%), deformities (%), and presence of pathogens.

In the pupal stage, the following parameters can be used: duration of pupal period, pupal weight at a fixed age (this type of observation is important since there is a narrow correlation between pupal weight and the capacity to lay eggs), pupal viability (%), sexual ratio (sr), and deformities. Pupae can be sensitive to desiccation and they should be kept in places with high relative humidities, 75% to 85%. Humidities greater than 85% favor pathogen development, especially fungi and bacteria. Pupae lose water over time. That is the reason why weighings should be done at a fixed age (e.g., 24 h, 48 h). To facilitate pupation, sand should be placed in the bottom of the rearing container or a mixture of one part vermiculite to two parts sand. In some holometabolous insects, the prepupal stage can be studied. According to Torre-Bueno (1978), the prepupa can be considered as "a quiescent instar between the end of the larval stage and the pupal stage or an active larval stage but which does not feed." Therefore, the duration (generally short) and the viability of the prepupal stage can be determined. In practice, the prepupal stage begins as soon as the last larval stage stops feeding.

In the adult stage (period from preoviposition; number of matings and behavior during mating) in Lepidoptera the following parameters can be used: the number of matings can be determined based on counting the number of spermatophores in the female bursa copulatrix, fecundity (number of eggs per female), total and daily numbers, aspects of the reproductive organs, mated and virgin male and female longevity (to evaluate adult survival, many researchers use Weibull's model), and the sex ratio. In general, for good reproductive performance, adults need high humidities, near to saturation. The papers that line the cages, for example, should be moistened daily. The food for Lepidoptera (sugary solutions) should be renewed periodically to avoid that fermentations (common in sugary liquids) harm the insect.

In general, there are no rules for rearing insects in the laboratory because of the enormous diversity in habits. However, the microclimate requirements of temperature, humidity, light, and ventilation should be considered for any rearing. In general terms, insects can be reared with temperatures of around 25°C; humidity requirements vary depending on the stage. In general, tropical insects develop in a photoperiod of 14 h (photo phase):10 h (scot phase); for oviposition, many adults need ventilation and hermetically closed cages should be avoided. When a biological laboratory study is started there are problems principally associated with mating, oviposition, and adult feeding (Parra 2002).

Regarding the nutritional criteria, the nutritional indices most used are those proposed by Waldbauer (1968) and revised by Kogan and Cope (1974), Scriber and Slansky (1981), with analytical considerations by Raubenheimer and Simpson (1992), Horton and Redak (1993), Simpson and Raubenheimer (1995), and Beaupre and Dunham (1995) (see Chapter 2).

Fertility life tables are normally used to compare diets (Silveira Neto et al. 1976; Gutierrez 1996). For elaboration of these tables the following biological parameters should be evaluated: duration of egg–adult period, viability of the immature stages, preoviposition period, sex ratio, daily mortality of males and females, and daily egg-laying capacity. At least 20 pairs per diet analyzed should be observed for this type of comparison. Salvadori and Parra (1990) compared four diets for *Pseudaletia sequax* Franclemont, based on a fertility life table observing values for the net reproductive rate (Ro) (number of times that a population increases at each generation) and the finite increase ratio (λ) (number of individuals added to the population by females that produce females), very different but sufficient to identify the two most suitable artificial diets (1 and 2) compared to the natural food (wheat) (4) (Table 3.12). Nalim (1991) used the same criterion to choose a diet for *S. frugiperda*. Nowadays, tests like those of Jackknife and Bootstrap (Gutierrez 1996), allow comparison of the results obtained with fertility life tables. There are cases in which the four criteria (morphological, biometric, nutritional, and life table) are insufficient for defining the best diet for an insect species. In this case, another type of analysis is used to arrive at a conclusion, the most common one being the cluster analysis; the method of principal components is also used (Figure 3.6).

TABLE 3.12

Net Reproduction Rate (Ro), Duration of Each Generation (T), Innate Capacity to Increase in Number (Rm), and Finite Increase Ratio (λ) of *Pseudaletia sequax* in Different Diets

Diet	Ro	T (days)	Rm	λ
1	282.63	57.9	0.097481	1.10239
2	310.20	56.1	0.102268	1.10768
3	54.14	59.3	0.067314	1.06963
4	346.84	52.5	0.111409	1.11785

Source: Salvadori, J. R., and J. R. P. Parra, *Pesq. Agropec. Bras.*, 25, 1701–13, 1990.
Note: 1, 2, and 3 = artificial diets; 4 = natural diet (wheat).

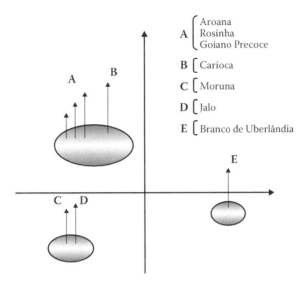

A { Aroana
    Rosinha
    Goiano Precoce

B [Carioca

C [Moruna

D [Jalo

E [Branco de Uberlândia

FIGURE 3.6 Analysis of principal components for identifying the best dry bean varieties as components of diets for *Spodoptera frugiperda*. (Modified from Parra, J. R. P., et al., *An. Soc. Entomol. Brasil*, 28, 49–57, 1999.)

3.17 Causes of Failures and the Advantages of an Artificial Diet for Insects

The failures of a diet for insects include the following: small food consumption because the diet has no phagostimulants (physical and chemical), the food is insufficiently digested because a digestive enzyme is not secreted or because an ingested antimetabolite inhibits the digestive enzyme, nutrient absorption and transport through the intestinal wall are inhibited by an ingested antimetabolite, and nutrient assimilation is retarded by a deficiency or excess of essential nutrients or by an antimetabolite that acts as an enzymatic inhibitor. The advantages of using artificial diets for insect rearing include the following: permits a constant supply of insects, uniform nutrition and biology, pathogens can be better controlled, and the possibility of automatization for mass rearings.

3.18 The Future of Artificial Diets

There have been no big advances with artificial diets over the last few years and most are still based on artificial media developed during the 1960s. This stagnation may be due to the frustration of poor performance with parasitoids and predators with artificial diets (rearings *in vitro*). But artificial diets are fundamental for advances in modern entomology, as detailed in Section 3.1.

As already mentioned, when dealing with a small-size rearing (rearing for research), the problems will be fewer. As the number of insects to be reared grows (mass rearing), problems with facilities, sanitation, cost, need for automatization, storage, and production forecasts also increase (Parra 2007). To increase the number of insects produced, the scale must often be adjusted, from a rearing for research (small number of insects produced) to a mass rearing (millions of insects produced). The quality is a further much-studied problem with evaluations that should be made periodically and considering the objective of the rearing. Among the processes that can result in genetic deterioration, the random and the nonrandom or adaptive ones should be considered. Genetic drift (or foundation effect), inbreeding and selection seem to be the most relevant (Bueno 2000; van Lenteren 2003; Prezotti and Parra 2002; Prezotti et al. 2004; Parra 2006; Parra and Cônsoli 2008).

Quality standards already exist for various insects, from which the competitiveness of the laboratory-reared insect can be compared to that of the wild insect (van Lenteren 2003). In biological control, this problem increases since because two insect species are reared (the pest and the natural enemy), quality control must be practiced for both.

With mass rearing, the problems change from being entomological to being technological, with the rearing locality acquiring the status of a factory. Then problems can arise with the storage of diet components (which can deteriorate over time), the maintenance of controlled environmental temperatures, humidities, and photoperiods, the availability and stocking of spare parts, and so forth. In this case, the problems increase. The search for better diets should favor studies involving food science, including nutritional requirements, with the analysis of the insect's food matrix but also considering the technology and the equipment used for large-scale production. More refined biotrials using microscopical techniques, nanotechnology, and molecular, biochemical, and fermentation processes can produce advances in the area. More sophisticated bioassays can improve the control of microorganisms in large rearings and clarify the role of symbionts (e.g., *Wolbachia, Buchnera*) in insect nutrition. It is also of fundamental importance that those who work with insect rearing, both graduates and postgraduates, should be valued since the process of using natural enemies to control pests, for example, is a cultural process that demands high-quality workers to avoid discrediting an option that is extremely important for the environment (Cohen 2004).

REFERENCES

Alverson, J., and A. C. Cohen. 2002. Effect of antifungal agents on biological fitness of *Lygus hesperus* Knight (Heteroptera: Miridae). *J. Econ. Entomol.* 95:256–60.

Anderson, T. E., and N. C. Leppla. 1992. *Advances in Insect Rearing for Research and Pest Management.* Boulder, CO: Westview Press.

Ashby, M. D., and P. Singh. 1987. *A Glossary of Insect Rearing Terms.* Bulletin 239 Entomology Division, Department of Scientific and Industrial Research, Wellington, New Zealand.

Ave, D. A. 1995. Stimulation of feeding: Insect control agents in regulatory mechanisms in insect feeding. In *Regulatory Mechanisms in Insect Feeding*, ed. R. F. Chapman, and G. de Boer, 345–63. New York: Chapman & Hall.

Beaupre, S. J., and A. E. Dunham. 1995. A comparison of ratio-based and covariance analyses of a nutritional data set. *Functional Ecol.* 9:876–80.

Bellows, T. S., and T. W. Fisher. 1999. *Handbook of Biological Control.* San Diego, CA: Academic Press.

Bento, F. M. M., S. R. Magro, P. Fortes, N. G. Zério, and J. R. P. Parra. 2007. Biologia e tabela de vida de fertilidade de *Agrotis ipsilon* em dieta artificial. *Pesq. Agropec. Bras.* 42:1369–72.

Bleicher, E., and J. R. P. Parra. 1991. Efeito do hospedeiro de substituição e da alimentação na longevidade de *Trichogramma* sp. *Pesq. Agropec. Bras.* 26:1845–50.

Bourtzis, K., and T. A. Miller. 2003. *Insect Symbiosis.* Boca Raton, FL: CRC Press.

Brues, C. T. 1946. *Insect Dietary: An Account of the Food Habits of Insects.* Cambridge, MA: Harvard University Press.

Bueno, V. H. P. 2000. *Controle Biológico de Pragas: Produção Massal e Controle de Qualidade.* Lavras, Brazil: Federal University of Lavras Press.

Campo, C. B. H., E. B. Oliveira, and F. Moscardi. 1985. *Criação Massal da Lagarta da Soja (Anticaria gemmatalis).* Embrapa, CNPSo, Documentos 10, Londrina, PR, Brazil.

Chagas, M. C. M., J. R. P. Parra, P. Milano, A. M. Nascimento, A. L. G. C. Parra, and P. T. Yamamoto. 2002. *Ageniaspis citricola*: Criação e estabelecimento no Brasil. In *Controle Biológico no Brasil: Parasitóides e Predadores*, ed. J. R. P. Parra, P. S. M. Botelho, B. S. Corrêa-Ferreira, and J. M. S. Bento, 377–94. São Paulo, Brazil: Manole.

Chambers, D. L. 1977. Quality control in mass rearing. *Annu. Rev. Entomol.* 22:289–308.

Chapman, R. F. 1982. *The Insects: Structure and Function.* Cambridge: Harvard University Press.

Cohen, A. C. 1985. Simple method for rearing the insect predator *Geocoris punctipes* (Heteroptera: Lygaeidae) on a meat diet. *J. Econ. Entomol.* 78:1173–5.

Cohen, A. C. 1995. Extra-oral digestion in predatory arthropods. *Annu. Rev. Entomol.* 40:85–103.

Cohen, A. C. 1998. Solid-to-liquid feeding: Story of extra-oral digestion in predaceous Artrhopoda. *Am. Entomol.* 44:103–16.

Cohen, A. C. 2004. *Insect Diets: Science and Technology.* Boca Raton, FL: CRC Press.

Cohen, A. C., and P. Crittenden. 2004. Deliberately added and "cryptic" antioxidants in three artificial diets for insects. *J. Econ. Entomol.* 97:265–72.

Cônsoli, F. L., and S. Grenier. 2010. In vitro rearing of egg parasitoids. In *Egg Parasitoids in Agroecosystems with Emphasis on Trichogramma*, ed. F. L. Cônsoli, J. R. P. Parra, and R. A. Zucchi, 293–314. Dordrecht, the Netherlands: Springer.

Cônsoli, F. L., and J. R. P. Parra. 1996a. Comparison of hemolymph and holotissues of different species of insect as diet components for in vitro rearing of *Trichogramma galloi* Zucchi and *T. pretiosum* Riley. *Biol. Contr.* 6:401–6.

Cônsoli, F. L., and J. R. P. Parra. 1996b. Biology and parasitism of *T. galloi* Zucchi and *T. pretiosum* Riley reared "in vitro" and "in vivo." *Ann. Entomol. Soc. Am.* 89:828–34.

Cônsoli, F. L., and J. R. P. Parra. 1997a. Development of an oligidic diet for in vitro rearing of *Trichogramma galloi* Zucchi and *Trichogramma pretiosum* Riley. *Biol. Contr.* 8:172–6.

Cônsoli, F. L., and J. R. P. Parra. 1997b. Produção "in vitro" de parasitóides: Criação de *Trichogramma galloi* Zucchi and *T. pretiosum* Riley, no Brasil. In *Trichogramma e o Controle Biológico Aplicado*, ed. J. R. P. Parra, and R. A. Zucchi, 259–302. Piracicaba, Brazil: FEALQ.

Cônsoli, F. L., and J. R. P. Parra. 1999a. Development of an artificial host egg for *in vitro* egg laying *Trichogramma galloi* and *T. pretiosum* using plastic membranes. *Entomol. Exp. Appl.* 91:327–36.

Cônsoli, F. L., and J. R. P. Parra. 1999b. "In vitro" rearing of parasitoids: Constraints and perspectives. *Trends Entomol.* 2:19–32.

Cônsoli, F. L., and J. R. P. Parra. 2002. Criação *in vitro* de parasitóides e predadores. In *Controle Biológico no Brasil: Parasitóides e Predadores*, ed. J. R. P. Parra, P. S. M. Botelho, B. S. Corrêa-Ferreira, and J. M. S. Bento, 249–71. São Paulo, Brazil: Manole.

Copersucar. 1987. Guia prático ilustrado para identificação e controle de contaminantes em insetários. Copersucar, Bol. Téc., 31 p.

Costa, A. S. 1980. Criação de insetos vetores para estudos de transmissão de fitoviroses, 157–183. In *Congresso Brasileiro de Entomologia 6*, Campinas, Brazil, SP.

Costa, M. M. L. 1991. Técnicas de criação de *Nezara viridula* (L., 1758) e sua relação com o parasitóide *Eutrichopodopsis nitens* Blanchard 1966. Ms. Sc. Dissertation, University of São Paulo, Piracicaba, SP, Brazil.

Dadd, R. H. 1977. Qualitative requirements and utilization of nutrients: Insects. In *Handbook Series in Nutrition and Food*. Section D, Vol. 1, ed. M. Rechcigl Jr., 305–346. Boca Raton, FL: CRC Press.

Dadd, R. H. 1985. Nutrition organisms. In *Comprehensive Insect Physiology, Biochemistry and Pharmacology*, ed. G. A. Kerkut, and L. I. Gilbert, Vol. 4, 313–90. Oxford, UK: Pergamon Press.

Dai, X. J., Z. J. Ma, L. W. Zhang, and A. H. Cao. 1992. Utilization of artificial host egg-wasps for control of insect pest in agriculture and forestry. *Proc. International Congress of Entomology 19*, Beijing. p. 191.

Degaspari, N., P. S. M. Botelho, L. C. Almeida, and H. J. Castilho. 1983. Biologia de *Sphenophorus levis* Vaurie, 1987 (Col: Curculionidae) em dieta artificial e no campo. *Rel. Na. Coord. Reg. Sul. Entomol. Planalsucar*, Vol. 2, 291–309.

Dias, N. S., J. R. P. Parra, and F. L. Cônsoli. 2010. Egg laying and development of neotropical Trichogrammatidae species in artificial eggs. *Entomol. Exp. Appl.* 137:126–31.

Dickerson, W. A., and N. C. Leppla. 1992. The insect rearing group and the development of insect rearing as a profession. In *Advances in Insect Rearing for Research and Pest Management*, ed. T. E. Anderson, and N. C. Leppla, 3–7. Boulder, CO: Westview Press.

Dickerson, W. A., J. D. Hoffman, E. G. King, N. C. Leppla, and J. M. Oddel. 1980. Arthropod species in culture in the United States and other countries. *Bull. Entomol. Soc. Am.*, Lanham.

Dougherty, E. C. 1959. Introduction to axenic culture of invertebrate metazoan: A goal. *Ann. N.Y. Acad. Sci.* 77:27–54.

Dunkel, F. V., and N. R. Read. 1991. Review of the effects of sorbic acid on insect survival in rearing diets with reference to other antimicrobials. *Amer. Entomol.* 37:172–3.

Edwards, D. R., N. C. Leppla, and W. A. Dickerson. 1987. Arthropod species in culture. *Bull. Entomol. Soc. Am.*, Lanham.

Etzel, L. K., and E. F. Legner. 1999. Culture and colonization. In *Handbook of Biological Control*, ed. T. S. Bellows, and T. W. Fisher. San Diego, CA: Academic Press.

Feeny, P. 1975. Biochemical coevolution between plants and their insect herbivores. In *Coevolution of Animals and Plants*, ed. L. E. Gilbert, and P. H. Raven, 3–19. Austin, TX: The University of Texas Press.

Finney, G. L., and T. W. Fisher. 1964. Culture of entomophagous insects and their host. In *Biological Control of Insect Pests and Weeds*, ed. P. De Bach, and E. I. Schlinger, 328–55. London: Chapman & Hall Ltd.

Fortes, P. S. R. Mayro, A. R. Panizzi, and J. R. P. Parra. 2006. Development of dry artificial diet for *Nezara viridula* (L.) and *Euschistus heros* (Fabricius) (Heteroptera: Pentatomidae). *Neotr. Entomol.* 35:567–72.

Fraenkel, G. 1953. *The Nutritional Value of Green Plants for Insects*. Symposia of the IXth International Congress of Entomology, Amsterdam, the Netherlands, pp. 90–100.

Frost, B. R. 1959. *Insect Life and Insect Natural History*. New York: Dover Press.

Funke, B. R. 1983. Mold control for insect-rearing media. *Bull. Entomol. Soc. Am.* 29:41–4.

Gallo, D., O. Nakano, S. Silveira Neto, R. P. L. Carvalho, G. C. de Baptista, E. Berti Filho, J. R. P. Parra, R. A. Zucchi, S. B. Alves, J. D. Vendramim, L. C. Marchini, J. R. S. Lopes, and C. Omoto. 2002. *Entomologia agrícola*, Piracicaba, Brazil: FEALQ.

Garcia, M. S., and J. R. P. Parra. 1999. Comparação de dietas artificiais, com fontes protéicas variáveis, para criação *Ecdytolopha aurantiana* (Lima) (Lepidoptera: Tortricidae). *An. Soc. Entomol. Brasil* 28:219–32.

Gomes, S. M., S. Grenier, J. Guillaud, and J. R. P. Parra. 2002. Desenvolvimento de *Trichogramma Pretiosum* Riley 1879 em meios artificiais. *Resumos Congresso Brasileiro de Entomologia XIX*, Manaus, p. 53.

Grenier, S. 1994. Rearing of *Trichogramma* and other egg parasitoids on artificial diets. In *Biological Control with Egg Parasitoids*, ed. E. Wajnberg, and S. A. Hassan, 73–92. Wallingford, UK: CAB International–IOBC.

Grenier, S. A. 1997. Desenvolvimento e produção "in vitro" de Trichogramma. In *Trichogramma e o Controle Biológico no Brasil*, ed. J. R. P. Parra, and R. A. Zucchi, 235–58. Piracicaba, Brazil: FEALQ/FAPESP.

Gutierrez, A. P. 1996. *Applied Population Ecology: A Supply Demand Approach*. New York: John Wiley & Sons, Inc.

Haddad, M. L., and J. R. P. Parra. 1984. Métodos para estimar os limites térmicos e a faixa ótima de desenvolvimento das diferentes fases do ciclo evolutivo dos insetos. *EMBRAPA, USP, FEALQ, Série Agricultura e Desenvolvimento*, Piracicaba, SP, Brazil.

Haddad, M. L., J. R. P. Parra, and R. C. Moraes. 1999. Métodos para estimar os limites térmicos inferior e superior de desenvolvimento de insetos. *Bol. Téc.*, FEALQ.

Heimpel, G. E., J. A. Rosenheim, and D. Kattari. 1997. Adult feeding and lifetime reproductive success in the parasitoid *Aphytis melinus*. *Entomol. Exp. Appl.* 83:305–15.

Hensley, S. D., and A. M. Hammond. 1968. Laboratory techniques for rearing the sugar cane borer on an artificial diet. *J. Econ. Entomol.* 61:1742–3.

Horton, D. R., and R. A. Redak. 1993. Further comments on analysis of covariance in insect dietary studies. *Entomol. Exp. Appl.* 69:263–75.

House, H. L. 1961. Insect nutrition. *Annu. Rev. Entomol.* 6:13–6.

House, H. L. 1963. Nutritional diseases. In *Insect Pathology: An Advanced Treatise*, ed. E. A. Steinhaus, 133–60, Vol. 1. New York: Academic Press.

House, H. L. 1966. The role of nutritional principles in biological control. *Can. Entomol.* 98:1121–34.

House, H. L. 1977. Nutritional of natural enemies. In *Biological Control by Augmentation of Natural Enemies*, ed. R. L. Ridgway, and S. B. Vinson, 151–82. New York: Plenum Press.

Hoy, M. A., P. Nowierski, M. W. Johnson, and J. L. Flexner. 1991. Issues and ethics in commercial releases of natural enemies. *Amer. Entomol.* 37:74–5.

Ignoffo, C. M., and O. P. Boening. 1970. Compartmented disposable plastic trays for rearing insects. *J. Econ. Entomol.* 63:1696–7.

Jones, Jr., W. A., and C. G. Jackson. 1990. Mass production of *Anaphes iole* for augmentation against *Lygus hesperus*: Effects of food on fecundity and longevity. *Southwest. Entomol.* 15:463–8.

Justi, Jr., J. 1993. Desenvolvimento de uma dieta artificial e técnicas de criação de *Helicoverpa zea* (Boddie, 1850) em laboratório. Ms. Sc. Dissertation, University of São Paulo, Piracicaba, SP, Brazil.

Kainoh, Y., and J. J. Bown. 1994. Amino acids as oviposition stimulants for the egg–larval parasitoid *Chelonus* sp. near *curvimoculatris* (Hymenoptera: Braconidae). *Biol. Control* 4:22–5.

Kasten, Jr., P., R. M. S. Molina, A. R. Iarossi, and J. R. P. Parra. 1998. Efeitos do pH no desenvolvimento de *Ecdytolopha aurantiana* (Lepidoptera, Tortricidae). *XVII Congr. Bras. Entomol.*, p. 658.

King, E. G., and N. C. Leppla, eds. 1984. *Advances and Challenges in Insect Rearing*. New Orleans, LA: Agricultural Research Service, Southern Region, U.S. Department of Agriculture.

Knipling, E. F. 1979. *The Basic Principles of Insect Population Suppression and Management*. Washington, DC: U.S. Department of Agriculture.

Kogan, N. 1980. Criação de insetos: Bases nutricionais e aplicações em programas de manejo de pragas. In *Anais VI Congr. Bras. Entomol.*, ed. Z. A. Ramiro, J. Grazia, and F. M. Lara, 45–75. Campinas, Brazil: Sociedade Entomológica do Brasil.

Kogan, M., and D. Cope. 1974. Feeding and nutrition associated with soybeans. 3. Food intake, utilization and growth in the soybean looper *Pseudoplusia includens*. *Ann. Entomol. Soc. Am.* 67:66–72.

Kunkel, H. 1977. Membrane feeding system in aphid research. In *Aphids as Virus Vectors*, ed. K. F. E. Harris, and K. Maramorosch, 311–38. New York: Academic Press.

Leppla, N. C. 1985. Gelling agents for insect diet: From mush to medium. *Frass Newsletter* 8:1–3.

Leppla, N. C., and F. Adams. 1987. *Insect Mass-Rearing Technology, Principles and Applications*. New York.

Leppla, N. C., and T. R. Ashley. 1978. *Facilities for Insect Research and Production*. United States Department of Agriculture Technical Bulletin 1576.

Li-Li-Ying, L. Wen-Hui, C. Chao-Shian, H. Shi-Tzou, S. Jia-Chi, D. Hansu, and F. Shu-Yi. 1986. In vitro rearing *Trichogramma* spp., *Anastatus* sp. in artificial "eggs" and the methods of mass production. *Second International Symposium on Trichogramma and other Eggs Parasites*, Guangzhou, China.

Lipke, H., and G. Fraenkel. 1956. Insect nutrition. *Annu. Rev. Entomol.* 1:17–44.

Liu, W., S. Yang Wen, and Li-Li-Ying. 1992. Lyophilized diets for in vitro rearing *Trichogramma dendrolimi* & *Bracon greeni*. *Proc. International Congress of Entomology 19*, Beijing, p. 321.

Mackauer, M. 1972. Genetic aspects of insect production. *Entomophaga* 17:27–48.

Magro, S. R., and J. R. P. Parra. 2004. Comparison of artificial diets for rearing *Bracon hebetor* Say (Hymenoptera: Braconidae). *Biol. Control* 29:341–7.

Magro, S. R., A. B. Dias, W. R. Terra, and J. R. P. Parra. 2006. Biological, nutritional and histochemical basis for improving an artificial diet for *Bracon hebetor* Say (Hymenoptera: Braconidae). *Neotr. Entomol.* 35:215–22.

McFarlane, J. E. 1978. Vitamins E and K in relation to growth of the house cricket (Orthoptera: Grylidae). *Can. Entomol.* 109:329–30.

Meneguim, A. M., J. R. P. Parra, and M. L. Haddad. 1997. Comparação de dietas artificiais, contendo diferentes fonts de ácidos graxos, para criação de *Elasmopalpus lignosellus* (Zeller) (Lepidoptera: Pyralidae). *An. Soc. Entomol. Brasil* 26:35–43.

Monnerat, R. G. 2002. Parâmetros bionômicos do bicudo-de-algodoeiro criado em dieta artificial para a realização de bioensaios. *Bol. Pesq. Desenvolvimento*, Brasília, v. 29, 12 p.

Nalim, D. M. 1991. Biologia, nutrição quantitativa e controle de qualidade de populações de *Spodoptera frugiperda* (J. E. Smith, 1797) (Lepidoptera: Noctuidae) em duas dietas artificiais. Ph.D. Thesis, University of São Paulo, Piracicaba, SP, Brazil.

Nantes, J. F. D., and J. R. P. Parra. 1978. Influência da alimentação sobre a biologia de *Perileucoptera coffeella* (Guérin Menèville, 1842) (Lepidoptera, Lyonetiidae). *Científica* 6:263–8.

Nava, D. E., and J. R. P. Parra. 2005. Biologia de *Stenoma catenifer* Walsingham (Lepidoptera: Elachistidae) em dieta natural e artificial e estabelecimento de um sistema de criação. *Neotr. Entomol.* 34:751–9.

Nava, D. E., P. Fortes, D. G. Oliveira, F. T. Vieira, T. M. Ibelli, J. V. C. Guedes, and J. R. P. Parra. 2006. *Plactynota rostrana* (Walker) (Tortridae) and *Phidotricha erigens* (Pyralidae) artificial diets effects on biological cycle. *Braz. J. Biol.* 66:29–41.

Nettles, W. C. 1986. Effects of soyflour, bovine serum albumin and three amino acid mixtures on growth and development of *Encelatoria bryani* (Diptera: Tachinidae) rearing on artificial diets. *Environ. Entomol.* 15:1111–5.

Panizzi, A. R., and J. R. P. Parra. 2009. *Bioecologia e Nutrição De Insetos: Base Para o Manejo Integrado de Pragas*. Brasília: Embrapa Informação Tecnológica.

Panizzi, A. R., J. R. P. Parra, C. H. Santos, and D. R. Carvalho. 2000. Rearing the southern green stink bug using an artificial dry diet and an artificial plant. *Pesq. Agropec. Bras.* 35:1709–15.

Parra, J. R. P. 1990. Técnicas de criação e produção massal de inimigos naturais. In *Manejo Integrado de Pragas*, ed. W. B. Crócomo, 146–177. Botucatu: UNESP.

Parra, J. R. P. 1991. Consumo e utilização de alimentos por insetos. In *Ecologia Nutricional de Insetos e Suas Aplicações no Manejo de Pragas*, ed. A. R. Panizzi, and J. R. P. Parra, 9–65. São Paulo, Brazil: Manole.

Parra, J. R. P. 1992a. Criação massal de inimigos naturais. In *II Ciclo de Palestras Sobre Controle Biológico de Pragas*, ed. P. B. Bastos, A. Batista Fo., and L. G. Leite, 5–20. Campinas, Brazil: Fundação Cargill.

Parra, J. R. P. 1992b. Situação atual e perspectives do controle biológico, através de liberações inundativas no Brasil. *Pesq. Agroc. Bras.* 27:271–9.

Parra, J. R. P. 1993. O controle biológico aplicado e o manejo integrado de pragas. In *Simpósio de Agricultura Ecológica I*, 116–139. Campinas, Brazil: Fundação Cargill.

Parra, J. R. P. 1997. O controle biológico aplicado como um componente do manejo de pragas. In *Interações Ecológicas e Biodiversidade*, ed. M. C. P. Araújo, G. C. Coelho, and L. Medeiros. Ijuí, Brazil. Ed. Unijuí.

Parra, J. R. P. 1998. Criação de insetos para estudos com patógenos. In *Controle Microbiano de Insetos*, ed. S. B. Alves, 1015–38. Piracicaba, Brazil: FEALQ.

Parra, J. R. P. 2000. A biologia de insetos e o manejo de pragas: da criação em laboratório à aplicação em campo. In *Bases e Técnicas do Manejo de Insetos*, ed. J. C. Guedes, I. D. Costa, and E. Castiglioni, 1–29. Santa Maria, Brazil: Federal University of Santa Maria Press.

Parra, J. R. P. 2002. Criação massal de inimigos naturais. In *Controle Biológico no Brasil: Parasitóides e Predadores*, ed., J. R. P. Parra, P. S. M. Botelho, B. S. Corrêa-Ferreira, and J. M. S. Bento, 143–64. São Paulo, Brazil: Manole.

Parra, J. R. P. 2006. A prática do controle biológico de pragas no Brasil. In *Controle Biológico de Pragas na Prática*, ed., A. S. Pinto, D. E. Nava, M. M. Rossi, and D. T. Malerbo-Souza, 11–24. Piracicaba, Brazil: FEALQ.

Parra, J. R. P. 2007. *Técnicas de Criação de Insetos Para Programas de Controle Biológico*. Piracicaba, Brazil: USP/ESALQ/FEALQ.

Parra, J. R. P. 2008. Mass rearing of natural enemies. In *Encyclopedia of Entomology*, 2nd Edition, Vol. 3, ed. J. L. Capinera, 2301–05. Springer.

Parra, J. R. P., and F. L. Cônsoli. 1992. In vitro rearing of *Trichogramma pretiosum* Riley, 1879. *Ciênc. Cultura* 44:407–9.

Parra, J. R. P., and F. L. Cônsoli. 2008. Criação massal e controle de qualidade de parasitóides de ovos. In *Controle Biológico de Pragas: Produção Massal e Controle de Qualidade*, ed. V. H. Bueno, 169–97. Lavras, Brazil: Federal University of Lavras Press.

Parra, J. R. P., and L. H. Mihsfeldt. 1992. Comparison of artificial diets for rearing the sugarcane borer. In *Advances in Insect Rearing for Research and Pest Management*, ed. T. E. Anderson, and N. C. Leppla. Boulder, CO: Westview Press.

Parra, J. R. P., and R. A. Zucchi. 2004. *Trichogramma* in Brazil: Feasibility of use after twenty years of research. *Neotrop. Entomol.* 33:271–81.

Parra, J. R. P., and S. M. Carvalho. 1984. Biologia e nutrição quantitativa de *Spodoptera frugiperda* (J. E. Smith, 1797) em meios artificiais compostos de diferentes variedades de feijão. *An. Soc. Entomol. Bras.* 13:305–19.

Parra, J. R. P., P. Kasten Jr., R. M. S. Molina, and M. L. Haddad. 2001. Efeito do pH no desenvolvimento do bicho-furão. *Laranja* 22:321–32.

Parra, J. R. P., P. S. M. Botelho, B. S. Corrêa-Ferreira, and J. M. S. Bento. 2002. *Controle Biológico no Brasil: Parasitóides e Predadores*. São Paulo, Brazil: Manole.

Parra, J. R. P., J. M. S. Bento, M. C. M. Chagas, and P. T. Yamamoto. 2004. O controle biológico da larva minadora-dos-citrus. *Visão Agric.* 2:64–7.

Parra, J. R. P., P. Milano, F. L. Cônsoli, N. G. Zério, and M. L. Haddad. 1999. Efeito de nutrição de adultos e da umidade na fecundidade de *Diatraea saccharalis* (Fabr.) (Lepidoptera: Crambiedae). *An. Soc. Entomol. Brasil* 28:49–57.

Parrella, M. P., K. M. Heinz, and L. Nunney. 1992. Biological control through augmentative releases of natural enemies: A strategy whose time has come. *Am. Entomol.* 38:172–8.

Pasini, A., J. R. P. Parra, and J. M. Lopes. 2007. Dieta artificial para criação de *Doru luteipes* (Scudder) (Dermaptera: Forficulidae), predador da lagarta-de-cartucho-do-milho, *Spodoptera frugiperda* (J. E. Smith) (Lepidoptera:Noctuidae). *Neotrop. Entomol.* 36:308–11.

Pedroso, A. S. 1972. Dados bionômicos de *Ceratitis capitata* Wied, 1824 (Diptera: Tephritidae) obtidos em laboratório em regime de dieta artificial. Ph.D. Thesis, University of São Paulo, Piracicaba, SP, Brazil.

Pereira, M. J. B., E. Berti-Filho, and J. R. P. Parra. 2004. Artificial diet rearing the annoma fruit borer *Cerconota anonella* spp. 1830 (Lepidoptera, Oecophoridae). *Insect Sci. Appl.* 23:137–41.

Prezoti, L., J. R. P. Parra, R. Vencovsky, C. T. S. Dias, I. Cruz, and M. C. M. Chagas. 2002. Teste de vôo como critério de avaliação da qualidade de *Trichogramma pretiosum* Riley (Hymenoptera: Trichogrammatidae): adaptação de metodologia. *Neotrop. Entomol.* 31:411–7.

Prezoti, L., J. R. P. Parra, R. Vencovsky, A. S. G. Coelho, and I. Cruz. 2004. Effect of the size of the founder population on the quality of sexual populations of *Trichogramma pretiosum* in laboratory. *Biol. Control* 30:174–80.

Raubenheimer, D., and S. T. Simpson. 1992. Analysis of covariance: An alternative to nutritional indices. *Entomol. Exp. Appl.* 62:221–31.

Reinecke, J. P. 1985. Nutrition: Artificial diets. In *Comprehensive Insect Physiology Biochemistry and Pharmacology*, ed. G. A. Kerkut, and L. I. Gilbert, Vol. 4, 391–419. Oxford, UK: Pergamon Press.

Ridgway, R. L., and M. N. Inscoe. 1998. Mass reared natural enemies for pest control: Trends and challenges. In *Mass Reared Natural Enemies: Application, Regulation, and Needs*, ed. R. L. Ridgway, M. P. Hoffmann, M. N. Inscoe, and C. S. Glenister, 1–26. Lanham, MD: Thomas Say Publications in Entomology.

Ridgway, R. L., and S. B. Vinson. 1977. *Biological Control by Augmentation of Natural Enemies*. New York: Plenum Press.

Rodrigues Fo., I. L. 1985. Comparação de dietas artificiais para *Heliothis virescens* (Fabr., 1781) (Lepidoptera: Noctuidae) através de estudos biométricos e nutricionais. Ms. Sc. Dissertation, University of São Paulo, Piracicaba, SP, Brazil.

Rodriguez, J. G. 1972. *Insect and Mite Nutrition: Significance and Implication in Ecology and Pest Management*. Amsterdam, the Netherlands: North Holland Publishing Company.

Salvadori, J. R., and J. R. P. Parra. 1990. Seleção de dietas artificiais para *Pseudaletia sequax* (Lep.: Noctuidae). *Pesq. Agropec. Bras.* 25:1701–13.

Schneider, J. C. 2009. *Principles and Procedures for Rearing High Quality Insects*. Stoneville, MS: Mississippi State University Press.

Scriber, J. M., and F. Slansky Jr. 1981. The nutritional ecology to immature insects. *Annu. Rev. Entomol.* 26:183–211.

Shorey, H. H. 1962. The biology of *Trichoplusia ni* (Lepidoptera: Noctuidae). II. Factors affecting adult fecundity and longevity. *Ann. Entomol. Soc. Am.* 56:476–80.

Signoretti, A. G. C., D. E. Nava, J. M. S. Bento, and J. R. P. Parra. 2008. Biology and thermal requirements of *Utetheisa ornatrix* (l.) (Lepidoptera: Arctiidae) reared on artificial diet. *Braz. Arch. Biol. Tech.* 51: 447–53.

Silveira-Neto, S., O. Nakano, D. Barbin, and N. A. Villa Nova. 1976. *Manual de Ecologia dos Insetos*. Piracicaba, Brazil: Ceres.

Simpson, S. J., and D. Raubenheimer. 1995. The geometric analysis of feeding and nutrition. *J. Insect Physiol.* 41:545–53.

Singh, P. 1977. *Artificial Diets for Insects, Mites, and Spiders*. New York: Plenum.

Singh, P. 1983. A general purpose laboratory diet mixture for rearing insects. *Insect Sci. Appl.* 4:357–62.

Singh, P. 1985. Multiple species rearing diets. In *Handbook on Insect Rearing*, ed. P. Singh, and R. F. Moore, Vol. I, 19–44. Amsterdam, the Netherlands: Elsevier.

Singh, P., and R. F. Moore. 1985. *Handbook of Insect Rearing*. Vols. 1 and 2. Amsterdam, the Netherlands: Elsevier.

Smith, C. N. 1966. *Insect Colonization and Mass Production*. New York: Academic Press.

Soares Jr., G. G. 1992. Problems with entomopathogens in insect rearing. In *Advances in Insect Rearing for Research and Pest Management*, ed., T. E. Anderson, and N. C. Leppla, 289–322. Boulder, CO: Westview Press.

Starler, N. H., and R. L. Ridgway. 1977. Economic and social considerations for the utilization of augmentation of natural enemies. In *Biological Control by Augmentation of Natural Enemies*, ed., R. L. Ridgway, and S. B. Vinson, 413–53. New York: Plenum Press.

Stimmann, M. W., R. Pangaldan, and B. S. Schureman. 1972. Improved method of rearing the beet armyworm. *J. Econ. Entomol.* 65:596–7.

Thompson, S. N. 1986. Nutritional and in vitro culture of insect parasitoids. *Annu. Rev. Entomol.* 31:197–219.

Thompson S. N., and K. S. Hagen. 1999. Nutrition of entomophagous insects and other arthropods. In *Handbook of Biological Control*, ed. T. S. Bellows, and T. W. Fisher, 594–652. New York: Academic Press.

Torre-Bueno, J. R. de la. 1978. *A Glossary of Entomology*. New York: New York Entomological Society.

Uvarov, B. P. 1928. Insect nutrition and metabolism. A summary of the literature. *Trans. R. Ent. Soc. Lond.* 74:255–343.

Van Drische, R. G., and T. S. Bellows Jr. 1996. *Biological Control*. London: Chapman & Hall.

Van Lenteren, J. C. 2000. Success in biological control of arthropods by augmentation of natural enemies. In *Biological Control Measures of Success*, ed., G. Gurr, and S. Wratten, 77–103. Dordrecht, the Netherlands: Kluwer Academic Publishers.

Van Lenteren, J. C. 2003. *Quality Control and Production of Biological Control Agents: Theory and Testing Procedures*. Cambridge, UK: CAB Publishing.

Van Lenteren, J. C., and J. Woets. 1988. Biological and integrated control in greenhouses. *Annu. Rev. Entomol.* 33:239–69.

Vanderzant, E. S. 1974. Development, significance and application of artificial diets for insects. *Annu. Rev. Entomol.* 19:139–54.

Vendramim, J. D., A. R. R. Souza, and J. R. P. Parra. 1982. Ciclo biológico de *Heliothis virescens* (Fabricius, 1781) (Lepidoptera, Noctuidae) em dietas com diferentes tipos de celulose. *An. Soc. Entomol. Brasil* 11:3–11.

Villacorta, A., and J. F. Barrera. 1993. Nova dieta merídica para criação de *Hypothenemus hampei* (Ferrari, 1867) (Coleoptera: Scolytidae) *An. Soc. Entomol. Brasil* 22:405–9.

Waage, J. K., R. P. Carl, J. Mills, and D. J. Greathead. 1985. Rearing entomophagous insects. In *Handbook of Insect Rearing*, ed., P. Singh, and R. F. Moore, Vol. 1, 45–66. Amsterdam, the Netherlands: Elsevier.

Waldbauer, G. P. 1968. The consumption and utilization of food by insects. *Adv. Insect Physiol.* 5:229–88.

Werren, J. H. 1997. Biology of *Wolbachia*. *Annu. Rev. Entomol.* 42:587–609.

4

*Molecular and Evolutionary Physiology of Insect Digestion**

Walter R. Terra and Clélia Ferreira

CONTENTS

4.1   Introduction 93
4.2   Gut Morphology and Function 94
4.3   Digestive Enzymes 98
     4.3.1   Digestion of Proteins 98
     4.3.2   Digestion of Carbohydrates 100
     4.3.3   Digestion of Lipids and Phosphates 101
4.4   Food Handling and Ingestion 102
     4.4.1   Preliminary Observations 102
     4.4.2   Solid Food 102
     4.4.3   Liquid Food 102
4.5   Overview of the Digestive Process 102
4.6   Role of Microorganisms in Digestion 104
4.7   Midgut Conditions Affecting Enzyme Activity 105
4.8   Basic Plans of the Digestive Process 106
     4.8.1   Evolutionary Trends of Insect Digestive Systems 106
     4.8.2   Blattodea 107
     4.8.3   Isoptera 108
     4.8.4   Orthoptera 108
     4.8.5   Hemiptera 108
     4.8.6   Coleoptera 111
     4.8.7   Hymenoptera 111
     4.8.8   Diptera 112
     4.8.9   Lepidoptera 113
4.9   Digestive Enzyme Secretion Mechanisms 113
4.10 Concluding Remarks 114
Acknowledgments 115
References 115

4.1 Introduction

Public and scientific awareness of the environmental problems caused by chemical insecticides led to the search for new approaches to insect control. Midgut studies were particularly stimulated after realization that the gut is a very large and relatively unprotected interface between the insect and its environment

---

* Part of this article was published in *Encyclopedia of Insects*, 2nd ed., 2009, edited by Resh V. H. and R. T. Cardé under the title "Digestion," 271–3, authored by W. R. Terra, and "Digestive System," 273–81, authored by W. R. Terra and C. Ferreira. © Elsevier 2009.

(Law et al. 1992). Accordingly, this chapter offers a wide and updated review of this research area, stressing the molecular processes underlying digestive phenomena, and highlighting, when possible, their potential in supporting the development of new control techniques. To provide a broad coverage while keeping the chapter within reasonable size limits, many details and original references were suppressed and the reader is led to appropriate reviews. Nevertheless, original references regarding the papers that support contemporary research trends were maintained.

Digestion is the process by which food molecules are broken down into smaller molecules that can be absorbed by the gut tissue. The digestive process occurs in the alimentary canal (gut) that is responsible for all steps in food processing: digestion, absorption, and feces delivery and elimination. The anterior (foregut) and posterior (hindgut) parts of the gut have cells covered by a cuticle, whereas in the midgut, cells are separated from the food by a filmlike anatomical structure referred to as peritrophic membrane. Salivary glands are associated with the foregut and may be important in food intake but usually not in digestion. Digestion is carried out by the insect digestive enzymes, apparently without the participation of symbiotic microorganisms. Nevertheless, microorganisms may have a role in allelochemical detoxification, pheromone production, and in making available nutrients for some insects.

4.2 Gut Morphology and Function

Figure 4.1 is a generalized diagram of the insect gut. The foregut begins at the mouth, includes the cibarium, the pharynx, the esophagus, and the crop (a dilated portion, as in Figure 4.2a, or a diverticulum, like Figure 4.2k). The crop is a storage organ in many insects and also serves as a site for digestion in others. The foregut is covered with a cuticle, which is nonpermeable to hydrophilic molecules and is reduced to a straight tube in some insects (Figure 4.2f). The proventriculus is a triturating (grinding into fine particles) organ in some insects and in most it provides a valve controlling the entry of food into the midgut, which is the main site of digestion and absorption of nutrients.

The midgut includes a simple tube (ventriculus) from which blind sacs (gastric or midgut ceca) may branch, usually from its anterior end (Figures 4.1 and 4.2a). Midgut ceca may also occur along the midgut in rings (Figure 4.2f) or not (Figure 4.2h) or in the posterior midgut (Figure 4.2q). In most insects, the midgut is lined with a filmlike anatomical structure (peritrophic membrane) that separates the luminal contents into two compartments: the endoperitrophic space and the ectoperitrophic space (Figure 4.1).

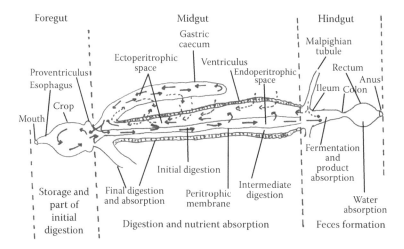

FIGURE 4.1 Generalized diagram of the insect gut. (Reprinted from *Encyclopedia of Insects*, 2nd Edition, Terra, W. R., and C. Ferreira, Digestive System, 273–81. Copyright 2009, with permission from Elsevier.)

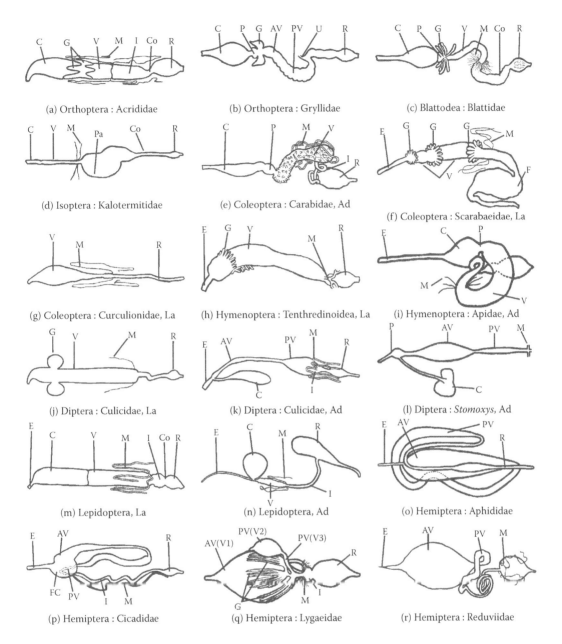

(a) Orthoptera : Acrididae

(b) Orthoptera : Gryllidae

(c) Blattodea : Blattidae

(d) Isoptera : Kalotermitidae

(e) Coleoptera : Carabidae, Ad

(f) Coleoptera : Scarabaeidae, La

(g) Coleoptera : Curculionidae, La

(h) Hymenoptera : Tenthredinoidea, La

(i) Hymenoptera : Apidae, Ad

(j) Diptera : Culicidae, La

(k) Diptera : Culicidae, Ad

(l) Diptera : *Stomoxys*, Ad

(m) Lepidoptera, La

(n) Lepidoptera, Ad

(o) Hemiptera : Aphididae

(p) Hemiptera : Cicadidae

(q) Hemiptera : Lygaeidae

(r) Hemiptera : Reduviidae

FIGURE 4.2 Major insect gut types: Ad = adult; AV = anterior ventriculus (midgut); C = crop; Co = colon; E = esophagus; F = fermentation chamber; FC = filter chamber; G = midgut (gastric) ceca; I = ileum; La = larva; M = Malpighian tubules; P = proventriculus; Pa = paunch; PV = posterior ventriculus (midgut); R = rectum; V = ventriculus. Not drawn to scale. (Reprinted from *Encyclopedia of Insects*, 2nd Edition, Terra, W. R., and C. Ferreira, Digestive System, 273–81. Copyright 2009, with permission from Elsevier; based partly on Terra, W. R., *Brazilian J. Med. Biol. Res.*, 21, 675–734, 1988.)

Some insects have a stomach, which is an enlargement of the midgut to store food (Figure 4.2r). In the region of the sphincter (pylorus) separating the midgut from the hindgut, Malpighian tubules branch off the gut. Malpighian tubules are excretory organs that may be joined to form a ureter (Figure 4.2b); in some species, however, they are absent (Figure 4.2o).

The hindgut includes the ileum, colon, and rectum and terminates with the anus (Figure 4.1). In some insects it is reduced to a straight tube (Figure 4.2g), in others it is modified in a fermentation chamber

(Figure 4.2f) or paunch (Figure 4.2d), with both structures storing ingested food and harboring microorganisms that have a controversial role in assisting cellulose digestion (see below).

The gut epithelium is always simple and rests on a basal lamina that is surrounded by conspicuous circular and a few longitudinal muscles. Wavelike contractions of the circular muscles cause peristalsis, propelling the food bolus along the gut. The gut is oxygenated by the tracheal system, and whereas the foregut and hindgut is well innervated, the same is not true for the midgut. The gut is also connected to the body wall through the extrinsic visceral muscles. These act as dilators of the midgut, mainly at the foregut, where they form a pump highly developed in fluid feeders (cibarium pump). However, it is also present in chewing insects (pharyngeal pump), which are then enabled to drink water and pump air into the gut during molts.

The epithelium of the midgut is composed of a major type of cell usually called the columnar cell (Figure 4.3a, c, f), although it may have other forms; it also contains regenerative cells (Figure 4.3g) that are often collected together in nests at the base of the epithelium, cells (Figure 4.3i) believed to have an endocrine function, and also specialized cells (goblet cells, Figure 4.3b, e; oxyntic cells, Figure 4.3d; hemipteran midgut cell, Figure 4.3h) (Terra and Ferreira 1994).

The peritrophic membrane (Figure 4.4a) is made up of proteins (peritrophins) and chitin to which other components (e.g., enzymes, food molecules) may associate (Hegedus et al. 2008). This anatomical structure is sometimes called the peritrophic matrix, but this term is best avoided because it does not convey the idea of a film and suggests that it is the fundamental substance of something, usually filling a space (as the mitochondrial matrix). The argument that "membrane" means a lipid bilayer does not hold here because the peritrophic membrane is an anatomical structure, not a cell part, as the tympanic membrane. Peritrophins have domains similar to mucins (gastrointestinal mucus proteins) and other domains able to bind chitin (Tellam et al. 1999). As mucins have a very early origin among animals (as confirmed by Lang et al. 2007), Terra (2001) proposed that the peritrophic membrane derived from the ancestral mucus. According to this hypothesis, the peritrophins evolved from mucins by acquiring chitin-binding domains. The parallel evolution of chitin secretion by midgut cells led to the formation of the chitin–protein network characteristic of the peritrophic membrane.

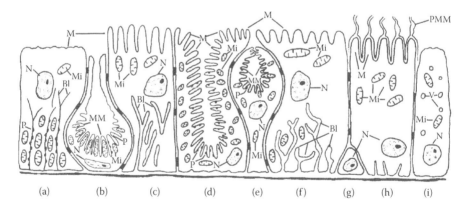

FIGURE 4.3 Diagrammatic representation of typical insect midgut cells: (a) columnar cell with plasma membrane infoldings arranged in long and narrow channels, usually occurring in fluid-absorbing tissues; (b) lepidopteran long-necked goblet cell; (c) columnar cell with highly-developed basal plasma membrane infoldings displaying few openings into the underlying space, usually occurring in fluid-absorbing tissue; (d) cyclorrhaphan dipteran oxyntic (cuprophilic) cell; (e) lepidopteran stalked goblet cell; (f) columnar cell with highly developed plasma membrane infoldings with numerous openings into the underlying space frequently present in fluid-secreting tissue; (g) regenerative cell; (h) hemipteran midgut cell; (I) endocrine cell. Note particles (portasomes) studding the cytoplasmic side of the apical membranes in b, d, and e and of the basal plasma membranes in a. Abbreviations: Bl = basal plasma membrane infoldings; M = microvilli; Mi = mitochondria; MM = modified microvilli; N = nucleus; P = portasomes; PMM = perimicrovillar membranes; V = vesicles. (Reprinted from *Encyclopedia of Insects*, 2nd Edition, Terra, W. R., and C. Ferreira, Digestive System, 273–81. Copyright 2009, with permission from Elsevier.)

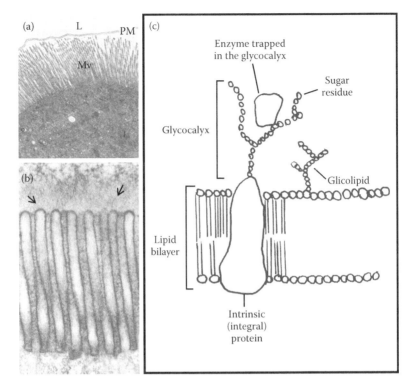

FIGURE 4.4 Midgut cell apexes. (a) Electron micrograph of *Musca domestica* posterior midgut cell. L = lumen; Mv = microvilli; PM = peritrophic membrane. Magnification: 7500×. (b) Electron micrograph of columnar cell of *Erinyis ello* anterior midgut. Detail of microvilli showing glycocalyx (arrows). Magnification: 52,000×. (c) Diagrammatic representation of the distribution of enzymes on the midgut cell surface. Glycocalyx: the carbohydrate moiety of intrinsic proteins and glycolipids occurring in the luminal face of microvillar membranes. ((a) and (c): Reprinted from *Encyclopedia of Insects*, 2nd Edition, Terra, W. R., and C. Ferreira, Digestive System, 273–81. Copyright 2009, with permission from Elsevier.)

The peritrophic membrane may be formed by a ring of cells (peritrophic membrane type 2, seen, for example, in *Aedes aegypti* L. larvae and *Drosophila melanogaster* Meigen) or by most midgut cells (peritrophic membrane type 1 found, for example, in *Ae. aegypti* adult and *Tribolium castaneum* Herbst). Larval *Ae. aegypti* and *D. melanogaster* peritrophins have domain structures more complex than those of adult *Ae. aegypti* and *T. castaneum*. Furthermore, mucin-like domains of peritrophins from *T. castaneum* (feeding on rough food) are lengthier than those of adult *Ae. aegypti* (blood-feeding). This suggests that type 1 and type 2 peritrophic membranes may have varied molecular architectures determined by different peritrophins, which may be partly modulated by diet (Venancio et al. 2009).

Although a peritrophic membrane is found in most insects, it does not occur in Hemiptera and Thysanoptera, which have perimicrovillar membranes in their cells (Figure 4.3h). The other insects that do not seem to have a peritrophic membrane are adult Lepidoptera, Phthiraptera, Psocoptera, Zoraptera, Strepsiptera, Raphidioptera, Megaloptera, and Siphonaptera as well as bruchid beetles and some adult ants (Hymenoptera). Some adults and the larvae and adults of Bruchidae have a peritrophic gel instead of a peritrophic membrane separating the midgut epithelium from the food.

Most of the pores of the peritrophic membrane are in the range of 7 to 9 nm, although some may be as large as 36 nm (Terra 2001). Thus, the peritrophic membrane hinders the free movement of molecules, dividing the midgut lumen into two compartments (Figure 4.1) with different molecules. The functions of this structure include those of the ancestral mucus (protection against food abrasion and microorganism invasion) and several roles associated with the compartmentalization of the midgut. These roles result in improvements in digestive efficiency and assist in decreasing digestive enzyme excretion, and in restricting the production of the final products of digestion close to their transporters, thus facilitating absorption (Terra 2001). Section 4.5 will examine this in detail.

4.3 Digestive Enzymes

4.3.1 Digestion of Proteins

Initial digestion of proteins is carried out by proteinases (endopeptidases), which are enzymes able to cleave the internal peptide bonds of proteins (Figure 4.5a). Different endopeptidases are necessary

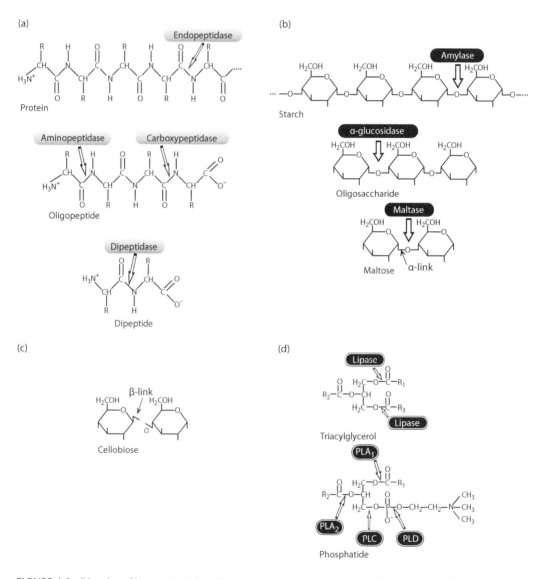

FIGURE 4.5 Digestion of important nutrient classes. Arrows point to bonds cleaved by enzymes. (a) Protein digestion; R, different amino acid moieties. (b) Starch digestion. (c) β-linked glucoside. (d) Lipid digestion; PL = phospholipase; R = fatty acyl moieties. (Reprinted from *Encyclopedia of Insects*, 2nd Edition, Terra, W. R., and C. Ferreira, Digestive System, 273–81. Copyright 2009, with permission from Elsevier.)

to do this because the amino acid residues vary along the peptide chain (R is a variable group in Figure 4.5a). Proteinases may differ in specificity toward the reactant protein (substrate) and are grouped according to their reaction mechanism into the subclasses serine, cysteine, and aspartic proteinases. Trypsin, chymotrypsin, and elastase are serine proteinases that are widely distributed in insects and have molecular masses in the range 20 to 35 kDa and alkaline pH optima. Trypsin preferentially hydrolyzes (its primary specificity) peptide bonds in the carboxyl end of amino acids with basic R groups (Arg, Lys); chymotrypsin is preferential toward large hydrophobic R groups (e.g., Phe, Tyr) and elastase, toward small hydrophobic R groups (e.g., Ala) (Terra and Ferreira 1994, 2005).

The activity of the trypsin also depends on the amino acid residues neighboring the bond to be cleaved. For example, lepidopteran trypsins better hydrolyze bonds at the carboxyl side of Lys inside a sequence of hydrophobic amino acids. Otherwise, trypsins from the other insects prefer to hydrolyze bonds at the carboxyl side of Arg with neighboring hydrophilic amino acids (Lopes et al. 2004). The reactive site of many plant protein inhibitors (PIs) are hydrophilic loops with a Lys residue in the sequence (Ryan 1960). As lepidopteran trypsins have hydrophobic subsites and prefer hydrolyzing Lys rather than Arg bonds, lepidopteran trypsins are more resistant to PIs than the other trypsins (Lopes et al. 2006). Also, insects fed on trypsin inhibitor-containing food may express new trypsin molecules insensitive to the inhibitors, due to changes in their primary specificities or binding properties of their subsites (Mazumdar-Leighton and Broadway 2001a,b; Brito et al. 2001). The details of the mechanism by which PIs in diet induce the synthesis of insensitive trypsins are unknown. Nevertheless, it was found that the first step in the process is the expression of the whole set of midgut trypsins (Brioschi et al. 2007). The evolutionary "arms race" between plants and insects regarding evolving new digestive proteinases and new PIs are reviewed in Christeller (2005).

Chymotrypsins are inactivated by synthetic ketones that react with a His residue at their active sites. However, chymotrypsins from polyphagous lepidopterans are resistant to chloromethyl ketone inactivation. Homology modeling and sequence alignment disclosed differences in the amino acids in the neighborhood of the chymotrypsin catalytic His residue that may affect its pKa value. This is proposed to decrease His reactivity toward chloromethyl ketones and thought to be an adaptation to the presence of dietary ketones (Lopes et al. 2009). Substrate subsite preferences from chymotrypsins pertaining to model insects are known, but in contrast to trypsins, no evolutionary trend was observed in them (Sato et al. 2008).

Cysteine (cathepsin L) and aspartic (cathepsin D) proteinases are the only midgut proteinases in hemipterans and they occur in addition to serine proteinases in cucujiformia beetles. Cathepsin L and cathepsin D have pH optima of 5.5 to 6.0 and 3.2 to 3.5 and molecular masses of 20 to 40 kDa and 60 to 80 kDa, respectively. Cathepsin L requires a mildly reducing midgut to maintain its active Cys residue able to react. Because of their pH optima, cathepsin Ds are not very active in the mildly acidic midguts of Hemiptera and cucujiformia beetles, but are very important in the middle midguts (pH 3.5) of cyclorrhaphous flies (Terra and Ferreira 1994, 2005). Digestive cathepsin D, like vertebrate pepsin, is homologous to lysosomal cathepsin Ds but lacks their characteristic proline loops (Padilha et al. 2009).

Intermediate digestion of proteins is accomplished by exopeptidases, enzymes that remove amino acids from the N-terminal (aminopeptidases) or C-terminal (carboxypeptidases) ends of oligopeptides (fragments of proteins) (Figure 4.4a). Insect aminopeptidases have molecular masses in the range 90 to 130 kDa, have pH optima of 7.2 to 9.0, have no marked specificity toward the N terminal amino acid, and are usually associated with the microvillar membranes of midgut cells. Because aminopeptidases are frequently active on dipeptides, they are also involved in protein-terminal digestion together with dipeptidases. Aminopeptidases may account for as much as 55% of the midgut microvillar proteins in larvae of the yellow mealworm, *Tenebrio molitor* L. Probably because of this, in many insects aminopeptidases are the preferred targets of *B. thuringiensis* endotoxins. These toxins, after binding to aminopeptidase (or other receptors), form channels through which cell contents leak, leading to insect death (Terra and Ferreira 1994, 2005).

The most important insect carboxypeptidases have alkaline pH optima, have molecular masses in the range 20 to 50 kDa, and require a divalent metal for activity. They are classified as carboxypeptidase

A or B depending on their activity upon neutral/acid or basic C-terminal amino acids, respectively. Dipeptidases hydrolyze dipeptides, thus completing the digestion of proteins, when they are aided by some aminopeptidases that also act on dipeptides (Terra and Ferreira 1994, 2005).

4.3.2 Digestion of Carbohydrates

Initial and intermediate digestion of starch (or glycogen) is accomplished by α-amylase. This enzyme cleaves internal bonds of the polysaccharide until it is reduced to small oligosaccharides or disaccharides (Figure 4.5b). Amylases are not very active on intact starch granules, making mastication necessary. Insect amylases depend on calcium ions for activity or stability, are activated by chloride ions (lepidopteran amylases are exceptions), have molecular masses in the range 48 to 68 kDa and their pH optima vary according to phylogeny; as described for trypsin, insects feeding on amylase inhibitor-containing food express new amylase molecules insensitive to the inhibitors (Terra and Ferreira 1994, 2005).

The final digestion of starch chains occurs under α-glucosidases, enzymes that sequentially remove glucosyl residues from the nonreducing ends of short oligomaltosaccharides. If the saccharide is a disaccharide, it is called maltose (Figure 4.5b). Because of that, α-glucosidase is also called maltase. As a rule, sucrose (glucose α1,β2-fructose) is hydrolyzed by α-glucosidase. Sucrose is found in large amounts in nectar, phloem sap, and in lesser amounts in some fruits and leaves (Terra and Ferreira 1994, 2005).

Polysaccharides are major constituents of cell walls. For phytophagous insects, disruption of plant cell walls is necessary in order to expose storage polymers in cell contents to polymer hydrolases. Cell wall breakdown may occur by mastication, but more frequently, it is the result of the action of digestive enzymes. Thus, even insects unable to obtain nourishment from the cellulosic and noncellulosic cell wall biochemical would profit from having enzymes active against these structural components. Cell walls are disrupted by β-glucanases, xylanases and pectinases (plant cells), lysozyme (bacterial cells) or chitinase and β-1,3-glucanase (fungal cells).

Although cellulose is abundant in plants, most plant-feeding insects such as caterpillars and grasshoppers do not use it (Terra et al. 1987; Ferreira et al. 1992). Cellulose is a nonramified chain of glucose units linked by β-1,4 bonds (Figure 4.5c) arranged in a crystalline structure that is difficult to disrupt. Thus, cellulose digestion is unlikely to be advantageous to an insect that can meet its dietary requirements using more easily digested food constituents. The cellulase activity found in some plant feeders facilitates the access of digestive enzymes to the plant cells ingested by insects. True cellulose digestion is restricted to insects that have, as a rule, nutritionally poor diets, as exemplified by termites, wood roaches, and cerambycid and scarabaeid beetles. There is growing evidence that insects secrete enzymes able to hydrolyze crystalline cellulose. This challenges the old view that cellulose digestion is carried out by symbiont bacteria and protozoa (Watanabe and Tokuda 2001). The end products of cellulase action are glucose and cellobiose (Figure 4.5c); the latter is hydrolyzed by a β-glucosidase, also called cellobiase.

The rupture of the bacterial cell wall catalyzed by a lysozyme active in very low pH value is important for insects like the housefly larvae. Because of this, the enzyme was isolated and characterized (Lemos et al. 1993), and after having its coding cDNA cloned, the 3-D structure of the recombinant lysozyme was resolved (Marana et al. 2006). Finally, site-directed mutagenesis confirmed the inference from 3-D studies regarding which residues around the catalytic groups lead to a decrease in the pH optimum. This decrease depends also in a less positively charged surface (Cançado et al. 2010).

Fungi are nutrients especially for detritivorous insects, although they are found contaminating stored products. The fungi wall is broken by digestive chitinases that are efficient in digestion, but harmless for the peritrophic membrane (Genta et al. 2006a). Some laminarinases (β-1,3-glucanases) are also able to digest the fungi wall (Genta et al. 2003, 2007, 2009).

Hemicellulose is a mixture of polysaccharides associated with cellulose in plant cell walls. They are β-1,4- and/or β-1,3-linked glycan chains made up mainly of glucose (glucans), xylose (xylans), and other monosaccharides. The polysaccharides are hydrolyzed by a variety of enzymes from which xylanases, laminarinases, and lichenases are the best known. Some laminarinases (β-1,3-glucanases) are processive;

that is, they perform multiple rounds of catalysis when the enzyme remained attached to the substrate. The exo-β-1,3-glucanase of *Abracris flavolineata* (De Geer) has a high-affinity accessory site that on substrate binding causes active site exposure, followed by the transference of the substrate to the active site. Processivity results in this case from consecutive transferences of substrate between accessory and active site (Genta et al. 2007).

Insect laminarinases are phylogenetically related to Gram-negative bacteria-binding proteins (GNBP) and other β-glucan-binding proteins that are active in the insect innate immune system. Both proteins are derived from the laminarinase of the ancestor of mollusks and arthropods. The insect lost an extended N-terminal region of the ancestral laminarinase, whereas the β-glucan binding proteins lost the catalytical residues (Bragatto et al. 2010).

The end products of the action of the above-mentioned hemicellulases are monosaccharides and β-linked oligosaccharides. The final digestion of those chains occurs under the action of β-glycosidases that sequentially remove glycosyl residues (glucose, galacose, or xylose) from the non-reducing end of the β-linked oligosaccharides. As these may be cellobiose, β-glycosidase is frequently also named cellobiase. Thus, β-glycosidases end the digestion of cellulose and hemicellulose (Terra and Ferreira 1994, 2005).

A special β-glycosidase (aryl β-glycosidase) acts on glycolipids and *in vivo* probably removes a galactose from monogalactosyldiacylglycerol that together with digalactosyldiacylglycerol is a major lipid of photosynthetic tissues. Digalactosyldiacylglycerol is converted into monogalactosyldiacylglycerol by the action of an α-galactosidase. The aryl β-glycosidase also acts on plant glycosides that are noxious after hydrolysis. Insects circumvent these problems by detoxifying the products of hydrolysis (Spencer 1988) or by repressing the synthesis and secretion of this enzyme while maintaining constant the synthesis and secretion of the other β-glycosidases (Ferreira et al. 1997; Azevedo et al. 2003).

Trehalase is the enzyme that hydrolyzes the disaccharide trehalose, the main sugar in insect hemolymph. As the sugar is used as an energy source, trehalase occurs in all insect tissues. In the midgut, trehalase may be found in a soluble form that is secreted into lumen or bound to the apical membranes (Terra and Ferreira 1994, 2005). Although Mitsumatu et al. (2005) state that the membrane-bound trehalase is present in the visceral muscles contaminating their preparations, their results do not seem conclusive.

At first, the midgut soluble trehalase was considered to be responsible for the hydrolysis of the trehalose diffusing from the hemolymph into midgut lumen. This would recover the resulting glucose, because glucose absorption follows a downward concentration gradient, as the concentration of glucose in the hemolymph is very low (Wyatt 1967). Since the soluble trehalase activity decreases during insect starvation and regains its normal level on feeding, whereas the hemolymph trehalose remains constant, soluble trehalase should be a true digestive enzyme (Terra and Ferreira 1981). All trehalases, including those of midgut, are inhibited by plant toxic β-glycosides and their aglycones. Insects may react to the ingestion of those substances by increasing trehalase activity (Silva et al. 2006).

A combination of chemical modification data (Silva et al. 2004) and site-directed mutagenesis (Silva et al. 2010) led to the finding that in addition to the catalytic residues (Asp322, Glu520) there are three other essential residues (Arg169, Arg227, Arg287) in insect trehalase.

4.3.3 Digestion of Lipids and Phosphates

Oils and fats are triacylglycerols and are hydrolyzed by a triacylglycerol lipase that preferentially removes the outer ester links of the substrate (Figure 4.1d) and acts only on the water–lipid interface. This interface is increased by surfactants that, in contrast to the bile salts of vertebrates, are mainly lysophosphatides. The resulting 2-monoacylglycerol may be absorbed or further hydrolyzed before absorption (Terra and Ferreira 1994, 2005).

Membrane lipids include glycolipids, such as galactosyldiacylglycerol and phosphatides. After the removal of galactose residues from mono- and digalactosyldiacylglycerol, which leaves diacylglycerol, it is hydrolyzed as described for triacylglycerols. Phospholipase A removes one fatty acid from the phosphatide, resulting in a lysophosphatide (Figure 4.5d) that forms micellar aggregates, causing the solubilization of cell membranes. Lysophosphatide seems to be absorbed intact by insects (Terra and Ferreira 1994, 2005).

Nonspecific phosphatases remove phosphate moieties from phosphorylated compounds to make their absorption easier. Phosphatases are active in an alkaline or acid medium (Terra and Ferreira 1994, 2005).

4.4 Food Handling and Ingestion

4.4.1 Preliminary Observations

Food preparation for ingestion varies with the kind of food, insect mouthparts (that depends on insect phylogeny), salivary glands, and strategies of development. In the following discussion, the subject will be treated according to the nature of the food and employed mouthparts. Only the major groups will be discussed.

4.4.2 Solid Food

Usually solid food is taken with the aid of chewing mouthparts lubricated with saliva. As a rule, saliva does not have enzymes, although in some cases, it may contain amylase and α-glucosidase (Walker 2003) and, in even rarer cases, cellulase and laminarinase may be found (Genta et al. 2003). The role of these enzymes in digestion is only subsidiary. Examples are given in Sections 4.8.2 and 4.8.5.

Other forms of ingesting solid food are observed in filterers (e.g., mosquito larvae), where saliva has no importance, and in insects with piercing-sucking mouthparts, such as hemipterans. If the hemipteran is a predator, digestion is usually preoral and depends on salivary enzymes injected into the host (Miles 1972). Otherwise, in seed suckers, saliva is usually devoid of enzymes and the ingested material corresponds to particles in suspension caused by movements of the mouthparts and saliva fluxes (e.g., *Dysdercus peruvianus* Guérin-Méneville, Section 4.8.5).

4.4.3 Liquid Food

Important differences are observed if the food is blood or plant sap, which are taken with piercing-sucking mouthparts and nectar, acquired with lapping (bees) or sucking (adult lepidopterans) mouthparts, respectively.

Blood intake should be fast and painless to avoid a host aggressive reaction. For this, the saliva of blood feeders has analgesics, vasodilators, and anticoagulants, but lacks digestive enzymes (Ribeiro 1987). This strategy is found both among the mosquitoes and blood-feeding hemipterans, despite the fact they have strikingly different digestive processes (see Sections 4.8.5 and 4.8.8).

Sap-sucking hemipterans have two types of saliva. One is responsible for the formation of a sheath that surrounds the stylets. The other contains digestive enzymes that aid in the penetration of the plant tissue to attain the conducting vessels. The enzymes include pectinases and others that break the intracellular cement (Walker 2003).

Finally, nectar feeders, like bees, have as a rule a salivary α-glucosidase that hydrolyzes sucrose from the nectar into glucose and fructose (see Section 4.8.7).

4.5 Overview of the Digestive Process

The proposal of models for the digestive process is based on the known relationships between the phases of digestion and the gut compartments (crop, endo- and ectoperitrophic space, midgut cells; Figure 4.1) where they occur. For this, samples of the ectoperitrophic space contents (Figure 4.1) are collected by puncturing the midgut ceca with a capillary or by washing the luminal face of midgut tissue. Midgut tissue enzymes are intracellular, glycocalyx-associated, or microvillar membrane-bound (Figure 4.4). Their location is determined by cell fractionation. In addition to the distribution of digestive enzymes, the spatial organization of digestion depends on midgut fluxes. Fluxes are inferred with the use of dyes. Secretory regions deposit injected dye onto the midgut hemal side, whereas absorbing regions accumulate orally fed dyes on the midgut luminal surface.

Frequently, initial digestion starts in the crop and goes on in the endoperitrophic space. Intermediate digestion takes place in the ectoperitrophic space (Figure 4.1) and the final digestion occurs at the midgut tissue surface, under the action of enzymes that are trapped into the glycocalyx or are integral proteins in microvillar membrane (Figure 4.4). Such studies (reviews: Terra and Ferreira 1994, 2005, 2009) led to the proposal of the endo-ectoperitrophic circulation of digestive enzymes. According to this recycling model (Figure 4.1), the food is moved inside the peritrophic membrane by peristalsis, whereas in the ectoperitrophic space there is a countercurrent flux of fluid caused by secretion of fluid at the end of the midgut and its absorption back in the ceca. As soon as the polymeric food molecules are digested to become sufficiently small to pass through the peritrophic membrane, they are displaced toward the ceca or anterior midgut, where intermediate and final digestion is completed and nutrient absorption occurs.

The origin and chemical nature of the peritrophic membrane were discussed in Section 4.2, whereas its functions will be analyzed here. As mentioned before, the peritrophic membrane protects the midgut tissues against food abrasion and invasion by microorganisms, but it also enhances digestive efficiency, as detailed below.

The first described function of the peritrophic membrane in enhancing digestive efficiency was that of permitting enzyme recycling (described above), which results in a decrease in digestive enzyme excretion. Since the first studies reporting that *Rhynchosciara americana* Wiedemann, and *Musca domestica* L. excreted less that 15% of the luminal trypsin at each midgut emptying (Terra and Ferreira 1994, 2005), the findings were now extended to include Lepidoptera (Borhegyi et al. 1999; Bolognesi et al. 2001, 2008), Coleoptera (Ferreira et al. 2002; Caldeira et al. 2007), and Orthoptera (Biagio et al. 2009).

The suggested increase in the efficiency of digestion of polymeric food, favored by oligomer (potentially inhibitors of initial digestion) removal (Terra 2001) was tested and confirmed. For this midgut contents from *S. frugiperda* larvae were placed into dialysis bags. Trypsin activities in stirred and unstirred bags were 210% and 160%, respectively, over the activities of similar samples maintained in a test tube (Bolognesi et al. 2008).

The hypothesis that the efficiency of oligomer digestion (intermediate digestion) increases if separated from the initial digestion (Terra 2001) was confirmed by the experiments of Bolognesi et al. (2008). They collected ectoperitrophic fluid from the large midgut ceca of *R. americana* and assayed several digestive enzymes restricted to the fluid. When those enzymes were put in the presence of peritrophic membrane contents, their activities decreased in relation to controls. These decreases in activity probably result from oligomer hydrolase competitive inhibition by luminal polymers.

Finally, the proposal that the peritrophic membrane avoids the unspecific binding of undigested material onto midgut cell surface, with benefic results (Terra 2001) was experimentally confirmed (Bolognesi et al. 2008). For this, purified midgut microvillar membranes were isolated. The activity of the microvillar enzymes decreased if peritrophic membrane contents were added to the assay media.

Upon studying the spatial organization of the digestive events in insects of different taxa and diets, it was realized that the insects may be grouped relative to their digestive physiology, assuming they have common ancestors. Those putative ancestors correspond to basic gut plans from which groups of insects may have evolved by adapting to different diets (Figure 4.6).

Midgut cell microvilli have, in addition to being the most frequent site of final digestion, a role in midgut protection in lepidopterans. This role includes protection against oxidative stress, detoxification of H_2O_2 and aldehydes, and against the action of the insect's own luminal serine proteinases (Ferreira et al. 2007).

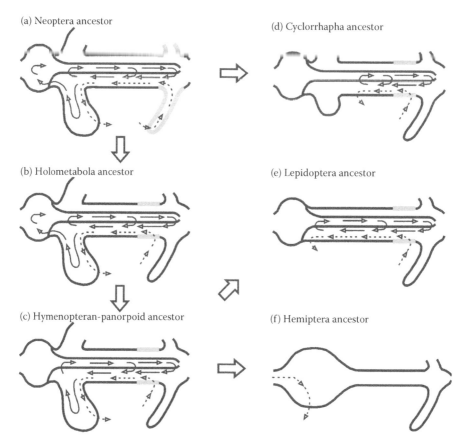

FIGURE 4.6 Diagrammatic representation of water fluxes (dashed arrows) and of the circulation of digestive enzymes (solid arrows) in putative insect ancestors that correspond to the major basic gut plans. In Neoptera ancestors (a), midgut digestive enzymes pass into the crop. Countercurrent fluxes depend on the secretion of fluid by the Malpighian tubules and its absorption by the ceca. Enzymes involved in initial, intermediate, and final digestion circulate freely among gut compartments. Holometabola ancestors (b) are similar except that secretion of fluid occurs in posterior ventriculus. Hymenopteran-Panorpoid (Lepidoptera and Diptera assemblage) ancestors (c) display countercurrent fluxes like Holometabola ancestors, midgut enzymes are not found in crop, and only the enzymes involved in initial digestion pass through the peritrophic membrane. Enzymes involved in intermediate digestion are restricted to the ectoperitrophic space and those responsible for terminal digestion are immobilized at the surface of midgut cells. Cyclorrhapha ancestors (d) have a reduction in ceca, absorption of fluid in middle midgut, and anterior midgut playing a storage role. Lepidoptera ancestors (e) are similar to panorpoid ancestors, except that anterior midgut replaced the ceca in fluid absorption. Hemiptera ancestors (f) lost crop, ceca, and fluid-secreting regions. Fluid is absorbed in anterior midgut. (Reprinted from *Encyclopedia of Insects*, 2nd Edition, Terra, W. R., and C. Ferreira, Digestive System, 273–81. Copyright 2009, with permission from Elsevier.)

4.6 Role of Microorganisms in Digestion

Most insects harbor a substantial microbiota including bacteria, yeast, and protozoa. Microorganisms might be symbiotic or fortuitous contaminants from the external environment. Microorganisms are found in the lumen, adhering to the peritrophic membrane, attached to the midgut surface, or intracellular. Intracellular bacteria are usually found in special cells, the mycetocytes that may be organized in groups, the mycetomes. Microorganisms produce and secrete their own hydrolases and cell death will result in the release of enzymes into the intestinal milieu. Any consideration of the spectrum of hydrolase activity in the midgut must include the possibility that some activities are derived from microorganisms. Despite the fact that digestive enzymes of some insects are thought to be derived from the microbiota, there are relatively few studies that show an unambiguous contribution of microbial hydrolases. Best examples are found among wood- and humus-feeding insects like termites, tipulid fly larvae, and scarabaeid beetle larvae. Although these insects may have their own cellulases (Section 4.3.2), only fungi and certain filamentous bacteria developed a strategy for the chemical breakdown of lignin. Lignin is a phenolic polymer that forms an amorphous resin in which the polysaccharides of the secondary plant cell wall are embedded, thus becoming hindered from enzymatic attack (Terra et al. 1996; Brune 1998; Dillon and Dillon 2004).

Microorganisms play a limited role in digestion of refractory materials, but they may enable phytophagous insects to overcome biochemical barriers to herbivory, for example detoxifying flavonoids, alkaloids, and the phenolic aglycones of plant glycosides. They may also provide complex-B vitamins for blood feeders and essential amino acids for phloem feeders, produce pheromone components, or withstand the colonization of the gut by nonindigenous species (including pathogens) (Dillon and Dillon 2004; Genta et al. 2006b).

4.7 Midgut Conditions Affecting Enzyme Activity

The pH of midgut contents is one of the important internal environmental properties that affect digestive enzymes. Although midgut pH is hypothesized to result from adaptation of an ancestral insect to a particular diet, its descendants may diverge, feeding on different diets, while still retaining the ancestral midgut pH condition. Thus, it is not necessary that there is a correlation between midgut pH and diet. Actually, midgut pH correlates well with insect phylogeny (Terra and Terra 1994; Clark 1999). The pH of insect midgut contents is usually in the 6 to 7.5 range. Major exceptions are the very alkaline midgut contents (pH 9–12) of Lepidoptera, scarab beetles, and nematoceran Diptera larvae, the very acid (pH 3.1–3.4) middle region of the midgut of cyclorrhaphous Diptera and the acid posterior region of the midgut of Hemiptera Heteroptera (Terra and Ferreira 1994; Clark 1999). pH values may not be equally buffered along the midgut.

The high alkalinity of lepidopteran midgut contents is thought to allow these insects to feed on plant material rich in tannins, which bind to proteins at lower pH, reducing the efficiency of digestion (Berenbaum 1980). This explanation may also hold for scarab beetles and for detritus-feeding nematoceran Diptera larvae that usually feed on refractory materials such as humus. Nevertheless, mechanisms other than high gut pH must account for resistance to tannin displayed by some locusts (Bernays et al. 1981) and beetles (Fox and Macauley 1977). One possibility is the effect of surfactants, such as lysolecithin that is formed in insect fluids due to the action of phospholipase A on cell membranes (Figure 4.5), and which occurs widely in insect digestive fluids (De Veau and Schulz 1992). Surfactants are known to prevent the precipitation of proteins by tannins even at pH as low as 6.5 (Martin and Martin 1984). In fact, the importance of a high midgut pH must be to free hemicelluloses that are digested even by insects unable to nourish from cellulose (Terra 1988).

Few papers have dealt with midgut pH buffering mechanisms. The best known is the alkaline buffer of lepidopterans midguts. Dow (1992) showed that the goblet cells from the lepidopteran larval midgut secrete a carbonate secretion that may be responsible for the high pH found in Lepidoptera midguts. It is not known if luminal midgut alkalinization in scarab beetles and nematoceran dipteran larvae occurs by a mechanism similar to that described for lepidopterans. Acidification of the midgut contents in *M. domestica* middle midgut apparently results from a proton pump, whereas neutralization at posterior midgut depends on ammonia secretion (Terra and Regel 1995).

Redox conditions in the midgut are regulated and may be the result of phylogeny, although data are scarce. Reducing conditions are important to open disulphide bonds in keratin ingested by some insects (clothes moths, dermestid beetles) (Appel and Martin 1990) to maintain the activity of the major proteinase in Hemiptera (see Section 4.3.1) and to reduce the impact of some plant allelochemicals, such as phenol, in some herbivores (Appel and Martin 1990).

Although several allelochemicals other than phenols may be present in the insect gut lumen, including alkaloids, terpene aldehydes, saponins, and hydroxamic acids (Appel 1994), data is lacking on their effect on digestion.

4.8 Basic Plans of the Digestive Process

4.8.1 Evolutionary Trends of Insect Digestive Systems

Neopteran (most of the winged forms) insects evolved along three lines: the Polyneoptera (which include Blattodea, Isoptera, and Orthoptera), the Paraneoptera (which include Hemiptera), and the Holometabola (which include Coleoptera, Hymenoptera, Diptera, and Lepidoptera).

About 86% of insect species are Holometabola, whereas the majority of the other species are Polyneoptera and Paraneoptera, and a few species are from nonneopteran lines. The success of Holometabola is probably related to the fact that their young forms (larvae) are adapted to ecological niches different from those of the adults, which do not allow them to compete with the adults for the same food, as is the case for the young forms (nymphs) and adults of the lines Polyneoptera and Paraneoptera (Kristensen 1999; Cranston and Gullan 2003).

Among Holometabola, the most successful orders are the Coleoptera, Hymenoptera, Diptera, and Lepidoptera. Perhaps Coleoptera insects are so numerous because they were the earliest group of Holometabola to evolve, occupying the relatively safe and abundant surface and subsurface niches initially available to insects (Evans 1975).

The evolution of the other Holometabola orders (the higher Holometabola) occurred through the occupation of ephemeral and mainly exposed niches. The occupation of less safe niches (or ephemeral ones) led to the appearance of several adaptations, the most effective probably is the reduction in life cycle, which makes the development of more generations possible within a fixed period of time, thus assuring the survival of numerous individuals, even if the mortality rate is high in each generation. Thus, whereas the life span of a beetle is about 12 months, that of a fly or a butterfly is about 6 weeks (Sehnal 1985). Associated with this decrease in life span, one would expect to find larger growth and food consumption rates in higher Holometabola than in Polyneoptera and Paraneoptera. Indeed, the relative growth rate (increase in biomass per initial biomass per day) is 0.07 (range, 0.01–0.16) for Coleoptera, 0.3 (range, 0.03–1.5) for Lepidoptera and 0.21 (range, 0.08–0.4) for Hymenoptera, whereas the relative food consumption rate (dry weight of food ingested per dry weight biomass) is 0.6 (range, 0.02–1.4) for Coleoptera, 1.8 (range, 0.27–6.9) for Lepidoptera, and 2.3 (range, 0.9–3.6) for Hymenoptera (Slansky and Scriber 1985).

It is possible that the remarkable increase in growth rate (which depends on the food consumption rate) of higher Holometabola in relation to Coleoptera is related to changes in the digestive physiology of Holometabola and even in the morphology of their gut. Otherwise, the fact that both the growth and food consumption rates of Polyneoptera (exemplified by Orthoptera) are similar to those of Coleoptera (Slansky and Scriber 1985) suggests that significant differences may not be found between their digestive physiology and gut morphology. Of course, these and the previous considerations refer to only the more generalized members of each group; specialized members may differ widely from the basic pattern of the group. It is interesting to note that the apparent digestibility [ratio between the mass of the absorbed food (dry mass of ingested food less the one of feces) and the mass of ingested food] depends more on the quality of the food than on the phylogenetic position of the insect (Slansky and Scriber 1985).

The organization of the digestive process in the different insect orders that correspond to the basic plans of the ancestral forms was reviewed several times (Terra 1988, 1990; Terra and Ferreira 1994, 2009). The following section is therefore an abridged version of those texts, highlighting new findings and trying to identify points that deserve more research, especially in relation to molecular aspects.

Polyneoptera and Paraneoptera evolved as external feeders occupying the ground surface, on vegetation, or in litter, and developed distinct feeding habits. Some of these habits are very specialized (e.g., feeding wood and sucking plant sap), implying adaptive changes of the digestive system. Major trends in the evolution of Holometabola were the divergence in food habits between larvae and adults and the exploitation of new food sources, exemplified by endoparasitism and by boring or mining living or dead wood, foliage, fruits, or seeds. This biological variation was accompanied by modifications in the digestive system. Among the hymenopteran and panorpoid (an assemblage that includes Diptera and

Lepidoptera) Holometabola, new selective pressures resulted from the occupation of more exposed or ephemeral ecological niches. Following this trend, those pressures led to shortening life spans, so that the insects may have more generations per year, thus ensuring species survival even if large mortality occurs at each generation. Associated with this trend, the digestive system evolved to become more efficient to support faster life cycles.

The basic plan of digestive physiology for most winged insects (Neoptera ancestors) is summarized in Figure 4.6a. In these ancestors, the major part of digestion is carried out in the crop by digestive enzymes propelled by antiperistalsis forward from the midgut. Saliva plays a variable role in carbohydrate digestion (see Section 4.4). After a while, following ingestion, the crop contracts, transferring digestive enzymes and partly digested food into the ventriculus. The anterior ventriculus is acid and has high carbohydrase activity, whereas the posterior ventriculus is alkaline and has high proteinase activity. The food bolus moves backward in the midgut of the insect by peristalsis. As soon as the polymeric food molecules have been digested to become small enough to pass through the peritrophic membrane, they diffuse with the digestive enzymes into the ectoperitrophic space (Figure 4.1). The enzymes and nutrients are then displaced toward the ceca with a countercurrent flux (mentioned in Section 4.5) caused by secretion of fluid at the Malpighian tubules and its absorption back by cells at the ceca (Figures 4.1 and 4.6), where final digestion is completed and nutrient absorption occurs. When the insect starts a new meal, the ceca contents are moved into the crop. As a consequence of the countercurrent flux, digestive enzymes occur as a decreasing gradient in the midgut, and lower amounts are excreted.

The Neoptera basic plan is the source of that of the Polyneoptera orders, which include Blattodea, Isoptera, and Orthoptera, and evolved to the basic plans of Paraneoptera and Holometabola. Lack of data limits the proposition of a basic plan to a single Paraneoptera order, Hemiptera.

The basic gut plan of the Holometabola (Figure 4.6b) (which include Coleoptera, Megaloptera, Hymenoptera, Diptera, and Lepidoptera) is similar to that of Neoptera, except that fluid secretion occurs by the posterior ventriculus instead of by the Malpighian tubules. The basic plan of Coleoptera did not evolve dramatically from the Holometabola ancestor, whereas the basic plan of Hymenoptera, Diptera, and Lepidoptera ancestor (hymenopteran-panorpoid ancestor; Figure 4.6c) presents important differences. Thus, hymenopteran-panorpoid ancestors have countercurrent fluxes like Holometabola ancestors, but differ from these in the lack of crop digestion, midgut differentiation in luminal pH, and in which compartment is responsible for each phase of digestion. In Holometabola ancestors, all phases of digestion occur in the endoperitrophic space (Figure 4.1), whereas in hymenopteran-panorpoid ancestors only initial digestion occurs in that region. In the latter ancestors, intermediate digestion is carried out by free enzymes in the ectoperitrophic space and final digestion occurs at the midgut cell surface by immobilized enzymes (Figure 4.4).

4.8.2 Blattodea

Cockroaches, which are among the first neopteran insects to appear in the fossil record, are extremely generalized in most morphological features. They are usually omnivorous. Digestion in cockroaches occurs as described for the Neoptera ancestor (Figure 4.6a), except that part of the final digestion of proteins occurs on the surface of midgut cells (Terra and Ferreira 1994). The differentiation of pH along the midgut is not conserved among cockroaches like *Periplaneta americana* (L.) (Blattidae), but is maintained in others exemplified by the blaberid *Nauphoeta cinerea* (Olivier) (Elpidina et al. 2001). Another difference observed is the enlargement of hindgut structures (Figure 4.2c), noted mainly in wood-feeding cockroaches. These hindgut structures harbor bacteria producing acetate and butyrate from ingested wood or other cellulose-containing materials. Acetate and butyrate are absorbed by the hindgut of all cockroaches, mainly of wood roaches (Terra and Ferreira 1994). Cellulose digestion may be accomplished in part by bacteria in the hindgut of *P. americana* or by protozoa in *Cryptocercus punctulatus* (Bignell 1981). Nevertheless, it is now clear that the saliva of *P. americana* has two cellulases and three laminarinases that open plant cells and lise fungi cells (Genta et al. 2003). This is in accordance with the detritivorous habit of this insect. The wood roach, *Panesthia cribata* Saussure, also has its own cellulose (Scrivener et al. 1989; Tokuda et al. 1999).

4.8.3 Isoptera

Termites are derived from and are more adapted than wood roaches in dealing with refractory materials such as wood and humus. Associated with this specialization, they lost the crop and midgut ceca and enlarged their hindgut structures (Figure 4.2d). Both lower and higher termites digest cellulose with their own cellulose (Watanabe et al. 1998; Tokuda et al. 1999), in spite of the presence of cellulase-producing protozoa in paunch, an enlarged hindgut region observed in lower termites. The products pass from the midgut into the hindgut, where they are converted into acetate and butyrate by hindgut bacteria as in wood roaches. Symbiotic bacteria are also responsible for nitrogen fixation in hindgut (Beneman 1973), resulting in bacterial protein. This is incorporated into the termite body mass after being expelled in feces by one individual and being ingested and digested by another. This explains the ability of termites to develop successfully in diets very poor in protein.

4.8.4 Orthoptera

Grasshoppers feed mainly on grasses, and their digestive physiology clearly evolved from the neopteran ancestor. Carbohydrate digestion occurs mainly in the crop, under the action of midgut enzymes, whereas protein digestion and final carbohydrate digestion take place at the anterior midgut ceca. The abundant saliva (devoid of significant enzymes) produced by grasshoppers saturate the absorbing sites in the midgut ceca, thus hindering the countercurrent flux of fluid. Starving grasshoppers present midgut countercurrent fluxes. Cellulase found in some grasshoppers is believed to facilitate the access of digestive enzymes to the plant cells ingested by the insects by degrading the cellulose framework of cell walls (Dow 1986; Terra and Ferreira 1994; Marana et al. 1997). Crickets are omnivorous or predatory insects with initial starch digestion occurring in their capacious crop (Figure 4.2b) and ending in caeca lumina. Regarding protein, initial trypsin digestion occurs mostly in caeca lumina, whereas final aminopeptidase digestion takes place in caeca and ventriculus. Differing from grasshoppers, the final digestion of both protein and carbohydrates depends on membrane-bound enzymes in addition to soluble ones. Both starving and feeding crickets present midgut countercurrent fluxes (Biagio et al. 2009).

4.8.5 Hemiptera

The Hemiptera comprise insects of the major infraorders Auchenorrhyncha (cicadas and cicadellids), and Sternorrhyncha (aphids) that feed almost exclusively on plant sap, and insects of the infraorder Heteroptera (e.g., assassin bugs, plant bugs, stinkbugs, and lygaeid bugs) that are adapted to different diets. The ancestor of the entire order is supposed to be a sapsucker similar to present-day Auchenorrhyncha.

The hemipteran ancestor (Figure 4.6f) differs remarkably from the neopteran ancestor, as a consequence of adaptations to feeding on plant sap. These differences consist of the lack of crop and anterior midgut ceca, loss of the enzymes involved in initial and intermediate digestion and loss of the peritrophic membrane associated with the lack of luminal digestion, and finally, the presence of hemipteran midgut cells (Figures 4.3h and 4.7). These cells have their microvilli ensheathed by an outer (perimicrovillar) membrane that extends toward the luminal compartment with a dead end enclosing a compartment, the perimicrovillar space (Figures 4.3h and 4.7).

Sap-sucking Hemiptera may suck phloem or xylem sap. Phloem sap is rich in sucrose (0.15–0.73 M) and relatively poor in free amino acids (15–65 mM) and minerals. Some rare phloems have considerable amounts of protein. Xylem fluid is poor in amino acids (3–10 mM) and contains monosaccharides (about 1.5 mM), organic acids, potassium ions (about 6 mM), and other minerals (Terra 1990). Thus, as a rule, no food digestion is necessary in sapsuckers except for dimer (sucrose) hydrolysis. The major problem facing a sap-sucking insect is to absorb nutrients, such as essential amino acids, that are present in very low concentrations in sap. Whichever mechanism is employed, xylem feeders may absorb up to 90% of the amino acids and carbohydrates from the sap (Andersen et al. 1989).

Amino acids are absorbed according to a hypothesized mechanism that depends on perimicrovillar membranes. Microvillar membranes actively transport potassium ion (the most important ion in sap) from perimicrovillar space (PMS, compartment between the microvillar and the perimicrovillar membrane)

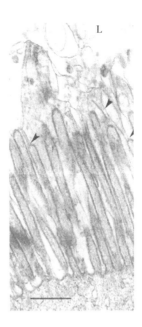

FIGURE 4.7   The microvillar border of midgut cells from Hemiptera. Electron micrograph of *Rhodnius prolixus* posterior midgut cell. Detail of microvilli showing the extension of the perimicrovillar membrane (PMM) in to the midgut lumen (arrowheads). Scale bar: 1 μm. (Reprinted from *Insect Biochem.*, 18, Ferreira, C., A. F. Ribeiro, E. S. Garcia, and W. R. Terra, Digestive enzymes trapped between and associated with the double plasma membranes of Rhodnius prolixus posterior midgut cells, 521–30. Copyright 1988, with permission from Elsevier.)

into the midgut cells, generating a concentration gradient between the gut luminal sap and the PMS. This concentration gradient may be a driving force for the active absorption of organic compounds (amino acids and sugars, for example) by appropriate carriers present in the perimicrovillar membrane. Organic compounds, once in the PMS, may diffuse up to specific carriers on the microvillar surface. This movement is probably enhanced by a transfer of water from midgut lumen to midgut cells, following (as solvation water) the transmembrane transport of compounds and ions by the putative carriers. The model assumes the presence of K⁺-amino acid symporter in the surface of the perimicrovillar membranes and amino acid uniporters and K⁺-pumps in the microvillar membranes (Terra 1988; Ferreira et al. 1988). Although amino acid transporters have been described in the midgut microvillar membranes of several insects (Wolfersberger 2000), no attempts have been made to study the other postulated proteins.

Organic compounds in xylem sap need to be concentrated before they can be absorbed by the perimicrovillar system. This occurs in the filter chamber (Figure 4.2p) of Cicadoidea and Cercopoidea, and Cicadelloidea, which concentrates the sap fluid by tenfold. The filter chamber consists of a thin-walled, dilated anterior midgut in close contact with the posterior midgut and the proximal ends of the Malpighian tubules. This arrangement enables water to pass directly from the anterior midgut to the Malpighian tubules through specific channels made up of aquaporin molecules, thus concentrating food in midgut (Le Cahérec et al. 1997).

Sternorrhyncha, as exemplified by aphids, may suck more or less continuously phloem sap of sucrose concentration up to 1.0 M and osmolarity up to three times that of the insect hemolymph. This results in a considerable hydrostatic pressure caused by the tendency of water to move from the hemolymph into midgut lumen. To withstand these high hydrostatic pressures, aphids developed several adaptations. Midgut stretching resistance is helped by the existence of links between apical lamellae (replacing usual midgut cell microvilli) that become less conspicuous along the midgut (Figure 4.8a). As a consequence of the links between the lamellae, the perimicrovillar membranes could no longer exist and were replaced by membranes seen associated with the tips of the lamellae: the modified perimicrovillar membranes (Figure 4.8b) (Ponsen 1991; Cristofoletti et al. 2003). A modified perimicrovillar membrane-associated α-glucosidase frees fructose from sucrose without increasing the osmolarity by

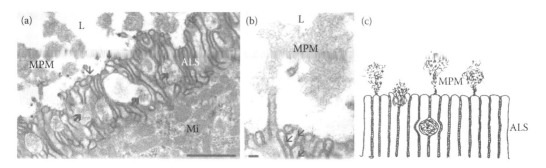

FIGURE 4.8   The apex of midgut cells from the aphid *A. pisum* (a) Electron micrograph of the apical surface of anterior midgut cells showing the lamellar system with associated modified perimicrovillar membranes (MPM) projecting into the lumen. Note trabecullae (small arrows) between lamellae and MPM masses moving among lamellae (large arrows). (b) Detail of modified perimicrovillar membranes associated with lamellae. Note trabecullae (arrows) between lamellae. (c) Model for the origin of membrane masses associated with the apical lamellar system, the modified perimicrovillar membranes. Abbreviations: ALS = apical lamellar system; L = lumen; Mi = mitochondria; MPM = modified perimicrovillar membranes. Bars, 1.0 μM (a), 0.1 μM (b). (Reprinted from *J. Insect Physiol.*, 49, Cristofoletti, P. T., A. F. Ribeiro, C. Deraison, Y. Rahbé, and W. R. Terra, Midgut adaptation and digestive enzyme distribution in a phloem feeding insect, the pea aphid *Acyrtosiphon pisum*, 11–24. Copyright 2003, with permission from Elsevier.)

promoting transglycosylations. As the fructose is quickly absorbed, the osmolarity decreases, resulting in a honeydew isoosmotic with hemolymph (Ashford et al. 2000; Cristofoletti et al. 2003). Another interesting adaptation is observed in whiteflies, where a trehalulose synthase forms trehalulose from sucrose, thus making available less substrate for an α-glucosidase that otherwise would increase the osmolarity of ingested fluid on hydrolyzing sucrose (Salvucci 2003).

A cathepsin L (see Section 4.3.1) bound to the modified perimicrovillar membranes of *A. pisum* (Cristofoletti et al. 2003) may explain the capacity of some phloem sap feeders to rely on protein found in some phloem saps (Salvucci et al. 1998) and the failure of other authors to find an active proteinase in sap feeders. They worked with homogenate supernatants or supernatants of Triton X-100-treated samples, under which conditions the cathepsin L would remain in the pellet. An aminopeptidase, also bound to the modified perimicrovillar membranes, is the major binding site of the lectin Concanavalin A. On binding, the lectin impairs the aphid development, in spite of the fact that the lectin does not affect aminopeptidase activity. It is thought that the aminopeptidase is located near the amino acid carriers responsible for amino acid absorption and that these are inhibited when the lectin binds to the aminopeptidase (Cristofolettti et al. 2006).

Amino acid absorption in *A. pisum* midguts is influenced by the presence of the bacteria *Buchnera* in the mycetocytes of the mycetomes occurring in the aphid hemocoel (Prosser et al. 1992). The molecular mechanisms underlying this phenomenon are not known, in spite of the fact that there is strong evidence showing that *Buchnera* uses the nonessential amino acids absorbed by the host in the synthesis of essential amino acids (Prosser and Douglas 1992; Shigenobu et al. 2000). It is likely that amino acid absorption through apical lamellar carriers depends on the amino acid concentration gradient between midgut lumen and hemolymph, whereas hemolymph titers vary widely according to *Buchnera* metabolic activity (Liadouze et al. 1995).

The evolution of Heteroptera was associated with regaining the ability to digest polymers. Because the appropriate digestive enzymes were lost, these insects instead used proteinases (cathepsins) derived from lysosomes (Houseman et al. 1985). Lysosomes are cell organelles involved in the intracellular digestion of proteins carried out by special proteinases, the cathepsins. Compartmentalization of digestion was maintained by the perimicrovillar membranes as a substitute for the lacking peritrophic membrane (Terra et al. 1988; Ferreira et al. 1988). Digestion in the two major Heteroptera taxa—Cimicomorpha, exemplified by the blood feeder *Rhodius prolixus,* and Pentatomorpha, exemplified by the seed sucker *Dysdercus peruvianus*—is similar (Terra et al. 1988a; Ferreira et al. 1988; Silva and Terra 1994; Silva et al. 1995). The dilated anterior midgut stores food and absorbs water, and, at least in *D. peruvianus,* also absorbs glucose that is transported with the aid of a uniporter (GLUT) and a K^+-glucose symporter (SGLT) (Bifano et al. 2010). Digestion of proteins and absorption of amino acids occur in the posterior ventriculus. Most protein digestion occurs in lumen with the aid of a cysteine proteinase (cathepsin L) and ends in the perimicrovillar space under the action of aminopeptidases and dipeptidases. Symbiont bacteria may occur in hematophagous insects, arguably to supply vitamins. Many Heteroptera feed on parenchymal tissues of plants. In some of these insects, excess water passes from the expanded anterior midgut to the closely associated midgut ceca, which protrude from the posterior midgut (Figure 4.2q). These ceca may also contain symbiotic bacteria (Goodchild 1966).

4.8.6 Coleoptera

Larvae and adults of Coleoptera usually display the same feeding habit; that is, both are plant feeders (although adults may feed on the aerial parts, whereas the larvae may feed on the roots of the same plant) or both are predatory. Coleoptera ancestors are like Holometabola ancestors except for the anterior midgut ceca, which were lost and replaced in function by the anterior midgut. Nevertheless, there are evolutionary trends leading to derived systems. Thus, in predatory Carabidae most of the digestive phases occur in the crop by means of midgut enzymes, whereas in predatory larvae of Elateridae initial digestion occurs extraorally by the action of enzymes regurgitated onto their prey. The preliquefied material is then ingested by the larvae, and its digestion is finished at the surface of midgut cells (Terra and Ferreira 1994).

Initial digestion of glycogen and proteins occur in the dermestid larval endoperitrophic space. Final digestion takes place at the midgut cell surface, in the anterior and posterior midgut in the case of glycogen and proteins, respectively. There is a decreasing gradient along the midgut of amylase and trypsin (major proteinase), suggesting the occurrence of digestive enzyme recycling (Caldeira et al. 2007).

Like dermestid beetles, the larvae of *Migdolus fryanus* Westwood (Cerambycidae) and *Sphenophorus levis* Vaurie (Curculionidae) have a peritrophic gel and a peritrophic membrane in the anterior and posterior midgut, respectively, and microvillar aminopeptidase and a decreasing gradient of amylase, maltase, and proteinase along the midgut (A. B. Dias and W. R. Terra, unpublished).

Tenebrionid larvae also have aminopeptidase as a microvillar enzyme and the distribution of enzymes in gut regions of adults is similar to the larvae (Terra and Ferreira 1994). This suggests that the overall pattern of digestion in larvae and adults of Coleoptera is similar, despite the fact that (in contrast to adults) beetle larvae usually lack a crop.

Insects of the series Cucujiformia (which includes Tenebrionidae, Chrysomelidae, Bruchidae, and Curculionidae) have cysteine proteinases in addition to (or in place of) serine proteinases as digestive enzymes, suggesting that the ancestors of the whole taxon were insects adapted to feed on seeds rich in serine proteinase inhibitors (Terra and Ferreira 1994). The finding of trypsin as the major digestive proteinase in *M. fryanus* (A. B. Dias and W. R. Terra, unpublished) confirms previous work (Murdock et al. 1987) that stated that Cerambycidae larvae reacquired the digestive serine proteinases

Scarabaeidae and related families are relatively isolated in the series Elateriformia. They evolved considerably from the Coleoptera ancestor. Scarabid larvae, exemplified by dung beetles, usually feed on cellulose materials undergoing degradation by a fungus-rich flora. Digestion occurs in the midgut, which has three rows of ceca (Figure 4.2f), with a ventral groove between the middle and posterior row. The alkalinity of gut contents increase to almost pH 12 along the midgut ventral groove. This high pH probably enhances cellulose digestion, which occurs mainly in the hindgut fermentation chamber (Figure 4.2f). The final product of cellulose degradation is mainly acetic acid, which is absorbed through the hindgut wall. It is not known with certainty if scarab larvae ingest feces to obtain nitrogen from the biomass, as described for termites (Section 4.8.3). Nevertheless, this is highly probable on the grounds that the microbial biomass in the fermentation chamber is incorporated into the larval biomass (Li and Brune 2005).

4.8.7 Hymenoptera

The organization of the digestive process is variable among hymenopterans and to understand its peculiarities it is necessary to review briefly their evolution. The hymenopteran basal lineages are phytophagous as larvae, feeding both ecto- and endophytically and include several superfamilies like Xyeloidea and Tenthredinoidea, all known as sawflies. Phylogenetically close to these are the Siricoidea (wood wasps) that are adapted to ingest fungus-infected wood. Wood wasplike ancestors gave rise to the Apocrita (wasp-waisted Hymenoptera) that are parasitoids of insects. They use their ovipositor to injure or kill their host that corresponds to a single meal for their complete development. A taxon sister of Ichneumonoidea in Apocrita gave rise to Aculeata (bees, ants, and wasps with thin waist) (Quicke 2003).

Hymenoptera ancestors are like panorpoid ancestor (Figure 4.6c), but there are trends leading to the loss of anterior midgut ceca, and in compartmentalization of digestion. These trends appear to be associated with the development of parasitic habits and were maintained in Aculeata, as detailed below.

The sawfly *Themos malaisei* Smith (Tenthredinoidea: Argidae) larva has a midgut with a ring of anterior caeca that forms a U at the ventral side (Figure 4.2h). Luminal pH is above 9.5 in the first two-thirds of the midgut. There is a recycling of enzymes involved in initial digestion and the final digestion occurs in the midgut cell surface (A. B. Dias, J. M. C. Ribeiro, and W. R. Terra, unpublished). These characteristics (except the presence of ceca) are similar to those of lepidopteran larvae.

Wood wasp larvae of the genus *Sirex* are believed to be able to digest and assimilate wood constituents by acquiring cellulase, xylanase, and possibly other enzymes from fungi present in wood on which they feed (Martin 1987). The first Apocrita were probably close to the ichneumon flies, whose larvae develop on the surface or inside the body of the host insect. Probably because of that, the larvae of Apocrita present a midgut that is closed at its rear end, and remains unconnected with the hindgut until the time of pupation.

In larval bees, most digestion occurs in the endoperitrophic space. Countercurrent fluxes seem to occur, but the midgut luminal pH gradient hypothetically present in the Hymenoptera-Panorpoidea ancestor was lost. Adult bees ingest nectar and pollen. Sucrose from nectar is hydrolyzed in the crop (Figure 4.2i) by the action of a sucrase from the hypopharyngeal glands. After ingestion, pollen grains extrude their protoplasm into the ventriculus, where digestion occurs. Adults seem to have midgut countercurrent fluxes as the larvae (Jimenez and Gilliam 1990; Terra and Ferreira 1994). Workers of leaf-cutting ants feed on nectar, honeydew, plant sap, or partly digested food regurgitated by their larvae. Because of this, it is frequently stated that they have no digestive enzymes or that they have only enzymes involved in intermediate and/or final digestion (Terra and Ferreira 1994). Although this seems to be true for leaf-cutting ants, which may depend entirely on monosaccharides produced by fungal enzymes acting on polysaccharides (Silva et al. 2003), this does not appear to be general among ants. Thus, adults of *Camponotus rufipes* (F.) (Formicinae) have as digestive enzymes amylase, trypsin (major proteinase), maltase, and aminopeptidase inside a type I peritrophic membrane. As only 14% of the amylase and less than 7% of the other digestive enzymes are excreted at each midgut emptying, these insects must have the usual enzyme recycling mechanism (A. B. Dias and W. R. Terra, unpublished).

4.8.8 Diptera

The Diptera evolved along two major lines: an assemblage of suborders corresponding to the mosquitoes, including the basal Diptera, and the suborder Brachycera, which includes the most evolved flies (Cyclorrhapha). The Diptera ancestor is similar to the hymenopteran-panorpoid ancestor (Figure 4.5c) in having the enzymes involved in intermediate digestion free in the ectoperitrophic fluid (mainly in the large ceca), whereas the enzymes of terminal digestion are membrane bound at the midgut cell microvilli (Terra and Ferreira 1994). Although these characteristics are observed in most nonbrachyceran larvae, the more evolved of these larvae may show reduction in size of midgut ceca (e.g., Culicidae, Figure 4.2k). Nonhematophagous adults store (nectar or decay products) in their crops and carried out digestion and absorption at the anterior midgut. Blood, which is sucked only by females, passes to the posterior midgut, where it is digested and absorbed (Billingsley 1990; Terra and Ferreira 1994).

The Cyclorrhapha ancestor (Figure 4.5d) evolved dramatically from the hymenopteran-panorpoid ancestor (Figure 4.5c), apparently as a result of adaptations to a diet consisting mainly of bacteria. Digestive events in Cyclorrhapha larvae are exemplified by larvae of the house fly *Musca domestica* (Espinoza-Fuentes and Terra 1987; Terra et al. 1988b). These larvae ingest food rich in bacteria. In the anterior midgut, there is a decrease in the starch content of the food bolus, facilitating bacteria death.

The bolus now passes into the middle midgut where bacteria are killed by the combined action of low pH, a special lysozyme, and a cathepsin D. Finally, the material released by bacteria is digested in the posterior midgut. Countercurrent fluxes occur in the posterior midgut powered by secretion of fluid in the distal part of the posterior midgut and its absorption back into the middle midgut. The middle midgut has specialized cells for buffering the luminal contents in the acidic zone (Figure 4.3d), in addition to those functioning in fluid absorption (Figure 4.3a). Except for a few bloodsuckers, Cyclorrhaphan adults feed mainly on liquids associated with decaying material (rich in bacteria) in a way similar to house fly adults. These salivate (or regurgitate the crop contents) on the food. After dispersion of the ingested material, starch digestion takes place at the crop under the action of a salivary amylase. Digestion is completed in the midgut as described for larvae (Terra and Ferreira 1994). The stable fly, *Stomoxys calcitrans* L., stores and concentrates the blood meal in the anterior midgut and gradually passes it to the posterior midgut, where digestion takes place, resembling what occurs in larvae. These adults lack the characteristic cyclorrhaphan middle midgut and the associated low luminal pH. Stable flies occasionally take nectar (Terra and Ferreira 1994).

4.8.9 Lepidoptera

Lepidopteran ancestors (Figure 4.5e) differ from hymenopteran-panorpoid ancestors because they lack midgut ceca, have all their digestive enzymes (except those of initial digestion) immobilized at the midgut cell surface, and present long-necked goblet cells (Figure 4.3b) and stalked goblet cells (Figure 4.3e) in the anterior and posterior larval midgut regions, respectively. Goblet cells excrete K^+ ions, which are absorbed from leaves ingested by larvae. Goblet cells also seem to assist the anterior columnar cells to absorb water and the posterior columnar cells to secrete water (Terra and Ferreira 1994; Ortego et al. 1996).

Although most lepidopteran larvae have a common pattern of digestion, species that feed on unique diets generally display some adaptations. *Tineola bisselliella* (Hummel) (Tineidae) larvae feed on wool and display a highly reducing midgut for cleaving the disulfide bonds in keratin to facilitate proteolytic hydrolysis of this otherwise insoluble protein (Terra and Ferreira 1994). Similar results were obtained with *Hofmannophila pseudospretella* (Stainton) (Christeller 1996). Wax moths (*Galleria mellonella* L.) infest beehives and digest and absorb wax. The participation of symbiotic bacteria in this process is controversial. The occurrence of the whole complement of digestive enzymes in nectar-feeding moths may explain, at least on enzymological grounds, the adaptation of some adult Lepidoptera to new feeding habits such as blood and pollen (Terra and Ferreira 1994).

4.9 Digestive Enzyme Secretion Mechanisms

Insects are continuous (e.g., Lepidoptera and Diptera larvae) or discontinuous (e.g., predators and hematophagous insects) feeders. Synthesis and secretion of digestive enzymes in continuous feeders seem to be constitutive; that is, these functions occur continuously, whereas in discontinuous feeders they are regulated (Lehane et al. 1996). It is widely believed (without clear evidence) that putative endocrine cells (Figure 4.3i) play a role in regulating midgut events. The presence of food in the midgut is necessary to stimulate synthesis and secretion of digestive enzyme. This was clearly shown in mosquitoes (Billingsley 1990).

Like all animal proteins, digestive enzymes are synthesized in the rough endoplasmic reticulum, processed in the Golgi complex, and packed into secretory vesicles (Figure 4.9). There are several mechanisms by which the contents of the secretory vesicles are freed in the midgut lumen. During exocytic secretion, secretory vesicles fuse with the midgut cell apical membrane, emptying their contents without any loss of cytoplasm (Figure 4.9a) (e.g., Graf et al. 1986). In contrast, apocrine secretion involves the loss of at least 10% of the apical cytoplasm following the release of secretory vesicles (Figure 4.9b). These have previously undergone fusions originating larger vesicles that after release eventually free their contents by solubilization (Figure 4.9b) (e.g., Cristofoletti et al. 2001). When the loss of cytoplasm is very small, the secretory mechanism is called microapocrine. Microapocrine secretion consists of releasing budding double-membrane vesicles (Figure 4.9c) or, at least in insect midguts, pinched-off

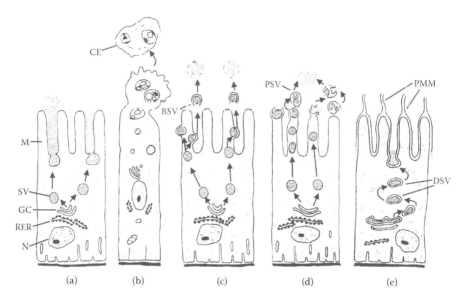

FIGURE 4.9 Models for secretory processes of insect digestive enzymes. (a) exocytic secretion, (b) apocrine secretion, (c) microapocrine secretion with budding vesicles, (d) microapocrine secretion with pinched-off vesicles, and (e) modified exocytic secretion in hemipteran midgut cell. Abbreviations: BSV = budding secretory vesicle; CE = cellular extrusion; DSV = double-membrane secretory vesicle; GC = Golgi complex; M = microvilli; N = nucleus; PMM = perimicrovillar membrane; PSV = pinched-off secretory vesicle; RER = rough endoplasmic reticulum; SV = secretory vesicle. (Reprinted from *Encyclopedia of Insects*, 2nd Edition, Terra, W. R., and C. Ferreira, Digestive System, 273–81. Copyright 2009, with permission from Elsevier.)

vesicles that may contain a single or several secretory vesicles (Figure 4.9d) (e.g., Jordão et al. 1999). In both apocrine and microapocrine secretion, the secretory vesicle contents are released by membrane fusion and/or by membrane solubilization due to high pH contents or to the presence of detergents.

Secretion by hemipteran midgut cells displays special features. Double-membrane vesicles bud from modified (double-membrane) Golgi structures (Figure 4.7e). The double-membrane vesicles move to the cell apex, their outer membranes fuse with the microvillar membrane, and their inner membranes fuse with the perimicrovillar membranes, emptying their contents (Figure 4.9e) (Silva et al. 1995). Aphids have a secretory mechanism that is derived from the one described and is depicted in Figure 4.8c.

Apocrine and microapocrine mechanisms waste membrane components and cytoplasm, and for this, they are employed only when they are advantageous relative to exocytic mechanisms. This happens when a fast delivery of digestive enzymes is necessary, as in blood feeders after a meal, and when the secretion occurs in a water-absorbing region. In most insects, water absorption is observed in the anterior midgut. An exocytic mechanism in an absorptive region is not efficient, since the movement of fluid toward the cell hinders the uniform dispersion of the secreted material. Fluid movement has a small effect on apocrine or microapocrine secretion because the enzymes are freed from budding or pinching-off vesicles far from the cells. Exocytosis from posterior midgut cells is efficient because, as a rule, those regions secrete fluids. Microapocrine mechanisms seem to be more advanced than apocrine secretion, since they waste less cell material. This agrees with the observation that apocrine mechanisms are found in the earlier evolved insects like grasshoppers and beetles, whereas the microapocrine mechanisms occur in more recently evolved insects, exemplified by lepidopterans.

4.10 Concluding Remarks

The molecular physiology of the digestive process is becoming a developed science and their methods are powerful enough to a steady progress. It is conceivable that, in the next few decades, knowledge on the structural biology and function of digestive enzymes, on the control of expression of alternate digestive enzymes and their secretory mechanisms as well as on microvillar biochemistry, will support the development of more effective and specific methods of insect control.

ACKNOWLEDGMENTS

Our work was supported by Brazilian research agencies FAPESP, INCT-Entomologia Molecular and CNPq. The authors are staff members of the Biochemistry Department and research fellows of CNPq.

REFERENCES

Andersen, P. C., B. V. Brodbeck, and R. F. Mizell. 1989. Metabolism of amino acids, organic acids and sugars extracted from the xylem fluid of four host plants by adult *Homalodisca coagulata. Ent. Exp. Appl.* 50:149–59.

Appel, H. M. 1994. The chewing herbivore gut lumen: Physicochemical conditions and their impact on plant nutrients, allelochemicals, and insect pathogens. In *Insect-Plant Interactions*, V.5, ed. E. A. Bernays, 203–23. Boca Raton, FL: CRC Press.

Appel, H. M., and M. M. Martin. 1990. Gut redox conditions in herbivorous lepidopteran larvae. *J. Chem. Ecol.* 16:3277–90.

Ashford, D. A., W. A. Smith, and A. E. Douglas. 2000. Living on a high sugar diet: The fate of sucrose ingested by a phloem-feeding insect, the pea aphid *Acyrthosiphon pisum. J. Insect Physiol.* 46:335–41.

Azevedo T. R., W. R. Terra, and C. Ferreira. 2003. Purification and characterization of three β-glycosidases from midgut of the sugar cane borer, *Diatraea saccharalis. Insect Biochem. Molec. Biol.* 33:81–92.

Benemann, J. R. 1973. Nitrogen fixation in termites. *Science* 181:247–62.

Berenbaum, M. 1980. Adaptive significance of midgut pH in larval Lepidoptera. *Am. Natur.* 115:138–46.

Bernays, E. A., D. J. Chamberlain, and E. M. Leather. 1981. Tolerance of acridids to ingested condensed tannin. *J. Chem. Ecol.* 7:247–56.

Biagio F. P., F. K. Tamaki, W. R. Terra, and A. F. Ribeiro. 2009. Digestive morphophysiology of *Gryllodes sigillatus* (Orthoptera: Gryllidae). *J. Insect Physiol.* 55:1125–33.

Bifano T. D., T. G. P. Alegria, and W. R. Terra. 2010. Transporters involved in glucose and water absorption in the *Dysdercus peruvianus* (Hemíptera: Pyrrhocoridae) anterior midgut. *Comp. Biochem. Physiol.* 157B:1–9.

Bignell, D. E. 1981. Nutrition and digestion. In *The American Cockroach*, ed. W. J. Bell and N. G. Adiyodi, 57–86. London: Chapman.

Billingsley, P. F. 1990. The midgut ultrastructure of hematophagous insects. *Annu. Rev. Entomol.* 35:219–48.

Bolognesi R., W. R. Terra, and C. Ferreira. 2008. Peritrophic membrane role in enhancing digestive efficiency. Theoretical and experimental models. *J. Insect Physiol.* 54:1413–22.

Bolognesi, R., A. F. Ribeiro, W. R. Terra, and C. Ferreira. 2001. The peritrophic membrane of *Spodoptera frugiperda*: Secretion of peritrophins and role in immobilization and recycling digestive enzymes. *Arch. Insect Biochem. Physiol.* 47:62–75.

Borhegyi, N. H., K. Molnár, G. Csikós, and M. Sass 1999. Isolation and characterization of an apically sorted 41-kDa protein from the midgut of tobacco hornworm (*Manduca sexta*). *Cell Tiss. Res.* 297: 513–25.

Bragatto I., F. A. Genta, A. F. Ribeiro, W. R. Terra, and C. Ferreira. 2010. Characterization of a β-1,3-glucanase active in the alkaline midgut of *Spodoptera frugiperda* larvae and its relation to β-glucan-binding proteins. *Insect Biochem. Molec. Biol.* 40(12):861–72.

Brioschi D., L. D. Nadalini, M. H. Bengtson, M. C. Sogayar, D. S. Moura, and M. C. Silva-Filho. 2007. General up regulation of *Spodoptera frugiperda* trypsins and chymotrypsins allows its adaptation to soybean proteinase inhibitor. *Insect Biochem. Molec. Biol.* 37:1283–90.

Brito L. O., A. R. Lopes, J. R. P. Parra, W. R. Terra, and M. C. Silva-Filho. 2001. Adaptation of tobacco budworm *Heliothis virescens* to proteinase inhibitors may be mediated by the synthesis of new proteinases. *Comp. Biochem. Physiol.* 128B:365–75.

Brune, A. 1998. Termite guts: The world's smallest bioreactors. *Trends Biotechnol.* 16:16–21.

Caldeira, W., A. B. Dias, W. R. Terra, and A. F. Ribeiro. 2007. Digestive enzyme compartmentalization and recycling and sites of absorption and secretion along the midgut of *Dermestes maculatus* (Coleoptera) larvae. *Arch. Insect Biochem. Physiol.* 64;1–18.

Cançado, F. C., A. R. G. Barbosa, and S. R. Marana. 2010. Role of the triad N46, S106 and T107 and the surface charges in the determination of the acidic pH optimum of digestive lysozyme from *Musca domestica.* *Comp. Biochem. Physiol.* 155B:387–95.

Christeller J. T. 2005. Evolutionary mechanisms acting on proteinase inhibitor variability. *FEBS J.* 272: 5710–22.

Christeller, J. T. 1996. Degradation of wool by *Hofmannophila pseudospretella* (Lepidoptera: Oecophoridae) larval midgut extracts under condition simulating the midgut environment. *Arch. Insect Biochem. Physiol.* 33:99–119.

Clark, T. M. 1999. Evolution and adaptive significance of larval midgut alkalinization in the insect superorder Mecopterida. *J. Chem. Ecol.* 25:1945–60.

Cranston, P. S., and P. J. Gullan. 2003. Phylogeny of insects. In *Encyclopedia of Insects*, ed. V. H. Resh, and R. T. Cardé, 882–98. New York: Academic Press.

Cristofoletti, P. T., A. F. Ribeiro, C. Deraison, Y. Rahbé, and W. R. Terra. 2003. Midgut adaptation and digestive enzyme distribution in a phloem feeding insect, the pea aphid *Acyrtosiphon pisum. J. Insect Physiol.* 49:11–24.

Cristofoletti, P. T., A. F. Ribeiro, and W. R. Terra. 2001. Apocrine secretion of amylase and exocytosis of trypsin along the midgut of *Tenebrio molitor. J. Insect Physiol.* 47:143–55.

Cristofoletti, P. T., F. D. Mendonça-de-Sousa, Y. Rahbé, and W. R. Terra. 2006. Characterization of a membrane-bound aminopeptidase purified from *Acyrtosiphon pisum* midgut cells. A major binding site for toxic mannose lectins. *FEBS J.* 273:5574–88.

De Veau, E. J. I., and J. C. Schultz. 1992. Reassessment of interaction between gut detergents and tannins in Lepidoptera and significance for gypsy moth larvae. *J. Chem. Ecol.* 18:1437–53.

Dillon, R. J., and V. M. Dillon. 2004. The insect gut bacteria: An overview. *Annu. Rev. Entomol.* 49:71–92.

Dow, J. A. T. 1986. Insect midgut function. *Adv. Insect Physiol.* 19:187–328.

Dow, J. A. T. 1992. pH gradients in lepidopteran midgut. *J. Exp. Biol.* 172:355–75.

Elpidina, E. N., K. S. Vinokurov, V. A. Gromenko, Y. A. Rudenskoya, Y. E. Dunaevsky, and D. P. Zhuzhikov. 2001. Compartmentalization of proteinases and amylases in *Nauphoeta cinerea* midgut. *Arch. Insect Biochem. Physiol.* 48:206–16.

Espinoza-Fuentes, F. P., and W. R. Terra. 1987. Physiological adaptations for digesting bacteria. Water fluxes and distribution of digestive enzymes in *Musca domestica* larval midgut. *Insect Biochem.* 17:809–17.

Evans, G. 1975. *The Life of Beetles*. London: George Allen and Unwin.

Ferreira, A. H. P., A. F. Ribeiro, W. R. Terra, and C. Ferreira. 2002. Secretion of β-glycosidase by middle midgut cells and its recycling in the midgut of *Tenebrio molitor* larvae. *J. Insect Physiol.* 48:113–8.

Ferreira, A. H. P., P. T. Cristofoletti, D. M. Lorenzini, L. O. Guerra, P. B. Paiva, M. R. S. Briones, W. R. Terra, and C. Ferreira. 2007. Identification of midgut microvillar proteins from *Tenebrio molitor* and *Spodoptera frugiperda* by cDNA library screenings with antibodies. *J. Insect Physiol.* 53:1112–24.

Ferreira, C., A. F. Ribeiro, E. S. Garcia, and W. R. Terra. 1988. Digestive enzymes trapped between and associated with the double plasma membranes of *Rhodnius prolixus* posterior midgut cells. *Insect Biochem.* 18:521–30.

Ferreira, C., J. R. P. Parra, and W. R. Terra. 1997. The effect of dietary plant glycosides on larval midgut β-glycosidases from *Spodoptera frugiperda* and *Diatraea saccharalis. Insect Biochem. Molec. Biol.* 27:55–9.

Ferreira, C., S. R. Marana, and W. R. Terra. 1992. Consumption of sugars, hemicellulose, starch, pectin, and cellulose by the grasshopper *Abracris flavolineata. Entomol. Exp. Appl.* 65:113–7.

Fox, L. R., and B. J. Macauley. 1977. Insect grazing on *Eucalyptus* in response to variation in leaf tannins and nitrogen. *Oecologia* 29:145–62.

Genta F. A., A. F. Dumont, S. R. Marana, W. R. Terra, and C. Ferreira. 2007. The interplay of processivity, substrate inhibition and a secondary substrate binding site of an insect exo-β-1,3-glucanase. *Biochim. Biophys. Acta* 1774:1070–91.

Genta F. A., I. Bragatto, W. R. Terra, and C. Ferreira. 2009. Purification, characterization and sequencing of the major β-1,3-glucanase from the midgut of *Tenebrio molitor* larvae. *Insect Biochem. Molec. Biol.* 39:864–74.

Genta, F. A., L. Blanes, P. T. Cristofoletti, C. L. do Lago, W. R. Terra, and C. Ferreira. 2006a. Purification, characterization and molecular cloning of the major chitinase from *Tenebrio molitor* larval midgut. *Insect Biochem. Molec. Biol.* 36:789–800.

Genta, F. A., R. J. Dillon, W. R. Terra, and C. Ferreira. 2006b. Potential role for gut microbiota in cell wall digestion and glucoside detoxification in *Tenebrio molitor. J. Insect Physiol.* 52:593–601.

Genta, F. A., W. R. Terra, and C. Ferreira. 2003. Action pattern, specificity, lytic activities, and physiological role of five digestive β-glucanases isolated from *Periplaneta americana. Insect Biochem. Molec. Biol.* 33:1085–97.

Goodchild, A. J. P. 1966. Evolution of the alimentary canal in the Hemiptera. *Biol. Rev.* 41:97–140.

Graf, R., A. S. Raikhel, M. R. Brown, A. O. Lea, and H. Briegel. 1986. Mosquito trypsin: Immunocytochemical localization in the midgut of blood-fed *Aedes aegyti* (L.). *Cell Tiss. Res.* 245:19–27.

Hegedus D., M. Erlandson, C. Gillott, and U. Toprak. 2008. New insights into peritrophic matrix synthesis, architecture, and function. *Annu. Rev. Entomol.* 54:285–302.

Houseman, J. G., P. E. Morrison, and A. E. R. Downe. 1985. Cathepsin B and aminopeptidase in the posterior midgut of *Phymata wolffii* (Hemiptera: Phymatidae). *Can. J. Zool.* 63:1288–91.

Jimenez, D. R., and M. Gilliam. 1990. Ultrastructure of ventriculus of the honey bee *Apis mellifera* (L): Cytochemical localization of the acid phosphatase, alkaline phosphatase, and non-specific esterase. *Cell Tiss. Res.* 261:431–43.

Jordão, B. P., A. N. Capella, W. R. Terra, A. F. Ribeiro, and C. Ferreira. 1999. Nature of the anchors of membrane-bound aminopeptidase, amylase, and trypsin and secretory mechanisms in *Spodoptera frugiperda* (Lepidoptera) midgut cells. *J. Insect Physiol.* 45:29–37.

Kristensen, N. P. 1999. Phylogeny of endopterygote insects, the most successful lineage of living organisms. *Eur. J. Entomol.* 96:237–353.

Lang T., G. C. Hansson, and T. Samuelsson. 2007. Gel-forming mucins appeared early in metazoan evolution. *Proc. Natl. Acad. Sci. USA* 104:16209–14.

Law, J. H., J. M. Ribeiro, and M. A. Wells. 1992. Biochemical insights derived from insect diversity. *Annu. Rev. Biochem.* 61:87–111.

Le Cahérec, F., M. T. Guillam, F. Beuron, et al. 1997. Aquaporin-related proteins in the filter chamber of homopteran insects. *Cell Tiss. Res.* 290:143–51.

Lehane, M. J., H. M. Müller, and A. Crisanti. 1996. Mechanisms controlling the synthesis and secretion of digestive enzymes in insects. In *Biology of The Insect Midgut,* ed. M. J. Lehane and P. F. Billingsley, 195–205. London: Chapman.

Lemos, F. J. A., A. F. Ribeiro, and W. R. Terra. 1993. A bacteria-digesting midgut-lysozyme from *Musca domestica* (Diptera) larvae. Purification, properties and secretory mechanism. *Insect Biochem. Molec. Biol.* 23:533–41.

Li, X., and A. Brune. 2005. Digestion of microbial mass, structural polysaccharides, and protein by the humivorous larva of *Pachnoda ephippiata* (Coleoptera: Scarabaeidae). *Soil Biol. Biochem.* 37:107–16.

Liadouze I., G. Febvay, J. Guillaud, and G. Bonnot. 1995. Effect of diet on free amino acid pools of symbiotic and aposymbiotic pea aphids, *Acyrtosiphon pisum. J. Insect Physiol.* 41:33–40.

Lopes A. R., M. A. Juliano, S. R. Marana, L. Juliano, and W. R. Terra. 2006. Substrate specificity of insect trypsins and the role of their subsites in catalysis. *Insect Biochem. Molec. Biol.* 36:130–40.

Lopes, A. R., M. A. Juliano, L. Juliano, and W. R. Terra. 2004. Coevolution of insect trypsins and inhibitors. *Arch. Insect Biochem. Physiol.* 55:140–52.

Lopes, A. R., P. M. Sato, and W. R. Terra. 2009. Insect chymotypsins: Chloromethyl ketone inactivation and substrate specificity relative to possible coevolutional adaptation of insects and plants. *Arch. Insect Biochem. Physiol.* 70:188–203.

Marana, S. R., A. F. Ribeiro, W. R. Terra, and C. Ferreira. 1997. Ultrastructure and secretory activity of *Abracris flavolineata* (Orthoptera: Acrididae) midguts. *J. Insect Physiol.* 43:465–73.

Marana, S. R., F. C. Cançado, A. A. Valerio, C. Ferreira, W. R. Terra, and J. A. R. G. Barbosa. 2006. Crystallization, data collection and phasing of two digestive lysozymes from *Musca domestica. Acta Cryst.* F62:750–2.

Martin, M. M. 1987. *Invertebrate–Microbial Interactions: Ingested Fungal Enzymes in Anthropod Biology.* Ithaca, NY: Cornell University Press.

Martin, M. M., and J. S. Martin. 1984. Surfactants: Their role in preventing the precipitation of proteins by tannins in insect guts. *Oecologia* 61:342–5.

Mazumdar-Leighton, S., and R. M. Broadway. 2001b. Transcriptional induction of diverse midgut trypsin in larval *Agrotis ipsilon* and *Helicoverpa zea* feeding on the soybean trypsin inhibitor. *Insect Biochem. Molec. Biol.* 31:645–57.

Mazumdar-Leighton, S., and R. M. Broadway. 2001a. Identification of six chymotrypsin cDNAs from larval midguts of *Helicoverpa zea* and *Agrotis ipsilon* feeding on soybean (Kunitz trypsin inhibitor). *Insect Biochem. Molec. Biol.* 31:633–44.

Miles, P. W. 1972. The saliva of Hemiptera. *Adv. Insect Physiol.* 9:183–255.

Mitsumasu, K., M. Azuma, T. Niimi, O. Yamashita, and T. Yaginuma. 2005. Membrane-penetrating trehalase from silkworm *Bombyx mori*. Molecular cloning and localization in larval midgut. *Insect Molec. Biol.* 14:501–8.

Murdock, L. L., G. Brookhart, P. E. Dunn, et al. 1987. Cysteine digestive proteinases in Coleoptera. *Comp. Biochem. Physiol.* 87B:783–7.

Ortego, F., C. Novillo, and P. Catañera. 1996. Characterization and distribution of digestive proteases of the stalk corn borer, *Sesamia nonagrioides* Lef. (Lepidoptera: Noctuidae). *Arch. Insect Biochem. Physiol.* 33:136–80.

Padilha, H. P., A. C. Pimentel, A. F. Ribeiro, and W. R. Terra. 2009. Sequence and function of lysosomal and digestive cathepsin D-like proteinases of *Musca domestica* midgut. *Insect Biochem. Molec. Biol.* 39:782–91.

Ponsen, M. B. 1991. Structure of the digestive system of aphids, in particular *Hyalopterus* and *Coloradoa*, and its bearing on the evolution of filter chambers in Aphidoidea. In *Wageningen Agricultural University Papers*, 91–95:3–61. Wageningen, the Netherlands: Agricultural University Press.

Prosser, W. A., and A. E. Douglas. 1992. A test of the hypotheses that nitrogen is upgraded and recycled in aphid (*Acyrtosiphon pisum*) symbiosis. *J. Insect Physiol.* 38:93–9.

Prosser, W. A., S. J. Simpson, and A. E. Douglas. 1992. How an aphid (*Acyrtosiphon pisum*) symbiosis responds to variation in dietary nitrogen. *J. Insect Physiol.* 38:301–7.

Quicke, D. L. J. 2003. Hymenoptera (ants, bees, wasps). In *Encyclopedia of Insects*, ed. V. H. Resh and R. T. Cardé, 534–46. San Diego, CA: Academic Press.

Ribeiro, J. M. C. 1987. Role of saliva in blood-feeding arthropods. *Annu. Rev. Entomol.* 32:463–78.

Ryan, C. A. 1990. Proteinase inhibitors in plants: Genes for improving defense against insects and pathogens. *Annu. Rev. Phytopathol.* 28:425–49.

Salvucci, M. E. 2003. Distinct sucrose isomerase catalyse trehalose synthesis in whiteflies, *Bernisia argentifolis*, and *Erwinia rhapontici*. *Comp. Biochem. Physiol.* 135B:385–95.

Salvucci, M. E., R. C. Rosell, and J. K. Brown. 1998. Uptake and metabolism of leaf proteins by the silverleaf whitefly. *Arch. Insect Biochem. Physiol.* 39:155–65.

Sato, P. M., A. R. Lopes, L. Juliano, M. A. Juliano, and W. R. Terra. 2008. Subsite substrate specificity of midgut insect chymotrypsins. *Insect Biochem. Molec. Biol.* 38:628–33.

Scrivener, A. M., M. Slaytor, and H. A. Rose. 1989. Symbiont-independent digestion of cellulose and starch in *Panesthia cribrata* Saussure, an Australian wood-eating cockroach. *J. Insect Physiol.* 35:935–41.

Sehnal, F. 1985. Growth and life cycles. In *Comprehensive Insect Physiology, Biochemistry and Pharmacology*, Vol. 2, ed. A. Kerkut and L. I. Gilbert, 1–86. Oxford, UK: Pergamon.

Shigenobu, S., H. Watanabe, M. Hattori, Y. Sasaki, and H. Ishikawa. 2000. Genome sequence of the endocellular bacterial symbiont of aphids *Buchnera* sp. APS. *Nature* 407:81–6.

Silva M. C. P., W. R. Terra, and C. Ferreira. 2004. The role of carboxyl, guanidine and imidazol groups in catalysis by a midgut trehalase purified from an insect larvae. *Insect Biochem. Molec. Biol.* 34:1089–99.

Silva M. C. P., W. R. Terra, and C. Ferreira. 2010. The catalytic and other residues essential for the activity of the midgut trehalase from *Spodoptera frugiperda*. *Insect Biochem. Molec. Biol.* 40(10):733–41.

Silva, A., M. Bacci, Jr., C. Q. Siqueira, O. L. Bueno, F. C. Pagnoca, and M. J. A. Hebling. 2003. Survival of *Atta sexdens* workers on different food sources. *J. Insect Physiol.* 49:307–13.

Silva, C. P., A. F. Ribeiro, S. Gulbenkian, and W. R. Terra. 1995. Organization, origin and function of the outer microvillar (perimicrovillar) membranes of *Dysdercus peruvianus* (Hemiptera) midgut cells. *J. Insect Physiol.* 41:1093–103.

Silva, C. P., and W. R. Terra. 1994. Digestive and absorptive sites along the midgut of the cotton seed sucker bug *Dysdercus peruvianus* (Hemiptera: Pyrrhocoridae). *Insect Biochem. Molec. Biol.* 24:493–505.

Silva, M. C. P., W. R. Terra, and C. Ferreira. 2006. Absorption of toxic β-glucosides produced by plants and their effect on tissue trehalases from insects. *Comp. Biochem. Physiol.* 143:367–73.

Slansky, F., and J. M. Scriber. 1985. Food consumption and utilization. In *Comprehensive Insect Physiology, Biochemistry and Pharmacology*, Vol. 4. ed. A. Kerkut and L. I. Gilbert, 87–163. Oxford, UK: Pergamon.

Spencer, K. C. 1988. Chemical mediation of coevolution in the *Passiflora–Heliconius* interaction. In *Chemical Mediation of Coevolution*, ed. K. C. Spencer, 167–240. San Diego, CA: Academic Press.

Tellam, R. L., G. Wijffels, and P. Willadsen. 1999. Peritrophic matrix proteins. *Insect Biochem. Molec. Biol.* 29:87–101.

Terra, W. R. 1988. Physiology and biochemistry of insect digestion: An evolutionary perspective. *Braz. J. Med. Biol. Res.* 21:675–734.

Terra, W. R. 1990. Evolution of digestive systems of insects. *Annu. Rev. Entomol.* 35:181–200.

Terra, W. R. 2001. The origin and functions of the insect peritrophic membrane and peritrophic gel. *Arch. Insect Biochem. Physiol.* 47:47–61.

Terra, W. R., A. Valentin, and C. D. Santos. 1987. Utilization of sugars, hemicellulose, starch, protein, fat and minerals by *Erinnyis ello* larvae and the digestive role of their midgut hydrolases. *Insect Biochem.* 17:1143–7.

Terra, W. R., and C. Ferreira. 2009. Digestive system. In *Encyclopedia of Insects*, 2nd Edition, ed. V. H. Resh and R. T. Cardé, 273–81. San Diego, CA: Academic Press.

Terra, W. R., and C. Ferreira.1981. The physiological role of the peritrophic membrane and trehalase: Digestive enzymes in the midgut and excreta of starved larvae of *Rhynchosciara*. *J. Insect Physiol.* 27:325–31.

Terra, W. R., and C. Ferreira. 1994. Insect digestive enzymes: Properties, compartmentalization and function. *Comp. Biochem. Physiol.* 109B:1–62.

Terra, W. R., and R. Regel. 1995. pH buffering in *Musca domestica* midguts. *Comp. Biochem. Physiol.* 112A:559–64.

Terra, W. R., C. Ferreira, and E. S. Garcia. 1988a. Origin, distribution, properties and functions of the major *Rhodnius prolixus* midgut hydrolases. *Insect Biochem.* 18:423–34.

Terra, W. R., C. Ferreira, B. P. Jordão, and R. J. Dillon. 1996. Digestive enzymes. In *Biology of the Insect Midgut*, ed. M. J. Lehane and P. F. Billingsley, 153–94. London: Chapman and Hall.

Terra, W. R., F. P. Espinoza-Fuentes, A. F. Ribeiro, and C. Ferreira. 1988b. The larval midgut of the housefly (Musca domestica)—Ultrastructure, fluid fluxes and ion secretion in relation to the organization of digestion. *J. Insect Physiol.* 34:463–72.

Terra, W. R., and C. Ferreira. 2005. Biochemistry of digestion. In *Comprehensive Molecular Insect Science*, ed. L. I. Gilbert, K. Iatrou, and S. S. Gill, 171–224. Oxford, UK: Elsevier.

Tokuda, G., N. Lo, H. Watanabe, M. Slaytor, T. Matsumoto, and H. Noda. 1999. Metazoan cellulase genes from termites: Intron/exon structures and sites of expression. *Biochim. Biophys. Acta* 1447:146–59.

Venâncio, T. M., P. T. Cristofoletti, C. Ferreira, S. Verjovski-Almeida, W. R. Terra. 2009. The *Aedes aegypti* larval transcriptome: A comparative perspective with emphasis on trypsins and the domain structure of peritrophins. *Insect Molec. Biol.* 18:33–44.

Walker, G. P. 2003. Salivary glands. In *Encyclopedia of Insects*. ed. V. H. Resh and R. T. Cardé, 1011–17. San Diego, CA: Academic Press.

Watanabe, H., and G. Tokuda. 2001. Animal cellulases. *Cell. Molec. Life Sci.* 58:1167–78.

Watanabe, H., H. Noda, G. Tokuda, and N. Lo. 1998. A cellulase gene of termite origin. *Nature* 394:330–31.

Wolfersberger, M. G. 2000. Amino acid transport in insects. *Annu. Rev. Entomol.* 45:111–20.

Wyatt, G. R. 1967. The biochemistry of sugars and polysaccharides in insects. *Adv. Insect Physiol.* 4:287–360.

5

Insect–Plant Interactions

Marina A. Pizzamiglio-Gutierrez

CONTENTS

5.1 Introduction...121
5.2 Development of Plants and Insects in Geologic Time.. 122
5.3 History of Plant–Insect Interactions and Theories on Evolution....................................... 123
    5.3.1 Theory of Coevolution... 124
5.4 Plant Perspective 125
    5.4.1 Factors Affecting Plant Defenses 128
    5.4.2 Cost of Defense in Plants.. 128
5.5 Herbivore Perspective 129
    5.5.1 Avoiding Host Plant Defenses 129
    5.5.2 Metabolizing and Sequestering Plant Toxins... 130
    5.5.3 Host Plant Manipulation 131
5.6 Herbivore Generalists and Specialists 131
5.7 The Tertiary Trophic Level... 132
    5.7.1 Effects of Abiotic Factors in Tritrophic Interactions 133
5.8 Final Considerations 135
Acknowledgments... 136
References... 136

5.1 Introduction

Terrestrial food chains and webs are composed of at least three trophic levels: plants, herbivores, and natural enemies (Price et al. 1980). Since the beginning of life on earth, plants and insects have evolved beneficial or detrimental interactions (Dethier 1976, Daly et al. 1978). The majority of ecological studies demonstrated that insects and plants do not simply live together, but rather interact, suffer the consequences of these interactions, and adapt interdependently (Schoonhoven 1990). Insects benefit plants through pollination or when they live in association with them, such as ants living on acacia plants protecting them against other insects and vertebrates herbivores and in return receiving food and shelter (Janzen 1966).

In the past decades, thousands of articles and reviews have been published on plant insect interactions, pollination, and coadaptation (Bernays 1982; Scriber 2002). Karban and Agrawal (2002) found that larger numbers of studies were conducted on plant defense mechanisms with a smaller number on the strategies insect use to overcome plant defenses. Price et al. (1980) suggested that the theory of plant–insect interactions cannot progress without considering the tertiary trophic level, as close observation reveals that plants exert direct and indirect, and positive and negative effects not only on herbivores, but also on herbivore natural enemies. Price (1980) views the tertiary trophic levels as part of the plant's battery of defenses (see De Moraes and Mescher 2007).

5.2 Development of Plants and Insects in Geologic Time

Angiosperms are the dominant plants today and first appeared in the early Cretaceous period about 135 million years ago. Over the last 30 million years, a wide range of vegetative and reproductive innovations evolved that changed the ecology and biogeochemistry of the planet (Feild and Arens 2007). Several authors have hypothesized on the origin of angiosperms and on the time and modes of their subsequent radiation. Barrett and Willis (2001) argued that radiation of angiosperms may have been influenced by the feeding behavior of herbivorous dinosaurs. However, direct evidence on this is scant, and probably arboreal mammals and insects had a far greater impact on angiosperms diversification than the herbivorous dinosaurs. In addition, high levels of CO_2 in the atmosphere may have played a considerable role in the early stages of angiosperms and herbivore radiation, and in the myriad of biological associations and feeding strategies that evolved in the different trophic levels. Figure 5.1, based on the work of Smart and Hughes (1973) and Gensel and Andrews (1987), summarizes some of the evolutionary events that occurred over geologic time.

In the Devonian, a significant diversification of vascular plants occurred including the emergence of seed plants, the production of spores, and other structures that provided food and shelter for insects. By the Carboniferous period, insects had become well diversified and distributed in various orders, some of which became extinct while others survive to the present (Ephemeroptera, Odonata, Orthoptera, Neuroptera). The plants present at this time were gymnosperms, calamites, and pteridofites.

In the Permian, fossil Hemiptera, Coleoptera, and others classified as Mecoptera were found, as well as the first evidence of leaf damage by insects. Primitive flowers and fossils of Diptera and Hymenoptera

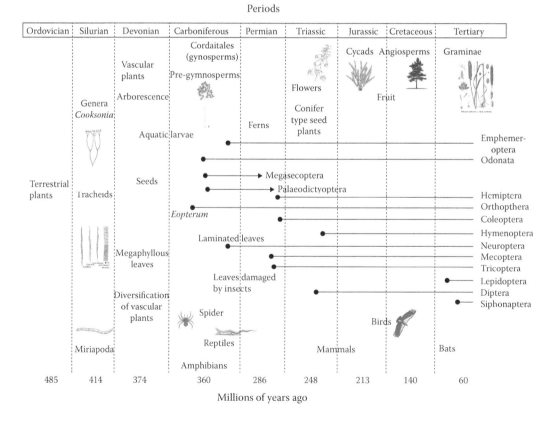

FIGURE 5.1 Chronology of some important events in the development of plants and insects across geologic time. (After Smart, J., and N. F. Hughes, In *Insect/Plant Relationships*, ed. H. F. Van Emden, 143–155, Oxford Blackwell, London, 1973; Gensel, P. G., and H. N. Andrews, *Amer. Scientist*, 75, 478–489, 1987. With permission.)

first appeared in the Triassic. Many groups of herbivorous insects and the first documented occurrence of herbivory associated with angiosperms occurred in the late Jurassic and early Cretaceous (Rasnitsyn and Krassilov 2000). During the Tertiary period, bark beetles (Coleoptera: Scolytidae), gall wasps (Hymenoptera: Cynipidae), and species of leaf miners (Diptera: Agromyzidae) (Zwölfer 1982) arose. Important groups of leaf miners that evolved during this period were Lyonetiidae, Gracillariidae, and Gelechiidae, and most fed on plants in the family Fagaceae (Opler 1973) comprising the first indications of true plant–insect interactions across geologic time.

Zwölfer (1975) (quoted in Prokopy and Owens 1983) argued that the emergence of a particular group of plants was often associated with the emergence of a parallel group of insects that exploited them. It is thought that interactions between plants and animal pollinators, mostly insects, were the driving force in the evolution of angiosperms (Stanton et al. 1986), and that the origins and much of the diversification of angiosperms was directly related to the coevolving behavior of insects (Daly et al. 1978; Price 1984). As a result of coevolution, insects and flowering plants became two of the largest taxonomic groups of organisms on the planet, with flowering plants achieving the highest level of organization in the plant kingdom (Takhtajan 1969).

The first insect pollinators were beetles that were inefficient pollinators destroying many flowers in the process (Smart and Hughes 1973). Primitive angiosperms such as the Magnoliaceae and Nymphaceae are still pollinated by beetles (Daly et al. 1978). These pollinators were common at a time when higher plants were evolving, well before the advent of currently important pollinators in the Lepidoptera and Hymenoptera orders (Faegri and Van der Pijl 1979). Bees (Apoidea) evolved from predatory wasps, with adaptations including feathery hair, changes in mouth parts to collect nectar and pollen, and an efficient system of communication in the genus *Apis* (Daly 1978).

5.3 History of Plant–Insect Interactions and Theories on Evolution

Early humans were hunters and nomads, and the food preferences of insects may have gone unnoticed except for disorders and diseases caused by flies, lice, fleas, and others (Flint and van den Bosch 1981). A rudimentary agriculture developed when humans began to live in permanent settlements, and as populations increased so did the need to increase food production, resulting in the domestication of plants about 10,000 years ago (Dethier 1976). Indications of the feeding habits of insects appear in early historical records, before biblical mention of famine and starvation caused by locusts and other insect pests (Berenbaum 1986). Before the Christian era, the Chinese reared the silkworm *Bombyx mori* (L.) (Lepidoptera: Bombycidae), observing that it fed exclusively on mulberry leaves *Morus alba* L. and *Morus nigra* (L.) (Harborne 1977a; Kogan 1986).

Fabre (1890), quoted in Kogan (1986), was one of the first to examine the feeding preferences of insects, which he called "botanical instinct." In 1910, Verschaffelt described the chemistry of the interaction between larvae and adults of Pieridae and cruciferous plants, noting that the insects were attracted to the plants due to the presence of a substance that came to be called sinigrin. These studies were classic contributions, but they did not stimulate significant research in the area of host plant–insect interactions for many years (Thorsteinson 1955, 1960). Stahl (1888) (cited in Rhoades 1979) and Errera (1886) (cited in Berenbaum 1986) suggested that some of these plant chemicals could have evolved to protect plants against attack from herbivores. In 1980, Mothes published a review of *secondary plant substances*, noting that the term was first used by Kossel (1891), and citing Czapek (1921), early elaboration on these substances. Brues (1920) published a study on the selection of host plants by insects, especially moths, and concluded that with exceptions, these insects showed marked preference in selecting plants in specific families or genera, suggesting the idea of a parallel development of deleterious characteristics in plants and adaptations by insects to overcome these barriers. Dethier (1941) suggested that some substances attracted insects to their preferred food, and Fraenkel (1959) published his pioneering work on the *raison d'etre* of secondary plant substances (see Schowalter et al. 1986).

The concept of *parallel evolution* was redefined by Fraenkel (1958) (quoted in Kogan 1986), suggesting that adaptive parallel and reciprocal evolution determined the patterns of host-plant use by insects. Fraenkel (1951) thought that primary substances were unimportant in insect host selection, and suggested that at the beginning of the coevolutionary process plants developed secondary substances to defend themselves from insects, and that later insects began to use the same substances in host plant finding (Fraenkel 1959). Kennedy and Booth (1951) presented the *theory of dual discrimination* in which primary and secondary substances are both important in food selection by the insects.

5.3.1 Theory of Coevolution

The concept of coevolution is relatively new, but the idea was present in the early observations of Darwin on pollination and adaptations between bees and flowers (Futuyma and Slatkin 1983). In 1964, Ehrlich and Raven published a classic paper on the coevolution of angiosperms and herbivorous insects, suggesting that through mutations and recombination, angiosperms produced secondary substances that altered their nutritional properties and formed new defenses against insect herbivores. Freed of insects attack, these plants colonized new areas; however, over time some groups of insects coevolved mechanisms to avoid or adapt to these substances and to exploit these plants without competition from nonadapted competitors. Plants developed substances to repel herbivores, and in turn, herbivores developed mechanisms to adapt or exploit these substances leading to a process wherein the plants became more toxic and the herbivores more specialized (Cornell and Hawkins 2003).

Edwards and Wratten (1981) credit the Ehrlich and Ravens (1964) theory of coevolution for helping understand the characteristics of the diversification of plants, insects, and their interactions, wherein plants evolved to use part of their metabolic budget for physical and or chemical protection, and insects evolved and invested in strategies to overcome plant defenses (Feeny 1975). Mello and Silva-Filho (2002) in a review of plant–insect interactions used the term *evolutionary race* to describe the escalation between the two processes.

Several authors have criticized the Ehrlich and Ravens (1964) theory of coevolution. Jermy (1976) proposed the *theory of sequential evolution* in which plant evolution was driven by selection factors such as climate, soil, and plant–plant interactions—influences more powerful than insect attack—and proposed that these factors produced the basic trophic diversification enabling the evolution of insects. Janzen (1980) and Futuyma (1983) also criticized Ehrlich and Raven's (1964) use of the term *coevolution*. According to Thompson (1986), plant chemicals certainly influenced the evolution of these interactions; however, intensive studies on the systematic, biogeography, and natural history of these insect groups would be necessary to explain the evolution of these interactions as proposed by the Ehrlich and Raven hypothesis (1964).

Research by Berenbaun and Feeny (1981) on the associations between insects and plants containing *coumarin* provided evidence that fit the various stages of the coevolutionary process described by Ehrlich and Raven (1964). Becerra (2005) studied beetles (Chrysomelidae) and their hosts in the family Burseracea and showed that plant defenses and counterdefenses by insects evolved roughly in synchrony and appear to confirm macroevolutionary coadaptation between plants and insects. Despite criticisms by Jermy (1976, 1980), Futuyma (1983), and support from Berenbaum (1983), the Ehrlich and Raven (1964) theory has stimulated considerable research on plant–insect interactions and coevolution (Futuyma and Slatkin 1983).

Feeny (1976) and Rhoades and Cates (1976), working independently, introduced the concept of *plant apparency* to explain the complex mechanisms involved in plant–insect interactions. The plant apparency hypothesis states that plants will invest heavily in broadly effective defenses if the plants are easily found by herbivores. Cornell and Hawkins (2003) reviewed the concept of plant apparency and aspects of the Ehrlich and Raven theory as concerns the existence of generalist herbivores. They suggested that nonapparent plants have defensive chemical substances and that groups of herbivores have adapted to feed on them and became specialized as proposed by the coevolutionary theory. In contrast, apparent plants have chemicals that reduce digestibility, and this suggests that both the coevolutionary and plant apparency theories agree in that the chemistry of plants has led to specialization in herbivores.

In higher trophic levels, Dietl and Kelley (2002) proposed that the evolution between predators and prey was induced by two related processes: *escalation* and *coevolution*. This type of escalation between predator and prey has been found in the fossil record of shellfish where evidence of predatory drilling or crushing of prey shows an increase in episodic pressure through time (Harper 2006). It is believed

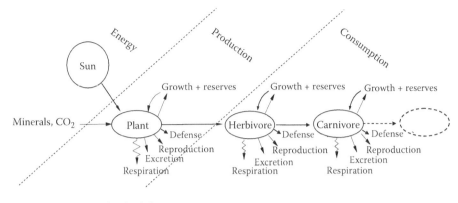

Tritrophic food chain

FIGURE 5.2 Acquisition and allocation of energy in a food chain with the same input and output flows in each trophic level. (Modified from Gutierrez, A. P., and G. L. Curry, In *Integrated Pest Management Systems for Cotton Production*, ed. R. F. Frisbie. 37–64. Copyright 1989, Wiley-VCH Verlag GmbH & Co. KGaA. Reproduced with permission.)

that a wide range of adaptations of defense by prey such as habits selection, morphology, physiology, and behavior evolved in response to increased pressure exerted by predation. The general consensus is that through the evolutionary process, in most cases escalation may be more important than coevolution, but the two hypotheses are difficult to separate using the fossil record (Harper 2006).

A special volume in the journal *Basic and Applied Ecology* published the following reviews on herbivore induced defenses: direct defenses of plants and the usefulness of molecular genetic techniques to investigate how this group of defenses function (Roda and Baldwin 2003); integration of functional strategies in indirect mechanical defenses by plants (Dicke et al. 2003a, b); interactions between microorganisms and insects and induced defenses (Rostas et al. 2003); induced defenses and the interactions that occur above and below ground (van Dam et al. 2003); the cost of induced plant defenses (Cipollini et al. 2003); and the evolution of induced defenses in plants (Zangerl 2003). The book edited by Dicke and Takken (2006) examines new directions that can be used in chemical ecology through molecular ecology (an *ecogenomic* approach).

In general, coevolution is important in all trophic levels, and while the details may vary, the general concepts apply because species of organisms at all levels face the same problems of resource acquisition and allocation, and all evolve in response to other organisms and the abiotic environment (Figure 5.2; Gutierrez and Curry 1989). In a bioeconomic sense, allocations to defense must be outweighed by increased fitness and adaptation (Gutierrez and Regev 2005).

5.4 Plant Perspective

Plants have evolved to survive assaults from organisms and environmental stresses for millennia through various mechanisms including physical barriers such as thorns and thick cuticles (Klein 2004), host free periods using short or long growth cycles, the dispersal of the progeny, association with other species, and tolerance to attack (Harris 1980). Plants can compensate for the loss of biomass caused by herbivore feeding with rapid regrowth, increased reproductive rates, and biochemical changes in response to herbivory (Giles et al. 2005). In addition to the chemistry of primary functions such as photosynthesis, respiration, and growth and reproduction, plants produce a variety of secondary substances that may be selectively concentrated in the reproductive organs responsible for the plant's existence (Price 1984). Whittaker (1972) suggested that some substances were initially expressed by plants in response to pressures exerted by herbivores, and when released into the environment accidentally became involved in interactions between plants. Due to the beneficial effects of reducing competition between plants, many of these substances continued to be synthesized and used by plants to protect against desiccation, salinity, and ultraviolet rays (Strong et al. 1984). For example, there is evidence that the phenolic resin in creosote bush, a desert plant, protects against desiccation and ultraviolet rays (Rhoades 1979). *Allelopathy*, the chemical interaction between plants, was first recorded by De Candolle in 1832 (quoted in Harborne 1977b, c) between cardo thistle (Compositae), oats, plants of *Euphorbia* spp., and flax (*Linum* sp.). Early research attempted to explain the presence of substances such as tannins and other phenolic compounds in primary metabolic processes of the plant and to demonstrate how they served to protect the plant from other plants, pathogens, and insects (Rhoades and Cates 1976; Rhoades 1985). Seigler and Price (1976) emphasized that the functions of some plant natural products can be multiple, involving both metabolic processes and primary defense. For example, primary metabolic compounds can act as direct defenses when the product occurs at high levels or at concentrations that prevent or impede herbivore growth, reproduction, and detoxification of metabolites (Slansky 1993; Simpson and Raubenheimer 2001).

Some plant secondary substances are similar to insect hormones and may alter the insect development and survival (Slama 1969). One of these substances, *juvabione*, was extracted from Canada balsam (*Abies balsamea*), while Kubo and Klocke (1983) found the same properties in extracts of the plant *Ajuga remota* (Labiatae).

Secondary chemical substances may affect the development of insects or act as chemical messengers. The term *semiochemical* was proposed by Law and Regnier in 1971 (quoted in Nordlund 1981) for volatile chemicals involved in the interactions between organisms. These chemicals may act as *allelochemicals* and include *allomone* and/or *kairomones*. Several terms were proposed to describe the effect of allelochemicals on the physiology and behavior of insects and these were summarized by Kogan (1986) (Table 5.1). Depending on circumstances, allomone and kairomones can repel the attack of insects or encourage others to feed (Daly et al. 1978). An example of an allelochemical having dual roles as allomone and kairomone is the substance *cucurbitacin* that may be an effective deterrent to most herbivores, but also act as a feeding stimulant for beetles of the genus *Diabrotica* (Coleoptera: Chrysomelidae) (Kogan 1986).

Plant phenotype may change due to abiotic conditions such as soil, water availability, nutrition, and other factors, as well as the interactions between plants and herbivores (Agrawal 2001). These changes may occur in the morphology, chemistry, and resource allocation to growth or defense, and other processes in plant parts above and below ground, and may alter the interactions of the plant with other members of the community (Dicke and Hilker 2003).

TABLE 5.1

Principal Classes of Chemical Plant Factors (Allelochemicals) and the Corresponding Behavior or Physiological Effects on Insects

Allelochemical Factors	Behavioral or Physiological Effects
Allomones	Gives adaptive advantage to the producing organism
Antixenotics	Disrupts normal host selection behavior
Repellents	Orients insects away from plants
Locomotory excitants	Starts or speeds up movement
Suppressants	Inhibits biting or piercing
Deterrents	Prevent maintenance of feeding or oviposition
Antibiotics	Disrupts normal growth and development of larvae; reduces longevity and fecundity of adults
Toxins	Produces chronic or acute intoxication syndromes
Digestibility reducing factor	Interferes with normal processes of food utilization
Kairomones	Gives advantage to the receiving organism
Attractants	Orients insect toward host plant
Arrestant	Slows down or stops movement
Feeding or oviposition excitant	Elicits biting, piercing, or oviposition; promotes continuation of feeding

Source: Kogan, M., In *Ecological Theory and Integrated Pest Management Practice*, ed. M. Kogan, 83–133, John Wiley & Sons, New York, 1986. With permission.

The defenses of plants against attack by herbivores or pathogens may be classified as *constitutive* or *induced* with research on the constitutive defenses being more developed. Constitutive defenses are always present in the plant and do not depend on the attack of herbivores or pathogens. They may include feeding inhibitors, toxins, and mechanical defenses. Induced defenses are triggered in response to attack by herbivores or pathogens and include the modification and accumulation of plant metabolites (Levin 1976). Induced indirect defenses in plants may interfere with herbivore feeding and development, or may cause the plant to emit volatile substances that attract natural enemies of herbivores (Walling 2000). Generally, induced defenses have been associated with the damage caused during feeding; however, recently, Hilker and Meiners (2002) demonstrated that oviposition by herbivores can also stimulate direct and indirect defenses in plants. Before this study, examples of information transfer via plant chemistry had been demonstrated only between healthy and attacked plants (Dicke and Bruin 2001). In this chapter, the constitutive and induced defenses are interpreted as direct defenses (Roda and Baldwin 2003).

Indirect plant defenses may increase the efficiency of natural enemies attacking herbivores by providing alternative food (Heil and McKey 2003), shelter (Grostal and O'Dowd 1994), or by releasing of volatiles that attract natural enemies (Dicke 1999; Hilker and Meiners 2002; Dicke et al. 2003a,b). An example of indirect defense is the maintenance of extrafloral nectaries in *Gossypium thurberi* (wild cotton) that feed ants protecting the plant against herbivores (Rudgers and Strauss 2004). Plant pathologists have long recognized the importance of constitutive or induced defenses wherein plants attacked by fungi induce resistance mechanisms involving the synthesis of *phytoalexins* in cells near the infection site to stop further spread of the disease (Rhoades 1985; Levin 1976; Ryan 1983). These interactions are illustrated in Figure 5.3.

Research by Feeny (1976) and Rhoades and Cates (1976) provided the basis for the *theory of optimal plant defense* (Rhoades 1979) that predicts defenses are produced and distributed in the plant tissues to obtain maximum benefits at the least costs. This means that tissues with a lower probability of being attacked have lower levels of constitutive defenses and high levels of induced defenses, while plant tissues more likely to be attacked contain high constitutive and less induced defenses (Zangerl and Rutledge 1996).

Rhoades and Cates (1976) and Coley (1980) postulated that some plants have developed quantitative chemical defenses such as tannins that reduce their digestibility. Nonapparent plants may also accumulate qualitative defenses (glucosides) but at lower concentrations and with associated reduced metabolic costs. In contrast, Bernays (1981) and Martin et al. (1987) argued that there was no evidence of tannins interfering with herbivore feeding. Zangerl and Rutledge (1996) investigated variations in defense

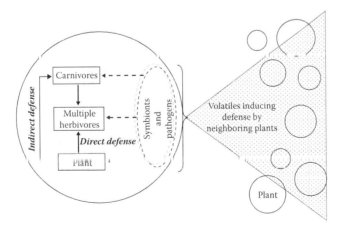

FIGURE 5.3 Illustration of direct defenses in a plant attacked by herbivores and/or pathogens and the stimulation of indirect defenses that maintain or attract predators that attack the herbivores. This figure also shows that damaged plants may release volatile compounds that induce defenses in neighboring healthy plants (circles). (Modified from Dicke, M., and M. Hilker, *Basic Appl. Ecol.*, 4, 3–14, 2003; Dicke, M., A. A. Agrawal, and J. Bruin, *Trends Plant Sci.*, 8, 403–405, 2003a. With permission.)

between plants of *Pastinaca sativa* L. and found results consistent with the predictions of the optimal plant defense theory. Kogan (1986) stated that while this theory is not complete, it helps explain the patterns of attack and defense in plant–insect interactions.

5.4.1 Factors Affecting Plant Defenses

Insect attack may act as inductors stimulating the accumulation of defenses in plants (Kogan and Paxton 1983). Environmental factors such as temperature, solar radiation, soil fertility, water stress, and pesticides also encourage resistance or susceptibility of plants to insect attack and affect the interactions between them (Kogan and Paxton 1983). Gallun and Khush (1980) proposed that levels of resistance and the ability of the plant to reduce infestation and damage caused by insects may be due to one or more mechanisms.

The age of the plant, leaves, fruits, and other organs infested by herbivores may influence the strength of the defense. For example, in temperate forests, Coley (1980) found defoliation by insects and herbivorous mammals was higher on young leaves. Holling et al. (1977) studied balsam fir forests and found that older trees were more susceptible to outbreaks of spruce budworm *Choristoneura fumiferana* (Clems). Old balsam fir leaves contain substances such as tannins or may be harder to digest, but there is no evidence that these effects are due to secondary substances or simply lower nutritional value. Fenny (1970) studied the relationship between the moth *Operophtera brumata* (L.) (Lepidoptera: Geometridae) and oak trees and found that seasonal variation in the quality and quantity of tannins in leaves influenced insect development, and higher concentration of tannin in older leaves significantly reduced larval growth and pupae weight. Moran and Hamilton (1980) proposed the hypothesis that low nutritional quality of plant tissues is an adaptation against herbivory, and while this seems plausible, its validity is uncertain. Zummo et al. (1984) studied the seasonal changes in the concentration of tannin and gossypol (terpenol aldehyde) in cotton and the damage caused by *Helicoverpa* (= *Heliothis*) *zea* (Boddie). They found that the amount and quality of tannin increased gradually from the cotyledon stage with a peak at the end of flowering, and decreased when the flower buds were one-third of the maximum size, increasing the vulnerability of the plants to caterpillar damage. In cotton, tannins have been shown to reduce larval size and survival (Chan et al. 1978). From a *fitness* point of view, Gutierrez and Regev (2005) posit that bioeconomically the loss of old leaves is less important because of lower photosynthetic rates, and hence low investment in protecting them is warranted. Similarly, the loss of surplus young cotton buds has a lower cost compared to the loss of older fruit in which greater time and energy has been invested.

The response of plants to herbivory may vary and depend on whether the attack is by a specialist or a generalist herbivore. In the medicinal plant *Hypericum perforatum* L., feeding by the specialist beetle *Chrysolina quadrigemina* (Suffrian) or simulation of physical damage results in a small accumulation of secondary substances in the tissues. However, when a generalist herbivore causes a little damage to the plant, an increase of 30% to 100% in the secondary compounds *hiperacins* and *hyperforin* were observed (Sirvent et al. 2003). These substances possess antimicrobial, antiviral, and antiherbivore properties.

Research on the weed *Silene latifolia* Poir, originally from Europe and accidentally introduced into North America about 200 years ago suggests that their escape from herbivore specialists favored natural selection. In their new range, the plants could invest more in growth and reproduction and less on defense than the same plants in Europe (Blair and Wolfe 2004). In another example, Zangerl and Berenbaum (2005) examined herbarium specimens of the European herb *Pastinaca sativa* (L.) collected over 152 years, and found an increase in phytochemicals after the accidental introduction of the herbivore *Depressaria pastinacella* (Duponchel) (Lepidoptera: Depressariidae).

5.4.2 Cost of Defense in Plants

The allocation of metabolic resources for the physical or chemical defenses against herbivores may represent a high energy and nutritive costs to plants (Chew and Rodman 1979). Direct and indirect defenses differ in their requirements for metabolic resources (Halitschke et al. 2001) that depending on the environment may have differing consequences on plant "fitness" (Turlings and Benrey 1998). Plants are under selection pressure to coordinate the metabolic processes and costs required for the direct induced defense and those needed for indirect defense or tolerance. The degree and type of defense in plant tissues or organs is related to the risk that the plant faces from herbivores, the importance of protecting the plant organs, and the costs involved (Rhoades 1983). It is often assumed that since reproductive organs are vital to the survival of the species and to individual fitness, it would seem better to protect them rather than vegetative parts that may be able to compensate for damage. In addition, it is assumed that perennial plants are better protected than the ephemeral ones (Price 1984; Kogan 1986).

Feeny (1976) and Kogan (1986) suggested that there was little data estimating the costs of defense because of experimental difficulties. Siemens et al. (2003) reviewed the evidence on the costs of induced defense in natural and agriculture systems reported in the literature, and concluded that it was difficult to estimate these costs because they are usually not measured in terms of fitness or fitness cost of the plant. Cipollini et al. (2003) posited that induced defenses are a form of adaptive phenotypic plasticity by which plants save metabolic costs by implementing the direct defenses only when needed and also to increase the indirect protection by natural enemies, while at the same time allowing constitutive defenses to remain active. This coordination would enable the plant to reduce the direct defense that could adversely affect the natural enemies if it were attacked by herbivores, and at the same time encourage indirect defenses (Cipollini et al. 2003). Hilker and Meiners (2002) attempted with little success to measure the costs and benefits of induced defenses using the reproductive parameters of plant fruit and seed production as the currency. Estimating these costs is complicated by the fact that the evidence of what are generally assumed to be direct defenses are difficult to isolate from the other changes that occur after insect attack (Roda and Baldwin 2003).

The ability of weeds to invade and colonize new areas is augmented when they escape adapted herbivores, allowing the weeds to reallocate resources to growth and reproduction that were previously used in chemical defense in its native range (Zangerl and Berenbaum 2005), and this change in allocation may be viewed as a measure of the fitness costs. Gutierrez and Regev (2005) provide a generalized economic framework for reviewing these trade-offs in any trophic level.

5.5 Herbivore Perspective

Insect herbivores are the most numerous organisms in the majority of natural ecosystems and may be responsible for about 80% of the plant material ingested annually (Thompson and Althof 1999). The ability of insects to feed involves a sequence of behavioral steps: *host habitat location*, *host finding*, *host acceptance*, and *host suitability* (Salt 1935; Kogan 1976; Matthews and Matthews 1978). These behavioral mechanisms allow herbivores to determine preferred plants and locations within plants that offer the best conditions for the development of progeny (Karban and Agrawal 2002) and are the result of coevolution.

Rhoades (1985) in a classic study examined the *offense–defense interactions* between herbivores and plants and how changes in chemical properties affected the development of herbivores. To feed on plants with high levels of secondary substances, adapted herbivores must invest resources (*costs*) in detoxification and this affects its growth and development (Roda and Baldwin 2003). Karban and Agrawal (2002) expanded the scheme proposed by Rhoades (1985) and presented three strategies employed by herbivores to exploit hosts (Table 5.2). The first strategy is considered the least aggressive and involves herbivores selecting certain plants and avoiding others. The second strategy involves changes in morphology and physiology of the insect that occurred through ecological and evolutionary time to exploit hosts. The third strategy is the most aggressive and occurs when herbivores actively modify the host plant, often before feeding, as, for example, by inducing the development of nutrient galls.

5.5.1 Avoiding Host Plant Defenses

Examples of insect that avoid plant defenses are abundant in the literature and the mechanisms are quite varied. Three examples are explored below. Ikeda et al. (1977) studied the larvae of pine sawflies

TABLE 5.2

Herbivore Offensive Strategies and Their Consequences

	Strategy	Tactic	Used by	Resulting Population Dynamics
Least aggressive	Choice	Avoidance, attraction	Opportunistic herbivores	Variable populations
	Change herbivore morphology, physiology	Metabolize, detoxify		
Most aggressive	Manipulate the host	Change host nutrition and defense	Stealthy herbivores	Low, invariant populations

Source: Modified from Rhoades, D. F. 1985. *Amer. Nat.* 125:205–38, by Karban, R., and A. A. Agrawal. 2002. *Annu. Rev. Ecol. Syst.* 33:641–64.

Neodiprion rugifrons Midd. and *N. Swainei* Midd. (Hymenoptera: Diprionidae) and found they fed only on old leaves and avoided new ones that contained acidic resin. The beetle *Ipilachna tredecimnotata* (Latreille) (Coleoptera: Coccinellidae) attacks a species of gourd that mobilizes toxic substances to the affected area, and this stimulates the beetle to cut a circular trench to isolate its feeding area from these substances (Carroll and Hoffman 1980). The beetle *E. borealis* (F.) (Coleoptera: Coccinellidae) also exhibits the same kind of adaptive behavior against plant defenses (Tallamy 1985). Dussourd (1999) investigated five species of leaf-feeding insects that cut trenches or ribs reducing the flow of resin, phloem fluid, or latex by 94% to the area where they fed.

Leaves, roots, and reproductive organs have very broad morphological diversification such as hairs, trichomes, spines, waxy layer, and hard tissues that evolved as defenses against insects and other herbivores. The caterpillar *Mechanitis isthnia* (Bates) (Lepidoptera: Ithomiidae) produces a web of silk on the leaf surface that enables it to avoid the trichomes as it feeds on the leaf margin (Rathcke and Pooler 1975).

Gregarious feeding behavior, wherein many individuals of the same species feed on the same host or plant parts, is a strategy that can help herbivores overcome plant defenses and also to defend against predators and parasitoids. This behavior is common in Aphidoidea, Coleoptera, Lepidoptera, and Orthoptera, and while benefits may accrue, the strategy may also increase intraspecific competition, attract more natural enemies, and may increase the probability of inducing plant defenses (Karban and Agrawal 2002).

5.5.2 Metabolizing and Sequestering Plant Toxins

Many herbivorous insects use plant toxins for their own benefit through detoxification mechanisms that may convert these substances into less toxic products. Herbivores adapted to feed on a specific plant often have an enzyme system able to metabolize toxic substances and use them as nutrients. This mechanism has been studied in detail by Rosenthal et al. (1977, 1983) in the bruchid beetle *Caryedes brasiliensis* Rolfe (Coleoptera: Bruchidae) that feeds on seeds of the tropical legume *Dioclea megacarpa* that are toxic to many organisms including other bruchid beetles. The seeds contain *canavanine*, a nonprotein amino acid that competes with the amino acid *arginine*, and causes protein deficiency in nonadapted herbivores. However, the larvae of *C. brasiliensis* possess a biochemical adaptation that enables them to discriminate between arginine and canavanine, and to degrade canavanine for use as a nitrogen source (Rosenthal et al. 1978). Larvae of *Utetheisa ornatrix* (L.) (Lepidoptera: Arctiidae) remove and accumulate alkaloids from host plants, and in the adult female the alkaloids are transferred to their eggs to protect them against predators (Dussourd et al. 1984).

Some insects store plant toxins to use against predators and parasitoids (Price 1984). Rothschild (1973) listed 43 species of insects that sequester plant substances. Eisner et al. (1974) studied the behavior of *Neodiprion sertifer* Midd. larvae that feed on pine resins, which it stores in its stomach diverticula. When disturbed by a predator, the larva regurgitates a drop of viscous fluid with chemical properties similar to pine resin. This fluid is unpleasant to vertebrate predators and sticks to the mouthparts of insects (Owen 1980a).

Sequestration of plant toxins by herbivores is often correlated with *aposematic coloration* and gregarious behavior (Muller 2003). Adults and nymphs of *Oncopeltus fasciatus* (Dallas) (Hemiptera: Lygaeidae) obtain glucosides from seeds of *Asclepias syriaca* (Asclepiadaceae) and store and metabolize them (Duffey and Scudder 1974), and in the process transform their hemolymph to repel predators. One of the best known examples of aposematic coloration and sequestration is the interaction between butterflies in the family Danaidae and plants in the family Asclepiadaceae. These butterflies have bright orange and black or white and black colors, and in the southwestern United States, *Dannaus plexippus* (L.) feed on *Asclepias curassavica* and *A. humictrata* that contain toxic *cardenolides*. These cardenolides are bitter and may cause vomiting or even death in birds and cattle when ingested in large amounts (Daly et al. 1978). Caterpillars of *D. plexippus* sequester these toxic substances and transfer them to the adult, where they are concentrated in the wings, the parts first attacked by birds. Insectivorous birds vomit after eating *D. plexippus* and learn to avoid them and other butterfly *mimics* such as the less common nontoxic species *Limenitis archippus* (Cramer) (Nymphalidae) that live in the same areas. Bates (1862) quoted in Owen (1980b) states that species of nontoxic *mimetic* butterflies are protected from predators because they look like the toxic species and that the phenotypic similarity between the species is a result of natural selection. Laboratory studies have shown that *L. archippus* reared on species of Asclepicidaceae that do not contain toxic substances are not toxic to predators (Huheey 1984).

5.5.3 Host Plant Manipulation

Insects may physically manipulate plant defenses in several ways. Galls may be formed by insects, mites, or other organisms through feeding or oviposition activity. Insects responsible for the induction of galls are host-specific and sometimes specific to a plant tissue. Due to physical injury or through the introduction of salivary secretions, growth hormone increase plant growth in the affected area, causing hypertrophy and hyperplasia that increases the size and number of cells resulting in an abnormal structure (Wawrzynski et al. 2005). Williams and Whitham (1986) investigated the patterns of leaf fall in poplar (*Populus*) in response to two gall-forming species of aphids. They found that aphid survival due to premature leaf fall induced by gall formation reduced aphid populations 25% and 53%.

Insects also reduce the efficiency of plant defenses in other ways. They may cause leaves to curl, reducing light interception that decreases leaf hardness and reduces the concentrations of tannin and other photoactive substances such as *hiperacin* (Berenbaum 1987; Berenbaun and Sandberg 1989; Sagers 1992). Herbivores may also produce substances that reduce direct defenses in plants (Roda and Baldwin 2003). For example, when *Manduca sexta* (L.) caterpillars feed on tobacco, they regurgitate components that cause a reduction in nicotine production (Halitschke et al. 2001).

Offense tactics used by herbivores to overcome plant defenses include morphological and physiological changes. For example, morphological adaptations of the mouthparts associated with feeding strategies may be countered by changes in the plant. Toju and Sota (2006) studied a beetle predator of camellia seed wherein the length of the beak (the strategy of offense) and the thickness of the seed pericarp (the plant defense) were positively correlated in field populations. The mouthparts of sucking insects in the Hemiptera and Homoptera enable them to avoid toxins by inserting their feeding stylets between cavities or ducts containing toxin (Slansky and Panizzi 1987).

5.6 Herbivore Generalists and Specialists

Dethier (1954) originated the concept that the evolution of phytophagy in insects has been from polyphagy toward monophagy. Rhoades (1979) argued that such selection does not withstand critical analysis, arguing that if polyphagous insects evolved toward a more restricted host diet, why are there still so many polyphagous species? In terms of the number of plant species available, the benefits point to polyphagy, but the relatively large number of insect specialists indicates that there are advantages to monophagy (Bernays and Graham 1988). Extreme monophagy seems disadvantageous from an evolutionary point of view, except in cases where the plants used are perennial and abundant; otherwise, fluctuations in plant populations could cause catastrophic effects on insect populations (Beck and Schoonhoven 1980).

In general, most herbivores are specialists and a smaller number of species such as the locust *Schistocerca gregaria* (Forsk) (Acrididae) are generalists (Bernays and Chapman 1994). An example of an insect specialist is the corn rootworm *Diabrotica longicornis* (Say) (Coleoptera: Chrysomelidae), which feeds only on specific parts of the root (Beck and Schoonhoven 1980). Due to specific requirements, specialists are generally less abundant than generalists that have more flexible requirements (Pianka 1978). Raubenheimer and Simpson (2003) examined the nutrition in *Locusta migratoria* (L.), a stenophagous specialist that feeds on grasses, and in *S. gregaria*, a true generalist herbivore, and found that the generalists had more flexibility in their behavioral and physiological responses to the imbalance of nutrients.

The generalist strategy is more adaptive, providing more alternatives for food and shelter (Price 1982), but herbivores must make choices between acceptable food plants that may vary in nutritional quality and may also have defenses to which they may be poorly adapted (Howard 1987). In relation to the theory of plant apparency (Feeny 1976), insect specialists will require more time and energy to find less apparent hosts that could also exposes them to factors that increase mortality, while in generalists the apparency of plants is less important because they can feed on a variety of plants (Rhoades and Cates 1976; Rhoades 1979). Research suggests that search in specialists can be more efficient due to the use of chemical signals.

Cornell and Hawkins (2003) examined four predictions of phytochemical coevolution found in the literature on the distribution and toxicity of phytochemicals and the level of specialization of herbivores. The predictions include the following: (1) herbivores may adapt to new chemicals and toxins and in the process become specialists; (2) herbivores can become generalists to feed on many hosts, but with less success; (3) the most widespread toxic substances are less toxic than those with a more restricted distribution; and (4) prediction (3) applies more to generalists than to specialists and depends on the presence or absence of the chemical in the normal plant host. Predictions (3) and (4) are related to the mechanisms of escape from herbivory and evolutionary radiation, positing that if a group of plant species with new chemicals becomes more widespread, the spread of the substances eventually lead to herbivores adapting to and disarming them.

5.7 The Tertiary Trophic Level

There are many examples where the characteristics of a plant such as secondary substances, trichomes, tissue hardness, and others factors may affect interactions between herbivores and their natural enemies by acting directly on the herbivore, the natural enemy, or both (Price et al. 1980). Hufbauer and Via (1999) suggested that the evolution between insect herbivores and their parasitoids may be influenced by the relationship between insect herbivores and their host plant. They demonstrated that populations of pea aphids may specialize on alfalfa, clover, or other hosts, but aphids specialized on alfalfa were parasitized less than those specialized on clover, regardless of whether the parasitoid was obtained from alfalfa or clover.

A vast amount of literature describes the effects of plant toxins on the primary consumers, but little is known about the impact of these toxins on natural enemies (Price et al. 1980; Price 1982). Survival of the parasitoid *Cotesia congregatus* (Say) (Hymenoptera: Braconidae) was reduced when tobacco hornworm caterpillars *Manduca sexta* (L.) and *Manduca quinquemaculata* (Haworth) (Lepidoptera: Sphingidae) were reared on tobacco plants with high concentrations of nicotine (Morgan 1910; Gilmore 1938a,b). Campbell and Duffey (1979) reported that when host larvae of *Helicoverta* (= *Heliothis*) *zea* (Boddie) (Lepidoptera: Noctuidae) eat plant with *tomatine*, the alkaloid may be toxic or even lethal to larvae of the parasitoid *Hyposoter exigua* (Viereck) (Hymenoptera: Ichneumonidae). Consequently, plants with high concentrations of tomatine may be better defended against herbivores, but they may in fact be more vulnerable because the tomatine has less effect on the caterpillar than on the parasitoid (Price 1986).

Parasitoids are attracted to volatile compounds released by plants in response to herbivore feeding (De Moraes et al. 2000). Such allelochemicals in plants have been shown to be beneficial to parasitoids in laboratory experiments. For example, when the parasitoid *Diaeretiella rapae* (McIntosh) (Hymenoptera: Braconidae) was offered aphids on sugar beet and cruciferous plants, it was more attracted to the cruciferous plants, resulting in more aphids being parasitized there (Read et al. 1970). Alborn et al. (1997) isolated the compound *volicitin* from oral secretions of *Spodoptera exigua* Hubner (Lepidoptera: Noctuidae), and demonstrated that when applied to maize seedlings, volatiles were released by the plant that attracted parasitic wasps. In contrast, simulated leaf damage without the application of *volicitin* did not attract the wasps.

Heliothis subflexa (Guenee) (Noctuidae: Heliothinae) larvae feed on fruits of *Physalis angulata* that lacks linoleic acid, a requirement for the development of many insects (De Moraes and Mescher 2004). However, *H. subflexa* can overcome this nutritional deficiency but its parasitoid *Cardiochiles nigriceps* Viereck cannot, thus reducing the vulnerability of *H. subflexa* to parasitism through a form of *cripsis* biochemistry.

5.7.1 Effects of Abiotic Factors in Tritrophic Interactions

In the long run, climate and biotic factors (e.g., competition, natural enemies) limit the distribution of plants and arthropods (Andrewartha and Birch 1954) and in the short run weather (e.g., temperature, solar radiation, rain, wind, and humidity) affects development, mortality, and abundance (Wellington et al. 1999). Organisms such as plants and insects have developed the ability to perceive environmental signals that warn of approaching changes and to respond to these signals through physiological, morphological, and behavioral changes that prepare them to face difficult conditions. Abiotic factors such as the characteristics of nutrition, photoperiod, and pH and other soil factors may have direct and indirect effects on natural populations, and may affect each species differently in the trophic cascade by affecting dormancy, migration, and polyphenism (Nechols et al. 1999). Plant growth is regulated by abiotic factors and soil nutrition, and these may affect the tertiary trophic levels via *bottom-up* effects, while herbivores and predation provide *top-down* regulation in the food chain (Hairston et al. 1960; Fretwell 1987). The sum of these abiotic and biotic interactions determines the regulation of species.

The effects of temperature on plant, poikilotherm herbivore, and predator growth rates are illustrated as growth indices (0–1) in Figure 5.4a, showing the range of temperatures (minimum and maximum) and the optimum for each species. The effects of temperature (TI) and humidity (MI) indices on growth rates of the three species are illustrated (Figure 5.4b) as ellipses representing the limits of favorability (Gutierrez 2001). (The effects of moisture are often defined in terms of vapor pressure deficit.) The weather experienced by the species and hence the values of the growth indices vary over time (i.e., the dashed line, Figure 5.4b). Hence, conditions favorable for the development of the species occur when the observed values (dashed line) fall within the limits of favorableness, but species have evolved various mechanisms to survive during unfavorable periods. The effects of some abiotic factors on the biology of the species are discussed below.

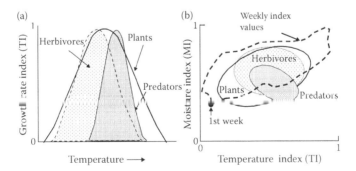

FIGURE 5.4 Effects of abiotic factors as physiological indices on species in a tritrophic food chain: (a) the effects of temperature on the growth rate (temperature index, $0 < TI < 1$) and (b) the effects of moisture (moisture index, $0 < MI < 1$) and temperature (TI). The dashed line represents the weekly MI and TI values starting week 1, while the three oblong shapes define the limits of favorability for the three species (see text). (From Gutierrez, A. P., In *Climate Change and Global Crop Productivity*, ed. K. R. Reddy and H. F. Hodges, CAB International, London, 2001. With permission.)

Insect survival depends on maintaining body water balance, and two important factors regulating this are humidity and temperature (Chapman 1982). The movement of air often causes a saturation deficit of water in plants (Ramsay et al. 1983). Insects living on exposed foliage may reduce the risk of desiccation by morphological adaptation and an efficient respiratory system (Daly et al. 1978), or through behavior, sheltering in the foliage, or becoming active at night when the risk of desiccation is lower. For example, Howard et al. (2002) found that swarms of the wasp *Apoica pallens* (Olivier) migrate only during a brief period before sunset. Costa and Varanda (2002) studied the construction of leaf shelters of *Stenoma scitiorella* Walker (Lepidoptera: Elachistidae) caterpillars and concluded that this behavior was likely selected as protection from desiccation and predation.

Plants also produce microclimates that can be very different from the weather in the area (i.e., the mesoclimate) (Edwards and Wratten 1981). Studies on oaks suggest that drought may be a factor reducing the abundance of leaf miners (Yarnes and Boecken 2005). Baumgaertner and Severini (1987) measured the temperature in the habitat of apple leaf miner *Phillonorycter blancardella* F.) (Lepidotpera: Gracillariidae) and found the temperature in the mines was higher than on leaves. Fennah (1963) found that thrip nymphs and adults normally feed on the underside of leaves, but when the bottom surface was exposed to the sun, the insects moved away even under conditions of high humidity. Many insects reduce their activity during periods of high winds or when the weather is overcast, such as adults of *Pieris rapae* (L.) (Lepidoptera: Pieridae) that do not fly or lay eggs in the field under such conditions (Gossard and Jones 1977).

Weather conditions influence trophic interactions and hence the success of biological control agents (Huffaker et al. 1971). A classic example of the effects of temperature on biological control is the case of the cottony cushion scale, *Icerya purchasi* Maskell that in areas of higher temperature is controlled by the predator *Rodolia cardinalis* Mulsant, while the dipterous parasitoid *Crytochaetum iceryae* (Will.) is active only in cooler regions (Quezada and DeBach 1973). Other examples are alfalfa aphid, *Therioaphis maculata* (Buckton) (Force and Messenger 1964), olive scale *Parlatoria oleae* (Colvée) (Huffaker and Kennett 1966; Rochat and Gutierrez 2001), and red scale *Aonidiella aurantii* (Maskell) (Murdoch et al. 2005).

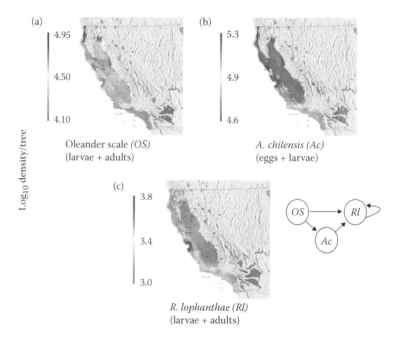

FIGURE 5.5 GIS maps of the annual cumulative log10 average number of larvae and adults of *A. nerii (OS)*, *A. chilensis (Ac)* eggs and larvae, and larval and adult of *R. lophanthae (Rl)* during the period 1995–2005 during the potential olive-growing regions of California at elevations below 750 m. (After Gutierrez, A. P., and M. A. Pizzamiglio, *Neotrop. Entomol.*, 36, 70–83, 2007. With permission.)

To examine the interactions between biotic and abiotic factors, it is necessary to develop a population dynamics model that incorporates the effects of the factors affecting the biology of each species. For example, field studies on the interactions between *Aspidiotus nerii* Bouché (Hemiptera: Aspididae) (oleander scale, *OS*) on California laurel (*Umbellularia californica*) and the parasitoid *Aphytis chilensis* (Howard) (Ac) (Hymenoptera: Aphelinidae) and the predator *Rhysobius lophanthae* (Blaisd.) (*Rl*) (Coleoptera: Coccinellidae) were conducted in the San Francisco Bay area of California and suggested that the beetle was the most important natural enemy. Studies in other climatic zones gave different results, and hence a mathematical model of this system was developed and embedded in a geographic information system (Grass GIS), and used to examine the effects of weather (temperature, rainfall, solar radiation) on these interactions throughout California (Gutierrez and Pizzamiglio 2007); see Figure 5.5. The analysis confirmed that the beetle was effective in the San Francisco Bay area (Figure 5.5c), but predicted the parasitoid should be more effective in controlling the oleander scale in areas with higher temperatures (Figure 5.5b). The model confirmed field studies in the Mediterranean basin where the parasitoid was the most important natural enemy. Several tritrophic systems as diverse as alfalfa, cotton, coffee, grape, and others have been modeled in this way (Gutierrez and Baumgärtner 2007), providing important ecological and economic information as well as information on the effect of climate change on them (Gutierrez et al. 2006a).

5.8 Final Considerations

Plants are the main sources of food and fiber for an increasing human population, and insects are their major competitors. The development of modern farming techniques and the use of herbicides and insecticides allowed the cultivation of large areas using a limited number of species that reduced plant diversity (Cromartie 1981). Pesticides have been used to protect these simplified systems, but they often cause the resurgence of target pests, outbreak of secondary pests, resistance of insects to these products, and the degradation of the environment (van den Bosch 1978). This has stimulated a search for environmentally friendly natural products such as botanical pesticides, semiochemicals, allelochemicals, and the genetic manipulation of plants.

The development of botanical pesticides has progressed over the past 50 years since the discovery that *Melia azedarach* and other plants contain substances that inhibit feeding in *Schistocerca gregaria* (Forsk) (Schoonhoven 1982). Another biopesticide is Avermectin, a natural product obtained by fermentation of the soil microorganism *Streptomyces avermectilis*. Avermectin is marketed as an effective treatment for controlling mites with little disruption of beneficial insects (Bull 1986; Dybas and Green 1984; Putter et al. 1981), and is used in more than 50 agricultural products in California. Despite being a natural product, it is toxic to fish and other aquatic invertebrates (Pesticides Action Network (PAN)). The entomopathogenic fungus *Metarhizium anisopliae var. acridum* that has been mass produced and used to control locusts in Africa (Arthurs and Thomas 2000) and other pests elsewhere.

Roitberg (2007) reviewed the area of behavior manipulation of insects in pest management and found that the response of insect behavior to various stimuli can vary greatly under different conditions. He warned that these factors must be understood before using attractant, kairomones, pheromones, and other tactics and concepts of behavioral ecology in pest management programs. Considerable progress has been made in the development and application of semiochemicals used as pheromones for detection and monitoring, and for the control of pests using mating disruption (Cardé and Millar 2004). Information on overwintering in the cotton boll weevil (*Anthonomus grandis* Boh.) and the emergence of the adults in spring were used in the development of an eradication/suppression program using pheromones and pesticides (Dickerson et al. 1987). The use of synthetic pheromones to prevent mating in *Cydia pomonella* (L), *Grapholita molesta* Busck, *Endopiza viteana* Clemens, and other species are common examples in the literature. Since 1970, over 548 research papers have been published on semiochemicals, but there has been no general review except for Cerambycidae and Scolytidae beetles in forest systems (Allison et al. 2004; Sun Xiao-Ling et al. 2006).

Bernays (1983) proposed that the best strategy would be to increase defenses in the plants whose actions are systemic and restricted to the crop. Insect control through resistant cultivars is an important defense strategy and involves the study of chemical defenses in plants, and the behavior, physiology, sense organs, and genetics of insects (Saxena and Barrion 1985). More recently, developments in molecular biology have produced a revolution in our knowledge of induced defenses in plants and refocused attention to the potential exploitation of the mechanisms of endogenous resistance in crop protection (Ferry et al. 2004).

Advances in biotechnology have enabled the introduction of genes into plants to protect them from pests (Horsch et al. 1985). Genes for the production of a protoxin (*Bt*) from the ubiquitous bacterium *Bacillus thuringiensis* have been inserted into numerous crops to protect them against herbivores and it is thought that *Bt* is nontoxic to the environment (Luttrell and Herzog 1994). The genetic basis of *Bt* protoxin production is relatively simple and this has facilitated its transfer into plants (Lindquist and Busch-Petersen 1987). This technology has been used in other ways to control pests, such as the use of genetically modified (GM) *Bt* maize as a "push–pull" trap crop to attract the pest *Eldana saccharina* Walker (Lepidoptera: Pyralidae) in sugar cane (Keeping et al. 2007). GM cotton, soybeans, and other crops have been developed for the control of Lepidoptera with considerable success against the pink bollworm (*Pectinophora gossypiella* Saunders), but less success has accrued against other cotton pests, and with some detrimental effects on the efficacy of natural enemies (Gutierrez et al. 2006b). Pemsl et al. (2005) analyzed the economic benefits of the use of transgenic *Bt* cotton and their role in outbreaks of secondary pests.

The use of transgenic herbicide-tolerant crops (HT) has increased herbicide use, resulting in the development of resistance in some weeds, additional pollution, and collateral deleterious effect on amphibian reproduction (Hayes 2003; Relyea 2005).

Finally, Lewis and Wilson (1980) questioned whether *ecological theories* can be applied in profoundly altered ecosystems (e.g., agriculture, forestry), but nevertheless, research on insect–plant interactions will continue to improve technologies in insect management, hopefully leading to environmental sustainability.

ACKNOWLEDGMENTS

The author gratefully acknowledges Prof. Andrew Paul Gutierrez for his assistance in the preparation of this chapter, and thanks editors Dr. A. R. Panizzi and Prof. J. R. P. Parra for the opportunity to contribute to this book.

REFERENCES

Agrawall, A. A. 2001. Phenotypic plasticity in the interactions and evolution of species. *Science* 294:321–26.

Allison, J. D., J. H. Borden, and S. J. Seybold. 2004. A review of the chemical ecology of the Cerambycidae (Coleoptera). *Chemoecology* 14 (3–4):123–50.

Alborn, H. T., T. C. Turlings, T. H. Jones, G. Stenhagen, J. H. Loughrin, and J. H. Tumlinson. 1997. An elicitor of plant volatiles from beet armyworm oral secretion. *Science* 276:945–49.

Andrewartha, H. G., and L. C. Birch. 1954. *The Distribution and Abundance of Animals*. Chicago: University of Chicago Press.

Arthurs, S., and M. B. Thomas. 2000. Effects of a mycoinsecticide on feeding and fecundity of the brown Locust *Locustana pardalina*. *Biocont. Sci. Tech.* 10:321–29.

Barrett, P. A., and K. J. Willis. 2001. Did dinosaurs invent flowers? Dinosaur and angiosperm coevolution revisited. *Biol. Rev.* 76:411–47.

Baumgärtner, J., and M. Severini. 1987. Microclimate and arthropod phenologies: The leaf miner *Phyllonoricter biancardella* F. (Lep.) as an example. In *Agrometeorology*, ed. F. Prodi, F. Rossi, and C. Cristoferi, 225–243. Cesena, Italy: Fondazione Cesena Agricultura Public.

Becerra, J. X. 2005. Synchronous coadaptation in an ancient case of herbivory. *PNAS* 100:12804–7.

Beck, S. D., and L. M. Schoonhoven. 1980. Insect behavior and plant resistance. In *Breeding Plants Resistant to Insects*, ed. F. G. Maxwell and R. A. Jennings, 115–135, New York: John Wiley & Sons.

Berenbaum, M. 1983. Coumarins and caterpillars: A case for coevolution. *Evolution* 37:163–79.

Berenbaum, M. 1986. Post-ingestive effects of phytochemicals on insects: On *Paracelsus* and plant products. In *Insect–Plant Interactions*, ed. T. A. Miller and J. Miller, 121–153. New York: Springer-Verlag.

Berenbaum, M. 1987. Charge of the light brigade; insect adaptations to phototoxins. In *Light-Activated Pesticides,* ed. J. R. Heitz and K. R. Downum, 206–216. Washington: ACS Symposium, Series 339.

Berenbaum, M., and P. Feeny. 1981. Toxicity of angular furanocoumarins to swallowtails: Escalation in the coevolutionary arms race? *Science* 212:927–9.

Bernays, E. A. 1981. Plant tannins and insect herbivores: An appraisal. *Ecol. Entomol.* 6:353–60.

Bernays, E. A. 1982. The insect on the plant: A closer look. *5th Int. Symp. Plant–Insect Relationships,* Wageningen, the Netherlands, 3–l7.

Bernays, E. A. 1983. Antifeedants in crop pest management. In *Natural Products for Innovative Pest Management,* ed. D. L. Whitehead and W. Bowers, 259–271. Oxford, UK: Pergamon Press.

Bernays, E. A., and R. F. Chapman. 1994. *Host-Plant Selection by Phytophagous Insects.* New York: Chapman and Hall.

Bernays, E. A., and M. Graham. 1988. On the evolution of host specificity in phytophagous arthropods. *Ecology* 69:886–92.

Blair, A. C., and L. M. Wolfe. 2004. The evolution of an invasive plant: An experimental study with *Silene latifolia. Ecology* 85:3035–42.

Brues, C. T. 1920. The selection of food-plants by insects, with special reference to lepidopterous larvae. *Amer. Nat.* 54:313–32.

Bull, D. L. 1986. Toxicity and pharmacodynamics of avermectin in the tobacco budworm, corn earworm, and fall armyworm (Noctuidae: Lepidoptera). *J. Agric. Food Chem.* 34:74–8.

Campbell, B. C., and S. S. Duffey. 1979. Tomatine and parasitic wasps: Potential incompatibility of plant antibiosis with biological control. *Science* 205:700–2.

De Candolle, A. M. P. 1832. Physiologie vegetale. Paris, Bechet Jenne, Lib. Fac. Med. v. III.

Cardé, R. T., and J. Millar. (eds.). 2004. *Advances in Insect Chemical Ecology,* Cambridge, UK: Cambridge University Press.

Carrol, C. R., and C. A. Hoffman. 1980. Chemical feeding deterrent mobilized in response to insect herbivory and counter adaptation by *Epilachna tredecimnotata. Science* 209:414–6.

Chan, B. G., A. C. Waiss, and M. Lukefahr. 1978. Condensed tannin, an anti-biotical chemical from *Gossypium hirsutum. J. Insect Physiol.* 24:113–8.

Chapman, R. F. 1982. *The Insects: Structure and Function.* Cambridge, MA: Harvard University Press.

Chew, F. S., and J. Rodman. 1979. Plant resources for chemical defense. In *Herbivores: Their Interaction with Secondary Plant Metabolites,* ed. G. A. Rosenthal and D. H. Janzen, 271–307. New York: Academic Press.

Cipollini, D., C. B. Purrington, and J. Bergelson. 2003. Costs of induced responses in plants. *Basic Appl. Ecol.* 4:79–89.

Coley, P. D. 1980. Effects of leaf age and plant life history patterns on herbivory. *Nature* 284:545–6.

Cornell, H. V., and B. A. Hawkins. 2003. A test of phytochemical coevolution theory. *Amer. Nat.* 161:507–22.

Costa, A. A., and E. M. Varanda. 2002. Building of leaf shelters by *Stenoma scitiorella* Walker (Lepidoptera: Elachistidae): Manipulation of host plant quality? *Neotrop. Entomol.* 31:537–40.

Cromartie, Jr., W. J. 1981. The environmental control of insects using crop diversity. In *Handbook of Pest Management,* v. 1, ed. D. Pimentel, 223–51. Boca Raton, FL: CRC Press, Inc.

Czapeck, F. 1921. *Biochemie der Pflanzen,* ed. G. Fisher, 3. Band, 2. Aufl. Jena.

Daly, H. V., J. T. Doyen, and P. Ehrlich. 1978. *Introduction to Insect Biology and Diversity.* New York: McGraw-Hill Book Co.

De Moraes, C. M., W. J. Lewis, and J. H. Tumlinson. 2000. Examining plant–parasitoid interactions in tritrophic systems. *An. Soc. Entomol. Brasil* 29:189–203.

De Moraes, C. M., and M. C. Mescher. 2007. Biochemical crypsis in the avoidance of natural enemies by an insect herbivore. *PNAS* 101:8993–7.

Dethier, V. G. 1941. Chemical factors determining the choice of plants by *Papilio* larvae. *Amer. Nat.* 75:61–73.

Dethier, V. G. 1954. Evolution of feeding preferences in phytophagous insects. *Evolution* 8:55–64.

Dethier, V. G. 1976. *Man's Plague? Insects and Agriculture.* Princeton, NJ: Darwin Press, Inc.

Dicke, M. 1999. Evolution of induced indirect defence of plants. In *The Ecology and Evolution of Inducible Defenses,* ed. R. Tollrian and C. D. Harvell, 62–88. Princeton, NJ: Princeton University Press.

Dicke, M., A. A. Agrawal, and J. Bruin. 2003a. Plants talk, but are they deaf? *Trends in Plant Science* 8:403–5.

Dicke, M., and J. Bruin. 2001. Chemical information transfer between plants: Back to the future. *Biochem. Syst. Ecol.* 29:981–94.

Dicke, M., and W. Takken. 2006. *Chemical Ecology: From Gene to Ecosystem,* Wageningen Ur Frontis Series: 16, Meeting Information: Spring School on Chemical Ecology—From Gene to Ecosystem, 2005, Wageningen, the Netherlands.

Dicke, M., R. M. P. van Poecke, and J. G. De Boer. 2003b. Inducible indirect defence of plants: From mechanisms to ecological functions. *Basic Appl. Ecol.* 4:27–42.

Dicke, M., and M. Hilker. 2003. Induced plant defenses: From molecular biology to evolutionary ecology. *Basic Appl. Ecol.* 4:3–14.

Dickerson, W. A., R. L. Ridgeway, and F. R. Planer. 1987. Southeastern boll weevil eradication program, improved pheromone trap, and program status. 335–337. *Proc. Beltwide Cotton Research and Production Conference, National Cotton Council*, Memphis, TN.

Dietl, G. P., and P. H. Kelley. 2002. The fossil record of predator–prey arms races: Coevolution and escalation hypotheses. *Paleontological Soc. Papers* 8:353–74.

Duffey, S. S., and G. G. E. Scudder. 1974. Cardiac glycosides in *Oncopeltus fasciatus* (Dallas) (Hemiptera: Lygaeidae). I. The uptake and distribution of natural cardenolids in the body. *Can. J. Zool.* 52:283–9.

Dussourd, D. E. 1999. Behavioral sabotage of plant defense: Do vein cuts and trenches reduce insect exposure to exudates? *J. Insect Behav.* 12:501–15.

Dussourd, D. E., K. Ubik, J. F. Resch, J. Meinwald, and T. Eisner. 1984. Egg protection by parental investment of plant alkaloids in Lepidoptera. *17th Int. Congr. Entomol.*, Hamburg, Germany, p. 840.

Dybas, R. A., and A. S. Green, Jr. 1984. Avermectins: Their chemistry and pesticidal activity. *British Crop Prot. Conf. Pests Dis.,* v. 3. Croydon, England.

Edwards, P. J., and S. D. Wratten. 1981. Ecologia das interações entre insetos e plantas. São Paulo, Brazil: EDUSP.

Ehrlich, P. R., and P. R. Raven. 1964. Butterflies and plants: A study in coevolution. *Evolution* 18:586–608.

Eisner, J. S., J. S. Johnessee, and J. Carrel. 1974. Defensive use by an insect of a plant resin. *Science* 184:996–9.

Errera, L. 1886. Rev. Sci. Soc. Bot. Real Belgique, 5. (citado em H.C.S. Abbott. (1887). *Comparative chemistry of higher and lower plants*).

Fabre, J. H. 1890. *Souvenirs entomologiques*. Paris, France: C. Delagrave.

Faegri, K., and L. Van Der Pijl. 1979. *The Principles of Pollination Ecology*. Oxford, UK: Pergamon Press.

Feeny, P. 1970. Seasonal changes in oak leaf tannins and nutrients as cause of spring feeding by wintermoth caterpillars. *Ecology* 51:565–81.

Feeny, P. 1975. Biochemical coevolution between plants and their insect herbivores. In *Coevolution of Animals and Plants,* ed. L. E. Gilbert and P. H. Raven, 3–19. Austin, TX: University of Texas Press.

Feeny, P. 1976. Plant apparency and chemical defense. In *Biochemical Interactions Between Plants and Insects, Rec. Adv. Phytochem.*, ed. J. W. Wallace and R. L. Mansell, 1–40. New York: Plenum Press.

Feild, T. S., and N. C. Arens. 2007. The ecophysiology of early angiosperms. *Plant Cell Environ.* 30:291–309.

Fennah, R. G. 1963. Nutritional factors associated with seasonal population increase of cacao thrips *Selenothrips rubrocinctus* (Giard.) (Thysanoptera) on cashew, *Anacardium occidentale. Bull. Entomol. Res.* 53:681–713.

Ferry, N., M. G. Edwards, J. A. Gatehouse, and A. M. R. Gatehouse. 2004. Plant–insect interactions: Molecular approaches to insect resistance. *Curr. Opin. Biotech.* 15:1–7.

Flint, M. L., and R. van den Bosch. 1981. *Introduction to Integrated Pest Management*. New York: Plenum Press.

Force, D. C., and P. S. Messenger. 1964. Fecundity, reproductive rates, and innate capacity for increase of three parasites of *Therioaphis maculata* (Buckton). *Ecology* 45:706–15.

Fraenkel, G. 1951. The nutritional value of green plants for insects. In *9th Int. Congr. Entomol.,* Amsterdam, the Netherlands.

Fraenkel, G. 1958. The basis of food selection in insects which feed on leaves. In *8th Ann. Meetings Entomol. Soc. Japan,* ed. G. Fraenkel, 5. Sapporo: Hokkaido University.

Fraenkel, G. 1959. The raison d'etre of secondary plant substances. *Science* 129:1966–70.

Fretwell, S. D. 1987. Food chain dynamics: The central theory of ecology? *Oikos* 50:291–301.

Futuyma, D. J. 1983. Evolutionary interactions among herbivorous and plants. In *Coevolution,* ed. D. J. Futuyma and M. Slatkin, 207–231. Sunderland, MA: Sinauer.

Futuyma, D. J., and M. Slatkin (eds.). 1983. In *Coevolution.* Sunderland, MA: Sinauer.

Gallun, R. L., and G. S. Khush. 1980. Genetic factors affecting expression and stability of resistance. In *Breeding Plants Resistant to Insects,* ed. F. G. Maxwell and P. R. Jennings, 63–85. New York: John Wiley & Sons.

Gensel, P. G., and H. N. Andrews. 1987. The evolution of early land plants. *Amer. Scientist* 75: 478–489.

Giles, C.T., J. M. Vivanco, B. Newingham, W. Good, H. P. Bais, P. Landrs, A. Caesar, and R. M. Caloway. 2005. Insect herbivory stimulates allelopathic exudation by an invasive plant and the suppression of natives. *Ecol. Lett.* 8:209–17.

Gilmore, J. V. 1983a. Observations on the hornworms attacking tobacco in Tennessee and Kentucky. *J. Econ. Entomol.* 31:706–12.

Gilmore, J. V. 1983b. Notes on *Apanteles congregatus* (Say) as a parasite of tobacco hornworms. *J. Econ. Entomol.* 31:712–15.

Gossard, T. W., and R. E. Jones. 1977. The effects of age and weather on egg-laying in *Pieris rapae* L. *J. Appl. Ecol.* 14:65–71.

Grostal, P., and D. J. O'Dowd. 1994. Plants, mites and mutualism: Leaf domatia and the abundance and reproduction of mites on *Viburnum tinus* (Caprifoliaceae). *Oecologia* 97:308–15.

Gutierrez, A. P., and G. L. Curry. 1989. Framework for studying crop–pest systems. In *Integrated Pest Management Systems for Cotton Production*, ed. R. F. Frisbie, 37–64. New York: John Wiley & Sons.

Gutierrez, A. P. 2001. Climate change: Effects on pest dynamics. In *Climate Change and Global Crop Productivity*, ed. K. R. Reddy and H. F. Hodges. London: CAB International.

Gutierrez, A. P., J. J. Adamcyzk, Jr., and S. Ponsard. 2006b. A physiologically based model of *Bt* cotton–pest interactions: II. Bollworm–defoliator–natural enemy interactions. *Ecol. Modelling* 191:360–82.

Gutierrez, A. P., and J. Baumgärtner. 2007. Modeling the dynamics of tritrophic population interactions. In *Perspectives in Ecology and Integrated Pest Management,* ed. M. Kogan and P. Jepson, 301–60. Cambridge, UK: Cambridge University Press.

Gutierrez, A. P., L. Ponti, C. K. Ellis, and T. D'oultremont. 2006a. Analysis of climate effects on agricultural systems: A report to the Governor of California sponsored by the California Climate Change Center. http://www.climatechange.ca.gov/climate_action team/reports/ index.html (accessed June 2, 2010).

Gutierrez, A. P., and M. A. Pizzamiglio. 2007. Physiologically-based GIS model of weather mediated model of competition between a parasitoid and a coccinellid predator of oleander scale. *Neotrop. Entomol.* 36:70–83.

Gutierrez, A. P., and U. Regev 2005. The bioeconomics of tritrophic systems: Applications to invasive species. *Ecol. Econ.* 52:382–96.

Hairston, N. G., F. E. Smith, and L. B. Slobodkin. 1960. Community structure, population control, and competition. *Am. Nat.* 44:421–5.

Halitschke, R., U. Schittko, G., Pohnert, W. Boland, and I. Baldwin. 2001. Molecular interaction between the specialist herbivore *Manduca sexta* (Lepidoptera, Sphingidae) and its natural host *Nicotiana attenuata*. III. Fatty acid–amino acid conjugated in herbivore oral secretions are necessary and sufficient for herbivore-specific plant responses. *Plant Physiol.* 125:711–7.

Harborne, J. B. 1977a. Insect feeding preferences. In *Introduction to Ecological Biochemistry*, ed. J. B. Harborne, 103–129. London: Academic Press.

Harborne, J. B. 1977b. Biochemical interaction between higher plants. In *Introduction to Ecological Biochemistry*, ed. J. B. Harborne, 178–195. London: Academic Press.

Harborne, J. B. 1977c. The plant and its biochemical adaptation to the environment. In *Introduction to Ecological Biochemistry,* ed. J. B. Harborne, 1–26. London: Academic Press.

Harper, E. M. 2006. Dissecting post-Palaeozoic arms races. *Palaeogeography, Palaeoclimatology, Palaeoecology* 232:322–43.

Harris, M. K. 1980. Arthropod–plant interaction related to agriculture emphasizing host plant resistance. In *Biology and Breeding for Resistance to Arthropods and Pathogens in Agricultural Plants*, ed. M. K. Harris, 23–51. College Station, TX: Texas A&M Press.

Hayes, T. 2003. Conservation physiology: The amphibian response to pesticide contamination. *Integ. Comp. Biol.* 43: 815.

Heil, M., and D. Mckey. 2003. Protective ant–plant interactions as model systems in ecological and evolutionary research. *Ann. Rev. Ecol. Evol. Syst.* 34:425–53.

Hilker, M., and T. Meiners. 2002. Induction of plant response to oviposition and feeding by herbivorous arthropods: A comparison. *Entomol. Exp. Appl.* 104:181–92.

Holling, C. S., D. Jones, and C. C. Clark. 1977. Ecological policy design: A case study of forest and pest management. In *Pest Management, International Institute for Appl. Syst. Analy.–Proc. Ser.*, ed. G. A. Norton and C. S. Holling, 13–90. Oxford, UK: Pergamon Press.

Horsch, R. B., J. E. Fry, N. L. Hoffmann, D. Eicholtz, S. G. Rogers, and R. T. Fraley. 1985. A simple and general method for transferring genes into plants. *Science* 227:1229–31.

Howard, J. J. 1987. Leaf-cutting and diet selection: The role of nutrients, water, and secondary chemistry. *Ecology* 68:503–15.

Howard, K. J., A. R. Smith, S. O'Donnell, and R. L. Jeanne. 2002. Novel method of swarm emigration by the epiponine wasp, *Apoica pallens* (Hymenoptera Vespidae). *Ethol. Ecol. Evol.* 14:365–71.

Hufbauer, R. A., and S. Via. 1999. Evolution of an aphid–parasitoid interaction: Variation in resistance to parasitism among aphid populations specialized on different plants. *Evolution* 53:1435–445.

Huffaker, C. B., and C. E. Kennett. 1966. Studies of two parasites of the olive scale, *Parlatoria oleae* (Colvée) in control of the olive scale, *Parlatoria oleae* (Colvée). IV. Biological control of *Parlatoria oleae* (Colvée) through the compensatory action of two introduced parasites. *Hilgardia* 37:283–334.

Huffaker, C. B., P. S. Messenger, and P. Debach. 1971. The natural enemy component in natural control and the theory of biological control. In *Biological Control*, ed. C. B. Huffaker, 16–69. New York: Plenum Press.

Huheey, J. E. 1984. Warning coloration and mimicry. In *Chemical Ecology of Insects*, ed. W. L. Bell and R. T. Cardé, 257–297. London: Chapman and Hall.

Ikeda, T., F. Matsumara, and D. M. Benjamin. 1977. Chemical basis for feeding adaptations of pine sawflies *Neodiprion rugifrons* and *Neodiprion swainei*. *Science* 197:497–9.

Janzen, D. H. 1980. When is it coevolution? *Evolution* 34: 611–12.

Jermy, T. 1976. Insect–host-plant relationship—coevolution or sequential evolution? *Symp. Biol. Hung.* 16:109–13.

Karban, R., and A. A. Agrawal. 2002. Herbivore offense. *Annu. Rev. Ecol. Syst.* 33:641–64.

Keeping, M. G., R. S. Rutherford, and D. E. Conlong. 2007. Bt-maize as a potential trap crop for management of *Eldana saccharina* Walker (Lep., Pyralidae) in sugarcane *J. Appl. Entomol.* 131:241–50.

Kennedy, J. S., and C. O. Booth. 1951. Host alternation in *Aphis fabae* Scop. I. Feeding preferences and fecundity in relation to the age and kind of leaves. *Ann. Appl. Biol.* 38:25–64.

Klein, R. 2004. Phytoecdysteroids. *J. Amer. Herbalist Guild.* 5:18–28.

Kogan, M. 1976. The role of chemicals factors in insect/plant relationships. *15th Int. Congr. Entomol.*, Washington, DC, 211–227.

Kogan, M. 1986. Plant defense strategies and host-plant resistance. In *Ecological Theory and Integrated Pest Management Practice*, ed. M. Kogan, 83–133. New York: John Wiley & Sons.

Kogan, M., and J. Paxton. 1983. Natural inducers of plant resistance to insects. In *Plant Resistance to Insects*, ed. P. A. Hedin, 52–171. Washington, DC: American Chemical Society.

Kossel, A. 1891. Ulber die *Chorda dorsalis*. *Zeitschrift für Physiologische Chemie*. 15:331–4.

Kubo, I., and J. A. Klocke. 1983. Isolation of phytoecdysones, insect ecdysis inhibitors and feeding deterrents. In *Plant Resistance to Insects*, ed. P. A. Hedin. Washington, DC: American Chemical Society.

Law, J. M., and E. Regnier. 1971. Pheromones. *Annu. Rev. Biochem.* 40: 533–48.

Levin, D. A. 1976. The chemical defenses of plants to pathogens and herbivores. *Annu. Rev. Ecol Syst.* 7:121–59.

Lewis, R., and N. Wilson. 1980. Ecological theory and pest management. *Annu. Rev. Entomol.* 25:287–308.

Lindquist, D. A., and E. Busch-Petersen. 1987. Applied insect genetics and IPM. In *Integrated Pest Management: Quo Vadis?*, ed. V. Deluchi, 2378–255. Geneva, Switzerland: Parasitis.

Luttrell, R. G., and G. A. Herzog. 1994. Potential effect of transgenic cotton expressing *Bt* cotton IPM programs. In *Proc. Beltwide Cotton Prod. Res. Conf.*, ed. D. J. Herber, 806–9. Memphis, TN: National Cotton Council.

Martin, J. S., M. M. Martin, and E. A. Bernays. 1987. Failure of tannic acid to inhibit digestion or reduce digestibility of plants protein in gut fluids of insect herbivores: Implications for theories of plant defense. *J. Chem. Ecol.* 13:605–21.

Matthews, R. W., and J. R. Matthews. 1978. *Insect Behavior.* New York: John Wiley & Sons.

Mello, M. O., and M. C. Silva-Filho. 2002. Plant–insect interactions: An evolutionary arms race between two distinct defense mechanisms. *Braz. J. Plant Physiol.* 14:71–81.

Moran, N., and W. D. Hamilton. 1980. Low nutritive quality as defense against herbivores. *J. Theor. Biol.* 89:247–54.

Morgan, A. C. 1910. Observations recorded at the 236th regular meeting of the Entomological Society of Washington. *Entomol. Soc. Wash. 12*, 72.

Mothes, K. 1980. Historical introduction. In *Secondary Plant Products. Encyclopedia of Plant Physiology*, v. 8, ed. E. A. Bell and B. V. Charlwood, 1–10. Berlin, Germany: Springer-Verlag.

Muller, C. 2003. Lack of sequestration of host plant glucosinolates in *Pieris rapae* and *P. brassicae*. *Chemoecology* 13: 47–54.

Murdoch, W. W., C. J. Briggs, and S. Swarbrick. 2005. Host suppression and stability in a parasitoid–host system: Experimental demonstration. *Science* 307: 610–613.

Nechols, J. R., M. J. Tauber, C. A. Tauber, and S. Masaki. 1999. Adaptations to hazardous seasonal conditions: Dormancy, migration, and polyphenism. In *Ecological Entomology*, Second Edition, ed., C. B. Huffaker and A. P. Gutierrez, 313–353. New York: John Wiley & Sons.

Nordlund, D. A. 1981. Semiochemicals: A review of the terminology in *Semiochemicals: Their Role in Pest Control*, ed. D. A. Nordlund, R. L. Jones, and W. J. Lewis, 13–24. New York: John Wiley & Sons.

Opler, P. A. 1973. Fossil lepidopterous leaf miners demonstrate the age of soma insect–plant relationships. *Science* 179:1321–3.

Owen, J. 1980a. *Feeding Strategy*. Chicago: University of Chicago Press.

Owen, D. 1980b. *Camouflage and Mimicry*. Chicago: University of Chicago Press.

Pemsl, D., H. Waibel, and A. P. Gutierrez. 2005. Why do some *Bt*-cotton farmers in China continue to use high levels of pesticides? *Intern. J. Agric. Sust.* 3:44–56.

Pianka, E. R. 1978. *Evolutionary Ecology*. New York: Harper and Row.

Price, P. W. 1982. Hypotheses on organization and evolution in herbivorous insect communities. In *Variable Plants and Herbivores in Natural and Managed Systems*, ed. R. F. Denno and M. S. McClure (eds.), 559–596. New York: Academic Press.

Price, P. W. 1984. *Insect Ecology*. New York: Wiley-Interscience.

Price, P. W. 1986. Ecological aspects of host plant resistance and biological control: Interactions among three trophic levels. In *Interactions of Plant Resistance and Parasitoids and Predators of Insects*, ed. D. J. Boethel and R. D. Eikenbary, 11–30. England: Ellis Horwood Ltd.

Price, P. W., C. E. Bouton, P. Gross, B. A. McPheron, J. N. Thompson, and A. E. Weis. 1980. Interactions among three trophic levels: Influence of plants on interactions between insect herbivores and natural enemies. *Ann. Rev. Ecol Syst.* 11:41–65.

Prokopy, R. J., and E. D. Owens. 1983. Visual detection of plants by herbivorous insects. *Annu. Rev. Entomol.* 28:337–64.

Putter, I., J. G. Maccornel, F. A. Preiser, A. A. Haidri, S. S. Ristich, and R. A. Dybas. 1981. Avermectins, novel insecticides, acaricides, and nematicides from a soil microorganism. *Experientia* 37:963–4.

Quezada, J. R., and P. Debach. 1973. Bioecological and population studies of the cottony scale, *Icerya purchasi* Mask. and its natural enemies. *Rodolia cardinalis* Mul. and *Cryptochaetum iceryae* Wil. in southern California. *Hilgardia* 41:631–88.

Ramsay, J. A., C. G. Butter, and J. H. Sang. 1983. The humidity gradient at the surface of a transpiring leaf. *J. Exp. Biol.* 15:255–65.

Rasnitsyn, A. P., and V. A. Kassilov. 2000. The first documented occurrence of phyllophagy in pre-Cretaceous insects: Leaf tissues in the gut of upper Jurassic insects from southern Kasakhstan. *Paleobotany J.* 34:301–9.

Rathcke, B. J., and R. W. Pooler. 1975. Coevolutionary race continues, butterfly larval adaptation to plant trichomes. *Science* 187:175–6.

Raubenheimer, D., and S. J. Simpson. 2003. Nutrient balancing in grasshoppers: Behavioural and physiological correlates of dietary breadth. *J. Exp. Biol.* 206:1669–81.

Read, D. P., P. Feeny, and R. B. Root. 1970. Habitat selection by the aphid parasite *Diaeretiella rapae* (Hymenoptera: Braconidae) and hyperparasite *Charips brassicae* (Hymenoptera: Cynipidae). *Can. Entomol.* 102:567–78.

Relyea, R. A. 2005. The lethal impacts of roundup and predatory stress on six species of North American tadpoles. *Arch. Environ. Contam. and Tox.* 48:351–7.

Rhoades, D. F. 1979. Evolution of plant chemical defenses against herbivores. In *Herbivores: Their Interactions with Secondary Plant Metabolites*, ed. G. A. Rosenthal and D. Janzen, 3–54. New York: Academic Press.

Rhoades, D. F. 1983. Herbivore population dynamics and plant chemistry. In *Variable Plants and Herbivores in Natural and Managed Systems*, ed. R. F. Denno and M. S. McClure, 155–220. New York: Academic Press.

Rhoades, D. F. 1985. Offensive–defensive interactions between herbivores and plants: Their relevance in herbivore population dynamics and ecological theory. *Amer. Nat.* 125:205–38.

Rhoades, D. F., and R. G. Cates. 1976. Toward a general theory of plant anti-herbivore chemistry. In *Biochemical Interactions between Plants and Insects*, ed. J. W. Wallace and R. L. Mansell, 168–213. New York: Plenum Press.

Rochat, J., and A. P. Gutierrez. 2001. Weather mediated regulation of olive scale by two parasitoids. *J. Anim. Ecol.* 70:476–90.

Roda, A., and I. T. Baldwin. 2003. Molecular technology reveals how the induced direct defenses of plants work. *Basic Appl. Ecol.* 4:15–26.

Roitberg, B. D. 2007. Why pest management needs behavioral ecology and vice versa. *Entomol. Res.* 37:14–8.

Rosenthal, G. A. 1983. A seed-eating beetle's adaptations to a poisonous seed. *Sci. Amer.* 249:164–71.

Rosenthal, G. A., D. L. Dahlman, and D. H. Janzen. 1978. L-canavanine detoxification: A seed predator's biochemical mechanism. *Science* 202:528–9.

Rosenthal, G. A., D. H. Janzen, and D. L. Dahlman. 1977. Degradation and detoxification of canavanine by a specialized seed predator. *Science* 196:658–60.

Rostás, M., M. Simon, and M. Hilker. 2003. Ecological cross-effects of induced plant responses towards herbivores and phytopathogenic fungi. *Basic Appl. Ecol.* 4:43–62.

Rothschild, M. 1973. Secondary plant substances and warning colouration in insects. In *Insect/Plant Relationships*, ed. H. F. Van Emden, 59–83. London: Oxford Blackwell.

Rudgers, J. A., and S. Y. Strauss. 2004. A selection mosaic in the facultative mutualism between ants and wild cotton. *Proc. R. Soc. Lond. Series B* 271:2481–88.

Ryan, C. A. 1983. Insect-induced chemical signals regulating natural plant protection responses. In *Variable Plants and Herbivores in Natural and Managed Systems*, ed. R. F. Denno and M. S. McClure, 43–60. New York: Academic Press.

Sagers, C. L. 1992. Manipulation of host plant quality: Herbivores keep leaves in the dark. *Functional Ecol.* 6:741–3.

Sandberg, S., and M. Berenbaum. 1989. Leaf-tying by tortricid larvae as an adaptation for feeding on phototoxic *Hypericum perforatum*. *J. Chem. Ecol.* 15:875–85.

Salt, G. 1935. Experimental studies on insect parasitism. III. Host selection. *Proc. R. Soc. Lond. B* 117:413–35.

Saxena, R. C., and A. A. Barrion. 1985. Biotypes of the brown planthopper *Nilaparvata lugens* (Stal) and strategies in deployment of host plant resistance. *Insect Sci. Appl.* 6:271–89.

Schoonhoven, L. M. 1982. Biological aspects of antifeedants. *Ent. Exp. Appl.* 3:57–69.

Schoonhoven, L. M. 1990. Host-marking pheromones in Lepidoptera with special reference to two *Pieris*-spp. *International Symposium on Semiochemicals and Pest Control: Prospects for New Applications*, Wageningen, the Netherlands, October 16–20, 1989. *J. Chem. Ecol.* 16:3043–52.

Schowalter, T. D., W. W. Hargrove, and D. A. Crossley, Jr. 1986. Herbivory in forested ecosystems. *Annu. Rev. Entomol.* 31:177–96.

Scriber, J. M. 2002. Evolution of insect–plant relationships: Chemical constraints, coadaptation, and concordance of insect/plant traits. *Ent. Exp. Appl.* 104:217–35.

Seigler, D., and P. W. Price. 1976. Secondary compounds in plants: Primary functions. *Amer. Nat.* 110:101–5.

Siemens, D. H., H. Lischke, N. Maggiulli, S. Church, and B. A. Roy. 2003. Cost of resistance and tolerance under competition: The defense–stress benefit hypothesis. *Evol. Ecol.* 17:247–63.

Simpson, S. J., and D. Raubenheimer. 2001. The geometric analysis of nutrient–allelochemical interactions: A case study using locusts. *Ecology* 82:422–39.

Sirvent, T. M., S. B. Krasnoff, and D. M. Gibson. 2003. Induction of hypericins and hyperforins in *Hypericum perforatum* in response to damage by herbivores. *J. Chem. Ecol.* 29: 2667–81.

Slama, K. 1969. Plants as a source of materials with insect hormone activity. *Ent. Exp. Appl.* 12:721–8.

Slansky, F. 1993. Nutritional ecology: The fundamental quest for nutrients. In *Ecological and Evolutionary Constraints on Foraging*, ed. N. E. Stamp and T. M. Casey, 29–91. New York: Chapman and Hall.

Slansky, Jr., F., and A. R. Panizzi. 1987. Nutritional ecology of seed-sucking insects. In *Nutritional Ecology of Insects, Mites, Spiders and Related Invertebrates*. ed. F. Slansky Jr. and J. G. Rodriguez, 283–320. New York: John Wiley & Sons.

Smart, J., and N. F. Hughes. 1973. The insect and the plant: Progressive palaecological integration. In *Insect/Plant Relationships*, ed. H. F. Van Emden, 143–155. London: Oxford Blackwell.

Stahl, E. 1888. Planzen und Schnechen. Biologische Studie ueber die Schutzmittel der Pflanzen gegen Schneckenfrass. *Jenaische Zeits. Naturwis* 22:555–684.

Stanton, M. L., A. A. Snow, and S. N. Handel. 1986. Floral evolution: Attractiveness to pollinators increases male fitness? *Science* 232:1625–27.

Strong, D. R., J. H. Lawton, and T. R. E. Southwood (eds.). 1984. *Insects on Plants. Community Patterns and Mechanisms.* Cambridge, MA: Harvard University Press.

Sun, X.-L., Q.-Y. Yang, J. D. Sweeney, and C.-Q. Gao. 2006. A review: Chemical ecology of *Ips typographus* (Coleoptera, Scolytidae). *J. For. Res.* 17:65–70.

Takhtajan, A. 1969. *Flowering Plants, Origin and Dispersal.* Edinburgh, Scotland: Oliver and Boyd Ltd.

Tallamy, D. W. 1985. *Epilachna borealis* squash beetle feeding behavior and adaptation against induced cucurbit defenses. *Ecology* 66:157.

Thompson, J. N. 1986. Patterns in coevolution and systematics. In *Coevolution and Systematics*, ed. A. R. Stone and D. L. Hawksworth, 119–143. Oxford, UK: Oxford University Press.

Thompson, J. N. and D. Althoff. 1999. Insect diversity and the trophic complexity of communities. In *Ecological Entomology*, Second Edition, ed. C. B. Huffaker and A. P. Gutierrez, 313–53. New York: John Wiley & Sons.

Thorsteinson, A. J. 1955. The experimental study of the chemotactic basis of specificity in phytophagous insects. *Can. Entomol.* 87:49–57.

Thorsteinson, A. J. 1960. Host selection in phytophagous insects. *Annu. Rev. Entomol.* 5:193–218.

Toju, H., and T. Sota. 2006. Imbalance in predator and prey armament: Geographic clines in phenotypic interface and natural selection. *Amer. Nat.* 167:105–17.

Turlings, T., and B. Benrey. 1998. The effects of plant metabolites on the behavior and development of parasitic wasps. *Ecoscience* 5:321–33.

van Dam, N. M., J. A. Harvey, F. L. Wäckers, T. M. Bezemer, W. H. van der Putten, and L. E. M. Vet. 2003. Interactions between aboveground and belowground induced responses against phytophages. *Basic Appl. Ecol.* 4:63–77.

van den Bosch, R. 1978. *The Pesticide Conspiracy.* 2nd Edition, Berkeley, CA: University of California Press.

Verschaffelt, E. 1910. The cause determining the selection of food in some herbivorous insects. *R. Acad. Amsterdam* 13:536–42.

Walling, L. L. 2000. The myriad plant responses to herbivores. *J. Plant Growth Regul.* 19:195–216.

Wawrzynski, R. P., D. J. Hahan, and M. E. Ascerno. 2005. University of Minnesota Horticulture Extension web page. http://www.extension.umn.edu/distribution/horticulture/DG1009.html (accessed September 10, 2007).

Wellington, W. G., D. L. Johnson, and D. J. Lactin. 1999. Weather and insects. In *Ecological Entomology*, 2nd Edition, ed. C. B. Huffaker and A. P. Gutierrez, 313–53. New York: John Wiley & Sons.

Whittaker, R. H. 1972. The biochemical ecology of higher plants. In *Chemical Ecology*, ed. E. Sondheimer and J. B. Simeone, 43–70. New York: Academic Press.

Williams, A. G., and T. G. Whitham. 1986. Premature leaf abscission: An induced plant defense against gall aphids. *Ecology* 67:1619–27.

Yarnes, C. T., and W. J. Boecklen. 2005. Abiotic factors promote plant heterogeneity and influence herbivore performance and mortality in Gambel's oak (*Quercus gambelii*). *Entomol. Exp. Appl.* 114:87–95.

Zangerl, A. R. 2003. Evolution of induced plant responses to herbivores. *Basic Appl. Ecol.* 4:91–103.

Zangerl, A. R., and M. R. Berenbaum. 2005. Increase in toxicity of an invasive weed after reassociation with its coevolved herbivore. *PNAS* 102:15529–32.

Zangerl, A. R., and C. E. Rutledge. 1996. The probability of attack and patterns of constitutive and induced defense: A test of optimal defense theory. *Amer. Nat.* 147:599–608.

Zummo, G. R., J. C. Segers, and J. H. Bennedict. 1984. Seasonal phenology of allelochemicals in cotton and resistance to bollworm (Lepidoptera: Noctuidae). *Environ. Entomol.* 13:1287–90.

Zwölfer, H. 1975. Mechanismen und ergebnisse der coevolution von phytophagen und entomophagen insekten und hoheren pflanzen. *20th Phylog. Symp.*, Hamburg, pp. 7–50.

Zwölfer, H. 1982. Patterns and driving forces in the evolution of plant–insect systems. *5th Int. Symp. Plant–Insect Relationships*, Wageningen, the Netherlands, pp. 287–96.

6

Symbionts and Nutrition of Insects

Edson Hirose, Antônio R. Panizzi, and Simone S. Prado

CONTENTS

6.1 Introduction 145
6.2 External Symbionts 146
    6.2.1 Fungus-Growing Insects 146
        6.2.1.1 Ambrosia Beetles—Subfamilies Scolytinae and Platypodinae 147
        6.2.1.2 Ant Subfamily Myrmicinae—Attine...147
        6.2.1.3 Termites Subfamily Macrotermitinae...148
6.3 Internal Symbionts 148
    6.3.1 Protozoa 149
    6.3.2 Secondary Symbionts 149
    6.3.3 Symbionts in Heteroptera 150
6.4 Primary or Essential Symbionts 153
    6.4.1 *Buchnera* 153
    6.4.2 *Wigglesworthia*... 154
    6.4.3 *Blochmannia* 155
    6.4.4 *Sitophilus Oryzae* Primary Endosymbiont (SOPE) 155
6.5 Nonnutritional Symbiotic Interactions...156
6.6 Conclusions 156
References... 156

6.1 Introduction

Insects are the most successful organisms on Earth; part of this success is due to their ability to feed on a wide variety of diets (Ishikawa 2003). Many of these foods have nutritional deficiencies that, in part, are supplied by microorganisms (Tamas et al. 2002). Therefore, microorganisms affected the development and survival of insects during millions of years of evolution, either being a direct food source or providing new metabolic pathways, which allowed the spread of these organisms (Berenbaum 1988; Wernegreen 2004; Schultz et al. 2005).

The term *symbiosis* was first coined by Anton de Bary in 1879 to define an intimate association between organisms of different species, usually a host and a microorganism (Rio et al. 2003). The potential interactions between hosts and symbionts microorganism may lie anywhere between parasitism and mutualism (Moran 2006; Haine 2008). Although the symbioses represent all relationship ranging from parasitism to mutualism, the term is usually used for a relationship where there are mutual benefits to the organisms involved. Many microorganisms are involved in insect digestion of food. Diverse insect groups that thrive on low nutrient diets depend on microorganisms to help them in the process of food digestion. These include insects that feed on diets with low digestibility due to complex molecules (Breznak and Brune 1994; Cazemier et al. 2003; Suh et al. 2003), diets with nutritional deficiencies such as phloem that lack lipids and essential amino acids, and blood, which is poor in several vitamins (Dadd 1985; Rainey et al. 1995; Adams and Douglas 1997; Byrne et al. 2003).

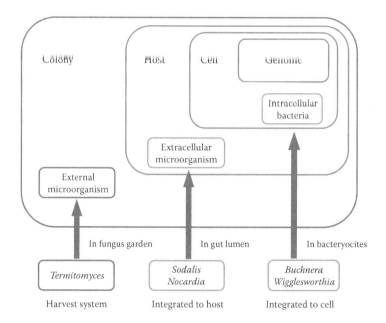

FIGURE 6.1 Relationships of insects and symbiotic microorganisms.

Other microorganisms are necessary to detoxification of plant material (Down 1989), and even in defense against parasitoid invasions and pathogen disease (Oliver et al. 2003; Dillon and Dillon 2004; Haine 2008). Microorganisms may be present inside or outside the insect body, and they maintain a complex and essential or casual relationship with the host; most of the ecological relationships between microorganisms and insects are constructive (Alves 1998).

In several orders of insects, the symbiotic nutritional relationships developed independently with different groups of microorganisms. Insects have developed farming systems where the symbiont fungus is maintained externally serving as food—ectosymbiosis. Other groups maintain close relationships carrying the symbionts—endosymbionts; these symbionts can be present in the lumen of the gut, the extracellular symbiont, or within specialized cells, the intracellular symbiont (Figure 6.1) (Douglas 1989; 1998; Stevens et al. 2001; Dillon and Dillon 2004; Wernegreen 2004).

The study of symbionts had a great momentum in the last two decades especially with the development of molecular techniques, which allowed better understanding of unknown interactions.

6.2 External Symbionts

6.2.1 Fungus-Growing Insects

Millions of years ago, insects from three different orders, Isoptera, Hymenoptera, and Coleoptera, developed the ability to cultivate specific fungi as food. Two of these groups of "farmer" insects became dependent of fungi crops and developed societies divided in castes that cooperate in complex cropping systems (Mueller and Gerardo 2002). These fungi are cultured under specific conditions, and insects regulate their growth under controlled conditions in gardens. In the absence of insects, gardens are quickly taken by microbial contaminants—this is characterized as an interdependent relationship between fungi and insects. Insects also prevent the occurrence of mites and nematodes, which are common invaders, and the contamination by spores of other fungi (Currie et al. 1999; Currie 2001; Farrell et al. 2001).

The symbiosis with fungi allowed ants, termites, and ambrosia beetles to occupy niches with abundant resources that had previously been inaccessible. With such complex interrelationships with their symbionts, these insects play an important role in their ecosystems, and in some cases are considered important pests in agricultural-forestry systems (Mueller and Gerardo 2002).

6.2.1.1 Ambrosia Beetles—Subfamilies Scolytinae and Platypodinae

A group of beetles, known as ambrosia beetles (Scolytinae and Platypodinae) bore long galleries in wood to feed and lay eggs (Cassier et al. 1996). About 3,400 ambrosia beetles species are known that developed strategies to feed and carry several types of fungal substrates. Some of these are forest species (Paine et al. 1997; van Zandt et al. 2003).

Ambrosia beetles have body compartments that vary from simple and shallow invaginations to complex structures associated with specialized glandular cells to maintain and carry fungi (Six 2003). The term *mycangia*, for example, has been applied to structures such as the double slots in thorax of *Dentroctonus frontalis* Zimmermann (Happ et al. 1971), scores on the head of *Scolytus ventralis* LeConte (Livingston and Berryman 1972), and paths in feathery arrows of *Pityoborus* spp. (Furniss et al. 1987). Some authors use the name pseudomycangia when these structures are not linked to glandular cells (Cassier et al. 1996).

The relationship between beetles and fungi is characterized as mutualistic when they feed directly from the fungus or when the fungus weakens the plant facilitating the feeding process. Insects provide proper development and efficient transmission of fungi; however, if the beetles are removed, gardens deteriorate quickly due to the excessive fungi growth that clog the galleries or spread contaminants (Wood and Thomas 1989). As termites and ants, ambrosia beetles protect fungal gardens from harmful contaminants and larvae develop on fungal diet (Beaver 1989).

Some ambrosia fungi mycelia are found only in galleries excavate by beetles, suggesting an essential association (Farrell et al. 2001). The interactions of beetles and their fungi are multifaceted and complex. They are related to insect developmental stage, vigor of the host, and composition of associated fungal flora (Paine et al. 1997). In general, this association is related to insect nutrition, with fungi changing the plant constituents and improving their assimilation.

Wood is a poor source of vitamins, sterols, and other nutrients, and fungi convert compounds into more digestible forms for the insects (Six 2003). Beetles spread fungi, which are protected from dissection (Beaver 1989). Coppedge et al. (1995) found that *D. frontalis* are larger and more fertile when they develop in the presence of symbiotic fungi, compared to insects reared in the absence of fungi. Ayres et al. (2000) showed evidence that support the theory that fungi concentrate nitrogen. When comparing two species of beetles, with and without mycangia, they demonstrated that insects without mycangia consume more phloem to get nitrogen needed for development. Other functions performed by fungal symbionts include mitigation of other fungi growth, and in some cases contribute to insect chemical communication (Hunt and Borden 1990).

6.2.1.2 Ant Subfamily Myrmicinae—Attine

About 50 to 60 million years ago, the leaf-cutting ants of the Attine tribe, which appeared in the Nearctic and Neotropical regions, acquired the ability to cultivate fungi (Bass and Cherrett 1994). Currently there are 190 known species, and each genus or species is associated to different fungi species (Mueller et al. 2001). This association is present from the beginning of the colony establishment, when the future queen takes from her parent colony a pellet of fungus inoculum, which serves as the starting point for a new garden (Mueller et al. 1998, 2001).

Although this behavior allows vertical vegetative propagation of fungi, genetic studies do not support a close correlation between ants' species and their fungi. This is because the ants occasionally replace their domesticated fungi with feral fungi and fungi from other colonies (Mueller et al. 1998; Green et al. 2002). The colony is dependent on fungi for food and the offspring is raised exclusively with fungal substrate. Therefore, ants developed the ability to cultivate fungi in underground chambers on culture substrates made of fragments of leaves and flowers. The genera *Atta* and *Acromyrmex* exclusively use leaves and fresh flowers that are transported to the nest. Recent studies increased considerably the understanding of the evolution of symbiosis between Attine ants and their fungi (Chapela et al. 1994; Mueller et al. 1998, 2001; Currie et al. 1999; Green et al. 2002).

To protect their gardens from parasitic fungi (e.g., *Escovopsis* sp.) that cause reduction in productivity and growth of the symbiotic fungus, Attine ants use antibiotics derived from bacteria, maintained in specialized regions of their own bodies (Currie et al. 1999; Currie 2001; Poulsen et al. 2003). Bacteria belong to the genus *Streptomyces*, a genus of soil bacteria that was used by the pharmaceutical industry for the discovery of modern antibiotics. Moreover, the worker ants perform the mechanical removal of contaminants and isolate areas of the gardens contaminated with other fungi (Bass and Cherrett 1994).

6.2.1.3 Termites Subfamily Macrotermitinae

There are over 2,600 species of termites, but only the subfamily Macrotermitinae family Termitidae, with approximately 330 known species, developed a symbiotic relationship with fungi of the genus *Termitomyces,* and became dependent on the fungus cultivation for food (Abe et al. 2000; Bignell and Eggleton 2000). Cultivation of fungi enabled this group of termites to turn the most important decomposers in the Old World. Termites cultivate fungi that can be found both in forests and mostly in savannas of this region (Aanen and Eggleton 2005).

The most common fungus cultivated by termites belongs to the genus *Termitomyces* (Basidiomicotina), which enable them to digest lignin. They grow on termites feces in special structures similar to a comb. This structure is maintained by the termites by continuous addition of predigested plant substrates, while the older material is consumed (Bignell and Eggleton 2000; Rouland-Lefevre 2000).

Termites forage in wood and other plant materials and eat fungi spores. These spores survive the gut passage and are deposited on the stool where the fungus grows and breaks down the plant material, allowing assimilation (Johnson et al. 1981).

According to Aanen et al. (2002), the origin of the symbiotic relationship between termites and fungi are symmetrical, with a single origin, both being dependent on that relationship. Aanen and Eggleton (2005) believe that the place of origin of this mutualism is in the rainforests of Africa, where termites of the subfamily Macrotermitinae and fungi of the genus *Termitomyces* are abundant due to high humidity and temperature. In savannas, this relationship proved to be essential—because of low humidity, the decomposition is slow and fungi are not able to develop. Termites found food in abundance, but that could only be exploited with the aid of fungi. Thus, the gardens would ensure optimal conditions for fungi development providing abundant food.

While the two main symbioses with fungi in social insects present many similar aspects, they are essentially different. The fungal symbionts of Attine ants rarely bear reproductive structures and are propagated in the vegetative form, spreading vertically by ant queens (Mueller et al. 2001; Green et al. 2002). In contrast, the symbionts of Macrotermitinae produce sexual structures that favor the horizontal acquisition of symbionts, although there are exceptions (Katoh et al. 2002).

Macrotermitini colonies of termites and Attine ants are among the most remarkable phenomena of nature. Some colonies have the volume of thousands of liters, with a complex system of chambers and galleries that can endure for decades. The study of these interrelationships shows how we can learn from these insects that feed on fungi, and this knowledge can give us new clues on how to maintain complex farming sustainable systems (Schultz et al. 2005).

6.3 Internal Symbionts

It is believed that most organisms of the class Insecta are involved in some kind of symbiosis, and most of these relationships are shared with bacteria (Rio et al. 2003); however, more complex organisms such as fungi and protozoa also may be present (Breznak and Brune 1994; Ohkuma and Kudo 1996; Brune 2003). Microbial symbionts are major evolutionary catalysts throughout the four billion years of life on Earth, shaping much of the evolution of complex organisms (McFall-Ngai 2002; Wernegreen 2004).

The primary habitat of bacteria is the digestive tract of their hosts, which contains a wide variety of nonpathogenic microorganisms that can be agents of mutualistic associations (Hackstein and Stumm 1994; Cazemier et al. 1997; Vries et al. 2001; Eichler and Schaub 2002). The study of microbial organisms is an important component in understanding the biology of insects (Dillon and Dillon 2004). Species of insects in several orders have structures modified in the digestive tract to contain and maintain these microorganisms (Douglas 1989).

The bacterial flora in the digestive tract of insect hosts most common are gram-negative and coliform (Dillon and Dillon 2004; Hirose et al. 2006). Many of these bacteria may multiply in culture media and are easily found in the environment, being casual inhabitants of the digestive tract (Hirose et al. 2006).

Culture of symbionts outside the host is one of the main factors limiting research with symbiotic associations (Wilkinson 1998). However, elimination of some of the microorganisms associated with insects has been used significantly to assess the effects of this association (Dale and Welburn 2001; Vries et al. 2001; Yusuf and Turner 2004). Although no method is used widely because of the particularities of each insect, several approaches such as heat, lysozyme, and antibiotics treatment have been successfully used to eliminate the microorganisms. For example, heat treatment is useful when the host is more tolerant to higher temperatures than the symbiont. Heat treatment works in aphids, several beetles, and stinkbugs (Montlor et al. 2002; Prado et al. 2009, 2010). However, antibiotic therapy administered orally or by injection is the most widely used method to disinfect the insects and kill their symbionts (Wilkinson 1998). On the other hand, lysozyme is in disuse due to its bad effects in host tissues (Douglas 1989).

With the advent of modern molecular techniques, the study of microbiota in insects has greatly improved, allowing the identification of species of bacteria without the need of growing in culture through the amplification of 16S ribosomal DNA (O'Neill et al. 1992; Brauman et al. 2001; Zchori-Fein and Brown 2002). The sequencing of the genome has shown similarities between the symbionts and the biochemistry machinery encoded in genes (Wernegreen 2002; Degnan et al. 2005).

6.3.1 Protozoa

In 1923, the American zoologist L. R. Cleveland first recognized that the cellulose-based food of termites was related to a mutualism association with intestinal protozoa (Slaytor 1992; Brune and Stingl 2005). The relationship is formed where the host takes advantage of the ability of the symbionts to produce enzymes that break down cellulose (O'Brien and Breznak 1984; Breznak and Brune 1994). It was believed that termites need only the enzymes produced by protozoa, but only 25% of termites (termites in the basal evolutionary scale) have protozoa in the hindgut; the other termites exhibit cellulolytic endogenous activity (Slaytor et al. 1997). Termites with protozoa also possess endogenous enzymatic activity (Slaytor 1992; Inoue et al. 1997; Watanabe and Tokuda 2001).

According to Nakashima et al. (2002), although basal termites present endogenous cellulases, enzymes from the symbiont are needed to support the metabolism of the host. This explains why basal termites, although producing endogenous enzymes, are dependent on protozoan for survival on a diet of cellulose. Some species of protozoan symbionts cannot survive when termites are fed a diet based on starch, which characterizes a dependent relationship. The diversity of protozoa found in the gut is perhaps due to the fact that different species of flagellates are specialized in other wood components in addition to cellulose (Inoue et al. 2000). Most of endoxylanases in *Reticulitermes speratus* (Kolbe) are located in the hindgut and are lost when the protozoa is removed by ultraviolet irradiation. The effects of an artificial diet in the protozoan community composition confirm that different species of flagellates are involved in the degradation of cellulose (Inoue et al. 1997).

6.3.2 Secondary Symbionts

Secondary or facultative symbionts are apparently recent habitants in insects (Chen and Purcell 1997). These microorganisms are transferred between species host and provide some benefits to the host biology, such as temperature tolerance (Chen et al. 2000; Sandström et al. 2001; Montllor et al. 2002), and increased resistance against development of parasitoids in aphids (Oliver et al. 2003). It was also suggested that these microorganisms might influence host characteristics such as susceptibility to disease, and transmission of other microorganisms such as infection by trypanosomes by *Glossina* spp. (Welburn et al. 1993). Table 6.1 shows some examples of these relationships.

These secondary or facultative symbionts may represent an intermediate stage between a free living style for a mandatory symbiosis, in which microorganisms are transmitted vertically and are essential to the host, and the parasite, in which optional mode transmission has typically been associated with

TABLE 6.1

Principal Examples of Internal Symbionts (Endosymbionts) Found in Insects

Bacteria	Host	Symbiont Function	Reference
Obligate primary symbionts			
Buchnera sp.	Acyrthosiphon pisum (Hemiptera: Aphidoidea)	Essential amino acid	Douglas 2006
	Schizaphis graminum (Hemiptera: Aphidoidea)		Tamas et al. 2002
	Baizongia pistacea (Hemiptera: Aphidoidea)		van Ham et al. 2003
Carsonella sp.	Psyllids	Amino acids	Thao et al. 2000
Trembalya sp.	Meal bugs		Baumann et al. 2005
Ascomycete fungi Clavicipitaceae	Planthoppers and a single tribe of aphids, Cerataphidini		Suh et al. 2001
Wigglesworthia	Glossina spp. (Diptera: Muscidae)	Vitamin B complex	Zientz et al. 2004
Blochmannia	Camponotus spp. (Hymenoptera: Formicidae)	Amino acids and fatty acids	Gil et al. 2003
Sitophilus oryzae Primary Endosymbiont (SOPE)	Sytophilus oryzae (Coleoptera: Curculionidae)	Vitamin and increase of enzymatic activity of mitochondria	Heddi et al. 1998
Baumannia	Homalodisca coagulate (Hemiptera: Cicadellidae)	Unknown	Moran et al. 2003
Facultative secondary symbionts			
Nocardia	Rhodillus spp. (Hemiptera: Triatomidae)	Vitamin B complex	Eichler and Schaub 2002
Sodalis	Glossina spp. (Diptera: Muscidae)	Unknown, probably nutritional	Aksoy et al. 1995
Symbiont R type	Aphids	Parasitoids resistance	Oliver et al. 2003

virulence (Fukatsu et al. 2000). Some secondary symbionts can employ similar mechanisms to intracellular parasites, overcoming the challenges to get in and share cells host, avoiding the host defense reactions, and multiplying within the cellular environment of the host (Hentschel et al. 2000). The sequencing of DNA from secondary endosymbionts identified genes that are required to pathogenicity (Dale et al. 2001, 2002). These pathways may have general utility for bacteria associated with host cells and may have evolved in the context of beneficial interactions.

6.3.3 Symbionts in Heteroptera

Many Heteroptera have appendices in the digestive tract called caeca or bacterial crypts. These are of various shapes and sizes and always house a large number of microorganisms. By now, symbionts are shown to be related to some insects of the families Plataspidae, Pentatomidae, Alydidae, Phyrrochoridae, Acanthosomatidae, Scutelleridae, Coreidae, and Parastrachiidae (Buchner 1965; Abe et al. 1995; Fukatsu and Hosokawa 2002; Kikuchi et al. 2005; Hirose et al. 2006; Prado et al. 2006; Kaltenpoth et al. 2009; Kikuchi et al. 2009; Prado and Almeida 2009a,b; Hosokawa et al. 2010; Kaiwa et al. 2010).

Buchner (1965) proposed that the symbionts present on the surface of the egg masses were vertically transmitted by the females and orally acquired by the first instars. After that, symbionts will reach and stay inside the gastric caeca. Goodchild (1978) studying *Piezosternum calidum* (F.) (Hemiptera: Pentatomidae) found evidence supporting Buchner's hypothesis that the caecum harbors symbiotic bacteria. In addition, Abe et al. (1995) confirmed the presence of symbionts in the pentatomid *Plautia stali* Scott, which were inhibited by the egg surface sterilization.

The genus *Triatoma* (Hemiptera), because of their restricted diet on blood, is dependent on symbiotic bacteria that are transmitted within the population via coprophagy. The first microorganism identified was the bacteria symbiont *Rhodococcus rhodnii*, an actinomycete, discovered in *Rhodnius prolixus* Stal (Erikson 1935). The species *Triatoma infestans* Klug, *T. sordida* (Stal), and *Panstrongylus megistus* (Burmeister) show, respectively, the symbiotic bacteria, *Nocardia* sp., *Gordini* sp., and *Rhodococcus equi* (Eichler and Schaub 2002). These endosymbionts allow their hosts to survive on restricted diets, which constitute their only food source. Thus, the symbionts provide their respective hosts with nutritional supplements such as amino acids and vitamins B complex (Buchner 1965; Nogge 1981). The loss of their symbiont result in damage to the host, such as sterility, reduced growth, and lower longevity (Nogge 1981). Aposymbiotic bugs present a series of deleterious effects such as elongation of the nymphal period, increase in mortality, and disturbances in digestion and excretion. These effects can be reduced by infection of these bugs with symbionts or by feeding on diets rich in vitamin B complex (Eichler and Schaub 1997).

Hirose et al. (2006) found that the region of the caecum in *Nezara viridula* (L.) (Hemiptera: Pentatomidae) (Figure 6.2) has low concentration of culturable bacteria; Prado and Almeida (2009a) also found a dominant caeca-associated bacterium on the egg mass surface of *N. viridula*, *Acrosternum hilare* (Say), *Murgantia histrionica* (Hahn), *Euschistus heros* (F.), *Chlorochroa ligata* (Say), *C. sayi* (Stal), *C. uhleri* (Stal), *Plautia stali* Scott, and *Thyanta pallidovirens* (Stal). Total number of bacterial colony forming units (CFU) in LB medium were at least 10^3, present in the ventricula 1 to 3 (V1 to V3) were 1,000× higher than in the caeca region of *Nezara viridula* (L.) (Hemiptera: Pentatomidae) (Figure 6.2) and presents intestinal low concentration of culturable bacteria. In many cases, the association among microorganisms and insects is casual and transient, in which microorganisms are probably derived from food ingested (Douglas 1989).

Hirose et al. (2006) found *Klebsiella pneumoniae* from *N. viridula* reared in laboratory, and it is possible that *K. pneumoniae* has been acquired by the insect through the food and has adapted to conditions of the insect in the laboratory rearing, not causing significant depletion of colony and helping to prevent the establishment of harmful microorganisms. For example, the colonization of grasshoppers (germ-free) by *Pantoea agglomerans* was favored by the presence of two native species, *K. pneumoniae* subsp. *pneumoniae* and *Enterococcus casseliflavus*. A simple inoculation with these three isolates was sufficient to establish a population that persisted for several weeks (Dillon and Dillon 2004).

By using a molecular technique, it was possible to detect the presence of a dominant bacterium closely related to *Pantoea* sp., restricted to the gastric caeca in a symbiotic relationship with insects of the Pentatomidae family and also on the egg mass surface (Hirose et al. 2006; Prado et al. 2006; Prado and Almeida 2009a).

Insects of the families Plataspidae and Acanthosomatidae present a dominant caeca-associated symbiont that form a monophyletic group, and have cospeciated with the host insects (Hosokawa et al. 2006;

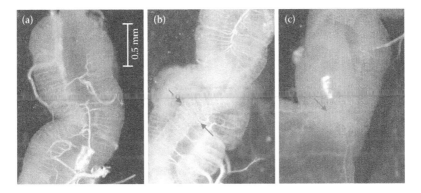

FIGURE 6.2 Details of the region of gastric caeca (V4) of the southern green stinkbug, *Nezara viridula*, formed by four rows of crypts and tracheae (silver staining tubules): (a) Proximal section; (b) Median section, arrows indicates one of the caeca row; and (c) Distal section, arrow indicates the beginning of the rectum and cecum.

Kikuchi et al. 2009). However, insects of the Pentatomidae family present a dominant bacterium that is polyphyletic, suggesting horizontal transmission and/or multiple introductions of the symbionts (Prado and Almeida 2009a).

Surface sterilization of *N. viridula*'s egg masses is able to negatively impact the maintenance of the symbionts by cleaning off the symbionts. Additionally, insects free of the symbionts, also called apo symbiotic insects, showed no impact on development and reproduction at 25°C for one generation; however, sterilized nymphs reared at 20°C had longer mean nymph developmental time and females never laid eggs (Prado et al. 2006; Prado et al. 2009). The impact of surface sterilization on the maintenance of the symbionts and in the development of *E. heros*, *Dichelops melacanthus* (Dallas), and *Pellaea stictica* (Dallas) is being evaluated. Additionally, comparisons between eggs of control insects and of surface-sterilized ones are shown by using scan electron microscopy in Figure 6.3 (Prado and Panizzi, unpublished).

Prado et al. (2010) tested the impact of temperature on the fitness of *Acrosternum hilare* and *Murgantia histrionica*, and their gut-associated symbionts showed that both stinkbug species lost their respective symbiont at 30°C. Data showed that decrease in host fitness was coupled with, and potentially mediated

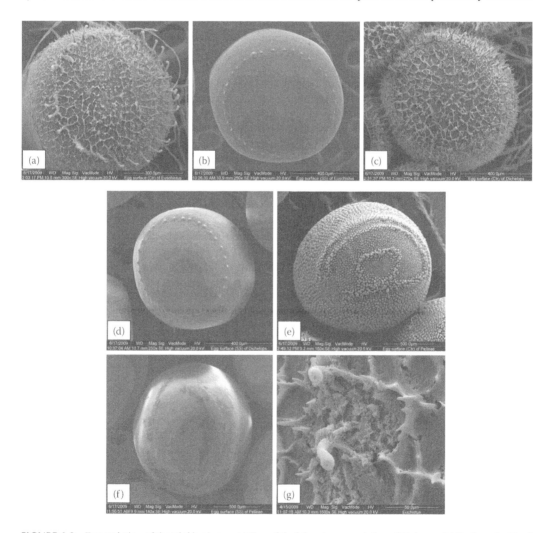

FIGURE 6.3 External view of the stinkbug's egg. (a) Egg of the laboratory population of *E. heros*. (b) Surface sterilized egg of *E. heros*. (c) Egg of the laboratory population of *Dichelops melacanthus*. (d) Surface sterilized egg of *D. melacanthus*. (e) Egg of the laboratory population of *Pellaea stictica*. (f) Surface sterilized egg of *P. stictica*. (g) Detail of the egg's surface of *E. heros*.

by, symbiont loss at 30°C, suggesting that not only egg mass sterilization but also climate changes may affect population performance of the insects directly or indirectly through mediated effects on their mutualists (Prado and Almeida 2009b; Prado et al. 2010).

While advances in our understanding of the biology of symbionts in heteropterans insects have been made, little is known about the nature of the caeca-associated symbionts of insects, and this relationship still requires further study.

It is important to note that intestinal bacteria can help with the digestion of food and produce essential vitamins, keeping potential pathogens under control (Dillon and Dillon 2004). The digestive system of insects is particularly vulnerable to attack of pathogens, parasites, and opportunistic organisms ingested with food (Lehane et al. 1997). There are other aspects of the association between insects and micro-organisms that should be considered, and it is important to recognize that many relationships between insects and microbial communities are not a simple one-to-one interaction. Another aspect is that microbial communities are dynamic through the course of interactions (Kaufman et al. 2000).

6.4 Primary or Essential Symbionts

Intracellular symbionts are especially present in Blattodea, Hemiptera, and Coleoptera (Curculionidae) (Dasch et al. 1984). It is estimated that 10% of the insects need intracellular bacteria for their development and survival (Baumann et al. 2000). Primary symbionts are essential for survival and reproduction of the host that feeds on unbalanced diets such as plant sap or blood. These symbionts primarily are within specialized host cells called bacteriocytes or mycetocytes (Baumann et al. 2000; Moran and Baumann 2000). The term *mycetocyte* was created because the first symbiotic observed was a fungus, and thus the cells that contain bacterial symbionts are more properly called bacteriocytes. However, the term *mycetocyte* is still used regardless of the symbiont that the cell harbors (Ishikawa 2003). Examples of obligate intracellular symbionts are *Buchnera* in aphids, *Wigglesworthia* in *Glossina* flies (Dale and Welburn 2001), *Blochmannia* in ants (Schröder et al. 1996; Degnan et al. 2005; Cook and Davidson 2006), *Carsonella* in psyllids (Thao et al. 2000), and *Blattabacterium* in cockroaches (Sabree et al. 2009) (Table 6.1). These bacteria live exclusively within host cells and are vertically transmitted to descendants. Molecular phylogenetic analysis demonstrated the stability of these mutualistic for long evolutionary periods, ranging from tens to hundreds of millions of years, which allowed their hosts to exploit food sources and inadequate habitats. Thus, the acquisition of these organisms can be seen as a fundamental innovation in the evolution of the host (Moran and Telang 1998). Due to the stable transmission of these symbionts from generation to generation (vertical transmission) and for long periods of time, these cytoplasmic genomes are seen as analogous to organelles (Zientz et al. 2001; Moran 2002; Wilcox et al. 2003).

6.4.1 *Buchnera*

Only insects of the order Hemiptera use phloem sap as the main or only source of food. This lifestyle has evolved several times among the Hemiptera, in Sternorrhyncha, and in Auchenorrhyncha (Dolling 1991). Because of the unbalanced nutritional quality of the sap content, all the Hemiptera that feed only on sap need symbionts (Douglas 2006).

Buchnera aphidicola is a gram-negative protobacteria that dominates aphid microbiota and represents over 90% of all microbial cells in the insect tissues. This bacterium lives inside large polyploidy cells called bacteriocytes, which are grouped into structures called bacteriomes, located adjacent to ovarioles. *Buchnera* within each *Buchnera* cell are separated from cytoplasmic contents by a membrane originating from the host cell called symbiosome membrane (Douglas 2003). These bacteria are transferred, vertically, directly from female to offspring during the blastoderm (Buchner 1965; Miura et al. 2003). In some aspects, the sap is an excellent food source, with high sugar concentrations providing an abundant source of carbon and nitrogen such as free amino acids. Besides being free of toxins and feeding deterrents, secondary compounds tend to be located in the apoplast and vacuoles of the cells (Brudenell et al. 1999; Thompson and Schulz 1999).

Sap has two major nutritional problems: nitrogen and sugar barriers that insects must overcome in order to feed on this material. The growth and fecundity of phytophagous insects are generally limited by the amount of nitrogen, the total amount of available nitrogen, or the quality of its composition. This quality question arises because animals are unable to synthesize 9 of 20 essential amino acids needed for the synthesis of proteins. If the concentration of these amino acids is low, there is a loss in protein synthesis that affects the insect development (Douglas 1998).

The relationship between the essential and nonessential amino acids is around 1:4 to 1:20 in the phloem. This relationship is considered low when compared with the ratio of 1:1 in animal proteins; consequently, the content of essential amino acids in the phloem is insufficient to support the growth of aphids (Douglas 2006). The evidence that *Buchnera* provides essential amino acids can be proved in three ways: via nutritional studies, via physiological evidence, and using genomics. The nutritional and physiological evidence depend on the development of two groups of techniques: elimination of *Buchnera* aphids with antibiotics that will generate aposymbiotic insects free of bacterial symbionts, and creation of these insects with defined diets that can be manipulated (Dadd 1985; Wilkinson 1998). Through the use of diets with amino acid deficiencies, it is possible to identify which amino acids are synthesized by *Buchnera,* and that aposymbiotic insects will develop only on diets with all essential amino acids (Douglas 1998). Additional physiological studies show that aphids with *Buchnera* can synthesize essential amino acids as precursors through sacarose and aspartate (Douglas 1989; Febvay et al. 1999; Wilkinson et al. 2001; Birkle et al. 2002). Figure 6.4 shows the essential amino acids provided by the biochemical machinery of symbiotic bacteria. Genomic evidence also shows that *Buchnera* provides essential amino acids to *Acyrthosiphum pisum* (Harris) (Shigenobu et al. 2000), *Schizaphis graminum* (Rondani) (Tamas et al. 2002), and *Baizongia pistacia* (L.) (van Ham et al. 2003). *Buchnera* in all these insects have genomes from 0.62 to 0.64 Mb (million base pairs) with 553 to 630 genes. These studies suggest that aphids overlap the barrier imposed by nitrogen supplied by the phloem. Sequencing the genome of the endosymbiotic bacterium *Buchnera aphidicola* in several aphid species revealed an extensive loss of genome (Shigenobu et al. 2000; Tamas et al. 2002; van Ham et al. 2003); however, it did not reveal the genetic basis for the interaction between bacteria and host cells. The fundamental changes that allow the incorporation of bacteria into cells host may be encoded in the host genome.

6.4.2 *Wigglesworthia*

The tsetse fly *Glossina* spp. (Diptera: Glossinidae) is an important vector of protozoa that causes sleeping sickness in humans and other diseases in animals. These insects feed exclusively on blood during all developmental stages, and the nutritional deficiencies are supplied by symbionts. Several

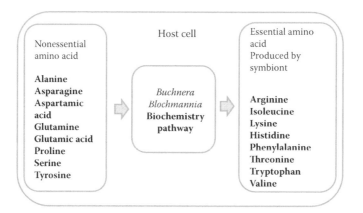

FIGURE 6.4 Interdependence between host (insect) and bacterium. The host needs the symbiont to synthesize essential amino acids, and the symbiont needs nonessential amino acids that are in the host cell cytoplasm.

microorganisms have been reported in tissues of the tsetse fly and recent discoveries have confirmed that these organisms represent three distinct associations. Aksoy (2003) confirmed that tsetse flies harbor three mainly endosymbionts representing three distinct associations. Two of these symbionts are members of Enterobacteriaceae: a symbiotic primary genus *Wigglesworthia* and a secondary symbiont of the genus *Sodalis* (Aksoy et al. 1995). The third association is reported with bacteria of the genus *Wolbachia* (Cheng et al. 2000).

It is difficult to study the functions of the symbionts in the tsetse fly; attempts to eliminate symbionts with antibiotics, lysozyme, and specific antibodies have resulted in delayed development and reduced egg production. The reproductive capacity is partially restored when aposymbiotic insects receive nutrient supplementation with vitamin B complex, suggesting that the endosymbionts are probably involved in the metabolism of these compounds (Nogge 1981). This evidence is reinforced by analyzing the genome of *Wigglesworthia;* it has the ability to synthesize various vitamins, including biotin, thiazole, lipoic acid, FAD (riboflavin, B2), folate, pantothenate, thiamine (B1), pirodixina (B6), proto-heme iron, and nicotianamine (Zientz et al. 2004). These findings reinforce the data that endosymbionts are involved in the metabolism of these compounds (Nogge 1981).

Wigglesworthia live intracellularly within specialized epithelial cells organized in bacteriomes located in the proctodeum. Those bacteria are found in the bacteriocytes cytoplasm (Chen et al. 1999). The primary symbiont does not infect the egg but is transmitted to the larva via secretion of milk glands. The tsetse fly has a viviparous reproduction; an adult female produces a single egg at a time, which develops internally. After a period of maturation and changes in the adult, the third instar larva pupates inside the mother and later the pupa is expelled (Cheng and Aksoy 1999).

6.4.3 *Blochmannia*

Blochmannia is an obligatory symbiotic bacterium associated exclusively with cells of the ants of the genera *Polyrhachis*, *Colobopsis*, and *Camponotus* (Dasch et al. 1984; Schröder et al. 1996; Sameshima et al. 1999; Sauer et al. 2002; Degnan et al. 2005). This symbiont has been studied extensively in *Camponotus*, a genus specialized in plant secretions and exudates from aphids (Davidson et al. 2004). Aphids, after digestion and assimilation of the phloem, eliminate waste rich in sugars and amino acids that are eaten by other insects. Among these, some ant species exploit this food source by collecting the drops of nectar directly from the secretions of aphids, creating a close relationship between ants and aphids (Douglas 2006).

Although the composition of nectar released by aphids is more balanced in essential amino acids compared with the phloem, it is possible that part of the amino acids acquired by ants that feed on nectar is provided by intracellular symbionts. *Blochmannia* genome retains genes that allow the biosynthesis of all nine essential amino acids and fatty acids, suggesting that the bacterium has a role in ant nutrition (Gil et al. 2003).

6.4.4 *Sitophilus Oryzae* Primary Endosymbiont (SOPE)

The primary intracellular symbiont of the genus *Sitophilus* (Coleoptera: Curculionidae) was called SOPE (*Sitophilus oryzae* primary endosymbiont) by Heddi et al. (1998). It is a gram-negative bacterium found in bacteriomes in larvae and in the ovaries of adults. This symbiont grows in number during the larval stage and accumulates in bacteriomes located at the apex of the gastric cacca. At adulthood, the number of bacteria decreases and disappears after three weeks. However, females retain these symbionts in an apical bacteriome in the ovaries and thereby transmit the symbiont to progeny. Thus, symbiosis in these weevils appears to be necessary only during the larval stage and early adulthood (Heddi 2003).

Sitophilus larvae feed mainly on albumen cereals, which have nutritional deficiencies such as pantothenic acid, biotin, and riboflavin, aromatic amino acids, phenylalanine, and tyrosine. Heddi et al. (1999) tested the possibility of SOPE to supply these nutrients; they noted differences in development between normal and aposymbiotic larvae fed diets supplemented with pantothenic acid and riboflavin. The deleterious effects were reduced in aposymbiotic larvae fed with the supplement diet.

6.5 Nonnutritional Symbiotic Interactions

An organism that has been extensively studied because of the series of changes that imposes to the host is *Wolbachia*. This protobacteria infects the reproductive organs of many arthropods; it may be transmitted horizontally and vertically by maternal transference (Noda et al. 2001; Hurst et al. 2005). Hosts infected with *Wolbachia* may suffer reproductive incompatibility, parthenogenesis, and feminization (Warren 1997; Ishikawa 2003). *Wolbachia* is generally not necessary for host survival, but in some hosts, it is an obligate symbiont (Dedeine et al. 2001). Recent estimates indicate that 66% of insect species are infected with *Wolbachia* (Hilgenboecker et al. 2008).

Another microorganism called CFB or CLO was discovered and also causes reproductive disturbances to hosts (Hunter et al. 2003). Despite the importance of these microorganisms in understanding the speciation of many arthropods (Bordenstein et al. 2001), apparently their presence does not guarantee a better performance of the host. Studies on the mechanisms of *Wolbachia* genome provided new information on how the integration occurs and the host–symbiont consequences. Since the hosts and symbionts often have different evolutionary interests, the different characteristics of insect–bacteria associations results in negotiation of genetic conflicts (Wernegreen 2004).

6.6 Conclusions

The study of symbiosis among microorganisms and insects can answer questions about the success of insects in diverse environments. Complete genomes of endosymbionts with a wide ecological and phylogenetic diversity will allow testing evolution models. A practical application in research with symbionts is the ability to manipulate bacterial symbionts in the insect vector of infectious diseases like malaria, Chagas disease, sleeping sickness, major causes of mortality mainly in developing countries (Alphey et al. 2002; Aksoy 2003). The knowledge of endosymbionts propitiates the use of genetically manipulated symbionts to control field populations of the vector insects. Additionally, there is the possibility of producing genetically modified symbionts that are incompatible with the pathogens that cause those diseases. Studies already show evidence that manipulation of the endosymbionts is a promising strategy to reduce the lifetime of the insect or limit transmission of parasites (Aksoy 2003).

To agricultural pests, the unveiling of the interrelationship of the insects with their symbionts lead to opportunities to develop sophisticated and efficient control techniques. Once the role of the symbionts on insect pests is known, their manipulation by genomic, biochemical, or conventional methods (for example, delete or select symbionts with the use of antibiotics) appears to be a real opportunity to mitigate the impact of pests on crops. Clearly, more advanced studies to achieve these goals are needed, but progress is being made. Recent studies of symbionts of insects suggest that this might be achieved in the not-too-distant future.

REFERENCES

Aanen, D. K., P. Eggleton, C. Rouland-Lefèvre, T. Guldberg-Frøslev, S. Rosendahl, and J. J. Boomsma. 2002. The evolution of fungus-growing termites and their mutualistic fungal symbionts. *Proc. Natl. Acad. Sci. USA* 99:14887–92.

Aanen, D. K., and P. Eggleton. 2005. Fungus-growing termites originated in African rain forest. *Curr. Biol.* 15:851–5.

Abe, Y., K. Mihiro, and M. Tanakashi. 1995. Symbiont of brown-winged green bug, *Plautia stali* Scott. *Jpn. J. Appl. Entomol. Zool.* 39:109–15.

Abe, T., D. E. Bignell, and M. Higashi. 2000. *Termites: Evolution, Sociality, Symbioses, Ecology.* Dordrecht, the Netherlands: Kluwer Academic Publications.

Adams, D., and A. E. Douglas. 1997. How symbiotic bacteria influence plant utilization by the polyphagous aphid, *Aphis fabae*. *Oecologia* 110:528–32.

Aksoy, S. 2003. Control of tsetse flies and trypanosomes using molecular genetics. *Vet. Parasitol.* 115: 125–45.

Aksoy, S., A. A. Pourhosseini, and A. Chow. 1995. Mycetome endosymbionts of tsetse flies constitute a distinct lineage related to *Enterobacteriaceae*. *Insect Mol. Biol.* 4:15–22.

Alphey, L., C. B. Beard, P. Billingsley, M. Coetzee, A. Crisanti, C. Curtis et al. 2002. Malaria control with genetically manipulated insect vectors. *Science* 298:119–21.

Alves, S. B. 1998. Microrganismos associados a insetos. In *Controle microbiano de insetos*, ed. S. B. Alves, 75–96. Piracicaba, Brazil: FEALQ.

Ayres, M. P., R. T. Wilkens, J. J. Ruel, M. J. Lombardero, and E. Vallery. 2000. Nitrogen budgets of phloem-feeding bark beetles with and without symbiotic fungi (Coleoptera: Scolytidae). *Ecology* 81:2198–210.

Bass, M., and J. M. Cherrett. 1994. The hole of leaf-cutting ant worker (Hymenoptera: Formicidae) in fungus garden maintenance. *Ecol. Entomol.* 19:215–20.

Baumann, P. 2005. Biology of bacteriocyte-associated endosymbionts of plant sap-sucking insect. *Annu. Rev. Microbiol.* 59:155–89.

Baumann, P., N. A. Moran, and L. Baumann. 2000. Bacteriocyte-associated endosymbionts of insects. In *The Prokaryotes*, ed. M. Dworkin. New York: Springer. Available online, http://link.springer.de/link/service/books/ 10125/.

Beaver, R. A. 1989. Insect–fungus relationships in the bark and ambrosia beetles. In *Insect–Fungus Interactions* ed. N. Wilding, N. M. Collins, P. M. Hammond, and J. F. Webber, 119–43. New York: Academic Press.

Berenbaum, M. R. 1988. Micro-organisms as mediators of intertrophic and intratrophic interactions. In *Novel Aspects of Insect–Plant Interactions*, ed. P. Barbosa, and D. K Letourneau, 91–123. New York: Wiley.

Bignell, D. E., and P. Eggleton. 2000. Termites in ecosystems. In *Termites: Evolution, Sociality, Symbiosis, Ecology*, ed. T. Abe, M. Higashi, and D. E. Bignell, 363–88. Dordrecht, the Netherlands: Kluwer Academic Publications.

Birkle, L. M., L. B. Minto, and A. E. Douglas. 2002. Relating genotype and phenotype for tryptophan synthesis in an aphid–bacterial symbiosis. *Physiol. Entomol.* 27:1–5.

Bordenstein, S. R., F. P. O'Hara, and J. H. Werren. 2001. *Wolbachia*-induced incompatibility precedes other hybrid incompatibilities in *Nasonia*. *Nature* 409:707–10.

Brauman, A., J. Doré, P. Eggleton, D. E. Bignell, J. A. Breznak, and M. D. Kane. 2001. Molecular phylogenetic profiling of prokaryotic communities in guts of termites with different feeding habits. *FEMS Microb. Ecol.* 35:27–36.

Breznak, J. A., and A. Brune. 1994. The role of microorganisms in the digestion of lignocellulose by Termitidae. *Annu. Rev. Entomol.* 39:453–87.

Brudenell, A. J. P., H. Griffiths, J. T. Rossiter, and D. A. Baker. 1999. The phloem mobility of glucosinolates. *J. Exp. Botany* 50:745–56.

Brune, A. 2003. Symbionts aiding digestion. In *Encyclopedia of Insects*, ed. V. H. Resh, and R. T Cardé, 1102–07. New York: Academic Press.

Brune, A., and U. Stingl. 2005. Prokaryotic symbionts of termite gut flagellates: Phylogenetic and metabolic implications of a tripartite symbiosis. In *Molecular Basis of Symbiosis*, ed. J. Overmann, 39–60. New York: Springer-Verlag.

Buchner, P. 1965. *Endosymbiosis of Animals with Plant Microorganisms*. New York: John Wiley.

Byrne, D. N., D. L. Hendrix, and L. H. Williams. 2003. Presence of trehalulose and other oligosaccharides in hemipteran honeydew, particularly Aleyrodidae. *Physiol. Entomol.* 28:144–9.

Cassier, P., J. Levieux, M. Morelet, and D. Rougon. 1996. The mycangia of *Platypus cylindrus* Fab., and *P. oxyurus* Dufour (Coleoptera:Platypodidae). Structure and associated fungi. *J. Insect Physiol.* 42:171–9.

Cazemier, A. E., J. H. P. Hackstein, H. J. M. Op den Camp, J. Rosenberg, and C. van der Drift. 1997. Bacteria in the intestinal tract of different species of arthropods. *Microb. Ecol.* 33:189–97.

Cazemier, A. E., J. C. Verdoes, F. A. G. Reubsaet, J. H. P. Hackstein, C. van der Drift, and H. J. M. O. den Camp. 2003. *Promicromonospora pachnodae* sp. nov., a member of the (hemi) cellulolytic hindgut flora of larvae of the scarab beetle *Pachnoda marginata*. *Anton. Leeuw.* 83:135–48.

Chapela, I. H., S. A. Rehner, T. R. Schultz, and U. G. Mueller. 1994. Evolutionary history of the symbiosis between fungus-growing ants and their fungi. *Science* 266:1691–4.

Chen, D. Q., and A. H. Purcell. 1997. Occurrence and transmission of facultative endosymbionts in aphids. *Curr. Microbiol.* 34:220–5.

Chen, X., L. Song, and S. Aksoy. 1999. Concordant evolution of a symbiont with its host insect species: Molecular phylogeny of genus *Glossina* and its bacteriome-associated endosymbiont, *Wigglesworthia glossinidia*. *J. Mol. Evol.* 48:49–58.

Chen, D. Q., C. B. Montllor, and A. H. Purcell. 2000. Fitness effects of two facultative endosymbiotic bacteria on the pea aphid, *Acyrthosiphon pisum*, and the blue alfalfa aphid, *A. kondoi*. *Entomol. Exp. Applic.* 95:315–23.

Cheng, Q., and S. Aksoy. 1999. Tissue tropism, transmission and expression of foreign genes *in vivo* in midgut symbionts of tsetse flies. *Insect Mol. Biol.* 8:125–32.

Cheng, Q., T. D. Ruel, W. Zhou, S. K. Moloo, P. Majiwa, S. L. O'Neill, and S. Aksoy. 2000. Tissue distribution and prevalence of *Wolbachia* infections in tsetse flies, *Glossina* spp. *Med. Vet Entomol.* 14:44–50.

Cook, S., and D. W. Davidson. 2006. Nutritional and functional biology of exudate-feeding ants. *Entomol. Exp. Appl.* 118:1–10.

Coppedge, B. R., F. M. Stephen, and G. W. Felton. 1995. Variation in female southern pine beetle size and lipid content in relation to fungal associates. *Can. Entomol.* 127:145–54.

Currie, C. R. 2001. A community of ants, fungi, and bacteria: A multilateral approach to studying symbiosis. *Annu. Rev. Microbiol.* 55:357–80.

Currie, C. R., J. A. Scott, R. C. Summerbell, and D. Malloch. 1999. Fungus-growing ants use antibiotic-producing bacteria to control garden parasites. *Nature* 398:701–4.

Currie, C. R., M. Poulsen, J. Mendenhall, J. J. Boomsma, and J. Billen. 2006. Coevolved crypts and exocrine glands support mutualistic bacteria in fungus-growing ants. *Science* 311:81–2.

Dadd, R. H. 1985. Nutrition: Organisms. In *Comparative Insect Physiology, Biochemistry and Pharmacology*, Vol. 4, ed. G. A. Kerkut, and L. I. Gilbert, 313–90. Oxford, UK: Pergamon.

Dale, C., and S. C. Welburn. 2001. The endosymbionts of tsetse flies: Manipulating host-parasite interactions. *Inter. J. Parasitol.* 31:627–30.

Dale, C., S. A. Young, D. T. Haydon, and S. C. Welburn. 2001. The insect endosymbiont *Sodalis glossinidius* utilizes a type III secretion system for cell invasion. *Proc. Natl. Acad. Sci. USA* 98:1883–8.

Dale, C., G. R. Plague, B. Wang, H. Ochman, and N. A. Moran. 2002. Type III secretion systems and the evolution of mutualistic endosymbiosis. *Proc. Natl. Acad. Sci. USA* 99:12397–402.

Dasch, G. A., E. Weiss, and K. P. Chang. 1984. Endosymbionts of insects. In *Bergey's Manual of Systematic Bacteriology*, ed. N. R. Krieg, 811–33. Baltimore: Williams and Wilkins.

Davidson, D. W., S. C. Cook, and R. R. Snelling. 2004. Liquid-feeding performances of ants (Formicidae): Ecological and evolutionary implications. *Oecologia* 139:255–66.

Dedeine, F., F. Vavre, F. Fleury, B. Loppin, M. E. Hochberg, and M. Bouletreau. 2001. Removing symbiotic *Wolbachia* bacteria specifically inhibits oogenesis in a parasitic wasp. *Proc. Natl. Acad. Sci. USA* 98:6247–52.

Degnan, P. H., A. B. Lazarus, and J. J. Wernegreen. 2005. Genome sequence of *Blochmannia pennsylvanicus* indicates parallel evolutionary trends among bacterial mutualists of insects. *Genome Res.* 15:1023–33.

Dolling, W. R. 1991. *The Hemiptera*. Oxford, UK: Oxford University Press.

Dillon, R. J., and V. M. Dillon. 2004. The gut bacteria of insects: Nonpathogenic interactions. *Annu. Rev. Entomol.* 49:71–92.

Douglas, A. E. 1989. Mycetocyte symbiosis in insects. *Biol. Rev.* 64:409–34.

Douglas, A. E. 1998. Nutritional interactions in insect–microbial symbioses: Aphids and their symbiotic bacteria *Buchnera*. *Annu. Rev. Entomol.* 43:17–37.

Douglas, A. E. 2003. *Buchnera* bacteria and other symbionts of aphids. In *Insect Symbiosis*, ed. K. Bourtzis, and T. A. Miller, 23–38. Boca Raton, FL: CRC Press.

Douglas, A. E. 2006. Phloem-sap feeding by animals: Problems and solution. *J. Exp. Bot.* 57:747–54.

Down, P. F. 1989. In situ production of hydrolytic detoxifying enzymes by symbiotic yeasts of cigarette beetle (Coleoptera: Anobiidae). *J. Econ. Entomol.* 82:396–400.

Eichler, S., and G. A. Schaub. 1997. The effects of aposymbiosis and of an infection with *Blastocrithidia triatomae* (Trypanosomatidae) on the tracheal system of the reduviid bugs *Rhodnius prolixus* and *Triatoma infestans*. *J. Insect Physiol.* 44:131–40.

Eichler, S., and G. A. Schaub. 2002. Development of symbionts in triatomine bugs and the effects of infections with trypanosomatids. *Exp. Parasit.* 100:17–27.

Erikson, D. 1935. The pathogenic aerobic organisms of the *Actinomyces* group. *Med. Res. Coun. Spec. Rep. Series* 203:5–61.

Farrell, B. D., A. S. Sesqueira, B. C. O'Meara, B. B. Normark, J. H. Chung, and B. H. Jordal. 2001. The evolution of agriculture in beetles (Curculionidae: Scolytinae and Platypodinae). *Evolution* 55:2011–27.

Febvay, G., Y. Rahbe, M. Rynkiewicz, J. Guillaud, and G. Bonnot. 1999. Fate of dietary sucrose and neo-synthesis of amino acids in the pea aphid, *Acyrthosiphon pisum*, reared on different diets. *J. Exp. Biol.* 202:2639–52.

Fukatsu, T., and T. Hosokawa. 2002. Capsule-transmitted gut symbiotic bacterium of the Japanese common plataspid stinkbug, *Megacopta punctatissima*. *Appl. Environ. Microbiol.* 68:389–96.

Fukatsu, T., N. Nikoh, R. Kawai, and R. Koga. 2000. The secondary endosymbiotic bacterium of the pea aphid *Acyrthosiphon pisum* (Insecta, Homoptera). *Appl. Environ. Microbiol.* 66:2748–58.

Furniss, M. M., Y. Woo, M. A. Deyrup, and T. H. Atkinson. 1987. Prothoracic mycangium on pine-infesting *Pityoborus* spp. (Coleoptera:Scolytidae). *Ann. Entomol. Soc. Am.* 80:692–6.

Gil, R., F. J. Silva, E. Zientz, F. Delmotte, and F. Gonzalez-Candelas. 2003. The genome sequence of *Blochmannia floridanus*: Comparative analysis of reduced genomes. *Proc. Natl. Acad. Sci. USA* 100:9388–93.

Green, A. M., U. G. Mueller, and M. M. Adams. 2002. Extensive exchange of fungal cultivars between sympatric species of fungus-growing ants. *Mol. Ecol.* 11:191–5.

Goodchild, A. J. P. 1978. The nature and origin of the mid-gut contents in a sap-sucking Heteropteran, *Piezosternum calidum* Fab. (Tessaratominae) and the role of symbiotic bacteria in its nutrition. *Entomol. Exp. Appl.* 23:177–88.

Hackstein, J. H. P., and C. K. Stumm. 1994. Methane production in terrestrial arthropods. *Proc. Natl. Acad. Sci. USA* 91:5441–5.

Haine, E. R. 2008. Symbiont-mediated protection. *Proc. R. Soc. B. Biol. Sci.* 275:353–61.

Happ, G. M., C. M. Happ, and S. J. Barra. 1971. Fine structure of the prothoracic mycangium, a chamber for the culture of symbiotic fungi, in the southern pine beetle *Dentroctonus frontalis*. *Tissue Cell* 3:295–308.

Heddi, A. 2003. Endosymbiosis in the weevil of the Genus *Sitophilus*: Genetic, physiological, and molecular interactions among associated genomes. In *Insect Symbiosis*, ed. K. Bourtzis, and T. A. Miller, 67–82. Boca Raton, FL: CRC Press.

Heddi, A., H. Charles, C. Khatchadourian, G. Bonnot, and P. Nardon. 1998. Molecular characterization of the principal symbiotic bacteria of the weevil *Sitophilus oryzae*: A peculiar G–C content of an endocytobiotic DNA *J. Mol. Evol.* 47:52–61.

Heddi, A., A. M. Grenier, C. Khatchadourian, H. Charles, and P. Nardon. 1999. Four intracellular genomes direct weevil biology: Nuclear, mitochondrial, principal endosymbiont, and *Wolbachia*. *Proc. Nat. Acad. Sci. USA* 96:6814–19.

Heddi, A., H. Charles, and C. Khatchadourian. 2001. Intracellular bacterial symbiosis in the genus *Sitophilus*: The "biological individual" concept revisited. *Res. Microbiol.* 152:431–7.

Hentschel, U., M. Steinert, and J. Hacker. 2000. Common molecular mechanisms of symbiosis and pathogenesis. *Trends Microbiol.* 8:226–31.

Hilgenboecker, K., P. Hammerstein, P. Schlattmann, A. Telschow, and J. H. Werren. 2008. How many species are infected with *Wolbachia*?—A statistical analysis of current data. *FEMS Microbiol. Lett.* 281:215–20.

Hirose, E., A. R. Panizzi, J. T. de Souza, A. J. Cattelan, and J. R. Aldrich. 2006. Bacteria in the gut of southern green stink bug (Heteroptera: Pentatomidae). *Ann. Entomol. Soc. Am.* 99:91–5.

Hosokawa, T., Y. Kikuchi, N. Nikoh, M. Shimada, and T. Fukatsu. 2006. Strict host–symbiont cospeciation and reductive genome evolution in insect gut bacteria. *PLoS Biol* 4:e337. doi:10.1371/journal.pbio.0040337.

Hosokawa, T., Y. Kikuchi, N. Nikoh, X.-Y. Meng, M. Hironaka, and T. Fukatsu. 2010. Phylogenetic position and peculiar genetic traits of a midgut bacterial symbiont of the stinkbug *Parastrachia japonensis*. *Appl. Environ. Microbiol.* 76: 4130–35.

Hunt, D. W. A., and J. H. Borden. 1990. Conversion of verbenols to verbenone by yeasts isolated from *Dendroctonus ponderosae* (Coleoptera: Scolytidae). *J. Chem. Ecol.* 16:1385–97.

Hunter, M. S., S. J. Perlman, and S. E. Kelly. 2003. A bacterial symbiont in the *Bacteroidetes* induces cytoplasmic incompatibility in the parasitoid wasp *Encarsia pergandiella*. *Proc. R. Soc. Lond. B. Biol. Sci.* 270:2185–90.

Hurst, G. D. D., M. Webberley, and R. Knell. 2005. The role of parasites of insect reproduction in the diversification of insect reproductive processes. In *Insect Evolutionary Ecology*, ed. M. D. E. Fellowes, G. J. Holloway, and J. Rolff, 205–29. Wallingford, UK: CABI Publishing.

Inoue, T., K. Murashima, J.-I. Azuma, A. Sugimoto, and M. Slaytor. 1997. Cellulose and xylan utilization in the lower termite *Reticulitermes speratus*. *J. Insect Physiol.* 43:235–42.

Inoue, T., O. Kitade, Y. Yoshimura, and I. Yamaoka. 2000. Symbiotic associations with protists. In *Termites: Evolution, Sociality, Symbiosis, Ecology*, ed. T. Abe, M. Higashi, and D. E. Bignell, 275–88. Dordrecht, the Netherlands: Kluwer Academic Publications.

Ishikawa, H. 2003. Insect symbiosis: An introduction. In *Insect Symbiosis*, ed. K. Bourtzis, and T. A. Miller, 1–21. Boca Raton, FL: CRC Press.

Johnson, R. A., R. J. Thomas, T. G. Wood, and M. J. Swift. 1981. The inoculation of the fungus comb in newly founded colonies of the Macrotermitinae (Isoptera) from Nigeria. *J. Nat. Hist.* 15:751–6.

Lehane, M. J., D. Wu, and S. M. Lehane. 1997. Midgut-specific immune molecules are produced by the blood-sucking insect calcitrans. *Proc. Natl. Acad. Sci. USA* 94:11502–7.

Livingston, R. L., and A. A. Berryman. 1972. Fungus transportation structures in the fir engraver, *Scolytus ventralis* (Coleoptera:Scolytidae). *Can. Entomol.* 114:174–86.

Kaiwa, N., T. Hosokawa, Y. Kikuchi, N. Nikoh, X. Y. Meng, N. Kimura, M. Ito, and T. Fukatsu. 2010. Primary gut symbiont and secondary, sodalis-allied symbiont of the scutellerid stinkbug *Cantao ocellatus*. *Appl. Environ. Microbiol.* 76: 3486–94.

Kaltenpoth, M., S. A. Winter, and A. Kleinhammer. 2009. Localization and transmission route of *Coriobacterium glomerans*, the endosymbiont of pyrrhocorid bugs. *FEMS Microbiol Ecol* 69:373–83.

Katoh, H., T. Miura, K. Maekawi, N. Shinzato, and T. Matsumoto. 2002. Genetic variation of symbiotic fungi cultivated by the macrotermitine termite *Odontotermes formosanus* (Isoptera: Termitidae) in the Ryukyu Archipelago. *Mol. Ecol.* 11:1565–72.

Kaufman, M. G., E. D. Walker, D. A. Odelson, and M. J. Klug. 2000. Microbial community ecology insect nutrition. *Am. Entomol.* 46:173–84.

Kikuchi, Y., X. Y. Meng, and T. Fukatsu. 2005. Gut symbiotic bacteria of the genus *Burkholderia* in the broad-headed bugs *Riptortus clavatus* and *Leptocorisa chinensis* (Heteroptera: Alydidae). *Appl. Environ. Microbiol.* 71:4035–43.

Kikuchi, Y., T. Hosokawa, N. Nikoh, X. Y. Meng, Y. Kamagata, and T. Fukatsu. 2009. Host–symbiont co-speciation and reductive genome evolution in gut symbiotic bacteria of acanthosomatid stinkbugs. *BMC Biol.* 7:2.

Kitano, H., and K. Oda. 2006. Self-extending symbiosis: A mechanism for increasing robustness through evolution. *Biol. Theory* 1:61–6.

McFall-Ngai, M. J. 2002. Unseen forces: The influence of bacteria on animal development. *Develop. Biol.* 242:1–14.

Miura, T., C. Braendle, A. Shingleton, G. Sisk, S. Kambhampati, and D. L. Stern. 2003. A comparison of parthenogenetic and sexual embryogenesis of the pea aphid *Acyrthosiphon pisum* (Hemiptera: Aphidoidea). *J. Exp. Zoolog. B Mol. Dev. Evol.* 295:59–81.

Montllor, C. B., A. Maxmen, and A. H. Purcell. 2002. Facultative bacterial endosymbionts benefit pea aphids *Acyrthosiphon pisum*, under heat stress. *Ecol. Entomol.* 27:189–95.

Moran, N. A. 2002. Microbial minimalism: Genome reduction in bacterial pathogens. *Cell* 108:583–6.

Moran, N. A. 2006. Symbiosis. *Curr. Biol.* 16: R866–R871.

Moran, N. A., and A. Telang. 1998. Bacteriocyte-associated symbionts of insects. *Bioscience* 48:295–304.

Moran, N. A., and P. Baumann. 2000. Bacterial endosymbionts in animals. *Curr. Opin. Microbiol.* 3:270–5.

Moran, N. A., C. Collin, H. Dunbar, W. A. Smith, and H. Ochman. 2003. Intracellular symbionts of sharpshooters (Insecta: Hemiptera: Cicadellinae) form a distinct clade with a small genome. *Environ. Microbiol.* 5:116–26.

Mueller, U. G., S. A. Rehner, and T. D. Schultz. 1998. The evolution of agriculture in ants. *Science* 281:2034–8.

Mueller, U. G., T. R. Schultz, C. Currie, R. Adams, and D. Malloch. 2001. The origin of the Attine ant–fungus symbiosis. *Quart. Rev. Biol.* 76:169–97.

Mueller, U. G., and N. Gerardo. 2002. Fungus-farming insects: Multiple origins and diverse evolutionary histories. *Proc. Natl. Acad. Sci. USA* 99:15247–9.

Nakashima, K., H. Watanabe, H. Saitoh, G. Tokuda, and J. I. Azuma. 2002. Dual cellulose-digesting system of the wood-feeding termite, *Coptotermes formosanus* Shiraki. *Insect Biochem. Mol. Biol.* 32:777–84.

Noda, H., T. Miyoshi, Q. Zhang, K. Watanabe, and K. Deng. 2001. *Wolbachia* infection shared among planthoppers (Homoptera: Delphacidae) and their endoparasite (Strepsiptera: Elenchidae): A probable case of interspecies transmission. *Mol. Ecol.* 10:2101–6.

Nogge, G. 1981. Significance of symbionts for the maintenance of an optional nutritional state for successful reproduction in hematophagous arthropods. *Parasitol.* 82:101–4.

O'Brien, R. W., and J. A. Breznak. 1984. Enzymes of acetate and glucose metabolism in termites. *Insect Biochem.* 14:639–43.

Ohkuma, M., and T. Kudo. 1996. Phylogenetic diversity of the intestinal bacterial community in the termite *Reticulitermes speratus*. *App. Envir. Microbiol.* 62:461–8.

Oliver, K. M., J. A. Russell, N. A. Moran, and M. S. Hunter. 2003. Facultative bacterial symbionts in aphids confer resistance to parasitic wasps. *Proc. Natl. Acad. Sci. USA* 100:1803–7.

O'Neill, S., R. Giordano, A. Colbert, T. Karr, and H. Robertson. 1992. 16S rRNA phylogenetic analysis of the bacterial endosymbionts associated with cytoplasmic incompatibility in insects. *Proc. Natl. Acad. Sci. USA* 89:2699–702.

Paine, T. D., K. F. Raffa, and T. C. Harrington. 1997. Interactions among scolytid bark beetles, their associated fungi, and live host conifers. *Annu. Rev. Entomol.* 42:179–206.

Poulsen, M., A. N. M. Bot, C. R. Currie, M. G. Nielsen, and J. J. Boomsma. 2003. Within-colony transmission and the cost of a mutualistic bacterium in the leaf-cutting ant *Acromyrmex octospinosus*. *Functional Ecol.* 17:260–9.

Prado, S. S., D. Rubinoff, and R. P. P. Almeida. 2006. Vertical transmission of a pentatomid caeca-associated symbiont. *Ann. Entomol. Soc. Am.* 99:577–85.

Prado, S. S., and R. P. P. Almeida. 2009a. Phylogenetic placement of pentatomid stink bug gut symbionts. *Curr. Microbiol.* 58:64–69.

Prado, S. S., and R. P. P. Almeida. 2009b. Role of symbiotic gut bacteria in the development of *Acrosterum hilare* and *Murgantia histrionica*. *Entomol. Exp. Appl.* 132:21–29.

Prado, S. S., M. Golden, P. A. Follet, M. P. Daugherty, and R. P. P. Almeida. 2009. Demography of gut symbiotic and aposymbiotic *Nezara viridula* (L.) (Hemiptera: Pentatomidae). *Environ. Entomol.* 38:103–9.

Prado, S. S., K. Y. Hung, M. P. Daugherty, and R. P. P. Almeida. 2010. Indirect effects of temperature on stink bug fitness, via maintenance of gut-associated symbionts. *App. Envir. Microbiol.* 76:1261–6.

Rainey, F. A., J. Burghardt, R. M. Kroppenstedt, S. Klatte, and E. Stackebrandt. 1995. Phylogenetic analysis of the genera *Rhodococcus* and *Nocardia* and evidence for the evolutionary origin of the genus *Nocardia* from within the radiation of *Rhodococcus* species. *Microbiol.* 141:523–8.

Rio, R. V., C. Lefevre, A. Heddi, and S. Aksoy. 2003. Comparative genomics of insect–symbiotic bacteria: Influence of host environment on microbial genome composition. *Appl. Environ. Microbiol.* 69:6825–32.

Rouland-Lefevre, C. 2000. Symbiosis with fungi. In *Termites: Evolution, Sociality, Symbiosis, Ecology*, ed. T. Abe, M. Higashi, and D. E. Bignell, 289–306. Dordrecht, the Netherlands: Kluwer Academic Publications.

Sabree, Z. L., S. Kambhampati, and N. A. Moran. 2009. Nitrogen recycling and nutritional provisioning by *Blattabacterium*, the cockroach endosymbiont. *Proc. Natl. Acad. Sci. USA* 106:19521–6.

Sameshima, S., E. Hasegawa, O. Kitade, N. Minaka, and T. Matsumoto. 1999. Phylogenetic comparisons of endosymbionts with their host ants based on molecular evidence. *Zool. Sci.* 16:993–1000.

Sandström, J. P., J. A. Russell, J. P. White, and N. A. Moran. 2001. Independent origins and horizontal transfer of bacterial symbionts of aphids. *Mol. Ecol.* 10:217–28.

Sauer, C., D. Dudaczek, B. Hölldobler, and R. Gross. 2002. Tissue localization of the endosymbiotic bacterium "*Candidatus Blochmannia floridanus*" in adults and larvae of the carpenter ant *Camponotus floridanus*. *Appl. Environ. Microbiol.* 68:4187–93.

Schröder, D., H. Deppisch, M. Obermayer, G. Krohne, and E. Stackebrandt et al. 1996. Intracellular endosymbiotic bacteria of *Camponotus* species (carpenter ants): Systematics, evolution and ultrastructural characterization. *Mol. Microbiol.* 21:479–89.

Schultz, T. R., U. G. Mueller, C. R. Curric, and S. A. Rehner. 2005. Reciprocal illumination: A comparison of agriculture of humans and in fungus-growing ants. In *Insect–Fungal Association: Ecology and Evolution*, ed. F. Vega, and M. Backwell, 149–190. New York: Oxford University Press.

Shigenobu, S., H. Watanabe, M. Hattori, Y. Sakaki, and H. Ishikawa. 2000. Genome sequence of the endocellular bacterial symbiont of aphids *Buchnera* sp. APS. *Nature* 407:81–6.

Six, D. L. 2003. Bark beetle–fungus symbiosis. In *Insect Symbiosis*, ed. K. Bourtzi, and T. A. Miller, 97–114. Boca Raton, FL: CRC Press.

Slaytor, M. 1992. Cellulose digestion in termites and cockroaches: What role do symbionts play? *Comp. Biochem. Physiol. B: Biochem. Mol. Biol.* 103:775–84.

Slaytor, M., P. C. Veivers, and N. Lo. 1997. Aerobic and anaerobic metabolism in the higher termite *Nasutitermes walkeri* (Hill). Insect Biochem. *Mol. Biol.* 27:291–303.

Stevens, L., R. Giordano, and R. F. Fialho. 2001. Male-killing, nematode infections, bacteriophage infection and virulence of cytoplasmic bacteria in the genus *Wolbachia*. *Annu. Rev. Ecol. Syst.* 32:519–45.

Suh, S.O., H. Noda, and M. Blackwell. 2001. Insect symbiosis: Derivation of yeast-like endosymbionts within an entomopathogenic filamentous lineage. *Mol. Biol. Evol.* 18:995–1000.

Suh, S. O., C. J. Marshall, J. V. McHugh, and M. Blackwell. 2003. Wood ingestion by passalid beetles in the presence of xylose-fermenting gut yeasts. *Mol. Ecol.* 12:3137–45.

Tamas, I., L. Klasson, B. Canbäck, A. K. Näslund, A. S. Eriksson, J. J. Wernegreen, J. P. Sandström, N. A. Moran, and S. G. E. Andersson. 2002. 50 million years of genomic stasis in endosymbiotic bacteria. *Science* 296:2376–9.

Thao, M. L., N. A. Moran, P. Abbot, E. B. Brennan, D. H. Burckhardt, and P. Baumann. 2000. Cospeciation of psyllids and their primary prokaryotic endosymbionts. *Appl. Environ. Microbiol.* 66:2898–905.

Thompson, G. A., and A. Schulz. 1999. Macromolecular trafficking in the phloem. *Trends Plant Sci.* 4:354–60.

van Ham, R. C. H. J., J. Kamerbeek, C. Palacios, C. Rausell, F. Abascal, U. Bastolla, J. M. Fernandez, L. Jimenez, M. Postigo, F. J. Silva, J. Tamames, E. Viguera, A. Latorre, A. Valencia, F. Moran, and A. Moya. 2003. Reductive genome evolution in *Buchnera aphidicola*. *Proc Natl. Acad. Sci. USA* 100:581–6.

van Zandt, P., V. Townsend, Jr., C. Carlton, M. Blackwell, and S. Mopper. 2003. *Loberus impressus* (LeConte) (Coleoptera: Erotylidae) fungal associations and presence in the seed capsules of *Iris hexagona*. *Coleop. Bull.* 57:281–8.

Vries, E. J., G. Jacobs, and J. A. J. Breeuwer. 2001. Growth and transmission of gut bacteria in the western flower thrips, *Frankliniella occidentalis*. *J. Inverteb. Pathol.* 77:129–37.

Warren, J. H. 1997. Biology of *Wolbachia*. *Annu. Rev. Entomol.* 42:587–609.

Watanabe, H., and G. Tokuda. 2001. Animal cellulases. *Cell. Mol. Life Sci.* 58:1167–78.

Welburn, S. C., K. Arnold, I. Maudlin, and G. W. Goday. 1993. *Rickettsia*-like organisms and chitinase production in relation to transmission of trypanosomes by tsetse flies. *Parasitology* 107:141–5.

Wernegreen, J. J. 2002. Genome evolution in bacterial endosymbionts of insects. *Nature Rev. Gen.* 3:850–61.

Wernegreen, J. J. 2004. Endosymbiosis: Lessons in conflict resolution. *PLoS Biol* 2(3):e68.

Wilcox, J. L., H. E. Dunbar, R. D. Wolfinger, and N. A. Moran. 2003. Consequences of reductive evolution for gene expression in an obligate endosymbiont. *Mol. Microbiol.* 48:1491–500.

Wilkinson, T. L. 1998. The elimination of intracellular micro-organisms from insects: An analysis of antibiotic-treatment in the pea aphid (*Acyrthosiphon pisum*). *Comp. Biochem. Physiol. A* 119:871–81.

Wilkinson, T. L., D. Adams, L. B. Minto, and A. E. Douglas. 2001. The impact of host plant on the abundance and function of symbiotic bacteria in an aphid. *J. Exp. Biol.* 204:3027–38.

Wood, T. G., and R. J. Thomas. 1989. The mutualistic association between Macrotermitinae and *Termitomyces*. In *Insect–Fungus Interactions*, eds. N. Wilding, N. M. Collins, P. M. Hammond, and J. F. Webber, 69–92. New York: Academic Press.

Yusuf, M., and B. Turner. 2004. Characterization of *Wolbachia*-like bacteria isolated from the parthenogenetic stored product pest psocid *Liposcelis bostrychophila* (Badonnel) (Psocoptera). *J. Stor. Prod. Res.* 40:207–25.

Zchori-Fein, E., and J. K. Brown. 2002. Diversity of prokaryotes associated with *Bemisia tabaci* (Gennadius) (Hemiptera: Aleyrodidae). *Ann. Entomol. Soc. Am.* 95:711–8.

Zientz, E., F. J. Silva, and R. Gross. 2001. Genome interdependence in insect–bacterium symbioses. *Genome Biol.* 2:1032.1–6.

Zientz, E., T. Dandekar, and R. Gross. 2004. Metabolic interdependence of obligate intracellular bacteria and their insect host. *Microb. Mol. Biol. Rev.* 68:745–70.

7

Bioecology and Nutrition versus Chemical Ecology: The Multitrophic Interactions Mediated by Chemical Signals

José M. S. Bento and Cristiane Nardi

CONTENTS

7.1 Introduction 163
7.2 Mechanisms for Host Search in Insects 164
7.3 Trophic Interactions Mediated by Semiochemicals 165
    7.3.1 Plant–Herbivore Interactions 165
        7.3.1.1 Constitutive Volatiles and Plant–Herbivore Interactions 166
        7.3.1.2 Induced Volatiles and Plant–Herbivore Interactions 167
        7.3.1.3 The Effect of Plant Volatiles on Insect Pheromone Emission 168
    7.3.2 Plant–Herbivore–Natural Enemy Interactions 169
        7.3.2.1 Host Searching Behavior in Parasitoids and Predators 169
        7.3.2.2 Induced Volatiles and Plant–Herbivore–Natural Enemies 169
        7.3.2.3 Extrafloral Nectar and Natural Enemy Attraction 171
7.4 Final Considerations 171
References 172

7.1 Introduction

To insects, the perception of chemical signals through long distance is highly important in the host localization process, since such signals provide precise information about them, such as developmental stage, physiological condition, and location. The efficiency in responding to such signals constitutes an important adaptive factor because besides providing access to food and supplying their nutritional needs, it may mean finding a place for mating, oviposition, and progeny survival (Dicke 2000; Bede et al. 2007).

Chemical signals used by insects are divided into allelochemicals, which include substances involved in interspecific communications and have as their main function food location and are used by both phytophagous and zoophagous species, and pheromones, which act as intraspecific signals. Allelochemicals can act as allomones, kairomones, synomones, or apneumones, depending on which organisms emit and respond to the signals (Nordlund and Lewis 1976; Dicke and Sabelis 1988). Pheromones can also play an important role in the search for hosts, acting as path markers through a food source, as aggregation stimuli, or as sexual engaging, favoring mate finding in suitable places for copulation and oviposition (Nordlund and Lewis 1976). In this context, some pheromones can act in association with allelochemicals (e.g., the synergistic action of aggregation pheromones and host plant compounds) or have their release influenced by them (Landolt and Phillips 1997; Reddy and Guerrero 2004).

Research in pheromones is extensive and includes the commercialization of innumerable synthetic components to manage several species worldwide. Over the last decades, studies have been focusing on the effects of plant allelochemicals volatiles over herbivorous, predatory, and parasitoid insects. Furthermore, researchers' interest on this subject is rising, whether regarding ecological matter and/

or new perspectives generated for the management of insects in agriculture (Karban and Baldwin 1997; Arab and Bento 2006; Turlings and Ton 2006; Cook et al. 2007).

In this chapter, we will discuss the trophic interactions mediated by volatile chemical signals emitted by plants and their role in the host searching by insects. Initially, we will analyze the mechanisms through which insects receive chemical signals and their orientation process toward the host organisms. Afterwards, we will discuss the influence of volatiles compounds over the host search in insects in the context of plant–herbivorous interaction, such as in the tritrophic and multitrophic relations involving natural enemies.

7.2 Mechanisms for Host Search in Insects

The search by insects for a nutritionally suitable host requires a sophisticated mechanism for detecting environmental signals, such as visual, auditory, tactile, and olfactory stimuli (Chapman 1998; Visser 1986).

Considering that the energy cost of traveling and searching over long distances is high, the ability and efficiency of the insect in recognizing the signal from its host are determining factors for successful localization (Dicke 2000; Bede et al. 2007). In this case, olfactory stimuli are very important due to their high capacity of being transmitted and last in the environment, and that they are relatively or highly specific and detectable in a precise way by insect receptors (Table 7.1) (Thornhill and Alcock 1983; Greenfield 2002).

Although the host searching process may vary along insect species, developmental stage, and available signals, well-defined behavioral sequences generally occur, beginning with random dispersal and localization over long distances, followed by recognition, selection, and acceptance/rejection after direct contact (Bernays and Chapman 1994; Schoonhoven et al. 2005). To perform such activities, insects must make nondirectional movements (kinesis) until the first contact is established with stimuli from the host and oriented movements (taxias) toward the food source. In this stage, the chemical molecules from the host, dispersed in the air as odor plumes, enter in contact with the sensillae of the antennae where they are absorbed and bind to the specific neurosensorial receptors. Insects immediately react by anemotaxis (i.e., maintaining a constant angle to the odor source), which promotes the activation and deactivation of receptors, orienting the insect toward the direction of the stimulus (Figure 7.1) (Chapman 1998; Greenfield 2002). In general, for this stimulus pattern to occur in the insect receptors, a zigzag movement is established by flying or walking. Other forms of signal reception may also be used such as clinotaxis and tropotaxis, based on the time or quantitative differences among signals received by the antennal sensillae (Schoonhoven et al. 2005).

After approaching the host, the insect starts to recognize and respond to concentration gradients of the chemical signals until the direct contact finally happens. In this stage, the sequence of behaviors is

TABLE 7.1

Characteristics of the Stimuli Received by Insects during Host Searching

Characteristic of Signal	Stimulus			
	Olfactive	Acoustic	Visual	Tactile/Gustatory
Distance	Long	Long	Medium	Very short
Transmission rate	Slow to rapid	Rapid	Rapid	Rapid
Power of deviation from barrier	Yes	Yes	No	No
Use in absence of light	Yes	Yes	No*	Yes
Localization of emitter	Difficult	Medium	Easy	Easy
Durability	Short to long	Instantaneous	Instantaneous	Short
Specificity	Very high	High	Low	Low to high

Source: Thornhill, R., and J. Alcock, *The Evolution of Insect Mating Systems*, Harvard University Press, Cambridge, MA, 1983.

*Except bioluminescent signals.

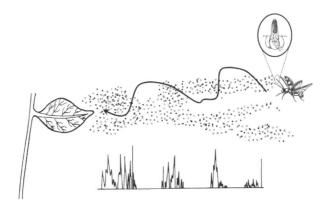

FIGURE 7.1 Schematic illustration showing the chromatographic profile of volatiles released by the plant, whose odor plume causes anemotaxia and the activation/inactivation of neurosensorial receptors in the insect. (Modified from Schoonhoven, L. M., J. J. A. Van Loon, and M. Dicke, *Insect–Plant Biology*, Oxford University Press, New York, 2005.)

defined by the chemical and physical characteristics of the host, with an exploratory evaluation through a superficial contact of the antennae, tarsi, mouthparts and/or ovipositor, before acceptance or rejection (Waldbauer and Friedman 1991; Schoonhoven et al. 2005).

7.3 Trophic Interactions Mediated by Semiochemicals

Ecosystems consist of complex trophic relationships among plants, herbivores, predators, and parasitoids. As organisms situated on the first trophic level, plants are important nutritional sources for a large number of consumers, including insects. However, herbivory can be avoided or reduced by several defense mechanisms, including the production of substances from secondary metabolism. Within the large complex of these substances produced in plants, volatile compounds are important in influencing host-searching behavior of insects, which, during their evolutionary process, have developed the ability of identifying such components and using them to perform their activities (Rhoades 1979).

Plant volatiles can be grouped into constitutive compounds, constantly produced and released by plants, and induced compounds, only synthesized after insect feeding or oviposition (Karban and Baldwin 1997). These compounds can be used by herbivores as signals for recognizing the host plant, its nutritional condition, the presence of cospecific species, competitors, or natural enemies (Bernasconi et al. 1998; Dicke and van Loon 2000; De Moraes et al. 2001; Randlkofer et al. 2007). On the other hand, organisms from other trophic levels can also recognize those chemical signals emitted by plants, orienting themselves toward the ones that signal the availability of food sources (nectar and pollen), shelter, or presence of prey, through the induced volatiles (Karban and Baldwin 1997).

Various studies have shown that plant volatiles are important sources for host searching on herbivores, predators, and parasites. These compounds have been indicated as mediators in the bitrophic, tritrophic, or multitrophic interactions, depending on the ecological context where they are analyzed (Figure 7.2) (Vet and Dicke 1992; Soler et al. 2007).

7.3.1 Plant–Herbivore Interactions

Plants secondary metabolic substances (SMSs) represent important components of their direct defense against herbivores, and include toxic substances, repellents, or deterrents (see Chapter 26). Over the time, however, herbivorous insects have developed mechanisms that allow them to overcome those barriers and exploit such compounds, increasing their adaptive success (Rhoades 1979). In this way, the recognition of plant volatiles is very important since it can determine the success of host searching and avoid contact with plants unsuitable for their needs.

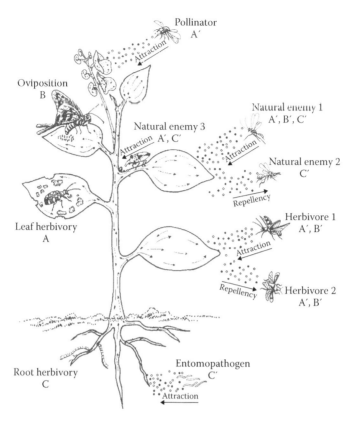

FIGURE 7.2 Interactions mediated by volatiles induced through insects leaf herbivory (A), oviposition (B), or root herbivory (C). Each induction type can cause an induced response and emission of volatiles by plants, which can interfere in
the behavior of pollinators, natural enemies, herbivores and/or entomopathogens (A', B', C'). (Modified from Kessler, A.,
and R. Halitschke, *Curr. Opin. Plant Biol.*, 10, 409–14, 2007.)

Such mechanisms for host searching and localization by herbivores can occur as a reaction to constitutive or induced volatiles, depending on the interactions involved and the signals available in the
environment.

7.3.1.1 Constitutive Volatiles and Plant–Herbivore Interactions

The constitutive plant volatiles have already been widely studied as signals for herbivore host searching.
Among these compounds are those that have a defensive action against some insect species (repellents)
and those acting as signals for the presence of nutritional resources (attractive) for phytophagous, frugivorous, poliniphagous, and so forth.

For herbivorous insects, the identification of the chemical signals of the host plant can be based
on isolated components or in compounds with distinct proportions (blend) common to various plant
groups. According to Bruce et al. (2005), the recognition of such volatiles blends occurs in most herbivores since the specific mixtures of various compounds are more efficient in their receptors. For *Cydia
molesta* (Busck) (Lepidoptera: Tortricidae), for example, a mixture of three components of the peach
tree ((Z)-acetate-3-hexen-1-ila; (Z)-3-hexen-1-ol; benzaldehyde) in a proportion of 4 : 1 : 1 was attractive
to females while the components tested individually were not (Natale et al. 2003).

In many cases, plants toxic to generalist insects emit volatiles that act as attractants to specialist herbivores that are capable of feeding on these plants and using them for their own benefit. For this, some
insects have the capacity to metabolize and transform toxins into nontoxic components or even sequester
such substances and use them to synthesize pheromones (Hartmann 1999; Nishida 2002). In this case,

specific volatiles from these plants are recognized and attract specialist insects but are avoided by the generalists. For *Tyria jacobaeae* L. (Lepidoptera: Arctiidae), for example, there is the recognition and orientation through the volatiles of *Senecio jacobea* L. (Asteraceae), which has pyrrolizidine alkaloids in its tissues, important food resources for the defense of the herbivore against natural enemies (van Dam et al. 1995; Hägele and Rowell-Rahier 1999). Adults of *Cisseps fulvicollis* Hubner (Lepidoptera: Arctiidae) are attracted to plant volatiles that contain pyrrolizidine alkaloids in their tissues, and after feeding sequester these compounds and use them as biochemical precursors for pheromone production (Hartmann and Ober 2000).

For generalists and those less adapted to toxic plant compound herbivores, the efficiency on recognizing and avoiding such compounds is extremely important to their survival. In this context, the plant volatiles act as allomones, warning the insects about the presence of components unsuitable for their survival or reproduction. An example is the isothiocyanates emitted by the Brassicaceae, which are highly repellent to the aphids *Phorodom humuli* (Schrank) and *Aphis fabae* Scopoli (Hemiptera: Aphididae), herbivores that do not use these plants as hosts (Notthinghan et al. 1991). Plants that are repellent to herbivores can also be used in agricultural situations as a manner of avoiding insect pests. Recent studies have demonstrated that *Diaphorina citri* Kuwayama (Hemiptera: Psyllidae), an important citrus pest, is repelled by volatiles from guava (Noronha and Bento, unpublished).

For herbivores, volatiles from flowers and extrafloral nectaries can also influence host searching since they indicate the presence of important nutritional resources, increasing herbivore longevity and reproductive potential (see revision in Wäckers et al. 2007). Several studies have been performed to characterize the interactions of plants, herbivores, and natural enemies mediated by the presence of pollen, floral, and extrafloral nectar in plants (Wäckers and Wunderlin 1999; Heil et al. 2001).

7.3.1.2 Induced Volatiles and Plant–Herbivore Interactions

In response to feeding or oviposition by phytophagous insects, plants activate metabolic pathways, resulting in the production of volatile chemical compounds qualitatively and quantitatively different from those released in the absence of any damage (Karban and Baldwin 1997). These responses induced by insects can be restricted to the damaged area (green leaf volatiles) or systemic, in which herbivore action on leaves, flowers, roots, or branches causes emission of volatiles from the whole plant (Dicke and van Loon 2000; van Dam et al. 2003; Turlings and Wäckers 2004). In the systemic response, induced components are transported by the phloem, altering the production and emission of compounds in all plant tissues (Mckey 1979; Karban and Baldwin 1997).

The induced plant volatiles include alcohols with six carbons, monoterpenes, sesquiterpenes, and other compounds derived from complex biochemical processes. Although some of these compounds are common to various plant species, the group of compounds released may vary according to genotypic characteristics, age, plant tissue, induction mechanisms, and herbivore species (Turlings et al. 1998; Ferry et al. 2004).

In the last few years, the induction by herbivores has been widely studied and the biochemical processes involved have already been clarified for various plant–insect systems (Turlings et al. 1990; Paré and Tumlinson 1997). According to these studies, the biochemical processes that trigger the induction are variable and depend on the organisms involved in the interaction, probably involving a considerable degree of plant–insect specificity. This way, there are already systems known in which the differential emission of compounds depends on the herbivore species that feeds on the plant (Paré and Tumlinson 1997; De Moraes et al. 1998). This specificity is due to enzymes present in the salivary secretion or oviposition fluids of some herbivore species, which enter into contact with specific receptors of the plant tissues and activate biochemical pathways for the production of specific compounds (Alborn et al. 2000). This process was first demonstrated by the action of the compound β-glucosinolate present in the salivary secretion of *Pieris brassicae* L. (Lepidoptera: Pieridae) and the fatty acid volicitin (N-(17-hidroxylinoleniol)-L-glutamine) in the salivary secretion of *Spodoptera exigua* (Hubner) (Lepidoptera: Noctuidae), which activate the production of specific volatiles in Brassicaceae and *Zea mays* L. (Poaceae), respectively (Mattiaci et al. 1995; Turlings et al. 2000). Later, other fatty acids were isolated from oral secretions of other species (Paré et al. 1998; Pohnert et al. 1999; Alborn et al. 2000, 2003; Halitschke et al. 2001).

The induction of volatile production in plants can also occur with less specificity, where herbivory or oviposition induces the action of hormones from the plant (e.g., jasmonic acid, salicylic acid, and ethylene), eliciting the production of the volatiles (Farmer and Ryan 1990; McConn et al. 1997; Ryan and Pearce 2001). In this case, although specificity has not been proved, studies demonstrate that the pattern of compounds released by induced plants is distinct from those emitted by healthy plants, and this difference distinctly affects plant interactions with associated insects (Hilker and Meiners 2002).

Studies involving induction of volatiles in plants have revealed the existence of even more complex interactions between organisms. Besides signaling the location of food for herbivores, the induced volatiles can also indicate the presence of competing organisms and the consequent reduction in plant quality (Bernasconi et al. 1998; Dicke 2000). The selection of an already induced plant can also means exposure to the harmful effects of defensive substances, either directly against the herbivore or indirectly by attracting natural enemies (Carrol et al. 2006).

The induced volatiles of plants can also attract and repel herbivore insects (Dicke and van Loon 2000). In some Aphidae, for example, the induction of plant volatiles is generally repellent, whereas in *Oreina cacaliae* (Schrank) and *Phyllotreta cruciferae* (Goeze) (Coleoptera: Chrysomelidae) they are attractive (Kalberer et al. 2001; Turlings and Wäckers 2004).

Up to now, studies have shown that insect response to induced volatiles can vary according to sex, physiological state, and time after induction, as well as the circadian rhythms of the plants and insects involved in the interaction. Arab et al. (2007) demonstrated that the induced volatiles of *Solanum tuberosum* L. (Solanaceae) are attractive to mated females of *P. operculella*. On the other hand, virgin females cannot differentiate volatiles from healthy or induced plants. These results indicate an alteration in searching behavior according to the insect physiological state. De Moraes et al. (2001) observed that induced *Nicotiana tabacum* L. (Solanaceae) plants emit specific volatiles during the night, which act as repellents for nocturnal females of *Heliothis virescens* F. (Lepidoptera: Noctuidae), preventing them of laying eggs in induced plants. The authors suggest that this repellence may be signaling the presence of competitors or natural enemies to *H. virescens*. Similarly, the response of many insects to induced volatiles can be used to identify oviposition sites since females tend to lay eggs in plants adequate for the survival of their offspring (Randlkofer et al. 2007).

7.3.1.3 The Effect of Plant Volatiles on Insect Pheromone Emission

Besides the isolated role of plant volatiles in the process of host searching in herbivores, these compounds can also influence insect pheromone emission and reception and, consequently, the combined action of these semiochemicals volatiles results in a behavioral response differentiated from that caused by isolated components (Dicke 2000; Reddy and Guerrero 2004).

In general, aggregation pheromones are among the most important pheromones for herbivores host searching and location because they are constantly associated with finding sites for feeding, mating, and oviposition (see revision in Landolt and Phillips 1997). There are insect species which are stimulated to release aggregation pheromones when they get in contact with the constitutive volatiles of the host plant, as happens with *Rynchophorus phoenicis* F. (Coleoptera: Curculionidae) and *Elaeis quineensis* (Jacq.) (Arecaceae) (Jaffé et al. 1993). Additionally, in many insect species, the synergistic effect of aggregation pheromones and induced plant volatiles increases the possibilities of herbivores successfully locate hosts. Dickens (1989) found an increase in the attraction of males and females of *Anthonomus grandis* Boheman (Coleoptera: Curculionidae) in response to the combined action of induced volatiles from the cotton plant (trans-2-hexenol, cis-3-hexenol or 1-hexenol) and aggregation pheromone compounds. A similar response was found in *Melolontha melolontha* L. (Coleoptera: Scarabaeidae), produced by the association of aggregation pheromone and induced plant volatiles (Reinecke et al. 2002). The increase in behavioral responses by herbivores was also verified as consequence of the association of induced plant volatiles and sexual pheromone, such as in *Plutella xylostella* (L.) (Lepidoptera: Plutellidae) and *Helicoverpa zea* (Boddie) (Lepidoptera: Noctuidae). In these species, the production of sexual pheromone by females occurs after contact with the host plants. Males are also more attracted by traps containing an association of sexual pheromone and plant allelochemicals volatiles than by those containing the compounds by themselves (Reddy et al. 2002). This increase on the attraction of cospecifics may be adaptively advantageous for herbivores since precision in locating food and individuals of the opposite sex are increased, mitigating the energy use for such activities. On the other hand, considering that the herbivores' natural enemies easily recognize these compounds, the significant increase in attraction may have negative implications for the final ability of the herbivore emitting the pheromone.

In contrast to allelochemicals, which increase the possibility of communication between cospecifics herbivores and the location of the host plant, there are plant volatiles whose action is antagonistic to pheromones, reducing or annulling their action. Haynes et al. (1994) demonstrated the reduction in pheromone responses by *Dendroctonus frontalis* Zimmermann (Coleoptera: Scolytidae), due to its association with constitutive compounds (4-alil anisol) emitted by *Pinus taeda* L. (Pinaceae).

Among the induced volatiles that inhibit pheromone action are green leaf volatiles. This effect was demonstrated in plants attacked by herbivores emitting compounds that reduce aggregation pheromone efficiency in various scolytid species (Dickens et al. 1992; De Groot and MacDonald 1999; Poland et al. 1998; Poland and Haack 2000; Huber and Borden 2001).

7.3.2 Plant–Herbivore–Natural Enemy Interactions

7.3.2.1 *Host Searching Behavior in Parasitoids and Predators*

Parasitoids and predators of herbivorous insects base their host searching on volatiles both from their own prey as well as from other sources associated with them, such as microorganisms or host plants. Considering that natural enemies are exposed to an enormous quantity of volatiles, the use of chemical signals that give a high degree of detectability and reliability is needed, and they influence the search and adaptation processes of these insects.

Chemical signals emitted by the prey are highly reliable because they provide precise information about location and abundance. However, these volatiles show low detectability over long distances. Therefore, to optimize the perception of chemical signals available in the environment, natural enemies use three distinct mechanisms: (1) the diversion of semiochemicals using volatiles indirectly related to the host (e.g., pupal parasitoid identifies larval volatiles); (2) associative learning, relating easy-to-detect stimuli to reliable stimuli but with low detectability; and (3) response to stimuli created by the specific interaction between the herbivore-prey and its host plant (Price et al. 1980; Vet and Dicke 1992).

Considering that plant volatiles are stimuli directly related to herbivores, these compounds acquire a significant importance in the searching process by natural enemies being detected at long distances. Also, volatiles emitted by plants after induction can increase the reliability of signals, supplying additional information to natural enemies, such as location, abundance, and developmental stage, among others (see revision in Vet and Dicke 1992; Karban and Baldwin 1997).

In parasitoids and predators, the degree of specificity in relation to the diet determines significant behavioral differences in searching for hosts and in the use of chemical signals (Figure 7.3). Specialist natural enemies may be associated with herbivores that exploit a single plant species, and thus both the constitutive and the induced plant volatiles can be used, being that the parasitoid responses to such compounds are maximized. Since parasitoids are more specific than predators, and this affects the differential exploitation of chemical signals, the response to the volatiles is also less specific for predators. Therefore, it is likely that these generalists use the information from their hosts more extensively.

7.3.2.2 *Induced Volatiles and Plant–Herbivore–Natural Enemies*

In the last few years, various studies have demonstrated that herbivory or oviposition of herbivorous insects induces the local or systemic production of plant volatiles that act as attractants to the herbivores' natural enemies (indirect defense) (De Moraes et al. 1998). These compounds are important for parasitoids and predators to locate hosts since they give specific and reliable information about the prey presence (Vet and Dicke 1992; Karban and Baldwin 1997).

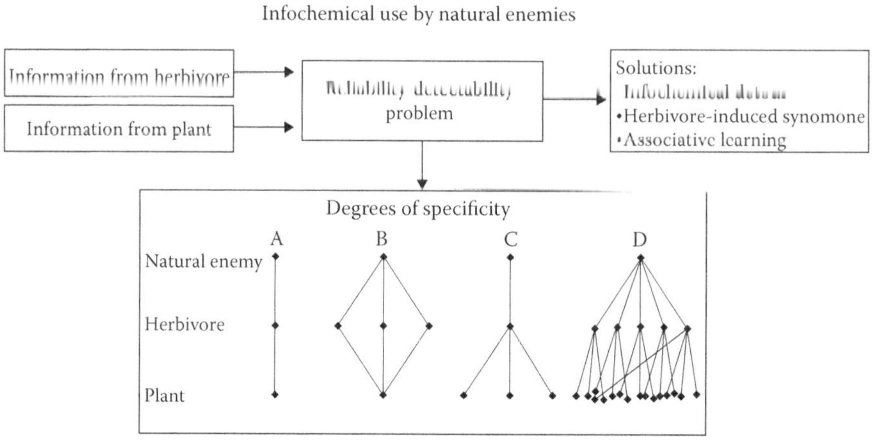

Infochemical use by natural enemies

FIGURE 7.3 Factors involved in the host searching process by parasitoids (A, B) and predators (C, D). (Modified from Vet, L. E. M., and M. Dicke, *Annu. Rev. Entomol.*, 37, 141–72, 1992.)

To predatory insects, the use of induced plant volatiles on host searching increase predation potential as well as reduce energy costs (Vet and Dicke 1992; Dicke and Vet 1999). However, studies on this are scarce for parasitoids.

The natural enemies' search for hosts consists of a series of behaviors that are affected by the chemical information available in the environment. Although many of these insects use compounds from herbivores or microorganisms associated with the prey or its habitat, plant volatiles are the main sources of information for host searching (Lewis et al. 1990; Vet et al. 1990; Karban and Baldwin 1997).

The identification of plant volatiles by parasitoids occurs through the recognition of specific blends, and many are able to distinguish between healthy and induced plants, orienting themselves toward those where their prey are. Studies by Turlings et al. (1991) demonstrated that volatiles emitted by maize plants in response to *S. exigua* feeding were used by its parasitoid *Cotesia marginiventris* (Cresson) (Hymenoptera: Braconidae) to find the host.

Induced plant volatiles are released where the damage occurred (green leaf volatiles), or in most cases, systemically. Thus, the damage caused to a plant part induces the production of volatiles in all its tissues (De Moraes et al. 1998). Besides this, plant responses to herbivores (feeding or oviposition) can be distinct, depending on the degree of specificity of the interactions involved and the biochemical routes elicited by induction (Hilker and Meiners 2002). Therefore, if plants respond in a differential manner to distinct forms of induction and herbivore species, blends are produced for each case and the chemical signals can relay specific information to natural enemies, such as the species or the developmental stage of the herbivore on the plant (Turlings and Wäckers 2004).

In *Cotesia kariyai* (Watanabe) (Hymenoptera: Braconidae), the attraction only occurs due to the volatiles induced by recently ecloded larvae of *Pseudaletia separata* Walker (Lepidoptera: Noctuidae) (Takabayashi et al. 1995) herbivory. Similarly, the egg parasitoid *Oomyzus gallerucae* (Fonscolombe) (Hymenoptera: Eulophidae) is attracted by volatiles emitted only after oviposition by *Xanthogaleruca luteola* Muller (Coleoptera: Chrysomelidae) on *Ulmus minor* (Ulmaceae).

In some cases, the high degree of specificity of the plant–herbivore–parasitoid interactions is due to the induction mechanism that elicits volatile emissions. Therefore, the presence of enzymes in the salivary secretion or oviposition fluid of herbivores can elicit specific pathways in the plant and the liberation of the resulting compounds exclusively from such interactions (Turlings and Wäckers 2004) (Figure 7.4). De Moraes et al. (1998) demonstrated that the parasitoid *Cardiochiles nigriceps* Viereck (Hymenoptera: Braconidae) can specifically recognize induced plant volatiles after herbivory by its prey, *H. virescens*. For the tritrophic interaction, *Microplitis croceipes* (Hymenoptera: Braconidae), *S. exigua,* and *Z. mays,* Turlings et al. (2000) proved that the substance volicitin, present in the salivary secretion of the herbivore, triggered the emission of specific volatiles, attracting the parasitoid.

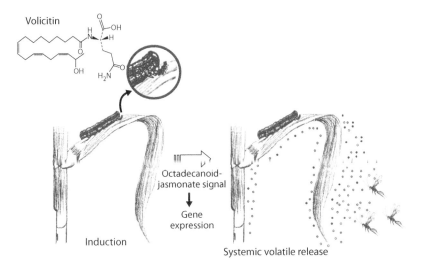

FIGURE 7.4 Induction in maize caused by the oral secretion of caterpillars and the systemic emission of volatiles that attract the specific parasitoids of these herbivores.

7.3.2.3 *Extrafloral Nectar and Natural Enemy Attraction*

The nutritional resources from extrafloral nectar are important for parasitoid and predator survival. Some systems are known in which the emission of extrafloral nectar increases visits to the plant by herbivore natural enemies (see revision in Turlings and Wäckers 2004). Such volatiles emitted by plants act as sinomones, supplying additional sources of nutrients to parasitoids and predators, which benefit from the easier location of their prey.

The production of extrafloral nectar is constitutive in plants but may be increased after its exploitation by consumers or in response to mechanical damage to the plant, caused or not by herbivory (Wäckers and Wunderlin 1999; Heil et al. 2001). These responses can trigger an increase in the volume of nectar produced or the alteration of its qualitative characteristics (e.g., increase in amino acid concentration) (Del-Claro and Oliveira 1993). Kost and Heil (2005) proved that an increase in extrafloral nectar availability attracted more predators (ants, wasps, and flies) and parasitoids (Chalcidoidea). For the parasitoid *Microplitis croceipes* (Cresson) (Hymenoptera: Braconidae), the resources offered by extrafloral nectar increase female longevity and reproduction, and it is the sole nutritional source for this insect when flower volatiles are unavailable. Behavioral observations by Röse et al. (2006) revealed that this parasitoid searches for hosts based on volatiles from the extrafloral nectar of cotton plants. The presence of this resource also increased the time the insects remained on the plant, which may result in an increase in the parasitism of the larvae of *H. virescens* and other noctuids.

7.4 Final Considerations

Over the last decades, the significant increase in studies on chemical ecology has made the use of semiochemicals such as pheromones and plant volatiles possible in behavioral insect control. The findings that plants attacked by herbivores can react by activating their indirect defenses and alerting predators and parasitoids about the presence of their prey, has resulted in a growing interest by researchers. They have investigated biochemical mechanisms and the ecological consequences of such interactions, as well as the implications and perspectives of using these compounds in agriculture (Turlings and Ton 2006). According to Karban and Baldwin (1997), artificial induction by applying inductive defense substances on plants may be one of the strategies for increasing the herbivore repellent potential or increasing the attraction to its natural enemies. Additionally, the molecular mechanisms involved in the induction of plant volatiles has indicated the possibility of developing varieties with a greater defensive potential, which can express such characteristic either constantly or inductively (Agelopoulos 1999; Turlings and Ton 2006). However, there is a consensus that many studies are still necessary for establishing an effective field strategy (Karban and Baldwin 1997; Turlings and Ton 2006; Arab and Bento 2006; Cook et al. 2007).

REFERENCES

Agelopoulos, N., M. A. Birkett, A. J. Hick, A. M. Hooper, J. A. Pickett, E. M. Pow, L. E. Smart, D. W. M. Smiley, L. J. Wadhams, and C. M. Woodcock. 1999. Exploiting semiochemical in insect control. *Pestic. Sci.* 55:225–35.

Alborn, H. T., M. M. Brennan, and J. H. Tumlinson. 2003. Differential activity and degradation of plant volatile elicitors in regurgitant of tobacco hornworm (*Manduca sexta*) larvae. *J. Chem. Ecol.* 29:1357–72.

Alborn, H. T., T. H. Jones, G. S. Stenhagen, and J. H. Tumlinson. 2000. Identification and synthesis of volicitin and related components from beet armyworm oral secretions. *J. Chem. Ecol.* 26:203–20.

Arab, A., and J. M. S. Bento. 2006. Plant volatiles: New perspectives for research in Brazil. *Neotropical Entomol.* 35:151–8.

Arab, A., J. R. Trigo, A. L. Lourenção, A. M. Peixoto, F. Ramos, and J. M. S. Bento. 2007. Differential attractiveness of potato tuber volatiles to *Phthorimaea operculella* (Gelechiidae) and the predator *Orius insidiosus* (Anthocoridae). *J. Chem. Ecol.* 33:1845–55.

Bede, J. C., J. N. McNeil, and S. S. Tobe. 2007. The role of neuropeptides in caterpillar nutritional ecology. *Peptides* 28:185–96.

Bernasconi, M. L., R. C. J. Turlings, L. Ambrosetti, P. Bassetti, and S. Dorn. 1998. Herbivore-induced emissions of maize volatiles repel the corn leaf aphid, *Rhopalosiphum maidis. Entomol. Exp. Appl.* 87:133–42.

Bernays, E. A., and R. F. Chapman. 1994. *Host-Plant Selection by Phytophagous Insects.* New York: Chapman & Hall.

Bruce, T. J. A., L. J. Wadhams, and C. M. Woodcock. 2005. Insect host localization: A volatile situation. *Trends Plant Science* 10:269–74.

Carrol, M. J., E. A. Schmetz, R. L. Meagher, and P. E. A. Teal. 2006. Attraction of *Spodoptera frugiperda* to volatiles from herbivore-damaged maize seedlings. *J. Chem. Ecol.* 32:1911–24.

Cook, A. M., Z. R. Kahan, and J. A. Pickett. 2007. The use of push–pull strategies in integrated pest management. *Annu. Rev. Entomol.* 52:57–80.

Chapman, R. F. 1998. *The Insects: Structure and Function.* Cambridge, UK: Cambridge University Press.

De Groot, P., and L. M. Macdonald. 1999. Green leaf volatiles inhibit response of red pine cone beetle *Conophthorus resinosae* (Coleoptera: Scolytidae) to a sex pheromone. *Naturwissenschaften* 86:81–5.

De Moraes, C. M., M. C. Mescher, and J. H. Tumlinson. 2001. Caterpillar-induced nocturnal plant volatiles repel conspecific females. *Nature* 410:577–80.

De Moraes, C. M., W. J. Lewis, P. W. Paré, H. T. Alborn, and J. H. Tumlinson. 1998. Herbivore-infested plants selectively attract parasitoids. *Nature* 393:570–3.

Del-Claro, K. and P. S. Oliveira. 1993. Ant–homoptera interaction: Do alternative sugar sources distract tending ants? *Oikos* 68:202–6.

Dicke, M., and M. W. Sabelis. 1988. Infochemical terminology: Based on cost–benefit analysis rather than origin of compounds? *Funct. Ecol.* 2:131–9.

Dicke, M., and L. E. M. Vet. 1999. Plant–carnivore interactions: Evolutionary and ecological consequences for plant, herbivore and carnivore. In *Herbivores: Between Plants and Predators,* ed. H. Olff, V. K. Brown, and R. H. Drent, 483–520. Oxford, UK: Blackwell Science.

Dicke, M. 2000. Chemical ecology of host-plant selection by herbivorous arthropods: A multitrophic perspective. *Biochem. Syst. Ecol.* 28:601–17.

Dicke, M., and J. A. Van Loon. 2000. Multitrophic effects of herbivore-induced plant volatiles in an evolutionary context. *Entomol. Exp. Appl.* 97:237–49.

Dickens, J. C., R. F. Billings, and T. L. Payne. 1992. Green leaf volatiles interrupt aggregation pheromone response in bark beetles infesting southern pines. *Experientia* 48:523–4.

Dickens, J. C. 1989. Green leaf volatiles enhance aggregation pheromone of boll weevil, *Anthonomus grandis. Entomol. Exp. Appl.* 52:191–203.

Farmer, E. E., and C. A. Ryan. 1990. Interplant communication: airborne methyl jasmonate induces synthesis of proteinase-inhibitors in plant-leaves. *Proc. Natl. Acad. Sci. USA* 87: 7713–6.

Ferry, N., M. G. Edwards, J. A. Gatehouse, and A. M. R. Gatehouse. 2004. Plant–insect interactions: Molecular approaches to insect resistance. *Curr. Opin. Biotechnol.* 15:155–61.

Greenfield, M. D. 2002. *Signalers and Receivers: Mechanisms and Evolution of Arthropod Communication*. New York: Oxford University Press.

Hägele, B. F., and M. Rowell-Rahier. 1999. Dietary mixing in three generalist herbivores: Nutrient complementation or toxin dilution? *Oecologia* 119:521–33.

Halitschke, R., U. Schittko, G. Pohnert, W. Boland, and I. T. Baldwin. 2001. Molecular interactions between the specialist herbivore *Manduca sexta* (Lepidoptera, Sphingidae) and its natural host *Nicotiana attenuata*. III. Fatty acid-amino acid conjugates in herbivore oral secretions are necessary and sufficient for herbivore-specific plant response. *Plant Physiol.* 125:711–7.

Hartmann, T. 1999. Chemical ecology of pyrrolizidine alkaloids. *Planta* 207:483–95.

Hartmann, T., and D. Ober. 2000. Biosynthesis and metabolism of pyrrolizidine alkaloids in plants and specialized insect herbivores. *Top. Curr. Chem.* 209:207–43.

Haynes, J. L., B. L. Strom, L. M. Roton, and L. L. Ingram. 1994. Repellent properties of the host compound 4-allylanisole to the southern pine beetle. *J. Chem. Ecol.* 20:1595–615.

Heil, M., T. Koch, A. Hilpert, B. Fiala, W. Boland, and K. E. Linsenmair. 2001. Extrafloral nectar production of the ant-associated plant, *Macaranga tanarius*, is an induced, indirect, defensive response elicited by jasmonic acid. *Proc. Natl. Acad. Sci. USA* 98:1083–8.

Hilker, M., and T. Meiners. 2002. Induction of plant responses towards oviposition and feeding of herbivorous arthropods: a comparison. *Entomol. Exp. Appl.* 104:181–92.

Huber, D. P. W., and J. H. Borden. 2001. Angiosperm bark volatiles disrupt response of Douglas-fir beetle, *Dendroctonus pseudotsugae*, to attractant-baited traps. *J. Chem. Ecol.* 27:217–33.

Jaffé, K., P. Sanchez, H. Cerda, R. Hernandez, N. Urdaneta, G. Guerra, R. Martinez, and B. Miras. 1993. Chemical ecology of the palm weevil *Rhynchophorus palmarum* (L.) (Coleoptera: Curculionidae): Attraction to host plants and to a male-produced aggregation pheromone. *J. Chem. Ecol.* 19:1703–20.

Kalberer, N. M., T. C. J. Turlings, and M. Rahier. 2001. Attraction of a leaf beetle (*Oreina cacaliae*) to damaged host plants. *J. Chem. Ecol.* 27:647–61.

Karban, R., and I. T. Baldwin. 1997. *Induced Responses to Herbivory*. Chicago: University of Chicago Press.

Kessler, A., and R. Halitschke. 2007. Specificity and complexity: The impact of herbivore-induced plant responses on arthropod community structure. *Curr. Opin. Plant Biol.* 10:409–14.

Kost, C., and M. Heil. 2005. Increased availability of extrafloral nectar reduces herbivory in lima bean plants (*Phaseolus lunatus*, Fabaceae). *Basic Appl. Ecol.* 6:237–48.

Landolt, P. J., and T. W. Phillips. 1997. Host plant influences on sex pheromone behavior of phytophagous insects. *Annu. Rev. Entomol.* 42:371–91.

Lewis, W. J., and W. R. Martin. 1990. Semiochemicals for use with parasitoids: Status and future. *J. Chem. Ecol.* 16:3067–89.

Mattiaci, L., M. Dicke, and M. A. Posthumus. 1995. Beta-glucosidase: An elicitor of herbivore induced plant odor that attracts host-searching parasitic wasps. *Proc. North Central Branch Entomol. Soc. Am.* 92:2036–40.

McConn, M., R. A. Creelman, E. Bell, J. E. Mullet, and J. Browse. 1997. Jasmonate is essential for insect defense in Arabidopsis. *Proc. Natl. Acad. Sci. USA* 94:5473–7.

McKey, D. 1979. The distribution of secondary compounds within plants. In *Herbivores: Their Interactions with Secondary Plant Metabolites*, ed. G. A. Rozenthal, and D. H. Janzen, 56–122. New York: Academic Press.

Natale, D., L. Mattiacci, E. Pasqualini, and S. Dorn. 2003. Response of female *Cydia molesta* (Lepidoptera: Tortricidae) to plant derived volatiles. *Bull. Entomol. Res.* 93:335–42.

Nishida, R. 2002. Sequestration of defensive substances from plants by Lepidoptera. *Annu. Rev. Entomol.* 47:57–92.

Nordlund, D. A., and W. J. Lewis. 1976. Terminology of chemical releasing stimuli in intraspecific and interspecific interactions. *J. Chem. Ecol.* 2:211–20.

Noronha, N. C. 2010. Efeito dos coespecíficos e voláteis das plantas *Murraya paniculata* (L.) Jack, *Psidium guajava* L. e *Citrus sinensis* (L.) Osbeck sobre o comportamento de *Diaphorina citri* Kuwayama (Hemiptera:Psyllidae). PhD thesis, Universidade de São Paulo.

Nottinghan, S. F., J. Hardie, G. W. Dawson, A. J. Hick, J. A. Pickett, L. J. Wadhams, and C. M. Woodcock. 1991. Behavioral and electrophysiological responses of aphids to host and nonhost plant volatiles. *J. Chem. Ecol.* 17:1231–42.

Paré, P., H. T. Alborn, and J. H. Tumlinson. 1998. Concerted biosynthesis of an insect elicitor of plant volatiles. *Proc. Natl. Acad. Sci. USA* 95:13971–5.

Paré, P. W., and J. H. Tumlinson. 1997. *De novo* biosynthesis of volatiles induced by insect herbivory in cotton plants. *Plant Physiol.* 114:1161–7.

Pohnert, G., V. Jung, E. Haukioja, K. Lempa, and W. Boland. 1999. New fatty acid amides from regurgitant of lepidopteran (Noctuidae, Geometridae) caterpillars. *Tetrahedron* 55:11275–80.

Poland, T. M., J. H. Borden, A. J. Stock, and L. J. Chong. 1998. Green leaf volatiles disrupt responses by the spruce beetle, *Dendroctonus rufipennis*, and the western pine beetle, *Dendroctonus brevicomis* (Coleoptera: Scolytidae) to attractant-baited traps. *J. Entomol. Soc. B.C.* 95:17–24.

Poland, T. M., and R. A. Haack. 2000. Pine shoot beetle, *Tomicus piniperda* (Col., Scolytidae), responses to common green leaf volatiles. *J. Appl. Entomol.* 124:63–9.

Price, P. W., C. E. Bouton, P. Gross, and A. E. McPheron. 1980. Interactions among three trophic levels: Influence of plant interactions between insect herbivores and natural enemies. *Annu. Rev. Ecol. Syst.* 11:41–65.

Randlkofer, B., E. Obermaier, and T. Meiners. 2007. Mother's choice of the oviposition site: Balancing risk of egg parasitism and need of food supply for the progeny with an infochemical shelter? *Chemoecology* 17:177–86.

Reddy, G. V. P., and A. Guerrero. 2004. Interactions of insect pheromones and plant semiochemicals. *Trends Plant Sci.* 9:253–61.

Reddy, G. V. P., J. K. Holopainen, and A. Guerrero. 2002. Olfactory responses of *Plutella xylostella* natural enemies to host pheromone, larval frass, and green leaf cabbage volatiles. *J. Chem. Ecol.* 28:131–43.

Reinecke, A., J. Ruther, and M. Hilker. 2002. The scent of food and defence: Green leaf volatiles and toluquinone as sex attractant mediate mate finding in the European cockchafer *Melolontha melolontha*. *Ecol. Lett.* 5:257–63.

Rhoades, D. F. 1979. Evolution of plant chemical defense against herbivores. In *Herbivores: Their Interactions with Secondary Plant Metabolites*, ed. G. A. Rozenthal, and D. H. Janzen, 4–48. New York: Academic Press.

Röse, U. S. R., J. Lewis, and J. H. Tumlinson. 2006. Extrafloral nectar from cotton (*Gossypium hirsutum*) as a food source for parasitic wasps. *Funct. Ecol.* 20:67–74.

Ryan, C. A., and G. L. Pearce. 2001. Polypeptide hormones. *Plant Physiol.* 125:65–8.

Schoonhoven, L. M., J. J. A. Van Loon, and M. Dicke. 2005. *Insect–Plant Biology*. New York: Oxford University Press.

Soler, R., J. A. Harvey, A. F. D. Kamp, L. E. M. Vet, W. H. Van Der Putten, N. M. Van Dam, J. F. Stuefer, R. Gols, C. A. Hordjk, and T. M. Bezemer. 2007. Root herbivores influence the behaviour of an above ground parasitoid through changes in plant-volatiles signals. *Oikos* 116:367–76.

Takabayashi, J., S. Takabayashi, M. Dicke, and M. A. Posthumus. 1995. Developmental stage of herbivore *Pseudaletia separata* affects production of herbivore-induced synomone by corn plants. *J. Chem. Ecol.* 21:273–87.

Thornhill, R., and J. Alcock. 1983. *The Evolution of Insect Mating Systems*. Cambridge, MA: Harvard University Press.

Turlings, T. C. J., and J. Ton. 2006. Exploiting scents of distress: The prospect of manipulating herbivore-induced plant odours to enhance the control of agricultutal pests. *Curr. Opin. Plant Biol.* 9:421–7.

Turlings, T. C. J., J. H. Tumlinson, R. R. Heath, A. T. Proveaux, and R. E. Doolittle. 1991. Isolation and identification of allelochemicals that attract the larval parasitoid, *Cotesia marginiventris* (Cresson), to the microhabitat of one of its hosts. *J. Chem. Ecol.* 17:2235–51.

Turlings, T. C. J., U. B. Lengwilwer, M. L. Bernasconi, and D. Wechsler. 1998. Timing of induced volatile emissions in maize seedlings. *Planta* 207:146–52.

Turlings, T. C. J., H. T. Alborn, J. H. Loughrin, and J. H. Tumlinson. 2000. Volicitin, an elicitor of corn volatiles in oral secretion of *Spodoptera exigua*: Isolation and bioactivity. *J. Chem. Ecol.* 26:189–202.

Turlings, T. C. J., and F. Wäckers. 2004. Recruitment of predators and parasitoids by herbivore-injured plants. In *Advances in Insect Chemical Ecology*, ed. R. T. Cardé, and J. G. Millar, 21–75. Cambridge, UK: Cambridge University Press.

Turlings, T. J. C., J. H. Tumlinson, and W. J. Lewis. 1990. Exploitation of herbivore-induced plant odors by host-seeking parasitic wasps. *Science* 250:1251–3.

Van Dam, N. M., L. W. M. Vuister, C. Bergshoeff, H. De Vos, and E. Van Der Meijden. 1995. The raison d'etre of pyrrolizidine alkaloids in *Cynoglossum officinale*-deterrent effects against generalist herbivores. *J. Chem. Ecol.* 21:507–23.

Van Dam, N. M., J. A. Harvey, F. L. Wäckers, T. M. Bezemer, W. H. Van Der Putten, and L. E. M. Vet. 2003. Interactions between above ground and below ground induced responses against phytophages. *Basic Appl. Ecol.* 4:63–77.

Vet, L. E. M., W. J. Lewis, D. R. Papaj, and J. C. Van Lenteren. 1990. A variable-response model for parasitoid foraging behavior. *J. Chem. Ecol.* 3:471–90.

Vet, L. E. M., and M. Dicke. 1992. Ecology of infochemical use by natural enemies in a tritrophic context. *Annu. Rev. Entomol.* 37:141–72.

Visser, J. H. 1986. Host odor perception in phytophagous insects. *Annu. Rev. Ecol. Syst.* 31:121–44.

Wäckers, F. L., and R. Wunderlin. 1999. Induction of cotton extrafloral nectar production in response to herbivory does not require a herbivore-specific elicitor. *Entomol. Exp. Appl.* 91:149–54.

Wäckers, F. L., R. Jörg, and P. Van Rijn. 2007. Nectar and pollen feeding by insect herbivores and implications for multitrophic interactions. *Annu. Rev. Ecol. Syst.* 52:301–23.

Waldbauer, G. P., and S. Friedman. 1991. Self-selection of optimal diets by insects. *Annu. Rev. Ecol. Syst.* 36:43–63.

8

Cannibalism in Insects

Alessandra F. K. Santana, Ana C. Roselino,
Fabrício A. Cappelari, and Fernando S. Zucoloto

CONTENTS

8.1 Introduction 177
8.2 Conditions for Cannibal Behavior Manifestation 179
    8.2.1 Genetic Bases 179
    8.2.2 Food Availability and Quality 179
    8.2.3 Population Density 182
    8.2.4 Victim Availability 183
        8.2.4.1 Sexual Cannibalism 184
    8.2.5 Other Factors 184
8.3 Food Impact and Ecological Significance 184
    8.3.1 Effects on the Cannibal Individual Performance 184
        8.3.1.1 Benefits 184
        8.3.1.2 Costs and Related Strategies 185
    8.3.2 Ecological Significance 186
        8.3.2.1 Effects on the Population Dynamics 186
8.4 Behavior Selection 187
8.5 Final Considerations 189
References 190

8.1 Introduction

When cannibalism is mentioned, an atypical and illogical behavior is imagined. When speaking of insects the image that comes to our minds is of a female praying mantis feeding on the male head during copulation, as well as of an animal feeding on its own species individual to follow the instinct of survival in an extreme situation of food shortage. Actually, these concepts and images do not represent entirely this fantastic and mysterious behavior.

Considered a laboratory phenomenon of little ecological and evolutionary significance (Fox 1975; Elgar and Crespi 1992; Polis 1981), cannibalism is a natural behavior that occurs with great frequency in nature. It is a differentiated prey–predator interaction since it occurs in an intraspecific level (Fox 1975; some anthropologists define cannibalism as the consumption of a part, parts or the entire co-specific. Elgar and Crespi (1992) consider as nonhuman cannibalism only the cases in which an individual is killed before being eaten; however, this definition must be considered with reservation since the cannibalized individual can be ingested alive.

About 1,300 animal species are considered cannibals (Polis 1981) with representatives, among others, in Platelmyntes (Armstrong 1964), Rotifera (Gilbert 1973), Copepoda (Anderson 1970), Centripeda (Eason 1964), mites (Somchoudhary and Mukherjee 1971), fish (Skurdal et al. 1985), anurans (Bragg 1965), snakes (Manzi 1970), birds and mammals (Yom-Tov 1974), and insects (Kirkpatrick 1957; Strawinski 1964). If there are cannibal individuals in so many different groups, what advantages would this kind of behavior offer? This chapter presents a review of the studies about cannibalism in insects aiming to clarify and deepen the knowledge about this intriguing subject.

Among the Arthropoda, cannibalism is usually related to carnivorous insects (Elgar 1992; Wise 2006) due to the existence of a structure adapted for the predator–prey relationship including mechanisms to localize the victim, strategies to attack, capture, and kill the prey, and a biochemical and physiological structure to digest and absorb animal tissue. Nevertheless, the distribution of cannibalism among non-carnivorous insects is more common than among the carnivorous, since they complement their protein-lacking diets with animal tissue by ingesting the egg chorion (Nielsen and Common 1991), exuviae, and other parts, and it is more common among generalist herbivorous than among the specialists (Bernays 1998; Barros-Bellanda and Zucoloto 2005). Cannibalism among noncarnivorous insects occurs in about 130 species in the orders Orthoptera, Blattodea, Hemiptera, Coleoptera, Hymenoptera, Lepidoptera, and Diptera (Richardson et al. 2009) (Figure 8.1); Coleoptera and Lepidoptera represent 75.3% of the cannibal species, largely because they are numerous groups that have been widely studied.

The practice of cannibalism may be advantageous to the individual by incrementing its fitness with the acquisition of essential nutrients (Polis 1981). Moreover, feeding on cospecifics can remove future competitors for food, space, or sexual partners (Stevens 1992; Wise 2006) and regulate population density (Fox 1975; Polis 1981). When consumption is oriented to sick or parasitized cospecifics, the practice of cannibalism can control some parasitic infestations and diseases in the population (Boots 1998).

Cannibalism is divided into (a) destructive cannibalism, when the cannibalized individual undergoes injuries or death, and (b) nondestructive cannibalism, when predation does not cause serious damage on the individual that suffered the action (Joyner and Gould 1987). The exchange of salivary secretions between colony members can exemplify the latter (Joyner and Gould 1985). Cannibalism can also be classified according to the kinship between cannibal and prey: filial when parents consume the offspring; fraternal when the individuals consume siblings; and heterocannibalism when no kinship exists (Smith and Reay 1991). Sexual cannibalism can be considered an example of heterocannibalism since copulating male and female are usually not kindred. Cannibalism by the female of the male whole body may occur during or soon after copulation and was registered in at least 16 Mantidae predator species, some Orthoptera, and in at least 25 Ceratopogonidea (Diptera) predatory species (Lawrence 1992). The study of cannibalism reveals a rich area for exploration of the bioecology and nutrition concepts (nutritional ecology), because the intraspecific feeding has a variety of ecological, nutritional, and interrelated consequences that differ from those of the interspecific feeding.

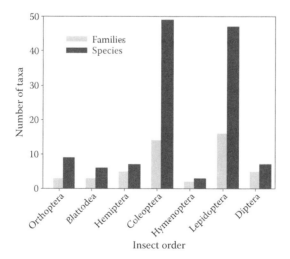

FIGURE 8.1 Number of families and species from seven typically noncarnivorous orders where cannibalism was registered. (Adapted from Richardson, M. L., P. F. Reagel, R. F. Mitchel, and M. W. Lawrence, *Annu. Rev. Entomol.*, 55, 39–53, 2009.)

8.2 Conditions for Cannibal Behavior Manifestation

As cannibalism is a very frequent behavior, what leads insects to practice it? The practice of cannibalism is often recognized by ecologists as a density-dependent mechanism that acts on the population only under intense environmental pressure. Notwithstanding, several other density-independent components can trigger this behavior manifestation (Fox 1975): asynchronous development with the host plant where the newly born emerge before the host plant is available (Dickinson 1992; Stam et al. 1987), poor nutritional quality of the host plant (O'Rourke and Hutchison 2004), asynchronous development with the population (Nakamura and Ohgushi 1981; Jasiénski and Jasiénska 1988; Kinoshita 1998), and encounter with cospecifics in vulnerable situations such as sick, parasitized individuals, or unable to move around (Bhatia and Singh 1959; Boots 1998), in addition to physical factors such as high temperature and variation of relative humidity (Ashall and Ellis 1962; Iqbal and Aziz 1976; Rojht et al. 2009).

8.2.1 Genetic Bases

The cannibal behavior expression is genetic but requires induction through environmental factors (Fox 1975). The existence of reproductive lineages with different cannibal tendencies constitutes the best evidence of the genetic component power for cannibalism; several pairs of genes act additively on the expression of the characteristic, which is a multifactorial quantitative polygenic heritage (Polis 1981). The smallest genetic change would be enough to determine a cannibal individual (Polis 1981) whose adaptations would inhibit or promote cannibalism by selecting genes that regulate the expression of the characteristic (Gould et al. 1980; Gould 1983; Tarpley et al. 1993). Two species of the genus *Tribolium* beetles present different cannibalism rates of eggs for males and for females (Stevens 1989). Females are more voracious than males; however, the contrary occurs in *T. confusum* Jacquelin du Val (Stevens 1989). As a result of egg ingestion, *T. castaneum* (Herbst) female shows increased egg production (Ho and Dawson 1996). Therefore, maintenance and expression of the characteristic in a species will depend on its reproductive success responses and on the selective pressure it is submitted to.

The geographic variation among insect populations provides a range of differences of morphological and behavioral characters (Tarpley et al. 1993). Aiming to examine the genetic base importance on the cannibal behavior, a laboratory experiment was carried out with two corn *Diatrea grandiosella* Dyar caterpillar populations, a subtropical insect that has expanded to North America; populations at 37°N and 19°N latitudes were tested (Tarpley et al. 1993). Under the same experimental conditions, marked differences in the cannibal behavior were found—cannibalism in the 37°N population was more intense probably as a result of the genetic differences between the tested populations.

Three behavioral hypotheses related to genetic lineages attempt to explain the existence of individuals relatively capable of obtaining success through the cannibal action: (a) locomotion activity, (b) search efficiency, and (c) appetite. The lineages with individuals with more ability to move around and find something to ingest with appetite will perform the cannibal act better than others not so skilled (Giray et al. 2001).

8.2.2 Food Availability and Quality

Food is fundamental for the living beings to survive. Partial or total alterations in its availability cause stress that in turn promotes behavioral, biochemical, and physiological changes that may eventually lead to cannibalism. Shortage of food causes the cotton *Spodoptera littoralis* (Boisd.) (Noctuidae) caterpillar to present a tendency to ingest cospecific caterpillars and return to the normal condition soon after food availability is restored (Abdel-Samea et al. 2006). The predator of aphids *Harmonia axyridis* Pallas (Coleoptera, Coccinelideae) shows cannibalism, which is reduced with high prey density (Burgio et al. 2002) (Figure 8.2). In *Chrysoperla carnea* (Stephens) (Neuroptera, Chrysopidae) green lacewing larvae, 100% cannibalism was observed when their main prey, aphids, were not available; when the prey were present the rate of cannibalism was negligible (Mochizuki et al. 2006; Rojht et al. 2009).

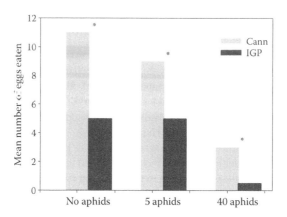

FIGURE 8.2 Comparison between intraguild predation (IGP) and cannibalism (CANN) in *Harmonia axyridis* adults in different aphids' densities. *: $P < 0.05$ (Kruskal–Wallis ANOVA test). (Modified from Burgio, G., F. Santi, and S. Maini, *Biol. Cont.*, 24, 110–6, 2002.)

The migration phenomenon is associated with the anticipated reduction of feeding resources after increase in population density, a situation observed in the North American *Anabrus simplex* Haldeman (Orthoptera) Mormon crickets (Simpson et al. 2006). During migration, crickets present deprivation of two essential nutrients: proteins and mineral salts (Figure 8.3). As the insects themselves are constituted by these nutrients, they practice cannibalism. Satiation of these two nutrients reduces the cannibalism rate; the protein satiation specifically inhibits migration. This suggests that the main reason for the migratory flock formation in this species is the nutritional protein and mineral salts deficiency (Simpson et al. 2006).

In the *Emblemasoma auditrix* Shewel parasitoid fly larvae, cannibalism is also manifested in situations of food deprivation. Females lay first instar larvae instead of eggs on the host, the *Okanagana rimosa* Say cicada; about 38 larvae eclode simultaneously in the uterus (De Vries and Lakes-Harlan 2007). Flies lay only one larva per host, and the development is completed in 5 days; however, the search for the host can take several weeks, and during that period, larvae cease their development and remain in the first instar. Surprisingly, some larvae inside the uterus gained weight and this was subsequently related to other larvae cannibalism (Figure 8.4). Dissecting the mother, dead and injured larvae were found (Figure 8.5a); larvae vestiges such as cuticle and mouth hooks were in the uterus (Figure 8.5b). This kind of cannibalism was described as prenatal cannibalism (De Vries and Lakes-Harlan 2007).

Not only the amount, but also the quality of the food that is influenced by the nutrients is determinant for cannibalism manifestation. The termite's food is markedly deficient in nutrients, particularly proteins; it is not surprising that most of the studied species show facultative carnivory usually in the form of cannibalism (Matthews and Matthews 1978). For instance, when the diet is protein rich, *Zootermopsis angusticollis* Hagen manifest almost zero cannibalism; however, when a diet of pure cellulose is experimentally provided, the colony becomes intensely cannibal (Cook and Scott 1933).

Social insects feed on dead individuals from the colony or on injured laborers (Whitman et al. 1994). Therefore, they adjust their cannibalism level according to their immediate nutritional needs. In ant colonies, the laborers routinely ingest damaged eggs, injured larvae, and pupae that have few chances to survive, but when the colony suffers periods of food scarcity healthy offspring can also be consumed (Wilson 1971; Joyner and Gould 1987). When the prey density in the foraging area of the red wood ant *Formica polyctena* Föster is low, the lack of nutrients stimulates the search for proteins in neighbor colonies where they fight, kill, and ingest cospecifics in the nest (Driessen et al. 1984).

Several authors found adaptive values for the cannibalism patterns in social insects. Kasuya et al. (1980) explained that the intercolony cannibalism in Japanese paper wasps *Polistes chinensis antennalis* Perez and *P. jadwigae* Dalla Torre is an effective way to accumulate food during the colony construction. In temperate climate species, food is scarce in autumn and due to reduction in the number of laborers

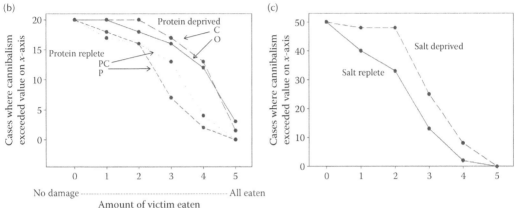

FIGURE 8.3 Changes in the cannibalism rates: (a) *Anabrus simplex* cricket nymphs collected in the field after pretreatment with the four artificial diets; (b) P: 42% proteins, no digestible carbohydrates, C: 42% carbohydrates, no proteins; PC: both nutrients; and O: no nutrients; (c) herbal seeds with or without 0.25 M NaCl. The tendency to cannibalism was reduced when the insects were exposed to diets rich in proteins and salts, increased with diets containing exclusively salts or proteins, and became much more evident with diets containing only carbohydrates. (Modified from Simpson, S. J., G. A. Sword, P. D. Lorch, and I. D. Couzin, *Proc. Natl. Acad. Sci. USA*, 103, 4152–6, 2006. With permission.)

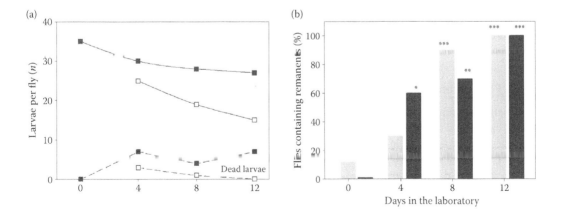

FIGURE 8.4 (a) Average number of larvae of *Emblemasoma auditrix* in the uterus per fly over time: flies that had no access to the host (black symbols, solid lines, group I) and consequently did not lay their larvae, and flies that had access to the host (open symbols, solid lines, group II) have shown significant reduction in the number of larvae over time. The number of dead larvae varied in both groups (dashed lines). (b) Percentage of flies with larvae vestiges (mouth hooks, cuticle parts) in both groups. Group I (open columns). Group II (black columns). Against day 0: *p = 0.02, **p = 0.002, ***p = <0.002. (Modified from De Vries, T., and R. Lakes-Harlan, *Naturwissenschaften*, 94, 477–82, 2007.)

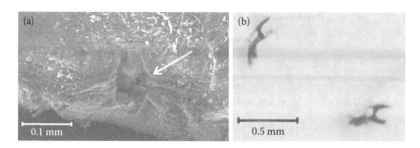

FIGURE 8.5 Micrography of *Emblemasoma auditrix* larvae. (a) detail of wounded larvae cuticle, (b) loose mouth hooks found in the mother uterus (From De Vries, T., and R. Lakes-Harlan, *Naturwissenschaften*, 94, 477–82, 2007. With permission.)

and in the proportion of reproductive individuals, intracolony cannibalism usually occurs (Wilson 1971; Joyner and Gould 1987).

In the omnivorous, flexibility in the protein source consumption is fundamental for cannibal behavior. This can reduce the cannibalism levels according to the feeding source availability and the animal or vegetal constitution (Coll and Guershon 2002; Leon-Beck and Coll 2007). Pollen in the diet can attenuate the cannibalism rates (Leon-Beck and Coll 2007). The *Coleomegilla maculata* De Geer lady beetle, abundant in corn fields, is a *Helicoverpa zea* (Boddie) larvae and eggs predator, in addition to using corn pollen as a feeding source (Cottrel and Yeargan 1998). When *C. maculata* populations are fed only on pollen, the cannibalism rates are lower than in populations that feed on *H. zea* with or without a pollen source.

Genetically modified vegetal species that express the *Bacillus thuringiensis* Berliner protein Cry1Ab are toxic to some insects. *Spodoptera frugiperda* J. E. Smith and *H. zea* caterpillars are important corn consumers and both practice cannibalism; in the United States, *H. zea* is one of the species that cause more losses to farmers (Hayes 1988; Mason et al. 1996; Buntin et al. 2004). When the corn MON810 that expresses the *B. thuringiensis* protein Cry1Ab is offered, both species tend to present more marked cannibal behavior compared to the same species in nongenetically modified corn (Horner and Dively 2003; O'Rourke and Hutchison 2004). This effect probably occurs due to reduction of available nutrients in plants that express the protein Bt, providing larvae a nutritionally negative effect (Chilcutt 2006).

Three factors can explain why starvation and reduction in food diversity promote cannibalism: (1) the feeding stress generally increases the foraging activity, increasing the probability of intraspecific contact, (2) food-deprived animals become weak and vulnerable to predation, and (3) consumers must expand their diet beyond their "limits" during the food deprivation period (Polis 1981).

8.2.3 Population Density

Cannibalism resulting from high population density is often erroneously attributed to food scarcity. In several cases, food is abundant or sufficient, and nevertheless the practice of cannibalism is observed. The ingestion of cospecifics in a situation of high birth rate in the population and the consequent higher density represent a nutritional increment for the cannibal, in addition to the elimination of future competitors (Fox 1975). *Lestes nympha* Selys (Odonata) adults feed on flies, mosquitoes, and other small insects (Fischer 1961). Environmental alterations such as food in excess and lack of predators make the population grow, causing an increase in the number of eggs and larvae; when the latter eclode they begin to compete for the same food (Fischer 1961). Therefore, the feeding target is changed and they start to eat cospecifics (Fischer 1961; Fox 1975).

Metahycus flavus Howard (Hymenoptera, Encyrtidae) females lay several eggs on the host with no conflicts among the gregariously developed larvae (Tena et al. 2009); occasionally, additional oviposition occurs on the same host, resources become limited, resulting in aggressive behavior of individuals, and cannibalism occurs.

Cannibalism resulting from high population density may stabilize population number and distribution (Dong and Polis 1992) as we discuss below. This behavior and other mortality factors modify the relationship size/distribution, acting as a selective agent regarding larvae and adults (Eisenberg 1966). The more species that use the same resource, the more interspecific events occur, and the relative amount of cannibalism and predation will depend on finding other individuals. In species that are always cannibals even when the interspecific competition is high, it is common that the cannibalism events are less frequent when there is less superposition in the use of resources (Fahy 1972).

8.2.4 Victim Availability

There are groups of insects that ingest their peers even in conditions of low population density and abundance of food. Some ladybug larvae can consume their siblings (Osawa 1992). The cannibal behavior can be triggered when an individual encounters another individual of the same species in vulnerable condition.

Cannibalism of eggs and newly emerged young can be determined by the dimension of the egg mass and by the time until eclosion. Egg cannibalism in *Ascia monuste* Godart (Lepidoptera), the kale caterpillar (Figure 8.6), occurs whenever the opportunity arises, regardless of food presence and/or availability; in other words, when the first caterpillars eclode or when the older caterpillars find eggs in the kale leaves, laboratory tests have shown that the caterpillars prefer cospecifics instead of the usual food (Barros-Bellanda and Zucoloto 2005). This can be explained because the egg protein content is higher than the vegetal tissue. This behavior is influenced by the development stage—the older the caterpillars, the higher their predation power (Zago-Braga and Zucoloto 2004) due to the mandible high rigidity in the final instars, as well as increased mobility facilitating predation of newly ecloded caterpillars that already present some mobility. This is a typical case of population asynchronous development.

When asynchrony occurs between insects and host plants, larvae eclode before the plant is available, either because the fruits are not ripe or because of some factor that makes the larval food unsuitable. *Nezara viridula* (L.) (Heteroptera: Pentatomidae) nymphs cannibalize in case they become active before the host plant is available (Stam et al. 1987).

Cospecifics infested by pathogens or parasites can also be consumed and the energy they contain is reused. In *Spodoptera frugiperda* (J. E. Smith) corn caterpillars, cannibalism of parasitized individuals reduces the parasite number in future generations (Holmes et al. 1963; Root and Chaplin 1976; Weaver et al. 2005). *Plodia interpunctella* Hübner (Lepidoptera) larvae kill and consume cospecifics parasitized by the wasp *Venturia canescens* Gravehorst (Reed et al. 1996). In *Oncopeltus fasciatus* Dallas (Hemiptera), cannibalism reduces a parasitoid presence in the eggs (Root and Chaplin 1976).

According to the bee colonies concept as superorganisms, a strong pressure for selection eliminates characteristics that reduce the inclusive aptitude (Moritz and Southwick 1992). As diploid males are reproductively inferior to haploid males, removal of larvae that will originate those males is considered an energy economy for the colony. Several arguments to support the evolutionary solution for the false male precocious cannibalism by adult laborer bees are in accordance with the kinship selection theory (Hamilton 1964). In *Apis mellifera* L., diploid male larvae are cannibalized by laborer bees in their first

FIGURE 8.6 *Ascia monuste* caterpillar (second instar) ingesting a cospecific egg. (Courtesy of Alessandra F. K. Santana.)

day of life; in *A. cerana* F., they survive longer, and some reach the fourth day of life (Woyke 1980). The removal would be related to the drone pheromone liberation ("cannibal pheromone"), indicating the diploid drone presence and its consequent elimination (Herrmann et al. 2005). Laborer bees usually remove the injured pupae by eating them instead of simply removing them (Gramacho and Gonçalves 2009).

8.2.4.1 Sexual Cannibalism

Sexual cannibalism can be started regardless of the food amount or population density. This behavior is triggered by the victim availability and/or behavior. In the Orthoptera *Hapithus agitator* Uhler and *Cyphoderris* sp., females feed on the males' wings during copulation. This behavior maintains males and females together during insemination and prevents females from feeding on the spermatophore before its total emptying (Vahed 1998). In males whose wings were surgically excised, the spermatophore transference was less successful than in those that remained winged (Vahed 1998).

Neurons in the Mantidae males heads inhibit copulation movements; when females ingest their heads during copulation, the neuronal inhibition disappears, the copulation movements are stimulated, and insemination is more efficient (Alcock 2001). Sexual cannibalism enhances offspring production in these cases: *Hierodula membranacea* Burmeister females maintained in poor diets but allowed to feed on males during copulation produced heavier oothecae than females that could not feed of males (Birkhead et al. 1988). In this species, there is a positive correlation between the ootheca mass and the number of juveniles.

Several authors proposed that the Mantidae sexual cannibalism appears only when the food is scarce, when the males lack space after copulation, or when observers cause disturbances: most observations of the Mantidae cannibalism are made in captured insects, and the consequent stress can initiate that behavior (Vahed 1998). *Mantis religiosa* L. sexual encounters, for instance, show 31% cannibalism rate (Lawrence 1992) and in *Torymus sinensis* Kamijo, 5% (Hurd et al. 1994). This data indicates that cannibalism is not necessarily a rule in the Mantidae sexual encounters.

8.2.5 Other Factors

Humidity and temperature somehow can stagnate, kill, or accelerate the population growth and this also happens regarding cannibalism. Nymphs of the Acrididae *Gastrimargus transverses* Thunberg cannibalize even with low population density and food abundance; and in the species *Spathosternum prasiniferum* Walker cannibalism is favored by high temperature and low humidity (Iqbal and Aziz 1976; Majeed and Aziz 1977). High temperatures heighten the cannibalism rate in *Blatta orientalis* L., *Blattella germanica* (L.), *Periplaneta americana* (L.) (Guthrie and Tindall 1968), and *C. carnea* (Rojht et al. 2009).

8.3 Food Impact and Ecological Significance

8.3.1 Effects on the Cannibal Individual Performance

8.3.1.1 Benefits

Perhaps the greatest benefit of cannibalism is nutritional (Whitman et al. 1994), since there is a direct influence on the insect fitness translated by prolonged survival, higher rate of development, and/or fecundity (Church and Sherratt 1996; Joyner and Gould 1985) in addition to probable increase in size and weight; the Coleoptera *Dorcus rectus* Motshulsky cannibal larvae gain more body mass than the noncannibal ones (Tanahashi and Togashi 2009).

The pierid *A. monuste* cannibal caterpillars show higher survival and weight rates (Table 8.1) as compared to the noncannibal ones, and the animal protein ingestion through the chorion is responsible for up to 50% of the newly ecloded caterpillar biomass (Barros-Bellanda and Zucoloto 2001). When *Ceratitis capitata* Wiedemann fruit flies larvae are food-deprived—either in quantity or in quality—they

TABLE 8.1

Survival and Dry Weight of *Ascia Monuste* Caterpillars That Ingested Chorion (Control Group C1) and Did Not Ingest Chorion (Experimental Group E1) Fed on Kale during the Immature Phase

	Number of Adults/Box (Survivors)	Dry Weight (G)		
		Newly Ecloded Caterpillars	First Instar Caterpillars	Fifth Instar Caterpillars
C1	5.1 ± 1.0 a	0.050 ± 0.008 a	0.100 ± 0.015 a	2.244 ± 0.310 a
E1	3.3 ± 1.2 b	0.035 ± 0.005 b	0.062 ± 0.010 b	2.0160 ± 0.250 a

Source: Barros-Bellanda, H. C. H., and F. S. Zucoloto, *Ecol. Entomol.*, 26, 557–61, 2001. With permission.

Note: $N = 6$ (survival); $n = 10$ (weight); means ± SD followed by different letters differ significantly (*t*-test, $P < 0.005$).

cannibalize smaller eggs and larvae, and this propitiates them a better development (Zucoloto 1993). On the other hand, there is no evidence that cannibalism benefits performance in some species. Survival of caterpillars that cannibalize the fall armyworm *S. frugiperda*, for instance, is significantly reduced in spite of food availability (Chapman et al. 1999). Cannibalism is clearly beneficial for *A. monuste* in the beginning of development; however, if practiced intensely it brings about developmental delay and reduction of fecundity (A.F.K. Santana, unpublished). In final instars, it could also result in adults with reduced sizes (Santana et al. 2011). This demonstrates that although cannibalism is clearly beneficial for some organisms, the intake of atypical foods may be detrimental.

Regarding nonnutritional benefits, the individual that cannibalizes eliminates a potential competitor and a possible cospecific predator (Fox 1975; Polis 1981). Therefore, the population size is reduced, the food *per capita* is more abundant, and the chances to survive and grow more rapidly increase. Damages caused in the plant due to herbivory and feces produced by the noctuid *Spodoptera* sp. caterpillars liberating semiochemicals can favor predatory Hemiptera and/or parasitoid Hymenoptera to find these larvae (Turlings et al. 1990; Yasuda 1997). Reduction of larval density reduces the liberation of odor hints and possibly predation and the risk of parasitism (Chapman et al. 1999). Demonstrably, predators have been more abundant in plants with higher herbivory rates (Chapman et al. 2000). Therefore, reduction of cospecifics in feeding can reduce considerably the chances of natural enemies to approach, favoring the cannibal survival.

8.3.1.2 Costs and Related Strategies

Cannibalism is clearly beneficial to the insect when food availability is low. It is not so when food availability is high and the victims are related (Burgio et al. 2002). *H. axyridis* females, which are aphid predators, present reduced fitness with a high prey density and the sibling cannibalism intensity indicates that this kind of cannibalism is not adaptive for females when the larval food is abundant (Osawa 1992).

Trying to eat the victim, in addition to the imminent predation risk, there is the risk of "role reversal": the cannibal may be injured or even cannibalized (Polis 1981). In migratory groups such as the grasshopper *Schistocerca gregaria* (Forskal), one individual bites another and in this situation the risk of being bitten is high (Bazazi et al. 2008).

The lack of predatory adaptation can be a problem faced by preferentially phytophagous cannibals. Regarding behavior, it is necessary that the insect have receptors to perceive, capture, accept, and ingest the animal source. The lack of adequate sensillas and neural programs to detect the prey and the absence of physical structure and adequate mouthparts for capture and manipulation of it make contact with the prey difficult. The exoskeletons and sharp extremities of the prey can damage the gut internal walls and make the passage by the digestive tract more difficult (Whitman et al. 1994). Considering digestion, insects must release different amounts of enzymes (mainly proteases). This is necessary because the animal feeding sources are richer in proteins than the vegetal sources. *C. capitata* cannibal fly larvae digestive enzymes consist of high trypsin secretion that increases protein digestion efficiency and lessens the aminopeptidase secretion, probably reducing deleterious effects due to the excess of free amino acids; in addition, salt accumulation was detected in the Malpighian tubules indicating an adaptation to

FIGURE 8.7 Green lacewing larvae cannibalism. Bigger and older larvae consume smaller cospecific larvae. (From Rojht, H., F. Budija, and S. Trdan, *Acta Agri. Slov.*, 93, 5–9, 2009. With permission.)

the excess of nutrients present in the food (Lemos et al. 1992). At least in theory, the facts mentioned above may have been a decisive step toward the appearance of exclusively carnivorous insect species.

As a strategy to reduce the risks related to intraspecific predation, insects often are opportunists; that is, preferentially feed on injured, incapacitated, dead, or fragile prey (Whitman et al. 1994). They cannibalize eggs, newly ecloded caterpillars, or when the potential prey ecdysis is about to occur; in the coccinelid *H. axyridis* and *A. bipunctata*, for instance, cannibalism only occurs in nonviable eggs (Santi and Maini 2007). *A. monuste* females avoid ovipositing in leaves where there are cospecific caterpillars because the eggs may be cannibalized (Barros-Bellanda and Zucoloto 2005). Predation of individuals in these conditions can bring about another risk to the cannibals: transmission of pathogens and parasites by ingesting a contaminated individual (Polis 1981; Boots 1998). In *P. interpuctella*, virus transmission and infection occurred via cannibalism (Boots 1998). This fact suggests that some cannibals do not prevent infection to the detriment of cannibalism, and the risk is consistently present.

Cannibals are often bigger and heavier than their victims (Boots 1998; Rojht et al. 2009). Bigger individuals probably have stronger mandibles and more adequate muscles for predation compared to smaller individuals (Tanahashi and Togashi 2009). In *C. carnea* green lacewing larvae, for instance, the difference in size between cannibals and victims is highly perceptible (Figure 8.7). There are exceptions to this generalization: some species are very cannibalistic when small (Polis 1981) or no differences exist between them. *Asynarchus nigriculus* Banks (Trichoptera) larvae, for instance, cannibalize larvae of the same size, often involving victim mobilization by cospecific groups (Wissinger et al. 1996).

Cannibalism is also disadvantageous when the cannibal becomes very aggressive, destroying its progeny or eliminating possible sexual partners; the cannibal that ingests relatives can reduce its inclusive fitness (Polis 1981; Burgio et al. 2002; Dobler and Kölliker 2009). The cannibalism advantages and disadvantages must be balanced against other factors that affect survival. In some cases, the consequences of cannibalism may be less severe than starvation and inadequate reproduction due to lack of nutrients.

8.3.2 Ecological Significance

Cannibalism mechanisms and ecological meanings may differ considerably among herbivores and carnivores. For the carnivores, the main benefit appears to be a quantitative compensation due to lack of prey (Dong and Polis 1992); the coccinelid *C. septempunctata*, for instance, cannibalize significantly more eggs in the absence than in the presence of aphids (Khan et al. 2003). When the herbivores manifest the cannibal behavior, they profit both, in better food quality by increasing the protein acquisition and in food quantity due to expansion of the essentially vegetal feeding. The herbivorous cannibals heighten their nitrogen indexes, possibly resolving nutritional deficiencies caused by the relative low nutritional quality of the plants they usually ingest.

8.3.2.1 Effects on the Population Dynamics

As mentioned above, the occurrence and intensity of cannibalism varies largely among species. A great deal of that variability is due to factors that influence the population. Some species respond to resource limitation with dispersion, diapause, alterations in the physiological characteristics, or interference of competitors, while other species may be unable to cannibalize because they do not succeed in capturing the prey or do not have adequate mouth morphology to ingest it.

TABLE 8.2

Ascia monuste Cannibalism Data Considering the Number of Cospecifics in the Groups

Number of Cospecifics	Number of Ingested Caterpillars	Cannibalized Caterpillars (%)
Group I	—	—
Group II	4.83 ±1.57 a	69.04 ± 22.46 a
Group III	8.17 ± 2.91 a	54.44 ± 19.40 a
Group IV	19.33 ± 4.78 b	64.40 ± 15.95 a

Source: Zago-Braga, R. C., and F. S. Zucoloto, *Rev. Bras. Entomol.*, 48, 415–20, 2004. With permission.

Note: $N = 6$, Mean ± standard deviation. Means followed by different letters differ significantly. Kruskal–Wallis ANOVA test, $P < 0.05$. Group I: control, fed only on *Brassica oleraceae* (kale); group II: fed on kale and 7 newly ecloded caterpillars; group III: fed on kale and 15 newly ecloded caterpillars; group IV: fed on kale and 30 newly ecloded caterpillars.

Usually, the number of individuals and the age of the population experience seasonal oscillations in response to variations in the environment's available resources. As mentioned above, restriction and/or reduction in the feeding resources quality can stimulate the practice of cannibalism. In this context, it may represent a variable in the population dynamics, acting drastically or irrelevantly in reducing its size (Hastings and Constantino 1991).

8.3.2.1.1 Population Control

Recently, cannibalism was considered a stimulating mechanism to form desert grasshopper migratory bands; when density is high, cannibal interactions among the population individuals increase the injury risk, intensifying the group flight behavior and resulting in a highly coordinated and mobile group (Bazazi et al. 2008). In dragonflies and ephemerides nymphs, cannibalism is not considered an important factor regarding population control since it is a rare behavior (Fox 1975). However, in the ladybugs *Plagiodera versicolora* Laicharting and *H. axyridis*, cannibalism is considered one of the main regulatory mechanisms responsible for up to 50% of the newly ecloded individuals' mortality (Osawa 1993), and the main cause of some Orthoptera species mortality in the field (Bazazi et al. 2008).

On the other hand, several authors suggest that cannibalism increases the number of survivors since the food in the environment is better used by a lower number of successful individuals. When food is scarce, some individuals may survive without cannibalism but cannibalism will propitiate a higher number of survivors (White 2005). It is the case of *T. castaneum* cannibal larvae: larvae from lineages that practice egg cannibalism are able to colonize the environment better than competitor lineages that do not cannibalize, which produces more survivals, shorter development, and more fertile females (Via 1999).

The number of newly ecloded caterpillars cannibalized in the laboratory is greater the larger the number of cospecifics for *A. monuste* caterpillars in the penultimate larval stage, indicating that cannibalism in this species can act as population control (Zago-Braga and Zucoloto 2004) (Table 8.2). In the field, egg cannibalism by newly ecloded caterpillars from the same oviposition also increases as the oviposition size increases, and it can also be observed between different ovipositions on the same leaf (Barros-Bellanda and Zucoloto 2005). In addition to affecting the number of individuals in a population, cannibalism can also affect its structure. As mentioned above, the cannibals can present high survival rates to the detriment of others, generating an oscillation in the age class distribution inside the population.

8.4 Behavior Selection

The genes responsible for the origin of the behavior probably promote a reproductive success response; they must then be distributed in the species as a result of natural selection in favor of the characteristic. Regarding diet, the cannibal individuals expand their gamma of resources through inclusion of previously ignored items due either to high cost or low energetic gain (Polis 1981). The feeding resources expansion is an evolutionary benefit for the cannibal organism (Crump 1990; Majerus 1994). On the other hand, the cannibal behavior evolution can be inhibited by feeding specialization (Richardson et al. 2009).

Several omnivorous insects have high rates of survival and fecundity when the vegetal diets are complemented with animal tissue (Coll 1998). Some studies report that cannibalism among cospecifics is avoided (Snyder et al. 2000) and that generalist predators eat preferentially heterospecifics, avoiding cannibalism whenever possible (Schausberger and Croft 2000).

The kinship selection theory suggests that cannibalism most probably spreads and fixes in a population when the cannibal act between close relatives is avoided (Hamilton 1970; Sherratt et al. 1999). Under intense competition, aggressive interactions among siblings are stimulated, causing death (Mock and Parker 1998). Fratricide is a common phenomenon in several taxa, including insects (Grbic et al. 1992; Van Buskirk 1992; Fincke 1994; Osawa 2002; Ohba et al. 2006). The examples below show that distinction in cannibalizing relatives may not occur (Sherratt et al. 1999): the grade of cannibalism of relatives and nonrelatives was compared between two mosquito species whose larvae are regularly found in the water: one detritivorous, *Trichoprosopon digitatum* Rondani, and one predator, *Toxorhynchites moctezuma* Dyar & Knab. Neither of these mosquitoes has shown preference to cannibalize nonrelatives (Sherratt et al. 1999). Apparently, cannibalism was determined by the relationship victim size × cannibal size, and this has made the selectivity to prefer consumption of nonrelatives; notwithstanding, observing *T. moctezuma*, the nutritional benefits have compensated the nondiscrimination cost between relatives and nonrelatives (Sherratt et al. 1999).

Protecting the eggs confers the male guardians some benefits. *Rhinocoris tristis* Stal (Hemiptera) males rely on eggs laid by females they copulated with, defending them from parasitoids and predators until the nymphs emerge (Thomas and Manica 2003). Males that copulate with females before oviposition are probably the fathers of at least some of the eggs. During this period, males do not abandon the eggs and the chances to obtain food are drastically reduced. Therefore, some of these eggs are consumed by the "fathers" that suck the egg content, leaving the wall and the operculum intact. The ingestion of their own offspring is evidence that confirms the link between food availability and incidence of cannibalism. There are cases of alloparental care resulting from direct competition between males regarding the offspring (Thomas 1994); in an experiment to investigate whether males discriminate between their own offspring and that of other males, no significant differences were found in the percentage of eggs cannibalized by the fathers or by the stepfathers (Thomas and Manica 2003).

There are three hypotheses to explain the occurrence of parental cannibalism: (1) nutritional gain for the cannibal, (2) gain for the relatives (kin selection), and (3) gain for the cannibal and for the victim, through close kinship with the cannibal and parental manipulation (Polis 1981). The insects that take care of the offspring but consume some of the eggs guarantee protection for the offspring as a whole, sacrificing part of it; abandonment of eggs by the males in search of food is frequently detrimental for the inclusive fitness—without the father protection, predation and/or egg parasitism is a certainty. Therefore, filial/parental cannibalism generates phenotypic (nutritional status) and genotypic (contribution for the following generations) benefits (Polis 1981). Parental cannibalism can also prevent damages or diseases when the peripheral eggs, which are more susceptible to parasitism, are ingested (Figure 8.8). Emptied but externally intact eggs may attract other females and aggregate more eggs, stimulating the males to protect them or use them as a feeding resource. Therefore, the males are investing in the care of present and future generations (Thomas and Manica 2003).

All models consider that sexual cannibalism has adaptive value for both genders. As previously mentioned, feeding on males' head offers a nutritional advantage for females, but what would be the advantage for males? Is sexual cannibalism adaptive under the male point of view? A supposition is that sexual cannibalism evolved and is maintained because of the reproductive benefits it confers to the adults.

Offering the head, the males can transfer more spermatozoa; the females receive a nutritional increment and guarantee egg fertilization by not ingesting the spermatophore (Fox 1975). In spite of not knowing for sure whether the males participate actively in being cannibalized, sexual cannibalism can be theoretically adaptive for the males if: (1) the opportunity to copulate with other females is limited; and (2) the nutrients available to the females contribute to the cannibalized males reproductive success (Buskirk et al. 1984). Alcock (2001) considers that the ability for copulation of headless mantid males has its origin in the reproductive consequence: the acephalous males increase posthumously their fitness

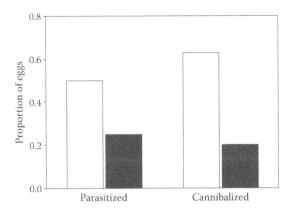

FIGURE 8.8   Proportion of peripheral (white columns) and internal (black columns) eggs that were parasitized by wasps and cannibalized by the guardian males; columns followed by different letters differ significantly. (Modified from Thomas, L. K., and A. Manica, *Anim. Behav.*, 66, 205–10, 2003.)

and this contributes for the positive selection of the characteristic, but any mutation that interferes in this ability can drive the selection pressure against the characteristic.

Sexual cannibalism in adult organisms can be the indirect result of adaptive behavior in previous stages of the organism life history (Arnqvist and Henriksson 1997; Suttle 1999). The reproductive success of some females is associated with the size they reach at the end of the juvenile phase when they become more fertile. Therefore, there is selection favoring greater female aggressiveness and this can be translated in aggressive behaviors resulting in sexual cannibalism. It is possible that the female behavior is only a subproduct of a selective process that favors other set of behaviors—in this case, high juvenile aggressiveness.

8.5 Final Considerations

The studies and research on cannibalism in insects is important not only regarding evolutionary aspects but also for those interested in applying the knowledge in practice. Even though there are some theories that disagree about insect behavior evolution and feeding habits, it is often accepted that the first exclusively terrestrial insects were saprophagous, followed by the phytophagous and the carnivorous. Nevertheless, several phytophagous species practice cannibal behavior that in several nonexclusive aspects brings advantages for the development, reproduction, and/or population control that ultimately can avoid competition and select the most capable individuals.

From an applied perspective, understanding the cannibal behavior and its consequences may help its use as an interference tool in the life history of insects that consume agricultural species and in other possible applications for biological control. (Mally 1892; Joyner and Gould 1985) proposed to associate corn and cotton in rows in order to reduce the *H. zea* caterpillars' damage. Moths prefer ovipositing in corn so few eggs would be laid in cotton. The high number of eggs oviposited in corn can promote cannibalism among the larvae and reduce the target population (Mally 1892; Joyner and Gould 1985).

Another important point regarding cannibalism is to situate the insect species that display that behavior in the adequate classification. For a long time, opportunistically cannibal species were classified as exclusively phytophagous. Today, it is known that cannibal behavior can improve these species' performance. Studies focusing on behavior should include descriptive, and whenever possible, quantified aspects. This will allow us to know the possible causes for the insects to practice cannibalism; therefore, populations should be studied in their habitats quantifying available foods and strategies used in cannibal behavior. At the same time, laboratory studies could help provide a better understanding of cannibalism, since manipulations in the laboratory are often impossible to do in the field.

REFERENCES

Abdel-Samea, S. A., H. A. Al-Kady, and G. A. Kherd. 2006. Cannibalism phenomenon between *Spodoptera littoralis* (Boisd.) (Lepidoptera: Noctuidae) larvae in the laboratory. *Egyptian J. Agric. Res.* 84:725–31.

Alcock, J. 2001. *Animal Behavior: An Evolutionary Approach,* 7th Edition. Sunderland, MA: Sinauer Associates, Inc.

Anderson, R. S. 1970. Predator–prey relationships and predation rates for crustacean zooplankters from some lakes in western Canada. *Can. J. Zool.* 48:1229–40.

Armstrong, J. T. 1964. The population dynamics of the planarian, *Dugesia tigrina. Ecology* 45:361–5.

Arnqvist, G., and S. Henriksson. 1997. Sexual cannibalism in the fishing spider and a model for the evolution of sexual cannibalism based on genetic constraints. *Evol. Ecol.* 11:255–73.

Ashall, C., and P. Ellis. 1962. Studies on numbers and mortality in field populations of the desert locust (*Schistocerca gregaria* Forskål). *Anti-Locust Bull.* 38:1–59.

Barros-Bellanda, H. C. H., and F. S. Zucoloto. 2005. Egg cannibalism in *Ascia monuste* in the field: opportunistic, preferential and very frequent. *J. Ethol.* 23:133–8.

Barros-Bellanda, H. C. H., and F. S. Zucoloto. 2001. Influence of chorion ingestion on the performance of *Ascia monuste* and its association with cannibalism. *Ecol. Entomol.* 26:557–61.

Bazazi, S., J. Buhl, J. J. Hale, M. L. Anstey, G. A. Sword, S. J. Simpson, and I. D. Couzin. 2008. Collective motion and cannibalism in locust migratory bands. *Curr. Biol.* 18:735–9.

Bernays, E. A. 1998. Evolution of feeding behavior in insect herbivores: success seen as different ways to eat without being eaten. *BioScience* 48:35–44.

Birkhead, T. R. 1988. Sexual cannibalism in the praying mantis. *Hierodula membranacea. Behaviour* 106:112–8.

Bhatia, D. R., and C. Singh. 1959. Cannibalism in locusts. *Indian J. Entomol.* 21:210–3.

Boots, M. 1998. Cannibalism and the stage-dependent transmission of a viral pathogen of the Indian meal moth, *Plodia interpunctella. Ecol. Entomol.* 23:118–22.

Bragg, A. N. 1965. *Gnomes of the Night.* University of Pennsylvania Press.

Buntin, G. D., J. N. All, R. D. Lee, and D. M. Wilson. 2004. Plant-incorporated *Bacillus thuringiensis* resistance for control of fall armyworm and corn earworm (Lepidoptera: Noctuidae) in corn. *J. Econ. Entomol.* 97:1603–11.

Burgio, G., F. Santi, and S. Maini. 2002. On intra-guild predation and cannibalism in *Harmonia axyridis* (Pallas) and *Adalia bipunctata* L. (Coleoptera: Coccinellidae). *Biol. Cont.* 24:110–6.

Buskirk, R. E., C. Frohlich, and K. G. Ross. 1984. The natural-selection of sexual cannibalism. *Am. Nat.* 123:612–25.

Chapman, J. W., T. Williams, A. M. Martínez, J. Cisneros, P. Caballero, R. D. Cave, and D. Goulson. 2000. Does cannibalism in *Spodoptera frugiperda* (Lepidoptera: Noctuidae) reduce the risk of predation? *Behav. Ecol. Sociobiol.* 48:321–7.

Chapman, J. W., T. Williams, A. Escribano, P. Caballero, R. D. Cave, and D. Goulson. 1999. Fitness consequences of cannibalism in the fall armyworm, *Spodoptera frugiperda. Behav. Ecol.* 10:298–303.

Chilcutt, F. F. 2006. Cannibalism of *Helicoverpa zea* (Lepidoptera: Noctuidae) from *Bacillus thuringiensis* (Bt) transgenic corn versus non-Bt corn. *J. Econ. Entomol.* 99:728–32.

Church, S. C., and T. N. Sherratt. 1996. The selective advantages of cannibalism in a Neotropical mosquito. *Behav. Ecol. Sociobiol.* 39:117–23.

Coll, M. 1998. Living and feeding on plants in predatory Heteroptera. In: *Predatory Heteroptera: Their Ecology and Use in Biological Control*, ed. M. Coll, and J. R. Ruberson, 89–130. Lanham, MD: Entomological Society of America.

Coll, M., and M. Guershon. 2002. Omnivory in terrestrial arthropods: Mixing plant and prey diets. *Annu. Rev. Entomol.* 47:267–97.

Cook, S. F., and K. G. Scott. 1933. The nutritional requirements of *Zootermopsis* (Termopsis) *angusticollis. J. Cell. Comp. Physiol.* 4:95–110.

Cottrell, T. E., and K. V. Yeargan. 1998. Effect of pollen on *Coleomegilla maculata* (Coleoptera: Coccinellidae), population density, predation, and cannibalism in sweet corn. *Environ. Entomol.* 27:1375–85.

Crump, M. L. 1990. Possible enhancement of growth in tadpoles through cannibalism. *Copeia* 1990:560–4.

De Vries, T., and R. Lakes-Harlan. 2007. Prenatal cannibalism in an insect. *Naturwissenschaften* 94:477–82.

Dickinson, J. L. 1992. Egg cannibalism by larvae and adults of the milkweed leaf beetle (*Labidomera clivicollis*, Coleoptera: Chrysomelidae). *Ecol. Entomol.* 17:209–18.

Dobler, R., and M. Kölliker. 2009. Kin-selected siblicide and cannibalism in the European earwig. *Behav. Ecol.* DOI: 10.1093/beheco/arp184.

Dong, Q., and G. A. Polis. 1992. The dynamics of cannibalistic populations: A foraging perspective. In: *Cannibalism: Ecology and Evolution among Diverse Taxa*, ed. M. A. Elgar, and B. J. Crespi, 13–7. Oxford, UK: Oxford Science.

Driessen, G. J. J., A. T. Van Raalte, and G. J. De Bruyn. 1984. Cannibalism in the red wood ant, *Formica polyctena* (Hymenoptera: Formicidae). *Oecologia* 63:13–22.

Eason, E. H. 1964. *Centipedes of the British Isles,* London: Frederick Warne & Co.

Eisenberg, R. M. 1966. The regulation of density in a natural population of the pond snail, *Lymnaea elodes*. *Ecology* 47:889–905.

Elgar, M. A. 1992. Sexual cannibalism in spiders and other invertebrates. In: *Cannibalism: Ecology and Evolution among Diverse Taxa*, ed. M. A. Elgar, and B. J. Crespi, 128–55. Oxford, UK: Oxford Science.

Elgar, M. A., and B. J. Crespi. 1992. *Cannibalism: Ecology and Evolution among Diverse Taxa*. Oxford, UK: Oxford University Press.

Fahy, E. 1972. The finding behaviour of some common lotic species in two streams of differing dentrital content. *J. Zool.* 167:337–50.

Fincke, O. M. 1994. Population regulation of a tropical damselfly in the larval stage by food limitation, cannibalism, intraguild predation and habitat drying. *Oecologia* 100:118–27.

Fischer, Z. 1961. Cannibalism among larvae of the dragonfly *Lestes nympha* Selys. *Ekol. Pol., Series B* 7:33–9.

Fox, L. R. 1975. Cannibalism in natural populations. *Annu. Rev. Ecol. Syst.* 6:87–106.

Gilbert, J. J. 1973. The adaptive significance of polymorphism in the rotifer *Asplanchna*. Humps in males and females *Oecologia* 13:135–46.

Giray, T., Y. A. Luyten, M. MacPherson, and L. Steven. 2001. Physiological bases of genetic differences in cannibalism behavior of the confused flour beetle *Tribolium confusum*. *Evolution* 55:797–806.

Gramacho, K., and L. S. Gonçalves. 2009. Sequential hygienic behavior in Carniolan honey bees (*Apis mellifera carnica*). *Gen. Mol. Res.* 8:655–63.

Grbic, M., P. J. Ode, and M. R. Strand. 1992. Sibling rivalry and brood sex-ratios in polyembryonic wasps. *Nature* 360:254–6.

Gould, F., G. Holtzman, R. L. Rabb, and M. Smith. 1980. Genetic variation in predatory and cannibalistic tendencies of *Heliothis virescens* strains. *Ann. Entomol. Soc. Am.* 73:243–50.

Gould, F. 1983. Genetic constraints on the evolution of cannibalism in *Heliothis virescens*. In: *Evolutionary Genetics of Invertebrate Behavior: Progress and Prospects*, ed. M. D. Huettel, 55–62. New York: Plenum Press.

Guthrie, D. M., and A. R. Tindall. 1968. *The Biology of the Cockroach*. New York: St. Martin's Press.

Hamilton, W. D. 1964. The genetic evolution of social behavior. *J. Theor. Biol.* 7:17–8.

Hastings, A., and R. F. Costantino. 1991. Oscillations in population numbers: Age-dependent cannibalism. *J. Anim. Ecol.* 60:471–82.

Hayes, J. L. 1988. A comparative study of adult emergence phenologies of *Heliothis virescens* (F.) and *H. zea* (Boddie) (Lepidoptera: Noctuidae) on various hosts in field cages. *Environ. Entomol.* 17:344–9.

Herrmann, M., T. Trenzcek, H. Fahrenhorst, and W. Engels. 2005. Characters that differ between diploid and haploid honey bee (*Apis mellifera*) drones. *Genet. Mol. Res.* 4:624–41.

Ho, F. K., and P. S. Dawson.1966. Egg cannibalism by *Tribolium* larvae. *Ecology* 47:318–22.

Holmes, N. D., W. A. Nelson, L. K. Peterson, and C. W. Farstad. 1963. Causes of variations in effectiveness of *Bracon cephi* (Gahan) (Hymenoptera: Braconidae) as a parasite of the wheat stem sawfly. *Can. Entomol.* 95:113–26.

Horner, T. A., and G. P. Dively. 2003. Effect of MON810 Bt field corn on *Helicoverpa zea* (Lepidoptera: Noctuidae) cannibalism and its implications to resistance development. *J. Econ. Entomol.* 96:931–4.

Iqbal, M., and S. A. Aziz. 1976. Cannibalism in *Spathosterum parasiniferum* (Walker) (Orthoptera: Acridoidea). *Indian J. Zool.* 4:43–5.

Jasieński, M., and G. Jasieńska. 1988. Parental correlates of the variability in body mass at hatching in *Arctia caja* larvae (Lepidoptera: Arctiidae). *Entomol. Gen.* 13:87–93.

Joyner, K., and F. Gould. 1987. Conspecific tissues and secretions as sources of nutrition. In: *Nutritional Ecology of Insects, Mites, Spiders and Related Invertebrates*, ed. F. Slansky Jr. and J. G. Rodrigues, 697–719. New York: Wiley-Interscience.

Joyner, K., and F. Gould. 1985. Developmental consequences of cannibalism in *Heliothis zea* (Lepidoptera, Noctuidae). *Ann. Entomol. Soc. Am.* 78:24–8.

Kasuya, F., Y. Hibino, and Y. Ito. 1980. On "intercolonial" cannibalism in Japanese paper wasps, *Polistes chinensis antennalis* Perez and *P. jadwigae* Dalla Torre (Hymenoptera: Vespidae). *Res. Popul. Ecol.* 23:255–62.

Khan, M. R., M. R. Khan, and M. Y. Hussein. 2003. Cannibalism and intersectific predation in ladybird beetle *Coccinella septempunctata* (L.) (Coleoptera: Coccinellidae) in laboratory. *Pak. J. Bio. Sci.* 6:2013–6.

Kinoshita, M. 1998. Effects of time-dependent intraspecific competition on offspring survival in the butterfly *Anthocharis scolymus* (L.) (Lepidoptera: Pieridae). *Oecologia* 114:31–6.

Kirkpatrick, T. W. 1957. *Insect Life in the Tropics.* Longmans, Green and Co.

Lawrence, S. E. 1992. Sexual cannibalism in the praying mantis, *Mantis religiosa*: A field study. *Anim. Behav.* 43:569–83.

Lemos, F. J. A., F. S. Zucoloto, and W. R. Terra. 1992. Enzymological and excretory adaptations of *Ceratitis capitata* (Diptera: Tephritidae) larvae to high protein and high salt diets. *Comp. Biochem. Physiol.* 102A:775–9.

Leon-Beck, M., and M. Coll. 2007. Plant and prey consumption cause a similar reductions in cannibalism by an omnivorous bug. *J. Insect Behav.* 20:67–76.

Majeed, Q., and S. A. Aziz. 1977. Cannibalism in *Gastrimargus transverses* (Orthoptera: Acrididae). *J. Entomol. Res.* 1:164–7.

Majerus, M. E. N. 1994. *Ladybirds.* New Naturalist Series No. 81. London: HarperCollins.

Mally, F. W. 1892. Report of progress in the investigation of the cotton bollworm. *U.S. Dep. Agric., Div. Entomol. Bull.* 26:45–56.

Manzi, J. J. 1970. Combined effects of salinity and temperature on the feeding, reproductive, and survival rates of *Eupleura caudata* (Say) and *Urosalpinx cinerea* (Say) (Prosobranchia: Muricidae). *Biol. Bull.* 138:35–46.

Mason, C. E., M. E. Rice, D. D. Calvin, J. W. Van Duyn, W. B. Showers, W. D. Hutchison, J. F. Witkowski, R. A. Higgins, D. W. Onstad, and G. P. Dively. 1996. European corn borer ecology and management. North Central Regional Publication, Ames, IA: Iowa State University Press.

Matthews, R. W., and J. R. Matthews. 1978. *Insect Behavior.* New York: John Wiley & Sons.

Mock, D. W., and G. A. Parker. 1998. Siblicide, family conflict and the evolutionary limits of selfishness. *Anim. Behav.* 56:1–10.

Mochizuki, A., H. Naka, K. Hamasaki, and T. Mitsunaga. 2006. Larval cannibalism and intraguild predation between the introduced green lacewing, *Chrysoperla carnea*, and the indigenous trash-carrying green lacewing, *Mallada desjardinsi* (Neuroptera: Chrysopidae), as a case study of potential nontarget effect assessment. *Environ. Entomol.* 35:1298–303.

Moritz, R. F. A., and E. E. Southwick. 1992. *Bees as Superorganisms: An Evolutionary Reality.* New York: Springer.

Nakamura, K., and T. Ohgushi. 1981. Studies on the population dynamics of a thistle feeding lady beetle *Henosepilachna pustulosa* in a cool temperate climax forest. II. Life tables, key factor analysis, and detection of regulatory mechanisms. *Res. Popul. Ecol.* 23:210–31.

Nielsen, E. S., and I. F. B. Common. 1991. Lepidoptera (moths and butterflies). In: *The Insects of Australia*, ed. I. D. Naumann, and CSIRO, 33–67. Melbourne, Australia: Melbourne University Press and Cornell University Press.

Ohba, S. Y., K. Hidaka, and M. Sasaki. 2006. Notes on paternal care and sibling cannibalism in the giant water bug, *Lethocerus deyrolli* (Heteroptera: Belostomatidae). *Entomol. Sci.* 9:1–5.

O'Rourke, P. K., and W. D. Hutchison. 2004. Developmental delay and evidence of for reduced cannibalism in corn earworm (Lepidoptera: Noctuidae) larvae feeding on transgenic Bt sweet corn. *J. Entomol. Sci.* 39:294–7.

Osawa, N. 2002. Sex-dependent effects of sibling cannibalism on life history traits of the ladybird beetle *Harmonia axyridis* (Coleoptera: Coccinellidae). *Biol. J. Linn. Soc.* 76:349–60.

Osawa, N. 1993. Population field studies of the aphidophagous lady beetle *Harmonia axyridis* Pallas (Coleoptera: Coccinellidae): Life tables and key factor analysis. *Res. Popul. Ecol.* 35:335–48.

Osawa, N. 1992. Sibling cannibalism in the lady beetle *Harmonia axyridis*: Fitness consequences for mother and offspring. *Res. Pop. Ecol.* 34:45–55.

Pexton, J. J., and P. J. Mayhew. 2004. Competitive interactions between parasitoid larvae and the evolution of gregarious development. *Oecologia* 141:179–90.

Polis, G. A. 1981. The evolution and dynamics of intraspecific predation. *Annu. Rev. Ecol. Syst.* 2:225–51.

Reed, D. J., M. Begon, and D. J. Thompson. 1996. Differential cannibalism and population-dynamics in a host–parasitoid system. *Oecologia* 105:189–93.

Richardson, M. L., P. F. Reagel, R. F. Mitchel, and M. W. Lawrence. 2009. Causes and consequences of cannibalism in non carnivorous insects. *Annu. Rev. Entomol.* 55:39–53.

Rojht, H., F. Budija, and S. Trdan. 2009. Effect of temperature on cannibalism rate between green lacewing larvae (*Chrysoperla carnea* [Stephens], Neuroptera, Chrysopidae). *Acta Agri. Slov.* 93:5–9. DOI:10.2478/v10014-009-0001-5.

Root, R. B., and J. Chaplin. 1976. The life styles of tropical milkweed bugs, *Oncopeltus* (Hemiptera: Lygaeidae), utilizing the same hosts. *Ecology* 57:132–40.

Santana, A. F. K., R. C. Zago, and F. S. Zucoloto. 2011. Effects of sex, host-plant deprivation and presence of conspecific immatures on the cannibalistic behavior of wild *Ascia monuste orseis* (Godart) (Lepidoptera, Pieridae). *Rev. Bras. Entomol.* 55: 95–101.

Santi, F., and S. Maini. 2007. Ladybirds eating their eggs: Is it cannibalism? *B. Insectol.* 69:89–91.

Schausberger, P., and B. A. Croft. 2000. Cannibalism and intraguild predation among phytoseiid mites: Are aggressiveness and prey preference related to diet specialization? *Exp. Appl. Acarol.* 24:709–25.

Sherrat, T. N., S. E. Ruff, and S. C. Church. 1999. No evidence for kin discrimination in cannibalistic tree-hole mosquitoes (Diptera: Culicidae). *J. Ins. Behav.* 12:123–32.

Simpson, S. J., G. A. Sword, P. D. Lorch, and I. D. Couzin. 2006. Cannibal crickets on a forced march for protein and salt. *Proc. Natl. Acad. Sci. USA* 103:4152–6.

Skurdal, J., E. Blegken, and N. C. Stenseth. 1985. Cannibalism in whitefish (*Coregonus lavaretus*). *Oecologia* 67:566–71.

Smith, C., and P. Reay. 1991. Cannibalism in teleost fish. *Rev. Fish Biol. Fish.* 1:41–64.

Snyder, W. E., S. B. Joseph, R. F. Preziosi, and A. J. Moore. 2000. Nutritional benefits of cannibalism for the lady beetle *Harmonia axyridis* (Coleoptera: Coccinellidae) when prey quality is poor. *Environ. Entomol.* 29:1173–9.

Somchoudhary, A. K., and A. B. Mukherjee. 1971. Biology of *Acaropsis docta* (Berlese) a predator on eggs of pests of stored grains. *Bull. Grain Technol.* 9:203–6.

Stam, P. A., L. D. Newsom, and E. N. Lambremont. 1987. Predation and food as factors affecting survival of *Nezara viridula* L. (Hemiptera: Pentatomidae) in a soybean ecosystem. *Environ. Entomol.* 16:1211–6.

Stevens, L. 1992. Cannibalism in beetles. In: *Cannibalism: Ecology and Evolution among Diverse Taxa*, ed. M. A. Elgar, and B. J. Crespi, 156–75. Oxford, UK: Oxford Science.

Stevens, L. 1989. The genetics and evolution of cannibalism in flour beetles (Genus *Tribolium*). *Evolution* 43:169–79.

Strawinski, K. 1964. Zoophagism of terrestrial hemiptera-heteroptera in Poland. *Ekol. Pol. Ser. A.* 12:428–52.

Suttle, K. B. The evolution of sexual cannibalism, 1999, Available at: http://ib.berkeley.edu/courses/ib160/pastpapers/suttle.html.

Tanahashi, M., and K. Togashi. 2009. Interference competition and cannibalism by *Dorcus rectus* (Motschulsky) (Coleoptera: Lucanidae) larvae in the laboratory and field. *Coleopts Bull.* 63:301–10.

Tarpley, M. D., F. Breden, and G. M. Chippendale. 1993. Genetic control of geographic variation for cannibalism in the southwestern corn borer, *Diatraea grandiosella*. *Entomol. Exp. Appl.* 66:145–52.

Tena, A., A. Kapranas, F. Garcia-Mari, and R. F. Luck. 2009. Larval cannibalism during the late developmental stages of a facultatively gregarious encyrtid endoparasitoid. *Ecol. Entomol.* 34:669–76.

Thomas, L. K., and A. Manica. 2003. Filial cannibalism in an assassin bug. *Anim. Behav.* 66:205–10.

Turlings, T. C. J., J. H. Tumlinson, and W. J. Lewis. 1990. Exploitation of herbivore-induced plant odors by host-seeking parasitic wasps. *Science* 250:1251–3.

Vahed, K. 1998. The function of nuptial feeding in insects: Review of empirical studies. *Biol. Rev.* 73:43–78.

Van Buskirk, J. 1992. Competition, cannibalism, and size class dominance in a dragonfly. *Oikos* 65:455–64.

Via, S. 1999. Cannibalism facilitates the use of a novel environment in the flour beetle, *Tribolium castaneum*. *Heredity* 82:267–75.

Weaver, D. K., C. Nansen, J. B. Runyon, S. E. Sing, and W. L. Morril. 2005. Spatial distributions of *Cephus cinctus* (Hymenoptera: Cephidae) and its braconid parasitoids in Montana wheat fields. *Biol. Control.* 34:1–11.

White, T. C. R. 2005. *Why Does the World Stay Green? Nutrition and Survival of Plant Eaters*. Sydney, Australia: CSIRO Publishing.

Whitman, D. W., M. S. Blum, and F. Slansky Jr. 1994. Carnivory in phytophagous insects. In: *Functional Dynamics of Phytophagous Insects*. ed. T. N. Anathakrisnan, 161–205. Lebanon, NH: Science Publishers.

Wilson, E. O. 1971. *The Insect Societies*. Cambridge, MA: Belknap Press of Harvard University Press.

Wise, D. H. 2006. Cannibalism, food limitation, intraspecific competition and the regulation of spider populations. *Annu. Rev. Entomol.* 51:441–65.

Wissinger, S. A., G. B. Sparks, G. L. Rouse, W. S. Brown, and H. Sterltzer. 1996. Intraguild predation and cannibalism among larvae of detritivorous caddisflies in subalpine wetlands. *Ecology* 77:2421–30.

Woyke, J. 1980. Evidence and action cannibalism substance in *Apis cerana indica*. *J. Apic. Res.* 19:6–16.

Yasuda, T. 1997. Chemical cues from *Spodoptera litura* larvae elicit prey locating behavior by the predatory stink bug, *Eocanthecona furcellata*. *Entomol. Exp. Appl.* 82:349–54.

Yom-Tov, Y. 1974. The effect of food and predation on breeding density and success, clutch size and laying date of the crow *Corvus corone*. *J. Anim. Ecol.* 43:479–98.

Zago-Braga, R. C., and F. S. Zucoloto. 2004. Cannibalism studies on eggs and newly hatched caterpillars in a wild population of *Ascia monuste* (Godart) (Lepidoptera, Pieridae). *Rev. Bras. Entomol.* 48:415–20.

Zucoloto, F. S. 1993. Acceptability of different Brazilian fruits to *Ceratitis capitata* (Diptera, Tephritidae) and fly performance on each species. *Braz. J. Med. Biol. Res.* 26:291–8.

9

Implications of Plant Hosts and Insect Nutrition on Entomopathogenic Diseases

Daniel Ricardo Sosa-Gómez

CONTENTS

9.1   Introduction 195
9.2   Starvation and Dietary Stress Effects on Entomopathogenic Diseases.................................... 196
9.3   Host Plant Effects on Bacterial Diseases... 197
      9.3.1   Preingestion Interactions between the Host Plant and Bacterial Agents 197
      9.3.2   The Gut Environment and Bacterial Diseases 197
      9.3.3   Host Plant Effects on Resistance to Bacterial Diseases... 199
9.4   Host Plant Effects on Viral Diseases 199
      9.4.1   Preingestion Interactions between the Host Plant and Viral Agents 199
      9.4.2   The Gut Environment and Viral Disease Interactions.. 199
9.5   Host Plant Effects on Mycoses...200
      9.5.1   Preinfection Interactions between the Host Plant and Fungal Mycopathogens.............200
      9.5.2   Postinfection Interactions between the Host Plant and Mycopathogens.......................201
9.6   Host Plant Effects on Diseases Caused by Nematodes...201
9.7   Assimilated Compounds and Disease Interactions 202
9.8   The Impact of Symbionts on Entomopathogenic Diseases..203
9.9   Interaction between Host Plant Pathogens and Entomopathogens 203
9.10 Nutritional Implications on Entomopathogenic Diseases in Insect Mass Rearing 203
9.11 Conclusions 204
Acknowledgment 204
References...204

9.1 Introduction

The susceptibility of phytophagous arthropods to their respective pathogens is influenced by an array of plant-associated factors. Both macro- and microenvironmental conditions associated with the plant may influence intrinsic properties of the pathogen and host. The plant microtopography may influence entomopathogen persistence and either favor or interfere with the survival of inoculum deposited on the plant surface. Foliar exudates and/or induced plant volatiles caused by herbivory may be antagonistic to entomopathogens on the phylloplane. The ingestion of entomopathogens with plant material and salivary secretions may stimulate or suppress the disease process. For example, there may be interactions between insect gut characteristics, leaf traits (i.e., pH and buffer capacity), entomopathogens, and their toxins (Figure 9.1). At this stage, gut-associated flora may also play an important role in disease expression. Insect-assimilated secondary plant components translocated into either the hemocoel or deposited onto the cuticle may further interface with the insect pathogen.

The impacts of insect nutrition on diseases are complex and often involve multitrophic interactions. Nutrition affects susceptibility to a certain pathogen or toxin, and food inputs affect the production of progeny inocula from diseased hosts (Cory and Myers 2004). Additionally, nutrition by altering either

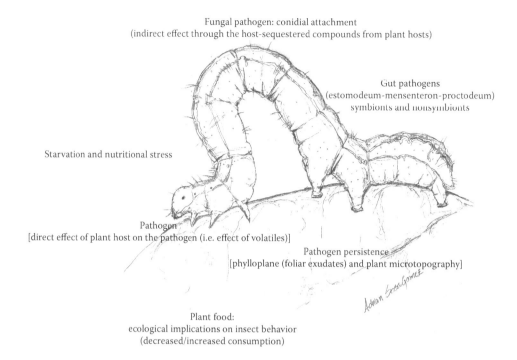

FIGURE 9.1 Illustration of potential host plant factors affecting causal agents of entomopathogenic diseases in a leaf-chewing insect.

the expression of a resistant trait or the functional dominance of resistant alleles strongly influences evolutionary trends between the insect host and entomopathogens (Janmaat and Myers 2006b). This chapter focuses on the direct and indirect effects of food (host plants, blood meal) and chemical compounds on the disease process of agriculturally important arthropods (insects and mites). Direct effects of volatiles and plant structure on entomopathogens, and the impact of symbionts on the nutritional aspects and health of insect diseases are also discussed.

9.2 Starvation and Dietary Stress Effects on Entomopathogenic Diseases

The current paradigm is that "stress" will increase host susceptibility to disease. Presently, the physiological, biochemical, and metabolic routes underlying stressed induced changes to disease are poorly understood. Studies on insect diseases have demonstrated that impacts of starvation, food quality, or dietary stress have unpredictable consequences on the infection process. In certain cases, starvation significantly increases disease-induced morbidity whereas in others starvation stress has no impact on host survival. For example, no starvation effect is observed for *Triatoma infestans* Klug nymphs by *Beauveria bassiana* conidia (Luz et al. 2003), whereas starvation of *Chrysoperla carnea* (Stephens) for 48 h prior to *B. bassiana* conidia exposure significantly increases mortality (Donegan and Lighthart 1989). Starvation of *Leptinotarsa decemlineata* (Say) larvae increased susceptibility, reducing the mean number of days before death from 7 (fed) to 5 (starved) in second instar larvae (Furlong and Groden 2003). In this case, application of *B. bassiana* delayed larval molting. The molting process, involving the shedding of the old cuticle, will remove entomopathogenic propagules from the cuticle, which may disrupt the penetration process and lower infection. Reduced susceptibility to *Pseudomonas fluorescens* infection is observed in the larval stage of the aphid predator, *Hippodamia convergens* Guérin-Méneville, when fed only water compared to larvae fed *Acyrthosiphon pisum* (Harris); starvation increased tolerance to *P. fluorescens* 150-fold (James and Lighthart 1992). Larvae of the Indian meal moth, *Plodia interpunctella* (Hubner), fed a high-quality diet were more susceptible to viral infection by *Plodia interpunctella* granulosis virus (*Pi*GV) than larvae fed on a low-quality diet (McVean et al. 2002). These effects should be considered in microbial control programs to improve the efficacy of insect pathogens or to understand epizootiological phenomena in the field and in artificial mass rearing.

9.3 Host Plant Effects on Bacterial Diseases

9.3.1 Preingestion Interactions between the Host Plant and Bacterial Agents

Exudates produced on the plant surface affect microorganisms inhabiting the phylloplane (McCormack et al. 1994). However, their interactions with entomopathogenic organisms present in this microhabitat need to be determined. The insect toxin-producing strains of *Bacillus thuringiensis* (Bt) are able to colonize plant surfaces, but compared to various *Pseudomonas* species, Bt seem to be a poor colonizers (Maduell et al. 2008). This study suggests that nutrient levels on common bean *Phaseolus vulgaris* leaves are not sufficient to enable Bt substantial growth.

9.3.2 The Gut Environment and Bacterial Diseases

Most studies on the impact of plant hosts on bacterial diseases are concerned with Bt, its toxins, and Lepidoptera (Hwang et al. 1995; Kleiner et al. 1998; Kouassi et al. 2001; Broderick et al. 2003). Physiological and biological traits of phytophagous arthropods can be highly influenced by compounds present in foliar tissues. Usually, allelochemicals that provide protection against herbivores interact with pathogenic microorganisms and regulate disease. However, in some circumstances, food quality may not alter the susceptibility of host. High mortality is simply due to an indirect interaction, whereby the ingestion of a larger amount of a preferred food results in the uptake of higher levels of pathogen or toxin. For example, Salama and Abdel-Razek (1992) found that larvae of *P. interpunctella* and *Sitotroga cereallela* (Olivier) suffered higher mortalities from Bt δ-endotoxin applied to crushed corn than to either whole grains or whole corn kernels; larvae preferred crushed corn and hence obtained more toxins with the preferred ingested food.

In contrast, phenolic glycosides from the quaking aspen, *Populus tremuloides* Michaux directly increased larval development time and decreased growth rates in both the gypsy moth, *Lymantria dispar* L. and the forest tent caterpillar, *Malacosoma disstria* Hübner (Hemming and Lindroth 2000). Survival of the first instar of *L. dispar* was reduced to 76% when fed with phenolic glycosides, to 89% when inoculated with Bt var. *kurstaki*, and to 66% when fed a combination of these agents. These results suggested an additive effect; potentially both agents caused by lesions in the midgut (Arteel and Lindroth 1992).

Orthoquinones, produced in tomato plants in response to herbivory and by the action of oxidative enzymes (polyphenol oxidases and chlorogenic acid), interact with Bt crystal proteins, leading to their solubilization and enhancing their toxicity to *Helicoverpa zea* (Boddie) (Ludlum et al. 1991). Increased mortality for Bt is also observed when the amino acid L-canavanine is added to the artificial diet of *Manduca sexta* (L) (Felton and Dahlman 1984). L-canavanine is synthesized by members of Lotoidea, a subfamily of Leguminosae, and could be toxic for sensitive insects such as *M. sexta*, but is less toxic to tolerant insects such as *Heliothis virescens* (F). Cinnamic acid also increases mortality of the sunflower moth, *Homoeosoma electellum* (Hulst) (Brewer and Anderson 1990).

Condensed tannin is considered an antagonist to Bt toxicity because it reduces *H. virescens* larval mortality when compared to Bt-induced mortality. Plant tannins may either act as a larval feeding deterrent or interfere with Bt's mode of action (Navon 1992). However, Kleiner et al. (1998) did not find any relationship between condensed tannin concentration in hybrid poplar (*Populus*) leaves and Bt efficacy against gypsy moth. Linear furanocoumarins, compounds found in Apiacea (Umbellifera) and Rutaceae, also has deterrent properties for *Spodoptera exigua* (Hübner) larvae. When added to a commercial formulation of Bt, these can reduce *S. exigua* feeding by acting independently, and can cause a mildly antagonistic effect on preference of diet amended with furanocoumarin (Berdegué and Trumble 1997). Other nonallelochemical compounds (such as ascorbic acid) present in food in suboptimal or supraoptimal concentrations may also affect susceptibility to bacterial infections. Prohemocyte and phagocyte counts are reduced in the hemolymph when ascorbic acid is present at unsuitable levels, correlating with a high mortality caused by Bt (Pristavko and Dovzhenok 1974).

Interactions with host plant compounds can reduce or accelerate different phases of the infection process. In *L. dispar*, the pH of the midgut is alkaline and is maintained independent of the ingested food. The foregut and hindgut pH and the oxidizing capacity of the entire gut are host-plant dependent (Appel and Maines 1995). The high alkalinity of the midgut significantly enhances protein extraction and the solubilization of cellular proteins from host plants, as well as affecting Bt protoxin cleavage into active toxins. Presently, less is known about the *in vivo* processing of the protoxin than the downstream steps such as toxin binding to brush border membrane of midgut cells, membrane insertion, pore formation, and osmotic shock of the midgut mesenteron cells (Ferré and Van Rie 2002).

An additional factor that may modulate bacterial disease expression is the insect intestinal microbiota. It may protect against pathogen attack (Raymond et al. 2009) or have no effect (Johnston and Crickmore 2009). Sequence analysis conducted on 16S rRNA and terminal restriction fragment length polymorphism analysis (T-RFLP) of the midgut microflora associated with gypsy moth larvae, *L. dispar*, showed that diet significantly impacted microbial diversity (Broderick et al. 2004). The capacity of these microorganisms to exclude harmful organisms will dictate a range of responses to entomopathogenic diseases. Gypsy moth mortality caused by CryIA(a) and CryIA(c) insecticidal proteins may be enhanced by feeding host insects various red maple epiphytic bacterial species. Spores from *Bacillus cereus, B. megaterium*, and from an acrystalliferous strain of Bt (nonepiphytic, HD-73 cry−) act as synergizers of CryIA proteins. Also, *Klebsiella* sp., *K. pneumonia, P. fluorescens, Xanthomonas* sp., *Actinomyces* sp., *Corynebacterium* sp., and *Flavobacterium* sp. are synergizers of at least one of the CryIA toxins. In the absence of the CryIA toxins, none of the identified bacterial or spore synergists are toxic or inhibit larval growth or molting (Dubois and Dean 1995).

Controversial pathogenic mechanisms have been proposed for the interaction between the gut flora of some species of Lepidoptera and Bt (Broderick et al. 2009; Johnston and Crickmore 2009; Raymond et al. 2009). According to Broderick et al. (2009), antibiotics administered *per os* reduce larval mortality to Bt (Dipel®) in *M. sexta, Vanessa cardui* L., *Pieris rapae* (L.), *L. dispar*, and *H. virescens*, possibly due to a reduction in gut bacteria prior to treatment. Reestablishment of the gut flora with *L. dispar* indigenous *Enterobacter* sp. restores larval susceptibility to Bt. *H. virescens* is the exception, where detectable gut bacteria are not observed before treatment, but antibiotic treatment also delays the kill time required by Bt. The reduction of mortality caused by antibiotics can be proved because mixtures of spores and toxins are virulent to *Plutella xylostella* (L.) in the absence of midgut bacteria in aseptically reared hosts, in which the residual effect of antibiotics is excluded (Raymond et al. 2009). Johnston and Crickmore (2009) obtained similar results with the tobacco hornworm, *M. sexta*; in which the insecticidal activity of Bt or the CryIAc toxin does not depend on the presence of gut bacteria.

Anopheline mosquitoes can harbor bacteria in their midguts. The most frequent species are Gram-negative rods, from the genera *Pseudomonas, Cedecea, Pantoea, Flavobacterium* (Pumpuni et al. 1996; Gonzalez-Cerón et al. 2003). In mosquitoes, blood feeding causes a pronounced increase in gut microbiota; bacterial populations may increase 11- to 40-fold, reaching 10^7 colony-forming units per milliliter. This significant increase in bacteria has been reported to elicit antibacterial and anti-*Plasmodium* immune responses (Pumpuni et al. 1996; Cirimotich et al. 2010). Simultaneous feeding of live or inactivated bacteria with these parasites decreased *Plasmodium falciparum* abundance. Mosquitoes with nondetectable bacteria in the midgut are more susceptible to *P. falciparum* infection. The resistance to bacterial and plasmodial challenges is related to proteins that are not related to specific immune responses (Cirimotich et al. 2010). According to Gonzalez-Cerón et al. (2003) midgut bacterial presence in *A. albimanus* Weidemann field populations may interfere on *P. vivax* transmission, and contribute to explain variations in malaria incidence in human populations.

The ingestion of nonnutritive particles, such as charcoal and carmine, offers protection against Bt. Median lethal concentrations of Bt subsp. *israelensis* against *Aedes aegypti* L. and subsp. *kenyae* against *S. littoralis* are approximately 20–217 and 2.3–44 times higher, respectively, in the presence of nutritional or nonnutritional particles (Ben-Dov et al. 2003). The ingested particles protect the midgut epithelial cells, thus reducing the larvicidal effect of Bt toxins and preventing their binding to receptors.

9.3.3 Host Plant Effects on Resistance to Bacterial Diseases

Lepidoptera resistance to Bt transgenic plants has been a primary concern in with the wide-scale use of recombinant plants containing insect resistant transgenes. Recently, resistance to Cry1F toxins in populations of *S. frugiperda* has been reported in Puerto Rico (Storer et al. 2010). Studies on the possible costs of resistance are important for understanding how the host plant may interfere with insect–pest biology. Janmaat and Myers (2005) compared the growth rate, pupal size, and survival of Bt-resistant, susceptible, and hybrids of cabbage looper *T. ni* larvae fed on tomato, bell pepper, and cucumber, and concluded that the fitness cost associated with Bt resistance increases with a reduction in host plant suitability. The least suitable food source for *T. ni* was pepper and the fitness cost of the Bt-resistant population was associated with low or no survivorship to pupation on pepper and reduced survival, fecundity, as well as pupal weight on pepper and tomato (Janmaat and Myers 2005, 2006a). Reduction in fecundity of the Bt-resistant population is followed by a reduction in offspring growth. Thus, food and particularly the environment experienced by the parents is reflected in progeny phenotypic variation, and consequently, in their response to pathogenic diseases (Janmaat and Myers 2006a).

9.4 Host Plant Effects on Viral Diseases

9.4.1 Preingestion Interactions between the Host Plant and Viral Agents

The host plant influences both the environmental persistence of viral pathogens and the viral infection process *per se*. A nucleopolyhedrovirus of the winter moth, *Operophtera brumata* (L.), persists more on Sitka spruce, *Picea sitchensis*, and oak, *Quercus robur*, than on heather, *Calluna vulgaris*, due to the shady environment that provides protection against ultraviolet irradiation (Raymond et al. 2005). Also, insect viruses are more persistent in stems and older trees with a more pitted bark than on leaves, due to the protection afforded by the more complex surface microtopography. Insects and mites living on leaves are exposed to the infections of phylloplane-inhabiting pathogens, for that reason foliar exudates can have profound influences modulating disease. The activity of *Heliothis armigera* (Hübner) nuclear polyhedrosis virus on cotton is reduced in four days but when applied to sorghum the inoculum remains active for one month (Roome and Daoust 1976). Coincidently, reduced mortality of *H. zea* has been observed in bioassays with HzNPV inoculated with cotton leaf discs compared to tomato leaf discs (Forschler et al. 1992; Farrar and Ridgway 2000). One of the best examples of the impact of plant surface on viral pathogens is the reported detrimental effect of cotton dew alkalinity (pH 8.5 to 9.1) on the biological activity of *Heliothis* nuclear polyhedrosis virus. After 7 days, polyhedral occlusion bodies from cotton and soybean dew resulted in 20% and 88.7% mortality, respectively (Young et al. 1977).

9.4.2 The Gut Environment and Viral Disease Interactions

Various intrinsic factors associated with the gut lumen, including pH, composition of food bolus, rate of food passage, influence the viral contact with the midgut microvilli, where primary viral infection occurs in the gut cells (Adams and McClintock 1991; Peng et al. 1997). Susceptibility of insects to viral infection can be influenced significantly by the plant material in the food bolus. For example, *H. zea* larvae are less susceptible to HzNPV when fed on the reproductive tissues of soybeans, velvetleaf, *Abutilon teophrasti*, crimson clover, *Trifolium incarnatum*, and Carolina geranium, *Geranium carolinium*, whereas *H. zea* larvae are more susceptible when fed on cotton reproductive structures compared to foliage (Ali et al. 1998).

Assays involving the velvetbean caterpillar, *Anticarsia gemmatalis* (Hübner), fed host plant leaves until the end of the second instar then challenged with AgNPV occlusion bodies as newly molted third instar, showed that the type of host plant affects host susceptibility to the virus. Larvae fed deer pea, *Vigna luteola*, are more resistant to AgNPV than those fed snout bean, *Rhynchosia minima* or soybean, *Glycine max*. However, control larvae (without viral infection) fed on *G. max* and *V. luteola* have a shorter development time and a greater pupal weight, indicating that they are the most suitable plants for *A. gemmatalis* larval development. Therefore, Peng et al. (1997) concluded that insect nutritional stress or host plant suitability does not affect velvetbean caterpillar susceptibility to the virus. However, Richter et al. (1987) did not observe this for the polyphagous fall armyworm larvae. Third instar larvae were less susceptible to NPV infection when fed on corn, brown top-top millet, and signalgrass than when fed on soybean, Bermuda grass, ryegrass, or sorghum. Differences in susceptibility were related to nutritional stress, but the possibility of plant antiviral components could not be excluded (Richter et al. 1987). Another rarely reported effect is the influence host plant on the yield of occlusion bodies produced by diseased insects. Western tent caterpillar, *Malacosoma californicum pluvial* (Dyar) larvae inoculated with the *Mcpl*NPV on alder, *Alnus rubra,* produced more progeny virus than those inoculated on wild rose, *Rosa nutkana,* or apple, *Pyrus malus.* The expectation was that insects that lived longer, providing the virus more biomass, would result in increased numbers of viral progeny. However, in the case of western tent caterpillar, larvae that died fastest on alder also produce more viruses (Cory and Myers 2004).

Biochemical conditions in the midgut may inhibit or favor ingested microbes. These conditions are highly variable depending on the food, colonizing organisms, developmental stage, and species. The pH in Lepidoptera usually ranges from 8 to 12. Even in the same species, the pH varies considerably in different parts of the gut, but is usually higher in the medial midgut (Dow 1992; Schmidt et al. 2009). Protease activity and diet pH may affect occlusion body solubilization and alter the rate of virion release in the midgut lumen (Wood 1980; Keating et al. 1990b). The pH in the *L. dispar* midgut could be influenced by the foliage type, and despite the midgut buffer capacity, larval susceptibility to nuclear polyhedrosis virus can be lower with acidic diets (Keating et al. 1988, 1990a). Scanning electron microscopy showed that a plant defense cysteine protease of 33 kDa severely damaged the peritrophic membrane, impairing the normal growth of *S. frugiperda* caterpillars feeding on resistant maize lines (Pechan et al. 2002). The peritrophic membrane is an important barrier in the midgut and is believed to restrict the ingress of microbes from accessing the mircovilli midgut layer (Lehane 1997). Certain properties of the peritrophic membrane can be influenced by the food ingested. For example, the thickness of the peritrophic membrane is influenced by the quality of the food ingested; it is thicker in larvae fed on cotton, oakleaf lettuce, or iceberg lettuce foliage than in larvae fed an artificial diet (Plymale et al. 2008). However, larvae fed either tobacco or artificial diet had the peritrophic membranes of similar widths. The number of layers comprising the peritrophic membrane was greater in plant-fed than in artificial diet–fed larvae.

9.5 Host Plant Effects on Mycoses

9.5.1 Preinfection Interactions between the Host Plant and Fungal Mycopathogens

The available nutritional sources have a profound impact on the disease determinants of entomopathogenic fungi (Shah et al. 2005), including conidial germination speed, attachment to the cuticle, penetration peg formation, and production of mucilage. However, little is known about the host characteristics that influence either the disease determinants or the disease dynamics within arthropod populations. Several species of fungi are important causal agents of epizootic diseases (Sosa-Gómez et al. 2010). Notwithstanding, the effect of the plant on the epizootic expression is not well understood. The influence of food plant on disease has been observed in populations of the pea aphid *Acyrthosiphon pisum* Harris. Pea aphids are killed by the entomophthoralean *Pandora neoaphidis* at higher proportions (approximately four times) on pea varieties that have a reduced surface wax bloom (Duetting et al. 2003). The reduced wax layer is believed to increase both adhesion and germination of *P. neoaphidis* conidia on the leaf surfaces.

Leaf discs of cassava, *Manihot esculenta* infested with the cassava green mite, *Mononychellus tanajoa* (Bondar), release plant volatiles that favor the production of primary conidia of the acaropathogenic fungus *Neozygites tanajoae* on mite mummies (Houndtondji et al. 2005). However, other studies indicate that tobacco plant volatiles induced by aphid feeding negatively impacted the germination of *P. neoaphidis* (Brown et al. 1995). Alternatively, no significant effects were observed on *in vivo* sporulation, conidia size, or *in vitro* growth rate of *P. neoaphidis* when the fungus was exposed to volatiles of *Vicia faba* damaged by *A. pisum* (Baverstok et al. 2005).

In some pathogen–arthropod systems, the physical (topography) and chemical factors that interact with fungal inoculum have been identified. *Metarhizium anisopliae* conidia deposition is higher along the junctions between epidermal cells of Chinese cabbage and in the leaf surface depressions of turnip leaves (Inyang et al. 1998; Inyang et al. 1999). Conidia appear to germinate faster on older compared to younger leaves, and conidia are more easily washed off from older leaves. Conidial attachment to the thorax and abdomen of the mustard beetle, *Phaedon cochleariae* (F.), is also influenced by the host plant, with attachment rates of 63% (oilseed rape), 53% (Chinese cabbage), and 43% (turnip). Similarly, *Frankliniella occidentalis* (Pergande) may acquire more *B. bassiana* conidia deposited on bean, *P. vulgaris,* leaves than on impatiens, *Impatiens wallerana.* The leaf veins of beans create ridges that acquire and accumulate conidia and that are favored by the thigmotactic behavior of thrips (Ugine et al. 2007). Phytochemicals excreted by plants may directly interfere with conidial viability and germination. The glycoalkaloids solanine and tomatine reduce *in vitro* growth of *B. bassiana*, although solanine is less toxic to *B. bassiana* than tomatine (Costa and Gaugler 1989).

9.5.2 Postinfection Interactions between the Host Plant and Mycopathogens

Infective units of acaropathogenic and entomopathogenic fungi show different strategies in the first steps of the infection process. On the outer layer of the exoskeleton, epicuticular compounds mediate the attachment through hydrophobic interactions of dry conidia, such as those produced by *B. bassiana*, *M. anisopliae*, and *N. rileyi* (Boucias et al. 1998). Little is known about the factors that modulate interactions in the attachment of hydrophilic conidia that are covered with a mucus layer as in *Hirsutella thompsonii*, *Lecanicillium lecanii*, or adhesive papillae as in *Neozygites* species (Boucias et al. 1998). However, there have been no studies to determine the effects of insect nutritional stress or host plant traits on these initial interactions, although one can assume that nutrition impacts epicuticle chemistry.

Early steps in the fungal infection process (i.e., conidial attachment and germination) and appressorium formation on the cuticle may also be affected by exocuticle exudates. Most published information refers to the effect of resistant host plant factors favoring mortality by an entomopathogen. Under experimental conditions, resistant tomatoes, *Lycopersicon hirsutum* f. *glabratum*, genotypes such as PI 134417 favor more mycoses caused by *B. bassiana* on *Tuta absoluta* (Meyrick) than the Santa Clara cultivar, *L. esculentum* (Giustolin et al. 2001). These factors also affect the kill time, spore or conidia production, and the effect on disease abundance. A number of reports mention the indirect influence of host plants on the kinetics of infection and on the degree of sporulation. For example, at the same concentration, *B. bassiana* conidia will kill *Bemisia tabaci* (Gennadius) nymphs within 4.7 and 6.9 days reared on cucumber or green pepper, respectively. Some members of the Cucurbitaceae, such as melon, cucumber, and marrow, favor *B. bassiana* sporulation on cadavers of *B. tabaci*, whereas hosts such as cotton, cabbage, and pepper do not (Santiago-Alvarez et al. 2006).

The gut flora of the insects can also influence mycopathogens. The subterranean termite, *Reticulitermes flavipes* (Kollar), produces volatile fatty acids with antimycotic activity. Microorganisms present in regurgitants and fecal pellets used during the construction of soil tunnels act as barriers against soil microbiota (Boucias et al. 1996; Boucias and Pendland 1998). Antimycotic compounds are also produced by the bacterium *Pantoea agglomerans* in the desert locust's (*Schistocerca gregaria* Forskal) gut and adversely affect pathogenic insect fungi (Dillon and Charnley 2002).

9.6 Host Plant Effects on Diseases Caused by Nematodes

Plant roots exude biologically active compounds into the rhizosphere that significantly influence several organisms by increasing or reducing herbivore populations and by attracting parasites and predators (Bais et al. 2006). Volatiles emitted by a damaged host plant are known to attract nematodes. Rasmann et al. (2005) identified (E)-β-m caryophyllene, a sesquiterpene plant signal emitted by maize plants below ground, which attracts *Heterorhabditis megidis,* an entomopathogenic nematode that causes disease in the corn rootworm *Diabrotica virgifera virgifera* LeConte. This signal, produced by European lines of maize, is not released by North American lines. Several species of mutualistic bacteria from the genera *Photorhabdus* and *Xenorhabdus* that are associated to *Heterorhabditis* and *Steinernema* nematodes are responsible for the pathogenic process and for killing the host insect. Most of the papers published about nematodes consider the indirect effects of host plants on the virulence of entomopathogenic nematodes and/or their bacterial symbionts (Epsky and Capinera 1994; Barbercheck et al. 1995; Barbercheck et al. 2003) and on their reproductive capacity (Barbercheck 1993; Barbercheck et al, 2003; Shapiro-Ilan et al. 2008). Barbercheck (1993) observed the harmful effect on rootworm larvae infected with nematodes and fed on different host plants. Mortality caused by *Steinernema carpocapsae* Weiser and *Heterorhabditis bacteriophora* Poinar is higher in larvae of *Diabrotica undecimpunctata howardi* Barber reared on the roots of squash, *Cucurbita maxima* Duchesne. However, when the rootworms are fed on peanuts, *Arachis hypogea*, mortality caused by *S. carpocapsae* is lower than for insects challenged with *H. bacteriophora*. On the other hand, larvae fed on corn suffer low mortality by *H. bacteriophora* and high mortality by *S. carpocapsae*. The nematode progeny of both species are also negatively affected by squash roots.

The antibiotic effect of cucurbitacin D has been observed for the symbiotic bacteria, *Xenorhabdus* and *Photorhabdus*, associated with the entomopathogenic nematode families, Steinernematidae and Heterorhabditidae, respectively. Cucurbitacin D inhibits the growth of bacteria from *S. carpocapsae* and one strain from *H. bacteriophora*, but does not inhibit other bacterial strains isolated from the nematodes *Steinernema glaseri*, *S. riobravis*, *H. bacteriophora*, or *Heterorhabditis* sp. However, cucurbitacin enhances *in vitro* growth of the bacterial strain isolated from *S. glaseri*. Cucurbitacin may indirectly affect nematode progeny and its negative effect on bacterial symbionts may be critical to nematode fitness (Barbercheck and Wang 1996).

Variable responses have been observed in experiments examining how different isolates of *Steinernema* behave after being continuously cultured (25 passages) in corn-fed or squash-fed southern corn rootworm *D. undecimpunctata howardi*. Two nematode isolates propagated in corn-fed rootworms were able to kill corn-fed rootworms more efficiently than the same isolates propagated in squash-fed rootworms. A squash-selected Mexican isolate became less virulent than nematodes reared on *Galleria mellonella* L. to kill squash-fed rootworms. Therefore, virulence changes and offspring (number of infective juvenile per rootworm) may be modified by the food plants on the insect develop and by the nematode isolate involved (Barbercheck et al. 2003).

9.7 Assimilated Compounds and Disease Interactions

Various stages of the disease process can be affected indirectly by the same plant compounds. For example, biologically active phytochemicals can act as phagostimulants for specialized luperini beetles, which may use them as a cue for host plant recognition. Cucumber beetles may gain protection against predators and parasitoids by sequestering highly bitter cucurbitacin triterpenes (Tallamy et al. 1998). This compound also seems to act against entomopathogens. Larvae of *Diabrotica undecimpunctata howardi* Barber fed on a diet rich in cucurbitacin express increased resistance to infection by the mycopathogen, *M. anisopliae*. Sequestered host-derived cucurbitacins can be transferred transovarially from females to the next generation and during mating through spermatophores. Thus, eggs produced by adults fed on *Cucurbita andreana* Naud. are more resistant to fungal infection. Larvae fed *C. maxima* "Blue Hubbard" cultivar exhibit a better chance of survival than those fed *C. pepo* "Yellow Crookneck," which contains traces of cucurbitacins in its roots (Tallamy et al. 1998; Tallamy et al. 2000). *Longitarsus melanocephalus* (DeGeer) flea beetles sequester iridoid glycoside compounds (up to 2% of their dry weight) from their host plant, *Plantago lanceolata*. These compounds show biological activity against bacteria and fungi. *In vitro* studies with iridoid glycosides extracted from *P. lanceolata* demonstrate antibacterial activity to Bt subsp. *kurstaki* and *Bt. tenebrionis,* but no activity was detected against the entomopathogenic fungi *M. anisopliae* or *B. bassiana* (Baden and Dobler 2009).

Arthropods can regulate their body temperature through behavior such as searching for warmer microenvironments or resting at the top of plants. These behaviors contribute to observed resistance to microbial infections caused by ricketssia and fungi (Louis et al. 1986; Ouedraogo et al. 2003). Because thermogenesis capacity depends on the ingested diet (Trier and Mattson 2003), diet may impact the immune response that is related to thermoregulation, as observed by Ouedraogo et al. (2003). However, the significance of the food for thermoregulation in each diet–arthropod–entomopathogen model remains to be determined.

9.8 The Impact of Symbionts on Entomopathogenic Diseases

Besides their nutritional role, symbionts are also important as protective agents against parasitoids and microbial infections (Haine 2008). Heddi et al. (2005) reported *Sitophilus zeamais* Motschulsky immune molecular responses toward endosymbiont invasion. They found six expressed sequence tags encoding a single peptidoglycan recognition protein (*PGRP*) gene. The *PGRP* gene family is known to trigger the immune defense system against fungi and bacteria.

The bacterium *Regiella insecticola*, a facultative endosymbiont of the pea aphid, *A. pisum*, confers host resistance to the widely distributed *P. neoaphidis* fungus. Also, the presence of *Regiella* in the aphids reduces the probability of successful sporulation by this mycopathogen (Scarborough et al. 2005). This symbiont is found most frequently in aphids feeding on *Trifolium*. When strains of *R. insecticola* are injected into aphid clones, the bacteria reduced the rate of acceptance of *V. faba*. However, in acceptance and performance assays on *Trifolium pretense*, the strain of bacteria seemed important in the specialization process (Ferrari et al. 2007). In contrast, no effects of *Wolbachia* infection, an intracellular symbiotic α-proteobacteria, have been observed in *Drosophila simulans* Sturtevant flies when inoculated with *B. bassiana*. However, Fytrou et al. (2006) implied that the conidial concentration was too high for determining subtle symbiont effects. On the other hand, Teixeira et al. (2008) found that *Wolbachia* increases the resistance of *Drosophila melanogaster* Meigen to RNA viral infections, such as a *Drosophila C* virus and Nora virus, which are both naturally occurring pathogens in *Drosophila*. *D. melanogaster* is also resistant to a Flock house virus, which is not its natural pathogen, but this resistance to infection does not extend to insect iridescent virus type 6, a DNA virus.

Bacterial diseases may be impacted by the presence/absence of microbial symbionts. Nonpathogenic *B. cereus* strains seem to serve as symbionts in cockroaches, termites, coreids, and isopod crustaceans (Singh 1974; Margulis et al. 1998). Raymond et al. (2007) reported that nonpathogenic *B. cereus* competes with Bt var. *kurstaki* in mixed infections of both bacteria in the diamondback moth. They observed that Bt replication is reduced when the nonpathogenic strain is present; *B. cereus* grows faster than Bt var. *kurstaki* in the host. The consequences of mixed infection on Bt virulence have not been reported. Some *B. cereus* strains, a normal constituent of gut flora and a soil-inhabiting bacterium can produce antibiotic compounds such as zwittermicin A (Stabb et al. 1994). This antibiotic has synergistic properties when inoculated with Bt var. *kurstaki* on *L. dispar*.

9.9 Interaction between Host Plant Pathogens and Entomopathogens

In terms of insect pathology, relatively few studies have addressed the impact of plant disease on host inset fitness. Plant host tissues attacked by phytopathogenic diseases undergo biochemical changes that likely have nutritional effects on arthropod herbivores. Larvae of the mustard leaf beetle, *P. cochleariae*, reared on Chinese cabbage leaves, *Brassica rapa* ssp. *Pekinensis*, and infected by the *Alternaria brassica* (Berk.) Sacc. fungus, suffer higher mortality when exposed to conidia of the *M. anisopliae* compared to larvae fed on healthy plants (Rostás and Hilker 2003). They attributed the lower susceptibility of larger larvae to a more effective immune response. Small larvae produce fewer defensive secretions from their exocrine glands than larger insects, and these secretions have antimicrobial properties. They also considered that the phytopathogenic agents induce phytochemicals in the foliar tissues

9.10 Nutritional Implications on Entomopathogenic Diseases in Insect Mass Rearing

Insect mass rearing is used in the silk industry, in factories producing microbial agents, for producing pharmacoproteins using baculoviruses as expression vectors, and in mass release insect programs that use sterile insects to compete with wild insect populations (Maeda et al. 1985). Therefore, cost/benefit relationships, efficiency, and competitive individuals are highly desirable for program success. Nutrition directly impacts all these aspects, and therefore care must be taken to obtain healthy and competitive insects. Stressor agents can be important predisposing factors for high disease levels in mass rearing laboratories (Fuxa et al. 1999). A spectrum of viral, bacterial, fungal, and microsporial agents impacts mass rearing facilities (Sikorowski and Lawrence 1994). Control of these diseases can be achieved by identifying the entomopathogenic organisms and determining and controlling the contributing stress factors. The most important stress factors are the changes in diet, unsuitable humidity conditions, high carbon dioxide levels, low aeration, excessive handling, the presence of contaminants (fungi, bacteria) usually caused by fecal accumulation and crowding that occur in mass rearing facilities. The choice of mulberry varieties can be critical for reducing viral diseases in the mass rearing of silkworm, *Bombyx mori* L. Mulberry genotypes HN-64 and Miura favor a high incidence of grasserie, which is caused by *Bm*NPV in silkworm populations. However, silkworms fed on the Calabreza genotype have a low or absent disease prevalence (Sosa-Gómez et al. 1991). Entomopathogenic nematodes are cultured *in vitro* or *in vivo* for large-scale commercial production. When production is *in vivo*, several insect hosts can be used. Yield and nematode quality are important parameters for achieving a desirable result and efficient microbial control. The response of *Heterorhabditis indica* and *Steinernema riobrave* to the nutritional quality of the host *Tenebrio molitor* L. is species dependent (Shapiro-Ilan et al. 2008). *T. molitor* fed with a higher starch:lipid ratio is more susceptible to *H. indica* than when fed with a lower starch:lipid ratio. In contrast, lipid supplements did not affect the host susceptibility to *S. riobrave* infection, and protein supplements did not affect the susceptibility of *T. molitor* to either nematode species.

Diseases in mass-reared insects have serious implications for mortality, life cycle, fecundity, fertility, and pheromone production, all potentially resulting in low body weight and altered longevity and insecticide susceptibility (Sikorowski and Lawrence 1994). Understanding these interactions is also important for explaining the epizootiology of entomopathogenic diseases in natural and agricultural settings and for handling and improving mass rearing of insects by eliminating plant stressors from artificial diets or selecting the appropriate cultivar or variety when reared on natural diets.

9.11 Conclusions

The understanding of nutritional aspects and their implications on arthropod diseases may serve multifunctional roles both for disease prevention in beneficial insect populations and for enhancing the incidence of disease in pest populations. To be able to achieve these goals, many questions still remain to be answered. What effect does the nutrition have on their arthropod defense systems? Can specific dietary factors be used either to upregulate or to suppress these defense barriers? What are the implications of nutrition on diseases as pertaining to reproductive performance? How can nutrition impact evolution/ selection for disease resistance? Can nutritional inputs such as plant secondary compounds negate the onset of resistance to microbial control agents? Nutritional studies are also important to predict how susceptible and resistant arthropod populations will respond to different genotype background of modified plants with insecticidal genes. Long-term management of arthropod pests lay in part on these principles and increasing our knowledge on plant/insect/microbe interactions will contribute to improved IPM tactics.

ACKNOWLEDGMENT

The author would like to acknowledge Dr. Drion G. Boucias, Entomology and Nematology Department, University of Florida, for his valuable comments on a draft of this chapter.

REFERENCES

Adams, J. R., and J. T. McClintock. 1991. Baculoviridae: Nuclear polyhedrosis viruses. Part I. Nuclear polyhedrosis viruses of insects. In *Atlas of Invertebrate Viruses*, ed. J. R. Adams, and J. R. Bonami, 87–204. Boca Raton, FL: CRC Press.

Ali, M. I., J. L. Bi, S. Y. Young, and G. W. Felton. 1999. Do foliar phenolics provide protection to *Heliothis virescens* from a baculovirus? *J. Chem. Ecol.* 25:2193–204.

Ali, M. I., G. W. Felton, T. Meade, and S. Y. Young. 1998. Influence of interspecific and intraspecific host plant variation on the susceptibility of heliothines to a baculovirus. *Biological Control.* 12:42–9.

Appel, H. M., and L. W. Maines. 1995. The influence of host plant on gut conditions of gypsy moth (*Lymantria dispar*) caterpillars. *Science* 41:241–6.

Arteel, G. E., and R. L. Lindroth. 1992. Effects of aspen phenolic on gypsy moth (Lepidoptera: Lymantriidae) susceptibility to *Bacillus thuringiensis*. *Great Lakes Entomol.* 25:239–44.

Baden, C. U., and S. Dobler. 2009. Potential benefits of iridoid glycoside sequestration in *Longitarsus melanocephalus* (Coleoptera, Chrysomelidae). *Basic Appl. Ecol.* 10:27–33.

Bais, H. P., T. L. Weir, L. G. Perry, S. Gilroy, and J. M. Vivanco. 2006. The role of root exudates in rhizosphere interactions with plants and other organisms. *Annu. Rev. Plant Biol.* 57:233–66.

Barbercheck, M. E. 1993. Tritrophic level effects on entomopathogenic nematodes. *Environ. Entomol.* 22:1166–71.

Barbercheck, M. E., and J. Wang. 1996. Effect of cucurbitacin D on in vitro growth of *Xenorhabdus* and *Photorhabdus* spp., symbiotic bacteria of entomopathogenic nematodes. *J. Invertebr. Pathol.* 68:141–5.

Barbercheck, M., J. Wang, and C. Brownie. 2003. Adaptation of the entomopathogenic nematode, *Steinernema carpocapsae*, to insect food plant. *Biol. Control* 27:81–94.

Barbercheck, M., J. Wang, and I. S. Hirsh. 1995. Host plant effects on entomopathogenic nematodes. *J. Invertebr. Pathol.* 66:169–77.

Baverstock, J., S. L. Elliot, P. G. Alderson, and J. K. Pell. 2005. Response of the entomopathogenic fungus *Pandora neoaphidis* to aphid-induced plant volatiles. *J. Invertebr. Pathol.* 89:157–64.

Ben-Dov, E., D. Saxena, Q. Wang, R. Manasherob, S. Boussiba, and A. Zaritsky. 2003. Ingested particles reduce susceptibility of insect larvae to *Bacillus thuringiensis*. *J. Appl. Entomol.* 127:146–52.

Berdegué, M., and J. T. Trumble. 1997. Interaction between linear furanocoumarins found in celery and a commercial *Bacillus thuringiensis* formulation on *Spodoptera exigua* (Lepidoptera: Noctuidae) larval feeding behavior. *J. Econ. Entomol.* 90:961–6.

Boucias, D. G., and J. C. Pendland 1998. *Principles of Insect Pathology.* Norwell, MA: Kluwer Academic Publishers.

Boucias, D. G., J. C. Pendland, and J. P. Latge. 1988. Nonspecific factors involved in attachment of entomopathogenic deuteromycetes to host insect cuticle. *Appl. Environ. Microbiol.* 54:1795–805.

Boucias, D. G., C. Stokes, G. Storey, and J. Pendland. 1996. Effect of imidacloprid on both the termite, *Reticulitermes flavipes* and its interaction with insect pathogens. *Pfanzenshutz-Natrichten Bayer* 49:103–44.

Brewer, G. J., and M. D. Anderson. 1990. Modification of the effect of *Bacillus thuringiensis* on sunflower moth (Lepidoptera: Pyralidae) by dietary phenols. *J. Econ. Entomol.* 83:2219–24.

Broderick, N. A., R. M. Goodman, and J. Handelsman. 2003. Effect of host diet and insect source on synergy of gypsy moth (Lepidoptera: Lymantriidae) mortality to *Bacillus thuringiensis subsp. kurstaki* by zwittermicin A. *Environ. Entomol.* 32:387–91.

Broderick, N. A., K. F. Raffa, R. M. Goodman, and J. Handelsman. 2004. Census of the bacterial community of the gypsy moth larval midgut by using culturing and culture-independent methods. *Appl. Environ. Microbiol.* 70:293–300.

Broderick, N. A., C. J. Robinson, M. D. McMahon, J. Holt, J. Handelsman, and K. F. Raffa. 2009. Contributions of gut bacteria to *Bacillus thuringiensis*-induced mortality vary across a range of Lepidoptera. *BMC Biol.* 7:11.

Brown, G. C., G. L. Prochaska, D. F. Hildebrand, G. L. Nordin, and D. M. Jackson. 1995. Green leaf volatiles inhibit conidial germination of the entomopathogen *Pandora neoaphidis* (Entomopthorales: Entomophthoraceae). *Environ. Entomol.* 24:1637–43.

Chapman, R. F. 1974. The chemical inhibition of feeding by phytophagous insects: A review. *Bull. Entomol. Res.* 64:339–63.

Cirimotich, C. M., Y. Dong, L. S. Garver, S. Sim, and G. Dimopoulos. 2010. Mosquito immune defenses against *Plasmodium* infection. *Develop. Comp. Immunol.* 34:387–95.

Cory, J. S., and J. D. Ericsson. 2010. Fungal entomopathogens in a tritrophic context. *BioControl* 55:75–88.

Cory, J. S., and K. Hoover. 2006. Plant-mediated effects in insect–pathogen interactions. *Trends Ecol. Evol.* 21:278–86.

Cory, J. S., and J. H. Myers. 2004. Adaptation in an insect host-plant pathogen interaction. *Ecol. Letters* 7:632–9.

Costa, S. D., and R. R. Gaugler. 1989. Sensitivity of *Beauveria bassiana* to solanine and tomatine: Plant defensive chemicals inhibit an insect pathogen. *J. Chem. Ecol.* 15:697–705.

Dillon, R., and K. Charnley. 2002. Mutualism between the desert locust *Schistocerca gregaria* and its gut microbiota. *Res. Microbiol.* 153:503–9.

Donegan, K., and B. Lighthart. 1989. Effect of several stress factors on the susceptibility of the predatory insect, *Chrysoperla carnea* (Neuroptera: Chrysopidae), to the fungal pathogen *Beauveria bassiana*. *J. Invertebr. Pathol.* 54:79–84.

Dow, J. A. T. 1992. pH gradients in lepidopteran midgut. *J. Exp. Biol.* 172:355–75.

Dubois, N. R., and D. H. Dean. 1995. Synergism between CryIA insecticidal crystal proteins and spores of *Bacillus thuringiensis*, other bacterial spores, and vegetative cells against *Lymantria dispar* (Lepidoptera: Noctuidae) larvae. *Environ. Entomol.* 24:1741–47.

Duetting, P. S., H. Ding, J. Neufeld, and S. D. Eigenbrode. 2003. Plant waxy bloom on peas affects infection of pea aphids by *Pandora neoaphidis*. *J. Invertebr. Pathol.* 84:149–58.

Epsky, N. D., and J. L. Capinera. 1994. Influence of herbivore diet on pathogenesis of *Steinernema carpocapsae* (Nematoda: Steinernematidae). *Environ. Entomol.* 23:487–91.

Farrar, R. R., and R. L. Ridgway. 2000. Host plant effects on the activity of selected nuclear polyhedrosis viruses against the corn earworm and beet armyworm (Lepidoptera: Noctuidae). *Environ. Entomol.* 29:108–15.

Felton, G. W., and D. L. Dahlman. 1984. Allelochemical induced stress: Effects of L-canavanine on the pathogenicity of *Bacillus thuringiensis* in *Manduca sexta*. *J. Invertebr. Pathol.* 44:187–91.

Ferrari, J., C. L. Scarborough, and H. C. J. Godfray. 2007. Genetic variation in the effect of a facultative symbiont on host-plant use by pea aphids. *Oecologia* 153:323–9.

Ferré, J., and J. Van Rie. 2002. Biochemistry and genetics of insect resistance to *Bacillus thuringiensis*. *Annu. Rev. Entomol.* 47:501–33.

Forschler, B. T., S. Y. Young, and G. W. Felton. 1992. Diet and susceptibility of *Helicoverpa zea* (Noctuidae, Lepidoptera) to a nuclear polyhedrosis virus. *Environ. Entomol.* 21:1220–3.

Furlong, M., and E. Groden. 2003. Starvation induced stress and the susceptibility of the Colorado potato beetle, *Leptinotarsa decemlineata*, to infection by *Beauveria bassiana*. *J. Invertebr. Pathol.* 83:127–38.

Fuxa, J. R., J. Sun, E. H. Weidner, and L. R. Lamotte. 1999. Stressors and rearing diseases of *Trichoplusia ni*: Evidence of vertical transmission of NPV and CPV. 1999. *J. Invertebr. Pathol.* 155:149–55.

Fytrou, A., P. G. Schofield, A. R. Kraaijeveld, and S. F. Hubbard. 2006. *Wolbachia* infection suppresses both host defence and parasitoid counter-defence. *Proc. R. Soc. B.* 273:791–6.

Giustolin, T. A., J. D. Vendramim, S. B. Alves, and S. A. Vieira. 2001. Patogenicidade de *Beauveria bassiana* (Bals.) Vuill. Sobre *Tuta absoluta* (Meyrick) (Lepidoptera: Gelechiidae) criada em dois genótipos de tomateiro. *Neotrop. Entomol.* 30:417–21.

Gonzalez-Ceron, L., F. Santillan, M. H. Rodriguez, D. Mendez, and J. E. Hernandez-Avila. 2003. Bacteria in midguts of field-collected *Anopheles albimanus* block *Plasmodium vivax* sporogonic development. *J. Med. Entomol.* 40: 371–4.

Haine, E. R. 2008. Symbiont-mediated protection. *Proc. R. Soc. B* 275:353–61.

Heddi, A., A. Vallier, C. Anselme, H. Xin, Y. Rahbe, and F. Wäckers. 2005. Molecular and cellular profiles of insect bacteriocytes: Mutualism and harm at the initial evolutionary step of symbiogenesis. *Cell. Microbiol.* 7:293–305.

Hemming, J. D. C., and R. L. Lindroth. 2000. Effects of phenolic glycosides and protein on gypsy moth (Lepidoptera: Lymantriidae) and forest tent caterpillar (Lepidoptera: Lasiocampidae) performance and detoxication activities. *Environ. Entomol.* 29:1108–15.

Hountondji, F. C. C., M. W. Sabelis, R. Hanna, and A. Janssen. 2005. Herbivore-induced plant volatiles trigger sporulation in entomopathogenic fungi: The case of *Neozygites tanajoae* infecting the cassava green mite. *J. Chem. Ecol.* 31:1003–21.

Hwang, S. Y., R. L. Lindroth, M. E. Montgomery, and K. S. Kathleen. 1995. Aspen leaf quality affects gypsy moth (Lepidoptera: Limantriidae) susceptibility to *Bacillus thuringiensis*. *J. Econ. Entomol.* 88:278–82.

Inyang, E. N., T. M. Butt, L. Ibrahim, and S. J. Clark. 1998. The effect of plant growth and topography on the acquisition of conidia of the insect pathogen. *Mycol. Res.* 102:1365–74.

Inyang, E., T. M. Butt, A. Beckett, and S. Archer. 1999. The effect of crucifer-epicuticlar waxes and leaf extracts on the germination and virulence of *Metarhizium anisopliae* conidia. *Mycol. Res.* 103:419–26.

James, R. R., and B. Lighthart. 1992. The effect of temperature, diet, and larval instar on the susceptibility of an aphid predator, *Hippodamia convergens* (Coleoptera: Coccinellidae), to the weak bacterial pathogen *Pseudomonas fluorescens. J. Invertebr. Pathol.* 60:215–18.

Janmaat, A. F., and J. H. Myers. 2005. The cost of resistance to *Bacillus thuringiensis* varies with the host plant of *Trichoplusia ni. Proc. R. Soc. B* 272:1031–38.

Janmaat, A. F., and J. H. Myers. 2006a. The influences of host plant and genetic resistance to *Bacillus thuringiensis* on trade-offs between offspring number and growth rate in cabbage loopers. *Trichoplusia ni. Ecol. Entomol.* 31:172–8.

Janmaat, A. F., and J. H. Myers. 2006b. Host plant effect the expression of resistance to *Bacillus thuringiensis kurstaki* in *Trichoplusia ni* (Hubner): An important factor in resistance evolution. *J. Evol. Biol.* 20:62–9.

Johnston, P. R., and N. Crickmore. 2009. Gut bacteria are not required for the insecticidal activity of *Bacillus thuringiensis* toward the tobacco hornworm, *Manduca sexta. Appl. Environ. Microb.* 75:5094–9.

Keating, S. T., M. D. Hunter, and J. C. Schultz. 1990a. Leaf phenolic inhibition of gypsy moth nuclear polyhedrosis virus: Role of polyhedral inclusion body aggregation. *J. Chem. Ecol.* 16:1445–57.

Keating, S. T., J. C. Schultz, and W. G. Yendol. 1990b. The effect of diet on gypsy moth (*Lymantria dispar*) larval midgut pH, and its relationship with larval susceptibility to a baculovirus. *J. Invertebr. Pathol.* 56:317–26.

Keating, S. T., W. G. Yendol, and J. C. Schultz. 1988. Relationship between susceptibility of gypsy moth larvae (Lepidoptera, Lymantriidae) to a baculovirus and host plant foliage constituents. *Environ. Entomol.* 17:952–8.

Kleiner, K. W., K. F. Raffa, D. E. Ellis, and B. H. McCown. 1998. Effect of nitrogen availability on the growth and phytochemistry of hybrid poplar and the efficacy of the *Bacillus thuringiensis* cry1A(a) D-endotoxin on gypsy moth. *Can. J. Forest Res.* 28:1055–67.

Kouassi, K. C., F. Lorenzetti, C. Guertin, J. Cabana, and Y. Mauffette. 2001. Variation in the susceptibility of the forest tent caterpillar (Lepidoptera: Lasiocampidae) to *Bacillus thuringiensis* variety *kurstaki* HD-1: Effect of the host plant. *J. Econ. Entomol.* 94:1135–41.

Lehane, M. J. 1997. Peritrophic matrix structure. *Annu. Rev. Entomol.* 42:525–50.

Louis, C., M. Jourdan, and M. Cabanac 1986. Behavioral fever and therapy in a rickettsia-infected Orthoptera. *Am. J. Physiol. Regul. Integr. Comp. Physiol.* 250:991–5.

Ludlum, C. T., G. W. Feiton, and S. S. Duffey. 1991. Plant defenses: Chlorogenic acid and polyphenol oxidase enhance toxicity of *Bacillus-thuringiensis* subsp. *kurstaki* to *Heliothis zea. J. Chem. Ecol.* 17:217–37.

Luz, C., J. Fargues, and C. Romaña. 2003. Influence of starvation and blood meal-induced moult on the susceptibility of nymphs of *Rhodnius prolixus* Stal (Hem., Triatominae) to *Beauveria bassiana* (Bals.) Vuill. infection. *J. Appl. Entomol.* 127:153–6.

Maduell, P., G. Armengol, M. Llagostera, S. Orduz, and S. Lindow. 2008. *Bacillus thuringiensis* is a poor colonist of leaf surfaces. *Microbiol Ecol.* 55:212–9.

Maeda, S., T. Kawai, M. Obinata, H. Fujiwara, T. Horiuchi, Y. Saeki, Y. Sato, and M. Furusawa. 1985. Production of human-interferon in silkworm using a baculovirus vector. *Nature* 315:592–4.

Margulis, L., J. Z. Jorgensen, S. Dolan, R. Kolchinsky, F. A. Rainey, and S. C. Lo. 1998. The *Arthromitus* stage of *Bacillus cereus*: Intestinal symbionts of animals. *Proc. Natl. Acad. Sci. USA* 95:1236–41.

McCormack, P. J., H. G. Wildman, and P. Jeffries. 1994. Production of antibacterial compounds by phylloplane-inhabiting yeasts and yeastlike fungi. *Appl. Environ. Microb.* 60:927–31.

McVean, R., S. Sait, D. Thompson, and M. Begon. 2002. Dietary stress reduces the susceptibility of *Plodia interpunctella* to infection by a granulovirus. *Biol. Control* 25:81–4.

Navon, A. 1992. Interactions among herbivores, microbial insecticides and crop plants. *Phytoparasitica* 20:21–4.

Ouedraogo, R., M. Cusson, M. S. Goettel, and J. Brodeur. 2003. Inhibition of fungal growth in thermoregulating locusts, *Locusta migratoria*, infected by the fungus *Metarhizium anisopliae* var. *acridum. J. Invertebr. Pathol.* 82:103–9.

Pechan, T., A. Cohen, W. P. Williams, and D. S. Luthe. 2002. Insect feeding mobilizes a unique plant defense protease that disrupts the peritrophic matrix of caterpillars. *Proc. Natl. Acad. Sci. USA* 99:13319–23.

Peng, F., J. R. Fuxa, S. J. Johnson, and A. R. Richter. 1997. Susceptibility of *Anticarsia gemmatalis* (Lepidoptera: Noctuidae), reared on four host plants, to a nuclear polyhedrosis virus. *Environ. Entomol.* 26:973–7.

Plymale, R., M. J. Grove, D. Cox-Foster, N. Ostiguy, and K. Hoover. 2008. Plant-mediated alteration of the peritrophic matrix and baculovirus infection in lepidopteran larvae. *J. Insect Physiol.* 54:737–49.

Pristavko, V. P., and N. V. Dovzhenok. 1974. Ascorbic acid influence on larval blood cell number and susceptibility to bacterial and fungal infection in the codling moth, *Laspeyresia pomonella* (Lepidoptera: Tortricidae). *J. Invertebr. Pathol.* 24:165–8.

Pumpuni, C. B., J. Demaio, M. Kent, J. R. Davis, and J. C. Beier. 1996. Bacterial population dynamics in three Anopheline species: The impact on *Plasmodium* sporogonic development. *Am. J. Trop. Med. Hyg.* 54:214–8.

Rasmann, S., T. G. Köllner, J. Degenhardt, I. Hiltpold, S. Toepfer, U. Kuhlmann, J. Gershenzon, and T. C. Turlings. 2005. Recruitment of entomopathogenic nematodes by insect-damaged maize roots. *Nature* 434:732–7.

Raymond, B., D. Davis, and M. B. Bonsall. 2007. Competition and reproduction in mixed infections of pathogenic and non-pathogenic *Bacillus* spp. *J. Invertebr. Pathol.* 96:151–5.

Raymond, B., S. E. Hartley, J. S. Cory, and R. S. Hails. 2005. The role of food plant and pathogen-induced behaviour in the persistence of a nucleopolyhedrovirus. *J. Invertebr. Pathol.* 88:49–57.

Raymond, B., P. R. Johnston, D. J. Wright, R. J. Ellis, N. Crickmore, and M. B. Bonsall. 2009. A mid-gut microbiota is not required for the pathogenicity of *Bacillus thuringiensis* to diamondback moth larvae. *Environ. Microbiol.* 11:2556–63.

Richter, A. R., J. R. Fuxa, and M. Abdel-Fattah. 1987. Effect of host plant on the susceptibility of *Spodoptera frugiperda* (Lepidoptera: Noctuidae) to a nuclear polyhedrosis virus. *Environ. Entomol.* 16:1004–6.

Roome, R. E., and R. A. Daoust. 1976. Survival of the nuclear polyhedrosis virus of *Heliothis armigera* on crops and soil in Botswana. *J. Invertebr. Pathol.* 27:7–12.

Rostás, M., and M. Hilker. 2003. Indirect interactions between a phytopathogenic and an entomopathogenic fungus. *Naturwissenschaften* 90:63–7.

Salama, H. S., and A. Abedel-Razek. 1992. Effect of different kinds of food on susceptibility of some stored products insects to *Bacillus thuringiensis*. *J. Appl. Entomol.* 113:107–10.

Santiago-Álvarez, C., E. A. Maranhão, E. Maranhão, and E. Quesada-Moraga. 2006. Host plant influences pathogenicity of *Beauveria bassiana* to *Bemisia tabaci* and its sporulation on cadavers. *BioControl* 51:519–32.

Scarborough, C. L., J. Ferrari, and H. C. Godfray. 2005. Aphid protected from pathogen by endosymbiont. *Science* 310:1781.

Schmidt, N. R., J. M. Haywood, and B. C. Bonning. 2009. Toward the physiological basis for increased *Agrotis ipsilon* multiple nucleopolyhedrovirus infection following feeding of *Agrotis ipsilon* larvae on transgenic corn expressing Cry1Fa2. *J. Invertebr. Pathol.* 102:141–8.

Shah, F. A., C. S. Wang, and T. M. Butt. 2005. Nutrition influences growth and virulence of the insect-pathogenic fungus *Metarhizium anisopliae*. *FEMS Microbiol. Letters* 251:259–266.

Shapiro-Ilan, D., M. G. Rojas, J. A. Morales-Ramos, E. E. Lewis, and W. L. Tedders. 2008. Effects of host nutrition on virulence and fitness of entomopathogenic nematodes: Lipid- and protein-based supplements in *Tenebrio molitor* diets. *J. Nematol.* 40:13–9.

Sikorowski, P. P., and A. M. Lawrence. 1994. Microbial contamination and insect rearing. *Am. Entomol.* 40:240–53.

Singh, G. 1974. Endosymbiotic microorganisms in *Cletus signatus* Walker (Coreidae: Heteroptera). *J. Cell. Mol. Life Sci. Experientia* 30:1406–7.

Sorvari, J., H. Hakkarainen, and M. Rantala. 2008. Immune defense of ants is associated with changes in habitat characteristics. *Environ. Entomol.* 37:51–6.

Sosa-Gómez, D. R., S. B. Alves, and L. G. Marchini. 1991. Variation in the susceptibility of *Bombyx mori* L. to nuclear polyhedrosis virus when reared on different mulberry genotypes. *J. Appl. Entomol.* 111:318–20.

Sosa-Gómez, D. R., C. C. López Lastra, and R. A. Humber. 2010. An overview of arthropod-associated fungi from Argentina and Brazil. *Mycopathologia* 170:61–76.

Stabb, E. V., L. M. Jacobson, and J. Handelsman. 1994. Zwittermicin A-producing strains of *Bacillus cereus* from diverse soil. *Appl. Environ. Microbiol.* 60:4404–12.

Storer, N. P., J. M. Babcock, M. Schlenz, T. Meade, G. D. Thompson, J. W. Bing, and R. M. Huckaba. 2010. Discovery and characterization of field resistance to Bt maize: *Spodoptera frugiperda* (Lepidoptera: Noctuidae) in Puerto Rico. *J. Econ. Entomol.* 103:1031–8.

Tallamy, D. W., P. M. Gorski, and J. K. Burzon. 2000. Fate of male-derived cucurbitacins in spotted cucumber beetle females. *J. Chem. Ecol.* 26:413–27.

Tallamy, D. W., D. P. Whittington, F. Defurio, D. A. Fontaine, P. M. Gorski, and P. W. Gothro. 1998. Sequestered cucurbitacins and pathogenicity of *Metarhizium anisopliae* (Moniliales: Moniliaceae) on spotted cucumber beetle eggs and larvae (Coleoptera: Chrysomelidae). *Environ. Entomol.* 27:66–372.

Teixeira, L., A. Ferreira, and M. Ashburner. 2008. The bacterial symbiont *Wolbachia* induces resistance to RNA viral infections in *Drosophila melanogaster*. *PLoS Biology* 6:2.

Trier, T. M., and W. J. Mattson. 2003. Diet-induced thermogenesis in insects: A developing concept in nutritional ecology. *Environ. Entomol.* 32:1–8.

Ugine, T. A., S. P. Wraight, and J. P. Sanderson. 2007. A tritrophic effect of host plant on susceptibility of western flower thrips to the entomopathogenic fungus *Beauveria bassiana*. *J. Invertebr. Pathol.* 96:162–72.

Wood, H. A. 1980. Protease degradation of *Autographa californica* nuclear polyhedrosis virus proteins. *Virology* 103:392–9.

Young, S. Y., W. C. Yearian, and K. S. Kim. 1977. Effect of dew from cotton and soybean foliage on activity of *Heliothis* nuclear polyhedrosis virus *J. Invertebr. Pathol.* 29:105–11.

Part II

Specific Aspects

10

Neotropical Ants (Hymenoptera) Functional Groups: Nutritional and Applied Implications

Carlos R. F. Brandão, Rogério R. Silva, and Jacques H. C. Delabie

CONTENTS

10.1 Introduction 213
10.2 Ant Guilds and Aspects about Their Nutritional Biology 214
10.3 The Neotropical Ant Guilds 216
    10.3.1 Generalist Predators 216
        10.3.1.1 Epigaeic Generalist Predators 216
        10.3.1.2 Hypogaeic Generalist Predators 217
    10.3.2 Specialists 217
        10.3.2.1 Predation in Mass and/or Nomadism 218
        10.3.2.2 Dacetini Predators 218
    10.3.3 Arboreal Predator Ants 220
    10.3.4 Generalists 220
        10.3.4.1 Generalized Myrmicines 221
        10.3.4.2 Generalized Formicines, Dolichoderines, and Some Myrmicines 222
        10.3.4.3 Small-Sized Hypogaeic Generalist Foragers 222
    10.3.5 Fungus Growers 223
        10.3.5.1 Leaf Cutters 223
        10.3.5.2 Litter-Nesting Fungus Growers 223
    10.3.6 Legionary Ants 224
    10.3.7 Dominant Arboreal Ants Associated with Carbohydrate-Rich Resources
        or Domatia 224
    10.3.8 Pollen-Feeding Arboreal Ants 225
    10.3.9 Subterranean Ants 226
10.4 Concluding Remarks: From Trophic Guilds to Applied Myrmecology 226
Acknowledgments 228
References 228

10.1 Introduction

Ants are eusocial organisms; that is, they adopt an advanced level of colonial structure in which adult individuals belonging to two or more generations contribute to the maintenance of the colonies and in the offspring care, and female individuals can be reproductive or sterile (Wilson and Hölldobler 2005b).

The Hymenoptera fossil record suggests that the most recent common ancestor of all present ant species lived more than 120 million years ago (Grimaldi and Engel 2005), although estimates with molecular data extend this origin even further in the past (132 to 176 million years) (Moreau et al. 2006). Brady et al. (2006), however, suggested that the origin of the ancestor of ants occurred between 105 to 143 million years ago (see comments in Crozier 2006), while Wilson and Hölldobler (2005a) reinforce these assumptions based on ecological arguments.

Among eusocial insects, ants represent the most diverse and ecologically dominant group (Wilson and Hölldobler 2005a,b). Among all insects, ants constitute one of the most important taxa in terms of biomass or relative local abundance (Hölldobler and Wilson 1990; Davidson et al. 2003; Ellwood and Foster 2004; Wilson and Hölldobler 2005a,b). Along with termites, ants compose some 2% of the approximately one million insect species so far described, but they can represent more than 50% of the insect biomass in the world rainforests (Wilson and Hölldobler 2005a,b).

As an ecologically dominant group in all Earth's ecosystems, from the tundra to tropical forests (Kaspari 2005; Wilson and Hölldobler 2005a,b), ants engage in interactions with many other organisms and thus participate significantly in functional ecosystem processes (Hölldobler and Wilson 1990) such as regulating populations of numerous arthropods (Floren et al. 2002; Izzo and Vasconcelos 2005; Philpott and Armbrecht 2006), in seed dispersal (Beattie 1985; Cristianini and Oliveira 2009; Lengyel et al. 2009), and in fostering changes in the physical structure of ecosystems (Folgarait 1998; Frouz and Jilková 2008).

Studies on ant communities have been the basis for evaluation programs and conservation of ecosystems (Bromham et al. 1999; Andersen et al. 2002) and have been used as indicators of biodiversity of other invertebrates; ant studies are also essential to make reliable estimates of "hyperdiverse" groups (insects, mites and other arachnids, and nematodes) and of species richness (Silva and Brandão 1999; Brown et al. 2003). Studies focusing on communities of ants have also been employed in various programs of conservation biology, as in the assessment of the impact of invasive species, population behavior detection of endangered species or group considered as "key" for monitoring in assessing recovery programs of land use (for example, mine rehabilitation), and in the long-term follow-up of changes in ecosystems (Underwood and Fisher 2006; Crist 2008).

The study of local communities of ants offers enormous potential for hypotheses testing about local and regional species richness (Kaspari et al. 2000b, 2004), relative abundance (Kaspari 2001; Kaspari et al. 2000a; Kaspari and Valone 2002), body size and its influence on the ecology of organisms (Kaspari 2005; Kaspari et al. 2010), dynamics of local communities, and intraspecific interactions and their ecological consequences (Gotelli and Ellison 2002; Sanders et al. 2003), and moreover in defining trophic networks characteristics (Guimarães et al. 2006, 2007; Chamberlain and Holland 2009).

The understanding of the communities' structure of neotropical ants and the factors that determine their organization has advanced with the use functional groups analysis or trophic guilds. This type of classification reveals group of sympatric species occupying similar roles or niches that show a high degree of interaction or overlap in their ecology; this might be seen as groups of species that influence together and in a similar way the structure of the community (Simberloff and Dayan 1991; Wilson 1999; Blondel 2003).

The adoption of the functional groups model has been highly successful in the analysis of ant ecological communities in Australia (Andersen 1995) by its predictive power with respect to the impact of factors such as stress (limiting productivity) and disturbance (responsible for the removal of biomass); this has been frequently used in studies to identify environmental bioindicators (Andersen et al. 2002, 2004; Andersen and Majer 2004; Majer et al. 2004).

In this chapter, we describe the nutritional biology of ants of the neotropical ant guilds, summarizing, in the descriptions, results of several studies, in particular in the characterization of ant guilds in neotropical forests (Delabie et al. 2000), the organization of Cerrado ant guilds (Silvestre et al. 2003), and on the leaf litter myrmecofauna (Silva and Brandão 2010). Considering the information already accumulated about ant guilds, we suggest that new tests of hypotheses about the factors that determine the ecology of ant communities and populations have their predictive power increased when taking into account guild models in comparison with traditional analytical methods.

10.2 Ant Guilds and Aspects about Their Nutritional Biology

Most imagoes (adults) of ants, predatory wasps, and other insects that develop through complete metamorphosis adopt as their main food resource hemolymph of their prey or sugary substances produced by nectaries (floral or extrafloral) and exudates of hemipterans. They spend much of the energy derived from these sugary and fatty inputs in the search of food for the immatures in the colony (Wilson and Hölldobler 2005b). So, larvae depend on protein and other substances to complete their development, while adult workers need energy for foraging activities, construction and maintenance of their nest, and care of the offspring.

Several ecomorphological studies clearly demonstrate the association between morphology, ecology, and taxonomy (Miles et al. 1987; Juliano and Lawton 1990; Price 1991; Douglas and Matthews 1992). Morphological patterns shared by species coexisting in space and time have been often used to characterize the organization of communities (Stevens and Willing 2000).

Body size alone may condition the dimensions of the ecological niche of an organism (Ovadia and Schmitz 2002; Ness et al. 2004; Woodward et al. 2005), and consequently, the community structure (Ovadia and Schmitz 2002; Cohen et al. 2003; Kaspari 2005). In ant communities' studies, evidence suggests that the interaction between body size and structural complexity of the environment influences the species composition (Farji-Brener et al. 2004; Sarty et al. 2006). This allows a wide diversity of species to share resources and prevents monopolization of these resources by only one or two more abundant species. For example, relatively large-sized, particularly common species are unable to access all microhabitats (such as the interstices of the leaf litter or small cavities in soils with special structural complexity), leaving, therefore, free refuge areas and enabling acquisition of food by other species (Sarty et al. 2006).

To describe ant guilds, analyses were performed based on the importance of morphology in the characterization of ecological groups. The classification scheme presented below incorporates morphological characters with known functional importance and therefore related to the foraging biology of the ant species, such as size, shape of various structures (head, eyes, legs, trunk, petiole), and preferred foraging places (Kaspari and Weiser 1999; Weiser and Kaspari 2006).

When applied to Atlantic forest studies, one of the richest habitats in environments and ecological niches in the neotropics, the classification scheme indicates consistently the presence of nine guilds inhabiting the leaf litter. Summarizing information from the literature and our own observations and aggregating the experiences found in surveys conducted in other biomes suggests the existence of another five guilds, adding to the species that inhabit the leaf liter, those with arboreal, nomadic, and subterranean habits.

The adopted scheme here is hierarchical and recognizes a vertical stratification of the fauna as its main compartment. This habitat segregation between species that share the same space is well documented. Various systematic surveys conducted in different regions of the planet biomes showed that there are significant differences in composition between the subterranean, leaf litter, soil surface, and vegetation faunas (Longino and Nadkarni 1990; Brühl et al. 1998; Silvestre et al. 2003; Delabie et al. 2007). In a second place, analyses indicate that body and eyes size are the major variables that can be used to characterize the functional groups, followed by information on the form of some morphological structures, especially of the mandibles, the petiole, and the relative position of the eye in relation to other structures of the cephalic capsule (Weiser and Kaspari 2006). In addition, a clustering analysis reveals that some groups include a range of taxa exploring the same resources, while other groups are phylogenetically consistent, implying that the exploitation of some niches is taxonomically constrained. In such cases, the specialized form of mandibles, body size, and behavioral features are important elements in the characterization of these groups, and result in very well defined groupings, even in taxonomic terms. As well, in the studied biogeographical region, some niches are composed exclusively of taxa belonging to the same clade, with, in general, highly specialized biology and anatomy, besides a relatively small body size.

It is expected that the same 14 guilds described here that compose the structure of the ant communities is repeated in every other neotropical forest site, including savanna areas in climax stage such as "cerradões" (a kind of savannah of central Brazil). As environments simplify and lose habitat, they also lose components or even guilds; consequently it is not to be expected that richer communities pass unnoticed (i.e., guilds not revealed by our studies). Regions under temperate climate regimes share the presence of particular social parasite species of ants, rarer in tropical regions.

The absence of social parasite species in the guilds scheme reflects the lack of appropriate information on the biology of numerous neotropical species. In the neotropics, some species have been identified as inquilines or social parasites, particularly in the genus *Acromyrmex*, like *Acromyrmex ameliae* Souza,

Soares & Della Lucia (host: *A. subterraneus brunneus* Forel and *A. subterraneus subterraneus* Forel), *Acromyrmex insinuator* Schultz, Bekkevold and Boomsma (host: *A. echinatior* Forel), *Pseudoatta argentina* Gallardo (host: *Acromyrmex lundii* (Guérin-Méneville)), and *Pseudoatta* sp. (host: *Acromyrmex rugosus* F. Smith) (Summer et al. 2004; Souza et al. 2007). There are also several known social parasites in neotropical species of the genera *Pheidole* and *Ectatomma* (Hora et al. 2005, Feitosa et al. 2008). It would be perfectly legitimate to consider the set of neotropical ant species with social parasites' habits like the 15th guild in the proposed schema.

We describe below the guilds or nutritional groupings of neotropical ants and the relevant data on their biology, following the taxonomic classification scheme of Bolton et al. (2006), with illustrations.

10.3 The Neotropical Ant Guilds

10.3.1 Generalist Predators

10.3.1.1 Epigaeic Generalist Predators

10.3.1.1.1 Large-Sized

Epigaeic generalist predator ants (species that forage on the soil surface) show relatively large body size (1 cm or more), with long and linear or triangular mandible, eyes distant from each other and located approximately at the midpoint between the insertion of the mandible and the vertexal margin, very large eyes, with the largest number of ommatidea among guilds. The worker caste is always monomorphic.

The taxa included in this group are species of the genera *Dinoponera*, *Odontomachus* (Figure 10.1a), *Pachycondyla*, *Ectatomma*, and the larger species of *Anochetus*. *Anochetus* is considered the sister group of *Odontomachus* because both genera share a closing mechanism of the mandibles unique in Ponerinae, known as trap-jaw (Gronenberg and Ehmer 1996). In general, older workers of the guild forage alone looking for prey, especially arthropods of similar size, but hunt other invertebrates as well, such as small gastropods and earthworms. Opportunistically they can be saprophytes; the larger species (scattered observations on *Dinoponera*, *Ectatomma,* and *Pachycondyla*) being sometimes found on corpses of small mammals. They rarely employ nest mate recruitment or visit or guard nectaries (except those in the genus *Ectatomma* and some arboreal *Pachycondyla*); in general, they live in nests in the soil, plants, or in cavities, or are associated with epiphytes inhabited by populations of a few dozen to a few hundred individuals.

There is a large variation in body size among the taxa in this guild, especially because it includes the *Anochetus* species. The shape of the mandible (long and linear) has a strong influence on the characterization of the group. In addition, this observation suggests that one should wait sharing of resources based on the prey size among taxa belonging to the *Anochetus* + *Odontomachus* group.

10.3.1.1.2 Medium-Sized

The epigaeic medium-sized are ant species with a body size averaging 0.5 to 1 cm, triangular mandible, with developed eyes that are distant from the insertion of the mandible and distant from each

FIGURE 10.1   (a) Worker of *Odontomachus bauri* Emery approximing prey. (b) Worker of *Gnamptogenys striatula* Mayr inspecting prey. (Courtesy of Alex Wild.)

other (Figure 10.1b). The group includes some species of *Heteroponera, Gnamptogenys, Hylomyrma, Megalomyrmex, Oxyepoecus, Pheidole, Solenopsis,* and *Basiceros*. All *Pheidole* and a large part of *Solenopsis* of the subgenus *Solenopsis* are polymorphic, while the remaining taxa are monomorphic. They form colonies with relatively small populations with some tens to hundreds of workers to large colonies, with up to several thousand workers, as in *Solenopsis*.

Some of the species classified in this group are known epigaeic generalist predators as *Gnamptogenys striatula* Mayr (Lattke 1995). The group also includes species with different habits, such as species of *Megalomyrmex* of the Silvestrii group, trophic parasites of fungus grown by Attini ants of the genera *Apterostigma, Cyphomyrmex, Trachymyrmex,* and *Sericomyrmex* (Adams et al. 2000b; Brandão 2003), or *Oxyepoecus punctifrons* (Borgmeier).

Most *Heteroponera* species (Heteroponerinae) nest in decaying trunks or leaf litter and its interstices (Kempf 1962; Françoso 1995). *Heteroponera dentinodis* (Mayr) and *H. dolo* (Roger) preferably feed on larvae and adults of *Tenebrio molitor* L. (Coleoptera: Tenebrionidae), larvae of *Alphitobius* sp. (Coleoptera: Tenebrionidae), adult *Folsomia candida* Willem (Collembola: Isotomidae) and *Drosophila* sp. larvae (Diptera) in the laboratory (Françoso 1995).

Species of *Basiceros* that are relatively large in size are included in this guild. These species are found exclusively in the neotropic forests, with triangular mandibles and multidenticulate chewing margin, and compound eyes positioned posteriorly on the head. Species are predators of small arthropods (Wilson and Hölldobler 1986). They show extremely slow movements and special mechanisms employed in camouflage, particularly two special types of hairs that form a double layer on the dorsal surface of the occiput, scape, pro and mesonotum, petiole, postpetiole and gaster, which retains particles of soil of approximately 10 μm that prevent them from being visually and chemically located by predators.

10.3.1.2 Hypogaeic Generalist Predators

10.3.1.2.1 Medium-Sized

Hypogaeic ants (species that forage exclusively within the leaf litter) present average body size (0.5 to 1 cm) and are characterized by the reduction of the eyes set relatively very close to the insertion of the mandibles. This group includes comparatively small monomorphic species of *Gnamptogenys, Hypoponera,* and *Pachycondyla* (as *P. ferruginea* (F. Smith) and *P. stigma* (F.)). The biology of these species is poorly known but the reduction of eyes and predatory hypogaeic habits suggests that they usually capture their small arthropod prey within the interstices of the leaf litter.

10.3.1.2.2 Small-Sized

Considering the morphological criteria employed here, the group of small-sized ants is formed exclusively by species of the genus *Hypoponera*, bringing together relatively small bodied ant species (less than 0.5 cm) with small triangular mandibles with eyes reduced to one ommatidium and inserted close to the articulation of the mandibles. In the morphological space that defined the nine ant guilds of the leaf litter, this group is well separated from the guild of medium-sized hypogaeic generalist predators, discussed above. There is a great uniformity in the general shape of the body and in other morphological characters. There is no detailed information about the nutritional biology of this group, but all species are considered generalist predators (Brown 2000). However, it is common to find several *Hypoponera* species living in the same 1 m² sample of leaf litter. As our analyses suggest, all species share food items, and it is expected that these ants must somehow segregate behaviorally and ecologically among themselves, avoiding competition (temporally or in the choice of resources).

10.3.2 Specialists

Specialist ants have specialized morphology and biology and are rarely studied. They have mandibles spanning from the classic triangular type to a strongly differentiated shape. They includes species of relative medium to small size, narrow mandibles with well separated points of articulation and with differentiated dentition; eyes next to the insertion of the mandibles that are distant, reduced, or absent (Figures 10.2a,b). They present a wide variety of body shapes, grouping representatives of Amblyoponinae (*Amblyopone, Prionopelta*), Cerapachyinae (*Acanthostichus, Cerapachys, Sphinctomyrmex*), Myrmicinae (*Adelomyrmex, Cryptomyrmex, Stegomyrmex*), Ectatomminae (some species of *Gnamptogenys, Typhlomyrmex*), Ponerinae (*Centromyrmex, Thaumatomyrmex*), and Proceratiinae (*Discothyrea, Proceratium*). All these taxa live in leaf litter or are hypogaeic. Information and discussion on the morphology and nutritional biology of this specialized ant taxa can be found in Fowler et al. (1991) and Brandão et al. (2009).

10.3.2.1 Predation in Mass and/or Nomadism

In this specialized group are included ant species that have relatively elaborate hunting behavior involving mass organized groups of workers that hunt in columns during foraging (e.g., *Pachycondyla marginata* (Roger) that predates exclusively *Neocapritermes opacus* Holmgren termites (Leal and Oliveira 1995), and in some cases nomadic behavior, as in some Cerapachyinae (*Acanthostichus, Cylindromyrmex*) and Ponerinae (*Leptogenys, Simopelta*)). The morphology and biology of reproductive females is similar to that observed in Ecitoninae, a condition known as dicthadiigyny, characterized by the combination of a relatively large and subquadrate head, reduced eyes (slightly larger than those of workers), reduced trunk devoid of wings; swollen gaster, with a poorly developed or missing constriction after the postpetiole.

Unlike legionary or army ants, which always forage in relatively large columns of workers, predation in species of this group usually involves a small group led by a single worker. This worker uses pheromones to mark a trail to the food resource and then recruit a large number of colony nestmates. Another fundamental difference with the true legionary ants (Ecitoninae and Dorylinae) is that the colony's frequent migration does not follow the characteristic rhythm of legionaries and the development of immatures is also not synchronized (Maschwitz et al. 1989). Mass predators specialize in the exploitation of several niches. There are those specialized in predating *Pheidole* species (e.g., *Simopelta*), isopods, and earwigs (in *Leptogenys*). Most of the known Cerapachyinae species (*Acanthostichus, Cerapachys, Cylindromyrmex,* and *Sphinctomyrmex*) prey exclusively on ants.

10.3.2.2 Dacetini Predators

Dacetini, according Baroni-Urbani and De Andrade (2007), is a tribe composed by five genera in the neotropics (*Acanthognathus, Basiceros, Daceton, Phalacromyrmex,* and *Strumigenys*), considering *Creightonidris* as a junior synonym of *Basiceros* (Feitosa et al. 2007). The much-differentiated mandibles of many dacetine ants are one of the most striking features of this segment of the typical leaf litter myrmecofauna. The morphology and the mechanism of action of these mandibles are quite different from the overall Myrmicinae pattern (Wilson 1956; Dietz 2004).

Most members of the tribe live in monogynic colonies, foraging and nidifying in the leaf litter and in the soil superficial layers, or among superficial roots (Bolton 1998). All known species are predatory, mainly of Collembola, but several species also are known to hunt a wide variety of other small arthropods, such as Diplura, Symphyla, Chilopoda, pseudoscorpions, mites, Araneae, isopods, amphipods, and small insects and their larvae (Dejean 1987a,b). Structurally, the mandibles are notably modified; being employed in predation, most specializations reflect a special technique for prey capture (Bolton 1998, 1999). These species are very common in leaf litter samples of tropical and subtropical forests. Some species are locally relatively abundant (Fisher 1999; Dietz 2004).

In ants, specialization of the mandibles not only involves the shape but also depends on the speed and strength that they can generate (Gronenberg et al. 1997, 1998). According to the mode of action of the mandibles, the species can be classified into two main subgroups, discussed below.

10.3.2.2.1 Species with Static Pressure Mandibles

The species of ants with static pressure mandibles includes Dacetini species such as *Strumigenys* spp., *Strumigenys schmalzi* Emery, and small-size *Basiceros* species (total length about 0.20 cm). They present a smaller and shorter mandible of all groups of ants, called static pressure mandibles. Their eyes are set very close to the insertion of the mandible and are relatively small (Figure 10.2c).

FIGURE 10.2 (a) Specialized predators: Worker of *Cerapachys augustae* Wheeler. (b) Worker of *Thaumatomyrmex contumax* Kempf transporting Polyxenidae Myriapod prey before stripping it from its unpalatable setae. (c) Worker of *Basiceros procerum* (Emery): Static pressure. (d) Worker of *Strumigenys louisianae* Roger (kinetic action mandibles during prey approach. (Courtesy of Alex Wild.)

The static pressure mandible's shape is usually triangular to elongated triangular but sometimes narrow, linear to sublinear, or forceps-like. Dentition is usually composed of numerous teeth and denticles on the chewing margin, although some species are almost edentate while others show specialized teeth, and with more teeth than in species that present kinetic mandibles; apical and isolated teeth are rare. The maximum gape of mandibles is 60 to 90 degrees and the labrum shows no lateral projections. The primary function of the mandible is to capture and firmly hold the prey. The initial stroke of the mandibles aims to capture the prey, and right after closing, the static pressure keeps the prey still.

Short mandibulate species initially place the antennae in direct contact with the prey and then stalk the prey for a long time before closing the mandibles around the prey in a sudden snap. The approach is slow and the mandibles firmly hold the prey when caught. Workers smear small pieces of the substrate mainly on the trunk and the head, "tricking" the prey, as in this way they "hide" their own odor and adopt that of the substrate (Masuko 1984).

Based on information in the literature, we included in this group the genus *Tatuidris* (Myrmicinae). There is no record on its biology, but its morphology and the presence of a highly differentiated sting suggest the species are specialized predators (Brown and Kempf 1968). Recent analyses of the Dacetini phylogeny recognizes *Tatuidris* as the sister group of all other Dacetini genera (Baroni Urbani and De Andrade 2007).

10.3.2.2.2 Species with Kinetic Mandibles

The species of ants with kinetic mandibles includes relatively small species (about 0.30 cm), with long and linear mandible and developed eyes (*Acanthognathus* spp., *Strumigenys splendens* (Borgmeier), *S. rugithorax* (Kempf), *S. denticulata* Mayr, *S. subedentata* Mayr, along with other species of *Strumigenys* (Figure 10.2d)).

Mandibles with kinetic action mode, regardless of form, are present in all species of the genera *Daceton*, *Acanthognathus*, and in some *Strumigenys*. They are always relatively narrow, linear or sublinear, usually long and when fully closed, touch only at the apex. There are always a few apical teeth (distally located), some of which may be relatively large. The gape of the mandible can reach 180 degrees, kept opened through basal lateral projections of the specialized labrum, which prevent the mandibles to close before or during the prey approach. In *Acanthognathus*, apical basimandibular processes keep the mandibles open (Dietz and Brandão 1993). When prey comes into contact with specialized hairs (trigger hairs) located in the mouth parts, the basimandibular processes disengage, freeing the accumulated strength of the adductor muscles and closing tightly the mandible in an explosive stroke (Gronemberg et al. 1998). The main mandibular function is to kill or injure the prey. Therefore, in most long mandibulate dacetines, dissipation of the energy released during the closing kills the prey through massive shock and structural damage of the body, and stings may be even not used.

The predation behavior of species with relatively long mandible is based largely on the powerful trap-mandible mechanism, whereas in short mandibulate species, it first involves a slow approach, direct contact with the prey, and persistence after detection, followed by a less powerful mandible lock. In other words, predation is more dependent on the morphology in long mandible species, and more dependent on the behavior in species with short mandibles (Masuko 1984).

10.3.3 Arboreal Predator Ants

This group includes species of ants as *Paraponera clavata* (F.), *Daceton armigerum* (Latreille), *D. boltoni* Azorsa & Sosa-Calvo, *Ectatomma tuberculatum* (Olivier) (which employs a "stalking" strategy), *Acanthoponera*, some species of *Pachycondyla* and *Gnamptogenys concinna* (F. Smith) (Delabie et al. 2010), and most *Pseudomyrmex* species, that actively explore the vegetation, preying on a wide diversity of arthropods. In general, these species have solitary foraging behavior and live in colonies with fewer individuals compared to dominant generalists (arboreal species associated with carbohydrate-rich resources). Workers can reach relatively large sizes (>1 cm), live in colonies with up to a few thousand individuals, and display strong aggressiveness, with presence of powerful poisons, such as in *Paraponera clavata*.

10.3.4 Generalists

The group of generalist ants includes a significant proportion of species in several local ant communities. When species of a community are classified into trophic habits (e.g., fungivore, herbivore, predator omnivore, and carnivore), omnivore is the dominant category both by the relatively high density of their nests (Kaspari 2001), and by the density of species per sampled unit area.

The food of the generalist species is extremely rich, both in breadth of collected food items and for the repertoire of behaviors used in interactions. Many ecologically dominant species (in terms of biomass) are considered "cryptic" herbivorous (Hunt 2003; Davidson et al. 2003; Philpott and Armbrecht 2006) because they maintain mutualistic interactions with sap-sucking insects (mainly Hemiptera Auchenorryncha (Membracoidea) and Sternorrhyncha (Coccoidea and Aphidae)) collectively called trophobionts when cared for by ants (Delabie 2001; Styrshy and Eubanks 2007) (Figure 10.3a–c).

As well, a large number of species forage on the vegetation looking for resources in the form of extrafloral nectaries (Figure 10.3d–e) of the microflora living on the surface of the leaves, secretions discarded by sucking sap not cared for by ants, fungi secretions, particulate matter (pollen, spores, or hyphae), and other animals' feces (Baroni Urbani and De Andrade 1997; Davidson et al. 2003, 2004; Davidson 2005; Oliveira and Freitas 2004).

In addition to the use of different food items, the species grouped here employ behaviorally varied strategies in the search of food resources, such as opportunists (species specialized in discovering food quickly and exploiting it before other ants arrive), species that employ massive recruitment of workers and that dominate the food resources, and behaviorally subordinate species that are behaviorally subordinate in encounter competition (Davidson 1998). These species can coexist with dominants because they are able to live in environments where food is and may even visit the food resource in the presence of dominant species (and because of this are also called insinuators).

Such behavioral patterns have profound implications in the ant community organization (Andersen 1992; Andersen and Patel 1994), especially for those species that inhabit the vegetation. Ant communities can show a strongly hierarchical structure, in which territorially dominant species can competitively control, in part, subordinate species distribution, generating what has been sometimes called "mosaics" of dominant arboreal ants (Majer et al. 1994; Dejean and Corbara 2003; Blüthgen and Fiedler 2004a,b).

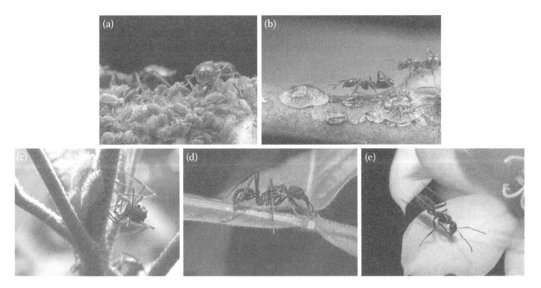

FIGURE 10.3   (a) *Crematogaster* sp. and aphids. (b) Workers of *Linepithema humile* (Mayr) tending an aggregation of scale insects (Hemiptera: Cocoidea). (c) Worker of *Ectatomma tuberculatum* (Olivier) with an aphid honeydew-drop between mandibles. (d) Worker of *Ectatomma goninion* Kugler & Brown visiting extrafloral nectary. (e) Worker of *Camponotus* sp. visiting an extrafloral nectary. (Courtesy of Alex Wild.)

For a critical discussion on hypotheses about mosaics of arboreal ants, see Ribas and Schoereder (2002), Blüthgen and Stork (2007), and Sanders et al. (2007).

There are two well-separated groups of generalist ant species in the morphometric space of the leaf litter ant fauna, which are described below.

10.3.4.1 Generalized Myrmicines

Generalized myrmicines are a group of ants that includes several species Myrmicinae with triangular and relatively short mandibles, widely separated well-developed eyes, as most *Pheidole* (Figure 10.4a) and *Wasmannia* species, some species of *Oxyepoecus* (*O. crassinodus* Kempf, *O. myops* Albuquerque and Brandão, *O. plaumanni* Kempf, *O. rastratus* (Mayr), *O. reticulatus* Kempf, *rosai* Albuquerque and Brandão), *Lachnomyrmex plaumanni* Borgmeier, *L. victori* Feitosa and Brandão, as well as *Solenopsis* species of relatively large body size.

Some species of this group are admittedly omnivorous and classified as generalists in other guilds' proposals, among which are those of the genera *Pheidole* and *Wasmannia*. Many studies have revealed the

FIGURE 10.4   (a) Workers and soldiers of *Pheidole rugulosa* Gregg collecting seeds (generalists: generalized myrmicines). (b) Worker of *Camponotus lespesii* Forel (generalists: generalized formicines). (c) Workers of *Paratrechina longicornis* (Latreille). (Courtesy of Alex Wild.)

food generalist and nesting habits of *W. auropunctata* Roger (Ulloa-Yashar and Cherix 1990). Colonies are relatively big, polydomic, and do not excavate deep nests, but exploit natural cavities preferably under stones or inside trunks and branches. All habitats are acceptable for them: dry, damp, open, or shaded. Colonies often move to more favorable sites when available (Way and Bolton 1997).

The genus *Pheidole* presents a rather uniform combination of anatomical features—almost all species are easily separated from those of all other genera (Wilson 2003), although there is a diversity of external morphological characters that allow identification of hundreds of species in the neotropics. It is the richest and hyperdiverse genus among all ants. The number of described species in the world reaches 900 and the overall richness is estimated at around 1,500 species (Wilson 2003). Moreover, it is locally abundant and is often the prevalent genus in most of the world's warm climate areas, especially in the soil and in the leaf litter. All species have reduced sting apparatuses; as a result, workers mainly use mandibles and toxic chemical repellents during interspecific interactions. Most studied species are predators and necrophagous (Wilson 2003), while some use seeds as a secondary food resource (Johnson 2000). A large proportion of *Pheidole* and *Solenopsis* nesting ground/leaf litter species use any available seed or fruit (Davidson et al. 1984; Kaspari 1993, 1996a; Kaspari and Byrne 1995; Passos and Oliveira 2003; Pizo and Oliveira 2000; Wilson 2003).

10.3.4.2 Generalized Formicines, Dolichoderines, and Some Myrmicines

The group of generalized formicines, dolichoderines, and some myrmicines comprises morphologically characterized ant species by their relatively average body size (0.30 cm), long legs and mandibles, quite narrow and short scapes, with developed eyes relatively close to one another. It includes several species of *Paratrechina* and *Brachymyrmex* (Formicinae), *Dorymyrmex* and *Linepithema* (Dolichoderinae), and a few *Pheidole* species (Myrmicinae) (Figure 10.4b,c).

Dolichoderinae and Formicinae species are omnivorous and particularly well adapted to liquid feeding (Eisner 1957). Along with an expansible esophagus, the modifications in the proventriculus allow an efficient storage of relatively large volumes of liquid. These key "innovations" in structure, connecting the crop (social stomach) to the individual stomach, evolved independently in the two subfamilies and differentiate them from all others, which require energetically expensive contractions to store liquid food (Davidson 1997; Davidson et al. 2004). The evolution of a more efficient way to store and process liquid food may have conditioned these taxa to a diet rich in carbohydrates but low in protein and amino acids (Davidson 1997).

Also some *Pheidole* species with relatively elongate bodies developed scapes and long legs were included with the generalist formicines and dolichoderines, especially when displaying long and narrow femora. These characteristics differentiate them from *Pheidole* species and from the generalist myrmicine guild. Some *Pheidole* and *Monomorium* (Myrmicinae) also have a liquid-rich diet.

10.3.4.3 Small-Sized Hypogaeic Generalist Foragers

The group of small-sized hypogaeic generalist foragers brings together the smaller known species of the leaf litter ants, with the smaller mandible among the guilds, as in *Solenopsis* spp. and *Carebara* spp. Eyes are small and placed very close to the insertion of mandibles.

This group includes predatory species of very small size (0.15 cm), with short escapes and mandibles, associated to vestigial eyes; among them there are numerous species of unknown biology such as those of *Carebara*. The species of reduced body size *Solenopsis* (often named *Diplorhoptrum* and called thief or robber ants) are extremely diverse, frequent, and abundant in leaf litter samples. Virtually all species are monogynic. Some are very characteristic and can be recognized by a combination of superficial sculptures and by the shape of the petiole and postpetiole. However, the most common species are difficult to separate due to a confusing character continuum. Although there is no information about the biology of this group, the ants are allegedly predatory thieves of other colonies of ant immatures, although its biomass apparently exceeds in many cases that of their supposed prey, which contradicts the ecological theory.

10.3.5 Fungus Growers

Ants of the Attini tribe (fungus growers) form a peculiar trophic category; studies on ant guilds employing classification analyses consistently separate them as a distinct group (Silvestre et al. 2003; Silva and Brandão 2010).

Most fungi grown by the Attini belong to *Leucoagaricus* and *Leucocoprinus* (Agaricales: Basidiomycota: respectively Lepiotaceae and Leucocoprinae), except for a few *Apterostigma* species that have secondarily adopted Tricholomataceae, that they grow and feed on. Leucocoprinaceous fungi are common decomposers in the leaf litter of neotropical forests and were probably repeatedly domesticated by ants from free-living populations (Adams et al. 2000a,b).

According the manner of collection and type of substrate used for the development of the fungus, the fungus grower ants can be divided into two subguilds, the leaf cutters and the cryptobiotic Attini (litter nesting fungus growers).

10.3.5.1 Leaf Cutters

Leaf cutter ants or "high" attines (the polymorphic *Atta* and *Acromyrmex* and some species of *Sericomyrmex* and *Trachymyrmex*) use live or dead plant substrate to grow their fungus (Figure 10.5a). Unlike cryptobiotic attine species, *Acromyrmex*, *Sericomyrmex,* and *Trachymyrmex* have colonies with hundreds to thousands of individuals; those of *Atta* reach millions of individuals. The fungus cultivated by the high monophyletic attines is probably transmitted only vertically (clonally) by founding queens (Chapela et al. 1994, but see Mikheyev et al. 2006). Additionally, this symbiotic fungus produces conspicuous nodules called staphylae, forming glycogen-rich vacuolized dilated ends of the hyphae (gongylidea).

Leaf cutting ants, especially *Atta*, are responsible for important ecological processes, through the excavation of large amounts of soil and herbivory in the understory of the vegetation. The colonies can deeply modify the environment near the nests, changing the physical structure of the soil, the distribution of nutrients in soil layers, as well as the composition, productivity, and distribution of plants (Weber 1972; Lofgren and Vander Meer 1986; Farji-Brener and Illes 2000).

10.3.5.2 Litter-Nesting Fungus Growers

The cryptobiotic Attini ants living in the leaf litter includes species of the monomorphic genera *Apterostigma, Cyphomyrmex, Mycetagroicus, Mycetosoritis, Mycocepurus, Myrmicocrypta, Sericomyrmex,* and *Trachymyrmex* (Figure 10.5b). The size of the colonies is always relatively small, with no more than few hundred individuals. The fungus grown by cryptobiotic Attini has likely polyphyletic origin, due to horizontal transmission mechanisms and independent and convergent domestication of living fungi.

The cryptobiotic Attini collect a wide variety of substrates for their fungus, such as leaves (rarely fresh, mostly already decayed), flowers, fruits, seeds, feces, lichen, moss, and carcasses of arthropods

FIGURE 10.5   (a) Attine (leaf-cutter ants): worker of *Atta texana* (Buckley) transporting a leaf fragment. (b) Cryptobiotic attines: worker of *Mycetosoritis hartmanni* (Wheeler) tending the fungus. (Courtesy of Alex Wild.)

(Leal and Oliveira 1998). The material is always collected on the soil, mostly within one to two meters from the nest entrance (Leal and Oliveira 2000).

10.3.6 Legionary Ants

The exclusively neotropical ant species of the Ecitoninae subfamily shows a combination of interrelated behavioral and morphological traits referred to as the legionary ant adaptive syndrome (Gotwald 1995; Brady 2003). The legionary ants are obligate collective foragers and group predators, they are nomadic, and have highly specialized, permanently wingless queens (Brady 2003; Brady and Ward 2007). Ecitoninae ants, together with the subfamilies Dorylinae and Aenictinae (tropical ants from Africa and Asia) are considered the dominant social hunters of invertebrates in tropical forests, affecting prey abundance and biodiversity (Roberts et al. 2000; Berghoff et al. 2003; Kaspari and O'Donnell 2003; Longino 2005; Kronauer 2008). Recent work suggests that responses of the leaf litter ant species to legionary ants predation may have determined life characteristics of the leaf litter ant communities, as small colony size, constant colony growth over time (Kaspari and Byrne 1995; Kaspari and Vargo 1995; Kaspari 1996b), and specialized nesting behavior (McGlynn et al. 2003, 2004; Longino 2005).

Several species included in this neotropical ant guild are mostly or exclusively carnivorous predators, and several species are specialized predators of other social insects (Gotwald 1995; O'Donnell et al. 2005). In general, a large part of the immature Ecitoninae diet consists of other species of ants (Gotwald 1995); *Nomamyrmex esenbeckii* (Westwood) preys on leaf-cutting ant colonies and attine brood may be an important food source (Swartz 1998).

According to preferred foraging place, legionary ants can be divided into two groups: epigaeic (foraging on the soil surface) or hypogaeic (foraging in the leaf litter and soil superficial layers). The genera *Eciton* and *Labidus* (Ecitoninae) and *Leptanilloides* (Leptanilloidinae, Brandão et al. 1999) mainly include epigaeic species (Figure 10.6a), while hypogaeic behavior (Figure 10.6b) occurs in *Neivamyrmex, Cheliomyrmex, Nomamyrmex* (Ecitoninae, Quiroz-Robledo et al. 2002), and *Asphinctanilloides* (Leptanilloidinae, Brandão et al. 1999). However, there is almost no information on the prey of these hypogaeic species and on their impact on leaf litter and soil surface invertebrate fauna (Berghoff et al. 2002, 2003).

10.3.7 Dominant Arboreal Ants Associated with Carbohydrate-Rich Resources or Domatia

In general, dominant arboreal ants associated with carbohydrate-rich resources or domatia are the ant species at the top of the prevalent dominance hierarchy in neotropical communities (Andersen et al. 2006; Blüthgen and Stork 2007). They predominantly eat liquid food resources widely distributed in vegetation, as nectar produced by floral or extrafloral nectaries of Angiosperms, carbohydrate-rich exudates produced by Auchenorrhyncha and Sternorrhyncha sucking hemipterans (improperly known as homopterans), and exudates of some Lepidoptera larvae (Blüthgen et al. 2003, 2004b; Davidson et al. 2003, 2004).

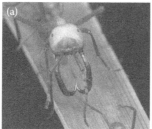

FIGURE 10.6 (a) Major worker of epigaeic forager *Eciton burchelli* (Westwood). (b) Workers of hypogaeic forager *Neivamyrmex californicus* (Mayr). (Courtesy of Alex Wild.)

The competition for these resources can be intense and asymmetric, resulting in hierarchically structured communities formed by competitively dominant and subordinate species (Blüthgen et al. 2000; Blüthgen and Fiedler 2004b; Blüthgen and Stork 2007), determining a distribution of the species in mosaic with the dominant species occupying exclusive territories, so that their home range does not overlap. Associated with the dominant species, coexist subdominant and subordinate species, also distributed in mosaics, since the association with the dominant species is species specific (Majer et al. 1994; Dejean et al. 2000).

Studies on the importance of mutualistic associations in the organization of ant communities indicate that the presence and type of exudate have a strong influence on the density, diversity, and distribution of ants, playing a key role in other arthropod community structures (Blüthgen et al. 2000, 2004b; Dejean et al. 2000). In addition, the quantity and quality of food resources, in particular the concentration of carbohydrates and composition of amino acids, influence the composition of species of ants that visit these resources (Blüthgen and Fiedler 2004a; Blüthgen et al. 2004a).

The arboreal species living in association with myrmecophyte plants (e.g., *Acacia*, *Cecropia*, and *Tococa*) are included here as well; they present specialized structures for the nesting of ants, as domatia (Figure 10.7), as well as structures that provide food (nectaries and Müllerian or food bodies). The term domatia has been applied to various types of cavities in plants that are used by ants as nests (Beattie 1985). Ant nesting in myrmecophytes can be active predators of herbivores associated with these plants, determining the structure of arthropods' communities (Yu et al. 2001; Izzo and Vasconcelos 2002, 2005). Among the ants associated with myrmecophytes, there are species of Dolichoderinae (*Azteca*), Formicinae (*Brachymyrmex*, *Camponotus*), and Myrmicinae (*Allomerus*, *Crematogaster*, *Pheidole*, and *Wasmannia*).

Some species can significantly influence the arthropod community structure by exerting strong predation pressure especially on Lepidoptera and Coleoptera larvae, expressed in the high values of beta diversity of herbivores observed in canopy samples (Floren et al. 2002; Philpott and Armbrecht 2006). Species of the following genera were recorded as having dominant or subdominant status in several studies on the structure of the arboreal fauna: Dolichoderinae (*Azteca*, *Dolichoderus*), Ectatomminae (*Ectatomma*), Formicinae (*Camponotus*, *Paratrechina*), and Myrmicinae (*Crematogaster*, *Monomorium*, *Pheidole*, *Solenopsis*, *Wasmannia*) (Majer et al. 1994; Dejean et al. 2000; Armbrecht et al. 2001).

10.3.8 Pollen-Feeding Arboreal Ants

In the exclusively neotropical Cephalotini tribe, the two genera *Cephalotes* and *Procryptocerus* represent a radiation of arboreal species that nest exclusively within live or dead branches (Figure 10.8). The nutritional biology of cephalotine ants is not well known yet, but the observed worker behavior in attractive baits (protein and sugar), carcasses, floral and extrafloral nectarines, and bird feces (rich in

FIGURE 10.7 Arboreal ants associated with domatia: *Pheidole melastomae* Wilson nesting in plant domatia of *Tococa* (Melastomataceae). (Courtesy of Alex Wild.)

FIGURE 10.8   Pollen eaten by arboreal ants: (a) Worker of *Cephalotes atratus* (L.) foraging on vegetation, (b) controlled gliding flight, and (c) worker of *C. clypeatus* (F.) in nest entrance. (Courtesy of Alex Wild.)

nitrogen bases) suggest widespread omnivory. In addition, workers constantly scrap pollen deposited on the vegetation surface by anemochory (Baroni-Urbani and De Andrade 1997), which is the most important component of the cephalotine ants' diet (De Andrade and Baroni-Urbani 1999). This is certainly one of the rare cases of pollen used as food by a terrestrial mesofauna component. The relatively slow Cephalotini behavior seems to be a foraging strategy: passive defense by virtue of spines covering in numerous species; in others, the generalized body flattening allows them to disguise themselves in the substrate (simulating the bark of trees), a cuticle that serves as a receptacle for filaments of algae in some species, allowing them to adopt the color of the substrate, and death feigning when threatened (De Andrade and Baroni-Urbani 1999).

10.3.9 Subterranean Ants

We employ the term subterranean here to describe the guild that encompass ant species living exclusively in the deeper layers of soil, including some who spend most of their life cycle in nests and cavities in the soil; only males and young queens come to the surface once or a few times a year (Silva and Silvestre 2004). They are considered as belonging to the specialized guild of predatory ant species, even if they have subterranean habits, whereas in this case the food habit is hierarchically more important than the occupied extract.

Species regarded as belonging to the subterranean ants guild include, for example, those of the genus *Tranopelta* that forage entirely underground (Delabie et al. 2000); they are characterized by the pale integument with little pigmentation, reduced scape length, antennae segmentation, and eyes. The genus *Acropyga*, of which *Tranopelta* are probably dependent (Delabie and Fowler 1993), also includes cryptobiotic and ground living species (although locally abundant) that maintain obligatory associations with scale insects (Hemiptera: Pseudococcidae) found in the roots of plants (LaPolla 2004). Mated *Acropyga* queens carry between the mandibles a symbiotic mated scale insect female during the nuptial flight (Johnson et al. 2001), in a behavior called trophoforesis (LaPolla et al. 2002) that allows the founding queens to start a new colony with a new generation of symbiotic Pseudococcidae.

The biology of most species that compose this group is unknown. The presence of subterranean species is often revealed only by records of males in light traps. Its richness seems to be greater than believed and has been regarded as a frontier in our knowledge about the ant fauna (Longino and Colwell 1997; Fisher and Robertson 2002; Silva and Silvestre 2004; Andersen and Brault 2010; Schmidt and Solar 2010). Systematic collection techniques for the assessment of the composition and abundance of species have not yet been properly tested (Esteves et al. 2008; Schmidt and Solar 2010); taxonomic novelties and more information about the feeding ecology of this group are expected from more intense surveys in different neotropical sites.

10.4 Concluding Remarks: From Trophic Guilds to Applied Myrmecology

The recognition that ants are major predators in many agro-ecosystems has led to generalist predator species being often used as insect pests and in phytopatogenic fungi control programs (Way and Khoo

1992; Philpott and Armbrecht 2006). Ants are well suited for pest control because they are capable of high offtake rates (food is stored in the nest, so individual foragers do not satiate even after killing many prey), and because the intensity and location of ant foraging efforts are comparatively easily manipulated (Agarwal et al. 2007; Ward-Fear et al. 2010). Several ant species prey on insect pests (Philpott and Armbrecht 2006; Philpott et al. 2008a,b). It has been already documented that several species of ants reduce densities of lepidopteran larvae in coffee plants (Perfecto and Vandermeer 2006). Ants also affect canopy arthropods in coffee plantations and prey on herbivores introduced into coffee farms (Philpott et al. 2004, 2008a). Ants have been traditionally installed by farmers in orchards or agroforestry systems to control pest populations in citrus, cocoa, and coconut plantations, banana groves, sweet potato fields, and in corn, bean, and squash systems because of the benefits ants provide to crop plants (Delabie et al. 2007; Philpott et al. 2008a).

Ants can control the abundance of herbivores via direct pest insect predation, or through interactions involving chemical repellents; they can even cause the demise of plant herbivores during their foraging activity, which reduces damage to plants, and increase cultures growth and production in agro-ecosystems (Symondson et al. 2002; Philpott and Armbrecht 2006). In addition, ants that use fungi as food reduce the presence of phytopatogenic fungi as they remove spores during foraging on the vegetation or restrict interactions between plants and disease vectors (Khoo and Ho 1992).

Recent studies on the ant fauna of cocoa plantations employed also the concept of trophic guilds to test the factors that determine the distribution of species in the vegetation and have helped to reveal the complex nature of the interactions in ant communities (Sanders et al. 2007), in particular in mosaics. This has profound implications in the use of predatory ants in pest control in agro-ecosystems (Philpott and Foster 2005; Armbrecht et al. 2006).

Using ants for pest control can be even more significant if the diversity and abundance of predatory species can be artificially manipulated. Increasing the diversity of predators can increase the likelihood that predatory species are included in the community and thus fosters the role of predators. In addition, it can increase the complementarity of prey, because broader diets and foraging behaviors will be effective, strongly influencing the density of herbivore populations (Hooper et al. 2005; Philpott and Armbrecht 2006).

The same idea applies to ants nesting on the ground, because the understanding of the factors that limit the diversity and density of trophic groups increases the potential for using ants as predators in agro-ecosystems (Philpott and Foster 2005). In experimental studies, litter ants have been recorded preying on fruit flies pupae (*Ceratitis capitata* Weidemann, Diptera: Tephritidae) and have the potential to control the coffee berry borer (*Hypothenemus hampei* Ferrari, Coleoptera: Scolytidae), the most important coffee pest (Armbrecht and Perfecto 2003).

Functional ant groups has been also used to control invasive species based on the concept of "mismatches" in which the characteristics of the invading species render them vulnerable to some mortality source operating within the introduced range (Ward-Fear et al. 2010). Globally, invasive species pose a major threat to biodiversity (Mack et al. 2000). The cane toads *Bufo marinus* L., for instance, are considered as one of the most significant invasive species in Australia. Cane toads are large anurans native to South and Central America, brought to Australia in 1935 in an unsuccessful attempt to control insect pests of sugar cane crops (Ward-Fear et al. 2010). Large, highly aggressive, and behavioral dominants ants (*Iridomyrmex* spp.) are ubiquitous in tropical Australia and can prey on cane toad metamorphs. Ants of this group occur across most continental Australia (Andersen 1995) and have the potential to contribute to cane toad control over a broad area (Ward-Fear et al. 2009, 2010).

On the other hand, there is a large number of ant species considered agricultural and public health pests. One important taxon is the *Solenopsis saevissima* species group of fire ants. This group consists of 13 described species of fire ants and their social parasites, all of which are native of South America. Two of the species, *S. invicta* Buren and *S. richteri* Forel occur also in the United States after unintentional introductions in the early 20th century, and *S. invicta* has been recently introduced to the West Indies islands and several Pacific Rim countries as well (Morrison et al. 2004). There is an impressive body of general knowledge on fire ant biology (Tschinkel 2006), although the development of a stable alpha-taxonomy for the *S. saevissima* species group has proven to be a difficult task, in large part because of the lack of informative invariant morphological characters among putative species (Shoemaker et al. 2006).

There is evidence that *S. invicta*, listed among the 100 worst invasive species in the world and largely restricted to disturbed habitats in its introduced range, can change or reduce the native nonant arthropod community, negatively affecting vertebrate populations, and disrupting mutualisms (Tschinkel 2006; King and Tschinkel 2008).

The Argentine ants (*Linepithema humile* (Mayr)) provide an example of a species that causes numerous direct and indirect impacts on communities. Native to South American, Argentine ants have been introduced nearly worldwide and eliminate nearly all epigaeic ants when they invade new habitats (Suarez et al. 1998). In South Africa, the displacement of native seed-dispersing ants by Argentine ants has resulted in reduced seedling recruitment in myrmecochorous shrub species. In addition, the displacement of native ants by *L. humile* has been implicated in the decline of an endangered vertebrate, the coastal horned lizard *Phrynosoma coronatum* (Blainville) (Suarez and Case 2002). A very similar situation can be exposed in parallel to the case of the neotropical little fire ant *Wasmannia auropunctata* (Roger) introduced worldwide (Foucaud et al. 2010).

The concept of functional groups has been explored to predict potential invasive species, based on analysis of convergent morphological and life history characteristics. Examining the functional group membership of invasive ants is useful to understand the life histories that are associated with infestations and invasions (McGlynn 1999). Therefore, using functional groups to examine convergent traits of invasive ants may lead to prediction of future invaders. In particular, information on queen number, unicoloniality, interspecific aggression, and generalized foraging and nesting, are important in the identification of future invaders (Brandão and Paiva 1994; McGlynn 1999). Predictive ecology may play an important role in the monitoring invasive species by focusing on the groups likely to contain invasive species (McGlynn 1999).

Finally, to understand and explain the role of ants in ecosystems is important in the context of the broader theme of environmental conservation. The functions of ant food chains have important consequences on the structure of the fauna and even on the structure of the vegetation in the tropical forests (Wilson 1987; Hunt 2003). Evidence suggests that the ants act as top predators in trophic chains (Floren et al. 2002) or as the main herbivores of tropical forests, because a significant part of energy resources used by the colonies (proportional to the enormous biomass of colonies in vegetation) is obtained from the exudates of Hemiptera and different types of nectaries (Davidson et al. 2003). As main predators and herbivorous throughout this long evolutionary history of more than 100 million years, ants had and still have considerable influence on the history of many other organisms and in the ecological dynamics of forests.

ACKNOWLEDGMENTS

We acknowledge support from the São Paulo State Science Foundation, within the Biota-FAPESP Program (grant 98/05083-0 to C.R.F. Brandão), a postdoctoral grant to R.R. Silva (grant 06/02190-8). C.R.F. Brandão and J.H.C. Delabie are fellows of the National Council for Science and Technology Development (CNPq) of Brazil. This is a contribution of the BIOTA-FAPESP, the Biodiversity Virtual Institute Program and of the Program PRONEX FAPESB CNPq, project PNX0011/2009.

REFERENCES

Adams, R. M. M., U. G. Mueller, A. M. Green, and J. M. Narozniak. 2000a. Garden sharing and garden stealing in fungus-growing ants. *Naturwissenschaften* 87:491–3.

Adams, R. M. M., U. G. Mueller, T. R. Schultz, and B. Norden. 2000b. Agropredation: Usurpation of attine gardens by *Megalomyrmex* ants. *Naturwissenschaften* 87:549–54.

Agarwal, V. M., N. Rastogi, and S. V. S. Raju. 2007. Impact of predatory ants on two lepidopteran insect pests in Indian cauliflower agroecosystems. *J. App. Entomol.* 131:493–500.

Andersen, A. N. 1992. Regulation of "momentary" diversity by dominant species in exceptionally rich ant communities of the Australian seasonal tropics. *Am. Nat.* 140:401–20.

Andersen, A. N. 1995. A classification of Australian ant communities, based on functional groups which parallel plant life forms in relation to stress and disturbance. *J. Biogeogr.* 22:15–29.

Andersen, A. N., and A. Brault. 2010. Exploring a new biodiversity frontier: Subterranean ants in northern Australia. *Biodivers. Conserv.* 19:2741–50.

Andersen, A. N., A. Fisher, B. D. Hoffmann, J. L. Read, and R. Richards. 2004. Use of terrestrial invertebrates for biodiversity monitoring in Australian rangelands, with particular reference to ants. *Aust. Ecol.* 29:87–92.

Andersen, A. N., T. Hertog, and J. C. Z. Woinarski. 2006. Long-term fire exclusion and ant community structure in an Australian tropical savanna: Congruence with vegetation succession. *J. Biogeogr.* 33:823–32.

Andersen, A. N., B. D. Hoffmann, W. J. Müller, and A. D. Griffiths. 2002. Using ants as bioindicators in land management: Simplifying assessment of ant community responses. *J. App. Ecol.* 39:8–17.

Andersen, A. N., and J. N. Majer. 2004. Ants show the way down under: Invertebrates as bioindicators in land management. *Front. Ecol. Environ.* 2:291–8.

Andersen, A. N., and A. D. Patel. 1994. Meat ants as dominant members of Australian ant communities: An experimental test of their influence on the foraging success and forager abundance of other species. *Oecologia* 98:15–24.

Armbrecht, I., and I. Perfecto. 2003. Litter-twig dwelling ant species richness and predation potential within a forest fragment and neighboring coffee plantations of contrasting habitat quality in Mexico. *Agric. Ecosyst. Environ.* 97:107–15.

Armbrecht, I., I. Perfecto, and E. Silverman. 2006. Limitation of nesting resources for ants in Colombian forests and coffee plantations. *Ecol. Entomol.* 31:403–10.

Armbrecht, I., E. Jiménez, G. Alvarez, P. Ulloa-Chacon, and H. Armbrecht. 2001. An ant mosaic in the Colombian rain forest of Chocó (Hymenoptera: Formicidae). *Sociobiology* 37:491–509.

Baroni-Urbani, C., and M. L. De Andrade. 1997. Pollen eating, storing, and spitting by ants. *Naturwissenschaften* 84:256–8.

Baroni-Urbani, C., and M. L. De Andrade. 2007. The ant tribe Dacetini: Limits and constituent genera, with descriptions of new species (Hymenoptera, Formicidae). *Annali di Museo Civico di Storia Naturale di Giacomo Doria* 99:1–191.

Beattie, A. J. 1985. *The Evolutionary Ecology of Ant–Plant Mutualism.* Cambridge, UK: Cambridge University Press.

Berghoff, S. M., U. Maschwitz, and K. E. Linsenmair. 2003. Influence of the hypogaeic army ant *Dorylus* (*Dichthadia*) *laevigatus* on tropical arthropod communities. *Oecologia* 135:149–57.

Berghoff, S. M., A. Weissflog, K. E. Linsenmair, R. Hashim, and U. Maschwitz. 2002. Foraging of a hypogaeic army ant: A long neglected majority. *Insectes Soc.* 49:133–41.

Blondel, J. 2003. Guilds or functional groups: Does it matter? *Oikos* 100:223–31.

Blüthgen, N., and K. Fiedler. 2004a. Preferences for sugars and amino acids and their conditionality in a diverse nectar-feeding ant community. *J. Anim. Ecol.* 73:155–66.

Blüthgen, N., and K. Fiedler. 2004b. Competition for composition: Lessons from nectar-feeding ant communities. *Ecology* 85:1479–85.

Blüthgen, N., G. Gebauer, and K. Fiedler. 2003. Disentangling a rainforest food web using stable isotopes: Dietary diversity in a species-rich ant community. *Oecologia* 137:426–35.

Blüthgen, N., G. Gottsberger, and K. Fiedler. 2004a. Sugar and amino acid composition of ant-attended nectar and honeydew sources from an Australian rainforest. *Aust. Ecol.* 29:418–29.

Blüthgen, N., and N. E. Stork. 2007. Ant mosaics in a tropical rainforest in Australia and elsewhere: A critical review. *Aust. Ecol.* 32:93–104.

Blüthgen, N., N. E. Stork, and K. Fiedler. 2004b. Bottom-up control and co-occurrence in complex communities: Honeydew and nectar determine a rainforest ant mosaic. *Oikos* 106:344–58.

Blüthgen, N., M. Verhaagh, W. Goitía, K. Jaffé, W. Morawetz, and W. Barthlott. 2000. How plants shape the ant community in the Amazonian rainforest canopy: The key role of extrafloral nectaries and homopteran honeydew. *Oecologia* 125:229–40.

Bolton, B. 1998. Monophyly of the dacetonine tribe-group and its component tribes (Hymenoptera: Formicidae). *Bull. Nat. Hist. Mus. Lond. (Ent.)* 67:65–78.

Bolton, B. 1999. Ant genera of the tribe Dacetonini (Hymenopera: Formicidae). *J. Nat. Hist.* 33:1639–89.

Bolton, B., G. Alpert, P. S. Ward, and P. Naskrecki. 2006. *Bolton's Catalogue of Ants of the World, 1758–2005.* Cambridge, MA: Harvard University Press.

Brady, S. G. 2003. Evolution of the army ant syndrome: The origin and long-term evolutionary stasis of a complex of behavioral and reproductive adaptations. *Proc. Natl. Acad. Sci. USA* 100:6575–9.

Brady, S. G., T. R. Schultz, B. Fisher, and P. S. Ward. 2006. Evaluating alternative hypotheses for the early evolution and diversification of ants. *Proc. Natl. Acad. Sci. USA* 103:18172–7.

Brady, S. G., and P. S. Ward. 2007. Morphological phylogeny of army ants and other dorylomorphs (Hymenoptera: Formicidae). *Syst. Entomol.* 30:593–618.

Brandão, C. R. F. 2003. Further revisionary studies of the ant genus *Megalomyrmex* Forel (Hymenoptera: Formicidae: Myrmicinae: Solenopsidini). *Pap. Avul. Zool.* 43:145–59.

Brandão, C. R. F., J. L. M. Diniz, D. Agosti, and J. H. C. Delabie. 1999. Revision of the neotropical ant subfamily Leptanilloidinae. *Syst. Entomol.* 24:17–36.

Brandão, C. R. F., and R. V. S. Paiva. 1994. The Galapagos ant fauna and the attributes of colonizing ant species. In *Exotic Ants*, ed. D. Willians, 1–10. Boulder, CO: Westview Press.

Brandão, C. R. F., R. R. Silva, and J. H. C. Delabie. 2009. Formigas (Hymenoptera). In *Bioecologia e Nutrição de Insetos: Base para o Manejo Integrado de Pragas*, ed. A. R. Panizzi, and J. R. P. Parra, 323–69. Brasília: Embrapa Informação Tecnológica.

Bromham, L., M. Cardillo, A. F. Bennett, and M. A. Elgar. 1999. Effects of stock grazing on the ground invertebrate fauna of woodland remnants. *Aust. J. Ecol.* 24:199–207.

Brown, K. S., A. V. L. Freitas, and R. B. Francini. 2003. Insetos como indicadores ambientais. In *Métodos de Estudos em Biologia da Conservação & Manejo da Vida Silvestre*, ed. L. Cullen Jr., R. Rudran, and C. Valladares-Padua, 125–51. Curitiba, Brazil: Editora da Universidade Federal do Paraná.

Brown, W. L., Jr. 2000. Diversity of ants. In *Ants: Standard Methods for Measuring and Monitoring Biodiversity*, ed. D. Agosti, J. D. Majer, L. E. Alonso, and T. Schultz, 45–79. Washington, DC: Smithsonian Institution Press.

Brown, W. L., Jr., and W. W. Kempf. 1968. *Tatuidris*, a remarkable new genus of Formicidae (Hymenoptera). *Psyche* 74:183–90.

Brühl, C. A., G. Gunsalam, and K. E. Linsenmair. 1998. Stratification of ants (Hymenoptera: Formicidae) in a primary rain forest in Sabah, Borneo. *J. Trop. Ecol.* 14:285–97.

Chamberlain, S. A., and J. N. Holland. 2009. Body size predicts degree in ant–plant mutualistic networks. *Funct. Ecol.* 23:196–202.

Chapela, I. H., S. A. Rehner, T. R. Schultz, and U. G. Mueller. 1994. Evolutionary history of the symbiosis between fungus-growing ants and their fungi. *Science* 266:1691–4.

Cohen, J. E., T. Jonsson, and S. R. Carpenter. 2003. Ecological community description using the food web, species abundance, and body size. *Proc. Natl. Acad. Sci. USA* 100:1781–6.

Crist, T. O. 2008. Biodiversity, species interactions, and functional roles of ants (Hymenoptera: Formicidae) in fragmented landscapes: A review. *Myrmecol. News* 12:3–13.

Cristianini, A. V., and P. S. Oliveira. 2009. The relevance of ants as seed rescuers of a primarily bird-dispersed tree in the neotropical cerrado savanna. *Oecologia* 160:735–45.

Crozier, R. H. 2006. Charting uncertainty about ant origins. *Proc. Natl. Acad. Sci. USA* 103:18029–30.

Davidson, D. W. 1997. The role of resources imbalances in the evolutionary ecology of tropical arboreal ants. *Biol. J. Lin. Soc.* 61:153–81.

Davidson, D. W. 1998. Resource discovery versus resource domination in ants: A functional mechanism for breaking the trade-off. *Ecol. Entomol.* 23:484–490.

Davidson, D. W. 2005. Ecological stoichiometry of ants in a new world rain forest. *Oecologia* 142:221–31.

Davidson, D. W., S. C. Cook, and R. R. Snelling. 2004. Liquid-feeding performance of ants (Formicidae): Ecological and evolutionary implications. *Oecologia* 139:255–66.

Davidson, D. W., S. C. Cook, R. R. Snelling, and T. H. Chua. 2003. Explaining the abundance of ants in lowland tropical rainforest canopies. *Science* 300:969–72.

Davidson, D. W., R. S. Inouye, and J. H. Brown. 1984. Granivory in a desert ecosystem: Experimental evidence for indirect facilitation of ants by rodents. *Ecology* 65:1780–6.

De Andrade, M. L., and C. Baroni-Urbani. 1999. Diversity and adaptation in the ant genus *Cephalotes*, past and present (Hymenoptera, Formicidae). *Stuttg. Beitr. Natkd. S. B (Geol. Palaeontol.)* 271:1–889.

Delabie, J. H. C. 2001. Trophobiosis between Formicidae and Hemiptera (Sternorrhyncha and Auchenorryncha): An overview. *Neotrop. Entomol.* 30:501–16.

Delabie, J. H. C., D. Agosti, and I. C. Nascimento. 2000. Litter ant communities of the Brazilian Atlantic rain forest region. In *Sampling Ground-Dwelling Ants: Case Studies from the World's Rain Forests*, ed. D. Agosti, J. D. Majer, L. T. Alonso, and T. R. Schultz, 1–17. Perth, Australia: Curtin University, School of Environmental Biology Bulletin No. 18.

Delabie, J. H. C., W. D. Da Rocha, R. M. Feitosa, P. Devienne, and D. Fresneau. 2010. *Gnamptogenys concinna* (F. Smith, 1858): nouvelles données sur sa distribution et commentaires sur ce cas de gigantisme dans le genre *Gnamptogenys* (Hymenoptera, Formicidae, Ectatomminae). *Bull. Soc. Entomol. Fr.* 115:269–77.

Delabie, J. H. C., and H. G. Fowler. 1993. Physical and biotic correlates of population fluctuations of dominant soil and litter ant species (Hymenoptera: Formicidae) in Brazilian cocoa plantations. *J. N. Y. Entomol. Soc.* 101:135–40.

Delabie, J. H. C., B. Jahyny, I. C. Nascimento, C. S. F. Mariano, S. Lacau, S. Campiolo, S. M. Philpott, and M. Leponce. 2007. Contribution of cocoa plantations to the conservation of native ants (Insecta: Hymenoptera: Formicidae) with a special emphasis on the Atlantic Forest fauna of southern Bahia, Brazil. *Biodiv. Conserv.* 16:2359–84.

Dejean, A. 1987a. Behavioral plasticity of hunting workers of *Serrastruma serrula* presented with different arthropods. *Sociobiology* 13:191–208.

Dejean, A. 1987b. Étude du comportement de prédation dans le genre *Strumigenys*. *Insectes Soc.* 33:388–405.

Dejean, A., and B. Corbara. 2003. A review of mosaics of dominant ants in rainforests and plantations. In *Arthropods of Tropical Forests: Spatio-Temporal Dynamics and Resource Use in the Canopy*, ed. Y. Basset, V. Novotny, S. E. Miller, and R. L. Kitching, 341–7. Cambridge, UK: Cambridge University Press.

Dejean, A., D. McKey, M. Gibernau, and M. Belin. 2000. The arboreal ant mosaic in a Cameroonian rainforest (Hymenoptera: Formicidae). *Sociobiology* 35:403–23.

Dietz, B. H. 2004. Uma revisão de Basicerotini Brown, 1948 (Formicidae: Myrmicinae), suas relações filogenéticas internas e com outras tribos dacetine (Dacetini e Phalacromyrmecini). Ph.D. thesis, Instituto de Biociências da Universidade de São Paulo, São Paulo, Brazil.

Dietz, B. H., and C. R. F. Brandão. 1993. Comportamento de caça e dieta de *Acanthognathus rudis* Brown & Kempf, com comentários sobre a evolução da predação em Dacetini (Hymenoptera, Formicidae, Myrmicinae). *Rev. Bras. Entomol.* 37:683–92.

Douglas, M. E., and W. J. Matthews. 1992. Does morphology predict ecology? Hypothesis testing within a freshwater stream fish assemblage. *Oikos* 65:213–24.

Ellwood, M. D. F., and W. A. Foster. 2004. Doubling the estimate of invertebrate biomass in a rainforest canopy. *Nature* 429:549–51.

Eisner, T. 1957. A comparative morphological study of the proventriculus of ants (Hymenoptera: Formicidae). *Bull. Mus. Comp. Zool.* 116:429–90.

Esteves, F. A., C. R. F. Brandão, and K. Viegas. 2008. Subterranean ants (Hymenoptera: Formicidae) as prey of fossorial reptiles (Reptilia, Squamata: Amphisbaenidae) in Central Brazil. *Pap. Avul. Zool.* 48:329–34.

Farji-Brener, A. G., G. Barrantes, and A. Ruggiero. 2004. Environmental rugosity, body size and access to food: A test of the size–grain hypothesis in tropical litter ants. *Oikos* 104:165–71.

Farji-Brener, A. G., and A. E. Illes. 2000. Do leaf-cutting ant nest make "bottom-up" gaps in neotropical rain forests?: A critical review of the evidence. *Ecol. Lett.* 3:219–27.

Feitosa, R. M., C. R. F. Brandão, and B. H. Dietz. 2007. *Basiceros scambognathus* (Brown, 1949) n. comb., with the first worker and male descriptions, and a revised generic diagnosis (Hymenoptera: Formicidae: Myrmicinae). *Pap. Avul. Zool.* 47:15–26.

Feitosa, R. M., R. R. Hora, J. H. C. Delabie, J. Valenzuela, and D. Fresneau. 2008. A new social parasite in the ant genus *Ectatomma* F. Smith (Hymenoptera, Formicidae, Ectatomminae). *Zootaxa* 1713:47–52.

Fisher, B. L. 1999. Improving inventory efficiency: A case study of leaf-litter ant diversity in Madagascar. *Ecol. Appl.* 9:714–31.

Fisher, B. L., and H. G. Robertson. 2002. Comparison and origin of forest and grassland ant assemblages in the high plateau of Madagascar (Hymenoptera: Formicidae). *Biotropica* 34:155–67.

Floren, A., A. Biun, and K. E. Linsenmair. 2002. Arboreal ants as key predators in tropical lowland rainforest trees. *Oecologia* 131:137–44.

Folgarait, P. J. 1998. Ant biodiversity and its relationship to ecosystem functioning: A review. *Biod. Cons.* 7:1221–44.

Foucaud, J., J. Orivel, A. Loiseau, J. H. C. Delabie, H. Jourdan, D. Konghouleux, M. Vonshak, M. Tindo, J.-L. Mercier, D. Fresneau, J.-B. Mikissa, T. McGlynn, A. S. Mikheyev, J. Oettler, and A. Estoup. 2010. Worldwide invasion by the little fire ant: Routes of introduction and eco-evolutionary pathways. *Evol. Appl.* 3:363–74.

Fowler, H. G., L. C. Forti, C. R. F. Brandão, J. H. C. Delabie, and H. L. Vasconcelos. 1991. Ecologia nutricional de formigas. In *Ecologia Nutricional de Insetos*, ed. A. R. Panizzi, and J. R. P. Parra, 131–223. São Paulo, Brazil: Editora Manole.

Françoso, M. F. L. 1995. Biologia e taxonomia de *Heteroponera* Mayr 1887 neotropicais (Hymenoptera: Formicidae). M.Sc. dissertation, Instituto de Biociências da Universidade de São Paulo.

Frouz, J., and V. Jilková. 2008. The effect of ants on soil properties and process (Hymenoptera: Formicidae). *Mirmecol. News* 11:191–9.

Gotelli, N. J., and A. M. Ellison. 2002. Biogeography at a regional scale: Determinants of ant species density in New England bogs and forests. *Ecology* 83:1604–9.

Gotwald, W. H., Jr. 1995. *Army Ants: The Biology of Social Predation*. New York: Cornell University Press.

Grimaldi, D., and M. S. Engel. 2005. *Evolution of the Insects*. New York: Cambridge University Press.

Gronenberg, W., C. R. F. Brandão, B. H. Dietz, and S. Just. 1998. Trap-mandibles revisited: The mandible mechanism of the ant *Acanthognathus*. *Physiol. Entomol.* 23:227–40.

Gronenberg, W., and B. Ehmer. 1996. The mandible mechanism of the ant genus *Anochetus* (Hymenoptera, Formicidae) and the possible evolution of trap-mandibles. *Zoology* 99:153–62.

Gronenberg, W., J. Paul, S. Just, and B. Hölldobler. 1997. Mandible muscle fibers in ants: Fast or powerful? *Cell Tiss. Res.* 289:347–61.

Guimarães, P. R., V. Rico-Gray, S. F. dos Reis, and J. N. Thompson. 2006. Asymmetries in specialization in ant–plant mutualistic networks. *Proc. R. Soc. B* 273:2041–7.

Guimarães, P. R., V. Rico-Gray, P. S. Oliveira, T. J. Izzo, S. F. Dos Reis, and J. N. Thompson. 2007. Interaction intimacy affects structure and coevolutionary dynamics in mutualistic networks. *Curr. Biol.* 17:1797–803.

Hölldobler, B., and E. O. Wilson. 1990. *The Ants*. Cambridge, MA: Harvard University Press.

Hooper, C., F. S. Chapin, III, J. J. Ewel, A. Hector, P. Inchausti, S. Lavorel, J. H. Lawton, D. M. Lodge, M. Loreau, S. Naeem, B. Schmid, H. Setälä, A. J. Symstad, J. Vandermeer, and D. A. Wardle. 2005. Effects of biodiversity on ecosystem functioning: A consensus of current knowledge. *Ecol. Monogr.* 75:3–35.

Hora, R. R., C. Doums, C. Poteaux, R. Fénéron, J. Venezuela, J. Heinze, and D. Fresneau. 2005. Small queens in the ant *Ectatomma tuberculatum*: A new case of social parasitism. *Behav. Ecol. Sociobiol.* 59:285–92.

Hunt, J. H. 2003. Cryptic herbivores of the rainforest canopy. *Science* 300:916–7.

Izzo, T. J., and H. L. Vasconcelos. 2002. Cheating the cheater: Domatia loss minimizes the effects of ant castration in an Amazonian ant-plant. *Oecologia* 133:200–5.

Izzo, T. J., and H. L. Vasconcelos. 2005. Ants and plant size shape the structure of the arthropod community of *Hirtella myrmecophila*, an Amazonian ant-plant. *Ecol. Entomol.* 30:650–6.

Johnson, C., D. Agosti, J. H. C. Delabie, K. Dumpert, D. J. Williams, M. Von Tschirnhaus, and U. Maschwitz. 2001. *Acropyga* and *Azteca* ants (Hymenoptera: Formicidae) with scale insects (Sternorrhyncha: Coccoidea): 20 million years of intimate symbiosis. *Am. Mus. Nov.* 3335:1–18.

Johnson, R. A. 2000. Seed-harvestor ants (Hymenoptera: Formicidae) of North America: An overview of ecology and biogeography. *Sociobiology* 36:89–122.

Juliano, S. A., and J. H. Lawton. 1990. The relationship between competition and morphology. I. Morphological patterns among co-occurring dytiscid beetles. *J. Anim. Ecol.* 59:403–9.

Kaspari, M. 1993. Body size and microclimate use in neotropical granivorous ants. *Oecologia* 96:500–7.

Kaspari, M. 1996a. Worker size and seed size selection by harvester ants in a neotropical forest. *Oecologia* 105:397–404.

Kaspari, M. 1996b. Litter ant patchiness at the 1-m^2 scale: Disturbance dynamics in three neotropical forests. *Oecologia* 107:265–73.

Kaspari, M. 2001. Taxonomic level, trophic biology and the regulation of local abundance. *Glob. Ecol. Biogeogr.* 10:229–44.

Kaspari, M. 2005. Global energy gradients and size in colonial organisms: Worker mass and worker number in ant colonies. *Proc. Natl. Acad. Sci. USA* 102:5079–83.

Kaspari, M., L. Alonso, and S. O'Donnell. 2000a. Three energy variables predict ant abundance at a geographical scale. *Proc. R. Soc. B* 267:485–9.

Kaspari, M., and M. M. Byrne. 1995. Caste allocation in litter *Pheidole*: Lessons from plant defense theory. *Behav. Ecol. Sociobiol.* 37:255–63.

Kaspari, M., and S. O'Donnell. 2003. High rates of army ant raids in the neotropics and implications for ant colony and community structure. *Evol. Ecol. Res.* 5:933–9.

Kaspari, M., S. O'Donnell, and J. R. Kercher. 2000b. Energy, density, and constraints to species richness: Ant assemblage along a productivity gradient. *Am. Nat.* 155:280–93.

Kaspari, M., B. S. Stevenson, J. Shik, and J. F. Kerekes. 2010. Scaling community structure: How bacteria, fungi, and ant taxocenes differentiate along a tropical forest floor. *Ecology* 91:2221–6.

Kaspari, M., and T. J. Valone. 2002. On ectotherm abundance in a seasonal environment—Studies of a desert ant assemblage. *Ecology* 83:2991–6.

Kaspari, M., and E. Vargo. 1995. Colony size as a buffer against seasonality: Bergmann's rule in social insects. *Am. Nat.* 145:610–32.

Kaspari, M., P. S. Ward, and M. Yuan. 2004. Energy gradients and the geographic distribution of local ant diversity. *Oecologia* 140:407–13.

Kaspari, M., and M. Weiser. 1999. The size–grain hypothesis and interspecific scaling in ants. *Funct. Ecol.* 13:530–8.

Kempf, W. W. 1962. Retoques à classificação das formigas Neotropicais do gênero *Heteroponera* Mayr. *Pap. Avul. Zool.* 15:29–47.

Khoo, K. C., and C. T. Ho. 1992. The influence of *Dolichoderus thoracicus* (Hymenoptera: Formicidae) on losses due to *Helopeltis theivora* (Heteroptera: Miridae), black pod disease, and mammalian pests in cocoa in Malaysia. *Bull. Entomol. Res.* 82:485–91.

King, J. R., and W. R. Tschinkel. 2008. Experimental evidence that human impacts drive fire ant invasions and ecological change. *Proc. Natl. Acad. Sci. USA* 105:20339–43.

Kronauer, D. J. C. 2008. Recent advances in army ant biology (Hymenoptera: Formicidae). *Mirmecol. News* 12:51–65.

LaPolla, J. S. 2004. *Acropyga* (Hymenoptera: Formicidae) of the world. *Contrib. Am. Entomol. Inst.* 33:1–130.

LaPolla, J. S., S. P. Cover, and U. G. Mueller. 2002. Natural history of the mealybug-tending ant *Acropyga epedana*, with descriptions of the male and queen castes. *Trans. Am. Entomol. Soc.* 128:367–76.

Lattke, J. E. 1995. Revision of the ant genus *Gnamptogenys* in the New World (Hymenoptera: Formicidae). *J. Hym. Res.* 4:137–93.

Leal, I., and P. S. Oliveira. 1995. Behavioral ecology of the neotropical termite-hunting ant *Pachycondyla* (=*Termitopone*) *marginata*: Colony founding, group-raiding and migratory patterns. *Behav. Ecol. Sociobiol.* 37:373–83.

Leal, I., and P. S. Oliveira. 1998. Interactions between fungus-growing ants (Attini), fruits and seeds in cerrado vegetation in Southeast Brazil. *Biotropica* 30:170–8.

Leal, I. R., and P. S. Oliveira 2000. Foraging ecology of attine ants in a neotropical savanna: Seasonal use of fungal substrate in the cerrado vegetation of Brazil. *Insectes Soc.* 47:376–82.

Lengyel, S., A. D. Gove, A. M. Latimer, J. D. Majer, and R. R. Dunn. 2009. Ants sow the seeds of global diversification in flowering plants. *PloS One* 4: e5480.

Lofgren, C. S., and R. K. Vander Meer. 1986. *Fire Ants and Leaf-Cutting Ants: Biology and Management.* Boulder, CO: Westview Press.

Longino, J. T. 2005. Complex nesting behavior by two neotropical species of the ant genus *Stenamma* (Hymenoptera: Formicidae). *Biotropica* 37:670–5.

Longino, J. T., and R. K. Colwell. 1997. Biodiversity assessment using structured inventory: Capturing the ant fauna of a tropical rain forest. *Ecol. Appl.* 7:1263–77.

Longino, J. T., and N. M. Nadkarni. 1990. A comparison of ground and canopy leaf litter ants (Hymenoptera: Formicidae) in a neotropical montane forest. *Psyche* 97:81–93.

Mack, R. N., D. Simberloff, W. M. Lonsdale, H. Evans, M. Clout, and F. Bazzaz. 2000. Biotic invasions: Causes, epidemiology, global consequences and control. *Ecol. Appl.* 10:689–710.

Majer, J. D., J. H. C. Delabie, and M. R. B. Smith. 1994. Arboreal ant community patterns in Brazilian cocoa farms. *Biotropica* 26:73 83.

Majer, J. D., S. O. Shattuck, A. N. Andersen, and A. J. Beattie. 2004. Australian ant research: Fabulous fauna, functional groups, pharmaceuticals, and the fatherhood. *Aust. J. Entomol.* 43:235–47.

Maschwitz, U., S. Steghaus, R. Gaube, and H. Hänel. 1989. A South East Asian ponerine ant of the genus *Leptogenys* (Hym., Form.) with army ant life habits. *Behav. Ecol. Sociobiol.* 24:305–16.

Masuko, K. 1984. Studies on the predatory biology of oriental Dacetine ants (Hymenoptera: Formicidae). I. Some Japanese species of *Strumigenys*, *Pentastruma*, and *Epitritus*, and a Malaysian *Labidogenys*, with special reference to hunting tactics in short-mandibulate forms. *Insectes Soc.* 31:429–51.

McGlynn, T. P. 1999. The worldwide transfer of ants: Geographical distribution and ecological invasions. *J. Biogeogr.* 26:535–48.

McGlynn, T. P., R. A. Carr, J. H. Carson, and J. Buma. 2004. Frequent nest relocation in the ant *Aphaenogaster aranoides*: Resources, competition, and natural enemies. *Oikos* 106:612–21.

McGlynn, T. P., M. D. Shotell, and M. S. Kelly. 2003. Responding to a variable environment: Home range, foraging behavior, and nest relocation in the Costa Rican rainforest ant *Aphaenogaster aranoides*. *J. Insect Behav.* 16:687–701.

Mikheyev, A. S., U. G. Mueller, and P. Abbot. 2006. Cryptic sex and many-to-one coevolution in the fungus-growing ant symbiosis. *Proc. Natl. Acad. Sci. USA* 103:10702–6.

Miles, D. B., R. E. Ricklefs, and J. Travis. 1987. Concordance of ecomorphological relationships in three assemblages of passerine birds. *Am. Nat.* 129:347–64.

Moreau, C. S., C. D. Bell, R. Vila, S. B. Archibald, and N. E. Pierce. 2006. Phylogeny of the ants: Diversification in the age of Angiosperms. *Science* 312:101–4.

Morrison, L. W., S. D. Porter, E. Daniels, and M. D. Korzhukin. 2004. Potential global range expansion of the invasive fire ant, *Solenopsis invicta*. *Biol. Invasions* 6:183–91.

Ness, J. H., J. L. Bronstein, A. N. Andersen, and N. Holland. 2004. Ant body size predicts dispersal distance of ant-adapted seeds: Implications of small-ant invasions. *Ecology* 85:1244–50.

O'Donnell, S., M. Kaspari, and J. Lattke. 2005. Extraordinary predation by the neotropical army ant *Cheliomyrmex andicola*: Implications for the evolution of the army ant syndrome. *Biotropica* 37:706–9.

Oliveira, P. S., and A.V. L. Freitas. 2004. Ant-plant–herbivore interactions in the neotropical cerrado savanna. *Naturwissenschaften* 91:557–70.

Ovadia, O. F., and O. J. Schmitz. 2002. Linking individuals with ecosystems: Experimentally identifying the relevant organizational scale for predicting trophic abundances. *Proc. Natl. Acad. Sci. USA* 99:12927–31.

Passos, L., and P. S. Oliveira. 2003. Interactions between ants, fruits and seeds in a restinga forest in south-eastern Brazil. *J. Trop. Ecol.* 19:261–70.

Perfecto, I., and J. Vandermeer. 2006. The effect of an ant–hemipteran mutualism on the coffee berry borer (*Hypothenemus hampei*) in southern Mexico. *Agric. Ecosyst. Environ.* 117:218–21.

Philpott, S. M., and I. Armbrecht. 2006. Biodiversity in tropical agroforests and the ecological role of ants and ant diversity in predatory function. *Ecol. Entomol.* 31:369–77.

Philpott, S. M., and P. F. Foster. 2005. Nest-site limitation in coffee agroecosystems: Artificial nests maintain diversity of arboreal ants. *Ecol. Appl.* 15:1478–85.

Philpott, S. M., R. Greenberg, P. Bichier, and I. Perfecto. 2004. Impacts of major predators on tropical agroforest arthropods: Comparisons within and across taxa. *Oecologia* 140:140–9.

Philpott, S. M., and I. Armbrecht. 2006. Biodiversity in tropical agroforests and the ecological role of ants and ant diversity in predatory function. *Ecol. Entomol.* 31:369–77.

Philpott, S. M., I. Perfecto, and J. Vandermeer. 2008a. Effects of predatory ants on lower trophic levels across a gradient of coffee management complexity. *J. Anim. Ecol.* 77:505–11.

Philpott, S. M., I. Perfecto, and J. Vandermeer. 2008b. Behavioral diversity of predatory ants in coffee agroecosystems. *Environ. Entomol.* 37:181–91.

Pizo, M. A., and P. S. Oliveira. 2000. The use of fruits and seeds by ants in the Atlantic Forest of Southeast Brazil. *Biotropica* 32:851–61.

Price, T. 1991. Morphology and ecology of breeding warblers along an altitudinal gradient in Kashmir, India. *J. Anim. Ecol.* 60:643–64.

Quiroz-Robledo, L., J. Valenzuela-González, and T. Suárez-Landa. 2002. Las hormigas ecitoninas (Formicidae: Ecitoninae) de la Estación de Biología Tropical Los Tuxtlas, Veracruz, México. *Folia Entomol. Mex.* 41:261–81.

Ribas, C. R., and J. H. Schoereder. 2002. Are all ant mosaics caused by competition? *Oecologia* 131:606–11.

Roberts, D. L., R. J. Cooper, and L. J. Petit. 2000. Use of premontane moist forest and shade coffee agroecosystems by army ants in Western Panama. *Cons. Biol.* 14:192–9.

Sanders, N. J., G. M. Crutsinger, R. R. Dunn, J. D. Majer, and J. H. C. Delabie. 2007. An ant mosaic revisited: Dominant ant species disassemble arboreal ant communities but co-occur randomly. *Biotropica* 39:422–7.

Sanders, N. J., N. J. Gotelli, N. E. Heller, and D. M. Gordon. 2003. Community disassembly by an invasive ant species. *Proc. Natl. Acad. Sci. USA* 100:2474–7.

Sarty, M., K. L. Abbott, and P. J. Lester. 2006. Habitat complexity facilitates coexistence in a tropical ant community. *Oecologia* 149:465–73.

Schmidt, F. A., and R. R. C. Solar. 2010. Hypogaeic pitfall traps: Methodological advances and remarks to improve the sampling of a hidden ant fauna. *Insectes Soc.* 57:261–6.

Schoemaker, D. D. W., M. E. Ahrens, and K. G. Ross. 2006. Molecular phylogeny of fire ants of the *Solenopsis saevissima* species-group based on mtDNA sequences. *Mol. Phylogenet. Evol.* 38:200–15.

Silva, R. R., and C. R. F. Brandão. 1999. Formigas (Hymenoptera: Formicidae) como indicadores da qualidade ambiental e da biodiversidade de outros invertebrados terrestres. *Biotemas* 12:55–73.

Silva, R. R., and C. R. F. Brandão. 2010. Morphological patterns and community organization in leaf-litter ant assemblages. *Ecol. Monogr.* 80:107–24.

Silva, R. R., and R. Silvestre. 2004. Riqueza da fauna de formigas subterrâneas (Hymenoptera: Formicidade) em Seara, Oeste de Santa Catarina, Brasil. *Pap. Avul. Zool.* 41:1–11.

Silvestre, R., C. R. F. Brandão, and R. R. Silva. 2003. Gremios funcionales de hormigas: el caso de los gremios del Cerrado. In *Introdución a las Hormigas de la Región Neotropical*, ed. F. Fernández, 113–48. Bogotá, Colombia: Fundación Humboldt.

Simberloff, D., and T. Dayan.1991. The guild concept and the structure of ecological communities. *Annu. Rev. Ecol. Syst.* 22:115–43.

Souza, D. J. De, I. M. F. Soareas, and T. M. C. Della Lucia. 2007. *Acromyrmex ameliae* s. n. (Hymenoptera: Formicidae): A new social parasite of leaf-cutting ants in Brazil. *Insect Sci.* 14:251–7.

Stevens, R. D., and M. R. Willig. 2000. Community structure, abundance, and morphology. *Oikos* 88:48–56.

Suarez, A. V., D. T. Bolger, and T. J. Case. 1998. The effect of fragmentation and invasion on the native ant community in coastal southern California. *Ecology* 79:2041–56.

Suarez, A. V., and T. J. Case. 2002. Bottom-up effects on persistence of a specialist predator: Ant invasions and horned lizards. *Ecol. Appl.* 12:291–8.

Styrsky, J. D., and M. D. Eubanks. 2007. Ecological consequences of interactions between ants and honeydew-producing insects. *Proc. R. Soc. B* 274:151–64.

Summer, S., D. K. Aanen, J. H. C. Delabie, and J. J. Boomsma. 2004. The evolution of social parasitism in *Acromyrmex* leaf-cutting ants: A test of Emery's rule. *Insectes Soc.* 51:37–42.

Swartz, M. B. 1998. Predation on an *Atta cephalotes* colony by an army ant *Nomamyrmex esenbecki*. *Biotropica* 30:682–4.

Symondson, W. O. C., K. D. Sunderland, and M. H. Greenstone. 2002. Can generalist predators be effective biocontrol agents? *Annu. Rev. Entomol.* 47 561–94.

Tschinkel, W. R. 2006. *The Fire Ants*. Cambridge, MA: Harvard University Press.

Ulloa-Chacón, P., and D. Cherix. 1990. The little fire ant *Wasmannia auropunctata* (R.) (Hymenoptera: Formicidae). In *Applied Myrmecology: A World Perspective*, ed. R. K. Vander Meer, K. Jaffe, and A. Cedeno, 281–9. Boulder, CO: Westview Press.

Underwood, E. C., and B. L. Fisher. 2006. The role of ants in conservation monitoring: If, when, and how. *Biol. Conserv.* 132:166–82.

Ward-Fear, G. P. Brown, M. J. Greenlees, and R. Shine. 2009. Maladaptive traits in invasive species: In Australia, cane toads are more vulnerable to predatory ants than are native frogs. *Funct. Ecol.* 23:559–68.

Ward-Fear, G., G. P. Brown, M. J. Greenlees, and R. Shine. 2010. Using a native predator (the meat ant, *Iridomyrmex reburrus*) to reduce the abundance of an invasive species (the cane toad, *Bufo marinus*) in tropical Australia. *J. Appl. Ecol.* 47:273–80.

Way, M. J., and B. Bolton. 1997. Competition between ants for coconut palm nesting sites. *J. Nat. Hist.* 31:439–55.

Way, M. J., and K. C. Khoo. 1992. Role of ants in pest-management. *Annu. Rev. Entomol.* 37:479–503.

Weber, N. A. 1972. Gardening ants: The attines. *Mem. Am. Phil. Soc.* 92:1–146.

Weiser, M. D., and M. Kaspari. 2006. Ecological morphospace of new world ants. *Ecol. Entomol.* 31:131–142.

Wilson, E. O. 1956. Feeding behavior in the ant *Rhopalhothrix biroi* Szabó. *Psyche* 63:21–3.

Wilson, E. O. 1987. The little things that run the world. *Conserv. Biol.* 1:344–6.

Wilson, E. O. 2003. *Pheidole in the New World: A Dominant, Hyperdiverse Ant Genus.* Cambridge, MA: Harvard University Press.

Wilson, E. O., and B. Hölldobler. 1986. Ecology and behavior of the neotropical cryptobiotic ant *Basiceros manni* (Hymenoptera: Formicidae: Basicerotini). *Insectes Soc.* 33:70–84.

Wilson, E. O., and B. Hölldobler. 2005a. The rise of the ants: A phylogenetic and ecological explanation. *Proc. Natl. Acad. Sci. USA* 102:7411–4.

Wilson, E. O., and B. Hölldobler. 2005b. Eusociality: Origin and consequences. *Proc. Natl. Acad. Sci. USA* 102:13367–71.

Wilson, J. B. 1999. Guilds, functional types and ecological groups. *Oikos* 86:507–522.

Woodward, G., B. Ebenman, M. Emmerson, J. M. Montoya, J. M. Olesen, A. Valido, and P. H. Warren. 2005. Body size in ecological networks. *Trends Ecol. Evol.* 20:402–9.

Yu, D. W., H.B. Wilson, and N. E. Pierce. 2001. An empirical model of species coexistence in a spatially structured environment. *Ecology* 82:1761–71.

11

Social Bees (Bombini, Apini, Meliponini)

Astrid M. P. Kleinert, Mauro Ramalho, Marilda Cortopassi-Laurino,
Márcia F. Ribeiro, and Vera L. Imperatriz-Fonseca

CONTENTS

11.1 Introduction 237
11.2 Resources Acquisition 238
    11.2.1 Physical Factors and Temporal Partitioning of Foraging Activity 238
    11.2.2 Niche Width and Floral Resources Allocation 239
    11.2.3 Floral Constancy, Load Capacity, and Foraging Strategies 243
11.3 Resource Utilization by Colonies 248
    11.3.1 Caste Determination and Differentiation in Bombini 248
    11.3.2 Caste Determination and Differentiation in Apini 249
    11.3.3 Caste Determination and Differentiation in Meliponini 250
11.4 Larval Food in Meliponini 251
11.5 Pollen 252
    11.5.1 Protein Value 252
11.6 Food Rich in Sugars Produced by Bees: Honey 252
    11.6.1 Honey Microscopy 252
    11.6.2 Honey in Apini 253
    11.6.3 Honey in Meliponini 254
        11.6.3.1 How to Exploit Meliponini Honey 254
        11.6.3.2 Antibacterial Activities 258
    11.6.4 Honey Microorganisms 259
11.7 Final Considerations 262
References 263

11.1 Introduction

In this chapter, we will address the role of food in the organization of social bees' colonies from foraging activity to its use on offspring feeding, emphasizing the characteristics of the two main resources collected and processed by them, pollen and honey.

In bees, eusociality emerged in Apinae and is present in the tribes Bombini, Apini, and Meliponini. One of the most obvious social conflicts is expressed in sexual production and in the existence of a specialized reproductive caste. Food has a strong influence on caste differentiation, and control of its amount or quality is one of the central mechanisms in social life. Foraging activity on flowers and food processing inside the colonies directly influence the social life of the colony. More emphasis will be given to the mechanisms described for Meliponini species or stingless bees.

Stingless bees are floral generalists, but they present a selective foraging activity, and as a rule, the annual economic budget of the colony depends on intensive foraging on few available pollen and nectar floral sources in each habitat. The choice of food sources is mediated by morphofunctional characteristics of the foragers, foraging strategies (solitary or collective), and social interactions in the colony and in the field. In communities, food overlap among species is a rule, but the role of competition in the organization of local assemblies is still quite controversial. Relationships among reproductive strategies of the colonies, diversity, and abundance distribution in the communities have just begun to be explored in recent studies.

11.2 Resources Acquisition

Although there are exceptions, social bees feed basically on pollen (protein source) and nectar (carbohydrate source) collected from flowers. Some species of stingless bees (Meliponini) are scavengers, feeding on decomposing organic matter (as *Trigona hypogea* Silvestri, *Trigona necrophaga* Camargo & Roubik, and *Trigona crassipes* (F.)); others also feed on honeydew, a sugar solution produced by membracids; and finally, some specialize in stealing food from other bees' nests (cleptobiotic bees, *Lestrimelitta* spp. in the Americas, and *Cleptotrigona* spp. in Africa).

11.2.1 Physical Factors and Temporal Partitioning of Foraging Activity

Stingless bees are found in tropical and subtropical regions. The most likely cause of this geographic distribution pattern is the sensitivity of both individuals and colonies to low temperatures. Although there are interspecific differences in relation to the ability of hive thermoregulation, bees of this tribe seem to depend more on structural characteristics of their nests than on physiological and behavioral responses for heat conservation (Sakagami 1982).

The main abiotic factors that singly or in combination influence stingless bees' flight activity are temperature, relative humidity, light intensity, and wind speed. According to Fowler (1979), extreme values would directly act on the bees, while moderate values would affect flight activity, as they reflect on food availability (for instance, nectar flow). Clearly, food availability only becomes important after bees meet favorable conditions for flight. Thus, species capable of flying in wider ranges of temperature, relative humidity, and so forth, eventually may have advantages over the others.

Temperature seems to be determinant to foragers, especially in small species, such as *Tetragonisca angustula* (Latreille) and *Plebeia* spp. that start foraging at temperatures above 16°C (Oliveira 1973; Iwama 1977; Kleinert-Giovannini 1982; Imperatriz-Fonseca et al. 1985). An exception, *Plebeia pugnax* Moure (in litt.) is capable to start foraging from 14°C (Hilário et al. 2001). All these species reduce flight activity at temperature below 20°C.

Larger species of Meliponini, with size between 8 and 12 mm, such as *Melipona*, begin flight activity in lower temperatures, starting from 11°C in *Melipona bicolor* Lepeletier (Hilário et al. 2000), and 13°C to 14°C in *M. quadrifasciata* and *M. marginata* Lepeletier (Guibu and Imperatriz-Fonseca 1984; Kleinert-Giovannini and Imperatriz-Fonseca 1986). Most species present optimal foraging activity between 20°C and 30°C, except *M. quadrifasciata* and *M. bicolor*, which preferentially collect food at 14°C to 16°C and 16°C to 26°C, respectively. Body biomass is the main variable in this relationship, since larger social bees such as *Apis mellifera* L. and *Bombus* spp. start foraging at lower temperatures, often well below 10°C (Gary 1967; Heinrich 1979). However, in Meliponini, relatively small species such as *Partamona helleri* (Friese) present an optimal foraging activity between 15°C to 24°C, similar to the range of large species such as *M. bicolor* (Azevedo 1997).

Optimum values of relative humidity (RH) for foraging range between 30% and 70 % for most species (Oliveira 1973; Iwama 1977; Kleinert-Giovannini 1982; Kleinert-Giovannini and Imperatriz-Fonseca 1986; Hilário et al. 2001). *Plebeia remota* (Holmberg), *Schwarziana quadripunctata* (Lepeletier), and *M. bicolor* present higher flight activity in higher ranges between 60% and 90% (Imperatriz-Fonseca et al. 1985; Imperatriz-Fonseca and Darakjian 1994; Hilário et al. 2000). *Plebeia emerina* (Friese) workers do not leave their nests when RH is above 70% (Kleinert-Giovannini 1982). With an optimal flight activity between 40% to 45% RH, *M. marginata* shows behavioral plasticity in relation to environmental conditions (Kleinert-Giovannini and Imperatriz-Fonseca 1986), intensifying foraging under extreme conditions (e.g., values above 80% RH), after prolonged periods of rainfall.

Light intensity seems to be important just for the start and the end of external activity. In other periods, it is difficult to dissociate it from changes in temperature, and therefore, its effects become secondary, or they are at least masked. Even so, some authors reported lower flight activity on cloudy days, when compared with sunny days with the same temperatures (Oliveira 1973; Kleinert-Giovannini 1982; Kleinert-Giovannini and Imperatriz-Fonseca 1986).

Wind speed between 2 and 3 m/s causes a decrease in the number of small foragers of *P. emerina* leaving the nest, leading to the interruption of foraging when it reaches 4 m/s (Kleinert-Giovannini 1982). In the same conditions, foragers of *T. angustula* continue to collect food (Iwama 1977), while other species such as *Plebeia droryana* (Friese), *P. saiqui* (Friese), and *M. marginata* are only slightly affected (Oliveira 1973; Kleinert-Giovannini and Imperatriz-Fonseca 1986).

The state of the colony also influences flight activity of stingless bees. Usually, weak colonies are more susceptible to variations in temperature (Kleinert-Giovannini and Imperatriz-Fonseca 1986) and have their activity shifted to later hours of the day, when compared to other colonies of the same species (Hilário et al. 2001). Nunes-Silva et al. (2010) also recorded distinct foraging patterns during two different reproductive phases in colonies of *P. remota*: during reproductive diapause, bees collected primarily nectar, while during the reproductive phase, they collected predominantly pollen.

Although many species present similar or overlapping optimum ranges of temperature and relative humidity, periods of increased flight activity tend to be different, allowing temporal partitioning of floral resources. Several species of the genus *Melipona* present higher flight activity at different times (e.g., *M. bicolor* and *M. quadrifasciata* are more active in the early morning hours, while *M. marginata* is more active between 11:00 a.m.–1:00 p.m.). Among the small species of *Plebeia*, *P. saiqui*, *P. remota*, and *P. pugnax* have higher activity concentrated between midmorning and midafternoon (Oliveira 1973; Imperatriz-Fonseca et al. 1985; Hilário et al. 2001); *P. emerina* and *Plebeia droryana* forage mainly in the afternoon.

Foragers leave the nest mainly to collect food (pollen and nectar), resin, water, and mud (for nest building), and these resources are sought at different times of the day. For instance, *P. pugnax* collects pollen preferentially in the morning until noon, while resin is collected throughout the day (Hilário et al. 2001). Although active all day, foragers of *M. rufiventris* and *Melipona bicolor* Lepeletier preferentially collect pollen in the early hours of the morning. While *M. bicolor* collects resin and mud preferably in late afternoon, foragers of *M. rufiventris* present, besides this peak, an additional peak of resin collection coincident with the morning pollen peak (Hilário et al. 2000; Fidalgo and Kleinert 2007). This temporal partitioning of resource gathering was also observed in other neotropical regions for other *Melipona* species (Bruijn and Sommeijer 1997) and increases the chances of coexistence of different species of stingless bees in a same place.

11.2.2 Niche Width and Floral Resources Allocation

Except for rare species that specialize in nest robbing (*Lestrimellita*) and in the use of animal protein from carcasses (*T. hypogea*), stingless bees feed on pollen and nectar. In this case, they are generalists foraging on a wide spectrum of floral types. However, few floral sources are heavily exploited in local communities. This foraging pattern was conceived from studies on pollen analysis of bee colonies and bee censuses on flowers (Imperatriz-Fonseca et al. 1984, 1987; Ramalho et al. 1985, 1989, 1990, 1991, 2007; Kleinert-Giovannini and Imperatriz-Fonseca 1987; Cortopassi-Laurino and Ramalho 1988; Ramalho 1990, 1995, 2004; Wilms et al. 1996, 1997). Both techniques generate extensive and independent data that depict the expression of this pattern in two hierarchical levels: colonies and local populations (Figure 11.1).

Although differences in the speed with which different species are able to muster a large number of individuals for a given food source (Lindauer and Kerr 1960; Hubbell and Johnson 1978; Johnson 1982; Johnson et al. 1987), there is no relationship between efficiency of communication and a colony's ability to concentrate foraging activity in few floral resources.

In local stingless bee communities, measures of niche width vary widely with temporal changes in the availability of floral sources and with the specific responses to the distribution of resources relative abundance (Ramalho et al. 1991). However, on average, niche width appears to be similar between groups of species with different foraging strategies (Biesmeijer and Slaa 2006). In general, this analysis confirms the pattern of concentrated use of floral sources, detected in local communities. The index often used as a measure of niche width (H', Shannon-Wiener) is extremely sensitive to the evenness of

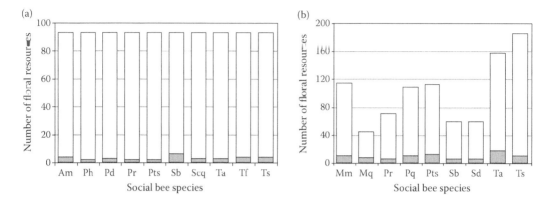

FIGURE 11.1   Floral sources allocation by stingless bees and Africanized honeybees (Am) in the Atlantic Forest domain: (a) Floral sources where at least 10% of foragers (black) were sampled in relation to all sources (white) visited by this bee group in the Atlantic Forest (State Park Cantareira, SP). (b) Floral sources with representation above 10% (black) in Meliponini diet—estimates by counting pollen grains in food stored in colonies at Universidade de São Paulo campus, SP. Meliponini: Ph = *Partamona helleri*; Pd = *Plebeia droryana*; Pq = *Plebeia saiqui*; Pr = *Plebeia remota*; Pts = *Paratrigona subnuda*; Sb = *Scaptotrigona bipunctata*; Sd = *Scaptotrigona depilis*; Scq = *Schwarziana quadripunctata*; Ta = *Tetragonisca angustula*, Ts = *Trigona spinipes*; Mm = *Melipona marginata*, Mq = *Melipona quadrifasciata*. (From Ramalho, M., A. Kleinert-Giovannni, and V. L. Imperatriz-Fonseca, In *Ecologia Nutricional de Insetos e Suas Implicações no Manejo de Pragas*, ed. A. R. Panizzi and J. R. P. Parra, 225–52, Editora Manole, São Paulo, Brazil, 1991; and Ramalho, M., Diversidade de abelhas (Apoidea, Hymenoptera) em um remanescente de Floresta Atlântica, em São Paulo, Ph.D. thesis, Universidade de São Paulo, Brazil, 1995. With permission.)

the most common sources in the diet, and only when foraging concentration becomes extreme, as for *Scaptotrigona* (Ramalho 1990), there is a marked reduction in its value.

Measures for niche width of these generalist consumers most likely reflect the effect and not the cause of ecological dominance. For instance, in local communities, as the number of foragers sampled on flowers increases, the number of sources visited by a species also increases (Ramalho 1995 and Figure 11.2). Similarly, greater colonies tend to use broader spectrum of floral sources (Cortopassi-Laurino and Ramalho 1988; Ramalho et al. 1991). Even species considered primitive and specialized in habitat types or nesting sites, such as *Mourella caerulea* (Friese) (Camargo and Wittmann 1989), are extremely generalist consumers of floral resources. Therefore, one should not expect a correlation between niche width (realized) and relative abundance of stingless bee species in ecological communities. For instance,

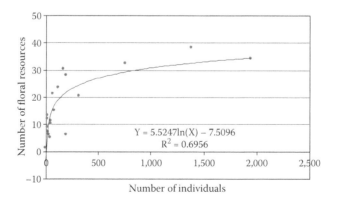

FIGURE 11.2   Relationship between number of individuals sampled on flowers (relative population size) and number of floral sources visited by 17 species of stingless bees, *Apis mellifera* and two species of *Bombus*. Linear correlation: $r = 0.72$, $p < 0.05$. Data regression curve is indicated. (From Ramalho, M., Diversidade de abelhas (Apoidea, Hymenoptera) em um remanescente de Floresta Atlântica, em São Paulo, Ph.D. thesis, Universidade de São Paulo, Brazil, 1995. With permission.)

Scaptotrigona bipunctata (Lepeletier) is dominant in the Atlantic forest at Serra da Cantareira, but has one of the smallest realized niche width (Ramalho 1990, 1995, 2004).

The premise of specific choices by modifying foraging pattern led to the hypothesis of floral preference in *Melipona* (Ramalho et al. 1989). The high relative pollen frequency from Solanaceae, Melastomataceae, and Myrtaceae in the diet of *Melipona* is related to pollen extraction capacity by vibration, especially from flowers with poricidal anthers, a skill that differentiates the group in relation to other Meliponini. Studies on feeding habits in different habitats as Cerrado, Atlantic forest, and Colombia Llanos, have been gradually giving empirical support to this hypothesis, with some reservations about preferred plant families (Silva and Schlindwein 2003; Antonini et al. 2006; Nates-Parra 2006).

Ramalho et al. (2007) showed that floral choices of *Melipona scutellaris* Latreille are not random. Using Africanized *A. mellifera* as control, they demonstrated that the diversity of floral sources in a colony's diet was dependent on species. In other words, independent of habitat type, colonies of *M. scutellaris* remained more similar to each other, forming clusters (Figure 11.3) significantly narrower than with *A. mellifera* colonies. The pattern also remains in line with more similar responses among colonies of *M. scutellaris* to local variations in blossoms supply.

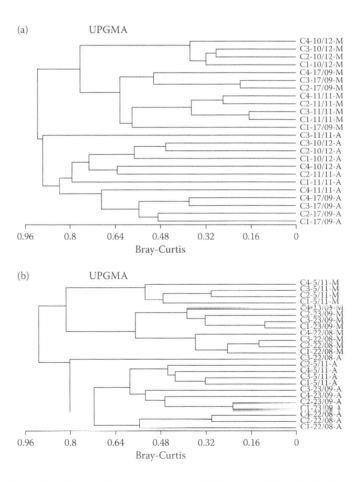

FIGURE 11.3 Analysis of trophic similarity between colonies of *Melipona scutellaris* (M1 to M4), with paired samples control from Africanized Apis mellifera colonies (A1 to A4) at two localities of the tropical Atlantic forest, in Bahia. Dissimilarity results (Bray-Curtis index, UPGMA method) of paired samples at two locations during three periods (months): (a) Alagoinhas, (b) Cruz das Almas. The significance test of similarity analysis (ANOSIM) supports the hypothesis that clusters are not random. (From Ramalho, M., M. D. Silva, and C. A. L. Carvalho, *Neotrop. Entomol.*, 36, 38–45, 2007. With permission.)

Secondary data analysis of 28 communities in several habitat types in Eastern Brazil indicates that the apparent competitive structure of Meliponini communities (Biesmeijer and Slaa 2006) is expressed in three trends: (1) retraction of niche width, with increasing number of species in communities, (2) with increasing number of plant species in communities, the ratio between number of stingless bee species and number of plant species decreases, that is, smaller niche packaging ("species packing"), and (3) local communities tend to be formed by species of distinct genera.

As more floral sources are explored, the number of sources shared among pairs of stingless bee species increases (Figure 11.4). Food overlap among these potentially generalist consumers tends to be very diffuse and extensive, and this is reflected mainly on measures based on presence–absence (e.g., Sorensen and Cody indexes). Clusters derived from these measures (Biesmeijer and Slaa 2006) poorly reflect the trophic distance among consumers and conceal the real community structures. But when measuring the intensity of use of shared floral sources (percentage of similarity, e.g., Schoener's index) (Ramalho 1995), we obtain the more functional ecological expression of species overlap (Figure 11.3), and clusters differ greatly. Comparing the percentage of food similarity in two nearby communities in the Atlantic forest from data generated with two independent techniques (pollen analysis and bee census on flowers), Ramalho et al. (1991) and Ramalho (1995) found that diets of *P. droryana* and *T. angustula* were closer to each other and to *Melipona* and *Scaptotrigona*, deflecting too much from clusters based on meta-analysis of presence-absence measures of species on flowers mentioned above.

Measures of diffuse exploitation of floral resources eventually shared (presence–absence) are certainly less informative than measures of Meliponini dependence in relation to a few productive flower sources in environment, for instance, mass flowering in the Atlantic forest (Ramalho 2004). An extreme case was observed in three species of *Scaptotrigona*, whose colonies concentrate their annual protein demands in one or few food sources, storing hundreds of surplus pollen grams for future offspring production (Ramalho 1990).

The increase in the average quality of a slightly higher number of floral sources in communities with more diverse flora would be enough to change sharing opportunities and to reduce niche width. For instance, in a local community in the Atlantic forest, less than two dozen trees with mass flowering attracted more than 70% of individuals sampled on flowers, and 100% of stingless bee species (Ramalho 1995, 2004). From an ecological point of view, that is, common in space and time, a recent

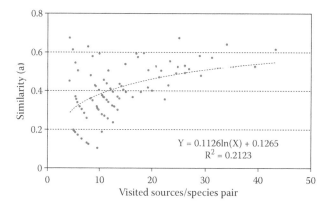

FIGURE 11.4 Variation in similarity between pairs of bee species (Apoidea) according to the number of foragers sampled on flowers (relative population size) in the Atlantic Forest (State Park Cantareira, SP). Comparison between species with proper number of individuals ($N \geq 50$) sampled over 18 months: 11 stingless bee species, *Apis mellifera*, and one species of each genus *Ceratina*, *Paratetrapedia*, and *Megachile*. Several points are overlapped: total number of pairs is equal to 105 (15!). Linear correlation: $r = 0.51$, $p < 0.05$. Data regression curve is indicated. Despite the consistent trend, there is a wide dispersion because most species forage with high intensity on few floral sources—when these few sources are also shared, overlapping (a = Cody index) is also high. (From Ramalho, M., Diversidade de abelhas (Apoidea, Hymenoptera) em um remanescente de Floresta Atlântica, em São Paulo, Ph.D. thesis, Universidade de São Paulo, Brazil, 1995. With permission.)

study supports the premise that there is a close (or predictable) relationship between Meliponini and mass flowering canopies (Monteiro and Ramalho 2010).

High overlap in abundant floral sources is not translated into higher competition. For instance, in the Atlantic forest of Cantareira state park, São Paulo, the vast majority of stingless bee species concentrated foraging activity in the canopy, where there are also more commonly canopies with huge mass flowering (Ramalho 2004). In these flowerings, spatial resources partitioning is easier and high values of similarity (above 50%) were observed among 22 species pairs that concentrated their foraging in the canopy, while only three pairs of species showed high similarity in the lower strata (Ramalho 1995). In a still more incisive way, *S. bipunctata* and *Paratrigona subnuda* Moure were extremely locally dominant and concentrated their visits in less than a dozen productive blossoms, with more than 50% overlap between them.

The mechanisms for floral sources sharing also need to be better contextualized in terms of costs and benefits. For instance, with their large relative size, species of *Melipona* should avoid floral resources whose supply is being depressed by exploitation; conversely, smaller bees such as *T. angustula* and some *Plebeia* and *Friesella* with larger pollen load capacity (Ramalho et al. 1994) could continue exploring floral resources in the process of local pollen depression, because they get profitable return rates with lower supply levels.

Considering random projections of taxonomic organization of 28 communities in Eastern Brazil, there is an overrepresentation in the number of genera in some cases (Biesmeijer and Slaa 2006). This situation seems consistent with the concept of limiting similarity. However, taxonomic proximity does not well represent functional similarity: there are common foraging strategies to several genera and cogeneric species with different strategies. Both pollen analysis from colonies (Ramalho et al. 1989, 1991; Ramalho 1990) and bee census on flowers (Ramalho 1995; Martins et al. 2003) have generated high percentage values of diet similarity between species of the same genus as *Melipona*, *Plebeia*, and *Scaptotrigona*.

In summary, the number of floral sources shared among stingless bee species depends on encounter chances and basically on the size of foragers' populations. However, the percentage of food similarity (i.e., the intensity of use of shared resources) is not related to niche width. Therefore, measures of food niche should be taken with caution in the analysis of Meliponini community structure because they cannot be translated into potential measures of competition between these generalist consumers.

11.2.3 Floral Constancy, Load Capacity, and Foraging Strategies

Under the adaptive logic, animals must be modeled to optimize their diet, and this means making the best possible choices in face of fluctuations in food supply, with appropriate adjustments in foraging (Pyke 1984). Models of foraging strategies explore the relationship between the total time of food intake and net energy gained (Schoener 1971). At the extremes, there are foragers that minimize food intake time and those that maximize energy gain. Animals with stable reproductive rates best fit to the first case, and those with variable offspring number to the second. Choices also depend on intrinsic factors such as body size, metabolic rate, and niche width. Furthermore, a set of environmental factors such as distribution, abundance, and risk exposure modify the access to food.

A significant part of experimental research on the economic foraging decisions in bees is included in the prediction analysis of optimal foraging theory and associated hypotheses (Pyke 1984). The basic premise is that the way an animal equates food acquisition, in terms of costs and benefits, determines its adaptive value, that is, the number of viable offspring it will leave for next generation (fitness). In the case of social bees with huge perennial colonies, this issue demands a solution and integration of decisions at different moments and in two hierarchical levels: in the field by forager, and in the colony, through social interactions that influence the long-term reproductive success. For these bees, there are two intrinsic constraints on foraging: displacement from a central point (i.e., the colony) and the short life of foragers. The need to return to the colony (central point) turns critical the equation of foraging cost and floral source distance. For these very small foragers with high-energy consumption during flight, autonomy is low and ultimately depends on the storage capacity of nectar in the honey pouch. Hence there is some common sense about the relationship between body size (honey pouch volume) and flight range in Meliponini (van Nieuwstadt and Iraheta 1996; Araújo et al. 2004).

Besides being short, forager longevity also has an inverse relationship with work intensity. This puts foraging decisions under the following perspective: forager should obtain and carry the largest possible load for each foraging trip or would have been modeled for maximizing net energy over lifetime, for the benefit of overall colony efficiency. Should it forage to maximize its own longevity, and thus indirectly increase long-term net returns for colonies? Studies with *A. mellifera* support this second alternative.

Foraging experimental studies have multiplied in recent decades, and in general confirm the expectations that foraging decisions are constrained by body size, metabolism, foragers' longevity, and social interactions. For instance, the selection process known as *majoring–minoring* seems to be typical of large foragers of the genus *Bombus* (Heinrich 1979). In this case, foragers modulate the intensity of use of several floral sources simultaneously in a single foraging trip. On successive trips they can make continuous adjustments intensifying visits to more profitable flowers (majoring), gradually reducing foraging on those that are depreciated (minoring). A *minor* source at a time becomes *major* in the other and vice versa.

Surveys on *A. mellifera* put into perspective foragers' ability to discriminate between net and gross foraging incomes (Seeley 1995). In the second case, foragers should focus on sources with a larger food supply, despite acquisition costs (collection and transport). In the case of net incomes, sources that offer a higher food return rate per unit of foraging time preferred to sources with a lower food return rate.

Foragers also carry information about foraging conditions to the colony. Their role as an information channel function on the frequency of food collecting trips (Nuñez 2000): the more times they get information in the field and delivery to the colony the higher their value. This would explain why foragers do not always completely fill the honey pouch. For instance, there is a gradual reduction of honey pouch load when nectar flow rate decreases. With this response, foragers reduce travel time and increase their value as information channel, contributing to speed collective colony responsiveness to changes in the relative value of floral sources.

Nuñez (2000) argues that the informational capacity of *M. quadrifasciata* is higher than that of the Africanized honeybee (*Apis mellifera scutellata* Lepeletier hybrid), which in turn is higher than that of European *A. mellifera*. When exposed to higher floral diversity, visits shorten and increased frequency of collecting trips explains higher foraging efficiency of Africanized bees. Stingless bees and Africanized honeybees are faster at choosing alternative sources, an ability correlated with higher floral diversity in tropical forests.

Stingless bees' foraging decisions are influenced by social interactions within colonies. For instance, some species of *Trigona* and *Scaptotrigona* use odor trails to communicate the location of attractive food sources; the expression of collective group foraging depends on the direct perception of stimuli in the field. Surprisingly, foragers' reaction to the presence of a conspecific on flowers is not related to the communication system (Slaa et al. 2003).

Honeybees are at the maximum extreme of colonial influence. Foragers bring to the colony profitability "expectations" of foraging in different floral sources, and through continuous exchange of information individuals compare sources' profitability and make joint decisions so the colony continuously redirects foraging effort to the most productive sources. Seeley (1985) named this process colonial thought to emphasize the emerging properties of the efficient integration of information.

The way a forager responds to the presence of individual conspecifics and heterospecific on flowers affects the spatial distribution pattern in food sources. Johnson and Hubbell (1974, 1975), Hubbell and Johnson (1978), Johnson (1983), Johnson et al. (1987) analyzed these responses in several species and proposed three forager categories: grouped, facultatively grouped, and opportunistic solitary. Considering the differences in aggressiveness among species, they recognized the existence of monopolistic and aggressive groups, nonaggressive group, and peaceful foragers. The strategy of aggressive groups also characterizes a "syndrome" called high-density floral specialists.

When a forager approaches a flower, its response to the presence of another individual of another species may be of repulsion or attraction. The answer is species specific but depends on the characteristics of the individual who is already on the flower: for instance, foragers avoid landing in the vicinity of individuals of larger or more aggressive species (Slaa et al. 2003).

Conspecific social interactions have an effect on spatial distribution and therefore affect foragers' activity (Slaa et al. 2003). On the approach to flowers, foragers of some stingless bee species react positively to the presence of a conspecific, while in others the reaction is negative. In the first case, there is a tendency for the distribution of foragers in groups. Also, there seem to be rules for individual decision making: inexperienced and experienced foragers react quite differently to the presence of a conspecific in *Trigona amalthea* Olivier, while the response is always positive in *Oxytrigona mellicolor* Packard. This difference explains satisfactorily why groups of foragers are less compact or more dispersed in the first species.

Foraging decisions and the role of communication in social bees were subjects of numerous experimental studies, especially in the last three decades. Even a brief review of this topic would be beyond the scope of this chapter. However, the reference is necessary to put into perspective the complexity and peculiarities of the economic functioning of large perennial colonies and also to better contextualize seemingly simple behaviors such as floral constancy of foragers.

When a worker visits just one floral source on each foraging trip, it displays floral constancy or fidelity. Foragers of stingless bees may also present fidelity to the same source during multiple trips for days (White et al. 2001). Bee floral fidelity has three basic causes: need, innate restriction, or preference (Faegri and van der Pijl 1979). The first two have no true relationship with floral constancy, because in the first case, environment offers no opportunity for choice, and in the second, choice is limited by morphophysiological constraints. In generalist species, as those of Meliponini, individuals have physical, physiological, and behavioral skills to visit several types of flowers, so that fidelity is expressed as preference (i.e., true floral constancy). Two nonmutually exclusive hypotheses were proposed for these learned responses: foraging efficiency (Levin 1978; Heinrich 1979) and memory constraints (Waser 1983).

Foraging efficiency hypothesis is based on the use of search images by bees: foragers discriminate between floral types and use this information from a distance before landing on the flower. The alternative hypothesis assumes that bees (e.g., *Bombus*) are able to use more than one search image simultaneously by memory constraints.

The basic problem of stingless bee colonies relies on equating the high food demands with temporal variation in floral resources availability in a small range area, given the constraints of foraging from and to a central point. In tropical environments with high floristic diversity, a perennial colony should be generalist, and a forager's floral constancy should be regarded as "behavioral specialization." As stingless bee foragers have a very short life, the learning cost of handling several flower types must have become an ecological constraint. An experimental approach with *Plebeia tobagoensis* Melo indicates that foragers avoid the trade-off between resource types, unless there are changes in food supply, due to the embedded cost of learning time (Hofstede and Sommeijer 2006).

The most widespread evidence of floral constancy by stingless bees resulted from analysis of foragers' pollen loads. This behavior was found in all studied species (Ramalho et al. 1994, 1998; Slaa et al. 1997, 1998; White et al. 2001). However, there are variations when comparing different species or habitats. Ramalho et al. (1994.) reported very high levels of floral constancy in nine species of stingless bees (Figure 11.5) foraging in gardens with high diversity of tree species on the Brazilian Atlantic coast. nearly 95% of the pollen loads came from one plant species. In gardens from Queensland, Australia, a high proportion of *Trigona carbonaria* Smith foragers (88%) also presented a high level of floral constancy even during successive foraging trips (White et al. 2001). On the other hand, in the Amazonic region, *Melipona* foragers carrying mixed pollen loads have often been recorded (Absy and Kerr 1977). This is not a peculiar pattern for this genus, for in three *Melipona* species studied in the Atlantic coast (Figure 11.5), floral constancy was close to 100%.

Floral constancy should represent the compromise between rate of change in floral resources supply and species-specific capabilities. For instance, the informational capacity (Nuñez 2000) and foraging speed (Slaa et al. 2003) should change the expression frequency of this behavior simply because there are differences in species responsiveness to fluctuations in floral resources supply. Floral constancy is expressed even when stingless bee foragers have many available alternative sources (Ramalho et al. 1994; White et al. 2001) and therefore it should be interpreted as part of a set of foraging strategies to maximize individual efficiency. Through floral constancy, a generalist visitor can become an efficient pollinator. There is huge interest in measuring the expression of this behavior in Meliponini, given their numerical dominance in mellitophilous flowers in most tropical habitats and biomes of the Americas, especially in the Atlantic and the Amazonic forests. Analysis of pollen loads from workers could be

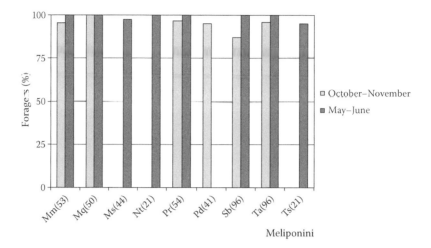

FIGURE 11.5 Floral constancy in stingless bee species. Percentage of foragers with unifloral pollen loads in two flowering periods: Mm = *Melipona marginata*; Mq = *M. quadrifasciata*; Ms = *M. scutellaris*; Nt = *Nannotrigona testaceicornis*; Pr = *Plebeia remota*; Pd = *P. droryana*; Sb = *Scaptotrigona bipunctata*; Ta = *Tetragonisca angustula*, Ts = *Trigona spinipes*. The numbers of sampled foragers are in parentheses. (From Ramalho, M., T. C. Giannini, K. S. Malagodi-Braga, and V. L. Imperatriz-Fonseca, *Grana*, 33, 239–244, 1994. With permission.)

widely used as an exploratory tool for choosing the more appropriate focal trees to analyze the effects of stingless bee foraging activity on plant reproduction (Ramalho 2004; Ramalho and Batista 2005).

Ramalho et al. (1994, 1998) focused on the relationship between pollen load capacity and workers size in Meliponini under standardized natural conditions, in the latter case comparing the transport of monofloral pollen (*Eucalyptus* pollen). They observed that pollen-carrying capacity/weight unit (load capacity) decreased as an exponential function of body weight or bee size (Figure 11.6a). Comparing pollen loads from different floral sources and pollen loads from *Eucalyptus* (Figure 11.6b), it was also evident that the fitting curve of body size becomes more accurate when comparing loads of the same pollen type.

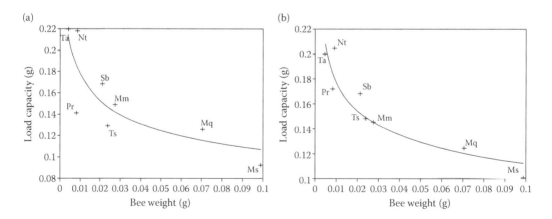

FIGURE 11.6 (a) Variation in pollen load per unit of body weight (load capacity) among stingless bee species. Pollen load capacity decreases with species size, independently of floral source ($N = 8$, $r = -0.77$, $p < 0.05$, and $Y = aXb$, and $a = 0.065$, $b = -0.218$). (b) Curve fits data better when comparing monofloral Eucalyptus sp pollen loads. ($N = 8$, $r = -0.90$, $p < 0.05$, $Y = aXb$ and $a = 0.073$, $b = -0.191$). Mm = *Melipona marginata*, Mq = *Melipona quadrifasciata*; Ms = *Melipona scutellaris*; Nt = *Nannotrigona testaceicornis*; Pr = *Plebeia remota*; Sb = *Scaptotrigona bipunctata*; Ta = *Tetragonisca angustula*, Ts = *Trigona spinipes*. (From Ramalho, M., T. C. Giannini, K. S. Malagodi-Braga, and V. L. Imperatriz-Fonseca, *Grana*, 33, 239–244, 1994. With permission.)

There are variations in the weight of foragers' loads that rely on their own pollen source and/or pollen type. Huge variations between individuals of a same species and same size category were also observed.

Load capacity decay is higher in the transition from small Meliponini, such as *T. angustula, P. remota, Nannotrigona testaceicornis* (Lepeletier), to those of medium size, such as *S. bipunctatata* and *Trigona spinipes* (F.). From one category to another, general differences were also observed concerning foraging strategies: from solitary opportunist, that avoids antagonistic interactions, to group foragers, sometimes aggressive and monopolists.

The variation pattern in workers' load capacity refers to theoretical questions about ecological constraints of body size, and considering hypotheses about foraging (Schoener 1971), two basic predictions arise: (1) it is expected that large bees are able to meet their energy needs more quickly than small bees when food is abundant, and more slowly when scarce, and (2) if competitors reduce the abundance of floral resources in a uniform way in several blossoms, size convergence should be favored, while differential depletion would promote divergence among species size. The first hypothesis leads to the following prediction: as the average floral resources supply changes, larger bees must answer to localized reduction, moving quickly to another site or another floral source. An experimental study with *M. quadrifasciata* (Nuñez 2000) suggests that foragers can behave according to this general prediction. In contrast, in ecological communities, the largest *Melipona* species would often avoid overlapping and antagonistic interactions with small Meliponini species. Both bee censuses on flowers as comparative analysis of pollen sources from colonies point in that direction. Among small stingless bees, there are extreme opportunistic strategies, such as presented by *Paratrigona subnuda* Moure, that often collect pollen remains on floral parts resulting from other visitors' activity. The second hypothesis serves as a starting point for a reflection about interactions among several midsized Meliponini. In particular, species that have more or less regular nest spacing (Hubbell and Johnson 1977; Breed et al. 1999) tend to homogenize spatial resources offered in the nearby habitat. These species would then present greater convergence of body size, as seems to be the case for a number of *Trigona* species.

Also common is the variation in worker size within the same colony and among colonies of the same species. Laboratory data suggested a slight trend toward smaller worker production by weak colonies. From the standpoint of foraging efficiency, reduction in size has survival colonial value (Ramalho et al. 1998). *M. quadrifasciata* small workers carried little more pollen/unit of body weight (Figure 11.7)

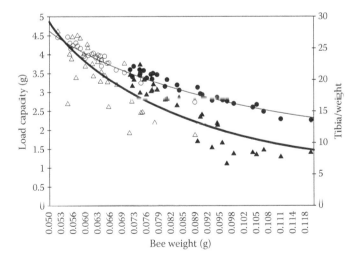

FIGURE 11.7 Relationship between pollen load capacity and forager body weight of *Melipona quadrifasciata*. Full and empty symbols represent workers of strong and weak colonies, respectively. Adjustment curve: $Y = aXb$. Triangles = relationship between load capacity and body weight ($a = 0.08$, $b = -1.37$, $r = -0.88$, $p < 0.05$). Circles = relationship between tibia surface (pollen-carrying structure) and body weight ($a = 2.61$, $b = 0.79$, $r = -0.97$, $p < 0.05$). Workers carry a little more pollen per unit of body weight (higher load capacity) and tibia allometric development explains most of observed variation. (From Ramalho, M., V. L. Imperatriz-Fonseca, and T. C. Giannini, *Apidologie*, 29, 221–8, 1998. With permission.)

(Ramalho et al. 1998). As pollen is essential for offspring production and smaller workers were also associated with weak colonies, the adaptability argument seemed to be supported. However, the primary cause of worker size variation is one of the basic problems of this apparently circular argument. When there are less floral resources in the environment, colonies need to reduce offspring production. There are fewer workers to forage, build, and provision the cells and thus a brood receives less food, and emerging bees are smaller. But why does the colony produce smaller bees rather than less offspring?

With the decrease in floral resource availability in a restricted range area, a colony has three options: reduce the amount of offspring, the size of offspring, or both. If the stability threshold of social functions in large perennial colonies were sensitive to the number of workers, colony survival would be less committed to the decrease of worker size: population fall is smaller and there is some efficiency gain in collecting pollen for production of the future offspring.

The inverse relationship between body size and pollen load capacity (Ramalho et al. 1994) means that food balance of colonies from different species can be achieved through different investment levels in foraging activity, with effects on life history. For instance, species with very small workers (*Plebeia, Tetragonisca, Paratrigona*) achieve a larger return of pollen biomass/foraging effort per capita and address more energy (and time) for offspring production. Variation in nests density and colonies longevity of *T. angustula* in disturbed forest habitats (Batista et al. 2003; Slaa 2006) supports this prediction. The opposite argument applies in very general lines to the large species of *Melipona*, whose foragers have lower pollen load capacity (Ramalho et al. 1994) and colonies invest more in longevity (Roubik 1989; Slaa 2006).

In summary, foraging economic decisions lead to floral constancy of stingless bee foragers. Allied to morphofunctional body size constraints, social interactions, and so forth, it can also be expressed as floral preference or realized niche narrowing, as has been observed in *Melipona* and *Scaptotrigona* (Ramalho 1990; Ramalho et al. 1989, 2007).

11.3 Resource Utilization by Colonies

In adult bees, food changes promote development of endocrine glands, determining workers skills. Bees in early adulthood participate in brood cell provisioning, producing larval food in their hypopharyngeal glands, which develop due to the consumption of large amounts of pollen. At a later age, the ingestion of pollen may stimulate ovarian development.

The queen receives a food rich in proteins that allows her to lay eggs continuously. In stingless bees, a queen may receive protein food through trophallaxis with workers or through ingestion of trophic eggs placed by workers. The queen occasionally feeds directly in the food pots or eats larval food from brood cells before oviposition. Studies with *P. remota* indicated that colony condition and food received by the queen determine egg size (M.F. Ribeiro, personal communication). During the larval period, food plays a key role in caste determination and/or differentiation.

11.3.1 Caste Determination and Differentiation in Bombini

Bombini are primitively eusocial bees (i.e., queens establish their nests and work at all tasks until the emergence of the first workers). Larval feeding is progressive or massal depending on the groups, which therefore are named pollen storers when larvae are fed slowly (*Bombus terrestris* (L.), *Bombus hypocrita* Perez) or pocket makers (present a food bag) when larvae obtain their food directly from a pollen mass (a bowl of wax with pollen, where eggs are laid by the queen) (Sladen 1912; Michener 1974).

In *Bombus,* mechanisms of caste determination differ among species. In *Bombus perplexus* Cresson, the size of female larvae is related to the amount of food in the colony, which is more abundant as the number of workers becomes larger in relation to the larvae (Plowright and Jay 1968). In *Bombus terricola* Kirby and *B. ternarius* Say, other mechanisms affect feeding rate and development of larvae into queens (Brian 1957; Plowright and Jay 1968). In some *Bombus hypnorum* (L.), *B. diversus* Smith, *B. ignitus,* and *B. hypocrita,* caste differentiation occurs later and larvae with longer developmental time eat more and become queens (Katayma 1966, 1973, 1975; Röseler 1970). A high feeding rate in the last phase of development in *B. rufocinctus* Cresson influences larval destiny, causing changes in growth rate and silk production. Larvae that are fed less often begin to produce silk earlier and soon weave their cocoons, becoming workers. Others, who receive food more frequently, spend less time in silk production, delaying pupation and achieving larger size to become queens (Plowright and Jay 1977). In *B. terricola*, the larval development period differs, although there are no differences in growth rates of queens and workers larvae (Pendrel and Plowright 1977). In pocket-maker species, reproductives and workers are distinctively fed. During most of their development, workers feed from pollen pockets. From an early age, males and larvae destined to be queens are fed regurgitated food by adult workers (Alford 1975).

Finally, in *B. terrestris*, the mother queen produces a pheromone that suppresses the endocrine system of female larvae, preventing them from becoming queens. With aging and the likely decrease in pheromone production by the queen and/or increase in colony size, some larvae escape this control and have their endocrine system activated and become queens (Röseler and Röseler 1974; Röseler 1976, 1991). The pheromone involved has not yet been identified but apparently acts in suppression of juvenile hormone production, leading larvae to suffer the first molt earlier and therefore become smaller.

Caste differentiation is expressed by the ingestion of different amounts of food. In turn, food availability in the colony depends on the ratio between workers that collect food and larvae that consume it. Efficiency of individual workers in foraging and offspring care is also important.

Another relevant aspect is food quality. Enzymes produced by workers' hypopharyngeal glands added to the food given to larvae apparently help in digestion (Free and Butler 1959; Röseler 1974). A protein source besides pollen was found in the food of *B. terrestris* larvae, not exclusively on the food of larvae that developed into queens (Pereboom and Shivashankar 1994; Pereboom 1996).

Growth rate of queen larvae is different from that of workers. Queens ingest proportionally less pollen than expected and probably accumulate more fat. This suggests that queen larvae make better use of ingested pollen, or receive an extra source of protein in their diet (Ribeiro 1994). During the second development period, feeding frequency is also higher in queen larvae (Ribeiro et al. 1999). As in *A. mellifera*, each feeding time is probably related to the presence of glandular material added to larval food (Browers et al. 1987). This is important in the final developmental phase of queen larvae, as they receive larger amounts of these nutritive substances, which promote higher growth even in the absence of adequate pollen supply to the colony (Ribeiro 1999).

Bombus species, like honeybees (Free et al. 1989; Huang and Otis 1991; Le Comte et al. 1995), signal a hunger state with pheromones modulating the larvae feeding pattern by workers. Comparing larvae that experienced food deprivation with a control group, Pereboom (1996, 1997) found that the former were fed before and with a higher initial rate than control larvae. Larval food composition induces females' caste development in *Bombus* (Pereboom 2000).

Approximately one million colonies of *B. terrestris* are sold each year for pollination in agriculture. This successful rearing trade, especially in Holland and Belgium (Velthuis and van Doorn 2006), brought contributions to the knowledge of food quality influence on nest development. For *Bombus*, nest development demands lots of pollen, obtained from colonies of *A. mellifera*. Ribeiro et al. (1996) found that pollen quality influences queens' production. Queens reared with dry pollen (which loses nutritional value in the drying process) in greenhouses were smaller, had higher mortality rates, and produced smaller colonies than those supplied fresh pollen. Qualitative and quantitative pollen variations influence colony development and reproductive success (Génissel et al. 2002).

11.3.2 Caste Determination and Differentiation in Apini

In honeybees, *A. mellifera*, caste determination occurs early in larval development. From the third day of life, food provided to worker and queen larvae changes quantitatively and qualitatively. Food for queen larvae, royal jelly, contains larger amounts of mandibular glands secretions than food for worker larvae. Three components were described in larval food: white (mandibular gland secretions), clear (hypopharyngeal gland secretions), and yellow (pollen) one. Worker larvae receive these components in the following proportions: 2:9:3, respectively, while a queen would receive 1:1 mainly from the first two components (Jung-Hoffman 1966).

Larvae feeding frequency varies. Queen larvae are fed >1,600 times while workers only 143 times during their development (Lindauer 1952). The amount of food seems to be relatively less important than quality. Larvae that were fed a queen's artificial diet *ad libitum* develop into queens, while those fed a worker's diet *ad libitum* do not (Moritz 1994). Queen larvae gain weight twice as fast than those of workers, weight gain 30 mg to >300 mg in just 2 days (Moritz 1994). Therefore, queen larvae are reared in larger cells, called royal cells.

Queen larvae have phagostimulant sugar present in 34% of royal jelly, while phagostimulant sugar is present in only 12% of workers' food (Beetsma 1979; Winston 2003). The type of sugar also differs in larval food: queens receive mainly glucose, whereas workers receive glucose in the first larval phases and fructose in the last ones (Browers 1984).

Finally, juvenile hormone (JH) produced by *corpora allata* exerts influence in caste differentiation. Queen larvae with 72 hours of age have JH levels 10× higher than worker larvae of the same age (Wirtz 1973). JH levels remain high throughout the remaining larval phases in queens (Beetsma 1979).

11.3.3 Caste Determination and Differentiation in Meliponini

In Meliponini, brood cells receive all food before egg laying by the queen, a feeding behavior known as mass provisioning. Individual brood cells are built by workers, following complex behavioral sequences (Sakagami and Zucchi 1963; Sakagami 1982). Workers provision cells with liquid larval food, and may place trophic eggs on this larval food that are consumed by the queen and more rarely by workers (Silva-Matos et al. 2000). The queen lays her egg in this cell, which is then closed by workers. This sequence of events called provisioning and oviposition process (POP) is variable (Sakagami and Zucchi 1963).

A basic question is whether the provisioned food differs in quality between cells that give rise to queens than in those of workers. In *Melipona* caste differentiation is genetic, although environmental influence may be also important (Kerr 1950b). Queens and workers are reared in identical cells; queen larvae are double heterozygous (AaBb) while worker larvae are homozygous; the amount of food is also important (Kerr 1950b, 1969; Kerr et al. 1966; Velthuis and Sommeijer 1991). Queens have four nodes in the ventral nerve cord, while workers have five (Kerr and Nielsen 1966). Another peculiarity of *Melipona* species is the large number of queens produced in the colonies, which can reach up to 25% of the offspring (Kerr 1946, 1948, 1950a,b; Santos-Filho et al. 2006).

A second hypothesis about *Melipona* caste determination was based on self-determination (Ratnieks 2001; Wenseleers et al. 2003). They consider that larvae "decide" their fate by choosing whether to be queens. Their model forecasts 14% of queens in the offspring, close to the 25% model suggested by Kerr (Santos-Filho et al. 2006). Caste determination in *Melipona* using molecular markers (Judice et al. 2004; Makert et al. 2006) provide a complete list of genes differently expressed in queens and workers of *M. quadrifasciata*, available at *http://www.lge.ibi.unicamp.br/abelha*. Hartfelder et al. (2006) put together a comprehensive review on caste determination in Meliponini.

In other genera of stingless bees, caste determination is essentially trophic, although several strategies have been developed for queen rearing in large cells, known as royal cells. Thus, consuming more food, female larvae become queens rather than workers (Engels and Imperatriz-Fonseca 1990). In *Frieseomelitta* and *Leurotrigona* species, where royal cells are not built by workers, one larva may consume all food of its neighboring cell and become a queen (Terada 1974; Faustino et al. 2002). This also happens in *Plebeia lucii* Moure, a species that builds bunch brood cells. In queenless colonies, these bees build cells for queen production (Teixeira and Campos 2005), and the larger amount of food determines the differentiation of larvae into queens.

Other studies suggest a greater complexity in trophic determination process. Giant workers can emerge from royal cells (a single observation in *P. remota*, Imperatriz-Fonseca 1975) and dwarf queens can arise from cells of equal size to those of workers (Ribeiro et al. 2006a), indicating that food amount alone is not enough to explain caste determination in royal cell builders. The emergence of dwarf queens from "normal" sized cells occurs in several genera on a regular *(Schwarziana, Cephalotrigona)* or occasional *(Plebeia, Nannotrigona)* basis. In general, some dwarf queens are viable, mate, and lay eggs normally, surviving for long time (Ribeiro and Alves 2001; Ribeiro et al. 2003; Wenseleers et al. 2005; Ribeiro et al. 2006a,b). Explanations for the existence of dwarf queens and their production mechanisms vary depending on the genus and on circumstances. Some larvae may escape the fate of becoming workers, using their ability of "self-determination" to become dwarf queens (Wenseleers et al. 2005; Ribeiro et al. 2006a). In this case, dwarf queen development is under the control of genetic mechanism (Wenseleers et al. 2004). Another possibility is the presence of larval food of better quality or in larger amounts in some cells. Castilho-Hyodo (2002) studied the quality of larval food in *S. quadripunctata*, showing the high variability in protein content of brood cells from the same comb.

In *Melipona beecheii* Bennet, colonies with reduced amount of food produced fewer queens than those that receive extra food. However, the latter did not produce a significantly higher number of queens, when compared with control colonies (Moo-Valle et al. 2001). In *P. remota*, however, there is no relationship between variation in the number of produced queens and colony food storage (Ribeiro et al. 2003).

11.4 Larval Food in Meliponini

In stingless bees, larval food seems to be species specific. Darchen and Delage-Darchen (1971) reared queens even with larval food of different species. Silva (1977) obtained queens in mixed colonies, made up of queens and workers from related species. Hartfelder and Engels (1989) studied the composition of larval food in stingless bees. They analyzed the water soluble constituents in larval food of seven species, and found that the variation in larval food proteins was consistent with phylogenetic trees. They also suggested that nurse workers of stingless bees would not control queen development, for instance, provisioning certain cells with a special diet. Instead, they would just place larger amounts of the same food type given to any other cell inside royal cells.

In stingless bees, the protein content of larval food is about 10 times lower than in *Apis* (Takenaka and Takahashi 1980), and that is the main difference between the two groups. The proportion of sugars and of free amino acids in larval food is similar in both (Shuel and Dixon 1959; Rembold and Lackner 1978).

Bionomic knowledge of the necrophagous bees (*T. crassipes, T. necrophaga,* and *T. hypogea*, Roubik 1982; Camargo and Roubik 1991) brought forth important issues on the quality of stingless bee larval food. These bees replaced pollen with animal protein. There is no pollen in their nests, but there are sugar solutions storages, probably obtained from extrafloral nectaries. Among the basic adaptations of these species to these new feeding habits are jaws with five teeth (maximum number found among Meliponini) and reduced corbicula on the third pair of legs (as they do not carry pollen).

Gilliam et al. (1985) studied the microbiology of larval food of *T. hypogea*, considered at that time an obligatory necrophagous. They mentioned these bees gathered food in a wide variety of freshly killed animals (frogs, toads, lizards, fish, birds, even monkeys). Later, Mateus and Noll (2004) found that this species fed on live wasp offspring caught in abandoned or unprotected nests. Once they find their food source, bees quickly recruit their nestmates, who monopolize the food source, excluding other insects. Workers place secretions on the organic matter for a predigestion (see review in Noll et al. 1997), then they ingest it and carry the thick liquefied material to the nest. There, that food is processed by other workers, probably adding large amounts of enzymes from hypopharyngeal glands. In *T. hypogea*, these secretory units are multicellular, while in Meliponini species that feed on pollen, they are unicellular (Cavasin-Oliveira and Cruz-Landim 1991). After being processed, the resulting viscous liquid has a pH between 3.0 and 4.0, very similar to *Apis* royal jelly, and it is stored in food pots. Several microorganisms transform and probably play an important role in the conservation of this protein food of animal origin. Gilliam et al. (1985) found in samples of larval food of this species *Bacillus pumilus, B. meggaterium, B. subtilis, B. circulans,* and *B. licheniformis*, which produce several enzymes. These microorganisms are responsible or have an important role in converting these reserves into a nutritious and metabolizable food for larvae and young bees. The same *Bacillus* species were found in pollen stored by *A. mellifera* (Gilliam and Morton 1978). Machado (1971) verified an association of a species similar to *B. pumilus* with pollen of *M. quadrifasciata*, which seemed to predigest pollen. It appeared in large amounts just in the glandular secretion placed between layers of pollen and nectar in brood cells. Machado (1971) also found *Bacillus* in larval food of 13 species of stingless bees: four species of Melipona, two of *Plebeia* and *Trigona*, and one of *Partamona, Frieseomelitta, Leurotrigona, Tetragona,* and *Nannotrigona*. Gilliam et al. (1985) argue that bees can add to larval food beneficial microorganisms, responsible for conversion, fermentation, and preservation of larval provisions, which also inhibit proliferation of other undesirable microorganisms, for instance, producing antibiotics and fat acids.

11.5 Pollen

Since plants cannot move to find reproductive partners, flowering plants developed a series of traits to overcome this difficulty: they attract insects or other animals to their flowers, favoring crossing among them. In flowers, plants provide food, nectar, and pollen, and use several features, such as vibrant colors, perfumes, and petals that serve as landing platforms, to attract floral visitors that carry pollen (male part) from one flower to the stigma (female part) of another, a phenomenon called pollination.

Pollen-collecting bees favor effective pollination of plants more than nectar gathers (Free 1966). Unlike nectar, which is available throughout the day, pollen from plants is a resource offered all at once. It is the main protein source for most bees and it is used for offspring development. Pollen is part of the diet of other insects and supplements the diets of bats, birds, and marsupials, and these animals, as well as bees, are pollinator agents.

11.5.1 Protein Value

Protein content of pollen grains varies from 2.5% up to 61% (Buchmann 1986). Pollen grain nutrients are found in their cytoplasm and are recovered after a digestive process. The grains' outer layers are not digested, because they are made of cellulose and sporopollenin, which are hard to decompose. As they retain their external structure, grains can be identified after passing through animals' digestive tract, which allows paleoecologists to reconstruct the original flora and climate of regions where they occurred.

Protein of pollen grains consists mainly of enzymes that act during pollen tube growth (Stanley and Linskens 1974). Roulston et al. (2000) showed that the protein content of pollen grains of 377 plants species is highly conservative within genera and families, with the exception of Cactaceae and Fabaceae. Plants taxa with buzz pollination are rich in proteins ($x = 47.8\%$), despite minute pollen grain size.

Anemophilous pollen grains have lower protein content than zoophiles, although anemophilous grains, such as those of Poaceae (maize) and Moraceae (*Cecropia*) are frequently collected by *Apis mellifera* and stingless bee species (Cortopassi-Laurino and Ramalho 1988). Protein content of pollen grains from corbiculae of some Meliponini of the Amazon region presented values between 15.7% and 23.8% (Souza et al. 2004).

11.6 Food Rich in Sugars Produced by Bees: Honey

Honey is still the main product of commercial rearing of honeybees and stingless bees. Flower nectar is the raw material used to create honey, which is produced and stored in large amounts inside the nests. As alternative food sources used by bees (Figure 11.8), there are plant secretions such as those from sugar cane, or excretions of insects that suck living parts of plant originating honeydew honey. In stingless bees, honey is stored in large oval pots, which vary in size according to species, whereas in *Apis* it is stored in hexagonal cells similar to those used to rear the brood.

11.6.1 Honey Microscopy

Bees visit flower nectaries mainly to collect nectar. In some cases, nectar gets contaminated with flower pollen. When observing honey under the microscope, pollen grains from flowers that were visited for nectar collection are identified (Figure 11.9). As a rule, most represented pollen grains indicate floral origin (i.e., the nectar that contributed most to the honey composition). Some pollen grains are considered geographical indicators, for they are only found at certain places.

FIGURE 11.8 Africanized honey bee (*Apis mellifera*) and *Nannotrigona testaceicornis* sucking secretions of scale insects. Wasps and ants also collect these secretions.

Melissopalynology, the study of honey pollen grains, depends mainly on data accumulation and knowledge of grain morphology. Pollen grains show typical shapes for each species, with different openings and ornamentation, and sizes ranging from 5 to 300 μm. Only the smallest grains are collected mostly by *Apis* and stingless bees (Barth 1989; Ramalho et al. 1990; Pirani and Cortopassi-Laurino 1993; Moreti et al. 2002).

Pollen analysis of food carried to nests has been used as an indirect method of assessment of bee visits to flowers. This has advantages and disadvantages in relation to field observations, which depend on several aspects, such as collection time, tree height, and "plant apparency." For beekeeping, pollen analysis allows identification of poorly known wild flora, supports planning honey annual production by migratory beekeeping, and allows control of floral and geographical origin of honey, information increasingly important for product credibility and for adoption of appropriate processing measures.

11.6.2 Honey in Apini

The most productive bees in Brazil are *A. mellifera*, or Africanized honeybees (Figure 11.10) as they are better known, frequently observed in urban centers. Africanized honeybees are not native to Brazil; they are a crossbreed between *A. mellifera*, brought from Portugal to Rio de Janeiro in 1839 by Father Antonio Carneiro and others (Nogueira-Neto 1997), and African *A. mellifera*, introduced in 1956, in order to increase honey production through selective breeding. Currently, it is estimated that domestic honey consumption in Brazil is around 40 to 60 thousand tons/year (C. Zara, personal communication).

Honeybee honey is composed mostly of water and sugars (99%). The remaining (1%) contains substances present in tiny amounts, but which are important in honey characterization, such as enzymes,

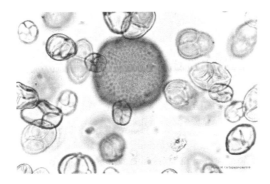

FIGURE 11.9 Pollen grains found in honey slides. The isolated central grain belongs to the family Euphorbiaceae, identified by the croton ornamentation pattern. The other grains are from Mimosaceae (*Mimosa bimucronata* and *M. taimbensis*).

FIGURE 11.10   Africanized honey bee (*Apis mellifera*) sucking nectar from *Citrus* sp. flower.

amino acids, and minerals. Its humidity is about 20% and it has approximately 80% of sugars such as glucose, fructose, and sucrose. Glucose is relatively insoluble and its amount determines honey crystallization tendency. Fructose is very sweet and hygroscopic that absorbs air humidity (Crane 1987). Color patterns, scent, and flavor vary with floral origin, geographical regions, and climatic conditions. Floral honeys can be separated from honeydew by morphological elements and physicochemical analyses (Barth 1989; Campos et al. 2003).

11.6.3 Honey in Meliponini

Honey production can reach just a few liters per hive per year. Nevertheless, the high market value turns stingless bee rearing into a profitable activity, at least in small scale. These bees' rearing, or meliponiculture, is based mainly on bees of the genus *Melipona*, which are large (15 mm) and store honey in big pots, which facilitates extraction. Since pre-Hispanic times in Mexico, rearing of *M. beecheii* testifies this long tradition; species of *Tetragonisca* and *Scaptotrigona* have also been widely reared. Traditionally, medicinal value is attributed to honey from the former genus, while the second are good producers because their colonies are very populous.

In Brazil, honey production from *Melipona* species is more expressive in the Northeast, where the product can be found in labeled packages, with details of the producer, origin, and collection date (Figure 11.11).

T. angustula (Figure 11.12) is the most popular stingless bee species, widely distributed throughout Latin America. Although having a small nest production, around one liter/year, its honey is considered medicinal and used in treating eye diseases by rural populations. Easiness of recognition and management have contributed to its popularity. Species of *Scaptotrigona* also have a wide distribution in Latin America. Usually, they have populous nests, are aggressive, and produce large amounts of honey. In Mexico and in Central and South America, several different species are reared for this purpose, as *Scapotrigona mexicana* (Guérin-Méneville), *S. depilis* (Moure), *S. nigrohirta* Moure, *S. polysticta* Moure, and *S. postica* (Latreille) (Cortopassi- Laurino et al. 2006).

11.6.3.1 How to Exploit Meliponini Honey

Compared with *A. mellifera* honey, stingless bee honey frequently has higher water percentage, higher acidity, and lower pH values (Cortopassi-Laurino and Gelli 1991). The high water percentage makes it more susceptible to fermentation, reducing storage time. The preparation of the Technical Regulation of Identity and Quality of Stingless Bee Honey faces two major basic problems: the lack of results of physicochemical analyses and the wide variety of bees.

FIGURE 11.11 Several honey packages of stingless bee honey. From left to right: honey from *M. scutellaris*; *Melipona* honey from the Amazon region; honey from *M. fasciculata* with glass coated with buriti fibers, which adds value to the product; honey from *M. rufiventris*; honey from *M. subnitida*, a unique honey with an annual registration label at the Agriculture Secretary from Rio Grande do Norte state; honey from *Scaptotrigona* sp, Belterra, PA; package and honey glass from *M. fasciculata* provided by the nongovernmental organization AMAVIDA from Maranhão state.

Technical studies of honey have focused on a few dozen species, especially *Melipona*. It has been suggested as a protocol for honey control from *Melipona, Trigona,* and *Scaptotrigona* (Vit et al. 2004). There is a technical foundation for a preliminary proposal on Meliponini honey legislation, considering that more than 1,100 samples of 18 species were already examined. Of these, there are a higher number of results for humidity, pH, acidity (free and total), ash, and HMF (hydroxy-methyl-furfural) parameters. However, as these physicochemical characteristics vary widely, there is need to enlarge the number of samples to obtain a consistent honey profile from most genera and species studied up to now (Bazlen 2000; Souza et al. 2004, 2006, 2009; Almeida and Marchini 2006; Carvalho et al. 2006; Cavalcante et al. 2006; Oliveira et al. 2006; Persano-Oddo et al. 2008, Anacleto et al. 2009, Rodrigués-Malaver et al. 2009). Table 11.1 summarizes the results of stingless bee honey analyses with at least five samples.

Until now, of the specified tests in the Technical Regulation for Identity and Quality of *A. mellifera* Honey, eight have been used in stingless bee honey analyses. This physicochemical test is applied with

FIGURE 11.12 Nest entrance of *Tetragonisca angustula* in a rational wooden box. This is one of the best-known stingless bee species that presents a wide geographical distribution, from Mexico to Misiones, Argentina. (From Nogueira-Neto, P., *Vida e Criação de Abelhas Indígenas sem Ferrão*, Editora Nogueirapis, São Paulo, Brazil, 1997. With permission.)

TABLE 11.1

Stingless Bee Honey: Physicochemical Characteristics

Species	Number	pH	Total Acidity	Humidity	HMF	Diastase Index	Invertase Index	Ashes	Locality	Reference
M. asilvai	11	3.3	41.6*	29.5	2.44				BA	Souza et al. 200?
M. asilvai	7	3.6	54.2	37.5	30.9			0.09	BR-BA	Souza et al. 200?
M. beecheii	5	4.2	59.4	27.0	5.4				Mexico	Santiesteban 199?
M. beecheii	7	3.7	23.2*	17.3	0.1	21.3		0.07	Guatemala	Dardón and Enríquez 2008
M. compressipes	35	3.3	91.1	25.6					MA	Oliveira et al. 20?6
M. compressipes	5		48.4	23.4	1.0	1.1		0.3	Venezuela	Vit et al. 1994
M. favosa	511			31.2					Trinidad/Tobago	Bijlsma et al. 20?6
M. favosa	14		62.9*	25.5	1.2	0.9		0.3	Venezuela	Vit et al. 1994
M. favosa favosa	6		36.8	24.2	17.1	2.9	90.1	0.2	Venezuela	Vit et al. 1994
M. grandis	5			27.5					Peru	Rodríguez-Malaver et al. 2009
M. mandacaia	20	3.3	43.5	28.8	5.8			0.4	BA	Alves et al. 200?
M. quadrifasciata	8	3.5	132.6*	32.2					SP	Cortopassi-Laurino 1997
M. quadrifasciata	6	4.0	38.5*	25.5	3.8	1.8		0.1	BA	Oliveira et al. 20?6
M. quadrifasciata	6					1.2–2.2			BA	Fonseca et al. 20?6
M. quadrifasciata anthidioides	9	4.0	40.6	32.1	16.0			0.1	BA	Souza et al. 20?4
M. scutellaris	20	4.1	31.1	28.6	2.7	4.7	201.9		BA	Bazlem 2000
M. scutellaris	7	3.6	39.8*	26.9	3.3	4.0		0.04	BA	Cavalcante et al. 20?6
M. scutellaris	7					0.7–19.8			BA	Fonseca et al. 20?6

Species	n								Country	Reference
M. scutellaris	15	4.4	19.9	29.1	2.0			0.19	BA	Souza et al. 2004
M. subnitida	27	4.4*	24.4*	24.0	8.7			0.5	PI	Camargo et al. 2006
M. trinitalis	62			33.0					Trinidad	Bijlsma et al. 2006
Plebeia wittmani	10	3.3	117.5					0.2	Argentina	Spariglia et al. 2010
Scaptotrigona pachysoma	7	3.9	6.6	26.9	1.0				Mexico	Santiesteban 1994
Tetragonisca angustula	12	4.3	7.9					0.07	Argentina	Spariglia et al. 2010
Tetragonisca angustula	261	4.2		27.7					SP	Iwama 1977
Tetragonisca angustula	20	4.0	5.1	27.9	5.7	22.0	38.9		SP/BA	Bazlem 2000
Tetragonisca angustula	10	4.4	20.6*	23.9	7.5	30.0		0.4	SP	Almeida and Marchini 2006
Tetragonisca angustula	14			24.9					Costa Rica	Demera and Angert 2004
Tetragonisca angustula	7	4.2	74.7	25.0				0.3	SP	Cortopassi-Laurino 1997
Tetragonisca angustula	20	4.1	45.2	24.4	9.4	7.2–54.1		0.39	SP	Anacleto et al. 2009
Trigona carbonaria	8	4.0	124.2*	26.5	1.2			0.48	Australia	Persano-Oddo et al. 2008

Note: Number of samples > 5. Most data from Brazil (BA = Bahia; MA = Maranhão; SP = São Paulo; PI = 0020 Piauí); other countries are identified.

* Free acidity.

reservations in the proposed legislation of stingless bee honey. Techniques adopted by the European Honey Commission (Bogdanov et al. 1997) can be adjusted to enhance technical control, and Souza et al. (2006) emphasized the need of obtaining additional data, such as sugar types, electric conductivity, and pollen analysis. Stingless bee honey collected in areas with different rainfall levels or from different nests in the same place shows variation in water content for the same species (Bijlsma et al. 2006).

In the Amazon region, the recent production of about three honey tons from *Melipona compressipes* F. and *M. seminigra* Friese (Villas-Boas and Malaspina 2004) shows that there is an underutilized potential production. Paradoxically, this "surplus" honey production is facing distribution and quality certificate problems. While this situation remains unsolved, stingless bee honey will continue to be sold as a natural product without official record, becoming more subject to adulteration. Table 11.2 presents a summary of physicochemical parameters that can be used for the Technical Regulations for the Quality of Stingless Bee Honey. They were compiled from analyses of 332 honey samples from *T. angustula* and 813 samples from *Melipona* spp.

11.6.3.2 Antibacterial Activities

Since ancient times, honey has been used as an antibacterial agent for wound and burn treatment. Initially it was thought that honey's antibacterial property was associated with high sugar concentration (± 80% to *Apis*) and low pH. However, some organisms that survive at low pH, such as *Staphylococcus aureus*, did not survive in honey, indicating that other substances were active against bacteria. This "inhibin" was later identified as hydrogen peroxide. This compound is produced by the action of a bee enzyme (glucoseoxidase) in honey sugar (glucose), resulting in gluconic acid plus hydrogen peroxide. The presence of H_2O_2 is higher in diluted honey.

Stingless bee honey still presents antibacterial activity even when hydrogen peroxide production is inhibited by catalase addition. Therefore, there are still other compounds that need to be chemically identified. Recently, antioxidant compounds such as polyphenols and flavonoids have been quantified in honey because they have bioactive properties that may be responsible for their biological and therapeutic properties (Vit and Tomaz-Barbéram 1998; Guerrini et al. 2009; Persanno-Oddo et al. 2009; Pitombeira et al. 2009; Rodrigués-Malaver et al. 2009).

Water activity available in honey has been quantified as it contributes to the development/inactivation of microorganisms. These values are between 0.59 and 0.82 (0.66) for *T. angustula* (Anacleto et al. 2009), 0.79 for *M. asilvai*, 0.75 for *M. mandaçaia*, 0.76 for *M. quadrifasciata anthidioides*, 0.71 for *M. scutellaris* (Souza et al. 2009), 0.74 for *T. carbonaria* (Persanno-Oddo et al. 2008). In honeybees, honey has lower humidity, with values between 0.48 and 0.65 (Schroeder et al. 2005).

Minimal inhibitory concentration (MIC) is another way to evaluate honey antibacterial value; this parameter identifies the minimum amount of honey with activity against certain bacteria strains. Rodrigués-Malaver et al. (2009) found in native bees of Peru an MIC of 50% (w/v) to inhibit *E. coli* and 12.5% to 50% to inhibit *S. aureus*. According to Boorn et al. (2009), *T. carbonaria* honey presented an

TABLE 11.2

Suggestions of Physical and Chemical Parameters for Stingless Bee Honey

Parameters	*Melipona*	*Tetragonisca angustula*
pH	3.3–4.4	4.0–4.4
Free acidity	<132.6	<71.9
Humidity	<37.5	<27.9
Ashes	<0.5	<0.4
HMF	<30.9	<9.4
Diastase	0.7–21.3	7.2–54.1
Invertase	90.1–201.9	38.9

MIC between 4% and >10% to inhibit Gram-positive bacteria, between 6% and >16% to inhibit those Gram-negative, and between 6% and >10% to inhibit *Candida* spp. Of nine species of stingless bees in Guatemala, except for *M. solani*, all presented an MIC between 2.5% and 10% against microorganisms, especially *N. perilampoides* with values between 2.5% and 5% (Dardón and Enriquéz 2008). Likewise, ethanolic fractions of honey from native bees of Ecuador presented inhibitory values against bacteria (Guerrini et al. 2009).

In relation to topical honey applications, Vit and Jacob (2008) found significant inhibition of induced cataracts in sheep when treated with flavonoids present in ethanolic fractions of honey, such as luteolin and orientin. Alves et al. (2008) verified that the application of *Melipona subnitida* honey in infected wounds of rats' skin stimulated immune response and reduced infection and healing time.

Unprocessed honeybee honey has been recommended as a topical agent in infected wounds, chronic ulcers, and burns, with excellent results in reducing infection and healing time (Tostes and Leite 1997). Similarly, honey from stingless bees has also been used as a topical agent in insect and snake bites and ocular inflammations in several Latin America countries. In the laboratory, stingless bee honey has shown bacteriostatic and bactericidal capacity equal to or greater than that of *A. mellifera*, against several bacteria strains, both Gram-positive and Gram-negative; however with less action against fungi and yeasts (Cortopassi-Laurino and Gelli 1991; Martins et al. 1997; Grajales-C. et al. 2001; Demera and Angert 2004; Gonçalves et al. 2005; Oliveira et al. 2005).

Tables 11.3 and 11.4 summarize current knowledge of stingless bee honey inhibiting power as compared to Africanized honeybee honey. In these tests, two methods have been used: dilution and application in a Petri dish (Anonymous 1977) and agar diffusion (Bauer et al. 1966). The most tested honeys were those of more productive bees such as *Melipona* and *T. angustula*. The most tested bacteria were *Staphylococcus aureus* and *Pseudomonas aeruginosa* as they are the major infectious agents of wounds and burns.

11.6.4 Honey Microorganisms

There is great interest in the characterization of microorganisms in honey, because it can be used as food or as a component of drugs and cosmetics. The microbial content of honey affects its "shelf life" and its validity for human use. Microorganisms associated with honey are fungi and spore-forming bacteria. Spores are present everywhere, even inside bee nests. They may come from external sources such as

TABLE 11.3

Antibiosis Values of Stingless Bees and Honeybees Honey by Dilution Methods[a] and Application on Petri Dishes

	Stingless Bees						Honeybees
Microorganisms	Msc = 5	Ms = 2	Pl = 1	Ta = 3	Mq = 2	Tc = 1	Am = 20
Bacillus subtilis	3.0	4.13	5.0	3.7	4.0	4.8	2.8
Bacillus subtilis Caron	3.3	3.9	5.0	3.7	4.0	4.0	2.7
Staphylococcus aureus	2.9	3.9	4.8	3.9	4.4	4.0	3.2
Klebisiella pneumoniae	3.1	4.3	5.0	3.3	5.0	5.0	3.0
Pseudomonas aeruginosa	3.0	3.8	5.0	3.8	4.6	5.0	3.1
Escherichia coli	1.7	3.8	5.0	3.3	4.3	4.8	2.0
Bacillus stearothermofilus	4.5	4.5	5.0	5.0	5.0	5.0	4.1

Source: Cortopassi-Laurino, M., and D. S. Gelli, *Apidologie*, 22, 61–73, 1991.

Methodology source: Anonymous, J. *Officiel de la République Française*, 22 avril, 3485–514, 1977.

Note: Species of bees: Msc = *Melipona scutellaris*; Ms = *M. subnitida*; Pl = *Plebeia pugnax*; Mq = *M. quadrifasciata*; Tc = *Tetragona clavipes*.

[a] 5%, 10%, 15%, 20%, and 25%, which correspond to notes 5, 4, 3, 2, 1, respectively.

TABLE 11.4

Antibiosis Value of Meliponini and *Apis* Honey by Agar Diffusion Method[a]

	Meliponini						Apis
Microorganisms	Msc = 1[b]	Ms = 1[b]	S.bip = 1[b]	Nt = 1[c]	Ta = 5[d]	Tl[e]	Am = 3[b]
B. subtilis	10.0	14.5	10.0				13.3
S. aureus	13	22	15.0			3.8	23.5
E. coli	10	28	10.0			2.9	24.0
S. cholerasuis	21	12	13.0				14.8
E. coli				19.0			
Proteus sp				10.0			
Pseudomonas aeruginosa				11.0			
Staphylococcus spp (coag-)				15.0			
Staphylococcus pyogenes				14.0			
							Am = +
Bacillus cereus					7.5		10.0
Pseudomonas aeruginosa					6.8		8.0
Saccharomice cerevisae					15.5	2.7	18
Candida albicans					20.4	2.7	18.0

Source: Bauer, A.W., W. M. M. Kirby, J. C. Sherris, and M. Turk, *Am. J. Clin. Pathol.*, 45, 493–6, 1966.

Note: Meliponini species: Msc = *Melipona scutellaris*; Ms = *M. subnitida*; S.bip = *Scaptotrigona bipunctata*; Nt = *Nannotrigona testaceicornis*; Ta = *Tetragonisca angustula*; Tl = *Trigona laeviceps*; Am = *Apis mellifera*.

[a] Inhibition zone size in 24 hours.

[b] Martins, S. C. S., L. M. B. Albuquerque, J. H. G. Matos, G. C. Silva, and A. I. B. Pereira, *Higiene Alimentar*, 52, 50–3, 1997.

[c] Gonçalves, A. L., A. Alves Filho, and H. Menezes, *Arq. Inst. Biol.*, 72, 455–9, 2005.

[d] Demera, J. H., and E. R. Angert, *Apidologie*, 35, 411–7, 2004.

[e] Chanchao, C., *Pak. J. Med. Sci.*, 25, 364–9, 2009.

pollen, nectar, air, and digestive tract of bees, and can survive in honey (Snowdon and Clever 1996). Secondary sources are those that can be incorporated into honey at any time after it is taken from the nest, but good handling practices and hygiene control these contaminants.

The greatest problem related to the presence of molds and yeasts in honey is fermentation, which results from sugar consumption by yeast, producing byproducts that change the final taste and flavor. The presence of yeasts in stingless bee honey is easily verified, since often they have a characteristic fermentation scent, besides physical identification in pollen grain slides (Barth 1989). Bacteria do not reproduce in honey and a large number of vegetative forms indicate recent contamination of honey is from secondary sources. As honey has antibacterial properties, it is expected to contain a low number and limited diversity of microorganisms. Tables 11.5 and 11.6 show the analyses of microorganism amounts in stingless bee honey. As there are no parameters for this honey, the results only indicate the number of colony forming units (CFU/g or ml). Parameters for honeybee honey in Brazil accept up to 100 CFU/g for fungi and yeasts. In all tested honeys, with a single exception, yeast amounts were higher than that of mold. Results indicate that those honeys from more humid areas tend to have higher values than those from dry regions such as Caatinga and Cerrado (*M. subnitida* and *M. quinquefasciata*, respectively), suggesting environmental influence. Standard bacteria counting (Table 11.6) revealed the same amount in all samples, regardless of bee species and geographic region. The value found, 10^2, regardless of bacteria type, indicates that stingless bee honey is not a sterile product. However, the National Agency for Sanitary Surveillance (Anvisa 2001) accepts the same value in products such as sweeteners, brown sugar, and molasses. Over time, a single honey sample from *T. angustula* exposed to different conditions and time periods showed no significant change in bacteria amount.

From a microbiological point of view, presence of *Bacillus*, yeasts, and molds in honey is considered a common occurrence, since these microorganisms are found in the intestinal microflora of solitary and

TABLE 11.5

Microbiological Analysis of Stingless Bee Honey Collected Aseptically

Species/Locality	Mold CFU/g	Yeast CFU/g	Total Coliform MPN/g	Fecal Streptococci MPN/g
M. fasciculata/MA[a]	1.5	<10.0	<0.18	<0.18[b]
M. fasciculata/PA[a]	2.5	23.5	<0.18	<0.18
M. quadrifasciata/SP	25	615	<0.18	<0.18
M. quinquefasciata/GO	1.5	55	<0.18	<0.18
M. rufiventris/SP	55.0	2.3×10^3	<0.18	<0.18
M. rufiventris/SP	70	255	<0.18	<0.18
M. rufiventris/SP	200	2.5×10^3	<0.18	<0.18
M. subnitida/RN	50	90	<0.18	<0.18
M. subnitida/RN	100	150	<0.18	<0.18
Tetragona clavipes/SP	<1	7.0×10^3	<0.18	<0.18
Tetragona clavipes/SP	50	3.3×10^3	<0.18	<0.18
Tetragona clavipes/SP	100	1.4×10^3	<0.54	<0.54
Tetragona clavipes/SP	<1	5.5	<0.18	<0.18
Melipona sp/AM[a]	2	3.0	<0.18	<0.18
S. depilis/Uruguay[a]	1.0×10^3	1.29×10^5	<0.18	<0.18
M. fuscopilosa/AC	<1.0	1.81×10^3	<0.1.8	<0.18
M. fuscopilosa/AC[c]	3	<1.0	<0.18	<0.18
M. crinita/AC	2×10^4	1.72×10^6	<0.18	<0.18

Note: CFU = colony forming unit according to Cetesb standard technique L5204; MPN = most probable number according to standard methods-APHA 2005; AC = Acre state; SP = São Paulo state; PA = Pará state; MA = Maranhão state; GO = Goiás state; RN = Rio Grande do Norte state; AM = Amazonas state. Technical advice: Elayse Maria Hachich from Microbiology and Parasitology Laboratory of Cetesb, São Paulo.

[a] Producer.

[b] <0.18 = absence of contamination within tests limits.

[c] Heated honey.

social bees, and its amount varies with bee age (function), seasons, food diets (deficient), and nest exposure to pesticides (Gilliam 1997). *Bacillus* species produce antimicrobial substances and enzymes, as do molds, and yeasts are the most important contributors of substances from a nutritional standpoint (Pain and Maugenet 1966). The questions here are which are those parameters limits and which are nonpathogenic and pathogenic microorganisms that can be found in stingless bee honey. Several species have already been studied in relation to pollen, honey, or larval food microflora: *Dactylurina staudingeri* (Gribodo), *T. hypogea*, *M. quadrifasciata*, *Melipona fasciata* Latreille, *T. angustula*, and *Frieseomelitta varia* (Lepeletier) (Machado 1971; Delage-Darchen and Darchen 1984; Gilliam et al. 1985, 1990; Rosa et al. 2003).

Of the 12 honey samples tested from Southeastern Brazil (Table 11.6), three indicated the presence of total coliforms (environmental), but not of fecal coliforms. More rigorous testing of presence/absence (P/A), which use samples ten times larger (10 g), showed one positive result for *E. coli* (fecal coliforms), three for *Enterococcus*, also of fecal origin, and six for *B. cereus*. *E. coli*, whose specific habitat is the gut of warm-blooded animals, does not multiply in nature and can be naturally found in honey if bees collect any material in creeping plants. In Table 11.5, samples analyzed with another method (NPM) also indicated the absence of contamination within the tests limits. Even in these samples, *Salmonella* sp., *S. aureus*, and *P. aeruginosa* were not found. These results open a perspective for the consumption of stingless bee honey, because some species have been observed visiting animal wastes and carcasses (Nogueira-Neto 1997), and therefore it was believed that their honey could contain large amount of fecal coliforms. If bees use this material in nests, it should be used in a restricted place, not in the food storage area, or honey eliminates these microorganisms with its antibacterial properties.

TABLE 11.6

Microbiological Analysis of Stingless Bee Honey Aseptically Collected from Southeastern Brazil

Species	Bacteria CFU/ml	Total Coliform MPN	Fecal Coliform MPN
Tetragonisca angustula	0.32×10^2	7.3×10^2	0
Tetragonisca angustula	0.51×10^2	39×10^2	0
Melipona bicolor	$>3 \times 10^2$	0	0
Melipona bicolor	$>3 \times 10^2$	0	0
Plebeia sp	0.2×10^2	0	0
Plebeia sp	$>3 \times 10^2$	0	0
Nannotrigona testaceicornis	$>3 \times 10^2$	0	0
Nannotrigona testaceicornis	$>3 \times 10^2$	0	0
Melipona subnitida	0.64×10^2	0	0
Melipona subnitida	0.18×10^2	0	0
Tetragonisca angustula	0.15×10^2	2.4×10^2	0
Tetragonisca angustula 1 day	5.6×10^2	—	—
Tetragonisca angustula 7 days in freezer	10×10^2	—	—
Tetragonisca angustula 7 days in environment	14×10^2	—	—

Note: CFU = colony forming unit according to Cetesb standard technique L5204; MPN = most probable number according to standard methods-ALPHA, 2005. Technical advice: Dilma S. Gelli and Harumi Sakuma from Microbiology Laboratory of Adolfo Lutz Institute, São Paulo, Brazil.

11.7 Final Considerations

Studies on the feeding habits of the social Apidae have contributed specifically to the understanding of energetics or foraging economy of these animals. Foragers of *Apis*, *Bombus*, and Meliponini are relatively easy to manage in field and laboratory, fitting well to the goals of controlled experiments where behaviors, benefits, and costs during foraging are analyzed. Information thus obtained refers to the discussion of an "optimal foraging theory," perhaps a controversy in itself (able to encompass the exceptions and dependent on them to explain the improvement of consumers in the evolutionary flow towards optimization) but without doubt, a biological paradigm.

Colonies of social Apidae are at the center of foraging economics both spatially (the fixed point for displacement) and behaviorally (changing foragers behavior). In a retrospective of the ecology of *A. mellifera*, Seeley (1985) notes that studies on colony functioning are well advanced, while investigations about historical conditions that favor emergence and establishment of specific responses (e.g., an elaborated communication system) began to appear only in the late 20th century. There is an intersection between physiological behavioral approaches (why a particular type of colony functions) and behavioral ecology (why a certain type of functioning was selected). In this new phase, intensification of studies in tropical regions is crucial, because these environments are the molds on which complex ecological mechanisms arose, and many geographic variants of *A. mellifera* and hundreds of species of stingless bees were differentiated.

When populations are isolated by any barriers, they start independent evolutionary histories. Among these barriers, genetic differentiation has often irreversible ecological and population effects. Thus, in Meliponini, hundreds of species with independent evolutionary histories share basic characteristics of the common ancestor, have a wide geographical distribution, and often occupy the same habitat. Given these facts, there is one basic question: what mechanisms regulate this coexistence?

In terms of feeding ecology, each species of stingless bees brings more or less altered "solutions" already encountered by its ancestors and that overlap with its own acquisitions, so that each colonial system works and acts on the environment, repeating in part the need to maintain foraging efficiency in different habitats or food sources, and differentiation of food habits to escape interspecific pressures represented by ancestral characteristics. The apparent contradiction between these two ecological goals was probably settled by morphological and functional diversification, often subtle, but still feasible in economic terms, allowing specific strategies for use of floral food sources and occupation of different habitats. Nevertheless, comparisons among most local Meliponini communities show relatively moderate variations in the number of coexisting species, indicating that there are also narrower limits for generalist social bees packaging in ecosystems.

In recent years, information on stingless bees' feeding habits have accumulated, but still with many basic gaps in view of the large number of species. In addition, there were few attempts to relate the expression of morphofunctional characteristics to food availability conditions. Thus, tracing parallels on how to allocate resources between colonies of closely related (e.g., same genus) and unrelated species is an open field for research that undoubtedly will help to understand the behavioral and ecological mechanisms that made coexistence possible, and therefore were important for any differences in feeding habits (e.g., floral preferences) and for finding specific solutions in colonial functioning (e.g., type of communication system).

Solving basic questions on how stingless bee species manage to coexist in the same locality will also be relevant for stingless bee management and utilization on applied fields, in terms of their use on crop pollination, since they are already known as effective pollinators of dozens of plant species.

REFERENCES

Absy, M. L., and W. E. Kerr. 1977. Algumas plantas visitadas para obtenção de pólen por operárias de *Melipona seminigra merrilae* em Manaus. *Acta Amazônica* 7:303–15.

Alford, D. V. 1975. *Bumblebees*. London: Davis-Poynter.

Almeida, D., and L. C. Marchini. 2006. Características físico-químicas de amostras de mel da abelha jataí (*Tetragonisca angustula*) provenientes do Município de Piracicaba-SP. Paper presented at the XVI Congesso Brasileiro de Apicultura, Aracaju.

Almeida-Muradian, L. B., A. H. Matsuda, and D. H. M. Bastos. 2007. Physicochemical parameters of Amazon Melipona honey. *Quim. Nova* 30:707–8.

Alves, D. F. S., F. C. J. Cabral, R. M. J. Oliveira, A. C. M. Rego, and A. C. Medeiros. 2008. Efeitos da aplicação tópica do mel de *Melipona subnitida* em feridas infectadas de ratos. *Rev. Col. Bras. Cir.* 35:188–93.

Alves, R. M. O., C. A. L. Carvalho, B. A. Souza, G. S. Sodré, and L. C. Marchini. 2005. Características físico-químicas de amostras de mel de *Melipona mandacaia* Smith (Hymenoptera: Apidae). *Ciênc. Tecnol. Aliment.* 25:644–50.

Anacleto, D. A., B. A. Souza, L. C. Marchini, and A. C. C. C. Moreti. 2009. Composição de amostras de mel da abelha Jataí (*Tetragonisca angustula* Latreille, 1811). *Ciênc. Tecnol. Aliment.* 29: 535–41.

Anonymous. 1977. Méthodes officielles d'analyse du miel. *J. Officiel de la République française 22 avril*, 3485–514.

Antonini, Y., R. G. Costa, and R. P. Martins. 2006. Floral preferences of a neotropical stingless bee, *Melipona quadrifasciata* Lepeletier (Apidae:Meliponina) in an urban forest fragment. *Braz. J. Biol.* 66:463–71.

Anvisa. 2001. Resolução nº 12, 2 de janeiro de 2001. Regulamento técnico sobre os padrões microbiológicos para alimentos. *Diário Oficial [da] República Federativa do Brasil*, Brasília, DF, 10 jan. 2001. Seção 1.

Araújo, E. D., M. Costa, J. Chaud-Netto, and H. G. Fowler. 2004. Body size and flight distance in stingless bees (Hymenoptera: Meliponini): inference of flight range and possible ecological implications. *Braz. J. Biol.* 64:563–8.

Azevedo, G. G. 1997. Atividade de vôo e determinação do número de ínstares larvais em *Partamona helleri* (Friese) (Hymenoptera, Apidae, Meliponinae). M.Sc. dissertation, Universidade Federal de Viçosa, Brazil.

Barth, M. O. 1989. *O Pólen no Mel Brasileiro*. Rio de Janeiro, Brazil: Editora Luxor.

Batista, M. A., M. Ramalho, and A. E. E. Soares. 2003. Nesting sites and abundance of Meliponini (Hymenoptera: Apidae) in heterogeneous habitats of the Atlantic Rain Forest, Bahia, Brazil. *Lundiana* 4:19–23.

Bauer, A. W., W. M. M. Kirby, J. C. Sherris, and M. Turk. 1966. Antibiotic susceptibility testing by a standardized single disk metodo. *Am. J. Clin. Pathol.* 45:493–6.

Bazlem, K. 2000. Charakterisierung von Honigen Stachelloser Bienen aus Brasilien. PhD thesis, Fakultäd für Biologie der Eberhard Karls, Universität Tubingen aus Stuttgart, Deutschland.

Beetsma, J. 1979. The process of queen–worker differentiation in the honeybee. *Bee World* 60:24–39.

Biesmeijer, J. C., and E. J. Slaa. 2006. The structure of eusocial bee assemblages in Brazil. *Apidologie* 37:240–58.

Bijlsma, L., L. L. M. Bruijn, E. P. Martens, and M. J. Sommeijer. 2006. Water content of stingless bees honeys (Apidae, Meliponini): Interspecific variation and comparison with honey of *Apis mellifera*. *Apidologie* 37:480–6.

Bogdanov, S., P. Martin, and C. Lullmann. 1997. Harmonised methods of the European honey commission. *Apidologie* (extra issue): 1–59.

Boorn, K. L., Y. Y. Khor, E. Sweetman, F. Tan, T. A. Heard, and K. A. Hammer. 2009. Antimicrobial activity of honey from the stingless bee *Trigona carbonaria* determined by agar diffusion, agar dilution, broth microdilution and time-kill methodology. *J. Appl. Microbiol.* 108:1534–1543.

Brasil. 2000. Regulamento técnico de identidade e qualidade de mel. *Ministério da Agricultura e do Abastecimento*. Instrução Normativa n° 11, 20 de outubro de 2000.

Breed, M. D., T. P. Mcglynn, M. D. Sanctuary, E. M. Stocker, and R. Cruz. 1999. Distribution and abundance of selected meliponine species in a Costa Rican tropical wet forest. *J. Trop. Ecol.* 15:765–77.

Brian, M. V. 1957. Caste determination in social insects. *Annu. Rev. Ent.* 2:107–20.

Browers, E. V. M. 1984. Glucose/fructose ratio in the food of honeybee larvae during caste differentiation. *J. Apic. Res.* 23:94–101.

Browers, E. V. M., R. Ebert, and J. Beetsma. 1987. Behavioral and physiological aspects of nurse bees in relation to the composition of larval food during caste differentiation in the honeybee. *J. Apic. Res.* 26:11–23.

Bruijn, de L. L. M., and M. J. Sommeijer. 1997. Colony foraging in different species of stingless bees (Apidae, Meliponinae) and the regulation of individual nectar foraging. *Insectes Soc.* 44:35–47.

Buchmann, S. L. 1986. Vibratile pollination in *Solanum* and *Lycopersicon:* A look at pollen chemistry. In *Solanaceae: Biology and Systematic*, ed. W.G. D'Arcy, 237–252. New York: Columbia University Press.

Camargo, J. M. F., and D. Roubik. 1991. Systematics and bionomics of the apoid obligate necrophages: The *Trigona hypogea* group. *Biol. J. Linean Soc.* 44:13–39.

Camargo, J. M. F., and D. Wittmann. 1989. Nest architecture and distribution of the primitive stingless bee, *Mourella caerulea* (Hymenoptera, Apidae, Meliponinae): Evidence for the origin of *Plebeia* (s.lat.) on the Gondwana Continent. *Stud. Neotrop. Fauna & Environm.* 24:213–29.

Camargo, R. C. R., M. S. Brito Neto, J. G. Ribeiro, M. C. Azevedo, A. L. H. Barreto, F. M. Pereira, and M. T. R. Lopes 2006. Avaliação da qualidade do mel de jandaíra (*Melipona subnitida* Ducke) produzido em área Resex do Delta do Parnaíba, por meio de Análises Físico-Química. Paper presented at the XVI Congresso Brasileiro de Apicultura, Aracaju, SE.

Campos, G., R. C. Della-Modesta, T. J. P. Silva, K. E. Baptista, M. F. Gomides, and R. L. Godoy. 2003. Classificação do mel em floral ou mel de melato. *Ciênc. Tecnol. Aliment.* 23:1–5.

Carvalho, C. A. L., G. S. Sodré, B. A. Souza, A. A. O. Fonseca, S. M. P. Cavalcante, G. A. Oliveira and L. C. Marchini. 2006. Composição físico-químicas de méis de diferentes espécies de abelhas sem ferrão provenientes da Ilha de Itaparica, Bahia. Paper presented at the XVI Congresso Brasileiro de Apicultura, Aracaju, SE.

Castilho-Hyodo, V. C. 2002. Rainha ou operária? Um ensaio sobre a determinação de castas em Schwarziana quadripunctata (Lepeletier, 1836) (Hymenoptera, Apidae, Meliponini). Ph.D. thesis, Universidade de São Paulo, Brazil.

Cavalcante, S. M. P., G. S. Sodré, C. A. L. Carvalho, A. A. O. Fonseca, B. A. Souza, G. A. Oliveira, and T. B. A. Santos. 2006. Características físico-químicas de méis de *Melipona scutellaris* de diferentes Municípios do Estado da Bahia. Paper presented at the XVI Congresso Brasileiro de Apicultura, Aracaju, SE.

Cavasin-Oliveira, G., and C. Cruz-Landim. 1991. Aspectos ultra-estruturais da glândula hipofaríngea de operária de *Trigona hypogea* (Hymenoptera, Apidae, Meliponinae). Paper presented at the XIII Colóquio Sociedade Brasileira de Microscoscopia Elctrônica, Caxambu, MG.

Chanchao, C. 2009. Antimicrobial activity by *Trigona laeviceps* (Stingless bee) honey from Thailand. *Pak. J. Med. Sci.* 25:364–9.

Cortopassi-Laurino, M. 1997. Comparing some physico-chemical parameters between stingless bee and Africanized Apis mellifera honeys from Brazil. p. 351. Paper presented at the XXXV International Apicultural Congress of Apimondia, Antwerp, Belgium.

Cortopassi-Laurino, M., and D. S. Gelli. 1991. Analyse pollinique, propriété physico-chimiques et action antibactérienne des miels d´abeille africanisées *Apis mellifera* et de Méliponinés du Brésil. *Apidologie* 22:61–73.

Cortopassi-Laurino, M., V. L. Imperatriz-Fonseca, D. W. Roubik, A. Dollin, T. Heard, I. Aguilar, C. Eardley, and P. Nogueira-Neto 2006. Global meliponiculture: Challenges and opportunities. *Apidologie* 37:275–92.

Cortopassi-Laurino, M., and M. Ramalho. 1988. Pollen harvest by Africanized *Apis mellifera* and *Trigona spinipes* in São Paulo: botanical and ecological views. *Apidologie* 19:1–24.

Crane, E. 1987. *O Livro do Mel*. São Paulo, Brazil: Editora Nobel.

Dardón, M. J., and E. Enríquez. 2008. Caracterización fisicoquímica y antimicrobiana de la miel de nueve especies de abejas sin aguijón (Meliponini) de Guatemala. *Interciência* 33:916–22.

Darchen, R., and B. Delage-Darchen. 1970. Facteur determinant les castes chez les trigones (Hym., Apidae). *C.R. Acad. Sci. Paris D* 270:1372–3.

Darchen, R., and B. Delage-Darchen. 1971. Le déterminisme des castes chez les Trigones (Hyménoptères Apidés). *Insectes Sociaux* 18:121–34.

Delage-Darchen, B., and R. Darchen. 1984. Les enzymes digestives de diverses abeilles socials et en particulier de *Dactylurina staudingeri*. *Publications Scientifiques Aceleres, Université René Descartes* 5:28.

Demera, J. H., and E. R. Angert. 2004. Comparison of the antimicrobial activity of honey produced by *Tetragonisca angustula* (Meliponinae) and *Apis mellifera* from different phytogeographic regions of Costa Rica. *Apidologie* 35:411–7.

Engels, W., and V. L. Imperatriz-Fonseca. 1990. Caste development, reproductive strategies and control of fertility in honeybees and in stingless bees. In *Social insects, An Evolutionary Approach to Caste Reproduction*, ed. W. Engels, 167–230. Berlin, Germany: Springer Verlag.

Faegri, K., and L. van der Pijl. 1979. *The Principles of Pollination Ecology*, 3rd ed. Oxford, UK: Pergamon Press.

Faustino, C. D., E. V. S. Matos, S. Mateus, and R. Zucchi. 2002. First record of emergency queen rearing in stingless bees. *Insectes Soc.* 49:111–13.

Fidalgo, A. O., and A. M. P. Kleinert. 2007. Foraging behavior of *Melipona rufiventris* Lepeletier (Apinae, Meliponini) in Ubatuba/SP, Brazil. *Braz. J. Biol.* 67:137–44.

Fowler, H. G. 1979. Responses by stingless bee to a subtropical environment. *Rev. Biol. Trop.* 27:111–8.

Free, J. B. 1966. The pollinating efficiency of honey-bee visits to apple flowers. *J. Hortic. Sci.* 41:91–94.

Free, J. B., A. W. Ferguson, and J. R. Simpkins. 1989. The effect of different periods of brood isolation on subsequent brood–cell visits by worker honeybees (*Apis mellifera* L.). *J. Apic. Res.* 28:22–5.

Gary, N. E. 1967. Diurnal variation in the intensity of flight activity from honeybee colonies *J. Apic. Res.* 6:65–8.

Génissel, A., P. Aupinel, C. Bressac, J. N. Tasei, and C. Chevrier. 2002. Influence of pollen origin on performance of *Bombus terrestris* micro-colonies. *EE et Applicata* 104:329–36.

Guerrini, A., R. Bruni, S. Maietti, F. Poli, D. Rossi, G. Paganetto, M. Muzzoli, L. Scalvenzi, and G. Sacchetti. 2009. Ecuadorian stingless bees (Meliponinae) honey: A chemical and functional profile of an ancient health product. *Food Chemistry* 114:1413–20.

Gilliam, M. 1997. Identification and roles of non-pathogenic microflora associated with honey bees. *FEMS Microbiology Letters* 155:1–10.

Gilliam, M., S. L. Buchmann, and B. J. Lorenz. 1985. Microbiology of the larval provisions of the stingless bee, *Trigona hypogea*, an obligate necrophage. *Biotropica* 17:28–31.

Gilliam, M., and H. L. Morton. 1978. Bacteria belonging to genus *Bacillus* isolated from honey bees, *Apis mellifera*, fed 2,4-d and antibiotics (1). *Apidologie* 9:213–21.

Gilliam, M., D. W. Roubik, and B. J. Lorenz. 1990. Microorganisms associated with pollen, honey, and brood provisions in the nest of a stingless bee, *Melipona fasciata*. *Apidologie* 21:89–97.

Gonçalves, A. L., A. Alves Filho, and H. Menezes. 2005. Atividade antimicrobiana do mel da abelha nativa sem ferrão *Nannotrigona testaceicornis* (Hymenoptera:Apidae, Meliponini). *Arq. Inst. Biol.* 72:455–9.

Grajales-C, J., M. Rincon, R. Vandame, A. Santiesteban, and M. Guzman. 2001. Caracteristicas físicas, químicas y efecto microbiológico de mieles de meliponinos y *Apis mellifera* de la Region Soconusco, Chiapas. Memórias II Seminário Mexicano sobre Abejas sin Aguijón. Mérida, Yucatán, México, p. 61–66.

Guibu, L. S., and V. L. Imperatriz-Fonseca. 1984. Atividade externa de *Melipona quadrifasciata quadrifasciata* Lepeletier (Hymenoptera, Apidae, Meliponinae). *Cienc. Cult.* 36:623.

Hartfelder, K., and W. Engels. 1989. The composition of larval food in stingless bees: Evaluating nutritional balance by chemosystematic methods. *Insectes Soc.* 36:1–14.

Hartfelder, K., G. R. Makert, C. C. Judice, G. A. G. Pereira, W. C. Santana, R. Dallacqua, and M. M. G. Bitondi. 2006. Physiological and genetic mechanisms underlying caste development, reproduction and division of labor in stingless bees. *Apidologie* 37:144–63.

Heinrich, B. 1979. *Bumblebee Economics*. Cambridge, MA: Harvard University Press.

Hilário, S. D., V. L. Imperatriz-Fonseca, and A. M. P. Kleinert. 2000. Flight activity and colony strength in the stingless bee *Melipona bicolor bicolor* (Apidae, Meliponinae). *Rev. Bras. Biol.* 60:299–306.

Hilário, S. D., V. L. Imperatriz-Fonseca, and A. M. P. Kleinert. 2001. Responses to climatic factors by foragers of *Plebeia pugnax* Moure (in litt.) (Apidae, Meliponinae). *Rev. Bras. Biol.* 61:191–6.

Hofstede, F. E., and M. J. Sommeijer. 2006. Effect of food availability on individual foraging specialization in the stingless bee *Plebeia tobagoensis* (Hymenoptera, Meliponini). *Apidologie* 37:387–97.

Huang, Z. Y., and G. W. Otis. 1991. Nonrandom visitation of brood cells by worker honeybees (Hymenoptera: Apidae). *J. Ins. Behav.* 4:177–84.

Hubbell, S. P., and L. K. Johnson. 1977. Competition and nest spacing in a tropical stingless bee community. *Ecology* 58:949–65.

Hubbell, S. P., and L. K. Johnson. 1978. Comparative foraging behavior of six stingless bees species exploiting a standardized resource. *Ecology* 59:1123–36.

Imperatriz-Fonseca, V. L., M. A. C. Oliveira, and S. Iwama. 1975. Notas sobre o comportamento de rainhas virgens de *Plebeia (Plebeia) remota* Holmberg (Apidae, Meliponinae). *Cienc. e Cult.* 27:665–9.

Imperatriz-Fonseca, V. L., A. Kleinert-Giovannini, M. Cortopassi-Laurino, and M. Ramalho. 1984. Hábitos de coleta de *Tetragonisca angustula angustula* Latreille (Hymenoptera, Apidae, Meliponinae). *Bol. Zool. Univ. S. Paulo* 8:115–31.

Imperatriz-Fonseca, V. L., A. Kleinert-Giovannini, and J. T. Pires. 1985. Climate variations influence on the flight activity of Plebeia remota Holmberg (Hym., Apidae, Meliponinae). *Rev. Bras. Ent.* 29:427–34.

Imperatriz-Fonseca, V. L., and P. Darakjian. 1994. Flight activity of *Schwarziana quadripunctata quadripunctata* (Apidae, Meliponinae): Influence of environmental factors. Paper presented at International Behavioural Ecology Congress, Nottingham (UK).

Iwama, S. 1977. Coleta de alimentos e qualidade do mel de *Tetragonisca angustula angustula* Latreille (Apidae, Meliponinae). M.Sc. dissertation, Universidade de São Paulo, Brazil.

Johnson, L. K., and S. P. Hubbell. 1974. Aggression and competition among stingless bees: Field studies. *Ecology* 55:120–7.

Johnson, L. K., and S. P. Hubbell. 1975. Contrasting foraging strategies and coexistence of two bee species on a single resource. *Ecology* 56:1398–1406.

Johnson, L. K. 1982. Patterns of communication and recruitment in stingless bees. In *The Biology of Social Insects*, ed. M. D. Breed, C. D. Michener, and H. E. Evans, 323–34. Boulder, CO: Westview Press.

Johnson, L. K. 1983. Foraging strategies and the structure of stingless bee communities in Costa Rica. In *Social Insects in the Tropics 2*, ed. P. Jaisson, 31–58. Paris, France: Université Paris-Nord.

Johnson, L. K., S. P. Hubbell, and D. H. Feener. 1987. Defense of food supply by eusocial colonies. *Am. Zool.* 27:347–58.

Judice, C., K. Hartfelder, and G. A. G. Pereira. 2004. Caste specific gene expression profile in the stingless bee *Melipona quadrifasciata*—Are there common patterns in highly eusocial bees. *Insectes Soc.* 51:352–8.

Jung-Hoffman, J. 1966. Die Determination von Königin und Arbeiterin der Honigbiene. *Z. Bienenforsch.* 8:296 322.

Katayma, E. 1966. Studies on the development of the broods of *Bombus diversus* Smith (Hymenoptera, Apidae). II. Brood development and feeding habits. *Kontyû* 34:8–17.

Katayma, E. 1973. Observations on the brood development in *Bombus ignitus* (Hymenoptera, Apidae). II. Brood development and feeding habits. *Kontyû* 41:203–16.

Katayma, E. 1975. Egg laying habits and brood development in *Bombus hypocrita* (Hymenoptera, Apidae). II. Brood development and feeding habits. *Kontyû* 43:478–96.

Kerr, W. E. 1946. Formação de castas no gênero *Melipona* (Illiger, 1806)–nota prévia. *Anais Esc. Sup. Agric. "Luiz de Queiroz"* 3:299–312.

Kerr, W. E. 1948. Estudos sobre o gênero *Melipona*. *Anais Esc. Sup. Agric. "Luiz de Queiroz"* 5:181–291.

Kerr, W. E. 1950a. Evolution of the mechanism of caste determination in the genus *Melipona*. *Evolution* 4:7–13.

Kerr, W. E. 1950b. Genetic determination of castes in the genus *Melipona*. *Genetics* 35:143–152.

Kerr, W. E., and R. A. Nielsen. 1966. Evidences that genetically determined *Melipona* queens can become workers. *Genetics* 54:859–66.

Kerr, W. E. 1969. Some aspects of the evolution of social bees (Apidae). In *Evolutionary Biology Vol. 3*, ed. T. Dobzhansky, M. K. Hecht, and W. C. Steere, 119–175. New York: Appleton-Century Crofts.

Kerr, W. E., A. C. Stort, and M. J. Montenegro. 1966. Importância de alguns fatores ambientais na determinação de castas do gênero *Melipona*. *An. Acad. Bras. Ciênc.* 38:149–68.

Kleinert-Giovannini, A. 1982. The influence of climatic factors on flight activity of Plebeia emerina (Hym., Apidae, Meliponinae) in winter. *Rev. Bras. Entomol* 26:1–13.

Kleinert-Giovannini, A., and V. L. Imperatriz-Fonseca. 1986. Flight activity and climatic conditions: responses by two subspecies of *Melipona marginata* Lepeletier (Apidae, Meliponinae). *J. Apic. Res.* 25:3–8.

Kleinert-Giovannini, A., and V. L. Imperatriz-Fonseca 1987. Aspects of the trophic niche of *Melipona marginata marginata* (Apidae, Meliponinae). *Apidologie* 18:69–100.

Le Comte, Y., L. Sreng, and S. H. Poitout. 1995. Brood pheromone can modulate the feeding behavior of *Apis mellifera* workers (Hymenoptera: Apidae). *J. Econ. Entomol.* 88:798–804.

Levin, D. A. 1978. Pollinator behavior and the feeding structure of plant-population. In *The Pollination of Flowers by Insects*, ed. A. J. Richards, 133–150. New York: Academic Press.

Lindauer, M. 1952. Ein Beitrag zur Frage der Arbeitsteilung im Bienenstaat. *Z. vergl. Physiol.* 34:299–345.

Lindauer, M., and W. E. Kerr. 1960. Communications between workers of stingless bees. *Bee World* 41:29–41.

Machado, J. O. 1971. Simbiose entre as abelhas sociais brasileiras (Meliponinae, Apidae) e uma espécie de bactéria. *Cienc. Cult.* 23:625–33.

Makert, G. R., R. J. Paxton, and K. Hartfelder 2006. An optimized method for the generation of AFLP markers in a stingless bee (*Melipona quadrifasciata*) reveals a high degree of genetic polymorphism. *Apidologie* 37:687–98.

Martins, C. F., A. C. A. Moura, and M. R. V. Barbosa. 2003. Bee plants and relative abundance of corbiculate Apidae species in a Brazilian Caatinga area. *Rev. Nordestina Biol.* 17:63–74.

Martins, S. C. S., L. M. B. Albuquerque, J. H. G. Matos, G. C. Silva, and A. I. B. Pereira. 1997. Atividade antibacteriana em méis de abelhas africanizadas (*Apis mellifera*) e nativas (*Melipona scutellaris, Melipona subnitida e Scaptotrigona bipunctata*) do Estado do Ceará. *Higiene Alimentar* 52:50–3.

Mateus, S., and F. B. Noll. 2004. Predatory behavior in a necrophagous bee, *Trigona hypogea* (Hymenoptera, Apidae, Meliponini). *Naturwiss.* 91:94–6.

Michener, C. D. 1974. *The Social Behavior of the Bees: A Comparative Study*. Cambridge, MA: Belknap Press of Harvard University Press.

Monteiro, D., and M. Ramalho. 2010. Abelhas generalistas (Meliponina) e sucesso reprodutivo de árvores com florada em massa de *Stryphnodendron pulcherrimum* (Willd.) Hochr., (Fabales-Mimosaceae) na Mata Atlântica (Bahia). *Neotrop. Entomol.* 39(4):519–26.

Moo-Valle, H., J. J. G. Quezada-Euán, and T. Wenseleers. 2001. The effect of food reserves on the production of sexual offspring in the stingless bee *Melipona beecheii* (Apidae, Meliponini). *Insectes Soc.* 48:398–403.

Moreti, A. C. C. C., L. C. Marchini, V. C. Souza, and R. R. Rodrigues. 2002. *Atlas de Pólen de Plantas Apícolas*. Rio de Janeiro, Brazil: Papel Virtual Editora.

Moritz, R. F. A. 1994. Nourishment and sociality in honeybees. In *Nourishment and Evolution in Insectes Societies*, ed. J. H. Hunt and C. A. Nalepa, 345–90. Boulder, CO: Westview Press.

Nates-Parra, G. 2006. Biodiversidad y meliponicultura en el Piedemonte Llanero, Meta, Colombia. In: VII Encontro sobre Abelhas, USP-Ribeirão Preto-SP. CD–ROM.

Nogueira-Neto, P. 1997. *Vida e Criação de Abelhas Indígenas sem Ferrão*. São Paulo, Brazil: Editora Nogueirapis.

Noll, F. B., R. Zucchi, J. A. Jorge, and S. Mateus. 1997. Food collection and maturation in the necrophagous stingless bee, *Trigona hypogea* (Hymenoptera, Meliponinae). *J. Kansas Ent. Soc.* 69:287–93.

Nuñez, J. A. 2000. Foraging efficiency and survival of African honeybees in the tropics. *Annals* IV Encontro Sobre Abelhas, Ribeirão Preto-SP, 9–16.

Oliveira, G. A., G. S. Sodré, C. A. L. Carvalho, B. A. Souza, S. M. P. Cavalcante, and A. A. O. Fonseca. 2006. Análises físico-químicas de méis de *Melipona quadrifasciata* do semi-árido da Bahia. Paper presented at the XVI Congresso Brasileiro de Apicultura, Aracaju, SE.

Oliveira, G. E., M. C. P. Costa, A. R. Nascimento, and V. M. Neto. 2005. Qualidade microbiológica do mel de tiúba (*Melipona compressipes fasciculata* Smith) produzido no Estado do Maranhão. *Higiene Alimentar* 19:92–133.

Oliveira, G. E., L. M. S. Silveira, A. R. Nascimento, and V. M. Neto. 2006. Avaliação de parâmetros físico-químicos do mel de tiúba (*Melipona compressipes fasciculata* Smith) produzido no Estado do Maranhão. *Higiene Alimentar* 20: 74–81.

Oliveira, M. A. C. 1973. Algumas observações sobre a atividade externa de *Plebeia saiqui* e *Plebeia droryana*. MSc dissertation, Universidade de São Paulo, Brazil.

Pain, J., and J. Maugenet. 1966. Recherches biochimiques et physiologiques sur le pollen emmaganisé para les abeilles. *Ann. Abeilles* 9:209–36.

Pereboom, J. J. M., and T. Shivashankar. 1994. Larval food processing by adult bumble workers. Paper presented at the XII Congress of IUSSI, Paris.

Pereboom, J. J. M. 1996. Food, feeding and caste differentiation in bumble bees. Paper presented at the XX Congress of Entomology, Firenze.

Pereboom, J. J. M. 1997. "… while they banquet splendidly the future mother…" The significance of trophogenic and social factors on caste determination and differentiation in the bumblebee *Bombus terrestris*. Ph.D. thesis, University of Utrecht.

Pereboom, J. J. M. 2000. The composition of larval food and the significance of exocrine secretions in the bumblebee *Bombus terrestris*. *Insectes Soc.* 47:11–20.

Persano-Oddo, L., T. A. Heard, A. Rodríguez-Malaver, R. A. Perez, M. Fernández-Muiño, M. T. Sancho, G. Sesta, L. Lusco, and P. Vit 2008. Composition and antioxidant activity of *Trigona carbonaria* honey from Australia. *J. Med. Food* 11:789–94.

Pirani, J. R., and M. Cortopassi-Laurino. 1993. *Flores e abelhas em São Paulo*. São Paulo, Brazil: Editora Universidade de São Paulo.

Pitombeira, J. S., M. C. Liberato, and N. F. Carvalho. 2009. Análise físico-química e fitoquímica de méis da abelha jandaíra (*Melipona subnitida* D.) oriundo da cidade de Mossoró-RN. Paper presented at the IX ENPPG, Ceará.

Plowright, R. C., and S. C. Jay. 1968. Caste differentiation in bumble bees (*Bombus* Latr.: Hym.). I. The determination of female size. *Insectes Soc.* 15:171–92.

Plowright, R. C., and S. C. Jay. 1977. On the size determination of bumble bee castes (Hymenoptera: Apidae). *Can. J. Zool.* 55:1133–8.

Plowright, R. C., and B. A. Pendrel. 1977. Larval growth in bumble bees (Hymenoptera: Apidae). *Can. Ent.* 109:967–73.

Pyke, G. H. 1984. Optimal foraging theory: A critical review. *Annu. Rev. Ecol. Syst.* 15:523–75.

Ramalho, M., V. L. Imperatriz-Fonseca, A. Kleinert-Giovannni, and M. Cortopassi-Laurino. 1985. Exploitation of floral resources by *Plebeia remota* Holmberg (Apidae, Meliponinae). *Apidologie* 16:305–30.

Ramalho, M., A. Kleinert-Giovannini, and V. L. Imperatriz-Fonseca. 1989. Utilization of floral resources by species of *Melipona* (Apidae, Meliponinae): Floral preferences. *Apidologie* 20:185–95.

Ramalho, M. 1990. Foraging by stingless bees of the genus *Scaptotrigona* (Apidae, Meliponinae). *J. Apic. Res.* 29:61–7.

Ramalho, M., A. Kleinert-Giovannini, and V. L. Imperatriz-Fonseca. 1990. Important bee plants for stingless bees (*Melipona* and Trigonini) and Africanized honeybees (*Apis mellifera*) in neotropical habitats: A review. *Apidologie* 21:469–88.

Ramalho, M., A. Kleinert-Giovannni, and V. L. Imperatriz-Fonseca. 1991. Ecologia alimentar de abelhas sociais. In *Ecologia nutricional de insetos e suas implicações no manejo de pragas*, ed. A. R. Panizzi and J. R. P. Parra, 225–52. São Paulo, Brazil: Editora Manole.

Ramalho, M., T. C. Giannini, K. S. Malagodi-Braga, and V. L. Imperatriz-Fonseca. 1994. Pollen harvest by stingless bee foragers (Hymenoptera, Apidae, Meliponinae). *Grana* 33:239–244.

Ramalho, M. 1995. Diversidade de abelhas (Apoidea, Hymenoptera) em um remanescente de Floresta Atlântica, em São Paulo. São Paulo, USP. Ph.D. thesis, Universidade de São Paulo, Brazil.

Ramalho, M., V. L. Imperatriz-Fonseca, and T. C. Giannini. 1998. Within colony-size variation of foragers and pollen load capacity in the stingless bee *Melipona quadrifasciata anthidioides* Lepeletier (Apidae, Hymenoptera). *Apidologie* 29:221–8.

Ramalho, M. 2004. Stingless bees and mass flowering trees in the canopy of Atlantic Forest: A tight relationship. *Acta Bot. Bras.* 18:37–47.

Ramalho, M., and M. A. Batista. 2005. Polinização na Mata Atlântica: perspectiva ecológica da fragmentação. In *Mata Atlântica e Biodiversidade*, ed. C. R. Franke, P. L. B. Rocha, W. Klein and S. L. Gomes, 93–142. Salvador: Editora Universidade Federal da Bahia.

Ramalho, M., M. D. Silva, and C. A. L. Carvalho. 2007. Dinâmica de uso de fontes de pólen por *Melipona scutellaris* Latreille (Hymenoptera, Apidae): uma análise comparativa com *Apis mellifera* L. (Hymenoptera, Apidae), no Domínio Tropical Atlântico. *Neotrop. Entomol.* 36:38–45.

Rembold, H., and B. Lackner. 1978. Vergleichenden Analyse von Weiselfuttersäften. *Mitt. dt. Ges. allg. angew. Ent.* 1:299–301.

Ribeiro, M. F. 1994. Growth in bumble bee larvae: Relation between development time, mass and amount of pollen ingested. *Can. J. Zool.* 72:1978–85.

Ribeiro, M. F., M. J. Duchateau, and H. H. W. Velthuis. 1996. Comparison of the effects of two kinds of commercially available pollen on colony development and queen production in the bumble bee *Bombus terrestris* L. (Hymenoptera, Apidae). *Apidologie* 27:133–44.

Ribeiro, M. F. 1999. Long duration feedings and caste differentiation in *Bombus terrestris* larvae. *Insectes Soc.* 46:315–22.

Ribeiro, M. F., H. H. W. Velthuis, M. J. Duchateau, and I. van der Tweel. 1999. Feeding frequency and caste differentiation in *Bombus terrestris* larvae. *Insectes Soc.* 46:306–14.

Ribeiro, M. F., and D. A. Alves. 2001. Size variation in *Schwarziana quadripunctata* queens (Hymenoptera, Apidae, Meliponini). *Rev. Etologia* 3:59–65.

Ribeiro, M. F., V. L. Imperatriz-Fonseca, and P. S. Santos-Filho. 2003. Exceptional high queen production in the Brazilian stingless bee *Plebeia remota*. *Stud. Neotrop. Fauna Env.* 38:111–4.

Ribeiro, M. F., P. S. Santos-Filho, and V. L. Imperatriz-Fonseca. 2006a. Size variation and egg laying performance in *Plebeia remota* queens (Hymenoptera, Apidae, Meliponini). *Apidologie* 37:1–12.

Ribeiro, M. F., T. Wenseleers, P. S. Santos-Filho, and D. A. Alves. 2006b. Miniature queens in stingless bees: Basic facts and evolutionary hypotheses. *Apidologie* 37:191–206.

Rodríguez-Malaver, A. J., C. Rasmussen, M. G. Gutiérrez, F. Gil, B. Nieves, and P. Vit. 2009. Properties of honey from ten species of Peruvian stingless bees. *Nat. Prod. Commun.* 4:1221–6.

Rosa, C. A., M. A. Lachance, J. O. C. Silva, A. C. P. Teixeira, M. M. Marini, Y. Antonini, and R. P. Martins. 2003. Yeast communities associated with stingless bees. *FEMS Yeast Res.* 4:271–5.

Röseler, P. F. 1970. Unterschied in der Kastendetermination zwischen den Hummelarten *Bombus hypnorum* und *Bombus terrestris*. *Z. Naturforsch.* 25b:543–8.

Röseler, P. F. 1974. Grössenpolymorphismus, Geschlechtsregulation und Stabilisierung der Kasten im Hummelvolk. In *Sozialpolymorphismus bei Insekten*, ed. G. H. Schmidt, 298–335. Stuttgart, Germany: Wissenschaliche Verlagsgesellschaft MBH.

Röseler, P. F. 1976. Juvenile hormone and queen rearing in bumble bees. In *Phase and Caste Determination in Insects*, ed. M. Lüscher, 55–61. Oxford, UK: Pergamon Press.

Röseler, P. F. 1991. Roles of morphogenetic hormones in caste polymorphism. In *Morphogenetic Hormones of Arthropods: Roles in Histogenesis, Organogenesis, and Morphogenesis—Part 3*, ed. A. P. Gupta, 384–99. New Brunswick, NJ: Rutgers University Press.

Röseler, P. F., and I. Röseler. 1974. Morphologische und physiologische Differenzierung der Kasten bei den Hummellarten *Bombus hypnorum* un *Bombus terrestris* (L). *Zool. Jb. Physiol.* 78:175–98.

Roubik, D. W. 1982. Obligate necrophagy in social bee. *Science* 217:1059–60.

Roubik, D. W. 1989. *Ecology and Natural History of Tropical Bees*. Cambridge, UK: Cambridge University Press.

Roulston, T. H., J. H. Cane, and S. L. Buchmann. 2000. What governs protein content of pollen: Pollinator preferences, pollen pistil interactions, or phylogeny? *Ecol. Monogr.* 70:617–43.

Sakagami, S. F., and R. Zucchi. 1963. Oviposition process in a stingless bee, *Trigona (Scaptotrigona) postica* Latreille. *Studia Entomol.* 6:497–510.

Sakagami, S. F. 1982. Stingless bees. In *Social Insects*, ed. H. R. Hermann, 361–423, v. 3. New York: Academic Press.

Santiesteban, H. A. 1994. *Características físicas y químicas de las mieles de cinco especies de abejas* Apis mellifera, Melipona beecheii, Scaptotrigona pachysoma, Tetragona jaty *y* Plebeia sp. *(Hymenoptera, Apidae) colectadas en el Municipio de Union Juarez, Chiapas, Mexico*. Escuela de Ciencias Químicas Chiapas: Universidade Autonoma de Chiapas.

Santos-Filho, P. S., D. A. Alves, A. Eterovic, V. L. Imperatriz-Fonseca, and A. M. P. Kleinert. 2006. Numerical investment in sex and caste by stingless bees (Apidae: Meliponini): A comparative analysis. *Apidologie* 37:207–21.

Schoeder, A., H. Horn, and H. J. Pieper. 2005. The correlation between moisture content and water activity (aw) in honey. _Deutsche Lebensmittel-Rundschau_ 101:139–42.

Schoener, T. W. 1971. Theory of feeding strategies. _Annu. Rev. Ecol. Syst._ 2:369–404.

Seeley, T. D. 1985. _Honeybee Ecology_. Princeton, New Jersey: Princeton University Press.

Seeley, T. D. 1995. _The Wisdom of the Hive_. Cambridge, MA: Harvard University Press.

Spariglia, M. A., M. A. Vattuone, M. M. S. Vattuone, J. R. Soberón, and D. A. Sampiero. 2010. Properties of honey from _Tetragonisca angustula fiebrigi_ and _Plebeia wittmanni_ of Argentina. _Apidologie_ (published online: 02 June 2010) DOI: 10.1051/apido/2010028.

Shuel, R. W., and S. E. Dixon. 1959. The early establishment of dimorphism in the female honeybee, _Apis mellifera_ L. _Insectes Soc._ 7:265–82.

Silva, C. E. P., and C. Schlindwein. 2003. Fidelidade floral e características polínicas das plantas relacionadas a _Melipona scutellaris_ (Hym., Apidae, Meliponini). Paper presented at the VI Congresso de Ecologia do Brasil, Fortaleza.

Silva, D. L. N. 1977. Estudos bionômicos em colônias mistas de Meliponinae. _Bol. Zool. Univ. São Paulo_ 2:7–106.

Silva-Matos, E. V., F. B. Noll, and R. Zucchi. 2000. Sistemas de regulação social encontrados em abelhas altamente eussociais (Hymenoptera; Apidae, Meliponinae). In: Encontro sobre Abelhas 4, Ribeirão Preto, 6–9 setembro, 2000. _Annals_ Riberão Preto: [s.n.], 2001. v. 4, 95–101.

Slaa, E. J., A. Cevaal, and M. J. Sommeijer. 1997. Flower constancy of three species of stingless bees (Apidae: Meliponinae) in Costa Rica. _Proc. Exper. Appl. Entomol. Soc._ 8:79–80.

Slaa, E. J., B. Ruiz, R. Salas, M. Zeiss, and M. J. Sommeijer. 1998. Foraging strategies in stingless bees: Flower constancy versus optimal foraging? _Proc. Exper. Appl. Entomol. Soc._ 9:185–90.

Slaa, E. J., J. Wassenberg, and J. C. Biesmeijer. 2003. The use of field-based social information in eusocial foragers: Local enhancement among nestmates and heterospecifics in stingless bees. _Ecol. Entomol._ 28:369–79.

Slaa, E. J. 2006. Population dynamics of a stingless bee community in the seasonal dry lowlands of Costa Rica. _Insectes Soc._ 53:70–9.

Sladen, F. W. L. 1912. _The Humble-Bee, Its Life-History and How to Domesticate It_. London: MacMillan & Co.

Snowdon, J. A., and D. O. Cliver. 1996. Microorganisms in honey. _Int. J. Food Microbiol._ 31:1–26.

Souza, B. A., C. A. L. Carvalho, G. S. Sodré, and L. C. Marchini. 2004. Características físico-químicas de amostras de mel de _Melipona asilvai_. Ciênc. Rural 34:1623–4.

Souza, B. A., D. W. Roubik, M. Barth, T. Heard, E. Enriquez, C. Carvalho, J. Villas-Boas, L. C. Marchini, J. Locatelli, L. Persano-Oddo, L. Almeida-Muradian, S. Bogdanov, and P. Vit 2006. Composition of stingless bee honey: Setting quality standards. _Interciência_ 31:867–75.

Souza, B. A., L. C. Marchini, M. Oda-Souza, C. A. L. Carvalho, and R. M. O. Alves. 2009. Caracterização do mel produzido por espécies de _Melipona_ Illiger, 1806 (Apidae:Meliponini) da região nordeste do Brasil: 1. Características físico-químicas. _Quím. Nova_ 32:303–8.

Stanley, R. G., and H. F. Linskens. 1974. _Pollen: Biology, Biochemistry, Management_. First edition, Heidelberg, Germany: Springer-Verlag.

Takenaka, T., and E. Takahashi. 1980. General chemical composition of the royal jelly. _Bull. Fac. Agric. Tamagawa. Univ._ 20:71–8.

Teixeira, L. V., and L. A. O. Campos. 2005. Produção de rainha de emergência em _Plebeia lucii_ (Hymenoptera, Apidae, Meliponina). Paper presented at the I Simpósio Brasileiro de Insetos Sociais, IUSSI/Brasil, Belo Horizonte.

Terada, Y. 1974. Contribuição ao estudo da regulação social em _Leurotrigona muelleri_ e _Frieseomelitta varia_ (Hymenoptera, Apidae). MSc dissertation, Universidade de São Paulo, Brazil.

Tostes, R. O. G., and F. E. P. Leite. 1994. Novas considerações sobre o uso tópico de açúcar e mel em feridas. _Rev. Med. Minas Gerais_ 4:35–9.

Van Nicuwstadt, M. G. L., and C. E. R. Iraheta. 1996. Relation between size and foraging range in stingless bees (Apidae, Meliponinae). _Apidologie_ 27:219–28.

Velthuis, H. H. W., and M. J. Sommeijer. 1991. Roles of morphogenetic hormones in caste polymorphism in stingless bees. In _Morphogenetic Hormones of Arthropods: Roles in Histogenesis, Organogenesis, and Morphogenesis—Part 3_, 346–83, ed. A. P. Gupta. New Brunswick, NJ: Rutgers University Press.

Velthuis, H. H. W., and A. van Doorn. 2006. A century of advances in bumblebee domestication and the economic and environmental aspects of its commercialization for pollination. _Apidologie_ 37:421–51.

Villas-Bôas, J. K., and O. Malaspina. 2004. Physical–chemical analysis of *Melipona compressipes* and *Melipona seminigra*: Honeys of Boa Vista do Ramos, Amazonas, Brazil. Paper presented at the 8th International Conference on Tropical Bees and VI Encontro sobre Abelhas.

Vit, P., S. Bogdanov, and V. Kilchenmann. 1994. Composition of Venezuelan honeys from stingless bees (Apidae, Meliponinae) and *Apis mellifera*. *Apidologie* 25:278–88.

Vit, P., M. Medina, and M. E. Enriquez. 2004. Quality standards for medicinal uses of Meliponinae honey in Guatemala, Mexico and Venezuela. *Bee World* 85:2–5.

Vit, P., A. Rodríguez-Malaver, D. Almeida, B. A. Souza, L. C. Marchini, C. F. Díaz, A. E. Tricio, J. K. Villas-Bôas, and T. A. Heard 2006. A scientific event to promote knowledge regarding honey from stingless bees: 1. Physical–chemical composition. *Magistra* 18:270–6.

Vit, P., and T. J. Jacob 2008. Putative anticataract properties of honey studied by the action of flavonoids on a lens culture model. *J. Health Sci.* 54:196–202.

Waser, N. M. 1983. The adaptive nature of floral traits: ideas and evidences. In *Pollination Biology*, ed. L. Real, 241–85. New York: Academic Press.

Wenseleers, T., F. L. Ratnieks, and J. Billen. 2003. Caste fate conflict in swarm-founding social hymenoptera: An inclusive fitness analysis. *J. Evol. Biol.* 16:647–58.

Wenseleers, T., A. G. Hart, and F. W. Ratnieks. 2004. When resistance is useless: Policing and the evolution of reproductive acquiescence in insects societies. *Am. Nat.* 164:154–67.

Wenseleers, T., F. W. Ratnieks, M. F. Ribeiro, D. A. Alves, and V. L. Imperatriz-Fonseca. 2005. Working-class royalty: Bees beat the caste system. *Biol. Letters* 1:125–8.

White, D., B. W. Cribb, and T. A. Heard. 2001. Flower constancy of the stingless bee *Trigona carbonaria* Smith (Hymenoptera: Apidae: Meliponini). *Aust. J. Entomol.* 40:61–4.

Wilms, W., V. L. Imperatriz-Fonseca, and W. Engels. 1996. Resources partitioning between highly eusocial bees and possible impact of the introduced Africanized honey bee on native stingless bees in the Brazilian Atlantic rainforest. *Stud. Neotrop. Fauna Environ.* 31:137–51.

Wilms, W., M. Ramalho, and L. Wendel. 1997. Stingless bees and Africanized honey bees in the Mata Atlântica rainforest of Brazil. XXXth Int. Apic. Congress of Apimondia, Antuérpia, 167–70.

Winston, M. L. 2003. *A Biologia da Abelha*. Porto Alegre: Editora Magister.

Wirtz, P. 1973. Differentiation in the honey bee larva. *Meded. Landbhoogesch.* 73–75:1–155.

Zucchi, R. 1993. Ritualized dominance, evolution of queen–worker interactions and related aspects in stingless bees (Hymenoptera, Apidae). In *Evolution of Insect Societies*, ed. T. Inoue, and S. Yamane, 207–49. Tokyo, Japan: Hakuhin-Sha Publishing Co.

12

Defoliators (Lepidoptera)

Alessandra F. K. Santana, Carla Cresoni-Pereira, and Fernando S. Zucoloto

CONTENTS

12.1 Introduction 273
12.2 Evolution of Feeding Habits 273
12.3 Morphology and General Biology of the Caterpillars 274
    12.3.1 Feeding and Digestion 276
    12.3.2 Food Perception 276
12.4 Caterpillar–Leaf Interaction 278
    12.4.1 Acceptability, Performance, and Preference 278
        12.4.1.1 Impact of Leaf Characteristics on Caterpillar Performance 280
    12.4.2 Competition and Food Deprivation 281
    12.4.3 Strategies for Food Utilization and Selection 282
    12.4.4 Dispersal 282
    12.4.5 Feeding on Alternative Sources (Nonvegetal) 283
    12.4.6 Feeding Periods 283
12.5 Tritrophic Relationships: Presence of Natural Enemies 285
12.6 Conclusions and Research Suggestions 287
References 287

12.1 Introduction

The order Lepidoptera includes the popular usually diurnal butterflies (Rhopalocera) and the usually nocturnal moths (Heterocera). The butterflies constitute about 5% of the order adult diversity but most of the representatives are the moths (Nielsen and Common 1991). Butterflies are one of the most popular groups of insects and are one of the largest animal taxa, with approximately 160,000 species divided in 47 superfamilies (Kristensen et al. 2007). The Lepidoptera, Coleoptera, and Hymenoptera immatures represent the defoliator insects (Bernays 1998); most of their caterpillars are phytophagous, predominantly specialists, feeding on only one or on a few families of related plants (Bernays 1998).

12.2 Evolution of Feeding Habits

The currently known species lineage is related to angiosperm radiation during the Cretaceous period (Whitfield and Kjer 2008). The feeding transition from larvae to angiosperm foliage may be related to the evolution of adult feeding on nectar and to the glossatan adaptive radiation (Stekolnikov and Korseev 2007).

There are four suborders in the group of Lepidoptera: Zeugloptera, Aglossata, Heterobathmiina, and Glossata. The suborder Glossata is divided in several infraorders and the series Ditrysia of the infraorder Heteroneura is the most abundant, containing approximately 98% of all species of the group (Kristensen 1984). The Ditrysia correspond to 29 superfamilies and one of the groups is monophyletic (Nielsen 1989). The superfamilies Papilonoidea and Hesperioidea constitute the butterfly group, Rhopalocera, and the monophyletism is supported by several synapomorphies (Kristensen 1976). Some authors consider the superfamily Papilonoidea as the most advanced Lepidoptera; undoubtedly, it is highly specialized.

12.3 Morphology and General Biology of the Caterpillars

The insect mandibular structure shows an adaptation to the food the species uses and can vary even inside the same group. The mouthparts in Lepidoptera are the most studied feeding organ under the anatomical, morphological, functional, and evolutionary aspects (Krenn et al. 2005).

Lepidoptera larvae and adults are related to vascular plants. The larvae have bite–chew mouthparts (Krenn 2010); the head general morphology of the heliconid *Dione moneta moneta* Hübner is shown (Figures 12.1 and 12.2), with structures following the general pattern in Lepidoptera (Kaminski et al. 2008). Most adults have a specialized proboscis to suck flower nectar and other liquid substances (Krenn 2010). When compared with other insects that use nectar, the Lepidoptera proboscis has unique characteristics (Krenn et al. 2005): the galea modification in the adult proboscis, widely found in this order, is not observed in other insect orders (Kristensen 1984). When at rest, it forms a flat vertical spiral with two components that primarily function as a hydraulic mechanism (Krenn 1990). Comparative studies show that the same movement mechanisms operate in all species, independently of the proboscis length, of the galea muscles arrangement, or of the behavioral adaptations for certain feeding resources (Krenn 1990; Krenn 2000; Wannenmacher and Wasserthal 2003).

Brown and Dewhurst (1975) illustrated mandible shapes in caterpillars of the genus *Spodoptera* (Noctuidae). Caterpillars that change their feeding habits during development show mandibles with different shapes throughout development (e.g., *Heterocampa obliqua* Packard) (Godfrey et al. 1989).

The functional importance of the mandible form was shown (Bernays and Janzen 1988). They show that in several species of Saturniidae caterpillars that feed on tough leaves, the mandible edge works against the surface of the other edge so a small disk is cut similarly at every bite. The fragments cut by the caterpillar show very regular size with low variation coefficient. On the other hand, Sphingidae caterpillars that feed on soft leaves produce variable leaf fragments, possibly due to the teeth complexity and salient conformation.

The ingestion mechanism of the food previously cut by the mandibles is not yet well studied. In most of the leaf-chewing insects, the maxilla is also well developed and plays an important role in food intake; the movement of the food into the mouth is a mechanical process that depends on the coordinated activity of the mouthparts (Chapman 1995a). Inside the mouth, the food presumably passes from the crop to the midgut through pharyngeal and esophageal peristaltic movements (Chapman 1995a).

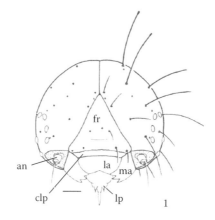

FIGURES 12.1   *Dione moneta moneta* Hübner, 1825, first instar: head capsule and mouthparts general morphology (front view) (right side setae omitted). fr = front; an = antenna; clp = clypeus; la = labro; ma = mandible; lp = labial palp; mp = maxillary palp; oc = ocellus. Bars = 100 μm. (Modified from Kaminski, L. A., R. Dell'Erba, and G. R. P. Moreira, *Rev. Bras. Entomol.*, 52, 13–23, 2008.)

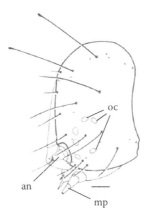

FIGURE 12.2 *Dione moneta moneta* Hübner, 1825, first instar: head capsule (lateral view). fr = front; an = antenna; clp = clypeus; la = labro; ma = mandible; lp = labial palp; mp = maxillary palp; oc = ocellus. Bars = 100 μm. (Modified from Kaminski, L. A., R. Dell'Erba, and G. R. P. Moreira, *Rev. Bras. Entomol.*, 52, 13–23, 2008.)

Morphologically, the caterpillar body is divided in three distinct parts: head, thorax, and abdomen. Hypognathism and prognathism are observed with very well developed mandibles and bites (Nielsen and Common 1991). The caterpillar head and body support the setae, whose distribution patterns are used in the caterpillar identification and classification (Carter 1987).

Using the proboscis, adults feed on nectar, honeydew, and fermented juices. Egg production is usually maintained by nutrients brought from the larval phase (Telang et al. 2001), since lepidopterans need a high content of proteins for egg production (Wheeler et al. 2000). To acquire proteins, newly hatched caterpillars usually ingest the chorions (Nielsen and Common 1991; Clark and Faeth 1998). Barros-Bellanda and Zucoloto (2001) investigating this behavior in *Ascia monuste* (Godart 1819) (Figure 12.3) found that chorion intake has a positive effect on the performance, and this behavior may have a relationship with egg cannibalism (see Chapter 8).

Caterpillars present intense intestinal capacity, quick digestion (Bernays 1998), and can cannibalize eggs and small caterpillars whether conspecific or not (Whitman et al. 1994). These adaptations were favored due to the diets with poor protein content of most phytophagous insects (Whitman et al. 1994).

Lepidopterans oviposit singly or in clusters; most of the butterflies lay solitary eggs (Stamp 1980). Egg clustering often results in larval aggregation in the beginning of the development (Figure 12.4a) and caterpillars that live in groups may or may not live isolated (Figure 12.4b) at the end of development (Clark and Faeth 1998; Barros-Bellanda and Zucoloto 2003). Solitary foraging is advantageous because there is less risk of spreading diseases (Hunter and Elkinton 2000; Wilson et al. 2003), intraspecific competition (Hunter and Elkinton 2000), cannibalism inside the same oviposition (Barros-Bellanda

FIGURE 12.3 *Ascia monuste* Godart newly hatched caterpillar ingesting chorion. (Courtesy of A. F. K. Santana.)

FIGURE 12.4   Different phases of *Ascia monuste* development in kale: (a) Egg clusters and larvae eclosion. (b) Fifth instar caterpillar foraging solitarily. (Courtesy of A. F. K. Santana.)

and Zucoloto 2005), and attack by predators (Bernays 1997; Hunter and Elkinton 2000). On the other hand, life in groups can offer advantages regarding performance such as facilitation of feeding due to microclimate modifications, intensification of thermoregulation (Ronnas et al. 2010), social stimulus for feeding (Stamp 1980; Clark and Faeth 1997), increased growth rate, adult weight, and larval survival (Le Masurier 1994; Bianchi and Moreira 2005; Allen 1010). Bianchi and Moreira (2005), working with *Dione juno juno* (Cramer) (Lepidoptera, Nymphalidae), found that immatures survival was affected by larval density: mortality was higher in groups with less than eight larvae. Larval aggregation can also increase the chances of survival of first instars when they feed on tough leaves (Kawasaki et al. 2009), and can cause quick depletion of resources triggering high rate of dispersal in the first instars (Rhainds et al. 2010).

12.3.1 Feeding and Digestion

Feeding is the caterpillar's main activity (Bernays 1997). As in other insects, the nutritional needs vary during the life cycle, and proteins are extremely important in the beginning of development (Simpson and Simpson 1990; Gaston et al. 1991). The relative growth rate, the food consumption rate, the metabolic rate, and the assimilation efficiency are higher in the initial than in the final instars (Scriber and Slansky 1981; Dix et al. 1996). However, the low availability of proteins during the first larval instars can reduce the capacity to transform nutrients in tissues in the postabsorption processes (Woods 1999) as conversion of digested food in animal biomass during that period is less efficient (Dix et al. 1996).

In the final instars, caterpillars tend to be >10,000 times heavier than the newly hatched ones (Simpson and Simpson 1990); the nutritional reserve concentration greatly increases as the proportion of metabolically active tissues is reduced, and most nutrients are deviated to conversion in biomass (Simpson and Simpson 1990).

Usually, the pH of the insect luminal content varies from 6 to 7.5, while lepidopterans larvae present alkaline pH from 8 to 12 (Dow 1984). These intestinal pH relatively high values probably are associated with the hemicelluloses release from the ingested plant cell walls. *Erinnyis ello* (L.) (Lepidoptera: Sphingidae) larvae are able, for instance, to efficiently digest hemicelluloses without affecting the ingested leaves cellulose (Terra et al. 1987).

12.3.2 Food Perception

Chemoreceptors associated with food ingestion are present in the caterpillar mouthparts and they are also found in tarsi and antennae of several insects; the insects use the antennae to monitor the food, vibrating them near or over the food surface (de Boer 1993). In several species, there are contact chemoreceptors in the top of the antennae, though olfactory receptors may also be present. In caterpillars, whose antennae are small and associated to the mouthparts, it is assumed that they also have direct involvement with the food mechanics and chemical selection (de Boer 1993).

When the caterpillars are in contact with potential feeding resources, it is difficult to make a distinction if the stimulus to promote feeding is olfactory or gustatory. According to Städler and Hanson (1975), the structurally and electrophysiologically gustatory sensilla also respond to odors.

Among caterpillars, the sensilla number is constant and it is not related to feeding habits or taxonomic position. According to Grimes and Neunzig (1986), all studied species have eight sensilla on the top of the maxillary palp and four on the galea; 41 species from 24 families of the order Lepidoptera with varied feeding habits (monophagous, oligophagous, and polyphagous) were analyzed. The Lepidoptera first larval instar has the same number of sensilla as the last instar, even when the mouthparts surface increases more than 100 times at the end of the caterpillar development (Chapman 1995).

The importance of nutrients and/or secondary compounds in lepidopterans' hierarchy of choices varies with different situations to which the insects are exposed. This presumably reflects variable sensitivity in different receptors. In a study about feeding behavior, using *Manduca sexta* (Cramer) (Lepidoptera: Sphingidae) caterpillars, it was detected that larvae reared on tomatoes rejected the acceptable nonhost *Vigna sinensis* (de Boer 1993). This behavior was primarily mediated by the galea lateral sensilla styloconica, since when these sensilla were removed caterpillars did not discriminate in favor of tomatoes. The lateral sensilla styloconica contains cells that detect deterrent substances (Peterson et al. 1993). Caterpillars without the sensilla styloconica preferred *V. sinensis* instead of moistened filter paper; however, when all sensilla were removed, they did not show any preferences and ingested pieces of paper. Studies with *Pieris brassicae* L. (Lepidoptera, Pieridae) demonstrated that the galea lateral sensilla styloconica is sensitive only to sucrose and glucose, while the medial sensilla respond to a greater variety of sugars (Ma 1972).

Cells that respond to amino acids are also present in some of the caterpillar sensilla, but there are several response variations in different species. Fourteen amino acids stimulate specific cells in *P. brassicae* caterpillars; the most effective are histidine, phenylalanine, and 4-hydroxyproline (Chapman 1995). The former two amino acids are those that stimulate *P. rapae* (L.) sensitive cells less. Eight amino acids do not produce responses in other species sensilla (Chapman 1995).

Sensitive cells respond to food-specific chemical components as the plants secondary compounds. This kind of response characterizes these components as attractant, unpalatable, and/or toxic, and the cells responsible for identification are called deterrents (Ma 1972; van Loon and Schoonhoven 1999). In caterpillars, these deterrent cells are restricted to four classes of chemosensilla, and these cells have receptors to varied molecules that may or not superpose and some can be completely distinct. In *M. sexta*, the sensilla styloconica deterrent cell contains at least two patterns of signs, one responding to phenolic glycosides and to methylxanthines, and the other responding to the aromatic nitroderivatives (Schoonhoven 1972). Notwithstanding, little is known about the signs' pattern nature in the deterrent cells (Glendinning et al. 2000).

Feeding inhibition due to reduction of the bite pattern or size occurs when the caterpillar finds unpalatable and/or toxic secondary compounds. The insect "dilemma" is that both the very toxic and the less toxic compounds activate the aversive response. Studies with *M. sexta* have shown that to solve this dilemma there are at least three mechanisms that reduce or inactivate the aversive response to the plant secondary compounds: (1) carbohydrates can mask the unpalatable taste of some compounds (Glendinning et al. 2000), (2) prolonged exposure to diets rich in some secondary compounds can initiate adaptive mechanisms in the peripheral and central gustatory system (Schoonhoven 1978; Usher et al. 1988), and (3) exposition to diets rich in toxic compounds can induce the production of P450 detoxification enzymes in the insect midgut (Brattsten et al. 1977).

As it is observed in other insects, the secondary compounds can also play an important role in lepidopteran learning. In the cabbage looper moth, *Trichoplusia ni* Hübner (Lepidoptera, Noctuidae), the last larval instar feeding on *Hoodia gordonii* latex in specific periods produced nondeterrence for oviposition by the adults arising from the experiment. Usually, the latex is highly deterrent for oviposition in this species when it is not submitted to a previous experience with this substance during the larval phase (Shikano and Isman 2009).

12.4 Caterpillar–Leaf Interaction

The interaction between caterpillars and their host plants has always intrigued naturalists. The consumption of vegetal biomass by herbivorous reaches 7% to 18% of the world vegetal biomass; nevertheless, "the world stays green" (White 2005). There are difficulties to overcome in order to detect and consume these plants: the production of chemical substances deterrent and/or toxic to the herbivores (Brattsten et al. 1977), surfaces that hamper fixation and feeding (Southwood 1978; Edwards and Wratten 1981), and the low protein content in vegetal tissues.

12.4.1 Acceptability, Performance, and Preference

To know whether a certain plant is suitable to the development of a species, the insect performance concept is used to measure survival of all immature stages (egg, larva, and pupa), larval growth, pupal mass, digestive efficiency indicated by nutritional indices (Waldbauer 1968; Scriber and Slansky Jr. 1981, see Chapter 2), fecundity, and adult longevity (Thompson 1988).

The plant nutritional value for a certain species is directly related to the insect performance in that host, and not necessarily related directly to the host chemical composition. The host plant with the apparent best nutritional content will not always yield the best performance for the insect, and the plant with the best nutritive value is not always chosen by the species in the field, since other variables influence the insect preference, such as predation and presence of deterrent substances. Monarch butterflies, for instance, demand high nitrogen levels to develop, but they do not choose necessarily the plants with higher nitrogen levels as these plants also present high concentrations of cardiac glycosides, which can be toxic (Awmack and Leather 2002).

Studying some aspects of the feeding habits of *A. monuste* it was demonstrated that certain hosts allow better performances (e.g., kale, cauliflower, rocket, and broccoli) than others (e.g., mustard and cabbage) (Felipe and Zucoloto 1993). To determine the performance and the preference for different hosts, more detailed experiments were conducted with *A. monuste* using kale and mustard (*B. juncea*). These experiments were carried out by Barros and Zucoloto (1999) and have shown that (1) these plants have different nitrogen contents and kale has the highest one, (2) the insect's performance was better with kale than with mustard (Table 12.1), (3) females chose kale in cage experiments regarding oviposition preferences, indicating a positive correlation with the performance (Table 12.2), (4) newly hatched caterpillars did not show significant preferences (Figure 12.5a), and (5) older caterpillars' preferences were not clear (Figure 12.5b).

The term preference is not a synonym of acceptability (van Loon 1996; Singer 2000). Acceptability is related to recognition of the host as part of the insect feeding menu, and this recognition is basically made analyzing the host chemical content, mainly regarding allelochemicals (van Loon 1996). The term

TABLE 12.1

Ascia monuste Performance Comparative Data after Feeding Exclusively on Kale or Mustard in the Laboratory (Eggs Collected in Kale Leaves)

Food	Emergence (%)	Number of Eggs/Female	Weight of the Pupa (mg)	AD (%)	ECI (%)	ECD (%)	Ingestion (mg)	Feces (mg)
Kale	92.2 a (±6.2)	38.5 a (±9.5)	77.7 a (±3.5)	53.4 a (±4.3)	25.8 a (±2.9)	48.3 a (±7.8)	230.3 a (±16.2)	107.2 a (±11.0)
Mustard	80.6 a (±6.9)	19.0 b (±4.2)	66.8 b (±3.3)	48.4 b (±4.8)	21.9 b (±3.4)	45.4 a (±8.5)	190.0 b (±13.3)	98.2 b (±7.9)

Note: The results of the first two parameters represent the mean ± SD of six groups with seven caterpillars each; the other indices were individually tested (10 replicates). Means followed by different letters differ from each other ($P < 0.05$, Mann–Whitney Test, $\alpha = 0.05$).

TABLE 12.2

Ascia monuste Butterflies Oviposition Regarding Preference for Kale or Mustard

Butterfly	Ovipositions (Kale)	Ovipositions (Mustard)	Eggs (Kale)	Eggs (Mustard)
1	1	0	27	0
2	1	0	36	0
3	0	1	0	22
4	1	0	34	0
5	2	0	28 + 52	0
6	0	1	0	30
7	1	1	16	29
8	1	1	79	19
9	2	2	16 + 14	23 + 20
10	1	0	23	0
11	0	1	0	28
12	1	0	33	0
13	3	0	12 + 27 + 21	0
14	2	0	20 + 27	0
15	2	0	21 + 19	0
16	1	0	39	0
Mean (± SD)	1.2 ± 0.83 a	0.4 ± 0.63 b	33.4 ± 23.4 a	10.7 ± 15.0 b

Note: The females were collected in the field and individualized in cages for 3 days. Means followed by different letters in each parameter differ from each other. ($P < 0.05$, Wilcoxon Test, $\alpha = 0.05$).

preference, on the other hand, involves a situation of choice where the insect establishes a hierarchy among plants (Thompson 1988); in addition to the allelochemicals, this implies other parameters such as nitrogen content (White 1984), water content (Scriber and Slansky 1981), plant physical characteristics (Roden et al. 1992), content of appealing volatile substances (de Boer 1993), amount of chemical defenses (Ehrlich and Raven 1964), absence of conspecific organisms (Miller and Strickler 1984), and the amount of resources, among others. The analysis of the mentioned characteristics is very important for the holometabolous insects, for instance, to prevent oviposition in hosts with unsuitable chemical content and/or that provide competition and/or high predation of the offspring.

Host plant selection by most of the Lepidoptera pregnant females is determinant for the immatures' survival and performance (van Loon 1996), mainly those in the first larval instar with very low mobility

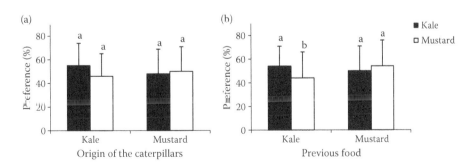

FIGURE 12.5 *Ascia monuste* caterpillars feeding preference for kale (dark columns) or mustard (white columns). (a) newly hatched caterpillars with two different origins (kale or mustard leaves in the field), and (b) fourth instar caterpillars with two different previous feedings (kale or mustard). The results represent the mean ± SD of 10 experiments. Different letters above the columns indicate significant differences ($P < 0.05$) among the means (Mann–Whitney test, $\alpha = 0.05$). (Modified from Barros, H. C. H., and F. S. Zucoloto, *J. Insect Physiol.*, 45, 7–14, 1999.)

and reserve of energy (Damman and Feeny 1988). Host preference by *A. monuste* is largely defined during oviposition; in addition, studies about immatures' feeding behavior indicate that the species first instars feed on the same site where the mother has laid the eggs (Catta-Preta and Zucoloto 2003). *A. monuste* adults, for instance, select hosts considering (1) plant quality (Barros and Zucoloto 1999), (2) leaf age (Bittencourt-Rodrigues and Zucoloto 2005), and (3) leaf part (Catta-Preta and Zucoloto 2003).

Experiments with three *Pieris* species have shown positive and negative correlation between preference regarding oviposition and the caterpillars' performance. In the laboratory, immatures of the three studied species presented better performance in the same plants even though only *P. melete* Menetries selected hosts coinciding exactly with the plants' best nutritive values for caterpillars: in *P. rapae* L. and *P. napi* L., the existence of other factors that influence the plant preference evolution is evident (Ohsaki and Sato 1994). In *P. napi*, the lowest risk of parasitism seems to influence host selection since the chosen plants guarantee low risk of parasitism for the immatures in spite of having lower nutritive value (Ohsaki and Sato 1994).

12.4.1.1 Impact of Leaf Characteristics on Caterpillar Performance

12.4.1.1.1 Nutrients and Allelochemicals

Food affects insect performance both in quality and in quantity (Slansky and Rodrigues 1987). Nitrogen and water contents are determinants in the caterpillar performance, particularly the newly ecloded ones (Mattson and Scriber 1987; Ojeda-Avila et al. 2003). In *Ostrinia nubilalis* Hübner young caterpillars, the high nitrogen concentration in corn plantations with low luminosity is more important regarding feeding rates than the chemical defenses concentration (Manuwoto and Scriber 1985). Similarly, the amino acids content is more important for the *S. frugiperda* first instar survival than the amount of toxins in the corn (Hedin et al. 1990). Laboratory studies have shown that the caterpillars appear to develop better with diets containing similar amounts of proteins and carbohydrates, or in some cases, with diets with high protein content (Waldbauer et al. 1984; Simpson and Raubenheimer 1993).

The secondary compounds importance regarding the caterpillar's success on the leaves is unquestionable (Bernays et al. 2002), but the level of sensibility to them may vary. In general, young caterpillars are more sensitive to the plant chemical defenses (Zalucki et al. 2002). In the specialist *Danaus plexippus* L., although no glycosides adverse effects were observed on caterpillars growth in the fourth and fifth instars and, subsequently, on the females' fecundity (Erickson 1973), in the first instar the results indicate physiological costs regarding this species feeding (Zalucki et al. 2001).

Variability in plant nutritional quality may also cause variation in the performance even in specialist species (Scriber and Slansky 1981). In general, the leaf nutritional quality changes with age; as the leaf ages, the contents of water and nitrogen usually are reduced and fiber content and toughness increase (Mattson 1980; Scriber and Slansky 1981; Slansky and Wheeler 1992; Bittencourt-Rodrigues and Zucoloto 2005). Consequently, the herbivorous insects develop better, survive in higher numbers, and weigh more when they feed on younger leaves (Dodds et al. 1996; Bittencourt-Rodrigues and Zucoloto 2005). In studies with *A. monuste* (Bittencourt-Rodrigues and Zucoloto 2009), the caterpillars that fed first on young leaves had better performance; however, in the second phase of larval development the performance did not vary according to the ingested food. Laboratory studies suggest a hierarchic differentiation in the immatures' degree of preference in this species; the first two instar caterpillars preferred young leaves, the third instar showed flexibility in the degree of preference, and the fourth and fifth instars did not show preference for young or old leaves.

12.4.1.1.2 Physical Structure and Associated Microfauna

The outer structure of the leaf also has a direct impact on caterpillar performance and behavior. Noctuidae caterpillars, for instance, prefer *Daphne laureola* leaves with shorter veins (Alonso and Herrera 1996).

Leaf toughness may extend the predator development time, affecting mainly newly hatched caterpillars that have relatively delicate mandibles; tougher leaves may increase the wear of the mandibles, hindering movements and food intake (Gaston et al. 1991; Lucas et al. 2000). In addition, they may interfere in the digestion processes (Scriber 1982). The high hemicellulose content in the corn, for instance, partially explains newly hatched *S. frugiperda* larvae resistance to ingest and adequately develop in this host (Hedin et al. 1990). Leaf toughness may also influence aggregation. Some studies (Young and Moffett 1979; Clark and Faeth 1997) demonstrated higher survival in tough leaves and higher success in finding feeding sites with larval clustering at the beginning of development (Kawasaki et al. 2009; Ronnas et al. 2010).

Another important barrier regarding larvae feeding is the difficulty to adhere to the leaf structure and to find the proper site to initiate feeding (Southwood 1978). The presence or absence of waxes has been considered a deterrent sign for larval feeding and adherence to the plant (Kantiki and Ampofo 1989; Yang et al. 1993). *Ilex aquifolium* leaves resistant to vertebrate herbivores have smooth cuticle and bristles along the margins, interfering with the *Lasiocampa quercus* L. caterpillars' feeding process (Edwards and Wratten 1981).

The nonglandular bristles and trichomes are mechanical defenses that prevent larval movement from one side to the other on the leaf or restrict their access to the leaf surface (Duffey 1986). The plant architectural obstacles can be overcome in part by foraging behavior; monarch butterfly caterpillars, for instance, pull out the *Asclepias syriaca* bristles by grazing before they feed on the leaves (Hulley 1988). In addition to the mechanical barrier, the glandular trichomes may contain toxic chemical compounds (Lin et al. 1987) or substances that stick to the insects (van Dam and Hare 1998). The latex of some plants, among them *Asclepias curassavica*, constitutes an effective defense against caterpillars (Dussourd 1999; Rodrigues et al. 2010), forming a mechanical and chemical barrier for the mandibular activity. The sabotaging behavior (i.e., inactivation of latex canals by cutting or trenching) is found in *D. erippus*, a specialist caterpillar that feeds on its leaves; in general, the sabotaging behavior does not cause damage to the larvae (Rodrigues et al. 2010).

Leaves are inhabited by a complex fauna of bacteria, fungi, and other microorganisms (Barbosa et al. 1991; Kinkel 1997). Feeding on the leaf surface also results in ingesting microorganisms and mortality is high when they are pathogenic. Alternatively, microorganisms may change positively the host quality for the caterpillars (Wilson et al. 2000). There are reasons to believe that the pathogens are also more dangerous for the younger caterpillars. *S. frugiperda* first instars prefer intact leaves of *Festuca arundinacea* instead of leaves infested by *Acremonium loliae* fungi (Hardy et al. 1985).

12.4.2 Competition and Food Deprivation

In some insects, migration from the original habitat is a response to competition, and that behavior originates a period of food deprivation and loss of energy; in some cases, migration entails the fatal risk of not finding food (Amano 1987).

In Lepidoptera, larvae usually eat continuously and their efficiency to use food can be affected by short periods of deprivation; however, if deprivation prolongs food permanence in the larvae gut, digestion and assimilation can be improved and growth efficiency can be increased (Waldbauer 1968). In *Calocalpe undulata* L. food deprivation for long periods of time (from 8 to 24 h) during the larval phase reduces growth, food intake, and fecal production; however, this situation results in relative increment in nutrients assimilation (Schroeder 1975).

Competition tends to be more intense within individuals that have common needs, and is more vigorous among members of the same species (Remmert 1982). It is expected that intra- and interspecific competitions make resource exploitation more efficient (Pianka 1983). According to Levot et al. (1979), species that tolerate better reduction in body size and develop in shorter periods have competitive advantages as compared to species with other strategies. According to Fretwell (1972), competitors adjust themselves to the amount of resources so that each individual enjoys the same rate of acquisition, although there may be individuals that develop better due to the lack of success of others. Issues such as higher mortality, smaller pupal volume or size, and reduction of adult size is common; however, other effects may appear, such as prolonged development period (Barros-Bellanda and Zucoloto 2002), differentiated sexual survival, or even reduction in size, as more evident either in the male or in the female (Schroeder 1975).

The *A. monuste* female oviposition prevents intraspecific larval competition for food during the first instars, and more developed caterpillars disperse to other plants (Barros-Bellanda and Zucoloto 2002). Some individuals do not adapt to long periods of food scarcity due to unviable reproductive strategies such as premature pupation, smaller adults, and adults with reduced reproductive potential (Barros-Bellanda and Zucoloto 2002). In *M. sexta*, a feeding pattern alteration was obtained increasing intake in deprivation periods exceeding 5h (Bowdan 1988). In long deprivation periods, the locomotor activity apparently intensifies according to the insect metabolic reserve level (Calow 1977); nevertheless, in periods of extreme deprivation, morbidity is observed (Bernays and Simpson 1982; Simpson and Simpson 1990).

Compensation behavior by intake is common among insects when the diet is qualitatively and/or quantitatively inadequate (Slansky and Scriber 1985; Simpson and Simpson 1990). *Anticarsia gemmatalis* Hübner (Lepidoptera, Noctuidae) and *M. sexta* (Timmins et al. 1988) caterpillars ingest higher diet amounts when it is more diluted. Frequently, increased food intake is linked to increased feeding time; this incurs important ecological consequences since feeding on the leaf surface is dangerous for caterpillars (Bernays 1997). According to Boggs and Freeman (2005), the last instar food is very important for the caterpillar's reserve: *Speyeria mormonia* (Boisduval) (Lepidoptera: Nymphalidae) butterfly females that were food deprived in the fifth instar have shown shorter survival in the adult phase than nondeprived butterflies, though effects on the final body mass were not observed.

12.4.3 Strategies for Food Utilization and Selection

In response to plant defenses, insects developed adaptations against the allelochemicals action; in a more radical attitude, some insects became addicted to them (White 2005). Certain secondary compounds are food markers for some Pieridae; the females only oviposit in the *Brassicaceae* family plants. Likewise, the caterpillars only ingest plants that contain a specific compound, the glucosinolates. Today, it is known that the mixture of allelochemicals as flavonoids (van Loon et al. 2002) can be more appealing to these caterpillars than the glucosinolates showing a higher degree of discriminatory ability than it was formerly thought. In addition, specialist caterpillars (e.g., *M. sexta* L., specialist in *Solanaceae*) are also able of feeding on plants of other families, even if they contain relatively toxic compounds (Campo and Renwick 1999).

Caterpillars can convert the plant secondary substances in nontoxic products and then use them for their own benefit. *Utetheisa ornatrix* L. (Lepidoptera, Arctiidae) caterpillars, for instance, sequestrate toxic alkaloids from the host plants and transfer them to adults that become chemically protected (Ferro et al. 2006). During mating, males transfer alkaloid reserves to females and to their eggs (Dussourd et al. 1991).

As the feeding site selection in the plant may favor caterpillar survival, protecting it against natural enemies and/or providing nutrients, the secondary compound concentration is also an important variable in selecting the feeding site. In general, larvae sensitivity to secondary compounds is reduced as they develop; *H. zea* first instars, for instance, avoid consuming gossypol, a polyphenol found in cotton plant glands (Chan et al. 1978). The same is observed in *H. virescens* young caterpillars; they become less selective only 48 to 72h after molt, also consuming the glands (Parrot et al. 1983).

12.4.4 Dispersal

Larval dispersal is an adaptive behavior that plays an important role in survival when food sources are limited. This behavior is well studied in *Chilo partellus* (Swinhoe), the corn caterpillar, and occurs because larval clustering exceeds what the plant can sustain, which is, among other things, a density-dependent behavior (Chapman et al. 1983). Dispersal behavior does not guarantee total success but does increase it. Desiccation and the difficulty in finding host plants are not the main issues for the dispersing caterpillars; however, predation risks are. Due to better mobility, older larvae have more chances to escape from predators than the smaller caterpillars (Berger 1992).

A. monuste apparently faces a weak selective pressure when immatures use kale as a feeding resource; although its nutritional quality is fairly reasonable, it presents high persistence and availability (Barros and Zucoloto 1999), and these three characteristics define the feeding relationships between insect and host (Tallamy and Wood 1986). In *A. monuste*, imperfections in the female egg distribution and reduced mobility at the end of larval phase were observed (Barros-Bellanda and Zucoloto 2003). Female oviposition indiscrimination may influence mobility of these caterpillars, and this in turn influences alteration of the female behavior. Kale is frequently found in big green gardens, influencing oviposition and larval dispersal when necessary. The dispersal behavior is successful in sites with high host density (Le Masurier 1994; Barros-Bellanda and Zucoloto 2003) such as large plantations (Le Masurier 1994).

Species with mobile larvae always show less discriminatory oviposition (e.g., Tammaru et al. 1995). *A priori*, it seems more logical that females characteristics precede and influence larvae characteristics because females are frequently better "equipped" to make choices among potential host plants (Price 1994), and the preference for oviposition seems to be ecologically and evolutionarily more plastic than the larval performance (Janz et al. 1994). Species such as *Charidryas harrissi* S., *Battus philenor* L., and *P. rapae* must be a few millimeters away from the host plant to detect it as food (Dethier 1959; Rausher 1979; Cain et al. 1985, respectively). In the evolutionary process, as less discriminatory ovipositions occur, natural selection favors more mobile larvae and also "allows" the survival of immatures from females that oviposit with less discrimination (Janz and Nylin 1997).

12.4.5 Feeding on Alternative Sources (Nonvegetal)

The herbivorous insect evolutionary success is surprising because plants have low protein levels in their tissues, making them a relatively poor feeding resource (Southwood 1978). Consequently, several biologists have suggested that omnivory precede generalized herbivory in insects that ingest a mixed diet (pteridophytes reproductive tissues, angiosperm floema, dead animal and vegetal tissues, and fungi) (Bernays 1998); afterwards, specialization by specific vegetal taxa has occurred (Dethier 1954).

Generalist insects ingest several feeding items, such as the *Schistocerca americana* (Drury) grasshopper that feeds on up to 20 different vegetal items in only one day (Bernays 1998). In these conditions, it seems probable that all essential nutrients for the insect to live can be acquired; that acquisition is more complicated in herbivorous specialists. As they feed on few or only one vegetal species, the strategy to obtain all the essential nutrients is based on the intake of alternative nonvegetal items as fungi, dead animal remains, exoskeletons, and spores (Whitman et al. 1994; Bernays 1998). Cannibalism may also be considered an important item in this discussion (Barros-Belanda and Zucoloto 2001, 2005; see Chapter 8).

Cannibalism is a common behavior in species other than Lepidoptera; this suggests basic tolerance for a diet based on animal proteins, both for species that cluster or do not cluster eggs, and for species with distinct feeding habits (i.e., generalist and specialist herbivorous). In *A. monuste orseis*, all instars exert cannibalism in the field (Table 12.3), and it is important to mention that ingested eggs are healthy and that egg ingestion occurs in the presence of abundant food (kale leaves). Egg cannibalism also occurs within the same oviposition (Barros-Belanda and Zucoloto 2005) (Table 12.4).

The main incentive for *U. ornatrix* (Lepidoptera, Arctiidae) cannibal behavior is alkaloid deficiency as this substance is important for chemical protection against potential predators and for the mating success (Bogner 1996). In this case, species specialty in recognizing plants due to a chemical marker presence is an incentive for the occurrence of cannibalism. Even so, it is believed that cannibalism is more related to the generalist herbivorous species than to the specialists (Bernays 1998).

Another interesting consequence of cannibalism is the reduced risk of attack by predators and parasitoids observed in *Spodoptera frugiperda* (J. E. Smith) immatures. The cannibal behavior reduces the immatures' density and the higher the number of existing immatures probably the greater is the predation and parasitism risk (Chapman et al. 2000). On the other hand, the conspecifics intake increases the risk of contracting diseases by ingesting contaminated individuals (Boots 1998; see Chapter 8).

12.4.6 Feeding Periods

The temporal analysis of insect feeding behavior has received little attention as compared to other aspects such as the intake regulation mechanisms (Bernays and Simpson 1982) or the host choice (Reynolds et al. 1986). Research with some caterpillars such as *M. sexta* (Reynolds et al. 1985; Bernays and Woods

TABLE 12.3

Eggs Ingested by *Ascia monuste* Caterpillars of Different Ages in the Field during 24 h

Replicate	L2	L3	L4	L5
1	4	11	25	24
2	20	10	31	28
3	12	6	25	40
4	17	7	30	39
5	7	12	22	40
6	11	21	29	30
7	15	12	26	40
8	11	24	33	40
9	3	11	25	40
10	8	19	32	40
11	13	20	31	32
12	15	12	25	34
Mean (± SD)	11.3 ± 5.1	13.7 ± 5.8	27.8 ± 3.6	35.5 ± 5.7

Note: L2 = second instar larvae with 30 available eggs; L3 = third instar larvae with 30 available eggs; L4 = fourth instar larvae with 40 available eggs; L5 = fifth instar larvae with 40 available eggs.

2000), *Helicoverpa armigera* (Hübner) (Raubenheimer and Barton-Browne 2000), and *Bombyx mori* L. (Nagata and Nagasawa 2006) have shown that feeding patterns (i.e., frequency and duration of meals) vary according to the type of food (Reynolds et al. 1986; Timmins et al. 1988; Bernays and Singer 1998), and can provide information about the caterpillar physiological responses as far as food is concerned.

The main larval need is to maximize growth rate while avoiding risks. In that sense, the feeding period is relatively more dangerous than other behaviors such as rest, since volatiles released by both the larvae movements and the consumed plant can catch predators' attention. The resting period is important to maximize the digestive efficiency (Bernays and Wood 2000).

TABLE 12.4

Ascia monuste Cannibalism inside Different Size Ovipositions

Oviposition	SP	AM	BG
1	0	0	16.4
2	0	9.1	8.1
3	12.9	0	14.1
4	23.6	9.1	18.4
5	0	0	10.0
6	0	17.5	15.3
7	0	10.0	10.0
8	0	17.5	15.3
9	22.8	0	8.1
10	0	9.1	14.1
11	0	16.4	10.0
12	12.9	10.0	11.5
13	0	0	18.4
14	24.3	0	12.9
15	0	10.0	14.1
Mean (± SD)	6.4 ± 9.9 a	7.2 ± 6.8 a	13.1 ± 3.4 b

Note: SP = small posture; AM = average posture; BG = big posture. Mean ± SD followed by different letters differ significantly in the groups (Student–Newman–Keuls Test, $P < 0.05$).

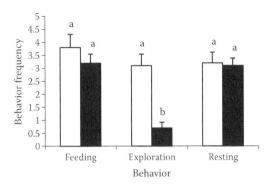

FIGURE 12.6 Behavioral events frequency (feeding, exploitation and rest) of fourth instar *A. monuste* caterpillars fed with strictly phytophagous (white bars) and mixed (leaves + eggs) (black bars) diets. Means ± SEM followed by different letters on the same behavioral event differed significantly (Mann–Whitney rank sum test, $P < 0.05$).

In most insects, the feeding and the resting periods are randomly distributed as well as their duration (Reynolds et al. 1986); ingestion of conspecific eggs (protein source) by *A. monuste*, for instance, did not influence feeding periods duration (Santana et al. 2011). *A. monuste* larvae feed during approximately 20% of the time, followed by exploration (31%), and rest (49%).

In other cases, there is a strong correlation between feeding period duration and the period that precedes feeding; *M. sexta* fifth instar caterpillars feed in turns and grow in the same rhythm with artificial and natural diets (tobacco leaves) although they spend proportionately more time feeding on tobacco than on the artificial diet (Reynolds et al. 1986). Feeding was constant also at night, and feeding periods became longer and resting periods shorter as development progressed (Reynolds et al. 1986).

The main difference between strictly phytophagous and mixed (leaves and eggs) feedings is the higher frequency of exploratory behaviors when there are no eggs in the diet (Figure 12.6). The exploratory behavior is important for the perception of predators and determinant for adequate food selection. The low frequency of caterpillar exploratory behavior with mixed feeding occurs because the available eggs are concentrated in the oviposition site, the area to be explored is smaller, and the homogeneity of forms, textures, and constitution of the same oviposition eggs is greater than that of the leaves (Santana et al. 2011). The long resting period in both groups can suggest more attention in selecting the resting site; in nature, that selection can increase the survival chances if the site chosen by the caterpillars is less obvious for predators and parasites.

Field studies conducted with *M. sexta* caterpillars in *Datura wrightii* plants shown that the insects move little, the intervals of feeding are regular, and they rest after beginning to feed, and each individual acts differently (Bernays and Woods 2000). As feeding and resting intervals did not show temperature influences, an endogenous neural oscillation may control the feeding rhythm and influence the entire feeding pattern. The meaning of the endogenous neural oscillation is discussed, particularly regarding the need for vigilance (Bernays and Woods 2000). Another field study conducted by Bernays et al. (2004) show differences in the foraging efficiencies between generalist and specialist Lepidoptera: the generalists spend more time moving, reject their potential host plants more often, take longer to initiate feeding after inspection, and have shorter feeding periods as compared to specialists. Shorter feeding periods can be a consequence of the generalists' reduced vigilance, thus reducing intake of possibly toxic plants (Bernays et al. 2004).

12.5 Tritrophic Relationships: Presence of Natural Enemies

Studies with Lepidoptera are important to understand interactions between insects and plants, though the exclusion of natural enemies from these studies does not permit a complete understanding of that relationship (Price et al. 1980).

In 1988, Bernays and Graham condemned the emphasis given to the host plant chemistry in detriment of the predators' influence in the herbivorous diet evolution. It is evident that the feeding resources used by the insects generate a very active selective force on their lives' historical features (Rhoades 1985); if these resources have low nutritional quality, low persistence, and/or availability (Tallamy and Wood 1986), their exposition to predators, parasitoids, and/or competitors may be affected (Price 1984) Consequently, the species' natural enemies and feeding resources intensely influence the population spatial distribution. About 70% to 85% of the studies provide evidence that food and/or oviposition site choices prioritize finding enemy free spaces (Berdegue et al. 1996).

Caterpillars present behaviors that can protect them against predator and parasite attacks such as a fall from the plant, regurgitation, defecation, and fight (Bernays and Woods 2000; Bernays et al. 2004). Caterpillar natural enemies include Odonata, several Hymenoptera and Diptera, as well as all kinds of vertebrates (Brown and Freitas 1999). Due to the difficulty to obtain direct predation determinations in small animals, their extinction has sometimes been used to determine predation. Following this objective, several studies focused on the importance of predation as a factor that affects the herbivorous insect mortality. The invertebrate predators are more important to small larvae while vertebrates became the main predators of bigger species. The predator evolutionary importance as an imminent and constant danger has been well discussed and it is known that they have always influenced the immature lepidopteran food utilization strategies (Heinrich 1993).

Clark and Faeth (1997) studied the effect of different predators on *Chlosyne lacinia* Geyer (Lepidoptera, Nymphalidae) caterpillars' survival (Figure 12.7); in the control group where no predator was excluded, mortality was significantly higher as compared to other groups, and ants were the most important predators. The movement of caterpillars is being associated to the predation risk offered by ants (Bergelson and Lawton 1988), pentatomids (Marston et al. 1978), spiders (de Boer 1971), and birds (Clark and Faeth 1997). Bernays's studies (1977) have shown that the feeding period was 100 times more dangerous than the resting period for *Uresiphita reversalis* Guenée caterpillars (aposematic coloration) and three times more dangerous for *M. sexta* (cryptic) caterpillars.

A clear example of the predation importance for the caterpillar–leaf interaction is described by Damman (1987): in the field, although the younger leaves provide better performances for *Omphalocera munroei* Martin (Lepidoptera, Pyralidae) larvae, this species preferentially feeds on *Asimina* spp. (Annonaceae) old leaves; the reasons for that choice are the better conditions to construct shelters in old leaves, an important step to protect against predators. In this case, protection against natural enemies was more important than efficient nutrition for these caterpillars. Predation also influences the females'

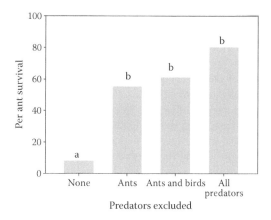

FIGURE 12.7 The *Chloyne lacinia* mean larval survival varied with the predator exclusion treatment: No predator was excluded, ants were excluded, ants and birds were excluded, and all the predators were excluded. Different letters above the columns indicate significant differences ($P < 0.05$) among the means (Tukey's HSD Test). (Modified from Clark, B. R., and S. H. Faeth, *Ecol. Entomol.*, 22, 408–15, 1998.)

oviposition behavior; in the field, *A. monuste* oviposition generally occurs in less nutritive but more protected parts of the plants (the leaf medial part) (Catta-Preta and Zucoloto 2003).

12.6 Conclusions and Research Suggestions

Herbivorous insect activity is very intense: first, they can reduce dramatically the plant fitness, directly and indirectly; second, they support an equal number of invertebrate predator and parasitoid species; and third, they are the largest feeding resource for a diversity of birds, lizards, and small mammals.

It is evident that many ecological, physiological, and behavioral insect processes are directly or indirectly related to feeding and nutritional contexts. Thus, it is extremely important to understand several aspects related to the feeding behavior as nutritional needs, habits, and preferences, in addition to the consequences of these aspects in adaptive value parameters. The knowledge of the insect preference by certain plants, for instance, permits the utilization of these plants as traps, enabling to control a certain population. Reducing the plant appearance to natural enemies by neighboring plants of different species is an important component of "associative resistance," a phenomenon frequently explored by organic gardeners (Feeny 1976). A clear and broad understanding of the caterpillar ecological nutrition is an important advancement not only regarding ecology and evolution but also for applied areas of economical interest, since the order Lepidoptera as a whole directly influences the vegetal losses suffered by agricultural crops (Nielsen and Common 1991).

In this chapter, basic ideas about the defoliator caterpillars' feeding biology were developed, making a relationship between nutrition (presence or absence of nutrients and/or secondary compounds) and influence of a varied environment, approaching several feeding behavioral aspects and the physiology of this group, taking into consideration ecological and evolutionary discussions. Suggestions for research related to the theme of this chapter include (1) The temporal analysis of the caterpillar feeding behavior in nature and the strategies to avoid predators and parasitoids during this process, and (2) the function of nonvegetal items and of egg cannibalism in feeding immatures of other Lepidoptera species. An important question to be answered is whether the specificity of a species would facilitate, turn difficult, or be irrelevant regarding caterpillar cannibal behavior.

REFERENCES

Ahman, I. 1985. Oviposition behavior of *Dasineura brassicae* on a high- versus a low-quality *Brassica* host. *Entomol. Exp. Appl.* 39:247–53.

Allen, P. E. 2010. Group size effects on survivorship and adult development in the gregarious larvae of *Euselasia chrysippe* (Lepidoptera, Riodinidae). *Insect. Soc.* 57:199–204.

Alonso, A., and C. M. Herrera. 1996. Variation in herbivory within and among plants of *Daphne laureola* (Thymelaceae): correlation with plant size and architecture. *J. Ecol.* 84:495–502.

Amano, K. 1987. Studies on the intraspecific competition in dung-breeding flies. III. Pupal size and mortality in immature stages under various larval density conditions in *Musca hevei* Villeneuve (Diptera, Muscidae). *Appl. Ent. Zool.* 22:59–67.

Awmack, C. S., and S. R. Leather. 2002. Host plant quality and fecundity in herbivorous insects. *Annu. Rev. Entomol.* 47:817–44.

Barbosa, P., V. A. Krischik, and C. G. Jones. 1991. *Microbial Mediation of Plant–Herbivore Interactions.* New York: Wiley & Sons.

Barros, H. C. H., and F. S. Zucoloto. 1999. Performance and host preference of *Ascia monuste* (Lepidoptera, Pieridae). *J. Insect Physiol.* 45:7–14.

Barros-Bellanda, H. C. H., and F. S. Zucoloto. 2001. Influence of chorion ingestion on the performance of *Ascia monuste* and its association with cannibalism. *Ecol. Entomol.* 26:557–61.

Barros-Bellanda, H. C. H., and F. S. Zucoloto. 2002. Effects of intraspecific competition and food deprivation on the immature phase of *Ascia monuste orseis* (Lepidoptera, Pieridae). *Iheringia* 92:93–8.

Barros-Bellanda, H. C. H., and F. S. Zucoloto. 2003. Importance of larval migration (dispersal) for the survival of *Ascia monuste* (Godart) (Lepidoptera: Pieridae). *Neotrop. Entomol.* 30:11–7.

Barros-Bellanda, H. C. H., and F. S. Zucoloto. 2005. Egg cannibalism in *Ascia monuste* in the field; opportunistic, preferential and very frequent. *J. Ethol.* 23:133–38.

Baumgartner, J., and M. Severini. 1987. Microclimate and arthropod phonologies: The leaf miner *Phyllonoricter blancardella* F. (Lep.) as an example. In *Agrometeorology,* ed. F. Prodi, F. Rossi, and C. Cristoferi, 225–43. Cesena, Italy: Fondazione Cesena Agricultura Public.

Beckwith, R. C. 1976. Influence of host foliage on the Douglas-fir tussock moth. *Environ. Entomol.* 5:73–7.

Berdegue, M., J. J. Trumble., D. Hare, and R. A. Redak. 1996. Is it enemy-free space? The evidence for terrestrial insects and fresh-water arthropods. *Ecol. Entomol.* 21:203–17.

Bergelson, J. M., and J. H. Lawton. 1988. Does foliage damage influence predation on the insect herbivores of bird? *Ecology* 69:434–45.

Berger, A. 1992. Larval movements of *Chilo partellus* (Lepidoptera: Pyralidae) within and between plants: Timing, density responses and survival. *Bull. Entomol Res.* 82:441–8.

Bernays, E. A. 1997. Feeding by caterpillars is dangerous. *Ecol. Entomol.* 22:121–3.

Bernays, E. A. 1998. Evolution of feeding behavior in insect herbivores. *Bioscience* 48:35–44.

Bernays, E. A. 1987. Evolutionary contrasts in insects: Nutritional advantages of holometabolous development. *Physiol. Entomol.* 11:377–82.

Bernays, E. A., and S. J. Simpson. 1982. Control of food intake. *Adv. Insect Physiol.* 16:59–118.

Bernays, E. A., and D. H. Janzen. 1988. Saturniid and sphingid caterpillars: Two ways to eat leaves. *Ecology* 69:1153–60.

Bernays, E. A., and M. S. Singer. 1998. A rhythm underlying feeding behaviour in a highly polyphagous caterpillar. *Physiol. Entomol.* 23:295–302.

Bernays, E. A., and M. Graham. 1988. On the evolution of host specificity in phytophagous arthropods. *Ecology* 69:886–92.

Bernays, E. A., R. F. Chapman, and T. Hartmann. 2002. A highly sensitive taste receptor cell for pyrrolizidine alkaloids in the lateral galeal sensillum of a polyphagous caterpillar. *Estigmene acrea. J. Comp. Physiol. A* 188:715–723.

Bernays, E. A., M. S. Singer, and D. Rodrigues. 2004. Foraging in nature: Foraging efficiency and oscillations in caterpillars with different diet breadths. *Ecol. Entomol.* 29:389–97.

Bernays, E. A., and H. A. Woods. 2000. Foraging in nature by larvae of *Manduca sexta*–influenced by an endogenous oscillation. *J. Insect Physiol.* 46:825–36.

Bianchi, V., and G. R. P. Moreira. 2005. Preferência alimentar, efeito da planta hospedeira e da densidade larval na sobrevivência e desenvolvimento de *Dione juno juno* (Cramer) (Lepidoptera, Nymphalidae). *Rev. Bras. Zool.* 22:43–50.

Bittencourt-Rodrigues, R. S., and F. S. Zucoloto. 2005. Effect of host age on the oviposition and performance of *Ascia monuste* Godart (Lepidoptera: Pieridae). *Neotrop. Entomol.* 34:169–75.

Bittencourt-Rodrigues, R., and F. S. Zucoloto. 2009. How feeding on young and old leaves affects the performance of *Ascia monuste orseis* (Godart) (Lepidoptera, Pieridae). *Rev. Bras. Entomol.* 53:102–6.

de Boer, M. H. 1971. A colour polymorphism in caterpillars of *Bupalus piniarius* (L.). *Netherlands J. Zool.* 21:611–16.

de Boer, G. 1993. Plasticity in food preference and diet-induced differential weighting of chemosensory information in larval *Manduca sexta. J. Insect Physiol.* 39:17–24.

Boggs, C. L., and K. D. Freeman. 2005. Larval food limitation in butterflies: Effects on adult resource allocation and fitness. *Oecologia* 144:253–61.

Bogner, F. X. 1996. Interspecific advantage results in intraspecific disadvantage: Chemical protection versus cannibalism in *Utetheisa ornatrix* (Lepdoptera: Arctiidae). *J. Chem. Ecol.* 22:1439–51.

Boots, M. 1998. Cannibalism and the stage-dependent transmission of a viral pathogen of the Indian meal moth, *Plodia interpunctella. Ecol. Entomol.* 23:118–22.

Bowdan, E. 1988. Microstructure of feeding by tobacco hornworm caterpillars, *Manduca sexta. Entomol. Exp. Appl.* 47:127–36.

Brattsten, L. B., C. F. Wilkinson, and T. Eisner. 1977. Herbivore plant interactions mixed function oxidases and secondary plant. *Science* 196:1349–52.

Brown, E. S., and C. Dewhurst. 1975. The genus *Spodoptera* (Lepidoptera, Noctuidae) in Africa and the Near East. *Bull. Entomol. Res.* 65:221–62.

Brown, Jr., K. S., and A. V. L. Freitas. 1999. Lepidoptera. In *Biodiversidade do Estado de São Paulo, Brasil: Síntese do Conhecimento ao Final do Século XX*, ed. C. R. F. Brandão, and E. M. Cancello, 225–43. São Paulo, Brazil: FAPESP.

Calow, P. 1977. Ecology, evolution and energetics: A study in metabolic adaptation. *Adv. Ecol. Res.* 10:1–62.

Cain, M. L., J. Eccleston, and P. M. Kareiva. 1985. The influence of food plant dispersion on caterpillar searching success. *Ecol. Entomol.* 10:1–7.

del Campo, M. L., and J. A. A. Renwick. 1999. Dependence on host constituents controlling food acceptance by *Manduca sexta* larvae. *Entomol. Exp. Appl.* 93:209–15.

Carter, D. J. 1987. Introduction to larvae. In *Lepidoptera. CIE Guides to Insects of Importance to Man*, ed. C. R. Betts, 191–224. London: Wallingford CAB International.

Catta-Preta, P. D., and F. S. Zucoloto. 2003. Oviposition behavior and performance aspects of *Ascia monuste* (Godart, 1919) (Lepidoptera, Pieridae) on kale (*Brassica oleracea* var. *acephala*). *Rev. Bras. Entomol.* 47:169–74.

Chan, B. G., A. C. Waiss, R. G. Binder, and C. A. Elliger. 1978. Inhibition of lepidopterous larval growth by cotton constituents. *Entomol. Exp. Appl.* 24:97–100.

Chapman, R. F., S. Woodhead, and E. A. Bernays. 1983. Survival and dispersal of young larvae of *Chilo partellus* (Swinhoe) (Lepidoptera: Pyralidae) in two cultivars of sorghum. *Bull. Entomol. Res.* 73:65–74.

Chapman, R. F. 1995a. Mechanics of food handling by chewing insects. In *Regulatory Mechanisms in Insect Feeding*, ed. R. F. Chapman, and G. de Boer, 3–31. New York: Chapman and Hall.

Chapman, R. F. 1995b. Chemosensory regulation of feeding. In *Regulatory Mechanisms in Insect Feeding*, ed. R. F. Chapman and G. de Boer, 101–36. New York: Chapman and Hall.

Chapman, W. J., T. Williams, A. M. Martinez, J. Cisneros, P. Caballero, R. D. Cave, and D. Goulson. 2000. Does cannibalism in *Spodoptera frugiperda* (Lepidoptera: Noctuidae) reduce the risk of predation? *Behav. Ecol. Sociobiol.* 48:321–7.

Clark, B. R., and S. H. Faeth. 1998. The consequences of larval aggregation in the butterfly *Chlosyne lacinia. Ecol. Entomol.* 22:408–15.

Damman, H. 1987. Leaf quality and enemy avoidance by the larvae of a pyralid moth. *Ecology* 68:88–97.

Damman, H., and P. Feeny. 1988. Mechanisms and consequences of selective oviposition by the zebra swallowtail butterfly. *Animal Beh.* 36:563–73.

van Dam, N. M., and J. D. Hare. 1998. Biological activity of *Datura wrightii* glandular trichome exudates against *Manduca sexta. Oecologia* 122:371–9.

Dethier, V. G. 1954. Evolution of feeding preferences in phytophagous insects. *Evolution* 8:33–54.

Dethier, V. 1959. Food-plant distribution and density and larval dispersal as factors affecting insects populations. *Can. Entomol.* 91:581–96.

Dix, M. E., R. A. Cunningham, and R. M. King. 1996. Evaluating spring cankerworm (Lepidoptera: Geometridae) preference for Siberian elm clones. *Environ. Entomol.* 25:56–62.

Dodds, K. A., K. M. Clancy, K. J. Leyva, D. Greenberg, and P. W. Price. 1996. Effects of foliage age class on Western spruce budworm oviposition choice and larval performance. *Great Basin Nat.* 56:135–41.

Dow, J. A. T. 1984. Extremely high pH in biological systems: A model for carbonate transport. *Am. J. Physiol.* 246:633–6.

Duffey, S. S. 1986. Plant glandular trichomes: Their partial role in defence against insects. In *Insect and the Plant Surface*, ed. B. E. Juniper and T. R. E. Southwood, 121–34. London: Arnold Publishers.

Dussourd, D. E., C. A. Harvis, J. Meinwald, and T. Eisner. 1991. Pheromonal advertisement of a nuptial gift by a male moth (*Utetheisa ornatrix*). *Proc. Natl. Acad. Sci. USA* 88:9224–7.

Dussourd, D. E. 1999. Behavioural sabotage of plant defence: Do vein cuts and trenchs reduce exposure to plant exudates? *J. Insect Behav.* 12:501–15.

Edwards, P. J., and S. D. Wratten. 1981. *Ecologia das interações entre insetos e plantas*. São Paulo, Brazil: EPU.

Ehrlich, P. R., and P. H. Raven. 1964. Butterflies and plants: A study in coevolution. *Evolution* 18:586–608.

Erickson, J. M. 1973. The utilization of various *Asclepias* species by larvae of the monarch butterfly, *Danaus plexippus. Psyche* 80:230–44.

Feeny, P. 1976. Plant apparency and chemical defense. *Rec. Adv. Phytochem.* 10:1–40.

Felipe, M. C., and F. S. Zucoloto. 1993. Estudos de alguns aspectos da alimentação em *Ascia monuste* Godart (Lepidoptera, Pieridae). *Rev. Bras. Zool.* 10:333–41.

Ferro, V. G., P. R. Guimarães, and J. R. Trigo. 2006. Why do larvae of *Utetheisa ornatrix* penetrate and feed in pods of *Crotalaria* species? Larval performance vs. chemical and physical constraints. *Entomol. Exp. Appl.* 121:23–9.

Fretwell, S. D. 1972. *Population in a Seasonal Environment.* Princeton, NJ: Princeton University Press.

Gaston, K. J., D. Reavey, and G. Valladares. 1991. Changes in feeding habit as caterpillars grow. *Ecol. Entomol.* 16:339–44.

Glendinning, J. I. N. Nelson, and E. A. Bernays. 2000. How do inositol and glucose modulate feeding in *Manduca sexta* caterpillars? *J. Exp. Biol.* 203:1299–315.

Godfrey, G. L., J. S. Miller, and D. J. Carter. 1989. Two mouthparts modifications in larval notodontidae (Lepidoptera)—Their taxonomic distribution and putative functions. *J. New York Entomol. Soc.* 97:455–70.

Grimes, L. R., and H. H. Neunzig. 1986. Morphological survey of the maxillae in last stage larvae of the suborder Ditrysia (Lepidoptera). *Ann. Entomol. Soc. Am.* 79:491–526.

Hardy, T. N., K. Clay, and A. M. J. Hammond. 1985. Fall armyworm (*Spodoptera frugiperda*) (Lepidoptera: Noctuidae): A laboratory bioassay and larval preference study for the fungal endophyte of perennial ryegrass (*Lolium perenne*). *J. Econ. Entomol.* 15:1083–9.

Heinrich, B. 1993. How avian predators constrain caterpillar foraging. In *Caterpillars: Ecological and Evolutionary Constraints on Foraging,* ed. N. E. Stamp, and T. M. Casey, 224–47. New York: Chapman and Hall.

Hedin, P. A., W. P. Williams, F. M. Davis, and P. M. Buckley. 1990. Roles of amino acids, protein, and fiber in leaf-feeding resistance of corn to the fall armyworm. *J. Chem. Ecol.* 16:1977–95.

Hulley, P. E. 1988. Caterpillar attacks plant mechanical defence by mowing trichomes before feeding. *Ecol. Entomol.* 13:239–41.

Hunter, A. F., and J. S. Elkinton. 2000. Effects of synchrony with host plant on populations of a spring feeding lepidopteran. *Ecology* 81:1248–61.

Janz, N., S. Nylin, and N. Wedell. 1994. Host plant utilization in the comma butterfly: Sources of variation and evolutionary implications. *Oecologia* 99:132–40.

Janz, N., and S. Nylin. 1997. The role of female search behaviour in determining host plant range in plant feeding insects: A test of the information processing hypothesis. *Proc. R. Entomol. Soc. Lond. B.* 264:701–07.

Kaminski, L. A., R. Dell'Erba, and G. R. P. Moreira. 2008. Morfologia externa dos estágios imaturos de heliconíneos neotropicais: VI. *Dione moneta moneta* Hübner (Lepidoptera, Nymphalidae, Heliconiinae). *Rev. Bras. Entomol.* 52:13–23.

Kantiki, L. M., and J. K. O. Ampofo. 1989. Larval establishment and feeding behavior of *Eldana saccharina* Walker (Lepidoptera: Pyralidae) on maize and sorghum plants. *Insect Sci. Appl.* 10:577–82.

Kawasaki, N., T. Miyashita, and Y. Kato. 2009. Leaf toughness changes the effectiveness of larval aggregation in the butterfly *Bryasa alcinous bradanus* (Lepidoptera: Papilionidae). *Entomol. Science* 12:135–40.

Kinkel, L. 1997. Microbial population dynamics on leaves. *Annu. Rev. Phytopathol.* 35:327–47.

Krenn, H. W. 1990. Functional morphology and movements of the proboscis of Lepidoptera (Insecta). *Zoomorphology* 110:105–14.

Krenn, H. W. 2000. Proboscis musculature in the butterfly *Vanessa cardui* (Nymphalidae, Lepidoptera): Settling the proboscis recoiling controversy. *Acta Zool.* 81:259–66.

Krenn, H. W., J. D. Plant, and N. U. Szucsich. 2005. Mouthparts of flower-visiting insects. *Arthropod Struct. Dev.* 34:1–40.

Krenn, H. W. 2010. Feeding mechanisms of adult lepidoptera: Structure, function, and evolution of the mouthparts. *Annu. Rev. Entomol.* 55:307–27.

Kristensen, N. P. 1976. Remarks on the family-level phylogeny of butterflies (Insecta, Lepidoptera, Rhopalocera). *Z. Zool. Syst. Evol.* 14:25–33.

Kristensen, N. P. 1984. The larval head of *Agathiphaga* (Lepidoptera, Agathiphagidae) and the lepidopteran ground plant. *Syst. Ent.* 9:63–81.

Kristensen, N. P., M. J. Scoble, and O. Karsholt. 2007. Lepidoptera phylogeny and systematics: The state of inventorying moth and butterfly diversity. *Zootaxa* 1668:699–747.

Le Masurier, A. D. 1994. Costs and benefits of egg clustering in *Pieris brassicae*. *J. Anim. Ecol.* 63:677–85.

Levot, G. W., K. R. Brown, and E. Shipp. 1979. Larval growth of some calliphorid and sarcophagid Diptera. *Bull. Entomol. Res.* 69:469–75.

Lin, S. Y. H., J. T. Trumble, and J. Kumamoto. 1987. Activity of volatile compounds in glandular trichomes of *Lycopersion* spp. against two insect herbivores. *J. Chem. Ecol.* 13:837–50.

Lucas, P. W., I. M. Turner, N. J. Dominy, and N. Yamashita. 2000. Mechanical defences to herbivory. *Ann. Bot.* 86:913–20.

Ma, W. C. 1972. Dynamics of feeding responses in *Pieris brassicae* Linn. as a function of chemosensory input: A behavioural and electrophysiological study. *Med. Land. Wag.* 72:1–162.

Manuwoto, S., and J. M. Scriber. 1985. Neonate larval survival of European corn borer, *Ostrinia nubilalis*, on high and low DIM-BOA genotypes of maize: Effects of light intensity and degree of insect inbreeding. *Agric. Ecosyst. Environ.* 14:21–36.

Marques, R. S. A., E. S. A. Marques, and P. W. Price. 1994. Female behavior and oviposition choices by an eruptive herbivore, *Disnycha pluriligata* (Coleoptera: Chysomelida) *Environ. Entomol.* 23:887–92.

Marston, N. L., G. T. Schmidt, K. D. Biever, and W. A. Dickerson. 1978. Reaction of five species of soybean caterpillars to attack by the predator, *Podisus maculiventris*. *Environ. Entomol.* 7:53–6.

Mattson, W. J. 1980. Herbivory in relation to plant nitrogen content. *Annu. Rev. Ecol. Syst.* 11:119–61.

Mattson, Jr., W. J., and J. M. Scriber. 1987. Nutritional ecology of insect folivores of woody plants: Nitrogen, water, fiber, and mineral considerations. In *Nutritional Ecology of Insects, Mites, Spiders and Related Invertebrates*, ed. F. Slansky and J. G. Rodriguez, 105–46. New York: Wiley & Sons.

Miller, J. R., and K. L. Strickler. 1984. Finding and accepting host plants. In *Chemical Ecology of Insects*, ed. W. J. Bell, and R. T. Cardé, 127–57. Sunderland, MA: Sinauer Associates.

Nagata, S., and H. Nagasawa. 2006. Effects of diet-deprivation and physical stimulation on the feeding behaviour of the larvae of the silkworm, *Bombyx mori*. *J. Insect Physiol.* 52:807–15.

Nielsen, E. S. 1989. Phylogeny of major lepidopteran groups. In *The Hierarchy of Life Molecules and Morphology in Phylogenetic Analysis*, ed. B. Fernholm, 281–94. Amsterdam, the Netherlands: Elsevier.

Ohsaki, N., and Y. Sato. 1994. Food plant choice of *Pieris* butterflies as a trade-off between parasitoid avoidance and quality of plants. *Ecology* 75:59–68.

Ojeda-Avila, T., H. A. Woods, and R. A. Raguso. 2003. Effects of dietary variation on growth, composition, and maturation of *Manduca sexta* (Sphingidae: Lepidoptera). *J. Insect Physiol.* 49:293–306.

Parry, D., J. R. Spence, and W. J. A. Volney. 1998. Budbreak phenology and natural enemies mediate survival of first-instar forest tent caterpillar (Lepidoptera: Lasiocampidae). *Environ. Entomol.* 27:1368–74.

Peterson, S. C., F. E. Hanson, and J. D. Warthen, Jr. 1993. Deterrence coding by a larval *Manduca* chemosensory neurone mediating rejection of a non-host plant *Canna generalis* L. *Physiol. Entomol.* 18:285–95.

Pianka, E. R. 1983. *Evolutionary Ecology*. New York: Harper and Row.

Price, P. W. 1984. *Insect Ecology*. New York: Wiley & Sons.

Price, P. W. 1994. Phylogenetic constraints, adaptive syndromes, and emergent properties: From individuals to population dynamics. *Res. Popul. Ecol.* 36:3–14.

Price, P. W., C. E. Bouton, P. Gross, B. A. McPherson, J. A. Thompson, and A. E. Weiss. 1980. Interaction among three trophic levels: Influence of plant interactions between insect herbivores and natural enemies. *Annu. Rev. Ecol. Syst.* 11:41–65.

Raubenheimer, D., and L. Barton-Browne. 2000. Developmental changes in the patterns of feeding in fourth- and fifth-instar *Helicoverpa armigera* caterpillars. *Physiol. Entomol.* 25:390–9.

Rausher, M. D. 1979. Larval habitat suitability and oviposition preference in three related butterflies. *Ecology* 60:503–11.

Remmert, H. 1982. *Ecologia*. São Paulo, Brazil: EPU.

Reynolds, S. E., M. R. Yeomans, and W. A. Timmins. 1986. The feeding behaviour of caterpillars (*Manduca sexta*) on tobacco and on artificial diet. *Physiol. Entomol.* 11:39–51.

Rhainds, M., C. Sadof, and C. Quesada. 2010. Dispersal and development of bagworm larvae (Lepidoptera: Psychidae) on three host plants. *J. Appl. Entomol.* 134:81–90.

Rhoades, A. F. 1985. Offensive–defensive interactions between herbivores and plants: Their relevance in herbivore population dynamics and ecological theory. *Am. Nat.* 125:205–38.

Roden, D. B., J. R. Miller, and G. A. Simmons. 1992. Visual stimuli influencing orientation by larval gypsy moth, *L. dispar* (L.). *Can. Entomol.* 134:284–304.

Rodrigues, D., P. H. S. Maia, and J. R. Trigo. 2010. Sabotaging behaviour and minimal latex of *Asclepias curassavica* incur no cost for larvae of the southern monarch butterfly *Danaus erippus*. *Ecol. Entomol.* 35:504–13.

Ronnås, C., S. Larsson, A. Pitacco, and A. Battisti. 2010. Effects of colony size on larval performance in a processionary moth. *Ecol. Entomol.* 35:436–45.

Santana, A. F. K., R. C. Zago, and F. S. Zucoloto. 2011. Effects of sex, host-plant deprivation and presence of conspecific immatures on the cannibalistic behavior of wild *Ascia monuste orseis* (Godart) (Lepidoptera, Pieridae). *Rev. Bras. Entomol.* 55:95–101.

Schoonhoven, L. M. 1972. Gustation and food-plant selection in some lepidopterous larvae. *Entomol. Exp. Appl.* 12:555–64.

Schoonhoven, L. M. 1978. Long-term sensitivity changes in some insect taste receptors. *Drug Res.* 28:2367–78.

Schroeder, L. A. 1975. Effect of food deprivation on the efficiency of utilization of dry matter, energy, and nitrogen by larvae of the *Calocalpe undulata*. *Ann. Entomol. Soc. Am.* 69:55–8.

Scriber, J. M., and F. Slansky, Jr. 1981. The nutritional ecology of immature insects. *Annu. Rev. Entomol.* 26:183–211.

Scriber, J. M. 1982. The behavior and nutritional physiology of southern armyworm larvae as a function of plant species consumed in earlier instars. *Entomol. Exp. Appl.* 31:359–69.

Simpson, S. J., and C. L. Simpson. 1990. The mechanisms of nutritional compensation by phytophagous insects. In *Insect–Plant Interactions*, ed. E. A. Bernays, 111–60. Boca Raton, FL: CRC Press.

Simpson, S. J., and D. Raubenheimer. 1993. The central role of the haemolymph in the regulation of nutrient uptake in insects. *Physiol. Entomol.* 18:395–403.

Singer, M. C. 2000. Reducing ambiguity in describing plant–insect interactions: "Preference," "acceptability" and "electivity." *Ecol. Lett.* 3:159–62.

Singer, M. S., E. A. Bernays, and Y. Carriére. 2002. The interplay between nutrient balancing and toxin dilution in foraging by a generalist insect herbivore. *Anim. Behav.* 64:629–43.

Slansky, Jr., F., and J. M. Scriber. 1985. Food consumption and utilization. In *Comprehensive Insect Physiology, Biochemistry and Pharmacology*, ed. G. A. Kerkut and L. I. Gilbert, 67–167. Oxford, UK: Pergamon Press.

Slansky, Jr., F., and J. G. Rodrigues. 1987. Nutritional ecology of insects, mites, spiders, and relates invertebrates: An overview. In *Nutritional Ecology of Insects, Mites, Spiders, and Related Invertebrates*, ed. F. Slansky, and J. G. Rodrigues, 1–69. New York: Wiley.

Slansky, Jr., F., and G. S. Wheeler. 1992. Caterpillar's compensatory feeding response to diluted nutrients leads to toxic allelochemical dose. *Entomol. Exp. Appl.* 65:171–86.

Southwood, T. R. E. 1978. The insect/plant relationship—An evolutionary perspective. In *Insect/Plant Relationships*, ed. H. F. van Emden, 3–32. Oxford, UK: Blackwell.

Städler, E., and F. E. Hanson. 1975. Olfactory capabilities of gustatory chemoreceptors of tobacco hornworm larvae. *J. Comp. Physiol.* 104:97–102.

Stamp, N. E. 1980. Egg deposition patterns in butterflies: Why do some species cluster their eggs rather than deposit them singly? *Am. Nat.* 115:367–80.

Stekolnikov, A. A., and A. I. Korzeev. 2007. The ecological scenario of Lepidopteran evolution. *Entomol. Rev.* 87:830–39.

Tallamy, D. W., and T. K. Wood. 1986. Convergence patterns in subsocial insects. *Annu. Rev. Entomol.* 31:369–90.

Tammaru, T., P. Kaitaniemi, and K. Ruohomaki. 1995. Oviposition choices of *Epirrita cutumnata* (Lepidoptera: Geometridae) in relation to its eruptive population dynamics. *Oikos* 74:296–304.

Telang, A., V. Booton, R. F. Chapman, and D. E. Wheeler. 2001. How female caterpillars accumulate their nutrient reserves. *J. Insect Physiol.* 47:1055–64.

Terra, W. R., A. Valentin, and C. D. Santos. 1987. Utilization of sugars, hemicellulose, starch, protein, fat and minerals by *Erinnyis ello* larvae and the digestion role of their midgut hydrolases. *Insect Biochem.* 17:1143–7.

Thompson, J. N. 1988. Evolutionary ecology of relation between oviposition preference and performance of offspring in phytophagous insects. *Entomol. Exp. Appl.* 47:3–14.

Timmins, W. A., K. Bellward, A. J. Stamp, and S. E. Reynolds. 1988. Food intake, conversion efficiency, and feeding behaviour of tobacco hornworm caterpillars given artificial diet of varying nutrient and water content. *Physiol. Entomol.* 13:303–14.

Usher, B. F., E. A. Bernays, and R. V. Barbehenn. 1988. Antifeedant tests with larvae of *Pseudaletia unipuncta*— Variability of behavioral response. *Entomol. Exp. Appl.* 48:203–12.

van Loon, J. J. A. 1996. Chemosensory basis of feeding and oviposition behaviour in herbivorous insects: A glance at the periphery. *Entomol. Exp. Appl.* 80:1–7.

van Loon, J. J. A., and L. M. Schoonhoven. 1999. Specialist deterrent chemoreceptors enable *Pieris* caterpillars to discriminate between chemically different deterrents. *Entomol. Exp. Appl.* 91:29–35.

van Loon, J. J. A., C. Z. Wang, J. K. Nielsen, R. Gold, and Y. T. Qiu. 2002. Flavonoids from cabbage are feeding stimulants for diamondback moth larvae additional to glucosinolates: Chemoreception and behaviour. *Entomol. Exp. Appl.* 104:27–34.

Visser, J. H. 1986. Host odor perception in phytophagous insects. *Annu. Rev. Entomol.* 31:121–44.

Waldbauer, G. P. 1968. The consumption and utilization of food by insects. *Adv. Insect Physiol.* 5:229–88.

Waldbauer, G. P., R. W. Cohen, and S. Friedman. 1984. Self-selection of and optimal nutrient mix from defined diets by larvae of the cornearworm, *Heliothis zea* (Boddie). *Physiol. Zool.* 57:590–7.

Wannenmacher, G., and L. T. Wasserthal. 2003. Contribution of the maxillary muscles to proboscis movement in hawkmoths (Lepidoptera: Sphingidae): An electrophysiological study. *J. Insect Physiol.* 49:765–76.

Wheeler, D. E., I. Tuchinskaya, N. A. Buck, and B. E. Tabashnik. 2000. Hexameric storage proteins during metamorphosis and egg production in the diamondback moth, *Plutella xylostella* (Lepidoptera). *J. Insect Physiol.* 46:951–8.

White, T. C. R. 1984. The abundance of invertebrate herbivores in relation to the availability of nitrogen in stressed food plant. *Oecologia* 63:90–105.

White, T. C. R. 2005. *Why Does the World Stay Green? Nutrition and Survival of Plant-Eaters*. Canberra, Australia: CSIRO Publishing.

Whitfield, J. B., and K. M. Kjer. 2008. Ancient rapid radiation of insects: Challenges for phylogenetic analysis. *Annu. Rev. Entomol.* 53:449–72.

Whitman, D. W., M. S. Blum, and F. Slansky Jr. 1994. Carnivory in phytophagous insects. In *Functional Dynamics of Phytophagous Insects*, ed. T. N. Ananthakrisnan, 161–205. New Delhi, India: Oxford & IBH Publishing Co. Pvt. Ltd.

Wilson, P. A., P. M. Room, M. P. Zalucki, and S. Chakraboty. 2000. Interaction between *Helicoverpa armigera* and *Colletotrichum gloeosporioides* on the tropical pasture legume *Stylosanthes scabra*. *Aust. J. Agric. Res.* 51:107–12.

Woods, H. A. 1999. Patterns and mechanisms of growth of fifth-instar *Manduca sexta* caterpillars following exposure to low or high-protein food during early instar. *Physiol. Biochem. Zool.* 72:445–54.

Yang, G., K. E. Espelie, B. R. Wiseman, and D. J. Isenhour. 1993. Effect of corn follicular lipids on the movement of fall armyworm (Lepidoptera: Noctuidae) neonate larvae. *Fla. Entomol.* 76:302–16.

Young, A. M., and M. W. Moffett. 1979. Studies on the population biology of the tropical butterfly *Mechanitis isthmian* in Costa Rica. *Am. Mid. Nat.* 101:309–19.

Zalucki, M. P., L. P. Brower, and A. M. Alfonso. 2001. Detrimental effects of latex and cardiac glycosides on survival and growth of first instar monarch butterfly larvae *Danaus plexippus* feeding on the sandhill milkweed *Asclepias humistrata*. *Ecol. Entomol.* 26:212–24.

Zalucki, M. P., A. R. Clarke, and S. B. Malcolm. 2002. Ecology and behavior of first instar larval lepidoptera. *Annu. Rev. Entomol.* 47:361–93.

13

Seed-Sucking Bugs (Heteroptera)

Antônio R. Panizzi and Flávia A. C. Silva

CONTENTS

13.1 Introduction 295
13.2 Food Characteristics (Seeds) 296
    13.2.1 Nutritional Composition 296
    13.2.2 Allelochemicals 297
    13.2.3 Physical and Structural Aspects 297
    13.2.4 Abundance 298
13.3 Biology of Seed-Sucking Heteropterans 298
    13.3.1 Feeding (Ingestion, Digestion, Excretion, and Food Utilization) 298
    13.3.2 Mating 302
    13.3.3 Oviposition 302
    13.3.4 Nymph Development 303
    13.3.5 Dispersal of Nymphs and Adults and Host Plant Choice 304
    13.3.6 Natural Enemies and Defense 307
13.4 Impact of Biotic Factors (Food) on Performance of Heteropterans 308
    13.4.1 Suitable Foods (Seeds/Fruits) 308
        13.4.1.1 Nymphs 308
        13.4.1.2 Adults 309
    13.4.2 Less Suitable Foods (Leaves, Branches, Trunks) 310
    13.4.3 Impact of Nymph-to-Adult Food Switch on Adult Performance 310
13.5 Impact of Abiotic Factors on Performance of Heteropterans 312
    13.5.1 Temperature and Light 312
    13.5.2 Humidity 312
    13.5.3 Rain and Wind 313
13.6 Adaptations and Responses of Heteropterans to Changes in Favorability of the Environment 313
13.7 Final Considerations 314
References 315

13.1 Introduction

The seed-sucking insects Hemiptera (Heteroptera—true bugs) include several families such as Alydidae, Corimelaenidae, Coreidae, Lygaeidae, Pentatomidae, Pyrrhocoridae, Rhopalidae, and Scutelleridae (Schuh and Slater 1995; Schaefer and Panizzi 2000). The majority of heteropterans prefer to feed on immature seeds, which are softer and therefore easier to penetrate than mature seeds; in addition they have greater water content. Species in Pyrrhocoridae and Lygaeidae feed on mature seeds (Janzen 1978). Pyrrhocorids include the cotton stainers (*Dysdercus* spp.), which are important pests (Schaefer and Ahmad 2000), and several species with no economic importance that inhabit tropical forests (Janzen 1972). Lygaeidae are known as "seed bugs" (Sweet 1960), although several species feed on sap from vegetative tissues (e.g., *Blissus* spp. and *Nysius* spp.) (Sweet 2000). Among the Alydidae, *Neomegalotomus parvus* (Westwood) have a better reproductive performance on mature than on immature seeds of legumes (Santos and Panizzi 1998).

Heteropteran feed by inserting their stylets (mandibles + maxillae) on plant tissues, causing damage to seeds and fruits as a result of stylet penetration and action of the saliva that cause tissue necrosis. Slansky and Panizzi (1987) revised the bioecology and nutrition of phytophagous heteropterans, and Hori (2000) revised the salivary secretions and tissue damage. In this chapter we will present food (seeds) characteristics, and the multiple interactions of seeds and sucking bugs, the impact of the biotic and abiotic environment to the insects' biology, and how they compensate for changes in favorability to achieve their maximum reproductive potential.

13.2 Food Characteristics (Seeds)

13.2.1 Nutritional Composition

Seeds present variable chemical composition depending on several factors, such as plant species, age, and cultivation management. Despite their similar chemical composition with that of other plant structures, seeds are "packs" of nutrients in high concentration (Slansky and Scriber 1985). Proteins and lipids present in seeds may differ from those in the rest of the plant and their concentration is defined genetically and/or by influence of the environment (Carvalho and Nakagawa 1983). For example, the percentage dry weight of protein and oil vary from 10% to 30% and from 10% to 40%, respectively, for seeds from various families (Earle and Jones 1962; Jones and Earle 1966). Proteins are the main components of leguminous seeds and vary from 20% to 40%, while cereals have 7% to 15% proteins (Vitale and Bollini 1995). Total protein and oil contents differ among plants in the same family. For example, soybean seeds have relatively higher protein (32.2% of the dry weight) and oil (21.8%) contents than other legumes such as green beans, *Phaseolus vulgaris* L., which have 24.2% and 1.2%, respectively (Earle and Jones 1962). In addition, the quality of soybean seed proteins, measured as the ratio of protein efficiency (i.e., weight gain/protein ingested), is higher (2.4%) that the one observed for green bean seeds (0.5%), as well as the digestibility of the proteins determined in rats (70.1–82.9% for soybean and 36.3–56.0% for green beans; Bressani and Elias 1980). Also, the percentage in dry weight of proteins vary from 11% to 22% among plants of the same species cultivated in different locations, which indicate the influence of the environment on the chemical composition of the seeds (Mayer and Poljakoff-Mayber 1982).

Proteins, lipids, and carbohydrates are the main chemical components of seeds. Considering their solubility, protein seeds are classified as albumins, globulins, glutelins, and prolamins. Glutelins and prolamins are abundant in cereals (80–90%), while albumins and globulins are less than 20% of the total proteins. In dicot plants, glutelins occur from low amounts up to 50% of total proteins, with prolamins occurring in low amounts or not at all. On the other hand, albumins and globulins are well defined in dicots (Duffus and Slaughter 1980; Mayer and Poljakoff-Mayber 1982; Carvalho and Nakagawa 1983). According to their function, proteins are ranked as storage proteins, structural and metabolic proteins, or protection proteins (Shewry and Halford 2002).

Lipids are the main reserve materials and are found in all seeds. They are present in the form of glycerides (tryglycerides), unsaturated fatty acids (e.g., oleic, linoleic, palmitic, and stearic acids), phospholipids, glycolipids, tocopherol, and others (Medcalf 1973; Mayer and Poljakoff-Mayber 1982). Carbohydrates are other major components of seeds, starch being the main reserve carbohydrate in cereals, as much as 65% in wheat seed and 79% in its endosperm (Medcalf 1973). Although sugars in general form a small part of carbohydrates in seeds, their percentage in dry weight can vary from 1% to 70% among plant species in different families (Mayer and Poljakoff-Mayber 1982).

Seeds also contain minerals, free amino acids, vitamins, and phytohormones (Duffus and Slaughter 1980; Carvalho and Nakagawara 1983). For the majority of plant families, elements such as phosphorous, potassium, and magnesium are present and their content may be positively correlated with the protein content in seeds (Lott et al. 1995).

Variations in dry weight and water (Figure 13.1) and in the chemical composition of seeds are observed with maturity. In peas, water content in seeds may decrease from 85% to 14% during seed development

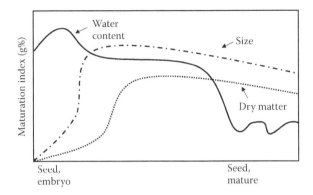

FIGURE 13.1 Modifications in some seed physiological traits during development and maturation. (Modified from Carvalho, N. M. and J. Nakagawa. *Sementes: Ciência, Tecnologia e Produção.* Campinas, Brazil: Fundação Cargil, 2ª ed., 1983.)

(Deunff 1989). In soybean, vitamin content decreases sharply with seed maturation and increases during germination (Bates and Matthews 1975). Protein and lipid contents when seeds reach maximum size and leaves turn yellow vary from 36.7% to 39.4%, decreasing to 20.5% and 21.5% at complete maturation, respectively (Bates et al. 1977). As seeds develop, starch and oil content increase, as well as oleic and linoleic acid contents; palmitic, stearic, and linoleic acids decrease and level off at the end of seed development (Rubel et al. 1972; Yazdi-Samadi et al. 1977).

13.2.2 Allelochemicals

In addition to nutrients, secondary compounds or allelochemicals are present in seeds and have toxic or repellent effects to insects. They include lectins (phytohemagglutinins), a group of glycoproteins present in seed cotyledons of legumes in high concentrations (Gatehouse and Gatehouse 2000). Lectins in bean seeds have severe toxic effects to vertebrates (Liener 1980) and insects (Janzen et al. 1976). Lectins present in soybean seeds are known to inhibit growth of *Manduca sexta* (L.) caterpillars (Shukle and Murdock 1983). Tannins are major chemical defense compounds against seed predators, forming a complex group derived from phenol and present all over the plants and abundant in seed teguments. They are considered antinutritional factors because they do not present direct effects but cause plant parts to be less digestible and difficult to be metabolized by microorganisms, insects, and vertebrates (Boeselwinkel and Bouman 1995).

Other allelochemicals common in seeds of legumes such as *Glycine* spp. and *Phaseolus* spp. include flavonoids, alkaloids, steroids, and phenolics (Kogan 1986). Yet, glycosides, nonproteic amino acids, trypsin inhibitors, antivitamins, and phytic acid are also antinutritional factors (Harbone et al. 1971, Janzen 1971; Liener 1979; Duffus and Slaughter 1980). Plants also produce proteins with antimetabolic activity against digestive enzymes (proteinases and amylases) in herbivores. Digestive proteinase inhibitors are small proteins that bind to digestive enzymes, preventing amino acid absorption resulting in insect death due to malnutrition (Gatehouse and Gatehouse 2000; Fontes et al. 2002). α-Amylase inhibitors present in legume seeds bind to amylases, forming inactive complexes that protect the seeds against seed borers (bruchids) (Shade et al. 1994, Shroeder et al. 1995).

13.2.3 Physical and Structural Aspects

Several physical and structural characteristics of seeds and/or pods are important with regard to feeding of seed-sucking insects. For instance, the seed tegument contains lignin that protects seeds against pathogens and predators (Boesewinkel and Bouman 1995). Pods exhibit pilosity or pubescence, and the toughness of pod walls and the air space between seeds and the pod walls may prevent sucking insects from feeding. In some plants, the aril or arillus (colorful and eatable layer that covers the seed tegument) attracts birds and mammals; however, its thickness may prevent feeding by sucking bugs (Boesewinkel and Bouman 1995).

In macadamia nuts, however, the husk and shell thickness is not always related to cultivar susceptibility to the pentatomid *Nezara viridula* (L.) (Follett et al. 2009). These characteristics may prevent the feeding activity of sucking bugs, mainly of young nymphs that possess mouthparts (stylets) that are shorter and more fragile than those of adults. The impact of these traits to the bugs' biology, despite the availability of some studies, should be further investigated. For example, adults of *Jadera haematoloma* Herrich-Schaeffer (Rhopalidae) that feed on seeds of *Cardiospermum corindum* (Sapindaceae) have stylets longer than those of individuals that feed on seeds of other species of Sapindaceae—this specialization allow bugs to reach the seeds that are protected by an air space between the seeds and the outer fruit wall (Carrol and Loye 1987). A similar case is observed for the southern green stink bug, *N. viridula*, the nymphs of which do not survive when exposed to pods of the bean *Sesbania vesicaria* (properly named bag pods) because they cannot reach the seeds, which are protected by the air space between the seeds and the pod walls (A.R. Panizzi, unpublished). The toughness of seed tegument and of the sorghum glume are greater on resistant cultivars to the mirid, *Calocoris angustatus* Leth., than on susceptible ones (Ramesh 1992). The pentatomid *Edessa medibunda* (F.) feeds on soybean stem (Galileo and Heinrichs 1979) and also on leaves (Rizzo 1971). The short mouthparts (stylets) of these bugs (Panizzi and Machado-Neto 1992) may explain why they prefer these plant structures compared with seeds protected by the pods.

13.2.4 Abundance

Seed abundance and availability to sucking insects are important for regulating the population dynamics of these insects in various ecosystems. In the case of annual crops, heteropterans need to colonize the plants fast—as soon as the seeds appear—because they are an ephemeral food source. There is great variability in the amount and periodicity of seed production due to climatic conditions (e.g., rain availability) and plant species in different habitats. Many times, these factors restrict seed availability and the finding of suitable food sources (references in Slansky and Panizzi 1987).

Seed size varies according to plant species and stage of development. Dramatic changes in seed size occur from seed formation to seed maturation and these changes affect the bugs' biology. For example, stink bugs (pentatomids) do not develop when feeding on pods without seeds—clearly because of lack of nutrients—and perform best when feeding on maturing seeds (Panizzi and Alves 1993). Seed size is critical to insects that live inside seeds, such as the chewing bruchids (Janzen 1969; Johnson and Kistler 1987; see also Chapter 14 in this book).

13.3 Biology of Seed-Sucking Heteropterans

13.3.1 Feeding (Ingestion, Digestion, Excretion, and Food Utilization)

Hemipterans (heteropterans) obtain nutrients and water through the stylets (mandibles + maxillae) that are inserted into the food source. This way of feeding probably evolved from a more primitive type of rasping–sucking mouthparts (Goodchild 1966). According to Hori (2000), the bugs will feed in one of the following ways: stylet-sheath feeding, lacerate-and-flush feeding, macerate-and-flush feeding, and osmotic pump feeding. In stylet-sheath feeding, the bugs insert their stylets into the tissue, mostly in the phloem, destroying a few cells; then, a stylet sheath is produced, which remains in the plant tissues and can be used to estimate the feeding frequency of these insects (Bowling 1979, 1980). The resulting damage is a minor mechanical damage (Miles and Taylor 1994). The external part of the stylet sheath is actually seen and recorded, and was called "flange" by Nault and Gyrisco (1966), who worked with other plant-sucking insects (aphids).

In lacerate-and-flush feeding, the bugs move their stylets vigorously back and forth, and several cells are lacerated. In the macerate-and-flush feeding type, the cells are macerated by the action of salivary pectinase. In both cases, cell contents are injected with saliva, damaging several cells. Finally, osmotic pump feeding occurs through the secretion of salivary sucrase injected into the plant tissue, which increase the osmotic concentration of intercellular fluids containing sugars and amino acids, which are then sucked, leaving empty cells around the stylets (Hori 2000).

The tips of outer stylets (mandibles) contain serrations that resemble teeth, which vary in size and shape. For example, the lengths and widths of mandible tips (areas holding serration) are larger for the pentatomids *N. viridula* and smaller for *Piezodorus guildinii* (Table 13.1). Another pentatomid, *Euschistus heros*, shows a mandible tip length greater than that of *Dichelops melacanthus*, but a similar mandible tip width. All four species of stink bugs have a similar pattern of serration (indentation) on the mandible tips, with a fixed number of central teeth and pairs of lateral teeth (Figures 13.2 and 13.3). Cohen (1990) observed mandible teeth ranging from 16 to 17 in reduviids, down to a few in pentatomids and lygaeids. The mandible tips in *D. melacanthus* and in *P. guildinii* are of similar border shape, and lack the squamous texture structures observed in *E. heros* and *N. viridula* (Figure 13.3a through d). The inner surface in the mandible of the four species of pentatomids mentioned above has a squamous texture. This groove accommodates the longitudinal external maxillary rib. Cobben (1978) reports that the orientation of this pavement is such that the forward thrust of one mandible will cause considerable friction against the outer surface of the adjacent maxillary stylet contributing to its inward deviation.

The saliva of heteropterans has been studied. It contains several enzymes and metabolites that vary among species, individuals, stage of development, and food source utilized (Miles 1972; Tingey and Pillemer 1977). When injected into plants, the salivary secretion cause deformation (e.g., galls; see Chapter 16) similar to those caused by excess growth hormones; indole acetic acid from the host plant or from the salivary gland is considered the most phytotoxic compound in the saliva of heteropterans (references in Hori 2000). Seeds damaged by the stylets may have greater incidence of pathogenic micro-organisms (e.g., Panizzi et al. 1979; Ragsdale et al. 1979).

Seed-sucking bugs require large amounts of water when feeding on dry (mature) seeds, and the watery saliva is produced in abundance during feeding. In general, water is obtained from other plants and/or from vegetative tissues of the host plant (e.g., Saxena 1963). Nymphs of the pyrrhocorid *Dysdercus bimaculatus* Stål feed on cotton seeds rich in water rather than dry seeds, and females tend to retard egg production under water stress (Derr 1980). Nutrient uptake is related to watery saliva production, and the rate food ingested/watery saliva indicates feeding efficiency (Eggermann and Bongers 1980).

Gregarism is an important component in the feeding activity of seed-sucking bugs (see references in Slansky and Panizzi 1987). The rhopalid, *Jadera choprai* Göllner-Scheiding is a good example of gregarism feeding on seeds, usually found on the soil (Figure 13.4); less nymph mortality and fasted nymph development was observed for nymphs raised in groups compared with nymphs raised in isolation (Panizzi et al. 2005b). This bug drags a seed with mouthparts for relatively long distances (up to 2 m in laboratory observation), and, while feeding, stays in a position forming an angle of about 45° relative to the soil surface, holding the seed with the first pair of legs (Panizzi and Hirose 2002).

The habit of carrying seeds has been reported for other species of heteropterans that live on the ground, such as cydnids (Sites and McPherson 1982; Tsukamoto and Tojo 1992; Takeuchi and Tamura 2000). In the case of *Parastrachia japonensis* Scott, males carry the fruits (drupes) of the host plant to niches or shelters (small shallow holes under the vegetation) to feed the nymphs (Tsukamoto and Tojo

TABLE 13.1

Mean (± SEM) Rostrum Length, Mandible Tips Length and Width, and Number of Teeth of Teneral (<1-day-old) Adult Females of Selected Species of Pentatomids

Species	Rostrum Length (mm)[a]	Mandible Tip Length (μm)[a]	Mandible Tip Width (μm)[a]	No. of Central Teeth	No. of Lateral Teeth Pair
Dichelops melacanthus	4.9 ± 0.029 b (10)	81.0 ± 0.75 c (8)	26.8 ± 0.35 b (8)	4	3
Euschistus heros	5.1 ± 0.030 b (10)	87.9 ± 0.89 b (6)	27.0 ± 0.36 b (6)	4	3
Nezara viridula	5.9 ± 0.046 a (10)	106.0 ± 1.11 a (4)	30.2 ± 0.50 a (4)	4	3
Piezodorus guildinii	3.5 ± 0.094 c (10)	71.1 ± 1.08 d (4)	23.7 ± 0.43 c (4)	4	3

Source: Data from Depieri, R. A. and A. R. Panizzi, *Rev. Bras. Entomol.*, in press, 2010.

Note: The numbers of observations are in parentheses.

[a] Means followed by the same letter in each column do not differ significantly using the Tukey test ($p \leq .05$).

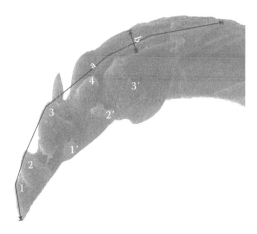

FIGURE 13.2 Scanning electron micrograph of mandible tip of adult pentatomid showing how mandible measurements were taken. Mandible tip length (line a); mandible tip width (line b). Numbers 1–4 indicate central teeth; 1'–3' indicate lateral teeth (augmentation 1200×). (From Depieri, R. A. and A. R. Panizzi, *Rev. Bras. Entomol.*, in press, 2010. With permission.)

1992). These drupes are, apparently, of better nutritional quality than those found in the soil at random since nymphs that feed on the latter show retarded development and higher mortality (Filippi et al. 2000). In some cases, females were found "stealing" drupes from other females' niches (kleptoparasitism), and this may influence niche location (Filippi et al. 2005).

The alydid *N. parvus* (Westwood) show gregarism on mature pods of pigeon pea, *Cajanus cajan* on which they feed on (Ventura and Panizzi 2003). However, this bug may feed on dead conspecifics. Second instars, without food (legume seeds), and feeding on dead nymphs reach the third instar. In the field, *N. parvus* adults are found feeding on carcasses and feces of mammals, such as dog drops (Ventura et al. 2000). This unusual feeding habit has been reported for alydids (Bromley 1937; Schaefer 1980).

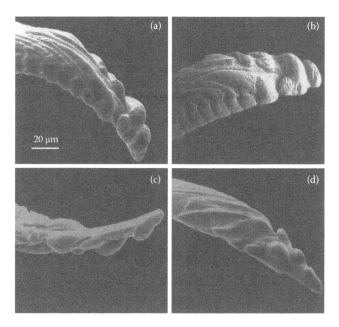

FIGURE 13.3 Scanning electron micrographs of mandibular stylets of pentatomids. (a) *Dichelops melacanthus.* (b) *Euschistus heros.* (c) *Nezara viridula.* (d) *Piezodorus guildinii* (augmentation 600×). (From Depieri, R. A. and A. R. Panizzi, *Rev. Bras. Entomol.*, in press, 2010. With permission.)

FIGURE 13.4 Nymphs of rhopalid, *Jadera choprai*, aggregated over balloon vine seed, *Cardiospermum halicacabum* (Sapindaceae), a common weed plant of soybean fields in the neotropics. (Courtesy of J. J. Silva.)

Aldrich (1995) associated the attraction of alydids to carcasses and feces with the production of rancid secretions (fatty acids of small chains) by the metathoracic glands of males and females.

Species of pentatomids when feeding on soybean pods prefer the seed closest to the pedicel, compared with seeds in the median or distal position (Panizzi et al. 1995). Apparently, the seed closest to the pedicel is the first to be reached by bugs walking on the plant. However, this also happens to detached pods from the plants, which suggests that other factors are influencing this choice. Usually, seed-sucking bugs are not adapted to utilize food sources other than seeds, although this happens (see Sections 13.4.1 and 13.4.2 for details).

Quantitiative studies on nutrition of seed-sucking insects, which were started 20+ years ago (Slansky and Panizzi 1987), remain tentative. This is probably due to the feeding habits of these insects that consume small amounts of liquid food and excrete liquid feces, making it difficult to estimate the different nutritional indexes. In general, nymphs and adults of heteropterans present low consumption rates, moderately high growth rates, and high efficiency of food assimilation, when compared with other feeding guilds such as leaf chewers (Table 13.2). The relative consumption and growth rates tend to decline with nymph development and vary with gender, age, and reproductive status (Slansky and Scriber 1985, Slansky and Panizzi 1987). Damage to seeds/fruits by late instars may be similar to that of adults (males or females), as reported for the pentatomid *N. viridula* (L.) feeding on cotton (Bommireddy et al. 2007).

TABLE 13.2

Quantitative Food Utilization of Seed-Sucking Heteroptera (Nymphs) and Leaf Chewers (Caterpillars of Lepidoptera)

Group of Insects and Limits	RCR		RGR		AD		NGE	
	X	Range	*X*	Range	*X*	Range	*X*	Range
Seed suckers	0.36	0.14–0.58	0.27	0.10–0.57	73	50–92	89	40–96
Leaf chewers[a]	1.46	0.31–6.60	0.17	0.03–0.80	41	12–98	37	2–93
Characterization of limits								
Low		<1		<0.1		<30		<40
Moderate		1–2		0.1–0.6		30–50		40–60
High		>2		>0.6		>50		>60

Source: Data from Slansky, Jr., F. and J. M. Scriber. 1985. In *Comprehensive Insect Physiology, Biochemistry, and Pharmacology*, ed. G. A. Kerkut and L. I. Gilbert, 87–163. Oxford: Pergamon Press.

Note: Note that, except for RCR, the values are greater for sucking insects compared with chewers. RCR, relative consumption rate; RGR, relative growth rate; AD, approximate digestibility; NGE, net growth efficiency. RCR and RGR are expressed as mg dry weight/day/mg of insect dry weight. AD and NGE are expressed in percentage values.

[a] Lepidoptera feeding on tree foliage.

13.3.2 Mating

Premating and mating behaviors have been studied for many species of seed-sucking heteropterans. These behaviors are influenced by several cues, including odors and sounds. In the southern green stink bug, *N. viridula* (Pentatomidae) males produce sex pheromones, which are important for attracting females, but they also attract parasitic flies (Tachinidae) (Harris and Todd 1980a; Borges et al. 1987; Borges 1995). For this bug, production of sound is an important component in mating, and vibrations are passed through the substrate (plant) (e.g., Harris et al. 1982; Čokl 1983; Ota and Čokl 1991; Čokl et al. 1999, 2000; Moraes et al. 2005). These vibratory signals interfere in the pheromone emission, and males emit more pheromones when stimulated by sounds produced by females (Miklas et al. 2003).

Male courtship to females occurs before mating; however, males of the lygaeid *Xyonysius* sp. perform courtship before and during mating. During mating, this behavior is more elaborate (see illustrations in Rodríguez and Eberhard 1994). For another species of lygaeid, *Ozophora baranowskii* Slater and O'Donnell, females touch males with their hind legs during mating, this being more intense during short-lasting copulae (Rodríguez 1998).

Copulae duration among seed-sucking heteropterans is variable and depends on temperature. For example, in the lygaeid *Oncopeltus fasciatus* (Dallas) it can last from 30 min (at 38°C) up to 2 days (at 24.5°C) (Andre 1935). For the pentatomid *N. viridula*, copula may last from 1 to 165 h (Harris and Todd 1980b); for another pentatomid, *Bathycoelia thalassina* (Herrich-Schaeffer), this period varies from 15 min to 8 h, and males move toward females to mate (Owusu-Manu 1980). Interestingly, for the bug *Corimelaena extensa* (Uhler) (Corimelaenidae), copulae last for only 12 s (Lung and Goeden 1982). For the pyrrhocorid *Dysdercus maurus* Distant, males also take initiative for mating, which may last up 70 h (Almeida et al. 1986). Longer copulae take place when males are more abundant than females, this being a strategy for avoiding sperm replacement. Prolonged mating prevents sperm replacement that occurs with multiple mating (McLain 1981; Carroll 1988). In some cases, males keep guarding females to prevent mating with other males during oviposition, as is the case of the rhopalid *J. haematoloma* (Herrich-Schaeffer) (Carroll 1993). Postmating refractory period is variable. For example, *Lygus hesperus* (Hemiptera: Miridae), not a seed-sucking bug and is specialized in feeding on vegetative tissues, takes from 1 day for males to 5 days to females to become sexually receptive again (Brent 2010).

Mating can be affected by the nutritional source. For instance, *O. fasciatus* mate two times more when feeding on seeds than when feeding on flowers or vegetative plant parts (Ralph 1976). *Dysdercus koenigii* (F.) copulates independently of its nutritional status; however, eggs are produced only if they feed on seeds of cotton (Shahi and Krishna 1981).

Oncopeltus fasciatus copulates more frequently during long photoperiods, and they do not mate under dark conditions (Walker 1979). The pentatomid *Euschistus conspersus* Uhler shows peaks of copulating activity at 11:00 p.m., with 80% of the bugs forming aggregations with copulatory activity (Krupke et al. 2006).

In general, heteropterans copulate with several individuals, such as the lygaeid *Lygaeus kalmii* (Stål) that copulates with up to six different partners (Evans 1987). In theory, multiple copulations with different males keep sperm provision viable and promote greater egg fertilization, resulting in greater genetic diversity of the progeny; females of *N. viridula* prefer polyandry (McLain 1992). The alydid *Riptortus clavatus* (Thunberg) present greater fecundity/fertility when females mate multiple times compared with females that mate only once (Sakurai 1996). The coreid *Leptoglossus clypealis* Heidemann can mate up to 17 times (Wang and Millar 2000). The Asian bambu coreid, *Notobitus meleagris* F., forms aggregation to mate, with a single male aggregating with several females; the male monitors the aggregation and shows aggressive behavior to intruder males (Miyatake 2002).

13.3.3 Oviposition

Several oviposition behaviors have been reported about seed-sucking bugs. The southern green stink bug, *N. viridula*, shows a curious oviposition behavior, previously described in detail (Panizzi 2006). During oviposition, soon after the egg is expelled, the female touches the egg mass with the last tarsomere. As she moves the leg, the tarsomere folds and the dorsal surface that will touch the egg is exposed. This movement occurs once, using one leg, each time an egg is deposited. For the next egg, the female slightly moves the tip of the abdomen to the side, and the process starts again, with the inflation of the genital plaques (gonocoxites and gonapophyses VIII), followed by expulsion of the egg and movement of one leg of the last pair, which touches the egg mass as described above. This oviposition behavior of *N. viridula* is speculated to help position and glue the eggs one to the other and to the substrate, soon after they are expelled.

The rhopalid *J. choprai* Göllner-Scheiding feeds on the ground on mature seeds of *Cardiospermum halicacabum* (L.) (Sapindaceae). Laboratory observations indicated that females dug a hole of ca. 0.5 cm in the soil with the forelegs, laid eggs, and covered them with lose soil. In artificial conditions, females buried the eggs in over 60% of the ovipositions, and nymphs were able to reach the soil surface from eggs buried 4 cm. This oviposition behavior is rare among seed-sucking heteropterans that usually oviposit on host plants (Panizzi et al. 2002b).

Seed-sucking heteropterans may lay eggs in groups or individually. These two patterns of egg laying show adaptive advantages and disadvantages (Panizzi 2004a). Apparently, no investigations have been conducted to compare the impact of natural enemies on eggs of bugs with different oviposition patterns, that is, those which lay eggs in masses or lay them singly. It is likely that predation/parasitism would greatly reduce the fitness of bugs that lay eggs in masses than those that lay eggs singly because, once located, eggs will be destroyed in greater number on the first case. Eggs laid singly and separated in time and space will, in theory, have a greater chance to escape from predators/parasitoids. Moreover, the egg mass guardian behavior shown by many species of heteropterans will facilitate the location of egg masses by natural enemies (Tallamy and Schaefer 1997), further reducing their fitness.

The uncommon oviposition on the body of conspecifics by the pentatomid *E. heros* (F.) and by the alydid *N. parvus* (Westwood) was observed in the laboratory (Panizzi and Santos 2001). This behavior has been reported about the coreids *Phyllomorpha laciniata* Vill. in Europe (Bolivar 1894), and *Plunentis porosus* Stål in South America (Costa Lima 1940). In the first case, females oviposit on the back of females and males a variable number of eggs (1–15) (Kaitala 1996); in the second case, eggs are laid on the ventral side of the male abdomen. Males either accept the egg deposition or reject or retard the process by making movements (Miettinen and Kaitala 2000).

The alydid *N. parvus* shows an interesting oviposition behavior on pods of pigeon pea, *C. cajan*, described and illustrated by Ventura and Panizzi (2000) (Figure 13.5). Female that initially is still moves the antenna alternately downward and upward. Dabbing/antennation is then accomplished, initially with the antenna and immediately after with antenna and labium. In the next step, the ovipositor is exposed and swept on the surface of the pod. After the female sweeps the ovipositor a few times, mechanoreceptors (sensilla on the ovipositor) are stimulated and eggs are laid on the depressions of pods between the seed loci. On soybean, this bug oviposits preferentially on the lower (abaxial) surface of leaves, near the central leaf vein (Panizzi et al. 1996a).

Oviposition rhythm is related to the food source. The pentatomid *N. viridula* fed on berries of privet, *Ligustrum lucidum* (Oleaceae), shows great peaks of oviposition, while when feeding on soybean, the oviposition rhythm did not peak during the ovipositing period (Figure 13.6) (Panizzi and Mourão 1999). Berries of privet are known to greatly increase fecundity of this and of other species of pentatomids (Panizzi et al. 1996b, 1998; Coombs 2004).

13.3.4 Nymph Development

As nymphs emerge, those originated from eggs laid in masses usually stay on top of around eggshells (corions). A mixture of visual, olfactory, and touch stimuli keep nymphs together as a group. For example, the pentatomid *N. viridula* utilize touch stimuli to remain aggregated during the first 2 days after emergence. After this period, chemical (*n*-tridecane) stimuli are used to keep nymphs together. However, depending on its concentration, this chemical compound may act to disperse the group (Lockwood and Story 1985).

During the first instar, nymphs do not feed. They are believed to ingest microorganisms (symbionts) and water at this age. For *N. viridula*, the bacteria *Klebsiella pneumoniae* (Schroeter), *Enterococcus faecalis* (Andrews & Horder) and *Pantoea* sp., were found in the gut possibly acting as symbionts (Hirose et al. 2006c). Also, bacteria were found on the eggshells after the nymphs' emergence and not in the females' ovarioles, indicating oral transmission of these symbionts (Prado et al. 2006).

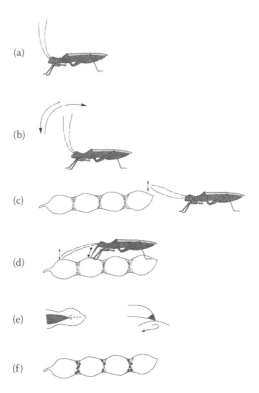

FIGURE 13.5 Behaviors related to the choice of the oviposition site by *Neomegalotomus parvus*. (a) Still female. (b) Antennae moving upward and downward. (c) Antennation–labium tip strikes pod surface. (d) Dabbing/antennation–labium and antennae tips strike pod surface. (e) Ovipositor is swept back and forward. (f) Eggs are laid and glued in the breaches of the pods. (From Ventura, M. U. and A. R. Panizzi, *An. Soc. Entomol. Bras.*, 29, 391, 2000. With permission.)

Apparently, the group functions as an organism, and the humidity is crucial at this early time of nymph development. Laboratory observations suggest that humidity keeps the colony united and when decreased, nymphs disperse and die (Hirose et al. 2006b; see Section 13.5.2). In general, nymphs in or around the eggshells become more susceptible to natural enemies (see Section 13.3.8).

Nymphs from isolated eggs leave the eggshells as they emerge and tend to feed and do not aggregate. Although there is no clear demonstration that these nymphs always feed, they show different survivorship in the presence of different foods. For instance, first instars of *N. parvus* do not die in the presence of mature soybean seeds; with immature soybean seeds, mortality was 16.7%; with soybean and green bean (*P. vulgaris*) immature pods and mature seeds of lupin (*Lupinus luteus*), mortality was <1.7%; with soybean stems and leaves, mortality ranged from 2.5% to 5.0% (Panizzi 1988). These data suggest these first instars ingest few nutrients and water, and with low nutritional food (vegetative tissues) they can reach the second instar. In fact, even without food they can do that using nutrients acquired during the embryonary development (A.R. Panizzi, unpublished). For another alydid, *Megalotomus quinquespinosus* Say, first instars do not feed (Yonke and Medler 1965). First instars of the rhopalid *J. choprai* feed on mature seeds of the balloon vine, *C. halicacabum* (Sapindaceae), a common weed in crop fields of southern Brazil (Panizzi and Hirose 2002).

13.3.5 Dispersal of Nymphs and Adults and Host Plant Choice

Dispersal of nymphs is limited since they do not fly nor disperse by walking. Not much data are available in that regard. Nymphs of the pentatomids *N. viridula* and *P. guildinii* move up to 12 m from the egg hatch point in soybean fields, and disperse more along than across rows; a greater distance is covered by fourth and fifth instars, when the gregarious behavior is mitigated (Panizzi et al. 1980).

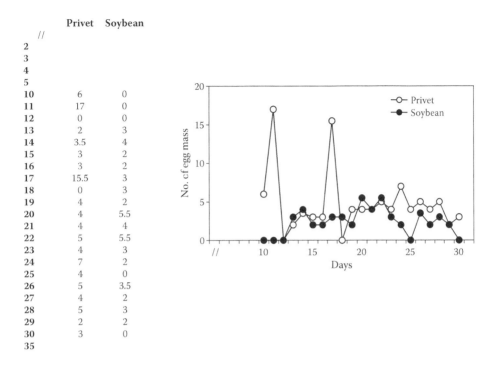

	Privet	Soybean
//		
2		
3		
4		
5		
10	6	0
11	17	0
12	0	0
13	2	3
14	3.5	4
15	3	2
16	3	2
17	15.5	3
18	0	3
19	4	2
20	4	5.5
21	4	4
22	5	5.5
23	4	3
24	7	2
25	4	0
26	5	3.5
27	4	2
28	5	3
29	2	2
30	3	0
35		

FIGURE 13.6 Oviposition rhythm of *Nezara viridula* fed with privet (*Ligustrum lucidum*) berries or soybean (*Glycine max*) pods. Note that with privet berries occur sharper peaks in oviposition rhythm, compared with those females fed with soybean pods, indicating greater fecundity of females on the former food. (From Panizzi, A. R. and A .P. M. Mourão, *An. Soc. Entomol. Bras.*, 28, 35, 1999. With permission.)

Adults of heteropterans are the main responsible for dispersion, and some species are known to migrate as the cereal bugs in the Middle East known as "Sunn pests" or "Soun pests"—pentatomids of the genus *Aelia* (references in Panizzi et al. 2000b), and scutellerids of the genus *Eurygaster* (references in Javahery et al. 2000). Other species disperse by flight among host plants (trees) such as the pentatomid *B. thalassina* (Herrich-Schaeffer), pest of cocoa in Africa (Owusu-Manu 1977). Females of *N. viridula* and of *P. guildinii* fly greater distances than males (Costa and Link 1982). Dispersal among host plants is mediated by the degree of pod and seed development for the coreid *Clavigralla tomentosicollis* Stål colonizing cowpea, *Vigna unguiculata* (Dreyer and Baumgärtner 1997).

As soon as stink bugs reach new areas, they start looking for preferred host plants (Panizzi 1997). Although polyphagous, there is some degree of preference for certain taxa. For example, members of Alydidae (Leptocorisinae) prefer to feed on grasses (Grammineae), while members of Alydidae (Alydinae) prefer legumes (Fabaceae) (Schaefer and Mitchell 1983). The pentatomid *N. viridula* prefer legumes and brassics (Brassicaceae) (Todd and Herzog 1980); another pentatomid, *Edessa meditabunda* (F.) prefer legumes and solanaceous (Solanaceae) plants (Silva et al. 1968); and the pentatomids of the genus *Chinavia* (*Acrosternum*) tend to associate with legumes, while those of the genera *Aelia*, *Mormidea*, and *Oebalus* prefer to feed on graminaceous plants (references in Panizzi et al. 2000b).

According to Tillman et al. (2009), the driving forces for dispersal of stink bugs and their distribution in space and time, such as in peanut–cotton farmscapes, seems to be food abundance and the structure of the landscape. Reeves et al. (2010) documented the influence of adjacent crops and uncultivated habitats on the distribution of stink bugs and boll injury in cotton field edges.

For locating and choosing the host plant, insects use their eyes, antennae, and palps. Plants' physical and chemical characteristics will influence the choice. A series of behaviors are shown by heteropterans that vary in intensity according to the food plant suitability. For instance the alydid *N. parvus* (Westwood) use sensilla on the antennae (Figure 13.7) and on the labium (Figure 13.8) to select the food

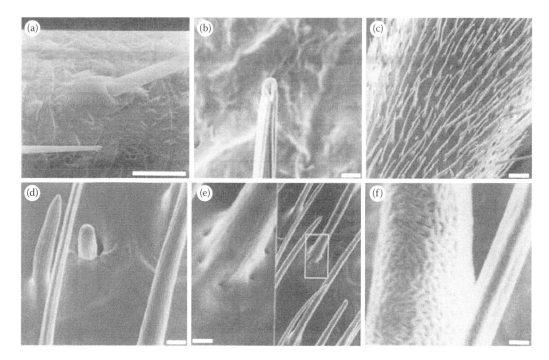

FIGURE 13.7 Sensilla on the *Neomegalotomus parvus* antennae. (a) Sensilla in flexible socket (bar, 20 μm). (b) Sensillum with a beveled point (bar, 2 μm). (c) Apical segment with several types of sensilla (bar, 20 μm). (d) Grooved peg sensilla on apical segment (bar, 2 μm). (e) Bristle sensilla with holes in the base on apical segment (bar, 20 μm). (f) Multiple pores sensilla on terminal segment (bar, 500 nm). (From Ventura, M. U. and A. R. Panizzi, *Braz. Arch. Biol. Technol.*, 48, 589, 2005. With permission.)

(Ventura et al. 2000, Ventura and Panizzi 2005). For the mirid, *Lygus rugulipennis* (Poppius), the presence of chemoreceptor sensilla on the mouthparts is directly related to host plant selection behavior (Romani et al. 2005). The antennal sensilla of pentatomid pest species have been described and their function compared and discussed (e.g., Silva et al. 2010). Semiochemically based monitoring and pheromone attraction and cross-attraction of seed/fruit-sucking hemipterans has gain momentum and seems to be a potential tool to manage these insects (e.g., Aldrich et al. 2009; Borges et al. 2010; Tillman et al. 2010).

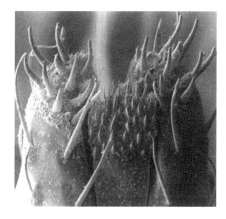

FIGURE 13.8 Sensilla on the extremity of the mouthparts (labium) of second instar nymph of the alydid, *Neomegalotomus parvus* (bar, 10 μm). (From Ventura, M. U. and A. R. Panizzi: Oviposition behavior of *Neomegalotomus parvus* (West.) (Heteroptera: Alydidae): Daily rhythm and site choice. *An. Soc. Entomol. Bras.* 2000. 29. 391–400. Copyright Wiley-VCH Verlag GmbH & Co. KGaA. Reproduced with permission.)

13.3.6 Natural Enemies and Defense

Seed-sucking insects are attacked by a great variety of natural enemies, including arthropod parasitoids and predators, reptiles, amphibians, birds, mammals, fungi, bacteria, and viruses (Slansky and Panizzi 1987).

Among the arthropods that attack seed-sucking heteropterans, egg parasitoids are very important. For example, stink bugs associated with soybean are attacked by at least 12 species of microhymenoptera, only in South America; *Trissolcus basalis* (Wollaston) and *Telenomus mormideae* Costa Lima are the main species. Among the adult parasitoids, tachinid flies are abundant and diverse. In North America at least 13 species of tachinids are found in soybean fields, the most common being *Trichopoda pennipes* (F.); in South America, *Eutrichopodopsis nitens* Blanchard is the most common adult parasitoid of the southern green stink bug *N. viridula* (Corrêa-Ferreira 1984, 1986; Panizzi and Slansky 1985a).

Plant architecture and smell, as well as insects' smell, are important cues used by parasitoids to locate their hosts. For instance, *T. pennipes* fly is attracted by the aggregation pheromone emitted by males of *N. viridula* (Harris and Todd 1980a). Also, the egg parasitoids utilize the host odor to locate it (e.g., Sales et al. 1978; Staddon 1986). *N. viridula* feeding on plants that grow straight upward and have their pods more greatly exposed are parasitized in greater proportion by the tachinid fly *T. pennipes*, compared with those feeding on lodged plants (Todd and Lewis 1976). *N. viridula* is also less abundant on lodged plants than on plants standing straight (Link and Storck 1978). This bug is less susceptible to the tachinid *E. nitens* when feeding on castor bean, *Ricinus communis* (Euphorbiaceae), than when it feeds on the weed Siberian motherwort, *Leonurus sibiricus* (Lamiaceae) (Panizzi 1989). Possible reasons to explain this difference in parasitism are as follows: castor beans are taller (usually 1–3 m) than the Siberian motherwort (<1 m), making the bugs on the former less "reachable" by the flies; castor bean plants form a habitat that is darker, cooler, and with higher relative humidity than that formed by a community of Siberian motherwort plants; and the long-lasting flowering period of Siberian motherwort might attract the flies, resulting in a more abundant insect population than in the castor bean habitat. The presence of multiple stink bug species on plant communities may lead to parasitism by flies on "wrong" hosts (Panizzi and Smith 1976; Panizzi and Slansky 1985c).

Not much data are found in the literature regarding predators of seed-sucking heteropterans. Carnivorous ants seem to be the main predators of the pentatomid *N. viridula* in soybean fields (Ragsdale et al. 1981; Krispyn and Todd 1982; Stam et al. 1987). Also, predatory heteropterans of the genera *Podisus* and *Tynacantha* are referred to as predators of phytophagous heteropterans (Panizzi and Smith 1976; Lockwood and Story 1986a).

The impact of natural enemies to seed-sucking heteropterans has not been evaluated in detail. It is well known that the remotion of natural enemies by pesticides cause pest resurgence, indicating their role in regulating pest populations. Moreira and Becker (1986) found 17% parasitism and 24% predation on *N. viridula* eggs from soybean fields.

Seed-sucking heteropterans show several defense mechanisms against natural enemies, including mimetic coloration, secretion defenses, parental care, aposematic coloration, gregarious behavior, and isolation of toxic compounds (allelochemicals) that make them distasteful (references in Slansky and Panizzi 1987).

Nymphs that show aposematic coloration (i.e., conspicuous color) as reported in the lygaeid *O. fasciatus* (Dallas) (Kutcher 1971) and in the coreid *Thasus acutangulus* Stål (Aldrich and Blum 1978) show gregarism, which is believed to increase the advertising color. In addition, gregarism increase survivorship by reducing the action of predators. Body size and the food effect are variable among aposematic species. For instance, the lygaeid *Lygaeus equestris* L. is less susceptible to predators when feeding on its preferred host plant than on an alternate plant; for another lygaeid, *Tropidothorax leucopterus* Goeze, this does not occur; however, both species are less attacked when nymphs are aggregated than when isolated (Tullberg et al. 2000).

Cryptic behavior or coloration is suggested to occur among seed-sucking heteropterans. Adults of the stink bug *Thyanta perditor* (F.) remain green when feeding on the weed plant *Bidens pilosa* or on plants of wheat, *Triticum aestivum*; however, they turn brown when feeding on maturing wheat, suggesting an adaptation to the color of the substrate (Panizzi and Herzog 1984). For *Thyanta calceata* (Say), a similar phenomenon was reported with green adults occurring during summer and brown adults occurring during fall, with adults able to change color according to the photoperiod (McPherson 1977a,b).

Defense secretions of heteropterans protect them against predators (see review by Aldrich 1988). Depending on the amount with which they are liberated, these substances can act as alarm or aggregation pheromones (Ishiwatari 1974, 1976). Members of Pentatomidae are known as "stink bugs" because of their unpleasant secretions. For example, *E. conspersus* Uhler when offered to birds cause an excitement behavior and a delayed and hesitant prey preparation, in contrast to the calm behavior of birds when feeding on worms. In addition to the bad taste, *E. conspersus* shows cryptic coloration and drops from the plant when disturbed (Alcock 1973). This last behavior is also attributed to nymphs of the lygaeid *L. kalmii* (Simanton and Andre 1936), and of the pentatomids *N. viridula* and *P. guildinii* (A.R. Panizzi, unpublished).

Another defense mechanism is shown by females that protect eggs or young nymphs (parental care). For instance, females of the pentatomid *Antiteuchus tripterus limbativentris* Ruckes protect eggs and first instars, but by so doing facilitate the finding of egg mass location by egg parasitoids that are able to attack eggs on the edge of the mass (Eberhard 1975). For *Elasmucha grisea* L. (Acanthosomatidae), egg protection is high against parasitoids and only predators can destroy the eggs. In addition to position on the top of the egg mass, females show several types of body movements and accelerated wing beat (Melber et al. 1980). *Elasmucha putoni* Scott also protects eggs and nymphs (Honbo and Nakamura 1985). The cydnid *P. japonensis* Scott keeps guard of eggs and young nymphs against predators (Nomakuchi et al. 2001). Furthermore, females of *D. maurus* Distant cover the egg mass with sand after oviposition, which suggests a defense behavior (Almeida et al. 1986). Similar behavior was observed for rhopalids that dig the soil, lay eggs, and cover them with soil particles (Carroll 1988; Panizzi et al. 2002a).

Nymphs of the pentatomid *N. viridula* are less predated by the heteropteran *Podisus maculiventris* Say and by the fire ant *Solenopsis invicta* Buren (Lockwood and Story 1986a). When captured, *Diactor bilineatus* (F.) (Coreidae), pest of passion fruit, *Passiflora* sp., releases the hind legs, which are conspicuous and colorful; this suggests a defense mechanism (J.C.M. Carvalho, personal communication to ARP). Certain heteropterans with migratory habits fly at night (McDonald and Farrow 1988), which suggests as a defense behavior to avoid daylight predators.

Coreids show several defense mechanisms (Mitchell 2000). *Leptoglossus zonatus* (Dallas) shows a curious behavior of landing on objects or persons that approach its habitat, suggesting territoriality. This behavior was studied in nearby corn fields, and results indicated that bugs fly in great numbers to objects placed on the edge of the crop field during the first day, but as time passes bugs "lose interest" and stop landing on the objects (Panizzi 2004b).

Some bugs overwinter underneath crop residues, and this offer them protection against natural enemies, as demonstrated by the neotropical brown stink bug, which is less attacked by tachinid flies than the southern green stink bug, *N. viridula*, which, in some latitudes, feeds year round and stay exposed to the flies (Panizzi and Oliveira 1999).

13.4 Impact of Biotic Factors (Food) on Performance of Heteropterans

13.4.1 Suitable Foods (Seeds/Fruits)

13.4.1.1 Nymphs

Food has a variable impact on nymph development and survivorship. For instance, developmental time of nymphs (second to fifth instar) of the alydid *N. parvus* (Westwood) varied from 17.3 to 34.1 days, and nymph mortality varied from 12.5% to 93.3% (see review in Panizzi 2007 and references therein). These two parameters are affected not only by the species of food plant explored but also by the stage of development of fruits and seeds, and whether seeds are exposed or not. In general, on exposed mature seeds, nymphs of *N. parvus* have a better performance than on immature seeds/fruits. For example, mature seeds of pigeon peas are commonly used to rear this bug in the laboratory, with high reproduction rates (Ventura and Panizzi 1997).

For the southern green stink bug, *N. viridula* (L.), time of nymphal development varied from 22.0 to 50.2 days, and nymph mortality ranged from 0% to 100%, with the majority of the values falling in the range of 22 to 26 days, and 15% to 30% mortality on its preferred food, soybean. For the red-banded stink bug, *P. guildinii*, nymph developmental time varied from 18.2 to 30.3 days and nymph mortality varied from 12.5% to 94.4%. Best results were obtained on fruits of the indigo legumes (*Indigofera endecaphylla* and *I. truxillensis*) and on sesbania, *Sesbania aculeata*. For another pentatomid, *Loxa deducta* (Walker) nymph developmental time and mortality varied from 35.8 to 56.6 days, and 17.1% to 82.6%, respectively, with better performance on fruits of privet, *L. lucidum* (Oleaceae) (references in Panizzi 2007). Survivorship of the neotropical pentatomid *D. melacanthus* was variable when nymphs feed on different diets. The mortality was smaller on immature seeds of corn compared with immature pods of soybean (Panizzi et al. 2007).

13.4.1.2 Adults

Fecundity (egg production) is highly variable and depends, basically, on the quality of the food ingested. For example, for the neotropical brown stink bug, *E. heros* (F.), fecundity varied from zero when feeding on star bristle, *Acanthospermum hispidum*, to 287.2 eggs per female when feeding on soybean pods, with intermediate values on other host plants (see review and references in Panizzi 2007; Medeiros and Megier 2009). For another pentatomid, *L. deducta* (Walker) fecundity varied from 27 eggs per female on soybean to almost 10-fold (236 eggs per female) on privet, *L. lucidum*. This last plant is known to be colonized by over 12 species of pentatomids in subtropical Brazil (Panizzi and Grazia 2001). For the alydid *N. parvus*, fecundity varied from 12 eggs per female on immature pods of lupin, *L. luteus*, to 118 eggs per female on mature pods of pigeon pea, *C. cajan*. For the majority of foods, fecundity was intermediate, with mature seeds/fruits yielding better results than immature ones. This means that not only the host plant but also the plant phenology affects fecundity.

The extremely polyphagous *N. viridula* also shows great variability on fecundity according to the food source utilized. It can vary from 0 to 298 eggs per female on sesame seeds. High fecundity was also observed in females feeding on sesbania, *Sesbania emerus* (274 eggs per female), and privet, *L. lucidum* (257 eggs per female). For the majority of foods, fecundity ranged from 50 to 100 eggs per female. The less polyphagous red banded stink bug, *P. guildinii*, laid from 11 eggs per female on pigeon pea up to 50 times more eggs (508 eggs per female) feeding on indigo, *Indigofera truxillensis*. This dramatic variability illustrates the importance of the quality of the food ingested for egg production.

The longevity of seed-sucking heteropterans has been studied for a great variety of species, and it varies according to food quality, gender, and sexual activity. In many studies, males are reported to live longer than females, as in the case of *Nysius vinitor* Bergroth (Lygaeidae), *O. fasciatus* (Dallas) (Lygaeidae), and *E. heros* (F.) (Pentatomidae) (Kehat and Wyndham 1972; Slansky 1980; Villas Bôas and Panizzi 1980; Malaguido and Panizzi 1999). Reduction in female's longevity seems to be due to reproduction caused by the drain of energy during oviposition (Slansky 1980). Lener (1967) observed that the mean longevity of virgin females and males of *O. fasciatus* was twice of that of adults that copulated, and suggested that sexual activity reduced longevity. However, in several studies with pentatomids such as *P. guildinii* (Panizzi and Smith 1977), *Acrosternum hilare* (Say) (Miner 1966), and *T. perditor* (F.) (Panizzi and Herzog 1984); mirids such as *L. hesperus* Knight (Al-Munshi et al. 1982); and other species of insects from different orders (Romoser 1973), either there are little differences in longevity between males and females or females live longer than males. Clearly, additional studies are needed to fully demonstrate the hypotheses of the impact of oviposition on mitigating or not female longevity.

Differences in longevity between genders are also affected by the quality of the food. For example, males of *P. guildinii* live longer than females when fed on pods of green beans, *P. vulgaris*, or on pods of soybeans. On raw shelled peanuts, *Arachis hypogaea*, longevity for both genders was similar, and on mature soybean seeds females lived twice as much as males (Panizzi and Slansky 1985b). Therefore, food quality influences the life of heteropterans both directly by reducing longevity due to low nutritional quality and indirectly by affecting reproduction (i.e., nutrients drain by egg production). However, on high-quality foods (e.g., *N. viridula* on sesame or privet, and *P. guildinii* on indigo) despite the high fecundity, females had longer longevity than males. This suggests that when feeding on very high quality foods, egg production seems not to affect longevity the way it happens with foods of low or moderate nutritional quality. For the lygaeid *Elasmolomus sordidus* (F.), the high fecundity of females on sesame did not cause substantial reduction in longevity, and males and females showed similar adult lifetime (Mukhopadhyay and Saha 1992).

13.4.2 Less Suitable Foods (Leaves, Branches, Trunks)

Phytophagous heteropterans are, in general, polyphagously feeding on an array of plants from different families. However, less preferred plants may be explored as food sources and, in some cases, bugs change their feeding habits on those plants; that is, they abandon the habit of feeding on seeds and/or fruits, and explore vegetative tissue of leaves or branches. This change in feeding habits has consequences to nymph and adult biologies. Quite often, the role of these less preferred food plants is underestimated.

The nutritional quality of plants is variable in space and time, and to compensate for changes in food quality heteropterans need to explore alternate, in general less preferred, plants. According to Simpson and Simpson (1990), there are three types of compensatory responses: alter consumption, select a different diet, and compensate after food ingestion.

When the preferred food (seeds and fruits as in the case of seed-sucking heteropterans) is not available, insects are able to obtain nutrients from buds or flowers; however, in general, these plant structures do not allow nymph development and adult reproduction. Although adults can fly in search of other foods, nymphs are seriously threatened because of their limited dispersal ability. Some species of heteropterans, however, feed preferably on stems such as the neotropical rice stink bug, *Tibraca limbativentris* Stål, which feeds on the lower plant stem, close to the soil (Rizzo 1976). The pentatomid *E. meditabunda* (F.) feeds preferably on stems and leaves of soybean (Rizzo 1971; Galileo and Heinrichs 1979), usually in an upside-down position. It is suggested that this position facilitate the penetration of its short stylets in the plant tissue (Panizzi and Machado-Neto 1992). Leaf feeding for the seed-sucking southern green stink bug, *N. viridula* (L.), on the main leaf vein of soybean and castor bean, *R. communis* L., has been observed (A.R. Panizzi, unpublished).

Several species of heteropterans feed on branches and trunks of trees, as the pentatomids *Antiteuchus mixtus* (F.) and *A. tripterus* (F.) on privet, *L. lucidum* (Oleaceae) (Panizzi and Grazia 2001). Attempts to raise these bugs in the laboratory on fruits of privet have failed, and, apparently, the bugs need the nutrients present on the tree bark or on the xylem/phloem. Other heteropterans such as the aradids (Aradidae) are specialized in feeding on micelia of fungi that grow under loose bark of trees; the species *Aradus cinnamomeus* Panzer, however, feed on the phloem and xylem of pine trees (Heliövaara 2000).

The majority of heteropterans spend only one-third of their lifetime feeding during spring/summer on preferred host plants. The rest of the time, they colonize alternate plants, generally of low nutritional quality, or occupy niches for overwintering. These alternate plants supply some nutrients and water, but sometimes bugs do not recognize them as toxic plants. For example, *N. viridula*, although extremely polyphagous, do not recognize star bristle, *A. hispidum* (Compositae), as an unsuitable and toxic plant. As soybean matures, bugs move to star bristle (a weed), and feed on its stems, become intoxicated, and have their longevity dramatically reduced (Panizzi and Rossi 1991). In several occasions, dead adults were found on the ground near the plant stalk, suggesting that they were probably feeding on stems and became intoxicated.

Certain heteropterans such as the pentatomid *D. melacanthus* (Dallas) and the alydid *N. parvus* (Westwood) feed on dropped mature seeds, and on seedlings of soybean, corn, and wheat, causing severe damage to the last two crops. On soybean they cause early yellowing of the cotyledons (Panizzi et al. 2005a). Damage to corn seedlings has been reported about pentatomids of the genus *Euschistus* (Sedlacek and Townsend 1988; Apryanto et al. 1989); in the United States, *N. viridula* feeds on corn, but this seems to be infrequent (Negron and Riley 1987). Species of *Chauliops* (Malcidae), which occur in Africa and in Asia, are known to feed on leaves of several plants species, and this feeding habit may explain their relatively small size (Sweet and Schaefer 1985).

13.4.3 Impact of Nymph-to-Adult Food Switch on Adult Performance

In general, adults of heteropterans disperse from their host plants to feed and reproduce on other plant species; that is, their progeny will feed on different food sources. These changes in food from nymph to adult, although a common event in their biology, have been little investigated and their importance overlooked.

Nymph-to-adult food switch may have a positive or negative effect on adult performance. Among heteropterans, there are few examples in the literature regarding this issue, and the southern green stink bug, *N. viridula*, perhaps is the most studied species in this regard. For example, when nymphs and adults fed on the same food, the low performance of nymphs on *Crotalaria lanceolata* and on mature seeds of soybean, *Glycine max*, caused low performance of adults on these foods; only one female oviposited, and it took twice as long (49 days) to produce a single egg mass on *C. lanceolata* and no nymph emerged (Table 13.3) (Panizzi and Slansky 1991). Mean longevity of females on both foods and of males on *C. lanceolata* was drastically reduced.

Performance of adults that fed on pods of bagpod, *S. vesicaria*, as nymphs and as adults was low (Table 13.3), with none reaching day 30. Seven out of 12 females were observed mating; however, only one female oviposited a single egg mass, and few nymphs were able to emerge (Table 13.3). In contrast, the high performance of nymphs on pods of another species of *Sesbania* (*S. emerus*) caused 85% of adults to mate and to lay several egg masses with a great number of eggs, and with >50% of nymphs emerging (Table 13.3) and a high adult survivorship up to day 40. Females that fed on pods of green bean, *P. vulgaris*, as nymphs and as adults, laid the second greatest number of egg masses and eggs; 75% of nymphs were able to emerge, 57% of females oviposited (Table 13.3), and adults showed great longevity. A high percentage of females raised as nymphs on soybean pods, seeds of peanuts (*A. hypogaea*), or on

TABLE 13.3

Reproductive Performance of *N. viridula* Females Feeding on Different Legume Foods (Immature Pods, Unless Indicated Otherwise) as Affected by Switching or Not Switching of Foods from Nymph to Adult

Food		Number		Number/♀ (X ± SEM)		% Egg Hatch, (X ± SEM)[b]
Nymph	Adult	Pairs	%♀ ovip.,[a]	Egg Masses	Eggs	
P. vulgaris	*P. vulgaris*	21	57.1	3.1 (0.5) ab	185.3 (33.0 ab)	75.6 (5.8) a
S. emerus	*S. emerus*	13	84.6	3.7 (0.5) a	273.9 (36.1) a	55.8 (6.4) ab
P. vulgaris	*S. emerus*	10	80.0	2.4 (0.6) B	172.1 (50.1) B	64.4 (6.8) A
A. hypogaea[c]	*A. hypogaea*[c]	5	60.0	3.0 (1.0) ab[d]	99.7 (50.4) bc*	26.1 (11.2)b*
P. vulgaris	*A. hypogaea*[c]	10	100.0	5.7 (1.0) A	446.4 (93.7) A[d]*	62.3 (10.3)A*
G. max	*G. max*	17	76.5	1.9 (0.2) bc	110.0 (11.8) bc	61.5 (10.2) a
P. vulgaris	*G. max*	10	90.0	2.4 (0.4) B	149.1 (20.4) B	70.0 (9.6) A
D. tortuosum	*D. tortuosum*	16	56.2	1.3 (0.2) c[d]	61.0 (15.0) c*	59.5 (11.3) ab
P. vulgaris	*D. tortuosum*	10	70.0	2.7 (0.3) B*	153.1 (17.8)B*	43.8 (16.2) A
G. max[c]	*G. max*[c]	10	10.0[e]	1.0	23.0	0
P. vulgaris	*G. max*[c]	10	70.0	2.6 (0.4) B	204.4 (28.3) B	72.5 (9.9) A
C. lanceolata	*C. lanceolata*	8	12.5	1.0	29.0	0
P. vulgaris	*C. lanceolata*	10	30.0	2.3 (0.3) B	122.7 (30.7) B	68.2 (18.5) A
S. vesicaria[f]	*S. vesicaria*	12	8.3 [e]	1.0	40.0	15.0
P. vulgaris	*S. vesicaria*	10	40.0	1.0 (0.0) B	87.5 (4.3) B	82.2 (6.6) A

Source: Data from Panizzi, A. R. and F. Slansky, Jr., *J. Econ. Entomol.*, 84, 103, 1991.

Note: Means in each columns followed by the same lowercase letter (nymphs and adult same food), and upper case letter (adults reared as nymphs on *P. vulgaris* and then switched to the various foods) are not significantly different (*p* = .05), Duncan's multiple range test. Asterisk indicates significant difference between the two series within each food (*p* = .05; *t* test).

[a] For both series, % females ovipositing was dependent on food (nymphs and adults, same food: *G* = 54.48; *df* = 7; *p* < .001; adults switched to different food: *G* = 12.24; *df* = 5; *p* < .05).

[b] Data transformed to arcsine for analysis.

[c] Mature seeds.

[d] Data were included in the analysis although the residuals were not normally distributed.

[e] Because only one female laid one egg mass in each of these treatments, data for these were excluded from the statistic analyses.

[f] Immature seeds.

pods of *Desmodium tortuosum* were observed to mate but fecundity was low (Table 13.3); longevity was similar on these three foods.

When nymphs of *N. viridula* were raised on food of moderate quality (i.e., green bean pods) and switched as adults to various foods, longevity was substantially increased for those adults fed on *C. lanceolata*, or for females fed mature soybean seeds, peanuts, or pods of *D. tortuosum*, compared with those fed on the same food as nymphs and as adults.

Reproductive performance of adults reared as nymphs on *P. vulgaris* and then switched to pods of *C. lanceolata*, *D. tortuosum*, or *S. vesicaria*, or mature seeds of *G. max* or *A. hypogaea*, was improved compared with the general poor performance of stink bugs fed as nymphs and adults on these foods (Table 13.3).

These results and others obtained with *N. viridula* (Kester and Smith 1984; Panizzi et al. 1989; Panizzi and Saraiva 1993; Velasco and Walter 1992, 1993), and with other species of pentatomids (Panizzi 1987; Panizzi and Slansky 1985b; Pinto and Panizzi 1994), reinforce the importance for the adult performance of heteropterans of the switch in food from nymph to adult.

13.5 Impact of Abiotic Factors on Performance of Heteropterans

13.5.1 Temperature and Light

The performance of seed-sucking heteropterans is affected by the variation of abiotic factors such as temperature and photoperiod. In general, biological parameters such as egg and nymph development, food ingestion, and egg production increase as temperature increase to a certain level, and then decrease thereafter (Slansky and Panizzi 1987).

The pentatomid *N. viridula* show different genetically determined biological types. Studies have shown that type O (f. *torquata*) has a better performance of nymphs and adults at lower temperatures than types G (f. *smaragdula*) and Y (f. *aurantiaca*) (Vivan and Panizzi 2005), which explains the greater abundance of type O in cooler areas of southern Brazil (Vivan and Panizzi 2006).

For another pentatomid, *D. melacanthus* (Dallas), a long photoperiod (14 h light/10 h dark) speeds development and reduces mortality of nymphs and increases fecundity of adults; in a similar way, different photophase lengths cause adult dimorphism—long and sharp pronotal spines and green abdomen on long photophase versus short and rounded pronotal spines and brownish abdomen on short photophase (Chocorosqui and Panizzi 2003). Relatively low temperatures (15°C) do not allow nymph survival, which is less than 5% at 20°C; clearly, higher temperatures are needed for nymphs of *D. melacanthus* to develop better (Chocorosqui and Panizzi 2002).

The neotropical brown stink bug, *E. heros*, enters in reproductive diapause under photophase of 12 h or less, showing partially or totally undeveloped reproductive organs, shorter pronotal spines, and reduced feeding activity (Mourão and Panizzi 2002). This pentatomid shows photosensibility from the first instar, which is more pronounced during the third instar; reduced photophase during nymph development cause adult reproductive diapause (Mourão and Panizzi 2000a). In north of Paraná state, Brazil, *E. heros* shows mature reproductive organs and long pronotal spines during summer (December–March) and colonize soybean and sunflower. During fall–winter (April–August), it shows immature reproductive organs and short pronotal spines, and is found on the soil under crop residues or in shelters (Panizzi and Niva 1994; Panizzi and Vivan 1997; Mourão and Panizzi 2000b, 2002).

The many interactions of photoperiod and food are well known among insects, including those feeding on seeds. For example, the lygaeid *Ochrimnus mimulus* (Stål) increase its fecundity in the presence of food (seeds) at a shorter (12 h light) photophase; with increased exposure to light conditions (14 h), the fecundity of females, in the presence or absence of seeds, was the same (Gould and Sweet 2000). The feeding activity of the pentatomid *E. conspersus* Uhler is greater during the scotophase (Krupke et al. 2006).

13.5.2 Humidity

The relative humidity of the air affects development and survival of insects, which must keep their body water content within certain limits. The water exchange is influenced by the degree of cuticle permeability (Raghu et al. 2004). Apparently the ability to keep body water content during the first instar is variable among heteropterans, once gregarism—important in this context during this age—may or may not occur (Panizzi 2004a). For those species that show gregarism, such as the pentatomid *N. viridula*, under low relative humidity, nymphs that aggregate together survive better and develop faster than those that remain isolated (Lockwood and Story 1986a). Nymphs reached about 90% survivorship with >80% relative humidity; with 60% relative humidity, 60% of the nymphs emerged and survived, while with 0% relative humidity only 15% of the nymphs emerged and the majority died. Nymphs that emerged and remained on the top of the corions (eggshells), during a period of 24 h, dispersed and regrouped 6.8 ± 0.67 times, and showed this behavior until the time they abandon the corions and moved toward the source of humidity. The duration of the rearrangement of the group (dispersion + regrouping) varied from 26 to 44 min. For each event, time decreased from about 102 min for the first rearrangement down to 24 min for the sixth and last rearrangement. These rearrangement behaviors of the nymphs on the top of the corions apparently compensate for the water lost of those nymphs on the outer position of the group, which are greatly exposed to desiccation (Hirose et al. 2006b). The impact of relative humidity seems to be more critical during the first instar than during the remaining instars since nymphs grow bigger and tend to become less susceptible to the change in relative humidity.

According to their habitat, seed-sucking heteropterans may show preference for different gradients of humidity. For example, the lygaeid *Nysius groenlandicus* (Zetterstedt), which lives in the Artic, prefers low humidity (xerophily), similar to those species of insects adapted to the desert conditions (Böcher and Nachman 2001).

13.5.3 Rain and Wind

Phytophagous insects that live on the foliage or on any other exposed part of plants are directly affected by rain and wind. These are much more harsh conditions that those experienced by insects that inhabit the soil (e.g., root feeders) or the interior of plants (e.g., borers). Edwards and Wratten (1980) discussed the great challenge that exposed phytophagous insects face—to keep themselves on plants during periods of severe rains and windy conditions. At this time, factors such as smooth leaf surfaces and waxy surfaces of other parts of plants make this task even harder.

The impact of rain and wind on the survivorship of seed-sucking heteropterans has, surprisingly, been very little investigated and data in the literature seems to be lacking. There is no doubt that heavy rains cause great disturbance and even death of young nymphs, either through the force of impact or due to drowness. In addition, the disruption of the colony itself may cause nymphs to fail to regroup, causing additional deaths. The wind fustigating the foliage disturbs the colonies, which also results in death of nymphs. These two abiotic factors, rain and wind, particular in the tropics where these conditions are accentuated during summer, should be better studied to improve our knowledge of heteropteran population growth, particularly on crops.

13.6 Adaptations and Responses of Heteropterans to Changes in Favorability of the Environment

The variable nature of abiotic (e.g., temperature, relative humidity, photoperiod) and biotic (e.g., food quality and availability, interspecific and intraspecific competition) factors is a constant challenge to seed-sucking heteropterans. Therefore, they need to adapt to the instability of the environment to achieve their maximal fitness. For example, with decrease in temperature, the pentatomid *N. viridula* change color, from green to russet (Harris et al. 1984). Bugs that reach adulthood during fall and enter diapause during winter show reduced reproductive performance after the winter diapause (Musolin and Numata 2004). Novak (1955) reported that the dark spots (melanization) on the body of adult *O. fasciatus* (Dallas) are bigger when bugs are raised under low temperatures. In a similar way, when the pentatomid *Plautia stali* Scott is transferred from a long to a short photoperiod, oviposition is inhibited and adults show darker coloration (Kotaki and Yagi 1987). This can be an adaptation to raise body temperature when exposed to the sun. This behavior of sun exposure (basking behavior) seems to be more pronounced from 7:00 to 9:00 a.m. for *N. viridula*, which can be extended during not so clear days. Pesticide applications to exposed bugs were suggested to increase their efficacy (Waite 1980; Lockwood and Story 1986b). Dark forms, however, may be associated with physiological age, independent of the photoperiod, as in the case of certain mirids (Wilborn and Ellington 1984). At low temperatures, aggregated nymphs of *N. viridula* may accelerate development, increase use of atmospheric water, and prevent desiccation (Lockwood and Story 1986a). Certain mirids show tolerance to desiccation (Cohen et al. 1984).

In heteropterans specialized in feeding on seeds/fruits, low food availability can be overcome by utilization of alternate plant tissues, but feeding on seeds/fruits is required for normal nymphal development and adult reproduction (Slansky and Panizzi 1987). Increased consumption by seed/fruit-sucking hemipterans of foods with low nutrient contents has been little investigated. Teneral adults of *O. fasciatus* that were fasted for 1 week showed 3-fold increased food consumption in the presence of abundant food (Slansky 1982). *N. viridula* fasted for 24 h gained 27 mg, while those not fasted gained 9 mg. Other responses shown by heteropterans include utilization of nutritional reserves (lipids), breakup of colony to increase the ability of food finding, and alteration of feeding habits, for example, utilizing unsuitable seeds or even practicing cannibalism (Slansky and Panizzi 1987).

Other adaptations and responses of seed-sucking heteropterans to variation in abiotic and biotic factors include induced responses such as migration, diapause, and seasonal polyphenism. In general, these are induced by photoperiod and temperature. For example, nymphs of *O. fasciatus* raised in short photoperiods originate adults with greater flight ability compared with those raised in long photoperiods; flight ability for long flights cease once reproduction starts (Dingle 1985). In general, there is a correlation between wing length, flight ability, and fecundity, which can be positive (in the case of migratory biotypes) or negative (in the case of nonmigratory biotypes) (Dingle and Evans 1987). Also, wing length can be influenced by the photoperiod, as is the case of *Phyrrocoris apterus* (L.) (Pyrrhocoridae) that show macroptery in long photoperiods and brachyptery in short photoperiods (Honek 1976). Yet, the species *Cavalerius saccharivorus* Okajima (Lygaeidae) shows polymorphism, with the proportion of individuals being long winged or short winged depending on the genetic variation of the population and the density of rearing conditions; long-winged individuals showed a better reproductive performance than short-winged ones in high-density conditions but not in low-density conditions; this mixed strategy allows better dispersion and exploration of the habitat of origin (Fujisaki 1985, 1986a,b). In low-temperature conditions and in the absence of food, *Cletus punctiger* Dallas (Coreidae) seeks for shelter in hibernacula (Ito 1988).

13.7 Final Considerations

Heteropterans that feed on seeds/fruits make up an important feeding group (guild), with several species pests of crops and fruit trees of economic importance worldwide (Schaefer and Panizzi 2000). Despite the information available on the impact of these insects on yield and quality of seeds/fruits, a lot remain to be done to better understand the many interactions of these insects with their host plants. For example, relatively few data exist on how the amounts and proportions of nutrient and alelochemicals in seeds/fruits and their physical attributes affect the prefeeding behavior and postfeeding performance. Moreover, interspecific and intraspecific competitions, in particular of those species that explore the same nutritional resources, await further investigations. The same can be said about the impact of natural enemies, such as parasitoids and predators, to the heteropterans' biology.

In the applied context, data that answer the many questions raised here and elsewhere (Slansky and Panizzi 1987) are necessary to allow full implementation of integrated pest management programs for heteropterans that feed on seeds/fruits. For example, utilization of tactics such as host plant resistance, manipulation of planting time, use of cultivars of different maturity groups, and use of attractive "trap" plants are all tactics that fit within the context of insect bioecology and nutrition. The myriad of interactions of seed/fruit-sucking heteropterans and their food certainly will be more understood taking in consideration the paradigm set by the bioecology and insect nutrition.

REFERENCES

Alcock, J. 1973. The feeding response of hand-reared red-winged blackbirds (*Agelaius phoeniceus*) to a stink bug (*Euschistus conspersus*). *Am. Midland Nat.* 89:307–13.

Aldrich, J. R. 1988. Chemical ecology of the Heteroptera. *Annu. Rev. Entomol.* 33:211–38.

Aldrich, J. R. 1995. Chemical communication in the true bugs and parasitoid exploitation. In *Chemical Ecology of Insects II*, ed. R. T. Carde and W. J. Bell, 318–63. New York: Chapman & Hall.

Aldrich, J. R., and M. S. Blum. 1978. Aposematic aggregation of a bug (Hemiptera: Coreidae): The defensive display and formation of aggregations. *Biotropica* 10:58–61.

Aldrich, J. R., A. Khrimian, X. Chen, and M. J. Camp. 2009. Semiochemically based monitoring of the invasion of the brown marmorated stink bug and unexpected attraction of the native green stink bug (Heteroptera: Pentatomidae) in Maryland. *Fla. Entomol.* 92:483–91.

Almeida, J. R., R. Xerez, and J. Jurberg. 1986. Comportamento de acasalamento de *Dysdercus maurus* Distant, 1901 (Hemiptera, Pyrrhocoridae). *An. Soc. Entomol. Bras.* 15:161–7.

Al-Munshi, D. M., D. R. Scott, and W. Smith. 1982. Some host plant effects on *Lygus hesperus* (Hemiptera: Miridae). *J. Econ. Entomol.* 75:813–15.

Andre, F. 1935. Notes on the biology of *Oncopeltus fasciatus* (Dallas). *Iowa J. Sci.* 9:73–87.

Apryanto, D., L. H. Townsend, and J. D. Sedlacek. 1989. Yield reduction from feeding by *Euschistus servus* and *E. variolarius* (Heteroptera: Pentatomidae) on stage V2 field corn. *J. Econ. Entomol.* 82:445–8.

Bates, R. P., F. W. Knapp, and P. E. Araújo. 1977. Protein quality of green-mature, dry mature and sprouted soybeans. *J. Food. Sci.* 42:271–2.

Bates, R. P., and R. F. Matthews. 1975. Ascorbic acid and B-carotene in soybeans as influenced by maturity, sprouting, processing, and storage. *Proc. Fla. State Hortic. Soc.* 88:266–71.

Böcher, J., and G. Nachman. 2001. Temperature and humidity responses of the arctic-alpine seed bug *Nysius groenlandicus*. *Entomol. Exp. Appl.* 99:319–30.

Boesewinkel, F. D., and F. Bouman. 1995. The seed: Structure and function. In *Seed Development and Germination*, ed. J. Kigel, and J. G. Galili, 1–24. New York: Marcel Dekker.

Bolivar, J. 1894. Observations sur la *Phyllomorpha*. *Feuille J. Natur.* 24:43.

Bommireddy, P. L., B. R. Leornard, and J. H. Temple. 2007. Influence of *Nezara viridula* feeding on cotton yield, fiber quality, and seed germination. *J. Econ. Entomol.* 100:1560–8.

Borges, M. 1995. Attractant compounds of the southern green stink bug, *Nezara viridula* (L.) (Heteroptera: Pentatomidae). *An. Soc. Entomol. Bras.* 24:215–25.

Borges, M., P. C. Jepson, and P. E. House. 1987. Long-range mate location and close range courtship behaviour of the green stink bug, *Nezara viridula* and its mediation by sex pheromones. *Entomol. Exp. Appl.* 44:205–12.

Borges, M., M. C. B. Moraes, M. F. Peixoto, C. S. S. Pires, E. R. Fujii, and R. A. Laumann. 2010. Monitoring the neotropical brown stink bug *Euschistus heros* (F.) (Hemiptera: Pentatomidae) with pheromone-baited traps in soybean fields. *J. Appl. Entomol.* 1:1–13.

Bowling, C. C. 1979. The stylet sheath as an indicator of feeding activity of the rice stink bug. *J. Econ. Entomol.* 72:259–60.

Bowling, C. C. 1980. The stylet sheath as an indicator of feeding activity by the southern green stink bug on soybeans. *J. Econ. Entomol.* 73:1–3.

Brent, C. S. 2010. Reproductive refractoriness in the western tarnished plant bug (Hemiptera: Miridae). *Ann. Entomol. Soc. Am.* 102:300–6.

Bressani, R., and L. G. Elias. 1980. Nutritional value of legume crops to humans and animals. In *Advances in Legume Science*, ed. R. J. Summerfield and A. H. Bunting, 135–55. Kew: Royal Botanic Gardens.

Bromley, S. W. 1937. Food habits of alydine bugs (Hemiptera: Coreidae). *Bull. Brooklyn Entomol. Soc.* 32:139.

Carroll, S. P. 1988. Contrasts in reproductive ecology between temperate and tropical populations of *Jadera haematoloma*, a mate-guarding hemipteran (Rhopalidae). *Ann. Entomol. Soc. Am.* 81:54–63.

Carroll, S. P. 1993. Divergence in male mating tactics between two populations of the soapberry bug: I. Guarding versus nonguarding. *Behav. Ecol.* 4:156–64.

Carroll, S. P., and J. E. Loye. 1987. Specialization of *Jadera* species (Hemiptera: Rhopalidae) on the seeds of Sapindaceae (Sapindales), and coevolutionary responses of defenses an attack. *Ann. Entomol. Soc. Am.* 80:373–8.

Carvalho, N. M., and J. Nakagawa. 1983. *Sementes: Ciência, Tecnologia e Produção*. Campinas, Brazil: Fundação Cargil, 2ª ed.

Chocorosqui, V. R., and A. R. Panizzi. 2002. Influência da temperature na biologia de ninfas de *Dichelops melacanthus* (Dallas, 1851) (Heteroptera: Pentatomidae). *Semina* 23:217–20.

Chocorosqui, V. R., and A. R. Panizzi. 2003. Photoperiod influence on the biology and phenological characteristics of *Dichelops melacanthus* (Dallas, 1851) (Heteroptera: Pentatomidae). *Braz. J. Biol.* 63:655–64.

Cobben, R. H. 1978. Evolutionary trends in Heteroptera. Part II. Mouthpart structures and feeding strategies. *Meded. Landbouwhogesch. Wageningen*, 78-5:407

Cohen, A. C. 1990. Feeding adaptations of some predaceous Hemiptera. *Ann. Entomol. Soc. Am.* 83:1215–23.

Cohen, A. C., S. L. Earl, and H. M. Graham. 1984. Rates of water loss and desiccation tolerance in four species of *Lygus*. *Southwest. Entomol.* 9:178–83.

Čokl, A. 1983. Functional properties of vibroreceptors in the legs of *Nezara viridula* (L.) (Heteroptera, Pentatomidae). *J. Comp. Physiol. A* 150:261–9.

Čokl, A., M. Virant-Doberlet, and A. McDowell. 1999. Vibrational directionality in the southern green stink bug, *Nezara viridula* (L.), is mediated by female song. *Animal Behav.* 58:1277–83.

Čokl, A., M. Virant-Doberlet, and N. Stritih. 2000. The structure and function of songs emitted by southern green stink bugs from Brazil, Florida, Italy and Slovenia. *Physiol. Entomol.* 25:196–205.

Coombs, M. 2004. Broadleaf privet, *Ligustrum lucidum* Aiton (Oleaceae), a late-season host for *Nezara viridula* (L.), *Plautia affinis* Dallas and *Glaucias amyoti* (Dallas) (Hemiptera: Pentatomidae) in northern New South Wales, Australia. *Aust. J. Entomol.* 43:335–9.

Corrêa-Ferreira, B. S. 1984. Incidência do parasitóide *Eutrichopodopsis nitens* Blanchard, 1966 em populações do percevejo verde *Nezara viridula* (Linnaeus, 1758). *An. Soc. Entomol. Bras.* 13:321–30.

Corrêa-Ferreira, B. S. 1986. Ocorrência natural do complexo de parasitóides de ovos de percevejos da soja no Paraná. *An. Soc. Entomol. Bras.* 15:189–99.

Costa, E. C., and D. Link. 1982. Dispersão de adultos de *Piezodorus guildinii* e *Nezara viridula* (Hemiptera: Pentatomidae) em soja. *Rev. Centr. Cien. Rur.* 12:51–7.

Costa Lima, A. M. 1940. *Insetos do Brasil. Hemípteros.* Série Didática Nº 3, 2º tomo. Rio de Janeiro: Escola Nacional de Agronomia.

Derr, J. A. 1980. Coevolution of the life history of a tropical-feeding insect and its food plants. *Ecology* 61:881–92.

Deunff, Y-le. 1989. Hydration of pea (*Pisum sativum*) seeds. *Seed Sci. Technol.* 17:471–83.

Dingle, H. 1985. Migration. In *Comprehensive Insect Physiology, Biochemistry and Pharmacology*, ed. G. A. Kerkut and L. I. Gilbert, 375–415, vol. 9. Oxford: Pergamon Press.

Dingle, H., and K. E. Evans. 1987. Responses in flight to selection on wing length in non-migratory milkweed bugs, *Oncopeltus fasciatus*. *Entomol. Exp. Appl.* 45:289–96.

Depieri, R. A., and A. R. Panizzi. 2010. Rostrum length and comparative morphology of mandible serration and of food and salivary canals of selected species of stink bugs (Heteroptera: Pentatomidae). *Rev. Bras. Entomol.* 54:584–7.

Dreyer, H., and J. Baumgärtner. 1997. Adult movement and dynamics of *Clavigralla tomentosicollis* (Heteroptera: Coreidae) populations in cowpea fields of Benin, West Africa. *J. Econ. Entomol.* 90:421–6.

Duffus, C. M., and J. C. Slaughter. 1980. *Seeds and Their Uses.* Chichester: J. Wiley & Sons.

Earle, F. R., and Q. Jones. 1962. Analyses of seed sample from 113 plants families. *Econ. Bot.* 16:221–50.

Eberhard, W. G. 1975. The ecology and behaviour of a subsocial pentatomid bug and two scelionid wasps: Strategy and counter strategy in a host and its parasites. *Smith. Contr. Zool.* 205, 39pp.

Edwards, P. J., and S. D. Wratten. 1980. *Ecology of Insect–Plant Interactions.* London: Edward Arnold Publishers.

Eggermann, W., and J. Bongers 1980. Die bedentung des wassering speilchelsekrets fur die nahrungsaufnahme von *Oncopeltus fasciatus* und *Dysdercus fasciatus*. *Entomol. Exp. Appl.* 27:169–78.

Evans, E. W. 1987. Dispersal of *Lygaeus kalmii* (Hemiptera: Lygaeidae) among prairie milkweeds: Population turnover as influenced by multiple mating. *J. Kansas Entomol. Soc.* 60:109–17.

Filippi, L., M. Hironaka, S. Nomakuchi, and S. Tojo. 2000. Provisioned *Parastrachia japonensis* (Hemiptera: Cydnidae) nymphs gain access to food and protection from predators. *Anim. Behav.* 60:757–63.

Filippi, L., M. Hironaka, and S. Nomakuchi. 2005. Kleptoparasitism and the effect of nest location in a subsocial shield bug *Parastrachia japonensis* (Hemiptera: Cydnidae). *Ann. Entomol. Soc. Am.* 98:134–42.

Follett, P. A., M. G. Wright, and M. Golden. 2009. *Nezara viridula* (Hemiptera: Pentatomidae) feeding patterns in macadamia nut in Hawaii: Nut maturity and cultivar effects. *Environ. Entomol.* 38:1168–73.

Fontes, E. M. G., C. S. S. Pires, E. R. Sujii, and A. R. Panizzi. 2002. The environmental effects of genetically modified crops resistant to insects. *Neotrop. Entomol.* 31:497–513.

Fujisaki, K. 1985. Ecological significance of the wing polymorphism of the oriental chinch bug, *Cavalerius saccharivorous* Okajima (Heteroptera: Lygaeidae). *Res. Pop. Ecol.* 27:125–36.

Fujisaki, K. 1986a. Reproductive properties of the oriental chinch bug, *Cavalerius saccharivorous* Okajima (Heteroptera: Lygaeidae) in relation to its wing polymorphism. *Res. Pop. Ecol.* 28:43–52.

Fujisaki, K. 1986b. Genetic variation of density responses in relation to wing polymorphism in the oriental chinch bug, *Cavalerius saccharivorous* Okajima (Heteroptera: Lygaeidae). *Res. Pop. Ecol.* 28:219–30.

Galileo, M. H. M., and E. A. Heinrichs. 1979. Danos causados à soja em diferentes níveis e épocas de infestação durante o crescimento. *Pesq. Agropec. Bras.* 14:279–82.

Gatehouse, J. A., and A. M. R. Gatehouse. 2000. Genetic engineering of plants for insect resistance. In *Biological and Biotechnological Control of Insect Pests*, ed. J. E. Rechcigl and N. A. Rechcigl, 211–41, Boca Raton: Lewis.

Goodchild, A. J. P. 1966. Evolution of the alimentary canal in Hemiptera. *Biol. Rev.* 41:97–140.

Gould, G. G., and M. H. Sweet, II. 2000. The host range and oviposition behavior of *Ochrimnus mimulus* (Hemiptera: Lygaeidae) in central Texas. *Southwest. Naturalist* 45:15–23.

Harborne, J. B., D. Boulter, and B. L. Turner. 1971. *Chemotaxonomy of the Leguminosae*. London: Academic Press.

Harris, V. E., and J. W. Todd. 1980a. Male-mediated aggregation of male, female and 5th-instar southern green stink bug and concomitant attraction of a tachinid parasite, *Trichopoda pennipes*. *Entomol. Exp. Appl.* 27:117–26.

Harris, V. E., and J. W. Todd. 1980b. Temporal and numerical patterns of reproductive behavior in the southern green stink bug, *Nezara viridula* (Hemiptera: Pentatomidae). *Entomol. Exp. Appl.* 27:105–16.

Harris, V. E., J. W. Todd, J. C. Webb, and J. C. Benner. 1982. Acoustical and behavioral analysis of the songs of the southern green stink bug, *Nezara viridula*. *Ann. Entomol. Soc. Am.* 75:234–49.

Harris, V. E., J. W. Todd, and B. G. Mullinix. 1984. Color change as an indicator of adult diapause in the southern green stink bug, *Nezara viridula*. *J. Agric. Entomol.* 1:82–91.

Heliövaara, K. 2000. Flat bugs (Aradidae). In *Heteroptera of Economic Importance*, ed. C. W. Schaefer and A. R. Panizzi, 513–7. Boca Raton: CRC Press.

Hirose, E., A. R. Panizzi, and A. J. Cattelan. 2006a. Potential use of antibiotic to improve performance of laboratory-reared *Nezara viridula* (L.) (Heteroptera: Pentatomidae). *Neotrop. Entomol.* 35:279–81.

Hirose, E., A. R. Panizzi, and A. J. Cattelan. 2006b. Effect of relative humidity on emergence and on dispersal and regrouping of first instar *Nezara viridula* (L.) (Hemiptera: Pentatomidae). *Neotrop Entomol.* 35:757–61.

Hirose, E., A. R. Panizzi, J. T. de Souza, A. J. Cattelan, and J. R. Aldrich. 2006c. Bacteria in the gut of southern green stink bug (Heteroptera: Pentatomidae). *Ann. Entomol. Soc. Am.* 99:91–5.

Honbo, Y., and K. Nakamura. 1985. Effectiveness of parental care in the bug *Elasmucha putoni* Scott (Hemiptera: Acanthosomatidae). *Jpn. J. Appl. Entomol. Zool.* 29:223–9.

Honek, A. 1976. The regulation of wing polymorphism in natural populations of *Pyrrhocoris apterus* (Heteroptera, Pyrrhocoridae). *Zool. Jb. Syst.* 103:547–70.

Hori, K. 2000. Possible causes of disease symptoms resulting from the feeding of phytophagous Heteroptera. In *Heteroptera of Economic Importance*, ed. C. W. Schaefer and A. R. Panizzi, 11–35, Boca Raton: CRC Press.

Ishiwatari, T. 1974. Studies on the scent of stink bugs (Hemiptera: Pentatomidae). I. Alarm pheromone activity. *Appl. Entomol. Zool.* 9:153–8.

Ishiwatari, T. 1976. Studies on the scent of stink bugs (Hemiptera: Pentatomidae). II. Aggregation pheromone activity. *Appl. Entomol. Zool.* 11:38–44.

Ito, K. 1988. Effects of feeding and temperature on the hiding behaviour of *Cletus punctiger* Dallas (Heteroptera: Coreidae) in hibernacula. *Jpn. J. Appl. Entomol. Zool.* 32:49–54.

Janzen, D. H. 1969. Seed eaters versus seed size, number, toxicity, and dispersal. *Evolution* 23:1–27.

Janzen, D. H. 1971. Seed predation by animals. *Annu. Rev. Ecol. Syst.* 2:465–92.

Janzen, D. H. 1972. Escape in space by *Sterculia apetala* seeds from the bug *Dysdercus fasciatus*, in a Costa Rican deciduous forest. *Ecology* 53:350–61.

Janzen, D. H. 1978. The ecology and evolutionary biology of seed chemistry as relates to seed predation. In *Biochemical Aspects of Plant and Animal Coevolution*, ed. J. B. Harborne, 163–206. New York: Academic Press.

Janzen, D. H., H. B. Juster, and I. E. Liener. 1976. Insecticidal action of the phytohemagglutinin in black beans on a bruchid beetle. *Science* 192:795–6.

Javahery, M., C. W. Schaefer, and J. D. Lattin. 2000. Shield bugs (Scutelleridae). In *Heteroptera of Economic Importance,* ed. C. W. Schaefer, and A. R. Panizzi, 475–503. Boca Raton: CRC Press.

Johnson, C. D., and R. A. Kistler. 1987. Nutritional ecology of bruchid beetles. In *Nutritional Ecology of Insects, Mites, Spiders, and Related Invertebrates*, ed. F. Slansky Jr. and J. G. Rodríguez, 259–82. New York: J. Wiley & Sons.

Jones, G., and F. R. Earle. 1966. Chemical analysis of seeds. II. Oil and protein content of 759 species. *Econ. Bot.* 20:127–55.

Kaitala, A. 1996. Oviposition on the back of conspecifics: An unusual reproductive tactic in a coreid bug. *Oikos* 77:381–9.

Kehat, M., and M. Wyndham. 1972. The influence of temperature on development, longevity, and fecundity in the rutherglen bug, *Nysius vinitor* (Hemiptera: Lygaeidae). *Aust. J. Zool.* 20:67–8.

Kester, K. M., and C. M. Smith. 1984. Effects of diet on growth, fecundity and duration of tethered flight of *Nezara viridula*. *Entomol. Exp. Appl.* 35:75–81.

Kogan, M. 1986. Natural chemicals in plant resistance to insects. *Iowa State J. Res.* 60:501–27.

Kotaki, T., and S. Yagi. 1987. Relationship between diapause development and coloration change in brown-winged green bug, *Plautia stali* Scott (Heteroptera: Pentatomidae). *Jpn. J. Appl. Entomol. Zool.* 31: 285–90.

Krispyn, J. W., and J. W. Todd. 1982. The red imported fire ant as a predator of the southern green stink bug on soybean in Georgia. *J. Ga. Entomol. Soc.* 17:19–26.

Krupke, C. H., V. P. Jones, and J. F. Brunner. 2006. Diel periodicity of *Euschistus conspersus* (Heteroptera: Pentatomidae) aggregation, mating, and feeding. *Ann. Entomol. Soc. Am.* 99:169–74.

Kutcher, S. R. 1971. Two types of aggregation grouping in the large milkweed bug, *Oncopeltus fasciatus* (Hemiptera: Lygaeidae). *Bull. Southern Calif. Acad. Sci.* 70:87–90.

Lener, W. 1967. Sexual activity and longevity in the large milkweed bug, *Oncopeltus fasciatus* (Hemiptera: Lygaeidae). *Ann. Entomol. Soc. Am.* 60:484–5.

Liener, I. E. 1979. Anti-nutritional factors as determinants of soybean quality. In *Proceedings World Soybean Research Conference II,* ed. F. T. Corbin, 703–12. Boulder: Westview Press.

Liener, I. E. 1980. Heat-labile antinutritional factors. In *Advances in Legume Science*, ed. R. J. Summerfield and A. H. Bunting, 157–70. Kew: Royal Botanic Gardens.

Link, D., and L. Storck. 1978. Correlação entre danos causados por pentatomídeos, acamamento e retenção foliar em soja. *Rev. Centr. Ciênc. Rur.* 8:297–301.

Lockwood, J. A., and R. N. Story. 1985. Bifunctional pheromone in the first instar of the southern green stink bug, *Nezara viridula* (L.) (Hemiptera: Pentatomidae): Its characterization and interaction with other stimuli. *Ann. Entomol. Soc. Am.* 78:474–9.

Lockwood, J. A., and R. N. Story. 1986a. Adaptive functions of nymphal aggregation in the southern green stink bug, *Nezara viridula* (L.) (Hemiptera: Pentatomidae). *Environ. Entomol.* 15:739–49.

Lockwood, J. A., and R. N. Story. 1986b. The diurnal ethology of the southern green stink bug, *Nezara viridula* (L.), in cowpeas. *J. Entomol. Sci.* 21:175–84.

Lott, J. N. A., J. S. Greenwood, and G. D. Batten. 1995. Mechanisms and regulation of mineral nutrient storage during seed development. In *Seed Development and Germination*, ed. J. Kigel, and G. Galili, 215–36. New York: Marcel Dekker.

Lung, K. Y. H., and R. D. Goeden. 1982. Biology of *Corimelaena extensa* on tree tobacco, *Nicotiana glauca*. *Ann. Entomol. Soc. Am.* 75:177–80.

Malaguido, A. B., and A. R. Panizzi. 1999. Nymph and adult biology of *Euschistus heros* (Hemiptera: Pentatomidae) and its abundance related to planting date and phenological stages of sunflower. *Ann. Entomol. Soc. Am.* 92:424–9.

Mayer, A. M., and A. Poljakoff-Mayber. 1982. *The Germination of Seeds*. Oxford: Pergamon Press, Oxford.

McDonald, G., and R. A. Farrow. 1988. Migration and dispersal of the rutherglen bug, *Nysius vinitor* Bergroth (Hemiptera: Lygaeidae), in eastern Australia. *Bull. Entomol. Res.* 76:493–509.

McLain, D. K. 1981. Sperm precedence and prolonged copulation in the southern green stink bug, *Nezara viridula*. *J. Ga. Entomol. Soc.* 16:70–7.

McLain, D. K. 1992. Preference for polyandry in female stink bugs, *Nezara viridula* (Hemiptera: Pentatomidae). *J. Insect Behavior* 5:403–10.

McPherson, J. E. 1977a. Notes on the biology of *Thyanta calceata* (Hemiptera: Pentatomidae) with information on adult seasonal dimorphism. *Ann. Entomol. Soc. Am.* 70:370–2.

McPherson, J. E. 1977b. Effects of developmental photoperiod on adult color and pubescence in *Thyanta calceata* (Hemiptera: Pentatomidae) with information on ability of adults to change color. *Ann. Entomol. Soc. Am.* 70:373–6.

Medcalf, D. G. 1973. Structure and composition of cereal component as related to their potential industrial utilization. In *Industrial Uses of Cereal*, ed. Y. Pomerans, 121–60. St. Paul: American Association of Cereal Chemistry.

Medeiros, L., and G. A. Megier. 2009. Ocorrência e desempenho de *Euschistus heros* (F.) (Heteroptera: Pentatomidae) em plantas hospedeiras alternativas no Rio Grande do Sul. *Neotrop. Entomol.* 38: 459–63.

Melber, A., L. Holscher, and G. H. Schmidt. 1980. Further studies on the social behavior and its ecological significance in *Elasmucha grisea* L. (Hem. Het. Acanthosomatidae). *Zool. Anz.* 205:27–38.

Miettinen, M., and A. Kaitala. 2000. Copulation is not a prerequisite to male reception of eggs in the golden egg bug *Phyllomorpha laciniata* (Coreidae; Heteroptera). *J. Insect Behav.* 13:731–40.

Miklas, N., T. Lasnier, and M. Renou. 2003. Male bugs modulate pheromone emission in response to vibratory signals from conspecifics. *J. Chem. Ecol.* 29:561–74.

Miles, P. W. 1972. The saliva of Hemiptera. *Adv. Insect Physiol.* 9:183–255.

Miles, P. W., and G. S. Taylor. 1994. "Osmotic pump" feeding by coreids. *Entomol. Exp. Appl.* 73:163–73.

Miner, F. D. 1966. Biology and control of stink bugs on soybeans. *Arkansas. Agric. Exp. Stn. Bull.* 708:1–40.

Mitchell, P. L. 2000. Leaf-footed bugs (Coreidae). In *Heteroptera of Economic Importance*, ed. C. W. Schaefer and A. R. Panizzi, 337–403. Boca Raton: CRC Press.

Miyatake, T. 2002. Multi-male mating aggregation in *Notobitus meleagris* (Hemiptera: Coreidae). *Ann. Entomol. Soc. Am.* 95:340–4.

Moraes, M. C. B., R. A. Laumann, A. Čokl, and M. Borges. 2005. Vibratory signals of four neotropical stink bug species. *Physiol. Entomol.* 30:175–88.

Moreira, G. R. P., and M. Becker. 1986. Mortalidade de *Nezara viridula* (Linnaeus, 1758) (Heteroptera: Pentatomidae) no estágio de ovo na cultura da soja. I—Todas as causas de mortalidade. *An. Soc. Entomol. Bras.* 15:271–90.

Mourão, A. P. M., and A. R. Panizzi. 2000a. Estágios ninfais fotossensíveis à indução da diapausa em *Euschistus heros* (Fabr.) (Hemiptera: Pentatomidae). *An. Soc. Entomol. Bras.* 29:219–25.

Mourão, A. P. M., and A. R. Panizzi. 2000b. Diapausa e diferentes formas sazonais em *Euschistus heros* (Fabr.) (Hemiptera: Pentatomidae) no norte do Paraná. *An. Soc. Entomol. Bras.* 29:205–18.

Mourão, A. P. M., and A. R. Panizzi. 2002. Photophase influence on the reproductive diapause, seasonal morphs, and feeding activity of *Euschistus heros* (Fabr., 1798) (Hemiptera: Pentatomidae). *Braz. J. Biol.* 62:231–8.

Mukhopadhyay, A., and B. Saha. 1992. Nutritional value of four host seeds and their relationship with adult and nymphal performance of seed bug (*Elasmolomus sordidus*) (Heteroptera: Lygaeidae). *Indian J. Agric. Sci.* 62:834–7.

Musolin, D. L., and H. Numata. 2004. Late-season induction of diapause in *Nezara viridula* and its effect on adult coloration and post-diapause reproductive performance. *Entomol. Exp. Appl.* 111:1–6.

Nault, L. R., and G. G. Gyrisco. 1966. Relation of the feeding process of the pea aphid to the inoculation of pea enation mosaic virus. *Ann. Entomol. Soc. Am.* 59:1185–97.

Negron, J. F., and T. J. Riley. 1987. Southern green stink bug, *Nezara viridula* (Heteroptera: Pentatomidae) feeding on corn. *J. Econ. Entomol.* 80:666–9.

Nomakuchi, S., L. Filippi, and M. Hironaka. 2001. Nymphal occurrence pattern and predation risk in the subsocial shield bug, *Parastrachia japonensis* (Heteroptera: Cydnidae). *Appl. Entomol. Zool.* 36:209–12.

Novak, V. 1955. Nektere zakonitosti vyvoje ventralnich cernych skvrn u plostice *Oncopeltus fasciatus*. *Mem. Soc. Zool. Tchekosl.* 19:233–46.

Ota, D., and A. Čokl. 1991. Mate location in the southern green stink bug, *Nezara viridula* (Heteroptera: Pentatomidae), mediated through substrate-borne signals on ivy. *J. Insect Behav.* 4:441–7.

Owusu-Manu, E. 1977. Flight activity and dispersal of *Bathycoelia thalassina* (Herrich-Schaeffer) Hemiptera: Pentatomidae. *Ghana J. Agric. Sci.* 10:23–6.

Owusu-Manu, E. 1980. Observations on mating and egg-laying behaviour of *Bathycoelia thalassina* (Herrich-Schaefer) (Hemiptera: Pentatomidae). *J. Nat. Hist.* 14:463–7.

Panizzi, A. R. 1987. Impacto de leguminosas na biologia de ninfas e efeito da troca de alimento no desempenho de adultos de *Piezodorus guildinii* (Hemiptera: Pentatomidae). *Rev. Bras. Biol.* 47:585–91.

Panizzi, A. R. 1988. Biology of *Megalotomus parvus* (Westwood) (Heteroptera: Alydidae) on selected leguminous food plants. *Insect Sci. Appl.* 9:279–85.

Panizzi, A. R. 1989. Parasitismo de *Eutrichopodopsis nitens* (Diptera: Tachinidae) em *Nezara viridula* (Hemiptera: Pentatomidae) observado em distintas plantas hospedeiras. *Pesq. Agropec. Bras.* 24:1555–8.

Panizzi, A. R. 1997. Wild hosts of pentatomids: Ecological significance and role in their pest status on crops. *Annu. Rev. Entomol.* 42:99–122.

Panizzi, A. R. 2004a. Adaptive advantages for egg and nymph survivorship by egg deposition in masses or singly in seed-sucking Heteroptera. In *Contemporary Trends in Insect Science*, ed. G. T. Gujar, 60–73. New Delhi: Campus Books International.

Panizzi, A. R. 2004b. A possible territorial or recognition behavior of *Leptoglossus zonatus* (Dallas) (Heteroptera, Coreidae). *Rev. Bras. Entomol.* 48:577–9.

Panizzi, A. R. 2006. Possible egg positioning and gluing behavior by ovipositing southern green stink bug, *Nezara viridula* (L.) (Heteroptera: Pentatomidae). *Neotrop. Entomol.* 35:149–51.

Panizzi, A. R. 2007. Nutritional ecology of plant feeding arthropods and IPM. In *Perspectives in Ecological Theory and Integrated Pest Management*, ed. M. Kogan, and P. Jepson, 170–222. Cambridge: Cambridge University Press.

Panizzi, A. R., and R. M. L. Alves. 1993. Performance of nymphs and adults of the southern green stink bug (Heteroptera: Pentatomidae) exposed to soybean pods at different phenological stages of development. *J. Econ. Entomol.* 86:1088–93.

Panizzi, A. R., S. R. Cardoso, and V. R. Chocorosqui. 2002a. Nymph and adult performance of the small green stink bug, *Piezodorus guildinii* (Westwood) on lanceleaf crotalaria and soybean. *Braz. Arch. Biol. Technol.* 45:53–8.

Panizzi, A. R., S. R. Cardoso, and E. D. M. Oliveira. 2000a. Status of pigeon pea as an alternative host of *Piezodorus guildinii* (Hemiptera: Pentatomidae), a pest of soybean. *Fla. Entomol.* 83:334–42.

Panizzi, A. R., V. R. Chocorosqui, J. J. Silva, and F. A. C. Silva. 2005a. Ataque de percevejos em plântulas de soja, *Folder* 12/2005, Embrapa Soja, Londrina, PR.

Panizzi, A. R., M. H. M. Galileo, H. A. O. Gastal, J. F. F. Toledo, and C. H. Wild. 1980. Dispersal of *Nezara viridula* and *Piezodorus guildinii* nymphs in soybeans. *Environ. Entomol.* 9:293–7.

Panizzi, A. R., and J. Grazia. 2001. Stink bugs (Pentatomidae) and a unique host plant in the Brazilian subtropics. *Iheringia, Ser. Zool.* 90:21–35.

Panizzi, A. R., and D. C. Herzog. 1984. Biology of *Thyanta perditor*. *Ann. Entomol. Soc. Am.* 77:646–50.

Panizzi, A. R., and E. Hirose. 2002. Seed-carrying and feeding behavior of *Jadera choprai* Göllner-Scheiding (Heteroptera: Rhopalidae). *Neotrop. Entomol.* 31:327–9.

Panizzi, A. R., E. Hirose, and V. R. Chocorosqui. 2002b. Unusual oviposition behavior by a seed feeding bug (Heteroptera: Rhopalidae). *Neotrop. Entomol.* 31:477–9.

Panizzi, A. R., E. Hirose, and E. D. M Oliveira. 1996a. Egg allocation by *Megalotomus parvus* (Westwood) (Heteroptera: Alydidae) on soybean. *An. Soc. Entomol. Bras.* 25:537–43.

Panizzi, A. R., and E. Machado-Neto. 1992. Development of nymphs and feeding habits of nymphal and adult *Edessa meditabunda* (Heteroptera: Pentatomidae) on soybean and sunflower. *Ann. Entomol. Soc. Am.* 85:477–81.

Panizzi, A. R., J. E. McPherson, D. G. James, M. Javahery, and R. M. McPherson. 2000b. Economic importance of stink bugs (Pentatomidae). In *Heteroptera of Economic Importance*, ed. C. W. Schaefer and A. R. Panizzi, 421–474. Boca Raton: CRC Press.

Panizzi, A. R., A. M. Meneguim, and M. C. Rossini. 1989. Impacto da troca de alimento da fase ninfal para a fase adulta e do estresse nutricional na fase adulta na biologia de *Nezara viridula* (Hemiptera: Pentatomidae). *Pesq. Agropec. Bras.* 24:945–54.

Panizzi, A. R., and A. P. M. Mourão. 1999. Mating, ovipositional rhythm and fecundity of *Nezara viridula* (L.) (Heteroptera: Pentatomidae) fed on privet, *Ligustrum lucidum* Thunb., and on soybean, *Glycine max* (L.) Merrill fruits. *An. Soc. Entomol. Bras.* 28:35–40.

Panizzi, A. R., A .P. M. Mourão, and E. D. M. Oliveira. 1998. Nymph and adult biology and seasonal abundance of *Loxa deducta* (Walker) on privet, *Ligustrum lucidum*. *An. Soc. Entomol. Bras.* 27:199–206.

Panizzi, A. R., and C. C. Niva. 1994. Overwintering strategy of the brown stink bug in northern Paraná. *Pesq. Agropec. Bras.* 29:509–11.

Panizzi, A. R., C. C. Niva, and E. Hirose. 1995. Feeding preference by stink bugs (Heteroptera: Pentatomidae) for seeds within soybean pods. *J. Entomol. Sci.* 30:333–41.

Panizzi, A. R., and E. D. M. Oliveira. 1999. Seasonal occurrence of tachinid parasitism on stink bugs with different overwintering strategies. *An. Soc. Entomol. Bras.* 28:169–72.

Panizzi, A. R., and C. E. Rossi. 1991. The role of *Acanthospermum hispidum* in the phenology of *Euschistus heros* and of *Nezara viridula. Entomol. Exp. Appl.* 59:67–74.

Panizzi, A. R., and C. H. Santos. 2001. Unusual oviposition on the body of conspecifics by phytophagous heteropterans. *Neotrop. Entomol.* 30:471–2.

Panizzi, A. R., and S. I. Saraiva. 1993. Performance of nymphal and adult southern green stink bug on an overwintering host and impact of nymph to adult food-switch. *Entomol. Exp. Appl.* 68:109–15.

Panizzi, A. R., C. W. Schaefer, and E. Hirose. 2005b. Biology and descriptions of nymphal and adult *Jadera choprai* (Hemiptera: Rhopalidae). *Ann. Entomol. Soc. Am.* 98:515–26.

Panizzi, A. R., L. J. Duo, N. M. Bortolato, and F. Siqueira. 2007. Nymph developmental time and survivorship, adult longevity, reproduction and body weight of *Dichelops melacanthus* (Dallas) feeding on natural and artificial diets. *Rev. Bras. Entomol.* 51:484–8.

Panizzi, A. R., and F. Slansky, Jr. 1985a. Review of phytophagous pentatomids (Hemiptera: Pentatomidae) associated with soybean in the Americas. *Fla. Entomol.* 68:184–214.

Panizzi, A. R., and F. Slansky, Jr. 1985b. Legume host impact on performance of adult *Piezodorus guildinii* (Westwood) (Hemiptera: Pentatomidae). *Environ. Entomol.* 14:237–42.

Panizzi, A. R., and F. Slansky, Jr. 1985c. *Piezodorus guildinii* (Hemiptera: Pentatomidae): An unusual host of the tachinid *Trichopoda pennipes. Fla. Entomol.* 68:485–6.

Panizzi, A. R., and F. Slansky, Jr. 1991. Suitability of selected legumes and the effect of nymphal and adult nutrition in the southern green stink bug (Hemiptera: Heteroptera: Pentatomidae). *J. Econ. Entomol.* 84:103–13.

Panizzi, A. R., and J. G. Smith. 1976. Observações sobre inimigos naturais de *Piezodorus guildinii* (Westwood, 1837) (Hemiptera: Pentatomidae) em soja. *An. Soc. Entomol. Bras.* 5:11–7.

Panizzi, A. R., and J. G. Smith. 1977. Biology of *Piezodorus guildinii*: Oviposition, development time, adult sex ratio, and longevity. *Ann. Entomol. Soc. Am.* 70:35–9.

Panizzi, A. R., J. G. Smith, L. A. G. Pereira, and J. Yamashita. 1979. Efeitos dos danos de *Piezodorus guildinii* (Westwood, 1837) no rendimento e qualidade da soja. *Anais I Semin. Nac. Pesq. Soja* 2:59–78.

Panizzi, A. R., and L. M. Vivan. 1997. Seasonal abundance of the neotropical brown stink bug, *Euschistus heros* in overwintering sites and the breaking of dormancy. *Entomol. Exp. Appl.* 82: 213–7.

Panizzi, A. R., L. M. Vivan, B. S. Corrêa-Ferreira, and L. A. Foerster. 1996b. Performance of southern green stink bug (Heteroptera: Pentatomidae) nymphs and adults on a novel food plant (Japanese privet) and other hosts. *Ann. Entomol. Soc. Am.* 89:822–7.

Pinto, S. B., and A. R. Panizzi. 1994. Performance of nymphal and adult *Euschistus heros* (F.) on milkweed and on soybean and effect of food switch on adult survivorship, reproduction and weight gain. *An. Soc. Entomol. Bras.* 23:549–55.

Prado, S. S., D. Rubinoff, and R. P. P. Almeida. 2006. Vertical transmission of a pentatomid caeca-associated symbiont. *Ann. Entomol. Soc. Am.* 99:577–85.

Raghu, S., R. A. I. Drew, and A. R. Clarke. 2004. Influence of host plant structure and microclimate on the abundance and behavior of a tephritid fly. *J. Insect Behav.* 17:179–90.

Ragsdale, D. W., A .D. Larson, and L. D. Newsom. 1979. Microorganisms associated with feeding and from various organs of *Nezara viridula. J. Econ. Entomol.* 72:725–7.

Ragsdale, D. W., A. D. Larson, and L. D. Newsom. 1981. Quantitative assessment of the predators of *Nezara viridula* eggs and nymphs within a soybean agroecosystem using an ELISA. *Environ. Entomol.* 10:402–5.

Ralph, C. P. 1976. Natural food requirements of the large milkweed bug, *Oncopeltus fasciatus* (Hemiptera: Lygaeidae), and their relation to gregariousness and host plant morphology. *Oecologia* 26:157–75.

Ramesh, P. 1992. Seed coat and glume toughness—a possible source of resistance in sorghum to the sorghum earhead bug, *Calocoris angustatus* Leth. (Hemiptera-Miridae). *Indian J. Entomol.* 54:266–71.

Reeves, R. B., J. K. Greene, F. P. F. Reay-Jones, M. D. Toews, and P. D. Gerards. 2010. Effects of adjacent habitat on populations of stink bugs (Heteroptera: Pentatomidae) in cotton as part of a variable agricultural landscape in South Carolina. *Environ. Entomol.* 39:1420–7.

Rizzo, H. F. E. 1971. Aspectos morfologicos y biologicos de *Edessa meditabunda* (F.) (Hemiptera, Pentatomidae). *Rev. Per. Entomol.* 14:272–81.

Rizzo, H. F. E. 1976. *Hemípteros de Interés Agrícola.* Buenos Aires: Editorial Hemisferio Sur.

Rodríguez, S. R. L. 1998. Possible female choice during copulation in *Ozophora baranowskii* (Heteroptera: Lygaeidae): Female behavior, multiple copulations, and sperm transfer. *J. Insect Behav.* 11:725–41.

Rodríguez, R. L., and W. G. Eberhard. 1994. Male courtship before and during copulation in two species of *Xyonysius* bugs (Hemiptera, Lygaeidae). *J. Kansas Entomol. Soc.* 67:37–45.

Romani, R., G. Salermo, F. Frati, E. Conti, N. Isidoro, and F. Bin. 2005. Oviposition behavior in *Lygus rugulipennis*: A morpho-functional study. *Entomol. Exp. Appl.* 115:17–25.

Romosei, W. S. 1973. *The Science of Entomology*. New York: MacMillan Publ. Co.

Rubel, A., R. W. Rinne, and D. T. Canvin. 1972. Protein, oil and fatty acids in developing soybean seeds. *Crop Sci.* 12:739–41.

Sakurai, T. 1996. Multiple mating and its effect on female reproductive output in the bean bug *Riptortus clavatus* (Heteroptera: Alydidae). *Ann. Entomol. Soc. Am.* 89:481–5.

Sales, F. M., J. H. Tumlinson, J. R. McLaughlin, and R. I. Sailer. 1978. Comportamento do parasitóide *Trissolcus basalis* (Wollaston) em resposta a queromônios produzidos pelo hospedeiro, *Nezara viridula* (L.). *Fitossanidade* 2:88.

Santos, C. H., and A. R. Panizzi. 1998. Nymphal and adult performance of *Neomegalotomus parvus* (Hemiptera: Alydidae) on wild and cultivated legumes. *Ann. Entomol. Soc. Am.* 91:445–51.

Saxena, K. N. 1963. Mode of ingestion in a heteropterous insect *Dysdercus koenigii* (F.) (Pyrrhocoridae). *J. Insect Physiol.* 9:47–71.

Schaefer, C. W. 1980. The host plants of the Alydinae, with a note on heterotypic feeding aggregations (Hemiptera: Coreoidea: Alydidae). *J. Kansas Entomol. Soc.* 53:115–22.

Schaefer, C. W., and I. Ahmad. 2000. Cotton stainers and their relatives (Pyrrhocoroidea: Pyrrhocoridae and Largidae). In *Heteroptera of Economic Importance*, ed. C. W. Schaefer and A. R. Panizzi, 271–307. Boca Raton: CRC Press.

Schaefer, C. W., and P. L. Mitchell. 1983. Food plants of the Coreoidea (Hemiptera: Heteroptera). *Ann. Entomol. Soc. Am.* 76:591–615.

Schaefer, C. W., and A. R. Panizzi. 2000. *Heteroptera of Economic Importance*. Boca Raton: CRC Press.

Schuh, R. T., and J. A. Slater. 1995. *True Bugs of the World (Hemiptera: Heteroptera). Classification and Natural History*. Ithaca: Cornell University Press.

Sedlacek, J. D., and L. H. Townsend. 1988. Impact of *Euschistus servus* and *E. variolarius* (Heteroptera: Pentatomidae) feeding on early growth stages of corn. *J. Econ. Entomol.* 81:840–4.

Shade, R. E., H. E. Schroeder, J. J. Pueyo, L. M. Tabe, L. L. Murdock, T. J. V. Higgins, and M. J. Chrispeels. 1994. Transgenic pea seeds expressing the α-amylase inhibitor of the common bean are resistant to bruchid beetle. *Biotechnology* 12:793–6.

Shahi, K. P., and S. S. Krishna. 1981. Influence of adult nutrition or some environmental factors on sexual activity and subsequence reproductive programming in the red cotton bug, *Dysdercus koenigii. J. Adv. Zool.* 2:25–31.

Shewry, P. R., and N. G. Halford. 2002. Cereal seed storage proteins: Structures, properties, and role in grain utilization. *J. Exp. Bot.* 53:947–58.

Shroeder, H. E., S. Gollasch, A. Moore, L. M. Tabe, S. Craig, D. C. Hardie, M. J. Chrispeels, D. Spencer, and T. J. V. Higgins. 1995. Bean alpha-amylase inhibitor confers resistance to pea weevil (*Bruchus pisorum*) in transgenic peas (*Pisum sativum* L.). *Plant Physiol.* 107:1233–9.

Shukle, R. H., and L. L. Murdock. 1983. Lipoxygenase, trypsin inhibitor, and lectin from soybeans: Effects on larval growth of *Manduca sexta* (Lepidoptera: Sphingidae). *Environ. Entomol.* 12:787–91.

Silva, A. G. D. A., C. R. Gonçalves, D. M. Galvão, A. J. L. Gonçalves, J. Gomes, M. N. Silva, and L. Simoni. 1968. *Quarto Catálogo dos Insetos que Vivem nas Plantas do Brasil—Seus Parasitas e Predadores*. Rio de Janeiro: Ministério da Agricultura.

Silva, C. C. A., G. de Capdeville, M. C. B. Moraes, R. Falcão, L. F. Solino, R. A. Laumann, J. P. Silva, and M. Borges. 2010. Morphology, distribution and abundance of antennal sensilla in three stink bug species (Hemíptera: Pentatomidae). *Micron* 41:289–300.

Simanton, W. A., and F. Andre. 1936. A biological study of *Lygaeus kalmii* Stal (Hemiptera-Lygaeidae). *Bull. Brooklyn Entomol. Soc.* 31:99–107.

Simpson, S. J., and C. L. Simpson. 1990. The mechanisms of nutritional compensation by phytophagous insects. In *Insect–Plant Interactions*. Volume II, ed. E. A. Bernays, 111–60. Boca Raton: CRC Press.

Sites, R. W., and J. E. McPherson. 1982. Life history and laboratory rearing of *Sehirus cinctus* (Hemiptera: Cydnidae), with description of immature stages. *Ann. Entomol. Soc. Am.* 75:210–5.

Slansky, Jr., F. 1980. Quantitative food utilization and reproductive allocation by adult milkweed bugs, *Oncopeltus fasciatus. Physiol. Entomol.* 5:73–86.

Slansky, Jr., F. 1982. Reproductive responses to food limitation in milkweed bugs: Sacrificing rate of egg production maintains egg quality and prolongs adult life. *Am. Zool.* 22:910.

Slansky, Jr., F., and A. R. Panizzi. 1987. Nutritional ecology of seed-sucking insects. In *Nutritional Ecology of Insects, Mites, Spiders and Related Invertebrates*, ed. F. Slansky, Jr. and J. G. Rodriguez, 283–320. New York: Wiley.

Slansky, Jr., F., and J. M. Scriber. 1985. Food consumption and utilization. In *Comprehensive Insect Physiology, Biochemistry, and Pharmacology*, ed. G. A. Kerkut and L. I. Gilbert, 87–163. Oxford: Pergamon Press.

Staddon, B. W. 1986. Biology of scent glands in the Hemiptera–Heteroptera. *Ann. Soc. Entomol. Fr.* 22:183–90.

Stam, P. A., L. D. Newsom, and E. N. Lambremont. 1987. Predation and food as factors affecting survival of *Nezara viridula* (L.) (Hemiptera: Pentatomidae) in a soybean ecosystem. *Environ. Entomol.* 16:1211–6.

Sweet, M. 1960. The seed bugs: A contribution to the feeding habits of the Lygaeidae (Hemiptera: Heteroptera). *Ann. Entomol. Soc. Am.* 53:317–21.

Sweet, M. 2000. Seed and chinch bugs (Lygaeoidea). In *Heteroptera of Economic Importance*, ed. C. W. Schaefer and A. R. Panizzi, 143–264. Boca Raton: CRC Press.

Sweet, M. H., and C. W. Schaefer. 1985. Systematics status and biology of *Chauliops fallax* Scott, with a discussion of the phylogenetic relationships of the Chauliopinae (Hemiptera: Malcidae). *Ann. Entomol. Soc. Am.* 78:526–36.

Takeuchi, M., and M. Tamura. 2000. Seed-carrying behavior of a stink bug, *Adrisa magna* Ueler (Hemiptera: Cydnidae). *J. Ethol.* 18:141–3.

Tallamy, D. W., and C. W. Schaefer. 1997. Maternal care in the Hemiptera: Ancestry, alternatives, and current adaptive value. In *The Evolution of Social Behavior in Insects and Arachnids*, ed. J. C. Choe and B. J. Crespi, 94–115. New York: Cambridge University Press.

Tillman, P. G., J. R. Aldrich, A. Khrimian, and T. E. Cottrell. 2010. Pheromone attraction and cross-attraction of *Nezara, Acrosternum*, and *Euschistus* spp. stink bugs (Heteroptera: Pentatomidae) in the field. *Environ. Entomol.* 39:610–17.

Tillman, P. G., T. D. Northfield, R. F. Mizell, and T. C. Riddle. 2009. Spatiotemporal patterns and dispersal of stink bugs (Heteroptera: Pentatomidae) in peanut–cotton farmscapes. *Environ. Entomol.* 38:1038–52.

Tingey, W. M., and E. A. Pillemer. 1977. Lygus bugs: Crop resistance and physiological nature of feeding injury. *Bull. Entomol. Soc. Am.* 23:277–87.

Todd, J. W., and D. C. Herzog. 1980. Sampling phytophagous Pentatomidae on soybean. In *Sampling Methods in Soybean Entomology*, ed. M. Kogan, and D. C. Herzog, 438–478. New York: Springer-Verlag.

Todd, J. W., and W. J. Lewis. 1976. Incidence and oviposition patterns of *Trichopoda pennipes* (F.), a parasite of the southern green stink bug, *Nezara viridula* (L.). *J. Ga. Entomol. Soc.* 11:50–4.

Tsukamoto, L., and M. S. Tojo. 1992. A report of progression provisioning in a stink bug, *Parastrachia japonensis* (Hemiptera: Cydnidae). *J. Ethol.* 10:21–9.

Tullberg, B. S., G. Gamberale-Stille, and C. Solbreck. 2000. Effects of food plant and group size on predator defence: Differences between two co-occurring aposematic Lygaeinae bugs. *Ecol. Entomol.* 25:220–5.

Velasco, L. R. I., and G. H. Walter. 1992. Availability of different host plant species and changing abundance of the polyphagous bug *Nezara viridula* (Hemiptera: Pentatomidae). *Environ. Entomol.* 21:751–9.

Velasco, L. R. I., and G. H. Walter. 1993. Potential of host-switching in *Nezara viridula* (Hemiptera: Pentatomidae) to enhance survival and reproduction. *Environ. Entomol.* 22:326–33.

Ventura, M. U., R. Montalván, and A. R. Panizzi. 2000. Feeding preferences and related types of behaviour of *Neomegalotomus parvus*. *Entomol. Exp. Appl.* 97:309–15.

Ventura, M. U., and A. R. Panizzi. 1997. *Megalotomus parvus* West. (Hemiptera: Alydidae): Inseto adequado para experimentação e didática entomológica. *An. Soc. Entomol. Bras.* 26:579–81.

Ventura, M. U., and A. R. Panizzi. 2000. Oviposition behavior of *Neomegalotomus parvus* (West.) (Heteroptera: Alydidae): Daily rhythm and site choice. *An. Soc. Entomol. Bras.* 29:391–400.

Ventura, M. U., and A. R. Panizzi. 2003. Population dynamics, gregarious behavior and oviposition preference of *Neomegalotomus parvus* (Westwood) (Heteroptera: Alydidae). *Braz. Arch. Biol. Technol.* 46:33–9.

Ventura, M. U., and A. R. Panizzi. 2005. Morphology of olfactory sensilla and its role in host plant recognition by *Neomegalotomus parvus* (Westwood) (Heteroptera: Alydidae). *Braz. Arch. Biol. Technol.* 48:589–97.

Villas Bôas, G. L., and A. R. Panizzi. 1980. Biologia de *Euschistus heros* (Fabricius 1789) em soja (*Glycine max* L. Merrill). *An. Soc. Entomol. Bras.* 9:105–13.

Vitale, A., and R. Bollini. 1995. Legume storage protein. In *Seed Development and Germination*, ed. J. Kigel and G. Galili, 73–102. New York: Marcel Dekker.

Vivan, L. M., and A. R. Panizzi. 2005. Nymphal and adult performance of genetically determined types of *Nezara viridula* (L.) (Heteroptera: Pentatomidae), under different temperature and photoperiodic conditions. *Neotrop. Entomol.* 34:911–5.

Vivan, L. M., and A. R. Panizzi. 2006. Geographical distribution of genetically determined types of *Nezara viridula* (L.) (Heteroptera: Pentatomidae) in Brazil. *Neotrop. Entomol.* 35:175 81.

Waite, G. K. 1980. The basking behavior of *Nezara viridula* (L.) (Pentatomidae: Hemiptera) on soybeans and its implication in control. *J. Aust. Entomol. Soc.* 19:157–9.

Walker, W. F. 1979. Mating behaviour in *Oncopeltus fasciatus*: Circadian rhythms of coupling, copulation duration and "rocking" behaviour. *Physiol. Entomol.* 4:275–83.

Wang, Q., and J. G. Millar. 2000. Mating behavior and evidence for male-produced sex pheromones in *Leptoglossus clypealis* (Heteroptera: Coreidae). *Ann. Entomol. Soc. Am.* 93:972–6.

Wilborn, R., and J. Ellington. 1984. The effect of temperature and photoperiod on the coloration of *Lygus hesperus desertinus* and *lineolaris*. *Southwest. Entomol.* 9:187–97.

Yazdi-Samadi, B., R. W. Rinne, and R. D. Seif. 1977. Components of developing soybean seeds oil, protein, sugars, starch, organic acids, and amino acids. *Agron. J.* 69:481–6.

Yonke, T. R., and J. T. Medler. 1965. Biology of *Megalotomus quinquespinosus* (Hemiptera: Alydidae). *Ann. Entomol. Soc. Am.* 58:222–4.

14

Seed-Chewing Beetles (Coleoptera: Chrysomelidae, Bruchinae)

Cibele S. Ribeiro-Costa and Lúcia M. Almeida

CONTENTS

14.1 Introduction...325
14.2 Distribution, Taxonomy, and Morphological Adaptations 327
14.3 Host Plant Specificity...329
14.4 Seed Availability over Time...332
14.5 Physical and Chemical Defenses of Fruits and Seeds 333
14.6 Obtaining Energy..334
14.7 Oviposition Behavior 335
14.8 Larval and Pupal Development 338
14.9 Intra- and Interspecific Competition..340
14.10 Predation Rate and Viability of Predated Seeds..340
14.11 Reproductive Performance, Diapause, and Dispersal..342
14.12 Natural Enemies..344
     14.12.1 Parasitoids..344
     14.12.2 Predators 346
14.13 Conclusions and Suggestions for Research...346
References...346

14.1 Introduction

Fruits and seeds are keys for plant propagation and survival. Seed production requires a high energetic investment, and the factors that act during this period are critical in plant life history and evolution. Seed consumption by insects is essential since seeds are rich sources of proteins, carbohydrates, and lipids and supply more nutrients than any other part of the plant. Some insects cause adverse effects since they can consume seeds intensely, thus limiting the seed supply and viability. However, effects can also be indirect, such as reducing seedling quality or causing signs of damage, resulting in rejection of fruit and/ or seed by dispersal agents.

Some groups of Coleoptera and Lepidoptera can be found mainly in the guild of insect seed consumers, and both orders can also be included in the seed-chewing guild if the type of larval mouthparts is considered. However, only members of Coleoptera have chewing mouthparts in both the larval and adult stages, representing the largest line of phytophagous insects. The most important seed-consuming Coleoptera belong to the Anthribidae, Chrysomelidae–Bruchinae, Cerambycidae, and Curculionidae.

Seed beetles are a monophyletic group with around 1700 species distributed throughout most of the world. They are considered either as a family of Coleoptera–Bruchidae, within the Chrysomeloidea superfamily (Vesperidae, Disteniidae, Oxypeltidae, Cerambycidae, Chrysomelidae, Orsodacnidae, Megalopodidae), or as a subfamily of the Chrysomelidae–Bruchinae, depending on the author. In this case, the group will be treated as a subfamily in accordance with present tendencies in phylogenetic studies (Reid 2000; Farrel and Sequeira 2004).

The food of bruchines depends on the developmental stage; larvae feed exclusively on seeds, whereas adults feed on pollen and nectar. There are records of bruchine larvae feeding on the seeds of 36 plant families (Table 14.1), but more than 80% feed on legumes, some of which have a high nutritional value and are economically important, such as dry beans, peas, and lentils. Bruchines even feed on seeds containing toxic compounds, which is another characteristic of the group.

The interaction between phytophagous insects and their host plants is one of the oldest and intriguing relationships and has been one of the most discussed in the Bruchinae as an insect–plant evolutionary model. This is because the Bruchinae are highly specific regarding the exploitation of plant tissue (seed endophages) and there is a high rate of specialization in the plants in which they occur. Generally, they are monophagous or oligophagous, and most subtribes are related to a certain plant family.

The bruchines stand out owing to their capacity to survive as adults and reproduce in stored grains for various generations without needing to feed. Species with these habits generally cause serious postharvest losses in economically important grains and include *Acanthoscelides obtectus* (Say), *Bruchus pisorum* (L.), *Callosobruchus chinensis* (L.), *Callosobruchus maculatus* (F.), *Caryedon serratus* (Olivier), and *Zabrotes subfasciatus* (Boheman).

Due to this endophagous feeding in seeds, the potential of bruchines as biological control agents has been evaluated for introduced beneficial plants or invasive plant species that have become weeds. A successful example has been the import of *Neltumius arizonensis* (Schaeffer), *Algarobius prosopis* (Le Conte), and *Algarobius bottimeri* Kingsolver from North America to South Africa to control *Prosopis* species. Another example is *Penthobruchus germaini* (Pic), which has been used to control *Parkinsonia aculeata*, introduced into Australia as an ornamental tree but which later became a weed.

In this chapter we characterize the bruchines within the seed consumer guild, describing their life cycle in the field and in grain storage conditions, and their morphological, biochemical, and behavioral specializations that permit them to obtain energy through seed consumption. Within the context of tritrophic relationships are the associations between plants and bruchines, and the potential of the group to interfere in seed germination and seedling viability with their complex interaction with parasitoids and mammals.

TABLE 14.1

Host Plant Families of Bruchinae According to the Angiosperm Phylogeny Group Classification (2003)

Acanthaceae	Humiriaceae
Anacardiaceae	Lythraceae
Apiaceae	Malphighiaceae
Arecaceae	Malvaceae
Asteraceae	Myrtaceae
Bignoniaceae	Nelumbonaceae
Bixaceae	Nitrariaceae
Boraginaceae	Nyctaginaceae
Casuarinaceae	Ochnaceae
Cistaceae	Oleaceae
Cochlospermaceae	Onagraceae
Combretaceae	Pandanaceae
Convolvulaceae	Poaceae
Dioscoreaceae	Putranjivaceae
Ebenaceae	Rhamnaceae
Euphorbiaceae	Sapindaceae
Fabaceae	Verbenaceae
Goodeniaceae	Vitaceae

14.2 Distribution, Taxonomy, and Morphological Adaptations

Bruchines can be found in almost all continents, but most are endemic to the Americas and their distribution generally match that of their host plants. One exception is *Stator generalis* Johnson & Kingsolver, which is restricted to an area of Panama while its only host plant is widely distributed throughout the Neotropical region. On the other hand, some species of *Acanthoscelides*, *Bruchus*, *Callosobruchus*, *Caryedon*, and *Zabrotes*, whose larvae develop in grains of economically important crops, have become cosmopolitan due to grain commercialization.

Bruchine species are divided into 67 genera and six tribes, Amblycerini, Bruchini, Eubaptini, Kytorhinini, Pachymerini, and Rhaebini. The largest tribe is Bruchini, with around 80% of the species, followed by Amblycerini (10%) and Pachymerini (9%); only 1% of species belong to the other three tribes (Johnson and Romero 2004). The fauna of the Neartic region is taxonomically much better known than that of the Neotropical region. This lack of knowledge, principally of the South American species, has been demonstrated in recent molecular studies that did not include these species (Alvarez et al. 2005; Tuda 2006; Kergoat et al. 2007; Kato et al. 2010).

Adults have compact body, almost square to oval, and a generally opistognathous head often with emarginate eyes. The dorsal side is covered with hairs, brown to black in color, forming different patters or is uniformly distributed, this being a noticeable characteristic of many species (Figures 14.1a and 14.1b).

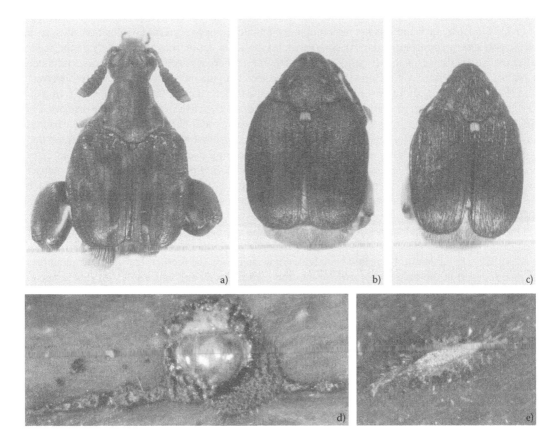

FIGURE 14.1 a) Dorsal view of *P. lineola*. b) and c) Intraspecific variation in *S. leptophyllicola*. d) Egg batch of *P. lineola* on immature fruits of *Cassia leptophylla*. e) Egg batch of *S. leptophyllicola* on immature fruits of *Cassia leptophylla*. (From Ribeiro-Costa, C. S. and A. S. Costa, *Rev. Bras. Zool.* 19, 305, 2002. With permission.)

The hind leg varies within the subfamily and is a source of generic and specific characters. The hind femur is often wide (Figure 14.1a) with teeth on the internal margin. The tibia generally has carinae and two spurs may be present. The pygidium is exposed (Figures 14.1a through 14.1c)

Sexual dimorphism can be observed on the last abdominal ventrite, which is slightly emarginate in most males whereas it is straight in females, and can also be seen in the pygidium, which is more convex in males when seen laterally. Other characteristics are the development of the eyes, body coloration, and the length and shape of the antennal segments.

Species identification, including some genera, is difficult and requires the study of male genitalia for correct identification. Many species are closely related, individuals are generally small (1.0–6.0 mm), and there may be intraspecific variation (Figures 14.1b and 14.1c). Both adults and larvae have chewing mouthparts. The adult mandible of some species is used to excavate the fruit wall for oviposition (Figure 14.1d); however, this type of behavior is not seen in the *Sennius* species (Figure 14.1e). The mandibular apex is sharp, has a prosteca, and the mola has teeth (Figure 14.2a). The mola teeth grind up pollen grains collected by the hairs of the galea and are specialized for this task. In *Caryedes brasiliensis* (Thunberg), for example, the teeth are uniformly distributed (Figure 14.2b), differing from those of *Ctenocolum tuberculatum* Motschoulsky, which are less differentiated and densely distributed (Figure 14.2c). The larval mandibles are short, robust, with a serrated or rounded apex, and rasp the seed integument and endosperm. The adult maxilla consists of the galea, the lacinia, and a four-segmented maxillary palp (Figure 14.2d). The galea setae are specialized for collecting pollen grains; they may be bifid or apically branched, pectinate, simple and pectinate or spatulate (Figure 14.2e), or just simple (Figure 14.2f) (Ribeiro-Costa and da Silva 2003, Silva and Ribeiro-Costa 2008).

Oviposition can vary and it is not uncommon for females to have preferred oviposition sites, such as the pod suture lines (Figure 14.3A). Some females, however, let their eggs fall freely on the seeds after the fruit wall has opened (Figure 14.3B,b); however, others, such as *Z. subfasciatus*, guarantee larval survival by directly attaching their eggs to the seed integument. Although the larva is destructive (Figure 14.3C,c,D,d), relatively few species have been described (Pfaffenberger 1985). In general, there are four larval instars; *Pachymerus cardo* Fåhraeus has five instars. With the exception of *Spermophagus* and some other species where the larvae are apodal during the larval stage, the first instar larva generally differs from the rest. This is the chrysomelid type with a sclerotinized and serrated prothoracic plate (Figure 14.3C,c), which helps during emergence from the egg and entry into the fruit or seed wall (Figure 14.3C), and legs that help seed penetration and that are compact or have a hard integument. Some species have an appendix on the 10th abdominal segment, which is attached to the surface for anchorage during entry into the fruit/seed (Pfaffenberger and Johnson 1976).

One of the most critical stages of the life cycle is from egg eclosion until entry into the seed. The first instar larva makes various movements, anchoring itself to the internal concave wall of the chorion with its legs, abdominal appendix, and body setae, and using the prothoracic plate and mandibles to rasp the wall of the flat chorion surface into the seed integument (Figure 14.4). At this moment, the egg can become unstuck because of the larva's efforts, resulting in its death. Therefore, species survival may be a function of the mechanisms used by the female to anchor its eggs to the fruit or seed wall.

After the first molt, the larva loses its dorsal plate and assumes a weevil-like larva appearance, with the legs reduced or even absent due to the endophagous mode of life, where movement is almost null since food is easily available (Figure 14.3D,d). In the last instar, the larva makes a mark on the internal seed wall (Figure 14.3E) and pupates (Figure 14.3F,f). The circular exit hole for the adult originates from this mark (Figure 14.3G,g). In some cases, in plants with dehiscent pods, the adults do not need to make the hole in the pod wall to exit and can reinfest available seeds afterward (Figure 14.3H). Under storage conditions, pest species reinfest grains without the adults needing to feed (Figure 14.3I,i). However, in the natural environment, flowers supply food and, therefore, bruchine populations start to appear during the floral phenophase (Figure 14.3J).

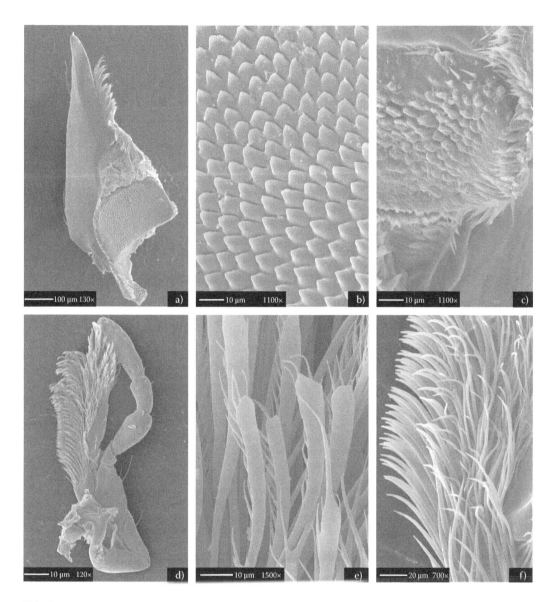

FIGURE 14.2 a) Mandible of *C. brasiliensis*. b) Mola teeth of *C. brasiliensis*. c) Mola teeth of *C. tuberculatum*. d) Maxilla of *C. brasiliensis*. e) Setae of the galea of *C. brasiliensis*. f) Setae of the galea of *P. lineola*. (From Silva, J. A. P. and C. S. Ribeiro-Costa, *Rev. Bras. Zool.*, 25, 802, 2008. With permission.)

14.3 Host Plant Specificity

In the Bruchinae, we observe associations of both larvae and adults with plants. Records of adults are few and should be carefully considered. The consumption of nectar and pollen was verified by Ott (1991) in males and females of *Acanthoscelides alboscutellatus* (Horn). However, *Althaeus hibisci* (Olivier) feeds on *Hibiscus moscheutus* pollen, which is also the host plant of its larvae (Shimamura et al. 2005); however, there is no information on nectar consumption. This specialized behavior is not the rule for the

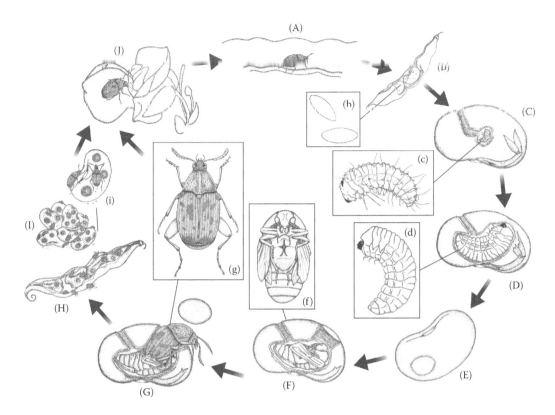

FIGURE 14.3 Generalized bruchine life cycle in pods and seeds of dry beans. (A) Oviposition on the ventral pod suture (or in adult exit holes). (B) Single eggs, freely deposited inside the pod (or attached to the seeds); b, aspect of egg. (C) Section of seed with entrance hole and tunnel excavated by a first instar larva; (c) generalized first instar larva showing specializations for penetrating the seed. (D) Section of seed showing larval growth and modifications after the first molt; (d) generalized final instar larva. (E) Seed showing demarcation of operculum by final instar larva. (F) Pupa inside the larval feeding chamber showing larval entrance hole and exit hole already prepared; (f) general view of pupa. (G) Emergence of adult showing operculum; (g) completely formed adult. (H) Emergence of adults with the possibility of reinfestation of seeds in partially opened fruits in the field. (I) Emergence of adults from stored grains with the possibility of reinfestation without the need of food for the adults; (i) aspect of infested grain with more than three circular holes. J. Adults in the field with a chance to feed on pollen and nectar. (From Pfaffenberger, G. S. and C. D. Johnson, *Tech. Bull. U.S. Dept. Agric.*, 1525, 1, 1976.)

Bruchinae. In *A. obtectus*, for example, pollen from 18 plant species was found in the gut with only 9% from its main host (Jarry 1987).

Records of larvae host plants are more relevant for coevolutionary studies since adult success depends on this stage of the life cycle. These records are also useful for identifying taxa at different taxonomic levels when considered with other conventional information.

Bruchine larvae have adapted to feeding on seeds by using specific mechanisms. Host selection is made by the female and involves the ability to find, recognize, and accept the plant, and more specifically, the choice of a certain fruit or seed for oviposition. Therefore, the success of larval development depends essentially on female choices (Johnson and Kistler 1987). For Janzen (1969), host divergence is more a function of nutritional rather than physical characteristics (e.g., type of seed, fruit structure, and form or nature of the seed integument). However, many plants produce toxins and these can be the reason why bruchines have specialized (Janzen 1969; Center and Johnson 1974).

Some species have specialized in feeding on one species or one plant genus (monophagous), whereas others are less specific (oligophagous) and still others can develop in the seeds of various genera (polyphagous). The most polyphagous bruchine species is *Stator limbatus* (Horn), which feeds

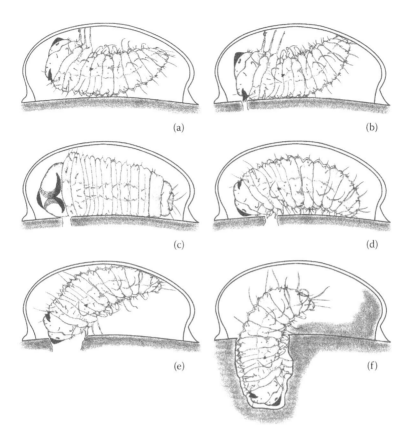

FIGURE 14.4 Movements of first instar larva to penetrate fruit/seed wall. (a) Larva supports itself on concave surface of the chorion with its legs; body is in contact with flat surface. (b) Abdominal appendix is supported on concave surface and with bodily contraction and distension movements, prothoracic plate reaches and damages the chorion in the flat region. (c) Larva turns its body with help of legs, abdominal appendix, and setae until it has reached a lateral position for a certain time. (d) Completes the turn, positioning itself with mandibles in contact with substrate. (e) Supporting itself with hind region of abdomen on internal wall of the chorion and with legs and mandibles on opposite side, it contracts and perforates the chorion and wall of seed/fruit. (f) During perforation, the remains, in the form of small spheres, accumulate in hind area and may be confused with feces. (Modified from Ramos, R. Y., *Bol. Asoc. Española Entomol.*, 31, 65, 2007. With permission.)

on more than 70 legume species in at least nine genera of the Caesalpinioideae, Mimosoideae, and Papilionoideae (Fox et al. 1997; Morse and Farrel 2005). However, the behavior of *Mimosestes amicus* Horn, which is a generalist, stands out. It develops better in seeds of *Cercidium floridium* than in seeds of *Prosopis velutina*, and adult metabolic rates are different in the two hosts (Kistler 1982). Thus, in some cases, the generalists can limit themselves to feeding on the seeds of only a few hosts just like the specialists.

Most Bruchinae tribes are associated with a certain plant family. The Pachymerini occur in the Arecaceae, the Spermophagini in the Convolvulacea and Malvacea, the Bruchini in the Fabacea, the Megacerini in the Convolvulacea, with the Fabaceae being the main host plant family. Some genera tend to be very specialized and are associated with only one subfamily, tribe, or with specific genera. Examples include *Bruchus*, which is mostly associated with plants of the Vicieae tribe of the Fabacea; *Sennius* with the subtribe Cassiinae; *Ctenocolum* with the Papilionoideae, principally *Lonchocarpus*; and *Gibbobruchus* with the Caesalpinioideae, principally *Bauhinia*.

Various studies have dealt with the taxonomic conservatism of the Bruchinae concerning their host plants (e.g., Farrell and Sequeira 2004; Morse and Farrel 2005; Kato et al. 2010). This is one of the most recognized patterns in insect–plant interactions where phylogenetically related species feed on plants that are also related (Kergoat et al. 2007). The influence of host plant secondary compounds is an important factor for conservatism since related plants generally share the same toxic compounds.

However, other characteristics, such as oviposition behavior, can affect plant use and host range. Recent molecular data have shown that different genera can evolve in parallel and independently colonize similar host plants in their respective areas of distribution (Kergoat et al. 2005). Molecular data have also provided evidence of specialization at the generic and species levels (Kergoat et al. 2005). Molecular phylogeography has shown that *Stator beali* Johnson, with its specialist habit, has evolved from *S. limbatus* with its generalist habit (Morse and Farrel 2005).

14.4 Seed Availability over Time

In general, the bruchines synchronize their life cycles with their host plants. The floral phenophase provides food for the adults and the fruiting phenophase, a substrate for oviposition and larval development (Figure 14.5). Fruit, and consequently seed availability, is not constant over time. The duration of the fruiting period is variable, both intra- and interspecifically, and dependent on abiotic factors. The bruchines have developed mechanisms for waiting for food, and when it is present they try to exploit it in the best way possible. Janzen (1975) discovered that there are species that prefer to deal with the extremes of seed availability of a given species within and between years instead of changing to different host species.

Depending on seed abundance over time, specialist species may prefer one or other hosts at different times during the year, acting like specialists. In this way, they synchronize their life cycle with that of different host plants and can complete more than one generation per year, known as bi- or multivoltine. Even species associated with a single plant can complete more than one cycle per year, the bi- or multivoltine specialists, with different oviposition behaviors (see Section 14.7 on oviposition guilds).

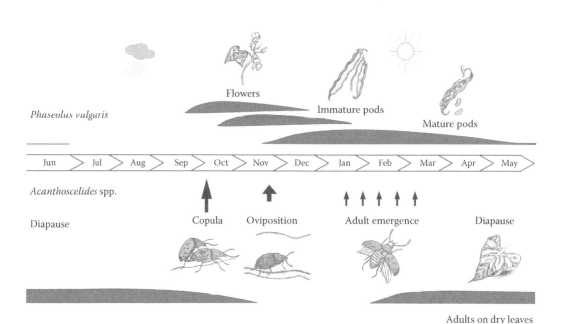

FIGURE 14.5 Cycle of *Acanthoscelides* spp. in wild *P. vulgaris* in Mexico. (Modified from Biemont, J. C. and A. Bonet, in *The Ecology of Bruchids Attacking Legumes (Pulses)*. Series Entomologica, vol. 19, ed. V. Labeyrie, 23–39. The Hague: W. Junk, 1980. With permission from Arturo Bonet.)

14.5 Physical and Chemical Defenses of Fruits and Seeds

The females lay eggs on immature or mature fruits or directly on seeds. However, according to the specific characteristics of these substrates, such as texture, hardness, curvature, pubescence, and size among others, oviposition may become more difficult or even stop. For example, the dense pubescence of *Astragalus utahensis* fruits is a physical barrier that makes oviposition by *Acanthoscelides fraterculus* (Horn) more difficult (Green and Palmbald 1975). In palms, where the fruit has a resistant exocarp and a succulent mesocarp, *Speciomerus giganteus* (Chevrolat) oviposits only in those parts that are eaten by frugivorous animals or that degrade in the soil.

The successful entry of the first instar larva into the seed depends, apart from other factors, on how the eggs are fixed to the integument. When they become unstuck from the surface before larval penetration of the seed, then larval death occurs. Studies with corrugated seeds of *Vigna sinensis* varieties showed higher first instar larval mortality of *Callosobruchus maculatus* than in smooth seed varieties (Nwanze and Horber 1976). Raina (1971) believes that the spiny integument of a variety of *Cicer arietinum* causes resistance to *Callosobruchus* spp. However, in *Phaseolus vulgaris*, the thick seed integument was the reason that Seifelnasr (1991) gave for the nondevelopment of *C. maculatus* in this host, although Silva et al. (2004) believed that the resistance of the seed to this species is due to the phaseolin of the integument, which is toxic. The high mortality of *Bruchidius sahlbergi* Schilsky in *Acacia erioloba* seeds and of *Bruchidius uberatus* in *Acacia nilotica* seeds may be due to larval difficulty in penetrating the hard integument, which requires a lot of energy during the perforation (Ernst 1992).

Seeds contain a wide variety of secondary compounds; however, the more toxic the food, the more specialized physiologically and biochemically the bruchine should be to exploit it. Similarly, the longer a population feeds on a species, the more it differentiates itself in its dependence on the secondary compounds (Janzen 1978). In this way, larval development is controlled mainly by the levels of seed secondary compounds and by the capacity of bruchines to detoxify them.

The chemical components of plant defenses include antibiotics, alkaloids, terpenes, cyanogenic glycosides, and proteins. The proteins associated with defense mechanisms include lectins, inhibitors of α-amylases, inhibitors of proteinases, protein inactivators of ribosomes, modified reserve proteins (vicilins), proteins to transport lipids, and glucanases and chitinases (Carlini and Grossi-de-Sá 2002). The reserve proteins of dry beans, *P. vulgaris*, are phaseolin, similar to vicilin 7S globulin and phytohemagglutinin. A third protein, arcelin, found in wild *P. vulgaris* in Mexico, has shown positive results for the control of *Z. subfasciatus* (Paes et al. 2000). Seven allelic variations have already been detected and according to Goossens et al. (2000), arcelins 1 to 5 are the most promising. In Brazil, Ribeiro-Costa et al. (2007) observed that the genotypes containing arcelins 1 and 2 suggest an antibiosis-type resistance, with a high mortality for immature stages, and arcelin 1 retards development (Table 14.2) and causes a drastic weight reduction in adults.

The so-called enzyme inhibitors are substances that stop the action of amylases and trypsin, which are essential for hydrolyzing the main constituents of insect diets, such as carbohydrates and proteins. Various types of α-amylase and proteinase inhibitors present in the seeds regulate bruchine development. However, not all these substances are commonly found in leguminous seeds, with trypsin inhibitors, for example, being present in only one of 5000 cowpea varieties studied for resistance to *C. maculatus* (Gatehouse and Boulter 1983).

One of the most well-known substances is the inhibitor of α-amylase, which occurs in *P. vulgaris*, and has been shown to efficiently control *Callosobruchus* spp. The growth of *C. maculatus* and *C. chinensis* larvae is inhibited when relatively low levels are added to the diets of these bruchines (Ishimoto and Kitamura 1989), and this has encouraged studies on the introduction of the gene into *Pisum sativum* (Shade et al. 1994). Another substance, vicilin, which is toxic (globulin of reserve 7S), isolated from *Vigna unguiculata*, also adversely affects the development and survival of *C. maculatus* (Mota et al. 2002). The saponins, which occur in various leguminous seeds (Applebaum and Birk 1972), are also interesting substances since they cause hormonal alterations, which stop pupal formation in *C. chinensis* (Johnson and Kistler 1987).

TABLE 14.2

Means (± SEM) of Development (Egg–Adult), in Days, and Mortality (%) of the
Immature Stages (Larva and/or Pupa) of *Z. subfasciatus* in Dry Bean Genotypes,
at a Temperature of 27 ± 2°C, RH 50 ± 10%, Photophase 12 h

Genotype	Development (Days)	Genotype	Mortality (%)
Arc 1	41.7 ± 1.17a	Arc 1	86.6 ± 1.78a
Arc 2	33.0 ± 0.46b	Arc 2	69.2 ± 1.24ab
IAPAR 44	32.7 ± 1.14b	IPR Uirapuru	45.5 ± 6.86bc
TPS Bionobre	32.6 ± 1.54b	IAC Una	45.0 ± 9.06cd
IPR Uirapuru	31.9 ± 0.21bc	TPS Bionobre	37.2 ± 5.08cd
IAC Una	31.4 ± 0.24bc	Carioca	35.0 ± 3.45cd
Pérola	30.3 ± 0.66bc	IAPAR 81	32.2 ± 5.31cde
Carioca	30.3 ± 0.91bc	IPR Juriti	27.0 ± 4.13cde
Bolinha	29.2 ± 1.73bc	IAPAR 44	25.8 ± 5.00cde
IAPAR 81	28.9 ± 0.56bc	Bolinha	18.1 ± 3.54de
IPR Juriti	27.8 ± 1.10c	Pérola	10.4 ± 1.77e
F	13.86	F	19.16[a]
CV%	6.98	CV%	28.78

Source: Modified from Ribeiro-Costa, C. S., et al., *Neotr. Entomol.* 36, 560, 2007.

Note: Means followed by the same letter, in the columns, do not differ statistically among
themselves by the Tukey test ($p \leq .05$).

The Bruchinae are especially known for the mechanisms they have developed for feeding on very toxic seeds. One classic example is *C. brasiliensis* (Thunberg), which feeds on *Dioclea megacarpa* seeds with >13% (dry weight) of L-canavanin, a lectin. Canavanin, like many other nonproteic amino acids, acts like a toxin. It has a similar structure to the amino acid arginine, and when incorporated into the protein, its physicochemical properties are altered and it becomes toxic. *C. brasilienis* larvae avoid the incorporation of canavanin into proteins owing to a specialized proteic synthesis system (arginil-tRNA synthase) that permits the distinction between arginine and canavanin. Arginase and urease degrade canavanin into ammonia, which is used as a nitrogen source for amino acids (Rosenthal 1983).

Other genera of Bruchinae also use toxic substances. Species of *Acanthoscelides* feed on *Astragalus* seeds, which contain selenium; one species may tolerate high levels of this substance, whereas others tolerate very low concentrations although these are enough to be toxic to mammals (Trelease and Treleae 1937; Johnson 1970). Species of *Megacerus* feed principally on Convolvulaceae seeds, which contain alkaloids (Janzen 1980), such as *Ipomoea pes-caprae* seeds, which contain ergotamine (Jirawongse et al. 1979). *A. obtectus* (=*A. obsoletus*) (Bridwell 1938) feeds on seeds of *Cracca virginiana*, which contain rotenone. Other bruchines feed on seeds of *Erythrina*, *Abrus*, *Dioclea*, and *Sarothamnus*, which also contain toxins (Janzen 1971). However, it is important to relate that in *Stator generalis*, which only feeds on the hard, toxic seeds of *Enterolobium cyclocarpum* (Johnson and Janzen 1982), survival is much lower (48%) than of other species of the genus bred in other hosts (75–81%) (Johnson 1982).

14.6 Obtaining Energy

The energy necessary for various adult activities, such as flight and reproduction, come principally from energy stored during larval feeding. Subsequent additions come from the consumption of pollen and nectar and, in the case of females, the nutritive ejaculatory secretions of the male. With age, the use of stored energy (lipids and glycogen) and water decreases. Females lose more water and weight than males owing to their higher metabolic activity, which includes egg laying (Sharma and Sharma 1984).

The first instar larva starts to feed when it reaches the seed endosperm. Before this, it moves only to perforate the surface and does not ingest any food. Thus, all the food sources are inside the seed and there is no possibility of changing seeds at this time; the food may only be complemented by more seeds in later instars. Although larval and adults foods are different (endosperm and pollen grains/nectar), the compositions of the carbohydrates in the food sources and glycosidase activity are similar in both adult and larval guts (Leroi et al. 1984).

Callosobruchus analis (F.) efficiently converts most of its diet into nutritional components, such as lipids (49% of the dry weight), and only a few of the seed components, such as cellulose and lignin, remain unused (Johnson and Kistler 1987). The conversion efficiency of food by females is higher than for the males. The dry weight consumption of *Phaseolus aureus* by female *C. maculatus* is greater and the caloric equivalents are also greater (7.2 cal/mg) than in males (6.99 cal/mg). Beans contain 4.45 cal/mg; females consume 64.5 cal of beans and convert this into 14.9 cal (Mitchell 1975). In *B. sahlbergi* Schilsky, the low efficiency of nitrogen use (34–41%) may be related to the concentration of the alkaloids, nonproteic amino acids, and cyanogenic glycosides of *A. erioloba*, which are not transformed but are eliminated in the feces (Ernst 1992).

Although recently emerged bruchines develop inside the seeds, they contain around 50% of water. During development, larvae may possibly use metabolic water and convert seed contents into lipids and other components of stored water (Johnson and Kistler 1987).

14.7 Oviposition Behavior

Bruchines show various oviposition behaviors that represent different ways of overcoming the barriers imposed by the host plants, strategies to stop egg mortality due to natural enemies, or even strategies to overcome intraspecific competition. Eggs are generally glued to the fruit or seed, or are left to fall into the fruit and reach the seed after perforations are made by females. They may also be laid in cracks or crevices in the fruit or even in old adult exit holes. Normally, oviposition occurs in the field when the seeds are completely developed, although some species lay their eggs on immature fruits whereas others wait until the seeds are exposed (Ribeiro-Costa and Costa 2002; Kingsolver 2004; Sari et al. 2005).

Johnson (1981) established three oviposition guilds: (a) the guild of mature pods—the bruchines of this group oviposit in the fruit wall when this is ripe and still attached to the plant; (b) the guild of mature seeds—oviposition occurs in the seed when the mature fruit is partially dehiscent and still fixed to the plant; and (c) the guild of dispersed seeds—when oviposition occurs in exposed seeds on the ground, after dispersal. He concluded that the legumes with dehiscent fruits are the most effective against attack because bruchines of guilds A and B are almost totally eliminated; in the plant species with indehiscent or late dehiscent fruits, the species of guild B are eliminated and in the case of partially dehiscent pods, bruchines belonging to all three guilds can predate their seeds. Examples of species belonging to guild A are *Merobruchus* spp., *Mimosestes* spp., *Acanthoscelides chiricahuae* (Fall), *Amblycerus hoffmannseggi* (Gyllenhal) (Figures 14.6a and 14.6b), *A. submaculatus* (Pic); guild B: *Sator limbatus*, *Sator pruininus* (Horn), *Sennius bondari* (Pic) (Figures 14.6c and 14.6d); and guild C: *Zabrotes* spp. and *Stator sordidus* (Horn) (Kingsolver 2004; Ribeiro-Costa 1998; Linzmeier et al. 2004). An example of a species belonging to more than one guild is *Megacerus baeri* (Pic), which can oviposit on the surface of mature fruits and also directly on the seeds of open fruits (guilds A and B).

Some species have developed special strategies to find an adequate substrate for oviposition and overcome host plant barriers. The behavior of female *Zabrotes interstitialis* (Chevrolat) is interesting since they use the exit holes left by *Pygiopachymerus lineola* (Chevrolat) in the wall of *Cassia grandis* fruits in order to gain access to the seeds because the pods of this species are indehiscent (Janzen 1978). Similarly, the females of *S. limbatus* also use the holes left by adult *Mimosestes* in *Cercidium floridum* fruits (Fox et al. 1997).

Larval survival and development depend directly on food availability—that is, fruits and seeds. Larger seeds supply more resources for larval development, with less intraspecific competition and, consequently, a higher fecundity and longevity of future adults due to stored energy. Female *C. maculatus* have specialized so they can oviposit in larger dry bean seeds and they also discriminate seeds on the basis of the number of eggs; that is, they avoid depositing the second egg in the same seed if there are still seeds

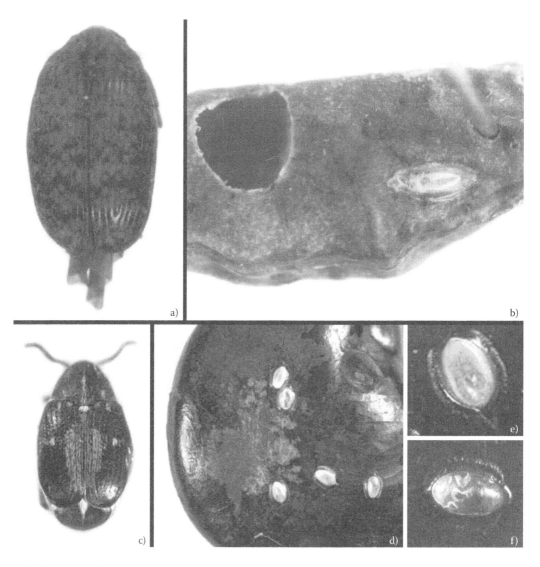

FIGURE 14.6 *Amblycerus hoffmanseggi.* a) Dorsal view of adult. b) Fruit of *Senna* cf. *bicapsularis* with egg and adult exit hole. (From Ribeiro-Costa, C. S., *Rev. Bras. Entomol.*, 36, 149, 1992. With permission.) *Sennius bondari.* c) Dorsal view of adult. (From Linzmeier, A. M., et al., *Rev. Bras. Zool.*, 21, 351, 2004. With permission.) d) Seed of *Senna macranthera* with eggs and demarcation of the adult exit hole. e) Viable egg. f) Nonviable egg.

without eggs (Mitchell 1975). The females of *Bruchidius villosus* (F.) select the larger fruits of *Cytisus scoparius* to lay their eggs since they contain more seeds (Redmon et al. 2000).

Variation in egg size is generally attributed to variations in nutrients or female age, but *S. limbatus* is a special case. Egg size is the result of the maternal effect that partly represents an adaptive species response to the chosen host in order to guarantee its success (Fox et al. 1997). Females lay smaller eggs in *Acacia greggii* with a lower larval mortality and larger eggs in *C. floridum* where there is a high larval mortality caused during seed penetration.

Eggs are opaque after larval emergence (Figure 14.6e) or translucid, generally when they are inviable (Figure 14.6f). They vary in form and sculpture and have glue that originates from the follicle epithelium (Snodgrass 1935) or the accessory glands (Wigglesworth 1947) and that hardens when in contact with the environment. This substance not only fixes the egg to the substrate but also protects it against adverse abiotic factors, strong sun, and/or low humidity, which result in desiccation (Figure 14.7). This covering may be smooth, reticulate (e.g., *Sennis bondari*, *Amblycerus submaculatus*, *A. hoffmanseggi*) (Caron et al. 2004; Ribeiro-Costa 1992, 1998) (Figures 14.6b and 14.7b), or have filaments [e.g., *Sennius lateapicalis* (Pic), *Sennius subdiversicolor* (Pic), *Sennius lamnifer* (Sharp), and *Sennius leptophyllicola* Ribeiro-Costa and Costa, *Sennius crudelis* Ribeiro-Costa and Reynaud] (Bondar 1937; Ribeiro-Costa and Costa 2002; Caron et al. 2004) (Figures 14.1e and 14.7d through 14.7f). Ribeiro-Costa and Costa (2002) discovered that the egg filaments of *S. leptophyllicola* laid on immature pods (Figure 14.1e) do not effectively fix the eggs, and they become unstuck when the fruit loses water and wrinkles with ripening, or by the entry of the first instar larva into the fruit wall. When eggs are laid on immature fruits, such as for *Sennius crudelis* and *Sennius puncticollis*, the seeds and the larvae develop simultaneously. A study of the dynamics of these species in *Senna multijuda* demonstrated a highly negative correlation between fruit length and the number of eggs. This indicated that in spite of a preference for ovipositing in immature fruits, the increase in fruit size over time due to ripening leads to the eggs falling off (Sari et al. 2005).

The morphological characteristics of the egg show specializations in bruchines and contribute to understanding the evolution of oviposition behavior within and between species groups within guilds. The eggs can be laid singly (Figures 14.1e and 14.6b) or in groups (Figure 14.1d). Tribes with many species, such as the Amblycerini and Bruchini, generally deposit single eggs. However, even within a genus such as *Amblycerus*, there are species that lay eggs singly on the fruit walls (Ribeiro-Costa 1998) (Figure 14.6b), while others lay eggs in groups of two or three (Ribeiro-Costa 1992). When the eggs are deposited in groups, partially overlapping, like those of *P. lineola* (Figure 14.2d) (Ribeiro-Costa and Costa 2002), this indicates a strategy against parasitism and desiccation since exposed eggs are more susceptible to

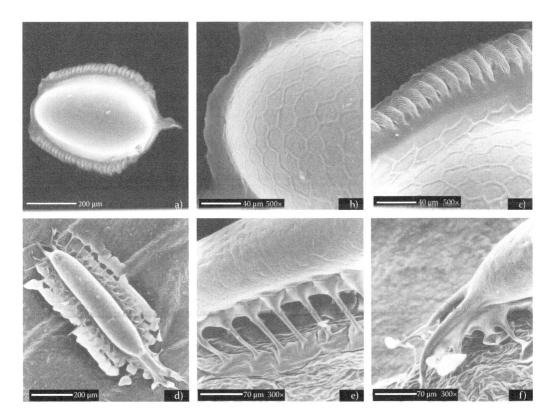

FIGURE 14.7 *Sennius bondari*. a) General aspect of egg. b) Details of reticulate covering. c) Details of undulated edge. *Sennius crudelis.* d) General aspect of egg. e) Details of lateral filaments. f) Details of terminal filaments. (From Caron, E., et al., *Zootaxa*, 556, 1, 2004. Reproduced with permission from Zootaxa.)

these factors. Other protection strategies are the eggs of *A. prosopis* (Le Conte), which are laid in cracks in the pods (Johnson 1983) and those of *Caryedon albonotatum* (Pic), which are covered with feces (Prevett 1966).

Some *Acanthoscelides* species do not attach their eggs to pods/seeds; they lay them in holes on the immature pod wall or inside the dehisced pod, or even spread them on to the seeds (Skaife 1926; Larson and Fisher 1938). An example is *A. obtectus*, where oviposition coincides with the start of fruiting by *Phaseolus* and finishes when the mature fruits drop off or are harvested. The females perforate the wall of the immature pods and lay their eggs, which drop into the pod. In the following generations, the eggs are laid in exit holes left by the previous generation (Kingsolver 2004). The female of *Z. subfasciatus* acts differently and lays single eggs directly on the seed integument after dehiscence of the *Phaseolus* pods or may infest the seeds while they are still inside the pods, using the holes made by other insects (Credland and Dendy 1992) or even cracks in the pod suture. Contrary to most other bruchines, the females need contact with the seed to stimulate ovigenesis (Pimbert and Pierre 1983).

14.8 Larval and Pupal Development

In general, the first instar larva emerges from the hole excavated in the flat surface of the egg chorion near the substrate, bores into the fruit and/or seed wall, toward the endosperm. These activities, as well as the molt, have a high energetic cost. Ernst (1992) calculated 5% to 10% larval weight loss for the first instar of *B. sahlbergi* over 1 to 2 days. After entering the seed, the larva generally molts another three times and should be ready to feed, assimilate and avoid, or detoxify, the food. In *B. sahlbergi*, which has five instars, the highest growth in *A. erioloba* seeds was reached between the third and fifth instars with a 78% to 90% increase in biomass (Ernst 1992).

Most times immediately before pupating, the larva makes a round hole in the seed (Figure 14.3E) or on the fruit wall, which is cut and removed during adult emergence (Figure 14.3F). After perforating, the larva returns to the feeding chamber to pupate. However, in some species, the pupal stage occurs partially or completely outside the seed.

One peculiar characteristic of bruchines, which uses up a considerable amount of energy, is the pre-emergence activity of the young adult to remove the operculum of the seed integument. Causes of mortality in this stage can be the unsuitable diameter of the hole, lack of energy for emergence, or low levels of relative humidity. Therefore, for successful rearing, a suitable humidity level should be maintained but without causing the seeds to develop fungus.

Most first instar larvae perforate the fruit wall and consume the first available seed, but larvae of some species are more selective, moving over various seeds before choosing one to feed on (Southgate 1979). *A. nilotica* fruits have internal septa that separate one seed from the next and only one seed is available per chamber in which the egg is laid (Southgate 1979). However, in *Amblycerus*, the septa are not physical barriers and its larvae perforate them during development to consume up to six *Senna* seeds (Ribeiro-Costa 1992, 1998). Bruchines generally consume all the seed contents and there are species that need more than one seed to complete their development, as discussed previously (Ribeiro-Costa 1998). In *Sennius morosus* (Sharp), which generally feeds on various seeds, when only one seed is available, the adult does not reach its normal size but it is unknown if these adults can produce offspring and compete like the largest ones (Johnson 1970; Center and Johnson 1974). Examples of species that consume many seeds are *Sennius morosus* and *Sennius simulans* (Horn) on *Cassia baubinioides* (Center and Johnson 1973), *Merobruchus julianus* (Horn) on *Acacia berlandieri* (Johnson 1967), and *Amblycerus submaculatus* on *Senna alata* (Figure 14.8a).

The quantity of ingested food depends on the available seed biomass. For *B. sahlbergi*, one larva's consumption varies from 9% to 38% depending on the seed weight of *A. erioloba*, reaching 100% in small seeds (Ernst 1992). The consumption of female *C. maculatus* on *P. aureus* is greater than for males, with 14.5 g and 9.5 mg of dry bean weight, respectively (Mitchell 1975). Generally, females are bigger than males and this difference is attributed to selection that favors larger females, which lay the eggs. However, contradicting this rule, males of *S. limbatus* are bigger than the females (Fox et al. 1995).

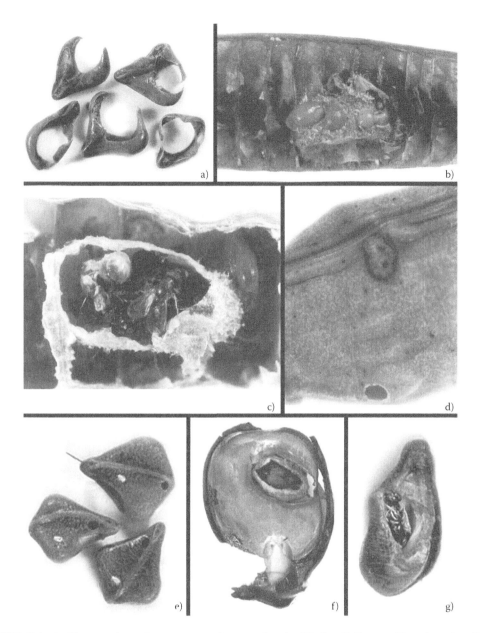

FIGURE 14.8 *Amblycerus submaculatus.* a) Aspect of consumed seeds of *S. alata.* b) Cocoon with seed remains. (From Ribeiro-Costa, C. S., *Coleopt. Bull.*, 52, 63, 1998. With permission.) c) Cocoon with cephalic capsule and remains of last instar larva and its parasitoid, *Horismenus* sp. d) Entrance hole (smaller) of first instar larva in fruit and exit hole (larger) of parasitoids. (From Ribeiro-Costa, C. S., *Rev. Bras. Entomol.*, 36, 149, 1992. With permission.) e) Seeds of *S. alata* with eggs and parasitoid exit holes. (From Ribeiro-Costa, C. S., *Coleopt. Bull.*, 52, 63, 1998. With permission.) f) Dissected seed of *S. macranthera* with an adult inside the pupal chamber. (From Linzmeier, A. M., et al., *Rev. Bras. Zool.*, 21, 351, 2004. With permission.) g) Dissected seed of *S. alata* with a parasitoid. (From Ribeiro-Costa, C. S., *Coleopt. Bull.*, 52, 63, 1998. With permission.)

It is interesting to observe that the size can vary depending on the host plant, as seen in *Acanthoscelides aureolus* (Horn), which is bigger when it feeds on *Astragalus* that has a toxic compound (selenium) and smaller when it feeds on *Lotus*, whose seeds are smaller.

The success of the pupal stage depends on the nutrients stored during the larval stage. Glycogen appears to be the biggest energy source for the first 3 days of the pupal stage and also supplies material for chitin synthesis; the lipids apparently supply energy to the pupa and adult (Johnson and Kistler 1987).

Bruchine larvae can pupate in different places, the most common being inside the seed, where they have fed (Figure 14.3F). An example of pupation inside the fruits is *S. morosus*, which builds a pupal chamber with the seeds of *Cassia bauhinioides* glued together with an adhesive substance and *S. simulans*, which pupates among fragments of *Cassia leptadenia* seeds remaining from larval development and when the fruits dehisce (Center and Johnson 1973). *Caryedon gonagra* (F.) pupates in a cocoon formed outside the fruit (Davey 1958) and *Caryedon interstinctus* (Fahraeus) pupates outside the fruit or in the soil (Skaife 1926). Species of *Amblycerus* pupate in a cocoon inside the fruit. The cocoons fixed to the internal fruit walls have a fibrous aspect and contain partially consumed seeds (Ribeiro-Costa 1992, 1998; Johnson et al. 2001). When the fruits dehisce, those with cocoons do not completely open, which stops seed dispersal but may be a pupal defense mechanism against parasitoids and adverse abiotic factors (Figures 14.8b and 14.8c). The larvae of *A. alboscutellatus* feed on various small seeds and also pupate to form a cocoon with various small seeds together (Ott 1991). Most adults emerge from the seeds a few months before their host plants have flowers or fruits. In economically important species, such as *A. obtectus*, the period is short, around 10 days (Skaife 1926). For *Caryedon palaestinicus* Southgate (=*C. serratus palaestinicus*), adults emerge from the cocoon after 150 days (26°C, 70%), some up to 120 days, and some enter diapause, emerging after more than 2 years (Donahaye et al. 1966).

14.9 Intra- and Interspecific Competition

Various seed-infesting species tolerate the presence of other larvae and complete their development. In a study with *Z. subfasciatus*, up to eight adults were observed emerging from the same dry bean seed (Sari et al. 2003), but other authors (Pajni and Jabbal 1986; Dendy and Credland 1991) have recorded more than 20 adults. In *C. brasiliensis*, more than 50 adults developed in a *D. megacarpa* seed (Rosenthal 1983). However, it is known that when food is limited during the larval stage, adults cannot reach their normal size. Female *C. chinensis* are around 3 mm long when developing under normal conditions but less than 1 mm long under adverse conditions (Skaife 1926). A size reduction due to food quantity or quality can adversely affect longevity, fecundity, and competiveness. When Poodler and Applebaum (1971) proposed a diet for *C. chinensis* using artificial beans with *C. arietinum* flour and substituting essential components, they emphasized that the proportion of carbohydrates: proteins should be high, mono-disaccharides should be present, and carbohydrates should have little amylase; in addition, a cowpea constituent, from the fraction soluble in methanol, was included.

In general, seeds infested with bruchines contain sufficient nutrients for the development of a single adult. The presence of two to six *B. pisorum* (L.) larvae per seed is common in *P. sativum*. All larvae feed and grow until a certain moment, but only one larva per seed will become an adult (Skaife 1926). Various authors have demonstrated intraspecific larval competition, such as Bradford and Smith (1977) for *S. giganteus* (Chevrolat) and Terán and L'Argentier (1979) for *Amblycerus dispar* (Sharp) (=*A. caryoboriformis*). Wang and Kok (1986) discovered that around 30% of seeds contain two larvae of *Megacerus discoidus* (Say) and proved that when there are various larvae, cannibalism occurs between the second and third instars.

Fruits of the same species can be consumed by more than one species of Bruchinae at the same time. *M. baeri* (Pic) and *Megacerus reticulatus* (Sharp) can be present in the same fruit of *I. pes-caprae* and each larva feeds on a seed (Castellani and Santos 2005).

14.10 Predation Rate and Viability of Predated Seeds

Bruchines can cause high infestation levels in their preferential hosts even after one or two generations. *Bruchidius villososus* (F.) damages more than 80% of seeds of the leguminous weed *C. scoparius*, which makes it a possible biological control candidate for this plant, whose control is made more difficult by a large and lasting seed bank (Redmon et al. 2000). *Lysiloma divaricata* is another plant that suffers from attacks at different stages: a bruchine occurs in the immature fruits, another at the beginning of dehiscence, and another when the seeds fall on to the soil; there is also evidence that rodents eat the seeds (Johnson and Romero 2004). The highest consumption of seeds by *Acanthoscelides* was observed in a *Mimosa* species with larger seeds, *Mimosa texana* var. *texana*, compared with *M. lacerata*, which is evidence of a direct relationship between seed size and percentage infestation (Orozco-Almanza et al. 2003). In the case of stored grain pests, the grains are eaten by various generations until they are completely consumed.

Seed predation under natural conditions can vary between months, years, and place and not only bruchines may be involved. It was observed that examples of *Senna multijuga* only 10 m away showed differential seed predation rates (Sari and Ribeiro-Costa 2005). Klips et al. (2005) characterized the predation levels of *H. moscheutus* seeds at four sites in two American states over 2 years. The percentage of seeds predated by the bruchine, *A. hibisci* (Olivier), varied from 4% to 27%, and of the curculionid, *Conotrachelus fissinguis* Le Conte, from 24% to 94%; the bruchine was present at the four sites and the curculionid at three.

Considering the different variables that influence the predation rate, studies should include sampling at various periods during the fruiting phenophase; that is, immature fruits and mature fruits still fixed to the plant or on the soil. For dehiscent fruits, evaluations should be made before and after dehiscence, when the fruit is fixed to the plant and in the dispersed seeds. The interaction with mammals attracted to feeding on the fruit can reduce fruit availability for oviposition, which is a further variable to be considered.

Janzen (1969) suggested that some plants had developed strategies to escape predation. According to this author, if a plant invests more energy in producing large, toxic seeds, the seedlings stand more chances of survival. Small seeds have less stored energy and toxins, with a lower chance of developing into viable seedlings. "Escape of predation" or "predator satiation," with the abundant production of small seeds for a short period of time, and which are rapidly dispersed, means that seeds are consumed by predators but some escape predation. The bruchines that feed on plants with this strategy would develop a tendency to be smaller in order to feed on small seeds or use more than one seed during development. An example of predator satiation was demonstrated by Raghu et al. (2005) for *Acanthoscelides macropthalmus* (Shaehher) predating on seeds of *Leucaena leucocephala*, a plant introduced from Australia. The fruits stay on the plant for a variable period, and when ripe split open and the seeds disperse. *A. macropthalmus* only lays eggs when the pod is ripe, and the number of damaged seeds increases the longer the fruit stays on the tree, varying from 11%, when the pods are on the plant for 1 month, to 53%, when pods remain for 4 months. The low population of bruchines in high fruit densities results in predator satiation. Another example is the low predation rate of *M. baeri* on *I. pes-caprae* compared with *Ipomoea imperati*, the result of different reproductive strategies by these plants. *I. pes-caprae* produces many fruits over 5 months, whereas *I. imperati* produces less fruit during 8 months. The low predation rate is due to the higher fruit and seed densities (Castellani 2003; Sherer and Romanovski 2005).

Seed damage depends principally on the consumption of the embryo by the larva and on the quantity of damaged cotyledons. Bruchine larvae can destroy a large part of the cotyledon, thereby reducing seed and seedling viability and also vigor. Legume grains cultivated when infestations are low have a large chance of survival. However, the chances for *V. unguiculata* seeds with more than three holes are minimal (Booker 1965). A *C. maculatus* larva removes about one-fourth of the cotyledon of a medium-sized seed of *V. unguiculata*; in smaller seeds, such as *Phaseolus radiatus*, the cotyledons are completely eaten, stopping germination (Southgate 1979). On the other hand, bruchines can also benefit plants by favoring germination. In *Acacia* seeds, which have a hard integument resistant to the entry of water, the entry holes of first instar larvae and emerging adults permit more hydration with a germination rate of 7% for infested *Acacia gerrardii* seeds and 17% for infested *Acacia sieberiana* seeds (Mucunguzi 1995).

An interesting relationship has been described between elephants, gazelles, and bruchines, which feed on *Acacia tortilis* fruits in Africa with a possible evolutionary interaction between these groups. When mammals feed on acacia fruits, the seeds are digested except for those with bruchine larvae, which are dispersed. Those seeds with holes and that have suffered the effects of digestive juices germinate faster than intact seeds (Johnson 1994; Baskin and Baskin 2001).

Another relationship with mammals was discovered for the palm *Attalea maripa*. Silvius and Fragoso (2002) observed different species of frugivorous vertebrates remove different amounts of the mesocarp. The intact fruits that fall on the soil without being picked up by primates, birds, or Brazilian agoutis escape from being eaten by bruchines since they do not oviposit on the exocarp. When the exocarp is completely removed and the mesocarp partially removed, there are more eggs. The fruits with the mesocarp intact or completely removed have low or intermediate numbers of eggs, respectively.

14.11 Reproductive Performance, Diapause, and Dispersal

Pollen, ripe fruits, and/or seeds in general stimulate mating and gamete production. Besides this, it has been observed that female bruchines produce a sexual pheromone to attract the males (Mbata et al. 2000). The number of matings varies, and a single mating is sufficient to fertilize the eggs of species that do not feed as adults (Pesho and van Houten 1982). Females that mate with many males lay more eggs than those that mate several times with the same male (Takakura 1999). Mating involves the deposition of male ejaculatory secretions on the female genitalia. These secretions contain heavy molecular weight proteins or mucopolysaccharides (Johnson and Kistler 1987). Ovogenesis and egg laying follows in female *A. obtectus* with a temporary inhibition in male receptivity (Huignard 1983). During the mating of *C. maculatus*, the sharp sclerites of the male genitalia evert and perforate the bursa (female genitalia), making the entry of ejaculatory secretions into the hemolymph easier and inducing oogenesis; females try to turn the males away during mating to reduce damage to their genitalia. Females that mate more have a reduced longevity that may be the result of perforations, which are seen 16 h after each mating (Crudgington and Silva-Jothy 2000). Males that mate for the first time have more spermatic liquid and the females receive more energy and more water, and tend to live longer (Paukku and Kotiaho 2005). The females of *Z. subfasciatus* can mate 1 h after emergence and lay eggs from 2 to 30 h after mating (Pajni and Jabbal 1986). Females of *Sennius bondari*, which develop in the ornamental species *Senna multijuga*, and oviposit on seeds, have a longer preoviposition period, which varies from 6 to 13 days. The periods of oviposition, postoviposition, and life cycle are comparatively longer than for *Z. subfasciatus* (Linzmeier et al. 2004) (Table 14.3).

Feeding on pollen and nectar influences longevity and induces gamete formation. Laboratory studies by Leroi (1978, 1981) on *A. obtectus* proved that solutions of saccharose, glucose, fructose, or a mixture of pollen, honey, and water, result in high fecundity and longevity. Adults fed with a mixture of honey and pollen can survive for more than 200 days and ovarian production was 50% higher when compared with unfed females. In the univoltine species *B. pisorum*, only the pollen of the host plant, *P. sativum*, promotes oocyte development and increases the probability and frequency of mating. Pollen from other hosts only keeps the species alive for long periods (Pesho and van Houten 1982). On feeding *C. chinensis*

TABLE 14.3

Means (± SEM) of Various Biological Parameters of *Sennius bondari* in *Senna macranthera* and *Z. subfasciatus* in *P. vulgaris* cv. Carioca, in the Laboratory

Parameters	*S. bondari*	*Z. subfasciatus*
Preoviposition	8.6 ± 1.92	1.2 ± 0.71
Oviposition	38.3 ± 4.77	5.9 ± 0.96
Postoviposition	52.6 ± 6.21	1.2 ± 1.10
Life cycle (days)	42.3 ± 0.34	28.9 ± 8.5
Fecundity	47.7 ± 4.13	38.1 ± 9.63
Longevity—male	94.3 ± 5.18	13.3 ± 2.51
Longevity—female	102.5 ± 2.66	9.4 ± 1.54

Source: Data from Sari, L. T., et al., *Rev. Bras. Entomol.*, 47, 621, 2003; Linzmeier, A. M., et al., *Rev. Bras. Zool.*, 21, 351, 2004.

with fungus (*Sphaerotheca fuliginea* or *Uromyces azukiola*), Shinoda and Yoshida (1987) obtained a longevity three times greater than for the control and double the number of eggs. It should be emphasized that for *Z. subfasciatus*, ovarian production and the beginning of oviposition are stimulated by ripe seeds (Pimbert and Pierre 1983).

A study of *Z. subfasciatus* on *P. vulgaris* c.v. Carioca, under excellent developmental conditions (30°C; 70 R.H.) and in the absence of food for the adults, found that the longer the oviposition period, the greater the number of eggs. Thus, although the females concentrate oviposition on the 3rd and 4th days after emergence, it is the duration of oviposition that determines how many adults emerge (Sari et al. 2003).

The energy spent is greater for reproduction and may be a limiting factor. In *A. obtectus*, the longer the preoviposition period, the fewer the eggs (Leroi 1980). The protein, lipid, and glycogen contents decrease with age in female *Z. subfasciatus* and *C. maculatus*, and are higher during the reproductive phase. These species also have a high lipid content (30–50% of dry weight), which may represent a survival strategy (Sharma and Sharma 1979a,b).

Besides the nutritional components, the genetic component also plays a significant role in reproduction. Huignard and Biemont (1978) found that the low altitude lines of *A. obtectus* with a high food availability showed a short longevity and reproduced early, mating soon after emergence and without needing any food stimulus. In the high altitude lines, in which the host plant is only available for a short time, longevity is longer and mating and oviposition only occur after food is available and fewer eggs are produced.

Bruchines complete one or a few generations a year (Johnson 1994). Those species from regions with cold or dry periods enter diapauses during the adult stage and are normally univoltine with the life cycle synchronized with their host plant. When pollen and nectar are available, diapauses terminate and mating occurs followed by oviposition when flowers/fruits appear. Species from more amenable climates where food resources are always available do not enter diapause.

Stored grains are a different environment (see Chapter 18), and the bruchines that develop under these conditions are multivoltine and do not diapause. Larval food is abundant, so population growth is continuous until all the food has been consumed.

Reproductive diapause has been registered in various species, such as *Bruchidius atrolineatus* (Pic) (Lenga et al. 1991) and *Bruchus pisorum* (Pesho and van Houten 1982; Annis and O'Keeffe 1984). A special case is *Bruchidius dorsalis* (Fahraeus), a multivoltine species that enters larval diapause or reproductive diapauses when the photoperiod is short (Kurota and Simada 2001, 2002). In *B. atrolineatus* (Pic), the reproductive diapause depends on the climatic conditions present at the beginning of the dry season and diapause is induced by a long photoperiod and high temperatures. The males terminate diapauses when exposed to a short photoperiod and high humidity; females produce mature oocytes only under similar climatic conditions and in the presence of inflorescences or pods of *V. unguiculata* (Monge et al. 1989; Lenga and Huignard 1992; Glitho et al. 1996). Similarly, the end of reproductive diapause in *Bruchus rufimanus* Boheman results from the interaction between an increase in the photophase and the ingestion of pollen from the host, *Vicia faba*, with the photoperiod being the most significant parameter (Tran and Huignard 1992). Biemont and Bonet (1980) observed in univoltine *Acanthoscelides* spp. on wild *P. vulgaris* in Mexico that diapausing adults stayed in dry, rolled-up leaves still attached to the plant. The adults move from one leaf to another, but ovarioles do not develop. Diapause ends when the first flowers appear at the end of September and beginning of October (Figure 14.5).

The number of generations of *Kytorhinus sharpianus* Bridwell varies geographically, from partially trivoltine to univoltine along a latitudinal gradient between 36°N and 41°N. The multivoltine and univoltine populations show a facultative diapause in response to photoperiod. The multivoltine populations show a seasonal variation between generations, which enter diapause or not, related to the number of eggs laid, longevity, and the preoviposition period (Ishihara 1999). The life cycle of *B. dorsalis* (Fahraeus) in hotter regions is trivoltine in *Gleditsia japonica*, and it passes winter in the larval or adult stages; in cold weather, the cycle is bivoltine, the phenology of the host plant is longer, and *B. dorsalis* enters diapause in the winter in the fourth larval instar or as an adult (Kurota and Shimada 2002).

Little is known about bruchine dispersal or their flight capacity. With herbivores attracted to eating infested fruits, it is possible that on dispersing the seeds, they also disperse the bruchine larvae inside the seeds (Or and Ward 2003). *Callosobruchus* shows polymorphism with a "normal" form, where individuals do not fly, and an "active" form, where they fly and disperse (Utida 1954; Caswell 1960). Active forms are more abundant when temperatures increase and the dry bean seeds have a higher humidity, resulting from larval metabolic activity. The individuals of the active form are larger, have more lipids, lay fewer eggs, take longer to emerge, and takes longer to have mature reproductive organs, which suggests a form of reproductive diapause (Gill et al. 1971; Utida 1972). Appleby and Credland (2007) confirmed reproductive diapause in active adults of *Callosobruchus subinnotatus*.

14.12 Natural Enemies

14.12.1 Parasitoids

Relevant studies on this subject include Whitehead (1975), Center and Johnson (1976), De Luca (1965, 1970), and Steffan (1981). Proving that eggs have been parasitized is simpler than for larvae or pupae. Generally, parasitized eggs are dark while nonparasitized ones are transparent when recently laid, turning to an opaque white when the first instar larva penetrates the seed and fills the egg with remains of the excavated seed wall (Figure 14.19); nonfertilized eggs are transparent. An indication of larval or pupal parasitism is when there are different-sized exit holes in the seeds. Parasitoids are normally smaller than bruchines and, consequently, the holes they make are smaller (Figures 14.8d and 14.8e). However, proof of parasitism is only possible after dissecting the seeds (Figures 14.8f and 14.8g).

In *Lonchocarpus muehlbergianus*, the exit holes left in the seeds by adult bruchines differ from those of parasitoids. The seeds with bigger holes are consumed by *Ctenocolum podagricus* (F.) (Figures 14.9a and 14.9b), while adult *Horismenus missouriensis* Ashmead (Eulophidae) (Figures 14.9c and 14.9d), a probable larval or pupal parasitoid, emerge from those with smaller holes (Sari et al. 2002). When the adult parasitoid is found inside the seed, it is impossible to associate it with the larval or pupal stage. The same thing occurs if there is a predator and a parasitoid, even considering the proportional sizes between predator and parasitoid. This was the case for *Horismenus* sp., found in *S. alata* seeds (Figure 14.8g), which were also infested by *A. submaculatus* and *S. bondari*. In the literature, there is an indication that from 18 to 30 individuals of *H. missouriensis* emerge from an *Amblycerus robiniae* (F.) larva (Bissel 1938), but this information does not invalidate the possibility of *Horismenus* also being a parasitoid of *S. bondari*. However, a small group of Hymenoptera is phytophagous and associated with fruit. Species of *Eurytoma* complete development by feeding on sap and *Bruchophagus* feeds on legume seeds and not on bruchines, as the name suggests (Steffan 1981).

Larval parasitism was found after dissecting the cocoon of *A. hoffmanseggi* when exuvia and adults of *Horismenus* sp., cephalic capsules, and the remains of bruchine larvae were found (Ribeiro-Costa 1992) (Figure 14.8c). When bruchines emerge, they perforate the fruit wall and the hole has a larger diameter than that left by first instar bruchine larvae and less than that of an adult during its emergence (Figure 14.8d).

The population cycles of bruchines and their parasitoids are little understood since various bruchine and parasitoid species occur in the same host plant. In the *S. multijuga* system, 3 bruchine and 11 parasitoid species were identified, with different population numbers, during 2 years of fruit collections. *S. crudelis* and *Eurytoma* sp. were the species that occurred together most, which is an indirect evidence of parasitism (Sari 2003; Sari et al. 2005).

Ott (1991) observed parasitism of the larvae and pupae of *A. alboscutellatus* by members of the Pteromalidae, Eupelmidae, and Eurytomidae, families commonly recognized as parasitoids of Bruchinae. The Eulophidae are also representative with around 20 bruchine parasitoids (Steffan 1981). The Trichogrammatidae include *Uscana*, which exclusively parasitize bruchine eggs. There are various hosts recorded under the name *Uscana semifumipennis*, including *Acanthoscelides alboscutellatus, C. maculatus, Althaeus hibisci*, and *S. limbatus* (Stephen 1981). Around 15 Braconidae species are bruchine parasitoids, most belonging to the genera *Triaspis, Heterospilus*, and *Urosigalphus* (Steffan 1981).

In the system composed of bruchines associated with *Prosopis*, Conway (1980) observed egg parasitism by *Trichogramma* sp. and estimated that *Horismeus productus* Ashmead (Eulophidae) parasitized from 1% to 4% of *M. amicus* (Horn) and *A. prosopis* (Le Conte) larvae. *Heterospilus prosopidis* Viereck

FIGURE 14.9 *Ctenocolum podagricus.* a) Dorsal view of adult. b) Seeds of *L. muehlbergianus* with exit holes. *Horismenus missouriensis.* c) Lateral view of adult. d) Seeds of *L. muehlbergianus* with exit holes of *H. missouriensis.* (From Sari, L. T., et al., *Neotr. Entomol.,* 31, 483, 2002. With permission.)

(Braconidae) parasitized from 9% to 17% of *M. amicus* and *A. prosopis* larvae and another braconid, *Urosigulphus bruchi* Crawford, destroyed from 4% to 7% of the bruchine larvae of *Prosopis,* with 17% to 25% of the larvae of Arizona being parasitized (Johnson 1983).

Parasitoids have a negative impact on natural populations, but effective control does not always occur. Parnell (1966), studying insect population dynamics in *C. scoparius,* observed parasitism of *Bruchidius ater* (Marsham) by *Habrocitus sequester* Kurdjumov, varying from 48% to 56% in 2 years. Many parasitoids have been introduced for bruchine biocontrol. *Uscana semifunipennis* was introduced into Hawaii to control *C. gonagra* and in Japan to control *B. rufimanus* Boheman. In 1989, *Uscana senex* was introduced into Chile to control *B. pisorum* and *A. obtectus* eggs have been used as alternative hosts for breeding this parasitoid (Rojas-Rousse et al. 1996). The braconid, *Triaspis thoracica* (Curtis), was introduced into the United States, Canada, and Australia to control *Bruchus* and *Tetrastichus bruchophagi* to control *Bruchus brachialis* Fahraeus (Van Huis 1991). *H. prosopidis* was introduced to control the larvae and pupae of *A. prosopis* Le Conte in Texas.

The biocontrol of bruchines in grain storage has given positive results. Examples include the egg parasitoid *Uscana lariophaga* Steffan, the larval and pupal parasitoids *Dinarmus basalis* (Rondani) (Van Huis 2002), and *Eupelmus orientalis* Crawford and *Eupelmus vuilleti* Crawford (Ndoutome et al.

2000) to control *C. maculatus*. Schmale et al. (2005) obtained a 48% to 75% reduction in an *A. obtectus* population during 16 weeks of storage of dry beans using *D. basalis* (Rondani). Other parasitoids of *A. obtectus*, which are also parasitoids of *Z. subfasciatus*, are *Stenocorse bruchivora* (Crawford) (Braconidae), *D. basalis* (Pteromalidae), and *Horismenus* sp. (Eulophidae). These three species are ectoparasitoids of the third and fourth larval instars and occasionally parasitize pupae. In *A. obtectus*, the species *Dinarmus laticollis*, *Eupelmus cushmani* (Crawford) and *Eupelmus cyaniceps* Ashmead (Eupelmidae), *Torymus atheatu* Grissel (Torymidae), and *Chryseida bennetti* Burks (Eurytomidae) are also known (Van Huis 1991). In *Z. subfasciatus*, *H. prosopidis* (Braconidae), *Anisopteromalus calandrae* (Howard) (Pteromalidae), *E. orientalis* (Eupelmidae), and *Dinarmus colemani* (Crawford) have been recorded (Kistler 1985, Van Huis 1991). Another economically important species in Brazil, *C. maculatus*, has *U. semifumipennis* Girault, *Uscana mukerjii* (Mani), *E. orientalis* Crawford, *Anisopteromalis calandrae* (Howard) (Pteromalidae), *Chaetopsila elegans* Westwood (Pteromalidae), *D. basalis*, *Dinarmus vagabundus* (Timberlake), *Lariophagus distinguendus* (F.), and *Lariophagus texanus* Crawford (Pteromalidae) recorded as parasitoids (Southgate 1979, Van Huis 1991).

14.12.2 Predators

A large number of bruchines in legume samples from tropical regions do not complete their life cycle owing to the action of mite predators of the genus *Pymotes*, which can feed on larvae, pupae, and adults. Various mammals also act as predators by consuming bruchine-infested seeds or fruits, which may or may not be attractive for consumption. The possible reasons for a preference for infested seeds is that they are more nutritious since the larvae synthesize fats and/or proteins or other nutrients, such as vitamins; they have a better flavor and are more easily opened and consumed (Gálvez and Jansen 2007).

14.13 Conclusions and Suggestions for Research

Bruchines constitute an interesting food group (guild) from the biological point of view, with innumerable species maintaining relationships with their host plants, which show specialized and sophisticated associations. By partially or totally feeding on seeds, they prejudice the reproductive potential of the plants and are a natural selection agent, influencing population sizes and the spatial distribution of plants. Their negative effect on plants is obvious, especially when the seedlings are subject to unfavorable climatic conditions and attacks by other animals and fungi. Studies on the germination and viability of predated seeds and the survival of seedlings from these seeds are necessary. The results from these studies would show the total effect of bruchine damage on seeds in the environment and the regeneration of reforested areas.

About 30 species are serious pests, and at least nine are cosmopolitan, principally due to commercial grain activities. Chemical insecticides are commonly used to control these pests, although objections to their application have increased because of toxic residues and the appearance of resistant insect populations. Studies focused on alternative control methods are recommended, including the use of resistant dry bean varieties, artificial cooling, postinert products, plant oils, and dusts, and also parasitoids as natural enemies. To develop new resistant varieties, research should be extended to plants with a certain degree of resistance to learn about the biosynthesis and regulation of chemical compounds associated with plant defenses. Significant studies on bruchines could elucidate the biochemical mechanisms that neutralize toxic compounds and could also contribute to explaining the evolution of the bruchines and their host plants.

REFERENCES

Alvarez, N., J. Romero-Napoles, K.-W. Anton, B. Benrey, and M. Hossaert-Mckey. 2005. Phylogenetic relationships in the neotropical bruchid genus *Acanthoscelides* (Bruchinae, Bruchidae, Coleoptera). *J. Zool. Syst. Evol. Res.* 44:63–74.

Annis, B., and L. E. O'Keeffe. 1984. Effect of pollen source on oogenesis in the pea weevil, *Bruchus pisorum* L. (Coleoptera: Bruchidae). *Prot. Ecol.* 6:257–66.

Angiosperm Phylogeny Group. 2003. An update of the angiosperm phylogeny group classification for the orders and families of flowering plants: APG II. *Bot. J. Linnean Soc.* 141:399–436.

Appleby, J. H., and P. F. Credland. 2007. Bionomics and polymorphism in *Callosobruchus subinnotatus* (Coleoptera: Bruchidae). *Bull. Entomol. Res.* 91:235–44.

Applebaum, S. W., and Y. Birk. 1972. Natural mechanisms of resistance to insects in legume seeds. In *Insect and Mite Nutrition*, ed. J. G. Rodriguez, 629–636. Amsterdam: North Holland.

Baskin, C. C., and J. M. Baskin. 2001. *Seeds: Ecology, Biogeography, and Evolution of Dormancy and Germination.* San Diego: Academic Press.

Biemont, J. C., and A. Bonet. 1980. The bean weevil populations from the *A. obtectus* Say group living on wild or subspontaneous *Phaseolus vulgaris* L. and *Phaseolus coccineus* L. and on *Phaseolus vulgaris* L. cultivated in the Tepotztlan region state of Morelos-Mexico. In *The Ecology of Bruchids Attacking Legumes (Pulses)*, Series Entomologica, ed. V. Labeyrie, 23–39. The Hague: W. Junk.

Bissel, T. 1938. The host plants and parasites of the cowpea curculio and other legume infesting weevils. *J. Econ. Entomol.* 31:531–7.

Bondar, C. 1937. Notas biológicas sobre bruquídeos observados no Brasil. *Arq. Inst. Biol. Vegetal* 3:7–44.

Booker, R. H. 1965. Pests of cowpea and their control in Northern Nigeria. *Bull. Entomol. Res.* 55:663–72.

Bradford, D. F., and C. C. Smith. 1977. Seed predation and seed number in *Scheelea* palm fruits. *Ecology* 58:667–73.

Bridwell, J. C. 1938. *Specularius erythrinae*, a new bruchid affecting seeds of *Erythrina* (Coleoptera). *J. Wash. Acad. Sci.* 28:69–76.

Carlini, C. R., and M. F. Grossi-de-Sá. 2002. Plant toxic proteins with insecticidal properties. A review on their potentialities as bioinsecticides. *Toxicon* 40:1515–39.

Caron, E., C. S. Ribeiro-Costa, and A. M. Linzmeier. 2004. The egg morphology of some species of *Sennius* Bridwell (Coleoptera: Chrysomelidae: Bruchinae) based on scanning electron micrographs. *Zootaxa* 556:1–10.

Castellani, T. T. 2003. Estrutura e dinâmica populacional de *Ipomoea pes-caprae* na Ilha de Santa Catarina. Tese de Doutorado, Universidade Estadual de Campinas, Campinas, SP.

Castellani, T., and F. A. M. Santos. 2005. Fatores de risco à produção de sementes de *Ipomoea pes-caprae. Rev. Bras. Bot.* 28:773–83.

Caswell, G. H. 1960. Observations on an abnormal form of *Callossobruchus maculatus* (F.). *Bull. Entomol. Res.* 50:671–80.

Center, T. D., and C. D. Johnson. 1973. Comparative life histories of *Sennius* (Coleoptera: Bruchidae). *Environ. Entomol.* 2:669–72.

Center, T., and C. D. Johnson. 1974. Coevolution of some seed beetles (Coleoptera: Bruchidae) and their hosts. *Ecology* 55:1096–103.

Center, T., and C. D. Johnson. 1976. Host plants of some Arizona seed-feeding insects. *Ann. Entomol. Soc. Am.* 69:195–201.

Conway, R. W. 1980. A comparative study of the bruchid-parasitoid complexes found in some Arizona legumes. Master of Science thesis, Northern Arizona University, Flagstaff, AZ.

Credland, P. F., and J. Dendy. 1992. Intraspecific variation in bionomic characters of the Mexican bean weevil, *Zabrotes subfasciatus. Entomol. Exp. Appl.* 65:39–47.

Crudgington, H. S., and M. T. Silva-Jothy. 2000. Genital damage, kicking and early death. The battle of the sexes takes a sinister turn in the bean weevil. *Nature* 407:855–6.

Davey, P. M. 1958. The groundnut bruchid, *Caryedon gonagra* (F.). *Bull. Entomol. Res.* 49:385–404.

De Luca, Y. 1965. Catalogue des métazoaires parasites et prédateurs de Bruchides (Coleoptera). *J. Stored Prod. Res.* 1:51–98.

De Luca, Y. 1970. Catalogue des métazoaires parasites et prédateurs de Bruchides (Coléoptères). *Ann. Soc. Hort. Hist. Nat. Hérault* 110:3–23.

Dendy, J., and P. F. Credland. 1991. Development, fecundity and egg dispersion of *Zabrotes subfasciatus. Entomol. Exp. Appl.* 59:9–17.

Donahaye, E., S. Navarro, and M. Calderón. 1966. Observations on the life-cycle of *Caryedon gonagra* (F.) on its natural hosts in Israel, *Acacia spriocarpa* and *A. tortilis. Trop. Sci.* 8:85–9.

Ernst, W. H. O. 1992. Nutritional aspects in the development of *Bruchidius sahlbergi* (Coleoptera: Bruchidae) in seeds of *Acacia erioloba. J. Insect Physiol.* 38:831–8.

Farrell, B. D., and A. S. Sequeira. 2004. Evolutionary rates in the adaptive radiation of beetles on plants. *Evolution* 58:1984–2001.

Fox, C. W., L. A. McLennan, and T. A. Mousseau. 1995. Male body size affects female lifetime reproductive success in a seed beetle. *Anim. Behav.* 50:281–4.

Fox, C. W., K. J. Waddell, and J. Deslauriers. 1997. Seed beetle survivorship, growth and egg size plasticity in a paloverde hybrid zone. *Ecol. Entomol.* 22:416–24.

Gálvez, D., and P. Jansen. 2007. Bruchid beetle infestation and the value of *Attalea butyracea* endocarps for neotropical rodents. *J. Trop. Ecol.* 23:381–4.

Gatehouse, A. M. R., and D. Boulter. 1983. Assessment of the antimetabolic effects of trypsin inhibitors on the development of the bruchid beetle, *Callosobruchus maculatus. J. Sci. Food Agric.* 34:345–50.

Gill, J., K. C. Kanwar, and S. R. Bawa. 1971. Abnormal "sterile" strain in *Callosobruchus maculatus. Ann. Entomol. Soc. Am.* 64:1186–88.

Glitho, I. A., A. Lenga, D. Pierre, and J. Huignard. 1996. Changes in the responsiveness during two phases of diapause termination in *Bruchidius atrolineatus* Pic (Coleoptera: Bruchidae). *J. Insect Physiol.* 42:953–60.

Goossens, A., C. Quintero, W. Dillen, R. de Rycke, J. F. Valor, J. de Clercq, M. V. Montagu, C. Cardona, and G. Angenon. 2000. Analysis of bruchid resistance in the wild common bean accession G02771: No evidence for insecticidal of arcelin 5. *J. Exp. Bot.* 51:1229–36.

Green, T. W., and I. G. Palmbald. 1975. Effects of insect seed predators on *Astragalus cibarius* and *Astragalus utahensis* (Leguminosae). *Ecology* 56:1435–40.

Huignard, J. 1983. Transfer and fate of male secretions deposited in the spermatophore of females of *Acanthoscelides obtectus* Say (Coleoptera: Bruchidae). *J. Insect Physiol.* 29:55–63.

Huignard, J., and J. C. Biemont. 1978. Comparison of four populations of *Acanthoscelides obtectus* (Coleopteran: Bruchidae) from different Colombian ecosystems. *Oecologia* 35:307–18.

Ishihara, M. 1999. Adaptive phenotypic plasticity and its difference between univoltine and multivoltine populations in a bruchid beetle, *Kythorhinus sharpianus. Evolution* 53:1979–86.

Ishimoto, M., and K. Kitamura. 1989. Growth inhibitory effects of an α-amylase inhibitor from kidney bean, *Phaseolus vulgaris* (L.) on three species of bruchids (Coleoptera: Bruchidae). *Appl. Ent. Zool.* 24:281–6.

Janzen, D. H. 1969. Seed-eaters versus seed size, number, toxicity and dispersal. *Evolution* 23:1–27.

Janzen, D. H. 1971. Escape of *Cassia grandis* L. beans from predators in time and space. *Ecology* 52:964–79.

Janzen, D. H. 1975. Interactions of seeds and their insect predators/parasitoids in a tropical deciduous forest. In *Evolutionary Strategies of Parasitic Insects and Mites*, ed. P. W. Price, 154–186. New York: Plenum.

Janzen, D. H. 1978. The ecology and evolutionary biology of seed chemistry as relates to seed predation. In *Biochemical Aspects of Plant and Animal Coevolution*, ed. J. B. Harborne, 163–206. London: Academic Press.

Janzen, D. H. 1980. Specificity of seed-attacking beetles in a Costa Rica deciduous forest. *J. Ecol.* 68:929–52.

Jarry, M. 1987. Diet of the adults of *Acanthoscelides obtectus* and its effect on the spatial pattern of the attacks in the fields of *Phaseolus vulgaris.* In *Insects–Plants. Proceedings of the 6th International Symposium on Insect–Plant Relationships (Pau 1986)*, eds. V. Labeyrie, V. G. Fabres, and D. Lachaise, 71–75. The Netherlands: W. Junk.

Jirawongse, V., T. Pharadai, and P. Tantivatana. 1979. The distribution of indole alkaloids in certain genera of Convolvulaceae growing in Thailand. *J. Nat. Res. Council Thai.* 9:17–24.

Johnson, C. D. 1967. Notes on the systematics, host plants, and bionomics of the bruchid genera *Merobruchus* and *Stator* (Coleoptera: Bruchidae). *Pan-Pacific Entomol.* 43:264–71.

Johnson, C. D. 1970. Biosystematics of the Arizona, California, and Oregon species of the seed beetle genus *Acanthoscelides* Schilsky (Coleoptera: Bruchidae). *Univ. Calif. Publ. Entomol.* 59:1–116.

Johnson, C. D. 1981. Interactions between bruchid (Coleoptera) feeding guilds and behavioral patterns of pods of the Leguminosae. *Environ. Entomol.* 10:249–253.

Johnson, C. D. 1982. Survival of *Stator generalis* (Coleoptera: Bruchidae) in host seeds from outside its geographical range. *J. Kans. Entomol. Soc.* 55:718–24.

Johnson, C. D. 1983. *Handbook on Seed Insects of Prosopis Species. Ecology, Control, and Identification of Seed-Infesting Insects of New World Prosopis (Leguminosae)*. Rome: Food and Agriculture Organization of the United Nations.

Johnson, C. D. 1994. The enigma of the relationships between seeds, seed beetles, elephants, cattle, and other organisms. *Aridus* 6:1–4.

Johnson, C. D., and D. H. Janzen. 1982. Why are the seeds of the Central American guanacaste tree (*Enterolobium cyclocarpum*) not attacked by bruchids except in Panama? *Environ. Entomol.* 11:373–7.

Johnson, C. D., and J. Romero. 2004. A review of evolution of oviposition guilds in the Bruchidae (Coleoptera). *Rev. Bras. Entomol.* 48:401–8.

Johnson, C. D., and R. A. Kistler. 1987. Nutritional ecology of bruchid beetles. In *Nutritional Ecology of Insects, Mites, Spider and Related Invertebrates*, eds. F. Slansky, Jr. and J. G. Rodriguez, 259–282. New York: John Wiley & Sons.

Johnson, C. D., J. R. Romero, and E. Raimúndez-Urrutia. 2001. Ecology of *Amblycerus crassipunctatus* Ribeiro-Costa (Coleoptera: Bruchidae) in seeds of Humiriaceae, a new host family for bruchids, with ecological comparisons to other species of *Amblycerus*. *Coleopt. Bull.* 55:37–48.

Kato T., A. Bonet, H. Yoshitake, J. Romero-Nápoles, U. Jinbo, M. Ito, and M. Shimada. 2010. Evolution of host utilization patterns in the seed beetle genus *Mimosestes* Bridwell (Coleoptera: Chrysomelidae: Bruchinae). *Mol. Phylogen. Evol.* 55:816–32.

Kergoat, G. J., N. Alvarez, M. Hossaert-Mckey, N. Faure, and F. Silvain. 2005. Parallels in the evolution of the two largest New and Old World seed-beetle genera (Coleoptera, Bruchidae). *Mol. Ecol.* 14: 4003–21.

Kergoat, G. J., J.-F. Silvain, A. Delobel, M. Tuda, and K.-W. Anton. 2007. Defining the limits of taxonomic conservatism in host-plant use for phytophagous insects: Molecular systematics and evolution of host–plant associations in the seed-beetle genus *Bruchus* Linnaeus (Coleoptera: Chrysomelidae: Bruchinae). *Mol. Phylogen. Evol.* 43:251–69.

Kingsolver, J. M. 2004. Handbook of the Bruchidae of the United States and Canada (Insecta, Coleoptera). *USDA Tech. Bull.* 1912:1–324.

Kistler, R. A. 1982. Effects of temperature on six species of seed beetles (Coleoptera: Bruchidae): An ecological perspective. *Ann. Entomol. Soc. Am.* 75:266–71.

Kistler, R. 1985. Host-age structure and parasitism in a laboratory system of two hymenopterous parasitoids and larvae of *Zabrotes subfasciatus* (Coleoptera: Bruchidae). *Environ. Entomol.* 14:507–11.

Klips, R. A., P. M. Sweeney, E. K. K Bauman, and A. A. Snow. 2005. Temporal and geographic variation in predispersal seed predation on *Hibiscus moscheutos* L. (Malvaceae) in Ohio and Maryland, USA. *Am. Midl. Nat.* 154:286–95.

Kurota, H., and M. Shimada. 2001. Photoperiod- and temperature-dependent induction of larval diapause in a multivoltine bruchid, *Bruchidius dorsalis*. *Entomol. Exp. Appl.* 99:361–9.

Kurota, H., and M. Shimada. 2002. Geographical variation in the seasonal population dynamics *Bruchidius dorsalis* (Coleoptera: Bruchidae): Constraints of temperature and host plant phenology. *Environ. Entomol.* 31:469–75.

Larson, A. O., and C. K. Fisher. 1938. The bean weevil and the southern cowpea weevil in California. *USDA Tech. Bull.* 593:1–70.

Lenga, A., and J. Huignard. 1992. Effect of changes in the thermoperiod on reproductive diapause in *Bruchidius atrolineatus* Pic (Coleoptera: Bruchidae). *Physiol. Entomol.* 17:247–54.

Lenga, A., C. Thibeaudeau, and J. Huignard. 1991. Influence of thermoperiod and photoperiod on reproductive diapause in *Bruchidius atrolineatus* (Pic) (Coleoptera: Bruchidae). *Physiol. Entomol.* 16:295–303.

Leroi, B. 1978. Alimentation des adultes d'*Acanthoscelides obtectus* Say (Coleopteran, Bruchidae): Influence sur la longévité et la production ovarienne des individus vierges. *Ann. Zoologie Ecologie Animale* 10:559–67.

Leroi, B. 1980. Regulation de la prodution ovarienne chez *Acanthoscelides obtectus* (Coleoptera: Bruchidae): Influence de l'age des femelles lors de la fecondation et de la presence des grains. *Entomol. Exp. Appl.* 28:132–44.

Leroi, B. 1981. Feeding, longevity and reproduction of adults of *Acanthoscelides obtectus* Say in laboratory conditions. In *The Ecology of Bruchids Attacking Legumes (Pulses)* Series Entomologica, ed. V. Labeyrie, 101–11. The Hague: W. Junk.

Leroi, B., C. Chararas, and J. M. Chipoulet. 1984. Etude des activités osidasiques du tube digestif des adultes et des larves de la bruche du haricot, *Acanthoscelides obtectus*. *Entomol. Exp. Appl.* 35:269–73.

Linzmeier, A. M., C. S. Ribeiro-Costa, and E. Caron. 2004. Comportamento e ciclo de vida de *Sennius bondari* (Pic, 1929) (Coleoptera, Chrysomelidae, Bruchinae) em *Senna macranthera* (Collad.) Irwin et Barn. (Caesalpinaceae). *Rev. Bras. Zool.* 21:351–6.

Mbata, G. N., S. Shu, and S. B. Ramaswamy. 2000. Sex pheromones of *Callosobruchus subinnotatus* and *C. maculatus* (Coleoptera: Bruchidae): Congeneric responses and role of air movement. *Bull. Entomol. Res.* 90:147–54.

Mitchell, R. 1975. The evolution of oviposition tactics in the bean weevil, *Callosobruchus maculatus* (F.). *Ecology* 56:696–702.

Monge, J. P., A. Lenga, and J. Huignard. 1989. Induction of reproductive diapause in *Bruchidius atrolineatus* during the dry season in a Sahelian desert. *Entomol. Exp. Appl.* 53.95–104.

Morse, G. E., and B. D. Farrel. 2005. Interspecific phylogeography of the *Stator limbatus* species complex: The geographic context of speciation and specialization. *Mol. Phylogen. Evol.* 36:201–13.

Mota, A. C., K. V. S. Fernandes, M. P. Sales, V. M. Q. Flores, and J. Xavier Filho. 2002. Cowpea vicilins: Fraction of urea denatured sub-units and effects on *Callosobruchus maculatus* F. (Coleoptera: Bruchidae) development. *Braz. Arch. Biol. Technol.* 25:1–5.

Mucunguzi, P. 1995. Effects of bruchid beetles on germination and establishment of *Acacia* species. *Afr. J. Ecol.* 33:64–70.

Ndoutome, A., R. Kalmes, and D. Rojas-Rousse. 2000. Reproductive potential of *Eupelmus orientalis* (Crawford) and *Eupelmus vuilleti* (Crawford) (Hymenoptera: Eupelmidae), two parasitoids of Bruchidae (Coleoptera) during the harvest and storage of cowpea pods (*Vigna unguiculata* (L.) Walp.). *Afr. Entomol.* 8:201–9.

Nwanze, K. F., and E. Horber. 1976. Seed coats of cowpeas affect oviposition and larval development of *Callosobruchus maculatus*. *Environ. Entomol.* 5:213–8.

Or, K., and D. Ward. 2003. Three-way interactions between *Acacia*, large mammalian herbivores and bruchid beetles—A review. *Afr. J. Ecol.* 41:257–65.

Orozco-Almanza, M. S., L. P. León-García, R. Grether, and E. García-Moya. 2003. Germination of four species of the genus *Mimosa* (Leguminosae) in a semi-arid zone of Central Mexico. *J. Arid Environ.* 55:75–92.

Ott, J. R. 1991. The biology of *Acanthoscelides alboscutellatus* (Coleoptera: Bruchidae) on its host plant, *Ludwigia alternifolia* (L.) (Onagraceae). *Proc. Entomol. Soc. Wash.* 93:641–51.

Paes, N. S., I. R. Gerhardt, M. V. Coutinho, M. Yokohama, E. Santana, N. Harris, M. J. Chrispeels, and M. F. G. de Sá. 2000. The effect of arcelin-1 on the structure of the midgut of bruchid larvae and immunolocalization of the arcelin protein. *J. Insect Physiol.* 46:393–402.

Pajni, H. R., and A. Jabbal. 1986. Some observations of *Zabrotes subfasciatus* (Boh.) (Bruchidae: Coleoptera). *Res. Bull. Panjab Univ. Sci.* 37:11–6.

Parnell, J. R. 1966. Observations on the population fluctuations and life histories of the beetles *Bruchidius ater* (Bruchidae) and *Apion fuscirostre* (Curculionidae) on broom (*Sarothamnus scoparius*). *J. Animal Ecol.* 35:157–88.

Paukku, S., and J. S. Kotiaho. 2005. Cost of reproduction in *Callosobruchus maculatus*: Effects of mating on male longevity and effect of male mating status on female longevity. *J. Insect Physiol.* 51:1220–6.

Pesho, G. R., and R. J. van Houten. 1982. Pollen and sexual maturation of the pea weevil (Coleoptera: Bruchidae). *Ann. Entomol. Soc. Am.* 75:439–43.

Pfaffenberger, G. S. 1985. Checklist of selected world species of described first and/or final larval instar (Coleopteran: Bruchidae). *Coleopt. Bull.* 39:1–6.

Pfaffenberger, G. S., and C. D. Johnson. 1976. Biosystematics of the first-stage larvae of some North American Bruchidae (Coleoptera). *Tech. Bull. U. S. Dept. Agric.* 1525:1–75.

Pimbert, M., and D. Pierre. 1983. Ecophysiological aspects of bruchid reproduction. I. The influence of pod maturity and seeds of *Phaseolus vulgaris* and the influence of insemination on the reproductive activity of *Zabrotes subfasciatus*. *Ecol. Entomol.* 8:87–94.

Poodler, H., and S. W. Applebaum. 1971. Basic nutritional requirements of larvae of the bruchid beetle, *Callosobruchus chinensis* L. *J. Stored Prod. Res.* 7:187–93.

Prevett, P. F. 1966. Observations on biology in the genus *Caryedon* Schoenherr in Northern Nigeria, with a list of parasitic Hymenoptera. *Proc. R. Entomol. Soc. Lond.* 41:9–16.

Raghu, S., C. Wiltshire, and K. Dhileepan. 2005. Intensity of pre-dispersal seed predation in the invasive legume *Leucaena leucocephala* limited by the duration of pod retention. *Austral. Ecol.* 30:310–318.

Raina, A. K. 1971. Comparative resistance to three species of *Callosobruchus* in a strain of chickpea (*Cicer arietinum*). *J. Stored Prod. Res.* 7:213–6.

Ramos, R. Y. 2007. Genera de Coleópteros de la Península Ibérica e Islas Baleares: Familia Bruchidae (Coleopteran, Chrysomeloidea). *Bol. Asoc. Española Entomol.* 31(1–2):65–114.

Redmon, S. G., T. G. Forrest, and G. P. Markin. 2000. Biology of *Bruchidius villosus* (Coleoptera: Bruchidae) on scotch broom in North Carolina. *Fla. Entomol.* 83:242–53.

Reid, C. A. M. 2000. Spilopyrinae Chapuis: A new subfamily in the Chrysomelidae and its systematic placement (Coleoptera). *Invertebr. Taxonomy* 14:837–62.

Ribeiro-Costa, C. S. 1992. Gênero *Amblycerus* Thunberg, 1815 (Coleoptera: Bruchidae). Grupo hoffmanseggi: II. Redescrições, chave e dados biológicos das espécies. *Rev. Bras. Entomol.* 36:149–75.

Ribeiro-Costa, C. S. 1998. Observations on the biology of *Amblycerus submaculatus* (Pic) and *Sennius bondari* (Pic) (Coleoptera: Bruchidae) in *Senna alata* (L.) Roxburgh (Caesalpinaceae). *Coleopt. Bull.* 52:63–9.

Ribeiro-Costa, C. S., and A. S. Costa. 2002. Comportamento de oviposição de bruquídeos (Coleoptera, Bruchidae) predadores de sementes de *Cassia leptophylla* Vogel (Caesalpinaceae), morfologia dos ovos e descrição de uma nova espécie. *Rev. Bras. Zool.* 19:305–16.

Ribeiro-Costa, C. S., and J. A. P. da Silva. 2003. Morphology of adult *Meibomeus cyanipennis* (Sharp) (Coleoptera: Bruchidae). *Coleopt. Bull.* 57:297–309.

Ribeiro-Costa, C. S., P. R. S. Pereira, and L. Zukovski. 2007. Desenvolvimento de *Zabrotes subfasciatus* (Boh.) (Coleoptera, Chrysomelidae, Bruchinae) em genótipos de *Phaseolus vulgaris* L. (Fabaceae) cultivados no Estado do Paraná e contendo arcelina. *Neotr. Entomol.* 36:560–4.

Rojas-Rousse, D., M. P. Gerding, and M. L. Céspedes. 1996. Caracterización de huevos parasitados por *Uscana senex* (Hymenoptera: Trichogrammatidae). *Agric. Téc. (Chile)* 56:211–3.

Rosenthal, G. A. 1983. A seed-eating beetle's adaptations to a poisonous seed. *Sci. Am.* 249:138–45.

Sari, L. T., C. S. Ribeiro-Costa, and A. C. Medeiros. 2002. Insects associated with seeds of *Lonchocarpus muehlbergianus* Hassl. (Fabaceae) in Tres Barras, Paraná, Brazil. *Neotr. Entomol.* 31:483–6.

Sari, L. T., C. S. Ribeiro-Costa, and P. R. V. S. Pereira. 2003. Aspectos biológicos de *Zabrotes subfasciatus* (Bohemann, 1833) (Coleoptera, Bruchidae) em *Phaseolus vulgaris* L., cv. Carioca (Fabaceae), sob condições de laboratório. *Rev. Bras. Entomol.* 47:621–4.

Sari, L. T., and C. S. Ribeiro-Costa. 2005. Predação de sementes de *Senna multijuga* L. C. Richard I. & B. (Caesalpinaceae) por bruquíneos (Coleoptera: Chrysomelidae). *Neotr. Entomol.* 34:521–24.

Sari, L. T., C. S. Ribeiro-Costa, and J. Roper. 2005. Dinâmica populacional de Bruchinae (Coleoptera: Chrysomelidae) em *Senna multijuga* L. C. Richard I. & B. (Caesalpinaceae). *Rev. Bras. Zool.* 22: 169–74.

Schmale, I., F. L. Wäckers, C. Cardona, and S. Dorn. 2005. How host larval age and density of the parasitoid *Dinarmus basalis* (Hymenoptera: Pteromalidae) influence control of *Acanthoscelides obtectus* (Coleoptera: Bruchidae). *Bull. Entomol. Res.* 95:145–50.

Seifelnasr, Y. E. 1991. The role of asparagine and seed coat thickness in resistance of *Phaseolus vulgaris* (L.) to *Callosobruchus maculatus* (F.) (Col., Bruchidae). *J. Appl. Entomol.* 111:412–7.

Shade, R. E., H. E. Shroeder, J. J. Pueyo, L. M. Tabe, L. L. Murdock, T. J. V. Higgins, and M. J. Chrispeels. 1994. Transgenic pea seeds expressing the α-amylase inhibitor of the common bean are resistant to bruchid beetles. *Bio-Technology* 12:793–6.

Sharma, S. P., and G. Sharma. 1979a. Age-related protein changes in bruchids *Zabrotes subfasciatus* and *Callosobruchus maculatus*. *Indian J. Exp. Biol.* 17:1197–200.

Sharma, S. P., and G. Sharma. 1979b. Age-related lipid studies in bruchids. *Curr. Sci.* 48:955–6.

Sharma, S. P., and G. Sharma. 1984. Changes in body weight, dry weight and water content with age in bruchids, *Callosobruchus maculatus* Fabr. and *Zabrotes subfasciatus* Boh. (Coleoptera: Bruchidae). *J. Entomol. Res.* 8:70–2.

Sherer, K. Z., and H. P. Romanowski. 2005. Predação de *Megacerus bueri* (Pic, 1934) (Coleoptera: Bruchidae) sobre sementes de *Ipomoea imperati* (Convolvulaceae), na praia da Joaquina, Florianópolis, sul do Brasil. *Biotemas* 18:39–55.

Shimamura, R., N. Kachi, H. Kudoh, and D. F. Whigham. 2005. Visitation of a specialist pollen feeder *Althaeus hibisci* Oliver (Coleoptera: Bruchidae) to flowers of *Hibiscus moscheutus* L. (Malvaceae). *J. Torrey Bot. Soc.* 132:197–203.

Shinoda, K., and T. Yoshida. 1987. Effect of fungal feeding on longevity and fecundity of the azuki bean weevil, *Callosobruchus chinensis* (L.) (Coleoptera: Bruchidae), in the azuki bean field. *Appl. Entomol. Zool.* 22:465–73.

Silva, J. A. P., and C. S. Ribeiro-Costa. 2008. Morfologia comparada dos gêneros do grupo *Merobruchus* (Coleoptera: Chrysomelidae: Bruchinae): Diagnoses e chave. *Rev. Bras. Zool.* 25:802–26.

Silva, L. B., M. P. Sales, A. E. A. Oliveira, O. L. T. Machado, K. V. S. Fernandes, and J. Xavier-Filho. 2004. The seed coat of *Phaseolus vulgaris* interferes with the development of the cowpea weevil *Callosobruchus maculatus* (F.) (Coleoptera: Bruchidae). *An. Acad. Bras. Cienc.* 66:57–65.

Silvius, K. M., and J. M. V. Fragoso. 2002. Pulp handling by vertebrate seed dispersers increases palm seed predation by bruchid beetles in northern Amazonian. *J. Ecol.* 90:1024–32.

Skaife, S. H. 1926. The bionomics of the Bruchidae. *South Afr. J. Sci.* 23:575–88.

Snodgrass, R. E. 1935. *Principles of Insect Morphology*. New York: McGraw-Hill.

Southgate, B. J. 1979. Biology of the Bruchidae. *Annu. Rev. Entomol.* 24:449–73.

Steffan, J. R. 1981. The parasitoids of bruchids. In *The Ecology of Bruchids Attacking Legumes (Pulses)*. Series Entomologica, ed. V. Labeyrie, 223–229. The Hague: W. Junk.

Takakura, K. 1999. Active female courtship behavior and male nutritional contribution to female fecundity in *Bruchidius dorsalis* (Fahraeus) (Coleoptera: Bruchidae). *Res. Pop. Ecol.* 41:269–73.

Terán, A. L., and S. M. de L'Argentier. 1979. Observaciones sobre Bruchidae (Coleoptera) del noroeste Argentino: II—Estudios morfológicos y biológicos de algunas especies de Amblycerinae y Bruchinae. *Acta Zool. Lilloana* 42:19–27.

Tran, B., and J. Huignard. 1992. Interactions between photoperiod and food affect the termination of reproductive diapause in *Bruchus rufimanus* (Boh.) (Coleoptera, Bruchidae). *J. Insect Physiol.* 38:633–7.

Trelease, S. F., and H. M. Trelease. 1937. Toxicity to insects and mammals of foods containing selenium. *Am. J. Bot.* 24:448–51.

Tuda, M., J. Rönn, S. Buranapanichpan, N. Wasano, and G. Arnqvist. 2006. Evolutionary diversification of the bean beetle genus *Callosobruchus* (Coleoptera: Bruchidae): Traits associated with stored-product pest status. *Mol. Ecol.* 15:3541–51.

Utida, S. 1954. "Phase" dimorphism observed in the laboratory population of the cowpea weevil, *Callosobruchus quadrimaculatus*. *Jpn. J. Appl. Zool.* 18:161–8.

Utida, S. 1972. Density dependent polymorphism in *Callosobruchus*. *Ser. Entomol.* 19:143–7.

Van Huis, A. 1991. Biological methods of bruchid control in the tropics: A review. *Insect Sci. Appl.* 12:87–102.

Van Huis, A. 2002. The wholesome effects of arthropods and their products in sub-Saharan Africa. *Entomol. Berichten* 62:8–13.

Wang, R., and L. T. Kok. 1986. Life history of *Megacerus discoidus* (Coleoptera: Bruchidae), a seedfeeder of hedge bindweed, in southwestern Virginia. *Ann. Entomol. Soc. Am.* 79:359–63.

Whitehead, D. R. 1975. Parasitic Hymenoptera associated with bruchid-infested fruits in Costa Rica. *J. Wash. Acad. Sci.* 65:108–16.

Wigglesworth, V. B. 1947. *The Principles of Insect Physiology*. London: Methuen.

15

Rhizophagous Beetles (Coleoptera: Melolonthidae)

Lenita J. Oliveira and José R. Salvadori

CONTENTS

15.1 Introduction...353
15.2 Roots as a Food Source...355
15.3 Morphological and Biological Features of Melolonthidians 356
15.4 Strategies Used by the Group to Explore Food..359
    15.4.1 Localization and Selection of Host Plant by Rhizophagous Insects.............................359
    15.4.2 Food Exploration by Rhizophagous Melolonthidians...360
15.5 Impact of Environmental Factors on Food Exploration and Performance of Larvae 361
15.6 Insect Adaptations and Responses to Variations in Abiotic and Biotic Factors.........................362
15.7 Conclusions and Suggestions for Research 364
References 364

15.1 Introduction

Considering the abundance and diversity of life habits among insects, relatively few species explore underground plant structures for feeding, including roots, shafts, rhizome, bulbs, and tubers. Although the agriculture literature contains many examples of production and product quality loss caused by insects associated with the soil, only 6 out of 26 orders of insects are well represented among herbivorous insects with underground habits (found in 11 orders). However, even in those six orders—Coleoptera, Diptera, Hemiptera (heteropterans and homopterans), Hymenoptera, Lepidoptera, and Orthoptera—the underground herbivores are restricted to a few families or subfamilies.

Insects that feed on underground plant parts are found in every continent, except Antarctica. Most orders, which include rhizophagous insects, have cosmopolitan distribution while families are more restricted and genders and species often show a high level of endemism in isolated habitats or islands (Brown and Gange 1990).

Coleoptera is the largest order and populates the most diverse ecosystems, with various roles on food chains, residue decomposition, and nutrients flow. The functional significance of coleopterans is due to the diversity of their eating behavior—acting like detritivores, herbivores, fungivores, or predators (Lawrence and Britton 1994). The underground species that are considered phytophagous feed mainly on live tissue from roots and phanerogamous underground stalks by chewing or absorbing juices. However, their habits can be really diverse; for example, some species behave like drills on roots, stalk, and tubers, creating galleries, while others cut the tissue from the outside, taking advantage of different parts of the root tissue, in accordance with their developing phases (Morón 2004).

Root feeding is widely spread among Coleoptera. In many groups, the larva is able to feed outside the roots, with high or low intensity, and therefore can be considered an underground species. The adults of most of these species feed from the aerial part of plants, not necessarily of the same species, whose roots nourish the larvae. In some cases, adults are adapted to live underground, but most of them deposit eggs on the superficial layer of the soil or in the base of the stalk of the host plant. Some species of rhizophagous coleopterans feed on nodules of leguminous plants or on mycorrhizae (Crowson 1981).

Many of the rhizophagous coleopteran species that are considered pests of cultivated plants and grazing in Brazil belong to the superfamily Lamellicornia or Scarabaeoidea. However, this group is very diverse regarding form, color, size, and eating habits, and there are thousands of species catalogued throughout the world (Morón et al. 1997). Species from this group can live in bird or insect (ants, termites) nests, rotten tree logs, soil associated with humus (decomposing animal or vegetable matter), on animal feces, or in the plant rhizosphere (Morón 1996). Adults and larvae are chewers, but in general, feed from different substrates, and the phytophagous species generally are polyphagous. Larvae can be phytophagous—feeding on roots (rhizophagous), underground stalk, bulbs, and tubers—or saprophagous, feeding on decomposing organic matter such as wood (xylophagous), feces (coprophagous), dead animals (necrophagous), humus, and straw. Larvae of some species were found hunting grasshopper eggs. Adults can feed from flowers, branches, leaves (phyllophagous), fruits (frugivorous), pollen, nectar, roots, excrement, corpses, and keratinized debris and decomposing matter. Male adults of some species do not feed (Oliveira et al. 2003).

The classification of rhizophagous coleopterans is controversial; however, according to Endrödi (1966), it is divided into five families: Melolonthidae, Scarabaeoidea, Trogidae, Passalidae, and Lucanidae. In the Brazilian ecosystems, the Melolonthidae family is one of the most common (Oliveira et al. 2003) and their larvae (white grubs), as well as other Scarabaeoidea species, are commonly known as *corós* or *bichos-bolo*.

The Melolonthidae species with edaphic larvae registered in Brazil are grouped into four, from six subfamilies of this family: 571 species in Melolonthinae (e.g., *Phyllophaga* spp., *Liogenys* spp., *Plectris* spp., and *Demodema* spp.), 210 species in Dynastinae (e.g., *Cyclocephala* spp., *Diloboderus* sp., *Eutheola* spp., *Dyscinetus* spp., *Ligyrus* spp., *Aegopsis* sp., *Bothynus* spp., and *Heterogomphus* spp.), 179 species in Rutelinae (e.g., *Analoma* spp.), and 49 species in Cetoniinae (Morón 2004). The different subfamilies have various eating habits. Rutelinae, Dynastinae, and Melolonthinae larvae generally have underground habits and can be saprophagous, phytosaprophagous, or phytophagous. On the other hand, adults are either phytophagous or do not feed. Most rhizophagous species, considered pests of cultivated plants in Brazil, belong to the subfamilies Melolonthinae and Dynastinae.

Under nonirrigated systems of grain production in the extreme south of Brazil, there are many species of Melolonthidae, in which the white grub-of-pastures (*Diloboderus abderus* Sturm) and the white grub-of-wheat (*Phyllophaga triticophaga* Morón & Salvadori) are the most important ones. This classification focuses on the possible damages on crops such as wheat, oat, rye, barley, triticale, corn, and soybean. Other cultivated plants such as Moorish wheat, colza, lupine, rye grass, vetch, and spontaneous vegetation can be hosts of *D. abderus* and *P. triticophaga* (Salvadori and Silva 2004; Silva and Salvadori 2004; Salvadori and Pereira 2006).

Despite the potential to cause damage, *D. abderus* can provide benefits, such as increasing the capacity of the soil to absorb water by opening soil galleries, and enhancing the physical, chemical, and biological features of the soil through incorporation and decomposition of crop residues (Gassen 1999). However, before this happens, it causes damages to cultures. The occurrence of the species *Demodema brevitarsis* Blanch., which cause damages on soybeans and in other cultures, restricted to a small area north of Rio Grande do Sul state, was also registered (Salvadori et al. 2006).

The small white grub (*Cyclocephala flavipennis* Burm.) is abundant and widely distributed in crops in the north region of Rio Grande do Sul state. In spite of consuming roots and damaging wheat plants in laboratory tests, in farming conditions under direct plantation, it does not cause considerable damage even in elevated populations (Salvadori 1999a; Salvadori and Pereira 2006). Besides the low potential of root consumption, it presents a facultative eating habit, with preference for decomposing organic matter.

In others regions of Brazil, *Phyllophaga cuyabana* (Moser), *Liogenys fuscus* Blanch., *Liogenys suturalis* Blanch, and *Plectris pexa* Germar frequently occur as pests in grain production systems, including soybean, corn, bean, and wheat in the states of Paraná, São Paulo, Minas Gerais, Mato Grosso do Sul, Mato Grosso, and Goiás (Corso et al. 1991; Nunes et al. 2000; Avila and Gomez 2001; Salvadori 2001; Salvadori and Oliveira 2001; Oliveira et al. 2004). Most of these species are neotropical and have wide distribution throughout Brazil, but species predominance and pest status varies according to the region.

Many melolonthidians, such as *Euetheola humilis* Burm., *Dyscinetus dubius* (Olivier), *Dyscinetus gagates* Burm., and *Ligyrus ebenus* (De Geer) feed on rice and some other cultivated plants. *E. humilis* is the most important species, and adults and larvae, known as black beetles and the white grubs-of-rice, respectively, cause severe damage and occur in every region in Brazil where rice is cultivated (Ferreira and Barrigossi 2006).

Aegopsis bolboceridus larvae (Thomson), the white grub-of-vegetables, already registered at the Federal District, Goiás, and Minas Gerais state of Brazil, can completely destroy the root system of greeneries (Solanaceae, Brassicaceae, and others). It was also recorded on bean, corn, sugarcane, brachiaria grass, ornamental, and weeds (Oliveira 2005).

Besides these rhizophagous species, other melolonthidians, considered beneficial, are common in Brazilian agroecosystems, especially in farms under direct sowing where species classified as "soil engineers" occur more often. These species build vertical tunnels on the soil (galleries), promote intense incorporation and decomposition of vegetable residues, and contribute to improve the physical and chemical features of the soil. This is the case for the white grub-of-straw (*Bothynus* spp.), named so owing to its feeding on plant remains, which does not cause direct damages to crops and builds vertical galleries of about 1.30 m deep. This species is found in southern Brazil in the Amazon region (Gassen 1999).

Many species of coprophagous white grubs are common in production systems that integrate farming and husbandry, promoting decomposition and incorporation of animal stool, as well as the biological control of pests of veterinary importance that are developed in fresh bovine feces (Honer et al. 1992).

15.2 Roots as a Food Source

Underground tissues contribute 50% to 90% to the plant biomass. Roots are the main biological component of the soil and represent an abundant supply of resources. However, the quality and distribution of this resource on the soil depend on several factors, from the morphology of the root system of different groups of plants to the longevity and specialization of the many types of roots (main or secondary), through the strategy of the roots for exploring water and nutritional resources in the soil.

Studies cited by Lavelle and Spain (2001) show that the root system of most plants develop relatively by chance; still roots tend to keep a minimum distance from each other to avoid overlapping and to optimize the extraction of water and nutrients available from the soil, especially in the case of perennial plants in arid environments. The depth of root distribution in the soil depends on the individual strategy of each species, and the physical and chemical conditions of the soil. Suberized and lignified main root can last the entire life period of the plant. Thin roots specialized in assimilating water and nutrients can last from a few days to months, and to many years when infected by mycorrhizae. Other biotic and abiotic factors, such as soil fertility, weather, root herbivory, and competition among plants, can affect root longevity. The production of thin roots is a highly seasonal process, spatially heterogeneous and apparently opportunistic, taking advantage of favorable conditions to develop new roots; when conditions become difficult, thin roots may mostly die (Lavelle and Spain 2001).

Besides being important pedogenic agents, roots keep a narrow interaction with the surrounding microflora and microfauna, and on natural systems and agroecosystems often form associations with symbionts such as nitrogen-fixating bacteria, fungus (mycorrhizae), or actinobacteria (actinorhizae). They provide energy and return nutrients to the soil by the production of organic matter below the surface and, while alive, by the production of exudates (Lavelle and Spain 2001).

The region of the soil that is under the immediate influence of the roots and in which there is propagation of microorganisms because of these roots is known as the rhizosphere (Paul and Clark 1996). Within the soil, roots are the main source of nutrients during a plant's life, but roots also play a role in the immobilization of nutrients during the initial phase of plant decomposition; subsequently, they are the main source of nutrients for the future plants and soil organisms (Van Noordwijk and Brouwer 1997), including insects.

It is likely that the rhizophagy of insects has been slowly developed millions of years ago, during the end of the Mesozoic era (Cretaceous) and the beginning of the Cenozoic era (Eocene). The evolution of the herbaceous angiosperm with abundant, fast-growing roots, between the Eocene and the Miocene, must have triggered the diversification of the strict and facultative rhizophagous insects that began to coexist with older saprophagous species (Morón 2004). However, besides rhizophagy being very common to many Coleoptera families, the first fossils from this order only appeared during the Permian period, at the beginning of the Paleozoic era (Futuyma 1992).

Studies cited by Brown and Gange (1990) show that the plant root system is the main mineral acquisition site and may serve as a place of synthesis of products involved in the growth and development of seedlings. It may also represent a place of storage of metabolic compounds, photosynthates, and carbohydrates, and may turn underground tissue of plants into a feeding source with high energetic content. Normally, the nitrogen content of roots is low relative to other plant parts, although this level may vary seasonally; in certain moments, it may be higher in the roots than in the rest of the plant. The long life cycle of some rhizophagous insects can be due to the relatively low content of nitrogen (Brown and Gange 1990) and other nutrients in the roots. However, larvae that attack the root nodules of leguminous plants (Jackai et al. 1990) have access to an extremely nitrogen-rich source.

Carbon dioxide seems to be one of the main chemical factors that determine the orientation of rhizophagous insects in the soil. However, as most roots produce CO_2, this is not enough to explain the ability of larvae to distinguish among roots from different species. Secondary volatile compounds important in the attraction of underground larvae have been identified with many species (Brown and Gange 1990). Once in the root, other chemical compounds can stimulate or inhibit feeding. The chemical substances that incite behavioral answers from rhizophagous insects may be attractive compounds, phagostimulants, or deterrents (Dethier 1970).

The metabolic compounds that often occur in the aerial part of the plant and act like feeding deterrents may also be found in the roots (Mckey 1979). The degree of specialization found among rhizophagous insects is probably the reflection of distribution of deterrents as well as attractants. Compounds with deterrent properties against rhizophagous insects that have been isolated from roots include alkaloids, fumitoric compounds, cyanogenic glycosides, glycosinates, isoflavonoids, phenolic acids, and saponins. However, chemical compounds from the roots that act as feeding deterrents to some species may not affect others (Brown and Gange 1990).

Exudates produced by roots are a mixture of carbohydrates and proteins that accelerate the activity and nutrient fixation in the rhizosphere (Lavelle and Spain 2001). The quality and quantity of exudates may vary among plant species (Curl and Truelove 1986), influencing the organisms associated with the roots (Bento et al. 2004).

Once the roots absorb minerals from the soil, the concentration of certain ions can be higher than on the leaves. Sodium, for example, is absorbed by the roots, although it is not required for plant development; however, all animals require sodium and this can represent an important component in the nutrition of rhizophagous insects (Brown and Gange 1990). Studies compiled by these authors show that roots can seclude HCO_3^- and OH^- ions, which tend to increase the pH of the rhizosphere. On the other hand, it is also known that the composition of root exudates involves a great variety of acids, which may have an overall effect on decreasing the soil pH around the root. However, the available data about the effect of soil pH on insects are conflicting.

The physical and chemical features of the soil as well as its temperature and humidity can affect the growth of roots and consequently the availability of this feeding resource to rhizophagous insects. Soil temperature and humidity may not only affect the feeding resource but may also interfere with the survival and abundance of rhizophagous insects, and therefore, may influence intra- and interspecific competition.

15.3 Morphological and Biological Features of Melolonthidians

The Scarabaeoidea adults are usually convex beetles, with oval or long body and lamellated antennae with 8–100 segments (Tashiro 1990). Larvae are usually white or yellowish, with an amber-yellow, brown, or black head.

FIGURE 15.1   Adults of *D. abderus* a) and *P. triticophaga* b). (Photos courtesy of Paulo R. V. S. Pereira.)

Melolonthidae adults (sensu Endrödi) have a proportionally small head in relation to the body that is usually oval and robust (rarely flat and thin). Their distinguishing features from other Scarabaeoidea are as follows: the antennal safe is much shorter than the blight; the antennae have three to seven long and flat articles, with small blades that are able to move, the surface of which presents a bright aspect scattered setae; the respiratory stigmas of the last three abdominal segments are placed at the lateral potion of the sternites and at least the last one is exposed when the elytra are at rest; tarsi are pentamerans, i.e., divided in five segments, and tarsal claws are well developed; the general coloration is variable; body length varies from 3 to 170 mm, and, often, present accentuated sexual dimorphism (Figure 15.1).

Melolonthidae larvae are typically scarabeiform with three pairs well-developed legs, with each leg having four differentiated articles, and with very apparent tarsungulus; antennae with four segments, the last one very conspicuous, maxillary feelers with four articles, and jaws with ventral process; epipharynx without epitome; one pair of thoracic respiratory stigmas and eight pairs of abdominal stigmas of the cribriform type (Morón et al. 1997; Morón 2004) (Figure 15.2).

In temperate weather, Melolonthidae species tend to be univoltine (having one generation per year) or have one generation every 2–4 years. On tropical areas there is a tendency to be multivoltines, but some may be univoltines (Luginbill and Painter 1953; Morón 1986). Generally the biological cycle of these insects is synchronized with environmental conditions.

In Brazil, the beginning of each generation of rhizophagous species varies depending on the weather. In regions where there is a dry season, such as north of Paraná, south of Mato Grosso do Sul, and Cerrado, it starts with the beginning of the rainy season, when adults leave the soil in flocks, usually at twilight or night time, for mating and, in some cases, feeding (Figure 15.3). Oviposition is done in the soil with larvae passing through three instars, with the last instar showing diapause or inactivity of variable duration, depending on the temperature and the water regimen. Pupation also occurs in the soil.

In south of Brazil, *D. abderus* and *P. triticophaga* are adapted to the temperate weather, with winters that can be restrictive. Usually, intense cold decrease larvae activity in the soil, which resume

FIGURE 15.2   Larvae of *D. abderus* a) *P. triticophaga* b) and *P. cuyabana* c). (Photos courtesy of Paulo R. V. S. Pereira and Crébio J. Ávila.)

FIGURE 15.3   Night flight of *P. cuyabana*. (Courtesy of Walter S. Leal.)

on periods of mild temperature, resulting in accentuated damage to plants. In *P. triticophaga*, which presents one generation every 2 years (Figure 15.4) (Salvadori 1997, 1999b, 2000), larvae of the third instar stop eating in November and go through the pupal phase after 60 days; adults complete the cycle over the fall, when temperature declines, and stay inactive in the soil during winter (Salvadori 1997, 2000).

In general, white grubs species in Brazilian agroecosystems are univoltine, such as *P. cuyabana* (Oliveira et al. 1996, 2004), *D. abderus* (Silva and Salvadori 2004) (Figure 15.5), *A. bolboceridus* (Oliveira 2005), and *Liogenys* spp. In Paraná state, for example, the flock of *P. cuyabana* begin usually at the end of October, after rains, and may occur until the beginning of December, with its peak in the middle of November; the active larvae can be found feeding on roots from November to April, but, from the end of April/beginning of May, all larvae start diapause, staying in soil chambers until the pupae begin to appear, normally in the middle of September/October (Oliveira et al. 1997). However, in Mato Grosso do Sul state, for example, the new cycle begins in September/beginning of October when adults of *P. cuyabana*, *L. fuscus*, and *L. suturalis* begin to leave the soil (Oliveira et al. 2004; Barbosa et al. 2006; Santos et al. 2006).

In the Cerrado region, adults of *A. bolboceridus* also leave the soil after the first rains of September and October, and its larvae, active on the rainy period (October to April), become inactive on the driest season (April to September) (Oliveira 2005).

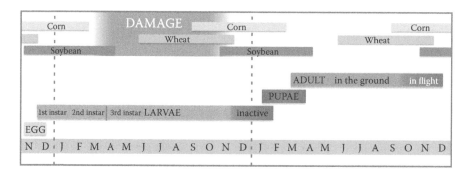

FIGURE 15.4   Biological cycle of *P. triticophaga*. (From Salvadori, J. R., and P. R. V. S. Pereira. 2006. Manejo Integrado de Corós em Trigo e em Culturas Associadas. Passo Fundo: Embrapa Trigo. http://www.cnpt.embrapa.br/publicacoes/p_co203.htm. With permission.)

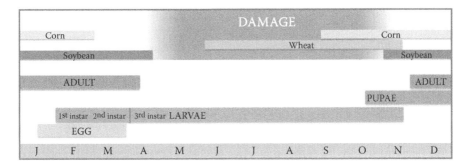

FIGURE 15.5   Biological cycle of *D. abderus.* (From Salvadori, J. R., and P. R. V. S. Pereira. 2006. Manejo Integrado de Corós em Trigo e em Culturas Associadas. Passo Fundo: Embrapa Trigo. http://www.cnpt.embrapa.br/publicacoes/ p_co203.htm. With permission.)

15.4 Strategies Used by the Group to Explore Food

The success of the holometabola group is due to the fact that immatures are better adapted to different ecological niches than adults (Terra 1991). Coleopteran adults and immatures, both chewers, in general explore distinct niches. The Scarabaeoidea adults are beetles that usually feed on plant tissues such as leaves, flowers, and fruits (Tashiro 1990). Larvae have strongly sclerotized and often asymmetrical jaws, allowing to explore different forms of food resources.

In general, edaphic melolonthidians not only share with other families of the order this strategy of exploring distinct niches at the adult and immature phases but also make use of two other strategies to explore food resources: polyphagy and a long biological cycle. However, the success of each species is closely linked to localization, selection, and utilization of hosts by adults and larvae, as well as to other biotic and abiotic factors. The diversity of eating habits of melolonthidians makes the group extremely important for Brazilian agroecosystems, whether for the damages caused to plants by the rhizophagous species or for the benefits to the soil quality promoted by the species classified as "soil engineers."

Melolonthidian larvae, saprophagous or phytophagous, may affect the chemical features of the soil and, therefore, indirectly, the availability of their own feeding resource. They need to consume 45–80 times their body weight from the food substrate to fully develop (Morón 1987), implying that for each gram of larva present in the soil, 63 g of substrate is needed. Hence, almost 60 g of stool enriched with bacteria and nitrogenated products that are easily assimilated are recycled per gram of larva (Morón and Rojas 2001). Larvae of *P. cuyabana* on the third instar, fed with soybean roots, sunflower, or *Crotalaria juncea*, weighing in average 0.8 to 1 g and can consume more than 30 times their biomass, return around 16–20% of this consumption to the soil in the form of feces (Oliveira 1997).

15.4.1 Localization and Selection of Host Plant by Rhizophagous Insects

Similar to other coleopterans, the behavior for host plant finding by adults, either for feeding or oviposition, can have a great influence on the distribution and survival of their progeny. This choice is important for determining what kind of feeding resource will be available to the larvae, which has limited moving ability. For instance, third instar larvae of *P. cuyabana*, with greater locomotion capability, prefer soybean roots as a more suitable food for their development, and avoid cotton roots that cause death of first instars (Oliveira 1997). However, farms with monocultures have limited food resource choices. In such situations, *P. cuyabana* females deposit fewer eggs near roots of hosts that are less appropriate for larval development, such as cotton and *Crotalaria spectabilis* (Oliveira et al. 2007).

During flight, female adults of *P. cuyabana* select the more conspicuous (attractive) plants to land on and attract males by the release of sexual pheromones. This behavior causes aggregation of adults in certain sites, and because females oviposit near the place of copulation, eggs and larval density decreases as the distance from the sites of aggregation of adults increases (Garcia et al. 2003).

Longevity and fertility of adults of *Popilia japonica* Newman (Japanese beetle) is affected by their feeding on different species or cultivars of host plants (Ladd 1987a). Adults are attracted by a wide range of plant species, and the acceptance or rejection of the hosts depends on the chemical stimulus from the leaf surface (Potter and Held 2002). One of the most effective compounds is a mixture of phenylpropionate, eugenol, and geraniol at 3:7:3 (Ladd and Mcgovern 1980). However, geraniol, phenylpropionate, and eugenol are not found in many of their favorite hosts (Loughrin et al. 1995, 1997). The Japanese beetle explores volatiles induced by feeding, such as aggregation kairomones; Potter and Held (2002) reported that plants whose leaves were damaged either by beetles or by caterpillars attracted more adults than intact or artificially damaged plants.

P. cuyabana, like *P. japonica*, is polyphagous and presents different levels of preference for some host plants. Males usually do not feed and some females eat leaves after copulation. The amount of leaves eaten and the proportion of females that feed vary according to the plant species (Oliveira et al. 1996; Oliveira and Garcia 2003) and, generally, are lower on less appropriate hosts for larvae development (e.g., cotton). Ingestion of leaves by females, even in small amounts, seems to be associated with the need for supplementary energy for reproduction. Around 52% of females of *P. cuyabana* that oviposit eat at least once after copulation, and most lay more eggs than the ones that never eat. However, females that never eat are able to produce fertile eggs (Oliveira and Garcia 2003).

Once on the soil, the search and exploration of food by rhizophagous larvae are influenced by the physical characteristics of the soil and the olfactory and gustatory stimulus coming from plants. Edaphic herbivores do not find roots by chance; rather, they orient themselves toward the roots by using semiochemicals that allow distinguishing between suitable and unsuitable plants. Secondary metabolic compounds released in the rhizosphere (e.g., alcohols, esters, and aldehydes) affect localization and recognition of host plants, with 80% having attractive properties. Insects that feed on a limited range of plants tend to explore specific metabolic compounds of the host, while nonspecialist herbivores seem to use more general semiochemicals (Johnson and Gregory 2006). Twenty studies cited by these authors concluded that CO_2 is the plant's main primary metabolic compound that allows insects to locate roots, although the emission of CO_2 by roots are too variable to allow a precise localization. Besides the lack of specificity, CO_2 gradients emitted by roots do not last for long periods, and vertical gradients tend to be stronger than horizontal gradients.

Many chemical compounds present in roots are not related to host localization in the rhizosphere, but they may stimulate insects to consume greater amounts of roots. When insects reach the roots, chemical substances act as phagostimulants (48% of compound are sugars) or feeding deterrents (mainly phenolic) (Johnson and Gregory 2006). For larvae of *P. japonica*, saccharose, maltose, fructose, glucose, and trehalose are important phagostimulants (Ladd 1988).

15.4.2 Food Exploration by Rhizophagous Melolonthidians

By exploring different types of feeding resources, adults and larvae of Melolonthinae decrease interspecific competition, increasing the chances of success for the species. However, despite this apparent separation, the performance of one phase is linked to the performance and behavior of the other. Adults of Cetoniinae and Rutelinae are common visitors of flowers where they consume nectar and pollen, while larvae of many species live in fallen logs or feed from humus and leaf litter (Berenbaum et al. 1998). Larvae of some Rutelinae and Dynastinae feed from decomposing matter and, rarely, from roots; larvae of Melolonthinae feed from roots, bulbs, tubers, and decomposing matter (Oliveira et al. 2003). Some species may change their eating habits during larval development, behaving as saprophagous during the first instar, and then consuming roots and gradually more fibrous and hard underground stalks at the last instars, finally behaving as strictly rhizophagous. Other species change in accordance to the available resource and are classified as facultative. For example, if eggs of some Dynastinae are deposited in soils rich in organic matter, their larvae will develop completely as saprophagous; if, however, the larvae begin their development in soil poor in humus, but with great offer of roots, they behave as rhizophagous during the three larval instars (Morón and Rojas 2001). Larvae of *A. bolboceridus* (Dynastinae) feed from roots of many plants at all stages; however, field observations have suggested that this species is able to survive long periods without eating or feeding on organic matter of vegetable origin (Oliveira 2005), most likely at the absence of its hosts.

Adults of Dynastinae usually attack stalks and roots in search of sap, while larvae have assorted habits, feeding from feces and decomposing vegetable matter, as well as roots of live plants (Tashiro 1990). Adults of *E. humilis* gnaw and dilacerate underground plant parts and their larvae feed on roots (Ferreira and Barrigossi 2004). Adults of other species of Dynastinae, such as *D. abderus*, do not feed, and females prefer to oviposit in areas with plant residues that supply feeding resources to the first instars. As they develop, larvae from this species start to behave like the phytophagous species, especially third instars, which feed on roots, but also consume seeds and the aerial part of small plants, which they pull within the soil after devouring the underground part of many wild and cultivated plants (Silva and Salvadori 2004). Larvae of *Cyclocephala* are root feeders, saprophagic, or have facultative eating habits (Morón 2010). With preference for decomposing organic matter, some species cause little or no damage to plants (Gassen 1999; Salvadori 1999a; Salvadori and Pereira 2006), while others appear associated with damage to several crops (Alzugaray et al. 1999; Morón et al. 2010; Villalobos-Hernández and Núñez-Valdez 2010). Some adults may be phyllophagous (Pérez-Domínguez et al. 2010) and also have an important role as pollinators (Cavalcante et al. 1999).

Adults of Melolonthinae and Rutelinae, known to be exclusively herbivorous, represent a different adaptive group, with a considerable body size. Larvae of Melolonthinae attack roots from grass, legumes, and other cultivated plants, as well as bushes and trees, while their adults devour leaves, flowers, and fruits (Tashiro 1990).

Members of *Phyllophaga* (Melolonthidae) are mainly associated with dicotyledons, and references on this genus with regard to monocotyledons and gymnosperms are rare (Morón 1986). However, in Brazil, many polyphagous species such as *P. triticophaga* and *P. cuyabana* occur in grass (Salvadori 2000; Oliveira et al. 2004). The Japanese beetle *P. japonica* (Rutelinae) feeds on leaves, flowers, and fruits of around 300 species of plants that belong to 79 families (Ladd 1987a,b, 1989), while its larvae feed on roots.

15.5 Impact of Environmental Factors on Food Exploration and Performance of Larvae

Because of their peculiar habits, rhizophagous melolonthidians are greatly affected by environmental factors, such as those that affect particularly the aerial part of plants on which adults feed on and those that affect the rhizosphere where larvae forage and live most of the time. The physical and chemical features of the soil influence the occurrence and abundance of rhizophagous insects: directly by influencing, for example, their survival, spatial distribution, and behavior, and indirectly when plants serve as food. Although there are similarities concerning how insects locate their host plants above or under the surface, the soil represents a much more complex environment and its nature (e.g., porosity, humidity, and density) is critical because it affects not only the mobility of the insect but also the diffusion of semiochemicals from roots (Johnson and Gregory 2006).

Besides abiotic factors inherent to the environment in which they live, such as temperature and soil humidity, survival of rhizophagous insects also depends on biotic factors such as natural enemies. To certain species, a significant part of larval mortality in the field is due to factors dependent on their density (Brown and Gange 1990). High densities of larvae cause significant reduction on growth rates because of the direct competition among larvae for the available food (Réginere et al. 1981b).

Soil structure is critical in determining the mobility and survival of rhizophagous insects. For scarabaeid larvae (Réginere et al. 1981a), survival is higher in soil of a thin texture, which retains humidity and reduces the risk of desiccation. It also has been suggested that the friction of sand particles may cause internal injury and reduce survival of digging larvae (Turpin and Peters 1971). Soil compacting may reduce survival, representing a physical barrier to the motion of larvae within the soil (Strnad and Bergman 1987), eventually impairing the availability and access to roots. The effect of the physical features of the soil over the edaphic larvae, however, depends on the behavior of each particular species. The white grub-of-pastures *D. abderus*, for example, seems not to be affected by compacted soils, probably because they build a permanent gallery that allows access to food (Torres et al. 1976). Besides, there are evidences that females prefer nonresolved soils for the construction of galleries that will serve as sites of oviposition and initial larval development (Silva et al. 1994). These features enabled this species of white grub, which originated from the native fields of the South American Pampas region, to perfectly adapt to the grain production systems where soil is not disturbed with the no-tillage cultivation system. On the other hand, the white grub-of-wheat, *P. triticophaga*, and the white grub-of-soybean, *P. cuyabana*, occur indistinctly in soils conventionally prepared for sowing, and in direct planting systems (Oliveira et al. 2000; Salvadori 2000). These species do not build permanent tunnels and live near the rhizosphere and the soil surface. In farms under direct planting, *P. triticophaga* moves preferentially along the line of sowing and concentrates in less crowded areas, along the terraces (Salvadori 2000; Salvadori and Silva 2004; Salvadori and Pereira 2006).

Many studies have shown that the growth rate of rhizophagous insects increases with the increase in soil temperature (Réginere et al. 1981c; Potter and Gordon 1984; Jackson and Elliot 1988). In species that grow during the hot season, soil temperature is important to determine the size of the larval populations and the probability of survival over winter (Brown and Gange 1990). The activity of larvae of *P. cuyabana* seems to be negatively affected by low temperatures and usually stay inactive in chambers in the soil, without feeding throughout the winter (Santos 1992; Oliveira 1997).

D. abderus and *P. triticophaga*, very common species in the south of Brazil, are adapted to low temperatures. Larvae feed from the end of fall until the beginning of spring, with the peak of food consuming matching the coldest season. However, during winter, it is normal that food consumption fluctuates according to temperature variability. Thus, in periods of extreme temperatures (close to 0°C), larvae decrease their activity to start over again, many times and more intensively, when the cold weather mitigates. Only exceptionally, when there are many consecutive days and nights with temperatures close to or below 0°C, larval death occurs (Salvadori and Silva 2004).

The consumption of food by third instars of *Sericesthis nigrolineata* Boisd. (Melolonthinae) is incremented by the increase in temperature within the range of 4–30°C; under the minimum limit, larvae stop eating; however, on the maximum limit, they perish. Larvae at the end of the third instar lose weight before entering the prepupal phase; mature larvae do not enter diapause (Ridsdill-Smith et al. 1974).

Humidity is the most important soil property for the survival and abundance of rhizophagous insects (Brown and Gange 1990). Larvae of *P. triticophaga*, in periods of water deficiency for the plants, search for more humid soil layers, deep in the soil profile, usually inside a chamber with smooth inner walls, possibly to avoid the effects of dehydration. When the unfavorable period is prolonged, larvae perish, and the performance of survivors is affected (Salvadori 2000; Salvadori and Silva 2004).

Information about the interaction of rhizophagous insects with the soil nutritional content is conflicting and, many times, the effect does not happen directly. The application of fertilizers, for example, apparently indirectly affects the rhizophagous herbivores through the root system, although the effects of the soil acidification should not be neglected (Brown and Gange 1990).

Prestidge et al. (1985) did not find any relation between the application of fertilizers and the feeding of scarabaeid larvae. However, for Spike and Tollefson (1988), the time of nitrogen application in relation to the establishment of larvae can be crucial. If the fertilizer is added before establishment of larvae, then the propagation of the root system may result in a greater feeding supply for the larvae, with greater survival and increase in damage to plants. If nitrogen addition happens after their establishment, then damages are lower.

15.6 Insect Adaptations and Responses to Variations in Abiotic and Biotic Factors

The pattern of answers from phytophagous insects to the spreading of resources in the field depend on: (a) the quality of this resource in relation to reproduction and survival (Bach 1988); (b) the manner of search for host plants (Ralph 1977; Bach 1988; Grez and Gonzalez 1995; Matter 1996); (c) the variation in resource concentration (Kareiva 1983); (d) the competition of individuals for resources (Adesiyun 1978; McLain 1981); and (e) the attack of natural enemies (Price et al. 1980).

Rhizophagous insects have a feeding source that, although abundant, may have exceptionally low quality. As an additional strategy to explore this resource, they often have long life cycles and thus tend to live in established plant communities. In these communities, they show highly aggregated distributions, usually corresponding to the negative binomial. This distribution results from the fact that the soil is a very heterogeneous environment and that insects are highly dependent on the texture, humidity, and temperature of the ambient soil, causing aggregation in favorable places. This grouping means that it is difficult to detect and quantify them by conventional sampling methods. Aggregated distributions may also result from the female's behavior concerning oviposition. The selection of the place for oviposition is critical; just-born insects are relatively immobile and must immediately find their source of food (Brown and Gange 1990).

The female preference for oviposition in some species, growth, survival, and reproduction of the offspring in those plants (performance) has been the central problems of the theory of insect–plant relationships (Thompson 1988). This author emphasizes that many studies suggest hypotheses that the relations between the preference for oviposition and the performance of the progeny can vary under ecological conditions and selective pressures. Larvae of *P. japonica* are polyphagous, but due to their restricted mobility, they stay in roots of plants near the place of oviposition (Potter and Held 2002); a similar thing happens with *P. cuyabana* (Garcia et al. 2003).

Species of Melolonthidae that feed from plants have in common the fact of usually being polyphagous, exploring plants of many families that favor their survival. However, this behavior of circumstantial monophagy or oligophagy at the larval phase, due to the adult behavior and the low mobility of larvae, is common among melolonthidians, especially on agricultural systems where the plant community tends to be less spatially variable and timely than natural systems.

These habits can also influence the physiology of larvae. The study by Potter and Held (2002) shows that the intestine of *P. japonica* larvae is alkaline and some of its enzymes, such as P450, are passively induced and have higher activities under facultative polyphagy than under monophagy. It also contains proteolytic enzymes that can be inhibited *in vitro*. Chronic ingestion of soybean trypsin results in elevated mortality of *P. japonica* larvae (Broadway and Villani 1995).

If the behavior of the adult influences the availability of larval food, then the quantity and quality of the diet of the insect at the larval phase can affect its survival and final weight, and thus the size of the adult. The adequacy of one plant to larval development is a result of many variables, including chemical and physical proprieties, microhabitat, and level of infestation (Jaenike 1978). The survival of larvae of *P. cuyabana*, for example, is affected by the diet, and their sensibility to allelochemicals in the food decreases as they develop. The ecological efficiency of larvae of *P. cuyabana* fed on a less appropriate host is reduced basically into two levels: exploration of the resource (consumption) and efficiency of its use. Usually, when they feed during the first instar on unsuitable hosts, their larvae die; however, third instars that consume roots of these plants are able to survive and reach diapause, although with lower final weight compared with those fed suitable hosts (Oliveira 1997).

Larvae of *A. bolboceridus* that had access to a greater quantity of food and/or nutritionally more appropriate food are able to accumulate more reserves and develop into bigger adults, with variations up to 80% in length (Oliveira 2005).

Many rhizophagous melolonthidians use the diapause strategy as a physiological answer to adverse conditions, such as unfavorable temperature and humidity. Diapause in larvae of third instars, within the *Phyllophaga* genus, is common during winter, an ecological strategy to survive unfavorable conditions (Ritcher 1958; Lim et al. 1980; Morón 1986). Diapause was demonstrated in species of Melolonthidae that occur in Brazil, such as *P. cuyabana* (Santos 1992) and *Phytalus sanctipauli* Blanch. (Redaelli et al. 1996), the last one possibly *P. triticophaga* (Salvadori 1999b).

Many Coleoptera do not feed during diapause or consume only small amounts of food occasionally (Guerra and Bishop 1962; Siew 1966; Hodek 1967), which reduces the chances for larval survival during winter. According to Tauber et al. (1986), part of the food eaten by insects before the beginning of diapause is accumulated as energy, in the form of fat, to be consumed during this period until reproduction.

Diapause in third instars of *Phyllophaga* is common over fall and winter and is an ecological survival strategy to face unfavorable conditions (Ritcher 1958; Lim et al. 1980; Morón 1986). Larvae gradually decrease their activities and cease eating despite food availability. At the beginning of this phase, larvae get deeper into the soil and prepare an individual and impermeable chamber, probably shaped with saliva (Morón 1986; Santos 1992), where they stay until adult sexual maturity. Small larvae of *P. cuyabana*, from feeding on unsuitable foods, are able to survive over the long period of diapause (Santos 1992; Oliveira 1997), developing into adults with low fertility, which will have an impact on the subsequent generation (Slansky and Scriber 1985; Honek 1993).

15.7 Conclusions and Suggestions for Research

Rhizophagous insects may cause considerable damage to agriculture by feeding on economically important crops, but also can be of use to control harmful plants (weeds) (Johnson and Gregory 2006). Among these insects, the melolonthidians are emphasized by the great number of species that occur in agroecosystems. Despite their importance, the biological aspects of most of the species that occur as pests of agricultural systems in Brazil continue to be little investigated.

The life strategies of these insects, including biological cycles (usually long) associated with a great diversity of eating habits of their immature and adult phases, make this group capable of exploring very distinct agroecosystems, increasing their chances of survival. Of particular importance are the feeding relations of rhizophagous insects in agricultural and cattle raising systems that involve rotation/succession of cultures, farm integration, and cattle raising and conservational handling of soil—all with ample and diversified nutritional possibilities that can determine the composition of the insect fauna. A more profound knowledge of the bioecology and nutrition of this group will certainly be of great value for defining managing strategies for rhizophagous insects in places where they often cause damage to crops.

REFERENCES

Adesiyun, A. A. 1978. Effects of seedling, density and spatial distribution of oat plants on colonization and development of *Oscinella frit* (Diptera Choloropidae). *J. Appl. Ecol.* 15:797–808.

Alzugaray, R., M. S. Zerbino, E. Morelli, E. Castiglioni, and A. Ribeiro. 1999. Manejo de gusanos blancos en cultivos cerealeros en Uruguay. In *Memórias 4ª Reunião Latino-Americana Scarabaeoidologia*, 83–92. Universidade Federal de Viçosa, Viçosa, MG, Brazil.

Ávila, C. J., and S. A. Gómez. 2001. Ocorrência de pragas de solo no Estado de Mato Grosso do Sul. In *Anais 8ª Reunião Sul-Brasileira Pragas Solo*, 34–61. Embrapa Soja, Londina, Londrina, PR, Brazil.

Bach, C. E. 1988. Patch size and herbivory: Mechanisms. *Ecology* 69:1103–17.

Barbosa, C. L., S. R. Rodrigues, A. Puker, and A. R. Abot. 2006. Estudo do comportamento reprodutivo de *Liogenys fuscus* (Coleoptera: Melolonthidae). Resumos 21º Cong. Bras. Entomol. Recife, PE, Brazil.

Bento, J. M. S., J. R. P. Parra, R. M. C. Muchovej, M. S. Araújo, and T. M. C. D. Lucia. 2004. Interações entre microrganismos edáficos e pragas de solo. In *Pragas de Solo*, ed. J. R. Salvadori, C. J. Ávila, and M. T. B. da Silva, 99–132. Embrapa Trigo, Passo Fundo, RS, Brazil.

Berenbaum, M. R., F. J. Radovsky, and V. H. Resh. 1998. Chemical ecology of phytophagous scarab beetles. *Annu. Rev. Entomol.* 43:39–61.

Broadway, R. M., and M. G. Villani. 1995. Does host range influence susceptibility of herbivorous insects to non-host proteinase inhibitors? *Entomol. Exp. Appl.* 76:303–12.

Brown, W. R., and A. C. Gange. 1990. Insect herbivory below ground. *Adv. Ecol. Res.* 20:1–58.

Cavalcante, T. R. M., M. F. Vieira, J. C. Zanúncio, and G. B. Freitas. 1999. Polinização da gravioleira (*Annona muricata* L. Annonaceae) por *Cyclocephala* spp. (Coleoptera: Scarabaeoidea) em Visconde do Rio Branco, MG, e Uma, BA. In *Memórias 4ª Reunião Latino-Americana Scarabaeoidologia*, 64–5. Universidade Federal de Viçosa, Viçosa, MG, Brazil.

Corso, I. C., L. J. Oliveira, and M. L. B. do Amaral. 1991. Ação de inseticidas sobre "coró da soja" (II) (Coleoptera: Scarabaeidae). In *Ata 3ª Reunião Sul-Brasileira Insetos Solo*, 10. Epagri-CPPP. Chapecó, SC, Brazil.

Crowson, R. A. 1981. *The Biology of Coleoptera*. London: Academic Press.

Curl, E. A., and B. Truelove. 1986. *The Rhizosphere*. Berlin: Springer-Verlag.

Dethier, V. G. 1970. Some general considerations of insects' responses to the chemicals in food plants. In *Control of Insect Behavior by Natural Products*, ed. D. L. Wood, R. M. Silverstein, and M. Nakajima, 21–28. London: Academic Press.

Endrödi, S. 1966. Monographie der Dynastinae (Coleoptera: Lamellicornia). I. Teil. *Entomol. Abh. Mus. Tierkunde* 33:1–457.

Ferreira, E., and J. A. F. Barrigossi. 2004. *Cultivo do Arroz Irrigado no Estado do Tocantins.* Goiânia, GO, Brazil: Embrapa Arroz e Feijão.

Ferreira, E., and J. A. F. Barrigossi. 2006. *Insetos Orizívoros da Parte Subterrânea.* Goiânia, GO, Brazil: Embrapa Arroz e Feijão.

Futuyma, D. J. 1992. *Biologia Evolutiva,* 2., ed. Ribeirão Preto, SP, Brazil: Sociedade Brasileira de Genética.

Garcia, M. A., L. J. Oliveira, and M. C. N. Oliveira. 2003. Aggregation behavior of *Phyllophaga cuyabana* (Moser) (Coleoptera: Melolonthidae): Relationships between sites chosen for mating and offspring distribution. *Neotrop. Entomol.* 3:537–42.

Gassen, D. N. 1999. Benefícios de escarabeídeos em lavouras sob plantio direto. *Memórias 4ª Reunião Latino-Americana Scarabaeoidologia,* 123–32. Viçosa, MG, Brazil.

Grez, A. A., and R. H. González. 1995. Resource concentration hypothesis: Effect of host plant patch size on density of herbivorous insects. *Oecologia* 103:471–4.

Guerra, A. A., and J. L. Bishop. 1962. The effect of aestivation on sexual maturation in female alfalfa weevil (*Hypera postica*). *J. Econ. Entomol.* 55:747–9.

Hodek, I. 1967. Bionomics and ecology of predaceous Coccinelidae. *Annu. Rev. Entomol.* 12:79–104.

Honek, A. 1993. Intraspecific variation in body size and fecundity in insects: A general relationship. *Oikos* 66:483–92.

Honer, M. R., I. Bianchin, and A. Gomes. 1992. Com besouro africano, controle rápido e eficiente. In *Manual de Controle Biológico,* 19–20. Rio de Janeiro: Sociedade Nacional de Agricultura.

Jackai, L. E. N., A. R. Panizzi, G. G. Kundu, and K. P. Sribastava. 1990. Insect pests of soybean in the tropics. In *Insect Pests of Foods Legumes,* ed. S. R. Singh, 91–156. New York: J. Wiley.

Jackson, J. J., and N. C. Elliott. 1988. Temperature-dependent development of immature stages of the western corn rootworm, *Diabrotica virgifera virgifera* (Coleoptera: Chrysomelidae). *Environ. Entomol.* 17:166–71.

Jaenike, J. 1978. On optimal oviposition behavior in phytophagous insects. *Theor. Popul. Biol.* 14:350–6.

Johnson, S. N., and P. J. Gregory. 2006. Chemically-mediated host-plant location and selection by rootfeeding insects. *Physiol. Entomol.* 31:1–13.

Kareiva, P. 1983. Influence of vegetational texture on herbivore populations resource concentration and herbivore movement. In *Variable Plants and Herbivores in Natural and Managed Systems,* ed. R. F. Denno and M. S. McClure, 259–89. New York: Academic.

Ladd, Jr., T. L. 1987a. Influence of food, age, and mating on production of fertile eggs by Japanese beetles (Coleoptera: Scarabaeidae). *J. Econ. Entomol.* 80:93–5.

Ladd, Jr., T. L. 1987b. Japanese beetles (Coleoptera: Scarabaeidae): Influence of favored food plants on feeding response. *J. Econ. Entomol.* 80:1014–7.

Ladd, Jr., T. L. 1988. Japanese beetles (Coleoptera: Scarabaeidae): Influence of sugars on feeding response of larvae. *J. Econ. Entomol.* 81:1390–3.

Ladd, Jr., T. L. 1989. Japanese beetles (Coleoptera: Scarabaeidae): Feeding by adults on minor host and non-host plants. *J. Econ. Entomol.* 82:1616–9.

Ladd, Jr., T. L., and T. P. McGovern. 1980. Japanese beetle: A superior attractant, phenethyl propionate + eugenol + geraniol 3:7:3. *J. Econ. Entomol.* 74:665–7.

Lavelle, P., and A. V. Spain. 2001. *Soil Ecology.* Dordrecht: Kluwer Academic.

Lawrence, J. F., and E. B. Britton. 1994. *Australian Beetles.* Melbourne: Melbourne University.

Lim, K. P., R. K. Stewart, and W. N. Yule. 1980. A historical review of the bionomics and control of *Phyllophaga anxia* (LeConte) (Coleoptera: Scarabaeidae) with special reference to Quebec. *Ann. Entomol. Soc. Am.* 25:163–78.

Loughrin, J. H., D. A. Potter, and T. R. Hamilton-Kemp. 1995. Volatile compounds induced by herbivory act as aggregation kairomones for Japanese beetle (*Popilia japonica* Newman). *J. Chem. Ecol.* 21:1457–67.

Loughrin, J. H., D. A. Potter, T. R. Hamilton-Kemp, and M. E. Byers. 1997. Response of Japanese beetles (Coleoptera: Scarabaeidae) of leaf volatiles of susceptible and resistant maple species. *Environ. Entomol.* 26:334–42.

Luginbill, P., and H. R. Painter. 1953. May beetle of the United States and Canada. *U.S. Dep. Agric. Tech. Bull.* 1060:1–103.

Matter, S. F. 1996. Interpatch movement of the red milkweed beetle, *Tetraopes tetraophthalmus*: Individual responses to patch size and isolation. *Oecologia* 105:447–53.

McKey, D. 1979. The distribution of secondary compounds within plants. In: *Herbivores—Their Interaction with Secondary Plant Metabolites*, ed. G. A. Rosenthal, and D. H. Janzen), 55–133. London: Academic Press.

McLain, D. K. 1981. Resource partitioning by three species of hemipteran herbivores on the basis of host plant density. *Oecologia* 48:414–7.

Morón, M. A. 2010. Diversidad y distribución del complejo "gallina ciega" (Coleoptera: Scarabaeoidea). In *Plagas del Suelo*, ed. L. A. Rodríguez del Bosque, and M. A. Morón, 41–63. México: Mundi-Prensa.

Morón, M. A. 1986. *El género Phyllophaga en México—Morfología, Distribución y Sistemática Supraespecifica (Insecta: Coleoptera)*. México: Instituto de Ecología.

Morón, M. A. 1987. Los estados inmaduros de *Dynastes hyllus* Chevrolat (Coleoptera: Melolonthidae, Dynastinae) con observaciones sobre su biología y el crecimiento alométrico del imago. *Folia Entomol. Mex.* 72:33–74.

Morón, M. A. 1996. *Los Coleoptera Melolonthidae Edafícolas en América Latina*. Puebla: Dica-IC-Ben. Univ. Aut. Puebla: Sociedad Mexicana de Entomología, México.

Morón, M. A. 2004. Melolontídeos edafícolas. In *Pragas de Solo*, ed. J. R. Salvadori, C. J. Ávila, and M. T. B. da Silva, 133–49. Embrapa Trigo, Passo Fundo, RS, Brazil.

Morón, M. A., and C. V. Rojas. 2001. Las especies de *Phyllophaga* en Brasil (Coleoptera: Melolonthidae; Melolonthinae). In *Anais 8ª Reunião Sul-Brasileira Pragas Solo*, 217–21. Embrapa Soja, Londrina, PR, Brazil.

Morón, M. A., B. C. Ratcliffe, and C. Deloya. 1997. *Atlas de los Escarabajos de México*. Xalapa: Sociedad Mexicana de Entomología-Conabio.

Morón, M. A., L. A. Rodríguez del Bosque, A. Aragón, and C. Ramírez-Salinas. 2010. Biología y hábitos de coleópteros escarabaeoideos. In *Plagas del Suelo*, ed. L. A. Rodríguez del Bosque and M. A. Morón, 65–82. México: Mundi-Prensa.

Nunes, Jr., J., L. J. Oliveira, I. C. Corso, and L. C. Farias. 2000. Controle químico de corós (Scarabaeoidea) em soja. In *Resumos 22ª Reunião Pesquisa Soja da Região Central do Brasil*, 58–9. Embrapa Soja, Londrina, PR, Brazil.

Oliveira, C. M. 2005. *Aspectos Bioecológicos do Coró-das-Hortaliças Aegopsis bolboceridus (Thomson) (Coleoptera: Melolonthidae) no Cerrado do Brasil Central*. Planaltina: Embrapa Cerrados.

Oliveira, L. J. 1997. Ecologia comportamental e de interações com plantas hospedeiras em Phyllophaga cuyabana (Moser) (Coleoptera: Melolonthidae, Melolonthinae) e implicações para o seu manejo em cultura de soja. PhD thesis, Campinas, SP, Brazil: Universidade de Campinas.

Oliveira, L. J., B. Santos, J. R. P. Parra, and C. B. Hoffmann-Campo. 2004. Coró-da-soja. In *Pragas de Solo*, ed. J. R. Salvadori, C. J. Ávila, and M. T. B. da Silva, 151–76. Embrapa Trigo, Passo Fundo, RS, Brazil.

Oliveira, L. J., B. Santos, J. R. P. Parra, M. L. B. Amaral, and D. C. Magri. 1996. Ciclo biológico de *Phyllophaga cuyabana* (Moser) (Scarabaeidae: Melolonthinae). *An. Soc. Entomol. Brasil* 25:433–9.

Oliveira, L. J., C. B. Hoffmann-Campo, and M. A. Garcia. 2000. Effect of soil management on the white grub population and damage in soybean. *Pesq. Agropec. Bras.* 35:887–94.

Oliveira, L. J., G. G. Brown, and J. R. Salvadori. 2003. Corós como pragas e engenheiros do solo em agroecossistemas. In *O Uso da Macrofauna Edáfica na Agricultura do Século XXI: A Importância dos Engenheiros do Solo*, 76–86. Londrina, PR, Brazil: Embrapa Soja.

Oliveira, L. J., M. A. Garcia, C. B. Hoffmann-Campo, and M. L. B. do Amaral. 2007. Feeding and oviposition preference of *Phyllophaga cuyabana* (Moser) (Coleoptera: Melolonthidae) on several crops. *Neotrop. Entomol.* 36:759–64.

Oliveira, L. J., M. A. Garcia, C. B. Hoffmann-Campo, D. R. Sosa-Gomez, J. R. B. Farias, and I. C. Corso. 1997. *Coró-da-Soja Phyllophaga cuyabana*. Londrina, PR, Brazil: Embrapa Soja.

Oliveira, L. J., and M. A. Garcia. 2003. Flight, feeding and reproductive behavior of *Phyllophaga cuyabana* (Moser) (Coleoptera: Melolonthidae) adults. *Pesq. Agropec. Bras.* 38:179–86.

Paul, E. A., and F. E. Clark. 1996. *Soil Microbiology and Biochemistry*, 2nd ed. San Diego: Academic.

Pérez-Domíngues, J. F., M. B. Nájera-Rincón, and R. Alvarez-Zagoya. 2010. Plagas del suelo en Jalisco. In *Plagas del Suelo*, ed. L. A. Rodríguez del Bosque and M. A. Morón, 251–61. México: Mundi-Prensa.

Potter, D. A., and D. W. Held. 2002. Biology and management of the Japanese beetle. *Annu. Rev. Entomol.* 47:175–205.

Potter, D. A., and F. C. Gordon. 1984. Susceptibility of *Cyclocephala immaculata* (Coleoptera: Scarabaeidae) eggs and immatures to heat and drought in turf grass. *Environ. Entomol.* 13:794–9.

Prestidge, R. A., S. van der Zijpp, and D. Badan. 1985. Effects of plant species and fertilizers on grass grub larvae, *Costelytra zealandica. N . Z . J . Agric. Res.* 28, 409–17.

Price, P., C. E. Bouton, P. Gross, B. A. McPheron, J. N. Thompson, and A. E. Weis. 1980. Interaction among three trophic levels: Influences of plants on interactions between insect herbivores and natural enemies. *Annu. Rev. Ecol. Syst.* 11:41–65.

Ralph, C. P. 1977. Search behavior of large milkweed bug, *Oncopeltus fasciatus* (Hemiptera: Lygaeidae). *Ann. Entomol. Soc. Am.* 70:337–42.

Redaelli, L., L. M. G. Diefenbach, and D. N. Gassen. 1996. Morfologia dos órgãos internos de reprodução de *Phytallus sanctipauli* Blanch., 1850 (Coleoptera: Scarabaeidae: Melolonthinae). *Resumos 21º Congresso Brasileiro Zoologia*, 129. Porto Alegre, RS, Brazil.

Réginere, J., R. L. Rabb, and R. E. Stinner. 1981a. *Popilia japonica*: Effect of soil moisture and texture on survival and development of eggs and first instar grubs. *Environ. Entomol.* 10:654–60.

Réginere, J., R. L. Rabb, and R. E. Stinner. 1981b. *Popilia japonica*: Intraspecific competition among grubs. *Environ. Entomol.* 10:661–2.

Réginere, J., R. L. Rabb, and R. E. Stinner. 1981c. *Popilia japonica*: Simulation of temperature-dependent development of immatures, and prediction of adult emergence. *Environ. Entomol.* 10:290–6.

Ridsdill-Smith, T. J., M. R. Porter, and A. G. Furnival. 1974. Effects of temperature and developmental stage on feeding by larvae of *Sericesthis nigrolineata* (Coleoptera: Scarabaeidae). *Entomol. Exp. Appl.* 18:244–54.

Ritcher, P. O. 1958. Biology of Scarabaeidae. *Annu. Rev. Entomol.* 3:311–33.

Salvadori, J. R. 1997. *Manejo de Corós em Cereais de Inverno.* Passo Fundo: EMBRAPA-CNPT, RS, Brazil.

Salvadori, J. R. 1999a. Efeito de níveis de infestação do coró *Cyclocephala flavipennis* em trigo. In *Anais 18ª Reunião Nacional Pesquisa Trigo*, 570–2, vol. 2, Embrapa Trigo, Passo Fundo, RS, Brazil.

Salvadori, J. R. 1999b. Manejo do coró-do-trigo (*Phyllophaga triticophaga*) no Brasil. In *Memórias 4ª Reunião Latino-Americana Scarabaeoidologia*, 106–12, Universidade Federal de Viçosa, Viçosa, MG, Brazil.

Salvadori, J. R. 2000. *Coró-do-Trigo.* Passo Fundo, RS, Brazil: Embrapa Trigo.

Salvadori, J. R. 2001. Influência do manejo de solo e de plantas sobre corós rizófagos, em trigo. In *Anais 8ª Reunião Sul-Brasileira Pragas Solo*, 79–89. Embrapa Soja, Londrina, PR, Brazil.

Salvadori, J. R., and L. J. Oliveira. 2001. *Manejo de Corós em Lavouras sob Plantio Direto.* Passo Fundo, RS, Brazil: Embrapa Trigo.

Salvadori, J. R., and M. T. B. da Silva. 2004. Coró-do-trigo. In *Pragas de Solo*, ed. J. R. Salvadori, C. J. Ávila, and M. T. B. da Silva, 211–32, Embrapa Trigo, Passo Fundo, RS, Brazil.

Salvadori, J. R., and P. R. V. S. Pereira. 2006. *Manejo Integrado de Corós em Trigo e em Culturas Associadas.* Embrapa Trigo, Passo Fundo, RS, Brazil. http://www.cnpt.embrapa.br/publicacoes/p_co.htm.

Salvadori, J. R., M. A. Morón, and P. R. V. S. Pereira. 2006. Ocorrência de *Demodema brevitarsis* (Coleoptera: Melolonthidae) em soja e em outras culturas, no sul do Brasil. *Resumos Congresso Brasileiro Entomologia*, Recife, PE, Brazil.

Santos, B. 1992. Bioecologia de Phyllophaga cuyabana (Moser 1918) (Coleoptera: Scarabaeidae), praga do sistema radicular da soja [Glycine max (L.) Merrill, 1917]. Master dissertation, Piracicaba: Universidade de São Paulo, SP, Brazil.

Santos, V., A. C. V. Portela, D. J. Salvador, and C. J. Ávila. 2006. Período de emergência e atividade diária de vôo de adultos de *Liogenys suturalis* (Blanchard, 1851) (Coleoptera: Melolonthidae). *Anais 21º Congresso Brasileiro Entomologia*, Recife, PE, Brazil.

Siew, Y. C. 1966. Some physiological aspects of adult reproductive diapause in *Galleruca taneceti* (L.) (Coleoptera: Chrysomelidae). *Trans. R. Entomol. Soc. Lond.* 118:59–374.

Silva, M. T. B. da, and J. R. Salvadori. 2004. Coró-das-pastagens. In *Pragas de Solo*, ed. J. R. Salvadori, C. J. Ávila, and M. T. B. da Silva, 191–210. Embrapa Trigo, Passo Fundo, RS, Brazil.

Silva, M. T. B. da, V. A. Klein, D. Link, and D. J. Reinert. 1994. Influência de sistemas de manejo de solos na oviposição de *Diloboderus abderus* (Sturm) (Coleoptera: Melolonthidae). *An. Soc. Entomol. Bras.* 23:543–8.

Slansky, Jr., F., and J. M. Scriber. 1985. Food consumption and utilization. In *Comprehensive Insect Physiology Biochemistry and Pharmacology*, ed. G. A. Kerkut and L. I. Gilbert, 87–163. New York: Pergamon.

Spike, B. P., and J. J. Tollefson. 1988. Western corn rootworm (Coleoptera: Chrysomelidae) larval survival and damage potential to corn subjected to nitrogen and plant density treatments. *J. Econ. Entomol.* 81:1450–5.

Strnad, S. P., and M. K. Bergman. 1987. Movement of first-instar western corn rootworms (Coleoptera: Chrysomelidae) in soil. *Environ. Entomol.* 16:975–8.

Tashiro, H. 1990. Insecta: Coleoptera Scarabaeidae (Larvae). In *Soil Biology Guide*, ed. D. L. Dindal, 1191–210. New York: J. Wiley.

Tauber, M. J., C. A. Tauber, and S. Masaki. 1986. *Seasonal Adaptations of Insects*. New York: Oxford.

Terra, W. R. 1991. Digestão do alimento e suas implicações na biologia dos insetos. In *Ecologia Nutricional de Insetos e Suas Implicações no Manejo de Pragas*, ed. A. R. Panizzi and J. R. P. Parra, 67–99. São Paulo: Manole.

Thompson, H. N. 1988. Evolutionary ecology of the relationship between oviposition preference and performance of offspring in phytophagous insects. *Entomol. Exp. Appl.* 47:3–14.

Torres, C., L. Alvarado, C. Senigagliesi, R. Rossi, and H. Tejo. 1976. Oviposición de *Diloboderus abderus* (Sturm) en relación a la roturación del suelo. *IDIA B* 32:124–5.

Turpin, F. T., and D. C. Peters. 1971. Survival of southern and western corn rootworm larvae in relation to soil texture. *Entomol. Exp. Appl.* 64:1448–51.

Van Noordwijk, M., and G. Brouwer. 1997. Roots as sinks and sources of nutrients and carbon in agricultural systems. In *Soil Ecology in Sustainable Agricultural Systems*, ed. L. Brussard and R. Ferrera-Cerrado, 71–89. Boca Raton: Lewis.

Villalobos-Hernández, F. J. and M. E. Núñez-Valdez. 2010. Manejo sustentable. In *Plagas del Suelo*, ed. L. A. Rodríguez del Bosque and M. A. Morón, 215–36. México: Mundi-Prensa.

16

Gall-Inducing Insects: From Anatomy to Biodiversity

G. Wilson Fernandes, Marco A. A. Carneiro, and Rosy M. S. Isaias

CONTENTS

16.1 Introduction 369
16.2 Herbivore Insect Guilds 370
16.3 Gall-Inducing Insect Taxa 372
    16.3.1 Hemiptera 372
    16.3.2 Thysanoptera 374
    16.3.3 Coleoptera 374
    16.3.4 Hymenoptera 375
    16.3.5 Lepidoptera 375
    16.3.6 Diptera 376
16.4 Host Plant Taxa 377
16.5 Location and Choice of the Host Plant 377
16.6 Gall Morphology 379
16.7 Gall Anatomy and Physiology 381
16.8 Gall Development 383
16.9 Gall Classification 385
16.10 Adaptive Significance 387
16.11 Concluding Remarks 389
References 389

16.1 Introduction

The earliest records of galls date to the times of Hippocrates (460–377 BC), Theophrastus (371–286 BC), and Pliny the Elder (23–79 AD). In his *Historia Naturalis XXVI*, published in the first century, Plinius, known as "the Merciful," was the first to use the word "gall" to designate the structure induced in oak trees by wasps from the family Cynipidae (Meyer 1987). Although the emergence of insects from these structures has been described by these ancient authors, it was only in the 17th century, with the works of Marcello Malpighi (1628–1694), Anthony van Leeuwenhoek (1632–1723), and Jan Schwammerdam (1630–1680) that gall development was linked to the oviposition of an insect.

Galls, or vegetative tumors, are plant tissues or organs formed by hyperplasia (increased cell number) and/or hypertrophy (increased cell size) induced by parasitic or pathogenic organisms (Mani 1964; Dreger-Jauffret and Shorthouse 1992). Galls can be induced by a wide variety of organisms (Figure 16.1), including viruses, bacteria, fungi, algae, nematodes, rotifers, copepods, and plants from the family Loranthaceae (popularly known as mistletoes), but are mainly caused by insects (Mani 1964; Raman et al. 2005).

Among all herbivorous insects, gall-forming insects are the most sophisticated because they are able to control and redirect the host plant for their own benefit. Galls represent a fascinating natural phenomenon reflecting intimate interactions between organisms that have been shaped by organic evolution throughout millions of years (see Larew 1992; Labandeira et al. 1994; Labandeira and Phillips 2002;

FIGURE 16.1   Galls induced by distinct organisms. a) Insect-induced gall. b) Acari-induced gall. c) Ambrosia gall (induced by an insect symbiont fungus). d) Fungus-induced gall (Witch's broom). e) Nematoid-induced gall. f) Loranthaceae-induced gall (*Struthantus flexicaulis*).

Stone et al. 2008). Gall-forming insects are able to modify the host plant's growth pattern, altering the structure of the vegetative tissue and driving the host to produce a food source that is rich in nutrients and free from chemical defenses together with a protective structure that is isolated from the environment (Price et al. 1986, 1987).

Galls are also known and used for their pharmacological properties, which have been recognized since ancient times. Aleppo galls (spherical galls formed on the twigs of *Quercus infectoria* by gall-wasp larvae) contain 50% to 60% galactonic acid and significant levels of gallic and ellagic acids. These substances are used to treat diarrhea, oral swelling, and hemorrhoids. The commercial exploitation of galls dates to the 17th century, when pigments extracted from galls were used to dye hair and other tissues, and as writing ink (Fernandes and Martins 1985). In China, galls have been extensively used for more than 1000 years in medicine, industrially, and as human food. In South America, the indigenous Aguaruna-Jívaro people of the Peruvian Amazon use leaf galls from *Licania cecidiophora* (Chrysobalanaceae) to make necklaces (Berlin and Prance 1978).

Recently, interest in galls has increased because of their potential uses as biological control agents for invasive plants and as bioindicators of environmental quality and health (Fernandes 1987; Julião et al. 2005; Moreira et al. 2007; Fernandes et al. 2010). Additionally, several authors have suggested that the interaction between plants and gall-forming insects is ideal for testing hypotheses about ecological relationships (Fernandes and Price 1988; Price 2003). Gall-forming insects present certain methodological advantages as model organisms, primarily due to their sessile habit. Gall-forming insect communities frequently include many species from different orders; galls are conspicuous structures that are persistent on the plant and can be easily observed and collected. Further, the interactions between the inducing insects and other organisms can be easily manipulated (Fernandes and Price 1988; Stone and Schönrogge 2003).

16.2 Herbivore Insect Guilds

Herbivorous or phytophagous insects are those that consume living parts of plants. They make up the largest portion of all extant species diversity. Nearly 50% of all herbivorous organisms are insects

(Gullan and Cranston 2005). Herbivorous insects are found in the orders Phasmatodea, Orthoptera, Thysanoptera, Hemiptera, Coleoptera, Diptera, Lepidoptera, and Hymenoptera (Triplehorn and Johnson 2005).

Considering the great diversity of species, different classifications can be used to differentiate the forms of use and distributions of insects and their host plants. In most cases, these classifications are useful only for didactic purposes because they do not encompass the full range of interactions among the organisms or because the different classes of interactions are not precisely delimited. Herbivorous insects can be grouped in terms of the variation in the number of host plants they utilize. Monophagous insects utilize a single plant taxon; oligophagous insects utilize a few plant taxa that are usually phylogenetically related (i.e., from the same genus or family); and polyphagous insects utilize a wide variety of host plant species that are not phylogenetically related (Price 1997).

Insects can also be separated into functional groups according to the type and form of utilization of a particular resource. These groups are called guilds (Root 1967); that is, they consist of species that exploit the same food class (or other type of resource) in a similar way. The species within a guild may or may not be phylogenetically related (generally, they are not). Herbivorous insects are divided into five principal guilds: chewers, suckers, miners, drillers, and gall makers (Price 1997). Chewers and suckers feed externally on the host plant and are therefore called free-living or exophytic herbivorous insects. Chewers possess mouthparts that are specialized for chewing and consume tissues from roots, stems, leaves, flowers, and fruits. They belong to the orders Orthoptera (grasshoppers, crickets), Coleoptera (beetles, weevils), Lepidoptera (butterflies and moths), and Hymenoptera (wasps). Sucking insects possess mouthparts that are modified to consume sap from plant vessels or the liquid contents of plant cells. These insects can feed on xylem sap, which is found in the xylem vessels (cells that carry nutrients and mineral salts from the soil to the plant); on phloem sap, which is found in phloem sieve tubes (cells that distribute carbohydrates and amino acids throughout the plant); or on the intracellular contents of vegetative cells in various organs of the host plant. Sucking insects are found in the order Hemiptera (true bugs, leafhoppers, and aphids; see Chapters 13 and 20). Many chewing and sucking insects are specialized feeders on seeds, which are nutrient-rich compared with other plant tissues. These insects are commonly referred to as seed predators. Seed-predating insects are found in the orders Hymenoptera, Coleoptera, Hemiptera, and Lepidoptera. Among coleopteran seed predators, members of the subfamily Bruchinae (Chrysomelidae), which mainly attack plant species from the family Fabaceae, are particularly important (see Chapter 14).

The three remaining guilds (miners, drillers, and gall makers) consist of insects whose larvae feed internally on plant tissues. Therefore, they are called endophytic insects. Mining insects are those whose larvae live in and feed on plant tissue between the epidermal layers (Dempewolf 2005). According to this definition, miners generally feed on parenchyma in leaves, fruits, and the cortex of branches but do not include insects that feed on pith or deep tissues. As a mining insect feeds, it forms a characteristic, externally visible tunnel called a mine, which often appears as a whitish track on the leaf. Mines are canals formed by insects feeding inside the parenchyma or epidermal tissue of a plant whose external walls remain intact. These canals can assume a variety of shapes depending on the species involved (DeClerck and Shorthouse 1985). The tissue that is most often consumed is the palisade parenchyma in the mesophyll, but many species preferentially consume other types of tissue (DeClerck and Shorthouse 1985). Mining insects are found in the orders Lepidoptera, Hymenoptera, Coleoptera, and Diptera (flies, midges) (Dempewolf 2005).

Drilling insects are differentiated from gall-making insects because they do not induce the formation of modified tissues, and from mining insects because they live and feed deeper within the plant tissue, forming cavities called galleries. Drilling insects can feed on living or dead tissue. Galleries are most often formed in stems but can also be formed in flower buds, roots, fruits, and seeds. Drilling insects are found in the orders Coleoptera, Lepidoptera, and Hymenoptera (Coulson and Witter 1984).

Gall makers are highly abundant, but their ecology and taxonomy remain poorly known; most gall-forming species have been described relatively recently or remain undescribed (Espírito-Santo and Fernandes 2007). In general, gall-forming insects are defined as herbivorous insects that, to complete their life cycles, obligatorily induce pathological modifications in the tissue of their host plants (galls). The interaction between the insect and the host plant results in hypertrophy and/or hyperplasia of

FIGURE 16.2   Orders of insects with gall-inducing species. a) Gall induced by a Diptera: Cecidomyiidae (*Paradasineura admirabilis* Maia) in the host plant *Erythroxylum suberosum* (Erythroxylaceae). b) Gall induced by a Hemiptera: Psyllidae (*Baccharopelma dracunculifoliae* Burkhardt) in the host plant *Baccharis dracunculifolia* (Asteraceae). c) Gall induced by a Lepidoptera (unknown species) in *Macairea radula* (Melastomataceae). d) Gall induced by a Hymenoptera: Cynipidae (unknown species) in the host plant *Quercus turbinela* (Fagaceae). e) Gall induced by a Thysanoptera (unknown species) in an unidentified the host plant. f) Gall induced by a Coleoptera: Brentidae: Apioninae in the host plant *Diospyrus hispida* (Ebenaceae).

the plant tissue (Weis et al. 1988). Gall-forming insects are found in all orders of herbivorous insects (Hemiptera, Thysanoptera [thrips], Coleoptera, Hymenoptera, Lepidoptera, and Diptera), with the exception of Orthoptera (Figure 16.2).

16.3 Gall-Inducing Insect Taxa

Around 13,000 species of gall-inducing insects are known worldwide, representing about 2% of the total number of insect species (Dreger-Jauffret and Shorthouse 1992; Raman et al. 2005). However, recent estimates have extrapolated this value to nearly 120,000 species of gall-forming insects (Espírito-Santo and Fernandes 2007). The habit of inducing galls in plants has evolved independently several times among the phytophagous insects (Roskam 1992; Gullan et al. 2005), occurring in at least 51 families distributed in six different orders (Figure 16.3), and is found in all biogeographic regions. Still, some groups are more species-rich in some regions than in others. Because of the great diversity of gall-forming insects and their host plants, and the great variability of the structures they form, we present some generalizations about the natural history, biology, and ecology of these organisms. More detailed information about each group can be found in a review by Raman et al. (2005).

16.3.1 Hemiptera

The order Hemiptera contains a large number of gall-forming insects distributed in 11 families, principally in the suborder Sternorrhyncha (Schaefer 2005). Less than a dozen species of gall-inducing insects are found in the suborder Heteroptera, all of them in the family Tingidae (Schaefer 2005).

    The superfamily Psylloidea includes around 3000 described species of gall-forming insects, which are found mainly in tropical and temperate regions of the southern hemisphere (especially in tropical Asia and the Australian region) (Gullan et al. 2005). This group remains poorly studied in tropical

1. Order Hemiptera
   Suborder Heteroptera
      Family Tingidae
   Suborder Sternorrhyncha
      Family Psyllidae
      Family Aleyrodidae
      Family Aphididae
      Family Phylloxeridae
      Family Adeligidae
      Family Eriococcidae
      Family Kermisidae
      Family Asterolecaniidae
      Family Coccidae
      Family Diapididae
2. Order Thysanoptera
   Suborder Tubulifera
      Family Phlaeotripidae
   Suborder Terebrantia
      Family Thripidae
3. Order Coleoptera
   Suborder Polyphaga
      Family Cerambycidae
      Family Chrysomelidae
      Family Brentidae
      Family Curculionidae
      Family Buprestidae
      Family Mordellidae
      Family Nitidulidae
      Family Scolytidae
4. Order Hymenoptera
   Suborder Symphyta
      Family Tenthredinidae
   Suborder Apocrita
      Family Agaonidae
      Family Pteromalidae
      Family Erytomidae
      Family Cynipidae

5. Order Lepidoptera
      Family Nepticulidae
      Family Heliozelidae
      Family Prodoxidae
      Family Cecidosidae
      Family Bucculatricidae
      Family Gracillariidae
      Family Tponomeutidae
      Family Ypsolophidae
      Family Glyphipterigidae
      Family Elachistidae
      Family Oecophoridae
      Family Coleophoridae
      Family Cosmopterigidae
      Family Gelechiidae
      Family Sesiidae
      Family Torticidae
      Family Alucitidae
      Family Pterophoridae
      Family Crambidae
      Family Thyrididae
6. Order Diptera
   Suborder Nematocera
      Family Cecidomyiidae
   Suborder Cyclorrhapha
      Family Tephritidae
      Family Chloropidae
      Family Agromyzidae
      Family Anthomyzidae
      Family Clythiidae

FIGURE 16.3 Families of gall-inducing insects are distributed into six different orders and found in all biogeographic regions.

areas, which probably contain the greatest species richness of gall-forming insects (Burckhardt 2005). Species in the superfamily Psylloidea induce galls of various shapes that are conspicuous in plant species from the families Asteraceae, Myrtaceae, Melastomataceae, Fabaceae, Lauraceae, Polygonaceae, Moraceae, and Salicaceae. For example, Ferreira et al. (1990) have described the biology and natural history of *Euphaleurus ostreoides* Crawford, which parasitizes a species of the family Fabaceae, while Lara and Fernandes (1994) and Espírito-Santo and Fernandes (2002) have described the natural history and ecology of *Baccharopelma dracunculifoliae* Burckhardt, which parasitizes *Baccharis dracunculifolia* (Asteraceae). Galls induced by species of the family Psylloidea are found in several plant genera, but they are particularly abundant in species of *Baccharis* (Burckhardt et al. 2004) and *Eucalyptus* (Burckhardt 2005).

The superfamily Coccoidea consists of plant parasites found in all biogeographic regions except the polar regions. They are classified into nearly 20 families, among which 230 gall-forming species (3% of known coccoidean species) are found in 10 families (Gullan et al. 2005). Coccoidean insects induce galls in 20 angiosperm families, principally in Myrtaceae (around 130 species), Fagaceae, Asteraceae, Ericaceae, and Verbenaceae. Records of galls formed by species of Coccoidea are rare in the Neotropics, although Gonçalves et al. (2005, 2009) have presented some biological and anatomical aspects of galls induced by *Pseudotectococcus rolliniae* Hodgson and Gonçalves (Eriococcidae) in *Rollinia laurifolia* (Annonaceae).

The superfamily Aphidoidea includes around 440 species of gall-forming aphids (Wool 2004). They exhibit complex life cycles, alternating between primary and secondary hosts and entering into sexual and parthenogenetic reproduction (holocycle). Each gall is induced by a single individual, the founder, which reproduces by parthenogenesis (Wool 2005). Thus, every individual within a gall is genetically identical. The individual insects within a gall obtain their food by sucking phloem contents from the vascular system of the plant inside the gall, but they are not capable of inducing gall formation. The number of nymphs per gall is highly variable but can reach thousands. For example, the host plant *Rhus glabra* (Anacardiaceae) can abscise its leaves in response to the galls induced by *Melaphis rhois* Ficht (Aphididae), which can contain more than 1700 nymphs in a single chamber (Fernandes et al. 1999).

16.3.2 Thysanoptera

The order Thysanoptera includes about 5500 species distributed in nine families, but gall-forming species are found mainly in the subfamily Phlaeothripinae Mound and Morris (2005). Gall-inducing Thysanopteran species are found in all biogeographic regions, especially in tropical Asia and the Australian region (Mound and Morris 2005). These insects live in colonies formed by multiple individuals. It is common to find more than one species associated with a single gall, thus making it difficult to determine the species responsible for inducing the gall. The galls are formed mainly on leaves, in flowers, or in fruits. Records of galls formed by species in the order Thysanoptera are rare in the Neotropics (Souza et al. 2000), although they are common in some species of the Brazilian Cerrado biome (GWF, personal observation).

16.3.3 Coleoptera

There are few gall-forming coleopteran species relative to the high species richness of beetles associated with plants. Gall-forming beetles are found mainly in the family Curculionidae. The habit of inducing galls is found exclusively in the derived superfamilies Chrysomeloidea and Curculionoidea. In these groups, the larvae are more sedentary, with reduced sensory (ocelli and antennae) and locomotor (legs) abilities, and present a lack of pigmentation on the body (Korotyaev et al. 2005).

Beetle larvae possess chewing mouthparts and cause considerable structural damage within galls, resulting in the rapid destruction of the tissues in contact with the larvae (Dreger-Jauffret and Shorthouse 1992). Galls induced by coleopterans can be recognized by the presence of large internal chambers. There may be one or multiple chambers within each gall, generally hosting only one larva per chamber. The pupal phase may occur inside the gall or in the soil; in the latter case, the larva pierces the wall of the gall and reaches the soil to initiate the pupal phase. The galls are primarily induced on branches and roots, but some insects from the superfamily Curculionoidea induce galls in leaves and flowers (Korotyaev et al. 2005). Galls induced by coleopterans vary from simple tumescences to structures that look like fruits, which are very different from the healthy organs of the plant (Souza et al. 1998; Korotyaev et al. 2005). There is no differentiation of nutritive tissue. Coleopterans induce galls in various plant families, including Asteraceae, Solanaceae, Brassicaceae, and Fabaceae. For example, *Collabismus clitellae* Boheman induces globular galls on the stems of *Solanum lycocarpum* (Solanaceae) in the cerrado (Souza et al. 1998, 2001), while *Apion* sp. (Brentidae) induces galls in sprouts of *Diospyros hispida* (Ebenaceae) (Araújo et al. 1995; Souza et al. 2006). In the Brazilian Cerrado biome, coleopteran galls are frequently used by large ant communities as shelter and for nest building (Craig et al. 1991; Araújo et al. 1995).

16.3.4 Hymenoptera

Along with the order Diptera, the order Hymenoptera presents the most complex entomogenous galls. Gall-inducing species of the order Hymenoptera are distributed into five families (Tenthredinidae, Cynipidae, Agaonidae, Tanaostigmatidae, and Eurytomidae) and are found in all biogeographic regions (Dreger-Jauffret and Shorthouse 1992; Stone et al. 2002).

The family Tenthredinidae (suborder Symphyta) consists of species that are primitively phytophagous. Their larvae are adapted to utilize a variety of resources, feeding externally or internally on plant tissues from branches, leaves, and fruits (Gauld and Bolton 1988). The distribution of gall-inducing species is restricted to the Northern Hemisphere, with records in the Palearctic, Nearctic, and Oriental regions (Roininen et al. 2005). Most wasps of the family Tenthredinidae are species specific; a few exceptions are known to induce galls in a few related host plant species. These wasps induce galls in leaves, branches, and flower buds in 11 genera in five angiosperm families (Salicaceae, Rosaceae, Caprifoliaceae, and Grossulariaceae) and one gymnosperm family (Pinaceae) (Price 2003).

There is an extensive literature concerning the biology and ecology of gall-inducing species that parasitize the genus *Salix* (Price 2003). Prominent among these insects are members of the family Cynipidae, which includes 1000 species in 41 genera that are mainly found in the Northern Hemisphere (Ronquist 1995; Liljeblad and Ronquist 1998). The largest number of known species occurs in the Nearctic region, particularly in Mexico, where 700 species of wasps in 29 genera are estimated to occur (Ronquist 1995; Liljeblad and Ronquist 1998). Species from the family Cynipidae are found on all continents except Australia. In terms of number of gall-forming species, this family is exceeded only by the family Cecidomyiidae; however, these families are equal in terms of their complexity and the great variety of families of host plants that they parasitize, especially Fagaceae, Fabaceae, Rosaceae, and Aceraceae (Csóka et al. 2005).

Chalcidoidea is a large superfamily of parasitoid wasps that attack numerous hosts. More than 20,000 species are known (Noyes 2002, 2003). Gall-inducing species in this superfamily are found in six families: Agaonidae, Eulophidae, Eurytomidae, Pteromalidae, Tanaostigmatidae, and Torymidae (La Salle 2005). Here, we comment on some aspects of the biology of the three largest families within the Neotropical region.

Wasps belonging to the family Agaonidae (Hymenoptera: Chalcidoidea) include many species that are intimately associated with the inflorescences of species of the genus *Ficus* (Moraceae) (Galil and Eisikowitch 1968; Wiebes 1979; Weiblen 2002). Species of the family Agaonidae can induce galls internally, penetrating figs as their pollinators, or externally (Kerdelhué et al. 2000; Kjellberg et al. 2005). This family contains more than 900 species and is found in tropical regions (Price 1997). The intimate and specific interactions between species of the family Agaonidae and their host plants may represent one of the clearest example of coevolution.

Tanaostigmatidae is a small family of wasps with a principally Neotropical distribution. Currently 92 species are known in nine genera worldwide (La Salle 2005). The great majority of species in this family induce galls or are inquilines in galls induced by other insects (Hardwick et al. 2005; La Salle 2005). These wasps induce galls in bushes and trees of the families Fabaceae, Polygonaceae, Lecythidaceae, and Rhamnaceae (La Salle 1987, 2005). Fernandes et al. (1987) have recorded the first occurrence of inquiline behavior in a species of this family in galls induced by a species of *Anadiplosis* (Diptera: Cecidomyiidae) on the legume *Machaerium aculeatum*.

The family Eurytomidae includes 1420 described species in 87 genera (Noyes 2002). Species of this family include parasitoid species, phytophagous species, gall inducing species and inquilines of galls. The gall-forming species are united in the subfamily Eurytominae. Galls are induced in species of the families Myrtaceae, Campanulaceae, Boraginaceae, Orchidaceae, and Pinaceae (2005). The number of galls induced by species in this family is likely to increase in tropical regions as more studies are conducted (Leite et al. 2007).

16.3.5 Lepidoptera

About 180 species of gall-forming lepidopterans have been identified. These insects parasitize members of 20 plant families. The lepidopteran families with the largest numbers of species are Gelechiidae and

Tortricidae (47 and 39 species, respectively). Gall-forming lepidopteran species occur in all biogeographic regions (Miller 2005).

Because of their feeding habits and chewing mouthparts, the larvae rapidly destroy tissues with which they come into contact. Lipid-rich nutritive tissues were detected in several lepidopteran galls by Vecchi (2004). Most galls formed by lepidopteran insects contain a single chamber hosting a single larva. The galls are induced by the larvae, except in the species *Heliozela stanleella* Fischer Von Röslerstamm (Heliozelidae), in which the female injects a gall-inducing substance during oviposition (Miller 2005). In addition to the identification of immature individuals, galls from species of Lepidoptera can be recognized by the large quantity of feces left by the larva.

Galls formed by lepidopteran species present a great variety of shapes, from simple tumescences to more complex structures that appear similar to fruits, which are very different from the healthy organs of the plant (Dreger-Jauffret and Shorthouse 1992). Galls are predominantly induced in the branches, although they also commonly develop in leaves of Melastomataceae (Gonçalves-Alvim et al. 1999). Lepidopteran species induce galls in at least 41 families of host plants, especially Asteraceae, Salicaceae, and Fabaceae (Miller 2005).

16.3.6 Diptera

Gall-forming species in the order Diptera occur in seven families, but mainly in the families Cecidomyiidae and Tephritidae. Species of the family Cecidomyiidae are the most important gall-forming insects and are widely distributed in all biogeographic regions, with 5451 described species in 598 genera (Gagné 2004). Their total number may exceed 100,000 species (Espírito-Santo and Fernandes 2007). Most of the species belonging to this family are associated with plants, inducing galls or living as inquilines therein, while a few species are predatory (Gagné 1994). Species of the subfamily Porrycondilinae feed on fungi, a condition considered ancestral with respect to the habit of inducing galls (Gagné 1994). Some species can induce galls in related plant species of the same genus or family. The existence of polyphagous species (using host plants from different families) is rare in the family Cecidomyiidae. Members of this family are particularly species-rich in certain plant families and genera, depending on the biogeographic region. In the Neotropical and Nearctic regions, they are most diverse in host plants from the genera *Baccharis* and *Solidago* (Asteraceae), respectively (Gagné 1989; Fernandes et al. 1996). In the Neotropical region, they are less numerous, with 500 species and 170 genera recorded (Maia 2005). In Brazil, 159 species and 75 genera have been described (Maia 2005). Many species described from Brazil are found in the restinga vegetation (a community characterized by shrubs and low forests growing on sandy dunes) in the state of Rio de Janeiro (Maia 2001a,b), where 95 species and 47 genera have been recorded (Maia 2005). However, records of species from the family Cecidomyiidae have increased considerably in recent years (Maia and Fernandes 2004, 2006).

Approximately 5% of the 4300 described species in the family Tephritidae are gall inducers, the majority of which belong to the subfamily Tephritinae (Freidberg 1998; Korneyev et al. 2005). The galls are induced principally in branches, flowers, leaves, and roots. More than 90% of the galls known to be induced by members of this family occur in host plants of the family Asteraceae (Freidberg 1998). For example, *Tomoplagia rudolphi* (Lutz & Lima) forms galls in *Vernonia polyanthes* (Asteraceae), which is widely distributed in southeastern Brazil (Silva et al. 1996). The families Melastomataceae, Aquifoliaceae, Acanthaceae, Fabaceae, and Onagraceae are also attacked by gall-forming species of this family.

Gall-forming insects of the family Chloropidae are apparently restricted to host plants of the family Poaceae, except for species of a genus that induce galls in species of the genus *Scirpus* (family Cyperaceae) (Dreger-Jauffret and Shorthouse 1992). As in other gall-forming cyclorrhaphan dipterans, the gall-forming process is not initiated at oviposition. The eggs are laid externally on branches or on the leaf surface. After hatching, the larva actively penetrates the branch, opening a hole with its mouthparts (Bruyn 2005). Once inside the branch, the larva begins to feed on the leaves that surround the meristem. Although the family is widely distributed, studies on gall-making species of Chloropidae are concentrated in the Palearctic and Nearctic regions.

16.4 Host Plant Taxa

Vascular plants, including gymnosperms (mainly conifers) and angiosperms, are the main hosts of gall-forming arthropods. In general, flowering plants (angiosperms) host more species of gall-forming insects. For example, in Brazil, plant families that are associated with a large number of gall-forming insects include Asteraceae, Myrtaceae, Malpighiaceae, Fabaceae, Rubiaceae, and Bignoniaceae (Fernandes 1987, 1992; Fernandes et al. 1988, 1996, 1997; Julião et al. 2002; Maia 2001b; Maia and Fernandes 2004). In one area of the Brazilian Cerrado, in Minas Gerais, the families Fabaceae, Myrtaceae, Malpighiaceae, Bignoniaceae, and Malvaceae accounted for 65% of the host plant species and hosted 70% of the gall-forming insect species (Gonçalves-Alvim and Fernandes 2001a,b). However, a brief analysis indicates substantial variation among biomes in the frequency of families attacked by gall-forming insects. This variation may be explained by the relative frequency of occurrence of the plant families. Broader studies across all Brazilian biomes are needed to better understand these patterns.

The species richness of gall-forming insects varies widely across biogeographic regions, and galls occur much more frequently in certain plant taxa. Species of the genus *Baccharis* (Fernandes et al. 1996), for example, have a large number of associated insect species (Table 16.1). In the region of Ouro Preto, *Baccharis pseudomyriocephala* (Figure 16.4) hosts 11 species of gall-forming organisms (Araújo et al. 2003). In addition to *Baccharis*, species of *Copaifera* (Neotropical region; GWF, personal observation; Oliveira et al. 2008), *Solidago*, and *Chrysothamnus* (Nearctic region; GWF, personal observation; Gagné 1994, Fernandes 1992) are rich in species of Cecidomyiidae; species of *Quercus* and *Rosa* (Nearctic region) and *Acacia* (Ethiopic region) are rich in species of Cynipidae (Shorthouse and Rohfritsch 1992; Stone et al. 2002); and species of *Eucalyptus* (Australian region) are rich in species of Chalcidoidea and Coccoidea (Blanche 1994). In the Sonoran desert, *Atriplex*, *Chrysothamnus*, and *Larrea* host a high diversity of gall-forming insects (McArthur 1986; McArthur et al. 1979; Fernandes and Price 1988; Waring and Price 1990). These data indicate the existence of super-hosts, that is, host plant taxa that sustain a large number of associated gall-forming insects (Fernandes and Price 1988; Veldtman and McGeoch 2003; Espírito-Santo et al. 2007). This conclusion is supported by the fact that few host plant taxa support a large number of insect species, independently of the sampling method (Hawkins and Compton 1992). However, the ecological mechanisms and selective pressures that influence these patterns within certain taxa remain unexplained or have not been adequately studied.

16.5 Location and Choice of the Host Plant

The free-living stage (adult stage) of gall-inducing insects is very short compared with the time they spend immersed in host plant tissues. In some cases, the larval stage may take several months. However, the adult stage is of extreme importance as it is at this stage of their life cycle that the galling herbivores must find their appropriate host plants and organs within all the available options. This is not an easy task as all plants may present a mosaic of resistance mechanisms to defend themselves against unbidden guests. The most studied ones are the physical and chemical defenses. Physical defenses include several types of trichomes that impair movement or even trap the adults, and tissue sclerophylly, which confer resistance to oviposition and feeding (Woodman and Fernandes 1991; Fernandes 1994; Lucas et al. 2000; Chen 2008). Further, as pointed out by Rasmann and Agrawal (2009), plant defenses are dependent on its genetics, ontogenesis, and also on environmental factors. Together, these features shape the multivariate defensive phenotype and outcome of the interaction. Also, the chemical defenses include the synthesis and accumulation of secondary metabolites (Gottlieb et al. 1996). By the time the host plant and organ are found, a succession of recognizing systems between both organisms are required to permit gall development (Rohfritsch and Shorthouse 1982). However, the elucidation of the plant's response to herbivore attack is much complex for it is difficult to establish the relevance of a particular trait for the interaction (Rasmann and Agrawal 2009).

For galling herbivores, host selection is vital because it is the offspring that stays most of their lifetime inside the host plant tissues, and most defenses are against the larval stage. On the basis of the crucial

TABLE 16.1

Number of Galling Species on Species of the Genus *Baccharis*

Host Plants	Galling Richness
Baccharis aphylla	1
Baccharis artemisioides	1
Baccharis bogotensis	2
Baccharis boyacensis	1
Baccharis capitalensis	1
Baccharis cf. bacchridastrum cabr.	1
Baccharis concinna	15
Baccharis confertifolia	1
Baccharis coridifolia	2
Baccharis dracunculifolia	17
Baccharis effusaaphylla	1
Baccharis elaegnoides	1
Baccharis eupatorioides	3
Baccharis genistelloides	1
Baccharis glutinosa	1
Baccharis latifolia	5
Baccharis lineares	1
Baccharis macrantha	2
Baccharis microphylla	1
Baccharis myrsinites	1
Baccharis nitida	2
Baccharis paucidentata	2
Baccharis platypoda	3
Baccharis poeppigiana	1
Baccharis prunifolia	1
Baccharis pseudomyriocephala	11[a]
Baccharis rosmarinifolia	7
Baccharis salicifolia	13
Baccharis schultzii	2
Baccharis serrulata	4
Baccharis spartioides	2
Baccharis subulata	2
Baccharis tricuneata	1
Baccharis trimera	1
Baccharis trinervis	2
Baccharis vulnerave	1
Total	125

[a] Fernandes, G. W., et al., *Trop. Zool.*, 9, 315, 1996; Araújo, A. P. A., et al., *Rev. Bras. Entomol.*, 47, 483, 2003.

FIGURE 16.4 Some galls induced by the insect community on a) *Chrysthamnus nauseosus hololeucus* and b) *Chrysthamnus nauseosus consimilis* in the Sonoran Desert, USA. On *C. n. hololeucus*: c) *Aciurina trixa* (Diptera: Tephritidade), d) *Rhopalomyia chrysothamni* (Diptera: Cecidomyiidae), e) unidentified Lepidoptera. On *C. n. consimilis*: (f–h) Cecidomyiidae (Diptera). (From Araújo, A. P. A., et al., *Rev. Bras. Entomol.*, 47, 483, 2003.)

events required to gall establishment, Moura et al. (2009) showed that the absence of *Aceria lantanae* (Acari) galls on sympatric varieties of *Lantana camara* (Verbenaceae) that present pink and white flowers could be related to their chemical contents and density of trichomes, which constitute part of the first line of resistance to herbivores, in general (Levin 1973, Woodman and Fernandes 1991, Lucas et al. 2000). Phytochemical profiles showed differences that could explain the selection of the group with red flowers as the host plants by the mite.

16.6 Gall Morphology

Galls can be formed on any organ of the host plant; nevertheless, the leaf is the most susceptible of the plant organs for the development of galls, and relatively few galls occur on branches, vegetative parts, or floral buds (Dreger-Jauffret and Shorthouse 1992). Although galls on fruits are not so numerous as galls on vegetative organs, they are interesting models to study sink and source relationships and of phenological synchronism. As an example, there is an unidentified species of wasp that oviposits through the pericarp of immature fruits of *Eugenia uniflora*. The mature galls are multichambered and no seeds develop, which constitute a serious damage to the host plant reproduction (Figure 16.5). There must be a synchronism between the life cycles of the wasp and the reproductive phase of *E. uniflora*; otherwise, the gall-inducing organism life cycle should be interrupted. Dorchin et al. (2006) affirm that the phenology and position of the gall on its host plant influence its ability to compete with other sinks. By the size and number of insects inside the galls on the fruits of *E. uniflora*, this affirmation seems to fit this system. In a Cerrado (savanna) vegetation reserve in Minas Gerais (Brazil), about 60% of the insect galls are formed on the leaves (Gonçalves-Alvim and Fernandes 2001a,b). In vegetation of rupestrian fields, the percentage of galls on the leaves was also similar, while in another Cerrado formation in Serra of San José, 70% of the galls were induced on leaves (Maia and Fernandes 2004). However, ratios may change as scale changes. For instance, in *Baccharis concinna*, *B. pseudomyriocephala*, and *B. dracunculifolia* most galls develop on stems (Fernandes et al. 1996; Araújo et al. 2003).

Morphological, biochemical, and phylogenetic studies on aphids (Stern 1995), cynipids (Stone and Cook 1998), sawflies (Nyman and Julkunen-Tiitto 2000), and thrips (Crespi and Worobey 1998) support the idea that the morphology of the gall is defined by the inducing insect and not by the host plant. Thus, the gall may be understood as an extended phenotype (sensu Dawkins 1982) of the inducing insect (Weis et al. 1988; Bailey et al. 2009). Besides the morphology, the inducing insect is also capable of controlling

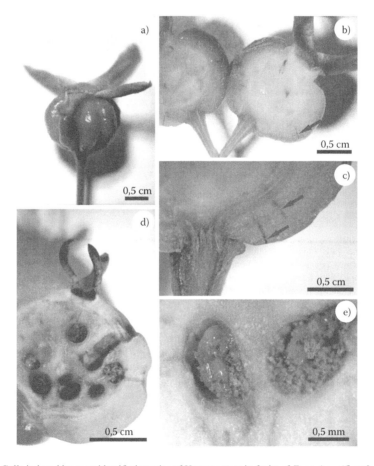

FIGURE 16.5 Galls induced by an unidentified species of Hymenoptera in fruits of *Eugenia uniflora* L. (Myrtaceae). a) Immature non-galled fruit. b) Immature fruit opened to show sites of oviposition through pericarp (arrow). c) Detail of oviposition site (arrows). d) Mature multichambered gall. e) Detail of two larval chambers with insect's excrement. (Courtesy of R. M. S. Isaias.)

the chemical properties of the galls (Nyman and Julkunen-Tiitto 2000), which may be regulated by the feeding activity of the insect, indicating that gall development depends on the behavior of its inducing insect. Insects that feed moving in a circle should produce round galls; insects that feed at one end of the gall (usually at the base) produce conical galls; while in lenticular galls, insects feed in the lateral margins (Rohfritsch and Shorthouse 1982). Thus, it is reasonable to assume that the gall has a function or adaptive significance for the inducing insect (see Price et al. 1986, 1987). The shape of the gall seems to be independent of the host plant, but as it is completely formed of plant tissues, its development must obey some constraints imposed by the cycle of the host plant cells at the site of oviposition. The gall is a phenotypic entity that represents the interaction between the genotype of the insect, that of the host plant, and the environment (Weis et al. 1988). Since most aspects of gall morphology are controlled by the galling insects, it is clear that the size and shape of the gall is crucial for their survival. Thus, if gall morphology results in differential survivorship, and the insect population shows a heritable variation in the ability of setting this feature, then the selection can act on it.

One of the best-studied galling insects is *Eurosta solidaginis* Fitch (Tephritidae), which induces galls on *Solidago altissima* (Asteraceae) and a few related species (Abrahamson and Weis 1997). In this system, gall size is important for the survival of the insect, although the genotype of the host plant has an important role in determining the characteristics of the gall. The variation in this gall morphology is explained by the genotype of the inducing insect. Thus, this system has proved that gall morphology should be understood as the extended phenotype of the insect and adaptive explanations can be related to the fitness of the insect.

16.7 Gall Anatomy and Physiology

The complexity in the structure of galls may vary by several degrees. The galls may vary from simple and isolated cytological transformations to a new arrangement of plant tissues. In these cases, the galls may be defined as new multicellular organs generated by coordinated cell division and expansion. The variation in morphological complexity is also followed by a variety of physiological traits.

In general, the galls induced by Cecidomyiidae and Cynipidae are the best studied from the structural point of view. The Cynipidae galls have two cortical regions, the inner cortex formed by a multilayered nutritive tissue located around the larval chamber and the outer cortex with a reserve tissue externally delimited by the epidermis. In some galls, the outer cortex is limited from the inner one by a thin layer of sclerenchyma. In fact, the high diversity of the outer cortex is said to be responsible for the great variety of gall morphotypes (Stone et al. 2002). Also, the number of larvae or nymphs per chamber may vary from one to hundreds, and may also be responsible for variations in the final size and shape of galls. Therefore, galling insects do not only control the developmental patterns of the host plant, so as to define the gall phenotype, but also its physiology. Moura et al. (2009) studied the ontogenesis of galls induced by an Acari and proposed that the cell divisions alter the leaf pattern first related to photosynthesis, and result in a new verrucous structure that guarantee an adequate microenvironment and nutrition source. This is also true for the majority of the gall morphotypes, from the simplest ones to the most complex.

That galls are sinks of photoassimilates is common sense (Larson and Whitham 1991; Larson 1998; Dorchin et al. 2006). Lemos Filho et al. (2007) presented data on the transpiration and photosynthetic performance in galls of two species on *Aspidosperma* (Apocynaceae) from southeast Brazil. These data showed that gall induction did not affect the photosystem II, and consequently there was no reduction of the relative electron transport rates. By establishing a relationship between the physiological and morphological features of the two gall systems, they concluded that the galls may produce photoassimilates, but in such low values that it seems improbable that they could guarantee gall maintenance without draining resources from their host organ. Another important feature are the physiological gradients inside and outside the gall tissues (Bronner 1992; Hartley 1998; Nyman and Julkunen-Tiitto 2000). These gradients have revealed specific enzymatic activities and are also accompanied by cytological peculiarities (Rehill and Schultz 2003; Oliveira and Isaias 2010a,b; Oliveira et al. 2010). The galls function as sinks of nutrients mobilized from the other host plant parts (Kirst and Rapp 1974; Fay et al. 1993; Whitham 1992). A large set of evidences support that the galling insect is able to manipulate the host plant, inducing the formation of nutritionally superior cells in comparison to the other healthy plant tissues, the nutritive tissue (Mani 1964; Shannon and Brewer 1980; Bronner 1992; Rohfritsch and Shorthouse 1982). The cells of this tissue have a high concentration of lipids, glucose, amino acids, and a high enzymatic activity, including phosphatases, proteases, and aminopeptidases rich in RNA and ribosomal RNA of the nucleolus (Bronner 1992). On the other hand, the parenchyma cells of the outer cortex form a reserve tissue characterized by a high concentration of starch, low concentration of lipids and glucose, and low enzyme activity. As the larva feeds on the cells of the nutritive tissue, there is a replacement of substances by the cells of the reserve tissue (Bronner 1992). The translocation of substances between the two tissue zones has been proven to need an intense enzymatic activity (Bronner 1992; Oliveira and Isaias 2010b; Oliveira et al. 2010). Also, the substances accumulated may be diverse, such as proteins, carbohydrates, and lipids (Figure 16.6).

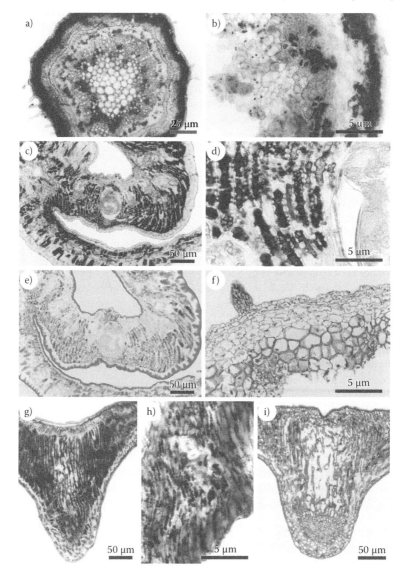

FIGURE 16.6   Histochemistry of galls. Flavonoid derivatives detected with 3,3′-diaminobenzidine in *Calliandra brevipes* Benth: a) non-galled stem; b) stem gall. *Machaerium uncinatum* (Vell.) Benth: c–d) Phenolic derivatives detected by ferric chloride in leaf galls; (e) lipids detected by Sudan Red B in cell wall and cuticle. f) *M. hirtum* (Vell.) Stellfeld. Proteins detected by Coomassie blue in leaf galls. g–i) Leaf galls of *M. aculeatum* Raddi. g) Proteins detected by Coomassie blue. h) Starch detected by Lugol's reagent. i) Carbohydrates detected by periodic acid–schiff reaction. (Courtesy of R. M. S. Isaias.)

Once the starch that accumulated in the reserve tissue cannot be directly used either for the gall-inducing larva or for the gall machinery, carbohydrate conversion is necessary. Oliveira et al. (2010c) detected the activity of the glucose-6-phosphatase, an enzyme responsible for the synthesis of intermediate compounds before the formation of sucrose in galls of Cecidomyiidae. Also, these authors detected a gradient of invertases, generally related to physiological sinks (Koch and Zeng 2002; Rehill and Schultz 2003), that fit the gradient of starch and sugars, and should provide resources to cell expansion and to the metabolism of the nutritive tissue. The detection of sucrose synthase was related to the maturation and formation of reserve tissues in gall systems. Thus, enzymatic gradients play key roles in maintaining the supply of nutrients for the developing galling larva (Bronner 1992), and also function in the maintenance of gall structure.

The nutritive tissue does not present defensive secondary compounds (Hartley and Lawton 1992; Hartley 1998; Nyman and Julkunen-Tiitto 2000), which may be detected in the outer cortical layers (Figure 16.6). Studies in galls induced by sawflies on willow (*Salix* spp.) showed that toxins are commonly accumulated in the outer cortex, suggesting that the insect can benefit from their defensive properties against other insects (Nyman and Julkunen-Tiitto 2000). These substances may have negative effects on the growth, development, or survival of another organisms (Wittstock and Gershenzon 2002), such as the natural enemies. In galls, the role of defense against natural enemies has been commonly attributed to phenolics. However, Formiga et al. (2009) did not find any relationship between the level of phenolics in *Aspidosperma spruceanum* and the degree of gall infestation by a galling Cecidomyiidae. Moreover, Abrahamson et al. (1991) affirmed that higher phenolic concentrations restricted to the gall tissues induced by *E. solidaginis* on *S. altissima* could also potentially play a role in gall formation by influencing the hormonal control of growth. Thus, the influence of phenolics in host plant–galling herbivore systems seems to be much more complex than just constitutive chemical defense.

In Brazil, few studies on the chemistry of galls have been developed, but the chemical analysis in a Lepidoptera–*Tibouchina pulchra* (Melastomataceae) system showed that defensive compounds were less abundant in the nutritive tissue, and more frequent in the outer cortex of the gall, corroborating the general premise. Moreover, carbohydrates and lipids were more abundant in the tissues of the gall than in the non-galled tissues of the host plant (Motta et al. 2005). However, in some galls, such as those induced by few species of Cecidomyiidae, there is no formation of a nutritive tissue (Bronner 1992). In these cases, two types of feeding strategies can be identified. The larvae feed directly from the contents of the cells (Gagné 1994) or from hyphae of fungi that line the larval chamber (Bronner 1992). This is the case of the ambrosia galls induced by three tribes of Cecidomyiidae: Asphondyliini, Alycaulini, and Lasiopterini (Meyer 1987; Yukawa and Rohfritsch 2005). The ambrosia galls received this name in reference to the similarities in food habits of these Cecidomyiidae and the ambrosia beetles (Meyer 1987). The hyphae of fungi are introduced into plant tissues during oviposition of the Asphondyliini, or by the first instar larvae in Alycaulini and Lasiopterini (Yukawa and Rohfritsch 2005). Up to the moment, ambrosia galls were found in *B. concinna*, *B. dracunculifolia* (Arduin and Kraus 2001), and in *Bauhinia brevipes* (Sá et al. 2009) in Brazil. These galls do not differ in external morphology from the galls where no association with fungi was detected.

16.8 Gall Development

The development of galls has four distinct phases: induction, growth and differentiation, maturation, and dehiscence (Dreger-Jauffret and Shorthouse 1992; Arduin et al. 2005). The induction phase is characterized by a sequence of events that define the recognition of the oviposition site (tissue, organ, and host plant), and the behavior of the inducing insect. It is a critical stage, and events during oviposition and/or feeding promote crucial changes in the tissues of the host plant. Generally, the galling larvae require a reactive, meristematic tissue for the formation of galls (Mani 1964; Weis et al. 1988; Dreger-Jauffret and Shorthouse 1992); however, there are cases of gall induction on non-meristematic tissues, as in the ambrosia galls on *B. concinna* and *B. dracunculifolia* (Arduin and Kraus 2001). Also, some insects may induce galls on immature or mature tissues, and in these cases, the mature galls may present anatomical and developmental features with distinct adaptive values (Oliveira and Isaias 2009). As a whole, the galling insect manipulates the potentialities of the host plant tissue to its own benefit. It conquers nutrition, protection, and an adequate microenvironment by generating patterns of cell redifferentiation (see Price et al. 1987). Moura et al. (2009) and Oliveira and Isaias (2010a) presented some ontogenetical analyses of gall development since the host leaf in its meristematic stage until gall maturation. These studies showed the correspondence of the tissue fates in non-galled organs and in galls of *L. camara* (Verbenaceae) and *Copaifera langsdorffii* (Fabaceae), respectively. Comparing these two gall systems, it is possible to set the ground meristem as the most plastic of the leaf tissues, capable of assuming several fates other than its primary one, the photosynthetic cells. In galls, the cells that originated from the ground meristem are redifferentiated into nutritive, protective, and also reserve tissues.

As the molecular mechanisms of gall induction and development remain mostly unknown about galls induced by insects, there is a great debate about the role of the insect and the host plant in the formation of the gall. In general, the gall-inducing stimuli originate during the feeding of the first instar larvae and more rarely during oviposition (body fluids of the female or the egg). In some groups, the role of the feeding activity of the larvae may have greater or lesser participation in gall development. For example, in Tenthredinidae, gall induction is initiated by the produced fluids of the accessory glands of the female reproductive system, injected into plant together with the eggs during oviposition (Meyer 1987). In the galls of Cynipidae, the induction process may have its origin in the fluids of the female egg or the larva (Bronner 1973; Rohfritsch and Shorthouse 1982). In Coleoptera, the galls can be induced by the larvae (e.g., Buprestidae) or during oviposition, when the eggs are laid in a cavity prepared by the female (Korotyaev et al. 2005). In some hemipteran galls, that is, Psylloidea (Burckhardt 2005) and Coccoidea (Gullan et al. 2005), the galls are usually initiated by the feeding activity of the first instars. In the galls induced by *P. rolliniae* (Eriococcidae) on *R. laurifolia* (Annonaceae), the second instar nymph induces sexually dimorphic galls (Gonçalves et al. 2005). Moreover, this species of Eriococcidae induces a stem gall for diapause to survive in the period when leaves of the host are shed (Gonçalves et al. 2009). Even though some new strategies of survival have been reported for Neotropical galling insects, the exact mechanism of the manipulation of host plant tissues remains unknown. There is evidence that the feeding action of aphids through the vascular system of the plant alters hormones and may be responsible for the initiation of the galls (Wool 2005). In Thysanoptera, gall formation is the result of the feeding activity of insects. By feeding on the plant cell content, one at a time, these insects alter the course of leaf expansion; the lamina becomes distorted because of distinct sites of hypoplasia and hyperplasia (Souza et al. 2000), and groups of necrotic cells (Mound and Kranz 1997). In the case of the Thysanoptera, the epidermal or mesophyll cells near the feeding sites (Ananthakrishnan and Raman 1989) are stimulated by an unknown mechanism (Mound and Kranz 1997) to return to their meristematic potentialities and produce a new structure.

The complexity of the gall systems may be higher when a third organism participate in the formation of the gall, as in the galls of ambrosia, or when the gall morphology is modified by inquilines and parasitoids, as is the case of many Cynipidae (Stone et al. 2002) and few Cecidomyiids (e.g., Fernandes et al. 1987). An example of this phenotype alteration was reported in the galls of *Anadiplosis* sp. whose larvae are parasitized by two plastygasterids, two eurytomids, and a tanaostigmatid (Hymenoptera). Galls due to this last parasitoid are distinguished from the others by their consistency, larger size, and different shape (Fernandes et al. 1987). In cases where the maintenance of gall development and its final size and shape are under the influence of the feeding activity of the gall maker, a third trophic level inside the gall may intermediate this feeding behavior or even block it. The first case is exemplified by ambrosia galls, in which the insects feed on the fungi and not on the plant cells. Thus, it seems plausible that the signaling molecules that trigger plant tissue transformation come from the fungi, whereas in the latter case, the end or the switch of an insect feeding activity inside the chamber alters cell fates.

The phase of growth and differentiation of the gall is the period in which its biomass increases remarkably due to the increased number of cells—hyperplasia (cell division) and/or hypertrophy (increase in cell size). As stated by Moura et al. (2009) and Oliveira and Isaias (2010a), these processes take place in all three plant tissue systems, but are more evident and crucial for gall functioning in the cells that originated from the ground meristem. Both hyperplasia and hypertrophy are defined by the feeding activity of larvae, whose saliva seems to modify the cell wall and dissolve cell contents. The activity then defines the form of the larval chamber, and possibly the external shape of the gall (Rohfritsch and Shorthouse 1982). However, in the Brazilian flora, some gall phenotypes are so peculiar and the larval chamber so small that it seems less difficult to assume just the feeding activity as responsible for the gall phenotype. This is the case of the horn-shaped galls on *C. langsdorffii* (Fabaceae) (Oliveira et al. 2008), the bivalve-shaped galls on *Lonchocarpus muehlbergianus* (Ferreira et al. 1990; Oliveira et al. 2006), and the bud galls of *Guapira opposita* (Nyctaginaceae) (Araújo 2008), for instance. Perhaps, the phenolics–auxin balance in these galls may be more important than the feeding sites for the definition of the gall size and shape by the time of its maturity.

The maturation phase of the gall occurs when the insect is in its last instar. This is the main trophic phase of the gall inducer, and that is the time when it eats a large mass of nutritive tissue. Then, the inner cortex will disappear under the control of the inducer, and the outer cortex of the gall, which is more under the influence of the plant (Dreger-Jauffret and Shorthouse 1992) will have its resources totally drained. Finally, the stage of dehiscence or the opening of the gall occurs at the end of the maturation phase, when the greatest physiological and chemical changes occur in the gall tissues. By the end of this phase, the flow of nutrients and water stops.

16.9 Gall Classification

Galls can be classically classified as either organoids or histioids, due to the developmental potentialities expressed by their cells (Meyer 1987; Dreger-Jauffret and Shorthouse 1992). The galls of the organoid type are those that differ little from the growth pattern of the host organ, which even galled, does not lose its identity. The organoid galls are represented by a swelling, callus-like growth, usually induced by insects and fungi. The galls of the histioid type exhibit a great variety of abnormal structures, the growth patterns of the host organ are changed, and the rearrangement or induction of new types of tissues occur. The histioid galls can be divided into cataplasmic and prosoplasmic. The cataplasmic galls

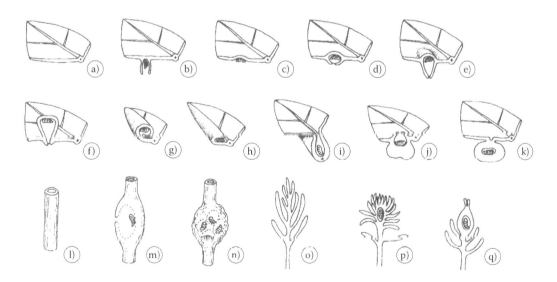

FIGURE 16.7 Morphological types of galls based on position of galling herbivores and gall development (Larew 1982): a) healthy leaf lamina; b) hairy galls; c) mark galls; d) discoid or vesicular gall; e and f) pouch galls; g) roll galls; h and i) fold galls; j and k) covering galls; l) healthy shoot; m) covering fall; n) typical gall with several chambers; o) healthy apical shoot; p) rosette gall (with increased number of leaves); and q) bud gall (reduced number of leaves). (From Dreger-Jauffret, F. and J. D. Shorthouse, Diversity of gall-inducing insects and their galls. In *Biology of Insect-Induced Galls*, ed. J. D. Shorthouse and O. Rohfritsch, 8–33. Oxford: Oxford University Press, 1992. Courtesy of Miriam Duarte.)

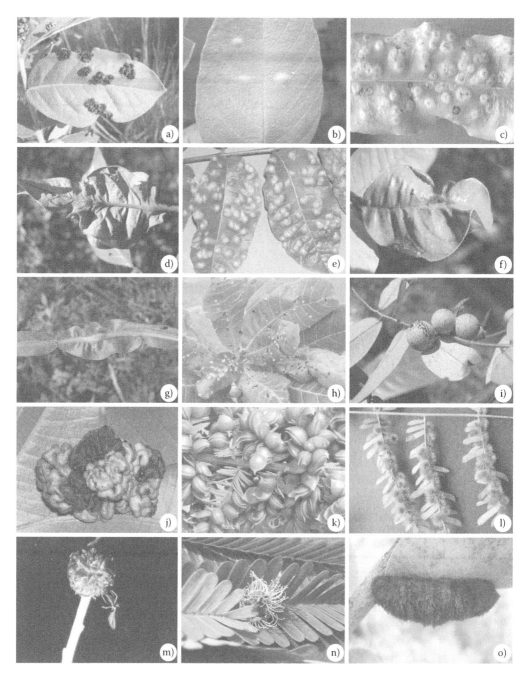

FIGURE 16.8 Several morphotypes of galls in different host plants: a) amorphous galls of thrips in Myrtaceae from Cerrado; b) discoid galls of Cecidomyiidae in *Davilla rugosa* from Cerrado; c) discoid galls of Cecidomyiidae in *Sacoglotis matogrossensis* from Amazonia; d) leaf gall of Cecidomyiidae in *Vismia latifolia* from Amazonia; e) insect galls in *Trattinickia rhoifolia* (Burseraceae) from Amazônia; f) elliptical gall at leaf margin in *Vismia latifolia* from Amazonia; g) elliptical galls in stem of *Baccharis* cf. *trimera* in Ouro Preto, MG; h) galls of a Cecidomyiidae in *Anacardium occidentale* from Amazonia; i) galls in a bud of an unidentified host plant species from Amazônia; j) cylindrical galls of a Cecidomyiidae in an unidentified host plant species from Cerrado; k) galls of Parkiamyia paraensis Maia in *Parkia pendula*; l) spheroid hairy gall of a Cecidomyiidae in Mimosa sp. from Amazônia; m) spherical or pineapple-like galls in *Chrysothmanus nauseosus* from Sonoran Desert, USA; n) galls of Hymenoptera in Mimosa sp. from Amazonia; and o) galls of *Paradasineura admirabilis* Maia in leaves of *Erythroxylum suberosum* (Erythroxylaceae) from Cerrado. (Courtesy of G. W. Fernandes.)

are amorphous and vary in volume and extent; they are less organized and differentiated from the host body, consisting mainly of different layers of histologically homogeneous parenchyma. Generally, the Hemiptera induce cataplasmic galls. The prosoplasmic galls are highly organized, with definite size and shape. Histologically, they are more complex and composed of differentiated tissues, although the degree of tissue differentiation is variable and dependent on the inducing insect. Also, independently of the gall type, the inducing insects are usually sedentary and therefore the site of stimulation and control of morphogenesis is restricted.

Different systems of classification of morphological types of galls are adopted. One of the first classification was proposed by E. Kuster in 1903 (Larew 1982) on the basis of the position of the gall and the type of gall development. Such classification includes just leaf galls and has been currently used. It proposes six main models of gall formation (Figure 16.7). *Covering galls*: in this gall type, the insect becomes encapsulated within the gall, which may present an opening (ostiole/operculum) or not. *Filz galls*: these galls are characterized primarily by dense pubescence in their outer walls (usually leaves), which houses the inducers. *Roll and fold galls*: these galls present differential growth caused by the feeding habits of the insect, resulting in winding, twisting, or folding of leaves and branches, which are often swollen. *Pouch galls*: in this gall type, tissue growth occurs in a restricted area around the larva, producing an invagination by differential growth on one side of the leaf blade. The gall tissues may have different degrees of differentiation, and the epidermis of the larval chamber originates from the epidermis of the plant body. *Mark galls*: these are the galls in which the eggs are oviposited on the plant surface; then, the first instar larvae penetrate the tissue that proliferates and surrounds it completely. *Pit galls*: these galls are characterized by a slight depression where the insect feeds, sometimes surrounding a protruding halo. Sometimes the epidermis forms a vesicle (discoid gall or blister galls). *Bud or Rosette* galls: these models cause the growth of buds or, sometimes, the proliferation and miniaturization of new leaves. There is a marked shortening of internodes. In Figure 16.8 some types of galls found on several Brazilian host plants are illustrated.

16.10 Adaptive Significance

The adaptive significance of the habit of inducing galls was currently revised (Price et al. 1986, 1987; Stone and Schönrogge 2003). At least one researcher hypothesized that galls do not have any adaptive value either for the insect or for the plant (Bequaert 1924), while another one proposed that they may have an adaptive value only for the host plant (Mani 1964).

According to the hypothesis of the adaptive value of galls for the plant, galls should limit the movement of the insect, restricting it in space and time, and thus, the gall structure is just a defensive structure. Most of the evidences do not support this hypothesis since galls act as sinks, translocating nutrients from other plant parts and limiting the growth and reproduction of host plants. These two hypotheses have few defenders today because the studies in recent decades have shown that galls probably both have an adaptive significance and are a detrimental structure to the host plant. Several lines of evidence illustrate the impact of galls on the fitness of their host plants (e.g., Fernandes 1987; Fernandes et al. 1993). Three other hypotheses advocate that the gall should present an adaptive value for the insect: the nutritional, the enemy-free space, and the microenvironmental hypotheses (reviewed by Price et al. 1987).

The nutritional hypothesis is supported by several studies that show that the galling insect is able to manipulate the host plant, inducing the formation of a nutritionally superior tissue (see Section 16.7) in comparison to the other non-galled tissues of the host plant (Shannon and Brewer 1980; Rohfritsch and Shorthouse 1982; Bronner 1992). This nutritive tissue is also free of defensive secondary compounds (Larew 1982; Price et al. 1986, 1987; Nyman and Julkunem-Tiitto 2000). Studies on galls induced by tenthredinids on species of willow (*Salix*) in the United States showed that defensive substances, mainly phenolic compounds, are common in the outer cortex of the galls, suggesting that the insect can benefit from their defensive properties against other insects (Larew 1982; Cornell 1983; Taper and Case 1987). The gall acts as a mobilizing sink of nutrients from the other tissues of the host plant (Fay et al. 1993; Larson and Whitham 1991). The enemy hypothesis argues that galling insects are less predated and/ or parasitized when compared with other phylogenetically close insects, but with a different feeding habit. For example, galls induced by tenthredinids are attacked by fewer parasitoid species and have lower mortality rates than free-living ones (Price and Pschorn-Walcher 1988). Nevertheless, according to Stone and Schönrogge (2003), some other galling herbivores are more attacked than their free-living relatives.

Some external features of galls may reduce the rates of attack by natural enemies. For example, increasing size (Stone et al. 2002; Weis et al. 1985; Rossi et al. 1992) or hardness (Weis 1982; Stone et al. 2002), or the presence of trichomes may reduce the attack by parasitoids and other natural enemies. The North American system *E. solidaginis–S. altissima* (Abrahamson and Weis 1997) has been widely studied in recent decades, and can help understand these different strategies. The success of *Eurytoma gigantea* Walsh in the parasite insect gall *E. solidaginis* depends on the ratio between the size of its ovipositor and the thickness of the gall. When the ratio exceeds 0.95 (a parasitoid with an ovipositor 10 mm in length can lay eggs on a branch with a wall of up to 9.5 mm), the parasitoid cannot successfully make the oviposition (Weis et al. 1985). Moreover, galls with greater diameter have greater chance of *Eurosta* being attacked by birds (e.g., woodpecker *Picoides pubescens*) than smaller galls (Weis et al. 1992). Thus, parasitoids and predators act as a selective force ("directional") about the size of the gall in different directions, first to increase the size of the gall and second in order to decrease it. Hence, there is a stabilizing selection favoring the differential reproduction and survival of galls of intermediate size. However, other studies did not statistically support this assertion (for a complete analysis, see Abrahamson and Weis 1997).

The beetle weevil *C. clitelae* Boheman, commonly found in the Brazilian Cerrado, induces galls with several chambers on *S. lycocarpum* (Solanaceae) (Souza et al. 1998, 2001). Although the beetle preferably attacks small plants, the size of their galls and the number of larvae per gall increase with the size of the branch. As in the case of *E. solidaginis*, the larger galls of *C. clitelae* are most often preyed on by the Cerrado woodpecker, *Colaptes campestre*.

The microenvironmental hypothesis states that because galls are sessile and protected by their structure, the galling larvae are less susceptible to abiotic environmental changes, particularly temperature and humidity (Fernandes and Martins 1985; Price et al. 1987). Hygrothermal and nutritional stress, defined here as high temperature and low humidity, and nutritional quality of the plants (Fernandes and Price 1988) should be the crucial environmental factors acting on the selective evolution of galling insects. The damage caused by herbivores on their host plants, preserved in the fossil record, showed its maximum in the Middle Eocene (a period characterized by a subtropical climate; less humidity; and a dry, defined, and cold weather), indicating a high diversity of galling organisms in xeric environments (Wilf et al. 2001). Recent studies support the assertion that galling insects are richer in species and more abundant in hygrothermal and nutritionally stressed habitats, with sclerophyllous vegetation in tropical and temperate regions (Price et al. 1998).

At the habitat or environmental scale, Fernandes and Price (1988, 1991, 1992) proposed the hygrothermal stress hypothesis that predicts that species richness and abundance of galling insects is higher in stressed hygrothermal habitats (i.e., in dry and sunny habitats) usually covered by sclerophyllous vegetation, with leaves of high phenolic compounds and low levels of nutrients (Turner 1994; Fernandes and Price 1991). The hypothesis of hygrothermal stress combines arguments of the three hypotheses about the adaptive nature of the habit of inducing galls to explain the distribution patterns of galling insects in ecological time (Fernandes et al. 2005). Also, Fernandes and Price (1991) observed that the negative relationship between altitude and the richness of the species of galling insects was dependent on the type of habitat. The richness of insect species is related to altitude in xeric habitats, but not in mesic ones at the same altitude, suggesting that the relationship between altitude and species richness is spurious and that hygrothermal stress is the key factor determining species richness of galling insects. This conclusion is supported by the latitudinal pattern: the richness of the species of galling insects is greatest in intermediate latitudes (25–40° north or south), coinciding with habitats submitted to water and nutrition stresses with sclerophyllous vegetation (e.g., Cerrado, Chaparral, and vegetation of the Mediterranean type) (Fernandes and Price 1988, 1991; Blanche and Westoby 1995; Lara and Fernandes 1996; Wright and Samways 1998; Price et al. 1998).

A few patterns have been proposed on the habit of inducing galls. First, that galls may confer effective protection against climatic variation (Price et al. 1987). Second, considering that some plant nutrients become toxic at high levels, and that the gall acts as a sink of nutrients from other plant parts (Nyman and Julkunen-Tiitto 2000), inducing insects may have more success in stressed habitats. This second pattern is based on the fact that in these habitats, plants tend to have low nutritional status (Fernandes and Price 1991), with low concentration of nutrients, and an excess of secondary compounds (Müller et al. 1987). Furthermore, gall-inducing insects are able to overcome these defensive substances, inducing a tissue free of phenolic compounds and high in nutrients (Larew 1982; Nyman and Julkunen-Tiitto 2000). The third factor that may modulate the pattern of species richness is a differential selective pressure inflicted by natural enemies, and plant resistance between xeric and mesic habitats on galling herbivores (Fernandes 1990, 1998; Fernandes and Price 1988, 1992). In summary, galls probably have an adaptive value for insects. The evolution of the habit of inducing galls can be explained by the action of different selective forces. The end result is the formation of a tissue rich in nutrients (according to the prediction of the nutritional hypothesis), and the development of galls with external structures and varying sizes in response to environmental pressures (according to the assumptions provided by the microenvironmental and enemy-free space hypotheses).

16.11 Concluding Remarks

In this chapter we discussed several aspects regarding insect galls and the main gall-inducing insect taxa, how insects select their host plants, how galls develop in plants, how they are classified, and their adaptive significance. In addition to these aspects, it should be mentioned that insects that induce galls can cause significant damage to wild and cultivated plants having a great economic impact. Just to mention one example, the coleopteran *Sternechus subsignathus* (Curculionidae) is known to feed on soybean plants and to cause gall formation on soybean stems; it is widely distributed, from the northeast to southern Brazil, and is regarded as a main pest of this important crop (e.g., Hoffmann-Campo et al. 1991, Silva 1998, Silva et al. 1998). The most recent review on the subject is almost 25 years old (Fernandes 1987); hence, a new revision is called for. We suggest that basic studies on bioecology and nutrition are important in order to support the managing strategies for species that has become pests.

REFERENCES

Abrahamson, W. G., K. D. McCrea, A. J. Whitwell, and L. A. Vernieri 1991. The role of phenolics in goldenrod ball gall resistance and formation. *Biochem. Syst. Ecol.* 19:615–22.

Abrahamson, W. G., and A. E. Weis. 1997. *Evolutionary Ecology across Three Trophic Levels: Goldenrods, Gallmakers and Natural Enemies.* New Jersey: Princeton University.

Ananthakrishnan, T. N., and R. Ananthanarayanan. 1989. *Thrips and Gall Dynamics.* Leiden: Brill Archive.

Araújo, L. M., A. C. F. Lara, and G. W. Fernandes. 1995. Utilization of *Apion* sp. (Coleoptera: Curculionidae) galls by an ant community in southeastern Brazil. *Trop. Zool.* 8:319–24.

Araújo, A. P. A., M. A. A. Carneiro, and G. W. Fernandes. 2003. Efeitos do sexo, do vigor e do tamanho da planta hospedeira sobre a distribuição de insetos indutores de galhas em *Baccharis pseudomyriocephala* Teodoro (Asteraceae). *Rev. Bras. Entomol.* 47:483–90.

Araújo, G. F. C. 2009. Reações estruturais, histoquímicas e fisiológicas de *Guapira opposita* à Cecidomyiidae galhadores. Ms. Sc. Dissertation, Federal University of Minas Gerais, Belo Horizonte, Brazil.

Arduin, M., and J. E. Kraus. 2001. Anatomia de galhas de ambrósia em folhas de *Baccharis concinna* e *B. dracunculifolia* (Asteraceae). *Rev. Bras. Bot.* 24:63–72.

Arduin, M., G. W. Fernandes, and J. E. Kraus. 2005. Morphogenesis of galls induced by *Baccharopelma dracunculifoliae* (Hemiptera: Psyllidae) on *Baccharis dracunculifolia* (Asteraceae) leaves. *Braz. J. Biol.* 65:559–71.

Bailey, R., K. Schönrogge, J. M. Cook, G. Melika, G. Csóka, C. Thuróczy, and C. N. Stone. 2009. Host niches and defensive extended phenotypes structure parasitoid wasp communities. *PLoS Biol.* 7:1–12.

Bequaert, J. 1924. Galls that secrete honeydew: A contribution to the problem as to whether galls are altruistic adaptations. *Bull. Brooklyn Entomol. Soc.* 19:101–24.

Berlin, B., and G. T. Prance. 1978. Insect galls and human ornamentation: The ethnobotanical significance of a new species of *Licania* from Amazonas, Peru. *Biotropica* 10:81–6.

Blanche, K. R. 1994. Insects induced galls on Australian vegetation. In *Gall-Forming Insects: Ecology, Physiology and Evolution*, ed. P. W. Price, W. J. Mattson, Y. N. Baranchikov, 49–55. New Hampshire: USDA Forest Service–North Central Research Station.

Blanche, K. R., and M. Westoby. 1995. Gall-forming insect diversity is linked to soil fertility via host plant taxon. *Ecology* 76:2334–7.

Bronner, R. 1973. Localisations ultrastructurales de l'activité phosphatasique acide (phosphomonoesterase acide) dans les cellule nourricieres de la galled du *Biorhiza pallida* sur le *Quercus penduculata* Ehrh. *Comptes Rendus de l'Académie des Sciences* 276:2677–80.

Bronner, R. 1992. The role of nutritive cells in the nutrition of cynipds and cecidomyiids. In *Biology of Insect-Induced Galls*, ed. J. D. Shorthouse, and O. Rohfritsch, 118–40. Oxford: Oxford University.

Bruyn, L. D. 2005. The biology, ecology, and evolution of shoot flies (Diptera: Chloropidae), In *Biology, Ecology, and Evolution of Gall-Inducing Arthropods*, ed. A. Raman, C. W. Schaefer, and T. M. Withers, v. 1, 373–405. New Hampshire: Science.

Burckhardt, D. 2005. Biology, ecology, and evolution of gall-inducing psyllids (Hemiptera: Psylloidea). In *Biology, Ecology, and Evolution of Gall-Inducing Arthropods*, ed. A. Raman, C. W. Schaefer, and T. M. Withers, v. 1, 143–57. New Hampshire: Science.

Burckhardt, D., M. M. Espírito-Santo, G. W. Fernandes, and I. Malenovský. 2004. Gall-inducing jumping plant-lice of the neotropical genus *Baccharopelma* (Hemiptera: Psylloidea) associated with *Baccharis* (Asteraceae). *J. Nat. Hist.* 38:2051–71.

Chen, M. 2008. Inducible direct plant defense against insect herbivores: A review. *Insect Sci.* 15:101–14.

Cornell, H. V. 1983. The secondary chemistry and complex morphology of galls formed by the Cynipidae (Hymenoptera): Why and how? *Am. Midl. Nat.* 136:581–97.

Coulson, R. N., and J. A. Witter. 1984. *Forest Entomology: Ecology and Management.* New York: J. Wiley.

Craig, T. P., J. Itami, L. M. Araújo, and G. W. Fernandes. 1991. Development of the insect community centered on a leaf-bud gall formed by a weevil (Coleoptera: Curculionidae) on *Xylopia aromatica* (Annonaceae). *Rev. Bras. Entomol.* 35:311–7.

Crespi, B. J., and M. Worobey. 1998. Comparative analysis of gall morphology in Australian gall thrips: The evolution of extended phenotypes. *Evolution* 52:1686–96.

Csóka, G., G. N. Stone, and G. Melika. 2005. Biology, ecology and evolution of gall-inducing Cynipidae. In *Biology, Ecology, and Evolution of Gall-Inducing Arthropods*, ed. A. Raman, C. W. Schaefer, and T. M. Withers v. 2, 573–642. New Hampshire: Science.

Dawkins, R. 1982. *The Extended Phenotype*. Oxford: Oxford University.

DeClerck, R. A., and J. D. Shorthouse. 1985. Tissue preference and damage by *Fenusa pusilla* and *Messa nana* (Hymenoptera: Tenthredinidae), leaf-mining sawflies in white birch (*Betula papyrifera*). *Can. Entomol.* 117:351–62.

Dempewolf, M. 2005. Dipteran leaf miners. In *Biology, Ecology, and Evolution of Gall-Inducing Arthropods*, ed. A. Raman, C. W. Schaefer, and T. M. Withers, v. 1, 407–29. New Hampshire: Science.

Dorchin, N., M. D. Cramer, and J. H. Hoffmann. 2006. Photosynthesis and sink activity of wasp-induced galls in *Acacia pycnantha*. *Ecology* 87:1781–91.

Dreger-Jauffret, F., and J. D. Shorthouse. 1992. Diversity of gall-inducing insects and their galls. In *Biology of Insect-Induced Galls*, ed. J. D. Shorthouse, and O. Rohfritsch, 8–33. Oxford: Oxford University Press.

Espírito-Santo, M. M., and G. W. Fernandes. 2002. Host plant effects on the development and survivorship of the galling insect *Neopelma baccharidis* (Homoptera: Psyllidae). *Austral. Ecol.* 27:249–57.

Espírito-Santo, M. M., and G. W. Fernandes. 2007. How many species of gall-inducing insects are there on earth, and where are they? *Ann. Entomol. Soc. Am.* 100:95–9.

Espírito-Santo, M. M., F. S. Neves, F. R. Andrade-Neto, and G. W. Fernandes. 2007. Plant architecture and meristem dynamics as the mechanisms determining the diversity of gall-inducing insects. *Oecologia* 153:353–64.

Fay, P. A., D. C. Hartnett, and A. K. Knapp. 1993. Increased photosynthesis and water potentials in *Silphium integrifolium* galled by cynipid wasps. *Oecologia* 93:114–20.

Fernandes, G. W. 1987. Gall forming insects: Their economic importance and control. *Rev. Bras. Biol.* 31:379–98.

Fernandes, G. W. 1990. Hypersensitivity: A neglected plant resistance mechanism against insect herbivores. *Environ. Entomol.* 19:1173–82.

Fernandes, G. W. 1992. Plant family size and age effects on insular gall forming species richness. *Global Ecol. Biogeogr. Lett.* 2:71–4.

Fernandes, G. W. 1994. Plant mechanical defenses against insect herbivory. *Rev. Bras. Entomol.* 38:421–33.

Fernandes, G. W. 1998. Hypersensitivity as a phenotypic basis of plant resistance against herbivory. *Environ. Entomol.* 27:260–7.

Fernandes, G. W., and R. P. Martins. 1985. As galhas. *Ciência Hoje* 19:58–64.

Fernandes, G. W., and P. W. Price. 1988. Biogeographical gradients in galling species richness: Tests of hypotheses. *Oecologia* 76:161–7.

Fernandes, G. W., and P. W. Price. 1991. Comparison of tropical and temperate galling species richness: The roles of environmental harshness and plant nutrient status. In *Plant–Animal Interactions: Evolutionary Ecology in Tropical and Temperate Regions*, ed. P. W. Price, T. M. Lewinsohn, G. W. Fernandes, and W. W. Benson, 91–115. New York: J. Wiley.

Fernandes, G. W., and P. W. Price. 1992. The adaptive significance of insect gall distribution: Survivorship of species in xeric and mesic habitats. *Oecologia* 90:14–20.

Fernandes, G. W., R. P. Martins, and E. Tameirão-Neto. 1987. Food web relationships involving *Anadiplosis* sp.—(Diptera: Cecidomyiidae) leaf galls on *Machaerium aculeatum* (Leguminosae). *Rev. Bras. Bot.* 10:117–23.

Fernandes, G. W., E. Tameirão-Neto, and R. P. Martins. 1988. Ocorrência e caracterização de galhas entomógenas na vegetação do Campus-Pampulha, UFMG, Belo Horizonte-MG. *Rev. Bras. Zool.* 5:11–29.

Fernandes, G. W., A. L. Souza, and C. F. Sacchi. 1993. Impact of a *Neolasioptera* (Cecidomyiidae) stem gall on its host plant *Mirabilis linearis* (Nyctaginaceae). *Phytophaga* 5:1–6.

Fernandes, G. W., M. A. A. Carneiro, A. C. F. Lara, L. A. Allain, G. I. Andrade, G. Giulião, T. C. Reis, and I. M. Silva. 1996. Galling insects on neotropical species of *Baccharis* (Asteraceae). *Trop. Zool.* 9:315–32.

Fernandes, G. W., F. M. C. Castro, and E. S. A. Marques. 1999. Leaflet abscission caused by a gall induced by *Melaphis rhois* (Aphididae) on *Rhus glabra* (Anarcadiaceae). *Int. J. of Ecol. Environ. Sci.* 25:63–9.

Fernandes, G. W., S. J. Gonçalves-Alvim, and M. A. A. Carneiro. 2005. Habitat-driven effects on the diversity of gall-inducing insects in the Brazilian Cerrado. In *Biology, Ecology, and Evolution of Gall-Inducing Arthropods*, ed. A. Raman, C. W. Schaefer, and T. M. Withers, v. 2, 693–708. New Hampshire: Science.

Fernandes, G. W., E. Almada, and M. A. A. Carneiro. 2010. Gall-inducing insect species richness as indicators of forest age and health. *Environ. Entomol.* 39:1134–40.

Ferreira, S. A., G. W. Fernandes, and L. G. Carvalho. 1990. Biologia e história natural de *Euphalerus ostreoides* (Homoptera: Psyllidae) cecidógeno de *Lonchocarpus guilleminianus* (Leguminosae). *Rev. Bras. Biol.* 50:417–23.

Formiga, A. T., S. J. M. R. Gonçalves, G. L. G. Soares, and R. M. S. Isaias. 2009. Relações entre o teor de fenóis totais e o ciclo das galhas de Cecidomyiidae em *Aspidosperma spruceanum* Müll. Arg. (Apocynaceae). *Acta Bot. Bras.* 23:93–9.

Freidberg, A. 1998. Tephritid galls and gall Tephritidae revisited, with special emphasis on myopitine galls. In *The Biology of Gall-Inducing Arthropods*, ed. G. Csóka, W. J. Mattson, G. N. Stone, and P. W. Price, 36–43. St Paul: USDA Forest Service.

Gagné, R. J. 1989. *The Plant-Feeding Midges of North America*. Ithaca: Comstock.

Gagné, R. J. 1994. *The Gall Midges of the Region Neotropical*. Ithaca: Comstock.

Gagné, R. J. 2004. A catalog of the Cecidomyiidae (Diptera) of the world. *Mem. Entomol. Soc. Wash.* 25:1–408.

Galil, J., and D. Eisikowich. 1968. Flowering cycles and fruit types of *Ficus sycomorus* in Israel. *New Phytol.* 67:745–58.

Gauld, I., and B. Bolton. 1988. *Hymenoptera*. Oxford: Oxford University.

Gonçalves, S. J. M. R., R. M. S. Isaias, F. H. A. Vale, and G. W. Fernandes. 2005. Sexual dimorphism of *Pseudotectococcus rolliniae* Hodgson; Gonçalves (Hemiptera Coccoidea Eriococcidae) influences gall morphology on *Rollinia laurifolia* (Annonaceae). *Trop. Zool.* 18:133–40.

Gonçalves, S. J. M. R., G. R. P. Moreira, and R. M. S. Isaias. 2009. A unique seasonal cycle in a leaf gall-inducing insect: The formation of stem galls for dormancy. *J. Natl. Hist.* 43:843–54.

Gonçalves-Alvim, S. J., M. L. Faria, and G. W. Fernandes. 1999. Relationships between four neotropical species of galling insects and shoot vigor. *An. Soc. Entomol. Bras.* 28:147–55.

Gonçalves-Alvim, S. J., and G. W. Fernandes. 2001a. Biodiversity of galling insects: Historical, community and habitat effects in four neotropical savannas. *Biodiversity Cons.* 10:79–98.

Gonçalves-Alvim, S. J., and G. W. Fernandes. 2010b. Comunidades de insetos galhadores (Insecta) em diferentes fisionomias do cerrado em Minas Gerais, Brasil. *Rev. Bras. Zool.* 18:289–305.

Gottlieb, O. R., M. A. C. Kaplan, and M. R. M. B. Borin. 1996. *Biodiversidade. Um Enfoque Químico-Biológico.* Rio de Janeiro: Ed. UFRJ.

Gullan, P. J., and P. S. Cranston. 2005. *The Insects: An Outline of Entomology.* Oxford: Blackwell.

Gullan, P. J., D. R. Miller, and L. G. Cook. 2005. Gall-inducing scale insects (Hemiptera: Sternorrhyncha: Coccoidea). In *Biology, Ecology, and Evolution of Gall-Inducing Arthropods,* ed. A. Raman, C. W. Schaefer, and T. M. Withers, v. 1, 159–230. New Hampshire: Science.

Hardwick, S., M. Harper, G. Houghton, A. La Salle, S. La Salle, M. Mullaney, and J. La Salle. 2005. The description of a new species of gall-inducing wasps: A learning activity for primary school students. *Austr. J. Ecol.* 44:409–14.

Hartley, S. E. 1998. The chemical composition of plant galls: Are levels of nutrients secondary compounds controlled by the gall-former? *Oecologia* 113:492–501.

Hartley, S. E., and J. H. Lawton. 1992. Host–plant manipulation by gall-insects: A test of the nutrition hypothesis. *J. Anim. Ecol.* 61:113–9.

Hawkins, B. A., and S. G. Compton. 1992. African fig wasp communities: Undersaturation and latitudinal gradients in species richness. *J. Anim. Ecol.* 61:361–72.

Hoffmann-Campo, C. B., J. R. P. Parra, and R. M. Mazzarin. 1991. Ciclo biológico, comportamento e distribuição estacional de *Sternechus subsignathus* Boheman, 1836 (Coleoptera: Curculionidae) em soja, no norte do Paraná. *Rev. Bras. Biol.* 51:615–21.

Julião, G. R., M. E. C. Amaral, and G. W. Fernandes. 2002. Galhas de insetos e suas plantas hospedeiras do Pantanal sul-mato-grossense. *Naturalia* 27:47–74.

Julião, G. R., E. M. Venticinque, G. W. Fernandes, and J. E. Kraus. 2005. Richness and abundance of gall-forming insects in the Mamirauá Varzea, a flooded Amazonian forest. *Uakari* 1:39–42.

Kerdelhué, C., J. P. Rossi, and J. Y. Rasplus. 2000. Comparative community ecology studies on old world figs and figs wasps. *Ecology* 81:2832–49.

Kirst, G. O., and H. Rapp. 1974. Zur Physiologie der Galle von *Mikiola fagi* Htg. auf Blättern von *Fagus sylvatica* L. 2. Transport 14C-markierter Assimilate aus dem befallenen Blatt und aus Nachbar-blättern in die Galle. *Biochem. Physiol. Pflanzen* 165:445–55.

Kjellberg, F., E. Jousselin, M. Hossaert-McKey, and J. Y. Rasplus. 2005. Biology, ecology, and evolution of fig-pollinating wasps (Chalcidoidea, Agaonidae). In *Biology, Ecology, and Evolution of Gall-Inducing Arthropods,* ed. A. Raman, C. W. Schaefer, and T. M. Withers, v. 2, 539–72. New Hampshire: Science.

Koch, K. E., and Y. Zeng. 2002. Molecular approaches to altered C-partitioning: Genes for sucrose metabolism. *J. Am. Soc. Horticult. Sci.* 127:474–83.

Korotyaev, B. A., A. S. Konstantinov, S. W. Lingafelter, M. Y. Mandelshtam, and M. G. Volkovitsh. 2005. Biology of gall inducers and evolution of gall induction in Chalcidoidea (Hymenoptera: Eulophidae, Eurytomidae, Pteromalidae, Tanaostigmatidae, Torymidae). In *Biology, Ecology, and Evolution of Gall-Inducing Arthropods,* ed. A. Raman, C. W. Schaefer, and T. M. Withers, v. 2, 507–538. New Hampshire: Science.

La Salle, J. 1987. New World Tanaostigmatidae (Hymenoptera, Chalcidoidea). *Contrib. Am. Entomol. Inst.* 23:1–181.

La Salle, J. 2005. Gall-inducing Coleoptera. In *Biology, Ecology, and Evolution of Gall-Inducing Arthropods,* ed. A. Raman, C. W. Schaefer, and T. M. Withers, v. 2, 239–71. New Hampshire: Science.

Labandeira, C., D. L. Dilcher, D. R. Davis, and D. L. Wagner. 1994. Ninety-seven million years of angiosperm–insect association: Paleobiological insights into the meaning of coevolution. *Proc. Natl. Acad. Sci. U. S. A.* 91:12278–82.

Labandeira, C., and T. L. Philips. 2002. Stem borings and petiole galls from Pennsylvanian tree ferns of Illinois, USA: Implications for the origin of the borer and galling functional-feeding-groups and holometabolous insects. *Palaeontographica* 264A:1–84.

Lara, A. C. F., and G. W. Fernandes. 1994. Distribuição de galhas de *Neopelma baccharidis* (Homoptera: Psyllidae) em *Baccharis dracunculifolia* (Asteraceae). *Rev. Bras. Biol.* 54:661–8.

Lara, A. C. F., and G. W. Fernandes. 1996. The highest diversity of galling insects: Serra do Cipó, Brazil. *Biol. Lett.* 3:111–14.

Larew, H. G. 1982. A comparative anatomical study of galls caused by the major cecidogenetic groups, with special emphasis on the nutritive tissue. PhD thesis, Oregon State University, Corvallis.

Larew, H. 1992. Fossil galls. In *Biology of Insect-Induced Galls,* ed. J. D. Shorthouse and O. Rohfritsch, 51–9. Oxford: Oxford University Press.

Larson, K. C. 1998. The impact of two gall-forming arthropods on the photosynthetic rates of their host. *Oecologia* 115:161–6.

Larson, K. C., and T. G. Whitham. 1991. Manipulation of food resources by a gall-forming aphid: The physiology of sink–source interactions. *Oecologia* 88:15–21.

Leite, G. L. D., R. V. S. Veloso, A. C. R. Castro, P. S. N. Lopes, and G. W. Fernandes. 2007. Efeito do AIB sobre a qualidade e fitossanidade dos alporques de influência da *Caryocar brasiliense* Camb. (Caryocaraceae). *Rev. Árvore* 31:315–20.

Lemos Filho, J. P., J. C. S. Christiano, and R. M. S. Isaias. 2007. Efeitos da infestação de insetos galhadores na condutância e taxa relativa de transporte de elétrons em folhas de *Aspidosperma australe* Mücll. Arg. e de *A. spruceanum* Benth. ex Müell. Arg. *Rev. Bras. Biociências* 5:1152–4.

Levin, D. A. 1973. The role of trichomes in plant defense. *Q. Rev. Biol.* 48:3–15.

Liljeblad, J., and F. A. Ronquist. 1998. Phylogenetic analysis of higher-level gall wasps relationships. *Syst. Entomol.* 23:229–52.

Lucas, P. W., I. M. Turner, N. J. Domuny, and N. Yamachita. 2000. Mechanical defences to herbivory. *Ann. Bot.* 86:913–20.

Maia, V. C. 2001a. The gall midges (Diptera, Cecidomyiidae) from three restingas of Rio de Janeiro State, Brazil. *Rev. Bras. Zool.* 18:583–629.

Maia, V. C. 2001b. New genera and species of gall midges (Diptera, Cecidomyiidae) from three restingas of Rio de Janeiro State, Brazil. *Rev. Bras. Zool.* 18:1–32.

Maia, V. C. 2005. Catálogo dos Cecidomyiidae (Diptera) do Estado do Rio de Janeiro. *Biota Neotropica* 5:1–15.

Maia, V. C., and G. W. Fernandes. 2004. Insect galls from serra de São José (Tiradentes, Mg, Brazil). *Braz. J. Biol.* 64:423–45.

Maia, V. C., and G. W. Fernandes. 2006. A new genus and species of gall midge (Diptera: Cecidomyiidae) associated with *Parkia pendula* (Fabaceae, Mimosoideae). *Rev. Bras. Entomol.* 50:1–5.

Mani, M. S. 1964. *Ecology of Plant Galls.* The Hague: W. Junk.

McArthur, E. D. 1986. Specificity of galls on *Chrysothamnus nauseosus* subspecies. In S*ymposium on the Biology of Artemisia and Chrysothmanus,* 205–10. Ogden, UT: Intermountain Research Station.

McArthur, E. D., C. F. Tiernan, and B. L. Welch. 1979. Subspecies specificity of galls forms on *Chrysothamnus nauseosus. Great Basin Nat.* 39:81–87.

Meyer, J. 1987. *Plant Galls and Gall Inducers.* Berlin: Gebruder Borntraeger.

Miller, W. E. 2005. Gall-inducing Lepidoptera. In *Ecology and Evolution of Plant-Feeding Insects in Natural and Man-Made Environments,* ed. A. Raman, 431–68. New Delhi: National Institute of Ecology.

Moreira, R. G., G. W. Fernandes, E. D. Almada, and J. C. Santos. 2007. Galling insects as bioindicators of land restoration in an area of Brazilian Atlantic Forest. *Lundiana* 8:107–12.

Motta, L. B., J. E. Kraus, A. Salatino, and M. L. F. Salatino. 2005. Distribution of metabolites in galled and non-galled foliar tissues of *Tibouchina pulchra. Biochem. Syst. Ecol.* 22:971–81.

Mound, L. A., and B. D. Kranz. 1997. Thysanoptera and plant galls: Towards a research programme. In *Ecology and Evolution of Plant-Feeding Insects in Natural and Man-Made Environments,* ed. A. Raman, 11–24. New Delhi: National Institute of Ecology.

Moura, M. Z. D., G. L. G. Soares, and R. M. S. Isaias. 2009. Ontogênese da folha e das galhas induzidas por *Aceria lantanae* Cook (Acarina: Eriophyidae) em *Lantana camar* L. (Verbenaceae). *Revista Brasileira Botanica* 32:271–82.

Müller, R. N., P. J. Kalisz, and T. W. Kimmerer. 1987. Intraspecific variation in production of astringent phenolics over a vegetation-resource availability gradient. *Oecologia* 72:211–5.

Nyman, T., and R. Julkunen-Tiitto. 2000. Manipulation of the phenolic chemistry of willows by gall-inducing sawflies. *PNAS* 97:13184–87.

Oliveira, D. C., and R. M. S. Isaias. 2009. Influence of leaflet age in anatomy and possible adaptive values of the midrib gall of *Copaifera langsdorffii* (Fabaceae: Caesalpinioideae). *Rev. Biol. Trop.* 57:293–302.

Oliveira, D. C., and R. M. S. Isaias. 2010a. Redifferentiation of leaflet tissues during midrib gall development in *Copaifera langsdorffii* (Fabaceae). *S. Afr. J. Bot.* 76:239–48.

Oliveira, D. C., and R. M. S. Isaias. 2010b. Cytological and histochemical gradients induced by a sucking insect in galls of *Aspidosperma australe* Arg. Muell (Apocynaceae). *Plant Sci.* 178:350–8.

Oliveira, D. C., J. C. S. Christiano, G. L. G. Soares, and R. M. S. Isaias. 2006. Reações de defesas químicas e estruturais de *Lonchocarpus muehlbergianus* Hassl. (Fabaceae) à ação do galhador *Euphalerus ostreoides* Crawf. (Hemiptera, Psyllidae). *Rev. Bras. Bot.* 29:657–67.

Oliveira, D. C., M. M. Drummond, A. S. F. P. Moreira, G. L. G. Soares, and R. M. S. Isaias. 2008. Potencialidades morfogênicas de *Copaifera langsdorffii* Desf. (Fabaceae): Super-hospedeira de herbívoros galhadores. *Rev. Biol. Neotrop.* 5:31–9

Oliveira, D. C., T. A. Magalhães, R. G. S. Carneiro, M. N. Alvim, and R. M. S. Isaias. 2010c. Do Cecidomyiidae galls of *Aspidosperma spruceanum* (Apocynaceae) fit the pre-established cytological and histochemical patterns? *Protoplasma* 242:81–93.

Price, P. W. 1997. *Insect Ecology.* New York: J. Wiley.

Price, P. W. 2003. *Macroevolutionary Theory on Macroecological Patterns.* Cambridge: Cambridge University Press.

Price, P. W., and H. Pschorn-Walcher. 1988. Are galling insects better protected against parasitoids than exposed feeders? A test using tenthredinid sawflies. *Ecol. Entomol.* 13:195–205.

Price, P. W., G. L. Waring, and G. W. Fernandes. 1986. Hypotheses on the adaptive nature of galls. *Proc. Entomol. Soc. Wash.* 88:361–3.

Price, P. W., G. W. Fernandes, and G. L. Waring. 1987. Adaptive nature of insect galls. *Environ. Entomol.* 16:15–24.

Price, P. W., G. W. Fernandes, A. C. F. Lara, J. Brawn, D. Gerling, H. Barrios, M. G. Wright, S. P. Ribeiro, and N. Rothcliff. 1998. Global patterns in local number of insect galling species. *Trop. Zool.* 25:581–91.

Raman, A., C. W. Schaefer, and T. M. Withers. 2005. *Biology, Ecology, and Evolution of Gall-Inducing Arthropods,* v. 2. New Hampshire: Science.

Rasmann, S., and A. A. Agrawal. 2009. Plant defense against herbivory: Progress in identifying synergism, redundancy, and antagonism between resistance traits. *Curr. Opin. Plant Biol.* 12:473–8.

Rehill, B. J., and J. C. Schultz. 2003. Enhanced invertase activities in the galls of *Hormaphis hamamelidis. J. Chem. Ecol.* 29:2703–20.

Rohfritsch, O., and J. D. Shorthouse. 1982. Insect galls. In *Molecular Biology of Plant Tumors,* ed. G. Kahl and J. S. Schell, 131–52. New York: Academic Press.

Roininen, H., T. Nyman, and A. Zinovjev. 2005. Biology, ecology, and evolution of gall-inducing sawflies (Hymenoptera: Tenthredinidae and Xyelidae). In *Biology, Ecology, and Evolution of Gall-Inducing Arthropods,* ed. A. Raman, C. W. Schaefer, and T. M. Withers, v. 2, 467–94. New Hampshire: Science.

Ronquist, F. 1995. Phylogeny, classification and evolution of the Cynipoidea. *Zool. Scr.* 28:139–64.

Root, R. B. 1967. The niche exploitation pattern of the blue-grey gnatcatcher. *Ecol. Monogr.* 37:317–50.

Roskam, J. C. 1992. Evolution of the gall-inducing insects guild. In *Biology of Insect-Induced Galls,* ed. J. D. Shorthouse, and O. Rohfritsch. 34–9. Oxford: Oxford University.

Rossi, A. M., P. D. Stiling, D. Strong, and D. M. Johnson. 1992. Does gall diameter affect the parasitism rate of *Asphondylia borrichiae* (Diptera: Cecydomiidae)? *Ecol. Entomol.* 17:149–54.

Sá, C., R. M. S. Isaias, F. A. O. Silveira, J. C. Santos, and G. W. Fernandes. 2009. Anatomical, histochemical and developmental aspects of an ambrosia leaf gall induced by *Schizomyia macrocapillata* (Diptera: Cecidomyiidae) on *Bauhinia brevipes* (Fabaceae). *Rev. Bras. Bot.* 32:319–27.

Schaefer, C. W. 2005. Gall-inducing heteropterans (Hemiptera) In *Biology, Ecology, and Evolution of Gall-Inducing Arthropods,* ed. A. Raman, C. W. Schaefer, and T. M. Withers, v. 1, 231–38. New Hampshire: Science.

Shannon, R. E., and J. W. Brewer. 1980. Starch and sugar levels in 3 coniferous insect galls. *Z. Angew. Entomol.* 89:526–33.

Shorthouse, J. D., and O. Rohfritsch. 1992. *Biology of Insect-Induced Galls.* New York: Oxford University Press.

Silva, I. M., G. I. Andrade, G. W. Fernandes, and J. P. Lemos Filho. 1996. Parasitic relationships between a gall forming insect *Tomoplagia rudolphi* (Diptera: Tephritidae) on its host plant (*Vernonia polyanthes,* Asteraceae). *Ann. Bot.* 78:45–8.

Silva, M. T. B. da. 1998. Aspectos ecológicos de *Sternechus subsignathus* Boheman (Coleoptera: Curculionidae) em soja no plantio direto. *An. Soc. Entomol. Brasil* 27:47–53.

Silva, M. T. B. da, N. Neto, and C. B. Hoffmann-Campo. 1998. Distribution of eggs, larvae and adult of *Sternechus subsignathus* Boheman on soybean plants a no-till system. *An. Soc. Entomol. Brasil* 27:513–18.

Souza, A. L. T., G. W. Fernandes, J. E. C. Figueira, and M. O. Tanaka. 1998. Natural history of a gall-inducing weevil *Collabismus clitellae* (Coleoptera: Curculionidae) and some effects on its host plant *Solanum lycocarpum* (Solanaceae) in southeastern Brazil. *Ann. Entomol. Soc. Am.* 91:404–9.

Souza, A. L. T., M. O. Tanaka, G. W. Fernandes, and J. E. C. Figueira. 2001. Host plant response and phenotypic plasticity of a galling weevil (*Collabismus clitellae*: Curculionidae). *Austral Ecol.* 26:173–8.

Souza, R. A., R. Nessim, J. C. Santos, and G. W. Fernandes. 2006. Influence of *Apion* sp. (Apionidae: Coleoptera) stems-galls on the induced resistance and leaf area of *Diospyros hispida* (Ebenaceae). *Rev. Bras. Entomol.* 50:433–5.

Souza, S. C. P. M., J. E. Kraus, R. M. S. Isaias, and L. J. Neves. 2000. Anatomical and ultrastructural aspects of leaf galls in *Ficus microcarpa* induced by *Gynaikothrips ficorum* Marshal (Thysanoptera) *Acta Bot. Bras.* 14:57–69.

Stern, D. L. 1995. Phylogenetic evidence that aphids, rather than plants, determine gall morphology. *Proc. R. Soc. Lond. B* 256:203–9.

Stone, G. N., and J. M. Cook. 1998. The structure of cynipid oak galls: Patterns in the evolution of an extended phenotype. *Proc. R. Soc. Lond. B* 265:979–88.

Stone, G. N., and K. Schönrogge. 2003. The adaptive significance of insect gall morphology. *Trends Ecol. Evol.* 18:512–22.

Stone, G. N., K. Schönrogge, R. J. Atkinson, D. Bellido, and J. Pujade-Villar. 2002. The population biology of oak gall wasps (Hymenoptera: Cynipidae). *Annu. Rev. Entomol.* 47:633–68.

Stone, G. N., R. W. J. M. van der Ham, and J. G. Brewer. 2008. Fossil oak galls preserve ancient multitrophic interactions. *Proc. R. Soc. Lond. B* 275: 2213–9.

Taper, M. L., and T. J. Case. 1987. Interactions between oak tannins and parasite community structure: Unexpected benefits of tannins to cynipid gall-wasps. *Oecologia* 71:254–61.

Triplehorn, C. A., and N. F. Johnson. 2005. *Borror and Delong's Introduction to the Study of Insects*. Belmont: Brooks Cole Thompson.

Turner, I. M. 1994. Sclerophylly: Primarily protective? *Funct. Ecol.* 8:669–75.

Vecchi, C. 2004. Reações diferenciais a herbivoros galhadores em espécies de Melastomataceae. PhD thesis, University of São Paulo, São Paulo.

Veldtman, R., and M. A. McGeoch. 2003. Gall-forming insect species richness along a non-scleromorphic vegetation rainfall gradient in South Africa: The importance of plant community composition. *Austr. J. Ecol.* 28:1–13.

Waring, G. L., and P. W. Price. 1990. Plant water stress and gall formation (Cecidomyiidae: *Asphondylia* spp.) on creosote bush. *Ecol. Entomol.* 15:87–95.

Weiblen, G. D. 2002. How to be a fig wasp. *Annu. Rev. Entomol.* 47:299–330.

Weis, A. E. 1982. Use of a symbiotic fungus by the gall maker *Asteromyia carbonifera* to inhibit attack by the parasitoid *Torymus capite*. *Ecology* 63:1602–5.

Weis, A. E., W. G. Abrahamson, and M. C. Andersen. 1992. Variable selection on *Eurostoma*'s gall size. I: The extent and nature of variation in phenotypic selection. *Evolution* 46:1674–97.

Weis, A. E., W. G. Abrahamson, and K. D. McCrea. 1985. Host gall size and oviposition success by the parasitoid *Eurytoma gigantea*. *Ecol. Entomol.* 10:341–8.

Weis, A. E., R. Walton, and C. L. Crego. 1988. Reactive plant tissue sites and the population biology of gall makers. *Annu. Rev. Entomol.* 33:467–86.

Whitham, T. G. 1992. Ecology of *Pemphigus* gall aphids. In *Biology of Insect-Induced Galls*, ed. J. D. Shorthouse and O. Rohfritsch, 225–37. Oxford: Oxford University.

Wiebes, J. T. 1979. Co-evolution of figs and their insect pollinators. *Annu. Rev. Ecol. Syst.* 10:1–12.

Wilf, P., C. C. La Bandeira, K. R. Johnson, P. D. Coley, and A. D. Cutter. 2001. Insect herbivory, plant defenses, and early Cenozoic climate change. *Proc. Natl. Acad. Sci. U. S. A.* 98:6221–6.

Wittstock, U., and J. Gershenzon. 2002. Constitutive plant toxins and their role in defense against herbivores and pathogens. *Curr. Opin. Plant Biol.* 5:1–8.

Woodman, R. L., and G. W. Fernandes. 1991. Differential mechanical defense: Herbivory, evapotranspiration and leaf-hairs. *Oikos* 60:11–9.

Wool, D. 2004. Galling aphids: Specialization, biological complexity, and variation. *Annu. Rev. Entomol.* 49:175–92.

Wool, D. 2005. Gall-inducing aphids: Biology, ecology, and evolution. In *Biology, Ecology, and Evolution of Gall-Inducing Arthropods*, ed. A. Raman, C. W. Schaeffer, and T. M. Withers. 73–132. New Hampshire: Science.

Wright, M. G., and M. J. Sanways. 1998. Insect species richness in a diverse flora: Gall-insects in the Cape Floristic region, South Africa. *Oecologia* 115:427–33.

Yukawa, J., and O. Rohfritsch. 2005. Biology of ecology of gall-inducing Cecidomyiidae (Diptera). In *Biology, Ecology, and Evolution of Gall-Inducing Arthropods*, ed. A. Raman, C. W. Schaeffer, and T. M. Withers, 273–304. New Hampshire: Science.

17

Detritivorous Insects

Julio Louzada and Elizabeth S. Nichols

CONTENTS

17.1 Introduction 397
17.2 Integrated View of Detritus as Food Resource 398
    17.2.1 Detritus Abundance 398
    17.2.2 Detritus Distribution 400
    17.2.3 Detritus Use 401
    17.2.4 Resource Allocation 402
    17.2.5 Population and Community Consequences of Detritus Use 402
17.3 Adaptations for Using Detritus as Food 403
    17.3.1 Adaptations to Access Nutrients in Low Availability 403
    17.3.2 Adaptations for Use of High-Availability Detritus That Is Unpredictable
        in Space and Time 405
17.4 Mutualisms between Insects and Microorganisms: Role of Coprophagy in Detritus Use 406
    17.4.1 External and Internal Rumen in Detritivores Insects 407
17.5 Ecological Functions of Detritivores Insects 407
    17.5.1 Leaf Litter Decomposition Rates 407
    17.5.2 Waste Removal and Related Functions 408
    17.5.3 Biological Control of Other Detritivores 409
17.6 Final Considerations 410
References 410

17.1 Introduction

The path traced by each chemical element within a given ecosystem is incredibly complex (Swift et al. 1979). Beginning with primary production—the process of transforming simple chemical elements, obtained through abiotic means, into complex molecules by autotrophic organisms via photosynthesis—approximately 1–5% of the available light energy that reaches the earth is transformed into plant tissue through photosynthesis (Begon et al. 2006). This energy and the nutrients obtained by these autotrophic organisms are subsequently consumed by a series of heterotrophic organisms, ranging from herbivores to predators and parasites, in a process termed "secondary production." Both autotrophic and heterotrophic organisms constantly excrete energy, nutrients, and materials into the environment over the course of their lives, in the form of leaves, hair, feces, urine, and ultimately, dead bodies (Swift et al. 1979; Begon et al. 2006). The detritivorous insects that consume these materials, and transfer the energy and nutrients they contain back into abiotic and biotic components of ecosystems, are the subject of this chapter.

In this chapter we discuss the principal points involved in insect use of detritus as a food resource. Their interactions with detritus are complex, and over time have resulted in a series of unique morphological, physiological, and behavioral adaptations that together have implications for the population and community structure of a vast number of species on Earth. We succinctly present the nutritional mechanisms and ecological aspects that affect insect use of these resources, as well as the consequences of this resource use for environmental services, and control of those detritivorous insects that can cause economic damage.

A central aim of this chapter is to expand this perspective, by providing information on several nutritional and ecological aspects of the use of detritus as food resources by insects, and the implications of detritus use for population and community ecology. Consequently, first it is necessary to define "resource," a term of some controversy across the ecological literature. For the purpose of this chapter, we will consider a resource as whatever substance or factor that can lead to an increase in population growth rates when its availability in the environment increases (Tilman 1982). This concept of resource is contingent on three factors. A resource must be (a) consumed, (b) limiting, and (c) have a direct effect on fitness (the ability to survive and reproduce).

17.2 Integrated View of Detritus as Food Resource

Detritus is a food resource that supports trophic food chains in practically every realm of the heterotrophic kingdom, among them innumerable species of insects. Its characteristics as a food resource are directly connected with the quality (the relative amount of available energy and nutrients) and its predictability in both space and time (Atinkson and Shorroks 1981; Hanski 1987). Certain types of detritus, such as fallen trees, twigs, and leaf litter, provide only minute amounts of available energy. Others, like feces and carcasses, represent enormous and rich concentrations of energy and nutrients. Similarly, some detritus resources are available nearly constantly in both space and time (e.g., leaf litter), while others are near impossible to predict in space and highly ephemeral in time (e.g., carcasses). Both of these aspects play an enormous role in the underlying ecology of detritivorous insects.

The distribution of detritus food resources in space and time interacts with the way these resources are used by organisms, particularly with how individuals allocate resources across functions and structures. These differences offer opportunities for natural selection, and ultimately have present-day consequences at the level of individual, populations, and communities. After consumption, resources acquired during feeding can be allocated across a diversity of activities and structures, such as movement, growth, reproduction, and competitive interactions. The integrated framework presented in Figure 17.1 allows a detailed assessment of consequences for populations and community structure of detritus use by insects.

17.2.1 Detritus Abundance

Resource abundance in the environment is key to studying the relationships between individuals, and consequently, of populations within any given biological community. Despite the constant production

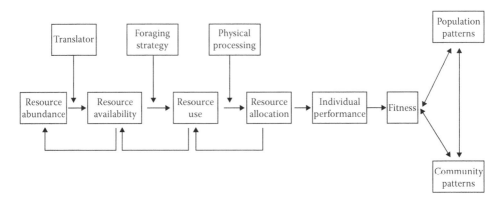

FIGURE 17.1 Steps involved in detritus-based food webs, with consequences for population and community structure. (Modified from Wiens, J. A. 1984. Resource systems, populations, and communities. In *A New Ecology: Novel Approaches to Interactive Systems*, ed. P. W. Price, C. N. Slobodchikoff, and W. S. Gand, 397–436. New York: John Wiley & Sons.)

of detritus in every ecosystem across the planet, empirical information on the availability of specific detritus types is surprisingly limited, and biased to those detritus types that are relatively straightforward to measure. For example, the production of leaf litter is fairly well understood across a range of ecosystems (Table 17.1) while the production dynamics of more ephemeral resources such as feces, vertebrate and invertebrate carcasses, or fungal fruiting bodies remain incognito for the majority of terrestrial ecosystems.

Given the difficulties in estimating the abundance of detritus in ecosystems, many authors have opted for indirect approaches to measuring detritus inputs (Bailey and Putman 2001; Laing et al. 2003). These methods associate, for example, vertebrate abundance with production of feces and carcasses over time. The abundance of detritus of animal origin can be presumed directly associated with overall animal biomass in any given area. The largest and most diverse mammal community on the planet, for example, can be found in the savannas of Africa, leading presumably to the largest known abundance of feces and certainly vertebrate carcasses for the copro/necrophagous insect community. Cambefort (1984) estimates that in East African savannas, detritivorous beetles in the subfamily Scarabaeinae incorporate an estimated 1000 kg/ha of herbivore feces per year into savanna soils. Alternatively, indirect measurements of detritus resources in an environment may be based on population or density estimates of those insects that feed on a given detritus (Nichols et al. 2009).

The abundance of detritus can vary in space as well as time. Detritus can be produced in quantities that are relatively uniform and homogenous throughout both space and time, or in highly ephemeral and concentrated patches that are difficult to predict in time. For example, mammalian feces, carcasses of small animals, and fruiting bodies of fungi in decomposition are often only available to scavengers for very short periods, being considered as temporally ephemeral resources (Hanski 1987). In this case, the resources behave like pulses in time, the insects that require a series of adjustments to its use. These resources often have unpredictable spatial and temporal abundance and availability (Hanski 1981). Several detritus types can behave like pulses of resources, but with one significant difference—the predictability of seasonal abundance or availability (Figure 17.2). A good example of this type are the detritus of decaying fruit, which are usually associated with a seasonal pattern of fruit and a territorial space limited to the size of the tree production (Muller-Landau and Hardesty 2005). The same can be said for flowers, which in some cases also represent important food resources for some scavengers. Alternatively, production of decomposing tree leaves and branches is a highly continuous pattern, making leaf litter

TABLE 17.1

Leaf Litter Amounts (Dry Weight/Day) Produced in Different Forest Ecosystems

Location	Latitude	Annual Precipitation (mm/m^2)	Forest Physiognomy	Leaf Litter (mg/ha)
Ivory Coast	6°N	1280	Gallery forest	6.2
Nigeria	7°N	1200	Mixed dry forest	5.6
Senegal	14°N	300	Savanna	1.2
Zaire	1°N	1700	Mixed forest	12.4
Zaire	11°S	1275	Evergreen tropical forest	9.1
Brazil	2°S	1700	Evergreen tropical forest	7.3
Colombia	4°N	8400	Lowlands tropical forest	8.5
Venezuela	2°N	3500	Amazonian savanna	5.6
Costa Rica	10°N	1500	Deciduous dry forest	7.8
Panama	9°N	2000	Moist tropical forest	11.4
Australia	17°S	2100	Moist tropical forest	10.4
Malaysia	3°N	2000	Tropical Dipterocarp forest	8.9
Papua New Guinea	8°S	1600	Semideciduous moist forest	8.8

Source: Modified from Vitousek, P., *Am. Nat.* 119, 553, 1982.

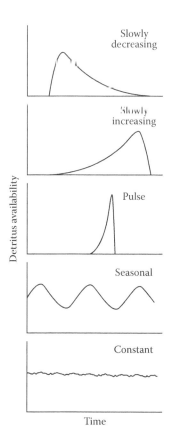

FIGURE 17.2   Patterns of abundance and temporal availability of detritus food resources.

one of the most temporally and spatially predictable detritus types in forest environments (Figure 17.3; Aerts and Caluwé 1997; González and Seastedt 2001).

17.2.2 Detritus Distribution

Generally, only part of the detritus produced and present in an area can actually be effectively used by scavengers. In this context several factors act as "translators" (Figure 17.1) that determine the relative proportion of detritus resources available to scavengers. These include characteristics of the abiotic

FIGURE 17.3   Forest litter, a highly predictable detritus.

environment, the detritus itself, and the composition of the scavenger community. We address these characteristics in detail in the sections ahead, but in the interests of illustration, we can take the simple example of high cellulose content in dead organic matter. Even large quantities of woody material or leaf litter represent a surprisingly low availability of food resources to detritivores (Berrie 1975). Interactions between different classes of translators help mediate the use of these cellulose fibers as food energy. Termites, for example, rely on the overall interaction between a consortium of gut-dwelling prokaryotic organisms and their cellulase enzymes; the wood can actually be a food resource for insect detritivores depending on its chemical nature (Smith and Douglas 1987).

Abiotic factors may also play a big role in the availability of detritus. The temperature and soil moisture, for example, can quickly change the nutritional aspects of detritus as to render it inaccessible to the community of detritivores (Swift et al. 1979; Jurgensen et al. 2004).

Temporal patterns in the abundance of detritus also greatly affect detritus-consuming insects, and the consequent development of foraging strategies (Figure 17.1). For example, the carcasses of large animals are a good example of a detritus type that appears suddenly (few hours after the death of the animal) and declines in its availability over time owing to the accumulation of toxic substances, consumption of matter itself, and competitive interactions that may limit the presence of some groups of organisms (Payne 1965). Other detritus types can slowly increase in abundance or availability with time. Decaying logs, for example, become available to scavengers in general when the tree falls to the ground. However, its availability as food source for insects depends on the slow process of colonization by fungi, decomposing bacteria, and gallery-burrowing insects galleries (e.g., passalid beetles) (Jurgensen et al. 2004).

17.2.3 Detritus Use

The spatiotemporal variations in detritus abundance can substantially affect populations of detritus-dependent insects, and have significant effects on the complex of adaptations that allow their use of detritus as a food resource. The organisms use some of the resources available to them aimed at meeting its energy demands for reproduction, use of space, growth, and other functions. A large variety of factors act as constraints on access to resources available, ranging from interspecific and intraspecific competition, foraging patterns, physiological resource needs, and the underlying quality of detritus as food sources. These may alter patterns of food preference and lead to extreme specializations for the use of detritus in the environment.

Scavenging insects have a range of morphological adaptations that enable the consumption of detritus, yet often at the cost of consuming other available resources in the environment. A good example is the beetle family Passalidae (Figure 17.4), with their body morphology specialized to dig and live deep inside galleries dug in rotting logs, and jaws highly adapted for wood chewing.

From an evolutionary point of view, one expects that insects are able to select their food on the basis of detectable correlates of characteristics that contribute positively to survival and reproduction. Valiela et al. (1979) and Valiela and Rietsima (1984) showed that detritivores utilize phenolic compounds and

FIGURE 17.4 A member of Passalidae insect family. Observe the body morphology and jaws adapted to chewing through decaying logs.

protein-based detritus and chemical properties that indicate both food quality and palatability. It has long been known that chemical changes in nitrogen and energy content of some detritus types can also affect the biomass (Tenore 1977, 1981), growth rate, and density of their respective detritivores (Findlay 1982). In other studies, feeding biology, phenolic compounds (Lincoln et al. 1982), and ATP content (Ward and Commins 1979) have been identified as important for invertebrate detritivores, resulting in patterns of resource selection expressed in both adults and larvae.

17.2.4 Resource Allocation

The resources acquired by scavengers are allocated across a variety of physiological functions within the body, such as metabolism, growth, movement, and reproduction. Energy and nutrients allocated to any given function are inherently unavailable for others, creating trade-offs at the level of individuals. Slight differences in the allocation patterns of these individuals ultimately can result in different individual performances (Wiens 1984). As for most taxa, heightened performance on one resource type or quantity typically comes at a cost in performance on other resource types (Hardin 1960) or quantities; these inter-specific ecological trade-offs promote the coexistence of competing species. The subsequent impacts on community structure and ecological function have been the subject of much ecological research (Levins and Culvert 1971; Vincent et al. 1996). For detritus-feeding insects, the spatial and temporal pattern of detritus availability interacts with the foraging behavior of individuals and their particular pattern of physiological functions related to resource allocation, to help structure the exact community composition at any single point in space and time.

17.2.5 Population and Community Consequences of Detritus Use

The ability to survive and reproduce greatly affect population and community patterns, through influencing the abundance of any given species and therefore its relationship with the rest of the community. Access to a particular resource can contribute to a greater ability to rapidly produce new individuals or to the accumulation of energy to be directed to reproductive events (Begon et al. 2006).

The detritivore community's relationship with detritus can be generally be called "noninteractive" (Monro 1967) because the use of waste by scavengers does not directly affect the rate of production of this resource by the system as a whole. Additionally, we can further distinguish between "reactive" detritus food webs, where changes in the rate of supply of the detritus of population responses are accompanied by changing abundances of scavengers, or "nonreactive," where changes in the rate of supply of detritus do not affect the status of the detritivore population (Caughley and Lawton 1981). Examples of reactive systems include the interaction between necrophagous flies and provision of carrion-feeding beetles and fungi in decomposition to the provision of fungi in a forest, while a nonreactive system may be the leaf litter and humus-eating insects, whose population fluctuations in time are relatively decoupled from their very steady input of food resources.

In communities associated with noninteractive food webs, the influence of competition as a structuring factor on community and population stability will be affected mainly by the quality of the detritus that the species used and by its relative unpredictability in space and time (Figure 17.5). Populations that behave in a reactive mode to the supply of detritus usually present population cycles and chaotic community dynamics, marked by lottery dynamics and the coexistence of competing species (Atinkson and Shorrocks 1981; Hanski 1981). These food webs tend to be based on resources that are ephemeral and spatially unpredictable and, at the same time, have a large amount of available energy (Figure 17.5). Several species that use resources with these characteristics developed evolutionary strategies for the location and rapid colonization of resource and its subsequent use as food or breeding substrate. Detritivorous Diptera and Scarabaeinae dung beetles are good examples of species with these types of adaptation.

On the other hand, nonreactive populations behave quite predictably in terms of population fluctuation and are affected mainly by factors of the physical environment rather than by competition. In this case the ecological and evolutionary context of a given species is much more focused on the establishment of strategies to access the available energy in the waste (or improve its nutritional quality) through

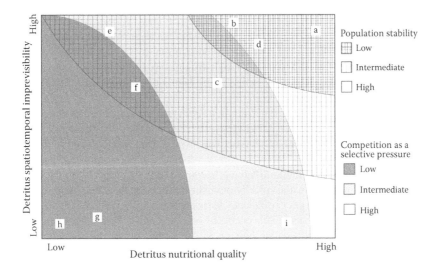

FIGURE 17.5 Distribution of detritus along the two axes of nutritional quality and predictability in space and time and their consequences on ecological community of scavengers: a) vertebrate carcasses, b) fruiting fungal bodies, c) herbivore feces, d) omnivore feces, e) rotting fruit, f) decaying flowers, g) leaf litter, h) decomposing tree trunks, and i) garbage nests of *Attini* ant species.

mutualistic partnerships, as is the case of passalid beetles (Figure 17.3) and most species of termites (Shellman-Reeve 1994), although always there are exceptions (Korb and Linsenmair 2001).

17.3 Adaptations for Using Detritus as Food

Detritivorous insects have evolved a number of unique adaptations. These can be grouped into along the same gradient of abundance versus availability. Here certain adaptations serve to either access the nutrients in those difficult-to-digest detritus present in constant abundance over space and time, or access the easily digestible and high-nutrition detritus resources that are unpredictable in space and ephemeral in time.

17.3.1 Adaptations to Access Nutrients in Low Availability

Some detritus types contain energy and nutrients in considerable quantity, although these are bound by macrostructural molecules and polymers that make them inaccessible without novel strategies to transform these molecules into smaller and more easily digestible forms. The most striking examples of detritus with these characteristics are those with large amounts of cellulose and lignin; leaf litter and decaying trunks in terrestrial ecosystems are good examples. Cellulose is by far the largest source of nonfossil carbon present on Earth, and much of the noncellulose, carbon-rich plant organic matter disappears within just weeks of a plant's death (Martin 1991). If we consider the number and diversity of insects that use plants as food resources, it is amazing how few are able to access the energy and matter contained in cellulose.

The digestion of cellulose is a complex process involving a series of enzymes with different modes of action (Coughlan and Ljungdahl 1988). The development of a complete cellulolytic enzyme system is common only in microorganisms (e.g., bacteria, fungi, and protozoa) and relatively uncommon in animals. The digestion of cellulose by animals is most often mediated by symbiotic cellulolytic microorganisms. Any discussion about the use of cellulose as food by the insects requires that two questions be addressed in an evolutionary ecology framework: (a) why is cellulose digestion in insects so rare, and (b) why is symbiont-dependent cellulose digestion more common than symbiont-independent cellulose digestion?

The digestion of cellulose by insects has been reported in at least 78 insect species from 20 families representing eight orders (Table 17.2) (Martin 1991). Orders where this digestion capacity is relatively cellulolytic are Thysanura (family Lepismatidae), Isoptera (termites), and three families of Coleoptera (Anobiidae, Buprestidae, and Cerambycidae). Cellulolytic capacity is probably also common in Scarabid beetles, cockroaches (family Blatoidea), as well as in the Tipulidae dipterans and Siricidae hymenopterans (Martin 1991).

The digestion of cellulose is relatively rare in detritivores in general, and digestive efficiency in this ecological group is usually low or moderate, ranging between 11% and 50% (Martin 1991). Among insects, termites are the most efficient digesters of cellulose, with assimilation efficiencies reaching 99% (Martin 1991). Wood-decaying (xylophagous) larvae of the families Siricidae, Anobiidae, Buprestidade, and Cerambicidae are less efficient, with assimilation efficiencies ranging from 12% to 68%. The digestion of cellulose also occurs in species with typically omnivorous diets, such as bookworms (Lepismatidae) and cockroaches, where the assimilation efficiency varies between 40% and 90%.

Cellulose digestion by detritivorous insects occurs by one of four different mechanisms (Martin 1991), including (a) the capacity to support cellulolytic protozoan symbionts in the hindgut; (b) exploitation of cellulolytic bacteria residing in the hindgut; (c) use of fungal cellulase enzymes, consumed along

TABLE 17.2

Distribution of Cellulose Digestion Capacity across Different Insect Orders

Order/Family	Number of Species Capable of Eating Cellulose
Thysanura	—
Lepismatidae	5
Orthoptera	—
Gryllidae	1
Cryptocericidae	1
Blattidae	1
Blaberidae	1
Isoptera	—
Mastotermitidae	1
Kalotermitidae	2
Hodotermitidae	1
Rhinotermitidae	6
Termitidae	8
Plecoptera	—
Pteronarcyidae	1
Coleoptera	—
Scarabaeidae	2
Buprestidae	6
Anobiidade	5
Coccinellidae	1
Cerambycidae	29
Curculionidae	1
Trichoptera	—
Limnephilidae	1
Diptera	—
Tipulidae	2
Hymenoptera	—
Siricidade	3

Source: Modified from Martin, M. M., *Philos. Trans. R. Soc. Lond. B*, 333, 281, 1991.

with food and that remain active after ingestion; and/or (d) secretions by the insect itself that complete a cellulase enzyme system. The existence of such different mechanisms of cellulose digestion probably implies that the appearance of lytic capacity occurred several times in insects and in different groups (Martin 1991).

The first of these four mechanisms was the earliest described, and early works documented the dependence of some groups of termites and cockroaches (families Kalotermitidae Rhinotermitidae) and wood roaches (family Cryptocercidae) on flagellated anaerobic protozoa in the alimentary canal, as early as 1924 (Cleveland 1924). The second mechanism, the exploitation of cellulolytic capacity, has been proposed as a strategy common in the termite family Termitidae, although the empirical evidence in support of this assumption remains generally tenuous (Breznak 1982; O'Brien and Slaytor 1982) outside of the domestic cockroach (*Periplaneta americana* L.) (Cruden and Markovetz 1979) and dung beetles of the genus *Oryctes* (Bayon 1981).

The intake of cellulase enzymes produced by microbial decomposers is the third possible mechanism of cellulose digestion by insects. This mechanism is apparently essential for the termite fungus growers (Abo-Khatwa 1978; Martin and Martin 1978; Rouland et al. 1988) and for larvae of wood borers (Kukor and Martin 1986; Kukor et al. 1988). It also seems to be common among the larvae of wood-eating wasps of the genus *Sirex* (Siricidae) (Kukor and Martin 1983). The fourth mechanism (production of insect enzyme cellulases) has been the subject of substantial controversy. While the existence of this mechanism has been proposed for a variety of insect groups ranging from Cerambycid beetles, termites, and cockroaches, it has been unequivocally demonstrated only in some species of cockroaches and termites (Scrivener et al. 1989; Martin 1991). Even in these species, the production of cellulolytic enzymes by cells in the intestinal wall of the insect does not necessarily imply the absence of additional mutual interactions between insects and microorganisms.

17.3.2 Adaptations for Use of High-Availability Detritus That Is Unpredictable in Space and Time

Highly nutritious food resources that are unpredictable in space or time create the conditions for intense interspecific and intraspecific competition for insect detritivores. Among the many adaptive strategies documented for such insects are (a) the ability to quickly and colonize detritus; (b) mechanisms that favor exploitative competition (resource domination); and (c) strategies to exclude other potential competitors (Table 17.3).

Flies are among the insects that exhibit sophisticated strategies for locating and rapid colonization detritus. The highly sensitive olfactory senses of flies allow them to perceive food deposits at great distances (up to tens of meters) (Cragg 1956; Shubeck 1975). Many flies and detritivorous beetles exhibit a "perching" behavior on leaves and stems in the forest understory, to maximize their ability to sense odor plumes and increase the chances of early colonization of potential food resources (Young 1982).

TABLE 17.3

Ecological, Behavioral, and Physiological Aspects Which Possibly Evolved as a Result of the Competitive Environment Existing in High-Quality and Spatio-Temporal Unpredictable Resources

Characteristic	Mechanism	Example
Exclusion of competitors from the detritus deposition	Relocation of detritus resources in locations distant from original deposition site, pheromone production	Subsocial scarabaeine and silphid beetles
Rapid colonization and use of detritus	Ovoviviparity, dial activity period, "endothermy"	Scarabaeine beetles in tropical forests, sarcophagid flies
Strong competitive ability during resource use	Elevated growth rate, group feeding, asymmetric competition	Various fly species
Post-feeding survival	Ability to pupate quickly	*Calliphora erythrocephala* Meigen

Source: Modified from Hanski, I., in *Nutritional Ecology of Insects, Mites, Spiders, and Related Invertebrates,* ed. F. Slansky, Jr. and J. G. Rodriguez , 837–85. New York: John Wiley & Sons, 1987.

Once food has been located, these detrivorous insects often demonstrate behaviors that lead to resource domination, often through the physical removal of the food resource from other potential competitors. This strategy is particularly conspicuous in beetles of the families Scarabaeidae and Silphidae. In these species, an individual or a male–female pair removes a portion of the carcass or feces, and buries it under the soil surface immediately, or prepares food balls that are then rolled some distance away from the deposition site, and then used as a feeding or breeding substrate (Halffter and Edmonds 1982).

Some additional techniques used by insect species to exclude other competitors include the use of repellent substances (Bellés and Favila 1983) or outright attack (Ridsdill-Smith 1981). The dung beetle *Canthon angustatus cyanellus* LeConte has chosen a largely chemical approach to controlling competition with dung-breeding flies, by producing a repellent substance that can easily "infect" the entire feces deposit, limiting its attractivity to flies and therefore the risk of competition with beetles (Bellés and Favila 1983).

When and where dung beetles and dung flies co-occur, fly survival also tends to decline as a consequence of increased fly mortality, from the combination of (a) direct mechanical damage to fly eggs and early instars caused during adult beetle feeding (Bishop et al. 2005; Ridsdill-Smith and Hayles 1990); (b) unfavorable microclimates for fly eggs and larvae caused by dung disturbance (Ridsdill-Smith and Hayles 1987); and (c) resource competition with older larvae, primarily from removal of dung for brood balls (Hughes 1975; Ridsdill-Smith and Hayles 1987, 1990).

17.4 Mutualisms between Insects and Microorganisms: Role of Coprophagy in Detritus Use

The use of detritus by insects involves a sort of physical and chemical mechanisms that evolved several independent times. Among these mechanisms the use of feces (or coprophagy) has a central place, by providing the insect with a rich food source and mainly as a component of the insect–microorganism mutualistic evolutive scenario.

Coprophagy offers at least three categories of nutritional benefits to dung-feeding insects. Fecal deposits are a tremendous source of (a) mutualistic fauna, (b) microbial protein, and (c) enzymes and secondary metabolites originating from the feces provider. From a purely nutritional perspective, feces and decaying vegetative matter are fundamentally similar food resources (Webb 1976; Cambefort 1991). In fact, over evolutionary time, multiple "food switching" events have occurred between predominantly coprophagous and saprophagous or mycophagous insect groups, such as the Scarabaeine dung beetles (Cambefort 1991) and butterflies of the genus *Telanepsia* (Common and Horak 1991). However, feces are unique in several respects from decomposed plant material (humus), often with (a) higher pH, (b) greater capacity for moisture retention, and (c) high area/volume ratio. These features further predispose feces to have a subsequently larger capacity for microbial growth (McBrayer 1973). Immediately following deposition, feces are typically colonized by succession of microorganisms that can very quickly become hyperabundant (Lodha 1974; Anderson and Bignell 1980; Bignell 1989). The increase in dung surface area caused by the removal and fragmentation of feces by coprophagous insects for feeding and brood balls is especially important for bacterial growth, which is largely confined to the surface of the deposit of feces. For those (often social) insects that additionally practice intraspecific coprophagy, metabolites from the feces provider can be further transferred across residents or individuals within the same insect group (Nalepa et al. 2001).

The microorganisms that colonize feces (both during digestion and after colonization) are rich in lipids, carbohydrates, and micronutrients (Martin and Kukor 1984). For many detritivores, these microorganisms are assimilated with high efficiency (Hargrave 1975; White 1993; Bignell and Eggleton 2000), and provide the largest, if not the sole source of protein food. Is not well understood whether some groups of coprophagous insects (e.g., non-Cryptocercidae cockroaches) that feed on dead plant material digest the substrate as a whole, or rather remove and consume only the microbial component, leaving the detritus relatively unaltered, in a process similar to the feeding strategies of millipedes (Vignelli 1989). The two processes are not mutually exclusive, and the importance of each component can be related to the specific characteristics of a given feces type and nutritional status of detritivorous (Berrie 1975).

17.4.1 External and Internal Rumen in Detritivores Insects

An important evolutive aspect of the detritus use by insects was the internalization of the microbiota responsible for detritus degradation. This process involves increasing the insect–microorganism interaction both in an ecological and an evolutive scenario, and expressed often as a sort of morphological and physiological modifications.

The process of degradation of dead plant material begins well before its consumption by detritus-feeding insects. This is known as conditioning and is accompanied by microbial detoxification of allelochemicals, softening of hard lignins in the detritus, and the progressive immobilization of N and P in the accumulating fungal and bacteria biomass. Consequently, when scavengers consume leaf litter after the microbial conditioning process, they interact with the microbial community in a process that can extend over several cycles if, as commonly happens, coprophagy is involved and the insect consume repeatedly its own feces. This process is known as "external rumen" (Lavelle et al. 1995; Shear and Selden 2001).

The evolution of a sophisticated community of intestinal microorganisms, such as those of termites, can be seen as a process of internalization of this external rumen, and is therefore referred to as an "internal" rumen. The essential difference between internal ruminant invertebrates and other invertebrate scavengers is that the former can feed on recently dead plant material, even in the absence of significant colonization and detritus conditioning by decomposing microorganisms (Wood 1976).

When an arthropod depends primarily on the external rumen to digest detritus, its association with the microbial community is both temporary and based largely outside the body, as occurs in passalid beetles and fungus-cultivating termites and ants. The interaction between insects and the microbial community is essentially the same, whether an insect feeds from organic material directly subject to microbial action, or when it secondarily feeds from organic material in the form of feces that are infected by microorganisms. In early evolutionary stages, coprophagy is indiscriminately directed at the fecal pellets of any detritivore (Hassall and Rushton 1985). The hindgut at this stage is relatively undifferentiated, and free-living microbes form the major part of the intestinal facultative mutualism. The isopods are a contemporary example of this evolutionary stage.

The sequence of environments, substrate–fecal pellet–hindgut, may represent a sequence of good–better–great environments for microbial growth. The stable environment and constant supply of food makes the hindgut of insects favorable to the growth of microorganisms if they are able to circumvent the digestive process of the host insect (Stevenson and Dindal 1987). Thus, it is expected that insect detritivores have arisen in several opportunities for mutualistic partnerships.

17.5 Ecological Functions of Detritivores Insects

Detritivorous insects cause significant alterations to the structure of habitats through breaking down available detritus into the soil of terrestrial ecosystems, principally through the physical disarticulation of hard plant structures and subsequent increase in decomposition rates (Harmon et al. 1986; Speight 1989). In the absence of mechanical damage to its tissue structure, some detritus types (e.g., fallen logs) may remain with relatively low microbial loads for years (Franklin et al. 1987).

17.5.1 Leaf Litter Decomposition Rates

Decaying woody material represents an important supply of nutrients and energy in forest ecosystems, and represents specialized habitat for certain decomposer organisms (Speight 1989; Key 1993; Grove 2000). Saproxylic scavengers assist in the mechanical breakdown of dead woody material, speeding the recycling of nutrients in forest ecosystems (Harmon et al. 1986; Speight 1989) and serving as important food sources for other organisms (Niwa et al. 2001). The action of insect detritivores can increase the contact surface for microbial attack by an order of magnitude or more, through their burrowing and feeding actions (Niwa et al. 2001).

17.5.2 Waste Removal and Related Functions

Dung beetles represent an ideal model system for understanding some of the ecological functions mediated by detritivorous insects. Dung beetles are a globally distributed insect group that reaches its highest diversity in tropical forests and savannas (Hanski and Cambefort 1991). Largely coprophagous dung beetle species feed on the microorganism-rich liquid component of mammalian dung (and less commonly that of other vertebrates, as well as rotting fruit, fungus, and carrion) and use the more fibrous material to brood their larvae (Halffter and Matthews 1966; Halffter and Edmonds 1982). As they feed on high-availability, low-predictability resources, dung beetle communities are characterized by high degrees of interspecific and intraspecific competition, and feeding and nesting behaviors that center on relocating resources to avoid competition. Most dung beetles use one of two broad nesting strategies to reduce the risks of competition, each with implications for ecological function. Tunneler species create brood balls that they then bury deep in vertical chambers, in close proximity to feces deposition site. Roller species create brood balls, then transport them some horizontal distance away before burial beneath the soil surface (Halffter and Edmonds 1982).

The extent to which nutrients in animal feces are returned to below the soil surface by dung beetles has strong implications for plant productivity. The transfer of freshly deposited waste below the soil surface by dung beetle species physically relocates nutrient-rich organic material and instigates microorganismal and chemical changes in the upper soil layers (Figure 17.6). By burying dung under the soil surface, dung beetles prevent the loss of N through ammonia (NH_3) volatilization (Gillard 1967), and enhance soil fertility by increasing the available labile N available for uptake by plants through mineralization (Yokoyama et al. 1991). Many dung beetle species move large quantities of earth to the soil surface during nesting (Mittal 1993), which may influence soil biota and plant productivity by increasing soil aeration and water porosity. Several experimental studies have reported a significant result of these combined bioperturbation and nutrient transfer activity of dung beetles on plant height (Kabir et al. 1985; Galbiati et al. 1995), increased biomass (Bang et al. 2005), significant gains in grain yield (Kabir et al. 1985), protein levels (Maqueen and Beirne 1975), and forage nitrogen content (Bang et al. 2005).

Another ecological function mediated by the use of feces by coprophagous dung beetles is the secondary dispersal of seeds. Vertebrate seed dispersal mechanisms are extremely widespread in tropical and temperate ecosystems (Howe and Smallwood 1982; Willson et al. 1990; Jordano 1992). For seeds, life between initial excretion in frugivorous animals dung and final seedling emergence is fraught with predators, pathogens, and a low probability of "landing" in an area suitable for future germination (Chambers and MacMahon 1994). Secondary seed dispersal is believed to play an important role in plant recruitment through lowering these post-primary dispersal risk factors (Chambers and MacMahon 1994).

FIGURE 17.6 Bioturbation action of dung beetles on cattle dung in highly compacted soil. Soil below feces deposit was brought to soil surface by the tunneling action of adult beetles, as they create tunnels where balls of animal feces are deposited, for adult and larval feeding.

From a dung beetle's perspective, most seeds present in dung simply represent contaminants since they occupy space in the dung and are not consumed by the larvae. However, with competition for dung usually intense and burial occurring rapidly, dung beetles often bury seeds, perhaps accidentally, as they bury dung for their larval brood balls. Dung beetles relocate seeds both horizontally and vertically from the point of deposition. The combined impact of this dispersal by tunneler and roller species benefits seed survival (and therefore plant recruitment) by (a) reducing seed predation and mortality due to seed predators and pathogens (Janzen 1983; Estrada and Coates-Estrada 1991; Chambers and MacMahon 1994; Shepherd and Chapman 1998; Andresen 1999; Feer 1999; Andresen and Levey 2004); (b) directing dispersal to favorable microclimates for germination and emergence (Andresen and Levey 2004); and (c) decreasing residual post-dispersal seed clumping (Andresen 1999, 2001), with potential effects on density dependent seed mortality, seedling competition, and predation risk (Andresen and Feer 2005). Dung beetle communities bury between 6% and 95% of the seeds excreted in any given fecal pile, although this percentage ranges widely across studies (47–95% [Shepherd and Chapman 1998]; 13–23% [Feer 1999]; 6–75% [Andresen 2002]; 35–48% [Andresen 2003]; 26–67% [Andresen and Levey 2004]). As they bury a disproportionate amount of dung, larger-bodied and nocturnal dung beetle species perform a disproportionate amount of the secondary seed dispersal function (Andresen 2002; Slade et al. 2007).

17.5.3 Biological Control of Other Detritivores

Another important ecological function that is mediated by detritivorous insects is the suppression of populations of harmful or pestiferous insects, particularly among those scavenger species that use ephemeral resources, and therefore are largely structured by interspecific and intraspecific competition dynamics. Scarabaeinae dung beetles again provide an ideal model system for further understanding these competitive interactions.

Through feeding and nesting, adult and larval dung beetle activity appears to interact with the abundance of dung-dispersed nematodes and protozoa. Much of our understanding of these relationships has arisen from the study of livestock parasites and pests. For example, a study in Australian cattle pastures found that manipulation of cattle dung by the dung beetle species *Digitonthophagus gazella* (Fabricius) resulted in significantly fewer emergent nematode larva (Bryan 1973), and that cattle feces deposits with dung beetles were excluded contained up to 50 times more helminth larvae than those with 10 or 30 *D. gazella* pairs (Bryan 1976). In an experimental manipulation of dung beetle abundances in cattle pastures in the southeastern United States, Fincher (1973) reported that a 5-fold increase in dung beetle abundance resulted in a nearly 15-fold reduction in the emergence in *Ostertagia ostertagi* relative to dung beetle free pastures, and a 3.7-fold reduction relative to pastures with natural dung beetle levels. Dung beetles have also been implicated in the reduction in abundance of the exploding fungus *Pilobolus sporangia*, which forcefully disperses nematodes in pasture systems along with its own spores (Gormally 1993). Laboratory studies additionally reveal that passage through certain dung beetle species significantly reduces the abundance of viable helminth eggs and protozoan cysts, including *Ascaris lumbricoides*, *Necator americanus*, *Trichuris trichiura*, *Entamoeba coli*, *Endolimax nana*, *Giardia lamblia* (Miller et al. 1961), and *Cryptosporidium parvum* (Mathison and Ditrich 1999). While dung beetles have been conjectured to be important suppressors of human endoparasites (Miller 1954), we know of no publication empirically relating dung beetles and human endoparasite transmission.

Fresh mammal dung is an important resource for a variety of dung-breeding flies as well as dung beetles. Several pestiferous, dung-dwelling fly species (principally *Musca autumnalis*, *M. vetustissima*, *Haematobia thirouxi potans*, *H. irritans exigua*, and *H. irritans irritans*) have followed the introduction of livestock globally, causing enormous reductions in livestock productivity (Haufe 1987) and hide quality (Guglielmone et al. 1999), and enormous financial costs to producers (Byford et al. 1992). A series of experimental manipulations of dung beetle and fly densities in artificial dung pats report elevated fly mortality in the presence of scarabaeine beetles, both in the laboratory and field (Bornemissza 1970; Wallace and Tyndale-Biscoe 1983; Bishop et al. 2005).

The relative impact of these dung beetle activities is modulated by several factors, including dung quality (Macqueen and Beirne 1975; Ridsdill-Smith et al. 1986; Ridsdill-Smith and Hayles 1990), beetle abundance (Bornemissza 1970; Hughes et al. 1978; Kirk and Ridsdill-Smith 1986; Ridsdill-Smith and

Hayles 1989; Ridsdill-Smith and Matthiessen 1988; Tyndale-Biscoe 1993), activity period (Fay et al. 1990), nesting strategy (Edwards and Aschenborn 1987), and importantly, arrival time (Hughes et al. 1978, Edwards and Aschenborn 1987, Ridsdill-Smith and Hayles 1987)

17.6 Final Considerations

Detrivorous insects are a fascinating, diverse, abundant, and critically important part of the animal kingdom. Understanding the relationship between insects and their food resources is the critical first step to any practical or theoretical understanding of insect population ecology, which subsequently underpins sustainable control efforts. While much of the published material on detritivorous insects exclusively focuses on their status as pests, or their potential for economic damage, detritivores form a crucial link in the nutrient cycling pathways of every ecosystem on Earth. While our knowledge of detritivorous insects (as with most insects) remains incipient in several areas, we hope this chapter can expand the perspectives of those entomologists interested in insect bioecology and nutrition, and those whose backgrounds have predominantly focused on the handful of species harmful to human well being and economy.

REFERENCES

Abo-Khatwa, N. 1978. Cellulase of fungus-growing termites: A new hypothesis on its origin. *Experientia* 34:559–60.

Aerts, R., and H. de Caluwé. 1997. Nutritional and mediated controls on leaf litter decomposition species. *Ecology* 78:244–60.

Anderson, J. M., and D. E. Bignell. 1980. Bacteria in the food, gut contents and feces of the litter-feeding millipede, *Glomeris marginata. Soil Biol. Biochem.* 12:251–4.

Andresen, E. 1999. Seed dispersal by monkeys and the fate of dispersed seeds in a Peruvian rain forest. *Biotropica* 31:145–58.

Andresen, E. 2001. Effects of dung presence, dung amount and secondary dispersal by dung beetles on the fate of *Micropholis guyanensis* (Sapotaceae) seeds in Central Amazonia. *J. Trop. Ecol.* 17:61–78.

Andresen, E. 2002. Dung beetles in a Central Amazonian rainforest and their ecological role as secondary seed dispersers. *Ecol. Entomol.* 27:257–70.

Andresen, E., and D. J. Levey. 2004. Effects of dung and seed size on secondary dispersal, seed predation, and seedling establishment of rain forest trees. *Oecologia* 139:45–54.

Andresen, E., and F. Feer. 2005. The role of dung beetles as secondary seed dispersers and their effect on plant regeneration in tropical rainforests. In *Seed Fate: Predation, Dispersal and Seedling Establishment*, ed P. M. Forget, J. E. Lambert, P. E. Hulme, and S. B. Vander Wall, 331–49. Wallingford: CABI International.

Atkinson, W. D., and B. Shorrocks. 1981. Competition on a divided and ephemeral resource: A simulation model. *J. Anim. Ecol.* 50:461–71.

Bailey, R. E., and R. J. Putman. 1981. Estimation of fallow deer *Dama dama* populations from faecal accumulation. *J. Appl. Ecol.* 18:697–702.

Bang, H. S., J. H. Lee, O. S. Kwon, Y. E. Na, Y. S. Jang, and W. H. Kim. 2005. Effects of paracoprid dung beetles (Coleoptera: Scarabaeidae) on the growth of pasture herbage and on the underlying soil. *Appl. Soil Ecol.* 29: 165–71.

Bayon, C. 1981. Modifications ultrastructurales des parois végétales dans le tube digestif d'une larva xylophage *Oryctes nasicornis* (Coleoptera, Scarabaeidae): Rôle des bactkries. *Can. J. Zool.* 59:2020–9.

Begon, M., C. R. Thownsend, and J. L. Harper. 2006. *Ecology: From Individuals to Ecosystems.* Oxford: Blackwell Scientific Publications.

Bellés, X., and M. E. Favila. 1983. Protection chimique du nid chez *Canthon cyanellus cyanellus* LeConte (Col. Scarabaeidae). *Bull. Soc. Entomol. Fr.* 88:602–7.

Berrie, A. D. 1975. Detritus, micro-organisms and animals in fresh water. In *The Role of Terrestrial and Aquatic Organisms in Decomposition Processes*, ed. J. M. Anderson and A. Macfadyen, 323–38. Oxford: Blackwell Scientific Publications.

Bignell, D. E. 1989. Relative assimilations of carbon-14-labeled microbial tissues and carbon-14-labeled plant fiber ingested with leaf litter by the millipede *Glomeris marginata* under experimental conditions. *Soil Biol. Biochem.* 21:819–28.

Bignell, D. E., and P. Eggleton. 2000. Termites in ecosystems. In *Termites: Evolution, Sociality, Symbioses, Ecology,* ed. T. Abe, D. E. Bignell, and M. Higashi, 363–87. Dordrecht: Kluwer Academic.

Bishop, A. L., H. J. McKenzie, L. J. Spohr, and I. M. Barchia. 2005. Interactions between dung beetles (Coleoptera: Scarabaeidae) and the arbovirus vector *Culicoides brevitarsis* Kieffer (Diptera: Ceratopogonidae). *Aust. J. Entomol.* 44:89–96.

Bornemissza, G. F. 1970. Insectary studies on the control of the dung breeding flies by the activity of the dung beetle, *Onthophagus gazella* F. (Coleoptera, Scarabaeidae). *J. Aust. Ent. Soc.* 9:31–41.

Breznak, J. A. 1982. Intestinal microbiota of termites and other xylophagous insects. *Annu. Rev. Microbiol.* 36:323–43.

Bryan, R. P. 1973. The effects of dung beetles activity on the numbers of parasitic gastrointestinal helmintic larvae recovered from pasture samples. *Aust. J. Agric. Res.* 24:161–8.

Bryan, R. P. 1976. The effects of dung beetle, *Onthophagus gazella*, on the ecology of the infective larvae of gastrointestinal nematodes of cattle. *Aust. J. Agric. Res.* 27:567–74.

Byford, R. L., M. E. Craig, and B. L. Crosby.1992. A review of ectoparasites and their effect on cattle production. *J. Anim. Sci.* 70:597–602.

Cambefort, Y. 1984. Etud écologique des Coléoptères Scarabaeidae de Côte d'Ivore. *Travaux des Chercheurs de la Station de LAMTO*, no. 3, Univ. d'Abidjan, 33p.

Cambefort, Y. 1991. From saprophagy to coprophagy. In *Dung Beetle Ecology*, ed. I. Hanski and Y. Cambefort, 22–35. Princeton: Princeton University Press.

Caugley, G., and J. H. Lawton. 1981. Plant–herbivore systems. In *Theoretical Ecology*, ed. R. M. May, 132–66. Oxford: Blackwell Scientific Publications.

Chambers, J. C., and J. A. MacMahon. 1994. A day in the life of a seed: Movements and fates of seeds and their implications for natural and managed systems. *Annu. Rev. Ecol. Evol. Syst.* 25:263–93.

Cleveland, L. R. 1924. The physiology and symbiotic relationships between the intestinal protozoa of termites and their host, with special reference to *Zeticulitermes laves* Kollar. *Biol. Bull.* 46:117–227.

Common, I. F. B., and M. Horak. 1994. Four new species of *Telanepsia* Turner (Lepidoptera: Oecophorideae) with larvae feeding on koala and possum scats. *Invert. Taxon.* 8:809–28.

Coughlan, M. P., and L. G. Ljungdahl. 1988. Comparative biochemistry of fungal and bacterial cellulolytic enzyme systems. In *Biochemistry and Energetics of Cellulose Degradation*, ed. J. P. Aubert, P. Beguin, and J. Millet, 11–30. London: Academic Press.

Cragg, J. B. 1956. The olfactory behaviour of *Lucillia* species (Diptera) under natural conditions. *Ann. Appl. Biol.* 44:467–77.

Cruden, D. L., and A. J. Markovetz. 1979. Carboxymethylcellulose decomposition by intestinal bacteria of cockroaches. *Appl. Environ. Microbiol.* 38:368–72.

Edwards, P. B., and H. H. Aschenborn. 1987. Patterns of nesting and dung burial in *Onitis* dung beetles: Implications for pasture productivity and fly control. *J. Appl. Ecol.* 24:837–51.

Estrada, A., and R. Coates-Estrada. 1991. Howler monkeys (*Alouatta palliata*), dung beetles (Scarabacidae) and seed dispersal: Ecological interactions in the tropical rain forest of Los Tuxtlas, Mexico. *J. Trop. Ecol.* 7:459–74.

Fay, H. A. C., A. Macqueen, B. M. Doube, and J. D. Kerr. 1990. Impact of fauna on mortality and size of *Haematobia* spp. (Diptera: Muscidae) in dung pads in Australia and South Africa. *Bull. Entomol. Res.* 80:385–392.

Feer, F. 1999. Effects of dung beetles (Scarabaeidae) on seeds dispersed by howler monkeys (*Alouatta seniculus*) in the French Guianan rain forest. *J. Trop. Ecol.* 15:129–42.

Fincher, G. T. 1973. Dung beetles as biological control agents for gastrointestinal parasites of livestock. *J. Parasitol.* 59:396–9.

Findlay, S. E. G. 1982. Effect of detrital nutritional quality on population dynamics of a marine nematode (*Diplomella chitwoodi*). *Mar. Biol.* 68:223–7.

Franklin, J. F., H. H. Shugart, and M. E. Harmon. 1987. Tree death as an ecological process. *BioScience* 37: 550–6.

Galbiati, C., C. Bensi, C. H. C. Conceição, J. L. Florcovski, M. H. Calafiori, and A. C. T. Tobias. 1995. Estudo comparativo entre besouros do esterco *Dichotomius analypticus* (Mann, 1829) *Ecossistema* 20:109–18.

Gillard, P. 1967. Coprophagous beetles in pasture ecosystems. *J. Aust. Inst. Agric. Sci.* 33:30–4.

González, G., and T. R. Seastedt. 2001. Soil fauna and plant litter decomposition in tropical and subalpine loreotor zoology 81:1151–64

Gormally, M. J. 1993. The effect of dung beetle activity on the discharge of *Pilobus sporangia* in cattle faeces. *Med. Vet. Entom.* 7:197–8.

Grove, S. J. 2000. Impacts of forest management on saproxylic beetles in the Australian lowland tropics and the development of appropriate indicators of sustainable forest management. PhD thesis. James Cook University.

Guglielmone, A. A., E. Gimeno, J. Idiart, W. F. Fisher, M. M. Volpogni, O. Quaino, O. S. Anziani, S. G. Flores, and O. Warnke. 1999. Skin lesions and cattle hide damage from *Haematobia irritans* infestations in cattle. *Med. Vet. Entom.* 13:323–8.

Halffter, G., and W. D. Edmonds. 1982. *The Nesting Behavior of Dung Beetles (Scarabaeinae): An Ecological and Evolutive Approach*. México, D.F.: Instituto de Ecología.

Hanski, 1981. Exploitative competition in transient habitat patches. In *Quantitative Population Dynamics*, ed. D. G. Chapman and V. F. Gallucci, 25–38. Fairland: International Cooperative Publishing House.

Hanski, I. 1987. Nutritional ecology of dung and carrion feeding insects. In *Nutritional Ecology of Insects, Mites, Spiders, and Related Invertebrates,* ed. F. Slansky Jr. and J. G. Rodriguez, 837–85. New York: John Wiley & Sons.

Hanski, I., and Y. Cambefort. 1991. *Dung Beetle Ecology*. Princeton University Press, Princeton. 520pp.

Hardin, G. 1960. The competitive exclusion principle. *Science* 131:1292–7.

Hargrave, B. T. 1975. The central role of invertebrate faeces in sediment decomposition. In *The Role of Terrestrial and Aquatic Organisms in Decomposition Processes*, ed. J. M. Anderson and A. MacFayden, 301–20. Oxford: Blackwell Scientific Publications.

Harmon, M. E., J. F. Franklin, F. J. Swanson, P. Sollins, S. V. Gregory, J. D. Lattin, N. H. Anderson, S. P. Cline, N. G. Aumen, J. R. Sedell, G. W. Lienkaemper, K. Cromack, and K. W. Cummins. 1986. Ecology of coarse woody debris in temperate ecosystems. *Adv. Ecol. Res.* 15:133–302.

Hassall, M., and S. P. Rushton. 1985. The adaptive significance of coprophagous behavior in the terrestrial isopod *Porcellio scaber. Pedobiologia* 28:169–75.

Haufe, W. O. 1987. Host–parasite interaction of blood feeding dipterans in health and productivity of mammals. *Int. J. Parasit.* 17:607–14.

Howe, H. F., and J. Smallwood. 1982. The ecology of seed dispersal. *Annu. Rev. Ecol. Syst.* 13:210–23.

Hughes, R. D., M. Tyndale-Biscoe, and J. Walker. 1978. Effects of introduced dung beetles (Coleoptera: Scarabaeinae) on the breeding and abundance of the Australian bushfly *Musca vetustissima* Walker (Diptera: Muscidae). *Bull. Entomol. Res.* 68:361–72.

Janzen, D. 1983. Insects at carrion and dung. In *Costa Rican Natural History*, ed. D. Janzen. Chicago: Chicago University Press.

Jordano, P. 1992. Fruits and frugivory. In *Seeds: The Ecology of Regeneration in Plant Communities*, ed. M. Fenner, 125–65. Wallingford: CAB International.

Jurgensen, M., P. Laks, D. Reed, A. Collins, D. Page-Dumroese, and D. Crawford. 2004. *Chemical, Physical and Biological Factors Affecting Wood Decomposition in Forest Soils*. Stockholm: IRG Documents Series.

Kabir, S. M. H., J. A. Howlader, and J. Begum. 1985. Effect of dung beetle activities on the growth and yield of wheat plants. *Bangladesh J. Agric.* 10:49–55.

Key, R. 1993. What are saproxylic invertebrates? In *Dead Wood Matters: The Ecology and Conservation of Saproxylic Invertebrates in Britain*, ed. K. J. Kirby and C. M. Drake, 5–6. Dunham Massey Park: English Nature Science.

Korb, J., and K. E. Linsenmair. 2001. Resource availability and distribution patterns, indicators of competition between *Macrotermes bellicosus* and other macro-detritivores in the Comoé National Park, Côte d'Ivoire. *Afr. J. Ecol.* 39:257–65.

Kukor, J. J., and M. M. Martin. 1983. Acquisition of digestive enzymes by Siricid wood wasps from their fungal symbiont. *Science* 220:1161–3.

Kukor, J. J., and M. M. Martin. 1986. Cellulose digestion in *Monochamus marmorator* Kby. (Coleoptera: Cerambycidae): The role of acquired fungal enzymes. *J. Chem. Ecol.* 12:1057–70.

Kukor, J. J., D. P. Cowan, and M. M. Martin. 1988. The role of ingested fungal enzymes in cellulose digestion in larvae of cerambycid beetles. *Physiol. Zool.* 61:364–71.

Laing, S. E., S. T. Buckland, R. W. Burn, D. Lambie, and A. Amphlett. 2003. Dung and nest surveys: Estimating decay rates. *J. Appl. Ecol.* 40:1102–11.

Lavelle, P., C. Lattaud, D. Trigi, and I. Barois. 1995. Mutualism and biodiversity in soils. *Plant Soil* 170:23–33.

Levins, R., and Culver, D. 1971. Regional coexistence of species and competition between rare species *Proc. Natl. Acad. Sci. U. S. A.* 68:1246–54.

Lincoln, D. E., T. S. Newton, P. R. Ehrich, and K. S. Williams. 1982. Coevolution of the checkerspot butterfly *Euphydryas clarcedona* and its larval food plant *Diplanscus aurantiacus*: Larval response to protein and leaf resin. *Oecologia* 52:216–23.

Lodha, B. C. 1974. Decomposition of digested litter. In *Biology of Plant Litter Decomposition*, ed. C. H. Dickinson and G. J. P. Pugh, 213–41. London: Academic Press.

MacQueen, A., and B. P. Beirne. 1975. Dung beetle activity and fly control potential of *Onthophagus nuchicornis* (Coleoptera: Scarabaeinae) in British Columbia. *Can. Entomol.* 107:1215–20.

Martin, M. M. 1991. The evolution of cellulose digestion in insects. *Philos. Trans. R. Soc. Lond. B* 333: 281–8.

Martin, M. M., and J. J. Kukor. 1984. Role of mycophagy and bacteriophagy in invertebrate nutrition. In *Current Perspectives in Microbial Ecology*, ed. M. J. Klug and C. A. Reddy, 257–63. Washington: American Society of Microbiology.

Martin, M. M., and J. S. Martin. 1978. Cellulose digestion in the midgut of the fungus-growing termite *Macrotermes natalensis*: The role of acquired digestive enzymes. *Science* 199:1453–5.

Mathison, B., and O. Ditrich. 1999. The fate of *Cryptosporidium parvum* oocysts ingested by dung beetles and their possible role in the dissemination of cryptosporidosis. *J. Parasitol.* 85:678–81.

McBrayer, J. F. 1973. Exploitation of deciduous leaf litter *by Apheloria montana* (Diplopoda: Eurydesmidae). *Pedobiologia* 13:90–8.

Miller, A. 1954. Dung beetles (Coleoptera, Scarabaeidae) and other insects in relation to human faeces in a hookworm area of southern Georgia. *Am. J. Trop. Med. Hyg.* 3:372–89.

Miller, A. 1961. The mouthparts and digestive tract of adult dung beetles (Coleoptera: Scarabeidae) with reference to the ingestion of helminth eggs. *J. Parasitol.* 47:735–44.

Miller, A., E. Chi-Rodriquez, and R. L. Nichols. 1961. The fate of Helminth eggs and protozoan cysts in human feces ingested by dung beetles (Coleoptera: Scarabeidae). *Am. J. Trop. Med. Hyg.* 10:748–54.

Mittal, I. 1993. Natural manuring and soil conditioning by dung beetles. *Trop. Ecol.* 34:150–9.

Monro, J. 1967. The exploitation and conservation of resources by populations of insects. *J. Anim. Ecol.* 36:531–47.

Muller-Landau, H. C., and B. D. Hardesty. 2005. Seed dispersal of woody plants in tropical forests: Concepts, examples, and future directions, In *Biotic Interactions in the Tropics*, ed. D. F. R. P. Burslem, M. A. Pinard, and S. Hartley, 435–56. Cambridge: Cambridge University Press.

Nalepa, C. A., D. E. Bignelland, and C. Bandi 2001. Detritivory, coprophagy, and the evolution of digestive mutualisms in Dictyoptera. *Insectes Soc.* 48:194–201.

Nichols, E., T. A. Gardner, C. A. Peres, and S. Spector 2009. Co-declining mammals and dung beetles: An impending ecological cascade. *Oikos* 118:481–7.

Niwa, C. G., R. W. Peck, and T. R. Torgersen 2001. Soil, litter, and coarse woody detritus habitats for arthropods in Eastern Oregon and Washington. *Northwest Sci.* 75:141–8.

O'Brien, R. W., and M. Slaytor. 1982. Role of microorganisms in the metabolism of termites. *Aust. J. Biol. Sci.* 35:239–62.

Payne, J. A. 1965. A summer carrion study of the baby pig (*Sus scrofa*). *Ecology* 46:592–602.

Ridsdill Smith, T. J. 1981. Some effects of three species of dung beetles (Coleoptera: Scarabaeidae) in south western Australia on the survival of the bush fly, *Musca vetustissima* Walker (Diptera: Muscidae), in dung pads. *Bull. Entomol. Res.* 71:425–33.

Ridsdill-Smith, T. J. 1986. The effects of seasonal changes in cattle dung on egg production by two species of dung beetles (Coleoptera: Scarabaeidae) in south-western Australia. *Bull. Entomol. Res.* 76:63–8.

Ridsdill-Smith, T. J., and L. Hayles. 1987. Mortality of eggs and larvae of the bush fly *Musca vetustissima* Walker (Diptera: Muscidae), caused by Scarabaeine dung beetles (Coleoptera: Scarabaeidae) in favourable cattle dung. *Bull. Ent. Res.* 77:731–6.

Ridsdill-Smith, T. J., and L. Hayles. 1989. Re-examination of competition between *Musca vetustissima* Walker (Diptera: Muscidae) larvae and seasonal changes in favourability of cattle dung. *J. Aust. Ent. Soc.* 28: 105–11.

Ridsdill-Smith, T. J., and L. Hayles. 1990. Stages of bush fly, *Musca vetustissima* (Diptera: Muscidae), killed by scarabaeine dung beetles (Coleoptera: Scarabaeidae) in unfavourable cattle dung. *Bull. Entomol. Res.* 80:477-8.

Ridsdill-Smith, T. J., L. Hayles, and M. L. Palmer. 1986. Competition between the bush fly and dung beetles in dung of differing characteristics. *Entomol. Exp. Appl.* 41:83–90.

Ridsdill-Smith, T. J., and J. N. Matthiessen. 1984. Field assessments of the impact of night-flying dung beetles (Coleoptera: Scarabaeidae) on the bush fly, *Musca vetustissima* Walker (Diptera: Muscidae), in south western Australia. *Bull. Ent. Res.* 74:191–5.

Ridsdill-Smith, T. J., and J. N. Matthiessen. 1988. Bush fly Musca vetustissima Walker (Diptera: Muscidae) control in relation to seasonal abundance of Scarabaeinae dung beetles (Coleoptera: Scarabaeidae) in south-western Australia. *Bull. Ent. Res.* 78:633–9.

Rouland, C., A. Civas, J. Rerioux, and F. Patek. 1988. Synergistic activities of the enzymes involved in cellulose degradation, purified from *Macrotermes mulleri* and from its symbiotic fungus *Termitomyces* sp. *Comp. Biochem. Physiol.* 91B:459–65.

Scrivener, A. M., M. Slaytor, and H. A. Rose. 1989. Symbiont-independent digestion of cellulose and starch in *Panesthia cribrata* Saussure, an Australian wood-eating cockroach. *J. Insect Physiol.* 35:935–41.

Shear, W. A., and P. A. Selden. 2001. Rustling in the undergrowth: Animals in early terrestrial ecosystems. In *Plants Invade the Land: Evolutionary and Environmental Perspectives*, ed. P. G. Gensel and D. Edwards, 29–51. New York: Columbia University Press.

Shellman-Reeve, J. 1994. Limited nutrients in a damp wood termite: Nest preference, competition and cooperative nest defense. *J. Anim. Ecol.* 63:921–32.

Shepherd, V. E., and C. A. Chapman. 1998. Dung beetles as secondary seed dispersers: Impact on seed predation and germination. *J. Trop. Ecol.* 14:199–215.

Shubeck, P. P. 1975. Do diurnal carrion beetles use sight, as an aid of olfaction, in locating carrion? *J. N. Y. Entomol. Soc.* 76:253–65.

Slade, E. M., D. J. Mann, J. F. Villanueva, and O. T. Lewis 2007. Experimental evidence for the effects of dung beetle functional group richness and composition on ecosystem function in a tropical forest. *J. Anim. Ecol.* 76:1094–104.

Smith, D. C., and A. E. Douglas. 1987. *The Biology of Symbiosis*. London: Edward Arnold.

Speight, M. C. D. 1989. Life in dead trees: A neglected part of Europe's wildlife heritage. *Environ. Conserv.* 16:354–6.

Swift, M. J., O. W. Heal, and J. M. Anderson. 1979. *Decomposition in Terrestrial Ecosystems*. Berkeley: University of California Press.

Tenore, K. R. 1977. Growth of *Capitella capitata* cultured on various levels of detritus derived from different sources. *Limnol. Oceonogr.* 22:936–41.

Tenore, K. R. 1981. Organic nitrogen and calorie content of detritus. I. Utilization by the deposit-feeding polychaete, *Capitella capitata. Estuar Coast Shelf Sci.* 12:39–47.

Tilman, D. 1982. *Resource Competition and Community Structure*. Princeton: Princeton University Press.

Tyndale-Biscoe, M. 1993. Impact of exotic dung beetles on native dung beetles, bush flies and on dung. In *Pest Control and Sustainable Agriculture*, ed. S. A. Corey, D. J. Dall, and W. M. Milne, 362–4. Melbourne: Commonwealth Scientific and Industrial Research Organization.

Valiela, I., L. Koumjian, T. Swain, J. M. Teal, and J. E. Hobbie. 1979. Cinnamic acid inhibition of detritus-feeding. *Nature* 280:55–7.

Vincent, T. L. S., D. Scheel, J. S. Brown, and T. L. Vicent. 1996. Trade-offs and coexistence in consumer-resource models: It all depends on what and where you eat. *Am. Nat.* 148:1038–58.

Vitousek, P. 1982. Nutrient cycling and nutrient use efficiency. *Am. Nat.* 119:553–72.

Wallace, M. M. H., and M. Tyndale-Biscoe. 1983. Attempts to measure the influence of dung beetles (Coleoptera, Scarabaeidae) on the field mortality of bush fly *Musca vetustissima* Walker (Diptera, Muscidae) in south-western Australia. *Bull. Entomol. Res.* 73:33–44.

Ward, G. M., and K. W. Cummins. 1979. Effects of food quality on growth of a stream detritivore, *Paratendipes albimanus* (Meigen) (Diptera: Chironomidae). *Ecology* 60:57–63.

Webb, D. P. 1976. Regulation of deciduous forest litter decomposition by soil arthropod feces. In *The Role of Arthropods in Forest Ecosystems,* ed. W. J. Mattson, 57–69. Berlin: Springer-Verlag.

White, T. C. R. 1993. *The Inadequate Environment: Nitrogen and the Abundance of Animals*. Berlin: Springer-Verlag.

Wiens, J. A. 1984. Resource systems, populations, and communities. In *A New Ecology: Novel Approaches to Interactive Systems*, ed. P. W. Price, C. N. Slobodchikoff, and W. S. Gand, 397–436. New York: John Wiley & Sons.

Willson, M. F., B. L. Rice, and M. Westoby. 1990. Seed dispersal spectra—A comparison of temperate plant communities. *J. Veg. Sci.* 1:547–62.

Wood, T. G. 1976. The role of termites (Isoptera) in decomposition processes. In *The Role of Terrestrial and Aquatic Organisms in Decomposition Processes*, ed. J. M. Anderson, 145–68. Oxford: Blackwell Scientific Publications.

Yokoyama, K., H. Kai, T. Koga, and T. Aibe 1991. Nitrogen mineralization and microbial populations in cow dung, dung balls and underlying soil affected by paracoprid dung beetles. *Soil Biol. Biochem.* 23:649–53.

Young, O. P. 1982. Perching behavior of *Canthon viridis* (Coleoptera, Scarabaeinae) in Maryland, USA. *J. N. Y. Entomol. Soc.* 90:161–5.

18

Insect Pests in Stored Grain

Sonia M. N. Lazzari and Flavio A. Lazzari

CONTENTS

18.1 Introduction...417
18.2 Grain Storage and Losses...418
18.3 Storage Ecosystem..418
18.4 Major Stored Grain Pests: Feeding Habits and Damage Caused 420
18.5 Mouthparts, Digestive and Excretory Systems..424
18.6 Food and Nutrition Characteristics..425
18.7 Search for Food and Its Utilization..426
     18.7.1 Stimuli for Oviposition..426
     18.7.2 Food Attractants and Gustatory Stimuli 427
     18.7.3 Nutritional Requirements 427
     18.7.4 Digestive Enzymes 430
     18.7.5 Nutrient Budget and Relative Growth Rate..432
     18.7.6 Microorganisms...434
18.8 Physiological and Behavioral Adaptations to Food and Environmental Changes 434
18.9 Applications and Perspectives for Stored Pest Management..435
     18.9.1 Monitoring and Food Baits..436
     18.9.2 Plant Resistance and Bioactive Compounds 436
          18.9.2.1 Grain Composition..437
          18.9.2.2 Enzyme Inhibitors...437
          18.9.2.3 Bioactive Compounds 438
     18.9.3 Biological Control..440
     18.9.4 Growth Regulators 442
     18.9.5 Lignocellulosic Biofuels 442
18.10 Final Considerations 443
References..443

18.1 Introduction

The chemical composition and nutritional quality of grains and seeds do not change substantially during storage; that is, insects that feed on stored grain have stable food source, without major changes in nutritional composition and their defense compounds over time. Despite the nutritional requirements of insects feeding on stored products being similar to those of other phytophagous species, the former exhibit an almost unique ability to grow and reproduce on relatively dry food. Early studies indicated that these insects use metabolic water for their development in such a dry environment (Fraenkel and Blewett 1944; Baker and Loschiavo 1987). Other studies, however, show that the passive diffusion of water vapor is also an important source of water for stored-product insects (Arlian 1979; Arlian and Veselica 1979). The wide availability of food coupled with adequate temperature and relative humidity (RH) favor population growth and distribution of these insects. This chapter explores the physiological and behavioral responses of insects to food and changes in storage environment—the bioecology and nutrition—focusing mainly on species that feed on grain and by-products.

Approximately 130 species of insects have been recorded in stored products in North America (Loschiavo and Okamura 1979; Barak and Harein 1981; Sinha 1995). Most insect species on stored products are cosmopolitan due to grain transportation around the world and the stability of the storage environment. The main insect pests of stored grains belong to the orders Coleoptera (beetles and weevils) and Lepidoptera (grain moths). There are also few species of Psocoptera (grain lice), parasitoids of the order Hymenoptera (wasps), and predators of the order Hemiptera. Many species of mites (Acarina) are associated with the storage environment.

Insects have qualitatively the same nutritional requirements as other animals, so they compete for food produced by man along the production chain. During storage, owing to the availability of food and protection offered by the environment, insects can significantly increase their population and cause considerable damage. In the context of this chapter, we will discuss the feeding behavior, physiology, and survival strategies adopted by the insect species present in the storage environment.

18.2 Grain Storage and Losses

The world population will reach 9.1 billion in 2050—a third more mouths to feed than there are today (Food and Agriculture Organization 2009). Thus, dependence on cereal grains and oilseeds will increase, not only for human consumption and animal feed, but also for ethanol and biodiesel production. The storage of agricultural production is needed to maintain the quality and quantity of seeds and/or grain until the time of their use and/or consumption. Storage is also necessary to balance stock fluctuation, thus preventing food and seed shortages. However, qualitative and quantitative losses may occur and their impact varies widely. Quality loss is more difficult to assess, and includes reduction in vigor and germination of seeds; changes in physical appearance (discoloration); nutrient loss; presence of insects, mites, and their fragments; fungal and mycotoxins contamination; and other impurities and foreign materials (Lazzari 1997). In tropical regions, weight loss and deterioration of products are more severe than in temperature regions, especially in subsistence farming. It is difficult to have an accurate figure for losses in storage due to insects alone, but we estimate that the total loss (insects, rodents, and other pests) is of the order of 0.5% to 10% of the volume of stored grain.

Among the factors causing losses in the storage system are insects, mites, and fungi associated with grain and other stored products. As the insect population increases, dry matter is lost and grain quality decreases due to increasing oil acidity and reduction of seed viability and germination (Sinha 1983). Heavy insect infestations can also reduce amino acid and protein content (Girish et al. 1975) and affect grain palatability (Khare et al. 1974). The presence of insect fragments and excreta in processed food results in quality loss and rejection of the product. Ladisch et al. (1968) associated the presence of quinones secreted by species of *Tribolium* spp. (Coleoptera: Tenebrionidae) with carcinomas in rats. Recent studies have shown that grain insects can harbor enteric bacteria resistant to antibiotics, which are potentially virulent when present in grains and by-products (Lakshmikantha et al. 2006).

18.3 Storage Ecosystem

The main components of the grain storage ecosystem are the storage structure (particularly the size and type of silos and bulk warehouse), environmental conditions (temperature and RH), condition of the grain (moisture or water content, time in storage, broken kernels, amount of impurities and foreign materials), organisms present (insects, fungi, mites, natural enemies), and management and control measures adopted (Sinha 1995).

Post-harvest loss is expressed as dry matter and nutrient loss. In tropical regions, weight loss and deterioration of products are more severe than in temperate regions, especially for subsistence farming. It is difficult to have an accurate figure for losses in storage due to insects alone, but it is estimated that the total loss (insects, rodents, and other causes) is of the order of 3% to 10% of the volume stored.

The type of storage structure has great influence on the ecological aspects of the grain mass, including the pest management practices necessary for maintaining the quality of stored products for long periods. The RH of the air and the temperature are the most relevant environmental factors because they affect the moisture balance of the grain mass. The moisture of the grain or seed, called moisture content (MC), is expressed on a wet basis and measured on samples that are taken to be representative of the grain lot. Insects and fungi require a minimum RH for their metabolic processes, as enzymes are inhibited or even destroyed when the available water content in the grain is below 10% (wet basis). According to Baker and Loschiavo (1987), the MC of stored products depends on temperature, type of grain or product, and mainly on the equilibrium RH. For whole wheat and wheat flour, 12% to 18% is the equilibrium MC for an RH of 40% to 80%. Cereal grains with MC below 12% are considered dry; on the other hand, MC above 15% favors the growth of several microorganisms and pests. The maximum water content (WC%) for corn kernels is 14.0 for storage up to 6 months; 13.0 for 6–12 months; and 12.5 for more than 12 months in storage, at 75% RH and 25°C (Lazzari 1997). When the grain MC is kept in a safe range, it can be stored for long periods without the development of microorganism and pest populations.

Storage insects are adapted to live in conditions of low humidity; however, moisture above that considered safe for storage of grain favors their development. The insects die when they lose about 60% of body water or 30% of total body weight (Ebeling 1971). Fungi are also very diversified and specialized concerning the moisture levels needed for their development and reproduction. Each species requires a specific grain moisture and optimal temperature for their survival.

The optimum temperature for the development of most insect species in stored grain ranges from 24°C to 32°C, as shown by Fields (1992) (Table 18.1). However, this range may vary for different species of insects and for the different stages within the same species. Below the low suboptimal or above the high supra-optimal levels, insect death depends on the exposure time. Acclimatization can also occur if the insect is subjected to gradual variations in temperature, extending survival in extreme temperatures. These considerations are fundamental for deciding on control measures by using cold or heat disinfestations in silos and flour mills (Burks et al. 2000).

Howe (1965) categorized some species according to their development on different temperature: those that grow best at high temperatures (optimum 30–34°C), including *Trogoderma granarium* Everts (Coleoptera: Dermestidae), *Oryzaephilus surinamensis* (L.) (Coleoptera: Silvanidae), and *Tribolium castaneum* (Herbst) (Coleoptera: Tenebrionidae), and those that do best in moderate temperatures (24–27°C), for example, *Anagasta kuehniella* (Zeller) [= *Ephestia kuehniella* (Zeller)] (Lepidoptera: Pyralidae), *Sitotroga cerealella* (Olivier) (Lepidoptera: Gelechiidae), and *Sitophilus granarius* (L.) (Coleoptera: Curculionidae).

Storage insects are less tolerant to freezing damage because body fluids crystallize, damaging membranes and affecting the osmolarity of tissues. However, some species can maintain their fluids in a state

TABLE 18.1

Response of Stored-Product Insects to Temperature, for Application of Cold and/or Heat Treatments

Condition for Development	Temperature Range (°C)	Insect Response to Temperature	Application
Lethal	Above 60	Death in seconds	Grain disinfestation with heat
	50 to 60	Death in minutes	Structural heat treatment
	43 to 46	Death in hours	Quarantine of perishable
Optimal	25 to 33	Maximum development	
Suboptimal	18 to 21	Reduced development	
	5 to 15	Arrested development	Grain protection and of durable commodities
	–1 to 3	Death in hours or days	Quarantine of certain products
Lethal	–16 to –22	Rapid death, frozen tissues	Rapid disinfestation of durable commodities

Source: Adapted from *J. Stor. Prod. Res.*, 28, P. G. Fields, 89–118, Copyright 1992, with permission from Elsevier.

of super-cooling (–10°C or less) without freezing, for varying periods (Burks and Hagstrum 1999). Heat can kill more quickly than cold, by disrupting the ionic balance across cell membranes and promoting the denaturation of DNA and enzymes. At high temperatures, the insects also lose water more rapidly due to the phase change of cuticular lipids (Edney 1977). Nerves and muscles are the tissues most susceptible to the deleterious effects of both cold and heat.

The development of insects inside the kernel is not directly affected by the physical condition of grains since they are able to break grain cuticles with their jaws and feed inside. The development of external pest populations, however, is favored when the amount of broken kernels or those damaged by primary pests or impurities is high. In addition, these conditions cause increase in temperature and humidity, favoring fungal development, which in turn facilitate the development of mycophagous species. Physical or chemical characteristics inherent to the grain or cultivar, such as cuticle hardness and presence of inhibitors of digestive enzymes, affect the nutrition and development of insect populations in stored grain or seed.

Due to the high biotic potential of insect species infesting stored grain, and the favorable conditions for their development, preventive and curative measures are usually needed, such as application of diatomaceous earth, aeration, cooling and bin transfers of grain, and chemical insecticides when necessary (Subramanyam and Hagstrum 2000; Lorini 2003). Preventive control measures should start at product reception, before storage, as it is not possible to improve the quality of a product during storage, although it is possible to maintain product quality (Lazzari 1997).

18.4 Major Stored Grain Pests: Feeding Habits and Damage Caused

Direct damage by insects in stored grain results from the feeding activity of larvae and/or adults that consume the endosperm and/or the embryo of intact or broken kernels or by-products, causing qualitative and quantitative losses. Indirect damage results from the contamination with live or dead insects, exuviae, feces, webs, and other wastes. Depending on the activity, insects generate heat and increase humidity, favoring the proliferation of other insects and microorganisms, causing deterioration of grain and risk of spontaneous combustion.

The economic damage caused by stored-product insects is difficult to establish. Silva et al. (2003) modeled the losses caused by *Sitophilus zeamais* Motschulsky (Coleoptera: Curculionidae) and *Rhyzopertha dominica* (F.) (Coleoptera: Bostrichidae), and determined that it takes 180 kg^{-1} adults of *S. zeamais* to cause 1.5% damaged grains. At this insect level, grain moisture increases to 0.13%, reducing the test weight by 0.4 kg hl^{-1} and causing a dry matter loss of 0.7%. *R. dominica* is comparatively more harmful since the presence of 64 kg^{-1} insects results in 1.5% damaged grains and a moisture increase of 0.07%, reducing the test weight by 0.5 kg hl^{-1} and causing 0.5% dry matter loss.

Stored-grain insects have a variety of eating habits; for example, chewing granivorous species or seed eaters feed on the endosperm and/or the embryo of grains and seeds; mycophagous species feed on fungi; and predators and parasitoids consume eggs or larvae of pests. According to the feeding habits and eating patterns, insects infesting stored grain are classified as primary and secondary pests or, internal and external pests, respectively. Primary or internal pests break the intact grain cuticle and enter the grain to complete their development, feeding on the endosperm and/or on the germ (embryo). In addition to the direct injury, they open entry points for other insects and microorganisms.

Hill (1990) grouped storage pests into six categories: (1) primary pests, which penetrate and grow inside the grain, consuming endosperm and germ; (2) secondary pests, which feed on broken grain or flour; (3) scavengers, which include cockroaches, crickets, moths, usually polyphagous and omnivorous, consuming waste products of animal and/or vegetable origin; (4) phytophagous species, specialized in oilseeds (various beetles), tobacco, chocolate, or dried fruits, which contain high levels of sugar and attract species of *Carpophilus* (Nitidulidae); (5) species that infest material of animal origin, utilizing protein (flies infesting meat—usually dry, dust mites that feed on cheese, ham, and bacon) and keratin (skin, wool, hair, hides, horns), including moths in closets, dermestid, psocids feeding on dried insect collections; and (6) predators and parasitoids of storage pests, mainly Hemiptera and some Hymenoptera and mites. Major insect species that infest stored products are listed in Table 18.2 according to their feeding habits (Baker and Loschiavo 1987).

TABLE 18.2

Feeding Habits of Some Stored-Product Insect Species

Type of Food	Species	Stage[a]
Whole, intact grains (cereals and oilseeds)	*Acanthoscelides obtectus*	Larva, Adult
	Corcyra cephalonica	Larva
	Rhyzopertha dominica	Larva, Adult
	Sitophilus granarius	Larva, Adult
	Sitophilus oryzae	Larva, Adult
	Sitophilus zeamais	Larva, Adult
	Sitotroga cereallela	Larva
	Trogoderma granarium	Larva
	Zabrotes subfasciatus	Larva, Adult
Broken or damaged kernels; feed	*Cryptolestes ferrugineus*	Larva, Adult
	Cryptolestes pusillus	Larva, Adult
	Oryzaephilus surinamensis	Larva, Adult
	Plodia interpunctella	Larva
	Tribolium castaneum	Larva, Adult
	Trogoderma spp.	Larva
Flour and processed foods; feed	*Anagasta küehniella*	Larva
	Tribolium confusum	Larva, Adult
	Tenebrio molitor	Larva
Moldy grain or flour	*Ahasverus advena*	Larva, Adult
	Pyralis farinalis	Larva
	Typhaea stercorea	Larva, Adult
Dried fruits	*Cadra cautella*	Larva
	Ephestia elutella	Larva
Nuts, cereals products with high oil content	*Oryzaephilus mercator*	Larva, Adult
Spices	*Lasioderma serricorne*	Larva
	Stegobium paniceum	Larva

Source: Adapted from Baker, J. E. and S. R. Loschiavo: In *Nutritional Ecology of Insects, Mites, Spiders and Related Invertebrates*, ed. F. Slansky, Jr. and J. G. Rodriguez, 321–44, 1987. Copyright Wiley-VCH Verlag GmbH & Co. KGaA. Reproduced with permission.

[a] Development stage responsible for significant damage.

Examples of internal pests are the weevils *Sitophilus oryzae* (L.) and *S. zeamais* (Coleoptera: Curculionidae); the lesser grain borer *R. dominica* and several species of bruchids; and the Angoumois grain moth *S. cerealella*. Secondary or external pests are those unable to break the grain cuticle and feed on damaged grain. Among these are the Coleoptera: rusty grain beetle *Cryptolestes ferrugineus* (Stephens) (Cucujidae); the saw-toothed grain beetle, *O. surinamensis*, and *Oryzaephilus mercator* (Fauvel) (Silvanidae); and the red flour beetle *T. castaneum* and the confused flour beetle *Tribolium confusum* Jaquelin du Val (Tenebrionidae). The larvae of the Indian meal moth *Plodia interpunctella* (Hübner) and several species of the genera *Ephestia*, *Cadra*, and *Corcyra* (Pyralidae) damage hulls and feed inside, however without developing inside the grain. Associated insect species, such as the psocids (Psocoptera), do not attack the grain itself, but are present in the storage environment and feed on fungus and grain residues.

The lesser grain borer, *R. dominica*, is considered the most serious pest in stored grain worldwide, damaging wheat, maize, rice, and other cereal grains (Figure 18.1). It is a very voracious species and its presence is characterized by abundant fine dust (result of feeding), mixed with sweet-smelling fecal material. The rice weevil *S. oryzae* and the maize weevil *S. zeamais* are key pests in stored grains, such as maize, rice, wheat, sorghum, rye, barley, oats, millet, and other grains and products. They consume

FIGURE 18.1 Damage caused by *R. dominica* in different grains: (a) rice; (b) wheat; (c) barley. (From Ceruti, F. C., Rastreabilidade de grãos: Conceito, desenvolvimento de software e estudos de casos de manejo de insetos no armazenamento. PhD dissertation, Universidade Federal do Paraná, PR, Brazil, 2007.)

both the endosperm and the embryo, causing qualitative and quantitative damage (Figure 18.2). The grain weevil larvae, *S. granarius*, develop within the grain and consume about 64% of its content, feeding especially on the germ (Campbell and Sinha 1976). The bean weevils, *Acanthoscelides obtectus* (Say) and *Zabrotes subfasciatus* Boheman (Chrysomelidae: Bruchinae), lay their eggs on the grain and one or more larvae penetrate the grain, consuming the endosperm and embryo. The mature larvae make circular exit holes before pupation and the adult pushes out the window when emerging (Figure 18.3). Adults are short-lived and normally do not feed, but may consume water and nectar. Bean weevils generate a characteristic odor in the infested beans. The flour beetles, *T. castaneum* and *Gnathocerus cornutus* (F.) (Coleoptera: Tenebrionidae), show preference for wheat flour and bran, but can attack a wide variety of cereal grains and animal feed, especially when these products have high MC and/or when they are moldy. As their chewing mouthparts are not able to break the intact grain cuticle, they feed on the germ and endosperm of cracked or grain damaged by primary insects.

Among the Lepidoptera, the Angoumois grain moth, *S. cerealella*, is a primary pest that attacks corn still in the field. Inside the silos and warehouses, its action is limited to the surface of the grain mass, 30 to 40 cm deep. This species attacks maize, wheat, paddy rice, barley, sorghum, and other cereals (Figure 18.4). The larva completes its cycle in a silken cocoon that is spun joining several kernels. Other species of Lepidoptera, such as the Indian meal moth, *P. interpunctella*, and *Ephestia* spp. infest grains (corn,

FIGURE 18.2 Damage caused by *Sitophilus* spp. in different grains: (a) *S. oryzae* in barley; (b) *S. zeamais* in millet; (c–d) *S. zeamais* in maize. (From Ceruti, F. C., Rastreabilidade de grãos: Conceito, desenvolvimento de software e estudos de casos de manejo de insetos no armazenamento. PhD dissertation, Universidade Federal do Paraná, PR, Brazil, 2007.)

FIGURE 18.3 Damage caused by bean weevils: (a) *A. obtectus*; (b) *Z. subfasciatus*. (From Ceruti, F. C., Rastreabilidade de grãos: Conceito, desenvolvimento de software e estudos de casos de manejo de insetos no armazenamento. PhD dissertation, Universidade Federal do Paraná, PR, Brazil, 2007.)

wheat, paddy rice, soybeans, and peanuts) and a variety of farinaceous products, dried fruits, nuts, and animal feed. In bulk warehouses, they stay on the surface, starting the attack of the grain preferably by the germ, as they cannot break the cuticle in other sites. The adult moth does not feed on the attacked products and lives only a few days. The Mediterranean flour moth, *A. kuehniella*, is a voracious species and weaves silk threads on the feeding sites, forming compact masses that may block machinery and pipes in flour mills. In bulk stores, despite their superficial attack, they weave a silk blanket on the grain mass, which serves as a refuge for other insects and makes insect control and grain management difficult.

With the exception of species that attack intact grains, others have a more diverse feeding habit (Levinson and Levinson 1978). For example, the cigarette beetle *Lasioderma serricorne* (F.) (Anobiidae) feed on more than 50 plant and animal products, including pepper and paprika (Howe 1957; LeCato 1978). LeCato and McGray (1973) observed populations of *O. surinamensis*, *O. mercator*, *T. castaneum*, and *T. confusum* on 15 different types of diets. These species, among others, may feed on fungi associated with high moisture in grains and flours (Sinha 1965, 1966, 1968; Loschiavo and Sinha 1966; Dolinski and Loschiavo 1973). Soybeans are the least damaged kernels during storage, but even so, Cox and Simms (1978) recorded 12 species of insect pests in soy flour, depending on temperature and humidity.

The presence of *Liposcelis* (Psocoptera: Liposcelididae) in stored grains indicates poor storage conditions, high moisture, mold growth, and broken kernels and fines. Although these insects are mycophagous and often ignored, infestations with psocids are becoming a great concern in storage facilities and flour mills.

Stored-product pests have been transported from one country or continent to another with the goods they infest, thus becoming cosmopolitan. Many of these species have geographical subspecies or races

FIGURE 18.4 Angoumois grain moth, *S. cerealella*, on corn, showing the characteristic emergence hole. (Courtesy of Clemson University, USDA Cooperative Extension, Slide Series, Bugwood.org, Tifton, GA.)

with different food preferences, climatic conditions, and susceptibilities to pesticides (Hill 1990). Grains and seeds are not good diets for insects when dry, as most species prefer soft and moist seeds. However, in the course of evolution, species that feed on seeds in the field have adapted to use older seeds, hard and dry, for which there is less competition. As man began to collect and store grain, storage areas became a niche for many species that had adapted over time, so the storage environment has become a habitat with abundant food and protection. Bruchids feed on legume seeds in the field and can continue the infestation in the warehouse. The scavengers are able to adapt easily to various waste and decaying material, both outside and inside the storage structure.

Mycophagous species (beetles and psocids) that originally fed on fungi and cellulose in tree bark became adapted to drier conditions of warehouses, where they consume fungi and grain dust. Dried fruits favor the adaptation of many species of Nitidulidae and Pyralidae. Some species that attack stored products have an unknown origin in nature, except around warehouses, as is the case of *Tribolium* spp. and *L. serricorne* (Hill 1990).

The lesser grain borer *R. dominica* has considerable longevity and disperses by flight to the periphery of storage units and in forested areas away from stores. This suggests a primitive habit under tree barks, now serving as temporary niche, and alternative food source or hibernation site in the absence of the preferred stored grain.

Each insect has, in their tissues or in their digestive tract, a combination and quantity of chemical elements characteristic of their diet and/or the environment. These elements may indicate the geographical origin of some populations (Bowden et al. 1984). Mahroof and Phillips (2006) traced the origin, movement, and feeding history of *R. dominica* using trace elements (isotopes $d^{13}C$ and $d^{15}N$). Adults of this insect were captured in warehouses and in forests, and some were reared on wheat, oak seeds (*Quercus* sp.), and corn. The values of carbon isotopes $d^{13}C$ in tissues were similar to the ones in the diets, indicating that they are more reliable markers than nitrogen isotopes for tracing food from the diet to the tissues. The analysis of carbon isotopes showed that the insects from the field and warehouses do not have significant differences in the values of $d^{13}C$, and zinc was the most promising marker. Insects collected around the storage units had similar zinc concentration to those inside the warehouse, but differed from those captured in the woods (oaks). This information can be useful in detecting movement of pests in and out of storage. Removal of trees and plants from the periphery of storage facilities is advised in order to avoid infestation from populations dwelling outside.

18.5 Mouthparts, Digestive and Excretory Systems

Larvae and adults of beetles have similar dietary habits, with chewing mouthparts and digestive tracts adapted to crush and process solid food. The crop is reduced or absent, cecum is also absent, and digestion is performed in the anterior midgut. In Tenebrionidae and Curculionidae, final digestion, especially of proteins, occurs on the surface of the midgut cells. In Dermestidae, the whole digestion process of the larvae occurs within the endoperitrophic space. Several families, such as Curculionidae, Tenebrionidae, and Chrysomelidae (Bruchinae), have cysteine proteinases in addition to or in place of serine proteinases as digestive enzymes, suggesting that their ancestors were adapted to feed on seeds rich in serine proteinase inhibitors (Terra 2003).

Most adult moths attacking stored products have atrophied sucking mouthparts, while larvae have chewing mouthparts. The gut of the larvae does not present ceca in the midgut, and all digestive enzymes (except those of the initial digestion) are distributed on the cell surface of the midgut. Goblet cells are found in the anterior midgut, have an elongated neck, and secrete dietary K^+ ions and may be involved in water absorption. Other types of cells with pedunculated neck are found in the posterior midgut. Although there is a widespread pattern of digestion in Lepidoptera larvae, those with a more specialized diet, such as the clothes moth (Tineidae), have peculiar digestive adaptations. These insects need to break the disulfide bridges of keratin to facilitate the proteolytic hydrolysis of this protein (Terra 2003).

Excretion is the physiological process that requires most water to function. It involves not only the availability of water to process urine but also recovering water before elimination of urine with feces. Due to the low MC of grain and low RH, water absorption in the hindgut is enhanced by the cryptonephric organ, which consists of the association of Malpighian tubules and rectal papillae. The reabsorption of water depends on the pumping of potassium chloride in the spaces of the basolateral folds of the absorptive cells, creating osmotic pressure that moves water to these areas, and then to the hemolymph (Terra 2003).

The recently acquired genome of the red flour beetle, *T. castaneum*, was an important step toward the process of profiling genes and proteins in the gut of this species. Data by Morris et al. (2009) show that about 17.6% of the genes represented in the array are predicted to be highly expressed in gut tissues. Their data provide the basis for comparative transcriptomic and proteomic studies related to the gut of coleopterans.

18.6 Food and Nutrition Characteristics

According to Hill (1990), the rate of development of storage pests is mediated by temperature, RH, water content of the food, and nutritional value of the diet. Diet preferences are often difficult to identify since many species can survive in a variety of foods but perform better and reproduce only in few kinds. In less suitable foods, reproduction may occur, but developmental time of immatures last longer and mortality is high (50–70%). Under optimal conditions, mortality of immatures and adults is low (1–2%). Thus, the distribution of pest species is usually the result of a combination of environmental conditions, availability and quality of food, and natural competition at various levels.

Grains and different parts of the grain or seed vary in composition and nutritional quality. For example, the corn kernel pericarp or cuticle has 3–10% starch, 1% oil, and 3.5% protein; the endosperm 86–89% starch, 0.8% oil, and 8% protein of low biological value; and the germ or embryo 5–10% starch, 31–35% oil, and 17–19% protein of high biological value (Lazzari and Lazzari 2002). For wheat, the endosperm consists of 70% starch, 8–13% protein, and a small amount of vitamins. The germ is nutritionally rich, with 25% protein, 20% sugars, and vitamins; other nutrients and trace elements vary with the type of grain and grain tissue (Waldbauer and Friedman 1991).

Each insect species has a particular capacity for food consumption. Demianyk and Sinha (1988) calculated the consumption rate for 10 species of stored-product insects, and transformed the consumption directly as a percentage of weight loss of the grain. The total is the sum of the weight loss caused by the larva and the adult; however, since the adult moths do not feed, the damage for most Lepidoptera is only the result of feeding by the larvae. In the case of beetles, damage by adults is higher because they live much longer than larvae. The equivalence is the weight loss caused by each species compared to the damage (1.00) of *Cynaeus angustus* (LeConte) (Tenebrionidae) (Table 18.3 with data from

TABLE 18.3

Relative Consumption Rate for Different Stored-Product Insect Species and Equivalence Value

Species	Diet	Consumption (mg)			Equivalence
		Larva	Adult	Total	
Cynaeus angustus	Maize	32	453	485	1.00
Tribolium castaneum	Wheat flour	13	315	328	0.68
Prostephanus truncatus	Maize	13	223	236	0.49
Rhyzopertha dominica	Wheat grain	5	149	154	0.32
Sitophilus granarius	Wheat grain	19	67	86	0.18
Cadra cautella	Wheat grain	36	—	36	0.07
Oryzaephilus surinamensis	Oats	2	33	35	0.07
Plodia interpunctella	Maize	34	—	34	0.07
Sitophilus oryzae	Wheat grain	7	25	32	0.07
Cryptolestes ferrugineus	Wheat grain	1	14	15	0.03

Source: Data from Hagstrum, D. W. and B. Subramanyam, *Fundamentals of Stored-Product Entomology.* St. Paul: AACC International, 2006. With permission.

Subramanyam and Hagstrum 2000; Hagstrum and Subramanyam 2006). They calculated that 32 adults of *C. ferrugineus*, 16 of *S. oryzae*, and 3 of *R. dominica* consume the same amount of food as 2 adults of *Prostephanus truncatus* (Horn) (Bostrichidae). However, it is important to take into account the feeding habits of specific species. For example, *C. ferrugineus* feeds on the embryo and reduces the grain weight, while the larvae and adults of *T. castaneum* and *O. surinamensis* do not reduce the total grain weight. Bull and Solomon (1958) determined that 0.214 g (wet weight) of adult *L. serricorne* could be fed 1 g of wheat from egg to adult. The yield of this species per gram of weight loss of food is 0.46 g (wet weight), which is comparable to 0.40 g for *T. confusum* and 0.43 g for *A. kuehniella* (Fraenkel and Blewett 1943).

The type of food, nutritional quality and quantity, and environmental conditions affect the developmental time, reproduction, and other biological parameters of storage insects (Subramanyam and Hagstrum 1991). *S. zeamais* feeding in grains of different sizes results in progenies with different body size. Adults fed since emergence on corn kernels were significantly larger (2.78 ± 0.18 mm long; 0.97 ± 0.14 mm wide) than those reared on pearl millet (0.98 ± 0.15 mm; 0.45 ± 0.08 mm). However, there was no significant difference in fertility of females (Ceruti and Lazzari, unpublished data).

According to Waldbauer and Friedman (1991), self-selection of food is a continuous regulation of food intake that involves frequent food changes. This selection, however, does not occur randomly, but allows the insect to take advantage of choice. In the case of stored-grain insects, when two or more types of food are present as a homogeneous mixture of small dry particles, the rate of selective ingestion of food can be compared to the rate of feed mixture. Selective feeding obviously benefits performance because food ingested selectively is used more efficiently compared to nonselective intake. From a strictly physiological aspect, increased efficiency alone offers little advantage because the insect may increase consumption instead of increasing efficiency. However, from the ecological point of view, this has vital value for the insect.

Food self-selection was first demonstrated for the larva of the flour beetle, *T. confusum*. Larvae fed whole wheat grain damaged around the germ consume the whole germ, and only a small portion of the endosperm around it (Fraenkel and Blewett 1943; Waldebauer and Bhattacharya 1973). When the larva was fed with a 1:1:1 mixture of small particles of wheat germ, bran, and endosperm, it did not feed randomly but selected a mixture of 81% of the germ, 17% of the endosperm, and 2% of bran. The selection from the mixture results in better growth than either one of the portions alone or fine grounded flour, which does not allow the larvae to feed selectively. Unlike a separate fraction or whole-wheat flour, the mixture provides a protein-to-carbohydrate ratio of 57:43, close to the optimal ratio of 50:50. This food balance is important for formulating artificial diets, as for the Mediterranean meal moth, *A. kuehniella*, used in biological control programs (Parra et al. 1989). Environmental conditions, diet, and protocols for rearing stored product pests and their parasitoids are presented in details by P. Flinn (http://ars.usda.gov/Research/docs.htm?docid=12885).

18.7 Search for Food and Its Utilization

18.7.1 Stimuli for Oviposition

Larval feeding depends largely on the choice of oviposition site by adults. *Cadra cautella* (Walker) (Lepidoptera: Pyralidae) is attracted to volatiles of wheat and oviposits near the source of the odor (Barrer and Jay 1980). According to Gomez et al. (1983b), the resistance of maize genotypes to *S. oryzae* may be partially explained by different levels of oviposition stimulants in the grain. Studies with *S. granarius* show that damage from feeding by this species occurs most frequently near the germ; however, about 70% of oviposition cavities are located at the opposite end. This would prevent first instar larva to get in contact with the tissue of the embryo that may be toxic (Gomez et al. 1982). Baker and Loschiavo (1987) mention that females of *S. granarius* make cavities for eggs in food pellets containing wheat extract, but do not lay eggs in them, suggesting that certain stimuli are necessary for oviposition. Females lay a substance that reduces the probability of more than one egg being placed in a single wheat kernel. Adults of *T. molitor* usually feed on the surface of grain or flour; however, they lay their eggs into the grain. Females tend to avoid low-quality food and lay fewer eggs when the medium is depleted; they also penetrate deeper into the mass when the population density is high or when there are many eggs and larvae on the food (Gerber and Sabourin 1984).

18.7.2 Food Attractants and Gustatory Stimuli

Milled cereals with high germ content are very attractive to storage insects. Wheat kernels contain 2–4% lipids (dry weight), while the germ contains 15% fat and 55–60% triglycerides, which elicits an aggregation response. Baker and Loschiavo (1987) present a review of the response of various insects to different compounds and foods (Table 18.4). Oil in the wheat germ and other volatiles in cereal grains act as an attractant. Some fatty acids can either prevent or induce aggregation and positive gustatory responses (Levinson and Levinson 1978). Other nutrients such as sucrose, fructose, glucose, maltose, and fatty acids combined with sucrose stimulate feeding of the larvae of *P. interpunctella* (Baker and Mabie 1973). Loschiavo (1975) demonstrated that maltose is a potent phagostimulant for *T. confusum.*

18.7.3 Nutritional Requirements

Nutrition affects development, size, color, reproduction, and other biological characteristics of insects. If the diet is qualitatively adequate but is only available in limited supply, the result will be small adults. The Mediterranean flour moth, *A. kuehniella,* for example, requires approximately 0.13 g of wheat flour for normal development. In smaller quantities, such as 0.04 g, moths usually emerge, but will be much

TABLE 18.4

Food Attractants, Phagostimulants, and Aggregants for Larvae and Adults of Stored-Product Insects

Species	Stage	Food or Food Extract	Activity
P. interpunctella	L_1	Extract of wheat, corn, and peanuts	Aggregants
P. interpunctella	L_4	Fatty acids + sucrose	Synergistic phagostimulants
S. cerealella	L	Wheat germ lipids and triglycerides	Aggregants; phagostimulants
C. ferrugineus	A	Wheat volatiles	Attractants
O. mercator and *O. surinamensis*	A	Volatiles of rolled oats and yeast	Attractants; arrestants
R. dominica	A	Wheat volatiles	Attractants
S. granarius	A	Wheat triglycerides	Aggregants
S. granarius	A	Aqueous extract of wheat	Arrestant and phagostimulants
S. granarius	A	Sesquiterpenes	Deterrents
S. oryzae	A	Ethanol extract of susceptible corn	Attractant
S. oryzae	A	Chloroform extract of corn	Aggregant
S. oryzae	A	Amylopectin	Phagostimulant
S. zeamais	A	Volatiles of wheat, corn and rice	Attractants
S. zeamais	A	Volatiles of corn; hexanoic acid	Attractants
T. castaneum	A	Fatty acids C_5–C_{11}	Repellents
T. castaneum	A	Fatty acids C_{13}, C_{15}, C_{18}	Aggregants
T. castaneum	A	Wheat volatiles	Attractants
T. confusum	A	Palmitic acid; maltose	Aggregants, phagostimulants
T. confusum	A	Wheat germ triglycerides; fungal triglycerides	Aggregants
T. granarium	A	Fatty acids C_5–C_8	Repellents
T. granarium	A	Fatty acids C_{12}–C_{16}	Aggregants

Source: Data from Baker, J. E. and S. R. Loschiavo: In *Nutritional Ecology of Insects, Mites, Spiders and Related Invertebrates,* ed. F. Slansky, Jr. and J. G. Rodriguez, 321–44. 1987. Copyright Wiley-VCH Verlag GmbH & Co. KGaA. Reproduced with permission.

Note: A, adult; L, larva and instar, if any.

smaller and may have the wing and body ratio changed (Norris 1933). Nutritional requirements, both qualitative and quantitative, vary between species and even within the same insect species, according to their stage of development. The following are the nutrients and their role in the nutrition and microorganism association of stored-product insects:

(a) *Carbohydrates.* They serve as energy source and can be converted into storage fat or contribute to the production of amino acids; they are essential and may be needed in large quantities. In the case of *Tenebrio* spp., development is optimum with 70% carbohydrates and growth is reduced when diet contains <40% carbohydrates. Use of different carbohydrates depends on hydrolysis of polysaccharides, on how fast substances are absorbed, and on the enzyme systems that introduce these substances into the metabolic processes (Chapman 1998). Stored-product insects are able to use a wide variety of carbohydrates. *Tenebrio*, for example, uses starch; mannitol; the trisaccharide raffinose; the disaccharides sucrose, maltose, and cellobiose; and the monosaccharides mannose and glucose. In many cases, carbohydrates are replaced by protein and fat, depending on the insect's ability to convert these compounds into intermediate products that can be used in energy cycles, and the speed at which these reactions occur (Dadd 1960).

(b) *Lipids.* Fat is the main form of energy storage. In *Ephestia*, the presence of linoleic acid in the diet is essential for molting; if the amount is suboptimal, wings are devoid of scales because they do not separate themselves from the pupal cuticle. The total absence of linoleic acid in the diet of the larva hinders the imago emergence. According to House (1974a), for *O. surinamensis*, oleic acid and palmitic acid are more efficient in promoting normal growth and development than linoleic acid, and for *T. granarium*, arachidonic acid accelerates larval growth. Lipogenic factors and sterols are needed in the diet of all insects. Cholesterol, for example, can be stored in older larvae of *Tenebrio*, reducing the need for greater amounts in the diet. Nutritional values of sterols in some species of stored product beetles are shown in Table 18.5.

(c) *Vitamins.* They are structural components of coenzymes and necessary in small amounts in the diet because they cannot be synthesized. B-vitamins thiamin, riboflavin, nicotinic acid, pyridoxine, and pantothenic acid are essential for most insects; also, biotin and folic acid are required for many insects. Other vitamins are more specific to certain species, as is the case of *Tenebrio* sp. that requires a source of carnitine. On the other hand, insects such as *Dermestes*

TABLE 18.5

Nutritive Value of Sterols in Stored-Product Beetles

Sterol	*Dermestes maculatus*	*Lasioderma serricorne*	*Oryzaephilus surinamensis*	*Pitinus tectus*	*Stegobium paniceum*	*Tenebrio molitor*	*Tribolium confusum*
Calciferol	–					–	
Cholesterol	+	+	+	+	+	+	+
7-Dehydrocholesterol	+	+	+	+	+		+
7-Dehydrocolesterol-monobenzoato	+				–		±
Dihydrocholesterol	–	±	+	±	±		±
Ergosterol	–	+	+	+	+	+	+
7-Hydroxycholesterol	–				–		±
7-Hydroxycholesterol-dibenzoato					–		±
β-Sitosterol	–	+	+	+	+	+	+
Zymosterol	–	±	+	±	–		±

Source: Data from House, H. L., In *Physiology of Insecta*, ed. M. Rockstein, 1–62. New York: Academic Press, Copyright 1974, with permission from Elsevier.

Note: +, Well-utilized; – not utilized; ± partially utilized; not demonstrated (no indication).

sp. (Coleoptera: Dermestidae) can synthesize this vitamin. For the drugstore beetle, *Stegobium* sp. (Coleoptera: Anobiidae), only thiamine and pyridoxine are necessary in the diet because riboflavin, nicotinic acid, pantothenic acid, folic acid, biotin, and choline are synthesized by intracellular symbionts. β-Carotene (provitamin A) is probably essential in the diet of all insects because it is a component of the visual pigment, and likely has other functions, especially concerning molt and development. There is evidence that α-tocopherol (vitamin E) is required by many insects and is related with fecundity in *P. interpunctella* (Dadd 1973). The vitamins required by the immature stages of three species of beetles are listed in Table 18.6 (House 1974a); two of the species feed on grains and/or flour (*O. surinamensis* and *T. confusum*) and one dwells on natural fibers and carpets (*Attagenus* sp., Dermestidae).

Since vitamins act as a constituent of enzyme systems in metabolic activities, their deficiency can affect the formation of various body structures and activities. House (1974a) mentions that in *Corcyra cephalonica* (Stainton) (Lepidoptera: Pyralidae), lack of thiamine causes degenerative changes in cell membranes, especially in muscle, fat tissue, and in the midgut epithelium, whereas in *T. confusum*, thiamine deficiency results in tissues with smaller cells. The lack of carnitine in *T. molitor* affects the system that controls water loss, with severe pathological effects in the oenocytes, Malpighian tubules, hemolymph and fat, but not in the muscular and nervous systems. Another deleterious effect of lack of carnitine is the occurrence of uric acid or its salts in the gut.

(d) *Amino acids.* They are required for the production of structural proteins and enzymes, usually present in the diet as protein and depend on the ability of the insect to digest it. The absence of any essential amino acid hinders growth. Even some of the nonessential amino acids are required in the diet for optimum growth because their synthesis from essential amino acids increases energy consumption and produces catabolites that need to be rapidly eliminated (Dadd 1973). In the saw-toothed grain beetle, *O. surinamensis*, alanine is necessary only in the absence of nucleic acids (Davis 1968). The amino acid requirements for other species of insects of grain and flour are presented by House (1974a) (Table 18.7). In general, the D-isomer of many of the nutritionally important amino acids is toxic to several species, but in larvae of *T. confusum* some of these compounds may be used (Fraenkel and Printy 1954). According to House (1974a), for *C. cephalonica*, some amino acids and iodine–protein complexes, although not essential, have a beneficial effect on growth and development.

TABLE 18.6

Vitamin Requirements for the Development of Immature Stored-Product Beetles

Vitamin	*Attagenus* sp.	*Oryzaephilus surinamensis*	*Tribolium confusum*
Ascorbic acid (C)		—	
Biotin	±	±	+
Carnitine (BT)		±	+
Choline	+	+	+
Folic acid	±	±	+
Inositol		—	—
p-Aminobenzoic acid	—		—
Nicotinic acid	+	+	+
Pantothenic acid	+	+	+
Pyridoxine	+	?	+
Riboflavin	+	+	+
Thiamine	+	—	+

Source: Data from House, H. L., In *Physiology of Insecta.* ed. M. Rockstein, 1–62. New York: Academic Press, Copyright 1974, with permission from Elsevier.

Note: +, essential (required); –, not required; ±, some growth promoting activity; ?, contradictory information; not demonstrated (no indication).

TABLE 18.7

Amino Acid Requirements of Some Immature Stored Grain Beetles

Amino Acid	*Oryzaephilus surinamensis*	*Tribolium confusum*	*Trogoderma granarium*
Arginine, histidine, isoleucine, leucine, lysine, methionine, phenylalanine, threonine, tryptophan, valine	+	+	+
Aspartic acid		±	−
Cysteine	+	−	−
Glutamic acid	−	±	−
Glycine	+	−	−
Hydroxyproline		−	−
Proline	−	−	−
Serine		−	−
Tyrosine	+	−	−

Source: Data from House, H. L., In *Physiology of Insecta*. ed. M. Rockstein, 1–62. New York: Academic Press, Copyright 1974, with permission from Elsevier.

Note: +, essential; −, not needed; ±, some growth promoting activity; not demonstrated (no indication).

(e) *Nucleic acids.* They can be synthesized by the insects; however, when present in the diet, they improve larval growth. RNA is usually synthesized in the adipose tissue of *T. molitor*, and is an essential component in the diet of *O. surinamensis*. In this species, RNA reduces mortality, and guanine (purines) and cytosine (pyrimidine), each at a dietary level of 0.04%, can replace the RNA to 0.5% (Davis 1966).

(f) *Inorganic salts.* They are essential for the ionic balance for cellular activities and co-factors, or can be an integral part of some enzymes, but these are usually present in trace amounts as impurities in various components of diet (Chapman 1998). Minerals are required in adult insects for reproductive activities of females, especially in vitellogenesis. *T. molitor* requires magnesium, calcium, and zinc (Fraenkel 1958), whereas in *C. cephalonica*, high levels of zinc are toxic because they reduce the activity of catalases in the tissues. *T. confusum*, which feeds primarily on flour, requires iron, magnesium, manganese, phosphorus, potassium, and zinc (Medici and Taylor 1966).

(g) *Water.* Its absorption is related to the active ion movement through the intercellular spaces of the midgut epithelium and rectal papillae, resulting in increased osmotic pressure in these spaces and in passive water flow in the intestinal lumen. The influx of water creates a positive hydrostatic pressure in the intercellular spaces, and water and ions pass into the hemolymph. During food digestion, absorption, and excretion, water is absorbed in various parts of the midgut and Malpighian tubules; however, because of the need for water conservation by stored-product insects, water in the urine is reabsorbed by the rectal papillae (Chapman 1998).

18.7.4 Digestive Enzymes

Insects generally have a broad spectrum of digestive proteinases that are expressed spatially and temporally in the midgut. Knowledge on the composition, arrangement, and operation of proteinases is essential for studies of control strategies based on proteinase inhibitors and toxins of *Bacillus thuringiensis*, for example (Terra and Ferreira 1994; Oppert 1999).

The pH of the midgut of *T. molitor* increases from 5.2–5.6 to 7.8–8.2 from anterior to posterior regions, and reflects the optimum pH for total proteolytic activity: 5.2 in the anterior region where 64% of the activity occurs, and almost 9.0 in the posterior portion, with 36% of the activity. Two thirds of proteolytic activity in the anterior midgut is due to cysteine proteinases and the rest is due to serine proteinases. In contrast, 76% of the activity in the posterior portion is due to serine proteinases; proteinases similar to chymotrypsin are also abundant in this region. Such diverse enzymatic activities indicate that the digestive system for protein in *T. molitor* is quite complex. The correlation of proteinase activity and pH indicates specialized physiological mechanisms for regulating intestinal enzymes (Vinokurov et al. 2006). Two soluble post-proline cleaving peptidases, PPCP1 and PPCP2, were detected in the midgut of *T. molitor* larvae. PPCP1 is located mainly in the more acidic anterior midgut lumen, at pH 5.3, and its activity is related with protein digestion. PPCP2 is a nondigestive tissue enzyme evenly distributed along the midgut, with maximal activity at pH 7.4 (Goptar et al. 2008).

Cysteine proteinases are important enzymes for digestion in beetles, while the vertebrates often use other classes of proteinases for digestion. According to Oppert et al. (2000), besides the cysteine proteinases, a complex pattern of proteinase activity occurs in the midgut of beetles. The rice weevil, *S. oryzae*, digests food using a combination of classes of cysteine and serine proteinases. Similarly, the combination of these inhibitors in the diet of larvae of the red flour beetle, *T. castaneum*, has a synergistic action in reducing insect growth. However, some insects have an adaptive phenotypic plasticity to compensate for the ingestion of digestive inhibitors, increasing the production of insensitive proteases. Thus, one explanation for the synergism of inhibitors of cysteine and serine proteinases observed in *T. castaneum* is that the combination of these inhibitors decreases the adaptive response of the insect.

The amylase-to-proteinase ratio in the midgut of four species of graminivorous beetles (*S. oryzae*, *S. granarius*, *T. molitor*, and *T. castaneum*), which feed primarily on cereal grains and by-products, are higher than in other species that feed and develop on animal diets or food with high protein content (Baker 1986). In the latter case, the general proteinase activity (caseinolytic activity) of aminopeptidase and, especially, the ratios of proteinase to amylase were much higher than in graminivorous species. The larvae of *A. kuehniella* and *P. interpunctella* showed lower levels of amylase but higher proteinase than the four beetle species previously mentioned. This is because these lepidopteran larvae, although feeding on cereals and by-products, have more varied eating habits than the beetles.

Several carbohydrases differ in their relative concentration in the gut of *T. castaneum*: amylase > invertase > β-glucosidade > α-galactosidase > β-galactosidase (House 1974b). Table 18.8 lists the main carbohydrases demonstrated for some species of Coleoptera of grain or flour. In the larvae of *T. molitor*, with the relative reduction of proteolytic activity during larval development, there is a relatively stable increase in amylase activity until both activities reach a constant level in the last larval instar (Applebaum et al. 1964a).

Studies by Cinco-Moroyoqui et al. (2006) demonstrate that higher activities of amylase are detected in populations of *R. dominica* reared on wheat kernels at low density than at high-density populations. As protein intake increases, reproductive rate also increases. However, the consumption of wheat protein is inversely correlated with the levels of amylase activity. Amylase activity in homogenates of *R. dominica* showed a variable degree of inhibition by protein extracts prepared from different wheat varieties, and those that had the lowest activities were more inhibited by extracts of wheat than those that had higher activities of amylase. The results suggest that the activity levels of α-amylase and composition

TABLE 18.8

Digestive Carbohydrases in the Gut of Some Stored-Products Insect Species

| Species | Glucosidase | | Galactosidase | | β-*h*-Fructosidase | Amylase |
	α	β	α	β		
Tenebrio molitor (larva)	+	+	+	+	+	+
Tribolium castaneum (larva)	+	+	+	+		+
Tribolium castaneum (adult)	+	+	+	+		+
Trogoderma sp.	+	−	+	+	+	+

Source: Data from House, H. L., In *Physiology of Insecta*, ed. M. Rockstein, 62–117. New York: Academic Press, 1974. With permission.

Note: +, present; −, absent; not demonstrated (no indication).

of isoamylases of populations of *R. dominica* are modulated by diet, and that the inhibitory activity of α-amylase of resistant and susceptible wheat varieties influences such variations.

Enzyme activity increases with temperature, and the highest rate occurs at 45–50°C, but only for short periods, because above 40°C the enzymes become denatured. Thus, for optimal enzymatic activity in the long term, there must be a balance between higher activity and faster denaturation at higher temperatures. In larvae of mealworms, there are changes in the activity of proteinases to compensate for temperature changes. If the larva is transferred from 23°C to 13°C, the activity of proteinase decreases and then increases, so that after 10 days the activity is twice as high as the original. Returning to 23°C, proteinase activity returns to its original level. Amylase activity does not present this kind of compensatory changes (Applebaum et al. 1964b).

Changes in enzyme activity suggest that both synthesis and secretion are regulated physiologically and may be induced by food, directly stimulating the secretion of midgut cells—a mechanism called secretagogue (Chapman 1998). The secretion of digestive enzymes can also be stimulated by neural mechanisms; that is, the presence of food stimulates a nerve reflex that, in turn, stimulates the activity of secretory cells in the intestinal epithelium. The stimulus may also be hormonal; that is, feeding results in the production of a hormone that reaches the gut via hemolymph. In *T. molitor*, proteinase secretion is endogenously induced, and at the time of molt and emergence, due to lack of food, there is no secretagogue activity, indicating that the secretion is an integral part of the hormonally regulated events of metamorphosis. The activity of proteinase in the midgut of *T. molitor* usually does not occur in beheaded adults the day before the emergence, but it works if decapitation occurs after emergence, indicating that the neurosecretory cells are the source of the hormones that control these secretions (Dadd 1961).

18.7.5 Nutrient Budget and Relative Growth Rate

Adults of several species of beetles of stored products consume significantly more food than their larvae. Campbell and Sinha (1978, 1990) estimated the efficiency of energy utilization from food for oviposition against the capacity of population increase, and obtained a high positive correlation.

According to Slansky and Scriber (1985), the efficiency of assimilation of graminivorous stored-product insects is generally higher than that of phyllophagous species. Baker and Loschiavo (1987) calculated the growth efficiency and relative growth rates of different insect species of stored products, on the basis of data submitted by various authors (Table 18.9). It can be observed that these values vary with the species and usually reflect the quality of food consumed. Shellenberger (1971) showed that corn has the lowest content of essential amino acids compared with other grains, which explains why growth efficiency and relative growth rate of *S. oryzae* are lower when the insect feeds on corn compared to wheat. Table 18.9 summarizes the results of tests performed by Gomez et al. (1982, 1983a), who studied the response of *S. oryzae* to different genotypes of corn. Larval development was faster on opaque maize Op2-conversion than on the normal Op2-counterpart. The efficiency of conversion of ingested food (ECI) and digested food (ECD) are lower in the Op2-conversion, probably due the higher concentration of lysine and softer endosperm, rendering the opaque maize more susceptible to attack by insects. Furthermore, the development time of the rice weevil is faster and the ECI is highest on the waxy maize-conversion line than on the waxy corn-counterpart because the rate of amylose–amylopectin is conversion higher in the former, which increases utilization efficiency.

Table 18.9 shows that the mixture of components of a diet (pelleted corn) offered to larvae of *S. oryzae* resulted in an increasing growth efficiency compared to a diet of whole kernels. In addition, the ECI and ECD values are highest when the insect is fed on an artificial diet, possibly because the diet is more homogeneous (Baker 1974). Tests with *T. confusum* (Waldbauer and Bhattacharya 1973) show that growth efficiency and other values are higher when the insect is fed with wheat germ. When a ground mixture with equal parts of wheat bran, germ, and endosperm was offered, larvae preferentially ate the lipid-rich germ. Tests with *T. castaneum* by Medrano and Gall (1976) proved that the utilization efficiency of food is affected not only by nutritional quality and by distribution of nutrients in the food, but that genetically controlled biochemical mechanisms also govern food conversion. Growth rates were faster and consumption rates and ECI were higher for the selected strain with higher pupal weight, when compared to the control strain.

TABLE 18.9

Relative Growth Rates and Food Utilization Indices for Stored-Product Insects Feeding on a Variety of Foods or Diets

Species	Food or Diet	Stage[a]	RGR[b]	AD[c,d]	ECI[e]	ECD[f]
C. cephalonica	Sorghum	L (20 days)	—	87 MB	11.4	13.1
	Maize		—	81	6.9	8.5
	Wheat		—	80	5.0	6.2
	Peanuts		—	86	9.0	10.8
C. cautella	Sorghum	L (20 days)	—	87 MB	5.9	6.7
	Maize		—	87	3.3	3.8
	Wheat		—	81	0.6	0.8
C. ferrugineus	Cut wheat kernel	E–P	0.13	66-79 EB	1–15	3–23
O. surinamensis	Rolled oats	E–P	0.20	89 EB	34.1	34.8
R. dominica	Cut wheat kernel	E–P	0.17	—	—	15–38
S. granarius	Artificial diet	L_1–P	0.20	88 MB	12.4	14.2
S. granarius	Wheat kernel	E–P	0.21	76 EB	10.5	14.1
S. oryzae	Artificial diet	L_1–P	0.27	96 MB	24.1	24.9
S. oryzae	Maize-Op2 conversion	E–P	0.15	76 MB	3.2	4.2
	Maize-Op2 counterpart		0.14	78	3.6	5.1
	Maize-waxy conversion		0.15	82	4.4	5.4
	Maize-waxy counterpart		0.13	80	3.9	5.0
S. oryzae	Maize—intact kernel	E–P	0.14	78 MB	4.5	3.4
	Maize—pelletized kernel		0.19	65	11.6	9.6
	Maize—germless pellets		0.15	51	7.5	4.8
S. oryzae	Wheat kernel	E–P	0.18	79 EB	13.4	16.9
T. castaneum	White flour + yeast	Control line[g]	0.14	66 MB	12.3	18.8
		Selected line[g]	0.18	69	16.6	24.1
T. confusum	Wheat bran	L_1–P	0.13	55 MB	6.5	11.7
	Wheat endosperm		0.11	66	4.7	6.5
	Wheat germ		0.21	65	9.7	14.9
	Mixture (1:1:1)		0.22	67	10.9	16.3

Source: Data from Baker, J. E. and S. R. Loschiavo: In *Nutritional Ecology of Insects, Mites, Spiders and Related Invertebrates*, ed. F. Slansky, Jr. and J. G. Rodriguez. 321–44. 1987. Copyright Wiley-VCH Verlag GmbH & Co. KGaA. Reproduced with permission.

[a] Development stages: E, egg; L_1, first instar larva; P, pupa.

[b] RGR, relative growth rate (mg dry weight gain/mg dry weight/day).

[c] AD, approximate digestibility or assimilation efficiency.

[d] Values based on mass budget (MB) or energy budget (EB).

[e] ECI, efficiency of conversion of ingested food into insect biomass and gross growth efficiency.

[f] ECD, efficiency conversion of digested food to insect biomass or gross growth efficiency.

[g] Medrano and Gall (1976): 12–14 days old *T. castaneum* larvae of a control line and selected line 9 based on pupal weight at 21 days.

Regarding the nutritional value of food, studies show that the highest level of lysine in different cultivars of barley is not the only factor responsible for the high development rate of insects. Lamb and Loschiavo (1981) established that the larval development rate of *T. confusum* is highly correlated with the lysine content in different cultivars of barley. However, there is a strong influence of temperature, shown by the logistic equation $y = K/(1 + \exp(a - bx))$, where y refers to the percentage of daily development, K is the maximum development rate, a and b are constants determined by the multiple curvilinear regression, and x is the rearing temperature. When the values of K for a particular barley cultivar are plotted against the lysine content of the same cultivar, a significant correlation is obtained, confirming the interaction between diet and environmental conditions (i.e., temperature).

18.7.6 Microorganisms

Several species of storage insects ingest microorganisms along with food, while others have a constant association with microorganisms present in the digestive tract or intracellularly in various tissues. The presence of microorganisms allows the use of diets that would otherwise be inappropriate. Several microorganisms present in the gut can supply the insect with essential vitamins or other dietary supplements. The obligate associations are particularly important for insects with a restricted diet deficient in certain essential nutrients, such as cereal grains, fur, feathers, wool, timber, and other stored products. In *Stegobium*, yeasts present in the gut produce B vitamins and sterols, which may be secreted into the intestinal lumen or released by the digestion of these microorganisms. The effect of the loss of microorganisms varies with the species and diet. In the case of *R. dominica*, apparently there is no negative consequence due to the absence of microorganisms (Chapman 1998).

18.8 Physiological and Behavioral Adaptations to Food and Environmental Changes

Behavioral and physiological responses affect the ability of population growth and efficiency of food utilization, even in species in the same genus. Baker (1974, 1986) and Baker and Woo (1985) observed that *S. oryzae* is more efficient in the use of food (wheat kernels) than *S. granarius* and *S. zeamais*. The rice weevil has a higher amylase-to-proteinase ratio, with a level of α-amylase three to eight times higher than the two other species. Since the endosperm of wheat is composed of about 55% starch and contains α-amylase inhibitors, the higher amylase levels in *S. oryzae* would act in the mechanism of detoxification because the substrate is provided in large quantities for both the inhibitor and for the normal digestive mechanism.

According to White and Sinha (1981), an increase in temperature from 25°C to 35°C accelerates the development of *O. surinamensis* and increases the food consumption rate; however, there is a reduction in the values of ECI and ECD. In the case of *S. granarius*, despite the development being faster and food consumption higher at 30°C than at 20°C, there is no reduction in efficiency of the ingested and consumed energy (Campbell et al. 1976), showing that *S. granarius* is more efficient in food utilization than *O. surinamensis*, independent of temperature.

Stored-product insects obtain water from the ingested food, by absorption from the air and from metabolism. Water can be lost through perspiration, respiration, excretion, feeding, and reproduction activities. Devine (1978) reported that an adult of *S. granarius* contains 1.6 mg of water in its body; the daily demand is 12%, of which 17% comes from food, 39% is metabolic water, and 44% is obtained from the air by diffusion. An adult of *O. surinamensis* contains 0.26 mg of water, with a daily demand of 34%, of which 10% comes from food, 15% from the metabolism and 75% from the environment.

Population growth is favored by successful compensatory responses to changes in water content of food and RH. According to Baker and Loschiavo (1987), insects respond to changes in ambient humidity and food MC through behavioral and physiological adaptations and population adjustments. The dispersal pattern of adults of the rusty grain beetle, *C. ferrugineus*, on wheat shows that this species responds to gradients of MC of the grain (Loschiavo 1983). About 90% of the adult population establishes in portions of high moisture, for example, in grain at 16% compared to areas with grain at 13.4%. This occurs because this species responds positively to high-moisture, moldy grain. In the case of the saw-toothed grain beetle, *O. surinamensis*, adults avoid areas of high RH (100%), but respond positively to gradients between 20% and 60% and even 10% and 50% RH. However, in situations of extreme dehydration and starvation, the insect can respond positively to areas with 100% RH or tend to aggregate in small amounts of free water or in points with grain at 18% humidity (Arbogast and Carthon 1972a,b; Stubbs and Griffin 1983).

Arlian (1979) studied the effects of wheat kernels with different MC on adults of the rice weevil. At low RH (22.5%), water loss by transpiration exceeds water obtained by all mechanisms. However, in the range from 65% to 85% RH, the water obtained from food intake alone exceeds the loss by transpiration, and water is still absorbed from the environment. When exposed to 99% RH, there is reduction in food consumption, but the water absorbed passively compensates for the low food intake. The author concluded that *S. oryzae* adjusts its physiology by reducing food consumption in conditions of extremely high or low humidity, so that the water balance and the potential for water gain is a factor that induces feeding.

Arbogast (1976) observed that the larval mortality and development time of *O. surinamensis* and *O. mercator* are reduced with increasing RH from 12% to 74%. The net reproductive rate and increase in the population size are also significantly higher at 74% RH for both species. Evans (1982) observed that the development time and survival of larvae and adults, and the net reproductive rate of *S. oryzae* reared on wheat kernels were positive and more favorable for the population increase at 14% grain moisture compared to 11.2%, independent of temperature.

Siva-Jothy and Thompson (2002) observed that the immune system of *T. molitor* is depressed after a short period of starvation, but it is rapidly activated to normal levels as soon as the insects feed. This immune response can manifest itself at the cellular level with the production of the enzyme phenol oxidase. This enzyme controls the melanization process to isolate pathogens present in the body, along with production of cytotoxic substances that kill the invading organism. Starvation leads to a reduction in the activity of phenoloxidase in the hemolymph of both males and females, regardless of the presence of reserves in the adipose tissue. This result suggests that high enzyme activity should have a relatively high-energy cost, so that it is reduced during starvation but quickly returns with the food supply because it has an important role in body defense. Thus, this rapid modulation of the immune function is affected not only genetically but also by the nutritional status of the insect.

Although insects in storage facilities are protected from extreme environmental conditions, some species may enter diapause. Among these are some species of Phycitinae (Lepidoptera) commonly present in the storage environment that enter diapause induced by low temperatures, short photoperiod, over-population, and diet (Cox et al. 1984). A condition similar to diapause was observed (Burges 1960) for *T. granarium*, characterized by discontinuous food intake, low respiration rate, and increased lipid and glycogen contents in the larvae. Because of the increased body weight, adults compensate for the slow larval development, laying more eggs that contribute to the population increase.

Psocoptera species may occur in large populations in warehouses and packed products (pasta, rice, corn, tea, and dried fruit). In storage, psocids feed on flour and grain germ, and on fungi and bacteria; however, in the external environment, they may consume dead insects, ascomycete fungi, organic matter, wood bark, and pollen. High infestations of Liposcelididae in certain products indicate high humidity and the presence of microflora. In laboratory tests, it was found that *Liposcelis bostrichophila* Badonnel, which is one of the most common psocid in grain warehouses, preferred crushed buckwheat and millet impregnated with glucose and fructose, among many other diets (Kalinovik et al. 2006). Investigation of the digestive tract and feces of several species of psocids indicates that they consume different species of fungi and bacteria. The feces of those fed with fungi present sporangiospores and conidia, which form colonies on medium plates. On the other hand, the specimens fed on intestinal flora bacteria completely absorb the vegetative forms and cellulose, by the action of cellulase, eliminating only spores in the feces, which are more resistant to fermentation in the digestive tract (Kalinovik et al. 2006).

18.9 Applications and Perspectives for Stored Pest Management

The fast development of insect populations in stored products is basically due to their high biotic potential, which takes into account the species reproductive capacity and resistance to the environment. Since the environmental resistance in silos and warehouses is virtually null because of great food availability, absence of parasitoid and predator pressure, and favorable abiotic conditions, the insect species in stored grain tend to express their full biotic potential. Considering the nutritional requirements of insects, their adaptive strategies to overcome low water availability and the strong interactions in the stored product ecosystem, one can define more cost-effective pest management measures, as discussed below.

18.9.1 Monitoring and Food Baits

The bioecology of insect nutrition provides essential information for pest monitoring using food baits, phagostimulants, and plant volatiles that can be combined with pheromones or kairomones in a variety of traps (Barak et al. 1990; Hagstrum and Subramanyam 2000). Specific foods, ground grain, vegetable oils, and volatile compounds derived from foods have been used as bait, increasing the attractiveness for several species of insects. Food-baited cage traps usually have a mesh covering the bait, allowing access of insects, but not rats and other animals (Figure 18.5). They are useful tools for simultaneously monitoring both sexes of several species. Capture data are used to determine critical infestation areas for cleaning measures. Baits should be removed frequently so they do not become an infestation source (Throne and Cline 1989, 1994).

Cage traps with bait consisting of a mixture of ground rice and other cereal grains plus wheat germ as an attractant were used to assess the presence and spatial distribution of insects in rice storage units (Trematerra et al. 2004). Food-baited traps can be placed on the floor of warehouses and probe traps, with or without food or pheromone in the grain mass, providing more complete and accurate information for decisions of control measures (Pinniger 1990; Pereira 1994; Pereira et al. 2000; Caneppele et al. 2003a; Ceruti 2003, 2007).

Mahroof and Phillips (2007) tested various types of food and other compounds (grains, nuts and ground spices, vegetable oils, and plant extracts) to monitor the cigarette beetle, *L. serricorne*. Traps with the pheromone bait serricornine combined with ground paprika or red chili pepper or leaf extracts of these spices had synergistic effect and captured significantly greater numbers of males and females compared with the other baits tested. Attractive and repellent food baits can also be used to keep insects out of industries and warehouses. The baits should be placed in traps with insecticide to eliminate the trapped insects, and be kept outside the storage unit.

18.9.2 Plant Resistance and Bioactive Compounds

Plant resistance to arthropods is one of the most important strategies in insect pest management (IPM) programs in field crops. Resistance can also be expressed in the grain or seed after being harvested and stored. Nevertheless, unlike plant tissues in development, the grains have a stable chemical composition and have no defense compounds such as alkaloids, saponins, non-protein amino acids, terpenoids, and phenolics that have antibiotic or deterrent effect on insects (Nawrot et al. 2006).

Several mechanisms may be involved in the resistance of grains and seeds to stored-product insects: chemical—lack of vital nutrients, presence of compounds that negatively affect their development, volatile repellents or deterrents that affect the feeding behavior, and inhibitors of digestive enzymes; and physical, represented by the resistance of grain cuticle and hardness of the endosperm, among other features.

The index of grain susceptibility, which includes the number of offspring produced and the development time, is a good measure to compare the nutritional quality of different varieties of grain, and the

FIGURE 18.5 Food-baited cage traps with bait made with ground cereal grains and wheat germ used for insect monitoring around silos. (Courtesy of F. Lazzari.)

hardness of the grain that limits its use as food for insects (Dobie 1974). Nawrot et al. (1985) found that albumin, globulins, and gliadins present in wheat grains negatively affect larval development, longevity, and fecundity of some species of storage insects.

18.9.2.1 Grain Composition

Some corn hybrids may present resistance against stored insects, as demonstrated by Caneppele et al. (2003b) and Marsaro et al. (2005b). They found that lipids present in some corn materials are positively correlated with resistance against the corn weevil, *S. zeamais*. Nawrot et al. (2006) found that a certain variety of hard wheat, despite having favorable conditions for the development of *S. granarius* (high amylolytic and low antiamylolytic activities), can show resistance against this insect due to the high protein and fiber content and hardness.

18.9.2.2 Enzyme Inhibitors

The presence of inhibitors of digestive enzymes in certain plants and genetic materials is a desirable feature for inhibiting the development of insect populations in stored grain. Cysteine proteinases are important digestive enzymes for beetles, while vertebrates often use other classes of proteinases for digestion. For this reason, the incorporation of genes that encode inhibitors of cysteine proteinases into genetically modified crops has been proposed as a method for preventing damage by beetles, without causing problems for vertebrates. Inhibitors of potato cysteine proteinase have been studied for *T. castaneum* (Oppert et al. 2003). Data by Oppert et al. (2005) suggest that the combination of cysteine and serine proteinase inhibitors exhibits a synergistic effect on midgut proteolytic activity and development of *T. castaneum* larvae. This is achieved by preventing the adaptive proteolytic response to overcome the activity of the inhibitors.

For *L. serricorne* (Oppert et al. 2002), the most potent inhibitors of caseinolytic activity in the intestinal lumen are inhibitors from soybeans, which also inhibit trypsin, chymotrypsin–trypsin, chymostatin, and *N*-tosyl-L-phenylalanine chloromethyl ketone. Leupeptin showed slight inhibition, while phenylmethylsulfonyl fluoride was inhibitory only at high concentrations (mM). The absence of cysteine, aspartic, and metallo-proteinase digestion of the cigarette beetle can be evidenced by the lack of activation by thiol reagents, at optimum alkaline pH (Figure 18.6).

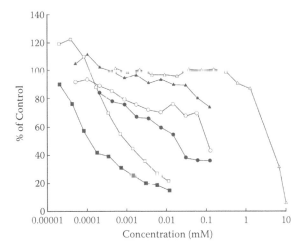

FIGURE 18.6 Inhibition of proteinases in the lumen of *L. serricorne* by selected inhibitors, as a percentage of the uninhibited activity (control): ■, soybean trypsin–chymotrypsin soybean inhibitor (Bowman Birk); □, soybean trypsin inhibitor (Kunitz); ●, chymostatin; ○, *N*-tosyl-L-phenylalanine chloromethyl ketone; ▲, leupeptin; △, phenylmethylsulfonyl fluoride. (Redrawn from Oppert, B., et al., *Bull. Entomol. Res.*, 92, 331, 2002. Cambridge University Press. With permission.)

Inhibitors that prevent the activity of α-amylase and proteinases in the gut of insects disrupt digestion because these enzymes play important roles in the digestion of starch and protein, respectively (Franco et al. 2002). Marsaro et al. (2005a) evaluated the presence of inhibitors of α-amylase in maize genotypes and found that their levels are negatively correlated with the index of susceptibility of corn hybrids against the corn weevil. The increase in α-amylase inhibitors can be obtained by crossing materials naturally endowed with this feature, or through genetic modification.

Arcelin is a protein that replaces phaseolin (normal protein stored in seeds) in some wild beans, and the arcelin-1 and arcelin-5 are the most active in antibiosis resistance against the Mexican bean weevil, *Z. subfasciatus*. Development slows down, and the number and weight of adults are lower in varieties rich in this compound, indicating that they are not nutritionally adequate for larval development. Since the beetles do not feed as adults, the effect of arcelin on this stage would be by antixenosis resulting in nonpreference for oviposition and not related with feeding (Lara 1997; Guzzo et al. 2006).

18.9.2.3 Bioactive Compounds

(a) *Avidin*. It is a biological material that has been tested for the control of several species of stored-product insects. It is a glycoprotein present in egg white, which prevents the absorption of biotin because it binds strongly to this compound ($K_d = 10^{-15}$ M). Biotin is a cofactor required for various types of carboxylase reactions, essential for all organisms (Morgan et al. 1993). According to Wright (1987), a concentration of about 2.5 ppm of avidin binds to approximately 38 ppb of biotin, which is about half the concentration of biotin in corn (70 ppb). Insects do not have a biosynthetic route for biotin and must obtain it from external sources. The avidin gene has been incorporated in corn and rice, rendering the grains resistant to the attack of insects, especially when the grains are ground with avidin and sprayed on the grain mass and eaten by insects.

When avidin is present in transgenic corn, a level of 100 ppm or more prevents the development of several species, including *S. zeamais*, *R. dominica*, *S. cerealella*, *O. surinamensis*, *T. castaneum*, *T. confusum*, *P. interpunctella*, and *A. kuehniella* (Kramer et al. 2000). However, only 50% of the harvested corn with avidin contains some level of recombinant avidin, as these transgenic plants are male sterile. However, in the flour and powder derived from this corn, there is a more homogeneous distribution of avidin, making them more toxic than the field corn avidin.

Avidin corn powder applied on stored maize showed an efficiency of 85% mortality of *T. castaneum*, compared with the control maize grain without avidin powder (Flinn et al. 2006). However, its effectiveness is lower for *C. ferrugineus* (40%), which is an external pest but tends to penetrate damaged grain and feed on the germ, reducing contact with avidin. In the case of *S. zeamais*, avidin has a reduced action (10%) because the insect feeds inside the grain, totally avoiding contact with avidin powder.

(b) *Protein-rich pea flour*. Seeds of legume plants contain toxic allelochemicals and deterrents that affect insects (Bell 1977), such as pea protein (from *Pisum sativum* L., Fabaceae), which disrupts the digestive processes. Pea flour is rich in protein that acts as an antifeedant and repellent to *S. oryzae*, with ingestion and contact actions (Hou et al. 2006). The median lethal time (LT_{50}) required to kill 50% of the population of adult rice weevils fed pea flour was 3 days (95% confidence interval, 2.8–3.2 days), compared with insects in total starvation (LT_{50} of 4 days, 95% confidence interval, 3.7–4.3 days) and those fed with other foods. The volume of bubbles in the midgut of insects fed the protein-rich pea flour increased rapidly (Figure 18.7b) compared with those fed wheat grain (Figure 18.7a) and other diets, but was similar to starving insects. It is possible that bubbles are produced by the action of symbiotic bacteria on the pea flour (Nardon and Grenier 1989) and because of gas pressure the midgut is distended and activates receptors of satiety, resulting in inhibition of feeding (Bernays and Simpson 1982). Protein-rich pea flour, the extract of the flour, and pea peptides all damaged and caused death of the midgut epithelium cells and of the insect (Figure 18.7c,d). The toxic effect of pea flour may

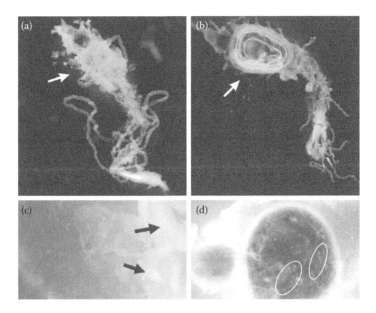

FIGURE 18.7 Midgut of adult *S. oryzae* fed wheat kernels (a, c) or protein-rich pea flour (b, d). Observe gas bubbles formed in midgut (indicated by arrow) only in insects fed pea flour (b). Midgut tissues stained with fluorescent dyes calcein AM and propidium iodide under a fluorescence microscope (c, d); dead tissue fluoresces—indicated by bright punctuations inside the ellipses (d), and live tissue—indicated by arrows (c). (From Hou, X., et al., *Can. Entomol.*, 138, 95, 2006. With permission.)

also be attributed to its direct action on the peritrophic matrix or epithelial cells of the midgut, similar to the action of neem (*Azadirachta indica* Adr. Juss., Meliaceae) (Nogueira et al. 1997).

The protein-rich pea flour contains albumin type PA1b, with peptides of 37 amino acids that are rich in sulfur (Taylor et al. 2004b). It has been shown that the peptides bind PA1b sites in the cell membrane (Gressent et al. 2003), affecting the functioning of ion channels. These insecticidal peptides have a molecular mass of approximately 3800 Da, which is much smaller than the toxic proteins of *B. thuringiensis* (Slaney et al. 1992) and can penetrate through the peritrophic matrix. It was found that the extracts contain a number of flour and soy saponins and lysolecithins (Taylor et al. 2004a) that, owing to their surface-active property, increase the absorption of toxic peptides across cell membranes. Thus, the full compound is more effective in controlling *S. oryzae* than just the pea peptides without saponins. The saponins of soybean have previously been considered as possible factors in insect repellence in legume seeds (Applebaum and Birk 1979).

A limitation of the use of this product in storage facilities is that *R. dominica*, one of the most destructive grain pests, is not controlled by the concentration of 0.1% protein-rich pea flour (Bodnaryk et al. 1997). However, as *A. calandrae* is a generalist parasitoid of lepidopteran and coleopteran pests, and is not affected by the product, the combination of pea protein with the parasitoids can help suppress the mixed populations of insects, including *R. dominica*

(c) *Plant extracts and essential oils.* Several essential plant oils have been tested and used as grain protectants against storage pests. These compounds play several roles in insect–plant interactions (repulsants and attractants for feeding and oviposition, feeding stimulation, ovicidals, antibiotics, and attractants to pollinators and parasitoids). They act in response to attacks by herbivores and in interactions between plants (allelopathy).

Hill (1990) recorded approximately 2000 plant species that produce substances with insecticidal activity against storage insects. Among them are ground black pepper (*Piper nigrum*), mint (*Mentha piperita*), basil (*Ocimum basilicum*), and *Eucalyptus* spp. Some essential oils have fumigant action, such as mustard oil from *Brassica rapa* L. There are many studies investigating these compounds or identifying their chemical nature, but there are few references to their mode of action and implications in food processing by insects. Many compounds act as attractants and phagostimulants and may be used as food baits in several traps (Section 18.9.1) (Ceruti et al. 2006).

It is desirable that plant breeding be concerned not only with the field stages but also with the development of materials that have resistance in the grain. Biochemical and technological properties that confer resistance to stored-product pests can be obtained by conventional cross breeding or by the insertion of genes, such as corn with the gene from *B. thuringiensis* (*Bt*) (Bacillaceae) or avidin, for example.

18.9.3 Biological Control

Biological control agents such as parasitoids, predators, and some microorganisms have been tested and used, particularly in the IPM, as possible alternatives to conventional pesticides used to control storage insects. All these agents may affect feeding behavior and eventually cause insect death due to several mechanisms.

The success of biological control depends on knowledge and selection of species that have the potential to be released into the storage environment and the interactions between them and the pest species. One of the advantages of releasing parasitoids and predators in stored grains is that they easily integrate the protocols for IPM, including sanitization and aeration (Flinn 1998), and also integrate with certain products used as grain protectants (Baker and Throne 1995). However, there is some concern about the presence of these insects or their fragments as contaminants in grain and flour, and with the specificity of the parasitoids to certain pest species, requiring supplementation with other measures to enhance the action of these control agents on primary species.

Most of the parasitoids that attack the primary pest beetles belong to the families Pteromalidae and Bethylidae, which occur naturally in the storage environment or can be released for a more effective control (Flinn and Hagstrum 2002). These hymenopterans do not feed on the grain and do not penetrate into it, and the adults can be easily removed by sieving before milling to reduce their fragments in flour (Flinn 1998). Some species of Hymenoptera in the families Ichneumonidae, Braconidae, and Trichogrammatidae may also occur in the storage environment, parasitizing larvae or eggs of Lepidoptera (Athié and Paula 2002).

Theocolax elegans (Westwood) (Hymenoptera: Pteromalidae) is an efficient parasitoid on primary pest larvae inside the grain, such as *Sitophilus* spp., *R. dominica*, *S. paniceum*, *Callosobruchus* spp., and *S. cerealella* (Burks 1979; Flinn et al. 1996; Flinn 1998; Flinn and Hagstrum 2001). A single female of *T. elegans* may parasitize up to six larvae of *R. dominica* per day (Flinn and Hagstrum 2001). However, this parasitoid species does not attack the secondary pests, whose immature stages develop out of grains, such as *Tribolium* spp. and *C. ferrugineus*. Temperature can affect the functional response of this parasitoid in controlling *R. dominica*. The highest parasitism rate occurs at 30°C (20 preys per day) and lowest at 20°C (2 preys per day). It is important to consider that temperatures above 32.5°C cause high mortality of *T. elegans* (Flinn and Hagstrum 2002).

Anisopteromalus calandrae (Howard) (Hymenoptera: Pteromalidae) is one of the most important parasitoids of the weevils *S. oryzae* and *S. granarius*, and should be released in sufficient numbers at the beginning of the storage period for a more effective control. The female detects the weevil larva inside the grain and usually deposits a single egg; the hatched larva will feed on the tissues of the host larva (Athié and Paula 2002). Other species, such as *Cephalonomia waterston* (Gahan) and *Cephalonomia tarsalis* (Ashmead) (Hymenoptera: Bethylidae) parasitize the species of beetles that develop out of the grain, *C. ferrugineus* and *O. surinamensis*, respectively. *Acarophenax lacunatus* (Cross and Krantz) (Prostigmata: Acarophenacidae) preferably parasitize eggs of *T. castaneum*, *C. ferrugineus*, and *R. dominica* (Oliveira et al. 2006).

The combination of two parasitoid species can result in more effective control of storage pests, such as the release of the egg parasitoid *Trichogramma deion* (Riley) (Hymenoptera: Trichogrammatidae) and the larval parasitoid *Bracon hebetor* (Say) (Hymenoptera: Braconidae) in the control of *P. interpunctella* in maize flour. The authors concluded that *B. hebetor* is efficient in controlling insects in packed and spilled grain, while *T. deion* was more effective against parasitism of larvae in packages. The combination of the two parasitoids can reduce population of *P. interpunctella* to almost 100% (Grieshop et al. 2006).

Another possible combination is the release of the parasitoid *T. elegans* with the spraying of avidin corn powder. The result is superior to other treatments for controlling mixed populations because *S. zeamais* develops within the grain and is controlled by the parasitoid. On the other hand, *T. castaneum* and *C. ferrugineus*, which feed and develop externally, are not parasitized by *T. elegans* and are subject to the action of avidin corn powder. Avidin does not exert negative effects on the parasitoid since the parasitoid does not feed on grain and flour (Flinn et al. 2006). Hou et al. (2004) found that avidin is not toxic and does not affect the progeny of *A. calandrae* and *C. waterston*, a parasitoid of *C. ferrugineus*. The simultaneous release of the two parasitoids plus avidin powder significantly reduced the population of *S. oryzae* and *C. ferrugineus*. Protein-rich pea flour can also supplement the action of parasitoids because it does not have negative effects on the parasitoids (Hou et al. 2004).

Several predacious species in the genus *Xylocoris* (Anthocoridae) feed especially on psocids, and on eggs and larvae of moths and beetles, being adapted to the high temperatures in the storage environment (Mound 1989). Some of these predator species also have cannibalistic habits (Arbogast 1979). It has been observed that finely ground flour prevents penetration of the predator to attack preys, reducing the impact of predators on this substrate.

Microbial pesticides can be used, stored, and manipulated to suppress populations of storage pests, particularly when the hosts are at high densities. However, there are some limitations, especially with regard to the application costs in low-value commodities or for cases where a quick action is needed.

(a) *Bacteria. B. thuringiensis* is a widely used entomopathogenic bacteria and is already formulated and approved as a protectant of several stored grains. It can be used for the treatment of empty structures in silos before filling, or applied on the grain surface to prevent or control infestations of lepidopterans. In the case of *Bt* corn, the grain can express its protein Cry 1AB and Cry 1F, exerting a negative impact on the emergence, development, and fecundity of the Indian meal moth, *P. interpunctella* (Sedlacek et al. 2001). However, resistance of populations of *P. interpunctella* to *Bt* has been detected in the laboratory (Oppert et al. 1997). Candas et al. (2003) consider that the increase in oxidative metabolism may be an adaptive response of the insect that has had its survival threatened. Both the detoxification and higher levels of generalized and localized mutations increase their resistance and would provide adaptive advantage.

(b) *Fungi. Metarhizium anisopliae* and *Beauveria bassiana* (Deuteromycota) present a satisfactory efficacy and ability to infect many insect species. However, their effectiveness is not enough to compete with chemical pesticides and in high doses can negatively affect the parasitoids. Conidia production requires atmospheric humidity close to saturation, but there are no major requirements for spore germination and early infectious process in the insect and fungi can easily germinate in the storage environment. These organisms may be associated with diatomaceous earth (DE) because this product acts in the epicuticle favoring spore germination (Akbar et al. 2004). *O. surinamensis* is more susceptible to infection by *B. bassiana* than *S. granarius* and *T. confusum* in the same concentration, mainly due to the adherence and germination of conidia on the insect's cuticle (Wakefield 2006). This author mentions that the production of quinone by *Tribolium* spp. can inhibit the germination of conidia of *B. bassiana* and growth of yeasts and bacteria. There are many studies on the adherence of conidia, including the combination with DE to facilitate this process, but there is little information about its mechanism of infection, and how it affects feeding behavior and kills the insect.

On the basis of data provided by the *T. castaneum* genome sequencing and quantitative real-time PCR, Lord et al. (2010) investigated the gene expression in this insect exposed to *B. bassiana*. Several protein genes were identified that can be used to clarify the tolerance of *T. castaneum* to fungi and other pathogens. These data can be used as a model for studies of insect immune defenses.

(c) *Viruses.* The baculoviruses (Baculoviridae) are the only viruses that show potential commercial use as pesticides. They are specific to a given species or closely related species, especially *P. interpunctella* in dried fruits and nuts, being easily transmitted from infected females to their offspring.

(d) *Protozoa.* The neogregarin *Mattese oryzaephili* (Neogregarinorida. Lipotrophidae) is pathogenic to several species of storage insects, including *O. surinamensis* and *C. ferrugineus* and for their parasitoids of the family Bethylidae. However, the parasitoids are able to inoculate and disperse the neogregarins in the act of oviposition in the host. Infection with this organism reduces the fitness of the target insect, and as a result, there is a reduction in pest population (Lord 2007).

(e) *Entomopathogenic nematodes.* These organisms have not been commonly used for insect control in storage silos, warehouses, and the food industry, but tests show they are effective in locating and infesting their hosts in hidden niches. *Steinernema riobrave* (Rhabditida: Steinernematidae) significantly reduces the survival of larvae, pupae, and adults of *T. castaneum* and *P. interpunctella,* and can be used in combination with other management measures (Ramos-Rodriguez et al. 2007). The infective forms of nematodes can penetrate through the seed coat, and spiracles, mouth, and anus of insects. Since it reaches the gastric ceca, Malpighian tubules and the spaces between the peritrophic matrix and epithelium of the digestive tract, expulsion in the feces becomes difficult. A general problem that occurs when penetration occurs via the digestive tract is the action of digestive enzymes that can kill up to 40% of the nematodes (Lewis et al. 2006).

(f) *Spinosad.* It is a product of fermentation of the actinomycete soil bacterium *Saccharopolyspora spinosa* Mertz and Yao (Actinomycetales) that has proven effective as a protector of bulk wheat, but has not been registered for this purpose (Subramanyam et al. 2002). Spinosad is a mixture of spinosyn A and D, which are toxic upon ingestion and contact, acting on the nicotinic acetylcholine receptors and γ-aminobutyric acid (Salgado 1998). Spinosad is particularly effective in controlling *R. dominica* and *P. interpunctella,* and its efficiency depends on the insect species, type or variety of the grain, exposure time, and temperature. Lorini et al. (2006) tested the effect of spinosad, growth regulators, and insecticides on two species of insects in wheat grain and found 100% mortality of *R. dominica* in the first day with the lowest dose. *S. zeamais* was more resistant to treatment with spinosad. Toews and Subramanyam (2003) and Flinn et al. (2004) found that spinosad is more effective when applied in whole grains, but effectiveness is reduced against the secondary species in broken grain, such as *T. castaneum* and *O. surinamensis.* The lethal time also varies according to species and stage/instar of the insect, but for all species tested, spinosad reduced reproduction rates. Even in low concentration of 1 mg/kg of grain, it significantly reduces the number of damaged kernels (Fang et al. 2002).

18.9.4 Growth Regulators

Insect growth regulators are pesticides that mimic hormones that regulate the developmental processes of insects, including molt. When added to the diet of larvae of Lepidoptera in the laboratory, insect growth regulators prevented these from reaching the adult stage (Mondal and Parween 2001). Methoprene can be applied on the bulk grain with effective control of *R. dominica.* As an aerosol, methoprene controls *T. castaneum* in warehouses and mills. Hydroprene applied on the grain surface and structures controls *P. interpunctella* (Arthur 2006). This technology can be combined with DE and other control measures, but is limited by the fact that is effective only against immature insects.

18.9.5 Lignocellulosic Biofuels

Even though cellulolytic activities were considered to be limited to plants, fungi, and bacteria, there is increasing evidence of the existence of animal cellulases, mainly in invertebrates. Insects are potential candidates for prospecting these enzymes, due to the diverse adaptation of several species to fibrous and lignocellulose-rich plant tissues. Oppert et al. (2010) prospected for cellulolytic activity in insect digestive fluids. They detected such activity in the gut of the red flour beetle *Tribolium castaneum* and *Tenebrio molitor*, among many other insects. They concluded that the origin of cellulolytic enzymes in insects and cellulase activity levels correlate with phylogenetic relationships, reflecting differences in host or feeding strategies.

18.10 Final Considerations

Several strategies, based on the paradigm of nutritional ecology (or bioecology and nutrition) proposed by Slansky (1982) and Slansky and Rodriguez (1987), can be employed in the IPM of stored products, as discussed in Section 18.9. On the basis of the relative difference in food consumption by different insect species, the damage equivalent of species present in the grain mass can be used to determine the economic action level of pests. This information has great value because grain and other stored products are frequently infested by several species of insects simultaneously or successively.

Because stored grain insects are confined to a protected environment and usually with an unlimited food supply, they can be used to model population growth of various organisms (Baker and Loschiavo 1987). Some species are easily reared and may be subjected to various conditions of food, temperature, humidity, and other factors, allowing the construction of models to predict the effect of interactions of biotic and abiotic factors on the development of infestations.

REFERENCES

Akbar, W., J. C. Lord, J. R. Nechols, and R. W. Howard. 2004. Diatomaceous earth increases the efficacy of *Beauveria bassiana* against *Tribolium castaneum* larvae and increases conidia attachment. *J. Econ. Entomol.* 97:273–80.

Applebaum, S. W., Y. Birk, L. Harpaz, and A. Bondi. 1964a. Comparative studies on proteolytic enzymes of *Tenebrio molitor* L. *Comp. Biochem. Physiol.* 11:85–103

Applebaum, S. W., M. Jankovic, J. Grozdanovic, and D. Marinkovic. 1964b. Compensation for temperature in the digestive metabolism of *Tenebrio molitor* larvae. *Physiol. Zool.* 37:90–5.

Applebaum, S. W., and Y. Birk. 1979. Saponins. In *Herbivores: Their Interaction with Secondary Plant Metabolites*, ed. G. A. Rosenthal and D. H. Janzen, 539–65. New York: Academic Press.

Arbogast, R. T. 1976. Suppression of *Oryzaephilus surinamensis* (L.) (Coleoptera: Cucujidae) on shelled corn by the predator *Xylocoris flavipes* (Reuter) (Hemiptera: Anthocoridae). *J. Ga. Entomol. Soc.* 11:67–71.

Arbogast, R. T. 1979. Functional response of *Xylocoris flavipes* to Angoumois grain moth and influence of predation on regulation of laboratory populations. *J. Econ. Entomol.* 72:847–9.

Arbogast, R. T., and M. Carthon. 1972a. Humidity response of adult *Oryzaephilus surinamensis* (Coleoptera: Cucujidae). *Environ. Entomol.* 1:221–7.

Arbogast, R. T., and M. Carthon. 1972b. Effect of starvation and desiccation on the water balance and humidity response of adult *Oryzaephilus surinamensis* (Coleoptera: Cucujidae). *Entomol. Exp. Appl.* 15:488–98.

Arlian, L. G. 1979. Significance of passive sorption of atmospheric water vapor and feeding in water balance of the rice weevil, *Sitophilus oryzae*. *Comp. Biochem. Physiol.* 62A:725–33.

Arlian, L. G., and M. M. Veselica. 1979. Water balance in insects and mites. *Comp. Biochem. Physiol.* 62A:191–200.

Arthur, F. H. 2006. Advances in integrating insect growth regulators into storage pest management. In *Proceedings of the 9th International Conference on Stored Product Protection*, ed. I. Lorini, B. Bacaltchuk, H. Beckel, D. Deckers, E. Sundfeld, J. Santos, J. Biagi, J. Celaro, L. Faroni, L. Bartolini, M. Sartori, M. Elias, R. Guedes, R. Fonseca, and V. Scussel, 217–23. Campinas: Abrapós.

Athié, I., and D. C. de Paula. 2002. *Insetos de Grãos Armazenados—Aspectos Biológicos e Identificação*. São Paulo: Varela Editora e Livraria Ltda.

Baker, J. E. 1974. Differential net food utilization by larvae of *Sitophilus oryzae* and *Sitophilus granarius*. *J. Insect Physiol.* 20:1937–42.

Baker, J. E. 1986. Amylase/proteinase ratios in larval midguts of ten stored-product insects. *Entomol. Exp. Appl.* 40:41–6.

Baker, J. E., and J. A. Mabie. 1973. Feeding behavior of larvae of *Plodia interpunctella*. *Environ. Entomol.* 2:627–32.

Baker, J. E., Hills M. et al. 1987. Purification, partial characterization and postembryonic levels of amylase from *Sitophilus oryzae* and *Sitophilus granarius*. *Arch. Insect Biochem. Physiol.* 2:415–28.

Baker, J. E., and S. R. Loschiavo 1987. Nutritional ecology of stored-product insects. In *Nutritional Ecology of Insects, Mites, Spiders and Related Invertebrates*, ed. F. Slansky Jr. and J. G. Rodriguez, 321–44. New York: John Wiley & Sons.

Baker, J. E., and J. E. Throne. 1995. Evaluation of a resistant parasitoid for biological control of weevils in insecticide treated wheat. *J. Econ. Entomol.* 88:1570–9.

Barak, A. V., and P. K. Harein. 1981. Insect infestation of farm-stored, shelled corn. *J. Econ. Entomol.* 74:197–202.

Barak, A. V., W. E. Burkholder, and D. L. Faustini. 1990. Factors affecting the design of traps for stored products insects. *J. Kans. Entomol. Soc.* 63:466–85.

Barrer, P. M., and E. G. Jay. 1980. Laboratory observations on the ability of *Ephestia cautella* (Walker) (Lepidoptera: Phycitidae) to locate, and to oviposit in response to a source of grain odour. *J. Stor. Prod. Res.* 16:1–7.

Bell, E. A. 1977. Toxins in seeds. In *Biochemical Aspects of Plant and Animal Coevolution*, ed. J. B. Harborne, 143–61. New York: Academic Press.

Bernays, E. A., and S. J. Simpson. 1982. Control of food intake. *Adv. Insect Physiol.* 16:59–118.

Bodnaryk, R. P., P. G. Fields, Y. S. Xie, and K. A. Fulcher. 1997. Insecticidal factors from peas. USA Patent 5.955:082.

Bowden, J., P. G. N. Digby, and P. L. Sherlock. 1984. Studies of elemental composition as a biological marker in insects. I. The influence of soil type and host plant on elemental composition of *Noctua pronuba* (L.) (Lepidoptera: Noctuidae). *Bull. Entomol. Res.* 74:207–25.

Bull, J. O., and E. Solomon. 1958. The yield of *Lasioderma serricorne* (F.) (Col., Anobiidae) from a given quantity of foodstuff. *Bull. Entomol. Res.* 49:193–200.

Burges, H. D. 1960. Studies on the dermestid beetle *Trogoderma granarium* Everts. IV. Feeding, growth, and respiration with particular reference to diapause larvae. *J. Insect Physiol.* 5:317–34.

Burks, D. B. 1979. Family Pteromalidae. In *Catalog of Hymenoptera in America North of Mexico*, ed. V. Krombein, P. D. Hurd, D. R. Smith, and D. B. Burks, 768–834. Washington, DC: Smithsonian Institution.

Burks, C. S., and D. W. Hagstrum. 1999. Rapid cold hardening capacity in five species of coleopteran pests of stored grain. *J. Stor. Prod. Res.* 35:65–75.

Burks, C. S., J. A. Johnson, D. E. Maier, and J. W. Heaps. 2000. Temperature. In *Alternative to Pesticides in Stored-Product IPM*, ed. B. Subramanyam and D. W. Hagstrum, 73–104. Boston: Kluwer Academic Publishers.

Campbell, A., and R. N. Sinha. 1976. Damage of wheat by feeding of some stored product beetles. *J. Econ. Entomol.* 69:11–3.

Campbell, A., and R. N. Sinha. 1978. Bioenergetics of granivorous beetles, *Cryptolestes ferrugineus* and *Rhyzopertha dominica* (Coleoptera: Cucujidae and Bostrichidae). *Can. J. Zool.* 56:624–33.

Campbell, A., and R. N. Sinha. 1990. Analysis and simulation modeling of population dynamics and bioenergetics of *Cryptolestes ferrugineus* (Coleoptera: Cucujidae) in stored wheat. *Res. Pop. Ecol.* 32:235–54.

Campbell, A., N. B. Singh, and R. N. Sinha. 1976. Bioenergetics of the granary weevil, *Sitophilus granarius* (L.) (Coleoptera: Curculionidae). *Can. J. Zool.* 54:786–98.

Candas, M., O. Loseva, B. Oppert, P. Kosaraju, and L. A. Bulla, Jr. 2003. Insect resistance to *Bacillus thuringiensis*—Alterations in the Indian meal moth larval gut proteome. *Mol. Cell. Proteomics* 2:19–28.

Caneppele, M. A. B., C. Caneppele, F. A. Lazzari, and S. M. N. Lazzari. 2003a. Correlation between the infestation level of *Sitophilus zeamais* Motschulsky, 1855 (Coleoptera, Curculionidae) and the quality factors of stored corn, *Zea mays* L. (Poaceae). *Rev. Brás. Entomol.* 47:625–30.

Caneppele, C., M. A. B. Caneppele, and S. M. N. Lazzari. 2003b. Resistência de híbridos de milho, *Zea mays* (L.), ao ataque de *Sitophilus zeamais* (Mots.). *Rev. Brás. Armazenamento* 28:51–8.

Ceruti, F. C. 2003. Técnicas de monitoramento e de controle de insetos em milho armazenado. MSc thesis, Universidade Federal do Paraná, PR, Brazil.

Ceruti, F. C. 2007. Rastreabilidade de grãos: Conceito, desenvolvimento de software e estudos de casos de manejo de insetos no armazenamento. PhD dissertation, Universidade Federal do Paraná, PR, Brazil.

Ceruti, F. C., A. R. Pinto, Jr., R. I. N. de Carvalho, and E. Vianna. 2006. Response of *Sitophilus zeamais* (Coleoptera: Curculionidae) to different volatiles of wheat grains. In *Proceedings of the 9th*

International Conference on Stored Product Protection, ed. I. Lorini, B. Bacaltchuk, H. Beckel, D. Deckers, E. Sundfeld, J. Santos, J. Biagi, J. Celaro, L. Faroni, L. Bartolini, M. Sartori, M. Elias, R. Guedes, R. Fonseca, and V. Scussel, 701–5. Campinas: Abrapós.

Chapman, R. F. 1998. *The Insects: Structure and Function.* Cambridge: Harvard University Press.

Cinco-Moroyoqui, F. J., E. C. Rosas-Burgos, J. Borboa-Flores, and M. O. Cortez-Rocha. 2006. α-amylase activity of *Rhyzopertha dominica* (Coleoptera: Bostrichidae) reared on several wheat varieties and its inhibition with kernel extracts. *J. Econ. Entomol.* 99:2146–50.

Cox, P. D., and J. A. Simms. 1978. The susceptibility of soya bean meal to infestation by some storage insects. *J. Stor. Prod. Res.* 14:103–9.

Cox, P. D., L. P. Allen, J. Pearson, and M. A. Beirne. 1984. The incidence of diapause in seventeen populations of the flour moth, *Ephestia küehniella* Zeller (Lepidoptera: Pyralidae). *J. Stor. Prod. Res.* 20:139–143.

Dadd, R. H. 1960. The nutritional requirements of locusts. I. Development of synthetic diets and lipid requirements. *J. Insect Physiol.* 4:319–47.

Dadd, R. H. 1961. The nutritional requirements of locusts—V. Observations on essential fatty acids, chlorophyll, nutritional salt mixtures, and the protein or amino acid components of synthetic diets. *J. Insect Physiol.* 6:126–45.

Dadd, R. H. 1973. Insect nutrition: Current developments and metabolic implications. *Annu. Rev. Entomol.* 18: 381–420.

Davis, G. R. F. 1966. Replacement of RNA in the diet of *Oryzaephilus surinamensis* (L.) (Coleoptera: Silvanidae) by purines, pyrimidines, and ribose. *Can. J. Zool.* 44:781–785.

Davis, G. R. F. 1968. Dietary alanine and proline requirements of the beetle, *Oryzaephilus surinamensis*. *J. Insect Physiol.* 14:1247–1250.

Demianyk, C. J., and R. N. Sinha. 1988. Bioenergetics of the larger grain borer, *Prostephanus truncatus* (Horn) (Coleoptera: Bostrichidae), feeding on corn. *Ann. Entomol. Soc. Am.* 81:449–59.

Devine, T. L. 1978. The turnover of the gut contents (traced with inulin-carboxyl-^{14}C), tritiated water and [22]Na in three stored product insects. *J. Stor. Prod. Res.* 14:189–211.

Dobie, P. 1974. The laboratory assessment of the inherent susceptibility of maize varieties to post-harvest infestation by *Sitophilus zeamais* Motsch. (Coleoptera: Curculionidae). *J. Stor. Prod. Res.* 10:183–197.

Dolinski, M. G., and S. R. Loschiavo. 1973. The effect of fungi and moisture on the locomotory behavior of the rusty grain beetle, *Cryptolestes ferrugineus* (Coleoptera: Cucujidae). *Can. Entomol.* 105:485–90.

Ebeling, W. 1971. Sorptive dusts for pest control. *Annu. Rev. Entomol.* 16:23–158.

Edney, E. B. 1977. *Water Balance in Land Arthropods.* Berlin: Springer-Verlag.

Evans, D. E. 1982. The influence of temperature and grain moisture content on the intrinsic rate of increase of *Sitophilus oryzae* (L.) (Coleoptera: Curculionidae). *J. Stor. Prod. Res.* 18:55–66.

Fang, L., B. Subramanyam, and F. H. Arthur. 2002. Effectiveness of spinosad on four classes of wheat against five stored-product insects. *J. Econ. Entomol.* 95:640–50.

Fields, P. G. 1992. The control of stored-product insects and mites with extreme temperatures. *J. Stor. Prod. Res.* 28:89–118.

Flinn, P. W. 1998. Temperature effects on efficacy of *Choetospila elegans* (Hymenoptera: Pteromalidae) to suppress *Rhyzopertha dominica* (Coleoptera: Bostrichidae) in stored wheat. *J. Econ. Entomol.* 91:320–3.

Flinn, P. W., and D. W. Hagstrum. 2001. Augumentative release of parasitoid wasps in stored wheat reduces insect fragment in flour. *J. Stor. Prod. Res.* 37:179–86.

Flinn, P. W., and D. W. Hagstrum. 2002. Temperature-mediated functional response of *Theocolax elegans* (Hymenoptera: Pteromalidae) parasitizing *Rhyzopertha dominica* (Coleoptera: Bostrichidae) in stored wheat. *J. Stor. Prod. Res.* 38:185–90.

Flinn, P. W., D. W. Hagstrum, and W. H. McGaughey. 1996. Suppression of beetles in stored wheat by augumentative releases of parasitic wasps. *Environ. Entomol.* 25:505–11.

Flinn, P. W., B. Subramanyam, and F. H. Arthur. 2004. Comparison of aeration and spinosad for suppressing insects in stored wheat. *J. Econ. Entomol.* 97:1465–73.

Flinn, P. W., K. J. Kramer, J. E. Throne, and T. D. Morgan. 2006. Protection of stored maize from insect pests using a two-component biological control method consisting of a hymenopteran parasitoid, *Theocolax elegans*, and transgenic avidin maize powder. *J. Stor. Prod. Res.* 42:218–25.

Food and Agriculture Organization. 2009. World agriculture 2030: Main findings. http://www.fao.org/english/newsroom/news/2002/7833-en.html (accessed June 14, 2009).

Fraenkel, G. 1958. The effect of zinc and potassium in the nutrition of *Tenebrio molitor*, with observations on the expression of a carnitine deficiency. *J. Nutr.* 65:361–395.

Fraenkel, G., and M. Blewett. 1943. The natural foods and the food requirements of several species of stored products insects. *Trans. R. Entomol. Soc. Lond.* 93:457–90.

Fraenkel, G., and M. Blewett. 1944. The utilisation of metabolic water in insects. *Bull. Entomol. Res.* 35:127–39.

Fraenkel, G., and G. E. Printy. 1954. The amino acid requirements of the confused flour beetle, *Tribolium confusum* Duval. *Physiol. Bull.* 106:149–157.

Franco, O. L., D. J. Rigden, F. R. Melo, and M. F. Grossi-de-Sá. 2002. Plant α-amylase inhibitors and their interaction with insects α-amylases: Structure, function and potential for crop protection. *Eur. J. Biochem.* 269:397–412.

Gerber, G. H., and D. U. Sabourin. 1984. Oviposition site selection in *Tenebrio molitor* (Coleoptera: Tenebrionidae). *Can. Entomol.* 116:27–139.

Girish, G. K., A. Kumar, and S. K. Jain. 1975. Assessment of the quality loss in wheat damaged by *Trogoderma granarium* Everts during storage. *Bull. Grain Technol.* 13:26–32.

Gomez, L. A., J. G. Rodriguez, C. G. Poneleit, and D. F. Blake. 1982. Preference and utilization of maize endosperm variants by the rice weevil. *J. Econ. Entomol.* 75:363–7.

Gomez, L. A., J. G. Rodriguez, C. G. Poneleit, and D. F. Blake. 1983a. Relationship between some characteristics of the corn kernel pericarp and resistance to the rice weevil (Coleoptera: Curculionidae). *J. Econ. Entomol.* 76:797–800.

Gomez, L. A., J. G. Rodriguez, C. G. Poneleit, D. F. Blake, and C. R. Smith Jr. 1983b. Influence of nutritional characteristics of selected corn genotypes on food utilization by the rice weevil (Coleoptera: Curculionidae). *J. Econ. Entomol.* 76:728–32.

Goptar, I. A., I. Y. Filippova, E. N. Lysogorskaya, E. S. Oksenoit, K. S. Vinokurov, D. P. Zhuzhikov, N. V. Bulushova, I. A. Zalunin, Y. E. Dunaevsky, M. A. Belozersky, B. Oppert, and E. N. Elpidina. 2008. Localization of post-proline cleaving peptidases in *Tenebrio molitor* larval midgut. *Biochimie* 90:508–14.

Gressent, F., I. Rahioui, and Y. Rahbe. 2003. Characterization of a high-affinity binding site for the pea albumin 1b entomotoxin in the weevil *Sitophilus*. *Eur. J. Biochem.* 270:2429–35.

Grieshop, M. J., P. W. Flinn, and J. R. Nechols. 2006. Biological control of Indian meal moth (Lepidoptera: Pyralidae) on finished stored products using egg and larval parasitoids. *J. Econ. Entomol.* 99:1080–4.

Guzzo, E. C., O. M. B. Corrêa, J. D. Vendramim, A. L. Lourenção, S. A. M. Carbonell, and A. F. Chiorato. 2006. Development of the bean weevil (Coleoptera: Bruchidae) on bean genotypes with and without arcelin over two generations. In *Proceedings of the 9th International Conference on Stored Product Protection*, ed. I. Lorini, B. Bacaltchuk, H. Beckel, D. Deckers, E. Sundfeld, J. Santos, J. Biagi, J. Celaro, L. Faroni, L. Bartolini, M. Sartori, M. Elias, R. Guedes, R. Fonseca, and V. Scussel, 914–9. Campinas: Abrapós.

Hagstrum, D. W., and B. Subramanyam. 2000. Monitoring and decision tools. In *Alternatives to Pesticides in Stored-Product IPM*, ed. B. Subramanyam and D. W. Hagstrum, 1–28. Boston: Kluwer Academic.

Hagstrum, D. W., and B. Subramanyam. 2006. *Fundamentals of Stored-Product Entomology*. St. Paul: AACC International.

Hill, D. S. 1990. *Pest of Stored Products and Their Control*. Boca Raton: CRC Press.

Hou, X., P. G. Fields, P. W. Flinn, J. Perez-Mendoza, and J. Baker. 2004. Control of stored-product beetles with combinations of protein-rich pea flour and parasitoids. *Environ. Entomol.* 33:671–80.

Hou, X., W. Taylor, and P. Fields. 2006. Effect of pea flour and pea flour extracts on *Sitophilus* oryzae. *Can. Entomol.* 138:95–103.

House, H. L. 1974a. Nutrition. In *Physiology of Insecta*, ed. M. Rockstein, 1–62. New York: Academic Press.

House, H. L. 1974b. Digestion. In *Physiology of Insecta*, ed. M. Rockstein, 62–117. New York: Academic Press.

Howe, R. W. 1957. A laboratory study of the cigarette beetle, *Lasioderma serricorne* (F.) (Col., Anobiidae) with a critical review of the literature on its biology. *Bull. Entomol. Res.* 48:9–56.

Howe, R. W. 1965. A summary of estimates of optimal and minimal conditions for population increase of some stored products insects. *J. Stor. Prod. Res.* 1:177–84.

Kalinovik, I., V. Rozman, and A. Liska. 2006. Significance and feeding of psocids (Liposcelidae, Psocoptera) with microorganisms. In *Proceedings of the 9th International Conference on Stored Product Protection*, ed. I. Lorini, B. Bacaltchuk, H. Beckel, D. Deckers, E. Sundfeld, J. Santos, J. Biagi, J. Celaro, L. Faroni, L. Bartolini, M. Sartori, M. Elias, R. Guedes, R. Fonseca, and V. Scussel, 1087–94. Campinas: Abrapós.

Khare, B. P., K. N. Singh, R. N. Chaudhary, C. S. Sengar, R. K. Agrawal, and P. N. Rai. 1974. Insect infestation and quality deterioration of grain. I. Germination, odour and palatability in wheat. *Ind. J. Entomol.* 36:194–9.

Kramer, K. J., T. D. Morgan, J. E. Throne, F. E. Dowell, M. Bailey, and J. A. Howard. 2000. Transgenic maize expressing avidin is resistant to storage insect pests. *Nat. Biotechnol.* 18:670–4.

Ladisch, R. K., H. Suter, and G. H. Froio. 1968. Sweat gland carcinoma produced in mice by insect quinones. *J. Penn. Acad. Sci.* 42:87–9.

Lakshmikantha, H. C., B. Subramanyam, Z. A. Larson, and L. Zurek. 2006. Association of *Enterococci* with stored products and stored-product insects: Medical importance and applications. In *Proceedings of the 9th International Conference on Stored Product Protection*, ed. I. Lorini, B. Bacaltchuk, H. Beckel, D. Deckers, E. Sundfeld, J. Santos, J. Biagi, J. Celaro, L. Faroni, L. Bartolini, M. Sartori, M. Elias, R. Guedes, R. Fonseca, and V. Scussel, 140–9. Campinas: Abrapós.

Lamb, R. J., and S. R. Loschiavo. 1981. Diet, temperature, and the logistic model of developmental rate for *Tribolium confusum* (Coleoptera: Tenebrionidae). *Can. Entomol.* 113:813–8.

Lara, F. M. 1997. Resistance of wild and near isogenic bean lines with arcelin variants to *Zabrotes subfasciatus* (Boheman). I—Winter crop. *An. Soc. Entomol. Bras.* 26:551–60.

Lazzari, F. A. 1997. Umidade, fungos e micotoxinas na qualidade de sementes, grãos e rações. Curitiba: Author's edition.

Lazzari, F. A., and S. M. N. Lazzari. 2002. Colheita, recebimento, secagem e armazenamento de milho. Apucarana: Abimilho.

LeCato, G. L. 1978. Infestation and development by the cigarette beetle in spices. *J. Ga. Entomol. Soc.* 13:100–5.

LeCato, G. L., and T. L. McGray. 1973. Multiplication of *Oryzaephilus* spp. and *Tribolium* spp. on 20 natural product diets. *Environ. Entomol.* 2:176–9.

Levinson, H. Z., and A. R. Levinson. 1978. Dried seeds, plant and animal tissues as food favoured by storage insect species. *Entomol. Exp. Appl.* 24:305–17.

Lewis, E. E., J. Campbell, C. Griffin, H. Kaya, and A. Peters. 2006. Behavioral ecology of entomopathogenic nematodes. *Biol. Contr.* 38:66–79.

Lord, J. C. 2007. Detection of *Mattesia oryzaephili* (Neogregarinorida: Lipotrophidae) in grain beetle laboratory colonies with an enzyme-linked immunosorbent assay. *J. Invertebr. Pathol.* 94:74–6.

Lord, J. C., K. Hartzer, M. Toutges, and B. Oppert. 2010. Evaluation of quantitative PCR reference genes for gene expression studies in *Tribolium castaneum* after fungal challenge. *J. Microbiol. Methods* 80:219–21.

Lorini, I. 2003. Manual técnico para o manejo integrado de pragas de grãos de cereais armazenados. Passo Fundo, RS, Brazil: Embrapa Trigo.

Lorini, I., H. Beckel, and S. Schneider. 2006. Efficacy of spinosad and IGR Plus to control the pests *Rhyzopertha dominica* and *Sitophilus zeamais* on stored wheat grain. In *Proceedings of the 9th International Conference on Stored Product Protection*, ed. I. Lorini, B. Bacaltchuk, H. Beckel, D. Deckers, E. Sundfeld, J. Santos, J. Biagi, J. Celaro, L. Faroni, L. Bartolini, M. Sartori, M. Elias, R. Guedes, R. Fonseca, and V. Scussel, 269–73. Campinas: Abrapós.

Loschiavo, S. R. 1975. Field tests of devices to detect insects in different kinds of grain storages. *Can. Entomol.* 107:385–9.

Loschiavo, S. R. 1983. Distribution of the rusty grain beetle (Coleoptera: Cucujidae) in columns of wheat stored dry or with localized high moisture content. *J. Econ. Entomol.* 76:881–4.

Loschiavo, S. R., and R. N. Sinha. 1966. Feeding, oviposition, and aggregation by the rusty grain beetle, *Cryptolestes ferrugineus* (Coleoptera: Cucujidae) on seed-borne fungi. *Ann. Entomol. Soc. Am.* 59:578–85.

Loschiavo, S. R., and G. T. Okamura. 1979. A survey of stored product insects in Hawaii. *Proc. Hawaii. Entomol. Soc.* 13:95–118.

Mahroof, R. M., and T. W. Phillips. 2006. Tracking the origins and feeding habitats of *Rhyzopertha dominica* (F.) (Coleoptera: Bostrichidae) using elemental markers. In *Proceedings of the 9th International Conference on Stored Product Protection*, ed. I. Lorini, B. Bacaltchuk, H. Beckel, D. Deckers, E. Sundfeld, J. Santos, J. Biagi, J. Celaro, L. Faroni, L. Bartolini, M. Sartori, M. Elias, R. Guedes, R. Fonseca, and V. Scussel, 433–40. Campinas: Abrapós.

Mahroof, R. M., and T. W. Phillips. 2007. Orientation of the cigarette beetle, *Lasioderma serricorne* (F.) (Coleoptera: Anobiidae) to plant-derived volatiles. *J. Insect Behav.* 20:99–115.

Marsaro, Jr., A. L., S. M. N. Lazzari, E. L. Z. Figueira, and E. Y. Hirooka. 2005a. Inibidores de amilase em híbridos de milho como fator de resistência a *Sitophilus zeamais* Motschulsky (Coleoptera: Curculionidae). *Neotr. Entomol.* 34:443–50.

Marsaro, Jr., A. L., S. M. N. Lazzari, P. Kadozawa, E. Y. Hirooka, and A. C. Gerage. 2005b. Avaliação da resistência de hibridos de milho ao ataque de *Sitophilus zeamais* Motschulsky (Coleoptera: Curculionidae) no grão armazenado. Summa Brism Agri Abril Al IIn

Medici, J. C., and W. Taylor. 1966. Mineral requirements of the confused flour beetle, *Tribolium confusum* (Duval). *J. Nutr.* 88:181–186.

Medrano, J. F., and G. A. E. Gall. 1976. Food consumption, feed efficiency, metabolic rate and utilization of glucose in lines of *Tribolium castaneum* selected for 21-day pupa weight. *Genetics* 83:393–407.

Mondal, K. A. M. S. H., and S. Parween. 2001. Insect growth regulators and their potential in the management of stored-product pests. *Integr. Pest Manag.* 5:255–95.

Morgan, T. D., B. Oppert, T. H. Czapla, and K. J. Kramer. 1993. Avidin and streptavidin as insecticidal and growth inhibiting dietary proteins. *Entomol. Exp. Appl.* 69:97–108.

Morris, K., M. D. Lorenzen, Y. Hiromasa, J. M. Tomich, C. Oppert, E. N. Elpidina, K. Vinokurov, J. L. Jurat-Fuentes, J. Fabrick and B. Oppert. 2009. *Tribolium castaneum* larval gut transcriptome and proteome: A resource for the study of the coleopteran gut. *J. Proteome Res.* 8:3889–98.

Mound, L. 1989. *Common Insect Pests of Stored Food Products. A Guide to their Identification.* London: Natural History Museum.

Nardon, P., and A. M. Grenier. 1989. Endocytobiosis in Coleoptera: Biological, biochemical, and genetic aspects. In *Insect Endocytobiosis: Morphology, Physiology, Genetics, Evolution*, ed. W. Schwemmler and G. Gassner, 175–216. Boca Raton: CRC Press.

Nawrot, J., J. R. Warchalewski, B. Stasinska, and K. Nowakowska. 1985. The effect of grain albumins, globulins and gliadins on larval development and longevity and fecundity of some stored product pests. *Entomol. Exp. Appl.* 37:187–92.

Nawrot, J., J. R. Warchalewski, D. Piasecka-Kwiatkowska, A. Niewiada, M. Gawlak, S. T. Grundas, and J. Fornal. 2006. The effect of some biochemical and technological properties of wheat grain on granary weevil (*Sitophilus granarius* L.) (Coleoptera: Curculionidae) development. In *Proceedings of the 9th International Conference on Stored Product Protection*, ed. I. Lorini, B. Bacaltchuk, H. Beckel, D. Deckers, E. Sundfeld, J. Santos, J. Biagi, J. Celaro, L. Faroni, L. Bartolini, M. Sartori, M. Elias, R. Guedes, R. Fonseca, and V. Scussel, 400–7. Campinas: Abrapós.

Nogueira, N. F., S. M. Gonzales, E. M. Garcia, and W. Souza. 1997. Effect of azadirachtin on the fine structure of the midgut of *Rhodnius prolixus*. *J. Invertebr. Pathol.* 69:58–63.

Norris, M. J. 1933. Contributions toward the study of insect fertility—II. Experiment on the factors influencing fertility in *Ephestia küehniella* Z. (Lepidoptera: Phycitidae). *Proc. Zool. Soc. Lond.* 2:902–4.

Oliveira, C. R. F., L. R. D. Faroni, A. H. Sousa, F. M. Garcia, and L. S. Souza. 2006. Preference of *Acarophenax lacunatus* (Cross and Krantz) (Prostigmata: Acarophenacidae) for eggs of different hosts. In *Proceedings of the 9th International Conference on Stored Product Protection*, ed. I. Lorini, B. Bacaltchuk, H. Beckel, D. Deckers, E. Sundfeld, J. Santos, J. Biagi, J. Celaro, L. Faroni, L. Bartolini, M. Sartori, M. Elias, R. Guedes, R. Fonseca, and V. Scussel, 711–8. Campinas: Abrapós.

Oppert, B. 1999. Protease interactions with *Bacillus thuringiensis* insecticidal toxins. *Arch. Insect Biochem. Physiol.* 42:1–12.

Oppert, B., K. J. Kramer, D. Johnson, and W. McGaughey. 1997. Proteinase-mediated insect resistance to *Bacillus thuringiensis* toxins. *J. Biol. Chem.* 272:23477–80.

Oppert, B., K. Hartzer, and C. M. Smith. 2000. Characterization of the digestive proteinases of *Hypera postica* (Gyllenhal) (Coleoptera: Curculionidae). *Trans. Kans. Acad. Sci.* 103:99–110.

Oppert, B., K. Hartzer, and M. Zuercher. 2002. Digestive proteinases in *Lasioderma serricorne* (Coleoptera: Anobiidae). *Bull. Entomol. Res.* 92:331–6.

Oppert, B., T. D. Morgan, K. Hartzer, B. Lenarcic, K. Galesa, J. Brzin, V. Turk, K. Yoza, K. Ohtsubo, and K. J. Kramer. 2003. Effects of proteinase inhibitors on growth and digestive proteolysis of the red flour beetle, *Tribolium castaneum* (Herbst) (Coleoptera: Tenebrionidae). *Comp. Biochem. Physiol. C* 134:481–90.

Oppert, B., T. D. Morgan, K. Hartzer, and K. J. Kramer. 2005. Compensatory proteolytic responses to dietary proteinase inhibitors in the red flour beetle, *Tribolium castaneum* (Coleoptera: Tenebrionidae). *Comp. Biochem. Physiol. C* 140:53–8.

Oppert, C., W. E. Klingeman, J. D. Willis, B. Oppert, and J. L. Jurat-Fuentes. 2010. Prospecting for cellulolytic activity in insect digestive fluids. *Comp. Biochem. Physiol. B* 155:145–54.

Parra, J. R. P., J. R. S. Lopes, H. J. P. Serra, and O. Sales Jr. 1989. Metodologia de criação de *Anagasta kuehniella* (Zeller, 1879) para produção massal de *Trichogramma* spp. *An. Soc. Entomol. Bras.* 18:403–15.

Pereira, P. R. V. S. 1994. Comparação entre métodos para detecção de coleópteros adultos (Insecta: Coleoptera) e ocorrência de fungos em trigo armazenado. MSc thesis, Universidade Federal do Paraná, PR, Brazil.

Pereira, P. R. V. S., F. A. Lazzari, and S. M. N Lazzari. 2000. Insect monitoring outside grain storage facilities in southern Brazil. In *Proceedings of the 7th International Working Conference on Stored-Product Protection*, ed. J. Zuxun, L. Quan et al., 1534–6. Beijing: Sichuan Publishing House of Science and Technology.

Pinniger, D. B. 1990. Food baited traps: Past, present and future. *J. Kans. Entomol. Soc.* 63:533–8.

Ramos-Rodriguez, O., J. F. Campbell, and S. B. Ramaswamy. 2007. Efficacy of the entomopathogenic nematoda *Steinernema riobrave* against the stored-product insect pests *Tribolium castaneum* and *Plodia interpunctella*. *Biol. Contr.* 40:15–21.

Salgado, V. L. 1998. Studies on the mode of action of spinosad: Insect symptoms and physiological correlates. *Pestic. Biochem. Physiol.* 60:91–102.

Sedlacek, J. D., P. A. Weston, B. D. Price, and P. M. Davis. 2001. Life history attributes of Indian meal moth (Lepidoptera: Pyralidae) and Angoumois grain moth (Lepidoptera: Gelechiidae) reared on transgenic corn kernels. *J. Econ. Entomol.* 94:586–92.

Shellenberger, J. A. 1971. Production and utilization of wheat. In *Wheat Chemistry and Technology*, ed. Y. Pomeranz, 1–18. St. Paul: American Association Cereal Chemists.

Silva, A. A. L., L. R. D'A Faroni, R. N. C. Guedes, J. H. Martins, and M. A. G. Pimentel. 2003. Modelagem das perdas causadas por *Sitophilus zeamais* e *Rhyzopertha dominica* em trigo armazenado. *Rev. Bras. Eng. Agríc. Ambiental* 7:292–6.

Sinha, R. N. 1965. Development of *Cryptolestes ferrugineus* (Stephens) and *Oryzaephilus mercator* (Fauvel) on seed-born fungi. *Entomol. Exp. Appl.* 8:309–13.

Sinha, R. N. 1966. Development and mortality of *Tribolium castaneum* and *T. confusum* on seed-born fungi. *Ann. Entomol. Soc. Am.* 59:192–201.

Sinha, R. N. 1968. Adaptive significance of mycophagy in stored-product Arthropoda. *Evolution* 22:785–98.

Sinha, R. N. 1983. Effects of stored-product beetle infestation on fat acidity, seed germination, and microflora of wheat. *J. Econ. Entomol.* 76:813–7.

Sinha, R. N. 1995. The stored grain ecosystem. In *Stored-Grain Ecosystems*, ed. D. S. Jayas, N. D. G. White, and W. E. Muir, 1–33. New York: Marcel Dekker Inc.

Siva-Jothy, M. T., and J. J. W. Thompson. 2002. Short-term nutrient deprivation affects immune function. *Physiol. Entomol.* 27:206–12.

Slaney, A. C., H. L. Robbins, and L. English. 1992. Mode of action of *Bacillus thuringiensis* toxin CryIIIA: An analysis of toxicity in *Leptinotarsa decemlineata* (Say) and *Diabrotica undecimpunctata howardi* Barber. *Insect Biochem. Mol. Biol.* 22:9–18.

Slansky, Jr., F. 1982. Insect nutrition: An adaptationist's perspective. *Fla. Entomol.* 65:45–71.

Slansky, Jr., F., and J. M. Scriber. 1985. Food consumption and utilization, In *Comprehensive Insect Physiology, Biochemistry and Pharmacology*, ed. G. A. Kerkut and L. I. Gilbert, 87–163. Oxford: Pergamon.

Slansky Jr. F. and J. G. Rodriguez. 1987. *Nutritional Ecology of Insects, Mites, Spiders and Related Invertebrates*. New York: John Wiley & Sons.

Stubbs, M., and R. Griffin. 1983. The response of *Oryzaephilus surinamensis* (L.) (Coleoptera: Silvanidae) to water. *Bull. Entomol. Res.* 73:587–95.

Subramanyam, B., and D. W. Hagstrum. 1991. Quantitative analysis of temperature, relative humidity, and diet influencing development of the larger grain borer, *Prostephanus truncatus* (Horn) (Coleoptera: Bostrichidae). *Trop. Pest Manage.* 37:195–202.

Subramanyam, B. and D. W. Hagstrum. 2000. *Alternative to Pesticides in Stored-Product IPM*. Boston: Kluwer Academic Publishers.

Subramanyam, B., L. Fang, and S. Dolder. 2002. Persistence and efficacy of spinosad in farm stored wheat. *J. Econ. Entomol.* 95:1102–09.

Taylor, W. G., P. G. Fields, and D. H. Sutherland. 2004a. Insecticidal components from field pea extracts: Soya saponins and lysolecithins. *J. Agric. Food Chem.* 52:7484–90.

Taylor, W. G., D. H. Sutherland, D. J. H. Olson, A. R. S. Ross, and P. G. Fields. 2004b. Insecticidal components from field pea extracts: Sequences of some variants of pea albumin 1b. *J. Agric. Food Chem.* 52:7499–7506.

Terra, W. R. 2003. Digestion. In *Encyclopedia of Insects*, ed. V. H. Resh and R. T. Cardé, 310–13. Amsterdam: Academic Press.

Terra, W. R., and C. Ferreira. 1994. Insect digestive enzymes: Properties, compartmentalization and function. *Comp. Biochem. Physiol. B* 109:1–62.

Throne, J. E., and L. D. Cline. 1989. Seasonal flight activity of the maize weevil, *Sitophilus zeamais* Motschulsky (Coleoptera: Curculionidae) and the rice weevil, *S. oryzae* (L.), in South Carolina. *J. Agr. Entomol.* 6:93–100.

Throne, J. E., and L. D. Cline. 1994. Seasonal flight activity and seasonal abundance of selected stored-product Coleoptera around grain storages in South America. *J. Agr. Entomol.* 11:321–38.

Toews, M. D., and B. Subramanyam. 2003. Contribution of contact toxicity and wheat condition to mortality of stored-product insects exposed to spinosad. *Pest Manage. Sci.* 59:538–44.

Trematerra, P., M. C. Z. Paula, A. Sciarretta, and S. M. N. Lazzari. 2004. Spatio-temporal analysis of insect pests infesting a paddy rice storage facility. *Neotr. Entomol.* 33:469–79.

Vinokurov, K. S., E. N. Elpidina, B. Oppert, S. Prabhakar, D. P. Zhuzhikov, Y. E. Dunaevsky, and M. A. Belozersky. 2006. Diversity of proteinases in *Tenebrio molitor* (Coleoptera: Tenebrionidae) larvae. *Comp. Biochem. Physiol. B* 1045:126–137.

Wakefield, M. E. 2006. Factors affecting storage insect susceptibility to the entomopathogenic fungus *Beauveria bassiana*. In *Proceedings of the 9th International Conference on Stored Product Protection*, ed. I. Lorini, B. Bacaltchuk, H. Beckel, D. Deckers, E. Sundfeld, J. Santos, J. Biagi, J. Celaro, L. Faroni, L. Bartolini, M. Sartori, M. Elias, R. Guedes, R. Fonseca, and V. Scussel, 855–62. Campinas: Abrapós.

Waldbauer, G. P., and A. K. Bhattacharya. 1973. Self-selection of an optimum diet from a mixture of wheat fractions by the larvae of *Tribolium confusum*. *J. Insect Physiol.* 19:407–18.

Waldbauer, G. P., and S. Friedman. 1991. Self-selection of optimal diets by insects. *Annu. Rev. Entomol.* 36: 43–63.

White, N. D. G., and R. N. Sinha. 1981. Energy budget for *Oryzaephilus surinamensis* (Coleoptera: Cucujidae) feeding on rolled oats. *Environ. Entomol.* 10:320–6.

Wright, K. N. 1987. Nutritional properties and feeding value of maize and its by-products, In *Maize: Chemistry and Technology,* ed. S. A. Watson and P. E. Ramstad, 447–78. St. Paul: American Association of Cereal Chemists.

19

Fruit Flies (Diptera)

Carla Cresoni-Pereira and Fernando S. Zucoloto

CONTENTS

19.1 Introduction 451
19.2 Fruit Flies Foodstuffs 452
19.3 Nutrition 453
 19.3.1 Proteins 454
 19.3.2 Carbohydrates 456
 19.3.3 Lipids 457
 19.3.4 Vitamins and Mineral Salts 457
 19.3.5 Symbionts 458
19.4 Abiotic Factors 458
19.5 Allelochemicals 459
19.6 Feeding 459
19.7 Behavior 461
19.8 Applicability and Conclusions 464
References 467

19.1 Introduction

Insects of the family Tephritidae, the true fruit flies, are known for having economic importance since they infest a series of host fruits of commercial interest. Their life cycle, particularly the larval phase, is closely related to the development of the host fruits because they start to damage them from the moment of oviposition. In addition, they are biologically important organisms, occurring in different habitats, exploring diverse feeding resources, and exhibiting a series of variable behavior.

Fruit flies belong to the order Diptera, suborder Brachycera, family Tephritidae. The genera comprising economically important species belong to the subfamily Trypetinae. The denomination "fruit flies" must be exclusively used for representatives of the family Tephritidae, and other flies that use fruits in their life cycle must not be included in this family (Zucchi 2000).

There is no satisfactory classification for this family (Silva 2000) probably because of the considerable size of the group (about 4200 species described), the regional nature of most of the taxonomical studies, and the intergradations of taxonomical signals among several superior taxa (Malacrida et al. 1996).

Usually, two large groups of tephritids are identified considering physiological, ecological, and behavioral differences (Selivon 2000). The number of annual generations, the exploration of resources, and the copulation behavior are characteristics often used in allocating groups.

Bateman (1972) suggested that fruit flies were initially divided according to the number of annual generations: multivoltine species (more than one annual generation, generally without diapause as the tropical and subtropical species of *Anastrepha*) and univoltine species (with only one annual generation, diapause in winter, and occurring in temperate climate regions as the species of *Rhagoletis*).

Another distinction concerns the kind of hosts used by the flies. Fletcher (1987) considers three main strategies of host utilization by the larval stage of fruit flies: monophagous species that explore only one vegetal species [e.g., *Bactrocera oleae* (Gmelin)]; oligophagous species that use few related species of

451

the same genus or same family (as *Anastrepha striata* Schiner); and polyphagous species that use a broad spectrum of hosts (as some species of *Anastrepha* and *Ceratitis*). Most species that damage fruit trees are included in these two last categories.

Oligophagous/polyphagous multivoltine species from tropical and subtropical regions do not present a life cycle closely linked to the phenology of only one host. In this case, females must select an oviposition site according to the availability of host fruits. Consequently, in the polyphagic regimen, diversification occurs in the exploration of resources (Selivon 2000).

Tephritids also present different sexual behaviors, and two patterns are essentially recognized regarding the courtship and copulation sites. In resource-based copulation system, copulation only occurs near or beside the host. Females search the fruit for oviposition and are compelled to copulate by males that have established territory on that site. In this system females do not choose males (Aluja et al. 2000). In the *leks* system, males form aggregates in the foliage, which may or not be of the host plant, and initiate a series of elaborate behavior—vibrating wings, emitting sounds, and liberating pheromones. Females, owing to the males' behavior, are attracted to those aggregates where they select the partner for copulation (Aluja et al. 2000).

Morgante et al. (1993) and Selivon and Morgante (1997) studied the copulation behavior of *Anastrepha striata* Schiner and *A. bistrigata* Bezzi, and verified that *A. striata* presents the typical behavior of a generalist species: males grouped in *leks* attract females emitting pheromones and producing sounds. *A. bistrigata* displays the typical behavior of monophagous temperate region species, such as *Rhagoletis*: the males choose the fruit as their territory, defending it from other males. Females search the fruit for oviposition, and males force copulation while the females are on the fruit.

By analyzing these parameters as a whole, some relationships can be established, recognizing two groups of fruit flies: the species that occur in temperate regions with stable populations whose successive generations remain in the same area and in the same host (Bateman 1972); in this case, a synchronization of high population density periods occurs with the available resources, and the host fruit itself is used as the copulation site. The other group, the multivoltine, polyphagous, tropical and subtropical region species are transient and establish themselves in regions where they find fruits in the process of ripening (Bateman 1972). The exploration of these resources leads to increased population density, which declines as the amount of available fruit mitigates. Therefore, contrariwise to the temperate species, which remain at the same site for several generations and are univoltine, the tropical and subtropical fruit flies disperse and form new populations where they find favorable conditions. In addition, the generations may superpose in the same host with several generations by period of fructification (Bateman 1972).

19.2 Fruit Flies Foodstuffs

Tephritidae are the guild of frugivorous insects. It is important to mention their life cycle and peculiar form of nourishment. Although members of the family are considered frugivorous, this classification is primarily related to the feeding habit of immatures. Actually, adults feed on a great variety of items, from exuded portions of host fruits to bird feces, organic matter in decomposition, nectar, pollen, and other substances (Christenson and Foote 1960). Feeding exclusively on fruit occurs only in the larval stage, although it is possible that larvae feed inside the fruit on their own exoskeletons (Zucoloto 1993b), on other small animals (larvae, worms, and other invertebrates), and on smaller co-specific larvae.

Sexually mature females search their hosts and deposit their eggs on the fruit skin. The larvae emerge directly inside the fruit where they feed continuously and complete development. At this point, the fruit is already ripe enough and has fallen from the tree. The last instar larvae leave the fruit and penetrate the ground where they will pupate until emergence as adults (Christenson and Foote 1960).

Nutritionally, not only adults but also immatures need carbohydrates, proteins, lipids, mineral salts, and vitamins to develop. These nutrients are called primary substances or compounds and are found in the food (Hsiao 1985) or can be acquired through symbiosis.

Another group of substances found in the food are secondary substances or allelochemicals that occur in plants (90% of the cases); they do not participate in the insect metabolism and are not directly used by animals that ingest or lay eggs on these plants (Hsiao 1985). Allelochemicals may be used by insects as signalers in several circumstances.

Initially, insects must accept the food for it to become part of their menu, and this acceptance occurs by recognition of the substances contained in the food that stimulate ingestion, and that are neither toxic nor deterrent. Among the foods, some have the insects' preference. The previous experience and the concentration of nutrients in the insects' hemolymph, in addition to other physiological and ecological factors, interfere in the food choice (Medeiros et al. 2008). Not only adults can choose their food but also immatures select more suitable food, or choose portions of a certain food that contains more quality and quantity of nutrients (Zucoloto 1987; Fernandes-da-Silva and Zucoloto 1993).

During oviposition, females spend time and energy to locate the host, besides being exposed to predators and parasitoids. Females must choose safe sites and suitable foods for the success of the offspring. Therefore, the choice of the oviposition site is an important and risky step. The search and host acceptance processes obey the same physiological factors that govern feeding, such as nourishment, egg availability, and age (Medeiros et al. 2008).

Medeiros-Santana and Zucoloto (2009) have shown that despite *A. obliqua* utilization of different hosts, differences in adults' performance were not related to the nutrient content of the host, and the percentage of adult emergence was due to different fruit sizes and not to their nutritional quality.

19.3 Nutrition

The nutritional needs of insects are qualitatively similar to those of the animals in general. They vary according to the stage of life cycle, and also vary with abiotic factors such as temperature, humidity, and luminosity.

When the insects' nutritional needs are not met by the food they consume, their performance will be negatively affected: expanded period of development, reduced fertility and fecundity, reduced adult size (Slansky and Scriber 1985), and this may interfere in the copulation and dispersal capacity, among other factors. Besides affecting the insects' performance, the nutritional aspects are determinants of food search and selection behavior, dispersal, choice, and acceptance of the sexual partners (Chapman 1998).

In the immature phase, quantity and quality of consumed nutrients affect weight, developmental period, survival, body chemical composition, adults' size and, in some cases, egg production. In the adult phase, nutrition is important regarding egg production, copulation, survival, dispersal, cuticle renewal, among other factors (Slansky and Scriber 1985).

One of the main difficulties in the study of fruit flies' nutrition and feeding behavior occurs in the immature phase. The feeding indexes considered for other insects (Slansky and Scriber 1985) cannot be applied to this particular group of insects, keeping in mind that feces collection of adults, and food ingestion and feces collection of larvae are also not feasible. The larval diet is extremely important for their adequate development as well as for the adults' performance, considering that the reproductive potential of the adults reflects the diet efficiency of the larvae.

Quality and quantity of the larvae food will interfere differently in the adults' size and in their reproductive potential (Zucoloto 1987). Although several studies have established relationships between adult size and reproductive success [copulation success by the males and egg production by the females (Liedo et al. 1992; Taylor and Yuval 1999)], that relationship does not always occur. Bigger males of *Ceratitis capitata* (Wiedeman) copulate more often than smaller males; however, their fecundity is similar (Blay and Yuval 1999). Bigger females of *C. capitata* did not show higher reproductive potential as compared with the smaller ones (Blay and Yuval 1999).

Aluja et al. (2009) studied the sexual behavior of *Anastrepha ludens* (Loew) and *A. obliqua* (Macquart) and found that factors considered important in partner choice and performance, such as adult size, played a minor role in male and female sexual responses. Under laboratory conditions, for these species, size was not important to achieve copulation, although it was important in inducing the females' refractory period. In a previous study with *A. ludens*, Aluja et al. (2008a) observed that the diet of adults was more important than the size in determining copulation success.

For individuals reared in the laboratory, genetic differences are more significant for the reproductive success than size differences. When larvae reach the critical weight, that is, the minimum weight for pupation and metamorphosis (Davidowitz et al. 2003), the performance of adults will depend more on the genetic quality than on the adults' size. Due to the genetic bottleneck of laboratory rearing, variability is reduced to a point where no differences were observed in the copulation success in experiments with *Bactrocera tryoni* (Hardy) (Meats et al. 2004), independently of the insect size. This is a characteristic effect of mass rearing that may interfere in behavioral studies. As suggested by Cayol (2000), behavioral and pheromonal characteristics may be lost in the rearing process and may be the key to explaining results found even in field studies with individuals reared in a large scale.

Resources that are essential for feeding adults for ovipositing and for larval development are unevenly available in quality and quantity during the different seasons of the year, and tephritids must adapt for foraging to meet their needs. The nutritional needs are not constant: they vary according to biotic (e.g., developmental, reproductive, or dispersal phase) and abiotic (e.g., temperature and humidity) factors. Flies must satisfy their minimum nutritional needs in order not to have their performance affected (Simpson and Simpson 1990).

Aluja et al. (2001a) compared *Anastrepha ludens*, *A. serpentina*, *A. striata*, and *A. obliqua* males at different ages and nutritional conditions. Results showed that the reproductive potential was affected by the diet quality; however, this effect was highly variable among related species with different ecological demands, and among different age groups in the same species.

Ovarian maturation and egg load are influenced particularly by feeding on protein. Although several studies have emphasized the presence of the host fruit or its odor as a stimulating factor for ovarian maturation (Fletcher and Kapatos 1983; Koveos and Tzanakakis 1990; Alonso-Pimentel et al. 1998), there is no evidence that the egg load has increased exclusively due to fruit stimulation (sensorial and/or olfactorial). The possibility that host stimulation triggers a feeding response that has favored ovarian maturation cannot be discarded (Papaj 2000).

The quality of the adult diet was the determinant factor for copulation success, duration of copulation, and induction of female refractory period in studies with *A. ludens* and *A. obliqua* (Aluja et al. 2009).

19.3.1 Proteins

Amino acids are the structural units of proteins. The basic functions of proteins in insects are tissue construction and maintenance; formation of enzymes, nucleotides, and chitin; maintenance of acid–base equilibrium; and, in the absence of primary sources, act as an energy source. Although carbohydrates are the classic phagostimulants, proteins can also function as phagostimulants. *A. obliqua* females intensify the artificial diet ingestion when the concentration of brewer's yeast, a protein source, is increased (Cresoni-Pereira and Zucoloto 2001a; Medeiros and Zucoloto 2006).

Larvae feed on the fruit pulp, but they may also eat their own exoskeletons, other invertebrates, and dead or smaller co-specific larvae inside the fruit. Zucoloto (1993b) has shown that *C. capitata* larvae can develop with diets containing meat powder and dead larvae of the same species, but the first generations have reduced performance. This suggests that cannibalism occurs when larvae are exposed to conditions of intraspecific competition and low nutrient contents, among others. Lemos et al. (1992) have shown that larvae fed on meat power discharge in the digestive tract more trypsin and less amino peptidases than larvae fed on yeast; this is a genetic adaptation found in wild larvae exposed to all kinds of environments that, occasionally, feed on co-specifics, exoskeletons, and other smaller invertebrates.

It is known that the natural food of larvae is poor in proteins or nitrogen compounds. Laboratory studies have demonstrated that larvae from several species of fruit flies prefer diets containing proteins over diets with a high sugar content. *A. obliqua* and *C. capitata* larvae do not survive with diets containing only carbohydrates, but complete development on diets containing exclusively proteins, although their performance is reduced (Message and Zucoloto 1980; Crisci and Zucoloto 2001). Canato and Zucoloto (1993) show that with concentrations of 6.5 to 25.0 g yeast per 100 ml diet, *C. capitata* larvae performed equally, suggesting that larvae do not regulate the amount of sugar ingested (which was constant) in the diet and successfully tolerated protein-based diets; with 30.0 g yeast per 100 ml diet concentration, performance was lightly reduced (Canato and Zucoloto 1993). In the immature phase, the need for protein of

C. capitata is similar both for males and females, as well as the discrimination threshold in the immature phase (Plácido-Silva et al. 2005).

C. capitata larvae need different amounts of protein and carbohydrate, depending on the age. Younger larvae need more proteins and older ones need more carbohydrates (Zucoloto 1987). This was also observed in choice tests where younger larvae chose diets with higher protein contents and older ones prefer diets with higher carbohydrate contents.

The main protein sources for adult fruit flies are pollen, bird feces, honeydew (Bateman 1972), and even carcasses of other dead insects, including co-specifics (Cresoni-Pereira, pers obs.).

Regarding protein utilization and need, female fruit flies are classified into two groups: those that need an exogenous protein source in the adult phase to reproduce (anautogenous, as the genus *Anastrepha* flies) and those that do not need such protein source (autogenous). Some authors rank *C. capitata* as an autogenous fruit fly; however, most studies are carried out with flies reared in the laboratory. Artificial diets for larvae used in the laboratory are rich in nutrients, so adults emerge with nutritional reserves sufficient to produce eggs even in the absence of a protein source in the adult diet. In these cases, egg production may intensify when adult females receive a protein source.

Proteins are also important regarding the mating processes. Females nourished with proteins signalize their nutritional status, influencing their reproductive potential. Non-protein-deprived *C. capitata* females in the adult phase were more receptive to copulation than the protein-deprived females (Cangussu and Zucoloto 1997). According to Trujillo (1998 cited by Aluja et al. 2000), 96% of *A. obliqua* females copulated at least once in spite of the diet; 91% of *A. ludens* females fed protein and sugar copulated with males fed the same diet, while only 50% of females fed with sugar only copulated with males also fed with sugar. *A. ludens* females fed with protein and sugar and maintained with males fed on the same diet produced more mature eggs than females and males fed on sugar (Mangan 2003).

The relationship between protein and ovarian maturation is a function of juvenile hormone and ecdisterone production, but interactions between these hormones are complex and poorly understood (Wheeler 1996). In general, some oocytes complete vitellogenesis, and that number is proportional to the protein ingestion rate (Chapman 1998). *B. tryoni* (Hardy) females feeding on proteins *ad libitum* peak when copulation occurs (Meats and Leighton 2004). A dramatic reduction in protein demand occurs when the flies remain noncopulated. There is variation in the number of eggs produced during the life of *B. tryoni* when females are submitted to different protein ingestion regimens (Meats and Leighton 2004).

B. tryoni flies sterilized by gamma rays in the pupal stage have no vitellogenesis, but protein consumption was similar to that of normal copulated females (Meats and Leighton 2004). This apparent lack of the effect on feeding rate contradicts data found for other species such as *C. capitata* (Galun et al. 1985).

Protein ingestion is important for male fruit flies. Non-protein-deprived males of *C. capitata* copulated significantly more than protein-deprived males, either wild or reared in the laboratory (Shelly and Kennely 2002). Drew (1987) observed that *B. tryoni* males do not need a protein source to produce sperm, but they need it to maintain sperm production throughout their lifetime.

Male copulation may be affected by ingestion of a protein source in the adult phase. For *A. serpentina*, *A. striata*, and *A. oblique*, the number of copulations was significantly influenced by adult diet. Age also influenced *A. obliqua* males, with those older (20 days) being more affected than those younger (12 days) (Aluja et al. 2001a). The addition of a protein hydrolysate to the diet conferred competitive advantage to males.

The adult diet can determine the male ability to inhibit the female re-copulation and induce a sexual refractory period during which females are not receptive, this can be particularly important for tephritids control, reducing the possibility of females re-copulating with nonsterile males. *C. capitata* and *B. tryoni* males with high-quality feeding have greater probability of inhibiting female re-copulation than males receiving low-quality feeding (Blay and Yuval 1997; Pérez-Staples et al. 2008).

C. capitata males fed high-quality diets were more capable of inhibiting re-copulation with wild females reared in the laboratory (Gavriel et al. 2009). In this case, male age was also important in reducing female receptiveness to mature males (11-day-old males), younger or older males being more successful (Gavriel et al. 2009).

The male physiological status determined by larval and adult feeding may influence re-copulation according to the amount of sperm ejaculation (Mossinson and Yuval 2003), and to the quality of accessory gland products (Radhakrishnan and Taylor 2007). Male physiological status also influences female behavior, promoting higher production of eggs and the tendency to re-copulate.

A. obliqua males maintained in different diet regimens promoted a change in utilization of ingested food by females favoring egg production, in addition to influencing other aspects of food selection (Cresoni-Pereira and Zucoloto 2005, 2006b). Another factor that may influence re-copulation is the availability of hosts for oviposition (Ringo 1996). As females lay eggs, reserves of sperm are reduced and females are prone to copulate again as shown for *Rhagoletis juglandis* Cresson and *Toxotrypana curvicauda* Gerstaecker (Landolt 1994, Carsten and Papaj 2005, Harano et al. 2006).

In general, protein addition to *C. capitata* adult diet intensifies copulation success when competition for partners among wild males is involved (Kaspi et al. 2000; Shelly et al. 2002; Shelly and Kennelly 2002), or competition among males reared in the laboratory (Kaspi and Yuval 2000; Shelly and Kennelly 2002); however, this was not observed for laboratory males competing with wild males (Shelly and Kennelly 2002; Shelly and McInnis 2003).

B. tryoni males fed continuously on hydrolyzed yeast after emergence had increased sexual performance, with more and longer copulation, and lower latency than males deprived of yeast (Pérez-Staples et al. 2007). Similar results were found by Shelly et al. (2005) for *B. dorsalis*. *C. capitata* males fed on brewer's yeast in the adult stage also had a higher number of copulations than males fed sucrose only, but copulation success and duration were not affected (Joachim-Bravo et al. 2009).

Duration of copulation is influenced in several ways by the ingestion of a protein source by adult fruit flies. *C. capitata* males fed on diets with high-quality protein perform short copulations (Taylor and Yuval 1999), while *B. tryoni* males perform long copulations under the same diet regimen (Pérez-Staples et al. 2007). Longer copulations may reduce the incidence of re-copulation in the species, which is a usual behavior; the duration of copulation is not always related to the amount of transferred ejaculation. According to Aluja et al. (2008b), shorter copulations may be advantageous to males by increasing the chances to meet other partners.

Salivary gland development responsible for pheromone production is highly influenced by the protein ingested during the larval phase (Ferro and Zucoloto 1989). Males with protein-deficient nutrition in the immature phase may have reduced production and emission of pheromones. Adult feeding can also affect the pheromones production to attract partners (Epsky and Heath 1993).

Females that participate in the *leks* system select their partners according to size, preferring larger ones (Aluja et al. 2000), but this selection apparently is not based only on male size (Mangan 2003). Other factors such as pheromone amount and constitution are also important. Wild *Ceratitis capitata* females (Wiedeman) select and discriminate males that liberate pheromones in the adequate composition (Heat et al. 1994) that effect courtship, sound, visual, and touch behaviors (Eberhard 2000).

Carey et al. (1998) with *C. capitata* and Jacome et al. (1999) with *A. serpentina* shown that females live longer and reproduce later when submitted to food deprivation, particularly lack of proteins. Mortality increases in reproducing females because it deviates somatic maintenance resources, particularly when protein content is low or absent in the diet (Carey et al. 1998).

19.3.2 Carbohydrates

Ordinarily, carbohydrates are of plant origin and, along with lipids, are the main sources of energy. They store energy, act as phagostimulants, and regulate the amount of ingested food. The great success of insects is due to their easy access to carbohydrates that are abundant in nature; most behavior and mouth structures of insects are adapted to explore these resources (Stoffolano 1995). Insects make better use of carbohydrates found in high proportion in nature such as glucose, fructose, and sucrose, followed by maltose, galactose, trehalose, raffinose, and starch.

Utilization of carbohydrates and feeding behavior of fruit flies are related to the abundance and distribution of these substances in nature (Hsiao 1985). In general, there must be little selective pressure on insects regarding mechanisms of carbohydrate acquisition and utilization since they are abundant in nature; this is not the case of proteins since they are limiting nutrients.

In general, fruit fly larvae are able to survive with artificial diets lacking carbohydrates or containing lipids as the energy source. The absence of carbohydrates in the diet reduces larval and adult performance of *C. capitata*, and adults develop compensation mechanisms (Canato and Zucoloto 1998). However, adults are unable to survive more than 3 days without carbohydrates in the diet (Fontellas and Zucoloto 1999, Manrakhan and Lux 2006).

Longevity of *A. obliqua* males on sucrose diets was reduced compared with males fed sucrose and protein (Cresoni-Pereira and Zucoloto 2006b). *C. capitata* (Chang et al. 2001, Manrakhan and Lux 2006), *C. cosyra* (Walker), and *C. fasciventris* (Bezzi) (Manrakhan and Lux 2006) males and females also showed low survival when fed with sucrose diets only.

The quality and ideal concentration of carbohydrates vary according to the species. *C. capitata* and *A. obliqua* females fed during the pre-oviposition period on different carbohydrates in different concentrations presented better performance when fed with glucose, sucrose, and fructose (12 g/100 ml diet); however, starch did not promote egg production (Zucoloto 1992; Fontellas and Zucoloto 1999). Carbohydrates are fundamental in *A. suspensa* male feeding for the development of sexual appeal (Landolt and Sivinski 1992).

A. serpentina females preferred carbohydrates instead of proteins or open fruits when in a choice situation; this kind of behavior was called "junk food." When flies find foods of questionable quality but are highly energetic, they may prefer them instead of other foods with higher sustenance. The ingestion of those highly energetic items may appease their hunger and block the ingestion of other items. This kind of behavior indicates that nutritional compensation, feeding autoselection, and other mechanisms of food selection are not a rule and that flies may commit metabolic errors (Jacome et al. 1999). Nigg et al. (2007) studied *A. suspensa* behavior and preferences regarding several sugars and observed that flies did not intensify sugar consumption as concentration was reduced.

Carbohydrate consumption by adult fruit flies is influenced in some species by gender and age. *Rhagoletis pomonella* (Walsh.) females consume more carbohydrates than males, possibly due to their more intense metabolism for egg production and host search for egg displacement (Webster et al. 1979). Wild *A. obliqua* females maintained in the laboratory consume a relatively constant amount of sucrose during the reproductive phase (15–60 days); afterward, there is a drastic reduction in carbohydrate ingestion until their death (Cresoni-Pereira and Zucoloto 2001b).

19.3.3 Lipids

Lipids mainly function as a source of energy, essential fatty acids, and steroids; structural components; and waterproofing agents (Dadd 1985). Several studies with fruit flies have shown that after emergence, either in the absence of adequate food (carbohydrates and proteins) or with carbohydrate deprivation, lipid levels fell dramatically and were re-established in subsequent days if adequate feeding was provided and lipid deprivation had not damaged performance (Cangussu and Zucoloto 1992; Jacome et al. 1995; Moreno et al. 1997; Warburg and Yuval 1997).

Studies carried out with lipids with fruit flies show that they transform carbohydrates and proteins into lipids; thus, there is no need, at least for adults, to supply lipids, except when the natural diet is rich in this nutrient as in the case of *B. oleae* (Manoukas 1977).

Studies with *C. capitata* showed that during metamorphosis, lipids (particularly triacylglycerols) and carbohydrates (especially glycogen and trehalose) are the main sources of energy (Nestel et al. 2003). However, little is known about the differential utilization of these resources and about the role of proteins in this process. For *A. fraterculus*, total lipids and triglycerides levels are constant during pupae development, while glycogen levels progressively mitigate with advancing pupae age (Dutra et al. 2007). Total protein in pupae peaks at larval tissue post-histolysis, providing evidence that in *A. fraterculus* proteins and glycogen are the main sources of energy for metamorphosis instead of lipids and glycogen as in *C. capitata* (Nestel et al. 2003).

19.3.4 Vitamins and Mineral Salts

Literature contains little information about the influence of vitamins on insect nutrition. The study of the species' natural food composition provides qualitative information about the necessary micronutrients; quantitative data are not available and are almost impossible to obtain using artificial diets. Nutritional studies can only be conducted in the laboratory, and artificial diet components and mineral salts are hardly vitamin free (Zucoloto 2000).

Current laboratory studies on insect nutrition use the deletion system: one nutrient at a time is taken from the diet. As mentioned above, the diet may be contaminated by traces of the nutrient under test. Furthermore, as the micronutrient needs are very small, deletion may not produce an immediate effect or these nutrients may pass from one generation to the next through the eggs. If the test is applied to only one generation, the nutrient need may not be detected (Zucoloto 2000). Several proportions of the same nutrient can also be tested in the diet, as carbohydrates, proteins, and lipids; however, this is unfeasible regarding micronutrients due to the very small amounts involved.

Canato et al. (1994) working with *C. capitata* observed that fecundity did not change when vitamins were withdrawn from the females' diet, while Moreno et al. (1997) demonstrated that vitamin addition to the *A. obliqua* adult females artificial diets improved performance. In both cases, passage of vitamins may have occurred from one generation to the other. According to Dadd (1985), all mineral salts important for other living organisms are also important for insects. Potassium, magnesium, and phosphate are important for growth in some species. Experimental elimination of symbionts from the digestive tract showed the need for iron, magnesium, and zinc (Dadd 1985).

19.3.5 Symbionts

The role of bacteria in nourishment and survival of tephritid larvae and adults is better known in *Bactrocera* and *Rhagoletis* (Fletcher 1987). These individuals have anatomical adaptations in order to host microorganisms: the gastric cecum in the larvae and the esophageal diverticulum (bulb) in the adults. Bacteria usually found in the digestive tube of fruit flies are members of Enterobacteriaceae (Fletcher 1987). The regurgitation behavior of *Bactrocera* and *Anastrepha* suggested that flies inoculate the fruit surface and use the resulting bacterial colonies as protein sources. The bacterial symbionts may also perform an important role in synthesis of essential amino acids, in breakage and digestion of fruit tissues, in detoxification of secondary substances, and in suppression of pathogenic microorganisms. Drew et al. (1983) show that bacteria derived from plant surfaces come from *Bactrocera* species, including *B. tryoni*, with a diet that enables immature flies to achieve sexual maturity and reproduction. These authors suggest that bacteria are digested since a great number is found in the flies' crop, but none reaches the hind gut or are found in the feces.

Drew and Lloyd (1987) show that *Bactrocera tryoni* (Hardy) and *B. neohumeralis* (Hardy) adults inoculate the fruit surface while foraging, thus attracting immature flies to feed on bacteria to reach sexual maturity. *Anastrepha ludens* (Loew) males did not show effect of the adults' diet on the sexual performance, while three other species (*A. serpentina*, *A. striata* and *A. obliqua*) did (Aluja et al. 2001a). A possible explanation for this difference is the presence of symbiont bacteria in the digestive tract of flies. Bacteria would be responsible for the individuals' adequate nourishment since *A. ludens* appears to have developed in a low nutritional quality environment, so that its success should be associated with the presence of these bacteria.

19.4 Abiotic Factors

The fruit flies' environment is extremely variable and there is no single component that can be considered determinant for their abundance or success. The main components include biotic (predation, competition, presence of co-specifics) and abiotic (humidity, temperature, light) factors (Bateman 1972). Light and temperature are particularly important for feeding because they trigger or inhibit daily activities, such as the feeding periods.

Biotic and abiotic factors interact in several ways to regulate the tephritid females' physiology and behavior. For instance, temperature, luminosity, and food quality may modulate the time of the day females lay eggs (Aluja et al. 1997) and influence copulation behavior (Aluja et al. 2000). In the field, *A. obliqua* oviposition behavior is influenced by luminosity, humidity, temperature, and the presence of hosts (Aluja and Birke 1993).

19.5 Allelochemicals

The role of plant allelochemicals is intensely discussed, but the most widespread theory is that they are related to defense against predation and, sometimes, competition with other plants, since due to their immobility plants could not "escape" from the environmental pressures (Hsiao 1985).

Many insect species developed physiological mechanisms through co-evolution to use feeding items containing allelochemicals as signalers. Although this relationship is more common among monophagous species, it may also occur in polyphagous species. Crisci and Zucoloto (2001) working with *C. capitata* found that secondary compounds in the diet can be tolerated by some generations, and this is important for the utilization of alternative hosts and expansion of the feeding niche. *A. obliqua* females are able to use a secondary compound that usually is not part of their "menu" as signaler of proteins in the diet, possibly through learning processes (Cresoni-Pereira and Zucoloto 2006a). Deleterious effects are observed due to the ingestion of allelochemicals by nonadapted species, such as abnormal utilization of nutrients, formation of unavailable complexes with essential nutrients, and digestive enzymes inhibition and phagoinhibition (Bernays 1985).

Some insects also use allelochemicals as pheromone precursors. In fruit flies, males produce pheromones, unlike most insects (Vilela and Kovaleski 2000). The only exception among tephritids is *B. oleae*, whose females produce the pheromone (Lima 2001). Females spend more energy for reproduction than for mating, compared with males. Signalizing with pheromones has lower energetic cost than searching for a sexual partner; females are less exposed to predators and unfavorable environmental conditions; and relatively little expenditure of energy and few risks are involved (Vilela and Kovaleski 2000).

Males of several *Bactrocera* species are intensely attracted by methyleugenol, a minor constituent of citronella oil that can be found in feeding sources of these species. This probably occurs because this compound is a pheromone precursor and also acts as an allomone that inhibits predation (Tan and Nishida 1998). The copulation success of *B. dorsalis* males was not intensified when methyleugenol was added to the larval diet, but it increased when this compound was added to the diet of adult males reared with larval diets lacking methyleugenol (Shelly and Nishida 2004). The great attraction of males of *Bactrocera* to methyleugenol is the basis of the control programs that use the annihilation technique (Malavasi 2000).

Allelochemicals are also involved in other behaviors and are often used in fruit fly control in commercial fruit farms. Some species change their oviposition behavior when exposed to the chemical information of natural enemies. This phenomenon was shown in *Rhagoletis basiola* (OS), which had reduced oviposition rate when exposed to chemical information of parasitoid eggs (Hoffmeister and Roitberg 1997), and in *Ceratitis cosyra* (Walker) and *Bactrocera invadens* (Drew, Tsuruta, and White), which change their behavior when exposed to hosts treated with ant secretions (Van Mele et al. 2009; Adandonon et al. 2009).

19.6 Feeding

The types of food holometabolous insects ingest depend on the insect mouthparts feeding mechanisms, nutritional value, presence of phagostimulants among others (see discussion below). Fruit flies use specific labellum modifications and positions to ingest diverse substances (Vijaysegaran et al. 1997). The form of ingestion as well as the types of food that can be ingested are influenced by a combination of structures and functions separated through three strictly associated components: (1) structure of the mouthparts, (2) flexion of the labellum during feeding, and (3) way of feeding centered in fluids through regurgitation and re-ingestion of the crop contents (Vijaysegaran et al. 1997).

Elzinga and Broce (1986) observed that the pseudotrachea has rings along the median line; these rings have numerous microteeth with spaces between them that result in numerous micropores that lead to the pseudotrachea lumen. These pseudotracheal rings are similar in some species of *Bactrocera* and *Anastrepha*. The pseudotracheal rings near the oral opening do not present microteeth; however, they are modified in wavy structures similar to bristles and are called prestomal thorns (Vihaysegaran et al. 1997). The opposite oral lobes near the distal scleritum flex to form the labellar cavity. Without that labellar flexion, the food would be introduced directly into the oral opening and filtration by the pseudo-tracheal micropores would not be possible. The oral lobes' flexion is fundamental for tephritids feeding since it protects the oral opening and prevents passing of food directly to the feeding canal. Liquid inges-tion and expulsion via pseudotrachea and numerous micropores that cover the labellar surface resemble the action of a sponge and are more effective than fluid ingestion or food liquefaction through a single wider opening as the oral opening (Vihaysegaran et al. 1997). Usually, flies ingest diluted liquids, but they also feed on solid dry items through regurgitation of the crop content, liquefaction of the food, and re-ingestion of the liquefied portion.

To ingest the food the fly must extend the proboscis, and this occurs after phagostimulation of mouth-part receptors and tarsal chemoreceptor bristles (Bernays 1985). If the sensorial input during ingestion declines below a certain threshold, or if a negative feedback promotes the end of the meal, the proboscis retracts slowly; retraction is slower when alimentary deterrents are found. Feeding continuity depends on continuous phagostimulation. This indicates that the feeding pattern is not continuous until satiation; it requires continuous positive feedback largely provided by the food chemical feedback (Bernays 1985). According to this author, four points are outstanding in starting and continuing feeding. First, continuous feeding demands continuous phagostimulation. Second, the ingested food amount in a meal depends on volumetric factors. Third, the regulation of the periods between the meals depends on the hemolymph composition. Finally, there are endogenous variations in the central nervous system that affect the feed-ing process as a whole.

The sequence of feeding events can be summarized as follows: grazing, suction, bubbling, and regurgi-tation. These events appear in the genus *Anastrepha* (Aluja et al. 1993), and on the species *R. pomonella* (Hendrichs et al. 1992, 1993) and *C. capitata* (Hendrichs et al. 1991). Grazing consists in repeated movements of extension and retraction of the proboscis to touch the food surface. This behavior was formerly described by Hendrichs et al. (1992), and is a way flies find and taste portions of nutrients in the explored surface. When that occurs in plant leaves, protein and carbohydrate acquisition of exudates occur. Proboscis extension allows suction of absorbed liquids, water droplets, bird fresh feces, other liq-uid foods, or items liquefied by the fly saliva. Adult flies can ingest dry foods provided they are liquefied by the saliva. It was demonstrated that performance is reduced when flies have to feed on solid items since a great deal of energy is spent to produce saliva in order to prepare the food for posterior suction (Hendrichs et al. 1993).

Bubbling is the formation of liquid drops of several sizes in the proboscis extremity while the fly is resting on the leaf. Regurgitation of drops deposited on leaf, fruit, or any other surface where the fly rests are re-ingested in variable intervals. Drop amounts and composition vary according to the kind of ingested food (Aluja et al. 2000). Re-ingestion of drops is also related to ingestion of symbiont bacte-ria, which may pass from one fly to another or may proliferate in the regurgitated material and then be ingested (Hendrichs et al. 1992). Bacteria are important for digestion and supply of nonsynthesized or acquired nutrients by adult flies.

Drew and Lloyd (1987) proposed that flies regurgitate to inoculate the host fruit surface with their intestinal bacteria. Hendrichs et al. (1992), describing bubbling in *R. pomonella*, proposed that flies regurgitate the ingested food to eliminate water excess through evaporation and to concentrate the nutri-ents suspended in the solution. Another alternative function for bubbling proposed by Vijaysegaran et al. (1997) is that flies regurgitate to reverse the flow in the labellum, removing particles accumulated in the micropores during feeding.

The flies' appetite for specific nutrients depends on a series of factors such as the concentration of each nutrient in the diet, the interval between the meals, and the amount of food ingested during each meal. The feeding pattern is quite variable, depending on the species and on the environmental condi-tions. Observations made with *A. suspensa* adults indicated that they prefer to feed during the morning period but also feed all day (Landolt and Davis-Hernandez 1993). *A. obliqua* in fruit farms under very high temperatures (45°C in areas without shade) preferred a bimodal pattern, with most of the activities occurring in cooler periods (Aluja and Birke 1993). With mild temperatures, feeding activity peak from 10 am to 3 pm.

Individualized *A. obliqua* females consume significantly more sucrose and brewer's yeast than females in groups (Cresoni-Pereira and Zucoloto, unpublished). Possibly females in groups have restricted access to the food. However, *A. serpentina* individual females do not feed daily even with sugar and protein sources available *ad libitum* (Aluja et al. 2001b).

Tephritids also feed by trophallaxis, the oral transference of a substance from males to females. The only reported case occurs with *A. striata* whose females circulate in front of courting males, touching males' labellum with their own, to drink the substance offered by the males (Aluja et al. 1993). It is believed that this substance contains nutrients or symbionts.

19.7 Behavior

All organisms exhibit a series of behaviors that link their physiological needs and the environment in which they live. These behaviors are highly adapted patterns that optimize the individuals' reproductive success (Yuval and Hendrichs 2000). Aluja et al. (2000) outlined the importance of quantifying and describing all environmental components when the feeding behavior is considered. Several factors influence tephritid behavior and their utilization of resources, such as social context (Zur et al. 2009), age (Carey et al. 2008), availability, food deprivation, and nutrient dilution in the diet, among others.

Fruit fly larvae are able to select the food or food portions more suitable for their development. Fernandes-da-Silva and Zucoloto (1993) working with *C. capitata* larvae demonstrated that they prefer the ripest portions of the fruit, and when fed exclusively with these portions they develop better than if they feed on other fruit parts. In the laboratory, Zucoloto (1987, 1988, 1993a) and Canato and Zucoloto (1993) have also shown this selection of more adequate parts or different foods. Dukas et al. (2001), also working with *C. capitata*, suggest that the most probable explanation for the different sizes of the same-age larvae found in kumquat fruits would be the nutritional variation inside the fruit and the fact that the larvae select certain portions; competition would be a weak argument considering the larval density per fruit and the relatively small size of the larvae as compared to fruit size.

The behavioral and physiological mechanisms that control the insects' food selection are not well known. Currently, the most accepted hypothesis is that the information about the food nutritional quality has its origin in the activity of several sensory receptors. The receptors would be integrated with information about the nutritional status provided by the nutrient content in the hemolymph (Simpson et al. 1995). According to Waldbauer and Friedman (1991), the self-selection feeding behavior is the continuous regulation of ingested food. It involves frequent changes of food to reach a favorable balance of nutrients through noncasual choices with consequent benefits for the individual.

Studies carried out with *C. capitata* (Cangussu and Zucoloto 1995) and *A. obliqua* (Cresoni-Pereira and Zucoloto 2001b; Medeiros and Zucoloto 2006) adult females show that they are able to select nutrient proportions that provide a balanced diet. However, results differ in a fundamental point (Figures 19.1 and 19.2). When the food was offered to *C. capitata* as one block containing sucrose and yeast, females' performance was better than when females selectively ingested sucrose or yeast from separated blocks (Cangussu and Zucoloto 1995). For *A. obliqua* females, total ingestion was lower and performance was better when these nutrients were offered separately than in a single block (Cresoni-Pereira and Zucoloto 2001b). This difference can be explained by the insects' origin and feeding antecedents. *C. capitata* females studied by Cangussu and Zucoloto (1995) were reared for several generations in the laboratory where the food consisted of nutrients in a single block, and this may have caused the population to more efficiently use the nutrients as a whole, probably by directional selection. The *A. obliqua* females studied by Cresoni-Pereira and Zucoloto (2001b) were wild; in nature nutrients are not found in one block but are dispersed, and flies need to search and select them adequately. This result match those of Simpson and Simpson (1990) who stated that insects compensate nutritional imbalance by selective feeding; the balance of ingested amounts from different feeding sources must meet their needs more efficiently than the balance obtained by feeding on a single food source.

In the natural environment, fruit flies select nutrients according to their feeding antecedents. As adults' feeding is extremely variable, flies select the most suitable items for that specific moment, considering other factors such as the presence of predators and competitors, the food availability, and the abiotic factors. Time spent feeding in the field is much longer than in the laboratory (Landolt and Davis-Hernandez 1993); since resources in the field are more dispersed, their nutritional quality is variable and they are

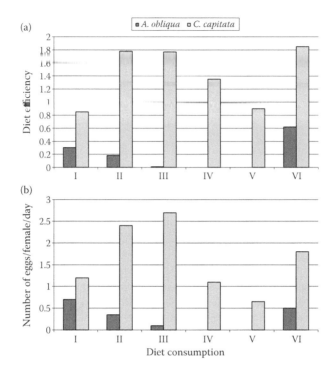

FIGURE 19.1 Diet efficiency (a) and number of eggs (b) produced by *Ceratitis capitata* and *Anastrepha obliqua* fed on one block of diet with the following yeast and sucrose proportions, respectively: diet I = 5.0 g : 6.5 g; diet II = 11.0 g : 6.5 g; diet III = 19.5 g : 6.5 g; diet IV = 27.0 g : 6.5 g; and diet V = 35.0 g : 6.5 g. Diet VI represents results found for females fed on 11.0 g sucrose and 6.5 g yeast separately. (Modified from *J. Insect Physiol.*, 41, J. A. Cangussu and F. S. Zucoloto, Self-selection and perception threshold in adult females of *Ceratitis capitata* (Diptera, Tephritidae), 223–7, Copyright 1995, with permission from Elsevier; and *J. Insect Physiol.*, 47, C. Cresoni-Pereira and F. S. Zucoloto, Dietary self-selection and discrimination threshold in *Anastrepha obliqua* females (Diptera, Tephritidae), 1127–32, Copyright 2001, with permission from Elsevier.)

not always accessible. In general, adults feed during the day, with feeding periods being concentrated in specific times of the day that vary from one species to another.

When insects are not able to balance their diet of essential nutrients offered in a single block, they eliminate the excess (Lemos et al. 1992; Zanotto et al. 1997); thus, this may damage performance and even cause death (Cresoni-Pereira and Zucoloto 2001a).

In *A. obliqua*, whose host fruits are highly ephemeral, the load of eggs is greatly influenced by the access to the resource than in *A. ludens*, which infests less ephemeral fruits. In this last case, egg load is greatly affected by the density of co-specific females and by age (Aluja et al. 2001b). They found that well-fed females of both species show a clear pattern of egg load related to aging, while protein-deprived females do not. Similarly, density of co-specifics only affects well-fed females.

Few studies focused on the effect of male nutritional status on female physiology and behavior. *A. obliqua* female discrimination threshold for a protein source is altered by the male presence and its nutritional status (Cresoni-Pereira and Zucoloto 2005). Diet utilization efficiency by *A. obliqua* females resulted in variable fecundity in male absence or when protein-deprived or nondeprived males were present (Cresoni-Pereira and Zucoloto 2006b). However, Fontellas-Brandalha and Zucoloto (2004) did not find effects of male presence on oviposition and on choice of substrates for oviposition. The female nutritional status was more important in determining the number of eggs and had no effects on the choice of the substrate. The male influence apparently seems to act on protein ingestion and on the physiological mechanisms of food utilization (Cresoni-Pereira and Zucoloto 2006b).

Oocyte numbers of *A. ludens* and *A. obliqua* was greater when females were submitted to fruit volatiles and male pheromones than when no chemical stimuli were present, and intermediate when one type

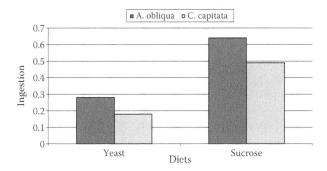

FIGURE 19.2 Sucrose (11.0 g) and brewer's yeast (6.5 g) diet ingestion (mg/female/day) by *Ceratitis capitata* and *Anastrepha obliqua* females from emergence day to day 60 of adult life. (Modified from Cangussu, J. A. and F. S. Zucoloto, *J. Insect Physiol.*, 41, 223, 1995, with permission from Elsevier; Cresoni-Pereira, C. and F. S. Zucoloto, *Iheringia*, 91, 53, 2001.)

of stimulus was present (Aluja et al. 2001b). Apparently, fruit volatiles and male pheromones independently affect egg production, and their effects may be additive or synergistic in nature.

A. ludens copulation with well-fed males resulted in higher fecundity than copulation with poorly fed males; females in the presence of males and fed protein and sugar laid more eggs than females and males fed on sugar only (Mangan 2003). Females without males during the maturation period show better egg maturation than females with males.

Fruit flies have highly refined mechanisms for food selection. The discrimination threshold is the lowest amount of nutrients that can be perceived by the insect in a certain volume of diet. *C. capitata* females show lower thresholds for proteins (Cangussu and Zucoloto 1995) than *A. obliqua* females (Cresoni-Pereira and Zucoloto 2001b). These data show variation of an extremely important physiological mechanism of food selection that reflects the insects' feeding antecedents. *C. capitata* flies used by Cangussu and Zucoloto (1995) were reared in the laboratory and fed on a protein-rich diet (brewer's yeast). As these females have larval reserve they do not need an exogenous protein source to lay the eggs, and consequently their threshold was low (Table 19.1). The wild *A. obliqua* females studied by Cresoni-Pereira and Zucoloto (2001b) needed exogenous protein sources in adequate concentrations in order to produce eggs. They should present a threshold inferior to that found for *C. capitata*, but that did not occur (Table 19.1). A possible explanation is that a minimum concentration of each nutrient is necessary to obtain adequate performances. Wild *A. obliqua* females needed higher protein amounts than *C. capitata* since they did not receive rich larval diet previously. Although there are several compensation mechanisms for nutrient dilution in the diet, compensation is limited by abdominal size. It would not be "advantageous" for *A. obliqua* to perceive very low protein amounts in the diet that could not be

TABLE 19.1

Discrimination Threshold of *Anastrepha obliqua* Wild Females and *Ceratitis capitata* Females Reared in the Laboratory and Maintained in Different Protein Source Deprivation Status

Nutritional Status	A. obliqua	C. capitata
	Yeast Perceived Amount (g)/100 ml Diet	
Newly emerged	0.7	—
Deprived	0.8	0.2
Nondeprived	1.6	0.6

Source: Modified from Cangussu, J. A. and F. S. Zucoloto, *J. Insect Physiol.*, 41, 223, 1995, with permission from Elsevier; Cresoni-Pereira, C. and F. S. Zucoloto, *J. Insect Physiol.* 47, 1127, 2001, with permission from Elsevier.

compensated by increasing ingestion. Other studies show the discrimination threshold for sucrose both in adults (Fontellas-Brandalha and Zucoloto 2003) and immatures (Zucoloto 1993c).

Choice of host for oviposition by *C. capitata* is not positively related to offspring performance (Joachim-Bravo and Zucoloto 1997). This has also been observed to species that explore new resources, a typical behavior of generalists. Apparently, the selection behavior aims nutrition of the individual adult, but it results in egg production and other reproductive behaviors of the progeny. Previous experience effects on behavior seem to be based on learning mechanisms (Bernays 1995; Carsten and Papaj 2005). Although these behavioral changes often reflect learning or imprinting, experience may also change behavior by altering the physiological and reproductive status (Carsten and Papaj 2005). Copulation of Caribbean fruit flies increases male juvenile hormone level and pheromone release, causing greater copulation success (Teal et al. 2000). The effects of the basic mechanisms of experience on copulation behavior may be difficult to distinguish particularly in systems where the experience mediates physiological changes (Carsten and Papaj 2005). Females of the papaya fly increase frequency of re-copulation when the host fruit is present; however, it is not clear if that is mediated by learning or by some effect of the physiological status (Landolt 1994).

Food selection is mediated by allelochemicals as signalers of quality feeding items (Bernays 1985), resulting in less searching time and less exposition to predators during the grazing behavior. This process occurs through positive associative learning. *A. obliqua* females are able to associate the presence of quinine sulfate, an unusual substance in this species menu, to the presence of brewer's yeast in the diet; however, they did not make this association regarding sucrose (Cresoni-Pereira and Zucoloto 2006a). That positive response to yeast may occur due to the need of an exogenous protein source for reproduction and because proteins are often less abundant than carbohydrates. As the latter are easily found in nature, the development of learning mechanisms for a nonlimiting nutrient would not be adaptive.

19.8 Applicability and Conclusions

All over the world, from about 500 *Bactrocera* species, 30 to 40 are known as potential pests. About 50 species of *Rhagoletis* have been described, and several are widely distributed. *Anastrepha* includes from 150 to 200 species. *Ceratitis* is notorious throughout the world mainly due to a single species, *C. capitata* (Lima 2001). Most economically important tephritids that occur in the neotropics belong to these genera (Zucchi 2000).

Fruit flies cause direct damage by oviposition and subsequent larvae development in fruits, and indirect damages when plant tissues are invaded by pathogenic microorganisms (Christenson and Foote 1960). Control programs of pest species have focus on population suppression and/or eradication of adults (Lima 2001).

Fruit fly monitoring estimate populations qualitatively and quantitatively to implement population control measures; monitoring efficiency of adult fruit flies depend on the quality of the attractant, on the kind of trap, and on its location in the field (Nascimento et al. 2000). Color traps may replace hydrolyzed protein, the most employed food attractant (Lima 2001). In addition to visual stimuli and feeding stimulants, plant semiochemicals have been investigated as attractants (Aluja 1994) and their synergistic effect with other odors (Aluja 1994; Aluja et al. 2001b). Use of traps based on sexual pheromone to monitor and control fruit flies is not yet possible. The pheromone system also depends on plant volatiles and food odors (Lima 2001). Traps associating host fruit and pheromone odors capture more flies due to the complementary effect of odors (Robacker et al. 1991). Also, information on how the physiological status affects smell and behavior is useful for understanding how fruit flies use semiochemicals (Lima 2001).

Integrated pest management programs in fruit crops have stimulated the use of biological control to reduce fruit fly population and to increase the abundance of natural enemies, parasitoids being the most effective (Carvalho et al. 2000). In general, the parasitoid life cycle is closely related to the fly life cycle. Therefore, to rear a great amount of parasitoids for liberation it is necessary to also rear the flies. The methodology to rear *Diachasmimorpha longicaudata* (Ashmead) uses as hosts third instar *C. capitata* larvae reared with artificial diets (Carvalho et al. 2000). Figure 19.3 illustrates techniques used for feeding, egg collecting, and maintaining *A. obliqua* in the laboratory.

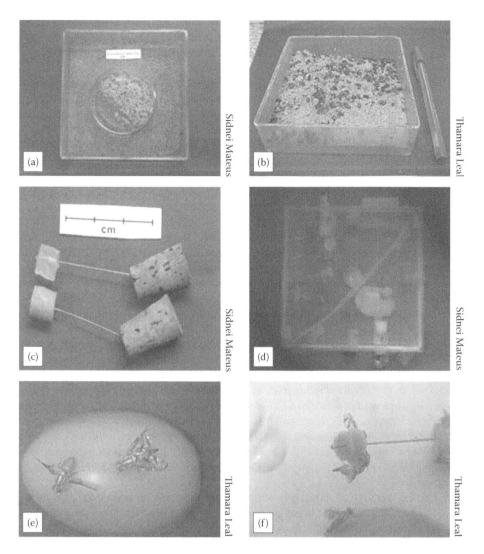

FIGURE 19.3 Laboratory techniques and materials for rearing *Anastrepha obliqua*. (a) Diet dish with larvae in sand boxes for pupation; (b) box with sand and pupae; (c) pins stuck in corks to offer the diet inside experimental boxes; (d) experimental box showing diet pins disposition, water sources (test tubes), and artificial substrate based on agar for oviposition; (e) male and female on the artificial oviposition substrate based on agar and yeast; and (f) flies on the diet and on oviposition substrate.

One of the techniques widely used as control method is the sterile male technique (Aluja 1994). This technique demands a high number of sterile males that should be reared in as close as possible to natural environmental conditions (Calkins and Parker 2005). This is important for males to find adequate feeding resources, reach sexual maturity, survive, find their co-specifics, and successfully copulate with wild females.

Studies on sterile male sexual compatibility and changes in the rearing environment suggest that the sterile male performance can be improved by changing the production process (Cayol 2000). The adult diet during the pre-liberation period has been a promising way to improve the sexual performance of sterile tephritid males (Yuval et al. 2007), with strong evidences that diet complementation with hydrolyzed yeast improves sexual performance. Studies with *Anastrepha obliqua* (Macquart), *A. serpentina* (Wiedemann), and *A. striata* (Schiner) males have shown that a protein source added to the adult male diet renders them more competitive when compared with males fed on other diets (Aluja et al. 2001a).

Bactrocera dorsalis (Shelly et al. 2005) and *C. capitata* (Blay and Yuval 1997) males fed on hydrolyzed yeast copulate more often than deprived males. However, the effect of hydrolyzed yeast on the copulation success was not evident when mass-reared males competed with wild males (Shelly and Kennelly 2002; Shelly and McInnis 2003; Shelly et al. 2005).

As mass-rearing conditions are artificial, there is a potential reduction of males to compete successfully with wild males for copulation with wild females (Meza et al. 2005). In addition to mass rearing, sterilization, shipment, and liberation methods can reduce the competitiveness of sterile males. The copulation performance of mass-reared *C. capitata* males was significantly worse than that of wild males (Pereira et al. 2007). Also in *C. capitata*, it was shown that females can develop behavioral resistance against sterile males over time (McInnis et al. 1996).

The integrated pest management system has been the most indicated alternative method for controlling fruit fly populations, avoiding the indiscriminate use of synthetic insecticides. Considering the tephritids' complexity, the fruit fly integrated pest management requires special attention regarding biological aspects, applied ecology, and use of available technology (Nascimento and Carvalho 2000). The utilization of any technique to control fruit flies demands deep knowledge of the insects' biology, physiology, behavior, and rearing. Feeding and nutrition are the determinant factors of tephritids' life cycle, and studies on those areas must be exhaustive, and comparisons between natural and laboratory populations must be constant. As diet offers nutrients in great amount, larvae mature earlier than wild ones, and this accelerates the succession of generations. These changes detected during the larval stage cause adult to emerge with high nutritive reserves and to reach sexual maturity earlier (Cayol 2000). This modification, associated with other behavioral alterations induced by the super-population of flies, may result in the reduction of sexual compatibility among reared and wild insects (McInnis et al. 1996; Cayol 2000). Several procedures are adopted to minimize the impact of mass rearing and sterilization on the quality of the insects reared in the large scale (Miyatake 1998; Taylor and Yuval 1999). Figure 19.4 shows the type of modifications that occurred in life traits of *B. cucurbitae* (Miyatake 1998).

The performance of a wild population of *C. capitata* was superior when flies fed on papaya, its natural food; while laboratory-reared flies show similar performance with both papaya and diet containing yeast and sucrose (Joachim-Bravo and Zucoloto 1998). Wild flies show preference for papaya and laboratory flies did not. Wild females only laid eggs on papaya while laboratory females oviposited indiscriminately. These data demonstrate the changes caused by mass rearing regarding feeding and oviposition behavior of these flies.

In conclusion, fruit flies are an important and unique feeding guild. To understand their lifestyle and to manage pest species, we should concentrate our efforts to investigate their bioecology and nutrition. This will allow increasing our present knowledge of these insects.

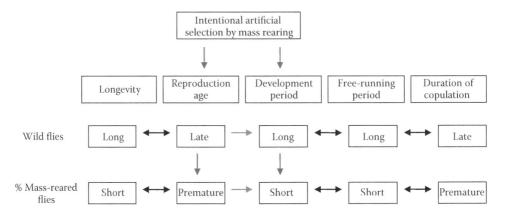

FIGURE 19.4 Genetic relationships between life antecedents and behavioral traits of *Bactrocera cucurbitae*, the melon flies. Blue arrows indicate artificial selection direction due to mass-rearing method. Black arrows indicate genetic correlation between two traits. Red arrows show "inadvertent" selection direction. (From Miyatake, T., *Res. Pop. Ecol.*, 40, 301, 1998.)

REFERENCES

Adandonon, A., J. F. Vayssiéres, A. Sinzogan, and P. Van Mele. 2009. Density of pheromone sources of the weaver ant *Oecophylla longinoda* affects oviposition behaviour and damage by mango fruit flies (Diptera: Tephritidae). *Int. J. Pest Manag*. 55:285–92.

Alonso-Pimentel, H., J. B. Korer, C. Nufio, and D. R. Papaj. 1998. Role of colour and shape stimuli in host-enhanced oogenesis in the walnut fly, *Rhagoletis juglandis*. *Physiol. Entomol*. 23:97–104.

Aluja, M. 1994. Bionomics and management of *Anastrepha*. *Annu. Rev. Entomol*. 39:155–78.

Aluja, M., and A. Birke. 1993. Habitat use by adults of *Anastrepha obliqua* (Diptera, Tephritidae) in a mixed mango and tropical plum orchard. *Ann. Entomol. Soc. Am*. 86:799–812.

Aluja, M., I. Jacome, A. Birke, N. Lozada, and G. Quintero. 1993. Basic patterns of behavior in *Anastrepha striata* (Diptera, Tephritidae) flies under field-cage conditions. *Ann. Entomol. Soc. Am*. 86:776–93.

Aluja, M., A. Jimenez, J. Piñero, M. Camino, L. Aldana, M. E. Valdés, V. Castrejón, I. Jacome, A. B. Dávila, and R. Figueroa. 1997. Daily activity patterns and within field distribution of papaya fruit flies (Diptera, Tephritidae) in Morelos and Veracruz, Mexico. *Ann. Entomol. Soc. Am*. 90:505–20.

Aluja, M., J. Piñero, I. Jacome, F. Diaz-Fleisher, and J. Sivinski. 2000. Behavior of flies in the genus *Anastrepha* (Trypetinae: Toxotrypanini). In *Fruit Flies (Tephritidae): Phylogeny and Evolution of Behavior*, ed. M. Aluja and A. L. Norrbom, 375–406. Boca Raton: CRC Press.

Aluja, M., I. Jacome, and R. Macias-Ordoñez. 2001a. Effect of adult nutrition on male sexual performance in four neotropical fruit flies species of the genus *Anastrepha* (Diptera, Tephritidae). *J. Insect Behav*. 14:759–75.

Aluja, M., F. Diaz-Fleischer, D. R. Papaj, G. Lagunes, and J. Sivinski. 2001b. Effects of age, diet, female density and the host resource on egg load in *Anastrepha ludens* and *Anastrepha obliqua* (Diptera, Tephritidae). *J. Insect Physiol*. 47:975–88.

Aluja, M., D. Pérez-Staples, J. Sivinski, A. Sánchez, and J. Piñero. 2008a. Effects of male condition on fitness in two tropical tephritid flies with contrasting life histories. *Anim. Behav*. 76:1997–2009.

Aluja, M., J. Rull, D. Perez-Staples, F. Díaz-Fleischer, and J. Sivinski. 2008b. Random mating among *Anastrepha ludens* (Diptera: Tephritidae) adults of geographically distant and ecologically distinct populations in Mexico. *Bull. Entomol. Res*. 99:207–14.

Aluja, M., J. Rull, J. Sivinski, G. Trujillo, and D. Pérez-Staples. 2009. Male and female condition influence mating performance and sexual receptivity in two tropical fruit flies (Diptera: Tephritidae) with contrasting life histories. *J. Insect Physiol*. 55:1091–8.

Bateman, M. A. 1972. The ecology of fruit flies. *Annu. Rev. Entomol*. 17:493–518.

Bernays, E. A. 1985. Regulation of feeding behavior. In *Comprehensive Insect Physiology, Biochemistry and Pharmacology*, ed. G. A. Kerkut and L. I. Gilbert, 390–467. London: Pergamon Press.

Bernays, E. A. 1995. Effects of feeding on experience. In *Regulatory Mechanisms in Insect Feeding*, ed. R. F. Chapman and G. de Boer, 279–306. New York: Chapman and Hall.

Blay, S., and B. Yuval. 1997. Nutritional correlates of reproductive success of male Mediterranean fruit flies (Diptera, Tephritidae). *Anim. Behav*. 54:59–66.

Blay, S., and B. Yuval. 1999. Oviposition and fertility in the Mediterranean fruit fly (Diptera, Tephritidae): Effects of male and female body size and the availability of sperm. *Ann. Entomol. Soc. Am*. 92:278–84.

Calkins, C. O., and A. G. Parker. 2005. Sterile insect quality. In *Sterile Insect Technique: Principles and Practice in Area-Wide Integrated Pest Management*, ed. V. A. Dyck, J. Hendrichs, and A. S. Robinson, 269–96. Netherlands: Springer-Dordrecht.

Canato, C. M., and F. S. Zucoloto. 1993. Influência da concentração de nutrientes no valor nutritivo e seleção de dietas em larvas de *Ceratitis capitata* Wied. (Diptera, Tephritidae). *An. Soc. Entomol. Bras*. 22:471–6.

Canato, C. M., G. Fernandes-da-Silva, J. A. Cangussu, and F. S. Zucoloto. 1994. Influência de sais e vitaminas na produção de óvulos por *Ceratitis capitata*. *Científica* 22:15–20.

Canato, C. M., and F. S. Zucoloto. 1998. Feeding behavior of *Ceratitis capitata* (Diptera, Tephritidae): Influence of carbohydrate ingestion. *J. Insect Physiol*. 44:149–55.

Cangussu, J. A., and F. S. Zucoloto. 1992. Nutritional value and selection of different diets by adult *Ceratitis capitata* flies (Diptera, Tephritidae). *J. Insect Physiol*. 38:485–91.

Cangussu, J. A., and F. S. Zucoloto. 1995. Self-selection and perception threshold in adult females of *Ceratitis capitata* (Diptera, Tephritidae). *J. Insect Physiol*. 41:223–7.

Cangussu, J. A., and F. S. Zucoloto. 1997. Effect of protein sources on fecundity, food acceptance and sexual choice by *Ceratitis capitata* (Diptera, Tephritidae). *Rev. Bras. Biol.* 57:611–8.

Carey, J. R., P. Liedo, H. G. Müller, J. L. Wang, and J. W. Vaupel. 1998. Dual modes of aging in Mediterranean fruit fly females. *Science* 28:1996–8.

Carey, J. R., L. G. Harshman, P. Liedo, H. G. Muller, J. L. Wang, and Z. Zhang. 2008. Longevity–fertility trade-offs in the tephritid fruit fly, *Anastrepha ludens*, across dietary-restriction gradients. *Aging Cell* 7:470–7.

Carsten, L. D., and D. R. Papaj. 2005. Effects of reproductive state and host resource experience on mating decisions in a walnut fly. *Behav. Ecol.* 16:528–33.

Carvalho, R. S., A. S. Nascimento, and W. J. R. Matrangolo. 2000. Controle biológico. In: *Moscas-das-Frutas de Importância Econômica no Brasil: Conhecimento Básico e Aplicado*, ed. A. Malavasi and R. A. Zucchi, 113–7. Ribeirão Preto: Holos Editora.

Cayol, J. P. 2000. Changes in sexual behavior and life history traits of tephritid species caused by mass-rearing process. In *Fruit Flies (Tephritidae): Phylogeny and Evolution of Behavior*, ed. M. Aluja and A. L. Norrbom, 843–60. Boca Raton: CRC Press.

Chapman, R. F. 1998. *The Insects: Structure and Function*. New York: Elsevier.

Chang, C. L., C. Albrecht, S. S. A. El-Shall, and R. Kurashima. 2001. Adult reproductive capacity of *Ceratitis capitata* (Diptera, Tephritidae) on chemically defined diet. *Ann. Entomol. Soc. Am.* 94:702–6.

Christenson, L. D., and R. H. Foote. 1960. Biology of fruit flies. *Annu. Rev. Entomol.* 5:171–92.

Cresoni-Pereira, C., and F. S. Zucoloto. 2001a. Influence of quantities of brewer yeast on the performance of *Anastrepha obliqua* wild females (Diptera, Tephritidae). *Iheringia* 91:53–60.

Cresoni-Pereira, C., and F. S. Zucoloto. 2001b. Dietary self-selection and discrimination threshold in *Anastrepha obliqua* females (Diptera, Tephritidae). *J. Insect Physiol.* 47:1127–32.

Cresoni-Pereira, C., and F. S. Zucoloto. 2005. The presence of the sexual partner and nutritional condition alter the *Anastrepha obliqua* Macquart (Diptera, Tephritidae) protein discrimination threshold. *Neotrop. Entomol.* 34:895–902.

Cresoni-Pereira, C., and F. S. Zucoloto. 2006a. Associative learning in wild *Anastrepha obliqua* females (Diptera, Tephritidae) related to a protein source. *Iheringia* 96:53–6.

Cresoni-Pereira, C., and F. S. Zucoloto. 2006b. Influence of male nutritional conditions on the performance and alimentary selection of wild females *Anastrepha obliqua* (Macquart) (Diptera, Tephritidae). *Rev. Bras. Entomol.* 50:287–92.

Crisci, V. L., and F. S. Zucoloto. 2001. Performance and selection of diets containing the allelochemical compound quinine sulphate by *Ceratitis capitata* (Diptera, Tephritidae). *Rev. Bras. Entomol.* 45:275–82.

Dadd, R. H. 1985. Nutrition: Organisms. In *Comprehensive Insect-Physiology, Biochemistry and Pharmacology*, ed. G. A. Kerkut and L. I. Gilbert, 313–89. London: Pergamon Press.

Davidowitz, G., L. J. D'Amico, and H. F. Nijhout. 2003. Critical weight in the development of insect body size. *Evol. Dev.* 5:188–97.

Drew, R. A. I. 1987. Behavioral strategies of fruit flies of the genus *Dacus* (Diptera, Tephritidae) significant in mating and host–plant relationship. *Bull. Entomol. Res.* 77:73–81.

Drew, R. A. I., A. C. Courtice, and D. S. Teakle. 1983. Bacteria as natural source of food for adult fruit flies (Diptera, Tephritidae). *Oecologia* 60:279–84.

Drew, R. A. I., and A. C. Lloyd. 1987. Relationship of fruit flies (Diptera: Tephritidae) and their bacteria to host plants. *Ann. Entomol. Soc. Am.* 80:629–36.

Dukas, R., R. Prokopy, and J. J. Duan. 2001. Effects of larval competition on survival and growth in Mediterranean fruit flies. *Ecol. Entomol.* 26:587–93.

Dutra, B. K., F. A. Fernandes, J. C. Nascimento, F. C. Quadros, and G. T. Oliveira. 2007. Intermediate metabolism during the ontogenetic development of *Anastrepha fraterculus* (Diptera: Tephritidae). *Comp. Biochem. Physiol.* 147A:594–9.

Eberhard, W. 2000. Sexual behavior and sexual selection in Mediterranean fruit fliy, *Ceratitis capitata* (Dacine, Ceratidini). In *Fruit Flies (Tephritidae): Phylogeny and Evolution of Behavior*, ed. M. Aluja and A. L. Norrbom, 459–89. Boca Raton: CRC Press.

Elzinga, R. J., and A. B. Broce. 1986. Labellar modifications of muscomorpha flies (Diptera). *Ann. Entomol. Soc. Am.* 79:150–209.

Epsky, N. D., and R. R. Heath. 1993. Food availability and food pheromone production of males of *Anastrepha suspensa* (Diptera: Tephritidae). *Environ. Entomol.* 22:942–7.

Fernandes-da-Silva, P. G., and F. S. Zucoloto. 1993. The influence of host nutritive value on the performance and food selection in *Ceratitis capitata* (Diptera, Tephritidae). *J. Insect Physiol.* 39:883–7.

Ferro, M. I. T., and F. S. Zucoloto. 1989. Influência da nutrição protéica no desenvolvimento da glândula salivar de machos de *Anastrepha obliqua*. *Científica* 17:1–5.

Fletcher, B. S. 1987. The biology of dacine fruit flies. *Annu. Rev. Entomol.* 32:115–44.

Fletcher, B. S., and E. T. Kapatos. 1983. The influence of temperature, diet and olive fruits on the maturation rates of female olive fruit flies at different times of the year. *Entomol. Exp. Appl.* 33:244–52.

Fontellas, T. M. L., and F. S. Zucoloto. 1999. Nutritive value of diets with different carbohydrates for adult *Anastrepha obliqua* (Macquart) (Diptera, Tephritidae). *Rev. Bras. Zool.* 16:1135–47.

Fontellas-Brandalha, T. M. L., and F. S. Zucoloto. 2003. Effect of sucrose ingestion on the performance of wild *Anastrepha obliqua* (Macquart) females (Diptera, Tephritidae). *Neotrop. Entomol.* 32:209–16.

Fontellas-Brandalha, T. M. L., and F. S. Zucoloto. 2004. Selection of oviposition sites by wild *Anastrepha obliqua* (Macquart) (Diptera, Tephritidae) based on the nutritional composition. *Neotrop. Entomol.* 33:557–62.

Galun, R., S. Gothilf, S. Blondheim, J. L. Sharp, M. Mazor, and A. Lachman. 1985. Comparison of aggregation and feeding responses by normal and irradiated fruit flies *Ceratitis capitata* and *Anastrepha suspensa* (Diptera, Tephritidae). *Environ. Entomol.* 14:726–32.

Gavriel, S., Y. Gazit, and B. Yuval. 2009. Remating by female Mediterranean fruit flies (*Ceratitis capitata*, Diptera: Tephritidae): Temporal patterns and modulation by male condition. *J. Insect Physiol.* 55:637–42.

Harano, T., M. Fujisawa, and T. Miyatake. 2006. Effect of oviposition substrate on female remating in *Callusobruchus chinensis* (Coleoptera: Bruchidae). *Appl. Entomol. Zool.* 41:569–72.

Heat, R. R., N. D. Epsky, B. D. Dueben, A. Guzman, and L. E. Andrade. 1994. Gamma radiation effects on production of four pheromonal components of male Mediterranean fruit flies (Diptera, Tephritidae). *J. Econ. Entomol.* 87:904–9.

Hendrichs J., B. I. Katsoyanos, D. R. Papaj, and R. J. Prokopy. 1991. Sex differences in movement between natural feeding and mating sites and tradeoffs between food consumption, mating success and predator evasion in Mediterranean fruit flies (Diptera, Tephritidae). *Oecologia* 86:223–31.

Hendrichs, J., S. S. Cooley, and R. J. Prokopy. 1992. Post-feeding bubbling behavior in fluid-feeding Diptera: Concentration of crop contents by oral evaporation of excess water. *Physiol. Entomol.* 17:153–61.

Hendrichs, J., B. S. Fletcher, and R. J. Prokopy. 1993. Feeding behavior of *Rhagoletis pomonella* flies (Diptera, Tephritidae): Effect of initial food quality and quantity on food foraging, handling cost and bubbling. *J. Insect Behav.* 6:43–64.

Hoffmeister, T. S., and B. D. Roitberg. 1997. To mark the host or the patch—Decisions of a parasitoid searching for concealed host larvae. *Evol Ecol.* 11:145–168.

Hsiao, T. H. 1985. Feeding behavior. In *Comprehensive Insect Physiology, Biochemistry and Pharmacology*, ed. G. A. Kerkut and L. I. Gilbert, 471–512. London: Pergamon Press.

Jacome, I., M. Aluja, P. Liedo, and D. Nestel. 1995. The influence of adult diet and age on lipids reserves in the tropical fruit fly *Anastrepha serpentina* (Diptera, Tephritidae). *J. Insect Physiol.* 41:1079–86.

Jacome, I., M. Aluja, and P. Liedo. 1999. Impact of adult diet on demographic and population parameters of the tropical fruit fly *Anastrepha serpentina* (Diptera, Tephritidae). *Bull. Entomol. Res.* 89:163–75.

Joachim-Bravo, I. S., and F. S. Zucoloto. 1997. Oviposition preference and larval performance in *Ceratitis capitata* (Diptera, Tephritidae). *Rev. Bras. Zool.* 14:795–802.

Joachim-Bravo, I. S., and F. S. Zucoloto. 1998. Performance and feeding behavior of *Ceratitis capitata*: Comparison of a wild population and laboratory population. *Entomol. Exp. Appl.* 87:67–72.

Joachim-Bravo, I. S., C. S. Anjos, and A. M. Costa. 2009. The role of protein in the sexual behaviour of males of *Ceratitis capitata* (Diptera, Tephritidae): Mating success, copula duration and number of copulation. *Zoologia* 26:407–12.

Kaspi, R., and B. Yuval. 2000. Post-teneral protein feeding improves sexual competitiveness but reduces longevity of mass-reared sterile male Mediterranean fruit flies (Diptera: Tephritidae). *Ann. Entomol. Soc. Am.* 93:949–55.

Kaspi, R., P. W. Taylor, and B. Yuval. 2000. Diet and size influence sexual advertisement and copulatory success of males in Mediterranean fruit fly leks. *Ecol. Entomol.* 25:1–6.

Koveos, D. S., and M. E. Tzanakakis. 1990. Effect of the presence of olive fruit on ovarian maturation in the olive fruit fly, *Dacus oleae*, under laboratory conditions. *Entomol. Exp. Appl.* 55:161–8.

Landolt, P. J. 1994. Mating frequency of the papaya fruit fly (Diptera, Tephritidae) with and without host fruit. _Fla. Entomol._ 77:305–12.

Landolt, P. J., and J. Sivinski. 1992. Effects of time of day, adult food and host fruit on the incidence of calling by male Caribbean fruit flies. _Environ. Entomol._ 21:382–7.

Landolt, P. J., and K. M. Davis Hernandez. 1993. Temporal patterns of feeding by Caribbean fruit flies (Diptera, Tephritidae) on sucrose and hydrolyzed yeast. _Ann. Entomol. Soc. Am._ 86:749–55

Lemos, F. J. A., F. S. Zucoloto, and W. R. Terra. 1992. Enzymological and excretory adaptations of _Ceratitis capitata_ (Diptera, Tephritidae) larvae to high-protein and high-salt diets. _Comp. Biochem. Physiol._ 102A:775–9.

Liedo, P., J. R. Carey, H. Celedonio, and J. Guillen. 1992. Size specific demography in three species of _Anastrepha_ fruit flies. _Entomol. Exp. Appl._ 63:135–42.

Lima, I. 2001. Semioquímicos das moscas-das-frutas. In _Feromônios de Insetos: Biologia, Química e Emprego no Manejo de Pragas_, ed. E. F. Vilela and T. M. C. Della-Lucia, 121–6. Ribeirão Preto: Holos Editora.

Malacrida, A. R., C. R. Guglielmino, P. D'Adamo, C. Torti, F. Marinoni, and G. Gasperi. 1996. Allozyme divergence and phylogenetic relationships among species of Tephritidae flies. _Heredity_ 76:592–602.

Malavasi, A. 2000. Técnica da aniquilação de machos. In _Moscas-das-Frutas de Importância Econômica no Brasil: Conhecimento Básico e Aplicado_, ed. A. Malavasi and R. A. Zucchi, 159–60. Ribeirão Preto: Holos Editora.

Mangan, R. L. 2003. Adult diet and male–female contact effects on female reproductive potential in Mexican fruit fly (_Anastrepha ludens_ Loew) (Diptera, Tephritidae). _J. Econ. Entomol._ 96:341–7.

Manoukas, A. G. 1977. Biological characteristics of _Dacus oleae_ larvae (Diptera, Tephritidae) reared on a basal diet with variable levels of ingredients. _Ann. Zool. Anim. Ecol._ 9:141–8.

Manrakhan, A., and S. A. Lux. 2006. Contribution of natural food sources to reproductive behaviour, fecundity and longevity of _Ceratitis cosyra, C. fasciventris_ and _C. capitata_ (Diptera, Tephritidae). _Bull. Entomol. Res._ 96:259–68.

McInnis, D. O., D. R. Lance, and C. G. Jackson. 1996. Behavioral resistance to the sterile insect technique by the Mediterranean fruit fly (Diptera, Tephritidae) in Hawaii. _Ann. Entomol. Soc. Am._ 89:739–44.

Meats, A., H. M. Holmes, and G. L. Kelly. 2004. Laboratory adaptation of _Bactrocera tryoni_ (Diptera, Tephritidae) decreases mating age and increases protein consumption and number of eggs produced per milligram of protein. _Bull. Entomol. Res._ 94:517–24.

Meats, A., and S. M. Leighton. 2004. Protein consumption by mated, unmated, sterile and fertile adults of the Queensland fruit fly, _Bactrocera tryoni_ and its relation to egg production. _Physiol. Entomol._ 29:176–82.

Medeiros, L., and F. S. Zucoloto. 2006. Nutritional balancing in fruit flies: Performance of wild adult females of _Anastrepha obliqua_ (Diptera, Tephritidae) fed on single-food or food-pair treatments. _J. Insect Physiol._ 52:1121–7.

Medeiros, L., C. Cresoni Pereira, and F. S. Zucoloto. 2008. Insects making choices: Physiology related to choices in feeding, mating and oviposition. In: _Insect Physiology: New Research_, ed. R. P. Maes, 143–83. New York: Nova Science.

Medeiros-Santana, L., and F. S. Zucoloto. 2009. Comparison of the performances of wild _Anastrepha obliqua_ (Diptera: Tephritidae) individuals proceeding from different hosts. _Ann. Entomol. Soc. Am._ 102:819–25.

Message, C. M., and F. S. Zucoloto. 1980. Valor nutritivo do levedo de cerveja para _Anastrepha obliqua_ (Diptera, Tephritidae). _Ciência Cultura_ 32:1091–4.

Meza, J. S., F. Diaz-Fleischer, and D. Orozco. 2005. Pupariation time as a source of variability in mating performance in mass-reared _Anastrepha ludens_ (Diptera, Tephritidae). _J. Econ. Entomol._ 98:1930–6.

Miyatake, T. 1998. Genetic changes of life history and behavioral traits during mass-rearing in the melon fly, _Bactrocera cucurbitae_ (Diptera, Tephritidae). _Res. Pop. Ecol._ 40:301–10.

Moreno, D. S., D. A. O. Zaleta, and R. L. Mangan. 1997. Development of artificial larval diets for West Indian fruit flies (Diptera, Tephritidae). _J. Econ. Entomol._ 90:427–34.

Morgante, J. S., D. Selivon, V. N. Solferini, and S. R. Matioli. 1993. Evolutionary patterns in specialist and generalist species of _Anastrepha_. In _Fruit Flies: Biology and Management_, ed. M. Aluja and P. Liedo, 15–20. New York: Springer-Verlag.

Mossinson, S., and B. Yuval. 2003. Regulation of sexual receptivity of female Mediterranean fruit flies: Old hypotheses revisited and a new synthesis proposed. _J. Insect Physiol._ 49:561–7.

Nascimento, A. S., R. S. Carvalho, and A. Malavasi. 2000. Monitoramento populacional. In *Moscas-das-Frutas de Importância Econômica no Brasil: Conhecimento Básico e Aplicado*, ed. A. Malavasi and R. A. Zucchi, 109–12. Ribeirão Preto: Holos Editora.

Nascimento, A. S., and R. S. Carvalho. 2000. Manejo integrado de moscas-das-frutas. In *Moscas-das-Frutas de Importância Econômica no Brasil: Conhecimento Básico e Aplicado*, ed. A. Malavasi and R. A. Zucchi, 169–173. Ribeirão Preto: Holos Editora.

Nestel, D., D. Tolmasky, A. Rabossi, and L. A. Quesada-Allué. 2003. Lipid, carbohydrate and proteins patterns during metamorphosis of the Mediterranean fruit fly, *Ceratitis capitata* (Diptera: Tephritidae). *Ann. Entomol. Soc. Am.* 96:237–44.

Nigg, H. N., R. A. Schumann, R. J. Stuart, E. Etxeberria, J. J. Yang, and S. Fraser. 2007. Consumption of sugars by *Anastrepha suspensa* (Diptera: Tephritidae). *J. Econ. Entomol.* 100:1938–44.

Papaj, D. R. 2000. Ovarian dynamics and host use. *Annu. Rev. Entomol.* 45:423–48.

Pereira, R., N. Silva, C. Quintal, R. Abreu, J. Andrade, and L. Dantas. 2007. Sexual performance of mass-reared and wild Mediterranean fruit flies (Diptera: Tephritidae) from various origins of the Madeira Islands. *Fla. Entomol.* 90:10–4.

Pérez-Staples, D., V. Prabhu, and P. W. Taylor. 2007. Post-teneral protein feeding enhances sexual performance of Queensland fruit flies. *Physiol. Entomol.* 32:225–32.

Pérez-Staples, D., A. M. T. Harmer, S. Collins, and P. W. Taylor. 2008. Potential for prerelease diet supplements to increase the sexual performance and longevity of male Queensland fruit flies. *Agric. Forest Entomol.* 10:255–62.

Plácido-Silva, M. C., F. S. Zucoloto, and I. S. Joachim-Bravo. 2005. Influence of protein on feeding behavior of *Ceratitis capitata* Wiedemann (Diptera, Tephritidae): Comparison between immature males and females. *Neotrop. Entomol.* 34:539–45.

Radhakrishnan, P., and P. W. Taylor. 2007. Seminal fluids mediate sexual inhibition and short copula duration in mated female Queensland fruit flies. *J. Insect Physiol.* 53:741–5.

Ringo, J. 1996. Sexual receptivity in insects. *Annu. Rev. Entomol.* 41:473–94.

Robacker, D. C., J. A. Garcia, and W. G. Hart. 1991. Attraction of laboratory strain of *Anastrepha ludens* (Diptera, Tephritidae) to the odor of fermented chapote fruit and pheromones in laboratory experiments. *Environ. Entomol.* 19:403–8.

Selivon, D. 2000. Biologia e padrões de especiação. In *Moscas-das-Frutas de Importância Econômica no Brasil: Conhecimento Básico e Aplicado*, ed. A. Malavasi and R. A. Zucchi, 25–8. Ribeirão Preto: Holos Editora.

Selivon, D., and J. S. Morgante. 1997. Reproductive isolation between *Anastrepha bistrigata* and *A. striata* (Diptera, Tephritidae). *Braz. J. Genet.* 20:583–5.

Shelly, T. E., and S. Kenelly. 2002. Influence of diet on male mating success and longevity and female remating in the Mediterranean fruit fly (Diptera, Tephritidae) under laboratory conditions. *Fla. Entomol.* 85:572–9.

Shelly, T. E., S. Kenelly, and D. O. McInnis. 2002. Effect of adult diet on signaling activity, mate attraction and mating success in male Mediterranean fruit flies (Diptera, Tephritidae). *Fla. Entomol.* 85:150–5.

Shelly, T. E., and D. O. McInnis. 2003. Influence of adult diet on the mating success and survival of male Mediterranean fruit flies (Diptera: Tephritidae) from two mass-rearing strains on field-caged host trees. *Fla. Entomol.* 86:340–4.

Shelly, T. E., and R. Nishida. 2004. Larval and adult feeding on methyl eugenol and the mating success of male oriental fruit flies, *Bactrocera dorsalis. Entomol. Exp. Appl.* 112:155–8.

Shelly, T. E., J. Edu, and E. Pahio. 2005. Influence of diet and methyl eugenol on the mating success of males of the Oriental fruit fly, *Bactrocera dorsalis* (Diptera: Tephritidae). *Fla. Entomol.* 88:308–13.

Silva, J. A. 2000. Estudos moleculares. In *Moscas-das-Frutas de Importância Econômica no Brasil: Conhecimento Básico e Aplicado*, ed. A. Malavasi and R. A. Zucchi, 29–39. Ribeirão Preto: Holos Editora.

Simpson, S. J., and C. L. Simpson. 1990. The mechanisms of nutritional compensation by phytophagous insects. In *Insect–Plant Interactions*, ed. E. A. Bernays, 111–60. Boca Raton: CRC Press.

Simpson, S. J., D. Raubenheimer, and P. G. Chambers. 1995. The mechanisms of nutritional homeostasis. In *Regulatory Mechanisms in Insect Feeding*, ed. R. F. Chapman and G. de Boer, 251–78. New York: Chapman and Hall.

Slansky, F., and J. M. Scriber. 1985. Food consumption and utilization. In *Comprehensive Insect Physiology, Biochemistry and Pharmacology*, ed. G. A. Kerkut and L. I. Gilbert, 87–163. Oxford: Pergamon Press.

Stoffolano, Jr., J. G. 1995. Regulation of a carbohydrate meal in the adult Diptera, Lepidoptera and Hymenoptera. In *Regulatory Mechanisms in Insect Feeding*, ed. R. F. Chapman and G. de Boer, 210–47. New York: Chapman and Hall.

Tan, K. H., and R. Nishida. 1998. Ecological significance of a male attractant in the defense and mating strategies of the fruit fly pest, *Bactrocera papayae*. *Entomol. Exp. Appl.* 89:155–8.

Taylor, P. W., and B. Yuval. 1999. Post-copulatory sexual selection in Mediterranean fruit flies: Advantages for large and protein-fed. *Anim. Behav.* 58:55–9.

Teal, P. E. A., Y. Gomez-Simuta, and A. T. Proveaux. 2000. Mating experience and juvenile hormone enhance sexual signaling and mating in male Caribbean fruit flies. *Proc. Nat. Acad. Sci. U. S. A.* 97:3708–12.

Van Mele, P., J. F. Vayssiéres, A. Adandonon, and A. Sinzogan. 2009. Ant cues affect the oviposition behaviour of fruit flies (Diptera: Tephritidae) in Africa. *Physiol. Entomol.* 34:256–61.

Vijaysegaran, S., G. H. Walter, and R. A. I. Drew. 1997. Mouthpart structure, feeding mechanisms and natural food sources of adult *Bactrocera* (Diptera, Tephritidae). *Ann. Entomol. Soc. Am.* 90:184–201.

Vilela, E. F., and A. Kovaleski. 2000. Feromônios. In *Moscas-das-Frutas de Importância Econômica no Brasil: Conhecimento Básico e Aplicado*, ed. A. Malavasi and R. A. Zucchi, 99–102. Ribeirão Preto: Holos Editora.

Waldbauer, G. P., and S. Friedman. 1991. Self-selection of optimal diets by insects. *Annu. Rev. Entomol.* 36:43–63.

Warburg, I., and B. Yuval. 1997. Effects of energetic reserves on behavioral patterns of Mediterranean fruit flies (Diptera, Tephritidae). *Oecologia.* 113:314–9.

Webster, R. P., J. G. Stoffolano, Jr., and R. J. Prokopy. 1979. Long-term intake of protein and sucrose in relation to reproductive behavior of wild and laboratory cultured *Rhagoletis pomonella*. *Ann. Entomol. Soc. Am.* 72:41–6.

Wheeler, D. 1996. The role of nourishment in oogenesis. *Annu. Rev. Entomol.* 41:407–31.

Yuval, B., and J. Hendrichs. 2000. Behavior of flies in the genus *Ceratitis* (Dacine: Ceratitidini). In *Fruit Flies (Tephritidae): Phylogeny and Evolution of Behavior*, ed. M. Aluja and A. L. Norrbom, 408–44. Boca Raton: CRC Press.

Yuval, B., M. Maor, K. Levy, R. Kaspi, P. W. Taylor, and T. E. Shelly. 2007. Breakfast of champions of kiss of death? Survival and sexual performance of protein-fed sterile Mediterranean fruit flies (Diptera: Tephritidae). *Fla. Entomol.* 90:115–22.

Zanotto, F. P., S. M. Gouveia, S. J. Simpson, D. Raubenheimer, and P. C. Calder. 1997. Nutritional homeostasis in locusts: Is there a mechanism for increased energy expenditure during carbohydrate overfeeding? *J. Exp. Biol.* 200:2437–48.

Zucchi, R. A. 2000. Taxonomia. In *Moscas-das-Frutas de Importância Econômica no Brasil: Conhecimento Básico e Aplicado*, ed. A. Malavasi and R. A. Zucchi, 13–24. Ribeirão Preto: Holos Editora.

Zucoloto, F. S. 1987. Feeding habitat of *Ceratitis capitata* (Diptera: Tephritidae): Can larvae recognize a nutritionally effective diet? *J. Insect Physiol.* 33:349–53.

Zucoloto, F. S. 1988. Qualitative and quantitative competition for food in *Ceratitis capitata* (Diptera, Tephritidae). *Rev. Bras. Biol.* 48:523–6.

Zucoloto, F. S. 1992. Egg production by *Ceratitis capitata* (Diptera, Tephritidae) fed with different carbohydrates. *Rev. Bras. Entomol.* 36:235–40.

Zucoloto, F. S. 1993a. Acceptability of different Brazilian fruits to *Ceratitis capitata* (Diptera, Tephritidae) and fly performance. *Braz. J. Med. Biol. Res.* 26:291–8.

Zucoloto, F. S. 1993b. Adaptation of a *Ceratitis capitata* population (Diptera, Tephritidae) to an animal protein base diet. *Entomol. Exp. Appl.* 67:119–27.

Zucoloto, F. S. 1993c. Nutritive value and selection of diets containing different carbohydrates by larvae of *Ceratitis capitata* (Diptera, Tephritidae). *Rev. Bras. Biol.* 53:611–8.

Zucoloto, F. S. 2000. Alimentação e nutrição de moscas-das-frutas. In *Moscas-das-Frutas de Importância Econômica no Brasil: Conhecimento Básico e Aplicado*, ed. A. Malavasi and R. A. Zucchi, 67–80. Ribeirão Preto: Holos Editora.

Zur, T., E. Nemny-Lavy, N. T. Papadopoulos, and D. Nestel. 2009. Social interactions regulate resource utilization in a Tephritidae fruit fly. *J. Insect Physiol.* 55:890–7.

20

Sap-Sucking Insects (Aphidoidea)

Sonia M. N. Lazzari and Regina C. Zonta-de-Carvalho

CONTENTS

20.1 Introduction 474
20.2 Evolutionary Aspects and Distribution 474
20.3 Structure of Mouthparts and Digestive Tract 475
    20.3.1 Structure and Function of Mouthparts 475
    20.3.2 Structure of Digestive Tract 476
20.4 Nutritional Physiology 477
    20.4.1 Host Location and Acceptance 478
    20.4.2 Feeding Strategies and Host Plant Condition 480
    20.4.3 Phloem Feeding 481
    20.4.4 Food Intake Mechanisms and Saliva Composition 482
    20.4.5 Food Constituents and Digestive Enzymes 484
    20.4.6 Nutritional Requirements 487
        20.4.6.1 Amino Acids 487
        20.4.6.2 Carbohydrates 488
        20.4.6.3 Vitamins 488
        20.4.6.4 Lipids, Minerals, and Trace Metals 488
    20.4.7 Feeding Rate and Nutritional Budget 488
    20.4.8 Intrinsic Rate of Natural Increase 492
        20.4.8.1 Development Rate 492
        20.4.8.2 Reproductive Rate 493
        20.4.8.3 Survival Rate 493
        20.4.8.4 Changes in Intrinsic Rate of Natural Increase 494
    20.4.9 Honeydew and Excretion 496
    20.4.10 Symbionts 496
20.5 Factors Affecting Feeding and Nutrition 498
    20.5.1 Physiological Status of Plant, Water Stress, and Aphid Performance 498
    20.5.2 Secondary Compounds 498
    20.5.3 Physical and Chemical Factors of Host Plant and Environment 499
20.6 Artificial Diets 499
20.7 Electrical Penetration Graph and Aphid Feeding 499
20.8 Perspectives for IPM 502
    20.8.1 Plant Nutrition 503
    20.8.2 Biological Control 503
    20.8.3 Physical and Chemical Traits in Plant Resistance 504
20.9 Final Considerations 505
References 506

20.1 Introduction

The relationship between sap-suckling insects and plant phloem is a highly specialized biotic interaction. Aphids, psyllids, whiteflies, scale insects, and mealy bugs (Hemiptera: Sternorrhyncha) are specialized phloem feeders, able to survive on a nutritionally unbalanced diet and minimize the defense responses of host plants. Phloem-sucking insects damage plants by reducing photosynthesis, affecting growth and distribution of nutrients, acting as the main vectors for plant viruses, and in some cases, injecting toxins into tissues (Thompson and Goggin 2006).

Aphids are small, soft-bodied, phloem-sucking insects. They have a very high reproductive potential, with parthenogenetic viviparous females for at least some generations, and occasionally oviparous. According to Dixon (1987a), aphid parthenogenesis and phloem sap feeding are the main factors that have shaped the evolution and ecology of the group, resulting in specificity, dependence, and adaptations of their life cycles to the host plant. More than 90% of aphid species exhibit considerable host plant specificity. However, some of the important pest species are extremely polyphagous. Host alternation allows many species to exploit new food resources so they may continue to grow and reproduce. Polymorphism, with aptera and winged morphs, is also a characteristic of aphids that confers them a highly specialized way of life (Risebrow and Dixon 1987).

Aphids are considered model organisms for studying speciation in animals and the mechanisms involving the utilization of plants by herbivores. Aphid–plant interactions comprise the selection of the host plant, penetration of plant tissues, phloem sap feeding, and plant response to insect attack. Aphid feeding can directly affect plant development, causing lesions or systemic symptoms, while the response of the plant affects the reproduction and feeding of the insect, and may also attract their natural enemies. Virus transmission by aphids is also a result of a specialized insect–plant interaction.

Excellent review papers and extensive studies on aphid feeding, nutrition, and related topics are presented by Auclair (1963, 1969), addressing physiological and biochemical aspects, and nutrition of some species in chemically defined diets; Miles (1972) deals with the saliva of sap-sucking insects; Miles (1987) discusses the effect of the feeding process of Aphidoidea in the host plant; Risebrow and Dixon (1987) address the nutritional ecology of phloem-sucking insects; Srivastava (1987), the nutritional physiology; Klingauf (1987) refers to the adaptive mechanisms for feeding and excretion; Pickett et al. (1992) review the chemical ecology of aphids; Powell et al. (2006), the behavior, evolution, and application perspectives on host plant selection by aphids; and van Emden and Harrington (2007) edited a book with several chapters by aphid authorities approaching aphid feeding and related topics, including integrated pest management (IPM) case studies. Many other general reviews and research papers on aphid nutrition are available, demonstrating the peculiarities and importance of this insect group.

Understanding the specialized aphid interaction with their host plants allows for improvement of management strategies for pest control. This chapter focuses on various aspects of bioecology and nutrition of Aphidoidea and their application, starting with aspects of evolution and biogeography; structure of mouthparts and digestive tract; nutritional physiology, including the insect–host plant interactions; composition of food; insect nutritional requirements; feeding rates; factors that affect nutrition and feeding strategies; followed by a brief discussion of artificial diets for aphids and the study of feeding behavior by an electronic monitoring technique, and concludes with the application of the available information on insect feeding and nutrition to the integrated management of aphid pest species. Much of the information presented herein is valid not only for aphids but also for other phloem feeders of the suborder Sternorrhyncha, such as Coccidae (mealy bugs and scale insects), Aleyrodidae (whiteflies), and Psyllidae (psylids).

20.2 Evolutionary Aspects and Distribution

The world's aphid fauna consists of about 4700 species classified into approximately 600 genera (Remaudière and Remaudière 1997). They are members of the suborder Sternorrhyncha, basal lineage of Order Hemiptera (Campbell et al. 1995; von Dohlen and Moran 1995).

The evolution of parthenogenesis and viviparity in combination with host alternation and polymorphism allowed this group to synchronize growth and reproduction with host phenology, resulting in wide variations in life cycles and within species. Within a parthenogenetic lineage, females may exhibit many phenotypes that differ in several attributes, including morphology, physiology, progeny size, development time, longevity, and utilization of preferred and alternate hosts (Hille Ris Lambers 1966; Heie 1987). Thus, a successful phenotype depends on a particular set of conditions, such as flight ability, tolerance to nutrient limitation and temperature, ability to locate suitable hosts, and potential fecundity (Moran 1992; Hales et al. 1997).

During spring the fundatrices develop, showing attributes that confer high fertility under favorable conditions. The summer generations are represented by females that reproduce parthenogenetically and may include one to several morphs with different attributes. In cycles with alternating hosts, migrant morphs fly from a primary tree host to a secondary herbaceous host. It generally occurs in response to overcrowding and/or deterioration of the nutritional status of the primary arboreous host plant. The large and fertile winged migrant females deposit their well-developed embryos on herbaceous plants in active growth (Dixon 1976). In autumn, low temperatures and short days stimulate the production of sexual morphs represented by a male and an oviparous female, which exhibits morphological and behavioral attributes associated with oogenesis, mating, and oviposition (Hille Ris Lambers 1966; Blackman and Eastop 1984, 1994; Miyazaki 1987; Moran 1992). On their primary host, the sexual morphs mate and produce the winter eggs. In regions where high temperatures and long days are prevalent throughout the year, many species remain in the same host, reproducing solely by parthenogenesis throughout the year (Blackman and Eastop 1984).

Several hypotheses have been proposed to explain host alternation in several aphid species. One proposes a selective advantage of the host alternating cycle based on the assumption of nutritional complementarity when switching between primary and secondary hosts (Dixon 1971). Several evidences indicate that woody plants are highly nutritious in the spring, when new leaves start to be produced, and less nutritious in the summer—explaining the migration to secondary herbaceous hosts as an alternative response to supplementation of nutrients (Dixon 1985a). The other hypothesis adopts a historical perspective, according to which, natural selection and adaptation are the forces involved in the origin of host alternating cycles (Moran 1988, 1990).

Continuous parthenogenesis is the most common form of aphid reproduction in the tropics and subtropics, different from what happens in temperate regions. The number of aphid species is inversely proportional to the number of plant species in different parts of the world. According to Eastop (1973), in temperate regions there are more species of aphids per thousand species of plants than in the tropics and subtropics. Dixon et al. (1987) demonstrated that most aphid species are host specific and that the host location is a random activity. Thus, where diversity of vegetation is high, few plant species are abundant enough to sustain a species of aphid. In this context, there are proportionally more polyphagous aphid species in the tropics than in temperate regions (Holman 1971; Eastop 1973, 1978; Heie 1994). Therefore, host specificity, random host location, and the little time aphids survive without food (Dixon 1985b) have limited the aphid species to commonly occurring plants, especially in temperate regions, where plant diversity is low (Dixon 1985b, 1987b).

20.3 Structure of Mouthparts and Digestive Tract

20.3.1 Structure and Function of Mouthparts

The sucking mouthparts of aphids are specialized for piercing and penetrating the plant tissue and sucking sap from the phloem sieve elements. Long and flexible chitinous stylets, formed by a pair of mandibles and a pair of maxillae, are lodged in a groove of a segmented labium, which forms a protective sheath. The labrum forms a short piece that holds the stylets lodged in the labium while they are at rest (Figure 20.1).

The average diameter of the stylet set along its length is 3.5 nm, narrowing abruptly to 2.5 nm at the apex (Miles 1987). Stylet length varies according to species, instar and insect morph, and they are not

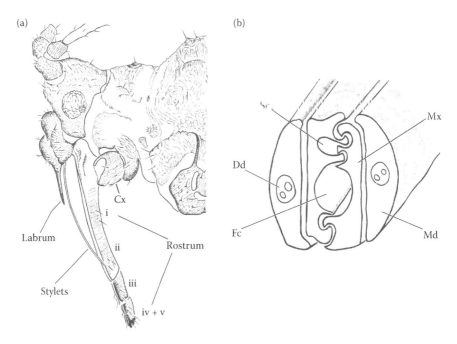

FIGURE 20.1 (a) Aphid rostrum forming a sheath that contains stylets: i to v, rostral segments; Cx, coxa. (b) Schematic cross section of stylets: Fc, food canal; Sc, salivary canal; Dd, dendrites; Md, mandible; Mx, maxilla. (Adapted from Miyazaki. In *Aphids: Their Biology, Natural Enemies and Control*, vol. 2A, ed. A. K. Minks and P. Harrewijn, 367–91, Copyright 1987, with permission from Elsevier.)

necessarily shorter in the immatures. The stylet length limits the access to certain parts or species/varieties of host plants, selecting even the caliber of the veins. Species that feed on trees branches and trunks tend to have longer stylets to reach a deeper phloem (Miles 1987; Risebrow and Dixon 1987). Nymphs have disproportionately long stylets relative to body length, and usually feed in the same site as adults (Klingauf 1987). The maxillae are the inner stylet pair, which have grooves that fit together to form one piece with two fine channels, one representing the food canal, used to ingest the sap, and the other, the salivary canal to inject saliva (Klingauf 1987). The two mandibles around the maxillae have the apex serrated to facilitate penetration and anchoring of the stylets in the plant tissue. The mandibles have an internal canal through which nerve fibers run, the function of which is not fully elucidated.

The penetration of stylets in plant tissues occurs by a protraction and retraction mechanism. The two pairs of stylets are withdrawn from the labium, which bends without penetrating the tissue; the mandibles move alternately, while the maxillae slide slightly ahead until they reach the phloem. The labium bears eight pairs of mechanoreceptors at the apex to sense the position of mouthparts. The stylet withdrawal can be quite fast, which is especially important in response to the presence of predators or parasitoids, to seek new feeding sites or during molting. Stylets can be inserted directly into the epidermis cells, intercellularly or through a stoma; aphids have control over stylet penetration, searching for less resistant path, through the middle lamella or the apoplasmic compartment, between adjacent cells, relying on salivary enzymes (Miles 1987).

20.3.2 Structure of Digestive Tract

Detailed descriptions of the structure, histology, innervation, and musculature of the digestive tract of Aphidoidea are presented by Ponsen (1987), summarized below. The digestive system of aphids is represented by an elongated tube, which is about three times the body length, with distinct compartments

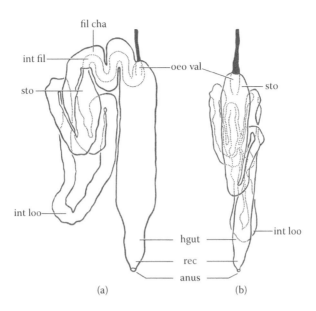

FIGURE 20.2 Semischematic representation of digestive tract of *Cryptomyzus ribis* (a) and *Myzus persicae* (b). anus, anus; fil cha, filter chamber; hgut, hindgut; int fil, intestinal filter; int loo, intestinal loop; oeo val, esophageal valve; rec, rectum; sto, stomach. (Redrawn from Ponsen, M. B., In *Aphids: Their Biology, Natural Enemies and Control*, vol. 2A, ed. A. K. Minks and P. Harrewijn, 79–97, Copyright 1987, with permission from Elsevier.)

and defined groups of specialized cells. The digestive tract begins in the thin alimentary canal formed between the maxillary stylets; continues in the pharynx, forming a pharyngeal pump; passes to the foregut, formed by an elongated esophagus; then to the midgut, divided into a dilated stomach and a tubular intestine; and finally passes into the hindgut, represented by the rectus, and ends at the anus (Figure 20.2). In some groups of Aphidoidea (Drepanosiphinae, Lachininae, and some Aphidinae), the digestive system has a filter chamber, encapsulating part of the midgut, so that the foregut connects almost directly to the hindgut. Aphids do not have Malpighian tubules, and the excretory function is performed partially or totally by the salivary glands.

The midgut is lined with an epithelium formed by striated border microvillous cells with deep folds. It is the longest portion of the digestive tract, consisting of either a dilated or tubular stomach, coiled or not, and a descendant intestine. The hindgut is a sac-like structure lined by inner circular and outer longitudinal muscles and has a distinct rectus. The anus is situated ventrally under the cauda, except in Adelgidae, in which it opens dorsally to the cauda and in Phylloxeridae, which lack an anal opening.

In general, it seems that all aphid species that have filter system also have the ectodermal hindgut, but there are some genera that have the ectodermal hindgut but no filter chamber. In some aphid species, there is a concentric filter system, in which the tubular anterior midgut is encapsulated by the anterior ectodermal hindgut chamber or filter. The tubular region or intestinal filter may be either curved, straight, slightly dilated, coiled, or has the appearance of an inverted V. The filter chamber is coated with an inner layer of endodermal epithelium and the outer layer is ectodermic.

20.4 Nutritional Physiology

The nutritional physiology of aphids includes the host plant location and acceptance, the mechanisms for phloem sap intake, its chemical composition, nutritional requirements, physiological and chemical processes to convert food into energy, and the role of nutrition in metabolic functions for growth, reproduction, and polymorphism (Srivastava 1987). However, to access the phloem sieve elements, aphids have to overcome plant defenses such as sieve tube occlusion, and phytohormone-signaling pathways to express anti-insect molecules. On the other hand, aphids affect plant primary metabolism, which could be a strategy to improve phloem sap composition to produce nutrients required for growth (Giordanengo et al. 2010). Symbiotic microorganisms may play an important role in nutritional physiology of aphids.

20.4.1 Host Location and Acceptance

Host plant selection by aphids is achieved by a sequence of clues and responses, including landscape, plant architecture and volatiles, and phloem sieve element properties. Initial visual clues are followed by olfactory and gustative inputs after plant contact. Plant penetration includes the intercellular pathway phase; the xylem phase (possibly for drinking water) and the phloem phase or effective feeding, preceded by sieve element salivation, presumably to suppress phloem wound responses. During feeding, the composition of the phloem is continuously monitored by the insect. Changes in phloem sap can be caused by plant ageing, daily and with seasonal changes, besides other conspecific and nonspecific interactions (Tjallingii 2006; Pettersson et al. 2007).

Aphids are poor flyers, but they can remain airborne for many hours and be transported through considerable distances by air movements (Pettersson et al. 2007). Visual and tactile surface characteristics of the plant can serve as reference for host recognition. During flight, before landing, the aphids seem to be able to make only a rough selection of the host, probably attracted by light waves more or less characteristic of the host plant or its physiological state. In the tropics, many aphid species, most of which are exotic, are polyphagous because of the difficulty of finding their preferred host in areas of high floristic diversity (Holman 1971; Eastop 1973).

There is great variation in plant structure, as well as in size and location of the phloem elements and organs in different plant species. Plant morphology may impose mechanical obstacles, such as hairs and wax that affect aphid movements. Despite and because of their small size (1–5 mm) and other adaptations, aphids can exploit phloem. If aphids were larger in size, their feeding rate would exceed the limit for dealing with plant response to feeding damage and to seal the phloem elements in the process (Dixon 1987b). Aphids are adapted to feed on leaves, branches, or trunks; the small species or immature morphs usually feed in the fine veins and large aphids in large veins (Dixon 1985a). The optimal plant part for settling combines quantity and quality of food, protection from natural enemies, and adequate microclimatic conditions. Aphids often show positive geotaxis and negative phototaxis after landing, and prefer the lower (abaxial) surface of the leaf (Pettersson et al. 2007).

Aphid feeding activity typically begins as soon as the insect lands on the plant. It begins probing the epidermis superficially, not penetrating the stylets beyond the epidermal cells. Depending on the sensory input, the aphid chooses to stop its feeding attempts, continue test probing, or probe deeper to start feeding (Miles 1987). Before inserting the stylets in the tissues, it secretes a droplet of saliva on the plant surface, apparently to test some of the material, using mechanoreceptors in the apex of the labium (Srivastava 1987). The nature of the sensory inputs received during probing is still unclear. The sensory hairs at the apex of the labium and the dendrites in the mandibles seem to be mechanoreceptors connected with the motor activities of the stylets. There is no evidence that they are chemoreceptors. Rather, the chemoreception function is assigned to the epipharynx gustatory organ that probes the ingested food from the feeding sites (Auclair 1963; Miles 1987).

To locate the phloematic vessels, aphids perceive the pH, osmotic gradients, or follow the cell walls with their stylets to reach the phloem elements (Figure 20.3 shows a schematic representation of the host selection process and stimuli involved). Aphids possess an array of olfactory organs on the antennae. Electrophysiological methods show that aphids respond to a variety of plant volatiles (Pickett et al. 1992). The major olfactory organs present in adult and nymphs are the primary rhinaria on the two last antennal segments (van Emden 1972).

The penetration and location of the phloem is a relatively slow process on herbaceous plants, from 30 min for *Aphis fabae* Scopoli to 60 min for *Myzus persicae* (Sulzer) (Pollard 1973). Penetration may be even longer for aphids on trees [12 h for *Cinara atlantica* (Wilson) on *Pinus taeda* (Pinaceae)] (Penteado 2007). Some amino acids and sucrose at certain concentration and/or combinations of these compounds

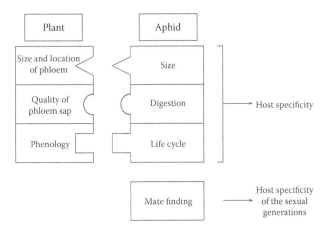

FIGURE 20.3 Scheme of plant and aphid characteristics that led to evolution of host specificity. (Redrawn from Dixon, A. F. G., In *Aphids: Their Biology, Natural Enemies and Control*, vol. 2A, ed. A. K. Minks and P. Harrewijn, 197–207, Copyright 1987, with permission from Elsevier.)

have phagostimulant action, whereas others are deterrents (Srivastava et al. 1983). The differential resistance of certain plant varieties to aphid feeding can be attributed to nutritional differences in the phloem sap. The susceptibility of cultivated tomato to the potato aphid *Macrosiphum euphorbiae* (Thomas) is associated with high sucrose concentration, free amino acids, and also high alanine and tyrosine content, when compared to a wild tomato variety (Risebrow and Dixon 1987).

The phloem sap feeders are affected not only by the primary nutrients but also by secondary compounds, which affect their feeding behavior, development, and survival. Plant-specific aphid species respond to qualitative and quantitative chemical signals, and to secondary compounds, selecting the most suitable host for feeding (Risebrow and Dixon 1987); these authors mentioned that the absence of certain secondary compounds hinders the establishment of monophagous aphids on artificial diets containing only primary nutrients.

Brevicoryne brassicae (L.) feeds readily on artificial diets and host plants containing the compound sinigrin, which is a mustard oil glycoside (alilisotiocianato), characteristic of the Brassicaceae hosting the cabbage aphid. On the other hand, *M. persicae* does not respond readily to this secondary compound, despite the performance of both species being positively correlated with nitrogen concentration and negatively correlated with some amino acids. It was observed that some amino acids in Brassicaceae are more suitable for the development of *B. brassicae* than for *M. persicae*, and that the content of these compounds directly correlates with that of sinigrin; thus, *B. brassicae* responds positively to mustard oil. *M. persicae*, however, responds negatively to sinigrin because it is not associated with the concentration of the group of amino acids more favorable for its development (van Emden 1978). Thus, when *M. persicae* feeds on Brassicaceae, it colonizes older leaves, where there is low concentration of sinigrin but acceptable levels of nutrients, whereas the colonies of *B. brassicae* are formed on young shoots, nutritionally rich and with high concentration of mustard oil glycosides. On the basis of these studies, van Emden (1978) concluded that aphid nutrition is less a matter of potential nutritional value and more a matter of the amount of nutrients that is effectively ingested per unit time. It is likely that the high capacity for population increase of *B. brassicae* on crucifers is due to its ability to detoxify the mustard oil glycosides, using glucosinolases, which have been detected in specialized aphid species colonizing Brassicaceae (MacGibbon and Beuzenberg 1978).

The few truly polyphagous species select the host according to their high nutritional quality in combination with low toxicity, thus reducing the amount of toxic compounds ingested and the cost to detoxify toxins. However, despite being a highly polyphagous species, some biotypes of *M. persicae* are able to select genotypes or varieties of plants with less active resistance factors, where they grow forming large populations (Weber 1982).

20.4.2 Feeding Strategies and Host Plant Condition

Many aphid species normally position themselves with the head down on vertical branches or leaves. When the colonies are large, the individuals may overlap one another. Some are positioned with the head toward the apex of the leaf or needle. It is evident that it is not only a matter of gravity, but also that intrinsic factors of the insect and characteristics of host plants affect the feeding positioning (Klingauf 1987). To cope with adverse microclimatic conditions, aphids settle near the large veins on the underside of leaves. This helps mitigate the action of wind, rain, and intense solar radiation (Dixon and McKay 1970). Many aphid species prefer to feed on the abaxial phloem of leaves, while others choose the adaxial and more internal phloematic elements, as is the case of *Aphis nerii* Boyer Fonscolombe on species of Asclepiadaceae. This is probably because different phloematic vessels translocate different nutritional and secondary compounds, or different concentrations of the same. Most prefer either new shoots or senescent parts; there is also greater preference for the leaves than the stem (Srivastava 1987).

Plant suitability varies according to factors that favor or hinder insect development, including circadian change in metabolism, plant growth, wax and secondary compound production, nutritional status, and resistance mechanisms (Risebrow and Dixon 1987). Aphids adapt to characteristics of food plants, and when transferred to a new host they need time to readjust. Acceptance will only occur after repeated exploration of the new host, with more individuals settling on the plant (Klingauf 1987). Auclair (1959) found that aphid species, feeding on nutritionally deficient resistant hosts, are able to assimilate much of the ingested sap. However, if this condition persists throughout the plant growth stages, it wanders and seeks new feeding sites, moving either to new shoots or senescent leaves.

Environmental and host plant conditions may determine the extent of population growth of different aphid genotypes, leading to a dramatic increase, reduction, or even extinction (Blackman 1985). In temperate zones, sexual reproduction in the fall produces a diversity of genotypes, which spend the winter as eggs. In the spring and summer some of these genotypes are eliminated, while others establish and form large clones of genetically identical individuals. Some of these survive until the fall, contributing with their gametes to form some generations in the following year (Figure 20.4).

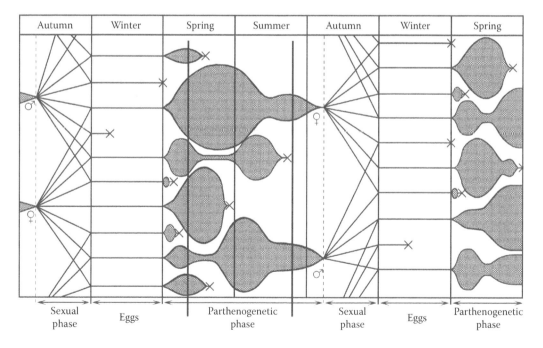

FIGURE 20.4 Diagram of alternation of parthenogenetic and sexual phases in life cycle of an aphid over 2 years. Width of each shaded area represents number of individuals sharing similar genotype—see text for explanation. (Redrawn from Blackman, R. L., In *Proceedings of the International Aphidological Symposium at Jablonna*, pp. 171–237. Warsaw: Polska Akademia Nauk, 1985. With permission.)

When the whole plant becomes unfavorable for aphid growth and reproduction, these phloematic feeders can adopt various strategies. Dixon (1963, 1975b,c) noticed that females of *Drepanosiphum plat-anoides* (Schrank) under high population density combined with low nutritional host quality spend less time feeding; are smaller in size; and have poorly developed gonads, small wings, and more adipose tissue. However, they do not fly readily, but enter reproductive diapause because other sycamore plants are under the same conditions during the summer. When favorable conditions return in the fall, many fly before reproducing. On the new host, they produce a progeny that will increase in size and weight and possess a higher reproductive rate, resulting in high population peaks during this period. Thus, the effect of population density results in changes in the potential rate of increase, obviously influenced by decreasing host qualitative condition induced by high aphid infestation.

Approximately 10% of aphid species exhibit host alternation, from trees to herbaceous plants, to avoid the low quality of host trees during summer (Mordvilko 1928; Shaposhnikov 1959), and to avoid increasing natural enemy populations on trees (Way and Banks 1968). This relatively rare condition of switching hosts are due to (a) difficulty of species to adapt to deep changes in phenology, quality, and architecture of a tree species versus herbaceous plant; (b) competition with aphids that are host specific and remain in one of the hosts; and (c) low availability of plants to sustain populations of a given species (Risebrow and Dixon 1987).

Aphids that do not alternate hosts (monoecia) exhibit less pronounced polymorphism than the heter-oecia that alternate hosts, but still the former produce alate migrants capable of colonizing other plants under unfavorable environmental conditions (Risebrow and Dixon 1987). Alate morphs are produced in response to overcrowding and adverse changes in host quality. Responses to overcrowding are under endocrine control and can act either before birth (on the embryos inside the mother) or after birth, pro-ducing winged morphs (Lees 1979). The winged females seek to colonize new hosts, possibly with better nutritional quality, where they produce a new generation of apterous morphs, for prompt and efficient exploitation of the host. Despite the lower reproductive rate, winged morphs maintain a high intrinsic growth rate for producing their offspring in early adulthood (Dixon and Wratten 1971).

Even in small groups of individuals, more gregarious aphid species perform better than isolated ones. The former has increased intrinsic growth rate, produce adults with higher weight, fecundity, and num-ber of embryos. This condition results from a change in plant metabolism caused by the injection of substances in the saliva; since they are aggregated, they act as a nutrient sink, exploring few sites to minimize the mechanical damage to the plant and its response. Even though the population growth rate decreases under food and space competition, there will be production of winged morphs that disperse before the total collapse of the population.

20.4.3 Phloem Feeding

In contrast to other species of Aphidoidea, species in the family Aphididae feed solely on phloem sap. Even though they probe other tissues, they reach and sustain suction in the phloem (van Emden 1969, Pollard 1973). The production and composition of the honeydew excreted by aphids indicates that the phloem sap is their primary food source; however, some authors discuss the fact that all Aphididae species have been considered phloem feeders. According to Lowe (1967), *A. fabae* usually feeds in the phloem of major veins of the leaves of *Vicia faba* (Fabaceae); however, the first instars of *M. persicae* and of *Myzus ornatus* Laing feed in the mesophyll of leaves. Saxena and Chada (1971), studying the trajectory of the stylet of two biotypes of *Schizaphis graminum* (Rondani), observed that biotype B, which is capable of feeding on the sorghum cultivar resistant to biotype A, made more penetrations in the intracellular mesophyll, causing more injuries than biotype A, that fed primarily in the phloem. The damage caused by biotype B included degeneration of protoplasts by the action of the toxicogenic saliva. Campbell et al. (1982) reinforce the fact that different *S. graminum* biotypes usually feed on the meso-phyll in more resistant cultivars of sorghum.

Although phloem sap is rich in carbohydrates and low in amino acids, the advantage of feeding in the phloem is that nutrients are in a soluble and readily assimilable and renewable form (Risebrow and Dixon 1987). Thus, even if there is more nitrogen in the leaves, sap ingestion by phloem feeders may be faster than the intake made by chewing insects (Mittler 1957). Moreover, the assimilation–consumption and production–assimilation relationships tend to be higher in sap-sucking insects (Llewellyn 1982). As a result, large amounts of soluble carbohydrates, especially sucrose, are ingested by aphids, although much of the excess is excreted in honeydew.

Aphids have a specific size in response to the structure and physiology of the phloem sieve elements, as well as complex life cycles in response to seasonal phenology of host plants. They also have intimate association with symbionts and a specialized digestive physiology for the chemistry of phloem sap (Figure 20.5). To optimize their life cycles, aphids tend to have a high degree of host specificity. The colonization of one or a few primary hosts is important for locating reproductive partners, an activity that would be hampered in holocyclic polyphagous species (Dixon 1987a). Monophagy represents a successful way of life for aphids, but it is important to consider that some truly polyphagous species, such as *A. fabae* and *M. persicae*, are examples of success. Polyphagy in aphids is usually restricted to the summer generations that alternate to herbaceous host plants. Herbaceous plants undergo constant changes in their development and abundance depending on the climate and region, so that monophagous aphids that exploit them may face lack of hosts and/or poor quality in certain situations, which can lead to population extinction. On the other hand, polyphagous species are able to exploit several plant species simultaneously, allowing them to choose the most nutritious parts and avoid certain parts or stages that contain high concentrations of secondary compounds. By doing so, they also avoid competition with monophagous species that eventually exploit the same resource, as discussed by Dixon (1987a).

20.4.4 Food Intake Mechanisms and Saliva Composition

The ingestion of phloem sap by aphids occurs by capillarity. Sap surface tension is reduced by the properties of the saliva, by the turgor pressure of the plant solution, and by active suction with the pharyngeal

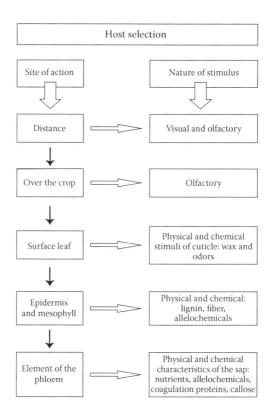

FIGURE 20.5 Host selection process of aphids. (Redrawn from Prado, F. 1997. Aphid–plant interactions at phloem level, a behavioural study. PhD thesis, Wageningen Agricultural University, 1997.)

pump. It is accepted that the normal phloem sap pressure is probably sufficient for food intake, but aphids exert control over suction and can stop it for the molting process and dispersal in the presence of natural enemies. However, aphids can feed in the absence of pressure, as occurs with artificial diets using parafilm membranes and with wilted plants, although reduced acceptability and development occur under these conditions (Klingauf 1987).

In addition to solutes, aphids ingest small particles (up to 1 μm in diameter; Miles et al. 1964), but if there is blockage of the alimentary canal, they retrieve the stylets and regurgitate since the pharyngeal pump can direct the flow in both directions. This regurgitation behavior has implications for the transmission of viruses (Harris and Bath 1973).

Aphid aggregation can affect food and nutrition because of the sink feeding effect, used to obtain a continuous and steady sap flow, without extending plant damage. For example, for *A. fabae*, reproduction increases in colonies with an average of eight individuals, as compared to smaller (2–4 individuals) or larger colonies (16–32 individuals), because the sink effect increases with the colony size up to a certain limit, after which intraspecific competition occurs, inhibiting growth and reproduction (Way 1967).

Two types of saliva are secreted by the salivary glands (Figure 20.6) and injected into the plant: watery saliva containing pectinases and cellulases (to break the cell walls), and gelatinous saliva, which forms a tubular sheath around the stylets (Miles et al. 1964). In addition to providing stiffness and protection for the stylets, the gelatinous saliva serves to minimize damage to cells in the stylet's pathway and to seal cell punctures, minimizing plant response to cell disruption. The presence of phenolases in the saliva of these insects may be involved in the detoxification mechanism of plant defense compounds (Miles 1969). Although the components of the saliva are primarily related to food and nutrition, the presence of amino acids and phytohormones (substances that regulate plant growth) in the saliva may be involved in the formation of plant galls and malformations (Klingauf 1987).

The Sternorrhyncha species secrete watery saliva during stylet penetration in plant tissues. The saliva is slightly alkaline (pH 8–9) (Miles 1972) and, in addition to enzymes, consists of several amino acids and amides (Miles 1987). Besides amino acids, other compounds were detected, such as phenolic compounds and indole-3 acetic acid, which is responsible for plant growth. Polyphenol oxidases and peroxidases have also been detected in the saliva of Sternorrhyncha and/or phytophagous Heteroptera (Miles 1964). A possible function of phenolases and other oxidizing enzyme systems in the saliva may be

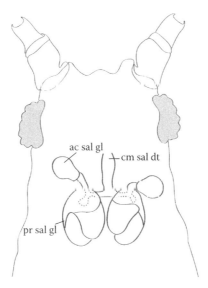

FIGURE 20.6 Salivary glands of *Myzus persicae*: ac sal gl, accessory salivary gland; cm sal dt, common salivary duct; pr sal gl, principal salivary gland. (Adapted from Ponsen, M. B., In *Aphids: Their Biology, Natural Enemies and Control*, vol. 2A, ed. A. K. Minks and P. Harrewijn, 79–97, Copyright 1987, with permission from Elsevier.)

the detoxification of deterrents or toxic substances in the plant. The polyphenoloxidases also have the function of promoting chemical bonds in the formation of gelatinous saliva (Miles 1965).

The presence of hydrolyzing enzymes in the watery saliva of Aphidoidea can be detected by tests in which insects probe a medium containing a specific substrate that causes hydrolysis. The presence of amylase (Staniland 1924), pectinases for intercellular penetration of stylets (Adams and McAllan 1958), and cellulases to penetrate cell walls, as well as the hydrolysis of carboxymethyl cellulose into oligosaccharides and glucose (Adams and McAllan 1956) were demonstrated. Some enzymes have been identified from crushing salivary glands, but in this case it is difficult to determine whether these are effectively secreted enzymes or endoenzymes. Adams and McAllan (1958) also determined that the saliva of many Aphidoidea contains polygalacturonase, which hydrolyzes pectin and determines the ability of the insect to achieve intercellular penetration.

The presence of cellulases in aphid saliva facilitates stylet insertion through the cell wall; however, neither cellulase nor pectinase seems to be essential to cell wall penetration. Many other hydrolyzing enzymes have not been detected in the saliva because substances can be processed in the gut (Miles 1987). For Aphidoidea species that feed on mesophyll and cortical tissue, there is no information on hydrolyzing enzymes in saliva. Anders (1961), cited by Miles (1987), refers to the presence of proteases in the saliva of *Viteus vitifoliae* (Fitch) [= *Daktulosphaira vitifoliae* (Fitch)] (Phylloxeridae). In addition to proteolytic enzymes, other compounds such as RNAase, DNAase, amylase, tryptophan, and even plant growth substances may occur (Zotov 1976, cited by Miles 1987). *V. vitifoliae* is capable of removing plant substances such as starch grains that accumulated in galls on leaves and roots of grape vines; however, there is no confirmation of saliva breakdown of plant reserves. There is evidence that saliva constituents are unused products from the diet, which are absorbed into the hemolymph and passed to the salivary glands and then excreted. The excess of water and amino acids absorbed from the diet and radioactive dye metabolites injected into the hemolymph recovered from the saliva indicate the excretory function of salivary glands in *V. vitifoliae* that lacks anus (Anders 1958, cited by Miles 1987).

The gelatinous sheath saliva formed around the stylet is secreted by secondary salivary glands and consists of lipoproteins and phospholipids, with pH around 6.0 (Miles 1965). The secretion gelatinizes immediately after being eliminated from the salivary canal, probably by the oxidation of sulfidric groups to form disulfide and hydrogen bonds. Some secretion is eliminated before the insertion of the stylets and continues to form a sheath as they penetrate into the tissues. The gelatinous saliva seals the puncture and damage caused to cells during penetration of the stylets, which trigger the release of proteinase inhibitors and phytoalexins.

Aphids are able to successfully puncture sieve elements and ingest phloem sap without eliciting normal calcium-triggered occlusion as the plant's response to injury. The watery saliva injected in the sieve elements immediately before sap ingestion and can sabotage plant defenses. It has been demonstrated that the saliva possesses anti-occlusion properties, provided by its effect on forisomes (Torsten et al. 2007). Forisomes are proteinaceous inclusions in sieve tubes of legumes that show calcium-regulated changes in conformation between a contracted state, which does not occlude the sieve tubes, and a dispersed state that occludes the sieve tubes. The authors demonstrated *in vitro* that aphid saliva induces dispersed forisomes to revert back to the contracted state because of molecular interactions between salivary proteins and calcium.

20.4.5 Food Constituents and Digestive Enzymes

The natural diet of aphids—the phloem sap—although apparently poor, is a fluid rich in amino acids and carbohydrates, and generally contains all other essential nutrients such as vitamins, sterols, minerals, and water needed for growth and normal reproduction. As soon as they locate and begin feeding on an appropriate site, aphids have access to a continuous source of nutrients with a high osmotic pressure (15–30 atm, according to Dixon 1975a), when the infestation is not too high and the plant is not wilted. The phloem sap contains a high and variable concentration of solutes, especially sucrose, which has an osmolality greater than that of the aphid hemolymph. Although lacking Malpighian tubules and possessing filter chamber, aphids do not dehydrate and do not absorb excessive amounts of sucrose (Ponsen 1979). Since the osmotic pressure of honeydew is comparable to that of the hemolymph, aphids are able to reduce the osmotic pressure of the sap, thus reducing osmotic water loss.

The quality of the phloem depends on water availability, temperature, photoperiod, age, and plant phenology. Miles et al. (1982) mention that the concentration of the phloem sap may change along the day, being lower at night; under water deficiency, free amino acids, proline, and carbohydrates increase in leaves and phloem sap.

Not all plants or plant parts are suitable for phloem feeders. The concentration of soluble nitrogen, in the form of free amino acids, can vary with plant species, variety, part, and other factors such as plant nutrition and soil conditions (Risebrow and Dixon 1987). Thus, aphids and other Sternorryncha tend to feed preferentially on young shoots, and occasionally on senescent leaves, which possess greater concentration of free amino acids. Aphids colonizing tree leaves switch to herbaceous plants during summer because trees have mature leaves during this period, while annual plants have variable age and physiological state suitable for feeding (Risebrow and Dixon 1987).

The nitrogen-to-carbon ratio may limit population development of phytophagous insects, both by quantity and quality of the compounds ingested. The most important nitrogen compounds are the nine essential amino acids that animals cannot synthesize: histidine, isoleucine, leucine, lysine, methionine, phenylalanine, threonine, tryptophan, and valine (Morris 1991). If the concentration of one of these amino acids is inadequate, development is hampered. Despite the variable amino acid composition in phloem sap, nitrogen quality is generally low; that is, the concentrations of essential amino acids are very low (Douglas 1993).

Mittler (1958a) evaluated the composition of sap ingested by *Tuberolachnus salignus* (Gmelin) feeding on *Salix acutifolia* (Salicicaceae), analyzing samples of sieve retained in the food canal within the stylets. The author found that the amount of protein or peptides, ammonia or uric acid was minimal. Nevertheless 12 amino acids were recovered, with concentrations varying from 0.03% to 0.2%. The amount varied with season and plant stage, being higher in the budding stage and the lowest at maturity, with intermediate values in young and senescent leaves. The free amino acid composition of the honeydew excreted by insects feeding on the branch was exactly the same as in the sap collected from the same branch.

Auclair (1963) detected 17 amino acids and amides, but no protein from the sap in the stylet of aphids feeding on different species of herbaceous plants. Barlow and Randolph (1978) recovered 18 free amino acids, nine in the form of proteins, from the sap in the stylets of *Acyrthosiphon pisum* (Harris), feeding on *Pisum sativum* (Fabaceae). Total amino acids made up 4.51% of the sap of initial growing plants, of which 98.9% were free amino acids (i.e., 4.46% of the total sap). Srivastava (1987) lists free and bound amino acids in the phloem sap of *P. sativum* var. Alaska, measured over a period of 5 days (Table 20.1).

Sucrose was the only sugar detected in the sap in the stylets of *T. salignus* feeding on *S. acutifolia*, in concentrations ranging from 5% to 10% (Mittler 1958a). However, other sugars have been recovered in low concentrations in sap from the stylets and phloem. There are reports of very high concentrations of sucrose (20–30%), along with 0.5–2% raffinose and stachyose, and traces of myo-inositol in the phloem exudates of *S. acutifolia* (Zimmermann and Ziegler 1975). On the basis of phloem exudates, Srivastava (1987) separated plants into three groups: (a) plants with sucrose as the predominant sugar, such as Fabaceae, occasionally with traces of raffinose-type oligosaccharides; (b) plants with similar quantities of raffinose-type oligosaccharides and sucrose, such as Myrtaceae and Tiliaceae; and (c) plants with considerable amounts of sugar alcohols, such as D-mannitol, sorbitol, and dulcitol, besides saccharose and raffinose-type oligosaccharides.

Phytosterols and cholesterol were detected in the sap of Brassicaceae and in the honeydew of *M. persicae* feeding on seedlings, indicating that these compounds are translocated by the phloem and ingested by the insect. Cholesterol and sitosterol are also translocated in the phloem of sorghum and ingested by *S. graminum* (Campbell and Nes 1982).

Water-soluble vitamins such as thiamin, niacin, pantothenic acid, and pyridoxine have been sampled in the sap of some trees, in addition to high concentrations of ascorbic acid and myo-inositol. Organic acids such as citric, tartaric, malic, fumaric, succinic, and pyruvic acids have been recovered from the sap in the stylets of aphids and/or in the phloem exudates of some plants, in minute quantities. Growth substances such as indolacetic acid, giberilins, auxins, and cytokinins have been detected in the phloem sap of many plants (Ziegler 1975). Nucleic acids and high concentrations of ATP were recovered, respectively, from the phloem sap and stylets of aphids on *Salix* (Gardner and Peel 1972). Hussain et al. (1974)

TABLE 20.1

Free and Bound Amino Acids in the Phloem Sap of *Pisum sativum* (L.) var. Alaska, Measured for 5 Days

	Free		Bound	
Amino Acids	Average nmol/mg Fresh Weight of Sap	Average % Fresh Weight of Sap	nmol/mg Fresh Weight of Sap	% Fresh Weight of Sap
Arginine	19.86	0.35	—	—
Aspartic acid	7.38	0.098	0.70	0.0093
Cysteine	0.13	0.0029	—	—
Glutamic acid	80.44	1.18	0.29	0.0034
Glycine	1.84	0.014	1.00	0.0075
Histidine	6.86	0.11	—	—
Isoleucine	8.82	0.12	0.06	0.00079
Leucine	6.10	0.080	—	—
Lysine	17.22	0.25	0.06	0.0088
Methionine	1.60	0.024	—	—
Methionine sulfone	5.48	0.099	—	—
Phenylalanine	12.70	0.21	0.19	0.0031
Proline	19.72	0.23	—	—
Serine	35.52	0.37	0.75	0.0079
Threonine	79.28	0.94	0.23	0.0027
Tyrosine	1.16	0.021	—	—
Valine	21.50	0.25	0.61	0.0072
Total% fresh weight of sap	—	4.46	—	0.051

Source: Data from Srivastava, P. N., In *Aphids: Their Biology, Natural Enemies and Control*, vol. 2A, ed. A. K. Minks and P. Harrewijn, 99–121, Copyright 1987, with permission from Elsevier.

demonstrated that phenolic compounds are translocated in the phloem: 15 substances were recovered from the honeydew of *M. persicae* and five from the sap of seedlings of radish, where they fed; the remaining substances resulted from the break down of the ingested compounds. Cations, inorganic anions, and heavy metals have been found in phloem exudates of several plants, and all of them can be translocated in the phloem, except for Ca ions, which are generally not translocated in the phloem but are present in the honeydew of many aphids (Dixon 1975a).

Because nutrients in the phloem sap have simple structure, enzymes of the digestive system of phloem feeders are basically polysaccharidases, invertase, and some proteases and/or peptidases. Since most aphids feed continuously, many nutrients are consumed in excess and are not absorbed; rather they are eventually eliminated as feces and/or honeydew, including amino acids and sugars that are not completely digested. The honeydew of *T. salignus* can contain all the amino acids present in the phloem sap of *Salix* sp., but always in lower concentration than in the plant sap (Mittler 1953).

Aphids feed primarily on phloem sap; thus, the only carbohydrate ingested in high concentration is sucrose, which is also present in the honeydew, indicating that this sugar is not completely hydrolyzed by α-glucosidases present in the digestive tract. Srivastava (1987) suggests that the partial hydrolysis of sucrose is due to the intestinal transit being too fast for the enzyme to complete its catalytic activity, or because pH is not optimal for a more effective action of this enzyme. Thus, sucrose present in the honeydew represents the excess of sugar intake that was not hydrolyzed. Glucose and fructose are the products of the digestion of sucrose since they are not ingested and their excess is excreted in the honeydew. Srivastava and Auclair (1962b) showed that invertase present in the gut of *A. pisum* hydrolyze sucrose, trehalose, melezitose, turanose, and maltose, but not melibiose and raffinose; in the digestive tract of this species, pH, temperature, and sucrose concentration for the optimal action of α-glucosidase is 6.2, 35°C, and 45 mg/ml, respectively.

Honeydew from aphids and many Coccoidea contains many oligosaccharides, such as malto-saccharose, malto-trisaccharose, melezitose, among others, indicating invertase activity in the digestive tract (Auclair 1963). Melezitose, which is excreted in high concentrations (>40%) in the honeydew, is apparently synthesized by the insect. Petelle (1980) suggests that the role of melezitose is to reduce the absorption of sugars by the intestinal wall. Kiss (1981) defends that the synthesis and excretion of this trisaccharide in the honeydew of aphids has the evolutionary function of attracting ants in mutualistic associations between these two organisms.

Peptidases occur in the saliva and digestive tract of aphids, but there are few references to proteases in these insects. Extracts from the stomach of *A. pisum* did not hydrolyze proteins such as casein, edestin, albumin, and hemoglobin, but hydrolyzed a series of peptides, suggesting the presence of peptidases, which were activated by Mn and Co, but not by Zn. The optimum condition for these enzymes is pH 7.0 at 40°C, which is much higher than the optimum temperature for development and reproduction of both the insect and host plant. These peptidases are active at low substrate concentrations (0.05–0.1%), suggesting an enzymatic adaptation for small amounts of peptides in the diet. The weak peptidase activity in freshly excreted honeydew indicates that this enzyme is secreted directly into the intestinal lumen (Srivastava and Auclair 1963). There is evidence that the proteolytic activity follows a circadian rhythm, with a significantly greater activity during the night, as for *A. fabae*, while for others, as *V. vitifoliae*, it is higher during the day (Vereshchagin 1980, cited by Srivastava 1987). Apparently, not all of the enzymatic activity described above is performed directly by aphids; rather it might be accomplished by microorganisms in the digestive tract or micetocytes in the mycetoma (Klingauf 1987).

20.4.6 Nutritional Requirements

In general, the nutritional requirements of aphids are very similar to those of other phytophagous insects. However, studies demonstrate that even strains of a given species may have different nutritional requirements for certain substances (Srivastava 1987). In most cases the nutritional requirements of aphids are demonstrated by withdrawing or adding substances in initially complete diets. Antibiotics can be used to eliminate intracellular symbionts and evaluate several biological performance parameters.

20.4.6.1 Amino Acids

Growth, reproduction, survival, polymorphism, and even the selection of feeding sites are influenced by the concentration of total amino acids in the diet (Srivastava and Auclair 1974). For *A. pisum* and *M. persicae*, the optimal concentration of amino acids in a chemically defined diet is between 2% and 4%. *A. pisum* requires cysteine and 10 essential amino acids (Markkula and Laurema 1967), whereas *Aphis gossypii* Glover also needs a source of tryptophan and phenylalanine (Turner 1971). For *M. persicae*, only histidine, isoleucine, and methionine are essential components in the diet for growth, for two generations (Dadd and Krieger 1968). However, the simultaneous absence of aspartic acid, glutamic acid, asparagine, and glutamine affects growth in this species.

Some amino acids can act in combination with sucrose as phagostimulants. To distinguish whether an amino acid has phagostimulant or nutritional function, one or more amino acids can be removed from a complete diet and the size of the nymphs evaluated. For *A. fabae*, alanine, proline, and serine function primarily as phagostimulants, whereas histidine and methionine were essential for protein synthesis (Leckstein and Llewellyn 1974). Srivastava and Auclair (1975) and Srivastava (1987) classified the functions of amino acids in the diet of *A. pisum*, offering each amino acid separately and with sucrose, and then correlating intake and growth. The nutritional role was played by arginine, aspartic acid, glutamic acid, glutamine, histidine, isoleucine, proline, and tyrosine. Glycine, leucine, methionine, threonine, and valine had both phagostimulant and nutritional functions. On the other hand, γ-aminobutyric acid, alanine, phenylalanine, tryptophan, glycine, homoserine, and leucine were mainly phagostimulants. Some amino acids may have no effect at all or may have a strong inhibitory effect of aphid feeding, acting as plant resistance factors (Srivastava 1987).

The utilization of phloem sap by aphid has been correlated with the presence of symbiotic bacteria of the genus *Buchnera*, involved with partial synthesis of essential amino acids (Douglas 1998). For example, different clones of *M. persicae* have differential nutritional requirement for methionine (Mittler

1971a). Srivastava et al. (1983) found that the amino acids essential for growth and reproduction differed significantly between biotypes of *A. pisum* due to differences in the composition of intracellular symbionts that supply amino acids.

20.4.6.2 Carbohydrates

Sucrose is the main carbohydrate present in the phloem sap and is a special requirement for aphids, besides acting as a potent phagostimulant. *M. persicae* requires 10–20% sucrose in the diet for optimal growth, reproduction, and survival (Mittler 1967) and for *A. pisum* >35% (Srivastava and Auclair 1971a). There is evidence that the requirement of relatively high sugar concentration by *M. persicae* is actually the need for a heavy metal contaminating sucrose (Srivastava and Auclair 1971b). When sucrose is replaced by the monosaccharides glucose and/or fructose, the survival of *A. pisum* and *A. gossypii* is drastically reduced, even in optimal concentrations, possibly because of the lack of phagostimulation provided by sucrose. Mittler et al. (1970) found that *M. persicae* can satisfactorily utilize diets with combinations of sugars, adding 1–2% sucrose, but not cellobiose or lactose because the pentoses act as feeding deterrents.

20.4.6.3 Vitamins

From the 10 water-soluble vitamins, only ascorbic acid, folic acid, niacin, calcium pantothenate, and thiamin are considered essential for growth and reproduction of *A. pisum*. Other vitamins may be relatively beneficial or have no appreciable effect on the performance (Auclair and Boisvert 1980). According to Dadd et al. (1967), *M. persicae* needs to ingest the following: ascorbic acid, calcium pantothenate, choline, folic acid, meso-inositol, pyridoxine, nicotinic acid, riboflavin, and thiamine. In *A. pisum* and *Neomyzus circunflexus* (Buckton), riboflavin has a detrimental effect due to the formation of stable complexes with metals present in the diet (Markkula and Laurema 1967); thus, the lack of this vitamin in the diet increases the performance of this species (Boisvert and Auclair 1981). Ascorbic acid, which forms chelates with minerals, is a requirement for *M. persicae*. It functions in the intake and absorption of minerals and is transmitted to the progeny (Mittler 1976).

20.4.6.4 Lipids, Minerals, and Trace Metals

Although cholesterol is necessary for normal insect development, insects do not have the ability to synthesize it and do not need to obtain it from the diet. Aphids can be reared normally for several generations in holidic diets without addition of cholesterol probably because symbionts are involved in cholesterol synthesis (Griffiths and Beck 1977), despite studies questioning this fact (Campbell and Nes 1983).

Studies with chemically defined diets show the need for K, P, and Mg and other ions and inorganic elements in the diet of *A. pisum* (Auclair 1965). The quantitative requirements of trace metals in the diet are quite restrictive, and concentrations slightly above or below dramatically affect growth. *M. persicae* requires very small quantities of Fe, Zn, Mn, and Cu, as chelants with ethylenediaminetetraacetic acid (EDTA) (Dadd 1967). Aphids grow best when minerals are added as chlorides in the diet rather than with metal complexes such as EDTA (Mittler 1976). Fe seems to be essential for reproduction, whereas Zn, Co, and Ca enhance growth and reduce mortality, and Ca is critical for adult development. Metallic ions, such as Mn and Co, are important in several enzyme systems, which activate peptidase in digestive tract of *A. pisum* (Srivastava and Auclair 1963). In contrast, these ions may block enzymatic reactions. Traces of Ca, Cu, Fe, Mn, and Zn are essential for the maintenance of intracellular symbionts, which degenerate when these ions are lacking, but when they are present in the diet symbionts look similar to those in aphids feeding on host plants (Ehrhardt 1968a, cited by Srivastava 1987).

20.4.7 Feeding Rate and Nutritional Budget

The feeding rates of aphids are affected by age and size, nutritional quality, and secondary compounds of host plant, presence of attending ants, and abiotic factors. Methods to calculate the consumption rate by sap-sucking insects differ from those used for chewing insects (Klingauf 1987). The method of Kennedy and Mittler (1953) consists in cutting the stylets of the insect while it is feeding and determining the rate of sap exuded from the apex of the stylet that remains in plant tissue. The sap exuded by the turgor pressure is supposed to represent the normal rate of aphid feeding. Besides determining the intake, this method allows for qualitative analysis of the phloem sap that is actually ingested. Another method to assess the intake of sap-sucking insects is the use of radioactive elements in the diet for quantitative and qualitative studies, as long as the radioactivity has been defined. An additional method consists in determining the sap ingestion by weight, especially when aphids are kept on an artificial diet with parafilm membrane (Klingauf 1987). He considers the better method to measure feeding rate of aphids is to radiolabeled diet with an isotope that is not absorbed or retained by the insect. Wright et al. (1985) used ^{3}H-inulin to measure the intake of an artificial diet by *M. persicae*; >99% of the substance is not absorbed by the insect, and its analysis in honeydew provides a quantitative measure of food intake. On the basis of the average rate of excretion of marked inulin between the second and last drop of honeydew, the ingestion rate of *M. persicae* was calculated as 24.8 ± 1.5 nl/h in a holidic diet.

The nutritional budget for *M. persicae* on holidic diets, recording body weight gain, amount of excrement, CO_2 production, and water loss through transpiration, yielded a respiratory quotient of 1.0, indicating that this species metabolizes carbohydrates, and nearly no amino acids (Kunkel and Hertel 1975, cited by Klingauf 1987). Apteriform third instar nymph accumulated 11% of ingested substances, lost 9% through transpiration, 5% CO_2-respired, and 73% eliminated through feces. Values for the alatiform nymph were similar, but the intake was higher with a weight gain of 15–20%. Apteriform nymph accumulated 70% of amino acids, while alatiform nymph accumulated 64%, and excretion was 30% and 36% of the ingested amino acids, respectively. The difference between the two morphs was even greater for the glucose budget: the apteriform nymph accumulated only 6%, breathed out 24%, and excreted 70%; alatiform accumulated 29%, breathed out 29%, and excreted 42%. The results show that alatiform nymph has a greater demand for carbohydrates and consume only 4% more energy during feeding than the apteriform; however, they accumulate 35% of the energy, whereas the apteriform nymph uses only 17% for growth. Thus, alatiform nymphs get their energy supply mainly from carbohydrates, and accumulate less amino acid. The high resorption of glucose results in a greater accumulation of fat for the adult to fly. In contrast, apteriform nymphs use more amino acids because they play a role in increasing the reproductive rate.

The nutritional budget of amino acids of *Rhopalosiphum padi* (L.), *S. graminum*, and *Diuraphis noxia* (Kurdjumov) on *Triticum aestivum* (Poaceae), measuring amino acid intake from phloem sap, their elimination in honeydew, and content in the insect tissues, indicated that *R. padi* had the highest rates for all variables, while *D. noxia* had the lowest intake rate due to the low growth rate and honeydew production (Sandström and Moran 2001). Both *D. noxia* and *S. graminum* induced increases in amino acid content in the phloem sap ingested. Many of the essential amino acids in the honeydew were present at levels lower than in the ingested sap, mainly methionine and lysine. However, arginine, cystine, histidine, and tryptophan were more abundant in the honeydew, suggesting an excess in supply. In the aphid tissues there were differences in the composition of free amino acids among the species, but the composition of proteins was similar, indicating that nutritional requirements are similar. In *R. padi* and *D. noxia*, the essential amino acids were ingested in insufficient quantities for growth, requiring provision by the symbionts.

Wilkinson et al. (2001) studied the fate of radioactive marked amino acids injected in the hemolymph of *A. fabae*. Part of the labeled amino acid was recovered quantitatively as carbon dioxide, but was not detected in the saliva or in honeydew. The loss of glutamic acid by the respiratory rate of aphids reared on chemically defined diets was twice as high as that of aphids reared on the host plant, *V. faba*. This indicates that glutamic acid and other amino acids are important respiratory substrates in *A. fabae*.

The energy budget of aphids can be calculated by the formula proposed by Petrusewicz and Macfadyen (1970): $C = P + R + U + F$, where C is energy intake (food consumption); P is the sum of the energy allocated for the nymphal growth (P_g), exuvial products (P_e), and reproduction (P_r); R is energy lost as heat during cellular respiration; U is energy allocated in nitrogeneous excreta resulting from catabolism; F is energy from food that passes through the digestive tract without absorption (feces). Ecologists refer to $P + R + U$ as assimilation (A), which is the energy that crosses the intestinal wall and represents the difference between the energy in food (C) and feces (F). On the other hand, physiologists refer to assimilation as the material incorporated and used by the organism, so that the U is excluded.

According to Llewellyn (1987), there are three classes of energy efficiency rates to describe the patterns of energy utilization (Figure 20.7): (a) the overall efficiency of energy transfer through the insect, which is given by the *P/C* ratio often called growth efficiency, and this ratio depends on two others: *A/C* and *P/A*; (b) the *A/C* value, also known as efficiency of assimilation, is a measure of food energy assimilation, which depends primarily on the composition of the ingested phloem sap—a low *A/C* indicates that a low energy content from food is absorbed through the intestinal wall and most is eliminated in the form of feces; and (c) *P/A*, a measure of cost of living, because the energy and materials released from food during digestion are used for growth and tissues maintenance and for providing energy for synthesis, active transport, and movement. The greater the portion of *A* required for the processes of energy demand, less energy is available for *P*. A low *P/A* ratio denotes that the *R* and *U* values are high, so that a smaller amount of energy is needed for living, providing more stored energy and increasing biomass.

Llewellyn (1987) described each component of the energy budget for aphids, referring also to the techniques that are used to estimate or measure consumption, production, and utilization of energy. For studies of energy flow of aphids, it is essential that the energy budget of the population be built for at least 1 year, considering environmental factors, such as temperature, as well as biological variables, such as changes in host plant and presence of ants. It is accepted that the insect in the field (cost of living in the real environment) extracts less energy from food, with an *A/C* ratio 12% lower than in the laboratory. The former uses only 19% of assimilated energy for production (*P*) compared with 45% in the laboratory. Thus, it is essential to be careful about extrapolating energy budget data from the laboratory to the field.

The energy budget for nine species of aphids is presented in Table 20.2. The energy budgets, in most cases, express parameters of the life cycle, by monitoring the energy flow from birth to death. It can be observed that energy consumption was lower for *M. euphorbiae* and higher for *A. pisum* (Llewellyn 1987). Comparison of energy ratios shows great consistency in the *P/A* ratio that averaged 78%, indicating that species have very similar energy requirements for maintenance. The phloem sap feeding is very economic, allowing the assimilated energy to be channeled into production. The author draws attention to the *P/A* ratio of the monophagous Drepanosiphinae *Eucallipterus tilia* (L.) and *Illinoia liriodendri* (Monell), calculated from the estimated energy budget in the field, which are not very different from those obtained for other species in the laboratory, suggesting that the aphids in the field were not subject to excessive environmental stresses. The *A/C* ratios are indicative of food quality and showed great

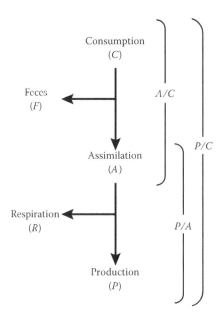

FIGURE 20.7 Outline of energy pattern of aphids. (Redrawn from Llewellyn, M., In *Aphids: Their Biology, Natural Enemies and Control*, vol. 2B, ed. A. K. Minks and P. Herrewijn, 109–17, Copyright 1987, with permission from Elsevier.).

TABLE 20.2

Energy Budgets and Energy Ratios for Some Aphid Species

Species	Conditions	Energy Requirements for the Life Cycle (J)				P/C[a] (%)	A/C[b] (%)	P/A[c] (%)	Adult Wet Weight (mg)
		Production	Respiration	Feces and Urine	Consumption				
Aphis fabae	Bean seedlings, stems, and leaves	23.2	2.7	19.9	45.7	51	57	89	0.78
Megoura viciae	Bean seedlings, stems, and leaves	56.8	9.3	30.9	97.0	59	68	86	3.07
Acyrthosiphon pisum	Pea seedling leaves	72.8	50.5	23.8	147.1	49	83	58	4.20
Eucallipterus salignus	Leaves and shoots	7.7	10.0	94.9	112.6	5	9	58	0.50
Tuberolachnus salignus	*Salignus*, twigs, stems	115.3	23.6	120.9	259.8	—	—	—	13.3
Illinoia liriodendra	Leaves of *Populus*	—	—	—	—	27	33	82	0.9
Macrosiphum euphorbiae	Apple seedlings, leaves, and galls	17.3	5.6	9.4	32.3	53	70	75	1.48
Dysaphis devecta	Apple seedlings, leaves, and galls	13.6	1.9	17.3	42.8	32	36	88	0.56
Aphis pomi	Apple seedlings, stems leaves, and galls	12.5	1.3	32.7	43.9	28	31	91	0.43

Source: Data from Llewellyn, M., In *Aphids: Their Biology, Natural Enemies and Control,* vol. 2B, ed. A. K. Minks and P. Harrewijn, 109–17, Copyright 1987, with permission from Elsevier.

a Total efficiency in energy transfer.

b Assimilation of energy from food (assimilation efficiency).

c Ratio of life maintenance.

variation for the species, ranging from 83% for *A. pisum* to 9% for *E. tilia*. The *A/C* value (48%) for the aphid species do not differ much from the value calculated for other invertebrate herbivores (45%), presented by Schroeder (1981). With *A/C* similar to other herbivores, but with a higher *P/A* rate, the *P/C* for aphids is 38%, considerably higher than that of herbivores (20%). Thus, aphids may be considered extremely efficient in converting their food into biomass; however, the type of plant from which sap is extracted plays an important role in energy budget

20.4.8 Intrinsic Rate of Natural Increase

The ability of most aphid species to become pests depends mainly on their remarkable growth rate. The intrinsic rate of natural increase (r_m), which is the innate ability of a species to increase in number, has been simplified for aphids (Wyatt and White 1977). These authors consider that 95% of the value of r_m is formed by the progeny produced in the period from birth until the end of reproduction (*d*), which corresponds to the effective fecundity (M_d). It is assumed, therefore, that the reproduction patterns are similar for all species and under any condition, and that the first progenies have the greatest influence on population growth. Moreover, the effective number of progeny (M_d) is assumed to be produced on a single date (T_d), equivalent to the duration of generation so that the equation to calculate r_m becomes greatly simplified: $r_m = [c\,(\ln M_d)]/d$, where *c* is a constant with value 0.738, and assuming that T_d is linearly related to the instantaneous mortality rate *d*. Since aphid populations rarely reach a stable age distribution (Carter et al. 1978), the value of r_m is not very useful for determining the increase of aphid populations in the field. However, it is useful to compare the potential growth rate and other parameters of different forms of a given species (Dixon 1987a). Thus, the value of r_m is determined largely by the reproductive rate in early adulthood, rather than by the number of nymphs born during a lifetime.

20.4.8.1 Development Rate

The time required for the development of an aphid, from birth to adult stage, depends on two extrinsic factors (food and temperature), and two intrinsic factors (birth weight and whether the form is winged or apterous) (Figure 20.8). Food and temperature also affect the birth weight due to the influence on the

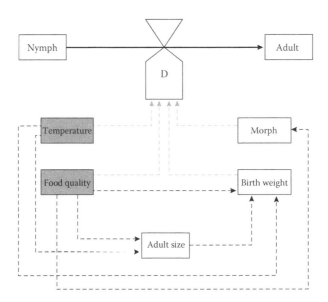

FIGURE 20.8 Diagram of interrelationship between factors that affect development (*D*) of aphids. (Redrawn from Dixon, A. F. G., In *Aphids: Their Biology, Natural Enemies and Control*, vol. 2A, ed. A. K. Minks and P. Harrewijn, 269–87, Copyright 1987, with permission from Elsevier.)

size of the mother and morph determination (Dixon 1987a). It is difficult to determine the quality of food ingested by the aphid, but it has been shown that aphids feeding on high-quality food develop faster and reach a larger size than those fed on poor food (Mittler 1958b). The relative growth rate (RGR) is the aphid growth per unit weight per unit time, and this rate represents a good indirect measure of food quality. Increase in the quality of food and temperature results in increased development rate and also affects the weight of the adult and, consequently, the weight of the individual at birth. If the aphid is small at birth, it will take a considerably longer time to reach maturity compared with an individual that is larger at birth. Also, the species that produces proportionally more offspring has a markedly shorter development time compared with one that produces few offspring. Dixon (1985a) considers that the interspecific correlation between birth weight and the weight of the mother is responsible for much of the variation.

20.4.8.2 Reproductive Rate

Aphids usually reach their highest reproductive rate in early adulthood, which, similarly to development rate, is also affected by food quality and temperature in addition to intrinsic factors, such as adult size, birth weight, number of ovarioles and aphid morph, either winged or apterous (Figure 20.9). The direct positive effect of food quality can be noticed when temperature and adult size are kept constant, and quality of food varies. On low-quality hosts, aphids with more developed gonads (higher number of ovarioles) have lower survival than those of the same clone with small gonads. However, those with larger gonads are able to achieve higher reproductive rates than those with small gonads on high-quality hosts. Species with a large number of ovarioles have a high initial reproductive rate and vice versa. It is interesting to note that morphs of *R. padi*, *Sitobion avenae* (F.), and *Metopolophium dirhodum* (Walker) have lower initial reproductive rates and fewer ovarioles when exploring grasses as secondary hosts. On the other hand, when they are on their primary hosts, they exhibit a greater number of ovarioles and longer development time (Dixon 1987a).

20.4.8.3 Survival Rate

Aphids respond to seasonal changes in food quality and produce morphs better adapted to survive in such conditions. However, unpredictable changes in climate or food quality result in low survival. Some species respond to abrupt changes in habitat quality by adjusting the proportion of progeny that is more

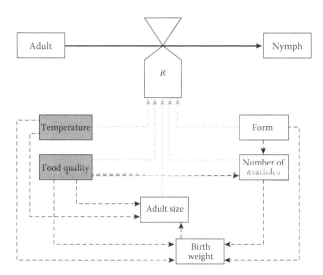

FIGURE 20.9 Diagram of interrelationship between factors that affect reproductive rate (*R*) of aphids. (Redrawn from Dixon, A. F. G., In *Aphids: Their Biology, Natural Enemies and Control*, vol. 2A, ed. A. K. Minks and P. Harrewijn, 269–87, Copyright 1987, with permission from Elsevier.)

adapted to a high or low level of reproductive investment. For example, the higher the nutritional stress experienced by *Megoura viciae* Buckton, the lower the number of its progeny committed to a high reproductive investment, and vice versa. Thus, some aphids try to adapt to habitat quality maintaining a high reproductive rate on nutritionally rich hosts and increasing their survival rate on poor hosts (Dixon 1987a). Because of its short life, aphids on a poor host do not expect improvement in habitat quality, and invest on development of mature embryos and absorb the smaller ones, sacrificing their fertility to maintain their full potential of reproductive rate and survival in the short term (Ward and Dixon 1982). In an optimal environment, females produce extra eggs and increase potential fertility. Thus, aphids are able to adjust their reproductive biomass and escape the restrictions imposed by physiological decisions made during the growth stages (Dixon 1987a).

20.4.8.4 Changes in Intrinsic Rate of Natural Increase

Fluctuations in both temperature and food quality lead to significant changes in the value of r_m; the intrinsic rate of increase is positively associated with the average growth rate, both within and between aphid species (Figure 20.10) (Leather and Dixon 1984; Dixon 1987). Based on this figure, $r_m = 0.86$ RGR, thus the relation between the intrinsic rate of increase and the mean relative growth rate is expressed in the equation RGR $= (0.86 \ln M_d)/d$. The equation states that the weight gain/unit weight at birth, from birth to adult stage, is equal to the number of nymphs born subsequently in a period equal to that from birth until the end of reproduction (d) multiplied by the conversion factor. This is expected because when they reach maturity, embryos represent a large proportion of the weight of adult females and the embryos are already developing their own embryos inside. Growth and reproduction occur simultaneously, with the effective number of offspring (M_d) being ovulated and reaching their embryonic development during the maternal development. At maturity, aphids cease growth and most of the assimilated food is channeled to embryo growth (Randolph et al. 1975). Regardless of its size as an adult, an aphid that reached a high growth rate during the nymphal stage achieves a high population increase rate (Figure 20.11) (Dixon 1987a).

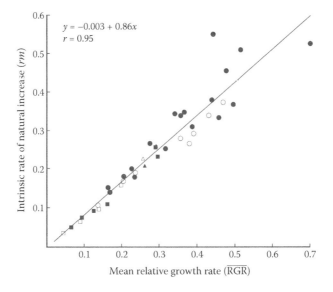

FIGURE 20.10   Intrinsic rate of natural increase (r_m) for the relative growth rate (RGR) of six species of aphids reared under various conditions. □ *Drepanosiphum acerinum*; ■ *Drepanosiphum platanoidis*; ○ *Sitobion avenae*; ● *Rhopalosiphum padi*; △ *Eucallipterus tiliae*; ▲ *Tuberolachnus salignus*. (Redrawn from Dixon, A. F. G., In *Aphids: Their Biology, Natural Enemies and Control*, vol. 2A, ed. A. K. Minks and P. Harrewijn, 269–87, Copyright 1987, with permission from Elsevier.)

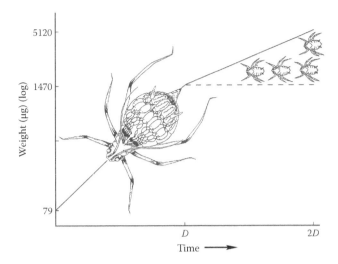

FIGURE 20.11 Biomass increase of an aphid species in terms of growth and progeny production as a function of time, where *D* is development time from birth to maturity. (Redrawn from Dixon, A. F. G., In *Aphids: Their Biology, Natural Enemies and Control*, vol. 2A, ed. A. K. Minks and P. Harrewijn, 269–87, Copyright 1987, with permission from Elsevier.)

Aphid pest species seem to channel proportionally more on reproduction and a shorter development time, and therefore, have a greater r_m than nonpest species (Llewellyn and Mohamed 1982). However, there are no indications that pest species allocate more resources to increase in number than nonpests, thus the value of r_m does not appear to be associated with their permanence or habitat condition. The variables that comprise the intrinsic rate of natural increase of aphids—development, reproduction, and survival rates—are affected by many factors. There is a direct and positive association between the relative average growth rate, which is determined primarily by food quality and temperature, and intrinsic rate of natural increase of the species. Thus the aphid's way of life determines its ability to achieve the maximum relative average growth rate under the given conditions (Figure 20.12).

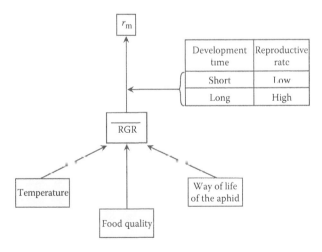

FIGURE 20.12 Diagram of interrelationship between factors affecting average relative growth rate (RGR) of an aphid, and exchange in development duration of life cycle and reproductive rate associated with a given RGR and intrinsic rate of natural increase (r_m). (Redrawn from Dixon, A. F. G., In *Aphids: Their Biology, Natural Enemies and Control*, vol. 2A, ed. A. K. Minks and P. Harrewijn, 269–87, Copyright 1987, with permission from Elsevier.)

20.4.9 Honeydew and Excretion

Aphids, unlike most terrestrial insects, have no Malpighian tubules and do not excrete uric acid (Kennedy and Fosbrooke 1972). However, due to the large amount of fluid intake, the excretion products may be removed by simple diffusion.

Honeydew—collection of excreta that aphids and other sap-sucking insects eliminate through the anus—is composed primarily of sugars and water, along with smaller quantities of amino acids and amides that compose the elaborate plant sap ingested. Aphids reach up to 55% efficiency in extracting amino acids from sap, and usually two-third of the ingested nitrogen is assimilated and one-third is excreted (Mittler 1958a) The predominant sugars in the honeydew are fructose, glucose, and sucrose; melezitose, trehalose, and some other oligosaccharides may also be present and are possibly related to the reduction of osmotic pressure of the gut. A large proportion of sugars present in honeydew are not in the same form as they are ingested. Melezitose, for example, is a common sugar in honeydew, but it is not a sap component. Melezitose in honeydew acts in osmoregulation and also has an important role in attracting ants associated with aphids (Kiss 1981). Because honeydew is rich in sugars, it can attract and serve as food for several species, including bees and other Hymenoptera, Diptera, and several species of predators and ants. The deposition of honeydew on the plant can also provide a favorable substrate for the development of fungi that cover the plant, forming the sooty mold that affects photosynthesis and leaf respiration.

Honeydew produced by myrmecophilous species is richer in amino acids than that produced by non-myrmecophilous aphids, especially in nonessential asparagine, glutamine, glutamic acid, and serine. It was found that species with higher amino acid concentrations also have higher concentrations of sugars attractive to ants, especially melezitose (Woodring et al. 2004).

The frequency of excretion can vary from 2 to 25 drops in 12 h, depending on the insect stage, host plant, and its physiological state, temperature, humidity, time of day, and presence of ants (Klingauf 1981). To determine the amount of honeydew excreted, one can measure the frequency of production, size, weight or volume of droplets. Heimbach (1985), cited by Klingauf (1987), used a device similar to a thermohygrograph to measure the frequency of excretion, and determined that the average volume of a droplet of honeydew is approximately 0.5 mm^3 and contains 5–15% dry matter and has specific gravity slightly above 1.

Excretion usually follows a diurnal rhythm of production, but within this period it may have an irregular pattern, without apparent changes in behavior, and being interrupted at the time of ecdysis and changes of feeding sites; the sticky honeydew can be a trap for the aphid itself. However, some morphological and behavioral adaptations minimize the possible adhesion, such as coating of the droplets of honeydew with wax filaments, spraying the droplets upon release, or removing it from the anus with the aid of the hind tibiae. In the case of Adelgidae and Pemphigidae, thin wax is used to coat and isolate the feces on the walls of the galls, which are also coated with wax (Klingauf 1987). The intensity of ant attendance is significantly lower in colonies of first and second instars compared with colonies of larger aphids. Ant attendance correlates with the amount of honeydew produced, and not with the total amino acid concentration (Fischer et al. 2002).

20.4.10 Symbionts

Even though some nutrients are found in low concentrations in the phloem, sap-sucking insects are capable of meeting their basic nutritional requirements, possibly due to the presence of symbiotic bacteria in their body. Buchner (1965) states that all aphids, with the exception of Phylloxeridae, have a mandatory association with microorganisms, whether bacteria or yeasts. The functions of these symbionts can range from an obligate nutritional role to a facultative role in protecting their hosts against environmental stresses (Burke et al. 2009).

In aphids, symbionts are confined to special cells (mycetocytes), which may be grouped forming the mycetoma that vary in form, location, and type of symbionts. In adults, the mycetoma are distributed in the abdominal region around the digestive tract, with some groups of mycetocytes free in the hemolymph and around the gut in embryos (Buchner 1965). The mycetocytes are hypertrophied cells, usually polyploids, with all the normal organelles, vesicles, and granular bodies (Houk and Griffiths 1980). Symbionts are found within vacuoles inside the mycetocytes, surrounded by a host cell membrane and its own plasmatic membrane and cell wall. These organisms also contain strands of DNA and ribosomal RNA. The symbionts are typically eubacteria, gram negative, but their taxonomic relationships are not well defined (Houk 1987).

Aphids deprived of their symbionts (aposymbiotics) have poor performance (Mittler 1971b, Houk and Griffiths 1980). Several functions have been attributed to symbiotic microorganisms, such as biosynthesis of amino acids, sterols, vitamins, and enzymes (polysaccharases); energy production; nitrogen fixation; and detoxification of catabolites and possibly of allelochemicals (Buchner 1965, Houk and Griffiths 1980).

The synthesis of some amino acids, such as tryptophan, has been attributed to the symbionts on the basis of analysis of plant sap and honeydew (Mittler 1953, 1958a). On the basis of radioactive ^{14}C- anthranilate tryptophan and on nymphal growth rate of *A. pisum* reared on tryptophan-free diet, Birkle et al. (2002) found that the production of tryptophan by *Buchnera* varies among parthenogenetic clones. The production values of this amino acid correlated significantly between the two methods, but not with the amplification level of the *Buchnera* gene *trpEG*, which codes for anthranilate synthase, a key enzyme in the tryptophan biosynthetic route. Methionine and cysteine are synthesized by microorganisms since the animals cannot incorporate inorganic sulfur in amino acids (Houk 1987). He mentions that *N. circumflexus* with its symbionts was able to incorporate radioactive sulfur into methionine, but aposymbiotic individuals failed to do so. Other studies with *A. gossypii*, however, show that the amino acids tryptophan, methionine, and cysteine are mandatory requirements of the diet and would not be supplied by symbionts (Turner 1971, 1977). For *A. fabae*, the key amino acids for embryo growth are phenylalanine and valine derived from the maternal tissues, and tryptophan derived from *Buchnera* (Bermingham and Wilkinson 2010).

Several studies based on metabolic pathways and histological techniques show that sterol is synthesized by symbionts in *A. pisum* (Houk 1987). Aphid rearing for several generations on defined diets show that there must be an endogenous source, corroborating the sterol synthesis by symbionts, discarding the possibility of fungal contamination that could produce phytosterols (Dadd and Mittler 1966). However, Campbell and Nes (1983) contradict these findings in their studies with *S. graminum*. They used intermediate metabolites and fractions of radioactive sterol to demonstrate that bioconversion of phytosterols into cholesterol is possible, as well as the use of acetate and mevalonic acid to synthesize other alcohols that compose the cuticular wax. The involvement of the symbionts in the synthesis of vitamins, especially B complex, has greater acceptance than sterol; nevertheless the evidence is dependent only with the survival of aphids on holidic diets devoid of exogenous source of vitamins (Buchner 1965, Houk and Griffiths 1980).

Symbionts have been implicated in the biosynthesis of polysaccharases (pectinase, cellulase, and hemicellulase) that degrade the cell wall and middle lamella of the matrix, used in the process of stylet penetration for probing (Campbell and Dreyer 1985). Their findings were based on substantial activity of these enzymes in homogenized aphids and also in prokaryotic organisms. Second, the activity of these enzymes in different aphid biotypes fed on different plant strains showed use of distinct substrates. Also, some of such biotypes can overcome plant resistance by hydrolyzing cell wall polysaccharides, while others cannot. *A. pisum* does not secrete amylase, but microorganisms in its digestive tract are capable of hydrolyzing starch (Srivastava and Auclair 1962a). *Sarcina, Micrococcus, Achromobacter,* and *Flavobacterium* bacteria were isolated from this aphid's gut (Srivastava and Rouatt 1963). The hypotheses of nitrogen fixation and energy production (mitochondrial function) by the symbionts have been discredited (Houk 1987).

Facultative bacterial symbionts, such as *Hamiltonella defensa, Regiella insecticola,* and *Serratia symbiotica* influence aphid's utilization of host plants and defense against parasitoids. Different clones of aposymbiotic *A. pisum* from *Buchnera aphidicola* may perform differently regarding the utilization of major nutrients (sucrose and amino acids). Yet there are no conclusive results on the impact of secondary symbionts on the nutrition of *A. pisum* (Douglas et al. 2006). The symbiont *S. symbiotica* has been involved in defense against heat and potentially also in aphid nutrition. It seems that their association with Lachninae aphids is a transition from facultative to obligate symbiosis. Their diversity in terms of morphology, distribution, and function is due to multiple independent origins of symbiosis from ancestors and possibly also to evolution within distinct symbiont clades (Burke et al. 2009).

The main problem in studying insect-symbiotic microorganisms is the difficulty in keeping the organisms separate from one another for long periods (Houk 1987). Obtaining aposymbiotic insects is also limited because the methods may change mitochondrial structures and affect their functions, masking the results (Griffiths and Beck 1974; Houk and Griffiths 1980). It is possible to clone DNA fragments in the symbiotic vectors to keep them continuously in culture for investigation of their functions. Also, the identification of genes responsible for synthesis of sterols, amino acids, vitamins, and enzymes can provide direct evidence of symbiont functions. Studies of cross-hybridization can also provide evidence of evolutionary relationships between aphids and eubacterial symbionts and help clarify such interactions. Genomic analysis shows that the pea aphid, *A. pisum*, lacks key purine recycling genes that code for purine nucleoside phosphorylase and adenosine deaminase. The symbiotic bacterium *Buchnera* possesses purine metabolism genes and can meet its nucleotide requirement from aphid-derived guanosine. The coupled purine metabolism of aphid and *Buchnera* could contribute to the dependence of the pea aphid on this symbiotic relationship (Ramsey et al. 2010).

20.5 Factors Affecting Feeding and Nutrition

20.5.1 Physiological Status of Plant, Water Stress, and Aphid Performance

The main factors controlling the dynamics of insect populations are as follows: (a) *Nitrogen content of the host plant*. The quality and quantity of food offered by a potential host in the form of nitrogen is crucial for the development of phytophagous insects. The proportions of essential nutrients in food play a more important role in nutrition than the absolute amounts of these nutrients (Slansky and Rodriguez 1987). Even though aphids usually have a high amount of nitrogen available in phloem sap, the behavior of these insects will vary with seasonal changes in the level of soluble nitrogen. Possibly, this is the reason for polymorphism and alternating of hosts observed in aphids. (b) *Age of the host plant*. Growth and development of the host plant directly affect feeding and nutrition of phloem sap–sucking insects due to changes in their physiology and nutrient availability. Klingauf (1987) refers to varieties of *V. faba* that show some degree of resistance to *A. fabae* during the vegetative growth, but become more susceptible during flowering, because of their high metabolic activity in floral organs that favor the aphids. Different strata of the same plant; leaf age and position; and availability of sunlight, moisture, and nutrient affect food quality or quantity. (c) *Biotic and abiotic factors of the host plant*. Feeding activity of aphids affect quality and quantity of food provided by the host plant. For example, infestations of *A. fabae* on sugar beet reduce the concentration of nonstructural carbohydrates in stem tissues and increase it in young and senescent leaves and roots (Capinera 1981).

Although aphids are able to control their feeding rate, regardless of the phloem sap pressure, deficiency of water in the soil can adversely affect their nutrition, reproduction, and survival. Wilting plants favor aphid infestation because it promotes senescence of older leaves while younger leaves remain turgid with greater availability of soluble nitrogen in the sap (Kennedy 1958). Risebrow and Dixon (1987) mentioned that the response of the insect depends on the nature and magnitude of plant stress, as well as on plant and insect species.

20.5.2 Secondary Compounds

Apparently, allelochemicals in plants do not significantly affect phloem feeders, as occurs with chewing phyllophagous insects. Schreiner et al. (1984) found that *M. euphorbiae* is distributed randomly on the common fern, *Pteridium aquilinum* (Hypolepidaceae), whereas chewing insects avoid cyanogenic parts of the plant and look for acyanogenic stems. This is probably because secondary compounds are synthesized in particular tissues and are not transported by the phloem, despite the evidence that some allelochemicals can be transported from one tissue to another by the phloem or xylem.

Several allelochemicals have deterrent effects on polyphagous aphids, such as *M. persicae*, while others may be tolerated at low concentration (Nault and Styer 1972; Dreyer and Jones 1981). Some allelochemicals are involved in plant resistance, such as hydroxamic acids and benzyl-alcohol in gramineous plants (Poaceae), alkaloids in species of *Lupinus* (Fabaceae) (Argondoña et al. 1980, Brusse 1962), and coumarins in different legumes (Mansour et al. 1982). Even though these compounds are not present in the phloem sap, aphids can come in contact with them during a probe test before feeding. Compounds present on the plant surface can also affect the behavior of aphids; the polyphagous aphid *A. fabae* is attracted by phenolic substances on the leaf surface (Jördens-Röttger 1979).

Nicotine is an alkaloid secreted by some species of *Nicotiana* (Solanaceae) that is lethal for most aphid species (Thurston et al. 1966). *Myzus nicotiana* Blackman, however, can feed and develop on tobacco plants because the phloem sap contains only traces of nicotine, while the leaf cells accumulate up to 10%. Gibson and Pickett (1983) demonstrated that glandular hairs of *Solanum berthaultii* (Solanaceae) release β-farnesene, which is a compound present in the alarm pheromone of aphids, thus preventing them from settling on the plant.

20.5.3 Physical and Chemical Factors of Host Plant and Environment

The presence of hairs on the leaf surface, glandular or not, represent a physical barrier against the small and delicate aphids, preventing them from reaching tissues with their stylets or walking due to the sticky secretions (Gibson and Pickett 1983). Thick cuticles and lignified vascular bundles can also act as physical barriers to phloematic suckers, making the reaching of the phloem via the stomata. Plant cell walls resist the action of aphid's digestive enzymes by forming a callus or necrosis preventing stylets penetration (Risebrow and Dixon 1987).

Aphids respond best when the pH of the diet is slightly alkaline (7.3–7.6), similar to that of the exudates from vessels of most plants (Auclair 1969). Thus, pH may be one of the factors used by aphids for initial acceptance of a feeding site (Pollard 1977). Aphids also respond to certain wavelengths of diets with colors close to their preferred natural hosts. Auclair (1969) found that *A. gossypii* is attracted to lengths from 610 to 570 nm (red–orange–yellow), but is repelled by 485 at 420 nm (blue–violet). The author also found that all morphs of *A. pisum* and *M. euphorbiae* prefer orange and/or yellow, instead of white, red, or blue; whereas some biotypes of *A. pisum* prefer yellow, others orange.

20.6 Artificial Diets

Nutritional requirements of aphids, with few exceptions, are similar to those of other insects, allowing their development on chemically defined diets for several generations. Auclair (1969) discusses and compares the nutritional requirements of aphids and other Hemiptera, considering the qualitative and quantitative aspects of the main classes of nutrients and certain compounds. He also discusses the role of symbionts and the influence of pH, light, and secondary compounds in aphid nutrition and metabolism, necessary for defining the diets.

20.7 Electrical Penetration Graph and Aphid Feeding

Studies on feeding behavior of aphids gained a new approach in the mid-1960s with the development of a technique that recorded the electrical waves produced by stylet penetration into plant tissues (McLean and Kinsey 1964, 1965). They developed a system based on an alternating current (AC) forming an electrical circuit, which included the aphid and the plant system. The waves recorded could be correlated with distinct activities of the insect during penetration of stylets in plant tissues. Subsequently, several modifications were made in the system. Tjallingii (1978, 1985, 1988) modified the system to DC, using a direct current, and named it the electrical penetration graph (EPG).

To obtain reliable recordings, aphids are starved for about 1 h before the test, and then fixed to a gold filament attached to an electrode, forming a probe connected to the amplifier equipment (GIGA 4–DC). The aphid is placed on the plant, while the other electrode is introduced into the soil. The circuit closes when the insect inserts the stylets into the plant (Figure 20.13). The technique has been used to study

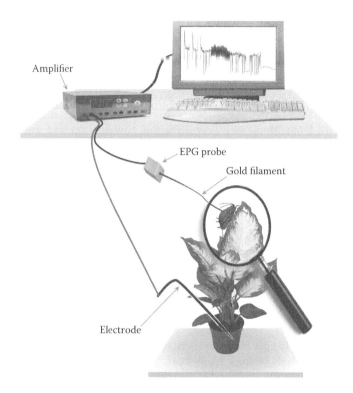

FIGURE 20.13   Electronic monitoring system for studying feeding behavior of aphids. (Illustration by Mirian N. Morales.)

the feeding behavior of aphids and others sap-sucking and blood-sucking insects. It has been used to evaluate the feeding behavior of aphids on resistant and susceptible hosts, to determine resistance factors (Dreyer and Campbell 1984; Mayoral et al. 1996) and to establish correlation between feeding and virus transmission mechanisms (Powell 1993; Prado and Tjallingii 1994).

The characterization and meaning of wave patterns (Figure 20.14) were defined by Tjallingii (1978, 1988, 1990). Backus (1994) reviewed the use of the EPG technique until 1990, establishing correlations of waves obtained by EPG and specific feeding behavior events. For aphids, it has been used to demonstrate the specialization of feeding mechanisms to prevent triggering plant responses that adversely affect food intake.

Successful phloem feeding depends on the insect's ability to interact with the host plant and to overcome various physical and chemical plant properties (Figure 20.15), according to Miles (1998). The plant responds to the insect when the cell membrane is disrupted by the stylet penetration, producing proteins that cause the coagulation of sap thus blocking the food canal (Prado 1997; Tjallingii 2006). To prevent protein clotting, aphids inject watery saliva. This activity corresponds to the E1 wave preceding sap ingestion, recorded on the EPG graphics. Biochemical characteristics of some plants prevent this aphid behavior, which might indicate their resistance against aphid feeding. A short duration of the pathway phase (waves A, B, and C) and a long phloem sap ingestion phase (waveform E2) are interpreted as host plant acceptance (Montllor and Tjallingii 1989). In contrast, long or short periods, or repetitive wave patterns for E1, without sap ingestion (E2), indicate the presence of plant defense mechanisms (Prado 1997). During phloem feeding, another regular activity occurs in the E2 phase, when the watery saliva injected into the plant is ingested passively with sap to prevent coagulation of phloem proteins within the fine food canal of the stylets (Figure 20.16).

In resistant genotypes of barley, *Hordeum vulgare* (Poaceae), the Russian wheat aphid, *D. noxia*, reaches the phloem in an average of 306 min, while in the susceptible genotype it takes 180 min (Brewer and Webster 2001). These authors observed that the corn leaf aphid, *Rhopalosiphum maidis* (Fitch), reaches the phloem of both susceptible and resistant barley plants faster than *D. noxia* (ca. 132 min). In

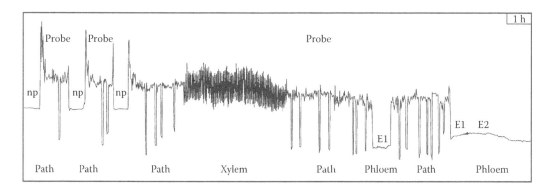

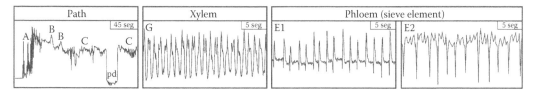

FIGURE 20.14 Electrical penetration graph (EPG) of feeding behavior of an aphid. Probe is the period of stylet penetration; np is the nonprobe. The first two probes contain only the pathway phase; the third probe includes also one phase in xylem and two in phloem: a short one with only E_1, the second with E_1 followed by E_2. Lower figure details each stage indicating the wave pattern; pd, potential drop (i.e., a brief intracellular puncture); A, B, C, pathway phases; G, xylem penetration. (Redrawn from Prado, E., Aphid–plant interactions at phloem level, a behavioural study. PhD thesis, Wageningen Agricultural University, 1997.)

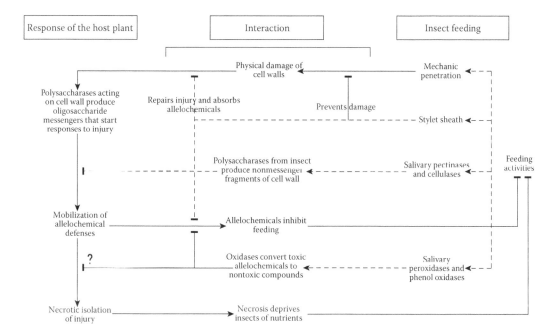

FIGURE 20.15 Interactions of feeding activities of Aphididae (lowercase and dashed lines) and response of host plant (capital letters and full lines), showing possible responses; arrows indicate increment; short lines indicate inhibition. (Redrawn from Miles, P. W., In *Aphids in Natural and Managed Ecosystems*, ed. J. M. Nieto-Nafría and A. F. G. Dixon, 255–63. International 5th Symposium on Aphids. León, 1997. León: Universidad de León, 1998. With permission.)

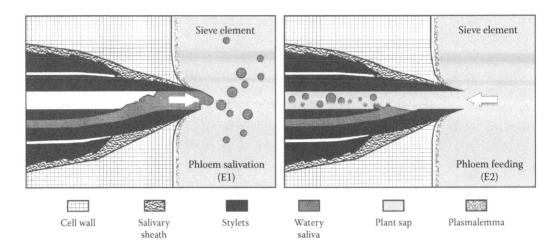

FIGURE 20.16 Diagram of a phloem-feeding aphid: E_1 corresponds to secretion of watery saliva in phloem element; during E_2, phloem sap is pumped up (high pressure) into food canal and saliva mixes with ingesting sap, preventing clotting of sap within the canal. Internal stylets are the maxillae and external stylets are the mandibles. (Redrawn from Prado, E., Aphid–plant interactions at phloem level, a behavioural study. PhD thesis, Wageningen Agricultural University, 1997.)

plant resistance studies, aphid feeding should be monitored for 24 h to record circadian activity patterns (Reese et al. 1994).

Penteado (2007) and Cardoso (2007) studied the feeding behavior of *C. atlantica* and *Pineus boerneri* Annand (Adelgidae), respectively, on *Pinus*. It was observed that the wave patterns for *C. atlantica* were similar to those recorded for other aphid species. However, the wave pattern G, which refers to xylem penetration, was not observed. The longer nonpenetration (np) phase indicates that *Pinus taeda* imposes some physical barriers to *C. atlantica* initial penetration compared with *Pinus elliotii*. However, after the insect penetrates the plant tissues, it rarely removes the stylets, which indicates host acceptance. In the case of the woolly pine aphid, *P. boerneri*, the EPG revealed two distinct wave forms: wave M, which represents extracellular stylet activity, with an irregular low frequency wave pattern; and the second pattern, named P, which represents intracellular activity and had a higher frequency. Honeydew excretion confirmed that the P pattern was associated with sap ingestion.

20.8 Perspectives for IPM

The ability of aphid species to become pests of many crops is attributed to their high population increase rate. This indicates that these insects have great ability to direct the resources obtained from their host plants for reproduction associated with short development time (Llewellyn 1982). Positive correlation between the relative growth rate and the intrinsic increase rate is observed for both pests and nonpest aphid species (Dixon 1987a). However, pest species achieve higher average increase rates because the crop plants represent an abundant and rich food source for aphids.

During compatible interactions, leading to successful feeding and reproduction, aphids cause alterations in their host plant, including morphological changes and various local and systemic symptoms (Giordanengo et al. 2010). Many aphid species exhibit the ability to enhance the nutritional quality of their host plants. Some do not induce any macroscopic changes in the plant; others, however, induce typical chlorotic lesions on plant tissues, such as *D. noxia* and *S. graminum*. Comparing the ingested phloem sap by *R. padi*, *S. graminum*, and *D. noxia* on wheat and barley, the later had a twofold higher concentration of amino acids, with higher proportion of essential amino acids. Changes in the phloem induced by both *S. graminum* and *D. noxia* appear to be systemic and are nutritionally advantageous for the aphids (Sandström et al. 2000).

Several principles and measures based on aphid bioecology and nutrition can be used for IPM of aphids on economic crops, as follows.

20.8.1 Plant Nutrition

Nitrogen, especially in the form of the amino acids glutamic acid and aspartic acid, and their amides asparagine and glutamine, promotes the successful establishment and development of aphid populations on plants (Klingauf 1987). As fertilizer use benefits production, it also favors development of pest populations. Treatments with nitrogen-rich fertilizers applied on annual crops affect not only the level of the soluble compound available for aphids but also the pH and tissue structure, promoting noticeable aphid population build up.

Phloem sap has a protein-to-carbohydrate ratio of ~0.1. Enhanced N fertilization increases the amino acid concentration in phloem sap and elevates the N/C ratio. *Uroleucon tanaceti* (Mordvilko) and *Macrosiphoniella tanacetaria* (Kaltenbach), specialist aphids feeding on tansy (*Tanacetum vulgare* L. (Asteraceae)), were reared hydroponically on this host plant under different N concentrations (Nowak and Komor 2010). Their feeding behavior was monitored by EPG and phloem sap was sampled by stylectomy. Both aphid species settled two to three times more frequently on plants fertilized with 6 or 12 mM NH_4NO_3, containing amino acid concentrations up to threefold higher, without a change in the proportion of essential amino acids. The duration of phloem feeding was two to three times longer in N-rich plants and the time spent in individual sieve tubes was up to tenfold longer. The authors concluded that aphids identified the nutritional quality of the host plant mainly by the amino acid concentration in the phloem sap, neither by leaf surface cues nor by the proportion of essential amino acids. Infestation by *U. tanaceti* also triggered a tenfold increase in the percentage of methionine and tryptophan, thus manipulating the plant nutritional quality, and causing premature leaf senescence.

M. persicae and *B. brassicae* respond positively to Brussels sprouts fertilized with high nitrogen and low potassium levels, increasing their reproductive rate. *M. persicae* prefers senescent leaves, which contain low potassium, thus concentrating the nitrogen content. Asparagine is the key compound that makes the senescent leaves more suitable for *M. persicae*. In contrast, *B. brassicae* usually feeds on younger leaves, where the nitrogen level is dependent on nitrogen availability for protein synthesis, with no influence of potassium levels (van Emden 1966). The differential behavior of both species is dependent on water pressure in the tissues. While *M. persicae* accepts greater variation in turgor pressure, including low turgidity of senescent leaves, *B. brassicae* has no ability to cope with the reduction in turgor and must remain in the shoots. Senescent leaves are in a state of water stress and increased proteolysis activity, which releases high amounts of nitrogen that is readily utilized by *M. persicae* (Wearing 1967; Wearing and van Emden 1967).

The complexity of the nutrient content for plant nutrition and aphid performance is not a function of the levels of each nutrient alone, but also of the rates and/or combinations of the different compounds (Jansson and Ekbom 2002). Phosphorous has a positive effect on various biological parameters of *M. euphorbiae*, and associated with potassium shortens the development time of this aphid. When reared on plants poor in nitrogen, the performance of *M. euphorbiae* is significantly reduced, demonstrating the positive effect of nitrogen nutrition on the plant and insect. However, high N/K fertilizers do not improve the performance of *M. euphorbiae*. Potassium deficiency in soybean fertilization can lead to higher populations of the soybean aphid because this deficiency improves the nitrogen content (Walter and Difonzo 2007). Thus, correct plant nutrition is essential not only for crop production but also to keep the pest population at an equilibrium level.

20.8.2 Biological Control

The Aphidoidea and other sap-sucking insects are subject to the action of parasitoids, predators, and pathogens, mainly because they stay longer on the feeding sites. Parasitized aphids experiment changes in nutritional behavior and physiology before death. Cloutier and Mackauer (1980) demonstrated this phenomena by comparing feeding parameters of parasitized and super-parasitized *A. pisum*. Super-parasitism occurs when the parasitoid lays more than one egg in the same host. During the early embryonic development of the parasitoid, the feeding rate of parasitized aphids increases, exceeding that of nonparasitized individuals. In super-parasitized aphids, the values are even greater than for a single parasitoid, with an increase in feeding rate of 133%, in honeydew excretion of 146%, in ingestion efficiency of 66%, and in food digestion of 86%. However, as soon as the parasitoid larvae complete their development, consuming the host tissues, the aphid stops feeding and dies. Some plant volatiles, such as kairomones, are used to attract natural enemies to enhance their action over aphid pests

20.8.3 Physical and Chemical Traits in Plant Resistance

Plant resistance to arthropods is the sum of constitutive qualities, which are genetically inherited and result in one cultivar or plant species that suffers less damage than other susceptible plant lacking such qualities (see Chapter 26 for details). The development of resistant cultivars to arthropods produces higher return per dollar invested than that for developing insecticides (Smith 2005).

The concept of plant defense states that plant species growing in similar biotic or abiotic constraints display convergent defensive traits. Volatile organic compounds may contribute to wild *Solanum* resistance, depending on *Solanum* accessions and aphid species. All species of *Solanum* tested by Le Roux et al. (2010) presented phloem-based antixenosis resistance against *M. persicae* and *M. euphorbiae*, determined by olfactometry and EPG tests.

Glandular hairs on leaves and petioles of plants represent a type of physical resistance against the establishment of aphid populations. These plant traits can be easily incorporated in strains of potato genes from wild species or varieties, as *Solanum berthaultii* (Solanaceae). This plant species has two types of glandular hairs, one secretes a sticky substance and the other releases substantial amounts of (E)-β-farnesene, which is the alarm pheromone of aphids, preventing their establishment on the plant (Auclair 1987).

Some nutrients, specifically amino acids, present in certain plant varieties or species, may negatively affect the interactions between aphids and their symbiotic microorganisms. The amino acid content in the phloem sap of *Lamium purpureum* (Lamiaceae) is very poor for *A. fabae* development. In addition, this plant promotes an exacerbated growth of the bacterial secondary symbionts *R. insecticola* and *H. defensa,* which disturbs aphid control over bacterial abundance (Chandler et al. 2008).

Another method of insect control based on physical barriers of plants is the use of silicon, which promotes silification or hardening of the cell walls. Ester 2-methyl-benzo (1,2,3)-thiadiazole-7-carbotioic, or acybenzolar-S-methyl, can lead to activation of genes coding for plant resistance and can be systemically translocated in the plant. This compound combined with silicic acid applied in the soil around the roots is effective in reducing insect fertility and consequent reduction in the growth rate of *S. graminum* on wheat plants (Costa and Moraes 2006).

Antinutritional factors may interfere with nutrient utilization, mainly of endogenous proteins. In some cases, antinutritional factors damage the gut epithelium and cause death of vertebrates and invertebrates. They are natural substances produced by the secondary metabolism of plants as a defense mechanism against the attack of several organisms, including insects, or under stress conditions. Hydroxamate 2,4-dihydroxy-7-methoxy-1,4-benzoxazines-3-one (DIMBOA) occurs in high levels in certain corn hybrids, especially in the seedlings, conferring resistance against *R. maidis*. The effect of DIMBOA in aphid suppression was observed by Long et al. (1977) and confirmed by Beck et al. (1983), and has been frequently used in plant breeding. Several glycoalkaloids from *Solanum* have deterrent, toxic, and antireproductive effects on *M. euphorbiae*. Their deleterious effects on the insect's basic biological processes are caused by the aglycone, while more generalized effects are caused by the carbohydrate monomer. Güntner et al. (2000) draw attention for the importance of considering that even subtle structural variations affect the bioactivity of these compounds. The alkaloid quinolizidine (tetracyclic), synthesized in the leaves of *Cytisus scoparius* and several other species of Fabaceae, has been investigated as a potential neurotoxic insecticide, producing signs of acute toxicity when ingested by aphids (Argondoña et al. 1980, Brusse 1962).

The pectin–pectinase combination seems to determine how quickly the aphid can reach the phloem, and this has been the chemical basis of plant resistance to *S. graminum* (Campbell and Dreyer 1985). Another resistance mechanism, incorporated in alfalfa cultivars to the spotted alfalfa aphid, *Therioaphis maculata* Buckton, may be the rapid formation of phytoalexins (polyphenols) in feeding sites. However, there are noticeable differences in the ability of some clones of *T. maculata* to detoxify these compounds (Nielson and Don 1974).

Transgenic plants expressing protease inhibitors have been considered as an alternative approach to pest control. However, just as the pests are affected by entomotoxins, beneficial insects such as parasitoids may also be exposed to the deleterious effects via host or directly from the plant. *Bacillus thuringiensis* entomotoxines are not effective against Sternorrhyncha insects, such as aphids. However, the gene for lectin (agglutinin) from the plant *Galanthus nivalis* (Amaryllidaceae), designated as *GNA*, has been introduced in several plants (wheat, rice, tobacco, and potato) and demonstrates a toxic effect on sap-sucking insects (Gatehouse et al. 1996). Lectins are proteins or glycoproteins that bind to specific carbohydrates, agglutinating cells or precipitating glycoconjugates, resulting in deleterious effects and even insect death. They can also act as antinutritional factors (antifeedants), affecting the conversion and assimilation of nutrients. *GNA* is naturally present in several plant species and can be incorporated into artificial diets and even in plants of economic importance. Potato plants expressing *GNA* have been shown to be partially resistant to *M. persicae* (Gatehouse et al. 1996) and *Aulacorthum solanum* (Kaltenbach) (Down et al. 1996). Lectin was incorporated in an artificial diet containing 0.1% *GNA* and was also incorporated in transgenic potatoes, and both caused reduction of body size and increased aphid mortality, among other deleterious effects. However, they had a negative effect on longevity, fecundity, and sex ratio of the parasitoid *Aphelinus abdominalis* (Dalmon) (Hymenoptera: Aphelinidae), reared on the potato aphid *M. euphorbiae* fed on artificial diet containing *GNA* (Couty et al. 2001). Thus, despite the potential of lectins when incorporated into plants for pest control, it is necessary to consider the deleterious effects of transgenic plants on natural enemies.

Smith and Boyko (2007) review the molecular basis for plant resistance and defense responses against aphid feeding. They consider that plant genes that participate in recognition of aphid herbivory together with genes involved in plant defense against herbivores mediate plant resistance to aphids. The rupture of the plant cell wall during aphid feeding triggers the activity of hundreds of genes that appear to be involved in the induction of defense responses in several plant species. Recent studies on the differential expression of the genes *Pto* and *Ptil* in wheat plants resistant to *D. noxia* provide evidence of the involvement of the *Pto* gene in plant resistance, suggesting that aphid feeding may trigger multiple signaling routes in plants. Early signs include gene-by-gene recognition and defense signaling in resistant plants, although recognition of cellular damage inflicted by *D. noxia* occurs in both susceptible and resistant plants. The signaling is mediated by several compounds, including jasmonic acid, salicylic acid, ethylene, abscisic acid, gibberellic acid, nitric oxide, and auxins. These signals lead to the production of chemical defenses that act directly on aphids. Despite differences in plant taxonomy, there are similarities in the types of plant genes that are expressed in response to feeding by different aphid species. Lazzari and Smith (unpublished) quantitatively investigated different phytohormones that may be involved with the resistance of wheat genotypes to *D. noxia*. They concluded that methyl salicylate, 12-oxophytodienoic acid, and *trans-* and methyl-jasmonate are produced at higher levels in resistant than in susceptible genotypes. The data, corroborated by microarray analysis, suggest that these phytohormones trigger the action of plant defense genes, and that these genes are upregulated on the resistant genotype.

Molecular techniques for gene manipulation may lead to new control methods. Turning out or silencing the genes that encode for the salivary enzymes may prevent the insect–plant interactions through the saliva, rendering the sap-sucking insect unable to feed normally and to deal with the plant defense responses. It should be considered that herbivores adapt or may exhibit compensatory mechanisms against the physical and chemical features of plants, taking advantage of their digestive and detoxification systems to counteract the effects of plant defenses. The study of these mechanisms and the dynamic chemistry in aphid–plant interactions is important to clarify and manipulate these processes both in the plants and insects for aphid pest management (Kessler and Baldwin 2002).

20.9 Final Considerations

Aphids are perfectly adapted to feed on the phloem sap, locating plant veins through physical and chemical stimuli. The saliva is the medium that promotes the interface with the host, carrying enzymes, probing the environment, and forming a gelatinous sheath to conduct the stylets and seal the injury in the plant cells. Different species and aphid morphs have different nutritional requirements, which can be met by their primary or secondary hosts in alternating host species or by seeking new food sources producing alate individuals. There is evidence that some symbionts supply nutritional requirements, but there is no evidence that they synthesize cholesterol in all aphid species. The relation between the sap-sucking insects and the plant phloem is a highly specialized biotic interaction. Aphids are able to survive on a nutritionally unbalanced diet because they are extremely efficient in converting food into biomass, and also because they minimize the plant defense responses. Thus, understanding the biology and nutritional interactions between aphids and their host plants, summarized in this chapter, is fundamental for management decisions for suppressing populations of aphid pests on several crops.

REFERENCES

Adams, J. B., and J. W. McAllan. 1956. Pectinase in the saliva of *Myzus persicae* (Sulz.) (Homoptera: Aphididae). *Can. J. Zool.* 34:541–3.

Adams, J. B., and J. W. McAllan. 1958. Pectinase in certain insects. *Can. J. Zool.* 36:305–8.

Argondoña, V. H., G. L. Juvenal, M. N. Hermann, and L. J. Corcuera. 1980. Role of hydroxamic acids in the resistance of cereals to aphids. *Phytochemistry* 19:1665–8.

Auclair, J. L. 1959. Feeding and excretion by the pea aphid, *Acyrthosiphon pisum* (Harr.) (Homoptera: Aphididae) reared on different varieties of peas. *Entomol. Exp. Appl.* 2:279–86.

Auclair, J. L. 1963. Aphid feeding and nutrition. *Annu. Rev. Entomol.* 8:439–90.

Auclair, J. L. 1965. Feeding and nutrition of the pea aphid, *Acyrthosiphon pisum* (Homoptera: Aphididae), on chemically defined diets of various pH and nutrient levels. *Ann. Entomol. Soc. Am.* 58:855–75.

Auclair, J. L. 1969. Nutrition of plant-sucking insects on chemically defined diets. *Entomol. Exp. Appl.* 12:623–41.

Auclair, J. L. 1987. Host plant resistance. In *Aphids: Their Biology, Natural Enemies and Control*, vol. 2C, ed. A. K. Minks and P. Harrewijn, 225–65. Amsterdam: Elsevier.

Auclair, J. L., and J. M. Boisvert. 1980. Besoins qualitatifs en vitamins hydrosolubles chez deux biotypes du puceron du pois, *Acyrthosiphon pisum*. *Entomol. Exp. Appl.* 28:233–46.

Backus, E. A. 1994. History, development, and applications of the AC electronic monitoring system for insect feeding. In *History, Development and Application of AC Electronic Insect Feeding Monitors*, ed. M. M. Ellsbury, E. A. Backus, and D. L. Ullman, 1–51. Lanham: Thomas Say Publications in Entomology.

Barlow, C. A., and P. A. Randolph. 1978. Quality and quantity of plant sap available to the pea aphid. *Ann. Entomol. Soc. Am.* 71:46–8.

Beck, D. L., G. M. Dunn, D. G. Routley, and J. S. Bowman. 1983. Biochemical basis of resistance in corn to the corn leaf aphid. *Crop Sci.* 23:995–8.

Bermingham, J., and T. L. Wilkinson. 2010. The role of intracellular symbiotic bacteria in the amino acid nutrition of embryos from the black bean aphid, *Aphis fabae*. *Entomol. Exp. Appl.* 134:272–9.

Birkle, L. M., L. B. Minto, and A. E. Douglas. 2002. Relating genotype and phenotype for tryptophan synthesis in aphid–bacterial symbiosis. *Physiol. Entomol.* 27:302–6.

Blackman, R. L. 1985. Aphid cytology and genetics. In *Proceedings of the International Aphidological Symposium at Jablonna*, pp. 171–237. Warsaw: Polska Akademia Nauk.

Blackman, R. L., and V. F. Eastop. 1984. *Aphids on the World's Crop: An Identification and Information Guide*. London: John Wilcy & Sons.

Blackman, R. L., and V. F. Eastop. 1994. *Aphids on the World's Trees: An Identification and Information Guide*. Wallingford: Cab Loxdale & H.D. International.

Boisvert, J. M., and J. L. Auclair. 1981. Influence de la riboflavin sur les besoins globaux en vitamins hydro-solubles chez le puceron du pois *Acyrthosiphon pisum*. *Can. J. Zool.* 59:164–9.

Brewer, M. J., and J. A. Webster. 2001. Probing behavior of *Diuraphis noxia* and *Rhopalosiphum maidis* (Homoptera: Aphididae) affected by barley resistance to *D. noxia* and plant water stress. *Environ. Entomol.* 30:1041–6.

Brusse, J. J. 1962. Alkaloid content and aphid infestation in *Lupinus angustifolius*. *N. Z. J. Agr. Res.* 5:188–9.

Buchner, P. 1965. *Endosymbiosis of Animals with Plant Microorganisms*. New York: Interscience.

Burke, G. R., B. B. Normark, C. Favret, and N. A. Moran. 2009. Evolution and diversity of facultative symbionts from the aphid subfamily Lachninae. *Appl. Environ. Microbiol.* 75:5328–35.

Campbell, B. C., and D. L. Dreyer. 1985. Host plant resistance of sorghum: Differential hydrolysis by sorghum pectic substances by polysaccharases of greenbug biotypes (*Schizaphis graminum*, Homoptera: Aphididae). *Arch. Insect Biochem. Physiol.* 2:203–15.

Campbell, B. C., and W. D. Nes. 1982. Translocation of sterols in sorghum and their uptake and metabolism by an aphid, *Schizaphis graminum* (Biotype C). Abstract of International Symposium on Biochemistry and Function of Isopentenoids in Plants. ARS, USA, Oakland, CA.

Campbell, B. C., and W. D. Nes. 1983. A reappraisal of sterol biosynthesis and metabolism in aphids. *J. Insect Physiol.* 29:149–56.

Campbell, B. C., D. L. McLean, M. G. Kinsey, K. C. Jones, and D. L. Dreyer. 1982. Probing behaviour of the greenbug (*Schizaphis graminum*, biotype C) on resistant and susceptible varieties of sorghum. *Entomol. Exp. Appl.* 31:140–6.

Campbell, B. C., J. D. Steffen-Campbell, J. T. Sorensen, and R. J. Gill. 1995. Paraphyly of Homoptera and Auchenorrhyncha inferred from 18S rDNA nucleotide sequences. *Syst. Entomol.* 20:175–94.

Capinera, J. L. 1981. Some effects of infestation by bean aphid, *Aphis fabae* Scopoli, on carbohydrate and protein levels in sugar beet plants, and procedures for estimating economic injury level. *Zeitsch. Ang. Entomol.* 92:374–84.

Cardoso, J. T. 2007. Morfologia, bioecologia e comportamento alimentar de *Pineus boerneri* Annand, 1928 (Hemiptera: Adelgidae) em *Pinus* spp. (Pinaceae). PhD thesis Universidade Federal do Paraná, Curitiba, PR, Brazil.

Carter, N., D. P. Aikman, and A. F. G. Dixon. 1978. An appraisal of Hughes' time-specific life table analysis for determining aphid reproduction and mortality rates. *J. Anim. Ecol.* 47:677–87.

Chandler, S. M., T. L. Wilkinson, and A. E. Douglas. 2008. Impact of plant nutrients on the relationship between an herbivorous insect and its symbiotic bacteria. *Proc. R. Soc. London B* 275:565–70.

Cloutier, C., and M. Mackauer. 1980. The effect of superparasitism by *Aphidius smithi* (Hymenoptera: Aphidiidae) on the food budget of the pea aphid, *Acyrthosiphon pisum* (Homoptera: Aphididae). *Can. J. Zool.* 58:241–4.

Costa, R. R., and J. C. Moraes. 2006. Efeitos do ácido silícico e do acibenzolar-*S*-methyl sobre *Schizaphis graminum* (Rondani) (Hemiptera: Aphididae) em plantas de trigo. *Neotr. Entomol.* 35:834–9.

Couty, A., S. J. Clark, and G. M. Poppy. 2001. Are fecundity and longevity of female *Aphelinus abdominalis* affected by development in GNA-dosed *Macrosiphum euphorbiae*? *Physiol. Entomol.* 26:287–93.

Dadd, R. H. 1967. Improvement of synthetic diet for the aphid *Myzus persicae* using plant juices, nucleic acids, or trace metals. *J. Insect Physiol.* 13:763–78.

Dadd, R. H., and T. E. Mittler. 1966. Permanent culture of an aphid on a totally synthetic diet. *Experientia* 22:832.

Dadd, R. H., and D. L. Krieger. 1968, Dietary amino acid requirements of the aphid *Myzus persicae*. *J. Insect Physiol.* 13:249–72.

Dadd, R. H., D. L. Krieger, and T. E. Mittler. 1967. Studies on the artificial feeding of the aphid *Myzus persicae* (Sulzer)—IV. Requirements of water-soluble vitamins and ascorbic acid. *J. Insect Physiol.* 13:249–72.

Dixon, A. F. G. 1963. Reproductive activity of the sycamore aphid, *Drepanosiphum platanoides* (Schr.) (Hemiptera, Aphididae). *J. Anim. Ecol.* 32:33–48.

Dixon, A. F. G. 1971. The life-cycle and host preferences of the bird cherry-oat aphid, *Rhopalosiphum padi* L. and their bearing on the theories of host alternation in aphids. *Ann. Appl. Biol.* 68:135–47.

Dixon, A. F. G. 1975a. Aphids and translocation. In *Transport in Plants. 1. Phloem Transport*, ed. M. H. Zimmerman and J. A. Milburn, 154–70. Berlin: Springer.

Dixon, A. F. G. 1975b. Seasonal changes in fat content, form, state of gonads and length of adult life in the sycamore aphid, *Drepanosiphum platanoides* (Schr.). *Trans. Entomol. Soc. Lond.* 127:87–99.

Dixon, A. F. G. 1975c. Effect of population density and food quality on autumnal reproductive activity in the sycamore aphid, *Drepanosiphum platanoides* (Schr.). *J. Anim. Ecol.* 44:297–304.

Dixon, A. F. G. 1976. Reproductive strategies of the alate morphs of the bird cherry-oat aphid *Rhopalosiphum padi* L. *J. Anim. Ecol.* 45:817–30.

Dixon, A. F. G. 1985a. *Aphid Ecology*. Glasgow: Blackie.

Dixon, A. F. G. 1985b. Structure of aphids populations. *Annu. Rev. Entomol.* 30:155–74.

Dixon, A. F. G. 1987a. Parthenogenetic reproduction and the intrinsic rate of increase of aphids. In *Aphids: Their Biology, Natural Enemies and Control*, vol. 2A, ed. A. K. Minks and P. Harrewijn, 269–87. Amsterdam: Elsevier.

Dixon, A. F. G. 1987b. The way of life of aphids: Host specificity, speciation and distribution. In *Aphids: Their Biology, Natural Enemies and Control*, vol. 2A, ed. A. K. Minks and P. Harrewijn, 197–207. Amsterdam: Elsevier.

Dixon, A. F. G., and S. McKay. 1970. Aggregation in the sycamore aphid *Drepanosiphum platanoides* (Schr.) (Hemiptera: Aphididae) and its relevance to the regulation of population growth. *J. Anim. Ecol.* 39:439–54.

Dixon, A. F. G., P. Kindlmann, J. Leps, and J. Holman. 1987. Why there are so few species of aphids, especially in the tropics. *Am. Nat.* 129:580–92.

Dixon, A. F. G., and S. D. Wratten. 1971. Laboratory studies on aggregation, size and fecundity in the black bean aphid, *Aphis fabae* Scop. *Bull. Entomol. Res.* 61:97–111.

Douglas, A. E. 1993. The nutritional quality of phloem sap utilized by natural aphid populations. *Ecol. Entomol.* 18:31–8.

Douglas, A. E. 1998. Nutritional interactions in insect–microbial symbioses: Aphids and their symbiotic bacteria *Buchnera*. *Annu. Rev. Entomol.* 43:17–37.

Douglas, A. E., C. L. M. J. François, and L. B. Minto. 2006. Facultative 'secondary' bacterial symbionts and the nutrition of the pea aphid, *Acyrthosiphon pisum*. *Physiol. Entomol.* 31:262–9.

Down, R. E., A. M. R. Gatehouse, W. D. O. Hamilton, and J. A. Gatehouse. 1996. Snowdrop lectin inhibits development and decreases fecundity of the glasshouse potato aphid (*Aulacorthum solani*) when administered in vitro and via transgenic plants both in laboratory and glasshouse trials. *J. Insect Physiol.* 42:1035–45.

Dreyer, D. L., and K. C. Jones. 1981. Feeding deterrency of flavonoids and related phenolics towards *Schizaphis graminum* and *Myzus persicae*: Aphid feeding deterrents in wheat. *Phytochemistry* 20:2489–93.

Dreyer, D. L., and B. C. Campbell. 1984. Chemical basis of host–plant resistance to aphids. *Plant Cell Environ.* 10:353–61.

Eastop, V. F. 1973. Deductions from the present day host plants of aphids and related insects. In *Insect–Plant Relationships*, ed. H. F van Emden, 157–78. Symposium of the Royal Entomological Society of London, No. 6. London: Blackwell.

Eastop, V. F. 1978. Diversity of the Sternorrhyncha within major climates zones. In *Diversity of Insect Faunas*, ed. L. A. Mound and N. Waloff, 71–88. Symposium of the Royal Entomological Society of London, No. 6. London: Blackwell.

Fischer, M. K., W. Völkl, R. Schopf, and K. H. Hoffmann. 2002. Age-specific patterns in honeydew production and honeydew composition in the aphid *Metopeurum fuscoviride*: Implications for ant-attendance. *J. Insect Physiol.* 48:319–26.

Gardner, D. C., and A. J. Peel. 1972. Some observations on the role of ATP in sieve-tube translocation. *Planta* 107:217–26.

Gatehouse, A. M. R., R. E. Down, K. S. Powell, N. Sauvion, Y. Rahbé, C. A. Newell, A. Merryweather, W. D. O. Hamilton, and J. A. Gatehouse. 1996. Transgenic potato plants with enhanced resistance to the peach–potato aphid *Myzus persicae*. *Entomol. Exp. Appl.* 79:295–307.

Gibson, R. W., and J. A. Pickett. 1983. Wild potato repels aphids by release of aphid alarm pheromone. *Nature* 302:608–9.

Giordanengo, P., L. Brunissen, C. Rusterucci, C. Vincent, A. van Bel, S. Dinant, C. Girousse, M. Faucher, and J. L. Bonnemain. 2010. Compatible plant–aphid interactions: How aphids manipulate plant responses. *C. R. Biol.* 333:516–23.

Griffiths, G. W., and S. D. Beck. 1974. Effects of antibiotics on intracellular symbionts in the pea aphid, *Acyrthosiphon pisum*. *Cell Tissue Res.* 148:287–300.

Griffiths, G. W., and S. D. Beck. 1977. In vivo sterol biosynthesis by pea aphid symbionts as determined by digitonin and electron microscopic autoradiography. *Cell Tissue Res.* 176:179–90.

Güntner, C., A. Vazquez, G. González, A. Usubillaga, F. Ferreira, and P. Moyna. 2000. Effect of *Solanum* glycoalkaloids on potato aphid, *Macrosiphum euphorbiae*: Part II. *J. Chem. Ecol.* 26:1113–21.

Hales, D. F., J. Tomiuk, K. Wöhrmann, and P. Sunnucks. 1997. Evolutionary and genetic aspects of aphid biology: A review. *Eur. J. Entomol.* 94:1–55.

Harris, K. F., and J. E. Bath. 1973. Regurgitation by *Myzus persicae* during membrane feeding: Its likely function in transmission of nonpersistent plant viruses. *Ann. Entomol. Soc. Am.* 66:793–6.

Heie, O. E. 1987. Paleontology and phylogeny. In *Aphids: Their Biology, Natural Enemies and Control*, vol. 2A, ed. A. K. Minks and P. Harrewijn, 367–91. Amsterdam: Elsevier.

Heie, O. E. 1994. Why are there so few aphid species in the temperate areas of the southern hemisphere? *Eur. J. Entomol.* 91:127–33.

Hille Ris Lambers, D. 1966. Polymorphism in Aphididae. *Annu. Rev. Entomol.* 11:47–78.

Holman, J. 1971. Factors influencing the host range of the aphids in the tropics. *Proc. 13th Int. Congr. Entomol.* 2:339–40.

Houk, E. J. 1987. Symbionts. In *Aphids: Their Biology, Natural Enemies and Control*, vol. 2A, ed, A. K. Minks and P. Harrewijn, 123–9. Amsterdam: Elsevier.

Houk, E. J., and G. W. Griffiths. 1980. Intracellular symbionts of the Homoptera. *Annu. Rev. Entomol.* 25:161–87.

Hussain, A., J. M. S. Forrest, and A. F. G. Dixon. 1974. Sugar, organic acid, phenolic acid and plant growth regulator content of extracts of honeydew of the aphid *Myzus persicae* and of its host plant *Raphanus sativus*. *Ann. Appl. Biol.* 78:65–73.

Jansson, J., and B. Ekbom. 2002. The effect of different plant nutrient regimes on the aphid *Macrosiphum euphorbiae* growing on petunia. *Entomol. Exp. Appl.* 104:109–16.

Jördens-Röttger, D. 1979. The role of phenolic substances for host-selection behaviour of the black aphid, *Aphis fabae*. *Entomol. Exp. Appl.* 26:49–54.

Kennedy, J. S. 1958. Physiological condition of the host plant and susceptibility to aphid attack. *Entomol. Exp. Appl.* 1:50–65.

Kennedy, J. S., and I. H. M. Fosbrooke. 1972. The plant in the life of an aphid. In *Insect–Plant Relationships*, ed. H. F. van Emden, 129–40. Symposium of the Royal Entomological Society of London. No. 6. London: Blackwell.

Kennedy, J. S., and T. E. Mittler. 1953. A method of obtaining phloem sap via the mouthparts of aphids. *Nature* 171:528.

Kessler, A., and I. T. Baldwin. 2002. Plant responses to insect herbivory: The emerging molecular analysis. *Annu. Rev. Plant Biol.* 53:299–328.

Kiss, A. 1981. Melezitose, aphids and ants. *Oikos* 37:382.

Klingauf, F. 1981. Inter-relations between pests and climatic factors. In *Food–Climate Interactions*, ed. W. Bach, J. Pankrath, and S. H. Schneider, 285–301. Dordrecht: D. Reidel Publishing Company.

Klingauf, F. A. 1987. Feeding, adaptation and excretion. In *Aphids: Their Biology, Natural Enemies and Control*, vol. 2A, ed. A. K. Minks and P. Harrewijn, 225–53. Amsterdam: Elsevier.

Leather, S. R., and A. F. G. Dixon. 1984. Aphid growth and reproductive rates. *Entomol. Exp. Appl.* 35:137–40.

Leckstein, P. M., and M. Llewellyn. 1974. The role of amino acids in diet intake and selection and the utilization of dipeptides by *Aphis fabae*. *J. Insect Physiol.* 20:877–85.

Lees, A. D. 1979. The maternal environment and the control of morphogenesis in insects. *Br. Soc. Dev. Biol. Symp.* 4:221–39.

Le Roux, V., S. Dugravot, L. Brunissen, C. Vincent, Y. Pelletier, and P. Giordanengo. 2010. Antixenosis phloem-based resistance to aphids: Is it the rule? *Ecol. Entomol.* 35:407–16.

Llewellyn, M. 1982. The energy economy of fluid-feeding herbivorous insects. In *Proceedings of the Fifth Symposium on Insect–Plant Relationship*, ed. J. H. Visser and A. K. Minks, 243–51. Wageningen, The Netherlands: Pudoc.

Llewellyn, M. 1987. Aphid energy budgets. In *Aphids: Their Biology, Natural Enemies and Control*, vol. 2B, ed. A. K. Minks and P. Herrewijn, 109–17. Amsterdam: Elsevier.

Llewellyn, M., and M. Mohamed. 1982. Inter- and intraspecific variation in the performance of eight species from the genus *Aphis*. *Proceedings of the 5th International Symposium on Insect–Plant Relationships*. Centre for Agricultural Publication and Documentation, Wageningen.

Long, B. J., G. M. Dunn, J. S. Bowman, and D. G. Routley. 1977. Relationship of hydroxamic acid content in corn and resistance to the corn leaf aphid. *Crop Sci.* 17:55–8.

Lowe, H. J. B. 1967. Interspecific differences in the biology of aphids (Homoptera: Aphididae) on leaves of *Vicia faba*. I. Feeding behaviour. *Entomol. Exp. Appl.* 10:347–57.

MacGibbon, D. B., and E. J. Beuzenberg. 1978. Location of glucosinolase in *Brevicoryne brassicae* and *Lipaphis erysimi* (Aphididae). *N. Z. J. Sci.* 21:389–92.

Mansour, M. H., N. Z. Dimetry, and I. S. Rofaeel. 1982. The role of coumarin as secondary plant substance in the food specificity of the cow pea aphid *Aphis craccivora* Koch. *Z. Angew. Entomol.* 93:151–7.

Markkula, M., and S. Laurema. 1967. The effect of amino acids, vitamins, and trace elements on the development of *Acyrthosiphon pisum* Harris (Hom. Aphididae). *Ann. Agr. Fenn.* 6:77–80.

Mayoral, A. M., W. F. Tjallingii, and P. Castañera. 1996. Probing behaviour of *Diuraphis noxia* on five cereal species with different hydroxamic acid levels. *Entomol. Exp. Appl.* 78:341–8.

McLean, D. L., and M. G. Kinsey. 1964. A technique for electronically recording aphid feeding and salivation. *Nature* 205:1358–9.

McLean, D. L., and M. G. Kinsey. 1965. Identification of electrically recorded curve patterns associated with aphid salivation and ingestion. *Nature* 205:1130–1.

Miles, P. W. 1964. Studies on the salivary physiology of plant-bugs: Oxidase activity in the salivary apparatus and saliva. *J. Insect Physiol.* 10:121–9.

Miles, P. W. 1965. Studies on the salivary physiology of plant-bugs: The salivary secretions of aphids. *J. Insect Physiol.* 11:1261–8.

Miles, P. W. 1969. Interaction of plant phenols and salivary phenolases in the relationship between plants and Hemiptera. *Entomol. Exp. Appl.* 12:736–44.

Miles, P. W. 1972. The saliva of the Hemiptera. *Annu. Rev. Phytopathol.* 9:183–255.

Miles, P. W. 1987. Feeding process of Aphidoidea in relation to effects on their food plants. In *Aphids: Their Biology, Natural Enemies and Control*, vol. 2B, ed. A. K. Minks and P. Harrewijn, 321–39. Amsterdam: Elsevier.

Miles, P. W. 1998. Aphid salivary functions: The physiology of deception. In *Aphids in Natural and Managed Ecosystems*, ed. J. M. Nieto-Nafría and A. F. G. Dixon, 255–63. International 5th Symposium on Aphids. León, 1997. León: Universidad de León.

Miles, P. W., D. L. McLean, and M. G. Kinsey. 1964. Evidence that two species of aphid ingest food through an open stylet sheath. *Experientia* 20:582.

Miles, P. W., D. Aspinall, and L. Rosenberg. 1982. Performance of the cabbage aphid, *Brevicoryne brassicae* (L.), on water-stressed rape plants, in relation to changes in their chemical composition. *Austr. J. Zool.* 30:337–45.

Mittler, T. E. 1953. Amino-acids in phloem sap and their excretion in aphids. *Nature* 172:207.

Mittler, T. E. 1957. Studies on the feeding and nutrition of *Tuberolachnus salignus* (Gmelin) (Homoptera: Aphididae). I. The uptake of phloem sap. *J. Exp. Biol.* 34:334–41.

Mittler, T. E. 1958a. Studies on the feeding and nutrition of *Tuberolachnus salignus* (Gmelin) (Homoptera: Aphididae). II. The nitrogen and sugar composition of ingested phloem sap and excreted honeydew. *J. Exp. Biol.* 35:74–84.

Mittler, T. E. 1958b. Studies on the feeding and nutrition of *Tuberolachnus salignus* (Gmelin) (Homoptera: Aphididae). III. The nitrogen economy. *J. Exp. Biol.* 35:626–38.

Mittler, T. E. 1967. Effect of amino acid and sugar concentration on the food uptake of the aphid *Myzus persicae*. *Entomol. Exp. Appl.* 10:39–51.

Mittler, T. E. 1971a. Dietary amino acid requirement of the aphid *Myzus persicae* affected by antibiotic uptake. *J. Nutr.* 101:1023–8.

Mittler, T. E. 1971b. Some effects on the aphid *Myzus persicae* of ingesting antibiotics incorporated into artificial diets. *J. Insect Physiol.* 17:1333–47.

Mittler, T. E. 1976. Ascorbic acid and other chelating agents in the trace-mineral nutrition of the aphid *Myzus persicae* on artificial diets. *Entomol. Exp. Appl.* 20:81–98.

Mittler, T. E., R. H. Dadd, and S. C. Daniels Jr. 1970. Utilization of different sugars by the aphid *Myzus persicae*. *J. Insect Physiol.* 16:1873–90.

Miyazaki, M. 1987. Forms and morphs of aphids. In *Aphids: Their Biology, Natural Enemies and Control*, vol. 2A, ed. A. K. Minks and P. Harrewijn, 367–91. Amsterdam: Elsevier.

Montllor, C. B., and W. F. Tjallingii. 1989. Stylet penetration by two aphids species on susceptible and resistant lettuce. *Entomol. Exp. Appl.* 52:103–11.

Moran, N. A. 1988. The evolution of host plant alternation in aphids: Evidence for specialization as a dead end. *Am. Nat.* 132:681–706.

Moran, N. A. 1990. Aphid's life cycles: Two evolutionary steps. *Am. Nat.* 136:135–8.

Moran, N. A. 1992. The evolution of aphid lifecycles. *Annu. Rev. Entomol.* 37:321–48.

Mordvilko, A. K. 1928. The evolution of cycles and the origin of heteroecy (migration) in plant-lice. *Ann. Mag. Nat. Hist.* 2:570–82.

Morris, J. G. 1991. Nutrition. In *Environmental and Metabolic Animal Physiology*, ed. C. L. Prosser, 231–76. New York: John Wiley & Sons.

Nault, L., and W. E. Styer. 1972. Effects of sinigrin on host selection by aphids. *Entomol. Exp. Appl.* 15:423–37.

Nielson, M. W., and H. Don. 1974. Probing behavior of biotypes of the spotted alfalfa aphid on resistant and susceptible alfalfa clones. *Entomol. Exp. Appl.* 17:477–86.

Nowak, H., and E. Komor. 2010. How aphids decide what is good for them: Experiments to test aphid feeding behaviour on *Tanacetum vulgare* (L.) using different nitrogen regimes. *Oecologia* 163:973–84.

Penteado, S. R. C. 2007. C*inara atlantica* (Wilson) (Hemiptera, Aphididae): Um estudo de biologia e associações. PhD thesis, Universidade Federal do Paraná, Curitiba, PR, Brazil.

Petelle, M. 1980. Aphids and melezitose: A test of Owen's 1978 hypothesis. *Oikos* 35:127–8.

Petrusewicz, K., and A. Macfadyen. 1970. *Productivity of Terrestrial Animals: Principles and Methods.* IBP Handbook No. 13. Oxford: Blackwell.

Pettersson, J., W. F. Tjallingii, and J. Hardie. 2007. Host-plant selection and feeding. In *Aphids as Crop Pests*, ed. H. F. van Emden and R. Harrington, 87–113. Wallingford: Cab International.

Pickett, J. A., L. J. Wadhams, C. M. Woodcock, and J. Hardie. 1992. The chemical ecology of aphids. *Annu. Rev. Entomol.* 37:67–90.

Pollard, D. G. 1973. Plant penetration by feeding aphids (Hemiptera, Aphidoidea): A review. *Bull. Entomol. Res.* 62:631–714.

Pollard, D. G. 1977. Aphid penetration of plant tissues. In *Aphids as Virus Vectors*, ed. K. F. Harris, and K. Maramorosch, 105–18. New York: Academic Press.

Ponsen, M. B. 1979. The digestive system of *Subsaltusaphis ornata* (Homoptera: Aphididae). *Meded. Landbouwhogeschool* 79–17:1–30.

Ponsen, M. B. 1987. Alimentary tract. In *Aphids: Their Biology, Natural Enemies and Control*, vol. 2A, ed. A. K. Minks and P. Harrewijn, 79–97. Amsterdam: Elsevier.

Powell, G. 1993. The effect of pre-acquisition starvation on aphid transmission of potyviruses during observed and electrically recorded stylet penetrations. *Entomol. Exp. Appl.* 66:255–60.

Powell, G., C. R. Tosh, and J. Hardie. 2006. Host plant selection by aphids: Behavioral, evolutionary, and applied perspectives. *Annu. Rev. Entomol.* 51:309–30.

Prado, E. 1997. Aphid–plant interactions at phloem level, a behavioural study. PhD thesis, Wageningen Agricultural University.

Prado, E., and W. F. Tjallingii. 1994. Aphid activities during sieve element punctures. *Entomol. Exp. Appl.* 72:157–65.

Ramsey, S. J. MacDonald, G. Jander, A. Nakabachi, G. H. Thomas, and A. E. Douglas. 2010. Genomic evidence for complementary purine metabolism in the pea aphid, *Acyrthosiphon pisum*, and its symbiotic bacterium *Buchnera aphidicola*. *Insect Mol. Biol.* 19:241–8.

Randolph, P. A., J. C. Randolph, and C. A. Barlow. 1975. Age-specific energetic of the pea aphid *Acyrthosiphon pisum*. *Ecology* 56:359–69.

Reese, J. C., D. C. Margolies, E. A. Backus, S. Noyes, P. Bramelcox, and A. G. O. Dixon. 1994. Characterization of aphid host plant resistance and feeding behavior through use of a computerized insect feeding monitor. In *History, Development and Application of AC Electronic Insect Feeding Monitors*, ed. M. M. Ellsbury, E. A. Backus, and D. L. Ullman, 52–72. Lanham: Thomas Say Publications in Entomology.

Remaudière, G., and M. Remaudière. 1997. *Catalogue of the World's Aphididae–Homoptera Aphidoidea.* Paris: INRA.

Risebrow, A., and A. F. G. Dixon. 1987. Nutritional ecology of phloem feeding insects. In *Nutritional Ecology of Insects, Mites, Spiders and Related Invertebrates,* ed. F. Slansky Jr. and J. G. Rodriguez, 421–48. New York: John Wiley & Sons.

Sandström, J. P., and N. A. Moran. 2001. Amino acid budgets in three aphid species using the same host plant. *Physiol. Entomol.* 26:202–11.

Sandström, J. P., A. Telang, and N. A. Moran. 2000. Nutritional enhancement of host plants by aphids—A comparison of three aphid species on grasses. *J. Insect Physiol.* 46:33–40.

Saxena, P. N., and H. L. Chada. 1971. The greenbug, *Schizaphis graminum*. 1. Mouth parts and feeding habits. *Ann. Entomol. Soc. Am.* 64:897–904.

Schreiner, I., D. Nafus, and D. Pimentel. 1984. Effects of cyanogenesis in bracken fern (*Pteridium aquilinum*) on associated insects. *Ecol. Entomol.* 9:69–79.

Schroeder, L. A. 1981. Consumer growth efficiencies; their levels and relationships to ecological energetics. *J. Theor. Biol.* 93:805–28.

Shaposhnikov, G. K. 1959. The initiation and evolution of the change of hosts and the diapause in plant-lice (Aphididae) in the course of the adaptation to the annual cycles of their host plants. *Entomol. Obozr.* 38:483–504.

Slansky, Jr., F., and J. G. Rodriguez. 1987. *Nutritional Ecology of Insects, Mites, Spiders and Related Invertebrates*. New York: John Wiley & Sons.

Smith, C. M. 2005. *Plant Resistance to Arthropods–Molecular and Conventional Approaches*. Netherlands: Springer.

Smith, C. M., and E. V. Boyko. 2007. The molecular bases of plant resistance and defense responses to aphid feeding: Current status. *Entomol. Exp. Appl.* 122:1–16.

Srivastava, P. N. 1987. Nutritional physiology. In *Aphids: Their Biology, Natural Enemies and Control*, vol. 2A, ed. A. K. Minks and P. Harrewijn, 99–121. Amsterdam: Elsevier.

Srivastava, P. N., and J. L. Auclair. 1962a. Amylase activity in the alimentary canal of the pea aphid, *Acyrthosiphon pisum* (Harr.) (Homoptera: Aphididae). *J. Insect Physiol.* 8:349–55.

Srivastava, P. N., and J. L. Auclair. 1962b. Characteristics of invertase from the alimentary canal of the pea aphid, *Acyrthosiphon pisum* (Harr.) (Homoptera: Aphididae). *J. Insect Physiol.* 8:527–35.

Srivastava, P. N., and J. L. Auclair. 1963. Characteristics and nature of proteases from the alimentary canal of the pea aphid, *Acyrthosiphon pisum* (Harr.) (Homoptera: Aphididae). *J. Insect Physiol.* 9:469–74.

Srivastava, P. N., and J. L. Auclair. 1971a. An improved chemically defined diet for the pea aphid, *Acyrthosiphon pisum. Ann. Entomol. Soc. Am.* 64:474–8.

Srivastava, P. N., and J. L. Auclair. 1971b. Influence of sucrose concentration on diet uptake and performance by the pea aphid, *Acyrthosiphon pisum. Ann. Entomol. Soc. Am.* 64:739–43.

Srivastava, P. N., and J. L. Auclair. 1974. Effect of amino acid concentration on diet uptake and performance by the pea aphid, *Acyrthosiphon pisum* (Homoptera: Aphididae). *Can. Entomol.* 106:149–56.

Srivastava, P. N., and J. L. Auclair. 1975. Role of single amino acids in phagostimulation, growth, and survival of *Acyrthosiphon pisum. J. Insect Physiol.* 21:1865–71.

Srivastava, P. N., J. L. Auclair, and U. Srivastava. 1983. Effect of non-essential amino acids on phagostimulation and maintenance of the pea aphid, *Acyrthosiphon pisum. Can. J. Zool.* 61:2224–9.

Srivastava, P. N., and J. W. Rouatt. 1963. Bacteria from the alimentary canal of the pea aphid, *Acyrthosiphon pisum* (Harr.) (Homoptera: Aphididae). *J. Insect Physiol.* 9:435–8.

Staniland, L. N. 1924. The immunity of apple stocks from attacks by woolly aphis (*Eriosoma lanigerum*, Hausmann). Part II. The causes of the relative resistance of the stocks. *Bull. Entomol. Res.* 15:157–70.

Thompson, G. A., and F. L. Goggin. 2006. Transcriptomics and functional genomics of plant defense induction by phloem-feeding insects. *J. Exp. Bot.* 57:755–66.

Thurston, R., W. T. Smith, and B. P. Cooper. 1966. Alkaloid secretion by trichomes of *Nicotiana* species and resistance to aphids. *Entomol. Exp. Appl.* 9:428–32.

Tjallingii, W. F. 1978. Electronic recording of penetration behaviour by aphids. *Entomol. Exp. Appl.* 24:721–30.

Tjallingii, W. F. 1985. Electrical nature of recorded signals during stylet penetration by aphids. *Entomol. Exp. Appl.* 38:177–86.

Tjallingii, W. F. 1988. Electrical recording of stylet penetration activities. In *Aphids: Their Biology, Natural Enemies and Control*, vol. 2B, ed. A. K. Minks and P. Harrewijn, 95–108. Amsterdam: Elsevier.

Tjallingii, W. F. 1990. Continuous recording of stylet penetration activity by aphids. In *Aphid–Plant Genotype Interactions*, ed. R. K. Campbell and R. D. Eikenbary, 89–99. Amsterdam: Elsevier.

Tjallingii, W. F. 2006. Salivary secretions by aphids interacting with proteins of phloem wound responses. *J. Exp. Bot.* 57:739–45.

Turner, R. B. 1971. Dietary amino acid requirements of the cotton aphid, *Aphis gossypii*: The sulphur-containing amino acids. *J. Insect Physiol.* 17:2451–6.

Turner, R. B. 1977. Quantitative requirements for tyrosine, phenylalanine and tryptophan by the cotton aphid, *Aphis gossypii* (Glover). *Comp. Biochem. Physiol.* 56A:203–5.

van Emden, H. F. 1966. Studies on the relations of insect and host plant. III. A comparison of reproduction of *Brevicoryne brassicae* and *Myzus persicae* (Hemiptera: Aphididae) on Brussels sprout plants supplied with different rates of nitrogen and potassium. *Entomol. Exp. Appl.* 9:444–60.

van Emden, H. F. 1969. Plant resistance to *Myzus persicae* induced by a plant regulator and measured by aphid relative growth rate. *Entomol. Exp. Appl.* 12:125–31.

van Emden, H. F. 1972. Aphids as phytochemists. In *Phytochemical Ecology*, ed. J. B. Harborne, 25–43. London: Academic Press.

van Emden, H. F. 1978. Insects and secondary plant substances—An alternative viewpoint with special reference to aphids. In *Biochemical Aspects of Plant and Animal Co-Evolution*, ed. J. B. Harborne, 309–23. London: Academic Press.

van Emden, H. F., and R. Harrington. 2007. *Aphids as Crop Pests*. Wallingford: CAB International.

von Dohlen, C. D., and N. A. Moran. 1995. Molecular phylogeny of the Homoptera: A paraphyletic taxon. *J. Mol. Evol.* 41:211–23.

Walter, A. J., and C. D. Difonzo. 2007. Soil potassium deficiency affects soybean phloem nitrogen and soybean aphid populations. *Environ. Entomol.* 36:26–33.

Ward, S. A., and A. F. G. Dixon. 1982. Selective resorption of aphid embryos and habitat changes relative to life-span. *J. Anim. Ecol.* 51:854–64.

Way, M. J. 1967. Intra-specific mechanisms with special references to aphid population. *Symp. R. Entomol. Soc. Lond.* 4:18–36.

Way, M. J., and C. J. Banks. 1968. Population studies on the active stages of the black bean aphid, *Aphis fabae* Scop., on its winter host *Euonymus europaeus* L. *Ann. Appl. Biol.* 62:177–97.

Wearing, C. H. 1967. Studies on the relations of insect and host plant. II. Effects of water stress in host plants in pots on the fecundity of *Myzus persicae* (Sulz.) and *Brevicoryne brassicae* (L.). *Nature* 213:1052–3.

Wearing, C. H., and H. F. van Emdem. 1967. Studies on the relations of insect and host plant. I. Effects of water stress in host plants in infestations by *Aphis fabae* Scop., *Myzus persicae* (Sulz.) and *Brevicoryne brassicae* (L.). *Nature* 213:1051–2.

Weber, G. 1982. Some ecological consequences of genetic variability in the polyphagous aphid *Myzus persicae*. In *Proceedings of the International Symposium on Insect–Plant Relationships*, 425–6. Wageningen, Netherlands: Pudoc.

Wilkinson, T. L., L. B. Minto, and A. E. Douglas. 2001. Amino acids as respiratory substrates in aphids: An analysis of *Aphis fabae* reared on plants and diets. *Physiol. Entomol.* 26:225–8.

Will, T., W. F. Tjallingii, A. Thönessen, and A. J. van Bel. 2007. Molecular sabotage of plant defense by aphid saliva. *Proc. Natl. Acad. Sci. U. S. A.* 104:10536–41.

Woodring, J., R. Wiedemann, M. K. Fischer, K. H. Hoffmann, and W. Völkl. 2004. Honeydew amino acids in relation to sugars and their role in the establishment of ant-attendance hierarchy in eight species of aphids feeding on tansy (*Tanacetum vulgare*). *Physiol. Entomol.* 29:311–9.

Wright, J. P., D. B. Fisher, and T. E. Mittler. 1985. Measurement of aphid feeding rates on artificial diets using ^{3}H-inulin. *Entomol. Exp. Appl.* 37:9–11.

Wyatt, I. J., and P. F. White. 1977. Simple estimation of intrinsic increase rates for aphids and Tetranychid mites. *J. Appl. Ecol.* 14:757–66.

Ziegler, H. 1975. Nature of transported substances. In *Encyclopedia of Plant Physiology,* vol. I, 59–100, ed. M. H. Zimmermann and J. A. Milburn. Berlin: Springer-Verlag.

Zimmermann, M. H., and H. Ziegler. 1975. List of sugars and sugar alcohols in sieve-tube exudates. In *Encyclopedia of Plant Physiology*, vol. I, 479–503, ed. M. H. Zimmermann and J. A. Milburn. Berlin: Springer-Verlag.

21

Parasitoids (Hymenoptera)

Fernando L. Cônsoli and S. Bradleigh Vinson

CONTENTS

21.1 Introduction...515
21.2 Parasitoid Development Strategies..516
21.3 Nutritional Requirements of Immature Parasitoids..517
    21.3.1 Koinobionts..517
    21.3.2 Idiobionts...518
21.4 Host as Nutritional Environment 519
    21.4.1 Egg..519
    21.4.2 Larva...519
    21.4.3 Pupa..520
    21.4.4 Adult...521
21.5 Effect of First Trophic Level in Host–Parasitoid Interactions 521
21.6 How Parasitoids Deal with Host Restrictions 522
21.7 Nutrition of Adult Parasitoids 530
21.8 Final Considerations 531
References...531

21.1 Introduction

Parasitoids are important regulators of insect populations, and they stand out as the main group of natural enemies in agricultural systems. Insect parasitoids are spread over a number of insect orders, but the adaptations to the parasitic way of life are more diverse and abundant in the Hymenoptera (Askew 1973, Vinson and Iwantsch 1980a, Pennacchio and Strand 2006). The efficiency of parasitic Hymenoptera in host exploitation is due to their long evolutionary process for overcoming the diverse restrictions imposed by the host and its habitat. The origin of parasitism in the Hymenoptera is still under debate, with some data indicating parasitism appeared as a way of life as early as the beginning of the Jurassic, around 200–205 million years ago (Grimaldi and Engel 2005), or only more recently, nearly 160 million years ago (Rasnitsyn 1988, Whitfield 1993). The adaptations of the Hymenoptera to parasitism, which have made this group one of the best-adapted insects to host exploitation, involve the integration of three processes: (i) the use of limited nutritional resources by the immature stage, since the latter should finish its development in just one host, (ii) the allocation of part of these resources to the adult stage; and (iii) the acquisition and use of nutrients during the adult stage. The use of limited resources by the immature stage involves a series of morphofunctional adaptations and the development of diverse strategies for handling the host, which can involve the regulation of several of the host's physiological processes aimed at optimizing nutrient acquisition and utilization (Vinson et al. 2001, Pennacchio and Strand 2006).

One of the alternatives found for integrating nutrient use to the other processes was the development of distinct reproductive strategies, such as proovigenesis and synovigenesis. In fact, most parasitoid species are distributed between these reproductive extremes, allowing them to maximize the use of resources obtained from the host by the immature parasitoid. Since the reproductive process can be sustained by nutrients obtained in the adult stage, immatures can regulate the amount of nutrients to be allocated for development of the soma (exoskeleton and tissues) and non-soma (germinative structures and nutrient reserves) for metamorphosis (see discussion in Jervis et al. 2008).

Therefore, in this chapter, we will discuss parasitoid bioecology and nutrition with emphasis on the Hymenoptera, focusing on aspects related to their way of life, their nutritional requirements for immature development, and their strategies for host exploitation and adult nutrition.

21.2 Parasitoid Development Strategies

Parasitoid development, like that of other entomophagous insects, depends on its host. However, in contrast to predators, which can reach their optimum development even by using suboptimum preys, since they can make use of several individual preys during their growth, parasitoids have their development constrained to a single host. Therefore, it is clear that the success of parasitism depends on the correct decisions parasitoid females make when selecting their hosts, as most immature parasitoids will be unable to exploit suboptimum hosts. However, both parasitoids and predators have to overcome the various challenges imposed by the hosts or prey to gain access to the nutritional resources they need. The first challenge to be overcome are the host/prey defensive barriers, which are more complex for parasitoids as compared to predators, as host defenses will include host immune defenses, besides all the

TABLE 21.1

Parasitoid Development Strategies for Host Exploitation

Idiobiont Parasitoids	
of eggs	Female parasitoids inject molecules to halt the host embryonic development and enzymes to aid in the digestion of the egg contents. Nutritional resources are relatively uniform in quantity, but conditions *iv* and *v* (see below) can apply.
of pupae	Females inject chemical molecules to paralyze host development and preserve host tissues. Similarly to egg parasitoids, enzymes can also be release into the host to help in host tissue digestion. Nutritional resources are relatively uniform in quantity, but conditions *iv* and *v* (see below) can apply.
Ectoparasitoids of protected hosts	A paralyzing venom is injected to avoid parasitoid elimination from the host cuticle due to active (defensive behavior) or passive (molt) behavior. However, once the host is paralyzed and eggs are laid, the eclosing parasitoid larvae will have to complete their development using the nutritional resources the host represents at parasitization. Under these conditions, there are five development alternatives: i) to only locate and attack large hosts ii) to attack hosts of different sizes, but regulate the clutch size allocated to each host or adjust the size of the developing parasitoid iii) to evaluate host quality and to regulate the sex ratio, laying male eggs in lower-quality hosts iv) to evaluate host quality and regulate clutch size according to host quality v) to lay several eggs and allow competition to adjust progeny size
Koinobiont Parasitoids	
Ectoparasitoids	Parasitoid female-derived molecules are injected into the host to avoid molt. Parasitoid females must host-feed without causing significant damage to the host.
Larval endoparasitoids	Immature parasitoids compete by nutritional resources with tissues of their hosts. Nitrogen availability is lower in young hosts, but will increase as the host ages. Parasitoids can use different strategies for host utilization either by synchronizing their own development to the host (conformers) or by regulating host development (regulators).
Adult endoparasitoids	The success of these parasitoids will depend on the host life span and on the competition the developing parasitoid will face with the host reproductive tissues. Adult parasitoids usually induce host castration or reduce host reproductive capacity.

Source: Modified from Vinson, S. B., F. Pennacchio, and F. L. Cônsoli, In *Endocrine Interactions of Insect Parasites and Pathogens*, ed. Edwards, J. P. and R. J. Weaver. Oxford: BIOS Scientific Publishers, 2001, 187–206.

behavioral, chemical, and physical barriers imposed by hosts/preys. Another challenge to be overcome is the maintenance of the host as a nutritional resource for suitable use. In this case, there will be distinct demands depending on the parasitoid's life history and the adaptive responses developed to overcome the restrictions imposed by the host for its suitable exploitation (see discussion in Section 21.6) (Table 21.1) (Vinson et al. 2001).

Parasitoids are characterized in different ways depending on the host stage exploited (parasitoid of egg, egg–larva, larva, larva–pupa, pupa, adult), its location in the host (ecto- or endoparasitoid), or the clutch size allocated to a same host (solitary or gregarious parasitoid) (Askew 1973). From the ecological point of view, hosts are grouped into *koinobionts* and *idiobionts* considering how the host development will progress after parasitism. Idiobionts are those parasitoids that paralyze the host or, by definition, exploit sessile hosts, such as eggs and pupae, whereas koinobionts are those developing on hosts that move and grow during parasitism. Koinobionts will exploit their hosts during their growth (larva) or reproductive (adult) stage, but parasitoids can initiate (egg and early larval stages) or finalize (later larval stages and pupal stage) their development in the hosts' immobile stages (egg or pupa), as observed with parasitoids that exploit two host stages for their complete development (egg–larval and larval–pupal parasitoids). Koinobionts can also attack exposed or protected hosts, adopting strategies of development aimed at adjusting their development to that of the host (*conformers*) or, inversely, at manipulating the host's physiology to their own requirements (*regulators*) (see discussion in Section 21.6) (Table 21.1) (Mackauer and Sequeira 1993, Vinson et al. 2001).

21.3 Nutritional Requirements of Immature Parasitoids

In general, the nutritional requirements of parasitoids are very similar to those of predators (House 1977, Thompson 1999). However, like zoophytophagous predators, some parasitoids have developed specific needs due to the co-evolution with their hosts. Parasitoids may require particular nutrients from their host owing to the loss of important biosynthetic pathways (Nettles 1990) in a similar way zoophytophagous predators can need nutrients derived from a host plant (Coll and Guershon 2002). Although there are clear implications that specific requirements for host-derived molecules may be required for parasitoid development, especially for those working on designing artificial rearing media, parasitoid development in their natural hosts is not limited by their needs.

The nutrients available for the developing parasitoid can be affected by the (i) host nutrition before and after parasitism; (ii) presence of substances harmful to parasitoids in the host's food substrate; (iii) modification, storage, and use of nutrients by the host; and (iv) the stage of development and endocrinal condition of the host (Vinson and Iwantsch 1980b, Vinson and Barbosa 1987, Barbosa 1988, Thompson and Redak 2001, 2005, 2008).

21.3.1 Koinobionts

Even solitary species of koinobionts will face severe competition for nutrient acquisition, as koinobionts will always be competing for the nutrients available with their hosts' own tissues. Therefore, in spite of the occasional limitation of certain nutrients in the host, amino acid and protein availability is certainly the most limiting factor for parasitoid growth. The requirements for such components are clearly noticed even for cuticle synthesis during parasitoid growth. Proteins are one of the main components of the cuticle and although the amino acid composition is complex, aromatic amino acids, such as phenylalanine and tyrosine, and the amino acid β-alanine, which are involved in the sclerotization and darkening of the cuticle, are relatively abundant (Andersen 1985, Chen 1985). However, some of these amino acids have a low solubility and should be available as components of peptide and/or complex proteins, which remain available to the growing parasitoid only during specific stages of development of the host, for example, as storage proteins (Rahbe et al. 2002). Another source rich in amino acids such as tyrosine is the cuticle, but the amino acids are cross-linked to other components of the cuticle, making their utilization difficult. The only condition where amino acids associated with the cuticle could be available to the parasitoid would be during ecdysis, when a partial digestion of the cuticle takes place and nutrients are resorbed by the insect. Thus, the nutrients available to the immature parasitoids vary both qualitatively and quantitatively, according to the physiological alterations inherent to the host's development (Vinson et al. 2001).

The nutrients derived from the host are used by many parasitoid species from the beginning of their embryonic development (Ferkovich and Dillard 1986, Côsnoli and Vinson 2004a). Uptake of nutrients from the host during the embryonic development is commonly observed in koinobionts that produce hydropic eggs, which are eggs characterized by their reduced yolk content. The amount of yolk produced and allocated to these eggs are insufficient to sustain the morphogenetic processes of the developing embryo, and nutrients must be obtained from the host hemolymph to sustain the parasitoid fully embryonic development (Le Ralec 1995). Nutrient absorption by the egg requires a very thin chorionic structure (Le Ralec 1995). This characteristic has several implications in the reproductive process and in the nutritional ecology of the adult female, allowing the latter to have a much lower energetic investment for egg development than species producing anhydropic eggs (Le Ralec 1995, Jervis et al. 2001). This characteristic also allows egg size to be much reduced, which appears to have been an evolutionary change required for the narrow ovipositor of parasitoids. Ovipositors are instrumental in allowing egg deposition directly into a nutritionally rich medium (hemolymph), resulting in rapid embryonic development (Schlinger and Hall 1940, Le Ralec 1995, Jervis et al. 2001).

As hydropic eggs will need to absorb small nutrients, such as amino acids, from the host hemolymph to sustain the embryonic development, there may be a need for host manipulation by the parasitoid, leading to a selective increase in the concentration of required amino acids, as observed in hosts parasitized by *Toxoneuron nigriceps* (Vierick). In this case, several amino acids that participate in the citric acid cycle, which is necessary for energy production, were shown to be upregulated early during parasitoid embryonic development (Côsnoli and Vinson 2004a). However, there are also signs that high molecular weight molecules can be necessary for embryonic development of endoparasitoid with hydropic eggs (Ferkovich and Dillard 1986, Greany et al. 1990).

Although the newly emerged larva gains direct access by oral ingestion to various nutrients available in the host's hemolymph, the acquisition of specific molecules through the tegument also appears to be necessary in this stage of the parasitoid's development (de Eguileor et al. 2001, Giordana et al. 2003).

Many endoparasitoids may actively feed on the host's tissues as their larvae grow, and some will assume a typical predatory behavior while others will need the action of enzymes produced by specific cells associated with them, so that the host's tissues are dissociated and the cell contents are made available for parasitoid consumption (Sequeira and Mackauer 1992, Hemerik and Harvey 1999). Normally, the destructive behavior of the immature parasitoid to the host's tissues occurs at advanced stages of parasitoid larval development, with the immature stage facing several changes in nutrient composition of the medium in which it develops (host). In some cases these changes are related to the stage-specific nutritional requirements of the developing parasitoid (Vinson et al. 2001).

The manipulation of the parasitoid nutritional environment (host hemolymph) can be dependent on the parasitoid host utilization strategy (*conformers* vs. *regulators*), and may require different levels of host manipulation and plasticity by the immature parasitoid (Vinson and Iwantsch 1980a, Lawrence 1986, Beckage and Kanost 1993). These aspects should also be considered due to the fact that newly eclosed parasitoid larvae will be limited regarding nutrient acquisition owing to the differences in their surface areas with the cellular surface area of the host tissues, which will be competing for the nutrients circulating in the host's hemolymph. Thus, the manipulation of nutrient levels is common in parasitized hosts and this often involves the endocrine system since the growth hormones participate in the gene regulation expression of innumerable proteins and in the nutrient levels in the hemolymph (Wyatt 1980, Arrese and Soulages 2010).

21.3.2 Idiobionts

The qualitative and quantitative requirements of idiobionts are basically the same of koinobionts. However, the relationships of immature idiobionts with their nutritive medium (the host) are very distinct from those established for koinobionts (Table 21.1). While koinobionts can either manipulate their own development (conformers) or regulate the host (regulators), idiobionts basically depend on the nutritional quality of the host at the moment of parasitism and on the size of the progeny allocated to the host to allow for their development under the best possible nutritional conditions (but see Vinson 2010 for a detailed discussion on egg parasitoids). Host nutritional quality is generally related to host size, which will in turn affect parasitoid progeny allocation, and the size of a clutch allocated to a host can alter nutrient availability due to the consumption of certain tissues being density-dependent (see discussion in Sections 21.4 and 21.6).

21.4 Host as Nutritional Environment

21.4.1 Egg

Most of the hosts of primary parasitoids produce anhydropic eggs (Flanders 1942). These eggs have a large maternal energetic investment, with high glycogen, lipid, and protein contents. Most of these metabolites are deposited in the eggs as proteins, such as the glycoproteins derived from vitellogenins, the vitellins (Kunkel and Nordin 1985, Ziegler and van Antwerpen 2006). These complex proteins serve as nutrient deposits, which are made available after processing, providing the necessary energy requirements to sustain the embryonic and morphogenic processes (Oliveira et al. 1989, Handley et al. 1998, Giorgi et al. 1999). As a result of embryonic development, the available nutrients in the vitellus are used for the construction of complex embryonic structures, which results in the reduction of their energetic value for egg parasitoids during the host's embryonic development (Strand 1986). Probably, the reduction in the energetic value of the host for these parasitoids is more likely to be associated with the costs of the digestive processes of the complex embryonic structures and to the existence of poorly digestible structures (e.g., cuticle), rather than to the energy losses from metabolic processes during host embryogenesis, as there is very little variation in the total content of some important metabolites during embryonic development (Constant et al. 1994).

21.4.2 Larva

In contrast to the egg stage, the nutritional quality of the larva as a host increases as it grows and develops. On eclosing, the larvae have used all the nutritional reserves that had been stored in the egg and can eat the egg chorion as the first "meal," using the nutrients stored in it (Barros-Bellanda and Zucoloto 2001). The newly eclosed larvae restrict their feeding to the soft, newer tissues, which have a lower nutritional value than the mature plant tissues. Thus, the hemolymph of young larvae has a reduced nutrient concentration, mainly amino acids and proteins, due in part to the low nutritional food quality, but also to the elevated metabolic requirements of the larval developing tissues (Scriber and Slansky 1981, Ellsbury et al. 1989).

Nutrient availability in the larval hemolymph will increase with larval growth as food protein content and food consumption also increases (Wyatt and Pan 1978). The major changes in the hemolymph metabolite levels, mainly proteins, occur in the later larval stages. These changes include nutrient mobilization for the synthesis and release of storage proteins by fat tissues in preparation for the pupal stage (Kanost et al. 1990, Haunerland 1996). Several groups of storage proteins are produced in preparation for pupation, and they differ in their amino acid composition and temporal requirement for release to and uptake from the hemolymph. Arylphorins are storage proteins with high content of aromatic amino acids, and become a major constituent of the hemolymph at later larval stages of most insects. In Diptera, arylphorins consist of more than 15% of aromatic amino acids (phenylalanine, tyrosine, tryptophan, among others) and 4% of methionine. On the other hand, in Lepidoptera, two classes of proteins are found, the arylphorins and the methionine-rich proteins, but the arylphorins of lepidoterans are only rich in aromatic amino acids. However, certain insect groups, such as the Hymenoptera, can have typical storage proteins (hexamerins), but also carry storage proteins rich in glutamine and glutamic acid as one the most abundant in the hemolymph (Wheeler and Martinez 1995, Hunt et al. 2003). Regardless of the group of storage proteins available, storage protein concentration in the hemolymph will reach 80% of the concentration of protein during the last two-thirds of the final instar, with some of these proteins being almost completely taken up by the fat body in the prepupal stage, where they are stored for further use for metamorphosis and reproduction (Haunerland 1996).

There are significant structural changes in various tissues and in the decision of nutrient allocation as insects approach metamorphosis. A large part of the nutrients that accumulate during the immature growth is used for the growth of imaginal discs, which display most of their growth in the late larval stage. Several structural changes occur in various tissues (digestive, muscular, nervous, among others) concomitantly with the metabolic changes, as tissues may go through partial or complete histolysis and are rearranged so as to suit the requirements of the adult stage. All these alterations are initiated and controlled by the endocrine system, which also fluctuates at this stage of insect development (Riddiford and Truman 2003, Hakim et al. 2010).

21.4.3 Pupa

Holometabolous insects suffer significant chemical and physical changes during the restructuring and synthesis of new tissues. Normally, the muscles are the first tissues to suffer intense degeneration, followed by the gut and the salivary glands, while there are fewer changes in the circulatory and nervous systems. At the same time as the degeneration of these tissues, there is much synthetic activity for building new structures from the imaginal discs. Obviously, all these changes involve the release and transport of a large quantity of nutrients, which are removed from the histolyzed structures and used in the synthesis of new ones, (Gilbert 2009, Merkey et al. 2011).

Tissue histolysis and synthesis result in a distinct metabolic activity in each of these processes. As an example, the low rates of CO_2 released at the prepupal and early pupal stages due to the histolysis of tissues that predominate at these stages leads to a steep reduction in the beginning of the curve of metabolic activity. However, the metabolic curve will later increase as the pupa develops, and the process of histogenesis and tissue growth predominate (Fink 1925).

In parallel with these structural changes at this development stage, there are also changes in the chemical composition of the pupa. The intensity of changes in nutrient levels varies according to the organism, being more drastic in Diptera than in Coleoptera. However, nutrients such as carbohydrates and soluble proteins will have a drop in their concentrations during metamorphosis, whereas insoluble proteins will have an increase (Figure 21.1) (Evans 1932, 1934). The accumulation of insoluble proteins at the end of the pupal development is compatible with the end of the histogenetic process. However, as with embryonic development (Evans 1932), nitrogen content availability during pupal development is kept constant but with changes in the way this nutrient is made available (Figure 21.2).

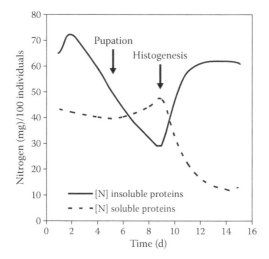

FIGURE 21.1 Nitrogen availability (mg/100 specimens) during histolysis and histogenesis of tissues in prepupal and pupal stages of *Lucilia sericata* (Meigen). (Modified from Evans, A. C., *J. Exp. Biol.*, 9, 314–321, 1932.)

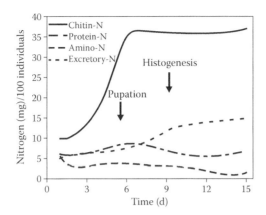

FIGURE 21.2 Nitrogen availability (mg/100 specimens) from different proteic sources during histolysis and histogenesis of tissues in prepupal and pupal stages of *L. sericata*. (Modified from Evans, A. C., *J. Exp. Biol.*, 9, 314–321, 1932.)

21.4.4 Adult

Although adult size correlates well with the individual energetic value, it is the physiological processes linked to reproduction and ageing that are responsible for changes in the internal environment, which could affect the individual's suitability as a host for the parasitoid. While the relationship between host size and energetic value to sustain the natural enemy development is obvious, as long as we disregard the existence of changes in the efficiency of specific defensive behaviors associated with insect size, it is the changes linked to the start of the reproductive stage, such as the synthesis and release of vitellogenins and other yolk proteins, and the use of nutrient reserves during ageing that affect the availability of nutrients in the host hemolymph.

The changes induced by the process of yolk protein synthesis are dependent on the adult reproductive strategy. The decisions on nutrient utilization and allocation during the immature and adult stages of insects are heavily dependent on their reproductive strategy (Boggs 1981, Jervis et al. 2007). Insects in which the reproductive capacity observed is very close or equivalent to the expected reproductive capacity (given by the number of mature oocytes at the beginning of the adult stage) do not suffer significant changes in their hemolymph protein levels since the yolk proteins are partially or totally produced during the pupal stage, using reserves accumulated during the larval stage.

Insects using this reproductive strategy are normally short lived and have no requirements for nutrients to sustain their egg development such as many lepidopterans species (Boggs 1997, Jervis et al. 2005). However, insects whose observed reproductive capacity is greater than expected show intense protein synthesis as adults, including the synthesis of yolk proteins. Adult nutrition in these species may or may not be necessary to sustain protein synthesis activity associated with reproduction since the necessary nutrients can also be derived from the reserves accumulated during the immature stage (Boggs 1981, Hamilton et al. 1990, Joern and Behmer 1997, Bauerfeind and Fischer 2005).

21.5 Effect of First Trophic Level in Host–Parasitoid Interactions

Parasitoids restrict their development to a single host and have their immature development constrained to this single food source. Once host size is normally correlated with the host food quality, and host size shows a direct correlation with host quality for parasitoid development (Hemerik and Harvey 1999, King 2002, Wang and Messing 2002), the quality of food hosts exploit has a direct impact in parasitoid development. However, recent data indicate the existence of other factors apart from host size that can affect the host nutritional quality (Häckermann et al. 2007). Changes in the nutritional quality of a host can be prejudicial to parasitism, and qualitative aspects of host nutrition can cause host modifications, which directly or indirectly affect host selection by parasitoids as well as their nutritional suitability to the natural enemy (Price et al. 1980).

Host nutritional quality can vary with changes resulting from exploitation of the host plant, as well as changes in the nitrogen-to-carbohydrate ratio (N:C) resulting from the ageing of the plant tissues being exploited due to plant physiological changes in response to seasonal changes or responses induced by environmental sources of stress. The quality of the plant used by the host herbivore is so significant that it can influence the spatial and temporal distribution of parasitism (Joern and Behmer 1997, Lill and Marquis 2001, Lill et al. 2002, Urrutia et al. 2007).

As well as the effect of food on the host nutritional quality, the food substrate exploited by the host can still directly and/or indirectly affect the natural enemy. Food substrates that are rich in secondary compounds involved in plant defense against herbivory can indirectly affect the natural enemy by influencing the metabolic rates of herbivores, reducing their growth and thus generating smaller individuals with a reduced nutritive value to the parasitoid. Secondary defense compounds can also directly affect the natural enemy because of their toxicity. This relationship is clear in herbivores able to sequester and store such compounds for their own defense, especially in nonspecific host–parasitoid associations or those which have a recent evolutionary history (Turlings and Benrey 1998, Kruse and Raffa 1999).

The evident relationship of the nutritional quality of the food of the host and the host suitability to parasitoids can be verified for natural enemies exploiting any stage of host development. Therefore, the nutritional quality of host eggs can depend on the nutrition obtained during the immature or adult stages, depending on the host's reproductive strategy (see Section 21.4.4). There are a number of cases in the literature that illustrate how the quality of the diet the host immature exploited can affect the host egg suitability for parasitoid development (van Huis and de Rooy 1998). However, information that allows us to understand the modifications in the biochemical composition of eggs leading to their reduced suitability as hosts for egg parasitoids is very limited. However, since the egg yolk has a high protein content, although also containing carbohydrates and lipids, it is possible that the diet could quantitatively or qualitatively affect the yolk protein composition. Another possible effect of nutrition in the production of lower quality eggs for parasitoids would be the production of smaller eggs. Since egg parasitoids are idiobionts, smaller hosts will provide a lower amount of food (especially for gregarious parasitoids) and consequently, be of a lower nutritional value. Egg size can be affected by various other environmental factors besides food. Responses induced by environmental conditions that favor the production of a large number of smaller eggs may be associated with the more successful exploitation of environments that have limiting conditions (Fox et al. 1997, Czesak and Fox 2003, Fischer et al. 2003, Bauerfeind and Fischer 2005).

The effect of the host's nutrition on pupal parasitoid development is similar to that observed for egg parasitoids. The quality of the pupa as a host will depend on the nutrients acquired and accumulated during the larval stage, with pupal size directly correlating to host quality (Greenblatt and Barbosa 1981, King 2002).

Parasitoids attacking actively feeding host stages are the most exposed to the effects of the host food quality (Zohdy and Zohdy 1976, Vinson and Iwantsch 1980b, Harvey et al. 1995, Kruse and Raffa 1999, Sarfraz et al. 2008). Yet, koinobiont parasitoids attacking hosts at different stages are certainly the most prone to such effects since they are exposed to the metabolic changes and growth rates imposed by the food being exploited by their host. The host's growth capacity on a certain food substrate is the most suitable parameter for indicating host quality for this parasitoid group (Harvey et al. 1994).

21.6 How Parasitoids Deal with Host Restrictions

Parasitoids have developed different strategies for overcoming the limitations imposed by hosts and by variables that affect their growth and development. These strategies allow for efficient host exploitation and are dependent on the life history of the natural enemy (Whitfield 1998, Harvey 2005).

Regardless of the parasitoid life history, whether koinobiont or idiobiont, parasitoid success relies on an efficient process of host selection. Host selection involves several stages, and for most parasitoid species they are all maternal, which include habitat and host location, and host evaluation and acceptance (Vinson and Iwantsch 1980a,b, Giraldeau and Boivin 2008, Colazza et al. 2010).

The role of female parasitoids in the host selection process is clear for most parasitoids that lay their eggs or larvae in the environment, such as those of the Coleoptera, Diptera, Lepidoptera, Neuroptera, and Trichoptera, including the parasitic Hymenoptera of Perilampidae, Eucharitidae, and Eucerotinae (Ichneumonidae). It is thought that the immatures of these species are unable to select for their host as it would be unlikely larvae would have a chance to locate several hosts and choose the most suitable among them. However, some species from these groups have particular traits that provide the newly hatched larvae to select for their hosts (reviewed in Brodeur and Boivin 2004). Obviously, there are cases in which the host selection will solely rely on the immature parasitoid, especially for those parasitoids with complex biology, such as Strepsiptera. Immatures of Strepsiptera abandon the mother while still in the host and actively search for their own hosts, with males and females exploiting hosts belonging to distinct species/groups (Kathirithamby 1989). However, given the complexity of these groups and the nonexistence of data on their bioecology, they will not be discussed in this chapter.

Since there are a number of factors (size, age, development stage, nutritional state) that can determine the biological characteristics of the natural enemy and the success of parasitism, parasitoids developed suitable sensillar structures and specific behaviors efficiently used in the process of host selection. These sensillae are distributed on the female antennae and ovipositor, and are used during an external (drumming, using the antennae) and internal (probing, using the ovipositor) assessment of the host (Vinson 1976, 1998, Cônsoli et al. 1999, Ochieng et al. 2000, Isidoro et al. 2001, Romani et al. 2010). Antennal sensillae are involved in host location and recognition even in those parasitoids in which no direct contact between the antenna and the host surface is observed, as for parasitoids attacking leafminers, borers, or protected pupae (Vinson 1998). However, there are cases in which host quality (=size) is assessed externally with the use of the antennae, as in the egg parasitoids of the genus *Trichogramma*. In these natural enemies, the curvature and external surface of the egg are evaluated as female parasitoids drums the host surface with their antennae, and will influence the size of the progeny to be allocated in the host (Schmidt and Smith 1985, 1987, Romani et al. 2010).

The stimuli perceived by the sensillae associated to the ovipositor, which stimulate oviposition and lead to host acceptance and oviposition, are still unknown for most parasitoid species. The compounds participating in host acceptance and oviposition by parasitoid females are produced by the host and should signal the host physiological and nutritional state. Normally, these compounds are proteins, amino acids, triglycerides, and salts, and several of them were identified from attempts to develop artificial rearing systems for natural enemies (Nettles et al. 1982, Kainoh et al. 1989, Rutledge 1996, Cônsoli and Grenier 2010). However, recent advances in the development of electrophysiological techniques for investigating the sensillae associated with the ovipositor of parasitoids allowed for the identification of compounds involved in host acceptance and oviposition by the larval parasitoid *Lepitopilina heteroma* (van Lenteren et al. 2007). Such techniques will perhaps allow the evaluation of factors that lead females to make wrong decisions in cases where the preference for oviposition superimposes the host suitability for parasitoid immature development. Female parasitoids of *Aphidius ervi* prefer to attack late stages (third and fourth instars) of the host *Aulacorthum solani*, even though hosts in the early stage (second instar) are the most suitable for immature development and result in a better reproductive performance (Henry et al. 2005).

The host evaluation and selection processes are key for the successful development of idiobiont parasitoids with a semigregarious or gregarious way of life (Vinson 2010). Since the host is a finite resource, cohorts exceeding the host support capacity from a single female oviposition or from superparasitism can result in unsuccessful development (although super-parasitism may be advantageous in certain cases; see van Alphen and Visser 1990, Dorn and Beckage 2007, for discussion). These parasitoids will usually display a certain level of plasticity in exploiting the food resource, resulting in the development of adults with distinct sizes and reproductive capacity (Vinson 2010). The plasticity of the development of semigregarious or gregarious parasitoids can be illustrated by the capacity of different sized cohorts of *Trichogramma* to develop in hosts of various sizes. In the case of *Trichogramma galloi* and *T. pretiosum,* it has been estimated that 1 µl of food would allow for the development of 9 to 87 individuals on the basis of the smallest and largest clutches laid in the smallest and largest hosts (Cônsoli and Parra 1999, Cônsoli et al. 1999).

In some parasitoids, the development plasticity in response to the amount and quality of the nutritional resources available to the immature may lead to the production of different morphological types and reproductive strategies. Females of the eulophid *Melittobia digitata* Dahms develop into two morphological types, a long- or a short-winged female, depending on the clutch size allocated to a host. Larvae developing from large clutches will emerge as long-winged females with a very low egg load (given by number of mature oocytes), while larvae developing from small clutches will emerge as short-winged females, but with a very high egg load. The plastic development observed in *M. digitata* is an adaptive strategy to maximize the exploitation of the nutritional resources available to larval development (Cônsoli and Vinson 2002a,b, 2004b).

Although the host of idiobiont parasitoids has a defined energetic value, nutrient release and use can vary in response to the envenomation by the female parasitoid and to the size of the clutch exploiting the host (Rivers and Denlinger 1995, Rivers et al. 1998). Gregarious or semigregarious larval or pupal idiobiont parasitoids can gain access to nutrients obtained from different tissues or only to nutrients available in the host hemolymph if attacking a host in small clutches. As the number of individuals exploiting the same host increases, nutrients stored in other tissues, such as the fat body, are released for consumption by the immature parasitoids. The use of nutrients stored in the fat body allows for the consumption of highly energetic nutrients, such as lipids and glycogen. These changes in availability of high-energy nutrients can influence the digestive physiology, growth, and development of the immature parasitoid.

Even when host size is almost always related to host nutritional quality, changes in nutrient availability can modify the energetic value of the host. For koinobiont parasitoids, host size at the time of parasitism is not directly correlated to host quality. Koinobionts develop in hosts that continue to grow, and the developing parasitoid must be prepared to adjust its development on the basis of the foreseen quality the host can reach during parasitism. To have a successful development according to future expectations of what the host quality would be for later stages of the larval development of parasitoids, these natural enemies can use a number of strategies for regulating the host development as well as had made their own larval development more flexible (Table 21.1) (Vinson and Iwantsh 1980a,b, Harvey et al. 1994, Vinson et al. 2001).

Therefore, the potential a host has to grow during the process of interaction with the parasitoid will depend on factors that can vary within and between host–parasitoid relationships. Obviously, as previously discussed, the first factors to influence the growth expectations of hosts are their own feeding rate and the nutritional quality of the food available to them. Both factors directly influence the development of the parasitoid itself (Guillot and Vinson 1973, Beckage and Riddiford 1983, Mackauer 1986, Croft and Copland 1995). The second aspect is specifically related to those parasitoids that attack hosts at different ages. In this case, the larval development of the parasitoid is very plastic and it is compatible with that of the host. When the host is too small, or is not in a suitable nutritional state to sustain successful development, the parasitoid remains as a first instar larva until the host reaches a suitable nutritional condition and only then does it continue to grow and consume the host (Smilowitz and Iwantsch 1973, Sato et al. 1986). Finally, the parasitoid can use strategies to manipulate host physiology, regulating its growth and development to support its establishment, when parasitoid larval growth and development will resume. These strategies are directly related to the parasitoid's nutritional ecology and are part of the process of host regulation, and we shall discuss it a little further.

The process of host regulation by koinobiont parasitoids has already been extensively revised in the literature, including discussions on the use of molecules involved in host regulation for the development of new strategies or applications for insect pest control (Vinson and Iwantsch 1980a, Beckage 1985, Vinson et al. 2001, Beckage and Gelman 2004). Parasitoids use a variety of chemical molecules to subdue and regulate their hosts for their successful development. Most of these molecules are proteins or small peptides produced by the ovary or venom glands associated with the female reproductive system that are injected into the host at egg laying (Asgari and Rivers 2011). Parasitoids of some subfamilies of Braconidae and Ichneumonidae are also associated with symbiotic viral particles fundamental to assure host colonization, as these viruses are largely involved in the regulation of the host immune responses to parasitism (Kroemer and Webb 2004, Bezier et al. 2009, Thézé et al. 2011). Braconids, platygasterids, and scelionids will also produce and release regulatory molecules in the host as the extraembryonic membrane of their eggs will develop into a particular cell type upon eclosion, the teratocytes (Dahlman and Vinson 1993). Teratocytes are not only involved in the synthesis and release of molecules for host regulation, but they can also assume other functions as they may also be involved in the synthesis and release of substances with a nutritive value to the developing parasitoid or of enzymes that will aid the parasitoid larvae to digest host tissues. Yet these cells are highly hypertrophic as they can accumulate nutrients during parasitoid development and serve as a source of stored nutrients for parasitoid consumption (Dahlman and Vinson 1993, Kadono-Okuda et al. 1998, Quin et al. 2000, Nakamatsu et al. 2002, Falabella et al. 2004, Gopalapillai et al. 2005, Cônsoli et al. 2007). Finally, the parasitoid larva itself can produce and release chemicals that will play a role in regulating host physiological processes involved with host metabolism, growth, and development (Führer and Willers 1986, Doury et al. 1997).

The first host physiological processes parasitoids have to deal with are concerned with the establishment of parasitism other than to the nutrition of the immature, with the exception of those parasitoids that lay hydropic eggs and requires host derived nutrients to sustain embryo development (see discussion above). Thus, the immune system is the first one to be targeted by parasitoids, and the molecules parasitoids inject into their hosts will be aimed at both the host cellular and humoral immune responses. The venom and/or the symbiotic viruses female parasitoids inject with the egg are the most common parasitoid-associated substances to affect the host's immune response. Both venom and virus affect the humoral response, altering biochemical processes involved with the phenoloxidase cascade or the cellular response by reducing the spreading capacity of plasmatocytes or by affecting the actin cytoskeletons of hemocytes (Strand and Pech 1995, Schmidt et al. 2001, Asgari and Rivers 2011).

Besides the nutrition of the host, the success of immature parasitoids in utilizing nutrients will also depend on the host's development stage and the competition parasitoid larvae will face with host tissues for nutrient utilization. In this way, the regulation of processes involved in nutrient consumption and utilization, development, metabolism, and allocation of nutritional resources is directly related to parasitoid nutrition. The intensity with which each of these processes is manipulated during parasitism depends, above all, on the parasitoid's development strategy (Table 21.1). The regulation of the host's endocrine system, for example, can happen only when the host reaches a suitable stage of development to sustain parasitoid development, even if the parasitism has been initiated in previous instars of the host, such as in the relationship *T. nigriceps—Heliothis virescens* (Pennacchio et al. 1993, Li et al. 2003). In other cases, the parasitoid can interrupt host development in the same instar in which parasitism occurred and activate the precocious expression of genes encoding for late proteins to adjust host quality to the parasitoid, like in the interaction *Euplectrus* sp.–*H. virescens* (Knop-Wright et al. 2001).

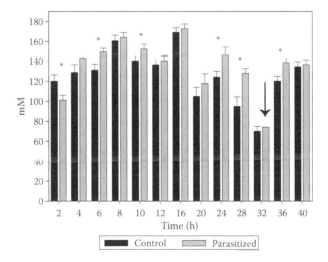

FIGURE 21.3 Amino acid concentration (mM) in larval hemolymph of *H. virescens* during embryonic development of the endoparasitoid *T. nigriceps*. *, differences between bars in each one of the sampling periods (*t* test, $P < .05$). Arrows, molt of the host to the fifth instar. (From Cônsoli, F. L. and S. B. Vinson, *Comp. Biochem. Physiol.*, 137B, 463–473, 2004a.)

The regulation of the endocrine system occurs through the direct or indirect manipulation of growth hormone levels (ecdysteroids and juvenile hormone), and the inactivation of biochemical processes. This results in the synthesis of ecdysteroids or in the cellular disruption of the prothoracic glands (Jones et al. 1992, Pennacchio et al. 1998, Cole et al. 2002), or even in the synthesis of substances that show similar activity to the juvenile hormone or in the reduction of the levels of enzymes responsible for its degradation in the hemolymph, the juvenile hormones esterases (Dover et al. 1995, Cusson et al. 2000). Independently of the endocrine regulation mechanism used by the natural enemy, control of host development is always related to maintaining it in the most suitable nutritional stage for sustaining parasitoid development (Vinson et al. 2001, Thompson and Redak 2008).

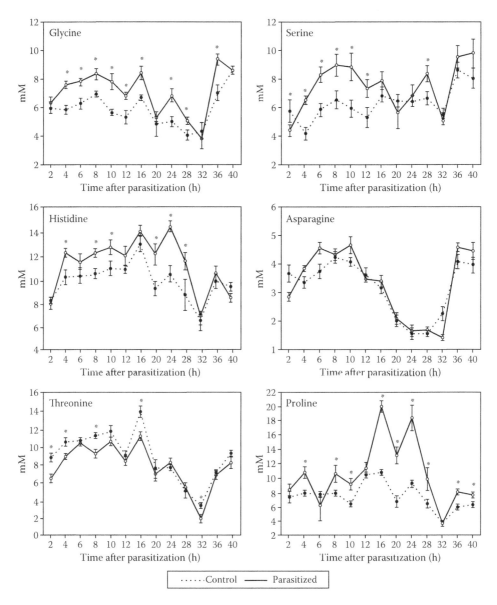

FIGURE 21.4 Changes in concentration of specific amino acids (mM) in larval hemolymph of *H. virescens* during embryonic development of the endoparasitoid *T. nigriceps*. *, differences between bars in each one of the sampling periods (*t* test, $P < .05$). (From Cônsoli, F. L. and S. B. Vinson, *Comp. Biochem. Physiol.*, 137B, 463–473, 2004a.)

The alterations that affect the synthesis, availability, and flow of nutrients are more easily related to natural enemy nutrition. One of the strategies used by parasitoids to mobilize host nutrients for their own development is the intervention in the host's nutrient distribution and allocation processes. One of the most easily observed events is the atrophy caused in imaginal discs and immature structures, such as the castration and interference in the growth of the wing imaginal discs (Digilio et al. 2000, Demmon et al. 2004). Therefore, instead of nutrients being absorbed by these developing tissues, they will stay available to the parasitoid larva(e) (Jones 1989, Falabella et al. 2000, Vinson et al. 2001, Rahbe et al. 2002). Interference can also be indirect through the action of molecular modulators produced by the natural enemy, which act on the symbionts responsible for specific nutrient production for sustaining the development of the host's reproductive apparatus. An example is the action of proteins produced by the teratocytes of *A. ervi* on the symbiont (*Buchnera aphidicola*) associated with the host *Acyrtosiphon pisum*

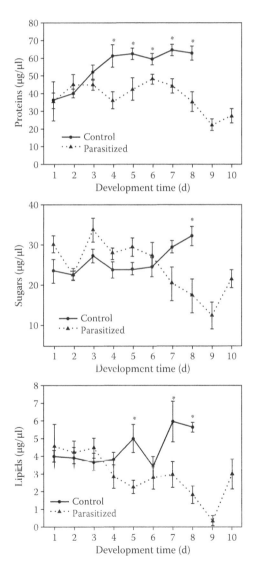

FIGURE 21.5 Metabolites composition (μg/μl) of hemolymph of last instar larvae of *Diatraea saccharalis* (F.) parasitized by *Cotesia flavipes* (Cameron) at different stages of parasitoid development. *, differences between treatments (*t* test, *P* < .05). (Modified from Salvador, G. and F. L. Cônsoli, *Biol. Control.*, 45, 103–110, 2008.)

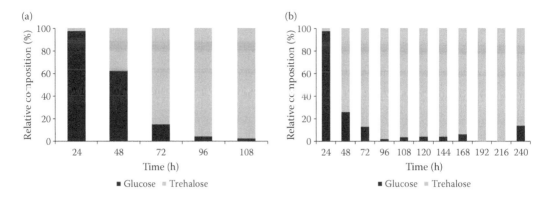

FIGURE 21.6 Relative composition of trehalose and glucose in hemolymph of *H. virescens* larvae during parasitoid development (b) as compared with control larvae (a). (Modified from Cônsoli, F. L., et al., *Comp. Biochem. Physiol.*, 142B, 181–191, 2005.)

(Harris) (Falabella et al. 2000, Rahbe et al. 2002). Another efficient way of regulating nutrient use by the host and, principally, of reducing its energetic costs, is the inhibition of the host's genetic expression, which permits primary molecules, such as amino acids, to remain available for parasitoid consumption (Dong et al. 1996, Kaeslin et al. 2005).

Alterations in the availability of primary molecules, such as amino acids, can be observed at the initial stages of parasitism and they are necessary for sustaining the embryonic development of the parasitoid. Endoparasitoids produce hydropic, yolk-poor eggs (see previous discussion), and the acquisition of

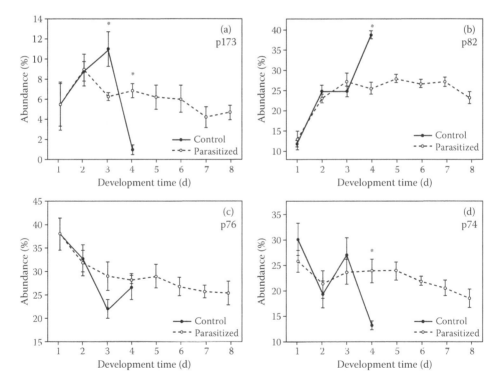

FIGURE 21.7 Abundance of selected proteins in larval hemolymph of *H. virescens* at different stages of *T. nigriceps* immature development (□) as compared with hemolymph of control larvae (●). p173, putative chromoprotein; p82, riboflavin monomer; p76 and p74, arylphorin monomers. *, differences between treatments (*t* test, $P < .05$). (Modified from Cônsoli, F. L., et al., *Comp. Biochem. Physiol.*, 142B, 181–191, 2005.)

nutrients from the host hemolymph is required to supply the energy for embryo development, as observed for *T. nigriceps*. Parasitized hosts have shown significant changes in the relative composition of free amino acids in the hemolymph particularly at the initial (tissue formation) and final (tissue differentiation) stages of the embryogenesis, especially for amino acids involved in the citric acid cycle (Figures 21.3 and 21.4) (Cônsoli and Vinson 2004a).

Other important changes in the host hemolymph composition during parasitism will occur late in the larval stage of parasitoid development (Figure 21.5). Amino acids such as tyrosine that play a key role in insect cuticle formation are also regulated during parasitism, making it available to the growing parasitoid either as a free amino acid or as part of easy-to-digest proteins (Rahbe et al. 2002). Changes in the levels/ratio of the carbohydrates glucose and trehalose in the host hemolymph can also occur as a result of the direct effect of parasitism on the process of glucogenesis (Pennacchio et al. 1993, Thompson and Dahlman 1998, Cônsoli and Vinson 2004a) (Figure 21.6).

Parasitoids also regulate the protein levels in the host hemolymph. Most of the proteins reported to be regulated by parasitoids are storage proteins, such as those carrying high aromatic amino acid contents. Regulation of these proteins can occur both at the transcription and/or the translation level (Shelby and Webb 1997, Knop-Wright et al. 2001, Cônsoli et al. 2005) (Figure 21.7). Finally, host composition can vary owing to the synthesis of specific proteins by the parasitoid or from the expression of genes derived from the symbionts associated with the natural enemy, which are located in host tissues and incorporated into its expression machinery (Kadono et al. 1998, Malva et al. 2004, Barat-Houari et al. 2006, Cônsoli et al. 2007). Unfortunately, despite the detection of various proteins in the hemolymph of parasitized hosts, often specifically associated with certain development periods of the natural enemy, very little is known on the function of these molecules, if nutritional, regulatory, or modulatory, in parasitism (Kadono-Okuda et al. 1998, Falabella et al. 2000, Hoy and Dahlman 2002, Cônsoli et al. 2005, 2007, Salvador and Cônsoli 2008) (Figure 21.8).

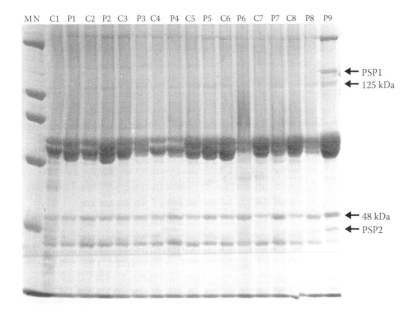

FIGURE 21.8 SDS-PAGE (7.5%) of proteins from the hemolymph of control (C) and parasitized (P) *D. saccharalis* larvae at different periods of development after parasitization by *C. flavipes*. Molecular weights (MW): myosin (200 kDa), β-galactosidase (116.25 kDa), bovine albumin (97.4 kDa), ovoalbumin (66.2 kDa), carbonic anydrase (45 kDa), trypsin inhibitor (31 kDa) (SDS-PAGE standards, Broad range; Bio-Rad, Hercules, CA, USA). PSP, parasitism-specific proteins; 125 and 48 kDa are host proteins that were found to be regulated during parasitism. (Modified from Salvador, G. and F. L. Cônsoli, *Biol. Control.*, 45, 103–110, 2008.)

21.7　Nutrition of Adult Parasitoids

Nutrition at the adult stage has implications in the insect reproduction and population ecology, but nutrition may affect adults differently depending on their reproductive physiology, and nutrient allocation and utilization strategies. The reproductive strategies of parasitoids range from fully synovigenic to fully proovigenic. It is clear that the decisions to be taken during the immature development stage on the allocation of nutritional resources, whether for somatic or nonsomatic tissues, will depend on the insect's reproductive strategy and the biotic and abiotic conditions to which they are exposed. Nutrition can serve the obvious purpose of supplying nutrients to sustain the metabolic activities of this stage, not only extending adult lifespan but it can also be essential for reproduction (Jervis et al. 2008).

Many parasitoids, especially synovigenic species, can acquire nutrients directly from the host through nondestructive or destructive feeding. More than 140 species in 17 families of Hymenoptera have already been listed as using host tissues for adult feeding, with an estimate that more than 100,000 parasitoid species have this habit (Jervis and Kidd 1986, Kidd and Jervis 1989). In nondestructive feeding, female parasitoids will feed on exudates of the host hemolymph after puncturing the host cuticle with the aid of their ovipositor. In some cases, they will not lay eggs after acquiring the host's nutrients even if feeding has not been harmful to the host, indicating the feeding activity may also be involved in the process of host selection.

In cases of destructive feeding, the host will become unsuitable for parasitism and, in many cases, will die. Destructive feeding can occur preferentially in smaller or larger instars and is dependent on the parasitoid species, on the different age classes, host distribution (isolated or grouped), and on host density (McGregor 1997, Zang and Liu 2007). Independently of the manner of feeding, there is clear indication that host feeding benefits parasitoid female fitness by increasing female reproductive capacity and longevity (Flanders 1953, Heimpel and Collier 1996, Giron et al. 2002, 2004, Burger et al. 2005). However, it should be mentioned that trehalose, the main sugar available in the hemolymph of insects, is not involved with the beneficial effects observed in host feeding. Very little or no gain at all in longevity were observed for females fed on trehalose solutions for most parasitoids feeding on host hemolymph, with trehalose being harmful to and reducing adult longevity of some parasitoid species (Jervis and Kidd 1986, Wäckers 2001).

There are also situations in which feeding on the host can reduce parasitoid fitness, such as with *Trichogramma turkestanica* Meyer, which often feeds on the first host it meets after parasitization. In this case, although there is a significant increase in fecundity (>70%), host-fed females will be short-lived as compared with females that do not host-feed. Besides, progenies produced from eggs in which hosts were also used as a food source by parasitoid females are much smaller than those produced on eggs in which females did not feed (Ferracini et al. 2006). It has also been argued that the reduction observed in female longevity in adults that host-feed is due to the allocation of nutrients to sustain the adult reproductive activities, such as egg development (Ferracini et al. 2006). Reduction in the size of the progeny from hosts that were also used as a food source for adults is related to the costs involved with the trade-offs of host feeding. At the same time females will benefit from the acquisition of nutrients from the host by increasing their fecundity, host feeding will affect the development of the progeny allocated to that host due to the reduction of the pool of nutrients that is left available to the parasitoid immature development (Rivero and West 2005, Ferracini et al. 2006).

However, even species that do not host-feed obtain nutrients from other sources during their adult life. Most adult parasitoids use carbohydrates as the primary source of energy to assure maximum longevity (Jervis et al. 1993). The main sources of carbohydrates for parasitoids are those available in floral and extrafloral nectaries and in honeydew, while pollen is used as a protein source. The nectar in floral and extrafloral nectaries contains high concentrations of sugars, but other substances required for oogenesis, such as amino acids, inorganic salts, and vitamins, are only present in very reduced concentrations (Baker and Baker 1983). However, some parasitoids that exploit hosts developing inside host plant tissues can use the available nutrients from the tissues attacked by their hosts to sustain their metabolic activities, resulting in increased longevity and fecundity (Sivinski et al. 2006, Hein and Dorn 2008). The eulophid ectoparasitoid, *Hyssopus pallidus* (Askew), uses nutrients taken directly from the fruit attacked by the host *Cydia pomonella* (L.) during the host selection process (Hein and Dorn 2008), while the braconid *Diachasmimorpha longicaudata* (Ashmead) feeds on juices released from fermented fruits attacked by its host (fruit flies) (Sivinsky et al.

2006). However, in general, carbohydrates present in the food exploited by adult parasitoids are nutrients that can be rapidly mobilized into sugars, which are especially suitable for supplying energy to sustain the basic metabolism and sudden activities, such as flight (Hoferer et al. 2000).

Honeydew is also a good source of nutrients for a number of parasitoid species and, like nectar, contains saccharose, glucose, and fructose as the main carbohydrates. However, various other carbohydrates can occur in significant concentrations and honeydew composition can vary qualitatively and quantitatively (Baker and Baker 1983, Koptur 1992). Despite the general recommendation on the use of these carbohydrate sources to improve parasitoid fitness, there are data indicating that the exploitation of these specific carbohydrates can vary considerably among parasitoids (Ferreira et al. 1998, Jacob and Evans 1998, Wäckers 2001). Thus, the composition of honeydew produced by various species of sucking insects can be unsuitable for the consumption of parasitoids and other consumers due to its low nutritional value. The production of "unsuitable" honeydew may be an evolutionary response of certain sucking insects to avoid the exploitation of their honeydew by insects other than their mutualistic predators (Wäckers 2000). Honeydew unsuitability is often associated with the presence of high concentrations of sugars derived from the insect, such as erlose, melezitose, trehalose, and raffinose. Nectar and honeydew are rich sources of saccharose, fructose, and glucose, but the possibility of poorly digestible sugars in honeydew yields honeydews that are nutritionally poor to parasitoids as compared to nectar (Wäckers 2001). Curiously, parasitoids of the genus *Diadegma* can produce some of the common sugars found in honeydew (melezitose, erlose, and maltose) in their own gut by using saccharose. However, there are no reports that these di- and trisaccharides can be absorbed and used by this parasitoid (Wäckers et al. 2006).

It is also assumed that these carbohydrate sources exploited by parasitoids can positively influence adult lipid reserves, which are an energy source used for maintaining the basic metabolic activities, egg production, and flight, and are strongly related to adult longevity (Eijs et al. 1998). However, there are strong indications that lipogenesis (*de novo* synthesis of lipids) from sugars obtained in the adult stage does not occur in adult parasitoids (Ellers 1996, Olson et al. 2000, Giron and Casas 2003). Thus, the effect of food on the lipid reserves of the adult parasitoid appears to be the possibility of their preservation when other nutrient sources are available during the adult stage.

The availability and/or use of food by the adult can influence the physiology and ecological behavior of parasitoids, inducing female wasps to forage for food or hosts (Jervis and Kidd 1995, Sirot and Bernstein 1996). These decisions involve questions related to the future probability of reproduction and adult life expectancy, and can directly affect parasitoid efficiency as biological control agents, changing, for example, the host searching time and the patch exploitation time of natural enemies in a certain area (Lewis et al. 1998).

21.8 Final Considerations

The biological diversity of parasitoids, their diverse strategies of development, and host interactions make broad generalizations on the bioecology and nutrition of this group of insects quite difficult. In spite of a considerable amount of literature on many subjects related to their bioecology and nutrition, such as host selection behavior, the effect of host nutrition on parasitoid development, the mechanisms by which the host is manipulated, and the effect of nutrition of the adult parasitoid, there are several issues still needing investigation to allow for a better understanding of this group of insects. Studies seeking to correlate the developmental aspects involved in parasitoid nutritional ecology with the first and second trophic levels are extremely important from the biological point of view, but they also have practical implications due to their involvement with the efficiency of these insects as biological control agents.

REFERENCES

Andersen, S. O. 1985. Sclerotization and tanning of the cuticle. In *Comprehensive Insect Physiology, Biochemistry and Pharmacology,* vol. 3, ed. G. A. Kerkut and L. I. Gilbert, 59–74. New York: Pergamon Press.

Arrese, E. L., and J. L. Soulages. 2010. Insect fat body: Energy, metabolism, and regulation. *Annu. Rev. Entomol.* 55:207–225.

Asgari, S., and D. B. Rivers. 2011. Venom proteins from endoparasitoid wasps and their role in host-parasite interactions. *Annu. Rev. Entomol.* 56:313–35.

Askew, R. R. 1973. Parasitic Hymenoptera. In *Parasitic Insects*, ed. R. R. Askew, 113–84. New York: American Elsevier Publishing Company, Inc.

Baker, H. G., and I. Baker. 1983. A brief historical review of the chemistry of floral nectar. In *The Biology of Nectaries*, ed. B. Bentley and T. Elias, 126–52. New York: Columbia University Press.

Barat-Houari, M., F. Hilliou, F. X. Jousset, L. Sofer, E. Deleury, J. Rocher, M. Ravallec, L. Galibert, P. Delobel, R. Feyereisen, P. Fournier, and A. N. Volkoff. 2006. Gene expression profiling *Spodoptera frugiperda* hemocytes and fat body using cDNA microarray reveals polydnavirus-associated variations in lepidopteran host genes transcript levels. *BMC Genomics* 7:160 (http://www.biomedcentral.com/1471-2164/7/160).

Barbosa, P. 1988. Natural enemies of herbivore–plant interactions: Influence of plant allelochemicals and host specificity. In *Novel Aspects of Insect–Plant Interactions*, ed. P. Barbosa and D. Letourneau, 201–29. New York: John Wiley & Sons, Inc.

Barros-Bellanda, C. H., and F. S. Zucoloto. 2001. Influence of chorion ingestion on the performance of *Ascia monuste* and its association with cannibalism. *Ecol. Entomol.* 26:557–61.

Bauerfeind, S. S., and K. Fischer. 2005. Effects of adult-derived carbohydrates, amino acids and micronutrients on female reproduction in a fruit-feeding butterfly. *J. Insect Physiol.* 51:545–54.

Beckage, N. E. 1985. Endocrine interactions between endoparasitic insects and their hosts. *Annu. Rev. Entomol.* 30:371–413.

Beckage, N. E., and L. M. Riddiford. 1983. Growth and development of the endoparasitic wasp *Apanteles congregatus:* Dependence on host nutritional status and parasite load. *Physiol. Entomol.* 8:231–41.

Beckage, N. E., and M. R. Kanost. 1993. Effects of parasitism by the braconid wasp *Cotesia congregata* on host hemolymph proteins of the tobacco hornworm, *Manduca sexta. Insect Biochem. Mol. Biol.* 23:643–53.

Beckage, N. E., and D. B. Gelman. 2004. Wasp parasitoid disruption of host development: Implications for new biologically based strategies for insect control. *Annu. Rev. Entomol.* 49:299–330.

Bezier, A., M. Annaheim, J. Herbiniere, C. Wetterwald, G. Gyapay, S. Bernard-Samain, P. Wincker, I. Roditi, M. Heller, M. Belzaghi, R. Pfister-Wilhem, G. Pirequet, C. Dupuy, E. Huguet, A. N. Volkoff, B. Lanzrein, and J. M. Drezen. 2009. Polydnavirus of Braconid wasps derive from an ancestral nudivirus. *Science* 323:926–30.

Boggs, C. L. 1981. Nutritional and life-history determinants of resource allocation in holometabolous insects. *Am. Nat.* 117:692–709.

Boggs, C. L. 1997. Reproductive allocation from reserves and income in butterfly species with differing adult diets. *Ecology* 78:181–91.

Brodeur, J., and G. Boivin. 2004. Functional ecology of immature parasitoids. *Annu. Rev. Entomol.* 49:27–49.

Burger, J. M. S., A. Kormany, J. C. van Lenteren, and L. E. M. Vet. 2005. Importance of host feeding for parasitoids that attack honeydew-producing hosts. *Entomol. Exp. Appl.* 117:147–54.

Chen, P. S. 1985. Amino acid and protein metabolism. In *Comprehensive Insect Physiology, Biochemistry and Pharmacology,* vol. 10, ed. Kerkut, G. A. and L. I. Gilbert, 177–217. New York: Pergamon Press.

Colazza, S., E. Peri, G. Salermo, and E. Conti. 2010. Host searching by egg parasitoids: Exploitation of host chemical cues. In *Egg Parasitoids in Agroecosystems with Emphasis on Trichogramma*, ed. Cônsoli, F. L., J. R. P. Parra, and R. A. Zucchi, 97–147. Dordrecht: Springer.

Cole, T. J., N. E. Beckage, F. F. Tan, A. Srinivasan, and S. B. Ramaswamy. 2002. Parasitoid–host endocrine relations: Self-reliance or co-optation? *Insect Biochem. Mol. Biol.* 32:1673–9.

Coll, M., and M. Guershon. 2002. Omnivory in terrestrial arthropods: Mixing plant and prey diets. *Annu. Rev. Entomol.* 47:267–97.

Cônsoli, F. L., and J. R. P. Parra. 1999. Development of an artificial host egg for in vitro egg laying of *Trichogramma galloi* and *T. pretiosum* using plastic membranes. *Entomol. Exp. Appl.* 91:327–36.

Cônsoli, F. L., and S. B. Vinson. 2002a. Larval development and feeding behavior of the wing dimorphics of *Melittobia digitata* Dahms (Hymenoptera: Eulophidae). *J. Hym. Res.* 11:188–96.

Cônsoli, F. L., and S. B. Vinson. 2002b. Clutch size, development and wing morph differentiation of *Melittobia digitata. Entomol. Exp. Appl.* 102:135–43.

Cônsoli, F. L., and S. B. Vinson. 2004a. Host regulation and the embryonic development of the endoparasitoid *Toxoneuron nigriceps* (Hymenoptera: Braconidae). *Comp. Biochem. Physiol.* 137B:463–73.

Cônsoli, F. L., and S. B. Vinson. 2004b. Wing morph development and reproduction of the ectoparasitoid *Melittobia digitata*: Nutritional and hormonal effects. *Entomol. Exp. Appl.* 112:47–55.

Cônsoli, F. L., E. W. Kitajima, and J. R. P. Parra. 1999. Sensilla on the antenna and ovipositor of the parasitic wasps *Trichogramma galloi* Zucchi and *T. pretiosum* Riley (Hym., Trichogrammatidae). *Microsc. Res. Tech.* 45:313–24.

Cônsoli, F. L., S. L. Brandt, T. A. Coudron, and S. B. Vinson. 2005. Host regulation and release of parasitism-specific proteins in the system *Toxoneuron nigriceps–Heliothis virescens. Comp. Biochem. Physiol.* 142B:181–91.

Cônsoli, F. L., D. Lewis, L. Keeley, and S. B. Vinson. 2007. Characterization of a cDNA encoding a putative chitinase from teratocytes of the endoparasitoid *Toxoneuron nigriceps. Entomol. Exp. Appl.* 122:271–8.

Constant, B., S. Grenier, and G. Bonnot. 1994. Analysis of some morphological and biochemical characteristics of the egg of the predaceous bug *Macrolophus caliginosus* (Het.: Miridae) during embryogenesis. *BioControl* 39:189–98.

Croft, P., and J. W. Copland. 1995. The effect of host instar on the size and sex ration of the endoparasitoid *Dacnusa sibirica. Entomol. Exp. Appl.* 74:121–4.

Cusson, M., M. Laforge, D. Miller, C. Cloutier, and D. Stoltz. 2000. Functional significance of parasitism-induced suppression of juvenile hormone esterase activity in developmental delayed *Choristoneura fumiferana* larvae. *Gen. Comp. Endocr.* 117:343–54.

Czesak, M. E., and C. W. Fox. 2003. Evolutionary ecology of egg size and number in a seed beetle: Genetic trade-off differs between environments. *Evolution* 57:1121–32.

Dahlman, D. L., and S. B. Vinson. 1993. Teratocytes: Developmental and biochemical characteristics. In *Parasites and Pathogens of Insects,* vol. 1, ed. Beckage, N. E., S. N. Thompson, and B. A. Federici, 145–65. San Diego: Academic Press.

de Eguileor, M., A. Grimaldi, G. Tettamanti, R. Valvassori, M. G. Leonardi, B. Giordana, E. Tremblay, M. C. Digilio, and F. Pennacchio. 2001. Larval anatomy and structure of absorbing epithelia in the aphid parasitoid *Aphidius ervi* Haliday (Hymenoptera, Braconidae). *Arthropod Struct. Dev.* 30:27–37.

Demmon, A. S., H. J. Nelson, P. J. Ryan, A. R. Ives, and W. E. Snyder. 2004. *Aphidius ervi* (Hymenoptera: Braconidae) increases its adult size by disrupting host wing development. *Environ. Entomol.* 33:1523–7.

Digilio, M. C., N. Isidoro, E. Tremblay, and F. Pennacchio. 2000. Host castration by *Aphidius ervi* venom proteins. *J. Insect Physiol.* 46:1041–50.

Dong, K., D. Zhang, and D. L. Dahlman. 1996. Down-regulation of juvenile hormone esterase and arylphorin production in *Heliothis virescens* larvae parasitized by *Microplitis croceipes. Arch. Insect Biochem. Physiol.* 32:237–48.

Dorn, S., and N. E. Beckage. 2007. Superparasitism in gregarious hymenopteran parasitoids: Ecological, behavioral and physiological perspectives. *Physiol. Entomol.* 32:199–211.

Doury, G., Y. Bigot, and G. Periquet. 1997. Physiological and biochemical analysis of factors in the female venom gland and larval salivary secretions of the ectoparasitoid wasp *Eupelmus orientalis. J. Insect Physiol.* 43:69–81.

Dover, B. A., A. Menon, R. C. Brown, and M. R. Strand. 1995. Suppression of juvenile hormone esterase in *Heliothis virescens* by *Microplitis demolitor* calyx fluid. *J. Insect Physiol.* 41:809–17.

Eijs, I. E. M., J. Ellers, and G. J. van Duinen. 1998. Feeding strategies in drosophilid parasitoids: The impact of natural food resources on energy reserves in females. *Ecol. Entomol.* 23:133–8.

Ellers, J. 1996. Fat and eggs: An alternative method to measure the trade-off between survival and reproduction in insect parasitoids. *Neth. J. Zool.* 46:227–35.

Ellsbury, M. M., G. A. Burkett, and F. M. Davis. 1989. Development and feeding behavior of *Heliothis zea* (Lepidoptera: Noctuidae) on leaves and flowers of crimson clover. *Environ. Entomol.* 18:323–7.

Evans, A. C. 1932. Some aspects of chemical changes during insect metamorphosis. *J. Exp. Biol.* 9:314–21.

Evans, A. C. 1934. On the chemical changes associated with metamorphosis in a beetle (*Tenebrio molitor* L.). *J. Exp. Biol.* 11:397–401.

Falabella, P., E. Trembaly, and F. Pennacchio. 2000. Host regulation by the aphid parasitoid *Aphidius ervi*: The role of teratocytes. *Entomol. Exp. Appl.* 97:1–9.

Ferkovich, S. M., and C. R. Dillard. 1986. A study uptake of radiolabeled host proteins and protein synthesis during development of eggs of the endoparasitoid, *Microplitis croceipes* (Cresson) (Braconidae). *Insect Biochem.* 16:337–45.

Ferracini, C., G. Boivin, and A. Alma. 2006. Costs and benefits of host feeding in the parasitoid wasp *Trichogramma turkestanica. Entomol. Exp. Appl.* 121:229–34.

Ferreira, C., B. B. Torres, and W. R. Terra. 1998. Substrate specificities of midgut β-glycosidases from insects of different orders. *Comp. Biochem. Physiol.* 119B:219–25.

Fink, D. E. 1925. Metabolism during embryonic and metamorphic development of insects. *J. Gen. Physiol.* 7:527–43.

Fischer, K., P. M. Brakefield, and B. J. Zwaan. 2003. Plasticity in butterfly egg size: Why larger offspring at lower temperatures? *Ecology* 84:3138–47.

Flanders, S. E. 1953. Predatism by the adult hymenopterous parasite and its role in biological control. *J. Econ. Entomol.* 46:541–4.

Fox, C. W., M. S. Thakar, and T. A. Mousseau. 1997. Egg size plasticity in a seed beetle: An adaptive maternal effect. *Am. Nat.* 149:149–63.

Führer, E., and D. Willers. 1986. The anal secretion of the endoparasitic larva *Pimpla turionellae*: Sites of production and effects. *J. Insect Physiol.* 32:361–7.

Gilbert, L. I. 2009. *Insect Development: Morphogenesis, Molting and Metamorphosis.* 778pp. New York: Elsevier BV.

Giordana, B., A. Milani, A. Grimaldi, R. Farneti, M. Casartelli, M. R. Ambroscecchio, M. C. Digilio, M. G. Leonardi, M. de Eguileor, and F. Pennacchio. 2003. Absorption of sugars and amino acids by the epidermis of *Aphidius ervi* larvae. *J. Insect Physiol.* 49:1115–24.

Giorgi, F., J. T. Bradley, and J. H. Nordin. 1999. Differential vitellin processing in insect embryos. *Micron* 30:579–96.

Giraldeau, L. A., and G. Boivin. 2008. Risk assessment and host exploitation strategies in insect parasitoids. In *Behavioral Ecology of Insect Parasitoids: From Theoretical Approaches to Field Applications*, ed. Wajnberg, E., C. Bernstein, and J. van Alphen, 212–27. New York: Blackwell Publishing.

Giron, D., and J. Casas. 2003. Lipogenesis in an adult parasitic wasp. *J. Insect Physiol.* 49:141–7.

Giron, D., A. Rivero, N. Mandon, E. Darrouzet, and J. Casas. 2002. The physiology of host feeding in parasitic wasps: Implications for survival. *Funct. Ecol.* 16:750–7.

Giron, D., S. Pincebourde, and J. Casas. 2004. Lifetime gains of host-feeding in a synovigenic parasitic wasp. *Physiol. Entomol.* 29:436–42.

Gopalapillai, R., K. Kadono-Okuda, and T. Okuda. 2005. Molecular cloning and analysis of a novel teratocyte-specific carboxylesterase from the parasitic wasp, *Dinocampus coccinellae*. *Insect Biochem. Mol. Biol.* 35:1171–80.

Greany, P., W. Clark, S. Ferkovich, J. Law, and R. Ryan. 1990. Isolation and characterization of a host hemolymph protein required for development of the eggs of the endoparasite *Microplitis croceipes*. In *Molecular Insect Science*, ed. H. H. Hagedorn, J. G. Hildebrandt, M. G. Caldwell, and J. H. Law, 306. New York: Plenum.

Greenblatt, J. A., and P. Barbosa. 1981. Effects of host's diet on two pupal parasitoids of the gypsy moth: *Brachymeria intermedia* (Nees) and *Coccygomimus turionellae* (L.). *J. Appl. Ecol.* 18:1–10.

Grimaldi, D., and M. S. Engel. 2005. Hymenoptera: Ants, bees, and other wasps. In *Evolution of the Insects*, ed. Grimaldi, D. and M. S. Engel, 407–67. New York: Cambridge University Press.

Guillot, F. S., and S. B. Vinson. 1973. Effect of parasitism by *Cardiochiles nigriceps* on food consumption and utilization by *Heliothis virescens*. *J. Insect Physiol.* 19:2073–82.

Häckermann, J., A. S. Rott, and S. Dorn. 2007. How two different host species influence the performance of a gregarious parasitoid: Host size is not equal to host quality. *J. Anim. Ecol.* 76:376–83.

Hakim, R. S., S. Baldwin, and G. Smagghe. 2010. Regulation of midgut growth, development, and metamorphosis. *Annu. Rev. Entomol.* 55:593–608.

Hamilton, R. L., R. A. Cooper, and C. Schal. 1990. The influence of nymphal and adult dietary protein on food intake and reproduction in female brown-banded cockroaches. *Entomol. Exp. Appl.* 55:23–31.

Handley, H. L., B. H. Estridge, and J. T. Bradley. 1998. Vitellin processing and protein synthesis during cricket embryogenesis. *Insect Biochem. Mol. Biol.* 28:875–85.

Harvey, J. A. 2005. Factors affecting the evolution of development strategies in parasitoid wasps: The importance of functional constraints and incorporating complexity. *Entomol. Exp. Appl.* 117:1–13.

Harvey, J. A., I. F. Harvey, and D. J. Thompson. 1994. Flexible larval growth allows use of a range of host sizes by a parasitoid wasp. *Ecology* 75:1420–8.

Harvey, J. A., I. F. Harvey, and D. J. Thompson. 1995. The effect of host nutrition on growth and development of the parasitoid wasp *Venturia canescens*. *Entomol. Exp. Appl.* 75:213–20.

Haunerland, N. H. 1996. Insect storage proteins: Gene families and receptors. *Insect Biochem. Mol. Biol.* 26:755–65.

Heimpel, G. E., and T. R. Collier. 1996. The evolution of host-feeding behaviour in insect parasitoids. *Biol. Rev.* 71:373–400.

Hein, S., and S. Dorn. 2008. The parasitoid of a fruit moth caterpillar utilizes fruit components as nutrient source to increase its longevity and fertility. *Biol. Control* 44:341–8.

Hemerik, L., and J. A. Harvey. 1999. Flexible larval development and the timing of destructive feeding by a solitary endoparasitoid: An optimal foraging problem in evolutionary perspective. *Ecol. Entomol.* 24:308–15.

Henry, L. M., D. R. Gillespie, and B. D. Roitberg. 2005. Does mother really know best? Oviposition preference reduces reproductive performance in the generalist parasitoid *Aphidius ervi. Entomol. Exp. Appl.* 116:167–74.

Hoferer, S., F. L. Wäckers, and S. Dorn. 2000. Measuring CO_2 respiration rates in the parasitoid *Cotesia glomerata. Mitt. Deuts. Ges allgem. Ang. Entomol.* 12:555–8.

House, H. L. 1977. Nutrition of natural enemies. In *Biological Control by Augmentation of Natural Enemies*, ed. R. L. Ridway and S. B. Vinson, 105–82. New York: Plenum Press.

Hoy, H. L., and D. L. Dahlman. 2002. Extended in vitro culture of *Microplitis croceipes* teratocytes and secretion of TSP14 protein. *J. Insect Physiol.* 48:401–9.

Hunt, J. H., N. A. Buck, and D. E. Wheeler. 2003. Storage proteins in vespid wasps: Characterization, developmental pattern, and occurrence in adults. *J. Insect Physiol.* 49:785–94.

Isidoro, N., R. Romani, and F. Bin. 2001. Antennal multiporous sensilla: Their gustatory features for host recognition in female parasitic wasps (Insecta, Hymenoptera: Platygastroidea). *Microsc. Res. Tech.* 55:350–8.

Jacob, H. S., and E. W. Evans. 1998. Effects of sugar spray and aphid honeydew on field populations of the parasitoid *Bathyplectes curculionis* (Hymenoptera Ichneumonidae). *Environ. Entomol.* 27:1563–8.

Jervis, M. A., and N. A. C. Kidd. 1986. Host-feeding strategies in hymenopteran parasitoids. *Biol. Rev.* 61:395–434.

Jervis, M. A., and N. A. C. Kidd. 1995. Incorporating physiological realism into models of parasitoid feeding behaviour. *Trends Ecol. Evol.* 10:434–6.

Jervis, M. A., N. A. C. Kidd, M. G. Fitton, T. Huddleston, and H. A. Dawah. 1993. Flower-visiting by hymenopteran parasitoids. *J. Nat. Hist.* 27:67–105.

Jervis, M. A., G. E. Heimpel, P. N. Ferns, J. A. Harvey, and N. A. C. Kidd. 2001. Life-history strategies in parasitoid wasps: A comparative analysis of "ovigeny." *J. Anim. Ecol.* 70:442–58.

Jervis, M. A., C. L. Boggs, and P. N. Ferns. 2005. Egg maturation strategy and its associated trade-offs: A synthesis focusing on Lepidoptera. *Ecol. Entomol.* 30:359–75.

Jervis, M. A., C. L. Boggs, and P. N. Ferns. 2007. Egg maturation strategy and survival trade-offs in holometabolous insects: A comparative approach. *Biol. J. Linn. Soc.* 90:293–302.

Jervis, M. A., J. Ellers, and J. A. Harvey. 2008. Resource acquisition, allocation, and utilization in parasitoid reproductive strategies. *Annu. Rev. Entomol.* 53:361–85.

Joern, A., and S. T. Behmer. 1997. Importance of dietary nitrogen and carbohydrates to survival, growth, and reproduction in adults of the grasshopper *Ageneotettix deorum* (Orthoptera: Acrididae). *Oeologia* 112:201–8.

Jones, D. 1989. Protein expression during parasite redirection of host (*Trichoplusia ni*) biochemistry. *Insect Biochem.* 19:445–55.

Jones, D., D. Gelman, and M. Loeb. 1992. Hemolymph concentrations of host ecdysteroids are strongly suppressed in precocious prepupae of *Trichoplusia ni* parasitized and pseudoparasitized by *Chelonus* near *curvimaculatus. Arch. Insect Biochem. Physiol.* 21:155–65.

Kadono-Okuda, K., F. Weyda, and T. Okuda. 1998. *Dinocampus* (=*Perilitus*) *coccinellae* teratocyte polypeptide: Accumulative property, localization and characterization. *J. Insect Physiol.* 44:1073–80.

Kaeslin, M., R. Pfister-Wilhelm, D. Molina, and B. Lanzrein. 2005. Changes in the haemolymph proteome of *Spodoptera littoralis* induced by the parasitoid *Chelonus inanitus* or its polydnavirus and physiological implications. *J. Insect Physiol.* 51:975–88.

Kainoh, Y., S. Tatsuki, H. Sugie, and Y. Tamaki. 1989. Host kairomones essential for egg-larval parasitoid, *Ascogaster reticulatus* Watanabe (Hymenoptera: Braconidae). *J. Chem. Ecol.* 15:1561–73.

Kanost, M. R., J. K. Kawooya, J. H. Law, R. O. Ryan, M. C. van Heusden, and R. Ziegler. 1990. Insect haemolymph proteins. *Adv. Insect Physiol.* 22:299–396.

Kathirithamby, J. 1989. Review of the order Strepsiptera. *Syst. Entomol.* 14:41–92.

Kidd, N. A. C., and M. A. Jervis. 1989. The effects of host-feeding behaviour on the dynamics of parasitoid–host interactions, and the implications for biological control. *Res. Popul. Ecol.* 31:235–74.

King, B. H. 2002. Offspring sex ratio and number in response to proportion of host sizes and ages in the parasitoid wasp *Spalangia cameroni. Environ. Entomol.* 31:505–8.

Knop-Wright, M., T. A. Coudron, and S. L. Brandt. 2001. Ecological and physiological relevance of biochemical changes in a host as a result of parasitism by *Euplectrus* spp.: A case study. In *Endocrine Interactions of Insect Parasites and Pathogens*, ed. Edwards, J. P., and R. J. Weaver, 153–86. Oxford: BIOS Scientific Publishers.

Koptur, S. 1992. Extrafloral nectary-mediated interactions between insects and plants. In *Insect–Plant Interactions*, ed. Bernays, E., 81–129. Boca Raton: CRC Press.

Kroemer, J. A., and B. A. Webb. 2004. Polydnavirus genes and genomes: Emerging gene families and new insights into polydnavirus replication. *Annu. Rev. Entomol.* 49:431–56.

Kruse, J. J., and K. F. Raffa. 1999. Effect of food plant switching by a herbivore and its parasitoid: *Cotesia melanoscela* development in *Lymantria dispar* exposed to reciprocal dietary crosses. *Ecol. Entomol.* 24:37–45.

Kunkel, J. G., and J. H. Nordin. 1985. Yolk proteins. In *Comprehensive Insect Physiology, Biochemistry, and Pharmacology*, ed. G. A. Kerkut and L. I. Gilbert, 83–111. Oxford: Bios Scientific Publishers.

Lawrence, P. O. 1986. Host–parasite hormonal interactions: An overview. *J. Insect Physiol.* 32:295–8.

Le Ralec, A. 1995. Egg contents in relation to host-feeding in some parasitic Hymenoptera. *Entomophaga* 40:87–93.

Lewis, W. J., J. O. Stapel, A. M. Cortesero, and K. Takasu. 1998. Understanding how parasitoids balance food and host needs: Importance to biological control. *Biol. Control* 11:175–83.

Li, S., P. Falabella, I. Kuriachan, S. B. Vinson, D. W. Borst, C. Malva, and F. Pennacchio. 2003. Juvenile hormone synthesis, metabolism, and resulting haemolymph titre in *Heliothis virescens* larvae parasitized by *Toxoneuron nigriceps. J. Insect Physiol.* 49:1021–30.

Lill, J. T., and R. J. Marquis. 2001. The effects of leaf quality on herbivore performance and attack from natural enemies. *Oecologia* 126:418–28.

Lill, J. T., R. J. Marquis, and R. E. Ricklefs. 2002. Host plants influence parasitism of forest caterpillars. *Nature* 417:170–3.

Mackauer, M. 1986. Growth and developmental interactions in some aphids and their hymenopteran parasites. *J. Insect Physiol.* 32:275–80.

Mackauer, M., and R. Sequeira. 1993. Patterns of development in insect parasites. In *Parasites and Pathogens of Insects vol. 1*, ed. Beckage, N. E., S. N. Thompson, and B. A. Federici, 1–23. New York: Academic Press.

Malva, C., P. Varricchio, P. Falabella, R. La Scaleia, F. Graziani, and F. Pennacchio. 2004. Physiological and molecular interaction in the host–parasitoid system *Heliothis virescens–Toxoneuron nigriceps. Insect Biochem. Mol. Biol.* 34:177–83.

McGregor, R. 1997. Host-feeding and oviposition by parasitoids on hosts of different fitness value: Influences of egg load and encounter rate. *J. Insect Behav.* 10:451–62.

Merkey, A. B., C. K. Wong, D. K. Hoshizaki, and A. G. Gibbs. 2011. Energetics of metamorphosis in *Drosophila melanogaster. J. Insect Physiol.* 57:1437–45.

Nakamatsu, Y., S. Fujii, and T. Tanaka. 2002. Larva of an endoparasitoid, *Cotesia kariyai* (Hymenoptera: Braconidae), feed on the host fat body directly in the second stadium with the help of teratocytes. *J. Insect Physiol.* 48:1041–52.

Nettles, W. C. Jr. 1990. *In vitro* rearing of parasitoids—Role of host factors in nutrition. *Arch. Insect Biochem. Physiol.* 13:167–75.

Nettles, W. C., Jr., R. K. Morrison, Z. N. Xie, D. Ball, C. A. Shenkir, and S. B. Vinson. 1982. Synergistic action of potassium chloride and magnesium sulfate on parasitoid wasp oviposition. *Science* 218:164–6.

Ochieng, S. A., K. C. Park, J. W. Zhu, and T. C. Baker. 2000. Functional morphology of antennal chemoreceptors of the parasitoid *Microplitis croceipes* (Hymenoptera: Braconidae). *Arthropod Struct. Dev.* 29:231–40.

Oliveira, P. L., M. D. de Alencar Petretski, and H. Masuda. 1989. Vitellin processing and degradation during embryogenesis in *Rhodnius prolixus. Insect Biochem.* 19:489–98.

Olson, D. M., H. Fadamiro, J. G. Lundgren, and G. E. Heimpel. 2000. Effects of sugar feeding on carbohydrate and lipid metabolism in a parasitoid wasp. *Physiol. Entomol.* 25:17–26.

Pennacchio, F., S. B. Vinson, and E. Tremblay. 1993. Growth and development of *Cardiochiles nigriceps* Viereck (Hymenoptera, Braconidae) larvae and their synchronization with some changes of the hemolymph composition of their host, *Heliothis virescens* (F.) (Lepidoptera, Noctuidae). *Arch. Insect Biochem. Physiol.* 24:65–7.

Pennacchio, F., P. Falabella, and S. B. Vinson. 1998. Regulation of *Heliothis virescens* prothoracic glands by *Cardiochiles nigriceps* polydnavirus. *Arch. Insect Biochem. Physiol.* 38:1–10.

Pennacchio, F., and M. R. Strand. 2006. Evolution of developmental strategies in parasitic Hymenoptera. *Annu. Rev. Entomol.* 51:233–58.

Price, P. W., C. E. Bouton, P. Gross, B. A. McPheron, J. N. Thompson, and A. E. Weis. 1980. Interactions among three trophic levels: Influence of plants on interactions between insect herbivores and natural enemies. *Annu. Rev. Ecol. Syst.* 11:41–65.

Quin, Q., H. Gong, and T. Ding. 2000. Two collagenases are secreted by teratocytes from *Microplitis mediator* (Hymenoptera: Braconidae) cultured in vitro. *J. Invert. Pathol.* 76:79–80.

Rahbe, Y., M. C. Digilio, G. Febvay, J. Guillaud, P. Fanti, and F. Pennacchio. 2002. Metabolic and symbiotic interactions in amino acid pools of the pea aphid, *Acyrthosiphon pisum*, parasitized by the braconid *Aphidius ervi*. *J. Insect Physiol.* 48:507–16.

Rasnitsyn, A. P. 1988. An outline of evolution of the hymenopterous insects (order Vespida). *Orient. Insects* 22:115–45.

Riddiford, L. M., and J. W. Truman. 1993. Hormone receptors and the regulation of insect metamorphosis. *Am. Zool.* 33:340–7.

Rivero, A., and S. A. West. 2005. The costs and benefits of host feeding in parasitoids. *Anim. Behav.* 69:1293–301.

Rivers, D. B., and D. L. Denlinger. 1995. Venom-induced alterations in fly lipid metabolism and its impact on larval development of the ectoparasitoid *Nasomia vitripennis* (Walker) (Hymenoptera: Pteromalidae). *J. Invertebr. Pathol.* 66:104–10.

Rivers, D. B., M. A. Pagnotta, and E. R. Huntington. 1998. Reproductive strategies of 3 species of ectoparasitic wasps are modulated by the response of the fly host *Sarcophaga bullata* (Diptera: Sarcophagidae) to parasitism. *Ann. Entomol. Soc. Am.* 91:458–65.

Romani, R., N. Isidoro, and F. Bin. 2010. Antennal structures used in communication by egg parasitoids. In *Egg Parasitoids in Agroecosystems with Emphasis on Trichogramma*, ed. Cônsoli, F. L., J. R. P. Parra, and R. A. Zucchi, 57–96. Dordrecht: Springer.

Rutledge, C. E. 1996. A survey of identified kairomones and synomones used by insect parasitoids to locate and accept their hosts. *Chemoecology* 7:121–31.

Salvador, G., and F. L. Cônsoli. 2008. Changes in the hemolymph and fat body metabolites of *Diatraea saccharalis* (Fabricius) (Lepidoptera: Crambidae) parasitized by *Cotesia flavipes* (Cameron) (Hymenoptera: Braconidae). *Biol. Control.* 45:103–10.

Sarfraz, M., L. M. Dosdall, and B. A. Keddie. 2008. Host plant genotypes of the herbivore *Plutella xylostella* (Lepidoptera: Plutellidae) affects the performance of its parasitoid *Diadegma insulare* (Hymenoptera: Ichneumonidae). *Biol. Control* 44:42–51.

Sato, Y., J. Tagawa, and T. Hidaka. 1986. Effects of the gregarious parasitoids *Apanteles ruficrus* and *A. kariyai* on host growth and development. *J. Insect Physiol.* 32:281–6.

Schlinger, E. I., and J. C. Hall. 1960. The biology, behavior, and morphology of *Praon palitans* Muesebeck, an internal parasite of the spotted alfafa aphid, *Therioaphis maculata* (Buckton) (Hymenoptera: Braconidae, Aphidiinae). *Ann. Entomol. Soc. Am.* 53:144–60.

Schmidt, J. M., and J. J. B. Smith. 1985. Host volume and measurement by the parasitoid wasp *Trichogramma minutum*: The roles of curvature and surface area. *Entomol. Exp. Appl.* 39:213–21.

Schmidt, J. M., and J. J. B. Smith. 1987. Measurement of host curvature by the parasitoid wasp *Trichogramma minutum*, and its effect on host examination and progeny allocation. *J. Exp. Biol.* 129:151–64.

Schmidt, O., U. Theopold, and M. Strand. 2001. Innate immunity and its evasion and suppression by hymenopteran endoparasitoids. *Bioessays* 23:344–51.

Scriber, J. M., and F. Slansky Jr. 1981. The nutritional ecology of immature insects. *Annu. Rev. Entomol.* 26:183–211.

Sequeira, R., and M. Mackauer. 1992. Nutritional ecology of an insect host-parasitoid association: The pea aphid-*Aphidius ervi* system. *Ecology* 73:183–9.

Shelby, K. S., and B. A. Webb. 1997. Polydnavirus infection inhibits translation of specific growth-associated host proteins. *Insect Biochem. Mol. Biol.* 27:263–70.

Sirot, E., and C. Bernstein. 1996. Time sharing between host searching and food searching in parasitoids: State dependent optimal strategies. *Behav. Ecol.* 7:189–94.

Sivinski, S., M. Aluja, and T. Holler. 2006. Food sources for adult *Diachasmimorpha longicaudata*, a parasitoid of tephritid fruit flies: Effects on longevity and fecundity. *Entom. Exp. Appl.* 118:193–202.

Smilowitz, Z., and G. F. Iwantsch. 1973. Relationships between the parasitoid *Hyposoter exiguae* and the cabbage looper *Trichoplusia ni*: Effects of host age on developmental rate of the parasitoid. *Environ. Entomol.* 2:759–63.

Strand, M. R., and L. L. Pech. 1995. Immunological basis for compatibility in parasitoid–host relationships. *Annu. Rev. Entomol.* 40:31–56.

Thézé, J., A. Bézier, G. Periquet, J. M. Drezen, and E. A. Herniou. 2011. Paleozoic origin of insect large dsDNA viruses. *Proc. Natl. Acad. Sciences USA* 38:15931–5.

Thompson, S. N. 1999. Nutrition and culture of entomophagous insects. *Annu. Rev. Entomol.* 44:561–92.

Thompson, S. N., and D. L. Dahlman. 1998. Aberrant nutritional regulation of carbohydrate synthesis by parasitized *Manduca Sexta* L. *J. Insect Physiol.* 44:745–53.

Thompson, S. N., and R. A. Redak. 2001. Altered dietary nutrient intake maintains metabolic homeostasis in parasitized larvae of the insect *Manduca sexta* L. *J. Exp. Biol.* 204:4065–80.

Thompson, S. N., and R. A. Redak. 2005. Nutrition interacts with parasitism to influence growth and physiology of the insect *Manduca sexta* L. *J. Exp. Biol.* 208:611–23.

Thompson, S. N., and R. A. Redak. 2008. Parasitism of an insect *Manduca sexta* L. alters feeding behaviour and nutrient utilization to influence developmental success of a parasitoid. *J. Comp. Physiol.* 178B:515–27.

Turlings, T. C. J., and B. Benrey. 1998. Effects of plant metabolites on the behavior and development of parasitic wasps. *Ecosciences* 5:321–33.

Urrutia, M. A. C., M. R. Wade, C. B. Phillips, and S. D. Wratten. 2007. Influence of host diet on parasitoid fitness: Unraveling the complexity of a temperate pastoral agroecosystem. *Entomol. Exp. Appl.* 123:63–71.

van Alphen, J. J. M., and M. E. Visser. 1990. Superparasitism as an adaptive strategy for insect parasitoids. *Annu. Rev. Entomol.* 35:59–79.

van Huis, A., and M. de Rooy. 1998. The effect of leguminous plant species on *Callosibruchus maculatus* (Coleoptera: Bruchidae) and its egg parasitoid *Uscana lariophaga* (Hymenoptera: Trichogrammatidae). *Bull. Entomol. Res.* 88:93–9.

van Lenteren, J. C., S. Ruschioni, R. Romani, J. J. A. van Loon, Y. T. Qiu, H. M. Smid, N. Isidoro, and F. Bin. 2007. Structure and electrophysiological responses of gustatory organs on the ovipositor of the parasitoid *Leptopilina heterotoma*. *Arthropod Struct. Dev.* 36:271–6.

Vinson, S. B. 1976. Host selection by insect parasitoids. *Annu. Rev. Entomol.* 21:109–33.

Vinson, S. B. 1998. The general host selection behavior of parasitoid Hymenoptera and a comparison of initial strategies utilized by larvaphagous and oophagous species. *Biol. Control* 11:79–96.

Vinson, S. B. 2010. Nutritional ecology of insect egg parasitoids. In *Egg Parasitoids in Agroecosystems with Emphasis on Trichogramma*, ed. Cônsoli, F. L., J. R. P. Parra, and R. A. Zucchi, 25–55. Dordrecht: Springer.

Vinson, S. B., and G. F. Iwantsch. 1980a. Host regulation by insect parasitoids. *Q. Rev. Biol.* 55:143–65.

Vinson, S. B., and G. F. Iwantsch. 1980b. Host suitability for insect parasitoids. *Annu. Rev. Entomol.* 25:397–419.

Vinson, S. B., and P. Barbosa. 1987. Interrelationships of nutritional ecology of parasitoids. In *Nutritional Ecology of Insects, Mites, and Spiders*, ed. Slansky, F. and J. G. Rodriguez, 673–95. New York: John Wiley & Sons, Inc.

Vinson, S. B., F. Pennacchio, and F. L. Cônsoli. 2001. The parasitoid–host endocrine interaction from a nutritional perspective. In *Endocrine Interactions of Insect Parasites and Pathogens*, ed. Edwards, J. P. and R. J. Weaver, 187–206. Oxford: BIOS Scientific Publishers.

Wäckers, F. L. 2000. Do oligosaccharides reduce the suitability of honeydew for predators and parasitoids? A further facet to the function of insect-synthesized honeydew sugars. *Oikos* 90:197–201.

Wackers, F. L. 2001. A comparison of nectar- and honeydew sugars with respect to their utilisation by the hymenopteran parasitoid *Cotesia glomerata*. *J. Insect Physiol.* 47:1077–84.

Wäckers, F. L., J. C. Lee, G. E. Heimpel, K. Winkler, and R. Wagenaar. 2006. Hymenopteran parasitoids synthesize 'honeydew-specific' oligosacchararides. *Funct. Ecol.* 20:790–8.

Wang, X. G., and R. H. Messing. 2004. Fitness consequences of body-size-dependent host species selection in a generalist ectoparasitoid. *Behav. Ecol. Sociobiol.* 56:513–22.

Wheeler, D. E., and T. Martínez. 1995. Storage proteins in ants (Hymenoptera: Formicidae). *Comp. Biochem. Physiol.* 112B:15–9.

Whitfield, J. B. 1998. Phylogeny and evolution of host–parasitoid interactions in Hymenoptera. *Annu. Rev. Entomol.* 43:129–51.

Wyatt, G. R. 1980. The fat body as a protein factory. In *Insect Biology in the Future*, ed. Locke, M. and D. S. Smith, 201–25. London: Academic Press.

Wyatt, G. R., and M. L. Pan. 1978. Insect plasma proteins. *Annu. Rev. Entomol.* 47:779–817.

Zang, L. S., and T. X. Liu. 2008. Host-feeding of three parasitoid species on *Bemisia tabaci* biotype B and implications for whitefly biological control. *Entomol. Exp. Appl.* 127:55–63.

Ziegler, R., and R. van Antwerpen. 2006. Lipid uptake by insect oocytes. *Insect Biochem. Mol. Biol.* 36:264–72.

Zohdy, N., and M. Zohdy. 1976. On the effect of the food of *Myzus persicae* Sulz. on the hymenopteran parasite *Aphelinus asychis* Walker. *Oecologia* 26:185–91.

22

Predatory Bugs (Heteroptera)

Vanda H. P. Bueno and Joop C. Van Lenteren

CONTENTS

22.1 Introduction 539
22.2 Taxonomy, Feeding Behavior, and Overview of Important Predatory Heteroptera 542
    22.2.1 Taxonomy 542
    22.2.2 Feeding Behavior and Prey Digestion 542
    22.2.3 Overview of Important Predatory Heteroptera 544
        22.2.3.1 Infraoder Cimicomorpha 544
        22.2.3.2 Infraorder Pentatomomorpha 546
22.3 Food Demands and Mass Rearing of Predatory Heteroptera 547
    22.3.1 Influence of Food on Development and Reproduction 547
    22.3.2 Influence of Food Quality on Mass Rearing of Predatory Heteroptera 548
22.4 Trophic Relationships in Predatory Heteroptera 551
    22.4.1 Within Trophic Level Relationships: Cannibalism 551
    22.4.2 Within Trophic Level Relationships: Intraguild Predation 552
    22.4.3 Predator–Plant Interactions 554
    22.4.4 Natural Enemies of Predatory Heteroptera 558
22.5 Habitat Choice and Distribution of Predatory Heteroptera 558
22.6 Predatory Heteroptera Used in Commercial Biological Control 559
22.7 Final Considerations 561
References 561

22.1 Introduction

Increased concern about human health and the environment, fast development of resistance by insects to pesticides, and the interest to utilize sustainable agricultural methods are important stimuli for the application of biological control of pests as an essential component in future crop protection programs. Nowadays, for several cropping systems, classical biological control (where natural enemies are collected in an exploration area—usually the area of origin of the pest—and introduced in new areas where the pest occurs) or augmentative biological control (where natural enemies are mass reared in biofactories and periodically released) can be more economical than conventional chemical pest control (Gurr and Wratten 2000). Also, conservation biological control (which consists of actions that protect and stimulate the performance of naturally occurring natural enemies) can be an important approach for developing more sustainable cropping systems.

Despite the fact that most of the research concerning predatory insects used in biological control has focused mainly on Coccinellidae and Chrysopidae, use of heteropteran species has strongly increased during the past two decades (Van Lenteren 2011). Both the number of species commercially available as well as the areas treated with Heteroptera increased. The recent interest in Heteroptera can be explained both from an ecological and a practical perspective because they form an important component of the arthropod predatory fauna in natural and managed ecosystems.

Within the context of biological control, predatory Heteroptera may perform an important role, although they are currently only forming a small portion of all biological control agents used. This is can be illustrated by recent data from Europe, where commercial use of natural enemies is largest and comprises 75% of all sales (Cock et al. 2010). In Europe, 15 heteropteran predatory species are marketed, while in total 176 invertebrate natural enemies are available (Figure 22.1). The commercial use of predatory bugs as biological control agents for the control of pests in agriculture and forestry has recently shown a strong increase, particularly in Europe, where several heteropteran species [*Orius laevigatus* (Fieber), *Macrolophus pygmaeus* (Rambur), and *Nesidiocoris tenuis* (Reuter)] are produced in very large numbers and released in greenhouses. In these protected situations, these predators are usually released as a second line of defense in addition to more specific natural enemy species, and mainly when pest populations reach high densities (Brodeur et al. 2002).

Particular characteristics of heteropteran predators, such as (facultative or obligatory) feeding on plants (i.e., phytophagy) and attacking other predators (i.e., intraguild predation), are very interesting in the light of population dynamics and community structures in terrestrial ecosystems. The habit of feeding on plants can sometimes result in plant damage (Figure 22.2) (Van Schelt et al. 1996; Albajes and Alomar 1999; Calvo et al. 2009; Arnó et al. 2010) and reduces the potential value of several species as biological control agents (Albajes and Alomar 1999). However, with well-planned releases, such problems can often be prevented and the phenomenon of plant feeding in heteropteran predators may have the beneficial effect that these predators can survive periods of prey scarcity (Albajes and Alomar 1999; Castane et al. 2011). Maybe there is no other insect order showing such a rich diversity of feeding habits as occurs in Heteroptera. The range in feeding from phytophagy to zoophagy can actually be considered as the extremes of one continuous feeding strategy.

Next to being of great scientific interest, many heteropterans are important for their contribution to natural control (the reduction of pest organisms by their naturally occurring enemies) of pests (Coll 1998; Wheeler 2001; Ingegno et al. 2009). According to Albajes and Alomar (1999), complexes of species of generalist predators may significantly reduce pest populations. The most generally found heteropteran predators in agroecosystems belong to the genera *Orius* (Anthocoridae), *Geocoris* (Lygaeidae), *Nabis* (Nabidae), *Macrolophus*, *Nesidiocoris* (Miridae), and *Podisus* (Pentatomidae). Although they are classified as polyphagous or generalist predators, many of the species may show a strong preference for a limited number of prey species, and thus can be considered oligophagous predators.

The knowledge of nutrition in entomophagous insects has greatly increased recently, which coincides with the emergence of an ecological view unifying ideas about nutrition, now called nutritional ecology. Nutritional ecology focuses on the interaction of nutrition, ecology, behavior, and physiology. The optimum nutritional situation can be accomplished by considering an ecological tritrophic interaction involving the entomophagous insect (third trophic level), the prey (second trophic level), and the food of the prey (first trophic level; often a plant). The co-occurrence of phytophagy and zoophagy in Heteroptera greatly influences the problem of best food provision for these predators. According to Thompson (1999), the application of this nutritional ecological approach to understand the feeding habits and nutritional requirements of entomophagous adults, has, for example, led to the use of supplementary food to increase the efficiency of beneficial insects in agroecosystems. The field of nutritional ecology

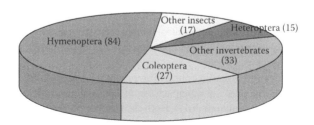

FIGURE 22.1 Commercially available invertebrate natural enemies in Europe. (After Van Lenteren, J. C. 2011, doi: 10.007/s10526-011-9395-1.)

FIGURE 22.2   Feeding injury on plant stem (A) caused by stings of *Nesidiocoris tenuis* (B). (Courtesy of V. H. P. Bueno.)

has also successfully been applied to obtain better understanding of the physiology and the type of food to be used to develop *in vitro* rearing of predators in the absence of their hosts and prey.

Nutritional demands do not differ greatly for the majority of predators, and thus most prey meet basic food composition demands. Consequently, prey choice seems to be determined mainly by prey capture costs, prey toxin content, and mortality risks during searching for prey. Generalist predators such as Heteroptera attack all potential prey of appropriate size and, initially, independent of food quality. They often have to learn which prey is of high or low quality (Sadeghi and Gilbert 1999; Gilbert 2005). Also, usually they first need to catch the prey before they are able to evaluate it properly. Generalist predators are initially not prey selective, but may become selective as a result of experience. Contrarily, specialist predators are usually able to determine prey quality before they have caught it, and they change prey preference only within a small range of prey species.

For heteropteran predators, little is known about food choice in natural ecosystems. According to Coll (1998) the majority of the studies on relationships between heteropteran predators, their prey, and host plants are limited to studies in managed agricultural systems and concern only a few genera belonging to the families Anthocoridae, Lygaeidae, Miridae, Nabidae, Pentatomidae, and Reduviidae. Ruberson and Coll (1998) suggested that there are three lines of research that need to be explored in heteropteran predators: (1) characterization of their predatory impact on the dynamics of agricultural and natural ecosystems; (2) elucidation of their phylogenetics and general systematics, and (3) development of efficient methods for mass rearing, commercialization, and release.

Because of the recently increased general interest to use biological control and the particular beneficial role that heteropteran predators can play in reducing pests in important cropping systems, there is a clear demand for knowledge about the nutritional ecology of these species. Such knowledge is expected to result in improved mass production and quality of heteropteran predators, and thus in optimization of their use in biological control.

In this chapter, first the taxonomic position of predatory Heteroptera will be summarized, as well as their feeding behavior. Next food demands and mass rearing of these predators will be described. This is followed by an overview of trophic relationships in Heteroptera. Finally, habitat choice, distribution, and use of predatory Heteroptera in commercial biological control will be discussed.

22.2 Taxonomy, Feeding Behavior, and Overview of Important Predatory Heteroptera

22.2.1 Taxonomy

The suborder Heteroptera (Order Hemiptera) currently consists of about 40,000 species worldwide. The suborder is divided into eight infraorders (Cimicomorpha, Dipsocoromorpha, Enicocephalomorpha, Gerromorpha, Leptopodomorpha, Nepomorpha, Peloridiomorpha, Pentatomomorpha) and about 75 families (http://bugguide.net).

Most species of Heteroptera are phytophagous, and some of these are considered serious agricultural pests. The suborder also contains a number of species that prey on insects, several of which are used in biological control pests (Table 22.1). Several of the predatory species also feed on plants (Table 22.2). Still other species are vectors of human or animal diseases, and some species are saprophagous. According the Borror et al. (1976), predatory species occur in the majority (69%) of the aquatic and semiaquatic families, and in only 29% of the terrestrial families. Predatory Heteroptera are thought to have evolved from litter-inhabiting omnivorous forms. Of the two major terrestrial infraorders with predatory species, Cimicomorpha are heteropteran predators that have a longer predation history than Pentatomomorpha. We will limit the discussion in the remainder of this chapter to species belonging to these two infraorders.

The best-known insect predators are represented in the families Anthocoridae (pirate bugs), Miridae (plant bugs), Nabidae (damsel bugs), Reduviidae (assassin bugs) belonging to the infraorder Cimicomorpha, and the Pentatomidae (stink bugs) and Lygaeidae (seed bugs) belonging to the infraorder Pentatomomorpha. An overview of important predatory families and species is given in Section 22.2.3.

22.2.2 Feeding Behavior and Prey Digestion

The mouthparts of the Heteroptera are of the piercing-sucking type and are in the form of a slender segmented beak (Figure 22.3). The segmented portion of the beak is the labium, which serves as a sheath for the four piercing stylets (two mandibles and two maxillae). The maxillae fit together in the beak to form two channels: a food channel and a salivary channel (Figure 22.3).

Feeding mechanisms of Heteroptera are among the most specialized in arthropods. The basic mechanism in predatory Heteroptera is the nonreflux type in which digestive enzymes originating from the salivary glands are injected into the prey to initiate a series of cycles. These cycles are then followed by a one-way flux of food coming from the prey to the gut of the predator where digestion is completed (Cohen 1995).

Usually, heteropteran predators ingest only a portion of the food present in the prey, a phenomenon known as partial prey consumption (Lucas 1985). Although the prey is only partially consumed,

TABLE 22.1

Infraorders of Heteroptera and Occurrence of Predatory Species in Terrestrial or Aquatic/Semiaquatic Environments

| | Order Hemiptera | |
Suborder Heteroptera	Infraorder Contains Predatory Species	Environment
Infraorders: Cimicomorpha	+	Terrestrial
Dipsocoromorpha	−	Terrestrial
Enicocephalomorpha	+	Terrestrial
Gerromorpha	+	Semiaquatic
Leptopodomorpha	+	Semiaquatic
Nepomorpha	+	Aquatic
Peloridiomorpha	−	Terrestrial
Pentatomomorpha	+	Terrestrial

TABLE 22.2

Importance of Plant Food for Predatory Heteroptera

Family	Plant Food Obligatory for Development	Complete Development Possible on Plant Food	Plant Food Improves Survival and/or Reproduction	Plant Food Important
Anthocoridae	−	+	−	−
Miridae	−	+	−	+
Pentatomidae	+	+	+	+
Lygaeidae	−	+	−	+
Nabidae	−	+	+	+

Note: +, important for species in this family; −, not important for species in this family.

heteropterans are very economical in extraction of nutrients present in the food. They show ingestion efficiencies of about 80% and absorption efficiencies of more than 90% (Cohen 1984). This is based on extraoral digestion, which allows the selection of prey-specific structures that are rich in nutrients. Cohen (1989a) reported 95% assimilation efficiency and 65% net food conversion efficiency for the predatory bug *Geocoris punctipes* (Say). Cohen (1998a,b, 2000a,b) observed that heteropteran predators not only ingest the body fluids of their prey but also use a "solid-to-liquid-feeding" method to attack soft organs of their prey. He concluded that they require diets with highly concentrated proteins (16–24% of total mass) and lipids (10–22%).

Heteropteran predators may increase their feeding efficiency by injecting digestive enzymes at specific sites in the prey. As a result, structures are liquefied, diluted, and sucked through the food channel into the digestive tract of the predator, a process called prey preparation by Kaspari (1990). The total estimated rate of digestion–ingestion in heteropteran predators is 25 mg/h (Cohen and Tang 1997). However, the real rate of ingestion will be much higher because the above estimate includes the time spent on liquefaction and digestion of the prey structures. The exact proportion of solid/water material during the ingestion is not known, but determination of the food composition of several feeding stages of heteropteran predators indicates that proportions of solids/water varying around 50% are common (Cohen 1998b).

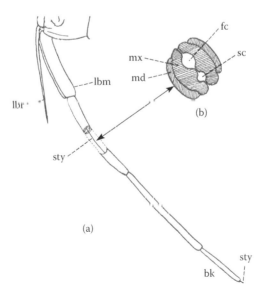

FIGURE 22.3 Basic structure of mouthparts of Heteroptera. (a) Lateral view of mouthparts: bk, beak; lbm, labium; lbr, labrum; sty, stylets. (b) Cross section of stylets: fc, food channel; md, mandibula; mx, maxilla; sc, salivary channel. (After Borror, D. J., D. M. Delong, and C. A. Triphlelorn, *An Introduction to the Study of Insects.* New York: Holt, Rinehart and Winston, 1976.)

Contrary to phytophagous heteropterans, predatory heteropterans apply the salivary secretion strictly to the external wall of the prey body to construct a salivary flange, which glues the labial tip to the external side of the prey cuticle. According to Cohen (1998a), the salivary flange was thought to be closely associated with plant feeding and strictly characteristic of Pentatomomorpha. However, the salivary flange has now also been found in predatory species in three families of Cimicomorpha (Anthocoridae, Nabidae, and Reduviidae; Cohen 1998a), and thus does not seem to be directly related to a history of plant feeding.

The time needed for feeding on a prey influences the risk of a predator of being attacked and the total number of prey a predator can kill per unit of time. Predators of the infraorder Cimicomorpha attack and consume prey quicker than the Pentatomomorpha because of their more potent poisons and salivary enzymes. The Cimicomorpha have an accessory salivary gland that makes a more efficient dilution and reduction of the viscosity of prey sap possible than the salivary gland of Pentatomomorpha. The Cimicomorpha seem to possess a more advanced biochemistry and physiology, resulting in a faster utilization of prey compared with the Pentatomomorpha (Cohen 1998a).

22.2.3 Overview of Important Predatory Heteroptera

The most important heteropteran predatory species used in biological control represented in the infraorder Cimicomorpha occur in the families Anthocoridae (pirate bugs), Miridae (plant bugs), Nabidae (damsel bugs), and Reduviidae (assassin bugs). Important species for biological control are also represented in the infraorder Pentatomomorpha and occur in the families Pentatomidae (stink bugs) and Lygaeidae (seed bugs).

22.2.3.1 Infraoder Cimicomorpha

22.2.3.1.1 Family Anthocoridae
The family Anthocoridae (pirate bugs) contains between 400 and 600 species distributed worldwide, and is composed of relatively small insects (1.4–4.5 mm) (Latin 2000). Literature about these insects is scarce. Most information concerns the original descriptions of species and a few notes about their distribution. Especially for South and Central America, Africa, and Southeast Asia, information is rudimentary (Latin 2000).

Species of the genus *Anthocoris* Fallén generally occur in habitats consisting of a variety of shrubs and trees, including fruit trees, and are highly polyphagous predators. Of this genus, the species *Anthocoris confusus* (Reuter), *Anthocoris nemoralis* (Fabricius), and *Anthocoris nemorum* L. are mentioned most often in the literature (Péricart 1972). *A. nemoralis* and *A. nemorum* are used for biological control of psyllids.

Pericart (1972) reports that the genus *Orius* Wolff consists of about 70 species distributed over all geographic regions that occupy a large variety of habitats. Because in the 1990s several *Orius* species [*Orius insidiosus* (Say), *O. laevigatus*, *Orius majusculus* (Reuter)] appeared to be effective predators of thrips, a large number of publications concerning their biology, mass rearing, and nutritional demands are now available (Salas-Aguillar and Ehler 1977; Ramakers and Rabasse 1995; Riudavets 1995; Van den Meiracker 1999; Tavella et al. 2000; Tommasini 2003; Silveira et al. 2003; Carvalho et al. 2004, 2005a,b; Mendes et al. 2005a,b,c; Bueno et al. 2007; Bueno 2009; Carvalho et al. 2011).

The popularity of *Orius* species for biological pest control has also led to better knowledge about their distribution and role in natural ecosystems (e.g., Lattin 1999, 2000; Tommasini 2003). Although being polyphagous, *Orius* predators often show a strong preference for particular prey species, such as thrips, particularly *Frankliniella occidentalis* (Pergande) (Thysanoptera: Thripidae) (Albajes and Alomar 1998). According to Van der Blom (2009), biological control has recently been implemented in about 50% of the most important greenhouse crops in Almeria (Spain), including virtually all sweet pepper, and *O. laevigatus* is one of the key beneficial species in this system. In greenhouse chrysanthemum crops in Brazil, thrips can be effectively controlled by the predator *O. insidiosus* (Bueno et al. 2003; Silveira et al. 2004). *O. insidiosus* (Figure 22.4) is the species most often used for thrips control in the Nearctic Region (Bueno 2009).

FIGURE 22.4 *Orius insidiosus* (Anthocoridae) adult. (Courtesy of V. H. P. Bueno.)

22.2.3.1.2 Family Miridae

The family Miridae (plant bugs) represents nearly one-third of the described species within the Heteroptera, with at least one third of those estimated to exhibit predatory habits (Wheeler 2000). This would mean that of the about 40,000 known heteropteran species, more than 4,000 species are predatory mirids. The size of mirid bugs varies mostly between 3 and 7 mm. The family contains predatory species of which the most studied are found in the subfamily Dicyphinae, in particular the species *Dicyphus tamaninii* Wagner and *Dicyphus errans* (Wolff), *Macrolophus caliginosus* (Wagner), and *N. tenuis* (Figure 22.5). All of these are polyphagous species, and currently of great interest for release in augmentative control programs or for their role in conservation biological control. Several papers have been published on their biology, mass rearing, and use as biological control agent in greenhouses (Riudavets 1995; Gabarra et al. 1995, 2008; Alomar and Albajes 1996; Albajes and Alomar 1999; Gabarra and Besri 1999; Arnó et al. 2009; Urbaneja et al. 2009). Of particular interest are *M. pygmaeus* and *N. tenuis* because they are both good control agents of whiteflies and *Tuta absoluta* (Meyrick) (Lepidoptera: Gelechiidae) (Arnó et al. 2009; Urbaneja et al. 2009; Mollá et al. 2009).

22.2.3.1.3 Family Nabidae

The family Nabidae consists of about 400 predatory species and their size is less than 12 mm in length. In the family Nabidae, the genus *Nabis* (Latreille) presents predatory species, such as *Nabis alternatus* (Parshley), *Nabis americaniformis* (Carayon), *Nabis ferus* (L.), and *Nabis pseudoferus* Remane (Figure 22.6). Nymphs and adults are predators, but to obtain water they need to suck plant saps that may damage the plant (Ridgway and Jones 1968). They are difficult to rear because of the occurrence of cannibalism between nymphs (Perkins and Watson 1972).

The biology and ecology of some species of Nabidae have been studied, although not very extensively (Lattin 1989; Braman 2000; Roth et al. 2008; Roth and Reinhard 2009), and only a few trials have been conducted concerning their usefulness as biological control agents (Elliot et al. 1998; Braman 2000; Cardinale et al. 2003; Cabello et al. 2009). Currently, *N. pseudoferus* is commercially available

FIGURE 22.5 *Nesidiocoris tenuis* (Miridae) preying on whitefly. (Courtesy of J. Belda, Koppert Biological Systems.)

FIGURE 22.6 *Nabis pseudoferus* (Nabidae). (Photo by J. Coelho. Accessed from http://www.flickr.com/photos/joaocoelho/3815800535/. Creative Commons license.)

in Europe for control of *T. absoluta* and *Spodoptera exigua* (Hubner) (Lepidoptera: Noctuidae) (Cabello et al. 2009).

22.2.3.2 Infraorder Pentatomomorpha

22.2.3.2.1 Family Pentatomidae

The family Pentatomidae consists of about 300 predatory species (Asopinae) and their size varies between 6 and 14 mm. Species of this family are associated with a wide range of natural and agricultural habitats; however, according to De Clercq (2000), many species appear to prefer shrubland and woods. In the family Pentatomidae, quite some species have been proposed or are actually used in biological control or integrated pest management programs. As a result, relatively many publications are available for species of this family, including an extensive review by De Clercq (2000). The biology, behavior, mass rearing, and nutritional demands have been studied for commercially important predatory species. *Brontocoris tabidus* (Signoret), *Podisus maculiventris* (Say), *Podisus nigrispinus* (Dallas), and *Perillus bioculatus* (F.) (Figure 22.7) are already commercially used in North and/or Latin America, or in Europe. Additionally, *Supputius cincticeps* (Stal) is evaluated for its pest control efficiency in Latin America (De Clercq 2000; Zanuncio et al. 2000; Lemos et al. 2003), and *Eocanthecona furcellata* (Wolff) in Asia (De Clercq 2000).

Several interesting studies have been done concerning evaluation of various preys for mass rearing of *S. cincticeps* (Zanuncio et al. 2002, 2005), *P. nigrispinus* (Lemos et al. 2003), *Podisus distinctus* (Dallas), and *Podisus sculptus* Distant (Nascimento et al. 1997, Lacerda et al. 2004). Pentatomid predatory species (*P. nigrispinus* and *B. tabidus*) are mass reared and released for control of caterpillar defoliators in *Eucalyptus* plantations in Brazil (Freitas et al. 1990, 2005; Zanuncio et al. 2002; Torres et al.

FIGURE 22.7 *Perillus bioculatus* adult and nymph feeding on a Colorado beetle larva. (Courtesy of Pascal De Rop and Patrick De Clerq.)

FIGURE 22.8 *Geocoris punctipes.* (Courtesy of J. C. Lins and V. H. P. Bueno.)

2006). Other pentatomid predatory species [*Picromerus bidens* (L.), *Podisus maculiventris*] are mass produced in Europe and North America mainly for biological control of the Colorado potato beetle in potatoes and of noctuid caterpillars in vegetables (De Clercq 2000; Anonymous 2001).

22.2.3.2.2 Family Lygaeidae

In the family Lygaeidae, predatory heteropterans occur in the genera *Geocoris* (Fallen) (Figure 22.8) with more than 120 species (Carayon 1961; Sweet 2000). Henry (1997) has divided the paraphyletic family Lygaeidae into smaller but monophyletic families; the subfamily Geocorinae becomes the family Geocoridae. The genus *Geocoris* belongs to the family Geocoridae (Sweet 2000). The size of predators in this genus varies between 2.7 and 5 mm. The amount of literature concerning predatory *Geocoris* species is limited. However, some information is available about their development and mass rearing (Butler 1966, Yokoyama 1980). A very interesting fact is that an artificial diet has been developed for *Geocoris* sp. (Cohen 1981, 1983). The diet seems to be so good that the species *G. punctipes* has not demonstrated any alteration in its prey selection behavior even after rearing it for about 50 generations on artificial diet (Hagler and Cohen 1991). At least one species, *G. punctipes*, is commercially produced for pest control (Yeargan and Allard 2002). Two other species, *Geocoris pallens* (Stal) and *Geocoris atricolor* Montadon, have been evaluated for control of thrips in Spain, but were not considered sufficiently effective (Riudavets 1995). Although *Geocoris* spp. are known to supplement their diets with plant material, they do not damage plants and are considered highly beneficial.

22.3 Food Demands and Mass Rearing of Predatory Heteroptera

22.3.1 Influence of Food on Development and Reproduction

The quantity and quality of available food influences the distribution, abundance, and biological parameters (development, fecundity, and longevity) of heteropteran predators (Molina-Rugama et al. 2001; Lundgren 2011). Although many factors affect development and fecundity of predatory bugs, the most important ones are food and temperature.

Survival of populations of these natural enemies in habitats with a shortage or lack of prey depends on their capacity to allocate energetic resources to specific activities (Legaspi and Legaspi 1998): at low prey densities, the energy reserved for reproduction may be limited and reproduction will reduce the survival capacity of these predators. The most generally occurring trade-offs in heteropteran predators are those between longevity and fecundity (Molina-Rugama et al. 2001; Mourão et al. 2003), and between oviposition and lipid content in the body (Legaspi et al. 1996; Legaspi and Legaspi Jr. 1998).

Some examples showing decreases in reproduction in favor of increased longevity are known in *Podisus rostralis* (Stål) and *P. maculiventris* (Wiedenmann and O'Neil 1990; Legaspi et al. 1996; Mourão et al. 2003). Females of the predator *P. maculiventris* also show trade-offs between oviposition and lipid content, as prey scarcity results in low oviposition and high lipid content in the fat body (Legaspi and O'Neil 1994; Legaspi et al. 1996).

Heteropteran predators can use a wide variety of prey species that may vary in quality. In general, the following "natural" prey species are considered to be of good quality: western flower thrips (*F. occidentalis*) for *O. insidiosus* and *O. laevigatus*; whiteflies (*Trialeurodes* and *Bemisia* spp.) for *M. caliginosus* and *N. tenuis*; and lepidopteran, coleopteran, and hymenopteran larvae for the pentatomids *S. cincticeps* and *P. nigrispinus*. In addition to the natural prey of heteropteran predators, eggs of *Anagasta (Ephestia) kuehniella* (Zeller) (Lepidoptera: Pyralidae) have shown to be of the good quality for their development. Development, survival, and reproduction of several species of *Orius* feeding on *A. kuehniella* have been studied (Van den Meiracker 1999; Henaut et al. 2000; Tommasini 2003; Carvalho et al. 2004, 2005a,b; Mendes et al. 2005a,b; Bueno et al. 2007). Blümel (1996) reported that individuals of *O. majusculus* and *O. laevigatus* each consumed approximately 210 *A. kuehniella* eggs during their nymphal development. Van den Meiracker (1999) found that nymphal survival and development rate of *O. insidiosus* increase with food supply until eight *A. kuehniella* eggs per individual per day were offered. Furthermore, each *A. kuehniella* egg consumed resulted in the production of approximately one *Orius* egg until a daily supply of eight *A. etc kuehniella* eggs per female.

The quality of various other preys has been studied as food for these predators. Mendes et al. (2002) found that different preys [*Caliothrips phaseoli* (Hood) (Thysanoptera: Thripidae), *Aphis gossypii* Glover (Hemiptera: Aphididae)] have a differential effect on the development time of the stages of *O. insidiosus*. According to Tommasini et al. (2004), *F. occidentalis* was an adequate prey for *Orius majusculus*, *O. laevigatus*, and *O. insidiosus*, but *Orius niger* (Wolff) was unable to develop and reproduce efficiently on *F. occidentalis*. *Podisus* bugs can be reared with relative ease on a variety of unnatural prey, as *Galleria mellonella* L. (Lepidoptera: Pyralidae) (De Clercq et al. 1998) and *Musca domestica* L. (Diptera: Muscidae) (Zanuncio et al. 2004). *Helicoverpa zea* (F.) (Lepidoptera: Noctuidae) eggs were nutritionally superior as prey to *G. punctipes* when compared with the aphid *Acyrthosiphum pisum* (Harris) (Hemiptera: Aphididae). *G. punctipes* survived four times longer when feeding on *H. zea* eggs than on aphids, and only individuals feeding on *H. zea* eggs completed their development (Eubanks and Denno 2000).

Several types of prey were used as food for various species of Pentatomidae predators, such as *Tenebrio molitor* L. (Coleoptera: Tenebrionidae) pupae, caterpillars of *Bombyx mori* L. (Lepidoptera: Bombycidae), larvae of *M. domestica* (Zanuncio et al. 2002), and caterpillars of *S. exígua* (Mohaghegh et al. 2001). *P. nigrispinus* showed a longer development time when feeding on *M. domestica* than having *Alabama argillacea* (Hueb.) (Lepidoptera: Noctuidae) or the third stage of *T. molitor* (Lemos et al. 2003) as prey.

Most heteropteran predators are able to use plant food (see Section 22.4.3) and are thus considered omnivores because they feed in more than one trophic level. However, several studies show that exclusive feeding on plant material does not allow many predator species to develop to the adult stage (Naranjo and Gibson 1996; Lemos et al. 2001), and this demonstrates that feeding on prey is essential to complete their life cycle (Coll 1998; Lalonde et al. 1990; Coll and Guershon 2002; Bueno 2009). When *O. insidiosus* feeds on corn plant material only, it may complete its development but it produces infertile females. *Orius vicinus* (Ribaut), when feeding only on plant material, produces smaller adults with structural alterations compared with the ones fed with animal food (Askari and Stern 1972).

22.3.2 Influence of Food Quality on Mass Rearing of Predatory Heteroptera

Food is an important factor that, besides determining the costs of mass rearing of a biological control agent, also has a strong influence on its development and reproduction (Lundgren 2011). A biologically effective predator will not become a successful biological control agent before a suitable type of prey is found for a mass-rearing system, so it can be produced at a reasonable price. The use of a natural or the target prey is not always possible due to rearing difficulties and/or high costs. Thus, in most of cases, a factitious prey has to be found. Lepidopteran eggs, especially eggs of *A. kuehniella*

(= *Ephestia kuehniella*), *Sitotroga cerealella* (Olivier) (Lepidoptera: Gelechiidae), and *Plodia interpunctella* (Hubner) (Lepidoptera: Pyralidae) have been used as factitious prey for *Orius* species (Arijs and De Clercq 2001, Tommasini et al. 2004) and for *Macrolophus* and *Nesidiocoris* species (Castané et al. 2007, Nannini and Souriau 2009). Of these prey species, *A. kuehniella* eggs appear to possess the best nutritional value for commercial production of several heteropteran predators (Van den Meiracker 1999; Tommasini et al. 2004; Bueno et al. 2007; Maselou et al. 2009; Bonte and De Clercq 2009, 2010). Also the cysts of the brine shrimp, *Artemia franciscana* (Kellogg), were tested as potential artificial food for *O. laevigatus* (Arijs and De Clercq 2001; De Clercq et al. 2005; Bonte and De Clercq 2008), *O. majusculus* (Riudavets et al. 2006), *M. caliginosus* (Castañé et al. 2006), and *O. insidiosus* (Carvalho et al. unpublished); however, these cysts did not approach the quality of *A. kuehniella* eggs.

Many insects have been successfully reared in small numbers in the laboratory, but large-scale rearing requires specific procedures and adaptations compared with small-scale laboratory rearing (Nordlund and Greensberg 1994). Particularly for heteropteran predators, cannibalism should be taken into account (Bueno et al. 2006). As the mass-rearing methods employed by commercial producers are usually not published, information on rearing systems is usually based on experience obtained from large-scale laboratory rearing. For *Orius* spp. different mass-rearing systems have been proposed (Isenhour and Yeargan 1981; Schmidt et al. 1995; Blümel 1996; Bueno 2000, 2009; Tommasini et al. 2004; Mendes et al. 2005a; Bueno et al. 2006), and two examples are given below.

De Clercq (2000) mentions that several authors in North America and Europe have described rearing procedures for *Podisus*. In Brazil, rearing methods have been published for *B. tabidus* and *P. nigrispinus* (Zanuncio et al. 2002). Richman and Whitcomb (1978) concluded that the key element for successful rearing of *Podisus* is provisioning the stinkbugs with more than one kind of prey.

The main factors that influence the rearing of heteropteran predators are food, temperature, microclimate, oviposition substrate, cannibalism, and type of container. According Mackauer (1976) and Van Lenteren (1991), the provision of variation in rearing conditions (food, microclimate, space) may enhance fitness and minimize selection during laboratory propagation of biological control agents.

Tommasini et al. (2004) describe a mass rearing of *O. laevigatus* starting from approximately 1000 wild predators (Figure 22.9). The rearing units used were transparent plastic boxes (3.6 dm³ in volume) with holes for aeration closed with fine steel netting on the sides and on the top. Each rearing unit started from approximately 1500 eggs of *O. laevigatus* laid into French bean pods, and the whole developmental cycle from egg to adult females laying the next generation of eggs was completed in the same box. As soon as females started to reproduce, the bean pods with eggs were collected twice per week, and fresh food and water were supplied. Adults were kept for oviposition in the same boxes for about 4 weeks. With this method, about 100,000 adults could be reared during a cycle. A similar rearing method was used for *O. laevigatus* and *O. majusculus* by Blümel (1996).

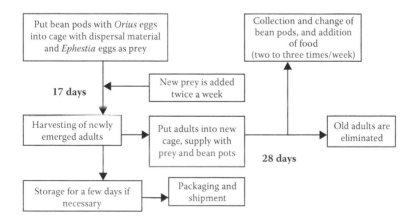

FIGURE 22.9 Production scheme for the thrips predator *Orius laevigatus*. (After Tommasini, M. G., et al., *Bull. Insectol.*, 57, 79, 2004. With permission.)

To develop a mass rearing method for *O. insidiosus*, Bueno et al. (2006) tested three types of containers: transparent plastic bags (4.0-liter capacity), Petri dishes (0.8-liter capacity), and glass jar (1.7-liter capacity). The development of densities of 100, 250, and 400 *O. insidiosus* eggs were evaluated for each type of container. The authors concluded that the best container for rearing *O. insidiosus* was the glass jar of 1.7 liters supplied with *A. kuehniella* eggs as food and farmer's friend inflorescence (*Bidens pilosa* L.) as oviposition substrate. Next, this method was improved by Bueno et al. (2007) and Bueno (2009), resulting in rearing of approximately 8000 adults per 1.7-liter jar from the initial 250 eggs. A scheme of this rearing process is shown in Figure 22.10.

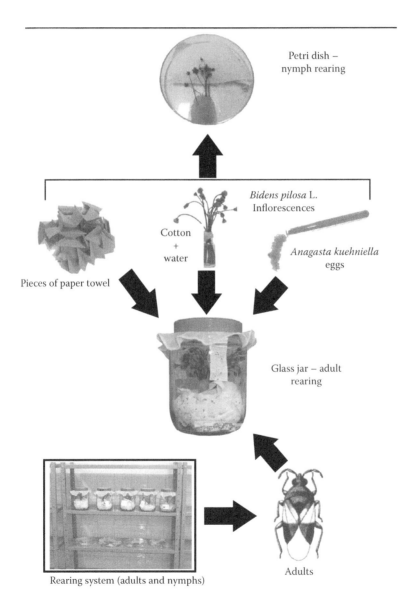

FIGURE 22.10 Production scheme for the predatory bug *Orius insidiosus*. (After Bueno, V. H. P., Desenvolvimento e criação massal de percevejos predadores *Orius*. In *Controle Biológico de Pragas: Produção Massal e Controle de Qualidade*, ed. V. H. P. Bueno, 33–76. Lavras: Editora UFLA, 2009.)

22.4 Trophic Relationships in Predatory Heteroptera

Predatory heteropterans exhibit many trophic relationships (Figure 22.11). Within their own—third— trophic level, they may prey on individuals of the same species (cannibalism) or on other predators (intraguild predation); they may prey on species that are predators of predators in the fourth trophic level; but usually they prey on herbivores that occur in the second trophic level, and many predatory heteropterans make use of plant food from the first trophic level. Also, predatory heteropterans themselves can be prey for other predators or host for other parasitoids. Several of these trophic level interactions will be illustrated below.

22.4.1 Within Trophic Level Relationships: Cannibalism

Polis and Holt (1992) distinguished intraspecific interactions, such as parental cannibalism or cannibalism among kin, from intraguild predation in which interspecific interactions are involved. The incidence and consequences of cannibalism and intraguild predation have been reviewed by several authors and are summarized by Schmidt et al. (1998) for predatory Heteroptera. Most of studies concerning cannibalism in Heteroptera concentrated on aquatic species from the Gerridae (Carcomo and Spence 1993) and Notonectidae (Streams 1992). However, cannibalism is also found in terrestrial predatory Heteroptera: cannibalism regularly occurs in nature or the laboratory in species of the families Nabidae (Braman and Yeargan 1989; Lattin 1989) and Lygaeidae (*Geocoris* spp.) (Crocker and Witcomb 1980). The predator *P. maculiventris* exhibits a high cannibalism rate in the laboratory, and there is at least one report of cannibalism between pentatomids under field conditions (Baker 1927). Also in Reduviidae, in the genera *Zelus*, *Sinea*, and *Apiomerus*, cannibalism is observed commonly between confined nymphs, but there are no reports about cannibalism between reduviids in nature. Further, cannibalism is often observed in the laboratory when immatures of *Sinea diadema* (F.) are confined in small arenas (Schmidt et al. 1998). Rates of cannibalism vary in response to several factors, including availability of alternative prey, starvation period, nymphal age, and size and complexity of the arena. Cannibalism is not common between adults of *S. diadema* and, if it occurs, it generally involves females preying on the smaller males (Schmidt 1994).

Cannibalism has been observed in several species of *Orius*, either in the laboratory (Askari and Stern 1972; Mituda and Calilung 1989; Bueno et al. 2007) or in the field (Nakata 1994). According to Malais and Ravensberg (2003), *Orius* spp. do not hesitate to feed on their own individuals and this is often observed in mass-rearing systems. The arrangement of heteropteran eggs, whether or not clustered, may be one factor that leads to cannibalism. According to Polis (1981), laying eggs in clusters may be an adaptation that allows some individuals to use their siblings as food when prey is scarce. However, this phenomenon of egg cannibalism is not often observed in predatory Heteroptera. Van den Meiracker (1999) reported that cannibalism on eggs does occur in anthocorids, although the author mentioned that in small stock colonies in the laboratory, the predator *Orius* was never found preying on conspecific

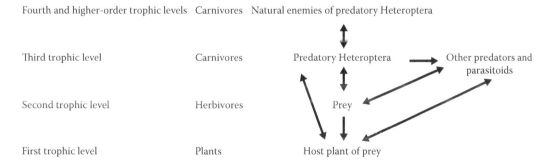

FIGURE 22.11 Trophic relationships in predatory Heteroptera.

eggs. Egg cannibalism was, however, observed in *A. nemorum* (Schmidt and Goyer 1983). The reduviid *S. diadema* lays its eggs in groups, but the nymphs disperse soon after hatching and the first instar only begins to feed 2 h after hatching (Schmidt et al. 1998). There is little evidence regarding the incidence of kin recognition in predatory Heteroptera. Schmidt et al. (1998) mentioned that even when forced encounters between conspecifics are created under artificial conditions, interactions rarely result in cannibalism, suggesting that conspecifics do not form an important component of the diet of Heteroptera.

Cannibalism may also be associated with periods of food shortage, low abundance, and or low quality of alternative prey (Fox 1975). Hunger can affect prey selection by increasing the range of prey size and prey species accepted by a predator (Molles and Pietruszka 1987). In addition, starvation can alter the predator's response to predation risk. Predators deprived of food may be more likely to intensify attacks on conspecifics capable of effective counterattacks, than satiated predators. Evans (1976) did not find cannibalism between nymphs of *A. confusus* when availability of non-heteropteran prey was adequate, but cannibalism occurred at low non-heteropteran prey densities. In general, cannibalism in Heteroptera will not frequently occur when alternative, non-heteropteran prey is abundant.

Next to prey shortage, lack of plant food or water may result in cannibalism. Schmidt et al. (1998) observed that nymphs of the reduviid *S. diadema* deprived of water showed cannibalism earlier than nymphs that had water or a glucose solution available.

Cannibalism is often mentioned as a serious obstacle for the mass rearing of predatory insects such as anthocorids, nabids, and reduviids (e.g., DeBach and Rosen 1991; Gilkeson et al. 1990). However, the above cited literature data suggest that cannibalism does not very frequently occur in predatory Heteroptera. Cannibalism might occur in those insects in mass-rearing systems where individuals that differ in age, size, or nutritional status are maintained together. According to Van den Meiracker (1999) and Bueno (2009), the age synchronization of a mass-rearing system will reduce cannibalism, as suggested earlier by Waage et al. (1985). Further, to prevent cannibalism of young nymphs, eggs should be collected frequently to prevent development of young nymphs in the mass rearing. Next, providing hiding places for nymphs and adults may further reduce cannibalism. Tommasini et al. (2004) stated that cannibalism during the juvenile instars of *O. laevigatus* can be reduced by adding buckwheat to the bottom of each rearing container. Finally, one should avoid accidental transfer of adults when the oviposition substrates with *Orius* eggs are collected.

In general, cannibalism in predatory Heteroptera seems to occur less frequently than in some other groups of predators, such as in Coleoptera. However, when prey is very scarce and plant food and/or water is not available, cannibalism may occur. As stated above, in mass-rearing systems cannibalism can be prevented or reduced by rearing in cohorts and by provision of sufficient space for hiding in rearing containers (e.g., using wrinkled wiping tissues) and by an adequate food supply.

22.4.2 Within Trophic Level Relationships: Intraguild Predation

Heteropteran predators are capable of preying on many species and, thus, it would be surprising if they would limit predation purely to herbivores. Diet composition is essentially determined by which prey are encountered and by the defense capabilities of the prey. If they meet and can master other predators, they will use them as prey. The literature demonstrates that intraguild predation, contrary to cannibalism (see Section 22.3.2), is very common in Heteroptera (e.g., Rosenheim et al. 1995, Rosenheim and Harmon 2006). Most of the cases show that the heteropteran species are the intraguild predators, but there are some examples demonstrating that heteropterans are eaten by other predators such as spiders or other heteropterans (Table 22.3) (Schmidt et al. 1998; Rosenheim and Harmon 2006).

Intraguild predation can occur as well between predators and parasitoids, primarily when predators consume insects that contain a developing parasitoid inside their body. Also parasitoids that already have completely consumed their host and are living within its cuticle can be used as prey by heteropterans (Ruberson and Kring 1990). The anthocorids *A. nemorum* and *O. insidiosus* prey on parasitized aphids by *Aphidius colemani* Viereck (Hymenoptera: Braconidae, Aphidiinae), even when unparasitized aphids are present. However, while *A. nemorum* may consume the pupa of the parasitoid inside the mummy, *O. insidiosus* does not prey on the pupa of the parasitoid (Meyling et al. 2002; Pierre et al. 2006). Like the two anthocorids mentioned above, the mirid *M. caliginosus* makes no distinction between parasitized

TABLE 22.3

Natural Enemies of Terrestrial Predatory Heteroptera

Natural Enemy Species	Heteropteran Prey	Stage Attacked	Reference
Hymenopteran Parasitoids			
Anastasus spp.	*Podisus nigrispinus*	Egg	Torres et al. 1996
Ooencyrtus sp.	*Podisus maculiventris*	Egg	Yeargan 1979
Ooencyrtus spp.	*Podisus nigrispinus*	Egg	Torres et al. 1996
Telenomus podisi	*Podisus nigrispinus*	Egg	Torres et al. 1996
Telenomus podisi	*Podisus maculiventris*	Egg	Okuda and Yeargan 1988
Telenomus calvus	*Podisus maculiventris*	Egg	Orr et al. 1986
Trissolcus brochymenae	*Podisus nigrispinus*	Egg	Torres et al. 1996
Trissolcus euschisti	*Podisus maculiventris*	Egg	Okuda and Yeargan 1988
Tachinid Parasitoids			
Cylindromiya euchenor	*Podisus maculiventris*	Adult	McPherson et al. 1982
Euthera tentatrix	*Podisus maculiventris*	Adult	McPherson et al. 1982
Hemyda aurata	*Podisus maculiventris*	Adult	McPherson et al. 1982
Predatory Spiders			
Araneae 9 spp.	*Nabis* spp.	Nymph, Adult	Schmidt et al. 1998
Lycosidae	*Nabis* spp.	Nymph, Adult	Rosenheim and Harmon 2006
Oxyopes salticus	*Geocoris punctipes*	Nymph, Adult	Schmidt et al. 1998
Peucetia viridans	*Zelus cervalicus*	Nymph, Adult	Schmidt et al. 1998
Peucetia viridans	*Tropiconabis capsiformis*	Nymph, Adult	Schmidt et al. 1998
Predatory Insects			
Geocoris punctipes	*Nabis alternatus*	Nymph	Schmidt et al. 1998
Geocoris spp.	*Orius tristicolor*	Nymph, Adult	Schmidt et al. 1998
Nabis alternatus	*Geocoris punctipes*	Nymph, Adult	Schmidt et al. 1998
Nabis roseipennis	*Geocoris punctipes*	Nymph, Adult	Schmidt et al. 1998
Nabis spp.	*Orius insidiosus*	Nymph, Adult	Schmidt et al. 1998
Podisus maculiventris	*Perillus bioculatus*	Nymph, Adult	Schmidt et al. 1998
Entomopathogens			
Entomophthera spp.	Miridae	Nymph, Adult	P. De Clercq pers. com. 2010

and unparasitized prey, in this case whitefly nymphs. However, as soon the whitefly pupae turn black as a result of parasitism by *Encarsia formosa* Gahan (Hymenoptera: Aphelinidae), or yellow if parasitized by *Eretmocerus eremicus* Rose & Zolnerowich (Hymenoptera: Aphelinidae), they are less often attacked by this mirid (Malais and Ravensberg 2003). Both ectoparasitoids (Press et al. 1974) and endoparasitoids that pupate outside the host (Jackson and Kester 1996) can be attacked by heteropteran predators. Even adult parasitoids can be the victim of intraguild predation as demonstrated by Wheeler (2000), who showed that a predatory *Nabis* spp. attacked adults of an *Aphidius* spp.

Several authors expressed concern that polyphagous heteropteran predators may reduce the total effect of biological control by preying on other natural enemies, which is often observed under laboratory conditions. However, field observations generally do not justify this concern (Daugherty et al. 2007; Jandricic et al. 2008). Christensen et al. (2002) showed that *O. majusculus* may eat eggs and larvae of the dipteran predator *Aphidoletes aphidimyza* (Rondani) (Diptera: Cecidomyiidae), but when the aphid *A. gossypii* was present predation of larvae of *A. aphidimyza* became significantly reduced.

Other experiments have indicated that some generalist predators are able to coexist with other biological control agents without negative effects on the control efficiency, and for a review, we refer to Rosenheim and Harmon (2006). For example, *Orius tristicolor* (White) may attack the predatory mite *Amblyseius cucumeris* (Oudemans) (Acari: Phytoseiidae), but use of the two species is compatible in greenhouses where thrips are present (Gillespie and Quiring 1992). Coexistence of potential intraguild predators can be influenced by presence of other prey. Coexistence of *P. maculiventris* and the predatory beetle *Harmonia axydiris* (Pallas) (Coleoptera: Coccinelidae) is possible in protected cultivation if other prey such as the noctuid *Spodoptera littoralis* (Boisduval) (Lepidoptera: Noctuidae) and the aphid *Myzus persicae* (Sulzer) (Hemiptera: Aphididae) are available, but coexistence is difficult in the absence of prey (De Clercq et al. 2003). In the latter situation *P. maculiventris* mainly attacks eggs and larvae of *H. axyridis*, and seldom the adults. The fact that presence of larvae of *S. littoralis* decreased larval predation of *H. axyridis* by *P. maculiventris* indicates that this coccinellid is either less preferred by, or less vulnerable to the attacks of *P. maculiventris* (De Clercq et al. 2003).

Intraguild predation can be bi- or unidirectional. For, example, interactions between *O. majusculus* and *M. caliginosus* are unidirectional and depend of the presence of alternative prey. Jakobsen et al. (2002) observed that *O. majusculus* was superior to *M. caliginosus*, and suggested this to be the result of its harder body structure and by being more vigorous. Direct observations revealed that adults of *O. majusculus* were the aggressor in all encounters between the two species. *M. caliginosus* only showed defensive reactions and never attempted attacks. An example of bidirectional intraguild predation is that of the two heteropteran predators, *G. punctipes* and *N. alternatus* (Schmidt et al. 1998).

As demonstrated above, heteropteran predators can be involved in intraguild predation as a prey and as a predator, and this may potentially lead to disruption of biological control programs. For biological control, maybe the most important question in the study of intraguild predation is how it affects pest population suppression (Brodeur and Rosenheim 2000; Brodeur et al. 2002), and this partly depends on whether we deal with field or greenhouse situations. For example, interference among heteropterans of the families Anthocoridae (*Orius*), Pentatomidae (*Podisus*), and Miridae (*Macrolophus* and *Dicyphus*) is less important in greenhouses than in the field where several species occur simultaneously. In greenhouses, these predators are released in a curative way, and mainly during periods of high pest infestation. Although they could interfere with other biocontrol agents, intraguild predation probability does not have significant consequences for biological control due to two reasons (Brodeur et al. 2002): (1) the predators will disappear from the system as soon as pest populations have decreased, as they show low survival and limited reproduction when present in food webs with a low diversity, and (2) the natural enemy fauna is regularly supplemented by periodic innundative releases of biological control agents.

Information in a recent review on intraguild predation by Rosenheim and Harmon (2006) leads to the conclusion that, contrary to earlier concerns, intraguild predation generally does not cause serious problems in biological control programs.

22.4.3 Predator–Plant Interactions

As mentioned earlier in this chapter, there might be no other insect order showing such a rich diversity of feeding habits as occurs in Heteroptera. Many species show cannibalistic behavior, may attack other predators (intraguild predation), or may feed facultatively or obligatorily on plants (phytophagy or herbivory). In this section we will discuss phytophagy and other predator–plant relationships in predatory Heteroptera.

Phytophagy. Field observations suggest that predators in their adult stage largely use plants either as their main source of nutrition (i.e., nonpredatory species in the adult stage) or as supplementary nutrition (i.e., predatory and phytophagous species in the adult stage) (Thompson 1999). Although phytophagy by predatory species is often mentioned in the literature, this phenomenon has only been studied with some detail for a few heteropteran species. In general, phytophagy is considered an important characteristic offering the ability to predators to colonize crops before the arrival of the pest, and also permitting survival during periods that prey is scarce (Albajes and Alomar 1999; Torres and Boyd 2009; Castane et al. 2011). Further, vegetable food can also provide useful complements to a carnivorous diet. Finally, it is tempting to speculate within an evolutionary perspective that the plant may benefit by providing food to predators and, thus, attract/arrest them to kill prey. There are several functional explanations why zoophytophagous insects use plant food: (1) equivalence—the plant provides enough nutrients to substitute the animal tissue when this is scarce; (2) facilitation—the plant provides some nutritional compounds that assists the carnivore in development or survival; and (3) independency—the plant provides essential nutrients that are not available in animal food (Gillespie and McGregor 2000). The benefits of phytophagy are specific for each species and depend on predator age, the type of prey, and the effects of plant compounds on the total diet (Table 22.2).

The ability to use plant material in addition to prey consumption represents an interesting aspect in the feeding habits of heteropteran predators (Eubanks and Denno 1999; Coll and Guershon 2002; Evangelista et al. 2003, 2004; Zanuncio et al. 2004; Castane et al. 2011). Because of feeding in more than one trophic level, these predators are considered omnivores. However, exclusive feeding on plants does not allow most heteropteran predators to develop to the adult stage (Naranjo and Gibson 1996; Lemos et al. 2001). This demonstrates that feeding on prey is essential to complete their life cycle, and because of this, they are often called zoophytophagous, and zoophytophagy is considered to be a special form of omnivory (Lalonde et al. 1990; Coll 1998; Coll and Guershon 2002). Several studies have demonstrated that individuals of the species *Orius* can survive on plant material only but cannot produce healthy offspring without animal food. When feeding on corn plants, *O. insidiosus* may complete its development but produces infertile females. *O. vicinus*, when only exposed to plant material, produces smaller adults with structural deformations compared with individuals exposed to prey (Askari and Stern 1972).

Naranjo and Gibson (1996) suggested that some species of Anthocoridae and Miridae are able to totally substitute carnivory for phytophagy, and showed that the predators *O. insidiosus* and *O. tristicolor* have the ability to complete nymphal development on plant material. The authors stressed, however, that addition of thrips, mites, or lepidopteran eggs to the diet consisting of bean or pollen decreased the development time with about 25%. The same two species of *Orius* can also complete their development on bean pods, but show high mortality (Salas-Aguilar and Ehler 1977; Richards and Schmidt 1996).

Which plant part can be used as nutrient for predators depends on the predatory species. Naranjo and Gibson (1996) demonstrated the omnivorous characteristic of the genus *Orius* and observed that they are able to feed on pollen of different plants and occasionally act as plant sap suckers. Plant sap sucking by *Orius* predators probably mainly serves for obtaining water, as they are not able to survive for long periods when only sucking plant sap.

Several predatory species in the families Lygaeidae, Miridae, and Pentatomidade can develop during the first instar or during several instars purely on plants food. Although some species of the families Nabidae and Reduviidade can extract useful moisture and nutrients from plants, their ability to develop only on plant food is limited.

The quality of plant food significantly influences if and how phytophagy can support development and survival. When feeding purely on nectar from cotton plants, the predator *Geocoris pallens* can develop till the fifth instar, and adult survival is about four times longer in a situation where nectar is available compared to a situation without nectar (Naranjo and Gibson 1996). When *G. pallens* is only exposed to the sap from cotton leaves, there is no development after the first instar. *G. punctipes* developed till the fifth instar when feeding purely on green beans or sunflower seeds, but only few individuals developed till the second instar when feeding on soybean leaves, and no development took place when feeding in cotton leaves. Sap from cotton leaves had no or only limited nutritional value for species of Nabidade and Reduviidade, although pollen, sunflower seeds, or green beans increased adult survival for weeks (Naranjo and Gibson 1996). Although plant saps stimulate better development of *M. caliginosus* populations, the sap is insufficient for complete development of individuals. When females are feeding only on plant sap they lay few eggs and show low survival (Malais and Ravensberg 2003). The pentatomid *B. tabidus* strongly benefits from plant supplements (Zanuncio et al. 2000). It appeared very important to provide plant material during the mass production of certain predators (Medeiros et al. 2003).

The capability to complete development on plant food only indicates the potential importance of phytophagy for predators. However, the inability to develop on one type of food from only one plant species may underestimate the capacity in which a predator can utilize phytophagy during its complete life cycle. Some predators may need a combination of compounds (nectar, pollen, seed, plant sap) from (more than one) plant species to obtain the essential nutrients for growth, development, and reproduction (Naranjo and Gibson 1996). Plant food supplements may be particularly essential for the development and reproduction of predators that are exposed to prey of low quality, but the effect of plant food may be of very limited value when the same predator can feed on high-quality prey. In many cases, the importance of phytophagy can be expressed best as a complement to food provided by the prey.

O. laeviagatus occurs frequently in flowers, and is known to use pollen. Kiman and Yeargan (1985) demonstrated that the number of eggs laid by *O. insidiosus* did not change when this predator was fed on pollen and on eggs of *Heliothis zirennium* (F.) (Lepidoptera: Noctuidae). However, Cocuzza et al. (1997) showed that females of this predator laid 40% more eggs on sweet pepper leaves when pollen were added to the diet of *A. kuehniella* eggs. According to Cocuzza et al. (1997), the presence of pollen increases the development rate and also improves the survival rate of some *Orius* species. Still, almost all *Orius* species show better development when an animal food source—prey—is available. Nymphs of *O. insidiosus* fed on pollen alone took about 1.3 to 1.4 times more time to develop to adults. Similarly, nymphs of *O. tristicolor* took 1.2 to 1.5 more times to complete their development when fed on green bean or pollen compared with nymphs fed on thrips, and they survived longer when fed on prey. Only about 2% of the nymphs of *G. punctipes* reached adulthood when feeding on green bean or oat seeds alone, and it took the individuals which completed their development two times longer to reach adulthood than those feeding on prey (Naranjo and Gibson 1996).

Studies have demonstrated that pollen can increase the oviposition capacity and decrease the dispersal rate of some predators occurring on crops in greenhouses. In cucumber crops where pollen are not available, populations of *O. laevigatus* invariably decline after their release for thrips control, making repeated releases necessary to reach sufficient pest control. However, when pollen are present in sweet pepper, the population of *O. insidiosus* remains constant even in the absence of thrips as a prey (Van Rijn and Sabelis 1993; Chambers et al. 1993; Van den Meiracker 1999). Also, *O. majusculus* establishes faster in sweet pepper crops than in other crops, and this is due the availability of the pollen as supplementary food source. This suggests that supplementing alternative food could facilitate preventive introductions of predators (Hulshof and Linnamaki 2002).

The effect of pollen on several biological characteristics of different heteropteran predators needs careful analysis because pollen of different plants species have distinct chemical–physical compositions. *O. insidiosus* completed its development on pollen of *Acer* spp. (McCaffrey and Horsburgh 1986), and the females laid eggs when fed on pollen of cultivated plants as corn (*Zea mays*) and sorghum (*Sorghum bicolor*), and pollen of weeds as farmer's friend (*Bidens pilosa*) and carthamus (*Amaranthus* sp.), but the number of eggs laid was low (Silva 2006). Females fed with pollen of corn and sorghum laid higher numbers of eggs (11.7 and 8.7 eggs, respectively) compared with females fed with pollen of farmer's friend and carthamus (2.3 and 3.5 eggs, respectively).

According to Albajes and Alomar (1999) the plant-feeding habits of pentatomid predators may be regarded as negative if feeding results in damage to the crop. Loomans et al. (1995) found that an occasional probe with the rostrum usually does not result in yield loss. There are, however, cases where excessive probing resulted in yield losses and damaged fruit. Thus, phytophagy may have positive (development, improved survival, and reproduction) and negative (fruit damage and yield loss) effects. Careful management and timing of releases may prevent such problems.

Other aspects of predator–plant interactions. The nutritional requirements of adult predators vary between heteropteran species. They all need prey for reproduction. Although detailed nutritional requirements of heteropteran predators have not been extensively investigated, the provision of optimal nutrition is often accomplished by a tritrophic ecological interaction involving the predators, its prey and the host plant (Figure 22.11) (Thompson 1999). Heteropteran predators may be affected directly and indirectly by plants, in the form of food relationships or by using it as a refuge and oviposition site. Plant characteristics can influence predators in many ways, such as their distribution on the plant, their oviposition, their foraging behavior, and their survival and abundance. The architecture of a plant, and its leaf and surface structure influence the diversity and the abundance of insects that occur on it (Lawton 1983). Larger and structurally more complex plants tend to host more insect species because they provide a greater variety of food, oviposition sites, refuges to hide or overwinter, as well as more diverse microclimates than plants that are structurally simpler. The distribution of heteropteran predators can further be influenced by different microclimates available on different parts of the plant. Nymphs of the anthocorid *A. confusus* remain 65% of their time immobile on the stems and in the petioles of bean plants (Evans 1976), and *O. tristicolor* use the junction between the veins of the leaves of cotton as resting sites. On corn plants, *O. insidiosus* are attracted to those structures on plants with good humidity conditions (Coll 1998).

The foraging behavior of heteropteran predators can be influenced by the leaf surface texture and by the presence of trichomes. The response of predatory bugs to leaf pubescence and leaf waxes appears to be species-specific. The anthocorid *A. nemorum* is less effective when searching on leaves with wax than on leaves with hairs (Lauenstein 1980). Adults of *O. insidiosus* move with great difficulties on leaf surfaces of soybean plants; the predators move with greater difficulty on lower than on upper leaf surfaces, presumably because of the presence of more and longer trichomes on the lower surface (Ysenhour and Yeargan 1981). A high trichome density was the cause for a low efficiency of *O. insidiosus* when foraging on tomato compared with foraging on bean (Coll and Ridgway 1995). The trichomes on tomato leaves also provide refuges for prey. These effects together may lead to a low predation capacity of *O. insidiosus* on tomato (Riudavets and Castañé 1994).

Studies conducted by Soglia et al. (2007) demonstrated that the consumption rates of nymphs of the aphid *A. gossypii* by *O. insidiosus* were influenced by difference in hairiness of two chrysanthemum cultivars. The aphid nymphs presents on the cultivar with most hairs (White Reagan) were eaten more frequently by the predator than nymphs present on the cultivar with fewer hairs (Yellow Snowdon). According to Malais and Ravensberg (2003), the predatory mirid *M. caliginosus* appears to have no problems with the glandular trichomes present on tomato plants. The adults are found in the developing shoots and along the petioles. Nymphs are mainly found on the upper side of the leaves; the adults are good flyers and easily disperse.

Another factor affecting the distribution of predators on plants is the presence of oviposition sites and egg survival possibilities. According to Isenhour and Yeargan (1981) and Coll (1998), species that lay their eggs in plant tissue show great specificity in their oviposition site selection. Egg viability depends on the oviposition sites, which are preferentially the meristematic regions of the plants. In addition, substrate characteristics such as hardness of the tissue and its humidity may influence acceptance of a site for oviposition (Van den Meiracker and Sabelis 1999).

The presence of plant food also influences predator distribution. Several heteropteran predatory species are more abundant in flowers that offer pollen and nectar, or on parts of the plant that offer extrafloral nectar (Shipp et al. 1992). *Orius* spp. live primarily on the central disk of the sunflower and on the panicle of sorghum, in flowers of soybean, in axils of corn leaves, and in chrysanthemum flowers. The importance of plant food for development and reproduction is described in the section on phytophagy above.

Extrafloral nectaries can present another important source of food for some species of predatory bugs. According to Scott et al. (1988), 50% fewer heteropteran predators were found in cotton cultivars without nectaries compared with cultivars with floral nectaries. *O. insidiosus* and some species of *Nabis* were less abundant in cotton without nectaries, but population size of *G. punctipes*, *Geocoris uliginosus* (Say), and most *Nabis* spp. do not differ between cultivars with and without floral nectaries. Bugg et al. (1987) found that several species of heteropteran predators survive for prolonged periods under field conditions when prey was scarce but plants with extrafloral nectaries were present (e.g., the weed *Polygonum aviculare*). Yokoyama (1978) observed *G. pallens* and *O. tristicolor* feeding on extrafloral nectaries in cotton plants and suggested that nectar probably is an important source of food only when prey is scarce. This observation is supported by studies showing that consumption of nectar by *G. punctipes* is limited when prey is available for the predator (Schuster and Calderon 1986).

Not only predators but also pests can use pollen and nectar as food. This does not necessarily corroborate the mutual interaction between plants and natural enemies of their herbivores (Bronstein 1994). If pollen not only attract predatory arthropods but also herbivores, then the presence of pollen increases the encounter rate between predator and prey, which can benefit the plant. Although several authors support the hypothesis about the importance of mutualistic relationships between plants and predators, a lot of experimental work is still needed to be able to make generalizations about this issue.

Changes in food availability during the season often lead to changes in distribution of heteropteran predators on the plant. Studies conducted by Dicke and Jarvis (1972) and Coll and Bottrell (1991) indicate that several species of *Orius* colonize the tassels of maize to feed on thrips and aphids. However, the predators feed on pollen on the axils of leaves when prey populations decline during the emission of the tassels. Later, with the appearance of corn cobs, predators colonize the corn cobs, where they reproduce and feed on hair fresh corn, where insect eggs and young larvae develop. Finally, they leave the old corn plant and colonize more attractive plants that have pollen. Soybean crops are also hosts to *O. insidiosus*, and the predator populations increase in size during the time of bud burst. Both in case of corn and soybean, Dicke and Jarvis (1972) and Isenhour and Yeargan (1981) concluded that the abundance of predators is due to the presence of pollen and thrips in the floral structures.

According to Coll (1998), information concerning effects of plants on heteropteran predators can be used to (1) increase the activity and efficiency of predator populations that naturally occur as a result of plant diversification; (2) maximize the compatibility of the activity of these predators with other strategies of pest control (particularly those concerning cultural practices and plant resistance); (3) develop effective protocols of surveys; (4) increase the establishment rate of mass-produced species and of those introduced by selection of release sites with appropriate resources for refuge and oviposition; (5) carry out mass rearing of heteropteran predators; and (6) evaluate the potential risks associated to augmentatives releases of heteropteran predators, such as damage to the plant and transmission of phytopathogens.

22.4.4 Natural Enemies of Predatory Heteroptera

Published information on natural enemies of predatory Heteroptera is scarce. There are records of egg parasitoids attacking *Podisus* species (Table 22.3) with one particular paper showing that high percentages (>80%) of parasitism could be found of egg masses of *P. nigrispinus*, which was released in *Eucalyptus cloeziana* forests for control of defoliating caterpillars (Torres et al. 1996). Also tachinid parasitoids are known to attack predatory Heteroptera (Table 22.3) (Yeargan 1979; McPherson et al. 1982). Further, several predators are known to attack heteropteran predators (Table 22.3).

However, as these data are often coming from field observations, it is difficult to determine the quantitative role of these predators in reduction of populations of heteropteran predators. Most of the known predators of Heteroptera are heteropterans as well. In addition, Araneae (spiders) are listed as predators of Heteroptera and they may attack species occurring in all predatory heteropteran families (Table 22.3).

We have not been able to find published information on entomopathogenic nematodes and microbial natural enemies of heteropteran predators, but we expect they will exist and those who rear heteropteran predators experience infection of mirids, for example, by *Entomophthera* spp. (P. De Clercq, personal communication).

22.5 Habitat Choice and Distribution of Predatory Heteroptera

The predatory heteropteran fauna in agroecosystems shows a large variation both in species composition and in relative abundance. Generally, these predators seem to be more abundant in annual crops such as cotton, soybean, and corn than in seasonal vegetable crops and perennial fruit orchards (Yeargan 1998). However, species of Miridae frequently occur in perennial crops, such as apples and nuts. Species of *Orius* occur in significant numbers in a large variety of crops, including annual crops, seasonal vegetable crops (tomato), in orchards (apple, peach, and citrus), as well as in several vegetable and ornamental crops under protected cultivation (Silveira et al. 2005). These anthocorids are also found in a large variety of layers in natural ecosystems. Latin (2000) mentions that *Orius* spp. occur mainly in *forb*, which, according to Lawton (1983) are layers of low structural complexity. The *forb* layer is composed of plants of simple structure and with annual flowering; they form the majority of cultivated plants and weeds. The diversity of habitats inside the *forb* where most *Orius* species occur probably reflects their adaptability. Lower numbers of *Orius* species are present in the arbustive category and even less in the trees.

Orius insidiosus (Say) has been collected in crops of corn (*Zea mays*), pearl millet (*Pennisetum glaucum*), sorghum (*Sorghum* spp.), beans (*Phaseolus vulgaris*), sunflower (*Helianthus annuus*), alfalfa (*Medicago sativa*), soybean (*Glycine max*), chrysanthemum (*Chrysanthemum* spp.), tango (*Solidago canadensis*), carthamus (*Carthamus tinctorius*), farmer's friend weed (*Bidens pilo*sa), amaranth (*Amaranthus* sp.), parthenium weed (*Parthenium hysterophorus*), and Joseph's coat (*Alternanthera ficoidea*). *Orius thyestes* Herring was found in farmer's friend weed, carthamus, and Joseph's coat. *Orius perpunctatus* (Reuter) and *Orius* sp. were collected mainly on farmer's friend weed, carthamus, and Joseph's coat and on corn plants (Silveira et al. 2005). Several of these plants are natural reservoirs for these predators, and the plants are used as habitat, refuge, food source (prey and pollen), and oviposition substrate. If the agricultural environment is well managed, the cultivated and wild plants can promote the conservation of several *Orius* species (Silveira et al. 2003). In Brazil, under field conditions, farmer's friend inflorescences function as habitat for several species of thrips, and provide prey and pollen (Silveira et al. 2005; Bueno 2009).

In greenhouses, several species of plants are used next to the crop species to provide alternative food for heteropteran predators and, thus, create better survival and reproduction possibilities. The presence of tobacco plants inside of greenhouses with tomato crops appeared important for survival and to stimulate dispersal of the predatory bug *M. caliginosus* (Miridae) in tomato plants for control of whitefly (*Bemisia* spp.) (Arnó 2000). Ornamental pepper varieties are used as banker plants to augment *Orius* establishment in ornamentals (Bio-Bulletin 2008). The plant *Tagetes erecta* L. showed potential for use as banker plant in ornamental crops, such as roses, to augment performance of *O. insidiosus*: the predator was observed both on rose and the banker plant mainly during the flowering period of the banker plant (Bueno et al. 2009).

Also the abundance of predators is often influenced by the presence of certain weeds that provide pollen, floral or extrafloral nectar, seeds, and plant sap. According to Wäckers (2008), the potential for using food supplements as a tool to improve the effectiveness of biological control agents is determined by the level in which key biological control agents depend on nonprey food, the level of nonprey food available in the system, and the suitability of existing food sources. However, it is often difficult to distinguish between the influence of prey abundance and availability of important plants resources offered by the associated plants. Elkassabany et al. (1996) mainly found *O. insidiosus* in several species of weeds when they were flowering and hosting high densities of thrips. Silveira et al. (2005) observed that different species of *Orius* and thrips occur simultaneously on several plants in the same habitat. Adults of the mirid *D. tamaninii* are found on the underside of bean and cucumber leaves, and on the upper surface of tomato and pepper leaves (Gesse Sole 1992). Adults of *M. pygmaeus* and *Engytatus nicotianae* (Koningsberger) occur on whitefly-infested vegetables and other plants (Eyles et al. 2008). Certain species of the Lamiaceae, Scrophulariacea, and Solanaceae serve as good host plants for the mirid *Dicyphus hesperus* Knight (Sanchez et al. 2004). Heteropteran predators, such as Anthocoridae and Geocoridae, are known to feed on and profit from extrafloral nectar on cotton plants. According to Butler et al. (1972), the extrafloral nectar of cotton is rich in sugars and contains certain amino acids that are essential for the growth and development of insects. Population numbers of *Nabis* spp., *G. punctipes*, and *G. uliginosus* did not differ on nectariless compared with necataried cultivars of cotton plants, whereas *O. insidiosus* and some *Nabis* spp. were less abundant in nectariless cotton (Schuster and Calderon 1986).

Heteropteran generalist natural enemies feed on a great variety of herbivore species; they also often use alternative food such as pollen and nectar and should, thus, profit from habitats with great plant diversity. Several heteropteran predators attack a large range of taxa and stages of prey and often feed on plant material. Thus, if the diversity of the plant community promotes the increase of predator populations, it is expected that this group of natural enemies is more abundant in rich habitats than in more simple habitats (Latin 1999, 2000; Bueno 2009).

Also the size of the heteropteran predatory species influences their distribution in agroecosystems. Heteropteran adults show variation in body size between 2 and 14 mm, with *Orius* species among the shortest ones (<2 mm), and Pentatomidae and Reduviidae among the largest ones (>10 mm). Larger species (>10 mm) are less abundant in agroecosystems than smaller ones such as anthocorids and nabids (Yeargan 1998).

Several ecological factors influence the choice of habitat and the relative abundance of heteropteran predators in different ecosystems, and particularly in agroecosystems. Proper knowledge of these factors could assist us in increasing endemic populations of these predators by strategies as habitat modification, thereby stimulating their role in biological control of pests.

22.6 Predatory Heteroptera Used in Commercial Biological Control

Twenty heteropteran predator species are commercially available worldwide (Table 22.4). With the exception of *O. insidiosus*, which is in use since 1985, all other species came to the market during the

TABLE 22.4

Families and Species of Heteropteran Predators Species Used Worldwide in Commercial Biological Control

Natural Enemy	Region Where Used	Year of First Use	Market Value
Anthocoridae (Pirate Bugs)			
Anthocoris nemoralis	Europe, North America	1990	S
Anthocoris nemorum	Europe	1992	S
Orius albidipennis	Europe	1993	S
Orius armatus	Australia	1990	S
Orius insidiosus	North and Latin America,	1985	M
Orius laevigatus	Europe, North Africa, Asia	1993	L
Orius majusculus	Europe	1993	M
Orius minutus	Europe	1993	S
Orius strigicollis	Asia	2000	S
Orius tristicolor	Europe	1995	S
Miridae (Plant Bugs)			
Dicyphus hesperus	North America	1995	M
Macrolophus caliginosus	Europe	2005	M
Macrolophus pygmaeus (nubilis)	Europe, North and South Africa	1994	L
Nesidiocoris tenuis	Europe, North Africa, Asia	2003	L
Pentatomidae (Stink Bugs)			
Brontocoris tabidus	Latin America	1990	S
Picromerus bidens	Europe	1990	S
Podisus maculiventris	Europe, North America	1996	S
Podisus nigrispinus	Latin America	1990	S
Lygaeidae (Seed Bugs)			
Geocoris punctipes	North America	2000	S
Nabidae (Damsel Bugs)			
Nabis pseudoferus ibericus	Europe	2009	S

Note: Market value: L, large (hundred thousands to millions of individuals sold per week); M, medium (10,000–100,000 individuals sold per week); S, small (hundreds to a few thousands of individuals sold per week). North Africa, north of Sahara; South Africa, south of Sahara; North America, Canada and United States of America; Aust, Australia.

Source: Based on Cock, M. J. W., et al., *Biocontrol*, 55, 199, 2010, and Van Lenteren, J. C. *BioControl*. http://www.springerlink.com/index/10.1007/s10526-011-9395-1, 2011.

last 20 years, so they are considered a relatively new group of biological control agents compared with other natural enemies. Three species are produced in very large numbers (*O. laevigatus*, *M. pygmaeus*, and *N. tenuis*; more than 100,000 per week), and four other species in relatively large numbers (*O. insidiosus*, *O. majusculus*, *D. hesperus*, and *M. caliginosus*; more than 10,000 per week).

Martinez-Cascales et al. (2006) reviewed the economic importance of *M. pygmaeus*, a well-known predator of small arthropod pests in vegetable crops in Europe. This heteropteran predator has been commercialized worldwide, mainly for the biological control of whitefly on a large scale in Europe, North Africa, and South Africa (Table 22.3). Eyles et al. (2008) reported the first record in New Zealand of *M. pygmaeus*, and stated that this may prove to be a fortuitous arrival of a beneficial insect, due to the importance of this heteropteran predator as biological control agent.

The mirid *N. tenuis* (Calvo et al. 2010) is used since 2003 on a large scale in Europe, North Africa, and Asia for control of *T. absoluta*, an important pest in tomato crops in greenhouses. The generalist predator *O. laevigatus* is mainly used to control thrips (Van der Blom 2009), and has been used since 1993 as an augmentative biological control agent in the protected cultivation of many European countries, North Africa, and Asia where hundred thousands to millions of individuals are sold per week (Table 22.4).

In Brazil, inoculative releases of predatory stinkbugs have been made on about 140,000 ha of *Eucalyptus* spp. forests. Between 1989 and 2005 about 3 million predatory stinkbug adults, mainly *P. nigrispinus* and *B. tabidus*, were released (Torres et al. 2006).

22.7 Final Considerations

Predatory Heteroptera form an interesting group for pure scientific studies, and also play a very important role in integrated pest management systems. About 20 species are currently commercially available worldwide, and various other species are under evaluation as biological control agents. Still the group is less intensively studied than other predatory insects, and various points need to be examined to be able to make better use of this group of predators. The following aspects should be considered for future studies: (1) basic biological and ecological aspects; (2) basic studies on particular characteristics, zoophytophagy, and intraguild predation; (3) applied research on the role they play in agroecosystems such as general predators and how they contribute to the reduction of pest populations; (4) applied studies to show their importance in augmentative biological control both in the field and in greenhouses; (5) basic studies to unravel the role of generalist predators in biological control; and (6) basic and applied research in the areas of nutrition, mass production, and predation capacity. Several species of Heteroptera have already been shown to play an important role in pest control, both in the field and in greenhouses. It is expected that studies as mentioned above will result in an even wider application of predatory Heteroptera in sustainable pest control today.

REFERENCES

Albajes, R., and O. Alomar. 1999. Current and potential use of polyphagous predators. In: *Integrated Pest and Disease Management in Greenhouse Crops*, ed. R. Albajes, M. L. Gullino, J. C. Van Lenteren, and Y. Elad, 265–75. Dordrecht: Kluwer Academic Publisher.

Alomar, O., and R. Albajes. 1996. Greenhouse whitefly (Homoptera: Aleyrodidae) predation and tomato fruit injury by the zoophytophagous predator *Dicyphus tamaninii* (Heteroptera: Miridae). In: *Zoophytophagous Heteroptera: Implications for Life History and Integrated Pest Management*, ed. O. Alomar and R. N. Wiedenmann, 155–77. Lanham, MD: Thomas Say Publications in Entomology.

Anonymous. 2001. Directory of least-toxic pest control products. *IPM Pract.* 22:1–38.

Arijs, Y., and P. De Clercq. 2001. Rearing *Orius laevigatus* on cysts of the brine shrimp *Artemia franciscana*. *Biol. Contr.* 21:79–83.

Arnó, J. 2000. Conservation of *Macrolophus caliginosus* Wagner (Het. Miridae) in commercial greenhouse during tomato crop-free periods. *IOBC/WPRS Bull.* 23:241–6.

Arnó, J., C. Castane, J. Riudavets, and R. Gabarra. 2010. Risk of damage to tomato crops by the generalist zoophytophagous predator *Nesidiocoris tenuis* (Rueter) (Hemiptera: Miridae). *Bull. Entomol. Res.* 100:105–15.

Arnó, J., R. Sorribas, M. Prat, M. Matas, C. Pozo, D. Rodrigues, A. Garreta, A. Gómez, and R. Gabarra. 2009. *Tuta absoluta*, a new pest in IPM tomatoes in the northeast of Spain. *IOBC/WPRS Bulletin* 49:203–8.

Askari, A., and V. M. Stern. 1972. Biology and feeding habitats of *Orius tristicolor* (Hemiptera: Anthocoridae). *Ann. Entomol. Soc. Am* 65:96–100.

Baker, A. D. 1927. Some remarks on the feeding process of the Pentatomidae (Hemiptera–Heteroptera). *Annu. Rep. Quebec Soc. Prot. Plants* 19:24–34.

Bio-Bulletin. 2008. *Biobest Biological Systems.* Belgium: BioBest.

Blümel, S. 1996. Effect of selected mass-rearing parameters on *O. majusculus* (Reuter) and *O. laevigatus* (Fieber). *IOBC/WPRS Bull.* 19:15–8.

Bonte, M., and P. De Clercq. 2008. Developmental and reproductive fitness of *Orius laevigatus* (Hemiptera: Anthocoridae) reared on factitious and artificial diets. *J. Econ. Entomol.* 101:1127–33.

Bonte, M., and P. De Clercq. 2009. Artificial rearing of the anthocorid predator *Orius laevigatus*. *IOBC/WPRS Bull.* 49:329–32.

Bonte, M., and P. De Clercq. 2010. Influence of diet on the predation rate of *Orius laevigatus* on *Frankliniella occidentalis*. *BioControl* 55:625–9.

Borror, D. J., D. M. Delong, and C. A. Triphlelorn. 1976. *An Introduction to the Study of Insects.* New York: Holt, Rinehart and Winston.

Braman, S. K. 2000. Damsel bugs (Nabidae). In *Heteroptera of Economic Importance*, ed. C. W. Shaefer and A. R. Panizzi, 639–56. Boca Raton, FL: CRC Press.

Braman, S., and K. V. Yeargan. 1989. Intraplant distribution of three *Nabis* species (Hemiptera: Nabidae), and impact of *N. roseipennis* on green cloverworm populations in soybean. *Environ. Entomol.* 18:240–44.

Brodeur, J., and J. A. Rosenheim. 2000. Intraguild interactions in aphid parasitoids. *Entomol. Exp. Appl.* 97:93–108.

Brodeur, J., C. Cloutier, and D. Gillespie. 2002. Higher-order predators in greenhouse systems. *IOBC/WPRS Bull.* 25:33–6.

Bronstein, J. L. 1994. Our current understanding on mutualism. *Quart. Rev. Biol.* 69:31–51.

Bueno, V. H. P. 2000. Desenvolvimento e multiplicação de percevejos predadores do gênero *Orius* Wolff. In *Controle Biológico de Pragas: Produção Massal e Controle de Qualidade*, ed. V. H. P. Bueno, 69–90. Lavras: Editora UFLA.

Bueno, V. H. P. 2009. Desenvolvimento e criação massal de percevejos predadores *Orius*. In *Controle Biológico de Pragas: Produção Massal e Controle de Qualidade*, ed. V. H. P. Bueno, 33–76. Lavras: Editora UFLA.

Bueno, V. H. P., J. C. Van Lenteren, L. C. P. Silveira, and S. M. M. Rodrigues. 2003. An overview of biological control in greenhouse chrysanthemums in Brazil. *IOBC/WPRS Bull.* 26:1–5.

Bueno, V. H. P., L. M. Carvalho, and N. Moura. 2007. Optimization of mass-rearing of the predator *Orius insidiosus* (Hemiptera: Anthocoridae): How far are we? *Global IOBC Bull.* 3:18–9.

Bueno, V. H. P., S. M. Mendes, and L. M. Carvalho. 2006. Evaluation of a rearing-method for the predator *Orius insidiosus*. *Bull. Insectol.* 59:1–6.

Bugg, R. L., L. E. Ehler, and L. T. Wilson. 1987. Effect of common knotweed (*Polygonum aviculare*) on abundance and efficiency of insect predators of crop pests. *Hilgardia* 55:1–52.

Butler, G. D. 1966. Development of several predaceous Hemiptera in relation to temperature. *J. Econ. Entomol.* 59:1306–7.

Butler, G. D., G. M. Loper, S. E. MacGregor, J. L. Webster, and H. Margolis. 1972. Amounts and kinds of sugars in the nectars of cotton (*Gossypium* spp.) and the time of their secretion. *Agron. J.* 64:364–8.

Cabello, T., J. R. Gallego, F. J. Fernandez-Maldonado, A. Soler, D. Beltran, A. Parra, and E. Vila. 2009. The damsel bug Nabis pseudoferus (Hem.: Nabidae) as a new biological control agent of the South American tomato pinworm, *Tuta absoluta* (Lep.: Gelechiidae), in tomato crops of Spain. *IOBC/WPRS Bulletin* 49: 219–23.

Calvo, J., J. E. Belda, and A. Gimenz. 2010. Una estrategia para el control biológico de mosca branca y *Tuta absoluta* em tomate. *Phytoma* 216:46–52.

Calvo, J., K. Bolckmans, P. Stansley, and A. Urbaneja. 2009. Predation by *Nesidiocoris tenuis* on *Bemisia tabaci* and injury to tomato. *Biocontrol* 54:237–46.

Carayon, J. 1961. Quelques remarques sur les Hémiptéres–Hétéroptères: Leer importante comme insectes auxiliares et les posibilites de leer utilisation dans la lutte biologique. *Entomophaga* 6:133–41.

Carcomo, H., and J. R. Spence. 1993. Kin discrimination and cannibalism in waterstriders (Gerridae). *Ann. Entomol. Soc. Am.* 35:35–6.

Cardinale, B. J., C. T. Harvey, K. Gross, and A. R. Yves. 2003. Biodiversity and biocontrol: Emergent impacts of a multi-enemy assemblage on pest suppression and crop yield in an agroecosystem. *Ecol. Lett.* 6:857–65.

Carvalho, L. M., V. H. P. Bueno, and C. Castane. 2011. Olfactory response towards its prey *Frankliniella occidentalis* of wild and laboratory-reared *Orius insidiosus* and *Orius laeviagatus*. *J. Appl. Entomol.* 135: 177–183 (doi 10.1111/j.1439-0418.2010.01527.x.).

Carvalho, L. M., V. H. P. Bueno, and S. M. Mendes. 2004. Response of two *Orius* species to temperature. *IOBC/WPRS Bull.* 28:43–6.

Carvalho, L. M., V. H. P. Bueno, and S. M. Mendes. 2005a. Desenvolvimento, consumo ninfal e exigências térmicas de *Orius thyestes* Herring (Hemíptera: Anthocoridae). *Neotrop. Entomol.* 34:607–12.

Carvalho, L. M., V. H. P. Bueno, and S. M. Mendes. 2005b. Influência da temperatura na reprodução e longevidade do predador *Orius thyestes* Herring (Hemiptera, Anthocoridae). *Rev. Bras. Entomol.* 49:409–14.

Castané, C., O. Alomar, J. Riudavets, and C. Gemeno. 2007. Reproductive biology of the predator *Macrolophus caliginosus*: Effect of age on sexual maturation and mating. *Biol. Contr.* 43:178–286.

Castané, C., J. Arnó, R. Gabarra, and O. Alomar. 2011. Plant damage to vegetable crops by zoophytophagous mirid predators. *Biol. Contr.* doi:10.1016/j.biocontrol.2011.03.007.

Castané, C., R. Quero, and J. Riudavets. 2006. The brine shrimp *Artemia* sp. as alternative prey for rearing the predatory bug *Macrolophus caliginosus*. *Biol. Contr.* 38:405–12.

Chambers, R. J., S. Long, and N. L. Helyer. 1993. Effectiveness of *Orius laeviagatus* for the control of *Frankliniella occidentalis* on cucumber and pepper in the UK. *Biocontr. Sci. Technol.* 3:295–307.

Christensen, R. K., A. Enkegaard, and H. F. Brodsgaard. 2002. Intraspecific interactions among the predators *Orius majusculus* and *Aphidoletes aphidimyza*. *IOBC/WPRS Bull.* 25:57–60.

Cock, M. J. W., J. C. Van Lenteren, J. Brodeur, B. I. P. Barratt, F. Bigler, K. Bolckmans, F.L. Cônsoli, F. Haas, P. G. Mason, and J. R. P. Parra. 2010. Do new access and benefit sharing procedures under the Convention on Biological Diversity threaten the future of biological control? *Biol. Contr.* 55: 199–218. doi:10.1007/s10526-009-9234-9.

Cocuzza, G. E., P. De Clercq, M. Van de Veire, A. De Cock, D. Degheele, and V. Vacante. 1997. Reproduction of *Orius laeviagatus* and *Orius albidipennis* on pollen and *Ephestia kuehniella* eggs. *Entomol. Exp. Appl.* 82:101–4.

Cohen, A. C. 1981. An artificial diet for *Geocoris punctipes*. *Southwest. Entomol.* 6:109–13.

Cohen, A. C. 1983. Improved method of encapsulating artificial diet for rearing predators of harmful insects. *J. Econ. Entomol.* 76:957–9.

Cohen, A. C. 1984. Food consumption, food utilization and metabolic rates of *Geocoris punctipes* (Het.: Lygaeidade) fed *Heliothis virescens* (Lep.: Noctuidae) eggs. *Entomophaga* 29:361–7.

Cohen, A. C. 1995. Extra-oral digestion in predatory Arthropoda. *Annu. Rev. Entomol.* 40:85–103.

Cohen, A. C. 1998a. Biochemical and morphological dynamics and predatory feeding habits in terrestrial Heteroptera. In *Predatory Heteroptera: Their Ecology and Use in Biological Control*, ed. M. Coll and J. R. Ruberson, 21–32. Lanham, MD: Thomas Say Publications in Entomology.

Cohen, A. C. 1998b. Solid-to-liquid feeding: The inside (s) story of extra-oral digestion in predaceous Arthropoda. *Am. Entomol.* 44:103–17.

Cohen, A. C. 2000a. How carnivorous bugs feed. In *Heteroptera of Economic Importance*, ed. C. W. Shaefer and A. R. Panizzi, 563–70. Boca Raton, FL: CRC Press.

Cohen, A. C. 2000b. A review of feeding studies of *Lygus* species with emphasis on artificial diets. *Southwest. Entomol.* 23:111–9.

Cohen, A. C., and R. Tang. 1997. Relative prey weight influences handling time and biomass extraction of *Sinea confusa* and *Zelus renardii* (Heteroptera: Reduviidae). *Environ. Entomol.* 26:559–65.

Coll, M. 1998. Living and feeding on plants in predatory Heteroptera. In: *Predatory Heteroptera: Their Ecology and Use in Biological Control*, ed. M. Coll and J. R. Ruberson, 89–129, Lanham, MD: Thomas Say Publications in Entomology.

Coll, M., and D. G. Bottrell. 1991. Microhabitat and resource selection of the European corn borer (Lepidoptera: Pyralidae) and its natural enemies in Maryland field corn. *Environ. Entomol.* 20:526–533.

Coll, M., and M. Guershon. 2002. Omnivory in terrestrial arthropods: Mixing plant and prey diets. *Annu. Rev. Entomol.* 47:267–97.

Coll, M., and R. L. Ridgway. 1995. Functional and numerical response of *Orius insidiosus* (Heteroptera: Anthocoridae) to its prey in different dynamics. *Ann. Entomol. Soc. Am.* 88:732–8.

Crocker, R. L., and W. H. Whitcomb. 1980. Feeding niches of the bigeyed bugs *Geocoris bullatus*, *G. punctipes* and *G. uliginosus* (Hemiptera: Lygaeidae: Geocorinae). *Environ. Entomol.* 9:508–13.

Daugherty, M., J. P. Harmon, and C. J. Briggs. 2007. Trophic supplements to intraguild predation. *Oikos* 116:662–77.

De Clercq, P. 2000. Predaceous stinkbugs (Pentatomidae: Asopinae). In *Heteroptera of Economic Importance*, ed. C. W. Shaefer and A. R. Panizzi, 737–86. Boca Raton, FL: CRC Press.

De Clercq, P., F. Merleved, and L. Tirry. 1998. Unnatural prey and artificial diets for rearing *Podisus maculiventris* (Heteroptera: Pentatomidae). *Biol. Contr.* 12:137–42.

De Clercq, P., I. Peeters, G. Vergauwe, and O. Thas. 2003. Interaction between *Podisus maculiventris* and *Harmonia axydiris*, two predators used in augmentative biological control in greenhouse crops. *Biol. Contr.* 48:39–55.

DeClercq, P., Y. Arijs, T. van Meir, G. van Stappen, P. Sorgeloos, K. Dewettinck, M. Rey, S. Grenier, and G. Febvay 2005. Nutritional value of brine shrimp cysts as a factitious food for *Orius laevigatus* (Heteroptera: Anthocoridae). *Biocontrol Science Technology* 15:467–79.

DeBach, P., and D. Rosen. 1991. *Biological Control by Natural Enemies*. Cambridge: Cambridge University Press.

Dicke, F. F., and J. L. Jarvis. 1972. The habitats and abundance of *Orius insidiosus* (Say) (Hemiptera: Heteroptera: Anthocoridae) on corn. *J. Kans. Entomol. Soc.* 35:339–44.

Elkassabany, E., J. R. Ruberson, and T. Kring. 1996. Seasonal distribution and overwintering of *Orius insidiosus* (Say) in Arkansas. J. Entomol. Sci. 31:76–88.

Elliot, N. C., R. W. Kieckhefer, J. H. Lee, and B. W. French. 1998. Influence of within-field and landscape factors on aphid predator populations in wheat. *Landsc. Ecol.* 14:239–52.

Eubanks, M. D., and R. F. Denno. 1999. The ecological consequences of variation in plants and prey for an omnivorous insect. *Ecology* 80:1253–66.

Eubanks, M. D., and R. F. Denno. 2000. Health food versus fast food: The effects of prey quality and mobility on prey selection by a generalist predator and indirect interactions among prey species. *Ecol. Entomol.* 25:140–6.

Evangelista Jr., W. S., M. G. C. Gondim Junior, J. B. Torres, and E. J. Marques. 2003. Efeito de plantas daninhas e do algodoeiro no desenvolvimento, reprodução e preferência para oviposição de *Podisus nigrispinus* (Dallas) (Heteroptera: Pentatomidae). *Neotrop. Entomol.* 32:677–84.

Evangelista Jr., W. S., M. G. C. Gondim Junior, J. B. Torres, and E. J. Marques. 2004. Fitofagia de *Podisus nigrispinus* em algodoeiro e plantas daninhas. *Pesq. Agropec. Bras.* 39:413–20.

Evans, H. F. 1976. The role of predator–prey size ratio in determining the efficiency of capture by *Anthocoris nemorum* and the escape reactions of its prey, *Acyrthosiphon pisum*. *Ecol. Entomol.* 1:85–90.

Eyles, A. C., T. Marais, and S. George. 2008. First New Zealand record of the genus *Macrolophus* Fieber, 1858 (Hemiptera: Miridae: Bryocorinae: Dicyphini): *Macrolophus pygmaeus* (Rambur, 1839), a beneficial predacious insect. *Zootaxa* 1779:33–7.

Fox, L. R. 1975. Cannibalism in natural populations. *Annu. Rev. Ecol. Syst.* 6:87–106.

Freitas, F. A., T. V. Zanuncio, J. C. Zanuncio, P. M. Conceição, M. C. Q. Fialho, and A. S. Bernardino. 2005. Effect of plant age, temperature and rainfall on Lepidoptera insect pests collected with light traps in a *Eucalyptus grandis* plantation in Brazil. *Ann. Forest Sci.* 62:85–90.

Freitas, G. D., A. C. Oliveira, E. J. Morais, and J. A. V. Barcelos. 1990. Utilização do hemíptero predador *Podisus connexivus* Bergroth, 1891 (Hemiptera: Pentatomidae) para o controle de lagartas desfolhadoras de *Eucalyptus* spp. *Boletim Interno da UFV*, Viçosa.

Gabarra, R., and M. Besri. 1999. Tomatoes. In: *Integrated Pest and Disease Management in Greenhouse Crops*, ed. R. Albajes, M. L. Gullino, J. C. Van Lenteren, and Y. Elad, 420–34. Dordrecht: Kluwer Academic Publisher.

Gabarra, R., C. Castané, and R. Albajes. 1995. The mirid bug *Dicyphus tamaninii* as a greenhouse whitefly and western flower thrips predator on cucumber. *Biocontr. Sci. Technol.* 5:475–88.

Gabarra, R., J. Arnó, and J. Riudavets. 2008. Tomate. In *Control Biologic de Plagas Agricolas*, ed. J. Jacas and A. Urbaneja, 410–422. Valencia: Phytoma.

Gesse Sole, F. 1992. Comportamiento alimenticio de *Dicyphus tamaninii* Wagner (Heteroptera: Miridae). *Bol. Sanid. Veg. Plagas* 18:685–91.

Gilbert, F. 2005. Syrphid aphidophagous predators in a food-web context. *Eur. J. Entomol.* 102:325–33.

Gilkeson, A. A., W. D. Morewood, and D. E. Elliot. 1990. Current status of biological control of thrips in Canadian greenhouses with *Amblyseius cucumeris* and *Orius tristicolor*. *IOBC/WPRS Bull.* 13:71–75.

Gillespie, D. R., and D. J. M. Quiring. 1992. Competition between *Orius tristicolor* (White) (Hemiptera: Anthocoridae) and *Amblyseius cucumeris* (Oudemans) (Acari: Phytoseiidae) feeding on *Frankliniella occidentalis* (Pergande) (Thysanoptera: Thripidae). *Can. Entomol.* 124:1123–8.

Gillespie, D. R., and R. R. McGregor. 2000. The functions of plant feeding in the omnivorous predator *Dicyphus hesperus*: Water places limits on predation. *Ecol. Entomol.* 25:380–6.

Gurr, G., and S. Wratten. 2000. *Measures of Success in Biological Control*. Dordrecht: Kluwer Academic.

Hagler, J. R., and A. C. Cohen. 1991. Prey selection by in vitro- and field-reared *Geocoris punctipes*. *Entomol. Exp. Appl.* 59:201–5.

Henaut, Y., C. Alauzet, A. Ferran, and T. William. 2000. Effect of nymphal diet on adult predation behavior in *Orius majusculus* (Heteroptera: Anthocoridae). *J. Econ. Entomol.* 93:252–5.

Henry, T. J. 1997. Phylogenetic analysis of family groups within the infraorder Pentatomomorpha (Hemiptera: Heteroptera) with emphasis on the Lygaeoidea. *Ann. Entomol. Soc. Am.* 90:275–301.

Hulshof, J., and M. Linnamaki. 2002. Predation and oviposition rate of the predatory bug *Orius laeviagatus* in the presence of alternative food. *IOBC/WPRS Bull.* 25:107–10.

Ingegno, B. L., M. G. Pansa, and L. Tavella. 2009. Tomato colonization by predatory bugs (Heteroptera: Miridae) in agroecosystems of NW Italy. *IOBC/WPRS Bull.* 49:287–91.

Isenhour, D. J., and K. V. Yeargan. 1981. Effect of crop phenology on *Orius insidiosus* populations on strip-cropped soybean and corn. *J. Ga. Entomol. Soc.* 16:310–22.

Jackson, D. M., and K. M. Kester. 1996. Effects of diet on longevity and fecundity of the spined stilt bug, *Jalysus wickhami. Entomol. Exp. Appl.* 80:421–5.

Jakobsen, D. R., A. Enkegaard, and H. F. Brodsgaard. 2002. Interactions between the two polyphagous predators *Orius majusculus* and *Macrolophus caliginosus. IOBC/WPRS Bull.* 25:115–8.

Jandricic, S., J. Sanderson, and S. Wraight. 2008. Intraguild predation among biological control agents used in greenhouse floriculture crops: A preliminary review. *IOBC/WPRS Bull.* 32:91–4.

Kaspari, M. 1990. Prey preparation and the determinants of handling time. *Anim. Behav.* 40:118–26.

Kiman, Z. B., and K. V. Yeargan. 1985. Development and reproduction of the predator *Orius insidiosus* (Hemiptera: Anthocoridae) reared on diets, selected plant material and arthropod prey. *Ann. Entomol. Soc. Am.* 78:464–7.

Lacerda, M. C., A. M. R. M. Ferreira, T. V. Zanuncio, J. C. Zanuncio, A. S. Bernardino, and M. Espíndula. 2004. Development and reproduction of *Podisus distinctus* (Heteroptera: Pentatomidae) fed on larva of *Bombyx mori* (Lepidoptera: Bombycidae). *Braz. J. Biol.* 64:237–42.

Lalonde, R. G., R. R. Mcgregor, and D. R. Gillespie. 1990. Plant-feeding by arthropod predators contributes to the stability of predator–prey population dynamics. *Oikos* 87:603–8.

Latin, J. D. 1989. Bionomics of the Nabidae. *Annu. Rev. Entomol.* 34:383–400.

Latin, J. D. 1999. Bionomics of the Anthocoridae. *Annu. Rev. Entomol.* 44:207–31.

Latin, J. D. 2000. Economic important of minute pirate bug (Anthocoridae). In *Heteroptera of Economic Importance*, ed. C. W. Schaefer and A. R. Panizzi, 607–37. Boca Raton, FL: CRC Press.

Lauenstein, V. G. 1980. Zum sucheverttraten von *Anthocoris nemorum* L. (Het.: Anthocoridae). *Z. Angew. Entomol.* 89:428–42.

Lawton, J. H. 1983. Plant architecture and the diversity of phytophagous insects. *Annu. Rev. Entomol.* 28:23–39.

Legaspi, J. C., and B. C. Legaspi Jr. 1998. Life history trade-offs in insects with emphasis on *Podisus maculiventris* (Heteroptera: Pentatomidae). In *Predatory Heteroptera: Their Ecology and Use in Biological Control*, ed. M. Coll and J. R. Ruberson, 71–87. Lanham, MD: Thomas Say Publications in Entomology.

Legaspi, J. C., and R. J. O'Neil. 1994. Lipids and egg production of *Podisus maculiventris* (Heteroptera: Pentatomidae) under low rates of predation. *Environ. Entomol.* 23:1254–9.

Legaspi, J. C., R. J. O'Neil, and B. C. Legaspi Jr. 1996. Trade-offs in body weights, egg loads, and fat reserves of field-collected *Podisus maculiventris* (Heteroptera: Pentatomidae). *Environ. Entomol.* 25:155–64.

Lemos, W. P., F. S. Ramalho, J. E. Serrão, and J. C. Zanuncio. 2003. Effects of diet on development of *Podisus nigrispinus* (Dallas) (Het., Pentatomidae), a predator of the cotton leafworm. *J. Appl. Entomol.* 127:389–95.

Lemos, W. P., R. S. Medeiros, F. S. Ramalho, J. C. Zanuncio. 2001. Effects of plant feeding on the development, survival and reproduction of *Podisus nigrispinus* (Dallas) (Heteroptera: Pentatomidae). *Int. J. Pest Manag.* 47:89–93.

Loomans, A. J. M., J. C. Van Lenteren, M. G. Tommasini, S. Maini, and J. Riudavets. 1995. *Biological Control of Thrips Pests.* Wageningen: Wageningen Agricultural University Press.

Lucas, J. R. 1985. Partial prey consumption by antilion larvae. *Anim. Behav.* 33:945–58.

Lundgren, J. G. 2011. Reproductive ecology of predaceous Heteroptera. *Biological Control.* doi: 10.1016/j.biocontrol.2011.02.009.

Mackauer, M. 1976. Genetic problems in the production of biological control agents. *Annu. Rev. Entomol.* 21:369–85.

Malais, M. H., and W. J. Ravensberg. 2003. *Knowing and Recognizing: The Biology of Glasshouse Pests and Their Natural Enemies.* Doetinchem: Koppert B.V.

Martinez-Cascales, J. I., J. L. Cenis, G. Cassis, and J. A. Sanchez. 2006. Species identity of *Macrolophus melanotoma* (Costa, 1853) and *Macrolophus pygmaeus* (Rambur, 1839) (Insecta: Heteroptera: Miridae) based on morphological and molecular data and bionomic implications. *Insecta Syst. Evol.* 37:385–404.

Maselou, D., D. Perdikis, and A. Fantinou. 2009. The effects of prey size and mobility on prey selection by the predatory bug *Macrolophus pygmaeus*. *IOBC/WPRS Bull.* 49:293–6.

McCaffrey, J. P., and R. L. Horsburgh. 1986. Biology of *Orius insidiosus* (Heteroptera: Anthocoridae): A predator in Virginia apple orchards. *Environ. Entomol.* 15:984–8.

McPherson, R. M., J. R. Pitis, L. D. Newsom, J. B. Chapin, and D. C. Herzog. 1982. Incidence of tachinid parasitism of several stink bug (Heteroptera: Pentatomidae) species associated with soybean. *J. Econ. Entomol.* 75:783–6.

Medeiros, R. S., F. S. Ramalho, J. C. Zanuncio, and J. E. Serrão. 2003. Effect of temperature on life table parameters of *Podisus nigrispinus* (Het., Pentatomidae) fed with *Alabama argillacea* (Lep., Noctuidae) larvae. *J. Appl. Entomol.* 127:209–13.

Mendes, S. M., V. H. P. Bueno, and L. M. Carvalho. 2005a. Adequabilidade de diferentes substratos à oviposição do predador *Orius insidiosus* (Say) (Hemiptera: Anthocoridae). *Neotrop. Entomol.* 34:415–21.

Mendes, S. M., V. H. P. Bueno, and L. M. Carvalho. 2005b. Reprodução e longevidade de *Orius insidiosus* (Say) (Hemíptera: Anthocoridae) em diferentes temperaturas. *Rev. Agric.* 80:87–101.

Mendes, S. M., V. H. P. Bueno, L. M. Carvalho, and R. P. Reis. 2005c. Custo da produção de *Orius insidiosus* como agente de controle biológico. *Pesq. Agropec. Bras.* 40:441–6.

Mendes, S. M., V. H. P. Bueno, V. M. Argolo, and L. C. P. Silveira. 2002. Type of prey influences biology and consumption rate of *Orius insidiosus* (Say) (Hemiptera: Anthocoridae). *Rev. Bras. Entomol.* 46:99–103.

Meyling, M. V., H. F. Brodsgaard, and A. Enkegaard. 2002. Intraguild predation between the predatory flower bug, *Anthocoris nemorum*, and the aphid parasitoid, *Aphidius colemani*. *IOBC/WPRS Bull.* 25:189–92.

Mituda, E. C., and V. J. Calilung. 1989. Biology of *Orius tantilus* (Motschulsky) (Hemíptera: Anthocoridae) and its predatory capacity against *Thrips palmi* Karny (Thysanoptera: Thripidae) on watermelon. *Philipp. Agric.* 72:165–184.

Mohaghegh, J., P. de Clercq, and L. Tirry. 2001. Functional response of the predators *Podisus maculiventris* (Say) and *Podisus nigrispinus* (Dallas) (Het., Pentatomidae) to the beet armyworm, *Spodoptera exigua* (Hübner) (Lep., Noctuidae): Effect of temperature. *J. Appl. Entomol.* 125:131–4.

Molina-Rugama, A. J., J. C. Zanuncio, E. Vinha, and F. S. Ramalho. 2001. Daily rate of egg laying of the predator *Podisus rostralis* (Stal, 1860) (Heteroptera, Pentatomidae) under different feeding intervals. *Rev. Bras. Entomol.* 45:1–5.

Mollá, O., H. Montón, P. Vanaclocha, F. Beitia, and A. Urbaneja. 2009. Predation by the mirids *Nesidiocoris tenuis* and *Macrolophus pygmaeus* on the tomato borer *Tuta absoluta*. *IOBC/WPRS Bull.* 49:209–14.

Molles, Jr., M. C., and R. D. Pietruszka. 1987. Prey selection by stonefly: The influence of hunger and prey size. *Oecologia* 72:473–8.

Mourão, S. A., J. C. Zanuncio, A. J. Molina-Rugama, E. F. Vilela, and M. C. Lacerda. 2003. Efeito da escassez de presa na sobrevivência e reprodução do predador *Supputius cincticeps* (Stal) (Heteroptera: Pentatomidae). *Neotrop. Entomol.* 32:469–73.

Nakata, T. 1994. Prey species of *Orius sauteri* (Poppius) (Heteroptera: Anthocoridae) in a potato field in Hokakaido. *Jpn. Soc. Appl. Entomol. Zool.* 29:614–6.

Nannini, M., and R. Souriau. 2009. Suitability of *Ceratitis capitata* (Diptera, Thephritidae) eggs as food source for *Macrolophus pygmaeus* (Heteroptera, Miridae). *IOBC/WPRS Bull.* 49:323–8.

Naranjo, S. E., and R. L. Gibson. 1996. Phytophagy in predaceous Heteroptera: Effects on life history and population dynamics. In *Zoophytophagous Heteroptera: Implications for Life History and Integrated Pest Management*, ed. O. Alomar and R. W. Wiedenmann, 57–93. Lanham, MD: Thomas Say Publications in Entomology.

Nascimento, E. C., J. C. Zanuncio, M. C. Picanço, and T. V. Zanuncio. 1997. Desenvolvimento de *Podisus sculptus* Distant, 1889 (Heteroptera: Pentatomidae) em *Bombyx mori* (Lepidoptera: Bombycidae) e *Tenebrio molitor* (Coleoptera: Tenebrionidae). *Rev. Bras. Biol.* 57:195–201.

Nordlund, D. A., and S. M. Greenberg. 1994. Facilities and automation for the mass production of arthropod predators and parasitoids. *Biocontrol News Inform.* 4:45–50.

Okuda, C. R., and K. V. Yeargan. 1988. Intra and interspecific host discrimination in *Telenomus podisi* and *Trissolcus euschisti* (Hymenoptera: Scelionidae). *Ann. Entomol. Soc. Am.* 81:1017–20.

Orr, D. B., J. S. Russin, and D. J. Boethel. 1986. Reproductive biology and behavior of *Telenomus calvus* (Hymenoptera: Scelionidae), a phoretic egg parasitoid of *Podisus maculiventris* (Hemiptera: Pentatomidae). *Can. Entomol.* 118:1063–72.

Péricart, J. 1972. Faune de l'Europe et du Bassin Méditerréen. Hémiptéres–Anhtocoridae, Cimicidae, Microphysidae de l'ouest-paleartique. Masson et Cie. 7.

Perkins, P. V., and T. F. Watson. 1972. Biology of *Nabis alternatus* (Hemiptera: Nabidae). *Ann. Entomol. Soc. Am.* 65:54–7.

Pierre, L. S. R., V. H. P. Bueno, M. V. Sampaio, J. C. Van Lenteren, B. F. de Conti, M. P. F. Silva and L. C. P. Silveira. 2006. Intraguild predation between *Orius insidiosus* (Say) and *Aphidius colemani* Viereck, and biological control of *Aphis gossypii* Glover. *IOBC/WPRS Bulletin* 29:219–22.

Polis, G. A. 1981. The evolution and dynamics of intraspecific predation. *Annu. Rev. Ecol. Syst.* 12:225–51.

Polis, G. A., and R. D. Holt. 1992. Intraguild predation: Dynamics of complex trophic interactions. *Trends Ecol. Evol.* 7:151–4.

Press, J. W., B. R. Flaherty, and R. T. Arbogast. 1974. Interactions among *Plodia interpunctella, Bracon hebetor*, and *Xylocoris flavipes*. *Environ. Entomol.* 3:183–4.

Ramakers, P. M. J., and J. M. Rabasse. 1995. Integrated pest management in protected cultivation. In *Novel Approaches to Integrated Pest Management*, ed. R. Reuveni, 199–229. Boca Raton, FL: CRC Press.

Richards, P. C., and J. M. Schmidt. 1996. The effects of selected dietary supplements on survival and reproduction of *Orius insidiosus* (Say) (Hemiptera: Anthocoridae). *Can. Entomol.* 128:171–6.

Richman, D. B., and W. H. Whitcomb. 1978. Comparative life cycles of four species of predatory stink bugs. *Fla. Entomol.* 61:113–9.

Ridgway, R. L., and S. L. Jones. 1968. Plant feeding by *Geocoris pallens* and *Nabis americoferus*. *Ann. Entomol. Soc. Am.* 61:232–3.

Riudavets, J. 1995. Predator of *Frankliniella occidentales* (Perg.) and *Thrips tabaci* Lind.: A review. In *Biological Control of Thrips Pests*, ed. A. J. M. Looman, J. C. Van Lenteren, M. G. Tommasini, S. Maini, and J. Riudavets, 43–87. Wageningen: Agricultural University Papers.

Riudavets, J., and C. Castañé. 1994. Abundance and host plant preferences for oviposition of *Orius* spp. (Heteroptera: Anthocoridae) along the Mediterranean coast of Spain. *IOBC/WPRS Bull.* 17:230–6.

Riudavets, J., J. Arnó, and C. Castané. 2006. Rearing predatory bug with the brine shrimp *Artemia* sp. as alternative prey food. *IOBC/WPRS Bull.* 29:235–40.

Rosenheim, J. A., and J. P. Harmon. 2006. The influence of intraguild predation on the suppression of a shared prey population: An empirical reassessment. In *Trophic and Guild Interactions in Biological Control*, ed. J. Brodeur and G. Boivin, 1–20. Dordrecht: Springer.

Rosenheim, J. A., H. K. Kaya, L. E. Ehler, J. J. Marois, and B. A. Jaffee. 1995. Intraguild predation among biological control agents: Theory and evidence. *Biol. Control* 5:303–35.

Roth, S., A. Janssen, and M. W. Sabelis. 2008. Odour-mediated sexual attraction in nabids (Heteroptera: Nabidae). *Eur. J. Entomol.* 105:159–62.

Roth, S., and K. Reinhardt. 2009. Sexual dimorphism in winter survival rate differs little between damsel bug species (Heteroptera: Nabidae). *Eur. J. Entomol.* 106:37–41.

Ruberson, J. R., and D. J. T. Kring. 1990. Predation of *Trichogramma pretiosum* by the anthocorid *Orius insidiosus*. *Les Coloques* 56:41–3.

Ruberson, J. R., and M. Coll. 1998. Research needs for the predaceous Heteroptera. In *Predatory Heteroptera: Their Ecology and Use in Biological Control*, ed. M. Coll and J. R. Ruberson, 225–33. Lanham, MD: Thomas Say Publications in Entomology.

Sadeghi, H., and F. Gilbert. 1999. Individual variation in oviposition preference, and its interaction with larval performance in an insect predator. *Oecologia* 118:405–11.

Salas-Aguilar, J., and L. E. Ehler. 1977. Feeding habits of *Orius tristicolor*. *Ann. Entomol. Soc. Am.* 70:60–2.

Sánchez, J. A., D. R. Gillespie, and R. R. MacGregor. 2004. Plant preference in relation to life history traits in the zoophytophagous predator *Dicyphus hesperus*. *Entomol. Exp. Appl.* 112:7–19.

Schmidt, J. M. 1994. Encounters between adult spined assassin bugs *Sinea diadema* (Fabr.) (Hemiptera: Reduviidae): The occurrence and consequences of stridulation. *J. Insect Behav.* 7:811–28.

Schmidt, J. M., J. R. Taylor, and J. A. Rosenheim. 1998. Cannibalism and intraguild predation in the predatory Heteroptera. In *Predatory Heteroptera: Their Ecology and Use in Biological Control*, ed. M. Coll and J. R. Ruberson, 131–69. Lanham, MD: Thomas Say Publications in Entomology.

Schmidt, J. M., P. C. Richards, H. Nadel, and G. Ferguson. 1995. A rearing method of the production of large numbers of the insidiosus flower bug, *Orius insidiosus* (Say) (Hemiptera: Anthocoridae). *Can. Entomol.* 127:445–7.

Schmidt, J., and R. A. Goyer. 1983. Consumption rates and predatory habits of *Scoloposcelis mississippensis* and *Lyctocoris elongates* (Hemiptera: Anthocoridae) on pine bark beetles. *Environ. Entomol.* 12:363–7.

Schuster, M. F., and M. Calderon. 1986. Interactions of host plant resistant genotypes and beneficial insects in cotton ecosystems. In *Interactions of Plant Resistance and Parasitoids and Predators of Insects*, ed. D. J. Boethel and R. D. Eikenbary, 84–97. Harvood: Chichester.

Scott, W. P., G. L. Snodgrass, and J. W. Smith. 1988. Tarnished plant bug (Hemiptera: Miridae) and predaceous arthropod populations in commercially produced selected nectaried and nectariless cultivars of cotton. *J. Entomol. Sci.* 23:280–6.

Shipp, J. L., N. Zariffa, and G. Ferguson. 1992. Spatial patterns of and sampling methods of *Orius* spp. (Hemiptera: Anthocoridae) on greenhouse sweet pepper. *Can. Entomol.* 124:887–94.

Silva, M. P. F. 2006. Influência do consumo de diferentes tipos de alimentos na longevidade e oviposição de fêmeas do predador *Orius insidiosus* (Say) (Hemíptera: Anthocoridae). Monografia, Univ. Federal Lavras, Brazil.

Silveira, L. C. P., V. H. P. Bueno, and J. C. Van Lenteren. 2004. *Orius insidiosus* as biological control agent of thrips in greenhouse chrysanthemums in the tropics. *Bull. Insectol.* 57:103–9.

Silveira, L. C. P., V. H. P. Bueno, J. N. C. Lousada, and L. M. Carvalho. 2005. Percevejos predadores (*Orius* spp.) (Hemiptera: Anthocoridae) e tripes (Thysanoptera): Interação no mesmo habitat? *Rev. Árvore* 29:767–73.

Silveira, L. C. P., V. H. P. Bueno, L. S. R. Pierre, and S. M. Mendes. 2003. Plantas cultivadas e invasoras como habitat para predadores do gênero *Orius* Wolff (Heteroptera: Anthocoridae). *Bragantia* 62:261–5.

Soglia, M. C. M, V. H. P. Bueno, and L. M. Carvalho. 2007. Efeito da presa alternativa no desenvolvimento e consumo de *Orius insidiosus* (Say) (Heteroptera, Anthocoridae) e comportamento de oviposição em cultivares de crisântemo. *Rev. Bras. Entomol.* 51:512–17.

Streams, F. A. 1992. Intrageneric predation by *Notonecta* (Hemiptera: Notonectidae) in the laboratory and in nature. *Ann. Entomol. Soc. Am.* 85:265–73.

Sweet II, M. H. 2000. Economic importance of predation by big-eyed bugs (Geocoridae). In *Heteroptera of Economic Importance*, ed. C. W. Shaefer and A. R. Panizzi, 713–25. Boca Raton, FL: CRC Press.

Tavella, L., R. Tedeschi, A. Arzone, and A. Alma. 2000. Predatory activity of two *Orius* species on the western flower trips in protected pepper crops. *IOBC/WPRS Bull.* 23:231–40.

Thompson, S. N. 1999. Nutrition and culture of entomophagous insects. *Annu. Rev. Entomol.* 44:561–92.

Tommasini, M. G. 2003. Evaluation of *Orius* species for biological control of *Frankliniella occidentalis* (Pergande) (Thysanoptera: Thripidae). PhD thesis, Wageningen University.

Tommasini, M. G., J. C. Van Lenteren, and G. Burgio. 2004. Biological traits and predation capacity of four *Orius* species on two prey species. *Bull. Insectol.* 57:79–93.

Torres, J. B., and D. W. Boyd. 2009. Zoophytophagy in predatory Hemiptera. *Braz. Arch. Biol. Technol.* 52:1199–208.

Torres, J. B., J. C. Zanuncio, and M. A. Moura. 2006. The predatory stinkbug *Podisus nigrispinus*: Biology, ecology and augmentative releases for lepidopteran larval control in *Eucalyptus* forests in Brazil. *CAB Rev., Perspect. Agric. Vet. Sci. Nutr. Nat. Resour.* 1:1–18.

Torres, J. B., J. C. Zanuncio, P. R. Cecon, and W. L. Gasperazzo. 1996. Mortalidade de *Podisus nigrispinus* (Dallas) por parasitóides de ovos em áreas de eucalipto. *An. Soc. Entomol. Bras.* 25: 463–71.

Urbaneja, A., H. Montón, and O. Mollá. 2009. Suitability of the tomato borer *Tuta absoluta* as prey for *Macrolophus caliginosus* and *Nesidiocoris tenuis*. *J. Appl. Entomol.* 133:292–6.

Van den Meiracker, R. A. F. 1999. Biocontrol of western flower thrips by heteropteran bugs, PhD thesis, Amsterdam University.

Van den Meiracker, R. A. F. and M. W. Sabelis. 1999. Do functional responses of predatory arthropods reach a plateau? A case study of *Orius insidiosus* with western flower thrips as prey. *Entomol. Exp. Appl.* 90:323–9.

Van der Blom, J., A. Robledo, S. Torres, and J. A. Sánchez. 2009. Consequences of the wide scale implementation of biological control in greenhouse horticulture in Almeria, Spain. *IOBC/WPRS Bull.* 49:9–13.

Van Lenteren, J. C. 1991. Harvesting safely from biodiversity: Natural enemies as sustainable and environmentally friendly solutions for pest control. In *Balancing Nature: Assessing the Impact of Importing Nonnative Biological Control Agents (An International Perspective)*, ed. J. A. Lockwood, F. G. Howarth, and M. F. Purcell, 15–30. Lanham, MD: Thomas Say Publication Entomology.

Van Lenteren, J. C. 2011. The state of commercial augmentative biological control: Plenty of natural enemies, but a frustrating lack of uptake. *BioControl* (on line first). doi:10.1007/s10526-011-9395-1.

Van Rijn, P. C. J., and M. W. Sabelis. 1993. Does alternative food always enhance biological control? The effect of pollen on the interaction between western flower thrips and its predators. *IOBC/WPRS Bull.* 17:123–5.

Van Schelt, J., J. Klapwijk, M. Letard, and C. Aucouturier. 1996. The use of *Macrolophus caliginosus* as a whitefly predator in protected crops. In *Bemisia 1995: Taxonomy, Biology, Damages, Control and Management*, ed. D. Gerling and R. T. Mayer, 515–21. Andover: Intercept.

Waage, J. K., K. P. Carl, N. J. Mills, and D. J. Greathead. 1985. Rearing entomophagous insects. In *Handbook of Insect Rearing*, ed. P. Singh, and R. F. Moore, 45–66. Amsterdam: Elsevier.

Wäckers, F. L. 2008. Food for thought: How to cater to the nutritional needs of biological control agents? *IOBC/WPRS Bull.* 32:253–60.

Wheeler, A. G. 2000. Predacious plant bugs (Miridae). In *Heteroptera of Economic Importance*, ed. C. W. Shaefer and A. R. Panizzi, 657–93. Boca Raton, FL: CRC Press.

Wheeler, A. G. 2001. *Biology of the Plant Bugs (Hemiptera: Miridae): Pest, Predators, Opportunists*. Cornell, NY: Cornell University Press.

Wiedenmann, R. N., and R. J. O'Neil. 1990. Effects of low rates of predation on selected life-history characteristics of *Podisus maculiventris* (Say) (Heteroptera: Pentatomidae). *Can. Entomol.* 122:271–83.

Yeargan, K. V. 1979. Parasitism and predation of stink bug eggs in soybean and alfalfa fields. *Environ. Entomol.* 8:715–9.

Yeargan, K. V. 1998. Predatory Heteroptera in North American agroecosystems: An overview. In *Predatory Heteroptera: Their Ecology and Use in Biological Control*, ed. M. Coll and J. R. Ruberson, 7–19. Lanham, MD: Thomas Say Publications in Entomology.

Yeargan. K. V., and C. M. Allard. 2002. Sensitive stage for photoperiod-induced reproductive diapause in the predator *Geocoris punctipes* (Heteroptera: Geocoridae). 2002. Esa Annual Meeting and Exhibition, Fort Lauderdale, Florida.

Yokoyama, V. Y. 1978. Relation of seasonal changes in extrafloral nectar and foliar protein and arthropod populations in cotton. *Environ. Entomol.* 7:799–802.

Yokoyama, V. Y. 1980. Method for rearing *G. pallens*, a predator in California cotton. *Can. Entomol.* 112:1–3.

Ysenhour, D. J., and K. V. Yeargan 1981.Effect of temperature on the development of *Orius insidiosus*, with notes on laboratory rearing. *Ann. Entomol. Soc. Am.* 74:114–6.

Zanuncio, J. C., E. B. Beserra, A. J. Molina-Rugama, T. V. Zanuncio, T. B. M. Pinon, and V. P. Maffia. 2005. Reproduction and longevity of *Supputius cincticeps* (Het.: Pentatomidae) fed with larvae of *Zophobas confusa*, *Tenebrio molitor* (Col.: Tenebrionidae) or *Musca domestica* (Dip.: Muscidae). *Braz. Arch. Biol. Technol.* 48:771–7.

Zanuncio, J. C., M. C. Lacerda, J. S. Zanuncio Junior, T. V. Zanuncio, A. M. C. Silva, and M. Espíndula. 2004. Fertility table and rate of population growth of the predator *Supputius cincticeps* (Heteroptera; Pentatomidae) on one plant of *Eucalyptus cloeziana* in the field. *Ann. Appl. Biol.* 144:357–61.

Zanuncio, J. C., R. N. Guedes, H. N. Oliveira, and T, V, Zanuncio. 2002. Uma década de estudos com perce vejos predadores: Conquistas e desafios. In *Controle Biológico no Brasil: Parasitóides e Predadores*, ed. J. R. P. Parra, P. S. M. Botellho, B. S. Corrêa Ferreira, and J. M. Bento, 495–509. São Paulo: Manole.

Zanuncio, J. C., T. V. Zanuncio, R. N. Guedes, and F. S. Ramalho. 2000. Effect of feeding on three *Eucalyptus* species on the development of *Brontocoris tabidus* (Het.: Pentatomidae) fed with *Tenebrio molitor* (Col.: Tenebrionidae). *Biocontr. Sci. Technol.* 10:443–50.

23

Predatory Beetles (Coccinellidae)

Lúcia M. Almeida and Cibele S. Ribeiro-Costa

CONTENTS

23.1 Introduction 571
23.2 Evolution, Taxonomy, and Morphology 572
23.3 Biology and Development 574
     23.3.1 Postembryonic Development 574
     23.3.2 Adult Development 576
23.4 Food Selection 576
     23.4.1 Food Specificity 577
     23.4.2 Food Quality 579
     23.4.3 Food Preferences 580
     23.4.4 Food Toxicity 581
23.5 Defense Strategies 581
23.6 Cannibalism 581
23.7 Intraguild Competition 582
23.8 Adaptations and Responses to Variations in Abiotic and Biotic Factors 583
23.9 Natural Enemies 584
23.10 Conclusions and Suggestions for Research 585
References 586

23.1 Introduction

Predators, together with parasites, parasitoids, and pathogens have received special attention, particularly from ecologists, due to their importance in biological control. Among the beetles (Coleoptera), coccinellids (Coccinellidae) are the most important predators.

The Coccinellidae family has more than 6000 described species distributed in 360 genera (Vandenberg 2002), with approximately 2000 in the Neotropical region. Most of these insects are important as efficient predators of aphids, coccids, psyllids, and other sucking insects, which are pests in agroforestry systems (Hodek and Honek 1996).

Among the members of the family, the small group of phytophagous species is also economically important because they are found feeding mainly on plants of the Cucurbitaceae and Solanaceae. In south Brazil, the species *Epilachna paenulata* (Germar), *Epilachna spreta* (Mulsant), and *Epilachna cacica* (Guèrin) are relatively common and feed on cultivated horticultural crops of *Cucurbita pepo* (squash), *Sechium edule* (chayote), and *Cucumis sativus* (cucumber) (Araújo-Siqueira and Almeida 2004).

Predatory coccinellids are very active searchers wherever prey can be found, as well as being very voracious, which characterizes them as efficient predators, principally of aphids (Hodek 1973). The natural occurrence of coccinellid larvae and adults during aphid infestations on crop plants is important for the latter's control, reducing field populations and potential damage.

Coccinellid aphid predators, both as larvae and adults, are generally well synchronized with pest populations and very sensitive to population changes of their prey. For this reason, they are considered more efficient natural enemies than predator species, which only act as larvae or as adults (Hagen and van den Bosch 1968). The presence of predators that control economically important pests is indispensable for the dynamic equilibrium of agroecosystems since this reduces human intervention for control and helps in regulating insect pests in many crops (Olkowskim et al. 1990, Obrycki and Kring 1998).

Coccinellids are considered efficient control agents of various aphids and other prey species. Their appetite, that is, the maximum number of prey individuals consumed by the predator; the functional response, which is the relationship between the number of prey captured and the number of prey available; and the preference, as well as the capacity to capture prey, are the main factors influencing their feeding process and predation efficiency as biological control agents. All these factors are intimately linked to temperature (Frazer 1988).

Biological control programs that use coccinellids started aiming to harmonize equilibrium processes, as well as avoiding the excessive use of agrochemicals in the environment. The first and greatest case of successful classical biological control was the introduction into California in 1888 of *Rodolia cardinalis* (Mulsant), an Australian species, to control the scale, *Icerya purchasi* Maskell, in citrus groves. Even after 100 years, these coccinellids are still important in control, maintaining scale populations below the economic injury level.

For insects, chemical and physical characteristics of the food, and interactions between substances and their harmful effects, may alter their performance. Therefore, in this chapter, aspects of coccinellid feeding behavior will be discussed, with emphasis on specificity, quality, preference, and toxicity of the foods exploited by these insects.

23.2 Evolution, Taxonomy, and Morphology

Coccinellids belong to an ancient and very successful group of beetles (Coleoptera), which originated in the Permian period, around 280 million years ago. Comparative morphological studies of present Coleoptera groups have shown that Coccinellidae are among the most advanced of the Coleoptera (Crowson 1981). However, more recent studies based on mitochondrial cytochrome oxidase indicate that these beetles may be linked to more primitive groups, such as the Carabidae, than to more recent ones (Howland and Hewitt 1995).

The Coccinellidae family is monophyletic, with species distributed throughout the world. The dorsal form of the body is extremely convexed and ventrally flat; the head is concealed by the pronotum, with antennae having 9 to 11 segments, with the last three to six segments in the form of a club and tarsi have four segments, rarely three. The family is divided into six subfamilies: Coccidulinae, Coccinellinae, Scyminae, Chilocorinae, Sticholotidinae, and Epilachninae (Vandenberg 2002). Except for a few members of the subfamily Coccinellinae (Psylloborini) that feed on fungi, and members of Epilachninae that feed on higher plants, all remaining coccinellids are predators of aphids, psyllids, scales, mites, and eventually other insect larvae (Dixon 2000).

Phylogenetic studies using 62 terminal taxa, and based on the ribosomal genes 18S and 28S, indicate that there appear to have been at least three transitions to aphidophagy and one to phytophagy originating from a common ancestor, and a second transition from phytophagy arose in the aphidiphagous/pollinivorous clade (Giorgi et al. 2009). Therefore, the authors conclude that modern coccinellid ancestors made a transition from mycophagy to predation, especially to coccidophagy.

The name Coccinellidae refers to the reddish color of the elytra of most species, principally those of the subfamily Coccinellinae, the first to be known and described, and the most typical. The standard dorsal color of the Coccinellidae is very variable, making identification difficult and it is necessary to look at the morphological characters on the ventral surface, especially the coloration of the pro-, meso-, and meta-epimeres, as well as the presence and shape of the postcoxal line and also the genitalia, principally of the male. According to Iperti (1999), field identification is possible using other characters, such as size, shape, and hairiness, which are sufficient for recognizing feeding preferences. It is possible to differentiate large coccinellids (3–9 mm), which are glabrous, from the small ones (<3 mm), which are pubescent, and the very small species, (<2 mm), which often feed on mites and aleyrodids. These three groups represent 60%, 39%, and 1% of species, respectively. Sometimes, the type of food can be forecasted from the elytral coloration (Iperti 1999). In Europe, for example, aphidophagous species have brilliantly red, yellow, or orange elytra and constitute 65% of the species of the family (Coccinellinae, Hippodamini; *Scymnus* spp. and *Pullus* spp.). Coccidophagous species are dark colored and represent 25% of the species of the family (Chilocorini, Hyperaspinae, *Sidis* spp., *Nephus* spp., and *Cryptolaemus* spp.). The mycophagous coccinellids are light brown, whitish, or yellowish and represent 8% of the species (Psylloborini; *Rhyzobius* sp.). Both the aphidophagous (Coccinellini, Hippodamiini) and the phytophagous species (Epilachninae) are yellowish brown, with the former not being pubescent. There are no large morphological variations among either larvae or adults of predatory coccinellids; however, species do differ in some aspects and this reflects, or is related to, their lifestyle.

The type of food eaten by an insect can be recognized from the morphology of the mandibles (Figure 23.1). Predators have mandibles with one or two sharp apical teeth. In coccid predators, the apical tooth is very sharp and serves to cut and remove the hard carapace that covers the prey (Samways and Wilson 1988). In some species, the inside tooth edge also has a cavity, so that when the sharp point perforates the prey, the food is directed to the buccal opening. Also, these predators generally inject a digestive enzyme into the prey to accelerate food digestion for which they use the same sharp tooth and the channel that functions like an intradermic needle. The whole apparatus will also serve to suck up the previously digested food.

Coccidophilus citricola Brèthes is a coccinellid that feeds on armored scales (Hemiptera, Diaspididae), abundant and common in Brazilian citrus orchards. First registered in Brazil in Rio de Janeiro and Pernambuco states, it is important in the natural control of *Diaspis echinocacti* (Bouché), the cactus scale; both the larvae and the adults have strong, pointed, and symmetrical mandibles (Silva et al. 2005).

Mandibles of phytophagous species have a series of apical teeth, generally five, for rasping and feeding on the leaf parenchyma. These species usually show an interesting behavior pattern, marking out the region of the leaf where they are going to feed and cutting with the mandibles each vein that nourishes the plant. This interrupts the sap flow, which stops any reaction by the plant of introducing toxic substances into the insect gut (Almeida and Marinoni 1986, Almeida and Ribeiro 1986, Ribeiro and Almeida 1989, Araújo-Siqueira and Almeida 2004).

In the mycophagous coccinellids, mandibles have a structure called prosteca, which consists of a series of sharp teeth forming a type of comb that is introduced into the fungal hyphae during feeding, pulling them out and placing them into the buccal opening. *Psyllobora gratiosa* Mader uses this form of feeding where mandibles are introduced into the fungal hyphae (Almeida and Milléo 1998). Like mandibles, maxillae and maxillary palps play a fundamental role in feeding and prey recognition (Kesten 1969, Nakamura 1985); amputation of the maxillary palps of *Coccinella septempunctata brucki* L. results in about 40% reduction of prey capture efficiency. The size and shape of the maxillary palps, as well as the presence and number of receptive sensillae at their tips, appear to influence the velocity and efficiency of prey searching. For example, aphid predator species should respond more rapidly to the presence of their prey, whereas for species that feed on scales, or even for the phytophages, this is unnecessary since their food can be more easily located due to its size and lack of mobility (Barbier et al. 1996).

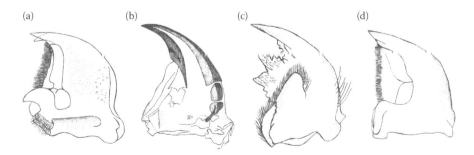

FIGURE 23.1 Representatives from mandibles of feeding guilds of Coccinellidae: (a) Aphid predator; (b) coccid predator; (c) phytophagous; and (d) mycophagous. (From Silva, R. A., et al., *Rev. Bras. Entomol.*, 49, 29, 2005. Creative Commons license.)

23.3 Biology and Development

Coccinellids are holometabalous; that is, their development include egg, four larval instars, prepupa, pupa, and adult. There are some rare species whose larval development has three or five instars. Only one known coccidophagous species has three instars (Hodek and Honek 1996); development cycle varies from less than 2 weeks to more than 2 months, depending on the size, thermal conditions, and trophic specificity.

23.3.1 Postembryonic Development

Eggs are small, elongated, and can be yellowish/orange at the beginning of oviposition, darkening just before emergence. The sculpturing of the chorion may be an important characteristic in phylogeny. Most species deposit their eggs in batches (Figures 23.2b and 23.3a), which stay glued to leaves, branches, or other solid substrates by the base. Predatory species lay their eggs on the ventral surface of leaves near their prey. Aphidophagous and phytophagous species lay their eggs in batches of 10 to 100; coccidophagous species deposit fewer egg batches or single eggs. There are exceptions, such as the aphidophagous species of *Platynaspis*, which deposit their eggs singly into slits or rolled-up leaves to protect them from ants (Völkl 1995). *Eupalea reinhardti* Crotch, which feeds on psyllids (Psyllidae), lays its eggs singly on the inside face of old and rolled-up leaves of *Caesalpinea peltophoroides* (Caesalpinacea) (Ferreira and Almeida 2000). *Zagloba beaumonti* Casey lays a single egg inside the carapace of the scale, *D. echinocacti* Bouché, probably as a strategy for larval protection and survival, since the larva has access to food to complete its development (Lima 1999). After eclosion, larvae stay on the chorion for another 1 or 2 days and feed on nonviable eggs.

Larvae that feed on aphids and psyllids hatch after 2–5 days; incubation time in coccidophagous species is much longer, around 7–9 days (Table 23.1). Before ecdysis, the larva stops feeding and uses its anal organ to fix itself to the substrate. Larvae of some species aggregate for this change. The prepupal phase is characterized by the fourth instar larva, which fixes itself to the substrate, remaining curved and does not feed, such as with *Olla v-nigrum* (Mulsant) and in *Harmonia axyridis* (Pallas) (Figure 23.2d). When larvae are fed *ad libitum* they grow exponentially. The fourth and final instars generally last the longest (Table 23.1), and the total amount of food consumed, as well as individual size, is determined during this period. The duration of larval instars may be influenced by temperature, but food quality and quantity are more important. Predator larvae are dark colored and very active, with a long, flattened

FIGURE 23.2 *Harmonia axyridis*: (a) Adults mating; (b) egg mass; (c) first instar; (d) fourth instar; (e) pupae; and (f) newly emerged adult.

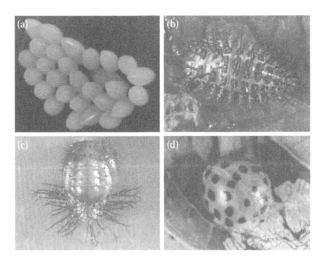

FIGURE 23.3 *Epilachna vigintioctopunctata*: (a) Egg mass; (b) fourth instar; (c) pupae; and (d) adult. (From Araújo-Siqueira, M. and L. M. Almeida, *Rev. Bras. Zool.* 21, 543, 2004. Creative Commons license.)

body, long thoracic legs, and characteristic bristles (Figure 23.2e). Larvae of the Scyminae have a thick layer of white wax similar in appearance to coccids. However, larvae of phytophagous species have a more globular body with short legs and are slower and have scoli (modified spines) distributed all over their body (Figure 23.3b).

Pupae of species that feed on aphids and psyllids, and species of the Sticholotidinae have no covering. In the coccidophagous species of Chilocorini and Noviini, pupae are partially covered and develop within a larval exuvia, whereas in the Hyperaspini and Scyminae, the larval exuvia completely covers the pupa (Iperti 1999). Pupae of phytophagous species are less well protected and remain with the exuvia of the fourth instar only on the hind region of the body, which is fixed to the substrate. Pupae are not totally immobile and if disturbed can move, pushing its body forward. Pupal color is influenced by temperature and humidity; *C. septempunctata* are orange when reared at 35°C and 55% relative humidity, turning dark brown at 15°C and 95% humidity (Hodek 1958).

TABLE 23.1

Development Time for Species of Coccinellidae Commonly Found in Brazil, Related to Different Types of Food

Types of Food	Species/Temperature	Egg	1° Instar	2° Instar	3° Instar	4° Instar	Pupa	References
Coccids	*Coccidophilus citricola*/24°C	9.54	4.22	2.85	2.94	3.22	5.7	Silva et al. 2004
	Zagloba beaumonti/25°C	8.0	3.6	3.0	2.9	3.2	5.5	Lima 1999
Psilids	*Eupalea reinhardti*/25°C	2.9	2.3	1.8	1.7	2.9	5.0	Almeida unpublished
	Olla v-nigrum/23°C	2.76	3.84	2.07	2.5	3.36	3.64	Kato et al. 1999
Aphids	*Hippodamia convergens*/23°C	3.9	2.9	2.2	2.6	3.1	6.58	Oliveira et al. 2004
	Cycloneda sanguinea/23°C	3.95	2.5	1.8	1.9	2.7	6,08	
	Eriopis connexa/23°C	3.96	3.1	2.2	2.5	3.0	5.74	
	Harmonia axyridis/17°C	4.42	7.0	6.57	8.28	15.14	6.42	Mise and Almeida unpublished
Plants	*Epilachna vigintioctopunctata*/24°C	7.14	5.88	4.62	5.88	9.81	8.19	Araújo-Siqueira and Almeida 2004

TABLE 23.2

Mean Developmental Stages (Days) and Viability of *Olla v-nigrum* at
Four Temperatures, Photoperiod of 12:12 h and 70% Relative Humidity

	Duration (days)			
Stage	17°C	21°C	25°C	29°C
Egg	6.26	3.78	2.76	2.00
1° Instar	7.33	4.78	3.84	2.00
2° Instar	4.79	3.14	2.07	1.62
3° Instar	6.24	2.31	2.50	1,50
4° Instar	10.22	3.42	3.36	3.26
Larval period	27.00	13.83	11.16	8.32
Prepupae	1.89	0.83	0.84	0.79
Pupa	10.83	4.83	3.64	2.53
Viability	77%	49%	78%	78%

23.3.2 Adult Development

Adult coccinellids disperse fast from the place of larval development, and mating occurs in less than a week after emergence, and about a week later females start laying eggs. In temperate regions, in summer and autumn, recently emerged adults enter dormancy (estivation and hibernation, respectively). For those species that live in temperate areas, the highest temperature limit for development is 32–33°C. For *O. v-nigrum* Mulsant, a cosmopolitan predator of psyllids (Hemiptera, Psyllidae), which develops on ornamental trees in south Brazil, the base temperature is 11.36°C, with a thermal constant of 240.93 degree days, but the most suitable temperature is 25°C. In colder temperatures, development is much longer and viability decreases (Table 23.2). In biological control programs a species is considered more efficient if it has incubation period and development time shorter than others (Nakajo 2006).

Adult longevity depends on voltinism and varies from a few months to years. The native coccinellids of temperate zones estivate or hibernate as adults and enter into quiescence or diapause. On the other hand, exotic species such as *Lindorus lophantae* Blaisdell, *Cryptolaemus mountrouzieri* Mulsant, and *Novius cardinalis* Mulsant do not estivate or hibernate, their larval stages resist drastic climate changes, reducing the speed of development during the winter but they never stop development; female ovaries mature soon after hibernation and after feeding on aphids; males gonads that have hibernated in autumn show certain degree of spermatogenesis (Dixon 2000).

At the beginning of the diapause *Semiadalia undecimnotata* (Schneider) males have active testicles, but this activity decreases with temperature reduction; females of this species, and of *Adalia bipunctata* (L.), have empty spermathecae at the beginning of dormancy until the spring (Hodek and Landa 1971, Hemptinne and Naisse 1987). In regions with severe winter, mating occurs after the end of dormancy in groups of individuals before their dispersal or migration; the preoviposition period lasts about a week. This behavior is important for survivorship of migrating species since only one mate is enough to fertilize a female during its lifetime.

The oviposition rate is proportional to the number of ovarioles, which varies considerably between trophic groups and species (i.e., from less than 4 to more than 50). Experiments demonstrate that polivoltine species generally have a high fecundity in the spring and monovoltine species have a high fecundity in the summer (Iperti et al. 1977).

23.4 Food Selection

The principal food of predatory coccinellids are aphids (Aphididae) and coccids (Coccidae), as well as other types of prey, such as mites (Putman 1955, Villanueva et al. 2004), Adelgidae (Delucchi 1954, Pope 1973), aleyrodids (Heinz and Zalom 1996), ants (Pope and Lawrence 1990), chrysomelid larvae (Elliot and Little 1980), as well as Hemiptera, Cicadellidae (Ghorpade 1979), Pentatomidae (Subramanyam 1925), and Phylloxeridae (Pope 1973).

The food of predatory coccinellids is dependent on prey abundance in their environment (Dixon 2000). In Europe, most of the species feed on aphids and coccids (Klausnitzer 1993); in Australia prey is variable (Hales 1979). Both adults and immatures consume the same food, and females help their progeny to find food by laying their eggs on plants that have prey colonies since coccinellids cannot detect their prey at a distance but only when in direct contact with it. Some species are attracted by plant volatiles colonized by aphids, such as *Anatis ocellata* (L.) attracted by volatiles from infested *Pinus* (Kesten 1969).

Phytophagous coccinellids such as *Epilachna borealis* Fabe and *Epilachna varivestis* Mulsant feed preferentially on soybean leaves, but can also feed on bean species, *Phaseolus vulgaris* and *Phaseolus lunatus*. As for the Brazilian *Epilachna* species, they feed on various species of Cucurbitaceae and Solanaceae, but their preference for certain plant species is clearly demonstrated; for example, *Epilachna paenulata* (Germar) that feeds on cucurbits does not develop beyond the first instar on *Secchium edule* (cucumber) (Marinoni and Ribeiro-Costa 1987).

Species of Psylloborini, in the genera *Halyzia*, *Vibidia*, and *Thea*, once thought predators, are exclusively mycophagous. *Tytthaspis sedecimpunctata* (L.) feeds on fungus in general (Turian 1969); *P. gratiosa* Mader appears to prefer *Oidium* sp., which occurs on *Hydrangea hortensis*, a common ornamental plant in south Brazil (Almeida and Milléo 1998).

23.4.1 Food Specificity

Coccinellid tend to be food specific. Among predators, some coccidophagous species are more specific than aphidophagous species (Kairo and Murphy 1995, Strand and Obrycky 1996). Among species that feed on psyllids, *E. reinhardti* is an example of specificity because it did not feed when offered aphids or coccids instead of psyllids (Ferreira and Almeida 2000).

Prey specificity in coccidophagous and aphidophagous species is a result of some mechanisms developed by these groups. Coccids, which are relatively immobile, invest more in defense, producing carapaces or toxins for their protection. Aphids, however, depend on their mobility to avoid capture (Dixon 1958). Therefore, the greater host specificity of coccidophagous species may be a response to the strong defense of their prey (Figure 23.4); however, in *Scymnus* spp., 23% of the species feed exclusively on aphids and 62% on coccids (Hatch 1961).

Host specificity have been studied by various authors and summarized by Hodek and Honek (1996) (Table 23.3). Coccidophagous coccinellids are generally smaller than aphidophagous species and tend

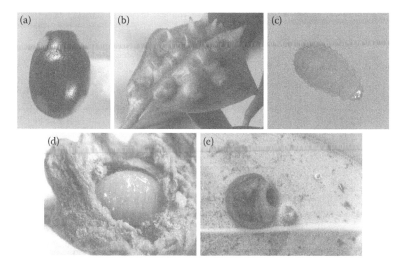

FIGURE 23.4 *Hyperaspis delicata*: (a) Adult; (b) leaf gall of guava; (c) larvae; (d) pupae; and (e) adult. (From Almeida L. M. and M. D. Vitorino, *Coleopt. Bull.* 51, 213, 1997. With permission.)

TABLE 23.3

Types of Food Used by Several Groups of Coccinellidae

Subfamily	Tribe	Prey
Sticholotidinae	Sukunahikonini	Coccids, Diaspidinae
	Serangiini	Aleirodids
	Sticholotidini	Coccids, Diaspidinae
	Pharini	Diaspidinae
	Microweiseini	Diaspidinae (*Aspidiotus*, *Chionaspis*)
Scyminae	Stethorini	Phytophagous mites, Tetranychidae
	Scymnillini	Aleyrodidae
	Scyminini	62% Coccids, 23% aphids
	Clitostethus, Lioscymnus	Aleirodids, aphids
	Diomus, Nephus	Pseudococcinae, coccids
	Sidis, Paradisis	*Pseudococcus*
	Cryptolaemus	Pseudococcinae
	Pseudoscymnus	Diaspidinae
	Platyorus	Aphids
	Scymnus (*Pullus*)	Aphids (shrubs and trees)
	Scymnus (*Scymnus*)	Aphids (grass)
	Aspidimerini	Aphids
	Hyperaspini	75% Coccids: Coccinae, Ortheziinae (*Pseudococcus*, *Phenacoccus*, *Ripersia*)
Ortaliinae	Ortaliini	Psyllids, Flatidae
Chilocorinae	Telsimiini	Coccids, Diaspidinae
	Platynaspini	Aphids
	Chilocorini	75% Coccids, aphids
Coccidulinae	Coccidulini	Coccids
	(Rhyzobiini)	51% Diaspidinae, 35% Coccinae, 14% Lecaniinae
	Exoplectrini	*Icerya* and related species
	Azyini	Diaspidinae
	Noviini	*Icerya* and closely related species
Coccinellinae	Coccinellini	85% Aphids, psyllids, Chrysomelidae
	(Hippodamiini)	75% Aphids
	(Synonychini)	Aphids
	Neda	Coccids
	Archaioneda	Coccids
	(Cheilomenini)	72% Aphids, coccids, Aleyrodidae
	(Veraniini)	Aphids
	Psylloborini	Mycophagous
Epilachninae		Phytophagous

Source: Modified from Hodek, I. and A. Honek, *Ecology of Coccinellidae*. Dordrecht: Kluwer Academic Publishers, 1996.

to feed on small prey. Species of *Stethorus*, which are very small, feed on small mites (Gordon 1985). Aphids are more mobile than coccids, and for that reason, the larger species of coccinellid that feed on aphids move faster.

Coccinellids guts vary in length. The gut length of herbivorous species can be twice that of predatory species. This reflects the need of herbivorous species to process large amounts of poor-quality food, of which they assimilate only 23% of the energy content, whereas carnivorous species process small amounts of high-quality food, of which 77% is assimilated (Brafield and Llewellyn 1982). Despite few studies, it seems that the gut length of coccidophagous species is less than that of aphidophagous species. This fact has been attributed to the high nutritional value of coccids and the low voracity of the coccidophagous coccinellids (Iperti et al. 1977).

Both larvae and adults of predatory coccinellids consume an enormous range of foods, and many species are considered polyphagous, making difficult to determine a correct predator–prey relationship for some groups. However, their essential foods must be available to complete larval development and to guarantee progeny (Hodek 1973), although adults can survive on alternative foods such as pollen grains and sugary substances (e.g., mixture of honey and water).

Many coccidophagous coccinellids deposit a single egg under the carapace of a scale or inside a gall where the larva develops, feeding on the prey; however, a single scale is not enough for its complete development. *Hyperaspis delicata* Almeida & Vitorino female deposits only one egg inside the gall produced by *Tectococcus ovatus* Hempel (Hemiptera, Eriococcidae), inside which live many nymphs feeding on *Psydium cattleianum* (Almeida and Vitorino 1997). Adults and larvae of *H. delicata* and *Hyperaspis vicinguerrae* feed on eggs and nymphs in the gall (Figure 23.4) (Hafez and El-Ziady 1952).

23.4.2 Food Quality

Predatory coccinellids accept a wide range of prey whose quality is suitable for their complete development (Hodek 1967). Most predators taste their food after making contact with the antennae and the special hairs (sensilla) found in their maxillary palps. In general, coccidophagous species take longer to develop than aphidophagous species, and this has been attributed to the poor food quality of the former. Both groups, however, must process a great volume of food.

Studies by Rana et al. (2002) with *A. bipunctata* (L.), reared on two different aphid species, *Acyrthosiphon pisum* (Harris) and *Aphis fabae* Scopoli, showed that predators are more adapted for exploiting their preferred prey and that this phenomenon may be generalized, indicating that the diet preference represents an evolutionary change, similar to what occurs with herbivorous insects. The nutritional quality of the food is an important factor in predator strategy; however, if their preferred prey is rare or absent, despite a poorer performance and increased mortality, these species use other foods and females lay their eggs on plants where these foods can be found. This indicates that predators can adapt to exploit less suitable prey in the absence of their preferred food.

Larvae of *Hippodamia convergens* Guérin-Méneville, fed with eggs of *Anagasta kuehniella* (Zeller) (Lepidoptera: Pyralidae) complete development to adult stage (Kato et al. 1999). *Coccinella septempunctata* L. and *Coccinella transversoguttata* Richardsoni fed with curculionid larvae grew and increased in weight but did not produce eggs (Richards and Evans 1998). However, the results found by Kalaskar and Evans (2001) with *C. septempunctata* and *H. axyridis* (Pallas) fed larvae of weevils showed that although these species prefer aphids, larvae and adults survived and completed development in alfalfa fields even when aphid population was sparse.

Coccinellids attack prey in isolation, and there is generally an adaptation between its size and that of the prey. First and second instar larvae of *H. convergens* and *Cycloneda sanguinea* (L.) offered different-sized aphids of *Pinus* sp., *Cinara atlantica* (Wilson) and *Cinara pinivora* (Wilson), consumed greater number of smaller nymphs, probably due to the easiness of prey manipulation. Temperature also influences the consumption of *H. convergens*, which increased at 25°C; *C. sanguinea* showed a similar consumption capacity between 15°C and 30°C (Cardoso and Lázzari 2003).

Coccinellid predators eventually eat pollen and nectar, which allows survival when food is unavailable. Feeding on pollen allows the accumulation of reserves during dormancy (Hagen 1962). Pollinivory has been registered for various species, such as *Hippodamia tredecimpuncta* (Goidanich). In *Chilocorus kuwanae* Silvestri, a coccinellid species introduced from Korea to the United States for biological control of scales, eat pollen and nectar (Nalepa et al. 1992). *Coleomegilla maculata* DeGeer can complete its development feeding on pollen of various plant species in the same way as it feeds on aphids (Smith 1960). In larvae of *Tytthaspis* (*Micraspis*) *sedecimpunctata* (L.) and *Tytthaspis trilineata* (Weise), mandibles have a ventral margin with 20–22 spiny processes, shaped like a comb, for the collection of pollen from *Lolium perenne* and *Lolium multiflorum* and also spores of *Alternaria* sp. (Ricci 1982). In Nova Friburgo, Rio de Janeiro, in December 2006, *Exoplectra miniata* (Germar) was observed feeding on

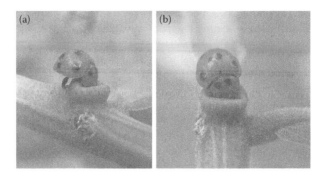

FIGURE 23.5   *Exoplectra miniata* feeding on nectar of *Inga edulis* (Leguminoseae, Mimosoideae), lateral (a) and front view (b).

nectar (Figure 23.5), in nectaries of ice-cream-bean, *Inga edulis* (Leguminoseae, Mimosoideae), where its preferred coccid food was absent (Almeida et al. 2011).

23.4.3 Food Preferences

Coccinellid predators accept a wide variety of foods. Besides feeding on aphids, coccids, and mites, they also feed on the young larvae of Lepidoptera, Coleoptera, Hymenoptera, small Diptera (Nematocera), and Thysanoptera. As already discussed in Section 23.4.1, there is specificity only in larger taxonomic groups, that is, within the same subfamily (Table 23.3).

Acceptability is often confused with suitability. Experiments have been developed to evaluate food nutritional quality, which analyzes the quantitative data of developmental parameters (e.g., developmental rate, survival, reproductive capacity). When prey is essential, it results in fast larval development, low mortality, high oviposition, and production of females. When the prey is an alternative food, it only serves as an energy source and increases survival. Various levels of both types of foods, essential and alternative, are found (Hodek 1962, 1993; Mills 1981; Hodek and Honek 1988).

Alternative food can vary from highly toxic to suitable, allowing survival when essential food is scarce and supplying energy sufficient to compensate metabolic losses or even accumulating reserves for dormancy. Coccinellids adopt a variety of foraging strategies to acquire resources for their survival when food is scarce, normally unused when their prey is abundant (Sloggett and Majerus 2000).

The presence of all developmental stages of the predator feeding on a certain prey is good evidence to evaluate predator specificity in the field. Such evidence may, however, be confusing since predators usually live with various insect species and any one may serve as food. There are isolated cases where the relationship between prey and predator is evident through methodic indirect observations over a long period. Eastop and Pope (1966, 1969) found strong coincidence over 5 years between the abundance of *Pulus auritus* Thunberg in oak trees infested with *Phylloxera glabra* (Heyden).

Larger species of coccinellids (aphidophages) are more prolific and lay larger eggs but have a shorter longevity and a short developmental period. The smaller species (coccidophages) are less prolific, have longer developmental period, lay smaller eggs, and live longer (Dixon 2000).

Another factor that interferes in the development of coccinellid predators is foraging success, which is influenced by the plant surface traits, such as the presence of trichomes and waxes. Trichomes, in the form of hooks on leaves of *Phaseolus coccineus* caused rapid death of *Stethorus punctillum* Weise larvae, and wounding of the delicate membranes of the adult abdominal segments (Putman 1955). Similarly, glandular trichomes on tobacco leaves significantly reduced the speed of *H. convergens* Guérin-Méneville larvae searching for prey. On the other hand, smooth surfaces may have a negative effect on larval development. *C. septempunctata* L. attacks its prey less efficiently and feeds on less *Acyrthosiphon pisum* Hart. on smooth leaves of *Pisum sativum* compared with the hairy leaves of *Vicia fabae* (Dixon 2000). Alternative food sources stabilize predator populations since individuals shift the type of food normally consumed in response to changes in prey abundance. Similarly, refuges for prey help avoid predation, keeping prey numbers at high levels, facilitating recovery of population cycles, and stabilizing preda-tor–prey relationships.

23.4.4 Food Toxicity

Some aphid species can be toxic, such as *Aphis nerii* Boyer de Fonscolombe, for example, which infests plants of Asclepiadaceae and Apocynaceae. These plants are toxic due to the high contents of oleandrin and nerrin, which are digested by aphids and sequestered and excreted in the honeydew (Rothschild et al. 1970, Malcom 1990). Various coccinellids do not survive after consuming *A. nerii* fed *Nerium oleander* (oleander) since it is a toxic plant containing the active ingredient oleandrin (Iperti 1966). Certain prey consumed by predators does not allow development, or may be toxic; others are rejected. These differ-ent relationships have been studied for aphidophages but also occur in coccidophages and acarophages. Some aphids are not accepted by some coccinellids, and this is often the result of palpal contact or trial tasting. *Macrosiphum aconitum* van der Goot feeds on *Aconitum*, which contains the toxic component aconitin. This allelochemical may be the reason why some coccinellids reject this prey. The unpalatabil-ity may also be attributed to the intense coloration of the aphid or the presence of surface wax (Hodek and Honek 1996).

An apparent case of acquired toxicity occurs in *R. cardinalis* (Mulsant), which does not feed on its essential host, *I. purchasi* Maskell, when this eats *Spartium* or *Genista*. After feeding on these plants, the coccid acquires the yellow pigment genistein and the alkaloid spartein, which makes it unpalatable; these substances may also be toxic (Dixon 2000).

23.5 Defense Strategies

Thanatosis is a type of defense, defined as the capacity of an animal to pretend it is dead in order to dis-courage predators. This behavior is very common in some vertebrates and invertebrates and also in coc-cinellids. Normally this phenomenon is characterized by showing an attractive color (aposematic), and the animal remains static pretending it is dead. When coccinellids are disturbed, they stop their move-ments, hide their legs and antennae, and exude a yellowish secretion from the femur–tibial articulation (adults) or from dorsal glands (larvae) in an attempt to stop their natural enemies from capturing them. However, thanatosis is often insufficient and other insects, such as wasps, ants, Mantidae, Chrysopidae, Asilidae, besides birds, rodents, and other mammals, manage to capture them. The bitter taste that the predator feels from this fluid has been attributed to alkaloids. Its smell is due to volatile repellent components, such as pirazines (Rothschild 1961). The substances coccinellin and precoccinellin were the first alkaloids to be extracted from *Coccinella septempunctata* and *Coccinella undecimpunctata* (Tursch et al. 1971a,b). Other alkaloids have been extracted from other species, such as propyleine in *Propylea quatuordecimpunctata* (L.) (Tursch et al. 1972), adaline in *A. bipunctata* L. (Tursch et al. 1973), and hippodamine in *H. convergens* Guérin-Méneville (Pasteels et al. 1973). Other alkaloids have been found in other species, but these compounds were not detected in species that are easily predated by birds.

23.6 Cannibalism

Cannibalism is one of the main problems when rearing Coccinellidae. The most vulnerable phases to cannibalism are those that are quiescent: egg, prepupa, pupa, or individuals that have recently changed skins and are, therefore, fragile (teneral). This behavior is of advantage because it preserves the species during periods of food scarcity (Osawa 1993). When larvae and pupae of *C. septempunctata bruckii* Mulsant are exposed to low aphid numbers, they cannibalize the eggs (Takahashi 1987). Larvae and adults of *Delphastus pusillus* (LeConte) feed on eggs when there are few of their favorite prey, *Bemisia tabaci* (Gennadius) (Hoelmer et al. 1993). Often, after dispersal of first instars toward aphids, conspecific egg batches may be located and cannibalism occur, which does not involve progeny. *H. axyridis* (Pallas) females tend to lay eggs far from the aphid colony to avoid cannibalism by individuals from another larval group (Osawa 1989).

Intraspecific predation is observed in a large variety of animals and generally occurs when preferential food is scarce; it is, therefore, a species survival strategy with an autoregulatory role (Agarwala and Dixon 1992). Coccinellid eggs are used as food by only a few predators compared with eggs of pest species, for example, some Lepidoptera species that coexist in the same habitat (Cottrell and Yeargan 1998a,b). This happens because eggs are protected by alkaloids, such as pirazines and quinolones (Ayer and Browne 1977, Agarwala and Yasuda 2001). The period of about 1 day, during which the larva remains on the chorion of recently ecloded eggs and also the occurrence of infertile eggs in a batch, facilitates both egg cannibalism of the same progeny and interspecific cannibalism.

The cannibalism of eggs of the native species *C. maculata* is much reduced during anthesis, when pollen is plentiful. However, for the exotic species *H. axyridis*, in the absence of prey, even with sufficient pollen, egg predation is high (Cottrell and Yeargan 1998b). In laboratory experiments of egg predation and cannibalism, Cottrell (2005) demonstrated that eggs of two native North American species, *C. maculata* DeGeer and *O. v-nigrum* Mulsant, were more predated than eggs of the exotic species *H. axyridis*. With the addition of alternative food, cannibalism and predation decreased; however, in the absence of food, the two native species predated less than the exotic species, which was more aggressive and consumed eggs of both species.

23.7 Intraguild Competition

Although there are few field studies with coccinellids that demonstrate intraguild competition, laboratory observations show that some aphidophagous species perform better if they exploit foods other than aphids. Behavioral studies on cannibalism and intraguild predation are used in trials to evaluate the possible impact of exotic coccinellid species on native ones (Burgio et al. 2005). *H. axyridis* has been shown to be a strong intraguild competitor (Takahashi 1989, Yasuda and Ohnuma 1999, Kajita et al. 2000, Yasuda et al. 2001). Larvae of *A. bipunctata* survive by feeding on eggs of *H. axyridis*, but not on eggs of their own (Sato and Dixon 2004); adults and larvae of *H. axyridis* feed on eggs of *A. bipunctata* (Burgio et al. 2002).

In pecan plantations in the United States, adult *H. axyridis* overlap in space and time with *O. v-nigrum* and eggs of both species are commonly found on leaves. When the preferential food is abundant, egg cannibalism (intraguild predation) is almost absent (Cottrell 2004).

Gardiner and Landis (2007) studied the intraguild impact on the population dynamics of soybean aphids to compare the impact of predation between *Aphidoletes aphidimyza* Rondani (Diptera, Cecidomyiidae) and *Chrysoperla carnea* Stephens (Neuroptera, Chrysopidae) in the presence or absence of *H. axyridis*. Results showed that the presence of *H. axyridis* contribute to the decline of *A. aphidimyza* and *C. carnea*, but the biological control of soybean aphids did not improve with the removal of *H. axyridis* from the system; that is, *H. axyridis* even as an intraguild predator contributes to the decline of aphid colonies. The Asiatic species *H. axyridis* was introduced into the United States for the first time in California in 1916, and into other states between 1978 and 1982. It does not appear to have become established until 1988 when it was collected again in various states. Since this time, it has shown enormous voracity for aphid pests, as well as competing with and dislodging native species. Various studies show that after the entry of *H. axyridis* into the United States, native predator densities decreased whereas *H. axyridis* density increased (Colunga-Garcia and Gage 1998, Brown and Miller 1998, Michaud 2002b, Alyokhin and Sewell 2004, Saini 2004), due to intraguild predation (Cottrell and Yeargan 1998, Michaud 2002b, Cottrell 2004, Yasuda et al. 2004). This species appears to be so aggressive that it can affect populations of the Monarch butterfly, *Danaus plexippus* L. (Koch et al. 2004c).

H. axyridis has the habit of massively invading residences and buildings; entering filing cabinets, computers, and machinery; and being a nuisance to people (Nalepa et al. 2004, 2005). It can occasionally feed on grapes and damage them (Koch et al. 2004a) and also contaminate wine production (Pickering et al. 2004, Galvan et al. 2006).

In Argentina, *H. axyridis* was introduced into Mendoza at the end of the 1990s, and it was registered in Buenos Aires for the first time at the end of 2001 (Saini 2004). This author also observed that the percentage of *C. sanguinea*, *O. v-nigrum*, *Eriopis connexa* L., *Coleomegilla quadrifasciata* (Schoenherr), and *A. bipunctata* L. decreased significantly between 2001 and 2004, suggesting that this exotic species has dislodged these once common predators.

In Brazil, *H. axyridis* was apparently accidentally introduced to Curitiba, Paraná state, in April 2002 and observed feeding on the aphid *Tinocallis kahawaluokalani* (Kirkaldy), on crape myrtle *Lagerstroemia indica*, and a species that is widely used as ornamental in the city. In the same year, adults and larvae were observed feeding on *C. atlantica* (Wilson) and *C. pinivora* (Wilson) (pine aphids). The larvae were found on the lower parts of young plants (Almeida and Silva 2002).

Since 1999, various species of coccinellids have been registered feeding on aphids, psyllids, and scales in Curitiba. After 2002, when *H. axyridis* was introduced, both species variety and population numbers decreased, probably due to the voracity of this species, which has shown the capacity to dislodge native species wherever it has been introduced. The most abundant species in 1999/2000, *C. sanguinea* and *H. convergens*, were significantly affected by the presence of *H. axyridis*, constituting less than 3% of the coccinellids collected in 2006/2007 (Martins et al. 2009). The authors also discovered that at least 20 aphid species were serving as a food source for *H. axyridis* and that 7 years after its introduction, this species was collected in the central-north part of the country (Brasília), more than 1000 km away, demonstrating its large capacity for dispersal. According to Koch et al. (2006), the invasion of Brazil and South America by *H. axyridis* appears to be established and its eradication difficult; although it has potential as a biological control agent, it is linked to adverse effects, including threats to fruit production and nontarget organisms. Future studies should be conducted to monitor the interaction of *H. axyridis* with other coccinellid species to evaluate its potential for dislodging native species in South America (Martins et al. 2009).

23.8 Adaptations and Responses to Variations in Abiotic and Biotic Factors

The specificity in behavior and food occurs within the limit of adult spatial distribution and depends on the preferential vegetation stratum, although microclimate conditions affect coccinellid habitat specificity. The wide range of host plants infested with *A. fabae* Scopoli attracts different coccinellid species. For example, *A. bipunctata* is found on the shrub *Evonymus europaeus*; *C. septempunctata*, on the native tree *Chenopodium album*; *S. undecimnotata*, on the annual legume *Vicia faba*; and *Adonia variegata* (=Hippodamia Adonia variegata) on dry beans, *P. vulgaris*. Certain types of vegetation are preferred by some coccinellid species that show a seasonal preference for habitat strata. This may be seen with some common European aphidophages, such as *C. septempunctata* and *S. undecimnotata*, which deposit eggs on low plants (0–50 cm) infested with aphids. Other aphidophages, such as *P. quatuordecimpunctata* and *A. variegata*, often occur on shrubs (0.50–2 m in height) and *A. bipunctata*, *S. conglobata*, and *A. decempunctata* depend on aphids that live in trees taller than 2 m (Iperti 1965).

Predators always search for suitable microclimate, the stratum of their preferred plant, and sufficient food resources. This is why the study of habit specificity is essential for understanding the behavior of aphidophagous predators. It is also necessary to differentiate the climate conditions in spring and summer. In spring, the young branches of many plants are infested by aphids, offering an excellent habitat for predators to complete their life cycle. In summer, infestations are much reduced and predator behavior depends more on the presence of the aphids and less on the microclimate conditions and food quality. Besides the influence of the synchronization of predators and prey, changes in seasonal climatic conditions affect coccinellid distribution by altering habitat microclimate characteristics and by influencing the growth of aphid populations due to the plant physiology.

In temperate climates, coccinellid predators generally reproduce in the spring when prey is abundant and become quiescent in the summer. Some species show some activity in the autumn, and all species show various levels of dormancy in the winter. The populations of a certain coccinellid species will react differently within the same geographic area, and no species produces the same number of generations throughout its area of distribution (Hagen 1962).

In most temperate climates, aphidophagous coccinellid species are univoltine and migrate. Migration occurs as soon as the adults emerge during periods of warmer weather (Iperti 1999). In Brazil, most species are multivoltine, principally in the warmer regions. When conditions are unfavorable, such as low temperatures and short photophase, coccinellids enter diapause, migrate, or exploit other food sources. Some species feed on alternative prey, practice cannibalism, or use other resources, such as pollen. Univoltinism is common in many aphidophagous species of Coccinellinae and Hippodamiinae (Banks 1954, Delucchi 1954, Hodek 1959, Hagen 1962) and may occur in coccidophagous (Katsoyannos 1983) and mycophagous species (Evans 1936). Bivoltinism is observed in aphidophagous species, principally the Hippodamiini (Hagen 1962) and Coccinellini (Hagen 1962, Ongagna et al. 1993). Bivoltinism with intervals of estivation (response to certain climatic adversities) is characteristic of some aphidophagous species of Hippodamiini and Coccinellini (Ibrahim 1955a,b, Hagen 1962, Quilici 1981). Multivoltinism with three generations per year is standard for all coccidophagous species of Chilocorini (Iperti et al. 1970, Katsoyannos 1983), some aphidophagous species of Coccinellini, and for *Scymnus apetzi* Mulsant and *Scymnus subvillosus* Goeze (Iperti 1986). Some coccinellids have successive generations without adult dormancy. Many are species from Australia and the Pacific region, introduced into California and Europe, especially the coccidophagous species of Coccidulinae (Sezeer 1970) and Scymninae. In Europe, coccinellids show all types of voltinism. Large species normally have one generation per year, and coccidophages produce at least three generations per year. Small coccinellids reproduce principally in the summer when temperatures are high. However, this is not the rule since *H. convergens* Guérin-Méneville, for example, complete up to five generations per year if its preferred food is available (Hagen 1962).

23.9 Natural Enemies

The toxic substances exuded by coccinellids protect them from many large predators, such as mammals, birds, reptiles, and amphibians. However, some species are predated principally by birds, which feed on them in flight and appear to be more resistant to the toxic effects of alkaloids because they do not have time to recognize their prey. Large aggregations of coccinellids during hibernations may serve as food for mammals (Majerus 1994).

Among the invertebrates, the arthropods are the most common natural enemies of coccinellids. There are various spider species that feed on *C. septempunctata*, *A. ocellata* L., and *Exochomus quadripustulatus* (L.), captured in the webs of *Araneus diadematus* and *Araneus quadratus* (Majerus 1994). Ants can kill coccinellid larvae and adults that enter their nests or interfere with the supply of the aphid honeydew. Groups of *A. fabae* aphids were defended by *Myrmica ruginodis* Nylander ants that drove off coccinellids close to the colony but ignored them on leaves distant from the aphid colonies (Jiggins et al. 1993). Richerson (1970) listed almost 100 parasites, including mites, nematodes, and insects, hosts of coccinellids, but information on their behavior is still scarce. There are few data on egg parasitism and most parasites develop in larvae, pupae, or adults. Among the Diptera, species of the Phoridae are the most important parasites. Richerson (1970), Klausnitzer (1976), and Disney et al. (1994) listed 25 species of coccinellids parasitized by *Phalacrotophora berolinensis* Schmitz and *Phalacrotophora fasciata* (Fallén) from Europe, Asia, and part of Russia.

The parasitoids (Hymenoptera) are perhaps the most important natural enemies of coccinellids. The main parasitoid species belong to the Braconidae, Encyrtidae, and Eulophidae families. *Dinocampus coccinellae* (Schrank) is a cosmopolitan braconid, endoparasite, of the subfamily Euphorinae, which has been studied in detail because it parasitizes more than 40 coccinellid species, especially adults of the subfamily Coccinellinae (Figure 23.6). This parasitoid species prefers larger species. Other parasitoid species of the same subfamily oviposit in young stages of coccinellids. In the subfamily Encyrtidae, the genus *Homalotylus* has more than 30 known species of coccinellid parasitoids. Among the Eulophidae, the Tetrastichinae are the most well-known parasitoids and most belong to the genus *Tetrastichus*, mostly attacking eggs, with others as hyperparasites. Also among the Eulophidae, a species of Entedontinae, *Pediobius foveolatus* (Crawford), is a parasitoid of larvae of the Epilachninae of the Ethiopian, Oriental, and Australian regions (Hodek and Honek 1996).

FIGURE 23.6 *Cycloneda sanguinea* in *Pinus* sp. parasitized by *Dinocampus coccinellae* (Hymenoptera).

Coccinellids are attacked by phoretic mites and also by parasitic mites of the family, Podapolipidae. Nematodes also attack coccinellids. Species of Allantonematidae; *Parasitilenchus coccinellinae* Iperti & van Waerebeke, a parasite of adults; *Howardula* sp., which parasitizes the gonads; and a solitary endoparasite of the family Mermitidae are the main pathogenic nematodes described in the literature (Hodek and Honek 1996). Shapiro-Ilan and Cottrell (2005) compared the susceptibility of two native coccinellid species, *C. maculata* and *O. v-nigrum*, and of two exotic species, *H. axyridis* and *C. septempunctata*, to two nematode species, *Heterorhabditis bacteriophora* and *Steinernema carpocapsae*. They concluded that the exotic species showed less susceptibility to nematode infection, which may contribute to its greater success.

There are few studies regarding fungus attack to coccinellids, but they may be attacked by *Beauveria* during dormancy. Prolonged hibernation of aggregating adults greatly increases the risk of fungal infection, as observed in *S. undecimnotata* inactive adults from large clusters (Hodek and Honek 1996).

23.10 Conclusions and Suggestions for Research

The relationships between the Coccinellidae and their preferred food have been focused mainly on economically important species. However, the limiting factor for understanding the evolution of predator intraguild feeding remains the absence, or in some cases, the complete lack of knowledge of the feeding habits, even of the commonest species. A few authors have discussed this topic but only for European or American species (Dixon 2000).

In this chapter, we have tried to bring together in a condensed form the main information on species biology and their food relationships. There are presently many alternatives for using predators in pest biological control, including the use of mathematical models, and it has become clear that a large range of factors can determine the ideal quantity of insects necessary for efficient control. However, it should be remembered that to be successful it is important to know the specific name of the study organism. This means that basic taxonomic and systematic studies are fundamental for fully understanding the group. The importance of identification becomes clear as this is the key word for the exchange of all information on the subject. Identification is most relevant for determining the biological cycle, habits, hosts, and even the control of a new species, by referring to previously known related forms. Therefore, in practice, the identification tells us if a certain specimen is important or not, and if it is potentially beneficial or harmful from a certain point of view. In concrete terms, the identification provides us with the opportunity to decide, for example, between an efficient predator, its place of origin, and the best way to manage it compared with an insect that does not show this pest control potential. Therefore, it is no exaggeration to say that the development of biological control is based on the taxonomy of the group (Zucchi 2002).

Basic biological studies of native species are necessary to understand the tri-trophic relationships. Understanding the dynamics of the plant–pest–predator relationship is essential for management and biological control. Although predators have not merited the same attention as parasitoids, a literature analysis shows that predators can exercise an important role in pest control. Studies on predator–prey relationships demonstrate that biological parameters are the most important characteristics for understanding the development rates of both predator and prey. These data should be compared with the response of the plant since these relationships are interdependent.

An interesting approach, which has been little explored in the case of predators, is the use of artificial diets or complimentary foods, which can contribute to greater predator efficiency in the field. Without doubt, this would be an aspect of biological studies that would contribute to the development of new technologies for the mass rearing of coccinellid species, as well as of other potentially important natural enemies of agricultural pests.

Another interesting aspect is the potential of semiochemicals for the manipulation of natural enemies. The use of synthetic kairomones and sinomones can improve the foraging capacity of predators by orienting their responses to the target prey. In the case of artificially reared natural enemies, their liberation in agroecosystems results in an uncontrolled dispersion and the success of this initiative may be compromised. Thus, specific chemical stimuli can be supplied to natural enemies to guide them to the target prey (Vilela and Pallini 2002).

There is presently an enormous effort under way to study exotic species whose introduction may have been accidental or planned. This subject has become a central focus of ecology, evolutionary biology, and conservation biology. However, the development of this type of study often suffers from a lack of information and the species identity. Therefore, a research network is necessary to gather together all the possible information, geographic and historical, of introductions, using museum data, so that the story of the species introduction into the country is known. The organization of a data bank and the availability of tools for communication, for the dissemination of information, are fundamental for supporting a system of exotic pest monitoring and detection. More importance should also be given to obtaining genetic data that could help trace the route of accidentally introduced exotic species.

REFERENCES

Agarwala, B. K., and A. F. G. Dixon. 1992. Laboratory study of cannibalism and interspecific predation in ladybirds. *Ecol. Entomol.* 17:303–9.

Agarwala, B. K., and H. Yasuda. 2001. Overlapping oviposition and chemical defense of eggs in two co-occurring species of ladybird predators of aphids. *J. Ethol.* 19:47–53.

Almeida, L. M., and J. Milléo. 1998. The immature stages of *Psyllobora gratiosa* Mader, 1958 (Coleoptera: Coccinellidae) with some biological aspects. *J. N. Y. Entomol. Soc.* 106:170–6.

Almeida, L. M., and R. C. Marinoni. 1986. Desenvolvimento de três espécies de *Epilachna* (Coleoptera: Coccinellidae) em três combinações de temperatura e fotoperíodo). *Pesq. Agropec. Bras.* 21:927–39.

Almeida, L. M., and C. S. Ribeiro. 1986. Morfologia dos estágios imaturos de *Epilachna cacica* Guérin, 1844 (Coleoptera, Coccinellidae). *Rev. Bras. Entomol.* 30:43–9.

Almeida, L. M., and M. D. Vitorino. 1997. A new species of *Hyperaspis* Redtenbacher (Coleoptera: Coccinellidae) and notes about the life habits. *Coleopt. Bull.* 51:213–6.

Almeida, L. M., and J. Milléo. 1998. The immature stages of *Psyllobora gratiosa* Mader, 1958 (Coleoptera: Coccinellidae) with some biological aspects. *J. N. Y. Entomol. Soc.* 106:170–6.

Almeida, L. M., and V. B. Silva. 2002. Primeiro registro de *Harmonia axyridis* (Pallas) (Coleoptera, Coccinellidae): Um coccinelídeo da região Paleártica. *Rev. Bras. Zool.* 19:941–4.

Almeida, L. M., G. H. Corrêa, J. A. Giorgi, and P. C. Grossi. 2011. New record of predatory ladybird beetle (Coleptera, Coccinellidae) feeding on extrafloral nectaries. *Rev. Bras. Entomol.* 55(3):447–50.

Alyokhin, A., and G. Sewell. 2004. Changes in a lady beetle community following the establishment of three alien species. *Biolog. Invas.* 6:463–71.

Araújo-Siqueira, M., and L. M. Almeida. 2004. Comportamento e ciclo de vida de *Epilachna vigintiopunctata* (Fabricius) (Coleoptera, Coccinellidae) em *Lycopersicum esculentum* Mill. (Solanaceae). *Rev. Bras. Zool.* 21:543–50.

Ayer, W. A., and L. M. Browne. 1977. The ladybug alkaloids including synthesis and biosynthesis. *Heterocycles* 7:685–707.

Banks, C. J. 1954. The searching behaviour of coccinellid larvae. *Anim. Behav.* 2:37–8.

Barbier, R., J. Le Lannic, and J. Brun. 1996. Récepteurs sensoriels des palpes maxilaires e Coccinellidae adultes aphidophages, coccidophages et phytophages. *Bull. Soc. Zool. Fr.* 121:255–68.

Brafield, A. E., and M. J. Llewellyn. 1982. *Animal Energetics*. Glasgow: Blackie.

Brown, M. W., and S. S. Miller. 1998. Coccinellidae (Coleoptera) in apple orchards of eastern West Virginia and the impact of invasion by Harmonia axyridis. *Entomol. News.* 102:136–42.

Burgio, G., F. Santi, and S. Maini. 2002. On intra-guild predation and cannibalism in *Harmonia axyridis* (Pallas) and *Adalia bipunctata* L. (Coleoptera: Coccinellidae). *Biol. Contr.* 24:110–6.

Burgio, G., F. Santi, and S. Maini. 2005. Intra-guild predation and cannibalism between Harmonia axyridis and Adalia bipunctata adults and larvae: Laboratory experiments. *Bull. Insectol.* 58:135–40.

Cardoso, J. T., and S. M. N. Lázzari. 2003. Consumption of *Cinara* spp. (Hemiptera, Aphididae) by *Cycloneda sanguinea* (Linnaeus, 1763) and *Hippodamia convergens* Guérin-Méneville, 1842 (Coleoptera, Coccinellidae). *Rev. Bras. Entomol.* 47:559–62.

Colunga-Garcia, M., and S. H. Gage. 1998. Arrival, establishment, and habitat use of the multicolored Asian lady beetle (Coleoptera: Coccinellidae) in a Michigan landscape. *Environ. Entomol.* 27:1574–80.

Cottrell, T. E. 2004. Suitability of exotic and native lady beetle eggs (Coleoptera: Coccinellidae) for development of lady beetle larvae. *Biol. Contr.* 31:362–71.

Cottrell, T. E. 2005. Predation and cannibalism of lady beetle eggs by adult lady beetles. *Biol. Contr.* 34:159–64.

Cottrell, T. E., and K. V. Yeargan. 1998a. Influence of a native weed, *Acalypha ostryaefolia* (Euphorbiaceae), on *Coleomegilla maculata* (Coleoptera: Coccinellidae) population density, predation, and cannibalism in sweet corn. *Environ. Entomol.* 27:1375–85.

Cottrell, T. E., and K. V. Yeargan. 1998b. Effect of pollen on *Coleomegilla maculata* (Coleoptera: Coccinellidae) population density, predation, and cannibalism in sweet corn. *Environ. Entomol.* 27:1402–10.

Crowson, R. A. 1981. *The Biology of Coleoptera*. London: Academic Press.

Delucchi, V. 1954. *Pullus impexus* (Muls.) (Coleoptera: Coccinellidae), a predator of *Adelgea picea* (Ratz.) (Hemiptera, Adelgidae) with notes on its parasites. *Bull. Entomol. Res.* 45:243–78.

Disney, R. H. L., M. E. N. Majerus, and M. J. Walpole. 1994. Phoridae (Diptera) parasitizing Coccinellidae (Coleoptera). *Entomologist* 113:28–42.

Dixon, A. F. G. 1958. The escape response shown by certain Aphids to the presence of the coccinellid *Adalia decempunctata* (L.). *Trans. R. Entomol. Soc. Lond.* 110:319–34.

Dixon, A. F. G. 2000. *Insect Predator–Prey Dynamics. Ladybird Beetles & Biological Control*. London: Cambridge University Press.

Eastop, V. F., and R. D. Pope. 1966. Notes on the ecology and phenology of some British Coccinellidae. *Entomologist* 99:287–9.

Eastop, V. F., and R. D. Pope. 1969. Notes on the biology of some British Coccinellidae. *Entomologist* 102:162–4.

Elliot, H. J., and D. W. Little. 1980. Laboratory studies on predation of *Chrysophtharta bimaculata* (Olivier) (Coleoptera: Chrysomelidae) eggs by the coccinellids *Cleobora mellyi* (Mulsant) and *Harmonia conformis* (Boisduval). *Gen. Appl. Entomol.* 12:33–6.

Evans, A. C. 1936. A note on the hibernation of *Micraspis sedecimpunctata* L. (*var. 12-punctata* L.) (Coccinell.) at Rothamsted Experimental Station. *Proc. R. Entomol. Soc. Lond.* 11:116–9.

Ferreira, F. A. S., and L. M. Almeida. 2000. Morfologia dos estágios imaturos de *Eupalea reinhardti* Crotch (Coleoptera, Coccinellidae) e alguns aspectos biológicos. *Rev. Bras. Zool.* 17:315–22.

Frazer, B. D. 1988. Predators. In *Aphids, Their Biology, Natural Enemies and Control*, ed. A. K. Minks and P. Harrewijn, 217–30. Vol. 2B. Amsterdam: Elsevier.

Galvan, T. L., R. L. Koch, and W. D. Hutchison. 2006. Toxicity of indoxacarb and spinosad to the multicolored Asian lady beetle, Harmonia axyridis (Coleoptera: Coccinellidae), via three routes of exposure. *Pest Manag. Sci.* 62:797–804.

Gardiner, M. M., and D. A. Landis. 2007. Impact of intraguild predation by adult *Harmonia axyridis* (Coleoptera: Coccinellidae) on *Aphis glycines* (Hemiptera: Aphididae) biological control in cage studies. *Biol. Contr.* 40:386–95.

Ghorpade, K. D. 1979. *Ballia eucharis* (Coleoptera: Coccinellidae) breeding on Cicadellidae (Homoptera) at Shillong. *Curr. Res.* 8:1–113.

Giorgi, J. A., N. J. Vandenberg, J. V. McHugh, J. A. Forrester, A. Ślipiński, K. B. Miller, L. R. Shapiro, and M. F. Whiting. 2009. The evolution of food preferences in Coccinellidae. *Biol. Contr.* 51:215–31.

Gordon, R. D. 1985. The Coccinellidae (Coleoptera) of America North of Mexico. *J. N. Y. Entomol. Soc.* 93:352–599.

Hafez, M., and S. El-Ziady. 1952. Studies on the biology of *Hyperaspis vinciguerrae* Capra, with a full description of the anatomy of the fourth stage larva. *Bull. Soc. Fouad 1er d'Entomol.* 36:211–46.

Hagen, K. S. 1962. Biology and ecology of predaceous Coccinellidae. *Annu. Rev. Entomol.* 7:289–326.

Hagen, K. S., and R. van Den Bosch. 1968. Impact of pathogens, parasites, and predators on aphids. *Annu. Rev. Entomol.* 13:325–84.

Hales, D. 1979. Population dynamics of *Harmonia conformis* (Boisd.) (Coleoptera: Coccinellidae) on acacia. *Gen. Appl. Entomol.* 11:3–8.

Hatch, M. H. 1961. *The Beetles of the Pacific Northwest Part III. Pselaphidae and Diversicornia.* Seattle: University of Washington Press.

Heinz, K. M., and F. G. Zalom. 1996. Performance of the predator *Delphastus pusillus* on Bemisia resistant and susceptible tomato lines. *Entomol. Exp. Appl.* 81:345–52.

Hemptinne, J. L., and J. Naisse. 1987. Ecophysiology of the reproductive activity of *Adalia bipunctata* (Coleoptera, Coccinellidae). *Med. Fac. Landbouww. Rijsksuniv. Gent.* 52:225–33.

Hodek, I. 1958. Influence of temperature, relative humidity and photoperiodicity on the speed of development of *Coccinella 7-punctata* L. *Cas. Cs. Entomol.* 55:121–41.

Hodek, I. 1959. Ecology of aphidophagous Coccinellidae. In *International Conference on Insect Pathology and Biological Control.* Prague: [s.n.] (1958), 543–7.

Hodek, I. 1962. Essential and alternative food in insects. In *11th International Congress of Entomology,* Vienna: [s.n.] 2:696–7.

Hodek, I. 1973. *Biology of Coccinellidae.* Academia, Prague & Dr. W. Junk, The Hague.

Hodek, I. 1993. Prey and habit specificity in aphidophagous predators (a review). *Biocontr. Sci. Technol.* 3:91–100.

Hodek, I., and A. Honek. 1988. Sampling, rearing and handling of aphid predators. In *Aphids, Their Biology, Natural Enemies and Control,* Vol. 2B, ed. A. K. Minks and P. Harrewijn, 311–21. Amsterdam: Elsevier.

Hodek, I., and A. Honek. 1996. *Ecology of Coccinellidae.* Dordrecht: Kluwer Academic Publishers.

Hodek, I., and V. Landa. 1971. Anatomical and histological changes during dormancy in two Coccinellidae. *Entomophaga* 16:239–51.

Hoelmer, K. A., L. S. Osborne, and R. K. Yokomi. 1993. Reproduction and feeding behaviour of *Delphastus pusillus* (Coleoptera: Coccinellidae), a predator of *Bemisia tabaci* (Homoptera: Aleyrodidae). *J. Econ. Entomol.* 86:322–9.

Howland, D. E., and G. M. Hewitt. 1995. Phylogeny of Coleoptera based on mitochondrial cytochrome oxidase 1 sequence data. *Insect Mol. Biol.* 4:203–15.

Ibrahim, M. M. 1955a. Studies on *Coccinella undecimpunctata aegyptica* Reiche. 1. Preliminary notes and morphology on the early stages. *Bull. Soc. Entomol. Egypte* 39:251–74.

Ibrahim, M. M. 1955b. Studies on *Coccinella undecimpunctata aegyptica* Reiche. 2. Biology and life-story. *Bull. Soc. Entomol. Egypte* 39:395–423.

Iperti, G. 1965. Contribution à l'étude de la specificité chez lês principales coccinelles aphidiphages dês Alpes Maritimes et dês Basses-Alpes. *Entomophaga* 10:159–78.

Iperti, G. 1986. Ecobiologie des coccinelles aphidiphages: Les migrations. In "Impact de la structure des paysages agricoles sur la protection des cultures." *Colloques de I.N.R.A.* 36:107–20.

Iperti, G. 1999. Biodiversity of predaceous Coccinellidae in relation to bioindication and economic importance. *Agric. Ecosyst. Environ.* 74:323–42.

Iperti, G., Y. Laudeho, J. Brun, and E. C. Janvry. 1970. Les entomophages de *Parlatoria blanchardi* Targ. Dans les palmerales de l'Adrar mauritanien III. Introduction, acclimatation et efficacité d'un nouveau prédateur Coccinellidae. *Ann. Zool. Ecol. Anim.* 2:617–38.

Iperti, G., P. Katsoyannos, and Y. Laudeho. 1977. Etude comparative de l'anatomie des coccinelles aphidi-phages et coccidiphages et appartenance d'*Exochomus quadripustulatus* L. à l'un de ces groupes ento-mophages (Col., Coccinellidae). *Ann. Soc. Ent. Fr.* 13:427–37.

Jiggins, C., M. Majerus, and U. Gough. 1993. Ant defense of colonies of *Aphids fabae* Scopoli (Hemiptera: Aphididae) against predation by ladybirds. *Br. J. Ent. Nat. Hist.* 6:129–37.

Kairo, M. T. K., and S. T. Murphy. 1995. The life history of *Rodolia iceryae* Janson (Col., Coccinellidae) and the potential for use in inoculative releases against *Icerya pattersoni* Newstead (Hom., Margarodidae) on coffee. *J. Appl. Entomol.* 119:487–91.

Kajita, Y., F. Takano, H. Yasuda, and B. K. Agarwala. 2000. Effects of indigenous ladybird species (Coleoptera: Coccinellidae) on the survival of an exotic species in relation to prey abundance. *Appl. Entomol. Zool.* 35:473–9.

Kalaskar, A., and E. W. Evans. 2001. Larval responses of aphidophagous lady beetles (Coleoptera: Coccinellidae) to weevil larvae versus aphids as prey. *Ann. Entomol. Soc. Am.* 94:76–81.

Kato, C. M., V. H. P. Bueno, J. C. Moraes, and A. M. Auad. 1999. Criação de *Hippodamia convergens* Guérin-Meneville (Coleoptera: Coccinellidae) em ovos de *Anagasta kuehniella* (Zeller) (Lepidoptera: Pyralidae). *An. Soc. Entomol. Bras.* 28:455–9.

Katsoyannos, P. 1983. Recherches sur la biologie de quelques espèces de Coccinellidae (Coleoptera) prédateurs de cochonilles (Homoptera, Coccoidea) dans la région méditerranéenne orientale. Thèse de Docteur d'Etat de l'Univ. P. Sabatier de Toulouse.

Kesten, U. 1969. Zur Morphologie und Biologie von Anatis ocellata (L.). (Coleoptera, Coccinellidae). *Z. Angew. Entomol.* 63:412–45.

Klausnitzer, B. 1976. Katalog der Entomoparasiten der mitteleuropäischen Coccinellidac (Col.). *Stud. Entomol. Forest.* 2:121–30.

Klausnitzer, B. 1993. Zur Nahrungsökologie der mitteleuropäischen Coccinellidae (Col.). *Jahrb. Naturwiss. Ver. Wuppertal* 46:15–22.

Koch, R. L., E. C. Burkness, S. J. Burkness, and W. D. Hutchison. 2004. Phytophagous preferences of the multicolored Asian lady beetle (Coleoptera, Coccinellidae) for autumn-ripening fruit. *J. Econ. Entomol.* 97:539–44.

Koch, R. L., R. C. Venette, and W. Hutchison. 2006. Invasions by *Harmonia axyridis* (Pallas) (Coleoptera: Coccinellidae) in the Western Hemisphere: Implications for South America. *Neotrop. Entomol.* 35:421–34.

Lima, I. M. M. 1999. Ciclo de vida de Zagloba Beaumonti Casey, 1899 (Coleoptera: Coccinellidae) como predador de Diaspis echinocacti (Bouché, 1833) (Hemiptera: Diaspididae): Duração, sobrevivência e fertilidade. PhD thesis, Universidade Federal do Paraná, Curitiba, PR, Brazil.

Majerus, M. E. N. 1994. *Ladybirds.* Somerset: Harper Collins Publishers. 367 p.

Malcom, S. B. 1990. Chemical defense in chewing and sucking insect herbivores: Plant-derived cardenolides in the monarch butterfly and oleander aphid. *Chemoecology* 1:12–21.

Marinoni, R. C., and C. S. Ribeiro-Costa. 1987. Aspectos bionômicos de *Epilachna paenulata* (Germar, 1824) (Coleoptera: Coccinellidae) em quatro diferentes plantas hospedeiras (Cucurbitaceae). *Rev. Bras. Entomol.* 31:421–30.

Martins, C. B. C., L. M. Almeida, R. C. Zonta-de-Carvalho, C. F. Castro, and R. A. Pereira. 2009. *Harmonia axyridis*: A threat to Brazilian Coccinellidae? *Rev. Bras. Entomol.* 53:663–71.

Michaud, J. P. 2002. Invasion of the Florida citrus ecosystem by Harmonia axyridis (Coleoptera: Coccinellidae) and asymmetric competition with a native species, Cycloneda sanguinea. *Environ. Entomol.* 31:827–35.

Mills, N. J. 1981. Essential and alternative foods for some British Coccinellidae (Coleoptera). *Entomol. Gaz.* 32:197–202.

Nakajo, J. C. 2006. Aspectos morfológicos e biológicos de Olla v-nigrum (Mulsant, 1866) (Coleoptera, Coccinellidae) alimentados com *Platycorypha nigrivirga* Burckhardt, 1987 (Hemiptera, Psyllidae). MSc dissertation, Universidade Federal do Paraná, Curitba, PR, Brazil.

Nakamura, K. 1985. Mechanism of the switchover from extensive to area-concentrated search behaviour of the ladybird beetle, *Coccinella septempunctata*. *J. Insect Physiol.* 31:849–56.

Nalepa, C. A., S. B. Bambara, and A. M. Burroughs. 1992. Pollen and nectar feeding by *Chilocorus kuwanae* (Silvestri) (Coleoptera: Coccinellidae). *Proc. Entomol. Soc. Wash.* 94:596–7.

Nalepa, C. A., G. G. Kennedy, and C. Brownie. 2004. Orientation of multicolored Asian lady beetles to build-ings. *Am. Entomol.* 50:165–66.

Nalepa, C. A., G. G. Kennedy, and C. Brownie. 2005. Role of visual contrast in the alighting behavior of Harmonia axyridis (Coleoptera: Coccinellidae) at overwintering sites. *Environ. Entomol.* 34:425–31.

Obrycki, J. J., and T. J. Kring. 1998. Predaceous Coccinellidae in biological control. *Annu. Rev. Entomol.* 43:295–321.

Oliveira, N. C., C. F. Wilcken, and C. A. O. Matos. 2004. Ciclo biológico e predação de três espécies de coccinelídeos (Coleoptera, Coccinellidae) sobre o pulgão-gigante-do pinus Cinara atlantica (Wilson) (Hemiptera, Aphididae). *Rev. Bras. Entomol.* 48:529–33.

Olkowskim, W., A. Zhang, and P. Siers. 1990. Improved biocontrol techniques with lady beetles. *IPM Monit. Field Pest Manage.* 12:1–12.

Ongagna, P., L. Giupe, G. Iperti, and A. Ferran. 1993. Cycle de développement d'*Harmonia axyridis* (Coleoptera, Coccinellidae) dans son aire d'introduction le sud-est de la France. *Entomophaga* 38:125–8.

Osawa, N. 1989. Sibling and non-sibling cannibalism by larvae of a lady beetle Harmonia axyridis Pallas (Coleoptera, Coccinellidae) in the field. *Res. Popul. Ecol.* 31:153–60.

Osawa, N. 1993. Population field studies of the aphidophagous ladybird beetle Harmonia axyridis (Coleoptera: Coccinellidae): Life tables and key factor analysis. *Res. Popul. Ecol.* 35:335–48.

Pasteels, J. M., C. Deroe, B. Tursch, J. C. Braekman, D. Daloze, and C. Hootele. 1973. Distribution et activités des alcaloides defensifs des Coccinellidae. *J. Insect Physiol.* 19:1771–84.

Pickering, G., J. Lin, R. Riesen, A. Reynolds, I. Brindle, and G. Soleas. 2004. Influence of Harmonia axyridis on the sensory properties of white and red wine. *Am. J. Enol. Vitic.* 55:153–9.

Pope, R. D. 1973. The species of *Scymnus* (s.str.), *Scymnus* (Pullus) and *Nephus* (Col., Coccinellidae) occurring in the British Isles. *Entomol. Month. Mag.*109:3–39.

Pope, R. D., and J. F. Lawrence. 1990. A review of *Scymnodes* Blackburn, with the description of a new Australian species and its larva (Coleoptera: Coccinellidae). *Syst. Entomol.* 15:241–52.

Putman, W. I. 1955. Bionomics of *Stethorus punctillum* Weise (Coleoptera: Cocinellidae) in Ontario. *Can. Entomol.* 87:9–33.

Quilici, S. 1981. Etude biologique de Propylea quatuordecimpunctata I. (Coleoptera, Coccinellidae). Efficacité prédatrice comparée de tríos types de coccinelles aphidiphages em lutte biologique contre les pucerons sous serres. Thèse de doctorat de 3ème Cycle Univ. P. et M. Curie, Paris VI.

Rana, J. S., A. F. G. Dixon, and V. Jarosik. 2002. Costs and benefits of prey specialization in a generalist insect predator. *J. Anim. Ecol.* 71:15–22.

Ribeiro, C. S., and L. M. Almeida. 1989. Descrição dos estágios imaturos de *Epilachna spreta* (Muls., 1850) (Coleoptera, Coccinellidae), com redescrição, comentários e chave para três outras espécies. *Rev. Bras. Zool.* 6:99–110.

Ricci, C. 1982. Sulla costituzione e funzione delle mandible delle larve di Tytthaspis sedecimpunctata (L.) e Tytthaspis trineleata (Weise). *Frustula Entomologica* 3:205–12.

Richards, D. R., and E. W. Evans. 1998. Reproductive responses of aphidophagous lady beetles (Coleoptera: Coccinellidae) to nonaphid diets: An example from alfafa. *Ann. Entomol. Soc. Am.* 91:632–40.

Richerson, J. V. 1970. A world list of parasites of Coccinellidae. *J. Entomol. Soc. Br. Columbia* 67:33–48.

Rothschild, M. 1961. Defensive odours and Müllerian mimicry among insects. *Trans. R. Entomol. Soc. Lond.* 113:101–21.

Rothschild, M., J. von Euw, and T. Reichstein. 1970. Cardiac glycosides in the oleander aphid Aphis nerii. *J. Insect Physiol.* 16:1191–5.

Saini, E. D. 2004. Presencia de *Harmonia axyridis* (Pallas) (Coleoptera: Coccinellidae) en la Provincia de Buenos Aires. Aspectos biológicos y morfológicos. *RIA* 31:151–60.

Samways, M. J., and S. J. Wilson. 1988. Aspects of the feeding behaviour of *Chilocorus nigritus* (F.). (Col., Coccinellidae) relative to its effectiveness as a biocontrol agent. *J. Appl. Entomol.* 106:177–82.

Sato, S., and A. F. G. Dixon. 2004. Effect of intraguild predation on the survival and development of three species of aphidophagous ladybirds: Consequences for invasive species. *Agric. For. Entomol.* 6:21–4.

Sezer, S. 1970. Etude morphologique, biologique et écologique de *Lindorus lophantae* Blaisd. et Scymnus (S.) apetzi Mulsant (Coleoptera, Coccinellidae) dans lê départment dês Alpes-Maritimes afin de préciser leur efficacité prédatrice à l'egard dês populations de cochonilles et aphides. Thèse de docteur-ingénieur de la Faculte dês Sciences de l'Université de Paris.

Shapiro-Ilan, D. I., and T. E. Cottrell. 2005. Susceptibility of lady beetles Coleoptera: Coccinellidae) to entomopathogenic nematodes. *J. Invert. Pathol.* 89:150–6.

Silva, R. A., L. M. Almeida, and A. C. Busoli. 2005. Morfologia dos imaturos e do adulto de *Coccidophilus citricola* Brèthes (Coleoptera, Coccinellidae, Sticholotidinae), predador de cochonilhas-de-carapaça (Hemiptera, Diaspididae) de citros. *Rev. Bras. Entomol.* 49:29–35.

Sloggett, J. J., and M. E. N. Majerus. 2000. Habitat preferences and diet in the predatory Coccinellidae (Coleoptera): An evolutionary perspective. *Biol. J. Linnean Soc.* 70:63–88.

Smith, B. C. 1960. A technique for rearing Coccinellid beetles on dry foods, and influence of various pollens on the development of *Coleomegilla maculate lengi* Timb. (Coleoptera: Coccinellidae). *Can. J. Zool.* 38:1047–9.

Strand, M. R., and J. J. Obrycky. 1996. Host specificity of insects parasitoids and predators. *Bioscience* 46:422–9.

Subramanyam, T. V. 1925. *Coptosoma ostensum* Dist. and its enemy *Synia melanaria* Mul. *J. Bombay Nat. Hist. Soc.* 30:924–5.

Takahashi, K. 1987. Cannibalism by the larvae of *Coccinella septempunctata bruckii* Mulsant (Coleoptera: Coccinellidae) in mass-rearing experiments. *Jpn. J. Appl. Entomol. Zool.* 31:201–5.

Takahashi, K. 1989. Intra- and interspecific predations of lady beetles in spring alfalfa fields. *Jpn. J. Entomol.* 57:199–203.

Turian, G. 1969. Coccinelles micromycétophages (Col.). *Mit. Schweiz. Entomol. Ges.* 42:52–7.

Tursch, B., D. Daloze, M. Dupont, J. M. Pasteels, and M. C. Tricot. 1971a. A defense alkaloid in a carnivorous beetle. *Experimentia* 27:1380.

Tursch, B., D. Daloze, M. Dupont, C. Hootele, M. Kaisin, J. M. Pasteels, and D. Zimmermann. 1971b. Coccinellin, the defensive alkaloid of the beetle *Coccinella septempunctata*. *Chimia* 25:307.

Tursch, B., D. Daloze, and C. Hootele. 1972. The alkaloid of *Propylea quatuordecimpunctata* L. (Coleoptera, Coccinellidae). *Chimia* 26:74.

Tursch, B., J. C. Braekman, D. Daloze, C. Hootele, D. Losman, R. Karlsson, and J. M. Pasteels. 1973. Chemical ecology of arthropods, VI. *Tetrahedron Lett.* 3:201–2.

Vandenberg, N. J. 2002. Coccinellidae Latreille 1807. In *American Beetles 2*, ed. R. H. Arnett, Jr., M. C. Thomas, P. E. Skelley, and J. H. Frank, 1–19. Boca Raton, FL: CRC Press LLC.

Vilela, E. F., and A. Pallini. 2002. Uso de semioquímicos no controle biológico de pragas. In *Controle Biológico no Brasil: Parasitóides e Predadores*, ed. J. R. P. Parra, P. S. M. Botelho, B. S. Corrêa-Ferreira, and J. M. S. Bento, 529–42. São Paulo: Manole.

Villanueva, R. T., J. P. Michaud, and C. C. Childers. 2004. Lady beetles as predators of pest and predacious mites in citrus. *J. Entomol. Sci.* 39:23–9.

Völkl, W. 1995. Behavioral and morphological adaptations of the coccinellid, *Platynaspis luteorubra* for exploiting ant-attended resources (Coleoptera: Coccinellidae). *J. Insect Behav.* 8:653–70.

Yasuda, H., and N. Ohnuma. 1999. Effect of cannibalism and predation on the larval performance of two ladybird beetles. *Entomol. Exp. Appl.* 93:63–7.

Yasuda, H., T. Kikuchi, P. Kindlmann, and S. Sato. 2001. Relationships between attack and escape rates, cannibalism, and intraguild predation in larvae of two predatory ladybirds. *J. Insect Behav.* 14:373–84.

Zucchi, R. A. 2002. A taxonomia e o controle biológico de pragas. In *Controle Biológico no Brasil: Parasitóides e Predadores*, Ed. J. R. P. Parra, P. S. M. Botelho, B. S. Corrêa-Ferreira, and J. M. S. Bento. São Paulo: Manole.

24

Green Lacewings (Neuroptera: Chrysopidae): Predatory Lifestyle

Gilberto S. Albuquerque, Catherine A. Tauber, and Maurice J. Tauber

CONTENTS

24.1 Introduction 594
    24.1.1 Taxonomic Considerations 595
    24.1.2 Voucher Specimens 596
24.2 Chrysopid Food and Artificial Diets 596
    24.2.1 Natural Diet of Lacewings 596
        24.2.1.1 Nonpredaceous Adults 597
        24.2.1.2 Omnivory: Occasional or Usual? 598
        24.2.1.3 Cannibalism 598
        24.2.1.4 Larval Consumption of Prey and Efficiency of Food Conversion 598
    24.2.2 Nutritional Requirements and Artificial Diets 599
        24.2.2.1 Larval Nutritional Requirements 600
        24.2.2.2 Adult Nutritional Requirements 600
24.3 Digestive System of Chrysopidae: Anatomy, Physiology, and Feeding 601
    24.3.1 Larval Digestive System 601
    24.3.2 Larval Predatory Behavior 604
        24.3.2.1 Search, Contact, and Recognition of Prey 604
        24.3.2.2 Prey Capture 605
        24.3.2.3 Prey Consumption 605
        24.3.2.4 Cleaning and Resting 606
    24.3.3 Adult Digestive System 606
        24.3.3.1 Internal Anatomy 606
        24.3.3.2 Anatomical Modifications Associated with Glyco-Pollenophagous and
                Prey-Based Diets 607
        24.3.3.3 Association with Symbiotic Yeast 608
    24.3.4 Adult Feeding Behavior 609
        24.3.4.1 Postemergence Movement 610
        24.3.4.2 Attraction to Plant Volatiles 610
        24.3.4.3 Response to Prey-Associated Volatiles 611
        24.3.4.4 Habitat and Food Finding: An Integrative Process 612
24.4 Effects of Food on Lacewing Performance 612
    24.4.1 Food, Development, and Survival of Immature Stages 613
    24.4.2 Interactions with Host Plant of Prey 613
    24.4.3 Food and Reproduction 615
24.5 Prey Specificity—Its Stability, Underlying Mechanisms, and Evolution 617
    24.5.1 Degree of Prey Specificity 617
    24.5.2 Prey Specificity—Determinants and Stability 618
    24.5.3 Prey Specificity—Its Practical Importance 618

24.6   Future: Suggestions for Basic and Applied Research 618
      24.6.1   Expand the Arsenal of Chrysopid Species in Biological Control 618
            24.6.1.1   Recommendation 618
            24.6.1.2   Rationale 618
            24.6.1.3   Specific Targets 619
      24.6.2   Improve Chrysopid Biosystematics 619
            24.6.2.1   Recommendation 619
            24.6.2.2   Rationale 619
            24.6.2.3   Specific Targets 619
      24.6.3   Research Priority: Nutrition and Chemical Ecology (Implications for Rearing) 619
            24.6.3.1   Recommendation 619
            24.6.3.2   Rationale 620
            24.6.3.3   Specific Targets 620
      24.6.4   Research Priority: Seasonality 620
            24.6.4.1   Recommendation 620
            24.6.4.2   Rationale 620
            24.6.4.3   Specific Targets 621
Acknowledgments...621
References..621

24.1 Introduction

The family Chrysopidae is part of Neuroptera (or Planipennia), one of the oldest holometabolous orders (insects with complete metamorphosis); its fossil record extends back to the late Paleozoic—the Permian period, about 270 million years ago (Grimaldi and Engel 2005). Of the more than 6000 known neuropteran species, approximately 1200 belong to Chrysopidae; it is the order's second largest family (Myrmeleontidae, the largest, contains about 2100 species). Currently, Chrysopidae includes three subfamilies: Nothochrysinae, Apochrysinae, and Chrysopinae; the latter contains 97% of the known species (Tauber et al. 2009). Chrysopids occur on all continents except Antarctica; interestingly, native species are unknown in New Zealand (Duelli 2001). While some species have broad distributions, for example, *Chrysoperla externa* (Hagen), which is present throughout most of the Neotropical region, many are restricted to small areas of the planet (Zeleny 1984; Tauber et al. 2009).

Adults of this family are commonly known as green lacewings or aphid lions. The larvae of numerous species conceal themselves under packets of debris that they place on their backs; these larvae are also referred to as trash carriers (or *bicho-lixeiro* in Portuguese). Such packets camouflage the larvae, and they form a protective barrier to attack from natural enemies (Eisner and Silverglied 1988; Milbrath et al. 1994). Trash- or debris-carrying behavior occurs throughout the Chrysopidae, including most species in the Neotropical region.

In the laborious construction and constant reforming of their packets, the larvae use a variety of materials, such as the exoskeletons of their prey, arthropod exuviae (including their own), whole or parts of dead insects, fibers of vegetable or animal origin, bits of lichen and tree bark, spider webs, insect waxes, and other similar particles that they encounter (Smith 1926; Slocum and Lawrey 1976; Wilson and Methven 1997; Canard and Volkovich 2001). The debris adheres to the larva by means of numerous, long, smooth or serrated, pointed- or hooked-tipped setae on the larva's dorsal surface and on the lateral tubercles of the thorax and abdomen (Smith 1926; New 1969).

Aside from this larval habit, green lacewings have several additional characteristics that confer protection against natural enemies. (a) Eggs are deposited at the ends of long, filamentous stalks that sometimes bear droplets of repellent chemicals (Duelli and Johnson 1992; Eisner et al. 1996). (b) Larvae have long, sharp mandibles that can be used in defense (Smith 1926). (c) Larvae may secrete repellent fluid through the anus (LaMunyon and Adams 1987). (d) Pupae are protected within tough cocoons containing numerous layers of firmly connected silk threads (Gepp 1984). (e) Adult prothoracic glands can emit foul-smelling liquids that repel predators (e.g., Blum et al. 1973). (f) Adult coloration, predominantly green in most species, combined with a habit of remaining immobile during the day, appear to confer crypsis (Smith 1926; Canard and Volkovich 2001). (g) Flying adults can detect the ultrasonic pulses of insectivorous bats and respond by making quick evasive movements (Miller and Olesen 1979).

Although green lacewings can be inconspicuous, they are highly attractive. Adults are medium-sized, delicate insects, with two pairs of membranous, lacey wings (forewings, 6–34 mm in length), large iridescent eyes, and long, filiform antennae, sometimes longer than the wings. They are predominantly green, but some species can be dark brown or reddish; the wings of some are spectacularly marked. The larvae, which pass through three instars before spinning the cocoon and pupating, have two basic types of morphology and behavior. Some species—the trash carriers—are cryptic and move about slowly; their bodies are oval, hunchback, and covered with numerous long setae. The larvae of the remaining species—the naked larvae—are relatively active and do not carry trash; their bodies are somewhat flattened, elongated, and sparsely covered with short setae.

24.1.1 Taxonomic Considerations

Green lacewings often inhabit agroecosystems, sometimes in high numbers. They have attracted the attention of biological control specialists during the last five to six decades. As a result, the biology and ecology of some species are now relatively well studied (Canard et al. 1984; Tauber et al. 2000, 2009; McEwen et al. 2001). This situation, however, does not apply to the Neotropics, which has a very rich chrysopid fauna containing more than 300 described species (>25% of the world's total) in about 20 genera (Brooks and Barnard 1990). A poor knowledge of the systematics and biology of the Chrysopidae in this region has significantly hindered advances in using these natural enemies against agricultural pests (Albuquerque et al. 2001).

Many neotropical chrysopids were described during the first half of the 20th century by the Spanish priest Longinos Navás S. J. and the American entomologist Nathan Banks. However, their descriptions were often abbreviated and imprecise by modern standards; also, during their time, genitalic characters were not used to distinguish taxa. Subsequently, Tjeder (1966), Adams (1967), and others led the way in making genitalic characters essential for species identification and classification of the family worldwide (Brooks and Barnard 1990). The systematics of the group is further complicated because many of the type specimens associated with Navás' numerous species were destroyed or lost over the years (Monserrat 1985; Legrand et al. 2008).

Neotropical chrysopids began to receive modern treatment, with the inclusion of genitalic characters, in the late 1960s and the 1970s (e.g., Adams 1969, 1975; Tauber 1969); a consideration of the fauna's generic classification followed (Adams and Penny 1986). Additional studies treated regional faunae and described new taxa (Adams 1982a,b, 1987; Adams and Penny 1985, 1992a,b; Penny 1997, 1998, 2001, 2002; Freitas and Penny 2000, 2001; Freitas 2003, 2007; Tauber et al. 2006, 2008a,b; Viana and Albuquerque 2009; Sosa and Freitas 2010; revisions: Tauber 2007, 2010; Freitas et al. 2009). Recently, the phylogeny of the family has been addressed (Winterton and Freitas 2006).

Historically, larval characters have played a significant role in chrysopid systematics (e.g., Withycombe 1925; Killington 1936; Tauber 1969, 1974, 2003; Díaz-Aranda and Monserrat 1995). They have also proven very useful in both identification and classification of neotropical chrysopids (Tauber et al. 2001, 2006, 2008a,b; Mantoanelli et al. 2006, 2011). We expect that they will continue to contribute significantly to the systematics of the group because they provide excellent taxonomic characters.

With the above advances in chrysopid systematics, we are confident that the biological and ecological investigation of green lacewing species in Central and South America will expand considerably in the near future (e.g., Silva et al. 2007; Multani 2008; Ribas et al. 2012). Meanwhile, the information about the bioecology and nutrition of Chrysopidae, reported below, is based largely on studies of European and North American species, mainly *Chrysoperla carnea* (Stephens), *sensu lato*. However, whenever available, we used information from neotropical species, especially *C. externa*, which is widely distributed throughout southern Florida, Mexico, Central America, and South America. This species is a highly attractive candidate for use in two types of biological control—conservation and augmentation (Albuquerque et al. 1994, 2001).

In our chapter we use the nomenclature of Brooks and Barnard (1990), and whenever synonymy was involved, we cited the current species names, instead of those cited by the original authors. Also, we use the name *C. carnea, s. lat.* (= *C. carnea, sensu lato*) for any species referred to in the literature as *Chrysoperla carnea*. The *C. carnea* has long been the subject of considerable taxonomic discussion and confusion; the taxonomic problems in this group ultimately could cause serious detriments to biological control (see Henry and Wells 2007 and references therein). Currently, the "*Chrysoperla carnea* species group" consists of a number of Palearctic and Nearctic entities whose names are largely based on different courtship songs. All of these entities are very difficult, sometimes impossible to distinguish morphologically. Previously, most were called *C. carnea*. In some cases, the differentiated song morphs represent distinct species; in other cases, species differentiation is not clear. For most of the studies on *C. carnea* cited here, it is not known which taxon the published research reports refer to (e.g., see Section 24.3.4.3).

24.1.2 Voucher Specimens

The above taxonomic issues prompt us to highlight the crucial importance of voucher specimens associated with virtually all biological research. We state emphatically that for each published (basic or applied) biological study that deals with organisms, voucher specimens should be deposited in a stable, well-cared-for, preferably public, museum. This requirement should apply no matter how well known the species, and even when the species identification was confirmed by a reputable taxonomist.

Two issues underlie our call for voucher specimens. First, as illustrated above, it is often difficult, even for systematists, to obtain reliable identifications; in the case of chrysopids, especially neotropical chrysopids, errors have been numerous—historically and recently (e.g., see Tauber et al. 2008c, 2011; Tauber and Flint 2010). Second, taxonomy is a dynamic science: research leads to the discovery of new species, synonymization of old names, differentiation of "cryptic species" (as in *C. carnea, s. lat.*), and the elucidation of phenotypic and genotypic intraspecific variation. Thus, a name that is valid today may be obsolete tomorrow; the situation with *C. carnea, s. lat.* is an important example (see Section 24.3.4.3). In most cases, well-preserved voucher specimens are the only reliable means of confirming or re-confirming the identity of subject species.

The voucher specimens should be appropriately labeled so that they are easily associated with the published work, and the publication should indicate the museum where the vouchers are deposited. These vouchers should include specimens that are preserved via traditional methods (pinned, in alcohol, on slides, etc.) as well as frozen specimens for future DNA analysis. Furthermore, the depository museums themselves should (a) be well maintained by professional curatorial staff, and (b) have sustained administrative and fiscal support. It is no exaggeration that published research without associated voucher specimens, ultimately, cannot be verified or replicated with confidence. Thus, if procedures such as those above are ignored, important and costly studies may be rendered problematic in the future.

24.2 Chrysopid Food and Artificial Diets

Although chrysopids are known as predators *par excellence*, predation is ubiquitous only in the larval stage. Adults in only a few chrysopid genera are predators; most are primarily glyco-pollenophagous, that is, they feed on nectar, pollen, and/or honeydew (see reviews: Principi and Canard 1984; Canard 2001). Consequently, the type of food that a lacewing eats is a key factor in its biology, ecology, and use in pest management, particularly biological control.

24.2.1 Natural Diet of Lacewings

The prey of chrysopid larvae and predaceous adults generally consists of small arthropods that are slow moving and have penetrable integuments (New 1975). Among the most common prey are mites (Tetranychidae and Eriophyidae) and several groups of insects, such as hemipterans from the suborders Sternorrhyncha (scale insects of the families Coccidae, Monophlebidae, Pseudococcidae, Diaspididae, and Eriococcidae; aphids of virtually all families; whiteflies; and psyllids) and Auchenorrhyncha (families Cercopidae, Cicadellidae, Membracidae, and Fulgoridae), eggs and small larvae of lepidopterans (families Noctuidae, Pieridae, Plutellidae, Pyralidae, Tortricidae, and Yponomeutidae), psocids (Psocidae), and thysanopterans (thrips). All these groups of prey include major insect pests.

Less commonly attacked are eggs and small larvae of beetles, dipterans, hymenopterans, and other neuropterans. Large insects are rarely preyed on (Killington 1936; Principi and Canard 1984; Canard 2001). Albuquerque et al. (2001) provide a list of potential prey for two of the most commonly found species in the Neotropics, *C. externa* and *Ceraeochrysa cubana* (Hagen); this list includes scales, aphids, lepidopteran eggs and larvae, whiteflies, and mites; most are pests of agricultural crops and horticultural plants.

Numerous types of prey seem to offer a diet that is adequate nutritionally, that is, the prey have proteins, amino acids, lipids, carbohydrates, vitamins, minerals, and other compounds in their tissues and hemolymph. However, the concentrations of these constituents and their accessibility to predators may vary among prey species (Florkin and Jeuniaux 1974; Yazlovetsky 1992; Cohen 1998). Furthermore, herbivores often sequester allelochemicals (toxins) from their host plants for their own defense against natural enemies; the accumulation of these toxic compounds may alter the potential value of the prey to lacewings (Bowers 1990; Rowell-Rahier and Pasteels 1992).

In addition, plants may have other, nonchemical defensive characteristics (waxy surfaces, trichomes, complex architecture, etc.) that can influence chrysopid mobility and efficiency in finding prey (see Section 24.4.2). Thus, from the lacewing's perspective, the quality and accessibility of prey can vary considerably; this variation is expressed in the performance of individual lacewing species when faced with diverse habitats or prey (e.g., Thompson and Hagen 1999; also see Section 24.4).

24.2.1.1 Nonpredaceous Adults

As their primary source of nutrients, the adults of most nonpredaceous lacewing species use metabolites (sugars, amino acids, and lipids) that are present in plant products—pollen, nectar, and, indirectly, honeydew excreted by members of the suborder Sternorrhyncha (Hemiptera). Even larvae and predaceous adults occasionally feed on such plant products (Downes 1974; Principi and Canard 1984; Hagen 1986; Wäckers et al. 2005). The nutritional composition of nectar and pollen varies among plant species, whereas honeydew varies with the aphid or mealybug producer and, apparently, with the host plant of these herbivorous insects (Hagen 1986). In general (see below), none of these three plant products alone (nectar, pollen, or honeydew) provides all of the nutrients that adult lacewings require; lacewings typically include a combination of two or all three in their diets.

Floral and extrafloral nectar is a source of sugars (sucrose, glucose, and fructose), and also, proteins, amino acids, lipids, antioxidants, alkaloids, phenolics, vitamins, saponins, dextrins, and inorganic substances (Baker and Baker 1983). Sugars make up 15% to 75% of nectar weight, and amino acid concentrations range from 0.2 to 0.7 μmol/ml in nectar from trees and shrubs and 0.4 to 4.7 μmol/ml in nectar from herbaceous plants; however, nectar rarely contains all 10 essential amino acids. As an example of the importance of nectar in the nutrition of chrysopids, Adjei-Maafo and Wilson (1983) recorded much higher densities of these (and other) predators in cotton varieties with extrafloral nectaries than in those without them.

As energy sources, pollen contains up to 14 different carbohydrates (including common sugars), as well as a variety of other nutrients. Some pollen may contain fatty acids and essential sterols; proteins constitute 6–35% of pollen weight; and usually all amino acids, except tryptophan and phenylalanine, are found in high concentrations. Also, minerals and vitamins A, C, E, and several of the B complex may be present. Thus, the pollen of at least some plant species has most of the essential nutrients required for reproduction, but often pollenophagous lacewings supplement their diets with nectar or honeydew.

Honeydew is composed mainly of sugars (fructose, glucose, and sucrose); it may also include some vitamins (such as vitamin C and several of the B complex) and amino acids (but rarely all of the 10 essential ones). Its composition may vary with the species of hemipteran producer and with a variety of factors related to the host plant of the producer. Generally, the concentration of amino acids in honeydew is related to their occurrence in the phloem. Thus, concentrations may vary with the condition of the host plant of the producer and with the seasons. In some cases, tryptophan and histidine, which commonly are absent from floral and extrafloral nectaries, are present in honeydew. Like the other plant products that lacewings eat, honeydew usually does not form a complete diet, and nonpredaceous chrysopid adults may supplement such a diet with nectar or pollen.

24.2.1.2 Omnivory: Occasional or Usual?

It has long been known that in nature lacewing larvae and predaceous adults supplement their diet of prey with floral and extrafloral nectar, pollen, and honeydew (Kawecki 1932; Killington 1936; Principi 1940; Downes 1974). However, the importance of these supplements has begun to be quantified only recently.

Working with neonate larvae of *C. plorabunda* (= *C. carnea, s. lat.*) in cotton, Limburg and Rosenheim (2001) showed that extrafloral nectar is a major component of the diet; its consumption increases with a decrease in prey availability. Although nectar alone is not sufficient for development, it can extend the longevity of first instars considerably.

In another study, Patt et al. (2003) demonstrated that *C. carnea, s. lat.* larvae supplemented their intake of carbon and nitrogen by feeding on nectar and pollen, thus improving their growth and development. A similar response occurs when the larval diet is supplemented with artificial honeydew (McEwen et al. 1993). Despite this evidence, it is not known whether larvae are commonly omnivorous or if omnivory is displayed only when prey are absent, scarce, or of poor quality. Moreover, to what extent this behavior occurs among chrysopid species other than the two *Chrysoperla* species tested is an intriguing area for investigation (see Section 24.4.1).

Among the groups in which the adults are generally considered carnivorous, *Chrysopa* spp. are not restricted to eating prey; they may also feed on pollen, yeasts, fungal spores, and honeydew. Although these species sometimes are classified as omnivorous (Stelzl 1992), the proportion of plant products in their diet is not quantified (Canard 2001).

24.2.1.3 Cannibalism

Both larval and adult chrysopids engage in intraspecific predation (cannibalism) in the laboratory (usually when food is scarce or unsuitable) (Smith 1922; Canard and Duelli 1984). How frequently cannibalism occurs in nature is not established.

In the laboratory, the most intense cannibalism involves the egg stage. Although chrysopid eggs are deposited atop a long stalk, which is considered an efficient defense, lacewing larvae, especially young first instars, have been observed to climb stalks and feed on eggs. In addition, predaceous and nonpredaceous females may sometimes eat their own eggs after they are laid.

Another similar behavior that females exhibit is auto-oophagy, in which females pull their own eggs from the genital chamber with their jaws. In *Chrysopa perla* (L.), this behavior is performed mainly by virgin females to keep their genital ducts free as new oocytes are produced (Philippe 1971).

Cannibalism among larvae is less commonly observed. Satiated, uncrowded larvae usually do not attack other lacewing larvae; apparently the risks outweigh the benefits in this condition (Canard 2001).

24.2.1.4 Larval Consumption of Prey and Efficiency of Food Conversion

Measuring the number of prey that each instar can consume is difficult (see review: Principi and Canard 1984). Such studies are usually conducted with a surplus of prey relative to the feeding potential of larvae, and thus they can provide an estimate of prey consumption capacity. However, a rigorous experimental design is essential and the resulting information needs cautious interpretation. For example, (a) the size and stage of the prey should be considered and, if possible, held constant; (b) the degree to which larvae ingest prey should be evaluated (often prey are killed but the contents are rejected or only partially ingested); and (c) environmental conditions, such as temperature, can influence the number of prey consumed (higher temperatures tend to induce higher consumption). Thus, physical conditions should be controlled during the experiment and taken into account when the results are interpreted.

Despite some contrary studies (e.g., Burke and Martin 1956), for many species, more than 75% of total consumption and most of the increase in weight occur in the third instar (review: Principi and Canard 1984). In several investigations with the neotropical *C. externa*, in which prey size was standardized, the pattern of prey consumption per instar was similar to that reported above for other species. That is, depending on the type of prey, the larvae of *C. externa* consumed from 3% to 8% of their total intake during the first instar, 11% to 21% during the second instar, and 72% to 85% during the third instar (Table 24.1). The total number of prey consumed during the larval stage varies widely and often depends on the size of the prey (Table 24.1).

Green lacewings seem to convert their food relatively efficiently, but experimental evidence is still largely nonexistent. In one of the few studies, Zheng et al. (1993a) demonstrated that *C. carnea, s. lat.* shows a gross conversion efficiency (= proportion of prey ingested converted to body mass) between 40% and 60%, depending on the instar and the level of consumption. They also found that conversion efficiency increased with instar and was higher when the supply of prey diminished.

24.2.2 Nutritional Requirements and Artificial Diets

Chrysopid larvae and adults have qualitative nutritional requirements similar to those of other predators, parasitoids, and even phytophagous insects (Vanderzant 1973; Hagen 1987; Thompson and Hagen 1999). That is, they need about 30 chemical compounds, including proteins and/or the 10 essential amino acids (arginine, phenylalanine, histidine, isoleucine, leucine, lysine, methionine, threonine, tryptophan, and valine), B-complex vitamins (folic acid, nicotinic acid, pantothenic acid, biotin, pyridoxine, riboflavin, and thiamine), other water-soluble growth factors (including choline and inositol), some fat-soluble vitamins, cholesterol or phytosterol, a polyunsaturated fatty acid, minerals, and an energy source (usually

TABLE 24.1

Mean Consumption of Prey (Number of Individuals) by the Three Instars of *Chrysoperla externa* (25 ± 1°C)

Prey Species	1st	2nd	3rd	Total	Reference
Hem.: Aleyrodidae					
Bemisia tabaci (4th instar nymphs)	107.8	288.0	1006.3	1402.1	Auad et al. (2005)
Hem.: Aphididae					
Aphis gossypii (3rd and 4th instar nymphs)	17.4	73.3	453.8	544.5	Santos et al. (2003)
Cinara pinivora + *C. atlantica* (1st and 2nd instar nymphs)	16.8	31.3	167.0	215.1	Cardoso and Lazzari (2003)
Cinara pinivora + *C. atlantica* (3rd and 4th instar nymphs)	5.2	12.3	49.5	67.0	Cardoso and Lazzari (2003)
Rhopalosiphum maidis (3rd and 4th instar nymphs)	21.9	40.1	279.0	341.0	Maia et al. (2004)
Schizaphis graminum (3rd and 4th instar nymphs)	13.7	34.7	266.2	314.6	Fonseca et al. (2001)
Lep.: Gelechiidae					
Sitotroga cerealella (eggs)	55.3	97.4	777.9	930.6	de Bortoli et al. (2006)
Lep.: Noctuidae					
Alabama argillacea (eggs)	11.6	43.7	290.3	342.7	Figueira et al. (2002)
A. argillacea (1st instar larvae)	23.9	85.3	365.5	474.7	Silva et al. (2002)
Lep.: Pyralidae					
Anagasta kuehniella (eggs)	95.8	192.4	1264.9	1553.1	de Bortoli et al. (2006)
Diatraea saccharalis (eggs)	21.8	77.1	468.4	567.3	de Bortoli et al. (2006)

Note: Figures refer to the regimen of higher density of prey or best cultivar/species of the prey's host plant used in the studies in which more than one of these factors were tested.

simple or complex carbohydrate and/or lipids). However, the quantitative requirement of each compound may differ according to the dietary habits of the species (Thompson and Hagen 1999).

24.2.2.1 Larval Nutritional Requirements

Knowledge of the qualitative and quantitative nutritional needs of chrysopid larvae stems from studies with artificial diets. The current status of the studies is summarized in reviews by Cohen (1995), Nordlund et al. (2001), and Yaslovetsky (2001). Below, we present some highlights.

Hagen and Tassan (1965) were the first to develop an artificial diet (a meridic diet, without insect material) for rearing lacewing larvae (*C. carnea, s. lat.*); this diet was an aqueous formulation of fructose, hydrolyzed protein, choline chloride, and ascorbic acid, encapsulated in paraffin film. Although lacking in some respects, the diet yielded complete development and fertile oviposition. Following this breakthrough, 14 other chrysopid species were reared, with varying degrees of success, on different artificial diets. In most of the studies, the proportion of proteins, lipids, and cholesterol mimicked those in the prey's hemolymph, and in all cases the nutrients were presented in a liquid preparation (e.g., Vanderzant 1969, 1973; Hassan and Hagen 1978; Yazlovetsky 1992). Numerous articles on the subject were published (see review: Yazlovetsky 2001).

From a nutritional viewpoint, in one of the few studies with a chemically defined (holidic) diet, Niijima (1989, 1993a,b) examined the specific value of individual nutrients by removing them serially from a relatively complete, defined, larval diet that Hasegawa et al. (1989) developed. This diet consisted of 23 amino acids, 17 vitamins, 11 minerals, 5 organic acids, 6 fatty acids, 2 sugars, and cholesterol. Niijima found that the 10 essential amino acids were required for larval molting, but several other amino acids could be removed without causing adverse effects (at least in the short run–for one generation). A 40% reduction in the level of amino acids in the diet greatly lengthened the larval developmental time. Choline, ascorbic acid, and some B vitamins (such as nicotinic acid and pantothenic acid) were essential for development, whereas the absence of most other B vitamins individually allowed development, but at a slower rate and with reduced emergence. Absence of other water-soluble and fat-soluble vitamins did not result in noticeable negative effects.

Despite the above findings and apparent success in developing artificial diets for chrysopid larvae, none has been used in commercial mass rearing. Cohen and Smith (1998) attributed this failure to several causes, such as the complexity, cost of manufacturing, and, especially, the liquid nature of the diets. Most importantly, these diets tend to decrease developmental and reproductive performance compared with that obtained under a regimen that included prey; the effect was especially apparent after several generations of rearing.

To overcome the above problems, Cohen and Smith (1998) used a different approach. Given that lacewing larvae perform some extraoral digestion of semisolid prey components (see Section 24.3.1), they developed a highly concentrated, semisolid artificial diet similar in texture and composition to the interior of prey. This diet, which was based on beef, chicken eggs, sugar, honey, and yeast, contained proteins and lipids in proportions approaching those found in the tissues of insects (15–20% and 12–18%, respectively); these proportions are well above those found in the hemolymph (4–5% and 2–3%, respectively). On this diet, *C. rufilabris* larvae developed and reproduced equally well or better for at least 15 generations, relative to prey-reared larvae. This approach represents a significant advancement for rearing chrysopids and other predaceous insects, and it offers new insights for examining how chrysopid larvae interact with prey in the field.

24.2.2.2 Adult Nutritional Requirements

As stated above, chrysopid adults can be either predaceous or glyco-pollenophagous. Although adults in each of these categories have been shown to have specific food needs for various aspects of reproduction (e.g., female mating, male mating, initiation of oviposition), females of all species studied thus far require a protein source to sustain oviposition. However, very little work has been done to define the nutritional requirements of adults in either group.

Several considerations, apart from the composition of the adult diet itself, may confound studies of adult nutritional requirements (Hagen 1987). First, because some metabolites obtained by the larvae are transferred to the adult, reproductive performance, at least in its early stages, can be strongly dependent on the larval diet (Hagen and Tassan 1966; Zheng et al. 1993b; Osman and Selman 1996). Second, the timing when various types of nutrients (carbohydrates, protein) are received during the adult stage can strongly influence subsequent levels of reproduction. For example, adults of *C. carnea, s. lat.* that received a carbohydrate-rich diet during the prediapause period had significantly higher levels of post-diapause fecundity than those that received a high-protein prediapause diet (for sustained fecundity, adults from both treatments required a high-protein diet after diapause) (Chang et al. 1995). Although the effect of dietary timing has been demonstrated only for *C. carnea, s. lat.*, it may occur in other glyco-pollenophagous lacewing species. Third, the gut (crop diverticulum) contains yeast symbionts that may synthesize essential amino acids; these symbionts may vary with geographic population and time (Hagen et al. 1970; Woolfolk et al. 2004; see also Section 24.3.4). All of the above factors can be difficult to control, and all can lead to variable results when adult diets are assessed experimentally.

Unlike the artificial diets for larvae, artificial diets for glyco-pollenophagous adults, which were initially developed in the late 1940s (Hagen 1950), have replaced the natural diet in mass rearing. They are efficient, low-cost, and provide high rates of fecundity and fertility in the laboratory. In general, these diets contain a source of protein (hydrolyzed or autolyzed yeast, such as *Saccharomyces cerevisiae* or *S. fragilis*) and carbohydrate (honey, fructose, or sucrose) (Hagen and Tassan 1966, 1970). Although they were originally tested and refined for *C. carnea, s. lat.*, they have been equally effective in rearing a substantial number of other lacewing species whose adults are glyco-pollenophagous, for example, several species of *Chrysoperla* (Tauber and Tauber 1983; Albuquerque et al. 1994; Carvalho et al. 1996) and *Ceraeochrysa* (López-Arroyo et al. 1999; Barbosa et al. 2002).

Despite the reported successes, adults of some glyco-pollenophagous species show relatively low reproductive performance on the artificial diets. For example, several *Chrysopodes* species produced only a few eggs when provided such diets, while other species of *Chrysopodes* and several species of *Leucochrysa* produced none (Silva 2006; Silva et al. 2007; Albuquerque and Tauber, unpublished data). Although the above examples refer to neotropical genera, this situation is probably true for chrysopids in other regions as well, but specific reports were not found. The poor performance of these species during laboratory rearing may have causes other than a dietary deficiency. They illustrate that investigations into the nutritional requirements of adult lacewings should be expanded beyond the few species studied thus far. They also illustrate the need to investigate the reproductive responses of chrysopid adults to other chemical and nonchemical stimuli.

It is important to reemphasize a point that Nordlund et al. (2001) made concerning the nutritional value of pollen. Although various types of pollen have been shown to differ in their dietary value for honeybees, the nutritional qualities of pollen have not been examined relative to lacewing reproduction. Given that most chrysopid adults are glyco-pollenophagous, this situation is especially regrettable.

24.3 Digestive System of Chrysopidae: Anatomy, Physiology, and Feeding

Adults and larvae of lacewings have evolved different modes of feeding, some of which are reflected in the morphology of their mouthparts and gut. On the one hand, the larvae of all species are primarily predators, although they may occasionally feed on pollen, nectar, and honeydew. As a consequence, there is a degree of morphological and functional uniformity in larval digestive systems and also in larval predatory behavior. Regrettably, very little is known about the prey associations of chrysopid larvae in nature. On the other hand, adults may be either predatory or glyco-pollenophagous, depending on the genus; this divergence in adult feeding habits is accompanied by morphological and functional differences in their digestive systems. Whether and to what degree these two types of adults also differ in their foraging behavior are unknown.

24.3.1 Larval Digestive System

The ingestive and digestive organs of chrysopid larvae (Figure 24.1) have a variety of properties that are unique among insects; our discussion of these features is based mainly on the works of Withycombe (1925), Killington (1936), and Yazlovetsky (2001).

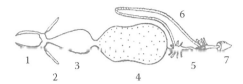

FIGURE 24.1 Schematic diagram of the digestive system of chrysopid larvae: (1) mouthparts (mandibles + maxillae); (2) salivary glands; (3) stomodaeum (pharynx, esophagus, and crop); (4) midgut; (5) proctodaeum; (6) Malpighian tubules (eight in total, but only one represented completely); (7) silk reservoir. (From Ermicheva, F. M., et al., *Izv. Acad. Sci. Moldavian SSR, Ser. Biol. Chem. Sci.,* 4, 49, 1987.)

Although lacewings are generally classified as chewing insects, a characteristic feature of their larval stage is the prominent, sucking-type feeding apparatus. The sickle-shaped jaws (mandibles and maxillae) are heavily sclerotized and grooved; they serve to capture, lacerate, and ingest prey. The mandibles and maxillae interlock via a small chitinized fold on the inner side of both; together they form a rigid tube with a sharp, piercing terminus and an internal feeding channel through which food is ingested (Figure 24.2a). The left and right feeding channels merge inside the head to form the pharynx; the pharynx is lined with a thick, chitinous layer, which extends through the esophagus. The jaws (mandibles and maxillae) are often as long as, or longer than the head, and they are always curved inward. The bases of the maxillae are enlarged and contain glandular material (Gaumont 1976); they are sometimes called "venom" glands, but we refer to them as "maxillary glands" (Figures 24.1 and 24.2b).

In addition to the mouthparts (mandibles, maxillae, labium, and labrum), which form the external parts of the digestive system and which contain the maxillary glands, there is a pair of unbranched, tubular salivary glands. These glands extend from the posterior region of the head or the anterior region of the prothorax to the base of the mouthparts, where they connect with the feeding channel.

There is no typical oral opening; indeed, the mouth is mechanically closed by the cephalic integument soon after larvae hatch or undergo molting (Killington 1936). Therefore, the route for ingestion is via the feeding tube (review: Canard 2001). Although ingestion is insufficiently studied, a number of factors are known to facilitate the process (Cohen 1998). First, larvae mechanically shred prey tissue into small pieces with the tips of their jaws; the mandibles have acute tips that lacerate tissue. Second, larvae inject secretions from the salivary glands and perhaps from the maxillary glands (Gaumont 1976; Cohen 1998;

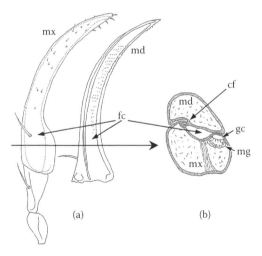

FIGURE 24.2 Mouthparts of the larvae of Chrysopidae: (a) Ventral view of maxilla (mx) and mandible (md), highlighting ridges on dorsal surface of the first and ventral surface of the second, which form the feeding channel (fc) when they join. (b) Cross section at the level of the arrow (gc, glandular channel; cf, chitinous folds that couple the mandible and maxilla together; mg, maxillary gland). (Modified from Canard, M. In *Lacewings in the Crop Environment,* ed. P. McEwen, T. R. New, and A. E. Whittington, 116–29. Cambridge: Cambridge Univ. Press, 2001.)

Yazlovetsky 2001). These secretions include hydrolytic enzymes that partially break down large molecules of nutrients, thereby facilitating movement of food through the feeding tube and its subsequent digestion in the midgut. Although the enzymes involved appear to be potent, the chemistry of extraoral digestion in chrysopids remains poorly known (Cohen 1995).

The suction of the partially digested fluids from the prey is accomplished by means of the synchronized action of muscles that control the mandibles and maxillae, as well as the pharynx. These muscles are attached to the dorsum and venter of the head and the arms of the tentorium. The motion of the pharyngeal muscles controls the expansion and contraction of the pharyngeal lumen, and the partially digested liquid is sucked up and pumped into the gut (Smith 1922; Killington 1936). This mechanism of extraoral digestion and ingestion fits the type that Cohen (1998) termed "intact without reflux": there is no destruction of the prey's cuticle, and enzymes only from the predator's extraintestinal glands (e.g., salivary and/or maxillary glands), not the gut, are injected into the prey. Later, the partially digested material is sucked into the predator's gut.

In the head, at the tentorium, there is a single valve formed by the projection of epithelial cells into the esophagus; this valve, combined with the peristaltic contractions of circular muscles of the esophagus, prevents the return of the fluids ingested during feeding. Within the prothorax, the esophagus enlarges to form a thin-walled crop, which is coated with a very delicate chitinous intima and ringed with bands of circular muscles. The crop occupies much of the internal cavity of the meso- and metathorax; there is no associated food reservoir (diverticulum) attached to it, as seen in the adults. Apparently, little or no digestion occurs in the crop.

The crop and mesenteron (midgut) are connected via a poorly developed esophageal valve, visible as a constriction between these two regions of the gut. The midgut is a large sac with a blind end; it occupies the anterior two thirds of the abdomen, and apparently, is without an outer layer of muscles. The midgut epithelium is composed of large cells and is lined with a thin peritrophic membrane—a continuation of the chitinous intima of the crop. Thus, food does not come into direct contact with the midgut epithelium. Most digestion and nutrient absorption occur within the midgut.

Recently, Chen et al. (2006) found numerous bacteria in the midgut of *C. carnea, s. lat.* larvae; they proposed that the bacteria may help decompose food waste in the midgut. When they are at rest, midgut epithelial cells are somewhat flattened, but when actively secreting, they become columnar and protrude into the mesenteric cavity. Their distal extremities swell with the secretions that are then released into the intestinal cavity.

After the midgut, the gut is closed for some distance, that is, the proctodaeum (hindgut) is composed of a string of solid cells that are thought to be nonfunctional. During larval development, little solid waste accumulates, and what does accumulate cannot be eliminated from the larva. Rather, waste material is stored at the midgut's posterior end and is excreted after adult emergence, in the form of a meconium—a small, hard, shiny, black or dark brown pellet, covered with peritrophic membrane.

At the distal end of the proctodaeum there is a small, thin-walled sac, the silk reservoir, which tapers toward the rectum; the rectum has a thick epithelium and is surrounded by circular muscles. The anus opens at the end of the 10th (last) abdominal segment. From the anterior end of the proctodaeum arise eight Malpighian tubules, which extend forward into the first few segments of the abdomen. Usually two of the tubules remain free in the lumen; the other six recurve posteriorly and reconnect with the proctodaeum immediately anterior to the silk reservoir, where their endings are surrounded by a group of epithelial cells. These six tubules are functional throughout larval development. They secrete a viscous, brown fluid, which is transferred to the silk reservoir and, thence, into the rectum. This brown fluid can be excreted through the anus as a defense against natural enemies or as an adherent, which helps the molting larva cling to the substrate or which gives traction to the false leg (tip of the abdomen) during locomotion. It may also have an excretory function (Spiegler 1962; LaMunyon and Adams 1987).

In the third instar, most of the cells at the posterior end of the recurrent Malpighian tubules increase in size and change their function: instead of producing the viscous fluid, they begin to secrete silk, which is stored in the silk reservoir. When the larva completes development, the silk is used to spin the cocoon, within which the larva metamorphoses into the pupal stage. The control of silk flow is accomplished via circular muscles around the rectum; the anal papilla functions as a spinner, and cocoon spinning is achieved by the active, multidirectional movement of the last abdominal segments.

24.3.2 Larval Predatory Behavior

Little is known about larval feeding behavior of green lacewings in nature. A few species have been studied in the laboratory, and most of this research involved *C. carnea, s. lat.* Therefore, generalizations in the chrysopid literature should be viewed with caution. For example, the naked larvae of the genus *Chrysoperla* are characterized by the agility of their movement, aggressiveness, and rapid growth, whereas trash-carrying larvae usually move and grow more slowly and are comparatively less aggressive. Their feeding behavior may differ from that of naked species.

As in many other groups of predators, the feeding behavior of lacewing larvae can be categorized into a sequence of steps (see reviews by Canard and Duelli 1984; Canard 2001; also see diagrammatic quantification of the behavior in Milbrath et al. 1993; Mantoanelli and Albuquerque 2007). First, the larva searches actively until it encounters a potential prey. Then the prey is contacted and recognized. If the prey is acceptable, it is captured with the jaws and then consumed. Finally, when feeding ends, the larva may clean its mouthparts and rest or resume searching for prey. These categories do not include the first step in the customary sequence of prey selection, that is, habitat location, because this function is largely performed by the adult female during the selection of oviposition sites (Greany and Hagen 1981) (see Section 24.3.5).

24.3.2.1 Search, Contact, and Recognition of Prey

Although lacewing larvae are able to withstand deprivation of food and water for relatively long periods after hatching (e.g., Tauber et al. 1991), their ultimate survival depends on searching for prey. Apparently, larvae require physical contact to recognize prey (Principi 1940; Fleschner 1950; Barnes 1975; Bond 1980). During the search, the larvae assume a characteristic posture, moving their heads from side to side, with the mouthparts partially open and parallel to the substrate, the labial palpi directed forward, and the antennae directed forward and slightly to the sides (Canard and Duelli 1984). Also, from time to time, larvae stop and the anterior part of the body sweeps from side to side, as they search for prey (Figure 24.3) (Bänsch 1964; New 1991). When they contact prey, their searching pattern becomes concentrated in the area of their encounter, and the frequency with which they change directions increases. Such behavior can increase searching efficiency because many types of lacewing prey (e.g., aphids, scales) occur in aggregations (Fleschner 1950; Bond 1980; New 1991).

The larvae of most chrysopid species are generally active, especially at night. Apparently, the greater the degree of hunger, the more intense their activity becomes (Sengonca et al. 1995); however, if no prey are found, the larvae gradually reduce their movements, become lethargic, and die.

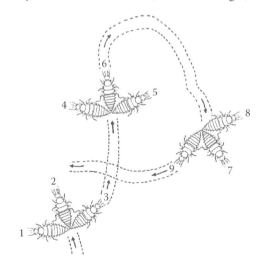

FIGURE 24.3 Lacewing larva exhibiting "casting behavior" in search of prey (numbers indicate successive positions of larva).

The direction of the light source has been shown to influence the direction of neonate movement, and the apparently species-specific response pattern may be associated with the location of preferred prey in relation to the oviposition site. Milbrath et al. (1994) found that newly hatched larvae of two species of *Chrysopa* show distinct phototactic responses, apparently related to the spatial distributions of their diverse prey. Neonate larvae of *Chrysopa slossonae* Banks exhibited negative phototaxis, a response that induced movement away from the leaves and branches in the upper part of the host plant, into the center of the plant, and toward colonies of their prey (aphids that live on branches and trunks in the lower part of the plant). On the other hand, the neonate larvae of *Chrysopa quadripunctata* Burmeister exhibited positive phototaxis, which tended to keep them on leaves in the tree canopy, where their primary prey are located.

Similarly, for older larvae that have fed, experimental evidence indicates that light and gravity may have a variety of species-specific influences on searching. Some species show positive phototaxis and geotaxis, while others show negative or mixed responses (see review: Hagen et al. 1976a).

Volatile chemicals (kairomones) associated with prey may aid larval searching. For example, volatiles emanating from honeydew, lepidopteran scales, and insect eggs were shown to increase searching efficiency in *C. carnea, s. lat.* larvae (Kawecki 1932; Lewis et al. 1977; Nordlund et al. 1977).

When prey is encountered, it can be recognized chemically, via sensory receptors on the terminal ends of the larval labial palpi and antennae. Visual stimuli may also aid in the initial identification of prey, as demonstrated in larvae of *C. carnea, s. lat.* and *Chrysopa oculata* Say (Lavallee and Shaw 1969; Ables et al. 1978). Then, the larva begins to examine the prey with its mouthparts, that is, via the sensory receptors at the tip of the maxillae; it then either accepts or rejects the prey (Hagen 1987; Canard 2001).

24.3.2.2 Prey Capture

After physical contact and recognition of prey, the larva stops and displays a characteristic posture—jaws open wide, parallel to the surface or directed slightly upward, and antennae and labial palpi directed laterally. Capture of prey follows a series of stereotyped movements: (a) slow approach; (b) stop; (c) attack with a rapid, forward thrust of the head and closing of the mouthparts; and (d) quick retraction of the head, and usually lifting of prey from the substrate (Canard and Duelli 1984). Prey lifting does not always occur and sometimes it is only partial, as in *C. carnea, s. lat.* (Bänsch 1964). Immobile prey, such as eggs and pupae of arthropods, as well as larvae of Coccidae and Diaspididae, are attacked differently, because initially the predator examines them slowly with the tips of its mouthparts, and then pierces the cuticle in several locations. Characteristically, only one of the jaws is inserted into the prey's body, while the other manipulates the prey (Canard and Duelli 1984; Milbrath et al. 1993).

Hagen et al. (1976a) suggest that the composition of the prey's cuticle or cuticular waxes may be important for inducing the insertion of the mouthparts. However, Hagen and Tassan (1965) showed that *C. carnea, s. lat.* larvae attempt to probe any projecting object.

24.3.2.3 Prey Consumption

As mentioned earlier (Section 24.3.1), secretions from the salivary and perhaps maxillary glands are injected into the body of the prey during and after capture, while their tissues are disrupted with the jaws; these actions help liquefy the internal contents of the prey and make them available for ingestion through the feeding channel (see Cohen 1998). Larvae tend to exhaust the contents of small prey before abandoning the carcass; however, after apparently consuming the entire contents, larvae may continue to manipulate the prey.

The duration of consumption depends on the size of the predator, size of the prey, and the level of larval hunger (e.g., Canard and Duelli 1984; Milbrath et al. 1993). For example, first, second, and third instars of *C. carnea, s. lat.* required 185, 130, and 80 sec to eat one egg and 13, 8, and 3 min to consume a caterpillar of *Prays oleae* Bern., respectively (Alrouechdi 1981). In a comparative study, first instars of *Ch. slossonae,* a fairly large-bodied species, consumed twice as many aphids in a 45-min period than did those of its close relative, *Ch. quadripunctata* (Milbrath et al. 1993). Moreover, it did so within an equal or shorter period of time. Notably, the third instars exhibited no interspecific differences in feeding efficiency.

24.3.2.4 Cleaning and Resting

After feeding, larvae often, but not always, clean their mouthparts by rubbing them against each other and/or the substrate. Among trash carriers that have been studied, feeding can frequently be followed by camouflaging behavior, that is, loading trash on their dorsa (Tauber et al. 1995b; Mantoanelli and Albuquerque 2007). Apparently, after ingesting sufficient food, the larvae tend to "rest"—they become inactive and assume a characteristic posture, with their jaws closed and perhaps touching the substrate, antennae and labial palpi extended forward, foretibiae nearly parallel to the axis of the body, and the anal papilla attached to the substrate (Canard and Duelli 1984). This posture is maintained until larvae resume searching for new prey or begin spinning a cocoon.

24.3.3 Adult Digestive System

The chrysopid digestive tract undergoes some significant changes during metamorphosis from larva to adult. For example, adults have a mouth opening through which food is ingested; the stomodaeum becomes differentiated into a crop diverticulum and proventriculus; and the proctodaeum opens for the passage of food residues. In general, the digestive systems of chrysopid adults share a common anatomical plan (shown schematically in Figure 24.4b). They also vary in certain features depending on the category of adult food habits (see Section 24.2.1). Below we discuss the commonalities; this discussion is based on the detailed work of Withycombe (1925), Killington (1936), Bitsch (1984), and Woolfolk et al. (2004). Subsequently, Sections 24.3.3.2 and 24.3.3.3 deal with the features of the predaceous and nonpredaceous adults.

24.3.3.1 Internal Anatomy

At the anterior end, the preoral cavity is flanked by a pair of mandibles; it receives secretions from two sets of salivary glands—(a) the labial glands, formed by several secretory tubes lying in the prothorax, united to a common duct within the head and which opens at the base of the labium (Figure 24.4a), and (b) the mandibular glands, which are located on the sides of the head and open at the base of the mandibles. The mouth entrance leads to a relatively large oral cavity, lined with chitinous tissue. The stomodaeum consists of a muscular pharynx in the head, coated with a chitinous layer (very similar to that of the larvae). The esophagus begins in the head, extends through the thorax, and expands considerably in

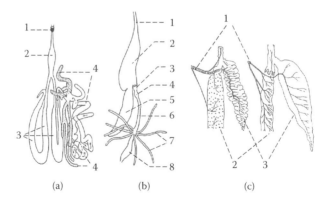

FIGURE 24.4 Digestive system of chrysopid adult. (a) Labial glands of *Chrysopa perla* (1, salivary pump; 2, median duct; 3, lateral duct; 4, secretory regions). (b) Schematic diagram of most of the digestive system of *Chrysoperla carnea s. lat.* (1, esophagus; 2, crop; 3, proventriculus; 4, valvula cardiaca; 5, crop diverticulum; 6, mesenteron; 7, Malpighian tubules; 8, proctodaeum). (c) Tracheal trunks and tracheoles associated with crop diverticulum in a glyco-pollenophagous species, *Anisochrysa prasina* (left) and a predaceous species, *Chrysopa walkeri* (1, tracheal trunk; 2, mesenteron; 3, crop diverticulum). (From Sulc, K., *Sber. K. Böhm. Ges. Wiss.*, 1914, 1, 1914 (a); Ickert, G., *Ent. Abh. Mus. Tierk. Dresden*, 36, 123, 1968 (b); With kind permission from Springer Science+Business Media. *Biology of Chrysopidae*, Feeding habits, 1984, 76–92, M. M. Principi and M. Canard, The Hague: Dr. W. Junk Publishers (c).)

the first and second abdominal segments, to form the crop. The entire foregut is lined with chitin, and the interior of the crop has large, sclerotized "teeth". At its distal end, the esophagus has an extension that occupies half the length of the abdomen—the crop diverticulum. Food passes through both the crop and its diverticulum, and as previously described (Section 24.2.2.2), the diverticulum (in glyco-pollenophagous species) may contain numerous yeast cells.

Apparently, some digestion occurs in the crop and diverticulum, but absorption is probably minimal (Ickert 1968; Woolfolk et al. 2004). The esophageal wall, including the crop, is composed of a thin epithelium and surrounded by circular muscles; the muscles are responsible for the peristaltic contractions that move the food posteriorly, into the diverticulum. The circular muscles of the diverticulum contract in the reverse direction and push the food anteriorly, back to the crop, or more frequently, to the proventriculus, which opens near the anteroventral end of the diverticulum (approximately at the third abdominal segment). The proventriculus is a complex, funnel-shaped, chitinous structure. It has a thick anterior margin from which long spines and "hairs" protrude into the lumen. Its inner wall has six chitinized grooves or eight lips with prominent, backward-directed spines, as well as other spines, hairs, and folds. The outer wall has circular and longitudinal muscles. The posterior, narrow end of the proventriculus protrudes into the midgut, forming the cardiac valve.

The midgut is a long and straight tube whose wall is formed by a single layer of epithelium. This layer is lined with the peritrophic membrane (a continuation of the chitinous intima of the proventriculus); food does not come into direct contact with the enteric epithelium. The existence of the peritrophic membrane in the midguts of adult lacewings, which was questioned by Bitsch (1984), was recently confirmed in *C. rufilabris* via scanning electron microscopy (Woolfolk et al. 2004). However, it was not found in the midguts of young (7-day-old) *C. carnea, s. lat.* adults (Chen et al. 2006). The columnar cells of the mesenteric epithelium are involved in both the secretion of enzymes for digestion and the absorption of nutrients, similar to that which occurs in the larvae.

The passage from the midgut to the proctodaeum is marked by a narrowing of the gut and by the openings of the eight Malpighian tubules, which are directed anteriorly and then flex backward (Withycombe 1925; Killington 1936; Woolfolk et al. 2004). Some of the tubules may have their posterior ends connected loosely with the proctodaeum, just anterior to the rectum. The proctodaeum is similar in structure to the esophagus, with its wall surrounded by circular muscles and a thin epithelium, but its inner cuticle is full of small spines (Woolfolk et al. 2004). At the end of the proctodaeum, in the dilated region of the rectum, there is a rectal chamber with a very thin epithelium, surrounded by circular and longitudinal muscles and six hemispherical rectal glands. The rectum is probably involved in the absorption of water and some small molecules. The posterior region of the rectum is narrow, and the anus, surrounded by circular muscles, opens in the membrane between the anal plates of the 10th tergite.

24.3.3.2 Anatomical Modifications Associated with Glyco-Pollenophagous and Prey-Based Diets

24.3.3.2.1 Nonpredaceous Adults

The digestive systems of nonpredaceous adults (those that eat pollen and/or sugary solutions) are generally considered to have a number of distinctive features—relatively small, symmetrical mandibles, without incisors (Figure 24.5a); spoon-shaped lacinia; and large tracheal trunks extending to their crop diverticula. The association of some of these attributes with glyco-pollenophagous feeding has support (Canard et al. 1990; Stelzl 1992; Canard 2001), but for others (especially the mouthparts), the evidence is not always strong. For example, the small size of the mandibles seems to be very consistent (Canard 2001). However, there are notable exceptions. That is, species in several glyco-pollenophagous genera have relatively large, asymmetric mandibles, the left one of which has a substantial left tooth similar to that found in prey-feeding adults (e.g., see Brooks and Barnard 1990; Canard 2001; Tauber 2010).

One trait that appears to be relatively distinctive for glyco-pollenophagous adults is an association with symbiotic yeasts. The symbionts typically inhabit the crop diverticulum, which, in glyco-pollenophagous adults, is enlarged, highly folded, and extensively tracheated (Hagen and Tassan 1966; Hagen et al. 1970; Canard et al. 1990; Woolfolk et al. 2004; Gibson and Hunter 2005). Most notably, the tracheal trunks are

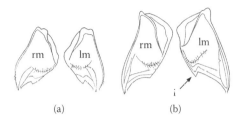

(a)   (b)

FIGURE 24.5   Mandibles of adult Chrysopidae: (a) glyco-pollenophagous habit, *Mallada prasinus* (Burmeister); (b) predatory habit, *Chrysopa perla* (L.) (i, incisor; rm, right mandible; lm, left mandible). (From Stelzl, M., In *Current Research in Neuropterology. Proceedings of the Fourth International Symposium on Neuropterology*, ed. M. Canard, H. Aspöck, and M. W. Mansell, 341–7. Toulouse, France, 1992. With permission.)

greatly enlarged (apparently to supply oxygen to the yeasts). These features are not found in predaceous adults (see Section 24.3.3.3).

In some nonpredaceous taxa, there are unusual exceptions to the generalizations above. For example, adults of *Hypochrysa elegans* (Burmeister) in the subfamily Nothochrysinae feed exclusively on pollen. The adults have large jaws, a diverticulum with small tracheal trunks, and no yeast symbionts. These attributes are consistent with a unique digestive process, not studied to date and well worth exploring (Canard et al. 1990). [Note: Adults in four other genera of Nothochrysinae, including *Nothochrysa*, are glyco-pollenophagous (Toschi 1965; Canard et al. 1990; Canard 2001), and at least two species of *Nothochrysa* take some prey (Principi and Canard 1984). Whether these species have the same anatomical and digestive traits as those of *H. elegans* is an intriguing question.]

24.3.3.2.2 Predaceous Adults

Lacewings whose adults are predaceous represent a small minority of the known chrysopid taxa. According to existing evidence, only 3 of the 75 currently recognized genera, that is, *Anomalochrysa*, *Atlantochrysa*, and *Chrysopa* (all in the tribe Chrysopini), show this kind of feeding behavior (Brooks and Barnard 1990). The first two of the above genera are restricted to islands, and their digestive systems have not been studied. In contrast, the agriculturally important genus *Chrysopa* is widely distributed in the Holarctic region, and the digestive systems of several of its species have been examined. Although generally considered carnivorous, *Chrysopa* spp. are not restricted to eating prey; they may also feed on pollen, yeasts, fungal spores, and honeydew.

The digestive systems of *Chrysopa* species exhibit several features that are assumed to be adaptations for a predaceous lifestyle. First, they have relatively large, asymmetric mandibles, with a robust tooth (incisor) on the left one, and a molar-like chewing surface (Figure 24.5b; Stelzl 1992; Canard 2001). Second, the lacinia is not modified (spoon-shaped) for holding pollen (Canard 2001). Third, although the diverticula are large, the associated tracheal trunks are much narrower and probably provide less oxygen than those of the nonpredaceous species (Figure 24.4c). And, fourth, no yeasts have been found inside the diverticula of the many *Chrysopa* species that have been examined (Hagen et al. 1970; Canard et al. 1990). Whether these attributes are adaptations associated with predation is not verified; some are also found in taxa that are believed to be glyco-pollenophagous (e.g., see Tauber 2010).

24.3.3.3 Association with Symbiotic Yeast

The pioneering study of Hagen et al. (1970) identified the symbiotic yeasts in the *C. carnea, s. lat.* diverticulum as *Torulopsis* sp. (now *Candida* sp.). Since then, several additional species in the genera *Metschnikowia* and *Candida* were identified from *Chrysoperla comanche* (Banks*)*, *C. carnea, s. lat.*, and *C. rufilabris* (Figure 24.6) (Woolfolk and Inglis 2004; Woolfolk et al. 2004; Suh et al. 2004). They probably are quite general among glyco-pollenophagous chrysopid adults (Johnson 1982 quoted by Gibson and Hunter 2005).

FIGURE 24.6 Convoluted inner surface of the crop diverticulum of *Chrysoperla rufilabris* adult, containing cells of the yeast *Metschnikowia pulcherrima* (arrows) (scanning electron micrograph; bar, 1000 μm). (From Woolfolk, S. W., et al., *Ann. Entomol. Soc. Am.*, 97, 796, 2004. With permission.)

The symbiosis between chrysopids and yeasts is intriguing, but not yet fully understood. It has long been held (and supported with strong morphological and experimental evidence) that they provide some essential nutrients that are absent or that occur at very low concentrations in the carbohydrate-rich diet (Hagen et al. 1970; Hagen and Tassan 1972; Woolfolk et al. 2004). On the basis of their detailed morphological study, Woolfolk et al. (2004) suggested that the most likely scenario is that yeast cells proliferate mainly within the diverticulum, and that yeast cells or the nutrients that they produce are transferred to the midgut where they are digested and absorbed. Apparently, the complex proventriculus is involved in controlling the transfer of the yeast cells to the midgut.

Recently, Gibson and Hunter (2005) questioned the nutritional role of yeasts, largely because they were unable to replicate some of the earlier experimental results of Hagen et al. (1970); these questions have not been resolved. [Note: Among other possibilities, Gibson and Hunter and Hagen et al. could have obtained their disparate results because they studied taxonomically different entities within the *C. carnea* species complex (see Section 24.1.1), and/or different species of yeasts. Hagen et al. collected their lacewings from the Central Valley of California, whereas Gibson and Hunter obtained theirs from an insectary culture in California. This uncertainty underscores the importance of pinned and frozen voucher specimens (see Section 24.1.2).]

The mode whereby lacewing adults transmit symbiotic yeasts is not well understood. Hagen et al. (1970) suggested that they are acquired only after adult emergence, via ingested food or trophallaxis. Recent discoveries indicate that vertical transmission (mother to offspring) also may occur, for example, active yeasts were found on chrysopid eggs and inside larvae (Woolfolk and Inglis 2004; Gibson and Hunter 2005). Nevertheless, it is not clear how chrysopid larvae might acquire yeast cells from the surface of the egg; it is unlikely that they ingest them during hatching because the egg is ruptured by a sclerotized structure, the "egg burster", and the larval mouth is closed at the time of hatching or very soon afterward (Withycombe 1925); larvae do not ingest food into the mouth and they do not "chew" their way out of the egg.

24.3.4 Adult Feeding Behavior

Knowledge of how adult lacewings find, recognize, and ingest food is largely derived from studies from very few species (mostly glyco-pollenophagous ones). We introduce the topic here with a discussion of chrysopid postemergence movement and then we consider how adults find food. It is noteworthy that when lacewings search for habitats and food, oftentimes they are simultaneously searching for suitable oviposition sites. In general, the studies include only the first two steps in the typical prey selection process (Hagen 1986), that is, habitat location and food location within the habitat.

24.3.4.1 Postemergence Movement

Chrysopid adults are mainly nocturnal creatures. During the day, they generally remain inactive on the lower surface of leaves (Smith 1922). In *C. carnea, s. lat.*, activity begins at dusk, reaches its peak during the early hours of darkness, and ceases at dawn. Sometimes flight occurs during the day, but such flights are usually of short duration, and only when individuals are disturbed (Duelli 1984a). There are some exceptions, such as *H. elegans*, which is active mainly during the day (Duelli 1986).

Lacewings, with their relatively large wings, may sometimes appear slow in flight. Nevertheless, some species are capable of swift, directed flight and quick maneuvers (e.g., Miller and Olesen 1979). During flight, they can be carried by wind, especially when they are more than 5 m above ground level. On the basis of laboratory and field studies, Duelli (1980a,b, 1984a,b) concluded that, during habitat finding and food finding, *C. carnea, s. lat.* displays three types of flight: migratory, appetitive downwind, and appetitive upwind. The first, migratory flight, involves long flights by newly emerged adults (the first two nights after emergence). During these flights, adults aided by wind can disperse up to 300 km (Chapman et al. 2006). According to Duelli (1980a), these migratory flights are closely associated with the postemergence period of sexual immaturity; they occur regardless of the presence of food nearby.

About 3 days after emergence, adults undertake short flights and begin responding to volatile chemicals (infochemicals) associated with food. Then, as they become sexually mature, adults tend to fly downwind approximately 1–5 m above the vegetation. When they enter an odor plume from a food source, they descend and land on the vegetation. If food is not discovered, they initiate upwind movement and approach the odor source slowly, in a series of short flights.

Even after mating (third and fourth nights, postemergence) and initiating oviposition (fifth night), *C. carnea, s. lat.* adults continue to move. The entire population proceeds downwind night after night. Duelli (1984b) interpreted this nomadic behavior as a means of dispersing offspring over a large area.

For species of lacewings other than *C. carnea, s. lat.*, there is very little information about flight behavior; what is known is largely inferred from trap samples. For example, in California (United States of America), large numbers of *Chrysopa nigricornis* Burmeister adults came to sticky traps in monocultures of alfalfa. This species usually inhabits trees and because no eggs, larvae, or adults were found in the alfalfa, such captures may indicate migratory flight (Duelli 1984a).

Infochemicals associated with the habitat and food itself (honeydew or prey) are known to influence chrysopid flight behavior (e.g., Hagen 1986, 1987; Szentkirályi 2001). Among these, volatile secondary metabolites from plants have an especially important role that has received considerable attention. These compounds may act directly on the predator (synomones) or indirectly, via the feces or honeydew from the prey (kairomones). In addition, pheromones produced by prey (acting as kairomones) and volatiles emanating from plants in response to prey damage (synomones) have also been identified as attractive to lacewings.

24.3.4.2 Attraction to Plant Volatiles

Although the attraction of adult lacewings to plant-based volatiles has been long known (Frost 1927, 1936), the compounds involved in the attraction were not isolated or identified until the 1960s. Currently, a large number of attractive compounds has been recognized, and they are perceived by both predaceous and nonpredaceous chrysopid adults in a wide variety of taxa: for example, neomatatabiol and other cyclic monoterpene alcohols, attractive to *Chrysopa formosa* Brauer and *Ch. pallens* (Ishii 1964; Hyeon et al. 1968; Sakan et al. 1970); terpenyl acetate, attractive to *Ch. nigricornis* and *C. carnea, s. lat.* (Caltagirone 1969); methyl eugenol, attractive to *Mallada basalis* (Walker), *Chrysopa* sp., and *Ankylopteryx exquisita* (Nakahara) (Suda and Cunningham 1970; Umeya and Hirao 1975; Pai et al. 2004); and a number of other compounds attractive to *C. carnea, s. lat.*, such as caryophyllene (Flint et al. 1979), eugenol (Wilkinson et al. 1980), 2-phenylethanol (Zhu et al. 1999, 2005; Zhu and Park 2005), (Z)-3-hexenil-acetate (Reddy et al. 2002), and phenylacetaldehyde (Tóth et al. 2006). These infochemicals appear to benefit not only the receiver (green lacewings) but also the producer (plants), because plant damage presumably is reduced when predators that feed on herbivores are attracted into the habitat.

The above plant compounds were regarded as synomones for lacewing habitat location. However, they also may be assimilated by phytophagous arthropods and eliminated in their feces, honeydew, or sex pheromones; in this sense they also may function as kairomones when they attract lacewings. Such a dual role has been confirmed only for neomatatabiol, a synomone produced by the plant *Actinidia polygama* (Ishii 1964; Hyeon et al. 1968; Sakan et al. 1970). This product was identified as a component of aphid sex pheromones and also shown to be attractive to males of several species of green lacewings (see below). [Note: The chemistry of the volatile was recently examined by Hooper et al. (2002); the main attractive ingredient for lacewings was (1R,4S,4aR,7S,7aR)-dihydronepetalactol.]

24.3.4.3 Response to Prey-Associated Volatiles

24.3.4.3.1 Long-Distance Attraction

Hagen et al. (1971) were pioneers in investigating the responses of chrysopids to volatiles presumably present in homopteran honeydew (kairomones for habitat location). They developed artificial honeydew containing hydrolyzed protein and used it to attract and arrest *C. carnea, s. lat.* in the field. In later studies, an active component of this artificial honeydew was identified as tryptophan, more specifically (but perhaps not exclusively) a volatile product resulting from the oxidation of this essential amino acid—3-indol-acetaldehyde (Hagen et al. 1976b; van Emden and Hagen 1976; Dean and Satasook 1983). However, in a subsequent study, *C. carnea, s. lat.* was more attracted to hydrolyzed or oxidized tryptophan, and less so to 3-indol-acetaldehyde (Dean and Satasook 1983). [Note: It is possible that Dean and Satasook (who worked with populations from England) and Hagen et al. (who collected their lacewings in California) obtained disparate results because the two studies examined different taxonomic entities within the *C. carnea* species complex (see Section 24.1.1). This uncertainty underscores, once again, the importance of pinned and frozen voucher specimens (see Section 24.1.2).]

Later, Harrison and McEwen (1998) claimed that volatiles are not produced in the acid hydrolysis of tryptophan; they discussed alternatives whereby spray applications of artificial honeydew could attract lacewings, including infochemicals emitted from the host plants, as a result of damage from the application of this artificial honeydew, or from the breakdown of tryptophan by microfauna.

Despite the above problems, protein hydrolysates (artificial honeydews) have been used to attract and arrest lacewings, especially those with a glyco-pollenophagous diet. In some cases, they resulted in larger lacewing populations and, consequently, pest reduction (Butler and Ritchie 1971; Hagen et al. 1971, 1976b; Ben Saad and Bishop 1976a,b; Neuenschwander et al. 1981; Liber and Niccoli 1988; Evans and Swallow 1993; McEwen et al. 1994). However, studies have not documented specific volatile compounds from natural honeydews that act alone as long-distance attractants for adult lacewings. Nevertheless, 3-indol-acetaldehyde is known to act in concert with plant-emitted synomones (see Section 24.3.4.4). Undoubtedly more work on the chemistry and mode of action of volatiles from honeydew would be very useful.

Adult lacewings in an array of diverse species respond to the components of aphid sex pheromones, such as (1R,4aS,7S,7aR)-nepetalactol (*Chrysopa phyllocroma* Wesmael, *Ch. pallens*, *Ch. formosa*, and *Ch. oculata*) and (1R,4S,4aR,7S,7aR)-dihydronepetalactol [*Ch. pallens*, *Nineta vittata* (Wesmael), and *Peyerimhoffina gracilis* (Schneider)] (Boo et al. 1998, 2003; Hooper et al. 2002; Zhu et al. 1999, 2005). These responses led Boo et al. (1998) and Zhu et al. (1999, 2005) to suggest that lacewings could use aphid pheromones as kairomones to locate their prey. This notion has been refuted (Hooper et al. 2002, Zhang et al. 2004), and it was concluded that lacewing attraction to the aphid pheromones results from a coincidence in the pheromone chemistry of the aphids and lacewings (Zhang et al. 2004). Although unexpected, the similarities in the lacewing and aphid pheromones are not completely surprising given the likely flow of the constituent plant compounds through the food chain. However, many questions remain about lacewings using aphid volatiles (e.g., pheromones) to locate prey in the field.

Finally, lacewing adults may be attracted to volatiles that plants produce in response to damage by phytophagous arthropods (synomones for habitat location). However, only one such compound has been identified, methyl salicylate. When attacked by phloem-feeding homopterans, a large number of plant species, including at least 13 cultivated crops, emit this volatile chemical (James 2003; James and Price

2004). Its attractiveness was demonstrated for both male and female *C. carnea, s. lat.* (Molleman et al. 1997), *Ch. nigricornis* (James 2003; James and Price 2004), and *Ch. oculata* (James 2006).

24.3.4.3.2 Short-Distance Responses

At short range, chrysopids perceive volatiles emanating from homopteran honeydew, and they respond by arresting their movement and intensifying their search behavior, feeding, or oviposition (Hagen 1986). These effects were demonstrated for *C. carnea, s. lat.*, before and after females established antennal contact with the honeydew of the scale *Saissetia oleae* (Olivier) (McEwen et al. 1993), and for *Chrysoperla nipponensis* (Okamoto) in the presence of honeydew of the aphid *Toxoptera aurantii* (Boyer) (Han and Chen 2002). Further studies are needed to determine whether the volatile compounds that are responsible for the short-distance effects are the same as those that attract lacewings at long distances.

Attraction of adult lacewings to volatiles found in arthropod feces has received little attention. In the only example found, Reddy et al. (2002) observed that the volatiles dipropyl-disulfide, dimethyl-disulfide, allyl-isothiocyanate, and dimethyl-trisulfide, in feces of *Plutella xylostella* (L.) caterpillars, attract both male and female *C. carnea, s. lat.* However, whether these kairomones act as stimuli to locate the habitat (long distance), food (short distance), or both, is unknown.

24.3.4.4 Habitat and Food Finding: An Integrative Process

Adult lacewings, predaceous or glyco-pollenophagous, locate their habitat and food by anemochemotactic attraction, induced by odor plumes of plant and prey origin (synomones or kairomones). Hagen (1986) showed that the attraction of *C. carnea, s. lat.* to artificial honeydew varies with the phenological status of the plant on which it is applied. He considered this finding, as well as lacewing attraction to plant volatiles, such as caryophyllene (Flint et al. 1979), and concluded that *C. carnea, s. lat.* must first receive a volatile signal (synomone) from the plant in order to respond to the chemical stimulus (3-indole-acet-aldehyde), which is associated with prey (honeydew). More recently, the discovery of several other plant synomones attractive to *C. carnea, s. lat.* [such as 2-phenylethanol emanating from leaves of corn and alfalfa (Zhu et al. 1999, 2005), (Z)-3-hexenil-acetate released by cabbage leaves (Reddy et al. 2002), and phenylacetaldehyde, a volatile present in various plants (Tóth et al. 2006)] appears to support Hagen's (1986) hypothesis.

Interactions and synergistic effects between classes of compounds have been shown for many species of insects, including chrysopids [e.g., between methyl salicylate (a synomone induced by herbivore feeding) and (1R,2S,5R,8R)-iridodial (a pheromone) (see Zhang et al. 2004)]. Therefore, it appears that lacewings may use a series of compounds from different categories (e.g., synomones and kairomones) and perhaps nonchemical, prey-related stimuli to find their habitats, food, oviposition sites (= larval feeding sites), and mates (Han and Chen 2002; Reddy et al. 2002; Zhang et al. 2004, 2006).

24.4 Effects of Food on Lacewing Performance

Universally, food is a major factor in shaping the survival, development, and reproductive success of animals. The role that food plays in these life history processes has been well studied in herbivorous insects; in contrast, predaceous insects are seldom examined, largely because of difficulties in observing them feed in nature, and in rearing them under experimental conditions. Nevertheless, there are some notable exceptions, including investigations with lacewings.

The effects of larval food on life history traits can be either overt (e.g., death versus survival) or subtle (e.g., altered rates of growth and/or development, size, and/or performance) (see review: Principi and Canard 1984). Moreover, these effects can be expressed in the short term (e.g., during the immature stages; covered in Section 24.4.1), in the intermediate term (during the subsequent adult stage; covered in Section 24.4.2), or perhaps even in a later generation (an unexplored, but potentially very fruitful topic for future research).

24.4.1 Food, Development, and Survival of Immature Stages

Numerous examples of dietary effects on lacewing immatures are in the literature. For example, when larvae of the neotropical species *Ce. cubana* were fed on five types of arthropods (species of mites, mealybugs, or whiteflies), larval and pupal developmental times ranged from 25 to 47 days and mortality varied from 2% to 80%, depending on the prey (Muma 1957). Another study, this one with *C. rufilabris*, showed similar ranges of variability (Chen and Liu 2001). One species of aphids, *Lipaphis erysimi* (Kaltenbach), was shown to be inadequate; all individuals that received it died before emergence. Two other species of aphids, *A. gossypii* and *Myzus persicae* (Sulzer), were equally favorable; development (hatching to adult emergence) required only 18 to 19 days and survival was 100%.

Even the performance of *C. carnea, s. lat.*, generally well recognized as polyphagous, is influenced by the prey species its larvae encounter; in some cases the impact was dramatic (Awadallah et al. 1976; Obrycki et al. 1989; Balasubramani and Swamiappan 1994; Osman and Selman 1996; Liu and Chen 2001; Matos and Obrycki 2006). Such a pattern does not appear to hold for *C. externa*, a neotropical species for which this topic has been well explored. The results, mainly from Brazil (see Table 24.2), show that a wide spectrum of prey is similarly adequate for *C. externa*'s development and survival.

The type of prey may also affect larval weight gain. For example, when *Ch. perla* larvae were fed 11 different species of aphids *ad libitum*, they produced cocoons weighing between 10 and 15 mg, depending on the prey (Principi and Canard 1984). Osman and Selman (1996) observed a less drastic but significant effect with *C. carnea, s. lat.*; in their study, the weight of cocoons varied between 9 and 12 mg, depending on which of six species of prey the larvae received.

The nutritional quality of prey not only varies among prey species, but notably, it can also show intraspecific variation. Prey may have high or low nutritional value for the predator, or even be toxic, depending on the prey's host plant. Such variation can be expressed in developmental and reproductive rates and also in rates of mortality. This aspect of chrysopid nutrition has received little attention, but Giles et al. (2000) demonstrated that when *C. rufilabris* larvae fed on the aphid *Acyrthosiphon pisum* (Harris), they had differential rates of development, depending on the aphids' host plant (alfalfa or fava bean).

In addition to the type and quality of prey, its quantity can also be decisive for larval development. Under conditions of low prey availability, the larvae often can spin cocoons, but they weigh less than usual and, consequently, the adults are small and reproductively deficient; in some cases, mortality within the cocoon can be considerable (Principi and Canard 1984).

It is noteworthy that chrysopid larvae can show some resiliency when prey levels are low during certain periods of their lifetimes. For example, when first and second instars of *C. carnea, s. lat.* were fed suboptimal amounts or prey and then third instars were provided optimal numbers of prey, there was considerable compensation for the low intake during early life. That is, although overall developmental times were extended by food deprivation early in life, total dry weight gains and overall conversion of food to body mass were not affected. In contrast, high intake during early instars, or by the adults, did not compensate for low intake during the third instar (Zheng et al. 1993a). Whether such attributes apply to species in other chrysopid taxa is an intriguing question.

24.4.2 Interactions with Host Plant of Prey

Plants can influence the performance of predators in multiple ways (Barbosa and Wratten 1998). Apart from serving as the primary source of nutrients for nonpredaceous adults or as an additional food source for larvae and predaceous adults (see Section 24.2.1), they also release volatile chemicals that help adults locate prey or oviposition sites (see Section 24.3.4). Other factors, related to the structure or morphology of the plant, may affect the efficiency of larval searching and prey capture.

The surface of the prey's host plants may have attributes that promote or hinder larval movement. In this respect, the presence, type, and density of trichomes are important. For example, first and second instars of *C. carnea, s. lat.* search for prey more rapidly on cotton than on the more hairy tobacco leaves. Moreover, on various types of tobacco leaves, the speed of larval movement is inversely proportional to the density of glandular trichomes (Elsey 1974). The same pattern seems to hold for *C. rufilabris* larvae on cotton plants; here, the lower the density of hairs on the leaves, the greater the efficiency of

TABLE 24.2

Developmental Time (Mean, Days) and Survival (%, between Parentheses) of the Three Instars, Prepupa (PP), and Pupa (P) of *Chrysoperla externa* under Different Prey Regimens (25 ± 1°C)

Prey Species	1st	2nd	3rd	PP	P	Total	Reference
Hem.: Aleyrodidae							
Bemisia tabaci (3rd and 4th instar nymphs)	4.1	3.7	6.0	3.5	6.4	23.7	Silva et al. (2004a)
	(100)	(100)	(97.2)	(97.2)	(97.2)		
Hem.: Aphididae							
Aphis gossypii (3rd and 4th instar nymphs)	3.5	3.0	3.8	3.0	7.0	20.3	Santos et al. (2003)
	(100)	(100)	(100)	(100)	(96.0)		
Cinara pinivora + C. atlantica (nymphs)	4.0	3.0	3.9	---- 11.1 ----		22.0	Cardoso and Lazzari (2003)
	(95.0)	(100)	(94.8)	(100)			
Rhopalosiphum maidis (3rd and 4th instar nymphs)	2.4	3.0	3.3	3.4	8.9	21.0	Maia et al. (2004)
Schizaphis graminum (3rd and 4th instar nymphs)	4.0	3.3	3.5	4.1	7.4	22.3	Fonseca et al. (2001)
	(100)	(100)	(100)	(100)	(100)		
Hem.: Pseudococcidae							
Dysmicoccus brevipes (nymphs and adults)	4.2	3.2	5.4	2.0	9.0	23.8	Gonçalves-Gervásio and Santa-Cecília (2001)
Lep.: Gelechiidae							
Sitotroga cereallela (eggs)	3.0	2.6	3.7	3.0	7.0	19.3	Costa et al. (2002)
	------ (90.2) ------		(90.2)	(90.6)			
Lep.: Noctuidae							
Alabama argillacea (eggs)	3.5	3.0	3.9	3.0	7.9	21.3	Figueira et al. (2000)
	(100)	(100)	(100)	(100)	(100)		
A. argillacea (1st instar larvae)	3.7	3.0	5.0	3.1	5.9	20.7	Silva et al. (2002)
	(100)	(100)	(90.0)	(100)	(88.9)		
Spodoptera frugiperda (eggs)	3.5	2.5	3.8	---- 10.1 ----		19.9	Auad et al. (2003)
	(55.0)	(100)	(95.0)	(100)			
S. frugiperda (1st instar larvae)	4.4	3.9	6.3	---- 9.9 ----		24.5	Auad et al. (2003)
	(93.0)	(86.0)	(100)	(100)			
Trichoplusia ni (eggs)	3.5	3.0	3.8	---- 9.9 ----		20.2	Ru et al. (1975)
Lep.: Pyralidae							
Anagasta kuehniella (eggs)	3.6	2.5	2.9	---- 11.2 ----		20.2	de Bortoli et al. (2006)
	(96.7)	(100)	(100)	(86.2)			
Diatraea saccharalis (eggs)	3.9	2.8	2.9	---- 11.3 ----		20.9	de Bortoli et al. (2006)
	(96.7)	(100)	(100)	(75.9)			
Lep.: Noctuidae + Hem.: Aphididae (mixed)							
S. cereallela (eggs) + *Myzus persicae* (nymphs)	3.4	2.8	4.0	3.3	7.6	21.1	Albuquerque et al. (1994)
	(100)	(100)	(100)	(100)	(100)		

Note: Figures refer to the regimen of higher density of prey or best cultivar/species of the prey's host plant used in the studies in which more than one of these factors were tested; information about survival are available only in some of these studies.

predation. However, in this case, the first instar was affected more than the second; thus the impact of the trichomes may vary with the predator's developmental state (Treacy et al. 1987). The state of the crop (e.g., the accumulation of dust throughout the season) may also affect the impact of trichomes on lacewings (Obrycki and Tauber 1985).

The presence or absence of epidermal waxes can also alter larval movement. For *C. carnea, s. lat.*, an abundant layer (a "bloom") of epicuticular waxes on the leaf surface reduces larval mobility on cabbage leaves relative to that on leaves with thin layers of wax (see review: Eigenbrode 2003). Another way that host plants may influence larval lacewings is through their architecture, that is, the degree of complexity in structure (see review: Barbosa and Wratten 1998; also see Messina et al. 1997; Clark and Messina 1998).

24.4.3 Food and Reproduction

The quality and quantity of food available during the larval and adult stages can strongly affect the reproductive performance of chrysopids (reviews by: Principi and Canard 1984; Rousset 1984; see also Section 24.2). The effects are noted in several reproductive traits, for example, the ability to mate, length of the preoviposition and oviposition periods, daily rates of oviposition, fecundity, and fertility. In general, females require large amounts of energy for oogenesis and sustained oviposition, and in some cases, both sexes require food before mating (see below).

According to Rousset (1984), in the small number of species that have been studied, previtellogenesis may begin before female emergence; this process uses nutritional reserves accumulated during larval life. In the male, spermatogenesis was shown to occur during the larval stage and inadequate prey during this stage can result in the sterilization of the adult (see Canard 1970 for *Ch. perla*). Moreover, a nutritious adult diet may not compensate for food deficiencies experienced during the larval stage (Figure 24.7). For example, when *C. carnea, s. lat.* larvae were provided low amounts of prey, they gave rise to smaller and less fecund adults, relative to those whose larvae received abundant prey; fecundity remained low even when the adults had unrestricted access to food (Zheng et al. 1993b). For the neotropical species *C. externa*, adults that were reared on different species of prey appear to vary in reproductive performance (see Table 24.3), but the specific effects of prey have not been explicitly examined.

Only some of the metabolites accumulated during the larval stage are transferred to the adults; therefore, for sustained fecundity, females are highly dependent on external sources of nutrients. As seen above (Section 24.2.3), the nutritional requirements for initiating oviposition may vary among species; however, all species seem to require a supply of proteins for continued egg production (Rousset 1984, note his Table 17). For example, *C. carnea, s. lat.*, which is nonpredaceous in the adult stage, is

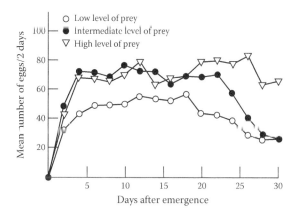

FIGURE 24.7 Mean oviposition, at intervals of 2 days, of *Chrysoperla carnea s. lat.* from larvae fed on three abundance levels of *Anagasta kuehniella* eggs. (From Zheng, Y., K. M. Daane, K. S. Hagen, and T. E. Mittler: Influence of larval food consumption on the fecundity of the lacewing *Chrysoperla carnea. Entomol. Exp. Appl.* 1993, 67, 9–14. Copyright Wiley-VCH Verlag GmbH & Co. KGaA. Reproduced with permission.)

TABLE 24.3

Reproductive Traits of *Chrysoperla externa* Females in Relation to the Prey Regimen Received during the Larval Stage (25 ± 1°C)

Prey Species	Preoviposition Period (Mean, Days)	Oviposition Period (Mean, Days)	Total Number of Eggs (Mean)	Reference
Hem.: Aleyrodidae				
Bemisia tabaci (3rd and 4th instar nymphs)	6.1	42.6	592.1	Silva et al. (2004b)
Hem.: Aphididae				
Aphis gossypii (3rd and 4th instar nymphs)	4.8	46.0	789.6	Santos et al. (2003)
Lep.: Gelechiidae				
Sitotroga cereallela (eggs)	4.6	46.0	667.5	Angelini and Freitas (2004)
Lep.: Noctuidae				
Alabama argillacea (eggs)	5.1	55.2	1020.3	Figueira et al. (2002)
Lep.: Pyralidae				
Anagasta kuehniella (eggs)	7.0	38.0	387.8	Boregas et al. (2003)

Note: Figures refer to the best prey regimen, including cultivar/species of host plant, used for feeding the larvae or the best adult diet in the studies in which more than one of these factors were tested.

autogenous; that is, females can mate and lay some eggs without postemergence feeding. However, they need to feed on pollen or honeydew to sustain a high level of egg production. Apparently, the neotropical *C. externa* (nonpredaceous adults) has the same requirements for sustained oviposition as *C. carnea, s. lat.*; females fed only carbohydrates (honey solution) oviposit a negligible number of eggs relative to the fecundity achieved when they have a protein source, such as pollen, soy, or yeast (Table 24.4). [Note: Under some circumstances, it appears that *C. externa* females may also require protein before mating (Tauber and Tauber 1974, as *Chrysoperla lanata* Banks).]

Chrysopa species, all of which have predaceous adults, show a much broader range of variation in postemergence dietary requirements for reproduction. Some are autogenous like the *Chrysoperla* species above; others differ and are anautogenous; that is, they need to ingest protein, for example, aphids, to start laying eggs (Tauber and Tauber 1974, 1987; Principi and Canard 1984; Jervis and Copland 1996). At one extreme, both females and males of *Ch. quadripunctata* mate without having fed on any protein, and some oviposition occurs when females receive a synthetic proteinaceous diet (Tauber and Tauber 1974). *Ch. oculata* and *Chrysopa coloradensis* Banks both produced eggs when provided a synthetic

TABLE 24.4

Reproductive Traits of *Chrysoperla externa* Females in Relation to the Diet Received by the Adult Stage (25 ± 2°C)

Diet	Preoviposition Period (Mean, Days)	Oviposition Period (Mean, Days)	Total Number of Eggs (Mean)	Egg Fertility (%)
Honey (40% solution)	8.8	59.8	22.3	98.9
Pollen	4.0	76.1	1742.4	95.6
Honey + pollen	3.2	100.5	1145.7	98.6
Honey + soybean protein	3.0	84.4	1985.4	98.9
Honey + yeast	3.0	81.2	2273.1	98.9

Source: Data from Ribeiro, M. J., et al., *Ciênc. Prát.*, 17, 120, 1993.
Note: Larval stage fed with eggs of *Anagasta kuehniella*.

proteinaceous diet, whereas *Ch. perla* and *Ch. nigricornis* require prey before ovipositing fertile eggs (Hagen and Tassan 1970; Philippe 1971).

Near the other extreme is *Ch. slossonae*, a specialist on a single aphid species (Tauber and Tauber 1987). Adults of this species reproduce successfully only if they have their specific prey. When provided other aphids, they will develop and lay eggs, but the eggs are not fertile. If these infertile pairs are then supplied their specific prey, fertile eggs are produced within 1 day. Whether the specific prey induces mating or some other crucial reproductive function (e.g., transfer of sperm to the spermatheca) is not known.

Females of *Ch. oculata* show geographic variation in their dietary requirements for reproduction. Those from western North America mated and had sustained oviposition when they received a synthetic proteinaceous diet (Hagen and Tassan 1970). In contrast, females from eastern North America required prey before mating and fertile oviposition (Tauber and Tauber 1974).

Although females have the larger protein requirement, males of some species may also require protein (or nutrients other than sugar) to achieve sexual maturation and successful mating (Principi and Canard 1984). For example, males of *Ch. perla* and *Ch. nigricornis*, both of which are predaceous in the adult stage, do not mate unless they feed on prey (Philippe 1971; Tauber and Tauber 1974). For *Ch. perla* males, proteinaceous food is necessary for the development of accessory glands, formation of spermatophores, and for the sperm to move to the seminal vesicles (Philippe 1971).

24.5 Prey Specificity—Its Stability, Underlying Mechanisms, and Evolution

As discussed above (Section 24.2.1), lacewings are often viewed as being generalist predators. They are very mobile, and they encounter many different prey species as they move, so polyphagy might be expected. However, studies have shown that lacewings vary greatly in the degree of their prey specificity. Furthermore, significant behavioral, physiological, and morphological characteristics are associated with diet range (Thompson 1951; Tauber and Tauber 1987; Canard 2001).

24.5.1 Degree of Prey Specificity

At one extreme, in terms of prey specificity, is *Ch. slossonae*, which in nature feeds on only one species of prey—*Prociphilus tesselatus* (Fitch), a robust woolly aphid that forms large, ant-tended, colonies on alder trees, *Alnus incana* ssp. *rugosa* (Eisner et al. 1978; Tauber and Tauber 1987; Tauber et al. 1993; Albuquerque et al. 1997). In contrast, *Ch. slossonae*'s sister species, *Ch. quadripunctata*, has a more general diet that is largely restricted to aphids. In nature it feeds on a large number of aphid species that occur on the leaves of a variety of tree species (Tauber et al. 1995b). Although these two sister species are reproductively isolated in nature, they hybridize under some conditions in the laboratory. Comparative studies of their ecophysiological and behavioral responses to food and studies of hybrid performance in relation to prey, demonstrated (a) the basis for a stable prey association and (b) how it can evolve in lacewings (see Sections 24.5.2 and 24.5.3).

Larvae of several species of *Italochrysa*, *Nacarina*, *Calochrysa*, and *Vieira* (tribe Belonopterygini) show similar types of morphological adaptations for feeding specialization (Principi 1946; New 1983, 1986; Tauber et al. 2006). This group of lacewings appears to be intimately associated with ants; the larvae have very dense coverings of hooked setae that retain protective packets on their dorsa. The range of ant species with which they are associated is completely unknown and an intriguing area for study.

Most *Chrysopa* species tend to feed only on aphids and could be termed "oligophagous", whereas species in other genera (e.g., *Chrysoperla*) seem to have broader prey ranges and take a taxonomically broad array of prey (i.e., prey from several arthropod orders). Among the species of lacewings that are considered "generalist predators" (e.g., *C. rufilabris*, *C. carnea, s. lat.*, and *Ch. oculata*), many show considerable variation in their performance according to the type of prey that the larvae feed on (Putman 1932; Hydorn and Whitcomb 1979; Obrycki et al. 1989). Some of the developmental and reproductive responses to various types of food were discussed above (Sections 24.4.1 and 24.4.3).

24.5.2 Prey Specificity—Determinants and Stability

Many specialized herbivores have been shown to have extensive suites of traits underlying their close associations with specific food (e.g., Tilmon 2008), but the *Ch. slossonae/Ch. quadripunctata* study was the first to demonstrate that food specialization in a predator can be based on a substantial number of morphological, behavioral, physiological, and phenological characteristics. Among the traits involved in the prey specialization are prey requirements for successful reproduction, female choice of oviposition site, egg and neonate size, larval mouthparts size, larval phototaxis, larval defensive behavior and morphology, and phenology (Tauber and Tauber 1987; Milbrath et al. 1993, 1994; Tauber et al. 1993, 1995a).

Most of the above traits were shown to have a genetic basis, and although phenotypic and genotypic variation was identified (Tauber et al. 1995a,b), the traits appear to be very stable. First, evolutionary change in these traits involves both benefits and costs; although the specialist and generalist can hybridize in the laboratory, hybrids performed poorly relative to the generalist and specialist (Albuquerque et al. 1997). Moreover, the specialist and generalist exhibit prezygotic and postzygotic reproductive isolating mechanisms (Tauber and Tauber 1987; Albuquerque et al. 1996). Indeed, a change in food association is believed to be the basis for the diversification and speciation of the generalist and specialist (Tauber and Tauber 1989; Tauber et al. 1993; Albuquerque et al. 1996).

24.5.3 Prey Specificity—Its Practical Importance

The evolution of specificity (whether food specificity, habitat specificity, or others) is considered a major factor in the evolutionary diversification and speciation of animals; a large body of literature deals with this topic (see Tilmon 2008; for chrysopids, see Tauber and Tauber 1989).

Apart from its significance to evolutionary biology, the finding of stability in insect prey specialization has particular importance for classical biological control, which involves the importation and release of nonnative species into a new environment. To protect vulnerable native, nonpest species, only natural enemies with very restricted host ranges are released for classical biological control. Moreover, before importation or release, it is required to demonstrate that the classical biological control agent will maintain its specific host association. The above studies, especially those with *Ch. slossonae*, provide a foundation for developing tests that assess the stability of food associations in predators.

24.6 Future: Suggestions for Basic and Applied Research

Despite much fine research on green lacewings around the world, the accumulated knowledge of their bioecology and nutrition continues to show large gaps. These gaps hinder the efficient use of the group in pest management. Below, we offer a few suggestions for future studies; with modest investment in effort, time, and resources, these research priorities would yield many benefits in the near future.

24.6.1 Expand the Arsenal of Chrysopid Species in Biological Control

24.6.1.1 Recommendation

In general, we recommend broadening the range of chrysopid taxa that are investigated in bioecological and nutritional studies.

24.6.1.2 Rationale

Most basic and applied research on the Chrysopidae has been based on a very small number of species in two or three genera—primarily *C. carnea, s. lat.* in the temperate regions and *C. externa* in the Neotropics. The vast majority of the approximately 1200 remaining chrysopid species worldwide remains totally unknown biologically. Of these, nearly 300 are in the Neotropics.

This overemphasis on a very restricted range of chrysopid species has two serious, negative consequences. First, much of the biological information that is now assumed to be applicable for green lacewings in general may need to undergo drastic changes. Second, the present narrow focus seriously limits the arsenal of potential natural enemies that can be used in suppressing arthropod pests in a broad array of agricultural and horticultural settings.

24.6.1.3 Specific Targets

Apart from sharing a common evolutionary history and a substantial suite of biological traits, both *C. carnea, s. lat.* and *C. externa* are excellent candidates for mass rearing and use in pest management (Albuquerque et al. 1994, 2001). However, these two species, although well suited for use in low-growing crops, such as vegetables and cotton, are not well adapted to orchards, parklands, or forests.

A large number of species in genera other than *Chrysoperla* (e.g., *Ceraeochrysa*, *Chrysopodes*, and *Leucochrysa*) occur naturally in forests and transitional habitats in the Neotropics; they appear better adapted to perennial plant communities. However, with few exceptions [mainly *Ce. cubana* and *Ceraeochrysa cincta* (Schneider)], the biology of these species is unexamined. Undoubtedly, among the remaining, largely unstudied genera, there are excellent candidates for use in pest management in a broad range of agroecosystems. Studies on the biology of species in these genera should be strongly encouraged and supported.

24.6.2 Improve Chrysopid Biosystematics

24.6.2.1 Recommendation

We recommend a strong emphasis on the biosystematics of the Chrysopidae. These studies should be broadly based, and include the following approaches: comparative adult and larval morphology, comparative biology, and molecular investigations. The systematics effort should (a) include the development of tools to aid in identification (e.g., illustrated keys, images of larvae and adults on the Internet) and (b) couple systematic studies with natural history observations wherever possible.

24.6.2.2 Rationale

As highlighted in Section 24.1.1, despite recent advances, the systematics and natural history of Chrysopidae continue to have serious gaps. The neotropical fauna is particularly problematic. It is difficult, if not impossible, to identify most neotropical species reliably, and the natural prey associations of most taxa are completely unknown. These deficiencies impose significant barriers to using chrysopids in basic and applied research (Tauber et al. 2000; Albuquerque et al. 2001; New 2001).

24.6.2.3 Specific Targets

Improved systematics is the key to broadening the arsenal of natural enemies that are available for use in biological control. Opportunities should be opened for young biosystematists to work with the group; jobs for systematists should be made available; and public collections should be ensured long-term administrative and fiscal support.

24.6.3 Research Priority: Nutrition and Chemical Ecology (Implications for Rearing)

24.6.3.1 Recommendation

We recommend comparative, in-depth research on several, well-targeted topics that deal with the nutrition and chemical ecology of chrysopids. Research on these topics could yield practical, short-term benefits.

24.6.3.2 Rationale

Given the dual goals of (a) increasing the arsenal of natural enemies for use in biological control and (b) basing biological control procedures on a reliable, predictive, and cost-effective footing, it is essential to increase the depth and breadth of research on chemical ecology and nutrition.

For example, the semisolid larval diet [developed by Cohen and Smith (1998)] and the research on which it was based represent important advances in understanding the nutritional physiology of *Chrysoperla*. It also provided standardized, economical procedures for rearing large numbers of vigorous larvae economically. Now, to benefit fully from these findings, it is essential to assess the diet's effectiveness for species in other taxa, and to explore in depth the biochemical processes involved in nutrient ingestion, digestion, and absorption.

As for the adult stage, although artificial diets have been used successfully for over half a century, many species do not reproduce under laboratory conditions. The examples are numerous (e.g., adults of *Chrysopodes*, *Leucochrysa*, and other smaller, more obscure neotropical genera). Moreover, the reasons for the problem are not clear. Thus, both fundamental and applied research on the nutritional and chemical ecology of adult chrysopids in diverse taxa is likely to yield excellent returns, for example, fine advances in expanding the arsenal of species available for use in biological control.

24.6.3.3 Specific Targets

Given that significant work, summarized in the sections above, has laid solid foundations, the following areas of research could lead to very practical findings in the near or intermediate term: (a) the biology and nutritional role of the symbiotic yeasts that occur in the adult digestive system; (b) studies on the nutritional value and neuroendocrine influence of pollen in chrysopid reproductive performance; and (c) biochemical research on chrysopid ingestion, digestion, and absorption, particularly the nature and role of extraoral secretions. Another, particularly difficult (and therefore longer-term) but potentially fruitful area of research is the chemical (nutritional and other) factors that stimulate reproduction (reproductive development, mating, and fertile oviposition) in chrysopids. All of the above areas have very strong implications for both basic and applied applications.

24.6.4 Research Priority: Seasonality

24.6.4.1 Recommendation

We recommend well-focused field and laboratory research to examine the seasonal cycles of neotropical chrysopids and to determine the underlying factors, especially the phenological association of chrysopids with prey and their host plants.

24.6.4.2 Rationale

In the temperate regions, where seasonal fluctuations in temperature and other environmental factors are large, the seasonality of insects has been studied extensively. These species have evolved adaptations that allow them to withstand long periods of dormancy (diapause and quiescence) (e.g., Tauber et al. 1986). In the tropics, the situation differs. Abiotic conditions in some, but not all, regions may be favorable for growth and development throughout the year (see Silva et al. 2007 for *Chrysopodes lineafrons* Adams and Penny), and seasonal variations in activity and abundance may be subtle. These fluctuations could have significant implications for biological control and pest management.

For example, in areas with seemingly favorable conditions all year round, field surveys show significant seasonal fluctuations in the abundance of some chrysopid species, such as *C. externa*, *Ceraeochrysa* spp., and *Leucochrysa* spp. (e.g., Gitirana-Neto et al. 2001; Souza and Carvalho 2002; Multani 2008). In some cases, the fluctuations over the year are quite dramatic. It is unknown whether these seasonal cycles involve the cessation or reduction of reproduction, short or long periods of dormancy, movement, or other altered behavior. Moreover, the factors responsible for the cycles remain unexplored.

24.6.4.3 Specific Targets

It is highly probable that seasonal changes in moisture, prey, or the prey's host plant are among the seasonal cues that tropical lacewings perceive. As seasonal factors, moisture and food may serve to delay or promote development or reproduction directly (Tauber and Tauber 1983; Tauber et al. 1998; Pires et al. 2000), or they may act as "token" stimuli for the induction of diapause leading to dormancy or movement (for *C. carnea, s. lat.*, see Tauber and Tauber 1992, and references therein).

Studies aimed at examining the role of these seasonal factors in the life histories of tropical chrysopids could have great basic and applied benefits in the near term. Although there are many others, two examples come to mind: (a) Understanding the seasonal synchrony between pests and their natural enemies is fundamental to developing reliable pest management procedures, and (b) knowledge of the seasonal changes in chrysopid dietary requirements can increase the efficiency of mass rearing and storage (e.g., Chang et al. 1995).

ACKNOWLEDGMENTS

Our work was supported, in part, by the National Science Foundation (DEB-0542373, MJT, CAT), the "Conselho Nacional de Desenvolvimento Científico e Tecnológico" (CNPq, Brazil, 475848/04-7 and 484497/07-3, GSA), Regional Project W-1185, Cornell University, and Universidade Estadual do Norte Fluminense. MJT and CAT also thank L. E. Ehler, L. Kimsey, M. Parella, and the Department of Entomology, University of California, Davis, for their help and cooperation in a variety of ways.

REFERENCES

Ables, J. R., S. L. Jones, and D. W. McCommas. 1978. Response of selected predator species to different densities of *Aphis gossypii* and *Heliothis virescens* eggs. *Environ. Entomol.* 7:402–4.

Adams, P. A. 1967. A review of the Mesochrysinae and Nothochrysinae (Neuroptera: Chrysopidae). *Bull. Mus. Comp. Zool.* 135:215–38.

Adams, P. A. 1969. New species and synonymy in the genus *Meleoma* (Neuroptera, Chrysopidae), with a discussion of genitalic homologies. *Postilla* 136:1–18.

Adams, P. A. 1975. Status of the genera *Ungla* and *Mallada* Navás (Neuroptera: Chrysopidae). *Psyche* 82:167–73.

Adams, P. A. 1982a. *Plesiochrysa*, a new subgenus of *Chrysopa* (Neuroptera) (Studies in New World Chrysopidae, Part I). *Neur. Int.* 2:27–32.

Adams, P. A. 1982b. *Ceraeochrysa*, a new genus of Chrysopinae (Neuroptera) (Studies in New World Chrysopidae, Part II). *Neur. Int.* 2:69–75.

Adams, P. A. 1987. Studies in Neotropical Chrysopidae (Neuroptera). III: Notes on *Nodita amazonica* Navás and *N. oenops*, n. sp. *Neur. Int.* 4:287–94.

Adams, P. A., and N. D. Penny. 1985. Neuroptera of the Amazon Basin. Part IIa. Introduction and Chrysopini. *Acta Amazonica* 15:413–79.

Adams, P. A., and N. D. Penny. 1986. Faunal relations of Amazonian Chrysopidae. In *Recent Research in Neuropterology. Proceedings of the Second International Symposium on Neuropterology*, ed. J. Gepp, H. Aspöck, and H. Hölzel, 119–24. Graz, Austria.

Adams, P. A., and N. D. Penny. 1992a. Review of the South American genera of Nothochrysinae (Insecta: Neuroptera: Chrysopidae). In *Current Research in Neuropterology. Proceedings of the Fourth International Symposium on Neuropterology*, ed. M. Canard, H. Aspöck, and M. W. Mansell, 35–41. Toulouse, France.

Adams, P. A., and N. D. Penny. 1992b. New genera of Nothochrysinae from South America (Neuroptera: Chrysopidae). *Pan-Pac. Entomol.* 68:216–21.

Adjei-Maafo, I. K., and L. T. Wilson. 1983. Factors affecting the relative abundance of arthropods on nectaried and nectariless cotton. *Environ. Entomol.* 12:349–52.

Albuquerque, G. S., C. A. Tauber, and M. J. Tauber. 1994. *Chrysoperla externa* (Neuroptera: Chrysopidae): Life history and potential for biological control in Central and South America. *Biol. Contr.* 4:8–13.

Albuquerque, G. S., C. A. Tauber, and M. J. Tauber. 1996. Postmating reproductive isolation between *Chrysopa quadripunctata* and *Chrysopa slossonae*: Mechanisms and geographic variation. *Evolution* 50:1598–606.

Albuquerque, G. S., M. J. Tauber, and C. A. Tauber. 1997. Life-history adaptations and reproductive costs associated with specialization in predacious insects. *J. Anim. Ecol.* 66:307–17.

Albuquerque, G. S., C. A. Tauber, and M. J. Tauber. 2001. *Chrysoperla externa* and *Ceraeochrysa* spp.: Potential for biological control in the New World tropics and subtropics. In *Lacewings in the Crop Environment*, ed. P. McEwen, T. R. New, and A. E. Whittington, 408–23. Cambridge: Cambridge Univ. Press.

Alrouechdi, K. 1981. Relations comportementales et trophiques entre *Chrysoperla carnea* (Stephens) (Neuroptera; Chrysopidae) et trois principaux ravageurs de l'olivier. I. La teigne de l'olivier *Prays oleae* Bern. (Lep. Hyponomeutidae). *Neur. Int.* 1:122–34.

Angelini, M. R., and S. Freitas. 2004. Desenvolvimento pós-embrionário e potencial reprodutivo de *Chrysoperla externa* (Hagen) (Neuroptera: Chrysopidae), alimentada com diferentes quantidades de ovos de *Sitotroga cerealella* (Lepidoptera: Gelechiidae). *Acta Scient. Agron.* 26:395–9.

Auad, A. M., C. F. Carvalho, B. Souza, and L. R. Barbosa. 2003. Duração e viabilidade das fases imaturas de *Chrysoperla externa* (Hagen, 1861) (Neuroptera: Chrysopidae) alimentada com ovos e lagartas de *Spodoptera frugiperda* (J. E. Smith, 1797) (Lepidoptera: Noctuidae). *Rev. Bras. Milho Sorgo* 2:106–11.

Auad, A. M., C. F. Carvalho, B. Souza, R. Trevizani, and C. M. F. R. Magalhães. 2005. Desenvolvimento das fases imaturas, aspectos reprodutivos e potencial de predação de *Chrysoperla externa* (Hagen) alimentada com ninfas de *Bemisia tabaci* (Gennadius) biótipo B em tomateiro. *Acta Scient. Agron.* 27:327–34.

Awadallah, K. T., N. A. Abou-Zeid, and M. F. S. Taufik. 1976. Development and fecundity of *Chrysopa carnea* Steph. *Bull. Soc. Ent. Égypte* 59:323–9.

Baker, H. G., and I. Baker. 1983. A brief historical review of the chemistry of floral nectar. In *The Biology of Nectaries*, ed. B. Bentley and T. Elias, 126–52. New York: Columbia Univ. Press.

Balasubramani, V., and M. Swamiappan. 1994. Development and feeding potential of the green lacewing *Chrysoperla carnea* Steph. (Neur. Chrysopidae) on different insect pests of cotton. *Anz. Schäd., Pflanzenschutz, Umweltschutz* 67:165–7.

Bänsch, R. 1964. Vergleichende Untersuchungen zur Biologie und zum Beutefangverhalten aphidivorer Coccinelliden, Chrysopiden und Syrphiden. *Zool. Jb., Syst. Ökol. Geogr. Tiere* 91:271–340.

Barbosa, L. R., S. Freitas, and A. M. Auad. 2002. Capacidade reprodutiva e viabilidade de ovos de *Ceraeochrysa everes* (Banks, 1920) (Neuroptera: Chrysopidae) em diferentes condições de acasalamento. *Ciênc. Agrotec.* 26:466–71.

Barbosa, P., and S. D. Wratten. 1998. Influence of plants on invertebrate predators: Implications to conservation biological control. In *Conservation Biological Control*, ed. P. Barbosa, 83–100. San Diego: Academic Press.

Barnes, B. N. 1975. The life history of *Chrysopa zastrowi* Esb.-Pet. (Neuroptera: Chrysopidae). *J. Entomol. Soc. S. Afr.* 38:47–53.

Ben Saad, A. A., and G. W. Bishop. 1976a. Effect of artificial honeydews on insect communities in potato fields. *Environ. Entomol.* 5:453–7.

Ben Saad, A. A., and G. W. Bishop. 1976b. Attraction of insects to potato plants through use of artificial honeydews and aphid juice. *Entomophaga* 21:49–57.

Bitsch, J. 1984. Anatomy of adult Chrysopidae. In *Biology of Chrysopidae*, ed. M. Canard, Y. Séméria, and T. R. New, 29–36. The Hague: Dr. W. Junk Publishers.

Blum, M. S., J. B. Wallace, and H. M. Fales. 1973. Skatole and tridecene: Identification and possible role in a chrysopid secretion. *Insect Biochem.* 3:353–7.

Bond, A. B. 1980. Optimal foraging in a uniform habitat: The search mechanism of the green lacewing. *Anim. Behav.* 28:10–9.

Boo, K. S., I. B. Chung, K. S. Han, J. A. Pickett, and L. J. Wadhams. 1998. Response of the lacewing *Chrysopa cognata* to pheromones of its aphid prey. *J. Chem. Ecol.* 24:631–43.

Boo, K. S., S. S. Kang, J. H. Park, J. A. Pickett, and L. J. Wadhams. 2003. Field trapping of *Chrysopa cognata* (Neuroptera: Chrysopidae) with aphid sex pheromone components in Korea. *J. Asia-Pacific Entomol.* 6:29–36.

Boregas, K. G. B., C. F. Carvalho, and B. Souza. 2003. Aspectos biológicos de *Chrysoperla externa* (Hagen, 1861) (Neuroptera: Chrysopidae) em casa-de-vegetação. *Ciênc. Agrotec.* 27:7–16.

Bowers, M. D. 1990. Recycling plant natural products for insect defense. In *Insect Defenses: Adaptive Mechanisms and Strategies of Prey and Predators*, ed. D. L. Evans and J. O. Schmidt, 353–86. Albany: State Univ. of New York Press.

Brooks, S. J., and P. C. Barnard. 1990. The green lacewings of the world: A generic review (Neuroptera: Chrysopidae). *Bull. Br. Mus. Nat. Hist. (Entomol.)* 59:117–286.

Burke, H. R., and D. F. Martin. 1956. The biology of three chrysopid predators of the cotton aphid. *J. Econ. Entomol.* 49:698–700.

Butler, G. D., Jr., and P. L. Ritchie, Jr. 1971. Feed Wheast and the abundance and fecundity of *Chrysopa carnea. J. Econ. Entomol.* 64:933–4.

Caltagirone, L. E. 1969. Terpenyl acetate bait attracts *Chrysopa* adults. *J. Econ. Entomol.* 62:1237.

Canard, M. 1970. Stérilité d'origine alimentaire chez le mâle d'un prédateur aphidiphage *Chrysopa perla* (L.) (Insectes, Névroptéres). *C. R. Hebd. Séanc. Acad. Sci.* 271:1097–9.

Canard, M. 2001. Natural food and feeding habits of lacewings. In *Lacewings in the Crop Environment*, ed. P. McEwen, T. R. New, and A. E. Whittington, 116–29. Cambridge: Cambridge Univ. Press.

Canard, M., and P. Duelli. 1984. Predatory behavior of larvae and cannibalism. In *Biology of Chrysopidae*, ed. M. Canard, Y. Séméria, and T. R. New, 92–100. The Hague: Dr. W. Junk Publishers.

Canard, M., and T. A. Volkovich. 2001. Outlines of lacewing development. In *Lacewings in the Crop Environment*, ed. P. McEwen, T. R. New, and A. E. Whittington, 130–53. Cambridge: Cambridge Univ. Press.

Canard, M., Y. Séméria, and T. R. New (ed.). 1984. *Biology of Chrysopidae*. The Hague: Dr. W. Junk Publishers.

Canard, M., H. Kokubu, and P. Duelli. 1990. Tracheal trunks supplying air to the foregut and feeding habits in adults of European green lacewing species (Insecta: Neuroptera: Chrysopidae). In *Advances in Neuropterology. Proceedings of the Third International Symposium on Neuropterology*, ed. M. W. Mansell and H. Aspöck, 277–86. Pretoria, Republic of South Africa.

Cardoso, J. T., and S. M. Lazzari. 2003. Development and consumption capacity of *Chrysoperla externa* (Hagen) (Neuroptera: Chrysopidae) fed with *Cinara* spp. (Hemiptera, Aphididae) under three temperatures. *Rev. Bras. Zool.* 20:573–6.

Carvalho, C. F., M. Canard, and C. Alauzet. 1996. Comparison of the fecundities of the Neotropical green lacewing *Chrysoperla externa* (Hagen) and the West-Palaearctic *Chrysoperla mediterranea* (Hölzel) (Insecta: Neuroptera: Chrysopidae). In *Pure and Applied Research in Neuropterology. Proceedings of the Fifth International Symposium on Neuropterology*, ed. M. Canard, H. Aspöck, and M. W. Mansell, 103–7. Toulouse, France.

Chang, Y.-F., M. J. Tauber, and C. A. Tauber. 1995. Storage of the mass-produced predator *Chrysoperla carnea* (Neuroptera: Chrysopidae): Influence of photoperiod, temperature, and diet. *Environ. Entomol.* 24:1365–74.

Chapman, J. W., D. R. Reynolds, S. J. Brooks, A. D. Smith, and I. P. Woiwod. 2006. Seasonal variation in the migration strategies of the green lacewing *Chrysoperla carnea* species complex. *Ecol. Ent.* 31:378–88.

Chen, T.-Y., and T.-X. Liu. 2001. Relative consumption of three aphid species by the lacewing, *Chrysoperla rufilabris*, and effects on its development and survival. *BioControl* 46:481–91.

Chen, T.-Y., C.-C. Chu, C. Hu, J.-Y. Mu, and T. J. Henneberry. 2006. Observations on midgut structure and content of *Chrysoperla carnea* (Neuroptera: Chrysopidae). *Ann. Entomol. Soc. Am.* 99:917–9.

Clark, T. L., and F. J. Messina. 1998. Foraging behavior of lacewing larvae (Neuroptera: Chrysopidae) on plants with divergent architectures. *J. Insect Behav.* 11:303–17.

Cohen, A. C. 1995. Extra-oral digestion in predacious terrestrial Arthropoda. *Annu. Rev. Entomol.* 40:85–103.

Cohen, A. C. 1998. Solid-to-liquid feeding: The inside(s) story of extra-oral digestion in predaceous Arthropoda. *Am. Entomol.* 44:103–17.

Cohen, A. C., and L. K. Smith. 1998. A new concept in artificial diets for *Chrysoperla rufilabris*: The efficacy of solid diets. *Biol. Contr.* 13:49–54.

Costa, R. I. F., C. C. Ecole, J. J. Soares, and L. P. M. Macedo. 2002. Duração e viabilidade das fases pré-imaginais de *Chrysoperla externa* (Hagen) alimentadas com *Aphis gossypii* Glover e *Sitotroga cerealella* (Olivier). *Acta Scient.* 24:353–7.

Dean, G. D., and C. Satasook. 1983. Response of *Chrysoperla carnea* (Stephens) (Neuroptera: Chrysopidae) to some potential attractants. *Bull. Ent. Res.* 73:619–24.

de Bortoli, S. A., A. C. Caetano, A. T. Murata, and J. E. M. Oliveira. 2006. Desenvolvimento e capacidade predatória de *Chrysoperla externa* (Hagen) (Neuroptera: Chrysopidae) em diferentes presas. *Rev. Biol. Ciênc. Terra* 6:145–52.

Díaz-Aranda, L. M., and V. J. Monserrat. 1995. Aphidophagous predator diagnosis: Key to genera of European chrysopid larvae (Neur.: Chrysopidae). *Entomophaga* 40:169–81.

Downes, J. A. 1974. Sugar feeding by the larva of *Chrysopa* (Neuroptera). *Can. Entomol.* 106:121–5.

Duelli, P. 1980a. Preovipository migration flights in the green lacewing, *Chrysopa carnea* (Planipennia, Chrysopidae). *Behav. Ecol. Sociobiol.* 7:239–46.

Duelli, P. 1980b. Adaptive dispersal and appetitive flight in the green lacewing, *Chrysopa carnea*. *Ecol. Ent.* 5:213–20.

Duelli, P. 1984a. Flight, dispersal, migration. In *Biology of Chrysopidae*, ed. M. Canard, Y. Séméria, and T. R. New, 110–6. The Hague: Dr. W. Junk Publishers.

Duelli, P. 1984b. Dispersal and oviposition strategies in *Chrysoperla carnea*. In *Progress in World's Neuropterology. Proceedings of the First International Symposium on Neuropterology*, ed. J. Gepp, H. Aspöck, and H. Hölzel, 133–45. Graz, Austria.

Duelli, P. 1986. Flight activity patterns in lacewings (Planipennia: Chrysopidae). In *Recent Research in Neuropterology. Proceedings of the Second International Symposium on Neuropterology*, ed. J. Gepp, H. Aspöck, and H. Hölzel, 165–70. Graz, Austria.

Duelli, P. 2001. Lacewings in field crops. In *Lacewings in the Crop Environment*, ed. P. McEwen, T. R. New, and A. E. Whittington, 158–71. Cambridge: Cambridge Univ. Press.

Duelli, P., and J. B. Johnson. 1992. Adaptive significance of the egg pedicel in green lacewings (Insecta: Neuroptera: Chrysopidae). In *Current Research in Neuropterology. Proceedings of the Fourth International Symposium on Neuropterology*, ed. M. Canard, H. Aspöck, and M. W. Mansell, 125–34. Toulouse, France.

Eigenbrode, S. D. 2003. The effects of plant epicuticular waxy blooms on attachment and effectiveness of predatory insects. *Arthropod Struct. Dev.* 33:91–102.

Eisner, T., and R. E. Silberglied. 1988. A chrysopid larva that cloaks itself in mealybug wax. *Psyche* 95:15–9.

Eisner, T., K. Hicks, M. Eisner, and D. S. Robson. 1978. "Wolf-in-sheep's-clothing" strategy of a predaceous insect larva. *Science* 199:790–4.

Eisner, T., A. B. Attygalle, W. E. Conner, M. Eisner, E. MacLeod, and J. Meinwald. 1996. Chemical egg defense in a green lacewing (*Ceraeochrysa smithi*). *Proc. Natl. Acad. Sci. U. S. A.* 93:3280–3.

Elsey, K. D. 1974. Influence of plant host on searching speed of two predators. *Entomophaga* 19:3–6.

Ermicheva, F. M., V. V. Sumenkova, and I. G. Yazlovetscky. 1987. Localization of protease in guts of *Chrysopa carnea* larvae. *Izv. Acad. Sci. Moldavian SSR, Ser. Biol. Chem. Sci.* 4: 49–52 (in Russian).

Evans, E. W., and J. G. Swallow. 1993. Numerical responses of natural enemies to artificial honeydew in Utah alfalfa. *Environ. Entomol.* 22:1392–401.

Figueira, L. K., C. F. Carvalho, and B. Souza. 2000. Biologia e exigências térmicas de *Chrysoperla externa* (Hagen, 1861) (Neuroptera: Chrysopidae) alimentada com ovos de *Alabama argillacea* (Hübner, 1818) (Lepidoptera: Noctuidae). *Ciênc. Agrotec.* 24:319–26.

Figueira, L. K., C. F. Carvalho, and B. Souza. 2002. Influência da temperatura sobre alguns aspectos biológicos de *Chrysoperla externa* (Hagen, 1861) (Neuroptera: Chrysopidae) alimentada com ovos de *Alabama argillacea* (Hübner, 1818) (Lepidoptera: Noctuidae). *Ciênc. Agrotec.* 26:1439–50.

Fleschner, C. A. 1950. Studies on searching capacity of the larvae of three predators of the citrus red mite. *Hilgardia* 20:233–65.

Flint, H. M., S. S. Salter, and S. Walters. 1979. Caryophyllene: An attractant for the green lacewing. *Environ. Entomol.* 8:1123–5.

Florkin, M., and C. Jeuniaux. 1974. Hemolymph: Composition. In *The Physiology of Insecta*, 2nd edition, v. 5, ed. M. Rockstein, 255–307. New York: Academic Press.

Fonseca, A. R., C. F. Carvalho, and B. Souza. 2001. Capacidade predatória e aspectos biológicos das fases imaturas de *Chrysoperla externa* (Hagen, 1861) (Neuroptera: Chrysopidae) alimentada com *Schizaphis graminum* (Rondani, 1852) (Hemiptera: Aphididae) em diferentes temperaturas. *Ciênc. Agrotec.* 25:251–63.

Freitas, S. 2003. *Chrysoperla* Steinmann, 1964 (Neuroptera, Chrysopidae): Descrição de uma nova espécie do Brasil. *Rev. Bras. Entomol.* 47:385–7.

Freitas, S. 2007. New species of Brazilian green lacewings genus *Leucochrysa* McLachlan, 1868 (Neuroptera Chrysopidae). *Ann. Mus. Civ. St. Nat. Ferrara* 8:49–54.

Freitas, S., and N. D. Penny. 2000. Two new genera of Neotropical Chrysopini (Neuroptera: Chrysopidae). *J. Kans. Entomol. Soc.* 73:164–70.

Freitas, S., and N. D. Penny. 2001. The green lacewings (Neuroptera: Chrysopidae) of Brazilian agro-ecosystems. *Proc. Calif. Acad. Sci.* 52:245–395.

Freitas, S., N. D. Penny, and P. A. Adams. 2009. A revision of the New World genus *Ceraeochrysa* (Neuroptera: Chrysopidae). *Proc. Calif. Acad. Sci.* 60:503–610.

Frost, S. W. 1927. Beneficial insects trapped in bait-pails. *Entomol. News* 38:153–6.

Frost, S. W. 1936. A summary of insects attracted to liquid baits. *Entomol. News* 47:89–92.

Gaumont, J. 1976. L'appareil digestif des larves de Planipennes. *Ann. Sci. Nat., Zool.* 18:145–250.

Gepp, J. 1984. Morphology and anatomy of the preimaginal stages of Chrysopidae: A short survey. In *Biology of Chrysopidae*, ed. M. Canard, Y. Séméria, and T. R. New, 9–19. The Hague: Dr. W. Junk Publishers.

Gibson, C. M., and M. S. Hunter. 2005. Reconsideration of the role of yeasts associated with *Chrysoperla* green lacewings. *Biol. Contr.* 32:57–64.

Giles, K. L., R. D. Madden, M. E. Payton, and J. W. Dillwith. 2000. Survival and development of *Chrysoperla rufilabris* (Neuroptera: Chrysopidae) supplied with pea aphids (Homoptera: Aphididae) reared on alfalfa and faba bean. *Environ. Entomol.* 29:304–11.

Gitirana-Neto, J., C. F. Carvalho, B. Souza, and L. V. C. Santa-Cecília. 2001. Flutuação populacional de espécies de *Ceraeochrysa* Adams, 1982 (Neuroptera: Chrysopidae) em citros, na região de Lavras-MG. *Ciênc. Agrotec.* 25:550–9.

Gonçalves-Gervásio, R. C., and L. V. C. Santa-Cecília. 2001. Consumo alimentar de *Chrysoperla externa* sobre as diferentes fases de desenvolvimento de *Dysmicoccus brevipes* em laboratório. *Pesq. Agropec. Bras.* 36:387–91.

Greany, P. D., and K. S. Hagen. 1981. Prey selection. In *Semiochemicals—Their Role in Pest Control*, ed. D. A. Nordlund, R. L. Jones, and W. J. Lewis, 121–35. New York: John Wiley & Sons.

Grimaldi, D., and M. S. Engel. 2005. *Evolution of the Insects.* New York: Cambridge Univ. Press.

Hagen, K. S. 1950. Fecundity of *Chrysopa californica* as affected by synthetic foods. *J. Econ. Entomol.* 43:101–4.

Hagen, K. S. 1986. Ecosystem analysis: Plant cultivars (HPR), entomophagous insects and food supplements. In *Interactions of Plant Resistance and Parasitoids and Predators of Insects*, ed. D. J. Boethel and R. D. Eikenbary, 151–97. Chichester: Ellis Horwood.

Hagen, K. S. 1987. Nutritional ecology of terrestrial insect predators. In *Nutritional Ecology of Insects, Mites, Spiders, and Related Invertebrates*, ed. F. Slansky Jr. and J. G. Rodriguez, 533–77. New York: John Wiley & Sons.

Hagen, K. S., and R. L. Tassan. 1965. A method of providing artificial diets to *Chrysopa* larvae. *J. Econ. Entomol.* 58:999–1000.

Hagen, K. S., and R. L. Tassan. 1966. The influence of protein hydrolysates of yeasts and chemically defined diets upon the fecundity of *Chrysopa carnea* Stephens (Neuroptera). *Vestnik Cs. Spol. Zool.* 30:219–27.

Hagen, K. S., and R. L. Tassan. 1970. The influence of food Wheast® and related *Saccharomyces fragilis* yeast products on the fecundity of *Chrysopa carnea* (Neuroptera: Chrysopidae). *Can. Entomol.* 102:806–11.

Hagen, K. S., and R. L. Tassan. 1972. Exploring nutritional roles of extracellular symbiotes on the reproduction of honeydew feeding adult chrysopids and tephritids. In *Insect and Mite Nutrition: Significance and Implications in Ecology and Pest Management*, ed. J. G. Rodriguez, 323–51. Amsterdam: North-Holland Publ.

Hagen, K. S., R. L. Tassan, and E. F. Sawall, Jr. 1970. Some ecophysiological relationships between certain *Chrysopa*, honeydews and yeasts. *Boll. Lab. Entomol. Agr. F. Silvestri Portici* 28:113–34.

Hagen, K. S., E. F. Sawall, Jr., and R. L. Tassan. 1971. The use of food sprays to increase effectiveness of entomophagous insects. *Proc. Tall Timb. Conf. Ecol. Anim. Contr. Habitat Manag.* 2:59–81.

Hagen, K. S., S. Bombosch, and J. A. McMurtry. 1976a. The biology and impact of predators. In *Theory and Practice of Biological Control*, ed. C. B. Huffaker and P. S. Messenger, 93–142. New York: Academic Press.

Hagen, K. S., P. Greany, E. F. Sawall, Jr., and R. L. Tassan. 1976b. Tryptophan in artificial honeydews as a source of an attractant for adult *Chrysopa carnea. Environ. Entomol.* 5:458–68.

Han, B., and Z. Chen. 2002. Behavioral and electrophysiological responses of natural enemies to synomones from tea shoots and kairomones from tea aphids, *Toxoptera aurantii. J. Chem. Ecol.* 28:2203–19.

Harrison, S. J., and P. K. McEwen. 1998. Acid hydrolysed L-tryptophan and its role in the attraction of the green lacewing *Chrysoperla carnea* (Stephens) (Neuropt., Chrysopidae). *J. Appl. Entomol.* 122:343–4.

Hasegawa, M., K. Niijima, and M. Matsuka. 1989. Rearing *Chrysoperla carnea* (Neuroptera: Chrysopidae) on chemically defined diets. *Appl. Entomol. Zool.* 24:96–102.

Hassan, S. A., and K. S. Hagen. 1978. A new artificial diet for rearing *Chrysopa carnea* larvae (Neuroptera: Chrysopidae). *Z. Ang. Ent.* 86:315–20.

Henry, C. S., and M. M. Wells. 2007. Can what we don't know about lacewing systematics hurt us? *Am. Entomol.* 53:42–7.

Hooper, A. M., B. Donato, C. M. Woodcock, J. H. Park, R. L. Paul, K. S. Boo, J. Hardie, and J. A. Pickett. 2002. Characterization of (1R,4S,4aR,7S,7aR)-dihydronepetalactol as a semiochemical for lacewings, including *Chrysopa* spp. and *Peyerimhoffina gracilis. J. Chem. Ecol.* 28:849–64.

Hydorn, S. B., and W. H. Whitcomb. 1979. Effects of larval diet on *Chrysopa rufilabris. Fla. Entomol.* 62:293–8.

Hyeon, S. B., S. Isoe, and T. Sakan. 1968. The structure of neomatatabiol, the potent attractant for *Chrysopa* from *Actinidia polygama. Tetrahedron Lett.* 51:5325–6.

Ickert, G. 1968. Beiträge zur Biologie einheimischer Chrysopiden (Planipennia, Chrysopidae). *Ent. Abh. Mus. Tierk. Dresden* 36:123–92.

Ishii, S. 1964. An attractant contained in *Actinidia polygama* Miq. for male lace wing, *Chrysopa septempunctata* Wesmael. *Jpn. J. Appl. Ent. Zool.* 8:334–7.

James, D. G. 2003. Field evaluation of herbivore-induced plant volatiles as attractants for beneficial insects: Methyl salicylate and the green lacewing, *Chrysopa nigricornis. J. Chem. Ecol.* 29:1601–9.

James, D. G. 2006. Methyl salicylate is a field attractant for the goldeneyed lacewing, *Chrysopa oculata. Biocontrol Sci. Technol.* 16:107–10.

James, D. G., and T. S. Price. 2004. Field-testing of methyl salicylate for recruitment and retention of beneficial insects in grapes and hops. *J. Chem. Ecol.* 30:1613–28.

Jervis, M., and M. J. W. Copland. 1996. The life cycle. In *Insect Natural Enemies: Practical Approaches to Their Study and Evaluation*, ed. M. Jervis and N. Kidd, 63–161. London: Chapman & Hall.

Johnson, J. B. 1982. Bionomics of some symbiote using Chrysopidae (Insecta: Neuroptera) from the Western United States. PhD diss., University of California, Berkeley.

Kawecki, Z. 1932. Beobachtungen über das Verhalten und die Sinnesorientierung der Florfliegenlarven. *Bull. Int. Acad. Pol. Sci. Lett.* 2:91–106.

Killington, F. J. 1936. *A Monograph of the British Neuroptera*, v. 1. London: Ray Society.

LaMunyon, C. W., and P. A. Adams. 1987. Use and effect of an anal defensive secretion in larval Chrysopidae (Neuroptera). *Ann. Entomol. Soc. Am.* 80:804–8.

Lavallee, A. G., and F. R. Shaw. 1969. Preferences of golden-eye lacewing larvae for pea aphids, leafhopper and plant bug nymphs, and alfalfa weevil larvae. *J. Econ. Entomol.* 62:1228–9.

Legrand, J., C. A. Tauber, G. S. Albuquerque, and M. J. Tauber. 2008. Navás' type and non-type specimens of Chrysopidae in the MNHN, Paris [Neuroptera]. *Rev. fr. Ent. (N.S.)* 30:103–83.

Lewis, W. J., D. A. Nordlund, H. R. Gross Jr., R. L. Jones, and S. L. Jones. 1977. Kairomones and their use for management of entomophagous insects. V. Moth scales as a stimulus for predation of *Heliothis zea* (Boddie) eggs by *Chrysopa carnea* Stephens larvae. *J. Chem. Ecol.* 3:483–7.

Liber, H., and A. Niccoli. 1988. Observations on the effectiveness of an attractant food spray in increasing chrysopid predation on *Prays oleae* (Berg.) eggs. *Redia* 71:467–82.

Limburg, D. D., and J. A. Rosenheim. 2001. Extrafloral nectar consumption and its influence on survival and development of an omnivorous predator, larval *Chrysoperla plorabunda* (Neuroptera: Chrysopidae). *Environ. Entomol.* 30:595–604.

Liu, T.-X., and T.-Y. Chen. 2001. Effects of three aphid species (Homoptera: Aphididae) on development, survival and predation of *Chrysoperla carnea* (Neuroptera: Chrysopidae). *Appl. Entomol. Zool.* 36:361–6.

López-Arroyo, J. I., C. A. Tauber, and M. J. Tauber. 1999. Comparative life histories of the predators *Ceraeochrysa cincta*, *C. cubana*, and *C. smithi* (Neuroptera: Chrysopidae). *Ann. Entomol. Soc. Am.* 92:208–17.

Maia, W. J. M. S., C. F. Carvalho, B. Souza, I. Cruz, and T. J. A. F. Maia. 2004. Capacidade predatória e aspectos biológicos de *Chrysoperla externa* (Hagen, 1861) (Neuroptera: Chrysopidae) alimentada com *Rhopalosiphum maidis* (Fitch, 1856) (Hemiptera: Aphididae). *Ciênc. Agrotec.* 28:1259–68.

Mantoanelli, E., and G. S. Albuquerque. 2007. Desenvolvimento e comportamento larval de *Leucochrysa* (*Leucochrysa*) *varia* (Schneider) (Neuroptera: Chrysopidae) em laboratório. *Rev. Bras. Zool.* 24:302–11.

Mantoanelli, E., G. S. Albuquerque, C. A. Tauber, and M. J. Tauber. 2006. *Leucochrysa* (*Leucochrysa*) *varia* (Neuroptera: Chrysopidae): Larval descriptions, developmental rates, and adult color variation. *Ann. Entomol. Soc. Am.* 99:7–18.

Mantoanelli, E., C. A. Tauber, G. S. Albuquerque, and M. J. Tauber. 2011. Larvae of four *Leucochrysa* (*Nodita*) species (Neuroptera: Chrysopidae: Leucochrysini) from Brazil's Atlantic coast. *Ann. Entomol. Soc. Am.* 104:1233–59.

Mantoanelli, E., C. A. Tauber, G. S. Albuquerque, and M. J. Tauber. 2011. Larvae of four species of *Leucochrysa* (*Nodita*) from Brazil's Atlantic coast (Neuroptera: Chrysopidae: Leucochrysini). Submitted.

Matos, B., and J. J. Obrycki. 2006. Prey suitability of *Galerucella calmariensis* L. (Coleoptera: Chrysomelidae) and *Myzus lythri* (Schrank) (Homoptera: Aphididae) for development of three predatory species. *Environ. Entomol.* 35:345–50.

McEwen, P. K., S. Clow, M. A. Jervis, and N. A. C. Kidd. 1993. Alteration in searching behaviour of adult female green lacewings *Chrysoperla carnea* (Neur.: Chrysopidae) following contact with honeydew of the black scale *Saissetia oleae* (Hom.: Coccidae) and solutions containing acid-hydrolysed L-tryptophan. *Entomophaga* 38:347–54.

McEwen, P. K., M. A. Jervis, and N. A. C. Kidd. 1994. Use of a sprayed L-tryptophan solution to concentrate numbers of the green lacewing *Chrysoperla carnea* in olive tree canopy. *Entomol. Exp. Appl.* 70:97–9.

McEwen, P., T. R. New, and A. E. Whittington (ed.). 2001. *Lacewings in the Crop Environment.* Cambridge: Cambridge Univ. Press.

Messina, F. J., T. A. Jones, and D. C. Nielson. 1997. Host-plant effects on the efficacy of two predators attacking Russian wheat aphids (Homoptera: Aphididae). *Environ. Entomol.* 26:1398–404.

Milbrath, L. R., M. J. Tauber, and C. A. Tauber. 1993. Prey specificity in *Chrysopa*: An interspecific comparison of larval feeding and defensive behavior. *Ecology* 74:1384–93.

Milbrath, L. R., M. J. Tauber, and C. A. Tauber. 1994. Larval behavior of predacious sister-species: Orientation, molting site, and survival in *Chrysopa. Behav. Ecol. Sociobiol.* 35:85–90.

Miller, L. A., and J. Olesen. 1979. Avoidance behavior in green lacewings. I. Behavior of free flying green lacewings to hunting bats and ultrasound. *J. Comp. Physiol.* 131:113–20.

Molleman, F., B. Drukker, and L. Blommers. 1997. A trap for monitoring pear psylla predators using dispensers with the synomone methylsalicylate. *Proc. Exp. Appl. Entomol., NEV Amsterdam* 8:177–82.

Monserrat, V. J. 1985. Lista de los tipos de Mecoptera y Neuroptera (Insecta) de la colección L. Navás, depositados en el Museo de Zoología de Barcelona. *Misc. Zool.* 9:233–43.

Multani, J. S. 2008. Diversidade e abundância de crisopídeos (Neuroptera, Chrysopidae) e interações com presas, parasitóides e fatores abióticos em pomares de goiaba em Campos dos Goytacazes, RJ. PhD thesis, State University of North Fluminense, Brazil.

Muma, M. H. 1957. Effects of larval nutrition on the life cycle, size, coloration, and longevity of *Chrysopa lateralis* Guer. *Fla. Entomol.* 40:5–9.

Neuenschwander, P., M. Canard, and S. Michelakis. 1981. The attractivity of protein hydrolysate baited McPhail traps to different chrysopid and hemerobiid species (Neuroptera) in a Cretan olive orchard. *Ann. Soc. Ent. Fr. (N.S.)* 17:213–20.

New, T. R. 1969. Notes on the debris-carrying habit in larvae of British Chrysopidae (Neuroptera). *Entomol. Gazette* 20:119–24.

New, T. R. 1975. The biology of Chrysopidae and Hemerobiidae (Neuroptera), with reference to their usage as biocontrol agents: A review. *Trans. R. Ent. Soc. Lond.* 127:115–40.

New, T. R. 1983. The egg and first instar larva of *Italochrysa insignis* (Neuroptera: Chrysopidae). *Aust. Entomol. Mag.* 10:29–32.

New, T. R. 1986. Some early stages of *Calochrysa* Banks (Neuroptera: Chrysopidae). *Aust. Entomol. Mag.* 13:11–4.

New, T. R. 1991. *Insects as Predators.* Kensington: The New South Wales Univ. Press.

New, T. R. 2001. Introduction to the systematics and distribution of Coniopterygidae, Hemerobiidae, and Chrysopidae used in pest management. In *Lacewings in the Crop Environment*, ed. P. McEwen, T. R. New, and A. E. Whittington, 6–28. Cambridge: Cambridge Univ. Press.

Niijima, K. 1989. Nutritional studies on an aphidophagous chrysopid, *Chrysopa septempunctata* Wesmael. I. Chemically-defined diets and general nutritional requirements. *Bull. Fac. Agric. Tamagawa Univ.* 29:22–30.

Niijima, K. 1993a. Nutritional studies on an aphidophagous chrysopid, *Chrysopa septempunctata* Wesmael (Neuroptera: Chrysopidae). II. Amino acid requirement for larval development. *Appl. Entomol. Zool.* 28:81–7.

Niijima, K. 1993b. Nutritional studies on an aphidophagous chrysopid, *Chrysopa septempunctata* Wesmael (Neuroptera: Chrysopidae). III. Vitamin requirement for larval development. *Appl. Entomol. Zool.* 28:89–95.

Nordlund, D. A., W. J. Lewis, R. L. Jones, H. R. Gross Jr., and K. S. Hagen. 1977. Kairomones and their use for management of entomophagous insects. VI. An examination of the kairomones for the predator *Chrysopa carnea* Steph. at the oviposition sites of *Heliothis zea* (Boddie). *J. Chem. Ecol.* 3:507–11.

Nordlund, D. A., A. C. Cohen, and R. A. Smith. 2001. Mass-rearing, release techniques, and augmentation. In *Lacewings in the Crop Environment*, ed. P. McEwen, T. R. New, and A. E. Whittington, 303–19. Cambridge: Cambridge Univ. Press.

Obrycki, J. J., and M. J. Tauber. 1985. Seasonal occurrence and relative abundance of aphid predators and parasitoids on pubescent potato plants. *Can. Entomol.* 117:1231–7.

Obrycki, J. J., M. N. Hamid, A. S. Sajap, and L. C. Lewis. 1989. Suitability of corn insect pests for development and survival of *Chrysoperla carnea* and *Chrysopa oculata* (Neuroptera: Chrysopidae). *Environ. Entomol.* 18:1126–30.

Osman, M. Z., and B. J. Selman. 1996. Effect of larval diet on the performance of the predator *Chrysoperla carnea* Stephens (Neuropt., Chrysopidae). *J. Appl. Ent.* 120:115–7.

Pai, K. F., C. J. Chen, J. T. Yang, and C. C. Chen. 2004. Green lacewing *Ankylopteryx exquisite* attracted to methyl eugenol. *Plant Prot. Bull.* 46:93–7.

Patt, J. M., S. C. Wainright, G. C. Hamilton, D. Whittinghill, K. Bosley, J. Dietrick, and J. H. Lashomb. 2003. Assimilation of carbon and nitrogen from pollen and nectar by a predaceous larva and its effects on growth and development. *Ecol. Ent.* 28:717–28.

Penny, N. D. 1997. Four new species of Costa Rican *Ceraeochrysa* (Neuroptera: Chrysopidae). *Pan-Pac. Entomol.* 73:61–9.

Penny, N. D. 1998. New Chrysopinae from Costa Rica (Neuroptera: Chrysopidae). *J. Neuropterol.* 1:55–78.

Penny, N. D. 2001. New species of Chrysopinae (Neuroptera: Chrysopidae) from Costa Rica, with selected taxonomic notes and a neotype designation. *Entomol. News* 112:1–14.

Penny, N. D. 2002. A guide to the lacewings (Neuroptera) of Costa Rica. *Proc. Cal. Acad. Sci.* 53:161–457.

Philippe, R. 1971. Influence de l'accouplement sur le comportement de ponte et la fécondité chez *Chrysopa perla* (L.) (Insectes–Planipennes). *Ann. Zool. Écol. Anim.* 3:443–8.

Pires, C. S. S., E. R. Sujii, E. M. G. Fontes, C. A. Tauber, and M. J. Tauber. 2000. Dry-season embryonic dormancy in *Deois flavopicta* (Homoptera: Cercopidae): Roles of temperature and moisture in nature. *Environ. Entomol.* 29:714–20.

Principi, M. M. 1940. Contributi allo studio dei neurotteri italiani. I. *Chrysopa septempunctata* Wesm. e *Chrysopa flavifrons* Brauer. *Boll. Ist. Ent. Univ. Bologna* 12:63–144.

Principi, M. M. 1946. Contributi allo studio dei neurotteri italiani. IV. *Nothochrysa italica* Rossi. *Boll. Ist. Ent. Univ. Bologna* 15:85–102.

Principi, M. M., and M. Canard. 1984. Feeding habits. In *Biology of Chrysopidae*, ed. M. Canard, Y. Séméria, and T. R. New, 76–92. The Hague: Dr. W. Junk Publishers.

Putman, W. L. 1932. Chrysopids as a factor in the natural control of the oriental fruit moth. *Can. Entomol.* 64:121–6.

Reddy, G. V. P., J. K. Holopainen, and A. Guerrero. 2002. Olfactory responses of *Plutella xylostella* natural enemies to host pheromone, larval frass, and green leaf cabbage volatiles. *J. Chem. Ecol.* 28:131–43.

Ribas, M. L., G. S. Albuquerque, M. J. Tauber, and C. A. Tauber. 2012. Life histories of two *Leucochrysa* species (Neuroptera: Chrysopidae), predators in Brazilian orchards. Submitted.

Ribeiro, M. J., C. F. Carvalho, and J. C. Matioli. 1993. Biologia de adultos de *Chrysoperla externa* (Hagen, 1861) (Neuroptera, Chrysopidae) em diferentes dietas artificiais. *Ciênc. Prát.* 17:120–30.

Rousset, A. 1984. Reproductive physiology and fecundity. In *Biology of Chrysopidae*, ed. M. Canard, Y. Séméria, and T. R. New, 116–29. The Hague: Dr. W. Junk Publishers.

Rowell-Rahier, M., and J. M. Pasteels. 1992. Third trophic level influences of plant allelochemicals. In *Herbivores: Their Interactions with Secondary Plant Metabolites. Evolutionary and Ecological Processes*, 2nd edition, v. 2, ed. G. A. Rosenthal and M. R. Berenbaum, 243–77. San Diego: Academic Press.

Ru, N., W. H. Whitcomb, M. Murphey, and T. C. Carlisle. 1975. Biology of *Chrysopa lanata* (Neuroptera: Chrysopidae). *Ann. Entomol. Soc. Am.* 68:187–90.

Sakan, T., S. Isoe, and S. B. Hyeon. 1970. The chemistry of attractants for Chrysopidae from *Actinidia polygama* Miq.. In *Control of Insect Behavior by Natural Products*, ed. D. L. Wood, R. M. Silverstein, and M. Nakajima, 237–47. New York: Academic Press.

Santos, T. M., A. L. Boiça Júnior, and J. J. Soares. 2003. Influência de tricomas do algodoeiro sobre os aspectos biológicos e capacidade predatória de *Chrysoperla externa* (Hagen) alimentada com *Aphis gossypii* Glover. *Bragantia* 62:243–54.

Sengonca, Ç., Y. K. Kotikal, and M. Schade. 1995. Olfactory reactions of *Cryptolaemus montrouzieri* Mulsant (Col., Coccinellidae) and *Chrysoperla carnea* (Stephens) (Neur., Chrysopidae) in relation to period of starvation. *J. Pest Sci.* 68:9–12.

Silva, C. G., B. Souza, A. M. Auad, J. P. Bonani, L. C. Torres, C. F. Carvalho, and C. C. Ecole. 2004a. Desenvolvimento das fases imaturas de *Chrysoperla externa* alimentadas com ninfas de *Bemisia tabaci* criadas em três hospedeiros. *Pesq. Agropec. Bras.* 39:1065–70.

Silva, C. G., A. M. Auad, B. Souza, C. F. Carvalho, and J. P. Bonani. 2004b. Aspectos biológicos de *Chrysoperla externa* (Hagen, 1861) (Neuroptera: Chrysopidae) alimentada com *Bemisia tabaci* (Gennadius, 1889) biótipo B (Hemiptera: Aleyrodidae) criada em três hospedeiros. *Ciênc. Agrotec.* 28:243–50.

Silva, G. A., C. F. Carvalho, and B. Souza. 2002. Aspectos biológicos de *Chrysoperla externa* (Hagen, 1861) (Neuroptera: Chrysopidae) alimentada com lagartas de *Alabama argillacea* (Hübner, 1818) (Lepidoptera: Noctuidae). *Ciênc. Agrotec.* 26:682–98.

Silva, P. S. 2006. Estudo comparativo da biologia e morfologia das espécies de *Chrysopodes* (Neuroptera, Chrysopidae) da região Norte Fluminense. PhD thesis, State University of North Fluminense, Brazil.

Silva, P. S., G. S. Albuquerque, C. A. Tauber, and M. J. Tauber. 2007. Life history of a widespread Neotropical predator, *Chrysopodes* (*Chrysopodes*) *lineafrons* (Neuroptera: Chrysopidae). *Biol. Contr.* 41:33–41.

Slocum, R. D., and J. D. Lawrey. 1976. Viability of the epizoic lichen flora carried and dispersed by green lacewing (*Nodita pavida*) larvae. *Can. J. Bot.* 54:1827–31.

Smith, R. C. 1922. The biology of the Chrysopidae. *Cornell Univ. Agric. Exp. Sta. Memoir* 58:1287–372.

Smith, R. C. 1926. The trash-carrying habit of certain lace wing larvae. *Sci. Monthly* 23:265–7.

Sosa, F., and S. Freitas. 2010. New Neotropical species of *Ceraeochrysa* Adams (Neuroptera: Chrysopidae). *Zootaxa* 2562:57–65.

Souza, B., and C. F. Carvalho. 2002. Population dynamics and seasonal occurrence of adults of *Chrysoperla externa* (Hagen, 1861) (Neuroptera: Chrysopidae) in a citrus orchard in southern Brazil. *Acta Zool. Acad. Sci. Hung.* 48(Suppl. 2):301–10.

Spiegler, P. E. 1962. The origin and nature of the adhesive substance in larvae of the genus *Chrysopa* (Neuroptera: Chrysopidae). *Ann. Entomol. Soc. Am.* 55:69–77.

Stelzl, M. 1992. Comparative studies on mouthparts and feeding habits of adult Raphidioptera and Neuroptera (Insecta: Neuropteroidea). In *Current Research in Neuropterology. Proceedings of the Fourth International Symposium on Neuropterology*, ed. M. Canard, H. Aspöck, and M. W. Mansell, 341–7. Toulouse, France.

Suda, D. Y., and R. T. Cunningham. 1970. *Chrysopa basalis* captured in plastic traps containing methyl eugenol. *J. Econ. Entomol.* 63:1706.

Suh, S.-O., C. M. Gibson, and M. Blackwell. 2004. *Metschnikowia chrysoperlae* sp. nov., *Candida picachoensis* sp. nov. and *Candida pimensis* sp. nov., isolated from the green lacewings *Chrysoperla comanche* and *Chrysoperla carnea* (Neuroptera: Chrysopidae). *Int. J. Syst. Evol. Microbiol.* 54:1883–90.

Sulc, K. 1914. Über die stinkdrüsen und speicheldrüsen der chrysopiden. *Sber. K. Böhm. Ges. Wiss.* 1914:1–50.

Szentkirályi, F. 2001. Ecology and habitat relationships. In *Lacewings in the Crop Environment*, ed. P. McEwen, T. R. New, and A. E. Whittington, 82–115, Cambridge: Cambridge Univ. Press.

Tauber, C. A. 1969. Taxonomy and biology of the lacewing genus *Meleoma* (Neuroptera: Chrysopidae). *Univ. Calif. Pub. Entomol.* 58:1–94.

Tauber, C. A. 1974. Systematics of North American chrysopid larvae: *Chrysopa carnea* group (Neuroptera). *Can. Entomol.* 106:1133–53.

Tauber, C. A. 2003. Generic characteristics of *Chrysopodes* (Neuroptera: Chrysopidae), with new larval descriptions and a review of species from the United States and Canada. *Ann. Entomol. Soc. Am.* 96:472–90.

Tauber, C. A. 2007. Review of *Berchmansus* and *Vieira* and description of two new species of *Leucochrysa* (Neuroptera: Chrysopidae). *Ann. Entomol. Soc. Am.* 100:110–38.

Tauber, C. A. 2010. Revision of *Neosuarius*, a subgenus of *Chrysopodes* (Neuroptera, Chrysopidae). *ZooKeys* 44:1–104.

Tauber, C. A., and M. J. Tauber. 1987. Food specificity in predacious insects: A comparative ecophysiological and genetic study. *Evol. Ecol.* 1:175–86.

Tauber, C. A., and M. J. Tauber. 1989. Sympatric speciation in insects: Perception and perspective. In *Speciation and Its Consequences*, ed. D. Otte and J. A. Endler, 307–44. Sunderland: Sinauer Associates.

Tauber, C. A., and M. J. Tauber. 1992. Phenotypic plasticity in *Chrysoperla*: Genetic variation in the sensory mechanism and in correlated reproductive traits. *Evolution* 46:1754–73.

Tauber, C. A., and O. S. Flint, Jr. 2010. Resolution of some taxonomic and nomenclatural issues in a recent revision of *Ceraeochrysa* (Neuroptera: Chrysopidae). *Zootaxa* 2565:55–67.

Tauber, C. A., M. J. Tauber, and M. J. Tauber. 1991. Egg size and taxon: Their influence on survival and development of chrysopid hatchlings after food and water deprivation. *Can. J. Zool.* 69:2644–50.

Tauber, C. A., J. R. Ruberson, and M. J. Tauber. 1995a. Size and morphological differences among the larvae of two predacious species and their hybrids. *Ann. Entomol. Soc. Am.* 88:502–11.

Tauber, C. A., M. J. Tauber, and L. R. Milbrath. 1995b. Individual repeatability and geographical variation in the larval behaviour of the generalist predator, *Chrysopa quadripunctata. Anim. Behav.* 50:1391–403.

Tauber, C. A., M. J. Tauber, and G. S. Albuquerque. 2001. *Plesiochrysa brasiliensis* (Neuroptera: Chrysopidae): Larval stages, biology, and taxonomic relationships. *Ann. Entomol. Soc. Am.* 94:858–65.

Tauber, C. A., G. S. Albuquerque, and M. J. Tauber. 2006. *Berchmansus elegans* (Neuroptera: Chrysopidae): Larval and adult characteristics and new tribal affiliation. *Eur. J. Entomol.* 103:221–31.

Tauber, C. A., G. S. Albuquerque, and M. J. Tauber. 2008a. *Gonzaga nigriceps* (McLachlan) (Neuroptera: Chrysopidae): Description of larvae and adults, biological notes, and generic affiliation. *Proc. Ent. Soc. Wash.* 110:417–38.

Tauber, C. A., M. J. Tauber, and G. S. Albuquerque. 2008b. A new genus and species of green lacewings from Brazil (Neuroptera: Chrysopidae: Leucochrysini). *Ann. Entomol. Soc. Am.* 101: 314–26.

Tauber, C. A., G. S. Albuquerque, and M. J. Tauber. 2008c. A new species of *Leucochrysa* and a redescription of *Leucochrysa (Nodita) clepsydra* Banks (Neuroptera: Chrysopidae). *Zootaxa* 1781:1–19.

Tauber, C. A., G. S. Albuquerque, and M. J. Tauber. 2011. Nomenclatural changes and redescriptions of three of Navás' baffling *Leucochrysa (Nodita)* species (Neuroptera: Chrysopidae). *ZooKeys* 92:9–33.

Tauber, M. J., and C. A. Tauber. 1974. Dietary influence on reproduction in both sexes of five predacious species (Neuroptera). *Can. Entomol.* 106:921–5.

Tauber, M. J., and C. A. Tauber. 1983. Life history traits of *Chrysopa carnea* and *Chrysopa rufilabris* (Neuroptera: Chrysopidae): Influence of humidity. *Ann. Entomol. Soc. Am.* 76:282–5.

Tauber, M. J., C. A. Tauber, and S. Masaki. 1986. *Seasonal Adaptations of Insects.* New York: Oxford Univ. Press.

Tauber, M. J., C. A. Tauber, J. R. Ruberson, L. R. Milbrath, and G. S. Albuquerque. 1993. Evolution of prey specificity via three steps. *Experientia* 49:113–7.

Tauber, M. J., C. A. Tauber, J. P. Nyrop, and M. G. Villani. 1998. Moisture, a vital but neglected factor in the seasonal ecology of insects: Hypotheses and tests of mechanisms. *Environ. Entomol.* 27:523–30.

Tauber, M. J., C. A. Tauber, K. M. Daane, and K. S. Hagen. 2000. Commercialization of predators: Recent lessons from green lacewings (Neuroptera: Chrysopidae: *Chrysoperla*). *Am. Entomol.* 46:26–37.

Tauber, M. J., C. A. Tauber, and G. S. Albuquerque. 2009. Neuroptera (Lacewings, Antlions). In *Encyclopedia of Insects*, 2nd edition, ed. V. H. Resh and R. Cardé, 695–707. San Diego: Academic Press.

Thompson, W. R. 1951. The specificity of host relations in predacious insects. *Can. Entomol.* 83:262–9.

Thompson, S. N., and K. S. Hagen. 1999. Nutrition of entomophagous insects and other arthropods. In *Handbook of Biological Control: Principles and Applications of Biological Control*, ed. T. S. Bellows and T. W. Fisher, 594–652. San Diego: Academic Press.

Tilmon, K. J. (ed.) 2008. *The Evolutionary Biology of Herbivorous Insects. Specialization, Speciation, and Radiation.* Berkeley: University of California Press.

Tjeder, B. 1966. Neuroptera–Planipennia. The lace-wings of Southern Africa, 5. Family Chrysopidae. *S. Afr. Anim. Life* 12:228–534.

Toschi, C. A. 1965. The taxonomy, life histories, and mating behavior of the green lacewings of Strawberry Canyon (Neuroptera, Chrysopidae). *Hilgardia* 36:391–433.

Tóth, M., A. Bozsik, F. Szentkirályi, A. Letardi, M. R. Tabilio, M. Verdinelli, P. Zandigiacomo, J. Jekisa, and I. Szarukán. 2006. Phenylacetaldehyde: A chemical attractant for common green lacewings (*Chrysoperla carnea* s.l., Neuroptera: Chrysopidae). *Eur. J. Entomol.* 103:267–71.

Treacy, M. F., J. H. Benedict, J. D. López, Jr., and R. K. Morrison. 1987. Functional response of a predator (Neuroptera: Chrysopidae) to bollworm (Lepidoptera: Noctuidae) eggs on smoothleaf, hirsute, and pilose cottons. *J. Econ. Entomol.* 80:376–9.

Umeya, K., and J. Hirao. 1975. Attraction of the jackfruit fly, *Dacus umbrosus* F. (Diptera: Tephritidae) and lacewing, *Chrysopa* sp. (Neuroptera: Chrysopidae) by lure traps baited with methyl eugenol and cue-lure in the Philippines. *Appl. Entomol. Zool.* 10:60–2.

Vanderzant, E. S. 1969. An artificial diet for larvae and adults of *Chrysopa carnea*, an insect predator of crop pests. *J. Econ. Entomol.* 62:256–7.

Vanderzant, E. S. 1973. Improvements in the rearing diet for *Chrysopa carnea* and the amino acid requirements for growth. *J. Econ. Entomol.* 66:336–8.

van Emden, H. F., and K. S. Hagen. 1976. Olfactory reactions of the green lacewing, *Chrysopa carnea*, to tryptophan and certain breakdown products. *Environ. Entomol.* 5:469–73.

Viana, G. G., and G. S. Albuquerque. 2009. Polimorfismo no padrão de manchas tegumentares de larvas e adultos de *Ceraeochrysa caligata* (Neuroptera: Chrysopidae) e redescrição dos instares larvais. *Zoologia* 26:166–74.

Wäckers, F. L., P. C. J. van Rijn, and J. Bruin (ed.). 2005. *Plant-Provided Food for Carnivorous Insects: A Protective Mutualism and Its Applications.* Cambridge: Cambridge Univ. Press.

Wilkinson, J. D., G. T. Schmidt, and K. D. Biever. 1980. Comparative efficiency of sticky and water traps for sampling beneficial arthropods in red clover and the attraction of clover head caterpillar adults to anisyl acetone. *J. Ga. Entomol. Soc.* 15:124–31.

Wilson, P. J., and A. S. Methven. 1997. Lichen use by larval *Leucochrysa pavida* (Neuroptera: Chrysopidae). *Bryologist* 100:448–53.

Winterton, S., and S. Freitas. 2006. Molecular phylogeny of the green lacewings (Neuroptera: Chrysopidae). *Austr. J. Entomol.* 45:235–43.

Withycombe, C. L. 1925. Some aspects of the biology and morphology of the Neuroptera. With special reference to the immature stages and their possible phylogenetic significance. *Trans. R. Ent. Soc. Lond.* 1924:303–411.

Woolfolk, S. W., and G. D. Inglis. 2004. Microorganisms associated with field-collected *Chrysoperla rufilabris* (Neuroptera: Chrysopidae) adults with emphasis on yeast symbionts. *Biol. Contr.* 29:155–68.

Woolfolk, S. W., A. C. Cohen, and G. D. Inglis. 2004. Morphology of the alimentary canal of *Chrysoperla rufilabris* (Neuroptera: Chrysopidae) adults in relation to microbial symbionts. *Ann. Entomol. Soc. Am.* 97:796–808.

Yazlovetsky, I. G. 1992. Development of artificial diets for entomophagous insects by understanding their nutrition and digestion. In *Advances in Insect Rearing for Research and Pest Management*, ed. T. E. Anderson and N. C. Leppla, 41–62. Boulder: Westview Press.

Yazlovetsky, I. G. 2001. Features of the nutrition of Chrysopidae larvae and larval artificial diets. In *Lacewings in the Crop Environment*, ed. P. McEwen, T. R. New, and A. E. Whittington, 320–37. Cambridge: Cambridge Univ. Press.

Zelený, J. 1984. Chrysopid occurrence in west palearctic temperate forests and derived biotopes. In *Biology of Chrysopidae*, ed. M. Canard, Y. Séméria, and T. R. New, 151–60. The Hague: Dr. W. Junk Publishers.

Zhang, Q.-H., K. R. Chauhan, E. F. Erbe, A. R. Vellore, and J. R. Aldrich. 2004. Semiochemistry of the golden-eyed lacewing *Chrysopa oculata*: Attraction of males to a male-produced pheromone. *J. Chem. Ecol.* 30:1849–70.

Zhang, Q.-H., M. Sheng, G. Chen, J. R. Aldrich, and K. R. Chauhan. 2006. Iridodial: A powerful attractant for the green lacewing, *Chrysopa septempunctata* (Neuroptera: Chrysopidae). *Naturwissenschaften* 93:461–5.

Zheng, Y., K. S. Hagen, K. M. Daane, and T. E. Mittler. 1993a. Influence of larval dietary supply on the food consumption, food utilization efficiency, growth and development of the lacewing *Chrysoperla carnea*. *Entomol. Exp. Appl.* 67:1–7.

Zheng, Y., K. M. Daane, K. S. Hagen, and T. E. Mittler. 1993b. Influence of larval food consumption on the fecundity of the lacewing *Chrysoperla carnea*. *Entomol. Exp. Appl.* 67:9–14.

Zhu, J., and K.-C. Park. 2005. Methyl salicylate, a soybean aphid-induced plant volatile attractive to the predator *Coccinella septempunctata*. *J. Chem. Ecol.* 31:1733–46.

Zhu, J., A. A. Cossé, J. J. Obrycki, K. S. Boo, and T. C. Baker. 1999. Olfactory reactions of the twelve-spotted lady beetle, *Coleomegilla maculata* and the green lacewing, *Chrysoperla carnea* to semiochemicals released from their prey and host plant: Electroantennogram and behavioral responses. *J. Chem. Ecol.* 25:1163–77.

Zhu, J., J. J. Obrycki, S. A. Ochieng, T. C. Baker, J. A. Pickett, and D. Smiley. 2005. Attraction of two lacewing species to volatiles produced by host plants and aphid prey. *Naturwissenschaften* 92:277–81.

25

Hematophages (Diptera, Siphonaptera, Hemiptera, Phthiraptera)

Mario A. Navarro-Silva and Ana C. D. Bona

CONTENTS

25.1 Introduction 633
25.2 Diptera 634
    25.2.1 Psychodidae (Phlebotominae) 634
    25.2.2 Culicidae 635
    25.2.3 Simuliidae 640
    25.2.4 Ceratopogonidae 641
    25.2.5 Tabanidae 642
    25.2.6 Sarcophagidae, Muscidae, and Calliphoridae 643
25.3 Siphonaptera 644
25.4 Hemiptera (Heteroptera) 645
    25.4.1 Cimicidae 645
    25.4.2 Triatominae 645
25.5 Phthiraptera 647
25.6 Final Considerations 647
References 648

25.1 Introduction

Insects of public health importance use different feeding strategies in the course of their life cycles. Strategies that allowed them to access new food sources made it possible for some insect species to become involved in the transmission of pathogenic agents, in different ways and with different degrees of efficiency.

More than 14,000 hematophagous arthropod species feed on the blood of different hosts, and in many cases can transmit pathogenic agents from one organism to another. Many species of insects, in the orders Diptera, Hemiptera, Phthiraptera, and Siphonaptera, for example, act as vectors of etiological agents that cause epidemics, including dengue, malaria, leishmaniasis, Chagas' disease, and bubonic plague, among others. In some species, all developmental stages and both sexes feed on blood; in others, only the adult females are hematophagous and seek out their hosts to obtain specific nutrients for egg production and for growth (Forattini 2002).

There are two basic ways of obtaining blood from the host: solenophagy, in which the blood is taken from the blood vessels, and telmophagy, in which blood is obtained by lacerating small vessels. Some insects, although they are not hematophagic and do not transmit etiological agents, may cause allergic reactions. These reactions are stimulated by compounds present in the saliva that is released during the bite, by inoculation of toxic substances (venom) through stingers or from urticating structures that are used for defense. Other species develop on the bodies of vertebrate hosts, where they feed on tissues and blood, and in this case, in addition to causing lesions that are sometimes serious, they can facilitate the development of secondary infections caused by microorganisms such as bacteria and fungi.

25.2 Diptera

The order Diptera includes a large number of insects that are involved in the transmission of agents harmful to humans and animals. This transmission is caused mainly by hematophagy, which attempts to disrupt the hemostasis of the host through a series of substances produced by the insect in glands associated with its digestive system.

25.2.1 Psychodidae (Phlebotominae)

The family Psychodidae belongs to the order Diptera and is divided into two subfamilies, Psychodinae and Phlebotominae, although only the latter includes hematophagic species. The phlebotomines have a cosmopolitan distribution and are represented by approximately 800 described species in 28 genera; 230 species occur in Brazil. Phlebotomines preferentially occupy forest environments, but in the course of the extensive historical deforestation, some species are now found in human-modified environments, in both rural and urban areas (Galati 1995; Andrade Filho et al. 2001; Azar and Nel 2003).

Females deposit their eggs in terrestrial microhabitats rich in organic matter that provides food for the larvae to develop. Breeding sites may be hollows and roots of trees, leaf axils, hiding places under rocks, or shelters of wild or domestic animals, and may be located in domestic, peridomestic, or wild environments. In rural areas, the accumulation of organic matter, including agricultural waste, leaves, and fallen fruit; pet droppings; food scraps; and the disposal of household water form peridomiciliar breeding sites. Precise information about the ecology of phlebotomines is scarce because of the difficulty in locating eggs, larvae, and pupae, and of monitoring their development (Forattini 1973; Alexander 2000; Ximenes et al. 2001; Massafera et al. 2005).

Although the majority of phlebotomine species do not transmit etiological agents, some become infected with and transmit arboviruses, bacteria, and protozoa. Many species of *Phlebotomus* and *Lutzomyia* are vectors of pathogenic agents. In the Americas, species of *Lutzomyia* are responsible for the transmission of *Leishmania*, which multiply in the cells of the mononuclear phagocyte system. When sucking blood, female *Lutzomyia* regurgitates from 10 to 100 promastigotes of *Leishmania* together with her saliva onto the skin of the host. Some components of the saliva with immunosuppressive and immunomodulatory activity inhibit the development of an immune-inflammatory response of the host against the parasite, aggravating the infection. Leishmaniasis is considered a zoonosis, and in humans can assume two main forms, cutaneous and visceral. Although humans are not important in the perpetuation of the transmission cycle, in certain areas they may play a leading role in maintaining the disease (Killick-Kendrick 1990; Marcondes 2001; Monteiro 2005).

Both male and female phlebotomines must supplement their diet with carbohydrates, which they acquire directly from the sap of plants, nectar, honeydew, and ripe fruit. These foods contain microorganisms, among them bacteria that are present in the digestive tracts of numerous species of insects. These carbohydrates are used as a complement to blood feeding. Prior feeding on a sugar solution is not necessary for hematophagy. Females of *Lutzomyia intermedia* (Lutz & Neiva) after feeding on a sugar solution are more stimulated to perform hematophagy, whereas females of *Lutzomyia longipalpis* (Lutz & Neiva) do not require sugar as a source of stimulation. Females of both species survive without a blood meal as long as 5 days after hatching, although *L. longipalpis* better tolerates the lack of a blood meal. Recently, high-performance liquid chromatography showed that honeydew is a natural source of energy for both sexes in phlebotomines. Honeydew is a liquid excreted by Aphididae or Coccoidea, which contains carbohydrates and amino acids, glucose, fructose, and sucrose, and is highly nutritious (Rangel et al. 1986; Tanada and Kaya 1993; Alexander and Usma 1994; Cameron et al. 1995; Souza et al. 1995). Following the blood meal, female initiates vitellogenesis, which requires the blood to be digested. The proteins in the blood are first digested by proteases secreted into the lumen of the digestive tract and that cross the peritrophic matrix. The peptides produced by the action of proteases are then digested by carboxy- and amino-dipeptidases, which may be free or attached to the gut epithelium. At the end of 4 days after a blood meal, the gut becomes empty (Telleria 2007).

Female sandfly can take their blood meals from a wide variety of vertebrates. The study of the stomach contents of blood-sucking insects is one alternative method to determine which animals are effectively used as blood hosts in the human environment. *L. longipalpis* (Lutz & Neiva), the vector of the main etiological agent causing visceral leishmaniasis in the Americas, sucks blood from birds, domestic mammals, and also from humans (Dias et al. 2003). Nery et al. (2004) observed that rodents are the main source of the blood meal for *Lutzomyia umbratilis* Ward & Fraiha and *Lutzomyia spathotrichia* Martins, Falcão & Silva, the principal vectors of leishmaniasis in the region of Manaus, Amazonas. Rangel et al. (1986), through laboratory analyses with *L. intermedia* and *L. longipalpis*, concluded that the best source of a blood meal is hamster, which is better accepted by females, and increase life cycle and fecundity. According to Muniz et al. (2006), *Nyssomyia whitmani* (Antunes & Coutinho) and *Pintomyia fischeri* (Pinto) arc opportunists, and females adjust their feeding habits to the availability of hosts. The feeding flexibility of certain species of sandflies, according to the availability of blood supplies in human-altered environments, suggests feeding eclecticism in these insects.

During the blood meal, the sandfly injects saliva into the host tissues. The molecules contained in the saliva assist in searching for the best place to bite and in overcoming the natural physiological consequences of the bite, to transmit the parasite. Sandfly saliva has chemotactic properties. This chemotaxy may be related to the saliva-induced release of cellular factors that attract certain cells of the immune system to the bite area. The chemoattractant action of extracts from the saliva of *L. longipalpis* has been demonstrated: in animals susceptible to infection by species of *Leishmania*, the number of macrophages in the liquid of the inflammation pocket was much greater than that of animals resistant to infection. The participation of chemoattractant molecules is induced by the sandfly's saliva, with consequent recruitment of the macrophage cells. This chemotactic factor is highly important for intracellular parasites such as species of *Leishmania*, which must enter the phagocytic cells and escape the elimination effects of the extracellular medium. The greater the number of macrophages present in the area of the vector's bite, the higher the chances of intracellular parasite to establish and initiate an infection (Teixeira et al. 2005; Silva 2009).

The presence of anti-saliva antibodies induced by the host and the role of the saliva in regulating the immune response show that salivary substances are strong candidates for the production of possible routes of immunization. Animals exposed to bite of *Phlebotomus* produce antibodies that are specifically directed toward the respective species of insect. Maxadilan, a vasodilator molecule found in the saliva of *L. longipalpis*, is also specifically immunogenic. In laboratory and domestic animals and in humans in an area endemic for *L. longipalpis*, exposure to this vector induced the production of anti-maxadilan antibodies. This provides valuable knowledge about the mechanisms of transmission involving pathogens, and constitutes a powerful target for interception of chains of transmission of the diseases spread by blood-sucking insects (Volf and Rohousová 2001; Milleron et al. 2004; Silva 2009).

25.2.2 Culicidae

The family Culicidae belongs to the order Diptera, suborder Nematocera, infraorder Culicomorpha, and is divided into two subfamilies: Anophelinae and Culicinae. The family contains 3490 described species in 44 genera and 145 subgenera (Forattini 2002). Culicids exploit a wide diversity of environments, from forests to areas heavily affected by human activities. The list of species important to human health in a particular region may change as a consequence of the introduction of new species or of behavioral changes, forming new epidemiological scenarios (Forattini 2002; Harbach 2007).

Culicids are holometabolic, and begin their life cycle in an aquatic environment, where the female deposits her clutch of eggs. Embryonic development generally takes only a few days and is influenced mainly by temperature and humidity. The eggs of some species resist dry conditions and remain viable for long periods. For example, females of *Aedes (Stegomyia) aegypti* L. deposit their eggs in water. These eggs are highly permeable; however, this permeability lasts for only a few hours, which is the time that the embryo takes to construct a serous cuticle. This cuticle makes the egg impermeable to the external medium, able to resist periods of desiccation for up to a year. Upon contact with water, the larva hatches and continues the life cycle (Rezende 2008).

We will discuss in detail two species that use similar resources in the larval and adult stages, with real and potential importance for public health. *Aedes aegypti* originated from the Afrotropical region, but now has a cosmotropical distribution. It is an urban species, capable of exploiting both natural and artificial breeding sites. Females seek their hosts during the day, and accomplish the bite quickly. This is the principal vector species of the four serotypes of dengue fever (Flaviviridae) and the yellow fever virus (rural and urban) (Forattini 2002). The other species, *Aedes albopictus* (Skuse), although found infected by the dengue virus in the state of Minas Gerais (Serufo et al. 1993), is not a vector in Brazil. However, it should be monitored because of its rapid expansion in the country (Forattini 1996; La Corte dos Santos 2003). *Ae. albopictus* shows the characteristics of an excellent invader because it is capable of using both natural and anthropic ecological niches (Gomes et al. 1992).

Aedes scapularis (Rondani) is the probable vector responsible for the epidemic of encephalitis caused by the Rocio virus in Vale do Ribeira in the state of São Paulo, Brazil (Forattini et al. 1981). It is also vector of *Dirofilaria immitis* (Nematoda, Onchocercidae), a parasite of dogs and cats. Immatures develop in temporary or semipermanent water accumulated on soil subject to periodic flooding and drying, such as pools in rivers, margins of swamps, and wetlands (Forattini 2002; Gomes et al. 2003; Paterno and Marcondes 2004).

Culex quinquefasciatus Say is the primary vector of *Wuchereria bancrofti*, the causative agent of human lymphatic filariasis (elephantiasis), a disease with socially stigmatizing sequelae. Filariasis occurs in the states of Pará and Pernambuco in Brazil, with a significant prevalence (Maciel et al. 1999). *Cx. quinquefasciatus* has been identified as a vector of *Di. immitis*. This species has a wide distribution and a high degree of association with domiciles, and is often found associated with human activity. The immatures can develop in artificial breeding sites such as plastic containers, pottery, discarded tires, and vases in cemeteries, as well as in water-treatment ponds or domestic reservoirs for gray (used) water. Another important site from the public health point of view is drainage ditches for improperly treated household wastes. These widely available breeding sites can generate significant numbers of adults, increasing the epidemiological risks.

Two other species, *Haemagogus janthinomys* Dyar and *Haemagogus leucocelaenus* (Dyar & Shannon), are found in forest environments; the latter species may also occur in woodlands such as parks, groves of trees, or patches of woods within the urban landscape. The former is the principal vector of the wild yellow fever virus, and the latter is considered the secondary vector (Forattini 2002). Immatures develop in treeholes, with peculiarities that affect the form of development of the immature stages. Both species prefer primates for blood feeding.

Culicids can develop in sites of natural or artificial origin, depending on the species' requirements. Natural sites include bromeliads, bamboo internodes, treeholes, ponds, and pools, among others. Artificial sites result from human activity, and may be discarded tires, water-storage tanks, cemetery vases, ceramic vases, cans, plastic or glass containers, or any other structure that accumulates water long enough to allow the immatures to develop into the emergence of the adults. In the Culicidae, some species are eclectic in the occupation of breeding sites, but the majority of them show preferences and specificity for the type of environment. *Aedes aegypti* and *Cx. quinquefasciatus* are capable of exploiting artificial and natural containers, and are easily found in the urban environment, although they show specificity concerning the water quality of sites for depositing their eggs. The immatures of *Anopheles (Kerteszia) cruzii* Dyar & Knab are found in bromeliads of different water-storage capacities, with a preference for shaded sites. Eggs may be placed individually, in a group forming "rafts" or in groups forming "rosettes" (*Mansonia*). The procedure adopted by the species for egg deposition determines the density of immatures in the containers as well as the dispersal of the species, although the number of eggs in a clutch may vary within a species. Egg deposition behavior also varies. Some species place their eggs directly on the water (*Culex* and *Anopheles*), others on wet surfaces next to water (*Aedes, Ochlerotatus, Psorophora*, and *Haemagogus*) or on leaves of aquatic macrophytes (*Mansonia* and *Coquillettidia*) (Veloso et al. 1956; Forattini 2002; Natal 2002).

Embryonic development generally occurs over a few days, and is influenced mainly by temperature and humidity. The eggs of some species can resist drying and remain viable for long periods. The eggs of *Ae. aegypti* and *Ae. albopictus*, for example, are resistant to lack of water and can remain out of water for several months, which makes it possible for them to be passively dispersed. The larvae take in oxygen directly from the air or from plant tissues, by means of a siphon located on the VIII abdominal segment. This siphon is modified in different species; the most profound changes are found in the genera *Mansonia* and *Coquilletidia*, where the respiratory siphon perforates the tissues of aquatic macrophytes, extracting oxygen from the aerenchyma of the plants. The larval stage is the period of feeding and growth of the insect; the immatures pass most of their time feeding, principally on organic matter and microplankton. The duration of the larval stage depends on the temperature, availability of food, and the density of larvae in the site. Any nutritional deficiencies of larval diets are reflected in lengthening of the development time of the immature stages, as well as increased mortality during the transition to the adult stage (Bergo et al.1990; Beyruth 1992).

The mouthparts of immature mosquitoes have toothed mandibles and oral brushes, which help in filtering foods. The larvae of Culicidae can be classified according to feeding behavior, as passive filterers or as active collectors using different methods such as scraping, biting, or predation (Forattini 2002). Access to oxygen tends to limit the life of immature culicids, since they have an open respiratory system that requires the larvae to remain at the water–air interface. Thus, the functionality of the respiratory siphon that facilitates contact with air is apparent. The immatures of Anophelinae do not possess this siphon, and they remain horizontally on the water interface (Forattini 2002). This behavior allows them to collect materials that accumulate on the surface film, richer in organic matter and microorganisms compared to the rest of the water column (Badii et al. 2006).

For certain species of culicids, such as those of the genera *Mansonia* and *Coquillettidia*, the larvae and pupae attach to roots of aquatic plants, where they extract oxygen for respiration (Forattini 1965). The breeding sites of these mosquitoes are rich in aquatic vegetation. Species of other genera generate their own feeding currents, and remove organic matter with their oral brush. *Cx. quinquefasciatus* uses the collecting–filtering feeding mode; this manner of feeding employs a well-developed filtration mechanism, with adaptations principally in the muscles of the pharynx. The food is efficiently selected and the stomach contents are small, even if the external medium is oversupplied with food substances (Morais et al. 2006). This characteristic allows *Cx. quinquefasciatus* to survive in polluted aquatic habitats.

Larvae of *Toxorhynchites* and *Psorophora* are highly voracious and predatory. Larvae of *Toxorhynchites* act as biological control agents of immatures of other culicids; for example, *Toxorhynchites splendens* Wiedemann predate larvae of *Ae. albopictus*. Normally the larvae of *Toxorhynchites* are positioned at a 45° angle to the water surface. At the moment of feeding, they move their body to the horizontal position, and when the prey approaches, the predator strikes laterally and seizes the prey with its mandibles (Steffan and Evenhuis 1981; Toma and Miyagi 1992; Collins and Blackwell 2000; Badii et al. 2006).

The liquid exploited by mosquitoes as breeding sites range in size from a large lake to the tiny dark interior of a bamboo internode. In the context of exploitation and colonization of different types of breeding site, specialization arises, which reduces competition for space and food. Evaluation of competition between *Ae. aegypti* and *Ae. albopictus* indicated that the former has a greater competitive capacity than the latter at an intermediate density; however, *Ae. albopictus* survived better at a high density (Forattini 2002; Nunes 2005).

Larvae that inhabit temporary water, or those that cohabit with predators must obtain energy fast in order to accelerate their development. Different intrinsic and extrinsic factors of the breeding site determine the success or failure of larvae in reaching the pupal stage. For *Ae. albopictus*, large and medium (permanent) breeding sites were more productive, each contributing 2.8 females/day. Small and medium-sized natural breeding sites produced a daily mean of 0.5 and 0.6 females/day, respectively (Brito and Forattini 2004).

Although brief, the pupal stage is a period of transition when profound tissue transformations take place, which lead to the formation of the adult and the change from the aquatic to the terrestrial habitat. During this period the pupae do not feed, although they continue to obtain oxygen from the atmosphere by means of structures called breathing tubes, located on the cephalothorax. After a brief period in the pupal stage, which can last up to 5 days, adults emerge. In *Ae. aegypti*, the period from egg to adult may take 10 days, depending on the environmental conditions. In *An. cruzii* under laboratory conditions, the observed duration of transition from egg to adult is approximately 30.71 ± 3.57 days (Forattini 2002; Chahad-Ehlers et al. 2007).

Male and female adults show different behavior in their nutritional requirements. Whereas adult females must feed on the blood of vertebrates and on plant sugars, males feed exclusively on plant carbohydrates. The single exception known to this type of behavior in Culicidae are females and males of *Toxorhynchites* that feed exclusively on carbohydrates. Sugars are ingested by Culicidae through the proboscis and are stored in the ventral diverticulum, which functions as a reservoir, resulting in slow digestion of sugars and absorption of water in the stomach, allowing the female to keep her stomach empty and ready to receive blood. In mosquitoes, the fat body is the principal organ of intermediary metabolism, functioning as a storage organ. These substances in the adults are derived from the energy reserves accumulated during the larval stage (Steffan and Evenhuis 1981; Foster 1995; Alves et al. 2004).

The nutrition of female culicids has physiological, reproductive, and epidemiological consequences. Females must activate neuroendocrine mechanisms to complete ovarian maturation by means of blood feeding. Nutrients in the blood complement the energy reserves acquired in the larval stage, initiating the process of yolk deposition in the ovarian follicles (vitellogenesis) (Zhou et al. 2004). The foregut contains two suction structures that serve to ingest food: (a) the cibarial pump, located beneath the clypeus and provided in its terminal part with a crest formed by sclerotized spicules, the teeth of the cibarium; and (b) the muscular pharyngeal pump, which generates the negative pressure to ingest food (Consoli and Oliveira 1994). This pump ruptures the red blood cells so that they do not obstruct the passage of blood through the gut.

Morphological studies with *Cx. quinquefasciatus* suggest that during blood digestion, the foregut participates in the initial stage of absorption and is probably related to the uptake of water, salts, and small molecules. This absorption activity reaches its peak 6 h after the blood is ingested, and ceases after about 18 h, when the peritrophic membrane is formed. Subsequently, absorption occurs only in the hindgut, with morphological and biochemical evidence of high synthetic activity, related to the secretion of proteases. Chymotrypsin, elastase, aminopeptidase, and trypsin reach their maximum activity at about 36 h. The products of digestion are apparently absorbed and transported to the basal labyrinth, from where they are released to the hemolymph; 72 h after the blood meal, proteases are used up and the protein levels return to those observed before the blood meal (Okuda et al. 2002).

Another important compound for maintenance is trealose. Mosquitoes utilize this disaccharide for energy metabolism during fasting and flight, mostly during long flights. They begin to fly using trealose, and, after a certain time, they obtain the required energy from lipids (Arrese and Soulages 2010).

Salivary glands of females play an important role in hemophagy. The injected saliva contains antihemostatic and anti-inflammatory substances that are used to overcome the hosts' barriers and allow blood sucking. The function of saliva differs when feeding on sugars or on blood; saccharose is hydrolyzed, trealose remains unaltered, and bacteriolytic factors are present (Forattini 2000).

The saliva of mosquitoes lubricates the stylets during hematophagy, and also serves to help locate the host's blood. With no saliva, *Ae. aegypti* takes longer to locate the host's blood, but feeding time does not change, which last less than 10 min (Mellink et al. 1981).

During blood feeding, in addition to bypassing physical defenses of the host, the insect also needs to bypass three efficient vertebrate defense systems: hemostasis, inflammation, and immunity. These three complex physiological responses potentially prevent hematophagy. In general, the saliva of blood-sucking insects contains at least one anticoagulant, one anti-platelet, and one vasodilator substance. Compounds such as adenosine and nitric oxide, which possess anti-platelet and vasodilator activity, are present. In *Ae. aegypti*, the vasodilator molecules consist of peptides belonging to the tachykinin family (sialokinin I and II). Tachykinins bind to receptors in the vessel wall and induce release of nitric oxide, causing vascular relaxation. Mosquitoes of the genus *Anopheles* use peroxidase as a vasodilator (Ribeiro and Francischetti 2003; Silva 2009).

Culex quinquefasciatus appears to have a recent relationship with humans since it is more efficient in sucking blood of birds compared to that of mammals (mouse and humans); its saliva apparently lacks efficient antiplatelet mechanisms. Other mosquitoes evaluated (*Anopheles albimanus* Weideman, *Ae. aegypti*), which feed on mammals, contain larger amounts of apyrase and better antiplatelet mechanisms in their saliva, supporting the idea that *Cx. quinquefasciatus* has only recently became in contact with humans (Ribeiro 1987, 2000; Silva 2009).

Some females do not require a blood meal to initiate vitellogenesis, and are termed autogenic. Studies with *Culex pipiens molestus* Forskål and *Ae. aegypti* indicate that the energy obtained from the larval diet (mainly lipids) has an important role in inducing ovarian development (Suwabe and Moribayashi 2000; Zhou et al. 2004). It is presumed that autogenesis is an adaptive strategy to shortage of blood supply (Forattini 2002). In *Ae. aegypti*, *Ae. albopictus*, and *An. cruzii* Dyar & Knab, multiple blood meals are necessary for eggs to mature (Dalla-Bona and Navarro-Silva 2006; Lima-Camara et al. 2007). The duration of the gonotrophic cycle is defined as the period between the blood meal and oviposition (Fernandez-Salas et al.1994). From the epidemiological point of view, the more blood meals females have during the same gonotrophic cycle, the greater the probability of getting infected and of transmitting pathogenic agents (Fernandez and Forattini 2003).

The parasite–vector physiology affects feeding by mosquitoes. As observed by Koella and Packer (1996), females of *Anopheles punctulatus* Doenitz complex infected by malarial plasmodia suck blood longer and more often. The short duration of the blood meal in *Ae. aegypti* is an adaptive strategy to mitigate interference of the host during feeding (Chadee et al. 2002). The number of egg clutches per female depends on the availability of blood to support ovarian development. A satisfactory meal is attained with 3.0 to 3.5 mg of blood, yielding about 120 eggs (Forattini 2002).

Food ingested by adult affects survival of culicid females. Female mosquitoes fed a sugar solution have longer life expectancy than females fed only blood (Harrington et al. 2001; Fernandez and Forattini 2003; Gary and Foster 2004). Probably an adaptive radiation occurred, which induced mosquitoes to take two evolutionary directions, one toward dependence on blood exclusively, and another toward dependence on sugar (Foster 1995). For instance, in the subfamily Toxorhynchitinae, where both sexes are restricted to feeding on carbohydrates, females do not require blood (Steffan and Evenhuis 1981). The option for sugars appears to have been retained in these mosquitoes, given the advantages that it offers for survival (Forattini 2002).

Most female mosquitoes are eclectic and opportunistic in their choice of host; however, the type of blood influences the mosquito's behavior. The high degree of synanthropy of some species is an adaptation for the strategy of feeding on human blood (Braks et al. 2001). Some anthropophilic species, such as *Ae. aegypti* and *Anopheles gambiae* Giles *sensu stricto*, obtain all the energy needed by the adult from human blood alone; their biting rate is increased, without affecting reproductive fitness (Gary and Foster 2001).

The biochemical composition of human blood, a rich cocktail of amino acids, has advantages for the reproductive capacity of mosquitoes and synthesis of energy reserves, compared to the blood of other vertebrates. The isoleucine present in the blood diet increases the fecundity of *Ae. aegypti*. Females fed rat blood (which has a high level of isoleucine) plus a sugar solution produce more eggs, followed by females fed with human blood (low level of isoleucine) plus sugar (Harrington et al. 2001). *Ae. aegypti* benefits from the accumulation of energy reserve and has reproductive success when it ingests both low and high levels of isoleucine. Comparison of the components of blood from different hosts demonstrated that *Ae. aegypti* females fed human blood lived as long as females fed rat blood. In females fed high levels of isoleucine (rats) with no access to carbohydrates, nutrients in the blood are used in vitellogenesis, resulting in death from depletion of their energy reserves. Triglycerides are the principal source of energy for mosquitoes, and plant carbohydrates and nectar are the most efficient substrates for synthesis of sugar and glycogen used for flight. *Ae. aegypti* synthesizes triglycerides from blood, remaining for long periods in the same house where the host, mating partners, and locations for oviposition are available within a few meters (Harrington et al. 2001).

In insect vectors of etiological agents, salivary glands play a fundamental role in the interaction between parasite and vector. When anophelines, which transmit the protozoan causing malaria, suck blood from an infected person, they ingest asexual forms of *Plasmodium*, which will then pass through a complex development within the mosquito. Physiological factors and strategies of the vector and the parasite are required, such as tissue specificity, so that all events of parasite differentiation can take place. The sporozoites, the infective forms of the parasite, possess specific compatible receptors, penetrate the salivary glands of the mosquito, and are injected directly into the circulating blood of a susceptible human (Lourenço-de-Oliveira 2005). These elements lead culicid females to enter into contact with human blood in order to assure their reproductive success, and in parallel to be disseminators of pathogens during blood meal.

Biological control strategies that are part of integrated pest management programs have emphasized the use of entomopathogenic bacteria to control mosquito larvae in different types of breeding sites. The bacterium *Bacillus thuringiensis israelensis* (Bti) possesses three different Cry toxins (toxic crystal) and one Cyt (toxin with cytolytic and homolytic activity). The high insecticidal activity is due to the toxic proteinases located in parasporal bodies (crystals). These are produced in the second stage of sporulation, during the formation of spores. After larvae ingest the crystals, these are dissolved in the acid or alkaline medium of the midgut and protoxins are released; protoxins do not yet show biological activity, and proteolytic activation is necessary. The proteases of the gut unfold the protoxins and produce a smaller activated protein. This toxin must pass through the peritrophic membrane to be recognized by the specific receptors present in the apical microvilli of the midgut. After binding to the receptor, the toxin creates pores that interfere with the system of ion transport across the tissue membrane. This process causes lysis of the midgut epithelium and/or interrupts normal secretion, lowering the pH of the lumen and favoring germination of the spores that will result in septicemia and death of the insect. The insect's feeding may be inhibited soon after it ingests the spores and the toxin, causing its death. Bti is marketed on a large scale for the control of Culicidae and Simuliidae, and a variety of efficient products are commercially available. Bti has been used in intensive campaigns in the United States and Germany for mosquito control, and in Africa to combat simuliid vectors of onchocercosis (Glare and O'Callagham 2000).

25.2.3 Simuliidae

The family Simuliidae belongs to the order Diptera, infraorder Culicomorpha, and superfamily Chironomoidea, and is more closely related to the families Ceratopogonidae and Chironomidae. The family has a wide distribution and contains approximately 1800 described species in 24 genera. In Brazil, approximately 78 species of simuliids are known, four of them belonging to the genus *Lutzsimulium* d'Andretta & Andretta, and the remainder included in *Simulium* Latreille (Crosskey and Howard 2004).

Dipterans of this family are called *borrachudos* or *piuns* in Brazil, and black flies in English. They are transmitters of protozoans, nematodes, viruses, and bacteria to domestic animals and to humans. Some species are intermediate hosts of *Onchocerca volvulus* and *Mansonella ozzardi*. Other species, although they do not have an important role as vectors, can bite humans and domestic animals, and cause economic damage to agriculture and tourism. In southern and southeastern Brazil, *Simulium pertinax* Kollar is the most common species and is an important pest: it is anthropophilic and can cause allergic reactions and great discomfort at the moment of the bite, principally when present in high densities (Oosterbroek and Courtney 1995).

Ornithophilic species frequently transmit *Leucocytozoon* (Apicomplexa: Plasmodiidae), parasites of birds. Some species that feed on mammals transmit the filarial nematodes *Dirofilaria*, *Mansonella*, and *Onchocerca* (Kinetoplastida: Onchocercidae). The allergic reaction to the bites of simuliids can be serious, and their saliva is suspected of being a cofactor in the transmission of the human herpes virus (Werner and Pont 2003). In the Neotropical region, several species of *Simulium* carry the filarial parasite *O. volvulus* that causes onchocercosis, a serious public health problem (Lozovei et al. 2004). Females oviposit on rocks, branches and leaves, substrates found in waterfalls, rivers, and streams. Each female deposits 200 to 300 eggs on average, which take 5 to 6 days to hatch, depending on the water temperature. The total number of instars ranges from 6 to 9 in the different species. The biological cycle from the incubation of the egg to the emergence of the adult in the genus *Simulium* can range from 30 to 49 days (Viviane and Araújo-Coutinho 1999).

Immatures of simuliids are restricted to river ecosystems. Larvae are filterers and require a water current for feeding; the nutrients are extracted by means of specialized structures named cephalic combs. These species are an important link between suspended particles and predators because larvae alter the size spectrum of organic matter particles. The dimensions of the food particles ingested by larvae of simuliids vary widely, depending on the availability in the biotope, since they do not select their food.

In general, they feed on a wide range of diatoms, algae, fungal spores and mycelia, rotifers, and bacteria. Larvae possess morphological adaptations such as suckers and false legs on the abdomen that aid in locomotion. They synthesize a silky substance from which they weave silk threads that are used for anchorage on solid substrates in the water. This allows constant move to new microhabitats that offer a renewed flow of food particles. Simuliids are ecological key species due to their ability to filter dissolved organic matter and make it available to the food chain (Thompson 1987; Lozovei 1994; Malmqvist et al. 1999; Alencar et al. 2001; Werner and Pont 2003).

In contaminated environments, simuliid larvae are potential subjects for study because they serve as food for many fishes and are considered an entry point for the uptake polluting substances in the food chain. The natural stages (larvae and pupae) are susceptible to organic and inorganic pollutants (Harding et al. 2006). The presence of organic matter in the water, in increasing amounts, is the principal cause of the growing populations of blackflies. In the absence of aquatic predators, fish, and other insects, larvae find ideal conditions for development. In some parts of the world, simuliids proliferate so uncontrolledly that they seriously influence the quality of life of the local human populations.

Adults, mostly males, feed on nectar, which satisfies their energy requirements. Females must feed on blood, which is digested, converted to reserve, and used mainly for maturation of the follicles. Females are telmophages, possess a short proboscis, and lacerate small vessels, producing hemorrhage from which they suck blood; they add saliva to the wound to prevent coagulation. The saliva of simuliids contains a diverse array of anticoagulant substances. For example, the salivary gland of *Simulium vittatum* (Zetterstedt) contains molecules that inhibit the coagulating elements of blood, such as thrombin, Xa, and factor V. In addition to the anticoagulant effect, the act of biting is rapid and painless as a consequence of the anesthetic properties of the saliva. Compounds present in the saliva prevent hemostasis in the vertebrate host, making the saliva a rich source of anti-hemostatic molecules (Basanova et al. 2002; Silva 2009; Chagas et al. 2010).

Some components in the saliva of *S. vittatum* show chemotactic properties. This chemotaxis may be related to the release of cellular factors with properties of attracting certain cells of the immune system to the area of the bite. For example, the anticoagulant molecule simulidin, extracted from the salivary contents of *S. vittatum*, is also chemotactic to macrophages. Naturally, these macrophage cells initiate an immune response at the location of the bite by releasing regulatory molecules that later define inflammatory occurrences in the host, exacerbating or inhibiting them. In addition, vasodilator substances are present, increasing the flow of blood to the location of the bite. The vasodilation of the capillaries increases the supply of blood, which is an important requirement for the transmission of pathogen. *Simulium ochraceum* (Walker) is a vector of the parasite *O. volvulus* (Abebe et al. 1995; Cupp and Cupp 1997; Silva 2009).

25.2.4 Ceratopogonidae

The family Ceratopogonidae belongs to the order Diptera, infraorder Culicomorpha, superfamily Chironomoidea. It has a worldwide distribution, with approximately 5500 described species (Borkent and Wirth 1997). Adults are only 1–4 mm in length, and are among the smallest insects of the order Diptera. In Brazil they are called *maruins*, *mosquitos do mangue*, or *mosquitos pólvora*; the English name is biting midges. They are of great importance because they are hematophages and potential transmitters of pathogens, such as *Onchocerca volvulus*, *O. gibsoni* (bovines) and *O. cervalis* (equines), *Haemoproteus* and *Leucocytozoon* (birds), *Hepatocystis* (monkeys), and also the bluetongue virus (sheep and cattle), African horse sickness (equines), bovine ephemeral fever (bovines), and Akabane disease (cattle, sheep, and goats) (Mellor et al. 2000). In Brazil, the hematophagous ceratopogonids belong to the genera *Culicoides* Latreille, *Forcipomyia* Meigen subgenus *Lasiohelea* Kieffer, and *Leptoconops* Skuse.

The genus *Culicoides* is the largest of the family, with a wide geographical distribution; more than 50 arboviruses have been isolated from members of the genus. According to (Mellor et al. 2000), more than 1400 species of *Culicoides* are known worldwide; 96% are obligate feeders on the blood of mammals and birds. Females are hematophagous and feed on the blood of vertebrates, including humans; they can become nuisances in areas of beaches, forests, mountains, and mangroves. *Culicoides* are mainly crepuscular, although few species take blood meal during the day. Females fly in search of males, a blood meal, and an oviposition site; males do not feed on blood (Mellor et al. 2000).

Eggs of *Culicoides* are generally placed in batches attached to the substrate, are not resistant to dry environments, and generally hatch within 2–7 days. Aquatic larvae inhabit lentic environments with fresh or brackish water, where they swim in search of detritus or microorganisms to feed on. The semi-aquatic larvae inhabit water-saturated plant detritus, swimming pools, rivers, swamps, marshes, beaches, treeholes, irrigation ditches, pipe leaks, puddles in the soil, animal manure, rotting fruit, and other vegetation. The duration of the four instars varies with the species and the temperature, from 4 to 5 days to several weeks. In countries with a temperate climate, this period may be extended because most species enter into diapause as fourth instar larvae. The food of the immatures consists of plant particles, but some species are predators, feeding on nematodes, rotifers, protozoans, and small arthropods; the salinity of seawater can strongly influence the nutrition of *Culicoides*, and high concentrations of seawater inhibit the survival and maturation of immatures of *Culicoides molestus* (Skuse) (Blanton and Wirth 1979; Mullen and Hribar 1988; Wirth and Hubert 1989; Meiswinkel et al. 1994; Mellor et al. 2000; Carrasquilla et al. 2010).

The pupal stage is short, generally lasting 2–3 days. Adults of *Culicoides* survive for 10 to 20 days, but may eventually live longer. Males and females meet in large swarms along riverbanks, where they form a swirl. Females ingest blood, sugar solution, water, and nectar; they are telmophages, lacerating vessels of the host producing hemorrhage to suck blood. The largest part of the food is deposited in acellular sacs in the midgut diverticulum; if the food is blood, the sphincter muscle contracts to direct the meal to the posterior part of the midgut.

The ovarian cycle in *Culicoides* is similar to that of the Culicidae, and the ovarian follicles do not mature past stage II without a blood meal. In general, adults of *Culicoides* take several blood meals. According to Brei et al. (2003), salinity can have a large impact on the nutrition of *Culicoides*. High concentrations of seawater inhibit the survival and maturation of immatures of *Cu. molestus* (Skuse), whereas lower concentrations are more appropriate for the survival of the adult. The ceratopogonid *Forcipomyia townsvillensis* (Taylor) feed on blood, carbohydrates, and water; it can survive up to 39 days, with 50% of the population living up to 2 weeks at 98% relative humidity (Megahed 1956; Kettle 1962; Cribb 2000; Mellor and Baylis 2000).

Culicoides variipennis (Coquillett) possesses a vasodilator substance in its saliva that produces a reddened halo (erythema) around the petechial hemorrhage caused by the bite. During feeding, this vasodilator increases the flow of blood from the blood vessels of the host near the location of the bite. In high densities, these insects cause serious dermatological problems. The harmful effect of the hematophagous females is apparent directly on the body; persons receiving multiple continuous bites develop bullous formations that are complicated by secondary infections resulting from injuries caused by scratching. Certain species of ceratopogonids are extremely annoying because of the insistence with which they seek to introduce themselves into the eyes, nostrils, and ears. They may cause inconvenience and harm, especially in sites with tourism potential (Perez De Leon et al. 1997; Marcondes 2009; Silva 2009).

25.2.5 Tabanidae

The family Tabanidae belongs to the order Diptera, suborder Brachycera, infraorder Tabanomorpha, superfamily Tabanoidea. The family includes 4300 described species in 137 genera. In the Neotropical region, there are 1172 species in 65 genera, belonging to three subfamilies: Chrysopsinae, Pangoniinae, and Tabaninae. The tabanids are popularly known as *mutucas* or *botucas* in Brazil, and horseflies or march flies in English-speaking countries (Fairchild and Burger 1994; Yeates and Wiegmann 1999).

Tabanids have medical and veterinary importance because of the blood spoliation by females. The wound opened by the bite allows bacterial invasion and the emergence of myiases. Tabanids are mechanical and biological transmitters of pathogens to domestic and wild animals, for example, infectious anemia of equids, vesicular stomatitis, encephalitis and swine fever, as well as *Anaplasma marginale*, *Trypanosoma evansi*, *Try. vivax*, and *Try. equiperdum*, and also carbuncles, brucellosis, tularemia, and bovine leukosis. Their economic importance is related to the spoliator and irritant effects they cause, mainly in equines and bovines, interfering with the feeding and rest of the animals, causing loss of milk production and weight, myiases, and depreciation of the value of the hide (Bassi et al. 2000; Prado 2004).

Oviposition occurs in aquatic or semiaquatic environments. Eggs are deposited in a mass on plants or muddy soil, but also on tree trunks, rocks, and bridge sides. The first instar larvae hatch after about 2–3 days. Larval development time can range from 1 year to more than 2 years, with up to 10 instars or more. Different aquatic systems are mentioned as breeding sites of immatures of Tabanidae, including lentic or lotic environments such as lakes, marshes, and shallow streams; on different natural substrates such as macrophytes, roots, leaf litter, or bromeliads; and even in artificial environments. Immatures of Tabanidae are terrestrial, aquatic, or semiaquatic predators, feeding on the body fluids of their prey. They frequently practice cannibalism in food-limited situations; thus, they require habitats that provide adequate food. When fully developed, larvae generally move to drier environments to pupate. The pupal stage lasts 2–3 weeks, depending on species and temperature (Wiegmann et al. 2000; Ferreira and Rafael 2006).

Adults live for about 2 months, and copulate soon after they emerge. Adult males feed only on carbohydrates; females are anautogenic, that is, they require a blood meal for maturation of the oocytes, and are telmophagic, making an incision in the skin and feeding on the extravasated blood. In addition to mammalian hosts, including equines, swine, bovines, tapirs, sloths, and humans, some species also feed on reptiles such as *Caiman crocodilus* (L.) and *Eunectes murinus* (L.), in Amazonia. Females lay from 100 to 1000 eggs for each meal taken. When they succeed in obtaining a full meal, they resume feeding at intervals of 6–10 days after oviposition. The majority of the hematophagous tabanids appear to rely on the abdominal stretch receptors as the preliminary mechanism to terminate the blood meal and search for the host. The energy used for flight in Tabanidae is derived from different sources of carbohydrates. Their behavior of flying above the host can reduce energy reserves, but can also be an evolutionary response to predation pressures or the activity profile of the female (Charlwood et al. 1980; Smith et al. 1994; Adams 1999; Ferreira et al. 2002).

According to Cilek and Schreiber (1996), 96% of *Chrysops celatus* Pechuman adults collected had fed on fructose, and 92% of those that sought food on a host were nulliparous. The relationship between mating and feeding of females of Tabanidae indicates that copulation precedes feeding on blood, since 90% of the females seeking hosts had previously copulated. The reproductive success of the males in flying over hosts of the females is directly related to sugared food (Yuval 2006).

The bite is painful because females show a high degree of hematophagy. They are able of ingesting up to 0.5 ml of blood per individual, and additional blood may be lost due to scraping that follows the bite. Other substances present in the salivary gland of hematophagous insects are linked to the feeding behavior of the vector. *Chrysops viduatus* (F.) possesses in its saliva compounds with enzymatic activity for hyaluronidase. This enzyme facilitates the enlargement of the feeding lesion and propagates other pharmacologically active compounds present in the saliva. The vasodilator effect is also observed in tabanids, the salivary extract causing arterial relaxation. Another important step for efficient blood feeding by hematophagous insects is the blocking of the coagulation cascade. Thrombin inhibitors were detected in the saliva of several species of tabanids (Kazimírová et al. 2001; Rajská et al. 2003; Takác et al. 2006; Krolow et al. 2007; Volfova et al. 2008; Silva 2009).

25.2.6 Sarcophagidae, Muscidae, and Calliphoridae

The dipterans of the families Muscidae, Calliphoridae, and Sarcophagidae are considered potential vectors of pathogenic agents, such as viruses, bacteria, cysts of protozoans, and eggs of helminths (Marchiori et al. 1999). Some species can cause discomfort and irritation at the moment of the bite or by remaining on the body of the host. With human population increase, abundance of these flies also increased, in domestic trash and feces of pets, which provide substrate for breeding and are a source of food (Oliveira et al. 2002).

During the larval stage, food resource limitation may occur and competition is increased, with each larva seeking to ingest the maximum amount of food before resources are exhausted (Gomes and Von Zuben 2003). Some larvae feed on carcasses (necrophages), and use decomposing organic matter as a source of proteins for ovarian development in the adult or for development of immatures (Oliveira-Costa 2003; Moura et al. 1997). Their activity accelerates the process of decomposition and disintegration of carcass, making them key elements in forensic entomology, serving as indicators decomposition time of human cadavers.

Among necrophagous insects, Sarcophagidae, Muscidae, and Calliphoridae are highly important in the succession pattern of the entomological fauna. Larvae of Calliphoridae are frequently collected from cadavers (Andrade et al. 2005) in different stages of decomposition. The species of *Ophyra* Robineau-Desvoidy (Muscidae, Azeliinae), in the third instar are facultative predators, and are generally associated with human feces, feces of chickens, and swines. In these substrates, larvae of *Ophyra* often predate larvae of *Musca domestica* L. and *Stomoxys calcitrans* L., in addition to sarcophagids and calliphorids (Krüger et al. 2003). These flies are also potential mechanical and biological transmitters of pathogens. In the family Muscidae, *Haematobia irritans* (L.) (horn fly) transmits the filarial nematode *Stephanofilaria stilesi* to bovines, and *Sto. calcitrans* (stable fly) transmits infectious anemia of equines (Prado 2004). Adults of the families Sarcophagidae and Calliphoridae are involved in the transmission of infectious agents such as *Escherichia coli* and the highly pathogenic avian influenza virus (Tachibana and Numata 2006).

Parasitism is also a food source for the larvae. Ectoparasites of medical and veterinary importance in Latin America include *Dermatobia hominis* (L.) (Diptera, Oestridae, Cuterebrinae), commonly known as the human bot fly. Larvae cause furunculous myiases in bovines and other domestic and wild animals, including humans. Parasitism by *De. hominis* larvae is called dermatobiosis, and is present in tropical and subtropical Americas (Pinto et al. 2002). Infestation occurs in certain types of fur and specific location on host body. Magalhães and Lima (1988), examining the frequency of larvae of *De. hominis* in cattle, noticed that the body sides are the most susceptible regions.

A peculiar characteristic of this fly is the need for another dipteran to transport its eggs. Several species of dipterans are known as vectors of *De. hominis* eggs, including members of Culicidae, Simuliidae, Anthomyiidae, Muscidae, Tabanidae, Fanniidae, Sarcophagidae, and Calliphoridae (Pinto et al. 2002). The number of specimens carrying eggs of *De. hominis* is higher at the end of the dry season, which explains the highest incidence of this parasite in cattle during this period (Gomes et al. 2002).

Losses caused by larvae of *De. hominis* in bovines include decreased in milk and meat production, and weight gain. The leather is most affected, reducing the commercial value. Larvae may be used in larval therapy, a type of biotherapy that involves the intentional introduction of disinfected live larvae into the tissues of wounded animals or humans (such as in the skin) for the purpose of selectively cleaning only the necrotic tissue of the wound, in order to accelerate the cure; this is possible because the larvae feed only on necrotic tissue (Marcondes 2006).

25.3 Siphonaptera

The name Siphonaptera derives from the Greek word "siphon," which means tube or pipe, regarding the mouthparts of fleas adapted to cut the skin and suck the blood. In the adult stage, hemophagy is carried out by the two sexes, during the day or night; fleas feed directly on the capillaries (solenophages) (Linardi 2004). Each blood meal lasts about 10 min, with 2–3 meals/day, and a female can ingest on average 14 µl of blood (Marcondes 2001). Studies with *Ctenocephalides felis felis* Bouche indicate that at the beginning of feeding on blood there is a reduction of proteins; the amount of proteins triples after the blood feeding, and then reaches a constant level (Hinkle et al. 1991). Larvae possess chewing mouthparts and live freely in burrows and nests of their hosts, feeding on the host blood expelled from the anus of the adult flea, and generally adhered to other organic detritus (Linardi and Guimarães 2000).

Fleas partake in different links of the epidemiological chain: invertebrate parasites per se, biological vectors, and invertebrate hosts (Linardi 2004). Pediculosis is caused by infestation by *Pediculus humanus corporis* De Geer (the body louse) or by *Pediculus humanus capitis* (De Geer) (head louse) (Heukelbach et al. 2003). Epidemic typhus, caused by *Rickettsia prowazekii*, trench fever, caused by

Bartonella quintana, and relapsing fever, caused by *Borrelia recurrentis*, are transmitted by body lice (Fournier et al. 2002).

This order also includes representatives of the family Tungidae such as the *bicho-de-pé* or chigger, which causes tungiasis, an ectoparasitic disease caused by the penetration of the female *Tunga penetrans* (L.) into the epidermis of its host; this flea subsequently hypertrophies to reach about 1 cm (Ariza et al. 2007).

25.4 Hemiptera (Heteroptera)

25.4.1 Cimicidae

The family Cimicidae belongs to the order Hemiptera, suborder Heteroptera, infraorder Cimicomorpha, superfamily Cimicoidea. The family is composed of 23 genera, and only three species are considered true ectoparasites of humans. They are known in Portuguese as *percevejos-de-cama*, and in English as bedbugs: *Cimex lectularius* L. occurs worldwide, *Cimex hemipterus* F. occurs mainly in the tropics, and *Leptocimex boueti* Brumpt is present in eastern Africa (Forattini 1990; Thomas et al. 2004).

Hematophagy probably developed only once in the Cimicidae since all species are obligate hematophages. Both sexes suck blood exclusively from vertebrates for their survival, growth, and reproduction (Reinhart and Silva-Jothy 2007). Cimicidae has a restricted choice of hosts; they were associated with bats, but adapted to other mammals and birds. The blood is sucked directly from the capillaries, and feeding is predominantly nocturnal (Marcondes 2001).

Nymphs of *Ci. lectularius* die within a few days after emerging if they do not feed on blood; blood meals account for 130% to 200% of the body weight of an unfed adult. A single full blood meal precedes eclosion for the next instar. Different host species or individuals generate meals of different sizes and amounts (Reinhart and Silva-Jothy 2007), due to variation in protein or micronutrients contents in the blood, such as calcium or vitamin B (DeMeillon and Hardy 1951). All cimicids shelter symbiotic microorganisms, usually *Rickettsia*, which aid in the digestion of vertebrate blood. These organisms, termed mycetones, are present in both sexes; they increase in number when the insect matures, but decrease with the age of the adult (Reinhart and Silva-Jothy 2007).

In *Ci. lectularius*, blood is sucked directly from capillaries, with injection of anticoagulant substances. They also feed on blood from damaged tissue. In general, bedbugs reach the state of repletion after 3–5 min for young nymphs and 10–15 min for adults. The frequency of blood meals is related to oviposition and molting, in addition to environmental factors such as temperature (Forattini 1990; Thomas et al. 2004).

Bedbugs cause great annoyance and loss of blood, which can lead to anemia in malnourished children (Marcondes 2001), disturbing sleep, since they inhabit the bed, in cracks and seams of the mattress fabric. Cimicids may carry the infectious agents of typhus, kala-azar (visceral leishmaniasis), anthrax, tularemia, hepatitis B, and HIV viruses. Silverman et al. (2001), in a review of the association between *Ci. lectularius* and the viruses of HIV and hepatitis, mentioned that both viruses persist inside the bedbug for several weeks, but no viral replication and no infectivity was detected.

25.4.2 Triatominae

The subfamily Triatominae belongs to the order Hemiptera, suborder Heteroptera, infraorder Cimicomorpha, family Reduviidae. The subfamily is constituted of five tribes, containing 15 genera. At present, 140 species are recognized, all of them hemophagous. Except for the genus *Lynchosteus* (found exclusively in India) and some species of the genus *Triatoma* Lap, all other triatomids are exclusive from the Americas. They are distributed from the United States to Argentina, the majority being neotropical. They are commonly known as *barbeiros, chupões, bicudos, vinchucas, chipos,* or *chinches* in Portuguese and Spanish, and as kissing bugs or assassin bugs in English (Forattini et al. 1982; Schofield et al. 1999; Schofield 2000; Galvão et al. 2003; Schofield and Galvão 2009).

Many species of triatomines are of epidemiological importance because they transmit *Trypanosoma cruzi*, the protozoan agent of Chagas' disease or American trypanosomiasis. Of these species, five are prominent: *Triatoma infestans* (Klug), *Rhodnius prolixus* (Stal.), *Panstrongylus megistus* (Burmeister), *Triatoma brasiliensis* (Neiva), and *Triatoma dimidiata* (Latreille). In these bugs, the protozoans multiply in the gut and the infectious forms are eliminated together with its feces and urine. Transmission of the infection occurs primarily through the deposition of the vector's feces on the cutaneous and mucosal tissues of humans (Coura 2003).

Triatomines probably evolved from various lineages of reduviids. It is possible that they were formerly predators and acquired hematophagous habits associated with a series of morphological, behavioral, and biochemical changes. These changes were mediated by the exploitation of the blood of vertebrate hosts as a source of food, by adaptation to the environment of the hosts, and by the dispersal of the hosts (Schofield et al. 1999).

After the egg stage, they pass through five immature stages (first to fifth instars) before reaching the adult stage. The majority of species of triatomines are wild, have nocturnal habits, and fly little. Females live longer and are more active than males, with a greater capacity for dispersal. In general, oviposition last 3–4 months, with a total production of 100–200 eggs/year. Eggs hatch 18–25 days after being laid (Coura 2005).

To molt from one stage to the next, at least one blood meal is necessary because it is the abdominal distension and protein factors obtained from the blood (hemoglobin) that activate the neurosecretory cells. These, in turn, determine a sequence of stimuli for the brain, which will produce hormones for molting (ecdysone) and growth (juvenile hormone) (Gonçalves and Costa 2010).

The nutritional state has little effect on the reproduction of males; however, lack of nutritional factors influence egg production by females. To compensate for smaller amounts of blood ingested, females seek food at briefer intervals and feed more often. From the epidemiological point of view, timing of meal is important since it is related to contact between vector and host (Gonçalves et al. 1997; Braga and Lima 2001).

Triatomines are voracious solenophagous hematophages in all developmental stages. They draw blood directly from the blood vessels and also take some extravascular blood, using the long flexible proboscis. They are able to resist long periods of fasting. For example, when *Pa. megistus* is starved, its reproductive potential is reduced but it is still able to maintain colonization. Virgin females of *Tri. brasiliensis* produce eggs even when they are not fed after the imaginal molt (Perondini et al. 1975; Forattini et al. 1981; Braga and Lima 2001; Gonçalves and Costa 2010).

A wide variety of biologically active substances have been detected in the saliva of triatomines: anticoagulants, molecules with an anesthetic effect, pore-forming molecules, complement inhibitor, inhibitors of collagen-induced platelet aggregation, arachidonic acid, ADP and thrombin, serotonin, epinephrine and norepinephrine, including vasodilators. Nitric oxide (NO) found in the saliva, functions as vasodilator, as in the case of *R. prolixus*. Because NO is an unstable gas, these insects store this compound in their saliva in the form of nitrophorines, proteins in the same class as the myoglobins and hemoglobins. NO injected in the form of nitrophorine induces vascular relaxation by elevating the intracellular levels of the mediator cyclic *guanosine* monophosphate in the muscle cells. That is, it produces vasodilation independently of the endothelium and inhibition of platelet aggregation, allowing continuous blood flow and creating a favorable feeding environment (Andersen and Montfort 2000; Silva 2009).

The source of the blood meal influences the life cycle and reproductive development in *Tri. infestans*, *Tri. brasiliensis*, *Tri. sordida*, and *Tri. pseudomaculata* Corrêa & Espínola. Studies on all these species found that the life cycle was shorter for the groups fed on mice than for those fed on pigeons. The mortality rate of nymphs tended to be higher in insects fed on pigeons than those fed on mice (Guarneri et al. 2000). Barbosa et al. (2007) found significant differences between two food sources (human and pigeon) for the total period of contact, frequency of the cibarial pump, speed of ingestion, and amount of liquid ingested through contraction of the cibarial pump.

Panstrongylus megistus is the main vector of Chagas' disease in Brazil, and information about its resistance to food limitation is import for control campaigns (Braga and Lima 2001). They concluded that when *Pa. megistus* is starved, its reproductive potential decreases, but the bug is still able to maintain colonization.

One factor to consider in morphological studies of the salivary glands is the state of nutrition of the insect. In starving bugs, glands are full of secretion, causing distension of the cell wall and consequently morphological modification (Lacombe 1999). Data on the feeding habits of Triatominae support epidemiological studies to perfect knowledge of local scenarios and guide control and monitoring activities (Forattini 1981).

25.5 Phthiraptera

Lice belong to the order Phthiraptera, the true parasites among the Exopterigota. Most species complete their life cycle in a single host, and transmission usually occurs opportunistically when hosts are in close contact. Host specificity led to numerous adaptations according to their niches, and consequently lice are diverse in body size and shape (Marcondes 2001). The dietary specializations of lice indicated their principal taxonomic divisions, and can be separated into those that feed on skin fragments, on feather and on fur, and those that specialize in feeding on blood. This order includes four superfamilies: Anoplura, Ischnocera, Amblycera, and Rhyncophthirina (Smith and Rod 1997).

The sucking lice show obligatory hemophagy for both sexes and all nymphal instars. As infesting agents, they are responsible for anoplurosis in domestic animals and for ptirosis and pediculoses in humans. Chewing lice feed on skin scales, secretions from sebaceous glands, and barbules of feathers. The buccal apparatus is modified primarily for chewing, but some species pierce the skin of the host, causing wounds. Little is known about blood in the diet of chewing lice. Ischnocera does not explore blood, and eat feathers and skin fragments. The diet of Amblycera is composed of feathers and secretions from epithelial tissue, together with blood. The Rhyncophthirina feed exclusively on blood (Marcondes 2001).

Morphological and behavioral adaptations allow lice to remain on with the host for long periods. The blood meal has the highest nutritional value and is easy to digest compared to skin, which result in greater fecundity. Sucking lice (Anoplura) are parasites of mammals. Of approximately 500 species, two-thirds are parasites of rodents. They feed exclusively on blood, and in high numbers cause anemia and weakness in the host (Price and Graham 1997). Only two superfamilies, Ischnocera and Amblycera, are found on birds. Lice belonging to Ischnocera live in the plumage or on the skin of their hosts and are highly host specific. Amblycera feed on epithelial tissue and blood, and are generally less host specific than the Ischnocera (Marshall 1981; Lehane 2005; Valim et al. 2005).

Bush et al. (2006) examined the role of melanin as a possible defense against lice. However, the infestation and feeding of lice was not interrupted by melanin. The human body louse *Pediculus humanus humanus* (L.) (Phthiraptera: Pediculidae) feeds exclusively on human blood. Proteins are the most abundant nutrients in the blood meal of this louse. The midgut of *Pe. humanus* contains leucine, an aminopeptidase. It is possible that ectoparasites that are highly specialized for a single host, such as the human body louse, are capable of digesting their blood meal with only one aminopeptidase; however, ectoparasites that are not specialized for a single host need a variety of aminopeptidases (Ochanda et al. 2000).

25.6 Final Considerations

The amount of information available on the importance of the food of the taxonomic groups treated here is highly asymmetrical. This imbalance is surprising in view of the actual and potential importance of some of the taxa discussed, since the majority of the species are vectors of etiological agents that cause maladies considered neglected diseases.

For the Phlebotominae, more detailed information about the food of immatures and the quality of resources used is needed to improve knowledge of the females' preferred environments for egg deposition. Even for the better-investigated groups, there still exist gaps that when filled will lead to significant advances in understanding the parasite–vector relationship. For example, chemical components in the saliva of hematophagous species show synergetic relationship with the parasite, facilitating its development or the processes of infection or contamination. Some nonhematophagous species that are phylogenetically related to taxa with greater impact on public health, such as in the Toxorhynchinae, are predatory in the first stages of development, which may help in designing methods to control species with predominantly wild habits. In this environment, the scale of action of natural predators can be more efficient than the traditional control methods employed in urban environments.

The recommendations by Forattini (1981) with regard to the Triatominae still appear true and valid for the diverse taxonomic groups treated here. It is needful to obtain data on the feeding habits of the various hematophagous species, particularly regarding ecological relationships and the biochemical processes involved in evaluating the host–parasite interaction. Information gained from such studies may illuminate these complex scenarios, with profound implications for the epidemiological studies that can aid in entomological monitoring and in control strategies.

REFERENCES

Abebe, M., M. S. Cupp, D. Champagne, and E. W. Cupp. 1995. Simulidin: A black fly (*Simulium vittatum*) salivary gland protein with anti-thrombin activity. *J. Insect Physiol.* 41:1001–6.

Adams, T. S. 1999. Hematophagy and hormone release. *Ann. Entomol. Soc. Am.* 92:1–13.

Alencar, Y. B., T. A. V. Ludwig, C. C. Soares, and N. Hamada. 2001. Stomach content analyses of *Simulium perflavum* Roubaud 1906 (Diptera: Simuliidae) larvae from streams in Central Amazônia, Brazil. *Mem. Inst. Oswaldo Cruz* 96:561–76.

Alexander, B., and M. C. Usma. 1994. Potential source of sugar for the phlebotomine sandfly *Lutzomyia youngi* (Diptera: Psychodidae) in a Colombian coffee plantation. *Ann. Trop. Med. Parasitol.* 88:543–9.

Alexander, B. 2000. Sampling methods for Phlebotomine sand flies (Diptera: Psychodidae). *Med. Vet. Entomol.* 14:109–22.

Alves, S. N., J. E. Serrão, G. Mocellin, and A. L. Mello. 2004. Effect of ivermectin on the life cycle and larval fat body of *Culex quinquefasciatus*. *Braz. Arch. Biol. Technol.* 47:433–9.

Andersen, J. F., and W. R. Montfort. 2000. The crystal structure of nitrophorin 2. *J. Biol. Chem.* 275:30496–503.

Andrade Filho, J. D., A. C. L. Silva, and A. L. Falcão. 2001. Phlebotomine sand flies in the state of Piauí, Brazil (Diptera: Psychodidae: Phlebotominae). *Mem. Inst. Oswaldo Cruz* 96:1085–7.

Andrade, H. T. A., A. A. Varela-Freire, M. J. A. Batista, and J. F. Medeiros 2005. Calliphoridae (Diptera) from human cadavers in Rio Grande do Norte State, northeastern Brazil. *Neotrop. Entomol.* 34:855–6.

Ariza, L., M. Seidenschwang, J. Buckendahl, M. Gomid, H. Feldmeie, and J. Heukelbach 2007. Tungiasis: A neglected disease causing severe morbidity in a shantytown in Fortaleza, State of Ceará. *Rev. Soc. Bras. Med. Trop.* 40:63–7.

Arrese, E. L., and J. L. Soulages. 2010. Insect fat body: Energy, metabolism, and regulation. *Annu. Rev. Entomol.* 55:207–25.

Azar, D., and A. Nel. 2003. Fossil psychodoid flies and their relation to parasitic disease. *Mem. Inst. Oswaldo Cruz* 98:35–7.

Badii, M., V. Garza, J. Landeros, H. Quiroz, N. L. San Nicolas, C. Juárez, and C. Saltillo 2006. Diversidad y relevancia de los mosquitos. *CUICYT//Bionomía* 13:4–16.

Barbosa, S. E., L. Diotaiuti, N. F. Gontijo, and M. H. Pereira. Comportamento alimentar de ninfas de 1° estádios de *Panstrongylus megistus* (Reduviidae, Triatominae) em hospedeiros humanos e pombos [online]: http://www.cpqrr.fiocruz.br/laboratorios/lab_triato/ comportamento%20alimentar%20de%20ninfas.htm. August 16, 2007.

Basanova, A. V., I. P. Baskova, and L. L. Zavalova. 2002. Vascular–platelet and plasma hemostasis regulators from bloodsucking animals. *Biochemistry* 67:143–50.

Bassi, R. M., M. C. I. Cunha, and S. Coscarón. 2000. Estudo do comportamento de tabanídeos (Diptera, Tabanidae) do Brasil. *Acta Biol. Par.* 29:101–15.

Bergo, E. S., G. M. Buralli, J. L. F. Santos, and S.M. Gurgel. 1990. Avaliação do desenvolvimento larval de *Anopheles darlingi* criado em laboratório sob diferentes dietas. *Rev. Saúde Pública* 24:95–100.

Beyruth, Z. 1992. Macrófitas aquáticas de um lago marginal ao rio Embu-mirim, São Paulo, Brasil. *Rev. Saúde Pública* 26:272–82.

Blanton, F. S., and W. W. Wirth. 1979. The sandflies (Culicoides) of Florida (Diptera: Ceratopogonidae). *Arthropods Fla. Neighb.* 10:1–204.

Borkent, A., and W. W. Wirth. 1997. World species of biting midges (Diptera: Ceratopogonidae). *Bull. Am. Mus. Nat. Hist.* 233:1–257.

Braga, M. V., and M. M. Lima. 2001. Efeitos de níveis de privação alimentar sobre a oogênese de *Panstrongylus megistus*. *Rev. Saúde Pública* 35:312–4.

Braks, M. A. H., S. A. Juliano, and L. P. Loubinos. 2006. Superior reproductive success on human blood without sugar is not limited to highly anthropophilic mosquito species. *Med. Vet. Entomol.* 20:53–9.

Brei, B., B. W. Cribb, and D. J. Merritt. 2003. Effects of seawater components on immature *Culicoides molestus* (Skuse) (Diptera: Ceratopogonidae). *Austr. J. Entomol.* 42:119–23.

Brito, M., and O. P. Forattini. 2004. Produtividade de criadouros de *Aedes albopictus* no Vale do Paraíba, SP, Brasil. *Rev. Saúde Pública* 38:209–15.

Bush, S. E., D. Kim, B. R. Moyer, J. Lever, and D. H. Clayton. 2006. Is melanin a defense against feather-feeding lice? *Auk* 123:153–61.

Cameron, M. M., F. A. Pessoa, A. W. Vasconcelos, and R. D. Ward. 1995. Sugar meal sources for the phlebotomine sandflies *Lutzomyia longipalpis* in Ceará State, Brazil. *Med. Vet. Entomol.* 9:263–72.

Carrasquilla, M. C., F. Guhl, Y. Zipa, C. Ferro, R. H. Pardo, O. L. Cabrera, and E. Santamaría. 2010. Breeding sites of *Culicoides pachymerus* Lutz in the Magdalena River basin, Colômbia. *Mem. Inst. Oswaldo Cruz* 105:216–9.

Chadee, D. D., J. C. Beber, and R. T. Mohammed. 2002. Fast and slow blood feeding durations of *Aedes aegypti* mosquitoes in Trinidad. *J. Vector Ecol.* 27:172–7.

Chagas, A. C., J. F. Medeiros, S. Astolfi-Filho, and V. PY-Daniel. 2010. Anticoagulant activity in salivary gland homogenates of *Thyrsopelma guianense* (Diptera: Simuliidae), the primary vector of onchocerciasis in the Brazilian Amazon. *Mem. Inst. Oswaldo Cruz* 105:174–8.

Chahad-Ehlers, S. S., A. L. Lozovei, and M. D. Marques. 2007. Reproductive and post embryonic daily rhythm patterns of the malaria vector *Anopheles (Kerteszia) cruzii*: Aspects of the life cycle. *Chronobiol. Int.* 24:289–304.

Charlwood, J. D., J. A. Rafael, and T. J. Wilkes. 1980. Métodos de determinar a idade fisiológica em Diptera de importância médica. Uma revisão com especial referência aos vetores de doenças na América do Sul. *Acta Amazonica* 10:311–33.

Cilek, J. E., and E. T. Schreiber. 1996. Diel host-seeking activity of *Chrysops celatus* (Diptera: Tabanidae) in Northwest Florida. *Fla. Entomol.* 4:520–5.

Collins, L. E., and A. Blackwell 2000. The biology of *Toxorhynchites* mosquitoes and their potential as biocontrol agents. *Biocontrol News Inf.* 2:105–16.

Consoli, R. A., and R. L. Oliveira. 1994. *Principais Espécies de Importância Sanitária no Brasil*. Rio de Janeiro: FIOCRUZ.

Coura, J. R. 2003. Tripanosomose, doença de chagas. *Ciênc. Cult.* 55:30–3.

Coura, J. R. 2005. *Dinâmica das Doenças Infecciosas e Parasitárias*. Rio de Janeiro: Editora Guanabara Koogan.

Cribb, B. W. 2000. Oviposition and maintenance of *Forcipomyia (Lasiohelea) townsvillensis* (Diptera: Ceratopogonidae) in the laboratory. *J. Med. Entomol.* 37:316–8.

Crosskey, R. W., and T. M. Howard. 2004. *A Revised Taxonomic and Geographical Inventory of World Black Flies (Diptera: Simuliidae)*. Washington, DC: The Natural History Museum.

Cupp, E. W., and M. S. Cupp. 1997. Black fly (Diptera: Simuliidae) salivary secretions: Importance in vector competence and disease. *J. Med. Entomol.* 34:87–94.

Dalla-Bona, A. C., and M. A. Navarro-Silva. 2006. Paridade de *Anopheles cruzii* em floresta ombrófila densa no Sul do Brasil. *Rev. Saúde Pública* 40:1118–23.

DeMeillon, B., and F. Hardy. 1951. Fate of *Cimex lectularius* on adult and on baby mice. *Nature* 167:151–2.

Dias, F. O. P., E. S. Lorosa, and J. M. M. Rebelo. 2003. Fonte alimentar sangüínea e a peridomiciliação de *Lutzomyia longipalpis* (Lutz & Neiva, 1912) (Psychodidae, Phlebotominae). *Cad. Saúde Pública* 19:1373–80.

Fairchild, G. B., and J. F. Burger. 1994. Catalog of the Tabanidae (Diptera) of the Americas South of the United States. *Mem. Am. Entomol. Inst.* 55:1–244.

Fernandez, Z., and O. P. Forattini. 2003. Sobrevivência de populações *de Aedes albopictus*; idade fisiológica e história reprodutiva. *Rev. Saúde Pública* 37:285–91.

Fernandez-Salas, I., M. H. Rodriguez, and D. R. Roberts. 1994. Gonotrophic cycle and survivorship of *Anopheles pseudopunctipennis* (Diptera: Culicidae) in the Tapachula foothills of Southern Mexico. *J. Med. Entomol.* 31:340–7.

Ferreira, R. L. M., L. H. Augusto, and J. A. Rafael. 2002. Activity of tabanids (Insecta: Diptera: Tabanidae) attacking the reptiles *Caiman crocodilus* (Linn.) (Alligatoridae) and *Eunectes murinus* (Linn.) (Boidae), in the Central Amazon, Brazil. *Mem. Inst. Oswaldo Cruz* 97:133–6.

Ferreira, R. L. M., and J. A. Rafael. 2006. Criação de imaturos de mutuca (Tabanidae: Diptera) utilizando briófitas e areia como substrato. *Neotrop. Entomol.* 35:141–4.

Forattini, O. P. 1965. *Entomologia Médica*. Vol. 3. São Paulo: Editora da Universidade de São Paulo.

Forattini, O. P. 1973. *Entomologia Médica*. Vol. 4. São Paulo: Editora Edgard Blucher Ltda./Editora da Universidade de São Paulo.

Forattini, O. P. 1996. *Culicidologia Médica*. Vol. 1. São Paulo: Editora da Universidade de São Paulo.

Forattini, O. P. 2002. *Culicidologia Médica*. Vol. 2. São Paulo: Editora da Universidade de São Paulo.

Forattini, O. P., J. M. S. Barata, J. L. F. Santos, and A. C. Silveira. 1981. Hábitos alimentares, infecção natural e distribuição de triatomíneos domiciliados na região nordeste do Brasil. *Rev. Saúde Pública* 15:113–64.

Forattini, O. P., J. M. S. Barata, J. L. F. Santos, and A. C. Silveira. 1982. Feeding habits, natural infection and distribution of domiciliary triatominae bugs in the central region of Brazil. *Rev. Saúde Pública* 16:171–204.

Foster, W. A. 1995. Mosquito sugar feeding and reproductive energetics. *Annu. Rev. Entomol.* 40:443–74.

Fournier, P. E., J. B. Ndihokubwayo, J. Guidran, P. J. Kelly, and D. Raoult. 2002. Human pathogens in body and head lice. *Emerg. Infect. Dis.* 8:1515–18.

Galati, E. A. B. 1995. Phylogenetic systematics of Phlebotominae (Diptera, Psychodidae) with emphasis on American groups. *Bol. Dir. Malariol. Saneamiento Ambiental* 35:133–42.

Galvão, C., R. Carcavallo, D. S. Rocha, and J. Jurberg. 2003. A checklist of the current valid species of the subfamily Triatominae Jeannel, 1919 (Hemiptera: Reduviidae) and their geographical distribution with nomenclatural and taxonomic notes. *Zootaxa* 202:1–36.

Gary, R. E., and W. A. Foster. 2001. Effects of available sugar on the reproductive fitness and vectorial capacity of the malaria vector *Anopheles gambiae* (Diptera: Culicidae). *J. Med. Entomol.* 38:22–8.

Glare, T. R., and M. O'Callagham. 2000. *Bacillus thuringiensis: Biology, Ecology and Safety*. Chichester: John Wiley & Sons.

Gonçalves, T. C. M., and J. Costa. Biologia dos vetores da doença de Chagas Disponível em: http://www .fiocruz.br/chagas/cgi/cgilua.exe/sys/start.htm?sid=23. June 9, 2010.

Gonçalves, T. C. M., V. Cunha, E. Oliveira, and J. Jurberg. 1997. Alguns aspectos da biologia de *Triatoma pseudomaculata* Corrêa & Espínola, 1964, em condições de laboratório (Hemiptera: Reduviidae: Triatominae). *Mem. Inst. Oswaldo Cruz* 92:275–80.

Gomes, P. R., W. W. Koller, A. Gomes, C. J. B. de Carvalho, and J. R. Zorzatto 2002. Dípetros fanídeos vetores de ovos de *Dermatobia hominis* em Campo Grande, Mato Grosso do Sul. Pesq. *Vet. Bras.* 22:114–8.

Gomes, L. and C. J. Von Zuben. 2003. Distribuição larval radial pós-alimentar em *Chrysomya albiceps* (Wied.) (Diptera: Calliphoridae): Profundidade, distância e peso de enterramento para pupação. *Entomologia y Vectores* 10:211–22.

Gomes, L., C. J. Von Zuben, and M. R. Sanches. 2003. Estudo da dispersão larval radial pós-alimentar em *Chrysomya megacephala* (Fabricius) (Diptera, Calliphoridae). *Rev. Bras. Entomol.* 47:229–34.

Guarneri, A. A., M. H. Pereira, and L. Diotaiuti. 2000. Influence of the blood meal source on the development ment of *Triatoma infestans, Triatoma brasiliensis, Triatoma sordida*, and *Triatoma pseudomaculata* (Heteroptera, Reduviidae). *J. Med. Entomol.* 37:373–9.

Harbach, R. E. 2007. The Culicidae (Diptera): A review of taxonomy, classification and phylogeny. *Zootaxa* 1668:591–638.

Harding, K. M., J. A. Gowland, and P. J. Dillon. 2006. Mercury concentration in black flies *Simulium* spp. (Diptera, Simuliidae) from soft-water streams in Ontario, Canada. *Environ. Pollut.* 143:529–35.

Harrington, L. C., J. D. Edman, and T. W. Scott. 2001. Why do female *Aedes aegypti* (Diptera: Culicidae) feed preferentially and frequently on human blood? *J. Med. Entomol.* 38:411–22.

Heukelbach, J., F. D. Oliveira, and H. Feldmeier. 2003. Ectoparasitoses e saúde pública no Brasil: Desafios para controle. *Cad. Saúde Pública* 19:1535–40.

Hinkle, N. C., P. G. Koehler, and W. H. Kern 1991. Hematophagous strategies of the cat flea (Siphonaptera: Pulicidae). *Fla. Entomol.* 74:377–85.

Kazimírová, M., M. Sulanová, M. Kozánek, P. Takac, M. Labuda, and P. A. Nuttall. 2001. Identification of anti-coagulant activities in salivary gland extracts of four horsefly species (Diptera, Tabanidae). *Haemostasis* 31:294–305.

Kettle, D. S. 1962. The bionomics and control of *Culicoides* and *Leptoconops* (Diptera, Ceratopogonidae, Heleidae). *Annu. Rev. Entomol.* 7:401–18.

Killick-Kendrick, R. 1990. Phlebotominae vectors of the leishmaniases: A review. *Med. Vet.* 4:1–24.

Koella, J. C., and M. J. Packer. 1996. Malaria parasites enhance blood-feeding of their naturally infected vector *Anopheles punctulatus*. *Parasitology* 113:105–9.

Krolow, T. K., R. F. Krueger, and P. B. Ribeiro. 2007. Chave pictórica para os gêneros de Tabanidae (Insecta: Diptera) do bioma Campos Sulinos, Rio Grande do Sul, Brasil. *Biota Neotrop.* 7:253–64.

Krüger, R. F., P. B. Ribeiro, and C. J. B. de Carvalho. 2003. Development of *Ophyra albuquerquei* Lopes (Diptera, Muscidae) in laboratory conditions. *Rev. Bras. Entomol.* 47:643–8.

Lacombe, D. 1999. Anatomia e histologia das glândulas salivares nos triatomíneos. *Mem. Inst. Oswaldo Cruz* 94:557–64.

La Corte dos Santos, R. 2003. Updating of the distribution of *Aedes albopictus* in Brazil (1997–2002). *Rev. Saúde Pública* 37:671–3.

Lehane, M. J. 2005. *The Biology of Blood-Sucking in Insects*. 2nd Ed. Liverpool: Liverpool School of Tropical Medicine.

Lima-Camara, T. N., N. A. Honório, and R. L. Oliveira 2007. Parity and ovarian development of *Aedes aegypti* and *Aedes albopictus* (Diptera: Culicidae) in Metropolitan Rio de Janeiro. *J. Vector Ecol.* 32:34–40.

Linardi, P. M. 2004. Biologia e epidemiologia das pulgas. *Rev. Bras. Parasitol. Vet.* 13:103–6.

Linardi, P. M., and L. R. Guimarães. 2000. *Sifonápteros do Brasil*. São Paulo: Museu de Zoologia USP/Fapesp.

Lourenço-de-Oliveira, R. 2005. Principais insetos vetores e mecanismos de transmissão das doenças infecciosas e parasitárias. In *Dinâmica de Doenças Infecciosas e Parasitárias*, ed. J. R. Coura, 75–97. Rio de Janeiro: Editora Guanabara Koogan S.A.

Lozovei, A. L. L. 1994. Microalgas na alimentação em larvas de simulídeos (Diptera) em criadouro natural em Curitiba, Paraná, Brasil. *Rev. Bras. Entomol.* 38:91–5.

Lozovei, A. L., F. Petry, L. G. S. Neto, and M. E. Ferraz. 2004. Levantamento das espécies de *Simulium* (Diptera, Simuliidae), Riacho dos Padres, município de Almirante Tamandaré, Paraná, Brasil. *Rev. Bras. Entomol.* 48:91–4.

Maciel, I. 1999. Avaliação epidemiológica do dengue no município de Goiânia no período de 1994 a 1997. *Mestrado. Instituto de Patologia Tropical e Saúde Pública*. Universidade Federal de Goiás: Goiânia, 119 pp.

Magalhães, F. E. P., and J. D. Lima. 1988. Freqüência de larvas de *Dermatobia hominis* (Linnaeus, 1781), em bovinos em Pedro Leopoldo, Minas Gerais. *Arq. Bras. Med. Vet. Zoot.* 40:361–7.

Malmqvist, B., Y. Zhang, and P. H. Adler. 1999. Diversity, distribution and larval habitats of North Swedish blackflies (Diptera: Simuliidae). *Freshw. Biol.* 42:301–14.

Marchiori, F. A. M., J. H. Guimarães, and E. Berti Fo. 1999. *A Mosca Doméstica e Algumas Outras Moscas Nocivas*. Piracicaba: FEALQ.

Marcondes, C. B. 2001. *Entomologia Médica e Veterinária*. São Paulo: Editora Atheneu.

Marcondes, C. B. 2006. *Terapia Larval de Lesões Cutâneas Causadas por Diabetes e Outras Doenças*. 1st ed Florianópolis: Editora da Universidade Federal de Santa Catarina.

Marcondes, C. B. 2009. *Doenças Transmitidas se Causadas por Artrópodes*. São Paulo: Editora Atheneu.

Marshall, A. G. 1981. *The Ecology of Ectoparasitic Insects*. London: Academic Press.

Massafera, R., A. M. Silva, A. P. Carvalho, D. R. Santos, E. A. B. Galati, and U. Teodoro. 2005. Fauna de flebotomíneos do município de Bandeirantes, no Estado do Paraná. *Rev. Saúde Pública* 39:571–7.

Megahed, M. M. 1956. Anatomy and histology of the alimentary tract of the female of the biting midge *Culicoides nubeculosus* Meigen. (Diptera: Ceratopogonidae). *Parasitology* 46:22–47.

Meiswinkel, R., E. M. Nevill, and G. J. Venter. 1994. Vectors: *Culicoides* spp. In: *Infectious Diseases of Livestock with Special Reference to Southern Africa*, ed. J. A. W. Coetzer, G. R. Thomson and R. C. Tustin, 68–89. Cape Town: Oxford University Press.

Mellink, J. J., and W. V. D. Bovenkamp. 1981. Functional aspects of mosquito salivation in blood feeding in *Aedes aegypti*. *Mosq. News* 41:115–9.

Mellor, P. S., J. Boorman, and M. Baylis. 2000. Culicoides biting midges: Their role as arbovirus vectors. *Annu. Rev. Entomol.* 45:307–40.

Milleron, R. S., J. P. Mutebi, S. Valle, A. Montoya, H. Yin, L. Soong, and G. C. Lanzaro. 2004. Antigenic diversity in maxadilan, a salivary protein from the sand fly vector of American visceral leishmaniasis. *Am. J. Trop. Med. Hyg.* 70:286–93.

Monteiro, M. C. 2005. Componentes antiflamatórios na saliva do *Lutzomyia longipalpis*, vetor da *Leishmania chagasi*. *Ambiência* 1:171–5.

Morais, S. A., M. T. Marelli, and D. Natal. 2006. Aspectos da distribuição de *Culex (Culex) quinquefasciatus* Say (Diptera, Culicidae) na região do rio Pinheiros, na cidade de São Paulo, Estado de São Paulo, Brasil. *Rev. Bras. Entomol* 50:413–8.

Moura, M. O., C J. B. de Carvalho, and E. L. A. Monteiro-Filho. 1997. Preliminary analysis of insects of medico-legal importance in Curitiba, State of Paraná. *Mem. Ins. Oswaldo Cruz* 92:269–74.

Mullen, G. R., and L. J. Hribar. 1988. Biology and feeding behavior of ceratopogonid larvae (Diptera: Ceratopogonidae) in North America. *Bull. Soc. Vect. Ecol.* 13:60–81.

Muniz, L. H. G., R. M. Rossi, H. C. Neitzke, W. M. Monteiro, and U. Teodoro. 2006. Estudo dos hábitos alimentares de flebotomíneos em área rural no sul do Brasil. *Rev. Saúde Pública*. 40:1087–93.

Natal, D. 2002. Bioecologia *Aedes aegypti*. *Biológico* 64:205–7.

Nery, L. C. R., E. S. Lorosa, and A. M. R. Franco. 2004. Feeding preference of the sand flies *Lutzomyia umbratilis* and *L. spathotrichia* (Diptera: Psychodidae, Phlebotominae) in an urbanf forest patch in the city of Manaus, Amazonas, Brazil. *Mem. Inst. Oswaldo Cruz* 99:571–4.

Nunes, S. L. L. 2005. Competição intra e interespecífica de larvas de *Aedes aegypti* e *Aedes albopictus* (Diptera: Culicidae) em condições laboratoriais. Master thesis, Public Health Faculty, São Paulo, SP.

Ochanda, J. O., E. A. C. Oduor, R. Galun, M. O. Imbuga, and K. Y. Mumcuoglu. 2000. Partial purification of the aminopeptidase from the midgut of the human body louse, *Pediculus humanus humanus*. *Physiol. Entomol.* 25:242–6.

Okuda, K., A. S. Caroci, P. E. Ribolla, A. G. Bianchi, and A. T. Bijovski 2002. Functional morphology of adult female *Culex quinquefasciatus* midgut during blood digestion. *Tissue Cell* 34:210–19.

Oliveira-Costa, J. 2003. *Entomologia Forense: Quando os Insetos são Vestígios*. Campinas: Editôra Millennium.

Oliveira, V. C., R. P. Mello, and J. M. d'Almeida. 2002. Dipteros muscóides como vetores mecânicos de ovos de helmintos em jardim zoológico, Brasil. *Rev. Saúde Pública* 36:614–20.

Oosterbroek, P., and G. Courtney. 1995. Phylogeny of the nematocerous families of Diptera (Insecta). *Zool. J. Linn. Soc.* 115:267–311.

Paterno, U., and C. B. Marcondes. 2004. Mosquitos antropofílicos de atividade matutina em trilha em Mata Atlântica em Florianópolis, Santa Catarina, Brasil (Diptera, Culicidae). *Rev. Saúde Pública* 38:133–5.

Perez De Leon, A. A., J. M. C. Ribeiro, W. J. Tabachnick, and J. G. Valenzuela. 1997. Identification of a salivary vasodilator in the primary North American vector of bluetongue viruses, *Culicoides variipennis*. *Am. J. Trop. Med. Hyg.* 57:375–81.

Perondini, A. L. P., M. J. Costa, and V. L. F. Brasileiro. 1975. Biology of *Triatoma brasiliensis*: II. Observations on the autogeny. *Rev. Saúde Pública* 9:363–70.

Pinto, S. B., T. Soccol, E. Vendruscolo, R. Rochadelli, P. B. Ribeiro, A. Freitag, C. Henemann, and M. Uemura 2002. Bioecologia de *Dermatobia hominis* (Linnaeus, 1781) em Palotina, Paraná, Brasil. *Ciênc. Rural* 32:821–7.

Prado, A. P. 2004. Dípteros de importância veterinária. *Rev. Bras. Parasitol. Vet.* 13:108.

Price, M. A., and O. H. Graham. 1997. Chewing and sucking lice as parasites of mammals and birds. *US Dept. Agric. Tech. Serv. Bull.* 1849:257.

Rajská, P., V. Knezl, M. Kazimirova, P. Takac, L. Roller, L. Vidlicka, F. Ciampor, M. Labuda, Weston-Davies, and P. A. Nutall. 2003. Vasodilatory activity in horsefly and deerfly salivary glands. *Med. Vet. Entomol.* 17:395–402.

Rangel, E. F., N. S. Souza, E. D. Wermelinger, A. F. Barbosa, C. A. A. Rangel. 1986. Biologia de *Lutzomyia intermedia* Lutz & Neiva, 1912 e *Lutzomyia longypalpis* Lutz & Neiva, 1912 (Diptera: Psychodidae), em condições experimentais. I. Aspectos da alimentação de larvas e adultos. *Mem. Inst. Oswaldo Cruz* 81:431–8.

Reinhardt, K., and M. T. S. Silva-Jothy. 2007. Biology of the bed bugs (Cimicidae). *Annu. Rev. Entomol.* 52: 351–74.

Rezende, G. L. 2008. Investigações sobre a aquisição de resistência à dessecação durante embriogênese de *Aedes aegypti* e *Anopheles gambiae*. PhD dissertation, Fiocruz, Rio de Janeiro.

Ribeiro, J. 2000. Blood-feeding in mosquitoes: Probing time and salivary gland anti-haemostatic activities in representatives of three genera (*Aedes, Anopheles, Culex*). *Med. Vet. Entomol.* 14:142–8.

Ribeiro, J. M. C. 1987. Role of saliva in blood feeding by arthropods. *Annu. Rev. Entomol* 32:463–78.

Ribeiro, J. M. C., and I. M. Francischetti. 2003. Role of arthropod saliva in blood feeding: Sialome and post-sialome perspectives. *Annu. Rev. Entomol* 48:73–88.

Serufo, J. C., H. M. Oca, V. A. Tavares, A. M. Souza, R. V. Rosa, M. C. Jamal, J. R. Lemos, M. A. Oliveira, R. M. R. Nogueira, and H. G. Schatzmayr. 1993. Isolation of dengue virus type 1 from larvae of *Aedes albopictus* in Campos Altos city, state of Minas Gerais, Brazil. *Mem. Inst.* Oswaldo Cruz 88:503–4.

Suwabe, K., and A. Moribayashi. 2000. Lipid utilization for ovarian development in an autogenous mosquito, *Culex pipiens molestus* (Diptera: Culicidae). *J. Med. Entomol.* 37:726–30.

Schofield, C. J. 2000. Biosystematics and evolution of the Triatominae. *Cad. Saúde Pública* 16:89–92.

Schofield, C. J., L. Diotaiuti, and J. P. Dujardin. 1999. The process of domestication in triatominae. *Mem. Inst. Oswaldo Cruz* 94:375–8.

Schofield, C. J., and C. Galvão. 2009. Classification, evolution and species groups within the Triatominae. *Acta Trop.* 110:88–100.

Silva, F. S. 2009. A importância hematofágica e parasitológica da saliva dos insetos hematófagos. *Rev. Trop. Ciênc. Agr. Biol.* 3:3–17.

Silverman A. L., L. H. Blow, J. B. Qu, I. M. Zitron, S. C Gordon, and E. D. Walker. 2001. Assessment of hepatitis B virus DNA and hepatitis C virus RNA in the common bedbug (*Cimex lectularius L.*) and kissing bug (*Rhodnius prolixus*). *Am. J. Gastroenterol.* 96:2194–8.

Smith, S. M., D. A. Turnbull, and P. D. Taylor. 1994. Assembly, mating, and energetics of *Hybomitra arpadi* (Diptera: Tabanidae) at Churchill, Manitoba. *J. Insect Behav.* 7:355–83.

Smith, V., and P. Rod. 1997. Phthiraptera lice. Version 07 March 1997 (under construction). http://tolweb.org/ Phthiraptera/8237/1997.03.07 in The Tree of Life Web Project, http://tolweb.org/.

Souza, N. A., C. A. Andrade-Coelho, A. F. Barbosa, M. I. Vilela, E. F. Rangel, and M. P. Deane. 1995. The influence of sugars and amino acids on the blood-feeding behaviour, oviposition and longevity of laboratory colony of *Lutzomyia longipalpis* (Lutz & Neiva, 1912) (Diptera: Psychodidae, Phlebotominae). *Mem. Inst. Oswaldo Cruz* 90:751–7.

Steffan, W. A., and N. L. Evenhuis. 1981. Biology of *Toxorhynchites*. *Annu. Rev. Entomol.* 26:159–81.

Tachibana, S. I., and H. Numata. 2006. Seasonal prevalence of blowflies and flesh flies in Osaka city. *Entomol. Sci.* 9:341–5.

Takác, P., M. A. Nunn, J. Mészaros, O. Pechánová, N. Vrbjar, P. Vlasáková, M. Kozánek, M. Kazimírovái, G. Hart, P. A. Nuttall, and M. Labuda. 2006. Vasotab, a vasoactive peptide from horse *Hybomitra bimaculata* (Diptera, Tabanidae) salivary glands. *J. Exp. Biol.* 209:343–52.

Tanada, Y., and H. K. Kaya. 1993. *Insect Pathology*. New York: Academic Press.

Teixeira, M. J., J. D. Fernandes, C. R. Teixeira, B. B. Andrade, M. L. Pompeu, J. S. Silva, C. I. Brodyskin, M. Barral-Neto, and A. Barral. 2005. Distinct *Leishmania braziliensis* isolados induce different paces of chemokine expressions patterns. *Infect. Immun.* 73:1191–5.

Telleria, E. L. 2007. Caracterização de tripsinas em *Lutzomyia longipalpis*–Principal vetor da leishmaniose visceral no Brasil. Mestrado. Instituto Oswaldo Cruz, Fiocruz, RJ, 92pp.

Thomas, I., G. G. Kihiczak, and R. A. Schwartz. 2004. Bedbug bites: A review. *Int. J. Dermatol.* 43:430–3.

Thompson, B. H. 1987. The use of algae as food by larval Simuliidae (Diptera) of Newfoundland streams. III. Growth of larvae reared on different algal and other foods. *Arch. Hydrobiol.* 76:459–66.

Toma, T., and I. Miyagi. 1992. Laboratory evaluation of *Toxorhynchites splendes* (Diptera: Culcidae) for predation of *Aedes albopictus* mosquito larvae. *Med. Vet. Entomol.* 6:281–9.

Valim, M. P., R. H. F. Teixeira, M. Amorim, and N. M. Serra-Freire. 2005. Malófagos (Phthiraptera) recolhidos de aves silvestres no Zoológico de São Paulo, SP, Brasil. *Rev. Bras. Entomol.* 49:584–7.

Veloso, H. P., J. V. Moura, and R. M. Klein. 1956. Delimitação ecológica dos anofelíneos do subgênero *Kerteszia* na região costeira do sul do Brasil. *Mem. Inst. Oswaldo Cruz* 54:517–49.

Viviane, A. B. P., and C. J. P. C. Araújo-Coutinho. 1999. Influência da temperatura no desenvolvimento embrionário de *Simulium pertinax* Kollar, 1832 (Diptera: Simuliidae). *Entomol. Vect.* 6:591–600.

Volf, P., and I. Rohousová. 2001. Species-specific antigens in salivary glands of phlebotomine sandflies. *Parasitology* 122:37–41.

Volfova, V., J. Hostomska, M. Cerny, J. Votypka, and P. Volf. 2008. Hyaluronidase of bloodsucking insects and its enhancing effect on *Leishmania* infection in mice. *PLoS Negl. Trop. Dis.* 2:e294.

Werner, D., and A. C. Pont. 2003. Dipteran predators of simuliid blackflies: A worldwide review. *Med. Vet. Entomol.* 17:115–32.

Wiegmann, B. M., S. C. Tsaur, D. W. Webb, D. K. Yeates, and B. K. Cassel. 2000. Monophyly and relationships of the Tabanomorpha (Diptera: Brachycera) based on 28S ribosomal gene sequences. *Ann. Entomol. Soc. Am.* 93:1031–8.

Wirth, W. W., and A. A. Hubert. 1989. The Culicoides of Southeast Asia (Diptera: Ceratopogonidae). *44th Mem. Am. Entomol. Inst.,* Gainesville, FL.

Ximenes, M. F. F. M., J. C. Maciel, and S. M .B. Jerônimo. 2001. Characteristics of the biological cycle of *Lutzomyia evandroi* Costa & Lima, 1936 (Diptera: Psychodidade) under experimental conditions. *Mem. Inst. Oswaldo Cruz* 96:883–6.

Yeates, D. K., and B. M. Wiegmann. 1999. Congruence and controversy: Toward a higher-level phylogeny of Diptera. *Annu. Rev. Entomol.* 44:397–428.

Yuval, B. 2006. Mating systems of blood-feeding flies. *Annu. Rev. Entomol.* 51:413–40.

Zhou, G., E. P. James, and M. A. Wells. 2004. Utilization of pre-existing energy stores of female *Aedes aegypti* mosquitoes during the first gonotrophic cycle. *Insect Biochem. Mol. Biol.* 34:919–25.

Part III

Applied Aspects

Plant Resistance and Insect Bioecology and Nutrition

José D. Vendramim and Elio C. Guzzo

CONTENTS

26.1 Introduction 657
26.2 Host Selection by Phytophagous Insects 658
    26.2.1 Stages in Host Selection 658
    26.2.2 Evolution of Insect–Plant Interactions 660
26.3 Morphological Resistance and Insect Bioecology and Nutrition 661
    26.3.1 Toughness and Thickness of Epidermis 662
        26.3.1.1 Silicon 662
        26.3.1.2 Lignin 664
    26.3.2 Epicuticular Waxes 665
    26.3.3 Pilosity of Epidermis 666
26.4 Chemical Causes and Insect Bioecology and Nutrition 667
    26.4.1 Antixenotic Factors 667
        26.4.1.1 Repellents 667
        26.4.1.2 Phagodeterrents 667
    26.4.2 Antibiotic Factors 669
26.5 Biotechnology and Resistance of Plants to Insects 670
    26.5.1 Lectins 671
    26.5.2 Enzymatic Inhibitors 673
        26.5.2.1 Protease Inhibitors 674
        26.5.2.2 α-Amylase Inhibitors 675
        26.5.2.3 Bifunctional Inhibitors 676
26.6 Final Considerations 677
References 677

26.1 Introduction

Methods promoting plant resistance, as alternatives to chemical pest control, have increased due to their advantages over conventional insecticides. These advantages include preventing environmental pollution, biological instability or toxicity to farm workers, and residues in foods. This technique is also cheap, has a continuous action on insects, and is compatible with other control methods and therefore can be included in any pest management program.

A resistant plant, due to its genotypic constitution, suffers less damage from insect attack than other plants under the same condition. Since resistance is the result of the relationship between plant and insect, identification of a resistant plant or variety can be done using parameters that consider the insect (e.g., difference in population, oviposition, consumption, duration of biological cycle, and fecundity) and the plant (e.g., difference in survival, destruction of different plant organs, production, and product quality). Observing the mechanisms by which a plant shows resistance to an insect, it can be verified that, in many cases, this involves changes in insect behavior and biology, while in other cases the plant itself reacts without having any effect on the insect. On the basis of these variations, plant resistance is classified into three types: nonpreference, antibiosis, and tolerance (Painter 1951).

A plant variety shows nonpreference (according to Kogan and Ortman 1978) when it is less used by insects for food, oviposition, or shelter, the most frequently observed being those concerning oviposition and feeding. The characterization of nonpreference for oviposition means less number of eggs, whereas nonpreference for feeding is characterized by less consumption or fewer insects on the plant. Antibiosis occurs when the insect feeds normally on the plant; however, this causes an adverse effect on its biology, such as mortality of immatures (often first instars), increased developmental period, and reduction in weight, fecundity, fertility, and oviposition period. Tolerance occurs when a plant suffers less damage than others with similar infestation levels, with no insect behavior/biology effects. Tolerant plants have greater capacity or speed to regenerate damaged areas, and greater vigor or leaf area.

These different types of resistance are due to complex interactions between phytophagous insects and their host plants, resulting from a long and continuous evolutionary process. These interactions developed basically in two ways: host selection by the insect and resistance of the plant to the insect (Lara 1991; Mello and Silva-Filho 2002); therefore, causes of resistance should consider both the plant and the insect. Protective mechanisms include physical, morphological, and chemical causes. Physical causes include color of the plant, which affects host selection for feeding and oviposition. Because color sensitivity differ for men and insects, it is difficult to work with this resistance mechanism; resistance caused by color are rarely documented, although cases of repellence caused by red color inhibiting insect oviposition have been reported (Lara 1991; Smith 2005). Chemical causes include substances that affect insect behavior and/or metabolism, and nutritional unsuitability of the plant. A change in insect behavior occurs mainly during host selection for feeding and oviposition (nonpreference or antixenosis). Metabolic effects caused by toxic compounds, enzymatic and reproductive inhibitors, or qualitative or quantitative nutritional deficiency, result in antibiosis. Finally, morphological causes are plant characteristics that affect insect locomotion, mating, host selection for feeding/oviposition, and ingestion/digestion of food. These characteristics include epidermis thickness, toughness, texture, waxiness, and pilosity.

26.2 Host Selection by Phytophagous Insects

26.2.1 Stages in Host Selection

Insects' preference for certain host plants has attracted man's attention over centuries. This has been speculated since the domestication of silkworm in China 5000 years ago, and owing to its specificity to feed on mulberry leaves. However, scientific investigation of host selection mechanisms dates back about a century when research on the ethological, ecological, and physiological bases of insect–plant interactions began.

Fabre (1890) cited by Kogan (1986) was one of the first researchers to question insect feeding preferences, when he unsuccessfully tried to rear *Bombyx mori* L. on hosts other than mulberry leaves. This preference was attributed to the insects' "botanical instinct." According to Kogan (1986), the term "chemical defense" was probably first used by Stahl, in 1888, but the unquestionable association of the "secondary substances" of plants with insect feeding and oviposition was first documented by Verschaffelt, in 1910, when he found that Pieridae butterflies were attracted by crucifers due to sinigrin, present in this plant family, which is a compound toxic to other insects unadapted to feeding on crucifers.

One of the oldest theories to explain host selection by phytophagous insects, the "Hopkins host selection principle," states that an insect species that feed on more than one host plant prefers to reproduce on the species it has become more adapted. The Hopkins principle was based on the beetle *Dendroctonus monticolae* Hopkins feeding on various *Pinus* species. After living for successive generations on a certain *Pinus* species, the beetle would prefer the species attacked previously, to which it became adapted (Hopkins 1917).

Brues (1920), studying host plant selection by phytophagous insects (Lepidoptera), concluded that, in general, these insects prefer plants of related families or genera. On the basis of these data, he suggested the hypothesis of parallel evolution in which plants will produce harmful substances and insects will adapt to overcome this barrier to feed on them. However, at the time, this study did not merit proper attention and was only discussed again by Fraenkel (1958) cited by Kogan (1986), who redefined the concept of parallel evolution, suggesting that a reciprocal parallel adaptive evolution determined insect host selection. The proper recognition of the importance of studies of parallel evolution only occurred, however, when Ehrlich and Raven (1964) established the basic principles of coevolution, although some advances in understanding the host selection process by insects had already been established before studies of Fraenkl, and Ehrlich and Raven. Dethier (1941), who had suggested that certain substances stimulate insects to seek their preferred host, later proposed the term "attractant" to define this group of substances and, in contrast to this, the term "repellent," to denominate the class of substances that stimulate insects to move away from a plant.

When Fraenkel (1953) cited by Lara (1991) chemically analyzed leaves of many plant species, he concluded that they contained the nutrients necessary for insect development as long as they were ingested in sufficient quantities. Thus, it could be proved that host selection could not be based on the nutrients because, in this case, all insects would be polyphagous or oligophagous. On the basis of these data, he proposed the "theory of secondary substances," which established that host selection would be regulated by the presence of substances whose occurrence was restricted to certain plant groups. According to this theory, these substances, whose occurrence was irregular among plants, would have no function in plant metabolism but would act only in defense against insects and microorganisms. However, with evolution, certain insects would become able to live with such substances and would later use them as a positive stimulus for host selection.

Despite clear demonstrations that secondary substances were important for host selection, the importance of nutrients in this process continued to be investigated. Kennedy (1958), working with aphids, proposed the "dual discrimination theory," according to which host selection would be governed by two types of stimuli: the exotic, conferred by secondary substances, and the nutritive, supplied by nutrients (essential and nonessential). The importance of nutritive substances in host plant selection by insects was definitively demonstrated by Cartier (1968), using artificial diets whose nutrient (sucrose and amino acids) concentrations varied.

Dethier et al. (1960) proposed the term "arrestant" for stimulus that causes insects to remain still or move slowly on the plant, and the term "stimulant" for that which causes insects to begin feeding. They further suggested the terms "locomotory stimulant" and "deterrent" for those stimuli that cause the opposite effects, respectively.

Beck (1965) reported three distinct stages in feeding preferences: (a) host plant recognition and orientation, (b) initiation of feeding, and (c) maintenance of feeding. These stages have a continuous sequence, and the insect's responses vary according to the plant's positive or negative stimuli. These stimuli can be chemical, physical, or morphological, and factors such as substance, color, or hairiness independently act as a stimulus. Besides this, insects will respond to the predominating stimuli. A chain of similar stimuli occurs in oviposition: insects need to orient themselves to the plant, locate it, and may oviposit or not, depending on the predominant stimulus.

Whittaker (1970) proposed the term, "allelochemical," a "nonnutritional chemical produced by organisms of one species that affect the growth, health, behavior, or population biology of organisms of another species." According to Whittaker and Feeny (1971), two types of allelochemicals are more important for insect–plant relationships: allomones, which give adaptive advantage to the producing organism (in this case, the host plant), and kairomones, which give adaptive advantage to the receiving organism (in this case, the phytophagous insect). Thus, in an insect–plant relationship, the allomones are unfavorable and the kairomones favorable to insects. Kogan (1986) adapted these concepts to the chains of stimuli proposed previously, classifying as allomones the repellents, locomotory excitants, suppressants, and deterrents, and as kairomones, the attractants, arrestants, and feeding and oviposition excitants. Also, he included in these groups, substances that affect insect metabolism and those that are generally responsible for antibiosis (Table 26.1). The classification of a substance, such as a kairomone or an allomone, should be carefully made, always considering the insect species involved, since substances that act as kairomones for some insects can act as allomones for others.

TABLE 26.1

Principal Classes of Alellochemicals and Their Corresponding Effects on Insects

Alellochemical Factors	Behavioral or Physiological Effects
Kairomones	Give adaptive advantage to the receiving organism
Attractants	Orient insects toward host plant
Arrestants	Slow down or stop movement
Feeding or oviposition excitants	Elicit biting, piercing, or oviposition; promote continuation of feeding
Allomones	Give adaptive advantage to the producing organism
Antixenotics	Disrupt normal host selection behavior
Repellents	Orient insects away from plant
Locomotory excitants	Start or speed up movement
Suppressants	Inhibit biting or piercing
Deterrents	Prevent maintenance of feeding or oviposition
Antibiotics	Disrupt normal growth and development of larvae; reduce longevity and fecundity of adults
Toxins	Produce chronic or acute intoxication syndromes
Digestibility reducing factors	Interfere with normal processes of food utilization

Source: Adapted from Kogan, M. In *Ecological Theory and Integrated Pest Management Practice*, ed. M. Kogan, 83–134. New York: John Wiley & Sons, 1986.

A new type of allelochemical was proposed by Nordlund and Lewis (1976), called synomone, which is defined as a chemical substance causing favorable behavioral or physiological reaction to both the producing and the receiving organism.

In insect–plant relationships, Price et al. (1980) mentioned the importance of natural enemies and that three trophic levels should be considered: plant, insect, and natural enemy. Considering plant resistance, Price (1986) highlighted two types of defense: intrinsic, where the plant defends itself against the insect by producing chemical substances and/or morphological characteristics, and extrinsic, in which the plant is benefitted by natural enemies that reduce the insect population; almost all plant's intrinsic defense mechanisms end up affecting natural enemies.

Observing the effects of these chemical substances at these three trophic levels, Whitman (1988) proposed a fourth type of allelochemical, antimone, defined as a chemical substance produced by an organism that causes an unfavorable reaction in both the producing and the receiving organism. Some authors include the apneumones, which are chemical substances produced by dead material, among the allelochemicals, causing favorable behavioral or physiological reaction to a receiving organism, in detriment to another organism present within or on the emitting material. Considering that the producing organism is not alive, many researchers prefer not to classify this type of substance among allelochemicals.

26.2.2 Evolution of Insect–Plant Interactions

The complex interactions between phytophagous insects and their hosts are the result of a long and continuing evolutionary process (Beck 1965). During this process, the feeding behavior of many insects became specialized, and they showed preference for certain plants or plant parts. Since this insect–plant relationship is a dynamic process, this preference can change over time, making certain common relationships more difficult and vice versa. The reasons for these changes are not clear in most cases, although there are various hypotheses on the evolution of these interactions.

The principles and evolutionary theories are divided into two hypotheses: coevolution and sequential evolution. The generally more accepted coevolution theory was explained in detail by Ehrlich and Raven (1964), who stated that the trophic relationships of phytophagous insects result from a very tight evolutionary interaction between plants and insects. In this interaction, the selection pressure represented by insect attacks induces the appearance of resistance mechanisms (mostly secondary substances) in the plants against the insects; while, on the other hand, some insects succeed in overcoming this resistance by adapting themselves to these substances that may become feeding stimulants. Thus, for each plant adaptation (evolution), an insect counter-adaptation (coevolution) tends to develop. Simply put, this process can be understood as being the production of new allomones by the plant through genetic recombination or random mutation (due to herbivore pressure) and the subsequent neutralization of these allomones by insects, who also manage to adapt to these allomones, transforming them into kairomones, by genetic recombination or random mutation.

An example of plant adaptation and insect coadaptation are the effects of L-canavanine on insect development. L-canavanine, a structural analog of L-arginine, is highly toxic to insects and when consumed by them makes up part of the structural proteins in place of L-arginine, resulting in the formation of defective and physiologically deficient proteins. In some legume seeds, L-canavanine is the commonest form of storing nitrogen, and its toxicity is a powerful allomonic barrier that gives protection against insects and other herbivore attack. However, the bruchid *Caryedes brasiliensis* (Thunberg) feeds exclusively on seeds containing L-canavanine, since its larvae have adapted physiologically, and consequently discriminate between L-arginine and L-canavanine and do not incorporate the latter amino acid into the insect's proteins. Besides this, in a second stage, the larvae started to use L-canavanine as a nitrogen source for other metabolic functions (Rosenthal et al. 1976, 1977).

The theory of sequential evolution proposed by Jermy (1976) was established as an alternative to the coevolution hypothesis, whose assumptions he considered inconsistent. Analyzing the aspects involved in the coevolution hypothesis, Jermy (1976) considered that (a) most phytophagous insects have very low population densities compared to the biomass of their host plants, and therefore they can hardly be important selection factors for the plant; (b) insect–plant interactions are not necessarily antagonistic: monophagous and oligophagous insects, if their number is fairly high, may ideally regulate the abundance of their host plants (which would bring mutual advantage); consequently, (c) resistance to insects is not a general necessity in plants and it cannot explain the presence of secondary plant substances; (d) parallel evolutionary lines of plants and insects that should result from coevolutionary interactions are rare, while many closely related insects feed on botanically very distant plant taxa, a relationship that cannot be related to coevolution. On the basis of these considerations, he proposed the hypothesis of sequential evolution to explain the insect–plant relationship: the evolution of flowering plants propelled by selection factor (e.g., climate, soil, plant–plant interactions), which are much more potent than insect attacks, creates the biochemically diversified trophic base for the evolution of phytophagous insects, while the latter do not appreciably influence the evolution of the plant. Other theories and speculations on these proposed theories have been discussed (Bernays and Graham 1988; Fox 1988; Thompson 1988), which seems to indicate that there is no a common hypothesis to explain all cases of insect–plant relationships, but that each case should be studied in detail.

26.3 Morphological Resistance and Insect Bioecology and Nutrition

As discussed previously (Section 26.2), the different strategies adopted by plants to make them resistant to insects are called causes or factors of resistance, and have been divided into physical, chemical, and morphological factors. Some authors consider the toughness of the epidermis, or some other plant tissue, and the presence of trichomes, as being physical factors (Beck 1965; Larsson 2002), while others consider them morphological factors (Lara 1991). Although these factors represent a physical barrier to penetration by insect stylets, mandibles, or the ovipositor, as well as to reduce their movements on the plant, thus making access to the desired resource difficult, here they will be treated as morphological factors, as agreed on by most researchers.

Although these morphological factors have a significant influence on insect behavior, adversely affecting its movements, mating, and host selection for feeding and/or oviposition, they can also have a big influence on the insect's physiology due to their chemical characteristics, by affecting food ingestion and digestion or by having a low nutritional quality. Morphological factors can be divided into epidermal and structural factors. The structural factors refer to the size of plant structures and how they are distributed on the plant. Although they contribute significantly to plant resistance to pests, they are not important from the nutritional point of view since they basically affect insect behavior. Epidermal factors include appendices or formations, such as the trichomes, as well as epidermal shape, texture, and consistency, which are determined by deposition of waxes, silica, lignin, and other compounds. These factors have a more direct influence on insect nutrition since they make up part of the plant tissue and are also ingested by the insect.

26.3.1 Toughness and Thickness of Epidermis

The epidermis is the first barrier the insect has to cross to enter the plant and/or feed on it, and several factors interfere in these processes. Epidermal toughness and thickness are generally caused by the deposition of silica or lignin, which act as resistance factors to insects in various crop plants.

26.3.1.1 Silicon

Silicon (Si) is the second most abundant element on the Earth's surface and although not considered essential for growth, development, and metabolism of higher plants, constitutes their quantitatively major inorganic component and acts as defense against insects (Epstein 1999). Absorbed by plants as monosilicic acid [$Si(OH)_4$], Si is concentrated in leaves, and first polymerizes into colloidal silicic acid and later into silica gel (SiO_2nH_2O), also called hydrated amorphous silica or opaline silica. Silicon accumulates in the lumen of epidermal cells, forming bodies called phytoliths, which once immobile are not redistributed in the plant, and this is why they occur in higher concentrations in older tissues (Barbosa Filho et al. 2000; Hunt et al. 2008; Keeping et al. 2009). The grass family (Poaceae) includes important crops, such as rice, sugarcane, corn, and wheat, which accumulate Si (Figure 26.1) that varies according to the genotype (Epstein 1999; Barbosa Filho et al. 2000). Although genotypes with high Si levels do not always show resistance to insects (Barbosa Filho et al. 2000), this element show positive effect on control of various insect pests, including xylem and phloem suckers, leaf chewers, and borers, but its mechanism of action has not yet been completely elucidated.

The most notorious effect of silica deposition in the plant epidermis is hardening of this tissue, which represents a mechanical barrier to insects. For chewing insects, silica wears out the mandibles, hindering or making chewing and plant tissue ingestion more difficult (Djamin and Pathak 1967; Goussain et al. 2002) (Figure 26.2).

Higher larval mortality of second and sixth instars of *Spodoptera frugiperda* (J. E. Smith) feeding on corn leaves with high silica content is in the first case due to small size and greater fragility of the mandibles, and in the second, because great food consumption cause higher wear and tear of mandibles due to abrasive action of silica deposited on leaf epidermis (Goussain et al. 2002). The harmful effect of silica was also observed for other leaf-feeding insects, such as *Spodoptera exempta* (Walker) and

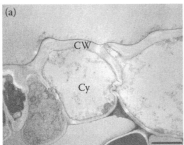
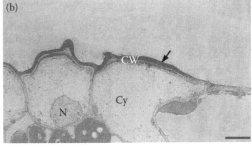

FIGURE 26.1 Transmission electron micrographs of rice leaves at four-leaf growth stage. (a) Epidermal cells of control plant. CW, cell wall; Cy, cytoplasm. Bar, 2 μm. (b) Epidermal cells of silicon-treated plant. Electron-dense layer (arrow) is evident in epidermal cell wall (CW). N, nucleus. Bar, 3 μm. (From Kim, S. G., et al., *Phytopathology*, 92, 1095, 2002. With permission.)

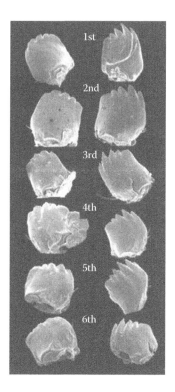

FIGURE 26.2 Mandibles of first to sixth instar larvae of *Spodoptera frugiperda*, fed with corn leaves with (left) and without (right) silicon application. The wearing down of incisor region of mandibles of larvae fed on leaves where silicon was applied is clearly seen. (From Goussain, M. M., et al., *Neotrop. Entomol.*, 31, 305, 2002. With permission.)

Schistocerca gregaria Forsk. (Massey et al. 2006). A similar effect has been verified for insect borers. The application of Si in sugarcane reduces the larval mass of the borer *Eldana saccharina* (Walker) by 19.8% and the length of the galleries it makes by 24.4% (Keeping and Meyer 2002). The application of Si in sugarcane also affects the borers *Chilo infuscatelus* Snellen (Rao 1967) and *Diatraea saccharalis* (F.) (Anderson and Sosa 2001). Besides the abrasive effect on insect mandibles, high quantities of Si in plant tissues cause nutritional imbalance since it is not used as nutrient, and is not absorbed by insects. For this reason, it is an ideal marker for measuring rates of food consumption and utilization by insects, which was proposed by Barbehenn (1993).

Among grasses, the highest quantities of Si are observed in rice plants (Epstein 1999) and the silica content (SiO_2) varies according to the genotype and may reach 13.9% of dry matter in genotypes considered resistant to *Chilo suppressalis* (Walker) (Djamin and Pathak 1967). Since insects do not digest and/ or absorb the Si from the diet, the higher the proportion of the mineral in the plant tissue, the lesser the values of approximate digestibility and conversion efficiency of ingested food.

Another indication of low nutritional value of food is increase in cannibalism, up to eight times among *S. frugiperda* larvae confined on corn leaves treated with Si compared with those confined on untreated leaves (Goussain et al. 2002). This behavior is more prevalent in situations of feeding stress, such as food scarcity (Raffa 1987; Nalim 1991). Some studies demonstrated that high Si content can cause feeding deterrence to *S. exempta* and *S. gregaria* (Massey et al. 2006) and the development of herbivores confined to plants with high Si content is adversely affected not only because they consume less food but also because they absorb less nitrogen (Massey and Hartley 2006) and carbohydrates (Massey et al. 2006) from the diet.

Hunt et al. (2008) verified that although the chlorophyll content in leaves of the grass *Lolium perenne* is the same with high and low silica, the chlorophyll content of *S. gregaria* feces was 38% greater when fed on leaves with high silica content compared with leaves with low silica content. This result suggests that more chlorenchyma cells remain intact in the high-silica plants, and that silica affects the disruption of cell walls. Since these cells contain high levels of starch and proteins, the mechanical protection provided by the silica could be partly responsible for the low digestibility of those high-silica grasses (Massey et al. 2006). However, Hunt et al. (2008) do not consider likely that silica reinforces the cell wall of chlorenchyma cells preventing their disruption, since significant silica levels are not observed in these cells. Thus, authors suggest that phytoliths act by keeping apart pestle, mortar, and cusps of the molar regions of the insect's jaws during chewing, preventing crushing of chlorenchymatic cells and reducing food digestibility.

In some cases, when confined on diets with low nutritional quality, insects compensate by ingesting large quantities of food. This has been observed for the grasshopper *S. gregaria* (Massey et al. 2006), feeding on plants with high silica content. In the specific case of silica, however, the increase in feeding activity can cause still greater wear and tear on the mandibles due to abrasive action. Si may not be a resistance factor against sucking insects, which insert their stylets into the plant between the phytoliths, avoiding the silica barrier of plant epidermis (Massey et al. 2006). However, it has been demonstrated that various sucking insects are affected by Si, for example, *Nilaparvata lugens* (Stal) (Yoshihara et al. 1979), *Sitobion avenae* and *Metopolophium dirhodum* (Walker) (Hanisch 1981), *Sogatella furcifera* (Horvath) (Kim and Heinrichs 1982; Salim and Saxena 1992), *Schizaphis graminum* (Rondani) (Carvalho et al. 1999; Basagli et al. 2003; Moraes et al. 2004; Goussain et al. 2005), *Rhopalosiphum maidis* (Fitch.) (Moraes et al. 2005), *Thrips palmi* Karny (Almeida et al. 2008), and *Frankliniella schultzei* Trybom (Almeida et al. 2009), whereas some chewing insect species are unaffected by high Si levels in the diet, for example, *Herpetogramma phaeopteralis* Guenee (Korndorfer et al. 2004) and *Agrotis ipsilon* (Hufnagel) (Redmond and Potter 2007).

Additionally, Gomes et al. (2005) verified that the application of Si on wheat plants activates the enzymes peroxidase, polyphenoloxidase, and phenylalanine ammonia-lyase, which act in the plant's defensive system against *S. graminum*, in the same way as a prior infestation with this aphid species. On the other hand, Correa et al. (2005) observed that the resistance of cucumber plants to *Bemisia tabaci* (Gennadius) induced by treatment with Si is identical to that induced by acibenzolar-*S*-methyl, a synthetic analog of salicylate, a natural plant elicitor. On the basis of all these examples of sucking insects that are adversely affected by high Si levels in the diet, allied to the fact already mentioned that silica does not constitute a mechanical barrier to stylet penetration, Keeping and Kvedaras (2008) stated that there is increasing evidence that soluble Si has an active role in the mechanism by which plants defend themselves from insect herbivores; this resistance mechanism, mediated by Si, includes one or a combination of induced and constitutive mechanical and chemical defense mechanisms.

Although most effects of Si on insects described in the literature suggest typical effects of antibiosis, such as reductions in survival and insect size, and in some cases, mandible wear and tear indicating low feeding efficiency (Keeping and Kvedaras 2008), antixenotic effects were also seen in *N. lugens* (Yoshihara et al. 1979), *S. gregaria*, and *S. exempta* (Massey et al. 2006), submitted to choice tests between their host plants with high and low Si content. Keeping et al. (2009) verified that the borer *E. saccharina* prefers to penetrate sugarcane at the leaf bud, the site with less Si accumulated, compared to the internode and the root band. In any event, plant defense involving Si has a very significant effect on feeding behavior, on performance, and finally, on the population dynamics of insect herbivores.

26.3.1.2 Lignin

Lignin, a polymer of phenylpropane units (monolignols), is located in the middle lamella and plays a fundamental role in cementing cell wall microfibrillae, maintaining the plant cells together (Esteban et al. 2003). The aromatic portions of the monolignols may show large variations between different plant groups, such as gymnosperms, woody angiosperms, and grasses. Besides these variations, the heterogeneity of lignin can also occur within the same group in different phenological stages of the plant (Lewis and Yamamoto 1990), and even at the subcellular level in the same phenological stage. These differences can alter plant vulnerability to insect attack, making it more or less resistant.

As with silica, lignin can also act as a physical barrier against certain insects. In sorghum, for example, the lignification and thickening of the walls of cells, which enclose the vascular bundle sheath within the central whorl of young leaves, stops penetration of larvae of the shoot fly, *Atherigona varia soccata* Rondani, in young leaves of resistant cultivars (Blum 1968).

The toughness of lignified structures can act as a physical deterrent to insects and, as with silica, can cause mandibular wear and tear, stopping or hindering the chewing and ingestion of plant tissue and reducing herbivore food consumption. Wainhouse et al. (1990) observed that in spruce trees with high lignin content, survival, growth rate, and larval weight of *Dendroctonus micans* (Kugelann) is lower. According to Swain (1979) cited by Wainhouse et al. (1990), lignin does not only show physical but also chemical effects due to hydrogen bonding to proteins and carbohydrates, reducing nutrient availability or being toxic to the insect. Lignin extracted from common beans tegument and incorporated into artificial diet of *Acanthoscelides obtectus* (Say) was toxic to early instar larvae, delayed development, and reduced fecundity of adults (Stamopoulos 1988).

The carbon–carbon links between monolignols make lignin very resistant to degradation/decomposition, and there is little evidence that insects can digest lignin. Even those that can, use microorganisms to help decompose the molecule and allow nutrient absorption. The beetle *Pselactus spadix* (Herbst), for example, can only digest lignin if the wood has been preconditioned by microbial decay (Oevering et al. 2003). Lignin concentration in insect feces is very high compared to other structural components and compared to its concentration in the wood ingested (Pitman et al. 2003), demonstrating its indigestibility.

26.3.2 Epicuticular Waxes

In most terrestrial plants, the cuticle forms a waxy layer that covers the apical wall of the epidermal cells of aerial organs (Jenks et al. 2002). This layer is composed of a lipid polymer, often covered with crystals that vary in shape and number, whose function is to protect the plant from dehydration (Eigenbrode and Espelie 1995). Among the principal classes of lipids that make up plant waxes are *n*-alkanes, wax esters, aldehydes, ketones, secondary alcohols, β-diketones, fatty alcohols, fatty acids, and triterpenoids. Their quantity and chemical composition vary among species (Figure 26.3), genotypes within species, parts within plants, and age of the plant part, affecting its ecological roles, here including interaction with herbivorous insects (Eigenbrode and Espelie 1995; Jenks et al. 2002).

Epicuticular waxes have been described as an important resistance factor to pest insects because their physical structure make permanence and maintenance of insects on plants more difficult. *Plutella xylostella* (L.) larvae show behavioral differences in host acceptance when exposed to waxes extracted from different varieties of *Brassica oleracea*, even though they possess the same surface morphology (Eigenbrode and Pillai 1998).

Varanda et al. (1992) observed that ursolic acid extracted from epicuticular waxes of *Jacaranda decurrens* acts as a feeding deterrent against the aphid *S. graminum*, when offered in an artificial diet and in high concentrations; however, it is toxic at lower concentrations. Although these results have little

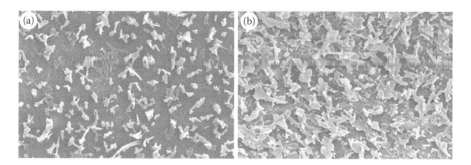

FIGURE 26.3 Surface morphology of air-dried adaxial leaf surfaces of *Thellungiella halophila* (a) and *T. parvula* (b) produced using scanning electron microscopy. Bar, 1 μm. (From Jenks, M. A., et al. Arabidopsis Book, 1, 1–22. With permission.)

practical value since aphids do not normally ingest plant epicuticular waxes, they suggest that these lipids act as allelochemicals in insects.

Heliocuerpa zea (Boddie) larvae fed artificial diet containing corn silks, from which epicuticular lipids were removed, develop better than those fed a diet containing silks plus epicuticular lipids (Yang et al. 1992). Similarly, *S. frugiperda* larvae fed artificial diet containing corn leaves without cuticular lipids develop better than those reared on a diet with leaves containing their lipids (Yang et al. 1991). When these lipids extracted from corn leaves were added to an artificial meridic diet, *S. frugiperda* development was also adversely affected (Yang et al. 1993). *S. frugiperda* fed on diet containing foliage of wild and cultivated species of peanuts without their cuticular lipids, showed increased weights and earlier pupation and adult emergence, compared with those fed on diet containing foliage with its lipids (Eigenbrode and Espelie 1995).

Shankaranarayana et al. (1980) described the extraction and the isolation of a triterpenoid from the epicuticular waxes of *Santalum album*, with activity on the lepidopteran *Atteva fabriciella* (Swedrus). Recently emerged adults fed with a solution of glucose containing a small quantity of the compound did not mate or oviposit.

Although plant epicuticular waxes show some toxic or antinutritional effect on insects, they do not allow postingestion activity to be clearly distinguished from a deterrent activity (Eigenbrode and Espelie 1995). On the other hand, waxes can indirectly contribute to insect development. Plants with a reduced or absent waxy layer, called glossy plants, are more subject to water stress and can suffer greater water loss, affecting the insect directly or indirectly by an increase in the concentration of some toxic compound or feeding deterrent.

26.3.3 Pilosity of Epidermis

Pilosity (presence of trichomes) is cited as one of the most important resistance factors in plants given their morphological variation and ways in which they can affect insects. Glandular trichomes, besides constituting the mechanical basis for resistance, are composed of specialized structures where certain chemical compounds, especially volatiles, are synthesized and stored.

In tomatoes (*Lycopersicon* spp. = *Solanum* spp.), whose plant–insect interactions are among the most deeply studied in crop plants, there are two main types of glandular trichomes: type IV with a single cell forming an apical gland, and type VI (corresponding to type A trichomes, which occur in the wild potato species *Solanum berthaultii*), with a gland consisting of four cells. In both types, the glands are supported by a multicellular peduncle on a monocellular base. The glandular secretion of the type IV trichomes has high levels of acylsugars, which are toxic to various insect species, such as *Macrosiphum euphorbiae* (Thomas), *Myzus persicae* (Sulzer), *B. tabaci* biotype B, *Trialeurodes vaporariorum* (Westwood), *H. zea*, *Spodoptera exigua* (Huebner), and *Liriomyza trifolii* (Burgess), whereas the type VI trichomes are more directly involved in the capture of small insects. Their glandular secretion is made up of phenolics (primarily rutin) as well as chlorogenic acid and conjugates of caffeic acid and polyphenoloxidase and peroxidase. When an insect damages a gland and the contents are mixed, the phenolics are oxidized to quinones, which polymerize, forming a sticky substance that fouls insect appendages and entangles them in exudate, glues them to the plant, or directly blocks up their mouthparts so they cannot feed (Kennedy 2003). Besides this mechanical effect, which stops the insect from feeding and starves it to death, there are also a series of postingestion effects observed with these compounds.

In *Lycopersicon hirsutum* f. *typicum* (= *Solanum hirsutum* f. *typicum* spp.), the trichome glands contain various sesquiterpenes, including γ-elemene, δ-elemene, α-curcumene, α-humulene, and zingiberene (Eigenbrode et al. 1994; Kennedy 2003), which are toxic to *Leptinotarsa decemlineata* (Say) larvae (Carter et al. 1989; Kennedy 2003). In *Lycopersicon hirsutum* f. *glabratum* (= *Solanum hirsutum* f. *glabratum*), the glands contain the methyl ketones 2-tridecanone and 2-undecanone, which are toxic to various insects, such as *H. zea*, *Keiferia lycopersicella* (Walsingham), *L. decemlineata*, *Manduca sexta* (L.), and *S. exigua* (Kennedy 2003). These examples demonstrate that trichomes show various interactions with insects, involving not only morphological but also chemical factors mainly present in the glandular exudates.

26.4 Chemical Causes and Insect Bioecology and Nutrition

Plants evolved resistance through the production of defense compounds that may be nonproteic (e.g., antibiotics, alkaloids, terpenes, cyanogenic glucosides) or proteic (e.g., lectins, arcelins, vicilins, systemins, chitinases, glucanases, and enzymatic inhibitors) (Franco et al. 2002). These compounds affect insect behavior (antixenosis or nonpreference) and/or insect biology and metabolism (antibiosis) (Table 26.1).

26.4.1 Antixenotic Factors

26.4.1.1 Repellents

In the host selection process by phytophagous insects, repellents cause a negative response by the insect, that is, to move away from the plant. Although at the early stage in the selection process, plant substrate color is important, in some cases selection is mediated mainly by volatile chemical substances.

To perceive odors emitted by potential host plants, insects rely on an olfactive guidance system controlled by particular sense organs known as sensila basiconica located on the antennae (Smith 2005). The number and arrangement of sensillae on the antennae are very diverse, which allows each insect to respond specifically to a certain amount of odors in a mixture. Most plants have specific composition of volatiles that differentiates them from other plants. Analyses of the air surrounding plants, the so-called headspace, yield mixtures of volatiles compounds that may number dozens or even a hundred volatile compounds. For example, the headspace odor of corn silk, which is attractive to certain moths, contains 30 volatile, and 40 compounds have been identified in the headspace odor of sunflower (Bernays and Chapman 1994).

Despite the large quantity of volatiles and their importance in host selection, the number of plant compounds that have been proven to be repellent to insects is still relatively low. Examples include monoterpenes present in the vapors from conifer resins, cited as repellents to scolytids (Bordasch and Berryman 1977; Werner 1995), and methyl salicylate and myrtenal found in various gymnosperms as repellents to the aphid *Aphis fabae* (Scop.) (Hardie et al. 1994).

The effect of various repellents present in diverse plant species makes possible the use of plant substances for insect pest control, especially those that attack stored products, including powder from leaves of the eucalyptus species *Corymbia citriodora*, repellent to *Sitophilus zeamais* Motschulsky, *A. obtectus*, and *Zabrotes subfasciatus* (Bohemann), and powder from leaves and fruits of *Chenopodium ambrosioides*, repellent to *Sitophilus oryzae* (L.), *Tribolium castaneum* (Herbst), *A. obtectus*, and *Z. subfasciatus* (Santos et al. 1984; Su 1991; Novo et al. 1997; Mazzonetto 2002; Mazzonetto and Vendramim 2003; Procópio et al. 2003). Tavares (2006), working with *p*-cimene and limonene compounds present in *C. ambrosioides* oil, observed repellence to *S. zeamais* but no effect on the behavior of *Rhyzopertha dominica* (F.). In practical terms, however, the use of repellents, such as citronella (*Cymbopogon* spp.), has been restricted almost only to the control of hematophagous mosquitoes and other insects of medical importance (Wasuwat et al. 1990; Suwonkerd and Tantrarongroj 1994; Tawatsin et al. 2001).

26.4.1.2 Phagodeterrents

Chapman (1974) suggested the term "feeding inhibitor" to describe allelochemicals that adversely affect insect feeding. Smith (2005) divided these allelochemicals into two groups, feeding deterrents and feeding inhibitors, without making a clear distinction between the two. Deterrents include substances that stop food intake, whereas the inhibitors include those that reduce feeding without stopping it. Schoonhoven (1982) referred to the term antifeedant to describe the function of these two groups of allelochemicals. Isman (2002) used the term antifeedant in a more restrictive way, considering it only a behavior-modifying substance that deters through a direct action on peripheral sensilla (= taste organs). He also mentioned that to determine an antifeedant action, bioassays should not exceed 6 h, when reduction in feeding can affect the insect's biology. Isman's (2002) definition of antifeedant excludes chemicals that suppress feeding by acting on the central nervous system after food ingestion and absorption, as well as substances that have sublethal toxicity. Since most studies do not separate substances that partially or totally inhibit food ingestion nor whether this reduction is related or not to the toxicity and/ or to the action on the insect's central nervous system, in this chapter the term phagodeterrent will be adopted to include allelochemicals that totally or partially inhibit insect food ingestion.

The ingestion of a smaller or larger amount of food depends on the palatability of plant tissues that insects perceive with gustatory receptors located in the maxillary palps and in the upper labium (Smith 2005). The variation in the types and number of gustatory receptors in different insect species results in a spectrum of different responses. Thus, in generalists (oligophagous), the response spectrum is wider than in the specialists (monophagous) (Visser 1983). Typical gustatory sensilla have selective receptors for phagodeterrents, such as sugars and amino acids (Isman 2002). Most allelochemicals inhibit feeding by stimulating a deterrent receptor, by blocking the perception of the feeding stimulants or by stopping the nervous impulses of gustatory information for selecting its food. This occurs with azadirachtin that stimulates deterrent receptors and also blocks sugar and inositol receptors (Schoonhoven 1988 cited by Koul 2005).

Phagodeterrence is most often caused by alkaloids, flavonoids, terpenoids, and phenols, which are produced and stored in the cell walls of leaves, vacuoles, or specialized structures, such as trichomes and waxes (Frazier 1986; Smith 2005). Terpenoids are the most diverse and potent phagodeterrents (Isman 2002). The triterpenoids, especially limonoids, include azadirachtin from *Azadirachta indica*, toosendanin from Chinaberry, *Melia azedarach*, limonin from *Citrus* spp., and various cardenolides, saponins, and withanolides from several plant species. Terpenoids include also diterpenes and sesquiterpenes. Among plant phenolics, the best known phagodeterrents are furanocoumarins and neolignans, and for the alkaloids, indoles and glycoalkaloids present in the Solanaceae (Isman 2002).

Lectins (see Section 26.5.1) can show phagodeterrence to insects, as shown by Czapla et al. (1992) cited by Czapla (1997), with *Ostrinia nubilalis* (Huebner) larvae feeding on artificial diets containing lectins from *Triticum aestivum* (WGA) or from *Bauhinia purpurea* (BPA) (Czapla et al. 1992 cited by Czapla 1997). Reduction in food consumption was also observed in spittlebug, *N. lugens*, fed with a diet containing lectin from *Galanthus nivalis* (GNA) (Powell et al. 1995). Through electrical penetration graphs, it was demonstrated that GNA added to artificial diet extends the probing period and reduces the ingestion period of this species (Rao et al. 1998). In all these cases, reduction in insect feeding could be due to lectin's harmful effects on the digestive tract, and not to the stimuli received from plant or diet. Sucking insects can be able to detect lectin during probing, and *N. lugens* nymphs avoid genetically modified plants expressing GNA in free choice tests with control plants, and this behavior is more pronounced in plants with a constitutive expression of lectin compared with those with tissue-specific expression (Foissac et al. 2000). The effect of lectins on insects is species specific; while GNA is phagodeterrent to *N. lugens* (Powell et al. 1995), concanavalin A (Con A; lectin from *Canavalia ensiformis*) causes increase in food consumption of *Lacanobia oleracea* (L.) larvae (Fitches and Gatehouse 1998).

Phagodeterrents used in pest control can be those present in resistant plants or those present in plants used as sources of bioactive vegetal extracts. Included in the first case are substances that are naturally present in resistant plants or that are transferred to them through traditional genetic breeding, as rutin, found in soybean that inhibits feeding by *Anticarsia gemmatalis* Huebner larvae (Hoffmann-Campo et al. 2006), and substances transferred by transgenic methods, as lectins (Vendramim and Nishikawa 2001). Phagodeterrents present in sources of bioactive vegetal extracts include the limonoid azadirachtin found in *A. indica*, which causes feeding inhibition in many insect species, and many other allelochemicals (Isman 2006). Lists of phagodeterrent allelochemicals are abundant (e.g., Panda and Khush 1995; Isman 2002; Smith 2005).

Many chemical substances present in plants, as the growth inhibitors as (or like) tannins, which bind to proteins to form difficult-to-digest complexes (Jãremo 1999), can also act as phagodeterrents (Karowe 1989). This was also verified with gossypol, which, besides affecting larval and pupal weights, development time, and larval survival, also inhibits feeding by *H. zea* (Stipanovic et al. 2006) and *Boarmia* (*Ascotis*) *selenaria* Schiffermüller larvae (Meisner et al. 1976). The alkaloid gramine, deterrent to *Rhopalosiphum padi* (L.) (Zuniga et al. 1988), is also considered toxic to this species (Zuniga and Corcuera 1986).

As a result of insect–plant coevolution, phagodeterrence is specific and one allelochemical that causes deterrence for one species may not affect or even stimulate another. One of the most well-known examples of this are the cucurbitacins, present in the roots and fruit of various cucurbits, which although repellent to various insects, such as *Apis mellifera* L. and *Vespula* sp. (Chambliss and Jones 1966), are also arrestant and phagostimulant to various other species, including chrysomelids of the genera *Diabrotica* and *Acalymma* (Metcalf et al. 1980; Eben et al. 1997). This characteristic has allowed the use of these compounds in baits containing insecticides for insect pests that are attracted (Arruda-Gatti and Ventura 2003).

In other situations, a highly deterrent compound to one species does not affect other closely related species. For example, azadirachtin phagodeterrence activity in relation to six noctuid species showed a variation of more than 30 times between the most sensitive, *Spodoptera litura* (F.), and the least sensitive species, *Actebia fennica* (Tauscher) (Isman 1993). Phagodeterrence of silphenene sesquiterpenes is high different for *Spodoptera littoralis* (Boisduval), *L. decemlineata*, and five aphid species (González-Coloma et al. 2002).

Prolonged exposure to the same allelochemical may mitigate the response, as observed in *P. xylostella* and *Pseudaletia unipuncta* (Haworth) larvae and *Epilachna varivestis* Mulsant adults to extracts of *Melia volkensii* and *Origanum vulgare* (Akhtar and Isman 2004), and this behavioral variation occurs for generalist and specialist species. Cases of rapid and total loss of activity have also been recorded, as observed with toosendanin, whose phagodeterrence to *S. litura* was totally blocked 4.5 h after the continuous exposure of the larvae to this allelochemical (Bomford and Isman 1996).

26.4.2 Antibiotic Factors

The main effects on insects of food with compounds that cause antibiosis are prolongation and mortality of immatures, reduction in size and weight of immatures and adults, reduction in fecundity and fertility, reduction in adult longevity, changes in sex ratio, and occurrence of abnormal pupae and adults. However, since some similar effects can be caused by deterrence, it may be difficult to discriminate between antibiosis and nonpreference or antixenosis (phagodeterrence). To characterize the type of resistance in this case, the first step is compare the test plant to the control plant. If there is no difference in consumption, it can be concluded that the effects on the biology are due to antibiosis. If there is a difference, more specific tests are needed using food consumption and utilization indices (Scriber and Slansky 1981; Slansky and Scriber 1985), and, in the case of caterpillars, removing the maxillae, which contain sensillae responsible for taste (Waldbauer and Fraenkel 1961). It is important to mention that there is not a strong relation between phagodeterrence and toxicity caused by chemical compounds of plants, which suggests that a behavioral rejection is not an adaptation to the effects of the compounds ingested but a consequence of the activity of deterrence receptors with wide chemical sensibility (Koul 2005).

Compounds that cause antibiotic effects are widely distributed throughout various genera of crop plants, including plants with different growth, fruiting, and propagation habits. Smith (2005), in an extensive bibliographic review, mentioned various examples of plants in cereals, forages, vegetables, fruit, and other tree species with antibiotic compounds against different insects.

Plant defense mechanisms against insects in tomatoes result from a series of interactive chemical characteristics that adversely affect nutrient acquisition and intoxicate insects (Duffey and Stout 1996). 2-Undecanone, which occurs together with 2-tridecanone in the trichome glands of tomatoes, *L. hirsutum* f. *glabratum* (= *S. hirsutum* f. *glabratum*), acts as a synergist, potentializing its toxicity to insects (Kennedy 2003). Herbivore performance depends both on the quantity and quality of the proteins that the plant contains. Protein quality change according to inter- and intraspecific genetic variations and to other phytochemicals ingested together with proteins, such as the alkylating agents. In other words, the nutritional value of a protein is not its inherent value but that obtained from the natural mixture (Duffey and Stout 1996).

Alkylating agents from plants are structurally very diverse, including quinones, phenolic compounds, aldehydes, pyrrolizidine alkaloids, lactone sesquiterpenes, and isothiocyanates. They form covalent bonds with amino acid chains, denaturing protein and limiting amino acid utilization by herbivores (Duffey and Stout 1996; Felton 1996; Felton and Duffey 1991). When, for example, a trichome's glandular content is discharged and mixed (which occurs on a large scale when a chewing insect feeds), the quinones resulting from the action of the polyphenoloxidases on the phenolic compounds (Duffey and Stout 1996; Felton 1996) do not always polymerize, and these quinones can be directly toxic to insects or they can react with the plant proteins, alkylating them and reducing or eliminating their nutritive value (Kennedy 2003).

Peroxidases oxidize mono- and dihydroxyphenols differently from polyphenoloxidases, but the effects on the proteins are essentially the same (Duffey and Stout 1996). Secondary products from these enzymatic reactions include hydrogen peroxide, the hydroxyl and the superoxide radicals, which can denature proteins and alter their digestibility (Felton 1996). Alkylated proteins reduce growth of *Pseudoplusia includens* (Walker) when supplied in an artificial diet; in soybean, hydrogen peroxide produced after tissue damage causes a quick cross link between the preexisting proteins of the cell walls. This link between the structural proteins of adjacent cell walls strengthens the walls and makes insoluble the proteins that may prove refractory for the herbivore's digestive enzymes (Felton 1996).

The intestinal pH significantly affects the activity of oxidative enzymes and the resulting chemical reactions (Duffey and Stout 1996). Most noctuid larvae have a very alkaline digestive tract, around pH 8.0 for *H. zea* and *S. exigua* (Felton and Duffey 1991), which favors the oxidation of chlorogenic acid into quinones and the production of free radicals and reactive hydrogen species (Felton and Duffey 1991; Duffey and Stout 1996). However, for *M. sexta*, which has an even higher pH (around 9.7), the alkalinity and detergency of the insect gut minimize the antinutritive effects of oxidized phenols. The solubility of tomato leaf proteins is significantly higher in a pH of 9.7, and this increase in solubility could compensate the loss in amino acid availability caused by the linking of the chlorogenic acid (Felton and Duffey 1991). The beetle *L. decemlineata*, which feeds on potato leaves having glandular trichomes containing chlorogenic acid and oxidizing enzymes, has an acidic digestive fluid, and even though the chlorogenic acid is alkylated, its effect is irrelevant since most amines are in the nonalkylable form. Low pH also disfavors the production of reactive oxygen species and organic free radicals (Duffey and Stout 1996).

26.5 Biotechnology and Resistance of Plants to Insects

Recent advances in biotechnology allowed the identification of genes, their functions, and their cloning, which has contributed significantly to insect plant resistance. Although classical breeding, made through crosses between resistant and susceptible varieties and based on the fundamentals of Mendelian genetics, is still important, biotechnology initiated a big revolution in the way of obtaining and evaluating cultivars resistant to insects. Its main advantages are specific introduction of the gene of interest into the plant, insertion of genes from phylogenetically unrelated organisms, control of the level of genetic expression, and the possibility of detecting the expression before plant maturation. Therefore, it has been possible to insert new genes into crop plants responsible for the expression of resistance factors or to increase the expression level of factors already present in the plant (Vendramim and Nishikawa 2001).

Although some secondary compounds, such as alkaloids, steroids, phenolic esters, terpenoids, cyanogenic glycosides, glucosinolates, saponins, flavonoids, pyrethrins, and nonprotein amino acids, are important in protecting plants against pests, their use in biotechnology is limited. Since they originate from complex biosynthetic processes with metabolic pathways involving various enzymes, these sorts of compounds are difficult to be inserted into plants; also, they add metabolic cost to the plant (Sharma et al. 2000). Proteins have a big advantage over phytochemicals since each protein is codified by a single gene that can be isolated and inserted into plants to increase their resistance to insect pests (Constabel 2000).

The technology presently available already permits the genetic transformation of plants to express peroxidases (Dowd and Lagrimini 1997), chitinases (Kramer et al. 1997), cholesterol oxidases (Greenplate et al. 1995; Purcell 1997), peptides from scorpion poison (Barton and Miller 1991 cited by Sharma et al. 2000; Wang et al. 2005), and proteins from parasitoids (Maiti et al. 2003); however, research has concentrated on two basic fronts: the insertion of bacterial genes (principally those codifying for the δ-endotoxins of *Bacillus thuringiensis*, the so-called Cry proteins) and the insertion of plant genes codifying for proteins that interfere in protein and carbohydrate metabolism (principally lectins and the inhibitors of proteases and amylases) (Bernal et al. 2004).

Discussions on the advantages and disadvantages of transgenic plants, or of each type of transformation, are not part of the scope of this chapter, but the proteins derived from plants are components normally used for feeding by man and animals and are considered to be safer strategies for insect pest control from the point of view of food safety (Vila et al. 2005). Besides this, since resistance based on plant genes does not show an acute toxicity, such as is shown by the *B. thuringiensis* proteins, it has been more widely accepted in integrated pest management (IPM) programs. The sublethal or chronic effects, such as reduction in growth or increasing on the development time, can normally be used together with biological control.

26.5.1 Lectins

Lectins are proteins of nonimmune origin that have the ability to bind to carbohydrates, glycoproteins, and substances that contain sugar, without altering the covalent structure of the glycosyl ligand (Peumans and Van Damme 1995). These proteins are present in diverse groups of organisms, such as animals, plants, fungi, bacteria, algae, protozoans, and even viruses, and are especially abundant in legume and cereal seeds (Constabel 2000).

Plant lectins are toxic to insect species from the orders Coleoptera, Diptera, Hemiptera, Hymenoptera, and Lepidoptera (Carlini and Grossi-de-Sá 2002), and this toxicity appears to be species specific and its effect cannot be generalized. The lectin from *Sphenostylis stenocarpa*, for example, affects the development of *Callosobruchus maculatus* (F.) at 0.2% concentration but this only occurs at 35% concentration for *Maruca vitrata* (F.). WGA is toxic to *Brevicoryne brassicae* (L.) (Cole 1994), but not to *Acyrthosiphon pisum* (Harris) (Rahbé and Febvay 1993).

The toxic action mechanism of lectins against insects is still not completely clear. Parameters such as survival, fecundity, fertility, food consumption, size, weight, color, and development period are used to measure their effects. This is a very complex issue because lectins show a wide range of effects on insects (Figure 26.4), considering the organism's complex physiology. Two basic requirements for lectin toxicity appear to be their resistance to digestion by proteolytic enzymes and the capacity to bind to glycoconjugates in some point of the insect's gut.

Harper et al. (1995), investigating the toxicity of 38 lectins to *O. nubilalis* and *Diabrotica virgifera virgifera* LeConte, found that all caused mortality or adverse effects to development due to binding to proteins in the insect midgut. They used the western blot technique, in which membrane destructuring expose lectin binding sites not exposed in living lectin-fed insects. Powell et al. (1998) used immuno-histochemical techniques to detect the binding of GNA to the gut of *N. lugens*. Transversal sections of the insect fed on a diet containing GNA were incubated with marked antibodies, preserving the tissue

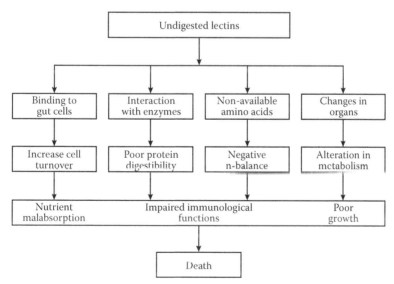

FIGURE 26.4 Spectrum of biological activities of plant lectins on insects. (Adapted from Vasconcelos, I. M., and J. T. A. Oliveira, *Toxicon*, 44, 385, 2004. With permission.)

structure and detecting potentially active binding sites *in vivo*. A combination of these two techniques was used by Bandyopadhyay et al. (2001) to identify the receptors of the lectin from *Allium sativum* in the midgut cells of *Dysdercus cingulatus* (F.) and *Lipaphis erysimi* (Kaltenbach), which were shown to be carbohydrate residues of the vesicle proteins of epithelial cells. In the midgut of *N. lugens*, where GNA has strong binding, thus causing toxic effects, the glycopeptide ferritin acts as the most abundant binding site for the lectin (Du et al. 2000).

Although not all lectins that bind to proteins show toxic effect, Zhu-Salzman et al. (1998) observed a correlation among insecticidal activity, binding, and resistance to proteolysis of the lectin from *Griffonia simplicifolia* (GSII) in *C. maculatus*. Zhu-Salzman and Salzman (2001) later proved that although the binding activity of GSII and its resistance to proteolysis contribute to lectin efficiency, they are independent activities.

Besides the feeding deterrent effect of lectins on insects, reducing food consumption (see Section 26.4.1.2), lectins have a low innate nutritional value since they have a very low content of half-cystine and have no methionine (Lajolo and Genovese 2002), the latter being one of the ten essential amino acids from which the others are synthesized (Parra 2001). In addition, in spite of lectins forming a very heterogeneous protein group, they have common structural similarity and similar capacity of not being degraded by digestive proteolytic enzymes. This has been demonstrated by Powell et al. (1998), investigating the proteolytic activity of GNA in the midgut and in the honeydew of *N. lugens* fed on an artificial diet containing lectin; they did not find any indication of this activity and concluded that GNA is resistant to proteolysis. This is to be expected with sap-sucking insects that have very low levels of proteases. However, this was also observed for Lepidoptera that show a high protease activity (Foissac et al. 2000). Larvae of *L. oleracea* fed on genetically modified plants expressing GNA had lectin in the feces (Fitches and Gatehouse 1998), indicating that GNA pass intact through the gut. When bound to the glycosylated digestive enzymes, lectins interfere in the insect's enzymatic function (Vasconcelos and Oliveira 2004). With the consequent loss in enzymatic activity, this binding may partly explain the resistance of the lectins to proteolysis.

The principal resistance mechanism of *N. lugens* to insecticides involves the production of high levels of glycosylated esterases, and GNA and the lectins from *Maackia amurensis* and *Dieffenbachia sequina* have the capacity to bind to the mannose of these enzymes, resulting in an alternative control of this pest (Vasconcelos and Oliveira 2004).

Even without directly binding to the insect's enzymes, lectins can, through their binding to other target sites, alter the normal enzymatic activity. Blakemore et al. (1995) verified increased secretion of trypsin in the gut of *Stomoxys calcitrans* (L.) larvae caused by WGA. Similarly, GNA and Con A cause increased amino peptide activity in the microvilosities of midgut cells of *L. oleracea* larvae (Fitches and Gatehouse 1998). Lectin activity on digestive enzymes, however, should be studied carefully because most of the assays are done with homogenized midgut tissues and intracellular and exogenous enzymes may be present (Gillot 2005).

Lectins can also have an effect on the insect's peritrophic matrix, which separates the food from the insect intestinal epithelium. This matrix is composed of a network of proteins and chitin with some proteins being highly glycosylated (Gillot 2005), providing linkage sites for diverse lectins. Chitin is a biopolymer of *N*-acetyl-D-glucosamine (GlcNac) units linked together, and the lectins with affinity to GlcNac (and in some cases to mannose and *N*-acetyl-D-galactosamine) are capable of binding to chitin, and consequently, to the insect peritrophic matrix (Raikhel et al. 1993). GSII, WGA, and the lectins from *Amaranthus caudatus*, *Bandeiraea simplicifolia*, *Phytolacca americana*, and *Talisia esculenta* are specific to GlcNac (Vasconcelos and Oliveira 2004) and their binding to chitin in the peritrophic matrix appears to be the principal cause of lectin toxicity to these insects. BPA and WGA cause the breaking of the peritrophic matrix of *O. nubilalis* 24 h after their ingestion, and it is possible to see pores in the fore region increasing progressively toward the hind region until the complete absence of the matrix 72 h after ingestion (Czapla et al. 1992 cited by Czapla 1997). This may cause consequences to the insect since, besides stopping the abrasive action of the food against the epithelial cells, the matrix also improves digestive efficiency by creating endo- and ectoperitrophic spaces (Gillot 2005).

The effect of lectins on the midgut epithelial cells occurs because these gut cells, whose microvilosities increase the organ's area of contact with food, have their membranes composed mainly by glycoproteins, which make the luminal surface of the intestine surrounded by potential linkage sites for the lectins present in the insect's diet (Peumans and Van Damme 1995). In insects where the peritrophic matrix has been damaged, or in species in which it is not present, there is no protective barrier against the action of these proteins.

Binding of lectins to the intestinal epithelium is often accompanied by breakage of the microvilosities and the disorganization of the principal absorptive cells, resulting in a decrease in the absorptive surface area with a consequent reduction in nutrient absorption (Vasconcelos and Oliveira 2004). Powell et al. (1998) observed that feeding with artificial diet containing GNA causes breakage of the midgut epithelium in adult *N. lugens*. Besides destruction of the cells due to their binding, lectins interfere in the normal functioning of the insect gut. Administered chronically, Con A caused hypertrophy in the gut of *L. oleracea* larvae, with an increase in mean intestine weight compared to total insect weight. As seen in some mammalian cell types, Con A acts as a mitogenic agent in the *L. oleracea* gut (Fitches and Gatehouse 1998). After destroying the gut epithelium, lectins reach the hemocoel and affect other insect organs, and the presence of GNA in the hemolymph, ovarioles, and fat body of *N. lugens* is attributed to the destruction of the microvilosities of the gut cells (Powell et al. 1998). GNA and Con A can also be found in the gut, hemolymph, and Malpighian tubules of *L. oleracea* larvae fed on artificial diets containing these lectins, demonstrating that the binding of these to the glycoproteins located along the insect gut can cause their internalization, resulting in systemic effects in the insect (Fitches et al. 2001). According to Vasconcelos and Oliveira (2004), lectin can be taken in by endocytosis and liberated into the intracellular space, being transported throughout the whole organism and causing harmful systemic effects in the internal tissues.

The effect of lectins on insect food proteins and on intestinal flora can happen owing to the potential binding of lectins to other glycosylated proteins in the food, stopping or slowing down their digestion (Czapla 1997). Lectins can also bind to the surface cell receptors of the microorganisms of the insect intestinal flora or indirectly interfere in their biology, structure, and population dynamics. Some lectins also show cytotoxic activity due to a catalytic site that, independent of the binding site to carbohydrate, acts intracellularly, cleaving an N-glycosidic bond in the rRNA adenosine. Therefore, these lectins are capable of inactivating the ribosomes of practically all eukaryotic organisms, interrupting cell protein synthesis and causing death (Peumans and Van Damme 1995; Carlini and Grossi-de-Sá 2002). These so-called ribosome-inactivating proteins have already been isolated from various different plants (Carlini and Grossi-de-Sá 2002).

The lectin family includes other proteins, such as the α-amylase inhibitors (see Section 26.5.2.2) and arcelins, and all these proteins are encoded by genes located at a single locus in the bean genome (Chrispeels 1997). Arcelins, which occur in wild accessions of *Phaseolus vulgaris* and are represented by seven different known isoforms, can constitute from 30% to 50% of seed proteins (Carlini and Grossi-de-Sá 2002). Genotypes containing arcelin have shown to be resistant to certain types of bruchids (Mazzonetto and Vendramim 2002; Guzzo et al. 2006, 2007). Arcelins are considered as "weak lectins" owing to some deletions or substitutions of essential amino acids at their active site (Chrispeels 1997; Carlini and Grossi-de-Sá 2002). If arcelins behave as lectins, they can have on insects the same effects as previously related for these proteins.

26.5.2 Enzymatic Inhibitors

Phytophagous insects use digestive enzymes, such as proteinases, amylases, lipases, glycosidases, and phosphatases, to process ingested food and obtain the necessary nutrients for their metabolism. During the evolutionary process, plants have adapted by using various defense mechanisms, including the production of insect digestive enzyme inhibitors. These inhibitors, acting in the digestive process, often prevent complex organic compounds from being degraded into simpler substances, such as amino acids, monosaccharides, and fatty acids, which are more easily absorbed. Therefore, the presence of these inhibitors can be an important resistance mechanism to insects. These compounds, however, are not always present in commercially exploited plants or, if present, do not always occur at levels sufficient to confer resistance. This problem can be overcome by using molecular biology techniques that permit the transfer of genes responsible for expressing these inhibitors to crop plants.

At present, the two most important groups of enzymatic inhibitors, and which have been extensively studied for use in transgenic-based plant improvement programs, are the protease and the α-amylase inhibitors, which inhibit insect digestive proteases and α-amylases, respectively. These enzymes are important in the digestion of proteins and starch, respectively (Carlini and Grossi-de-Sá 2002; Lajolo and Genovese 2002).

26.5.2.1 Protease Inhibitors

Insect proteases are an enzyme group responsible for the sequential hydrolysis of proteins into oligomers, dimers, and monomers inside the gut (Reeck et al. 1997), catalyzing the cleavage of peptide bonds (Carlini and Grossi-de-Sá 2002). Protease inhibitors (PIs), also consisting of proteins, form complexes that have a high affinity with proteases, inhibiting their hydrolytic activity (Outchkourov et al. 2004) and, although terms are used indiscriminately, originally, the "proteinases" only refer to endopeptidases (enzymes that hydrolyze internal peptide bonds), and the "proteases" include both the endopeptidases and the exopeptidases (enzymes that hydrolyze N-terminal or C-terminal bonds) (Ryan 1990).

According to the amino acids present at their active site, endopeptidases are classified as serine proteinases (containing serine and histidine), cysteine proteinases (containing cysteine), aspartic proteinases (containing an aspartate group), or metalloproteinases (containing a metallic ion) (Boulter 1993; Carlini and Grossi-de-Sá 2002). PIs are grouped into at least 24 distinct families (Ryan 1990; Reeck et al. 1997), of which at least 9 contain representatives found in plant tissues: soybean Kunitz trypsin inhibitor; Bowman–Birk inhibitor; potato inhibitor I; potato inhibitor II; squash inhibitor; barley trypsin inhibitor; cystatin, inhibitor of cysteine proteinase; carboxypeptidase inhibitor of potato; and bifunctional inhibitor of corn Ragi I-2. In general, PIs of the same family have specificity to a certain class of proteinases.

PIs affect water balance, molting, and the enzymatic and hormonal regulation of insects (Boulter 1993); the principal effect appears to be related to insect nutrition, interfering in the digestive processes. In *C. maculatus*, nutritional supplementation with methionine is capable of overcoming the effects of ingesting the cowpea trypsin inhibitor (Gatehouse and Boulter 1983). Besides mortality, insects subjected to diets containing PIs show delayed growth and development and reduction in individual weight (Ussuf et al. 2001). These effects are much more complex than the simple reduction in hydrolytic activity of digestive proteases (Ryan 1990).

The PIs stefin A and equistatin expressed in potato leaves show phagodeterrent action to adults of *Frankliniella occidentalis* (Pergande). The mechanism that senses PIs and affects the behavior of *F. occidentalis* utilizes a completely different signaling pathway from the well-known olfactory and gustatory signaling pathways (Outchkourov et al. 2004). Deterrent effect has also been observed with fragments of PI from pea seeds with anti-chymotrypsin activity, which interrupts *Acyrthosiphon pisum* (Harris) feeding (Rahbé et al. 2003). In general, aphids are insensitive to PIs because they feed on sap taken directly from the phloem, which contains high levels of free amino acids, and are not dependent on proteases to fulfill their nutritional needs. The fact that a PI with anti-chymotrypsin activity can cause deterrence in *A. pisum* and that no chymotrypsin activity has been observed in this insect's gut (Rahbé et al. 2003) reinforces the hypothesis that the insect satiation mechanism is regulated by hormones, and this explains the deterrence observed.

Degradation of PIs in insect guts potentially affect the efficacy of PIs (Outchkourov et al. 2004), and the active PIs against insects intrinsically show a relative resistance to digestion by their proteolytic enzymes. Therefore, the low digestibility of the PIs also results in their little contribution to insect nutritional needs.

In plant storage organs, such as tubers and seeds, PIs correspond to 10% of total protein (Ussuf et al. 2001). In bean seeds, with trypsin inhibitors of only around 2.5% of total protein, this value corresponds to 32% and 40% of its cystine content in *Phaseolus lunatus* and *P. vulgaris*, respectively (Kakade 1974 cited by Lajolo and Genovese 2002).

The effect of inhibitors on insect proteases occurs when, in contact with these enzymes, these inhibitors bind to their active site in a practically irreversible way, forming a complex with a very low dissociation constant and blocking the active site. In this way, the inhibitor acts as a pseudosubstrate for the enzyme, imitating the original substrate but not allowing the cleavage of the peptide bond. The enzyme often manages to hydrolyze the inhibitor; however, due to the binding site configuration, the hydrolyzed inhibitor can stay bonded to the enzyme in the same way as in its nonhydrolyzed form (Lawrence and Koundal 2002). By stopping the normal functioning of the proteases, the PIs make protein digestion by the insect more difficult. This provokes amino acid deficiency resulting in undernourished insects and possibly leading them to death.

A common response of insects to PIs is increase in levels of expression of digestive enzymes, by increasing the production of already existing proteolytic enzymes, by producing enzymes insensitive to the inhibitors, or by producing enzymes capable of inactivating or degrading the inhibitors (Silva-Filho and Falco 2000). *C. suppressalis* larvae, which use serine and cysteine proteinases, significantly increase the production of these enzymes as well as of exopeptidases, leucine aminopeptidase and carboxypeptidases A and B, after ingesting corn PI (Vila et al. 2005). *S. exigua* larvae reared on transgenic plants expressing the potato inhibitor II show only 18% of its proteases sensitive to the inhibitor, whereas in the control larvae it reaches 78% (Jongsma et al. 1995). The beetle *L. decemlineata* also synthesizes proteinases insensitive to the inhibitor expressed in potato leaves after chronic ingestion of leaves (Bolter and Jongsma 1995). This is a general adaptive response of the insect that secretes proteases insensitive to inhibitors and is able to digest them. However, the increase in the production of digestive enzymes and the induction of proteolytic activity in *C. suppressalis*, in response to corn PI, is not sufficient to avoid the harmful effects of the inhibitor (Vila et al. 2005). In any event, the presence of inhibitors in the insect gut results in the secretion of insensitive proteases. The hypersecretion of digestive enzymes, such as trypsin and chymotrypsin, which are rich in sulfurated amino acids, causes the loss of these endogenous amino acids (Shukle and Murdock 1983), and can be another nutritional problem for the insect, mainly when it feeds on a diet poor in this type of amino acid (Lajolo and Genovese 2002). To overcome the problems of adaptation of insects to enzymatic inhibitors, the solution is make plants express more than one type of digestive enzyme inhibitor and/or produce new, more potent, and specific inhibitors against insects' digestive enzymes (Marsaro et al. 2006).

26.5.2.2 α-Amylase Inhibitors

The α-amylases are wide-spectrum hydrolytic enzymes found in microorganisms, animals, and plants. They catalyze the first hydrolysis of the sugar polymers, such as starch and glycogen, into simpler units to allow their assimilation by the organism. These widely distributed molecules are the most important digestive enzymes of many insects that feed exclusively on seed products during larval and/or adult lives. When the action of the α-amylases is inhibited, nutrition of the organism is impaired, causing shortness in energy (Carlini and Grossi-de-Sá 2002).

The inhibitors of α-amylases (α-AIs) are divided into two groups: the non-protein or non-proteinaceous, and the protein or proteinaceous. The non-protein inhibitors contain diverse types of organic compounds, such as acarbose, iso-acarbose, acarviosine glucose, and cyclodextrins. These inhibitors have practically not been studied for insect control (Franco et al. 2002). However, the proteinaceous α-AIs, especially those found in plants as part of their natural defense mechanisms, constitute a potentially important tool in research for obtaining resistant varieties to insect pests, whether by classical improvement methods or by genetic engineering (Chrispeels et al. 1998; Gatehouse and Gatehouse 1998; Iulek et al. 2000; Carlini and Grossi-de-Sá 2002; Franco et al. 2002; Svensson et al. 2004).

α-AIs present in plants occur principally in cereals (Feng et al. 1996; Franco et al. 2000; Iulek et al. 2000) and legume seeds (Shade et al. 1994, Ishimoto et al. 1996, Grossi-de-Sá et al. 1997) but also in other botanical groups (Lu et al. 1999; Hansawasdi et al. 2000; Figueira et al. 2003; Marsaro et al. 2005). According to Franco et al. (2002), the α-AIs with potential use in insect pest control are classified into six classes on the basis of their tertiary structures: lectin type, knottin type, cereal type, Kunitz type, c-purothionin type, and thaumatin type. These classes of inhibitors show remarkable structural variety leading to different modes of inhibition and different specificity profiles against diverse α-amylases. The determination of the specificity of inhibition is very important and constitutes the first step for discovering an inhibitor that can be used to obtain a transgenic plant resistant to insects. In some cases, the α-AIs only act against the α-amylases of mammals or, only against the α-amylases of insects. However, in general, these inhibitors inhibit α-amylases from different sources. Franco et al. (2002) state that the specificity of inhibition is an important issue as the introduced inhibitor must not adversely affect the plant's own α-amylases, nor the nutritional value of the crop. In these cases, a better knowledge of the structural bases that determine the mode of inhibition will allow a rational design of mutants with the desirable characteristics.

Various insects, especially coleopterans that feed on seeds rich in starch during the larval and/or adult stages, depend on their α-amylases for survival. This has stimulated research on starch digestion aiming to control pests that depend on this nutrient (Franco et al. 2000). Particular attention has been given to the α-AIs from seeds of the common bean, *P. vulgaris*, which cause toxic effects to various insect pests (Carlini and Grossi-de-Sá 2002; Franco et al. 2002) and which, together with the α-AIs found in wheat, were initially the best characterized inhibitors (Hilder and Boulter 1999). The lectin-type α-AIs have been purified and characterized from different varieties of *P. vulgaris*, including the white, red, and black beans (Franco et al. 2000). Studies with the α-AIs of *P. vulgaris* intensified after the discovery that they adversely affect the development of *Callosobruchus chinensis* (L.) and *C. maculatus* (Ishimoto and Kitamura 1989; Shade et al. 1994).

The genus *Phaseolus* contains at least four phenotypes of α-AIs: α-AI-1, α-AI-2, α-AI-3, and a null type against all the α-amylases tested. Of particular interest is the specificity of the two isoforms α-AI-1 and α-AI-2 toward different α-amylases; α-AI-1, found in most cultivated common bean varieties, inhibits mammalian α-amylases such as porcine pancreatic amylase, and the insect larval α-amylases of *C. chinensis*, *C. maculates,* and *Bruchus pisorum* L., but is not active against the α-amylase of the Mexican bean weevil, *Zabrotes subfasciatus*, an important storage pest of the common bean (Grossi-de-Sá and Chrispeels 1997). This lack of activity against *Z. subfasciatus* may be due to the noninhibition of amylase by α-AI-1 or the presence of intestinal serine proteinase capable of digesting the inhibitor (Ishimoto et al. 1996; Silva et al. 2001). The second variant, α-AI-2, which shares 78% amino acid homology with α-AI-1, is found in few wild accessions of common bean and specifically inhibits the *Z. subfasciatus* larval α-amylase (Ishimoto and Kitamura 1993; Grossi-de-Sá and Chrispeels 1997; Grossi-de-Sá et al. 1997).

To validate α-amylases as a target for plant protection, it is important to know its variety and how the expression is controlled. Different forms of α-amylases were observed in the midgut lumen of *C. maculatus* and *Z. subfasciatus* (Campos et al. 1989; Silva et al. 1999). For *Z. subfasciatus*, the patterns of expression of the α-amylase vary according to this insect's diets, apparently more in response to antimetabolic proteins, such as the α-AIs, than as a response to structural differences in starch granules. This beetle regulates the concentration of α-glucosidases and α-amylases when reared on different diets (Silva et al. 1999).

These inhibitors from common beans have been used in transformation of other plant species. The inhibitor of α-amylase responsible for resistance of *P. vulgaris* to bruchid attack is used in pea plants with the aim of transferring the resistance character (Shade et al. 1994). After the transformation, a high level of α-AI in plants, which were then submitted to tests with *C. maculatus* and *C. chinensis*, was observed. All transformed plants showed resistance to these beetles, although *C. maculatus* was less susceptible to α-AI than *C. chinensis*. Resistance to *B. pisorum*, *C. maculates*, and *C. chinensis* was also found in pea and adzuki bean, *Vigna angularis* L., seeds transformed by the introduction of α-AI-1 (Shade et al. 1994; Schroeder et al. 1995; Ishimoto et al. 1996; Morton et al. 2000).

Because of their practical application in transgenic plant production, α-AIs should have the appropriate specificities and be effective against a range of insect pests. Moreover, they should not interfere with endogenous α-amylases, important for plant metabolism (Kadziola et al. 1998). Since there is a wide structural and functional variation among α-AIs, one should screen those with desirable characteristics. The rational redesign of known inhibitors to confer upon them the required specificity profile, faster than the screening approach, requires a full understanding of the basic structural interactions between the amylases and the inhibitors.

26.5.2.3 Bifunctional Inhibitors

PIs may have different reactive sites that can consequently act on different types of enzymes. This is what happens, for example, with the corn bifunctional inhibitor, which is active against trypsin and α-amylase (Boulter 1993). Such inhibitors have two types of independent reactive sites located in separate regions of the protein: one type specific for proteases and the other for α-amylases. Since they are capable of inactivating proteases and α-amylases simultaneously, these inhibitors are called bifunctional inhibitors (Ryan 1990; Ussuf et al. 2001). Since the reactive sites for proteases and α-amylases are independent in the bifunctional inhibitors, the effects on these types of enzymes are also independent and are the same as those already described for the PIs and the α-AIs. Franco et al. (2002) stated that there are compounds that, besides simultaneously inhibiting proteases and α-amylases, also show other activities, as for example, activity similar to the chitinases, which increases the potential for resistance in plants that contains this inhibitor.

26.6 Final Considerations

Utilizing plant resistance to insects is a control method that has been employed for more than a century, significantly reducing the need for conventional insecticide use in insect pest management, with little or no economic or ecological disadvantages. In many cases, the use of resistant varieties may be the only control method used, but even when only moderate resistance is obtained, this method can be used together with other control tactics in agreement with the IPM philosophy. Some resistant varieties make the crop unsuitable for the insect as a food source, and the limitation of this essential resource to animals, that is, of the energy necessary for the maintenance of vital functions, adversely affects survival and reproduction, which affects insect pest population size in the field.

To better understand the bioecology and nutrition of phytophagous insects, it is important to know the mechanisms of host plant resistance. For this, it is first necessary to characterize the resistant variety and this can be done using various parameters, which consider the effect of the plant on the insect, causing eventual changes in its behavior and biology, as well as the effect of the insect on the plant demonstrating how much its development and survival and, consequently, its production, are affected. On the basis of these aspects, the type of resistance involved can be identified. Once the types of resistance have been identified, it is important to consider that the resistance is the result of complex interactions between insects and plants through which, during the evolutionary process, the insect adapts to the plant and there is a subsequent development of resistance by the plant to this (counter-adaptation). Therefore, the plant should not be seen as a passive entity but as an active organism, which, by selection, develops defense mechanisms against insects. These mechanisms constitute the factors or causes of resistance, commonly divided into physical, chemical, and morphological. Knowledge of the types and causes involved in resistance of a plant to an insect pest is fundamental for the orientation of improvement programs aimed at obtaining resistant varieties.

The use of resistant varieties produced by classical crosses has been shown to be economically viable to growers and ecologically and socially acceptable to consumers. In the future, genetically modified varieties will play an important role in the sustainability of world agriculture. Despite the existence of hundreds of resistant genes available for insertion into crop plants, the continuous adaptation of insects to these, as well as the appearance of more virulent biotypes of insect pests, also demands the constant identification of new sources of resistance. The rational design of new more potent and more specific inhibitors against insects' digestive enzymes will assume an important role in the development of resistant plant varieties to insect pests.

REFERENCES

Akhtar, Y., and M. B. Isman. 2004. Feeding responses of specialist herbivores to plant extracts and pure allelochemicals: Effects of prolonged exposure. *Entomol. Exp. Appl.* 111:201–8.

Almeida, G. D., D. Pratissoli, J. C. Zanuncio, V. B. Vicentini, A. M. Holtz, and J. E. Serrão. 2008. Calcium silicate and organic mineral fertilizer applications reduce phytophagy by *Thrips palmi* Karny (Thysanoptera: Thripidae) on eggplants (*Solanum melongena* L.). *Interciencia* 33:835–8.

Almeida, G. D., D. Pratissoli, J. C. Zanuncio, V. B. Vicentini, A. M. Holtz, and J. E. Serrão. 2009. Calcium silicate and organic mineral fertilizer increase the resistance of tomato plants to *Frankliniella schultzei*. *Phytoparasitica* 37:225–30.

Anderson, D. L., and O. Sosa Jr. 2001. Effect of silicon on expression of resistance to sugarcane borer (*Diatraea saccharalis*). *J. Am. Soc. Sugar Cane Technol.* 21:43–50.

Arruda-Gatti, I. A., and M. U. Ventura. 2003. Iscas contendo cucurbitacinas para o manejo de *Diabrotica* spp. *Semin. Ciênc. Agrár.* 24:331–6.

Bandyopadhyay, S., A. Roy, and S. Das. 2001. Binding of garlic (*Allium sativum*) leaf lectin to the gut receptors of homopteran pests is correlated to its insecticidal activity. *Plant Sci.* 161:1025–33.

Barbehenn, R. V. 1993. Silicon: An indigestible marker for measuring food consumption and utilization by insects. *Entomol. Exp. Appl.* 67:247–51.

Barbosa Filho, M. P., G. H. Snyder, A. S. Prabhu, L. E. Datnoff, and G. H. Korndörfer. 2000. Importância do silício para a cultura do arroz. Informações agronômicas no. 89, Encarte técnico. Piracicaba, SP, Brazil.

Basagli, M. A. B., J. C. Moraes, G. A. Carvalho, C. C. Ecole, and R. C. R. Gonçalves-Gervásio. 2003. Effects of sodium silicate application on the resistance of wheat plants to the green-aphid *Schizaphis graminum* (Rond.) (Hemiptera: Aphididae). *Neotrop. Entomol.* 32:659–63.

Beck, S. D. 1965. Resistance of plants to insects. *Annu. Rev. Entomol.* 10:207–32.

Bernal, J. S., J. Prasifka, M. Sétamou, and K. M. Heinz. 2004. Transgenic insecticidal cultivars in integrated pest management: Challenges and opportunities. In *Integrated Pest Management: Potential, Constraints and Challenges*, ed. O. Koul, G. S. Dhaliwal, and G. W. Cuperus, 123–145. Cambridge: CABI Publishing.

Bernays, E. A., and M. Graham. 1988. On the evolution of host specificity in phytophagous arthropods. *Ecology* 69:886–92.

Bernays, E. A., and R. F. Chapman. 1994. *Host–Plant Selection by Phytophagous Insects. Contemporary Topics in Entomology.* New York: Chapman and Hall.

Blakemore, D., S. Williams, and M. J. Lehane. 1995. Protein stimulation of trypsin secretion from the opaque zone midgut cells of *Stomoxys calcitrans*. *Comp. Biochem. Physiol. B* 110:301–7.

Blum, A. 1968. Anatomical phenomena in seedlings of sorghum varieties resistant to the sorghum shoot fly (*Atherigona varia soccata*). *Crop Sci.* 8:388–91.

Bolter, C. J., and M. A. Jongsma. 1995. Colorado potato beetles (*Leptinotarsa decemlineata*) adapt to protein-ase inhibitors induced in potato leaves by methyl jasmonate. *J. Insect Physiol.* 41:1071–8.

Bomford, M. K., and M. B. Isman. 1996. Desensitization of fifth instar *Spodoptera litura* to azadirachtin and neem. *Entomol. Exp. Appl.* 81:307–13.

Bordasch, R. P., and A. A. Berryman. 1977. Host resistance to the fir engraver beetle, *Scolytus ventralis* (Coleoptera: Scolytidae) 2. Repellency of *Abies grandis* resins and some monoterpenes. *Can. Entomol.* 109:95–100.

Boulter, D. 1993. Insect pest control by copying nature using genetically engineered crops. *Phytochem.* 34:1453–66.

Brues, C. T. 1920. The selection of food-plants by insects, with special reference to lepidopterous larvae. *Am. Nat.* 54:313–32.

Campos, F. A. P., J. Xavier-Filho, C. P. Silva, and M. B. Ary. 1989. Resolution and partial characterization of proteinases and α amylases from midguts of larvae of the bruchid beetle *Callosobruchus maculatus* (F.). *Comp. Biochem. Physiol. B* 92:51–7.

Carlini, C. R., and M. F. Grossi-de-Sá. 2002. Plant toxic proteins with insecticidal properties. A review on their potentialities as bioinsecticides. *Toxicon* 40:1515–39.

Carter, C. D., J. N. Sacalis, and T. J. Gianfagna. 1989. Zingiberene and resistance to Colorado potato beetle in *Lycopersicon hirsutum* f. *hirsutum*. *J. Agric. Food Chem.* 37:206–10.

Cartier, J. J. 1968. Factors of host plant specificity and artificial diets. *Bull. Entomol. Soc. Am.* 14:18–21.

Carvalho, S. P., J. C. Moraes, and J. G. Carvalho. 1999. Efeito do silício na resistência do sorgo (*Sorghum bicolor*) ao pulgão-verde *Schizaphis graminum* (Rond.) (Homoptera: Aphididae). *An. Soc. Entomol. Bras.* 28:505–10.

Chambliss, O. L., and C. M. Jones. 1966. Cucurbitacins: Specific insect attractants in *Cucurbitaceae*. *Science* 153:1392–3.

Chapman, R. F. 1974. The chemical inhibition of feeding by phytophagous insects: A review. *Bull. Entomol. Res.* 64:339–63.

Chrispeels, M. J. 1997. Transfer of bruchid resistance from the common bean to other starchy grain legumes by genetic engineering with the α-amylase inhibitor gene. In *Advances in Insect Control: The Role of Transgenic Plants*, ed. N. Carozzi and M. Koziel, 139–56. London: Taylor & Francis.

Chrispeels, M. J., M. F. Grossi-de-Sá, and T. J. V. Higgins. 1998. Genetic engineering with α-amylase inhibitors makes seeds resistant to bruchids. *Seed Sci. Res.* 8:257–63.

Cole, R. A. 1994. Isolation of a chitin-binding lectin with insecticidal activity in chemically-defined synthetic diets from two wild *Brassica* species with resistance to cabbage aphid *Brevicoryne brassicae. Entomol. Exp. Appl.* 72:181–7.

Constabel, C. P. 2000. A survey of herbivore-inducible defensive proteins and phytochemicals. In *Induced Plant Defenses against Pathogens and Herbivores. Biochemistry, Ecology and Agriculture*, ed. A. A. Agrawal, S. Tuzun, and E. Bent, 137–66. St. Paul, MN: The American Phytopatological Society.

Correa, R. S. B., J. C. Moraes, A. M. Auad, and G. A. Carvalho. 2005. Silicon and acibenzolar-*s*-methyl as resistance inducers in cucumber, against the whitefly *Bemisia tabaci* (Gennadius) (Hemiptera: Aleyrodidae) biotype B. *Neotrop. Entomol.* 34:429–33.

Czapla, T. H. 1997. Plant lectins as insect control proteins in transgenic plants. In *Advances in Insect Control: The Role of Transgenic Plants*, ed. N. Carozzi and M. Koziel, 123–38. London: Taylor & Francis.

Dethier, V. G. 1941. Chemical factors determining the choice of plants by *Papillio* larvae. *Am. Nat.* 75:61–73.

Dethier, V. G., L. B. Browne, and C. N. Smith. 1960. The designation of chemicals in terms of the responses they elicit from insects. *J. Econ. Entomol.* 53:134–6.

Djamin, A., and M. D. Pathak. 1967. Role of silica in resistance to Asiatic rice borer, *Chilo suppressalis* (Walker), in rice varieties. *J. Econ. Entomol.* 60:347–51.

Dowd, P. F., and L. M. Lagrimini. 1997. The role of peroxidase in host insect defenses. In *Advances in Insect Control: The Role of Transgenic Plants*, ed. N. Carozzi and M. Koziel, 195–223. London: Taylor & Francis.

Du, J., X. Foissac, A. Carss, A. M. R. Gatehouse, and J. A. Gatehouse. 2000. Ferritin acts as the most abundant binding protein for snowdrop lectin in the midgut of rice brown planthoppers (*Nilaparvata lugens*). *Insect Biochem. Mol. Biol.* 30:297–305.

Duffey, S. S., and M. J. Stout. 1996. Antinutritive and toxic components of plant defense against insects. *Arch. Insect Biochem. Physiol.* 32:3–37.

Eben, A., M. E. Barbercheck, and M. S. Aluja. 1997. Mexican diabroticite beetles: II. Test for preference of cucurbit hosts by *Acalymma* and *Diabrotica* spp. *Entomol. Exp. Appl.* 82:63–72.

Ehrlich, P. R., and P. R. Raven. 1964. Butterflies and plants: A study in coevolution. *Evolution* 18:586–608.

Eigenbrode, S. D., and K. E. Espelie. 1995. Effects of plant epicuticular lipids on insect herbivores. *Annu. Rev. Entomol.* 40:171–94.

Eigenbrode, S. D., and S. K. Pillai. 1998. Neonate *Plutella xylostella* responses to surface wax components of a resistant cabbage (*Brassica oleracea*). *J. Chem. Ecol.* 24:1611–27.

Eigenbrode, S. D., J. T. Trumble, J. G. Millar, and K. K. White. 1994. Topical toxicity of tomato sesquiterpenes to the beet armyworm and the role of these compounds in resistance derived from an accession of *Lycopersicon hirsutum* f. *typicum. J. Agric. Food Chem.* 42:807–10.

Epstein, E. 1999. Silicon. *Annu. Rev. Plant Physiol. Plant Mol. Biol.* 50:641–64.

Esteban, L. G., A. G. Casasús, C. P. Oramas, and P. de Palacios. 2003. *La Madera y su Anatomía*. Madrid: Fundación Conde del Valle de Salazar, Ediciones Mundi-Prensa and AITIM.

Felton, G. W. 1996. Nutritive quality of plant protein: Sources of variation and insect herbivore responses. *Arch. Insect Biochem. Physiol.* 32:107–30.

Felton, G. W., and S. S. Duffey. 1991. Reassessment of the role of gut alkalinity and detergency in insect herbivory. *J. Chem. Ecol.* 17:1821–36.

Feng, G. H., M. Richardson, M. S. Chen, K. J. Kramer, T. D. Morgan, and G. R. Reeck. 1996. α-Amylase inhibitors from wheat: Amino acid sequences and patterns of inhibition of insect and human α-amylases. *Insect Biochem. Mol. Biol.* 26:419–26.

Figueira, E. L. Z., A. Blanco-Labra, A. C. Gerage, E. Y. S. Ono, E. Mendiola-Olaya, Y. Ueno, and E. Y. Hirooka. 2003. Amylase inhibitor present in corn seeds active in vitro against amylase from *Fusarium verticillioides. Plant Dis.* 87:233–40.

Fitches, E., and J. A. Gatehouse. 1998. A comparison of the short and long term effects of insecticidal lectins on the activities of soluble and brush border enzymes of tomato moth larvae (*Lacanobia oleracea*). *J. Insect Physiol.* 44:1213–24.

Fitches, E., S. D. Woodhouse, J. P. Edwards, and J. A. Gatehouse. 2001. In vitro and in vivo binding of snowdrop (*Galanthus nivalis* agglutinin; GNA) and jackbean (*Canavalia ensiformis*; Con A) lectins within tomato moth (*Lacanobia oleracea*) larvae; mechanisms of insecticidal action. *J. Insect Physiol.* 47:777–87.

Foissac, X., N. T. Loc, P. Christou, A. M. R. Gatehouse, and J. A. Gatehouse. 2000. Resistance to green leafhopper (*Nephotettix virescens*) and brown planthopper (*Nilaparvata lugens*) in transgenic rice expressing snowdrop lectin (*Galanthus nivalis* agglutinin; GNA). *J. Insect Physiol.* 46:573–83.

Fox, L. R. 1988. Diffuse coevolution within complex communities. *Ecology* 69:906–7.

Franco, O. L., D. J. Rigden, F. R. Melo, and M. F. Grossi-de-Sá. 2002. Plant α-amylase inhibitors and their interaction with insect α-amylases: Structure, function and potential for crop protection. *Eur. J. Biochem.* 269:397–412.

Franco, O. L., D. J. Rigden, F. R. Melo, C. Bloch Jr., C. P. Silva, and M. F. Grossi-de-Sá. 2000. Activity of wheat α-amylase inhibitors towards bruchid α-amylases and structural explanation of observed specificities. *Eur. J. Biochem.* 267:2166–73.

Frazier, J. L. 1986. The perception of plant allelochemicals that inhibit feeding. In *Molecular Aspects of Insect–Plant Associations*, ed. L. B. Brattsten and S. Ahmad, 1–42. New York: Plenum Press.

Gatehouse, A. M. R., and D. Boulter. 1983. Assessment of the anti-metabolic effects of trypsin inhibitors from cowpea (*Vigna unguiculata*) and other legumes on development of the bruchid beetle *Callosobruchus maculatus*. *J. Sci. Food Agric.* 34:345–50.

Gatehouse, A. M. R., and J. A. Gatehouse. 1998. Identifying proteins with insecticidal activity: Use of encoding genes to produce insect-resistant transgenic crops. *Pestic. Sci.* 52:165–75.

Gillot, C. 2005. *Entomology.* 3 ed. Dordrecht: Springer.

Gomes, F. B., J. C. Moraes, C. D. dos Santos, and M. M. Goussain. 2005. Resistance induction in wheat plants by silicon and aphids. *Sci. Agric.* 62:547–51.

González-Coloma, A., F. Valencia, N. Martín, J. J. Hoffmann, L. Hutter, J. A. Marco, and M. Reina. 2002. Silphinene sesquiterpenes as model insect antifeedants. *J. Chem. Ecol.* 28:117–29.

Goussain, M. M., E. Prado, and J. C. Moraes. 2005. Effect of silicon applied to wheat plants on the biology and probing behaviour of the greenbug *Schizaphis graminum* (Rond.) (Hemiptera: Aphididae). *Neotrop. Entomol.* 34:807–13.

Goussain, M. M., J. C. Moraes, J. G. Carvalho, N. L. Nogueira, and M. L. Rossi. 2002. Efeito da aplicação de silício em plantas de milho no desenvolvimento biológico da lagarta-do-cartucho *Spodoptera frugiperda* (J. E. Smith) (Lepidoptera: Noctuidae). *Neotrop. Entomol.* 31:305–10.

Greenplate, J. T., N. B. Duck, J. C. Pershing, and J. P. Purcell. 1995. Cholesterol oxidase: An oostatic and larvicidal agent active against the cotton boll weevil, *Anthonomus grandis. Entomol. Exp. Appl.* 74:253–8.

Grossi-de-Sá, M. F., and M. J. Chrispeels. 1997. Molecular cloning of bruchid (*Zabrotes subfasciatus*) α-amylase cDNA and interactions of the expressed enzyme with bean amylase inhibitors. *Insect Biochem. Mol. Biol.* 27:271–81.

Grossi-de-Sá, M. F., T. E. Mirkov, M. Ishimoto, G. Colucci, K. S. Bateman, and M. J. Chrispeels. 1997. Molecular characterisation of a bean α-amylase inhibitor that inhibits the α-amylase of the Mexican bean weevil *Zabrotes subfasciatus. Planta* 203:295–303.

Guzzo, E. C., O. M. B. Corrêa, J. D. Vendramim, A. F. Chiorato, S. A. M. Carbonell, and A. L. Lourenção. 2007. Searching for resistance sources against the Mexican bean weevil (*Zabrotes subfasciatus*) in common bean (*Phaseolus vulgaris*) genotypes. In *Proceedings of the 16th International Plant Protection Congress*, 720–1. Glasgow: British Crop Protection Council.

Guzzo, E. C., O. M. B. Corrêa, J. D. Vendramim, A. L. Lourenção, S. A. M. Carbonell, and A. F. Chiorato. 2006. Development of the Mexican bean weevil (Coleoptera: Bruchidae) on bean genotypes with and without arcelin over two generations. In *Proceedings of the 9th International Working Conference on Stored Product Protection*, ed. Lorini, I., B. Bacaltchuk, H. Beckel, D. Deckers, E. Sundfeld, J. P. dos Santos, J. D. Biagi, J. C. Celaro, L. R. D'A. Faroni, L. de O. F. Bortolini, M. R. Sartori, M. C. Elias, R. N. C. Guedes, R. G. da Fonseca, and V. M. Scusse, 914–9. Passo Fundo: Associação Brasileira de Pós-Colheita.

Hanisch, H. C. 1981. Populationsentwicklung von getreideblattlausen an weizenpflanzen nach verschieden hoher stickstoffdungung und vorbeugender applikation von kieselsaure zur wirtspflanze. *Mitt DGAAE* 3:308–11.

Hansawasdi, C., J. Kawabata, and T. Kasai. 2000. α-Amylase inhibitors from roselle (*Hibiscus sabdariffa* Linn.) tea. *Biosci. Biotechnol. Biochem.* 64:1041–3.

Hardie, J., R. Isaacs, J. A. Pickett, L. J. Wadhams, and C. M. Woodcock. 1994. Methyl salicylate and (–)-(1*R*,5*S*)-myrtenal are plant-derived repellents for black bean aphid, *Aphis fabae* Scop. (Homoptera: Aphididae). *Earth Environ. Sci.* 20:2847–55.

Harper, S. M., R. W. Crenshaw, M. A. Mullins, and L. S. Privalle. 1995. Lectin binding to insect brush border membranes. *J. Econ. Entomol.* 88:1197–202.

Hilder, V. A., and D. Boulter. 1999. Genetic engineering of crop plants for insect resistance—A critical review. *Crop Prot.* 18:177–91.

Hoffmann-Campo, C. B., J. A. Ramos Neto, M. C. N. Oliveira, and L. J. Oliveira. 2006. Detrimental effect of rutin on *Anticarsia gemmatalis*. *Pesq. Agropec. Bras.* 41:1453–9.

Hopkins, A. D. 1917. A discussion of C. G. Hewitt's paper on "Insect Behaviour." *J. Econ. Entomol.* 10:92–3.

Hunt, J. W., A. P. Dean, R. E. Webster, G. N. Johnson, and A. R. Ennos. 2008. A novel mechanism by which silica defends grasses against herbivory. *Ann. Bot.* 102:653–6.

Ishimoto, M., and K. Kitamura. 1989. Growth inhibitory effects of an α-amylase inhibitor from kidney bean, *Phaseolus vulgaris* (L.) on three species of bruchids (Coleoptera: Bruchidae). *Appl. Ent. Zool.* 24:281–6.

Ishimoto, M., and K. Kitamura. 1993. Specific inhibitory activity and inheritance of an α-amylase inhibitor in a wild common bean accession resistant to the Mexican bean weevil. *Jpn. J. Breed.* 43:69–73.

Ishimoto, M., T. Sato, M. J. Chrispeels, and K. Kitamura. 1996. Bruchid resistance of transgenic azuki bean expressing seed α-amylase inhibitor of common bean. *Entomol. Exp. Appl.* 79:309–15.

Isman, M. B. 1993. Growth inhibitory and antifeedant effects of azadirachtin on six noctuids of regional economic importance. *Pestic. Sci.* 38:57–63.

Isman, M. B. 2002. Insect antifeedants. *Pestic. Outlook* 13:152–7.

Isman, M. B. 2006. Botanical insecticides, deterrents, and repellents in modern agriculture and an increasingly regulated world. *Annu. Rev. Entomol.* 51:45–66.

Iulek, J., O. L. Franco, M. Silva, C. T. Slivinski, C. Bloch Jr., D. J. Rigden, and M. F. Grossi de Sa. 2000. Purification, biochemical characterization and partial primary structure of a new α-amylase inhibitor from *Secale cereale* (Rye). *Int. J. Biochem. Cell Physiol.* 32:1195–204.

Jāremo, J. 1999. Plant adaptations to herbivory: Mutualistic versus antagonistic coevolution. *Oikos* 84:313–20.

Jenks, M. A., S. D. Eigenbrode, and B. Lemieux. 2002. Cuticular waxes of Arabidopsis. *Arabidopsis Book* 1:1–22

Jermy, T. 1976. Insect–host–plant relationship—Coevolution or sequential evolution? *Symp. Biol. Hung.* 16:109–13.

Jongsma, M. A., P. L. Bakker, J. Peters, D. Bosch, and W. J. Stiekema. 1995. Adaptation of *Spodoptera exigua* larvae to plant proteinase inhibitors by induction of gut proteinase activity insensitive to inhibition. *Proc. Nat. Acad. Sci. U. S. A.* 92:8041–5.

Kadziola, A., M. Sogaard, B. Svensson, and R. Haser. 1998. Molecular structure of a barley α-amylase inhibitor complex: Implications for starch binding and catalysis. *J. Mol. Biol.* 278:205–17.

Karowe, D. N. 1989. Differential effect of tannic acid on two tree-feeding Lepidoptera: Implications for theories of plant antiherbivore chemistry. *Oecologia* 80:507–12.

Keeping, M. G., and J. H. Meyer. 2002. Calcium silicate enhances resistance of sugarcane to the African stalk borer *Eldana saccharina* Walker (Lepidoptera: Pyralidae). *Agr. Forest Entomol.* 4:265–74.

Keeping, M. G., and O. L. Kvedaras. 2008. Silicon as a plant defence against insect herbivory: Response to Massey, Ennos and Hartley. *J. Anim. Ecol.* 77:631–3.

Keeping, M. G., O.L. Kvedaras, and A. G. Bruton. 2009. Epidermal silicon in sugarcane: Cultivar differences and role in resistance to sugarcane borer *Eldana saccharina*. *Environ. Exp. Botany* 66:54–60.

Kennedy, G. G. 2003. Tomato, pests, parasitoids, and predators: Tritrophic interactions involving the genus *Lycopersicon*. *Annu. Rev. Entomol.* 48:51–72.

Kennedy, J. S. 1958. Physiological condition of the host plant and susceptibility to aphid attack. *Entomol. Exp. Appl.* 1:50–65.

Kim, H. S., and E. A. Heinrichs. 1982. Effects of silica level on whitebacked planthopper. *Int. Rice Res. Newsl.* 7:17.

Kim, S. G., K. W. Kim, E. W. Park, and D. Choi. 2002. Silicon-induced cell wall fortification of rice leaves: A possible cellular mechanism of enhanced host resistance to blast. *Phytopathology* 92:1095–103.

Kogan, M. 1986. Plant defense strategies and host-plant resistance. In *Ecological Theory and Integrated Pest Management Practice*, ed. M. Kogan, 83–134. New York: John Wiley & Sons.

Kogan, M., and E. F. Ortman. 1978. Antixenosis–a new term proposed to define Painter's "non-preference" modality of resistance. *Bull. Entomol. Soc. Am.* 24:175–6.

Korndorfer, A. P., R. Cherry, and R. Nagata. 2004. Effect of calcium silicate on feeding and development of tropical sod webworms (Lepidoptera: Pyralidae). *Fla. Entomol.* 87:393–5.

Koul, O. 2005. *Insect Antifeedants*. Boca Raton: CRC Press.

Kramer, K. J., S. Muthukrishnan, L. Johnson, and F. White. 1997. Chitinases for insect control. In *Advances in Insect Control: The Role of Transgenic Plants*, ed. N. Carozzi and M. Koziel, 185–93. London: Taylor & Francis.

Lajolo, F. M., and M. I. Genovese. 2002. Nutritional significance of lectins and enzyme inhibitors from legumes. *J. Agric. Food Chem.* 50:6592–8.

Lara, F. M. 1991. *Princípios de Resistência de Plantas a Insetos.* 2 ed. São Paulo: Ícone.

Larsson, S. 2002. Resistance in trees to insects—An overview of mechanisms and interactions. In *Mechanisms and Deployment of Resistance in Trees to Insects*, ed. M. R. Wagner, K. M. Clancy, F. Lieutier, and T. D. Paine, 1–29. New York: Kluwer Academic Publishers.

Lawrence, P. K., and K. R. Koundal. 2002. Plant protease inhibitors in control of phytophagous insects. *EJB Electron. J. Biotechnol.* 5:1–17.

Lewis, N. G., and E. Yamamoto. 1990. Lignin: Occurrence, biogenesis and biodegradation. *Annu. Rev. Plant Physiol. Plant Mol. Biol.* 41:455–96.

Lu, S., P. Deng, X. Liu, J. Luo, R. Han, X. Gu, S. Liangi, X. Wangi, F. Lii, V. Lozanov, A. Patthy, and S. Pongor. 1999. Solution structure of the major α-amylase inhibitor of the crop plant amaranth. *J. Biol. Chem.* 274:20473–8.

Maiti, I. B., N. Dey, S. Pattanaik, D. L. Dahlman, R. L. Rana, and B. A. Webb. 2003. Antibiosis-type insect resistance in transgenic plants expressing a teratocyte secretory protein (TSP14) gene from a hymenopteran endoparasite (*Microplitis croceipes*). *Plant Biotechnol. J.* 1:209–19.

Marsaro, Jr., A. L., S. M. N. Lazzari, and A. R. Pinto Jr. 2006. Inibidores de enzimas digestivas de insetos-praga. *Rev. Acad.* 4:57–61.

Marsaro, Jr., A. L., S. M. N. Lazzari, E. L. Z. Figueira, and E. Y. Hirooka. 2005. Inibidores de amilase em híbridos de milho como fator de resistência a *Sitophilus zeamais* (Coleoptera: Curculionidae). *Neotrop. Entomol.* 34:443–50.

Massey, F. P., A. R. Ennos, and S. E. Hartley. 2006. Silica in grasses as a defence against insect herbivores: Contrasting effects on folivores and a phloem feeder. *J. Anim. Ecol.* 75:595–603.

Massey, F. P., and S. E. Hartley. 2006. Experimental demonstration of the antiherbivore effects of silica in grasses: Impacts on foliage digestibility and vole growth rates. *Proc. R. Soc. Lond. B* 273:2299–304.

Mazzonetto, F. 2002. Efeito de genótipos de feijoeiro e de pós de origem vegetal sobre *Zabrotes subfasciatus* (Boh.) e *Acanthoscelides obtectus* (Say) (Col.: Bruchidae). PhD diss., Escola Superior de Agricultura "Luiz de Queiroz," Universidade de São Paulo, Brazil.

Mazzonetto, F., and J. D. Vendramim. 2002. Aspectos biológicos de *Zabrotes subfasciatus* (Boh.) (Coleoptera: Bruchidae) em genótipos de feijoeiro com e sem arcelina. *Neotrop. Entomol.* 31:435–9.

Mazzonetto, F., and J. D. Vendramim. 2003. Efeito de pós de origem vegetal sobre *Acanthoscelides obtectus* (Say) (Coleoptera: Bruchidae) em feijão armazenado. *Neotrop. Entomol.* 32:145–9.

Meisner, J., M. Wysoki, and L. Telzak. 1976. Gossypol as phagodeterrent for *Boarmia* (*Ascotis*) *selenaria* larvae. *J. Econ. Entomol.* 69:683–5.

Mello, M. O., and M. C. Silva-Filho. 2002. Plant-insect interactions: An evolutionary arms race between two distinct defense mechanisms. *Braz. J. Plant Physiol.* 14:71–81.

Metcalf, R. L., R. A. Metcalf, and A. M. Rhodes. 1980. Cucurbitacins as kairomones for diabroticite beetles. *Proc. Natl. Acad. Sci. U. S. A.* 77:3769–72.

Moraes, J. C., M. M. Goussain, M. A. B. Basagli, G. A. Carvalho, C. C. Ecole, and M. V. Sampaio. 2004. Silicon influence on the tritrophic interaction: Wheat plants, the greenbug *Schizaphis graminum* (Rondani) (Hemiptera: Aphididae), and its natural enemies, *Chrysoperla externa* (Hagen) (Neuroptera: Chrysopidae) and *Aphidius colemani* Viereck (Hymenoptera: Aphidiidae). *Neotrop. Entomol.* 33:619–24.

Moraes, J. C., M. M. Goussain, G. A. Carvalho, and R. R. Costa. 2005. Feeding non-preference of the corn leaf aphid *Rhopalosiphum maidis* (Fitch, 1856) (Hemiptera: Aphididae) to corn plants (*Zea mays* L.) treated with silicon. *Cienc. Agrotec.* 29:761–6.

Morton, R. L., H. E. Schroeder, K. S. Bateman, M. J. Chrispeels, E. Armstrong, and T. J. V. Higgins. 2000. Bean α-amylase inhibitor-1 in transgenic peas (*Pisum sativum*) provides complete protection from pea weevil (*Bruchus pisorum*) under field conditions. *Proc. Natl. Acad. Sci. U. S. A.* 97:3820–5.

Nalim, D. M. 1991. Biologia, nutrição quantitativa e controle de qualidade de populações de *Spodoptera frugiperda* (J. E. Smith, 1797) (Lepidoptera: Noctuidae) em duas dietas artificiais. MSc diss., Escola Superior de Agricultura "Luiz de Queiroz," Universidade de São Paulo, Brazil.

Nordlund, D. A., and W. J. Lewis. 1976. Terminology of chemical-releasing stimuli in intraspecific and interspecific interactions. *J. Chem. Ecol.* 2:211–20.

Novo, R. J., A. Viglianco, and M. Nasseta. 1997. Actividad repelente de diferentes extractos vegetales sobre *Tribolium castaneum* (Herbst). *Agriscientia* 14:31–6.

Oevering, P., A. J. Pitman, and K. K. Pandey. 2003. Wood digestion in *Pselactus spadix* Herbst-a weevil attacking marine timber structures. *Biofouling* 19:249–54.

Outchkourov, N. S., W. J. de Kogel, A. Schuurman-de Bruin, M. Abrahamson, and M. A. Jongsma. 2004. Specific cysteine protease inhibitors act as deterrents of western flower thrips, *Frankliniella occidentalis* (Pergande), in transgenic potato. *Plant Biotechnol. J.* 2:439–48.

Painter, R. H. 1951. *Insect Resistance in Crop Plants*. New York: McMillan.

Panda, N., and G. S. Khush. 1995. *Host Plant Resistance to Insects*. Guildford: Biddles.

Parra, J. R. P. 2001. *Técnicas de Criação de Insetos para Programas de Controle Biológico*. 6ª ed. Piracicaba: FEALQ.

Peumans, W. J., and E. J. M. Van Damme. 1995. Lectins as plant defense proteins. *Plant Physiol.* 109:347–52.

Pitman, A. J., E. B. G. Jones, M. A. Jones, and P. Oevering. 2003. An overview of the biology of the wharf borer beetle (*Nacerdes melanura* L., Oedemeridae) a pest of wood in marine structures. *Biofouling* 19:239–48.

Powell, K. S., A. M. R. Gatehouse, V. A. Hilder, and J. A. Gatehouse. 1995. Antifeedant effects of plant lectins and an enzyme on the adult stage of the rice brown planthopper, *Nilaparvata lugens. Entomol. Exp. Appl.* 75:51–9.

Powell, K. S., J. Spence, M. Bharathi, J. A. Gatehouse, and A. M. R. Gatehouse. 1998. Immunohistochemical and developmental studies to elucidate the mechanism of action of the snowdrop lectin on the rice brown planthopper, *Nilaparvata lugens* (Stal). *J. Insect Physiol.* 44:529–39.

Price, P. W. 1986. Ecological aspects of host plant resistance and biological control: Interactions among three trophic levels. In *Interactions of Plant Resistance and Parasitoids and Predators of Insects*, ed. D. J. Boethel and R. D. Eikenbary, 11–30. New York: John Wiley & Sons.

Price, P. W., C. E. Bouton, P. Gross, B. A. McPheron, J. N. Thompson, and A. E. Weis. 1980. Interactions among three trophic levels: Influence of plants on interactions between insect herbivores and natural enemies. *Annu. Rev. Ecol. Syst.* 1:41–65.

Procópio, S. O., J. D. Vendramim, J. I. Ribeiro Jr., and J. B. Santos. 2003. Bioatividade de diversos pós de origem vegetal em relação a *Sitophilus zeamais* Mots. (Coleoptera: Curculionidae). *Ciênc. Agrotec.* 27:1231–6.

Purcell, J. P. 1997. Cholesterol oxidase for the control of boll weevil. In *Advances in Insect Control: The Role of Transgenic Plants*, ed. N. Carozzi and M. Koziel, 95–108. London: Taylor & Francis.

Rahbé, Y., and G. Febvay. 1993. Protein toxicity to aphids-an in vitro test on *Acyrthosiphon pisum. Entomol. Exp. Appl.* 67:149–60.

Rahbé, Y., E. Ferrasson, H. Rabesona, and L. Quillien. 2003. Toxicity to the pea aphid *Acyrthosiphon pisum* of anti-chymotrypsin isoforms and fragments of Bowman-Birk protease inhibitors from pea seeds. *Insect Biochem. Mol. Biol.* 33:299–306.

Raffa, K. F. 1987. Effect of host plant on cannibalism rates by fall armyworm (Lepidoptera: Noctuidae) larvae. *Environ. Entomol.* 16:672–5.

Raikhel, N. V., H.-I. Lee, and W. F. Broekaert. 1993. Structure and function of chitin-binding proteins. *Annu. Rev. Plant Physiol. Plant Mol. Biol.* 44:591–615.

Rao, S. D. V. 1967. Hardness of sugarcane varieties in relation to shoot borer infestation. *Andhra Agric. J.* 14:99–105.

Rao, K. V., K. S. Rathore, T. K. Hodges, X. Fu, E. Stoger, D. Sudhakar, S. Williams, P. Christou, M. Bharathi, D. P. Bown, K. S. Powell, J. Spence, A. M. R. Gatehouse, and J. A. Gatehouse. 1998. Expression of snowdrop lectin (GNA) in transgenic rice plants confers resistance to rice brown planthopper. *Plant J.* 15:469–77.

Redmond, C. T., and D. A. Potter. 2007. Silicon fertilization does not enhance creeping bentgrass resistance to cutworms and white grubs. *Appl. Turfgrass Sci.* 6:1–7. http://www.plantmanagementnetwork.org/pub/ats/research/2006/grubs.

Reeck, G. R., K. J. Kramer, J. E. Baker, M. R. Kanost, J. A. Fabrick, and C. A. Behnke. 1997. Proteinase inhibitors and resistance of transgenic plants to insects. In *Advances in Insect Control: The Role of Transgenic Plants*, ed. N. Carozzi and M. Koziel, 157–83. London: Taylor & Francis.

Rosenthal, G. A., D. H. Janzen, and D. L. Dahlman. 1977. Degradation and detoxification of canavanine by a specialized seed predator. *Science* 196:658–60.

Rosenthal, G. A., D. L. Dahlman, and D. H. Janzen. 1976. A novel means for dealing with L-canavanine, a toxic metabolite. *Science* 192:256–8.

Ryan, C. A. 1990. Protease inhibitors in plants: Genes for improving defenses against insects and pathogens. *Annu. Rev. Phytopathol.* 28:425–49.

Salim, M., and R. C. Saxena. 1992. Iron, silica, and aluminum stresses and varietal resistance in rice: Effects on whitebacked planthopper. *Crop Sci.* 32:212–9.

Santos, J. P., J. Cruz, and R. A. Fontes. 1984. *Armazenamento e Controle de Pragas.* Documentos 1. Brasília: EMBRAPA/CNPMS.

Schoonhoven, L. M. 1982. Biological aspects of antifeedants. *Entomol. Exp. Appl.* 31:57–69.

Schroeder, H. E., S. Gollasch, A. Moore, L. M. Tabe, S. Craig, D. C. Hardie, M. J. Chrispeels, D. Spencer, and T. J. V. Higgins. 1995. Bean α-amylase inhibitor confers resistance to the pea weevil, *Bruchus pisorum*, in genetically engineered peas (*Pisum sativum* L.). *Plant Physiol.* 107:1233–9.

Scriber, J. M., and F. Slansky Jr. 1981. The nutritional ecology of immature insects. *Annu. Rev. Entomol.* 26:183–211.

Shade, R. E., H. E. Schroeder, J. J. Pueyo, L. M. Tabe, L. L. Murdock, T. J. V. Higgins, and M. J. Chrispeels. 1994. Transgenic pea seeds expressing the α-amylase inhibitor of the common bean are resistant to bruchid beetles. *Bio/Technol.* 12:793–6.

Shankaranarayana, K. H., K. S. Ayyar, and G. S. K. Rao. 1980. Insect growth inhibitor from the bark of *Santalum album. Phytochemistry* 19:1239–40.

Sharma, H. C., K. K. Sharma, N. Seetharama, and R. Ortiz. 2000. Prospects for using transgenic resistance to insects in crop improvement. *EJB Electron. J. Biotechnol.* 3:76–95.

Shukle, R. H., and L. L. Murdock. 1983. Lipoxygenase trypsin inhibitor and lectin from soybeans: Effects on larval growth of *Manduca sexta* (Lepidoptera: Sphingidae). *Environ. Entomol.* 12:787–91.

Silva, C. P., W. R. Terra, and R. M. Lima. 2001. Differences in midgut serine proteinases from larvae of the bruchid beetle *Callosobruchus maculatus* and *Zabrotes subfasciatus. Arch. Insect Biochem. Physiol.* 47:18–28.

Silva, C. P., W. R. Terra, J. Xavier-Filho, M. F. Grossi-de-Sá, A. R. Lopes, and E. G. Pontes. 1999. Digestion in larvae of *Callosobruchus maculatus* and *Zabrotes subfasciatus* (Coleoptera: Bruchidae) with emphasis on α-amylases and oligosaccharidases. *Insect Biochem. Mol. Biol.* 29:355–66.

Silva-Filho, M. C., and M. C. Falco. 2000. Interação planta-inseto. *Biotecnol. Ciênc. Desenv.* 12:38–42.

Slansky Jr., F., and J. M. Scriber. 1985. Food consumption and utilization. In *Comprehensive Insect Physiology, Biochemistry and Pharmacology*, ed. G. A. Kerkut and L. I. Gilbert, 87–164. Oxford: Pergamon Press.

Smith, C. M. 2005. *Plant Resistance to Arthropods: Molecular and Conventional Approaches.* Berlin: Springer.

Stamopoulos, D. C. 1988. Toxic effect of lignin extracted from the tegument of *Phaseolus vulgaris* seeds on the larvae of *Acanthoscelides obtectus* (Say) (Col., Bruchidae). *J. Appl. Entomol.* 105:317–20.

Stipanovic, R. D., J. Lopez, M. K. Dowd, L. S. Puckhaber, and S. E. Duke. 2006. Effect of racemic and (+) and (–)-gossypol on the survival and development of *Helicoverpa zea* larvae. *J. Chem. Ecol.* 32:959–68.

Su, H. C. F. 1991. Toxicity and repellency of *Chenopodium* oil to four species of stored product insects. *J. Entomol. Sci.* 26:178–82.

Suwonkerd, W., and K. Tantrarongroj. 1994. Efficacy of essential oil against mosquito biting. *Commun. Dis. J.* 20:4–11.

Svensson, B., K. Fukuda, P. K. Nielsen, and B. C. Bonsager. 2004. Proteinaceous α-amylase inhibitors. *BBA-Proteins Proteom.* 1696:145–56.

Tavares, M. A. G. C. 2006. Busca de compostos em *Chenopodium* spp. (Chenopodiaceae) com bioatividade em relação a pragas de grãos armazenados. PhD diss., Escola Superior de Agricultura "Luiz de Queiroz," Universidade de São Paulo, Brazil.

Tawatsin A, S. D. Wratten, R. R. Scott, U. Thavara, and Y. Techadamrongsin. 2001. Repellency of volatile oils from plants against three mosquito vectors. *J. Vector Ecol.* 26:76–82.

Thompson, J. N. 1988. Coevolution and alternative hypotheses on insect/plant interactions. *Ecology* 69: 893–5.

Ussuf, K. K., N. H. Laxmi, and R. Mitra. 2001. Proteinase inhibitors: Plant-derived genes of insecticidal protein for developing insect-resistant transgenic plants. *Curr. Sci.* 80:847–53.

Varanda, E. M., G. E. Zúñiga, A. Salatino, N. F. Roque, and L. J. Corcuera. 1992. Effect of ursolic acid from epicuticular waxes of *Jacaranda decurrens* on *Schizaphis graminum. J. Nat. Prod.* 55:800–3.

Vasconcelos, I. M., and J. T. A. Oliveira. 2004. Antinutritional properties of plant lectins. *Toxicon* 44:385–403.

Vendramim, J. D., and M. A. N. Nishikawa. 2001. Melhoramento para resistência a insetos. In *Recursos Genéticos e Melhoramento: Plantas*, ed. L. L. Nass, A. C. C. Valois, I. S. Melo, and M. C. Valadares-Inglis, 737–81. Rondonópolis: Fundação Mato Grosso.

Vila, L., J. Quilis, D. Meynard, J. C. Breitler, V. Marfà, I. Murillo, J. M. Vassal, J. Messeguer, E. Guiderdoni, and B. S. Segundo. 2005. Expression of the maize proteinase inhibitor (*mpi*) gene in rice plants enhances resistance against the striped stem borer (*Chilo suppressalis*): Effects on larval growth and insect gut proteinases. *Plant Biotechnol. J.* 3:187–202.

Visser, J. H. 1983. Differential sensory perceptions of plant compounds by insects. In *Plant Resistance to Insects*, ed. P. A. Hedin, 215–30. Washington, DC: American Chemical Society.

Wainhouse, D., D. J. Cross, and R. S. Howell. 1990. The role of lignin as a defence against the spruce bark beetle *Dendroctonus micans*: Effect on larvae and adults. *Oecologia* 85:257–65.

Waldbauer, G. P., and G. Fraenkel. 1961. Feeding on normally rejected plants by maxillectomized larvae of the tobacco hornworm, *Protoparce sexta* (Lepidoptera, Sphingidae). *Ann. Entomol. Soc. Am.* 54:477–85.

Wang, J., Z. Chen, J. Du, Y. Sun, and A. Liang. 2005. Novel insect resistance in *Brassica napus* developed by transformation of chitinase and scorpion toxin genes. *Plant Cell Rep.* 24:549–55.

Wasuwat, S., T. Sunthonthanasart, S. Jarikasem, N. Putsri, A. Phanrakwong, S. Janthorn, and I. Klongkarn-ngan. 1990. Mosquito repellent efficacy of citronella cream. *J. Sci. Tech.* 5:62–8.

Werner, R. A. 1995. Toxicity and repellency of 4-allylanisole and monoterpenes from white spruce and tama-rack to the spruce beetle and eastern larch beetle (Coleoptera: Scolytidae). *Environ. Entomol.* 24:372–9.

Whitman, D. W. 1988. Allelochemical interactions among plants, herbivores, and their predators. In *Novel Aspects of Insect–Plant Interactions*, ed. P. Barbosa and D. K. Letourneau, 11–64. New York: John Wiley & Sons.

Whittaker, R. H. 1970. The biochemical ecology of higher plants. In *Chemical Ecology*, ed. E. Sondheimer and J. B. Simeone, 43–70. New York: Academic Press.

Whittaker, R. H., and P. P. Feeny. 1971. Allelochemics: Chemical interactions between species. *Science* 171:757–70.

Yang, G., B. R. Wiseman, and K. E. Espelie. 1992. Cuticular lipids from silks of seven corn genotypes and their effect on development of corn earworm larvae [*Helicoverpa zea* (Boddie)]. *J. Agric. Food Chem.* 40:1058–61.

Yang, G., B. R. Wiseman, D. J. Isenhour, and K. E. Espelie. 1993. Chemical and ultrastructural analysis of corn cuticular lipids and their effect on feeding by fall armyworm larvae. *J. Chem. Ecol.* 19:2055–74.

Yang, G., D. J. Isenhour, and K. E. Espelie. 1991. Activity of maize leaf cuticular lipids in resistance to leaf-feeding by the fall armyworm. *Fla. Entomol.* 74:229–36.

Yoshihara, T., M. Sogawa, B. O. Pathak, B. O. Juliano, and S. Sakamura. 1979. Soluble silicic acid as a suck-ing inhibitory substance in rice against the rice brown planthopper (Delphacidae: Homoptera). *Entomol. Exp. Appl.* 26:314–22.

Zhu-Salzman, K., and R. A. Salzman. 2001. Functional mechanics of the plant defensive *Griffonia simplicifolia* lectin II: Resistance to proteolysis is independent of glycoconjugate binding in the insect gut. *J. Econ. Entomol.* 94:1280–4.

Zhu-Salzman, K., R. E. Shade, H. Koiwa, R. A. Salzman, M. Narasimhan, R. A. Bressan, P. M. Hasegawa, and L. L. Murdock. 1998. Carbohydrate binding and resistance to proteolysis control insecticidal activity of *Griffonia simplicifolia* lectin II. *Proc. Natl. Acad. Sci. U. S. A.* 95:15123–8.

Zuniga, G. E., and L. J. Corcuera. 1986. Effect of gramine in the resistance of barley seedlings to the aphid *Rhopalosiphum padi*. *Entomol. Exp. Appl.* 40:250–62.

Zuniga G. S., E. M. Varanda, and L. J. Corcuera. 1988. Effect of gramine on the feeding behavior of the aphids *Schizaphis graminum* and *Rhopalosiphum padi*. *Entomol. Exp. Appl.* 47:161–5.

27

Insect Bioecology and Nutrition for Integrated Pest Management (IPM)*

Antônio R. Panizzi, José R. P. Parra, and Flávia A. C. Silva

CONTENTS

27.1 Introduction 687
27.2 Bioecology and Nutrition of Phytophagous Arthropods and IPM 688
    27.2.1 Insect–Plant Interactions 688
    27.2.2 Plant Diversity and Stability 688
    27.2.3 IPM Tactics in Context of Bioecology and Insect Nutrition 689
        27.2.3.1 Host Plant Resistance 689
        27.2.3.2 Trap Crops 690
        27.2.3.3 Cultivation of Mixed Crops 691
        27.2.3.4 Functional Allelochemicals 691
27.3 Management of Pests within the Context of Insect Bioecology and Nutrition 692
    27.3.1 Managing Heteropterans on Soybean 692
        27.3.1.1 Host Plant Resistance 692
        27.3.1.2 Use of Trap Crops 693
        27.3.1.3 Managing Mixed Crops to Mitigate Heteropterans' Impact on Soybean 693
        27.3.1.4 Use of Substances That Interfere with Feeding Process to Reduce
                  Heteropterans' Impact on Soybean 694
    27.3.2 Managing Heteropterans on Host Plants 695
        27.3.2.1 Host Plant Sequences 695
        27.3.2.2 Local Populations with Specific Feeding Habits 696
        27.3.2.3 Manipulation of Preferred Host Plants as Traps 696
        27.3.2.4 Role of Less Preferred Plant Food Sources 697
    27.3.3 Managing Heteropterans in Overwintering Sites and Host Plants 698
        27.3.3.1 Managing Crop Residues 698
        27.3.3.2 Monitoring Bugs in Overwintering Niches to Determine Crop Colonization 699
27.4 Conclusions 699
References 700

27.1 Introduction

Insect bioecology and nutrition within the context of evolution (nutritional ecology) has been defined as an area of entomology that involves the integration of biochemical, physiological, and behavioral information (Slansky and Rodriguez 1987a). Such a broad view suggests the need for basic studies essential to understand the different life styles of insects.

---

* This chapter was modified and updated from Nutritional ecology of plant feeding arthropods and IPM by A. R. Panizzi, in *Perspectives in Ecological Theory and Integrated Pest Management*, edited by Marcos Kogan and Paul Jepson. Copyright 2007 Cambridge University Press. Reprinted with permission.

Considering the damage inflicted to plant structures by feeding arthropods, it is possible to identify several feeding guilds, from the more conspicuous foliage and fruit chewers to the less noticeable seed suckers, fruit borers, and root feeders. All of these, and many others, have been studied and reviewed under the paradigm of insect nutritional ecology (see chapters in Slansky and Rodriguez 1987b). In general, these reviews using the insect nutritional ecology model have focused primarily on basic aspects of the different insects (feeding guild biology), and have not dealt with applied aspects, despite the enormous importance of insects in these guilds as pests of major crops worldwide.

Within the context of integrated pest management (IPM) systems, several tactics taking into account insect bioecology and nutrition can be considered. They include host plant resistance, trap crops, asynchrony of foods and pests phenology, crops consortiums, and functional allelochemicals. These tactics, although considered in several IPM textbooks (e.g., Pimentel 1981; Kogan 1986a; Rechcigl and Rechcigl 2000; Flint and Gouveia 2001; Pedigo 2002; Pimentel 2002; Norris et al. 2003), still must be further explored within the context of the insect bioecology and nutrition (nutritional ecology) paradigm, considering each of the major feeding guilds associated with plants.

Attempts to stress IPM under the scope of insect bioecology and nutrition are rare (see chapters in Panizzi and Parra 1991a, 2009); in these chapters, greater attention was given to management tactics considering insect feeding activity, host plant preferences, host plant impact on pest populations, and feeding behavior.

In this chapter, we will touch on basic information for holistic IPM programs, including insect–plant interactions, plant diversity and stability, and IPM tactics in the context of insect bioecology and nutrition ecology. As a case study, we will present in greater detail a system with soybean as the major commodity and the guild of seed-sucking insects associated with it. This guild includes many severe pests of several crops worldwide (Schaefer and Panizzi 2000), and it is the most important pest complex of soybean in the Neotropics, a region that hosts the largest soybean production area in the world. Using this system, it will be shown how basic information on interactions of these pests with their entire host plant range may be used to mitigate impact on the main crop plant.

27.2 Bioecology and Nutrition of Phytophagous Arthropods and IPM

27.2.1 Insect–Plant Interactions

Insect–plant interactions have been explored in many ways, and the literature covering this subject has exploded in the last 20+ years (Ahmad 1983; Crawley 1983; Bernays 1989–1994; Bernays and Chapman 1994; Brackenbury 1995; Jolivet 1998; Finch and Collier 2000). Phytophagous insects plus the plants they feed on make up about 50% of all living species; members of Lepidoptera, Hemiptera, and Orthoptera are mostly phytophages (Strong et al. 1984).

Despite the gigantic biomass formed by plants, only nine orders of insects utilize plants as their main food, which suggests that plants may not be an ideal food. Owing to many physical (e.g., pilosity, toughness of tissues, and thorns) and chemical (e.g., non-nutritional compounds, imbalance of nutrients, and lack of water) attributes, insects cannot or do not explore plants fully as food sources (Edwards and Wratten 1980).

Because of the diversity of plant defenses, and insect adaptations to feed on the defended plants, studies on coevolution have proliferated during the past 30–40 years, since the publication of the paper by Erlich and Raven (1964) on the coevolution of butterflies and plants. Despite this and many other studies that followed, several authors do not consider coevolution to be the general mechanism driving insect–plant interactions or as the mechanism responsible for the structure of phytophagous insect communities (e.g., Janzen 1980; Fox 1981; Futuyma 1983; Jermy 1984; Strong et al. 1984). Insects and plants coexist, and considering the integrated management of pests on crops, the theoretical bases of insect–plant interactions provide subsidies for research on host plant resistance and the practice of IPM (Kogan 1986a).

27.2.2 Plant Diversity and Stability

The issue of species diversity and stability of biotic communities has been the object of considerable interest and debate. Although this seems to be the case for natural ecosystems, when considering agroecosystems, ecologists and pest management experts and practitioners still argue whether the "diversity–stability hypothesis" holds true. In general, studies suggest that with the increase of biodiversity, that is, all species of plants, animals, and microorganisms interacting in an ecosystem, it is possible to stabilize the community of insects and to enhance the management of pests (Altieri and Letourneau 1984; Andow 1991; Wratten et al. 2007). With expansion of monocultures plant biodiversity is reduced, with consequent habitat destruction, decrease in resource availability, and reduction in numbers of arthropod species; this leads to changes in the functioning of the ecosystem, affecting its productivity and sustainability (Altieri and Nicholls 1999, 2004; Nicholls and Altieri 2007).

The importance of habitat heterogeneity compared to pure habitat area for biodiversity should be strong when the focal insect groups show (a) high degrees of habitat specialization and (b) high densities, thereby requiring only a small area for persistence (Ricklefs and Lovette 1999).

Southwood and Way (1970) considered factors influencing the degree of biodiversity in agroecosystems: (a) the diversity of vegetation within and around the agroecosystem; (b) the temporal and spatial permanence of the various crops within the agroecosystem; (c) the intensity of management practices, such as tillage and pesticide applications; and (d) the degree of isolation of the agroecosystem from natural vegetation. The role of uncultivated land in the biology of crop pests and their natural enemies has been recognized (van Emden 1965).

In agroecosystems, biodiversity can be planned or associated, as suggested by Vandermeer (1995). In the first case, biodiversity consists of cultivated crops, livestock, and associated organisms, which are introduced into the system on purpose, for economic or aesthetic reasons, and are managed intensively. In the second case, biodiversity includes all organisms, from plants and higher animals to microbes, which naturally existed or moved into the system from surrounding areas. This associated biodiversity is important to maintain or mitigate the unbalance that usually is associated with the planned biodiversity.

It may be stated that the stability of ecosystems, in general, is a result of the addition of all interactions among the living organisms. Therefore, the more structured the agroecosystem, the greater the stability. Altieri (1994) reported that cropping systems with taller plants (e.g., corn) mixed with shorter plants (squash or beans) provide more niches, enhancing species biodiversity. In southern Brazil, small growers cultivate beans, cassava, and small grains, in areas surrounded by taller plants such as corn or pigeon pea. These latter plants not only provide increased species diversity but also function as barriers to insects' dispersion, preventing pest outbreaks. Producers of organic soybean plant the beans in relatively small areas surrounded by natural vegetation or corn to reduce the attack of pests.

27.2.3 IPM Tactics in Context of Bioecology and Insect Nutrition

The IPM tactics of host plant resistance, trap cropping, mixed cropping, and the allelochemicals associated with those systems can be profitably analyzed under the context of bioecology and insect nutrition.

27.2.3.1 Host Plant Resistance

The use of cultivars resistant to pests is one of the most effective, economical, and environmentally safe management tactics (Pedigo 2002) and should be a key component of any IPM system.

The development of host plant resistance within the context of bioecology and insect nutrition includes the interrelationships of food attributes, with the insect consumption and utilization of the food, and its consequences to the insect performance and fitness. These interrelationships between insect bioecology and nutrition (nutritional ecology) and host plant resistance were illustrated by Slansky (1990). In this diagram, studies on insect nutritional ecology focus on the understanding of the effect of food on the insect biology, while host plant resistance attempts to manipulate food attributes to manage insect pests. Therefore, the basic insect nutritional ecology approach supports the applied approach of host plant resistance, and the convergence of the two disciplines results in a better understanding of the whole process.

Of the three fundamental modalities of host plant resistance, that is, antibiosis, antixenosis, and plant tolerance, stated over 50 years ago by Painter (1951), the first component—antibiosis—greatly relates to insect nutrition. Plant attributes comprising nutrients, non-nutrients, and morphological features, will dictate the extent of the food's impact on the insect's biology. This impact may result in death of immatures, reduced growth rates, increased mortality of pupae, small adults with reduced fecundity, shortened adult life span, morphological malformations, restlessness, and other abnormal behaviors (Pedigo 2002).

With the introduction of genetically modified (GM) crops carrying toxins, host plant resistance is gaining a new momentum. This approach using modern biotechnology is being considered a new technological breakthrough in agriculture, comparable to the green revolution of the early 1970s. For example, transgenic plants expressing the bacterium *Bacillus thuringiensis* (*Bt*) Berliner, which produces toxins that confer pest resistance to plants, has been introduced in at least 18 crops; corn, cotton, and potato GM cultivars are already commercially available (Shelton et al. 2002). In 2001, about 13 million hectares were cultivated with *Bt* corn and *Bt* cotton, mainly in the United States and Canada (James 2001); other countries where these and other *Bt* crops are cultivated include China, India, South Africa, and Argentina (Carpenter et al. 2002), and nowadays the area covered with transgenic plants has increased substantially in many countries worldwide. Other toxins, such as inhibitors of digestive enzymes—proteinases, amylases—and lectins, are also being introduced to plants to give them protective effects (Gatehouse and Gatehouse 2000). These and other toxins, being introduced in cultivars of many crops, will certainly make the host plant resistance strategy a major component of IPM programs worldwide. However, concerns about the possible environmental effects of GM crops resistant to insects have risen, and this issue has been extensively debated (see reviews by Fontes et al. 2002; O'Callaghan et al. 2004; Kenkel 2007; Kennedy 2008). There are several advantages to the use of host plant resistance in IPM, as well as a number of disadvantages. Among the former, host plant resistance is generally compatible with other IPM tactics, is often easy and inexpensive for producer to implement, and is cumulative in its impact on herbivore populations. The disadvantages include the sometimes long, difficult, and expensive process of developing (breeding) resistant varieties; the instability inherent in some types of plant resistance; and, occasionally, incompatibility of plant resistance with other IPM tactics (Stout and Davis 2009).

27.2.3.2 Trap Crops

Trap crops are plants, usually preferred hosts, planted to attract insects and, in consequence, to divert their attack from the crops. This can be accomplished by deviating the pests from attacking the target crop, and concentrating them in great numbers in restricted areas where control measures can be taken; this method is usually much more economical than conventional control methods such as the use of pesticides (Hokkanen 1991, Shelton and Badenes-Perez 2006).

This tactic (trap crop) has strong components in the context of the nutritional ecology model. These components include, first, the feeding preference. Although most insects are polyphagous or oligophagous, they tend to show preferences for certain plant taxa, and one can use this preference to attract the insects. This preference will be dictated, at least in part, by the nutritional value of the plants. Apparently, insects can predict or evaluate the nutritional value of plants, and "choose" them for oviposition. Although less preferred host plants also have an important role in the insect's biology, the preferred hosts usually contribute more to the insect's fitness.

A second component of the trap crop tactic, considering the nutritional ecology paradigm, has to do with the impact of the trap crop on the performance of larvae/nymphs and adults. Usually, on these preferred hosts, the maximum potential contribution to the next generation is expected to be achieved, with production of the fittest individual. Survivorship of immatures and reproduction of adults are the greatest. Therefore, populations will tend to explode on these preferred hosts and exhaust them; an accurate estimate of the holding capacity of the trap crop should be determined so pest insects do not leave the trap crop because of interspecific competition and lack of food. Thus it is important to determine when to interfere with control measures to avoid dispersion of populations to the target crop.

A third component of trap crops, considering the nutritional ecology model, is that preferred, and therefore highly nutritional host plants, may allow pest species to store energy in their bodies to overcome unfavorable periods of food scarcity. Although not considered widespread, this is a very important event in the biology of those insects that accumulate energy. Feeding on a rich nutritional food source, such as trap crops, in particular at times that precede the winter, might be crucial for these insects' survivorship.

Shelton and Badenes-Perez (2005) consider that the potential success of a trap cropping system depends on the interaction of the characteristics of the trap crop and its deployment with the ecology and behavior of the targeted insect pest. However, the characteristics of the trap crop and insect alone are not sufficient to predict whether a trap crop will be successful. Ultimately, the combination of insect and trap crop characteristics and practical considerations determines the success of a trap cropping system.

27.2.3.3 Cultivation of Mixed Crops

The cultivation of mixed crops is another IPM tactic that fits in the context of insect bioecology and nutrition. In general, we may say that as the diversification of cropping systems is increased, or, as the number of cultivated plant species is increased in a particular system (polycultures), outbreaks of herbivores populations are decreased (Andow 1991; Altieri 1994; Altieri and Nicholls 2004).

There are several reasons why polycultures are less susceptible to pest attack. First, different species of plants that are intercropped may provide mutual protection by acting as a physical barrier, each for the other; second, one species of plant may camouflage another species of plant, forming a mosaic that will confound the behavior of pests; and third, the odor produced by a particular plant may repel or disrupt the searching ability of pest species (Altieri 1994; Altieri and Nicholls 2004). The various interactions between plants in intercrops have strong impact in many parameters in host plant quality, for example, nutrient status, plant morphology, and secondary compound content. The nature of these interplant interactions depends on the characteristics of components crop plants (Langer et al. 2007).

Another major point that makes polycultures less susceptible to pest outbreaks is the greater occurrence of natural enemies (predators and parasitoids) in such a system than in monocultures. An extensive body of literature demonstrates this fact (references in Altieri and Nicholls 1999, 2004; Horn 2002; Norris et al. 2003). Also, the dispersal of insects in response to vegetation diversity is greatly affected. These authors stated that the establishment of a system of corridors of natural vegetation linking crop fields may serve multiple purposes in implementing IPM at the landscape level. For example, it may block dispersion of plant inoculums; it may block pest movement; and it may produce biomass for soil fertility, among other effects. The fact is that by making cropping systems more diverse, we make them more sustainable with greater conservation of resources (Vandermeer 1995).

To function properly as a management tactic, the cultivation of mixed crop demands a very accurate study of the local conditions. In general, there is a need to get information on the population trends of the different pests *locally* before deciding on any type of polycultivation. Once the decision of establishing a system using several crops is taken, one should decide which crops and what percentage of the total area should be dedicated to each of them. As mentioned before, a certain amount of the area should be allocated to host the natural vegetation, to provide refuges and corridors linking the system to allow the balance of pests with their natural enemies. A strong program of monitoring these insects, that is, pests and their natural enemies, during the cropping season and after harvest is crucial to understand the flows of insects from one crop to the other and to the natural vegetation. There is a need to assess each agricultural system separately, to understand the many interactions of pests and natural enemies, which will depend on the size of the field, location, plant composition, surrounding vegetation, and cultural management (Altieri and Nicholls 1999, 2004; Nicholls and Altieri 2007).

27.2.3.4 Functional Allelochemicals

Allelochemicals are compounds that mediate behavioral or physiological interactions among organisms of different species. There are thousands of compounds mediating a myriad of interactions, within the three classical categories of allelochemicals: kairomones (i.e., allelochemicals that provide an adaptive advantage to the perceiver), allomones (i.e., allelochemicals that provide an adaptive advantage to the emitter), and synomones (i.e., allelochemicals that provide an adaptive advantage to both, the perceiver and the emitter). For the purpose of IPM, the classification proposed by Kogan (1986b) is a good example of how these compounds function: as kairomones they may function as attractants, driving the insects toward the host plant; as arrestants, slowing or stopping movement; and as feeding or oviposition excitants, provoking biting/piercing or oviposition. In the second case, as allomones, they may function as antixenotics, by orienting insects away from the plant (repellents), speeding up movement (locomotory excitants), inhibiting biting/piercing (suppressants), and preventing maintenance of feeding/oviposition (deterrents); or as antibiotic, causing intoxication (toxins) or reducing food utilization processes (digestibility reducing factors).

Most plants synthesize toxins that affect herbivores. Those toxins that increase the fitness of the plants have a metabolic cost. Studies indicate that there is a balance between the cost and the various ecological effects (Karban and Baldwin 1997), although it is often difficult to measure either the costs or the benefits associated with defensive substances. In addition, plants have also developed mechanisms to produce certain defense plant secondary metabolites only upon herbivore attacks or after receiving warnings from neighboring plants (Baldwin et al. 2006).

Plant toxins have played an important role in agricultural plants, and most crops contain one or more types of toxins (see Seigler 2002 for important groups of toxins in major crops). Some plants produce toxins in their roots with toxic and/or repellent effects to root feeders, such as nematodes. These plants are called antagonistic plants (see review by Owino 2002).

Resistance has been managed in crops either by using traditional plant breeding or new molecular techniques (Karban and Baldwin 1997). Despite the many successful examples of breeding for changes in secondary metabolite chemistry to enhance resistance in crop plants, undesirable side effects have been observed. For example, cotton lines with high contents of gossypol, a sesquiterpene toxin, show resistance to bollworm larvae and other herbivores, but also show detrimental effects to humans and livestock that use cotton products (Gershenzon and Croteau 1991). On the other hand, the elimination of cyanogenic glycosides from the tuber roots of cassava, mitigating the poisoning effects to humans, is highly desirable, but this might increase herbivory and fungal attacks on plants free of these compounds (Moeller and Siegler 1999). Therefore, there is a need to balance the cost/benefits of manipulating plant toxins. Many studies report a wide range of interactions of allelochemicals. Borden (2002) exemplifies these interactions among terrestrial plants, arthropods, and vertebrates. Despite these many studies and examples in the literature, the adoption of allelochemicals as pest management tools has been limited, for several reasons, some of them discussed above. Clearly, much remains to be done and there is no doubt that the management of insect pests through the many possible uses of allelochemicals will play a major role in IPM programs in the future.

27.3 Management of Pests within the Context of Insect Bioecology and Nutrition

To discuss the management of insect pests within the context of insect bioecology and nutrition, we will take as an example the feeding guild of seed suckers (heteropterans or true bugs) that feed on soybean. The basic studies on their biology and ecology were discussed previously in this book (see Chapter 13).

27.3.1 Managing Heteropterans on Soybean

27.3.1.1 Host Plant Resistance

Host plant resistance is an important IPM tactic in the context of bioecology and insect nutrition (see Chapter 26). In the case of heteropterans, many studies have been conducted over the years, including evaluation of commercial cultivars, evaluation of genotypes from germplasm banks, and development of new cultivars.

Early studies by McPherson et al. (1979) with soybean suggested the commercial cultivar "Lee 68" might possess some mechanism of tolerance to stink bug feeding. Similarly, Link et al. (1971, 1973) found a lower percentage of damaged seeds for the cultivar "Bienville," compared with cv. "Santa Rosa" and "Industrial," and that cv. "Serrana" were less affected by stink bugs than "Bienville." Jones and Sullivan (1978) found cv. "Essex," which matured earlier than other cultivars, to escape severe damage by stink bugs. This observation of early-maturity cultivars avoiding stink bug damage, which was also observed by other researchers elsewhere, was the basis for the development of massive breeding programs that led to the development of early-maturity cultivars that escape the damage by heteropterans, particularly, in Brazil.

The evaluation of germplasm led to the discovery of several plant introductions (PIs) with variable degrees of resistance to several insect pests, including stink bugs. For instance, Turnipseed and Sullivan (1975) reported adverse effects of PIs 229358, 227687, and 171451, and of the line ED 73–371 to *Nezara viridula* nymphs. Jones and Sullivan (1979) showed that PI 229358 was the most consistently resistant PI to *N. viridula* nymphs. Another germplasm, PI 17144, was also shown to be resistant to *N. viridula* due to antibiosis and antixenosis (Gilman et al. 1982; Kester et al. 1984). The lines IAC 74-2832 and Chi-Kei No. 1B showed less damage by stink bugs than did many cultivars and lines evaluated in the field (Panizzi et al. 1981).

Despite these many years of research on host (soybean) plant resistance to stink bugs, it was only in 1989 that the first variety was released, which presented antibiosis and tolerance types of resistance (Rossetto 1989). This variety was named IAC-100, which stands for Instituto Agronômico de Campinas (IAC) in São Paulo, Brazil, at its 100-year anniversary in 1989. It was cultivated by certain growers after its release, but was soon replaced by other cultivars with higher seed yield, despite their susceptibility to heteropterans.

As pointed out by Boethel (1999), soybean breeders and entomologists have discovered many obstacles in their attempts to develop soybean insect-resistant cultivars. The incorporation of the *Bt* gene into soybean against chewing insects raised hopes for the revitalization of host plant resistance. However, thus far, the discovery of an effective toxin against heteropterans that might be incorporated into soybean by those working on traditional breeding and in biotechnology remains a challenge.

27.3.1.2 Use of Trap Crops

The use of trap crops, in the context of using the same species of host plant in different phenological stage of development, to attract pest species that prefer to feed on plants during a certain time of plant development, has been used with some success on soybean to manage heteropterans in different parts of the world.

Apparently, the first study carried out on soybean was by Newsom and Herzog (1977), who reported on the attractiveness of soybean planted early to stink bugs in Louisiana, USA. The bugs concentrated in small areas of early planted soybeans, which, because planted earlier, were already in the reproductive stage, with pods filled with seeds that attracted the bugs. In the remaining area, which was planted later, the plants were still in the vegetative stage, and less attractive. Chemical control was applied to the plants in reproduction, avoiding the dispersal of the bugs to the surrounding soybean crop. Similar results were reported later on soybean in the United States by Ragsdale et al. (1981) and McPherson and Newsom (1984).

In Brazil, early-maturity and early-planted soybean, occupying about 10% of soybean fields, were reported to attract several species of stink bugs, reducing the degree of colonization of the main area (Panizzi 1980). This tactic to control stink bugs on soybean fields was used together with the release of egg parasitoids, such as *Trissolcus basalis* (Wollaston), in the trap area early in the season; this was effective in managing pest populations of stink bugs (Corrêa-Ferreira 1987). Additional studies on the use of the trap crop technique for stink bugs were also conducted in the expanding area of soybean cultivation in central Brazil (Kobayashi and Cosenza 1987).

Despite these many studies and favorable results, the trap crop technique for management of stink bugs on soybean and other crops has been limited to special situations, such as small isolated fields or organic fields where the use of pesticides is prohibited. There are several reasons why this technique is not widely adopted by growers: the polyphagous feeding habits of heteropterans that increase the difficulty in attracting the bugs to a specific trap crop more effectively; the limited knowledge on the host plant–bug interactions; and the lack of interest by growers that, in general, prefer more conventional methods of pest control (e.g., chemical control), which are considered dependable and easier to use.

27.3.1.3 Managing Mixed Crops to Mitigate Heteropterans' Impact on Soybean

Soybean is usually cultivated in large areas worldwide. However, in some regions of the world where the crop is expanding, such as in the tropics, a growing percentage of the total acreage is in small fields.

These fields are usually exploited by small farmers with specific purposes, such as the production of organic soybean or vegetable type soybean to be used for human consumption.

In these small fields, mixing crops in the same area or cultivation of several crops in adjacent areas are common events. For example, in some areas of southern Brazil, the landscape with relatively small soybean fields is surrounded by natural vegetation and other crops. In this scenario, soybean usually escapes the damage caused by stink bugs. It is known that in the tropics, soybean cultivated near other legume fields is much less damaged by heteropterans than when cultivated alone (Jackai 1984; Naito 1996).

For more than 15 years we have cultivated soybean for research in small fields surrounded by several other crops. We observed a much slower rate of plant colonization by heteropterans than in soybean fields in the so-called open areas. Even in these latter areas, which are usually more flat and large than the small fields surrounded by other crops, additions of different crops tend to mitigate the impact of heteropterans.

27.3.1.4 Use of Substances That Interfere with Feeding Process to Reduce Heteropterans' Impact on Soybean

The use of secondary compounds or allelochemicals that interfere with the feeding process of insects on plants was briefly discussed. With regard to heteropterans that feed on soybean, an example of a substance that interferes with their feeding behavior and is being used to manage these pests will be presented in detail.

Field observations in soybean fields in southern Brazil, of an apparent attraction of stink bugs to clothes or handles of tools, caused growers and extension entomologists to speculate that human sweat was attracting the bugs. Field trials were set using sodium chloride (NaCl) mixed with water and sprayed over soybean plants. Initial studies, in the greenhouse, with potted soybean plants in cages, indicated that the southern green stink bug, *Nezara viridula* (L.), preferred plants sprayed with NaCl over plants sprayed with water only (Corso 1989). Results of additional tests, using a mixture of NaCl (0.5%) with half the recommended dosage of conventional pesticides to control stink bugs, indicated a similar efficacy in control. The reduced dosage was promptly adopted by growers, for economic reasons (Corso 1990).

Additional field studies were conducted at the research station of the National Soybean Research Center of Embrapa, in Londrina, Paraná state, by Panizzi and Oliveira (1993), to test the "attraction" of stink bugs to NaCl. They selected field plots (32 m × 7 m) in nine different locations of soybean fields, and sprayed half the plots with NaCl (0.5%). The other half was sprayed with water only. The bugs were sampled about twice per week for 11 weeks (15 sampling dates), using the beat cloth method, and the number of nymphs and adults of the three major species of pentatomids (i.e., *N. viridula*, *Piezodorus guildinii*, and *Euschistus heros*) were recorded. The results indicated that nymphs and adults of all three species were consistently more abundant in areas treated with salt than in the untreated areas.

Because laboratory bioassays indicated that NaCl did not have a synergistic effect when mixed with insecticides (Sosa-Gómez et al. 1993), additional investigations were conducted by Niva and Panizzi (1996) to test the hypothesis that common salt was interfering in the stink bug feeding behavior. They compared the feeding behavior of adults of the southern green stink bug, *N. viridula* (L.), on soybean pods treated with NaCl (0.5%) and soybean pods treated with water only (control). They offered soybean pods with and without salt to bugs confined in arenas (Petri dish 14 cm × 2 cm), and recorded the time spent on food touching with mouthparts and feeding (i.e., insertion of the stylets into the soybean pods). The bugs spent considerably more time touching the pods treated with salt than the untreated pods, and the feeding time was similar in both treatments. Food touching behavior was greatly increased on soybean pods treated with salt, causing an arrestant behavior, which means the bugs stayed longer on treated soybean pods. These results might explain both the greater number of bugs found on treated soybean plants than on control plants referred by Corso (1989) and Panizzi and Oliveira (1993), and the greater insecticide efficacy at reduced dosages when mixed with pesticides reported by Corso (1990). This example illustrates the potential of using a substance that interferes with the feeding to manage a pest by taking advantage of basic studies on the feeding behavior driven by gustatory sensilla, which are present on the labial tips of heteropterans.

27.3.2 Managing Heteropterans on Host Plants

Most species of heteropterans spend only a third of their lifetimes feeding and breeding on cultivated spring/summer crops. The rest of the time these bugs are found feeding and reproduce on alternate (wild) host plants, or occupy overwintering sites provided by these hosts, such as under the trees' bark or underneath fallen dead leaves. Therefore, it is important to monitor the bugs' population while living on these wild plants or underneath debris, and to devise tactics to manipulate these pests before they colonize cultivated plants. This is, perhaps, one of the greatest challenge faced by entomologists because much knowledge is needed regarding the biology, ecology, behavior, and physiology of bugs' life, and because little has been investigated compared with what is known on these bugs while damaging economic cultivated plants.

To design strategies to manage pest species, one must know which wild host plants are used by heteropterans, how suitable they are for nymphal development and adult reproduction, what sequence of plants are used by sequential generations, and when dispersal occurs from crop plants to wild plants and vice versa (Panizzi 1997).

27.3.2.1 Host Plant Sequences

In general, heteropterans explore a variety of host plants within and between generations. Nymphs and adults move among the same or different plant species, which may be colonized in sequence. There are several examples of sequences of plants used by different species of heteropterans (references in Panizzi 1997).

In Paraná state (Brazil), the highly polyphagous southern green stink bug, *N. viridula*, colonizes soybean during late spring and summer, completing three generations on this crop, before it moves to alternate hosts such as *Crotalaria lanceolata*, where a fourth generation is completed. During this time it may feed on the weed star bristle, *Acanthospermum hispidum*, but no reproduction occurs on this plant. A fifth generation is completed during the mild winter of northern Paraná, on host plants such as wild radish, *Raphanus raphanistrum*; mustard, *Brassica campestris*; and pigeon pea, *Cajanus cajan*. During winter, *N. viridula* may feed on wheat, *Triticum aestivum*, but does not reproduce on this plant. A sixth generation is completed, during spring, on siberian motherworth, *Leonurus sibiricus*. During the entire year, the southern green stink bug is observed on castor bean, *Ricinus communis*, on which plant it may feed but not reproduce. The less polyphagous red-banded stink bug, *P. guildinii* (Westwood), also completes three generations on soybean. A fourth generation is completed on legumes such as crotalaria, pigeon pea, and several indigo species (*Indigofera hirsuta, I. truxillensis*, and *I. suffruticosa*). During the winter it feeds on indigo legumes, but, in contrast to *N. viridula*, it does not reproduce at this time. A neotropical species, *P. guildinii*, is less adapted to the somewhat cooler temperatures of the "winter." A fifth generation is completed on indigo legumes, before the bug starts colonizing soybean again during late spring. The neotropical brown stink bug *E. heros* (F.), like the previous two species, completes three generations on soybean. During late summer–early fall, a fourth generation is completed on pigeon pea, *C. cajan*. During the summer it may be found feeding on the euphorb *Euphorbia heterophylla*, but reproduction on this plant was observed to occur only under laboratory conditions (Pinto and Panizzi 1994). *E. heros* may feed, but will not reproduce, on the weed plant star bristle, *A. hispidum*. It is interesting to note that on this plant, this typical seed sucker feeds on the stems of the plant. At this time *E. heros* starts moving to shelters underneath leaf litter, where it remains until the next summer. During this time this bug accumulates lipids and does not feed, remaining in a state of partial hibernation (Panizzi and Niva 1994; Panizzi and Vivan 1997). Despite completing fewer generations than the former two species, *E. heros* is the most abundant species, particularly in the warmer regions. Its long time in shelters help avoid the attack of natural enemies, increasing its survivorship (Panizzi and Oliveira 1999).

In the study of host plant sequences used by heteropterans, it is important to determine which plants are used in sequence and how suitable they are to nymphal development and adult reproduction. By doing so, one will know which host plants are the most important to the bugs' biology, and on which plants studies should be concentrated in order to devise management tactics to mitigate their impact on crops.

27.3.2.2 Local Populations with Specific Feeding Habits

Heteropterans are, in general, polyphagous, feeding on an array of plant of different species belonging to different families. Despite this polyphagy, species have shown preference for certain plant taxa, such as legumes and brassicas, as in the case of the southern green stink bug, *N. viridula* (Todd and Herzog 1980); legumes and solanaceous plants, as in the case of *Edessa meditabunda* (Silva et al. 1968; Lopes et al. 1974); or grasses in general, as for species of *Aelia, Mormidea*, and *Oebalus* (references in Panizzi et al. 2000).

However, local populations of *N. viridula* in the southern United States may feed on Gramineae, such as corn (Negron and Riley 1987), which has not been reported as a food plant of this bug elsewhere. Moreover, in the southeastern United States, a farmscape can consist of wooded areas, weedy field edges, and a variety of field crops including corn, peanuts, and cotton (Turnipseed et al. 1995). In Hawaii, this stink bug is a serious pest of macadamia nuts, *Macadamia integrifolia* (Proteaceae) (Mitchell et al. 1965; Jones et al. 2001; Wright et al. 2007).

Local populations of the neotropical brown stink bug, *E. heros*, will feed on the euphorb, *E. heterophylla*; however, in general, this bug does not explore this plant as host (Pinto and Panizzi 1994). These and several other examples demonstrate that depending on time of exposure to restricted hosts and their availability, polyphagous species will act as monophagous or oligophagous (Fox and Morrow 1981).

The neotropical stink bug *Dichelops melancathus* (Dallas) was considered rare in southern Brazil and, because it overwinters in crop residues, with the increase of the no-tillage cultivation system it adapted to feed on corn and wheat seedlings, and populations dramatically increased (Chocorosqui and Panizzi 2004, see Section 27.3.2.4.).

This phenomenon of local populations with specific feeding habits makes clear the complexity of the biology of phytophagous heteropterans. What may be valid information in one place may not apply in another. This indicates that to devise management tactics that involve manipulation of host plants, studies should be done locally. The host plant sequences used by each species at each place should be determined and fully understood, considering such biotic factors as characteristics of the bug species, of the host plant species, and host plant sequences; and such abiotic factors as rain regimen, range of favorable temperatures that allow the bug reproduction, and photoperiod.

27.3.2.3 Manipulation of Preferred Host Plants as Traps

There have been several studies concerning preferred host plants as traps, as a tool to manage pest species (see references in Hokkanen 1991, and Shelton and Baldenes-Perez 2006). In the case of heteropterans, several studies have been conducted, manipulating plant phenology, as in the case of using the preference of stink bugs to feed on soybean plants with pods/seeds compared to plants in the vegetative period.

Because heteropterans are, in general, polyphagous, this makes the use of the trap crop technique, in the context of attraction by different plant species, a more complicated issue.

Despite their polyphagy, several attempts have being made to use the classical trap crop concept to manage heteropterans. An early report is by Watson (1924), who referred to the use of legumes (*Crotalaria*) to attract populations of the southern green stink bug, *N. viridula*, in citrus orchards in Florida, USA. The bugs concentrated on the legume plants and were killed before colonizing the citrus plants. Tillman (2006) shows that sorghum (*Sorghum bicolor* L.) can serve as a trap crop for *N. viridula* adults in cotton fields, in the southern United States. Other studies demonstrated that early-maturing soybeans could act as a potential trap for stink bugs during the highly susceptible periods of cotton development (Bundy and McPherson 2000; Gore et al. 2006). Ludwig and Kok (1998) evaluated trap crops to manage the harlequin bug, *Murgantia histrionica* (Hahn) (Pentatomidae), on broccoli.

In the tropics, Jackai (1984) reported on the attraction of cowpea, *Vigna unguiculata*, to stink bugs, mitigating their damage on soybean in Africa. Moreover, in Indonesia, the alydid *Riptortus linearis* (L.), also a soybean pest, was controlled using the legume *Sesbania rostrata* as a trap crop (Naito 1996).

In southern Brazil, there is good potential to use some legume host plants as traps for the heteropterans that feed on soybean. For instance, the close association of the red-banded stink bug, *P. guildinii*, with wild legumes (indigo, genus *Indigofera*) can be used to attract the bugs and concentrate them in particular areas where they can be eliminated. Similarly, pigeon pea, *C. cajan*, which is used on contour lines as a wind barrier, can be used as a trap crop for this and other pentatomids, such as *N. viridula* and *E. heros*, and for alydids, such as *Neomegalotomus parvus* (Westwood). This legume produces pods almost all year round, and is attractive to bugs when they leave their preferred host, soybean.

27.3.2.4 Role of Less Preferred Plant Food Sources

In general, during their lifetime, insects are faced with less preferred food sources and must adapt to explore alternate food sources when more preferred species are unavailable.

Most species of hemipterans spend only a third of their lifetimes feeding on spring/summer crops, usually their preferred hosts. The rest of the time they spend feeding and breeding on alternate hosts, some of them of low nutritional quality, or occupying overwintering sites. Therefore, the less preferred food plants are usually overlooked, and their roles in the life history of hemipterans are, in general, underestimated.

Although hemipterans do not breed on these plants (at least on some of these plants), they provide nutrients, to some extent, and water, as well. However, because bugs are not used to them, sometimes they may not recognize these "host plants" as potential toxic plants, despite their polyphagy and wide capacity to overcome toxic allelochemicals or lack of essential nutrients.

Among the less preferred host plants of hemipterans, some are cultivated and some are wild, uncultivated plants. Usually they occur nearby cultivated fields where the preferred hosts were harvested or ended their cycle and became mature. In some cases, weeds that remain green between mature plants of a certain crop are temporarily used as a source of nutrients and water. This situation is common in tropical or subtropical areas, where most bugs are active during the entire year—some species, however, enter diapause, underneath debris, without feeding, such as the neotropical brown stink bug, *E. heros* (Panizzi and Vivan 1997).

When phytophagous hemipterans are faced with a scarcity of preferred host plants, and environmental conditions are favorable, that is, temperatures and humidity are relatively high and photoperiod adequate, bugs will feed and remain active on less preferred plant food sources for several reasons: the less preferred plants possess seeds or fruits the bugs are not accustomed to feed on; the less preferred plants may be at a vegetative stage, and lacking seeds and fruits; or the less preferred plants may produce fruits and seeds that are suitable but inaccessible (out of reach, like seeds protected by thick pod walls, or by an empty space between the pod walls and the seeds). Faced with one of these conditions or others, bugs must change their feeding habits and feed on other plant structures, usually not explored as food sources.

For instance, the southern green stink bug, *N. viridula*, will feed on less preferred host plants in northern Paraná state, Brazil, such as star bristle, *A. hispidum*. Nymph mortality on this plant is high in the laboratory (in the field nymphs may not even feed on this plant), and adults will not reproduce on it, and their longevity is reduced; although a seed feeder, this bug strongly prefers feeding on stems of this plant (Panizzi and Rossi 1991). The stems are mostly filled with an aqueous tissue and the insects apparently detect this abundant source of water.

On castor bean, *R. communis*, late instars and adult *N. viridula* show an atypical feeding behavior by feeding on the leaf veins (Panizzi 2000). Eggs are not laid by females on castor bean leaves, unless accidentally. On wheat, *T. aestivum*, *N. viridula* adults have been observed feeding on reproductive plants during mild winters. Adults will feed on seedheads but will not lay eggs on plants. Attempts to raise nymphs in the laboratory using seedheads or mature seeds did not succeed.

N. viridula, although extremely polyphagous, does not breed on graminaceous plants. There are reports of its damage to wheat in Brazil (Maia 1973), and to wheat and corn in the United States (Viator et al. 1983; Negron and Riley 1987). However, these may be cases of local populations with specific feeding habits, as previously discussed. In northern Paraná state, *N. viridula* may eventually feed on corn, but on the stems, not on the ears, of seedling corn grown under no-tillage cultivation system. Bugs that stay in areas with weed plants or with scattered cultivated host plants will eventually feed on corn seedlings that are established in these areas. However, these events are uncommon.

Other species of hemipterans, such as *N. viridula,* will also feed on less preferred food plants. For instance, the neotropical brown stink bug *E. heros*, a typical seed sucker, will feed on star bristle stems

(Panizzi 2000). In Rio Grande do Sul state, adults were recorded feeding on seeds of *Amaranthus retro-flexus* (Amaranthaceae), and on fruits of three Solanaceae, mainly *Vassobia breviflora*. Immature survi-vorship on soybean (86.7%) and on *V. breviflora* (81.8%) was the same, but nymphs fed on *A. retroflexus* did not reach adulthood (Medeiros and Megier 2009).

Another pentatomid, *Dichelops melacanthus* (Dallas), previously reported as a pest of soybean, and feeding on pods (Galileo et al. 1977), has been observed feeding on corn, and on wheat. It is interesting that on these two graminaceous plants, the bugs feed on stems of young plants, causing substantial damage (Ávila and Panizzi 1995; Manfredi-Coimbra et al. 2005). This change in feeding habits, from reproductive structures of more preferred hosts, such as legumes (soybean), to vegetative tissues of less preferred hosts (graminaceous), is attributed to the low availability of preferred hosts. After soybean harvest, *D. melacanthus* stays on the ground underneath debris, and will feed on corn or wheat plants growing on areas under conservation tillage (Chocorosqui and Panizzi 2004). In these areas, bugs found shelter (straw) and food (dried seeds dropped on the ground) and will thrive, unlike what occurs on areas under conventional cultivation systems, where bugs are dislodged from their shelters and killed by plowing.

A similar situation occurs with the alydid *Neomegalotomus parvus* Westwood, a typical seed sucker that feeds on mature seeds of legumes. On areas under conservation tillage, this bug will feed on soybean seedlings. In areas under conservation tillage, it stays on the ground feeding on its preferred food (mature seeds) and will complement its diet with young plants (Panizzi and Chocorosqui 1999).

In conclusion, although many aspects of the biology of hemipterans have been investigated, perhaps, an aspect least studied is this subject of hemipterans on less preferred plant food sources. If we are to develop holistic IPM systems, more attention must be devoted to this subject.

27.3.3 Managing Heteropterans in Overwintering Sites and Host Plants

After colonizing spring/summer crops, heteropterans disperse to overwintering sites or, especially in the tropics, to alternate hosts. In general, bugs begin to disperse even before the crop they are feeding on completes maturation. For example, pentatomids that feed on soybean will start leaving the crop after reaching the population peak, during the time plants begin to senesce. This process of crop abandonment increases in intensity as the plants dry out completely and become mature.

In general, after leaving the summer crops, heteropterans feed on alternate hosts and may complete an extra generation before moving to diapause sites, or may continue to breed on these alternate hosts. This will depend not only on the favorability of abiotic factors, such as temperature and photoperiod, but also on the capability of certain species to reproduce on these alternate host plants. These bugs will emerge from diapause sites or alternate hosts to start colonizing spring/summer preferred hosts, such as soybean. In soybean, colonization begins during the vegetative period (V0 to Vn), increases with reproduction during blooming/early pod set (R1 to R3), and the numbers increase to reach the so-called critical period at pod filling (R4 to R5.n); at this stage the damage to the crop is crucial. At the end of the pod-filling period (R6), the bugs' population reaches its peak, and dispersal begins again.

27.3.3.1 Managing Crop Residues

The management of crop residues to mitigate the impact of pest species is becoming increasingly important, and is the main reason for the adoption of no-tillage or minimum-tillage cropping systems in many regions of the world. These systems provide favorable conditions for soil-inhabiting insects or those that live in or under debris.

At least three species of soybean heteropterans have been favored by no-tillage or minimum tillage cropping systems. These are the neotropical brown stink bug *E. heros* (Fabricius), the neotropical green-belly stink bug *Dichelops melacanthus* (Dallas) (Pentatomidae), and the brownish root bug *Scaptocoris castanea* Perty (Cydnidae).

E. heros, because of its habit of hiding underneath crop residues during more than 6 months of the year, particularly during late fall, winter, and early spring, has increased in abundance dramatically.

Considered a secondary pest in the 1970s, it is the most common stink bug pest of soybean in Brazil today.

D. melacanthus has also increased its abundance probably because the adoption of tillage cultivation systems. Once considered a minor pest of soybean, together with *D. furcatus* (F.), *D. melacanthus* now is a major pest of corn and wheat. It also remains on the ground in partial hibernation (oligopause), and, when corn or wheat are sown directly over the crop residues during the fall, in southern Brazil, it attacks the seedlings and the resulting plants show severe damage (Chocorosqui and Panizzi 2001). Similar damage by stink bugs to seedling corn has also been observed in the United States (Sedlacek and Townsend 1988; Apriyanto et al. 1989).

The third species, *S. castanea*, attacks the roots of many economically important plants, such as corn, cotton, rice, groundnuts, sugarcane, potato, peas, tomatoes, pimentos, and lucerne, as well as wild uncultivated plants in the Neotropics. It can also be devastating to soybean (Lis et al. 2000 and references therein). A root feeder, it spends most of its life underground.

To control these three species and others that live in the soil at least part of their life, management of crop residues is mandatory. Plowing or burning the residues is recommended.

27.3.3.2 Monitoring Bugs in Overwintering Niches to Determine Crop Colonization

Perhaps one of the most important steps toward implementing holistic IPM programs is to monitor overwintering niches and host plants to determine the abundance of pest populations and likely time of crop invasion. This "preventive" step is, in general, overlooked and its importance underestimated.

How can one estimate the intensity of stink bugs colonizing a soybean field by monitoring the bugs during the overwintering period? This depends on several factors. For example, temperature and humidity are crucial. If, after soybean harvest, the temperature falls below 5°C during the fall–winter and remains low for a certain period, a high mortality of bugs is expected. Similarly, if strong spring rains precede the cultivation period, the population of bugs on alternate host plants or in the soil under crop residues will be heavily affected. These two factors might mitigate the impact of bugs during the following soybean season.

Another important factor influencing the population dynamics of the bugs during overwintering is the cultivation system. As mentioned above, the no-tillage or minimum-tillage cultivation systems may promote a greater than expected population of bugs, particularly of *E. heros* and *D. melacanthus*. These two species overwinter on the soil under crop residues. Plowing eliminates a great portion of these two bugs' populations.

Finally, the presence of host plants may allow predicting which species of stink bugs are likely to predominate in the following soybean season. For example, the presence of indigo legumes as overwintering host plants will increase the population of the red-banded stink bug, *P. guildinii*. Similarly, the weed plant siberian motherwort, *L. sibiricus*, which grows before soybean during early spring, is a preferred host of the southern green stink bug, *N. viridula*, allowing reproduction and therefore population build-up. These and other examples illustrate that it is possible, to a certain degree, to predict both qualitatively and quantitatively the populations of stink bugs that colonize soybean.

27.4 Conclusions

As stated in the beginning of this chapter, research on insect bioecology and nutrition (nutritional ecology) has concentrated on basic aspects, relating food characteristics to food intake and utilization by insects, and its consequences on their performance. In a more applied field, as a support to pest management programs, this has been, in general, overlooked. An exception is the paper by Slansky (1990), which relates insect nutritional ecology to host plant resistance.

Several decades ago, during the 1960s and 1970s, several authors concentrated on pest management strategies, considering insect bioecology and nutrition aspects in a broad context (references in Panizzi and Parra 1991b, 2009). Today, after more than 40 years, these pest management strategies based on insect nutrition, such as host plant resistance, trap crops, polycultivation, and use of allelochemicals, remain a challenge to be fully implemented in IPM programs.

As new areas in biology gain momentum, such as the development of GM crops resistant to insects, insect bioecology and nutrition becomes a very important area of research in entomology, now within a more applied context. These cultivars of many important crops are being widely adopted by growers all over the world, and will certainly influence the insect pests causing dramatic changes. Many questions will arise, such as how these GM plants will fit into current IPM programs. Clearly, much research will be needed to change the traditional IPM programs to accommodate this new technological tool.

To conclude, it is reasonable to assume that as we develop new IPM programs that are more efficient and more ecologically sound, the tactics taking into account the interactions of insects with their food will play a growing role in achieving our goals.

REFERENCES

Ahmad, S. 1983. *Herbivorous Insects: Host-Seeking Behaviour and Mechanisms*. New York: Academic Press.

Altieri, M. A. 1994. *Biodiversity and Pest Management in Agroecosystems*. New York: Food Products Press.

Altieri, M. A., and D. L. Letourneau. 1984. Vegetation diversity and insect pest outbreaks. *CRC Crit. Rev. Plant Sci.* 2:131–69.

Altieri, M. A., and C. I. Nicholls. 1999. Biodiversity, ecosystem function, and insect pest management in agricultural systems. In *Biodiversity in Agroecosystems*, ed. W. W. Collins and C. O. Qualset, 69–84. Boca Raton, FL: CRC Press.

Altieri, M. A., and C. I. Nicholls. 2004. *Biodiversity and Pest Management in Agroecosystems*. New York: Food Products Press.

Andow, D. A. 1991. Vegetational diversity and arthropod population response. *Annu. Rev. Entomol.* 36:561–86.

Apriyanto, D., L. H. Townsend, and J. D. Sedlacek. 1989. Yield reduction from feeding by *Euschistus servus* and *E. variolarius* (Heteroptera: Pentatomidae) on stage V2 field corn. *J. Econ. Entomol.* 82:445–8.

Ávila, C. J., and A. R. Panizzi. 1995. Occurrence and damage by *Dichelops* (*Neodichelops*) *melacanthus* (Dallas) (Heteroptera: Pentatomidae) on corn. *An. Soc. Entomol. Bras.* 24:193–4.

Baldwin, I. T., R. Halitschke, A. Paschold, C. C. von Dahl, and C. A. Preston. 2006. Volatile signaling in plant–plant interactions: "Talking Trees" in the genomics era. *Science* 311:812–5.

Bernays, E. A. 1989–1994. *Insect–Plant Interactions*. Vols. I–V. Boca Raton, FL: CRC Press.

Bernays, E. A., and R. F. Chapman. 1994. *Host-Plant Selection by Phytophagous Insects*. New York: Chapman and Hall.

Boethel, D. J. 1999. Assessment of soybean germplasm for multiple insect resistance, p. In *Global Plant Genetic Resources for Insect-Resistant Crops*, ed. S. L. Clement and S. S. Quisenberry, 101–29. Boca Raton, FL: CRC Press.

Borden, J. H. 2002. Allelochemics. In *Encyclopedia of Pest Management*, ed. D. Pimentel, 14–7. New York: Marcel Dekker, Inc.

Brackenbury, J. 1995. *Insects and Flowers: A Biological Partnership*. London: Cassell Brandford.

Bundy, C. S., and R. M. McPherson. 2000. Dynamics and seasonal abundance of stink bugs (Heteroptera: Pentatomidae) in a cotton–soybean ecosystem. *J. Econ. Entomol.* 93:697–706.

Carpenter, J., A. Felsot, T. Goode, M. Hamming, D. Onstad, and S. Sankula. 2002. *Comparative Environmental Impacts of Biotechnology-Derived and Traditional Soybean, Corn, and Cotton Crops*. Ames, IA: Council for Agricultural Science and Technology.

Chocorosqui, V. R., and A. R. Panizzi. 2001. Evolução dos danos causados por *Dichelops melacanthus* (Dallas) (Heteroptera: Pentatomidae) a plântulas de milho e trigo. In *Resumos XIX Congresso Brasileiro de Entomologia, Manaus, AM*, 276.

Chocorosqui, V. R., and A. R. Panizzi. 2004. Impact of cultivation systems on *Dichelops melacanthus* (Dallas) (Heteroptera: Pentatomidae) populations and damage and its chemical control on wheat. *Neotrop. Entomol.* 33:487–92.

Corrêa-Ferreira, B. S. 1987. Liberação do parasitóide *Trissolcus basalis* em cultivar armadilha e seu efeito na população de percevejos da soja. *Res. Pesq. Soja.* 20:142–3.

Corso, I. C. 1989. Atratividade do sal de cozinha para espécimes de *Nezara viridula* (L., 1758). *Res. Pesq. Soja.* 45:78–9.

Corso, I. C. 1990. Uso de sal de cozinha na redução da dose de inseticida para controle de percevejos da soja. *Cent. Nac. Pesqui Soja, Londrina, PR, Com. Téc*. 45:1–7.

Crawley, M. J. 1983. *The Herbivory: The Dynamics of Animal–Plant Interactions*. Berkeley, CA: University of California Press.

Edwards, P. J., and S. D. Wratten. 1980. *Ecology of Insect–Plant Interactions*. London: Edward Arnold Publ. Limited.

Erlich, P. R., and P. H. Raven. 1964. Butterflies and plants: A study in coevolution. *Evolution*. 18:586–608.

Finch, S., and R. H. Collier. 2000. Host-plant selection by insects—A theory based on 'appropriate/inappropriate landings' by pest insects of cruciferous plants. *Entomol. Exp. Appl*. 96:91–102.

Flint, M. L., and P. Gouveia. 2001. *IPM in Practice—Principles and Methods of Integrated Pest Management*. Oakland, CA: University of California Press.

Fontes, E. M. G., C. S. S. Pires, E. R. Sujii, and A. R. Panizzi. 2002. The environmental effects of genetically modified crops resistant to insects. *Neotrop. Entomol*. 31:497–513.

Fox, L. R. 1981. Defense and dynamics in plant–herbivore systems. *Am. Zool*. 21:853–64.

Fox, L. R., and P. A. Morrow. 1981. Specialization: Species property of local phenomenon? *Science* 211:887–93.

Futuyma, D. J. 1983. Evolutionary interactions among herbivorous insects and plants. In *Coevolution*, ed. D.G. Futuyma and M. Slatkin, 207–31. Sunderland, MA: Sinauer Assoc. Inc.

Galileo, M. H. M., H. A. O. Gastal, and J. Grazia. 1977. Levantamento populacional de Pentatomidae (Hemiptera) em cultura de soja (*Glycine max* (L.) Merr.) no município de Guaíba, Rio Grande do Sul. *Rev. Bras. Biol*. 37:111–20.

Gatehouse, J. A., and A. M. R. Gatehouse. 2000. Genetic engineering of plants for insect resistance. In *Biological and Biotechnological Control of Insect Pests*, ed. J. E. Rechcigl and N. A. Rechcigl, 211–41. Boca Raton, FL: Lewis Publishers.

Gershenzon, J., and R. Croteau. 1991. Terpenoids. In *Herbivores: Their Interactions with Secondary Plant Metabolites,* ed. G. A. Rosenthal and M. R. Berenbaum, 165–219. San Diego, CA: Academic Press.

Gilman, D. F., R. M. McPherson, L. D. Newsom, D. C. Herzog, R. L. Jensen, and C. L. Williams. 1982. Resistance in soybeans to the southern green stink bug. *Crop Sci*. 22:573–6.

Gore, J., C. A. Abel, J. J. Adamczyk, and G. Snodgrass. 2006. Influence of soybean planting date and maturity group on stink bug (Hemiptera: Pentatomidae) populations. *Environ. Entomol*. 35:531–6.

Hokkanen, H. M. T. 1991. Trap cropping in pest management. *Annu. Rev. Entomol*. 36:119–38.

Horn, D. J. 2002. Ecological aspects of pest management. In *Encyclopedia of Pest Management*, ed. D. Pimentel, 211–3. New York: Marcel Dekker, Inc.

Jackai, L. E. N. 1984. Using trap plants in the control of insect pests of tropical legumes. In *Proceedings of the International Workshop in Integrated Pest Control of Grain Legumes,* ed. P. C. Matteson, 101–12. Brasília: Embrapa.

James, C. A. 2001. Global review of commercialized transgenic crops, 2001. ISAAA Briefs Number 24, ISAAA, Ithaca, New York.

Janzen, D. H. 1980. When is it coevolution? *Evolution* 84:611–2.

Jermy, T. 1984. Evolution of insect/host plant relationships. *Am. Nat*. 124:609–30.

Jolivet, P. 1998. *Interrelationship between Insects and Plants*. Boca Raton, FL: CRC Press.

Jones, Jr., W. A., and M. J. Sullivan. 1978. Susceptibility of certain soybean cultivars to damage by stink bugs. *J. Econ. Entomol*. 71:534–6.

Jones, Jr., W. A., and M. J. Sullivan. 1979. Soybean resistance to the southern green stink bug, *Nezara viridula*. *J. Econ. Entomol*. 72:628–32.

Jones, Jr., W. A., and M. J. Sullivan. 1981. Overwintering habitats, spring emergence patterns, and winter mortality of some South Carolina Hemiptera. *Environ. Entomol*. 10:409–14.

Jones, Jr., W. A., and M. J. Sullivan. 1982. Role of host plants in population dynamics of stink bug pests of soybean in South Carolina Hemiptera. *Environ. Entomol*. 11:867–75.

Jones, V. P., D. M. Westcott, N. M. Finson, and R. K. Nishimoto. 2001. Relationship between community structure and southern green stinkbug (Heteroptera: Pentatomidae) damage in macadamia nuts. *Environ. Entomol*. 30:1028–35.

Karban, R., and I. T. Baldwin. 1997. *Induced Responses to Herbivory*. Chicago: University of Chicago Press.

Kenkel, P. 2007. Economics of host plant resistance in integrated pest management systems. In *Ecologically Based Integrated Pest Management*, ed. O. Koul and G. W. Cuperus, 194–200. Cambridge, MA: CAB International.

Kennedy G. G. 2008. Integration of insect-resistant genetically modified crops within IPM programs. In *Progress In Biological Control—Integration of Insect-Resistant Genetically Modified Crops within IPM Programs.* ed. J. Romeis, A. M. Shelton and G. G. Kennedy, 1–26. New York: Springer.

Kester, K. M., C. M. Smith, and D. F. Gilman 1984. Mechanisms of resistance in soybean [*Glycine max* (L.) Merrill] genotype PI 171444 to the southern green stink bug *Nezara viridula* (L.). *Environ. Entomol.* 13:1208–15.

Kobayashi, T., and G. W. Cosenza. 1987. Integrated control of soybean stink bugs in the Cerrados. *Jpn. Agric. Res. Q.* 20:229–36.

Kogan, M. 1986a. *Ecological Theory and Integrated Pest Management Practice.* New York: Wiley.

Kogan, M. 1986b. Plant defense strategies and host-plant resistance. In *Ecological Theory and Integrated Pest Management Practice,* ed. M. Kogan, 83–134. New York: Wiley.

Langer, V., J. Kinane, and M. Lyngkjaer. 2007. Intercropping for pest management: The ecological concept. In *Ecological Based Integrated Pest Management,* ed. O. Koul and G. W. Cuperus, 74–110. Cambridge, MA: CAB International.

Link, D., V. Estefanel, and O. S. Santos. 1971. Danos causados por percevejos fitófagos em grãos de soja. *Rev. Centr. Cienc. Rur.* 1:9–13.

Link, D., V. Estefanel, O. S. Santos, M. C. Mezzomo, and L. E. V. Abreu. 1973. Influência do ataque de pentatomídeos nas características agronômicas do grão de soja, *Glycine max* (L.) Mer. *An. Soc. Entomol. Bras.* 2:59–65.

Lis, J. A., M. Becker, and C. W. Schaefer. 2000. Burrower bugs (Cydnidae), In *Heteroptera of Economic Importance,* ed. C. W. Schaefer, and A. R. Panizzi, 405–19. Boca Raton, FL: CRC Press.

Lopes, O. J., D. Link, and I. V. Basso. 1974. Pentatomídeos de Santa Maria—Lista preliminar de plantas hospedeiras. *Rev. Centr. Cien. Rur.* 4:317–22.

Ludwig, S. W., and L. T. Kok. 1998. Evaluation of trap crops to manage harlequin bugs, *Murgantia histrionica* (Hahn) (Hemiptera: Pentatomidae) on broccoli. *Crop Prot.* 17:123–8.

Maia, N. G. 1973. Ocorrência do percevejo da soja—*Nezara viridula* (L.) em espigas de trigo no Rio Grande do Sul. *Agron. Sulriogr.* 9:241–3.

Manfredi-Coimbra, S., J. J. Silva, V. R. Chocorosqui, and A. R. Panizzi. 2005. Danos do percevejo barriga-verde *Dichelops melacanthus* (Dallas) (Heteroptera: Pentatomidae) em trigo. *Ciênc. Rur.* 35:1243–7.

McPherson, R. M., and L. D. Newsom. 1984. Trap crops for control of stink bugs in soybean. *J. Ga. Entomol. Soc.* 19:470–80.

McPherson, R. M., L. D. Newsom, and B. F. Farthing. 1979. Evaluation of four stink bug species from three genera affecting soybean yield and quality in Louisiana. *J. Econ. Entomol.* 72:188–94.

Medeiros, L., and G. A. Megier. 2009. Ocorrência e desempenho de *Euschistus heros* (F.) (Heteroptera: Pentatomidae) em plantas hospedeiras alternativas no Rio Grande do Sul. *Neotrop. Entomol.* 38: 459–63.

Mitchell, W. C., R. M. Warner, and E. T. Fukunaga. 1965. Southern green stink bug, *Nezara viridula* (L.), injury to macadamia nut. *Proc. Hawaii. Entomol. Soc.* 19:103–9.

Moeller, B. L., and D. S. Seigler. 1999. Biosynthesis of cyanogenic glycosides, cyanolipids, and related compounds. In *Plant Amino Acids,* ed. B. J. Singh, 563–609. New York: Marcel Dekker, Inc.

Naito, A. 1996. Insect pest control through use of trap crops. *Agrochem. Jpn.* 68: 9–11.

Negron, J. F., and T. J. Riley. 1987. Southern green stink bug, *Nezara viridula* (Heteroptera: Pentatomidae) feeding on corn. *J. Econ. Entomol.* 80:666–9.

Newsom, L. D., and D. C. Herzog. 1977. Trap crops for control of soybean pests. *Louisiana Agric.* 20:14–5.

Nicholls, C. I., and M. A. Altieri. 2007. Agroecology: Contributions towards a renewed ecological foundation for pest management. In *Perspectives in Ecological Theory and Integrated Pest Management,* ed. M. Kogan and P. Jepson, 431–68. Oxford: Cambridge University Press.

Niva, C. C., and A. R. Panizzi. 1996. Efeitos do cloreto de sódio no comportamento de *Nezara viridula* (L.) (Heteroptera: Pentatomidae) em vagem de soja. *An. Soc. Entomol. Bras.* 25:251–7.

Norris, R. F., E. P. Caswell-Chen, and M. Kogan. 2003. *Concepts in Integrated Pest Management.* Upper Saddle River, NJ: Prentice Hall.

O'Callaghan, M., R. Travis, T. R. Glare, E. P. J. Burgess, and L. A. Malone. 2005. Effects of plants genetically modified for insect resistance on nontarget organisms. *Annu. Rev. Entomol.* 50:271–92.

Owino, P. O. 2002. Antagonistic plants. In *Encyclopedia of Pest Management,* ed. D. Pimentel, 21–3. New York: Marcel Dekker, Inc.

Painter, R. H. 1951. *Insect Resistance in Crop Plants*. New York: McMillan.

Panizzi, A. R. 1980. Uso de cultivar armadilha no controle de percevejos em soja. *Trigo Soja*. 47:11–4.

Panizzi, A. R. 1997. Wild hosts of pentatomids: Ecological significance and role in their pest status on crops. *Annu. Rev. Entomol*. 42:99–122.

Panizzi, A. R. 2000. Suboptimal nutrition and feeding behavior of hemipterans on less preferred plant food sources. *An. Soc. Entomol. Bras*. 29:1–12.

Panizzi, A. R., and V. R. Chocorosqui. 1999. Pragas—Elas vieram com tudo. *Cultivar* 11:8–10.

Panizzi, A. R., J. E. McPherson, D. G. James, M. Javahery, and R. M. McPherson. 2000. Economic importance of stink bugs (Pentatomidae). In *Heteroptera of Economic Importance*, ed. C. W. Schaefer and A. R. Panizzi, 421–74. Boca Raton, FL: CRC Press.

Panizzi, A. R., and C. C. Niva. 1994. Overwintering strategy of the brown stink bug in northern Paraná. *Pesq. Agropec. Bras*. 29:509–11.

Panizzi, A. R., and E. D. M. Oliveira. 1999. Seasonal occurrence of tachinid parasitism on stink bugs with different overwintering strategies. *An. Soc. Entomol. Bras*. 28:169–72.

Panizzi, A. R., and N. Oliveira. 1993. Atração do cloreto de sódio (sal de cozinha) aos percevejos-pragas da soja. *Res. Pesq. Soja*. 58:71–6.

Panizzi, A. R., and J. R. P. Parra. 1991a. *Ecologia Nutricional de Insetos e Suas Implicações no Manejo de Pragas*. São Paulo, Brazil: Manole.

Panizzi, A. R., and J. R. P. Parra. 1991b. A ecologia nutricional e o manejo integrado de pragas, In *Ecologia Nutricional de Insetos e Suas Implicações no Manejo de Pragas*, ed. A. R. Panizzi and J. R. P. Parra, 313–36. São Paulo, Brazil: Manole.

Panizzi, A. R., and J. R. P. Parra. 2009. *Bioecologia e Nutrição de Insetos: Base Para o Manejo Integrado de Pragas*. Brasília: Embrapa Informação Tecnológica.

Panizzi, A. R., and C. E. Rossi. 1991. The role of *Acanthospermum hispidum* in the phenology of *Euschistus heros* and of *Nezara viridula*. *Entomol. Exp. Appl*. 59:67–74.

Panizzi, A. R., and L. M. Vivan. 1997. Seasonal abundance of the neotropical brown stink bug, *Euschistus heros* in overwintering sites and the breaking of dormancy. *Entomol. Exp. Appl*. 82:213–7.

Panizzi, M. C. C., I. A. Bays, R. A. S. Kiihl, and M. P. Porto. 1981. Identificação de genótipos fontes de resistência a percevejos-pragas da soja. *Pesq. Agropec. Bras*. 16:33–7.

Pedigo, L. P. 2002. *Entomology and Pest Management*. 4th ed. Upper Saddle River, NJ: Prentice Hall.

Pimentel, D. 1981. *Handbook of Pest Management in Agriculture*. Vols. I–III. Boca Raton, FL: CRC Press.

Pimentel, D. 2002. *Encyclopedia of Pest Management*. New York: Marcel Dekker, Inc.

Pinto, S. B., and A. R. Panizzi. 1994. Performance of nymphal and adult *Euschistus heros* (F.) on milkweed and on soybean and effect of food switch on adult survivorship, reproduction and weight gain. *An. Soc. Entomol. Bras*. 23:549–55.

Ragsdale, D. W., A. D. Larson, and L. D. Newsom. 1981. Quantitative assessment of the predators of *Nezara viridula* eggs and nymphs within a soybean agroecosystem using an ELISA. *Environ. Entomol*. 10:402–5.

Rechcigl, J. E., and N. A. Rechcigl. 2000. *Insect Pest Management—Techniques for Environmental Protection*. Boca Raton, FL: Lewis Publishers.

Ricklefs, R. E., and I. J. Lovette. 1999. The roles of island area per se and habitat diversity in the species–area relationships of four lesser Antillean faunal groups. *J. Anim. Ecol*. 68:1142–60.

Rossetto, C. J. 1989. Breeding for resistance to stink bugs. In *Proceedings of the World Soybean Research Conference IV*, ed. A. J. Pascale, 20–46. Buenos Aires, Argentina: Impresiones Amawald S.A.

Schaefer, C. W., and A. R. Panizzi. 2000. *Heteroptera of Economic Importance*. Boca Raton, FL: CRC Press.

Sedlacek, J. D., and L. H. Townsend. 1988. Impact of *Euschistus servus* and *E. variolarius* (Heteroptera: Pentatomidae) feeding on early growth stages of corn. *J. Econ. Entomol*. 81:840–4.

Seigler, D. S. 2002. Toxins in plants. In *Encyclopedia of Pest Management*, ed. D. Pimentel, 840–2. New York: Marcel Dekker, Inc.

Shelton, A. M., J. Z. Zhao, and R. T. Roush. 2002. Economic, ecological, food safety, and social consequences of the deployment of *Bt* transgenic plants. *Annu. Rev. Entomol*. 47:845–81.

Shelton A. M., and F. R. Badenes-Perez. 2006. Concepts and applications of trap cropping in pest management. *Annu. Rev. Entomol*. 51:285–308.

Silva, A. G. D. A., C. R. Gonçalves, D. M. Galvão, A. J. L. Gonçalves, J. Gomes, M. N. Silva, and L. Simoni. 1968. *Quarto Catálogo dos Insetos que Vivem Nas Plantas do Brasil—Seus Parasitas e Predadores*. Part II, v. I, Rio de Janeiro: Ministério da Agricultura.

Slansky, Jr., F. 1990. Insect nutritional ecology as a basis for studying host plant resistance. *Fla. Entomol.* 73:359–78.

Slansky, Jr., F., and J. G. Rodriguez. 1987a. Nutritional ecology of insects, mites, spiders, and related invertebrates. In *Nutritional Ecology of Insects, Mites, Spiders and Related Invertebrates*, ed. F. Slansky, Jr. and J. G. Rodriguez, 1–69. New York. Wiley.

Slansky, Jr., F., and J. G. Rodriguez. 1987b. *Nutritional Ecology of Insects, Mites, Spiders and Related Invertebrates*. New York: Wiley.

Sosa-Gómez, D. R., C. Y. Takachi, and F. Moscardi. 1993. Determinação de sinergismo e suscetibilidade diferencial de *Nezara viridula* (L.) e *Euschistus heros* (F.) (Hemiptera: Pentatomidae) à inseticidas em mistura com cloreto de sódio. *An. Soc. Entomol. Bras.* 22:569–76.

Southwood, T. R. E., and M. J. Way. 1970. Ecological background to pest management. In *Concepts of Pest Management*, ed. R. L. Rabb and F. E. Guthrie, 6–29. Raleigh, NC: North Carolina State University Press.

Stout M., and J. Davis. 2009. Keys to the increased use of host plant resistance in integrated pest management. In *Integrated Pest Management: Innovation–Development Process*. Vol. 1, ed. R. Peshin and A. K. Dhawan, 163–81. New York: Springer.

Strong, D. R., J. H. Lawton, and T. R. E. Southwood. 1984. *Insects on Plants. Community Patterns and Mechanisms*. Cambridge, MA: Harvard University Press.

Tillman, P. G. 2006. Sorghum as a trap crop for *Nezara viridula* L. (Heteroptera: Pentatomidae) in cotton in the southern United States. *Environ. Entomol.* 35:771–83.

Todd, J. W., and D. C. Herzog. 1980. Sampling phytophagous Pentatomidae on soybean. In *Sampling Methods in Soybean Entomology*, ed. M. Kogan and D. C. Herzog, 438–78. New York: Springer-Verlag.

Turnipseed, S. G., and M. J. Sullivan. 1975. Plant resistance in soybean insect management. In *Proceedings of the World Soybean Research Conference I*, ed. L .D. Hill, 549–60. Urbana-Champaign, IL: Interstate Printers and Publishers.

Turnipseed, S. G., M. J. Sullivan, J. E. Mann, and M. E. Roof. 1995. Secondary pests in transgenic *Bt* cotton in South Carolina. *Proceedings of the Beltwide Cotton Conferences*, Memphis, TN.

Vandermeer, J. 1995. The ecological basis of alternative agriculture. *Annu. Rev. Ecol. Syst.* 26:201–24.

van Emden, H. F. 1965. The role of uncultivated land in the biology of crop pests and beneficial insects. *Sci. Hort.* 17:121–6.

Viator, H. P., A. Pantoja, and C. M. Smith. 1983. Damage to wheat seed quality and yield by the rice stink bug and southern green stink bug (Hemiptera: Pentatomidae). *J. Econ. Entomol.* 76:1410–3.

Watson, J. R. 1924. *Crotalaria* as a trap crop for pumpkin bugs in citrus groves. *Fla. Grow.* 29:6–7.

Wratten, S. D., D. F. Hochuli, G. M. Gurr, J. Tylianakis, and S. L. Scarratt. 2007. Conservation, biodiversity, and integrated pest management. In *Perspectives in Ecological Theory and Integrated Pest Management*, ed. M. Kogan and P. Jepson, 223–45. Oxford: Cambridge University Press.

Wright, M. W., P. A. Follett, and M. Golden. 2007. Longterm patterns and feeding sites of southern green stink bug (Hemiptera: Pentatomidae) in Hawaii macadamia orchards, and sampling for management decisions. *Bull. Entomol. Res.* 97:569–75.

Index

A

Abies balsamea, 126
Abiotic variation adoptions, in predatory beetles, 583–584
Abracris flavolineata, 101
Abrus, 334
Abutilon teophrasti, 199
Acacia, 225, 377
Acacia berlandieri, 338
Acacia erioloba, 333, 335, 338
Acacia gerrardii, 341
Acacia greggii, 336
Acacia nilotica, 333, 338
Acacia sieberiana, 341
Acacia tortilis, 341
Acalymma, 669
Acanthaceae, 376
Acanthocinus obsoletus, 334
Acanthognathus spp., 219
Acanthoponera, 220
Acanthoscelides, 327, 334, 338, 341
Acanthoscelides alboscutellatus, 329, 340, 344
Acanthoscelides aureolus, 339
Acanthoscelides chiricahuae, 335
Acanthoscelides fraterculus, 332–333
Acanthoscelides macropthalmus, 341
Acanthoscelides obtectus, 326, 330, 334, 338, 340, 342–343, 345–346, 421, 422–423, 665, 667
Acanthoscelides spp., 332, 343
Acanthospermum hispidum, 309–310, 695, 697
Acanthostichus, 218–219
Acari, 553
Acari-induced gall, 370
Acarophenacidae, 440
Acarophenax lacunatus, 440
Aceraceae, 375
Acheta domesticus, 35
Achromobacter, 497
Aciurina trixa, 379
Aconitum, 581
Acremonium loliae, 281
Acromyrmex, 223
Acromyrmex ameliae, 213
Acromyrmex echinatior, 216
Acromyrmex insinuator, 216
Acromyrmex lundii, 216
Acromyrmex rugosus, 216
Acromyrmex subterraneus brunneus, 216
Acromyrmex subterraneus subterraneus, 216
Acropyga, 226
Acrosternum hilare, 151–152, 309
Actebia fennica, 669
Actinomyces sp., 198

Acyrthosiphon pisum, 55, 68, 110, 150, 196, 200, 203, 485–488, 490–492, 497–499, 503, 527, 579–580, 613, 671, 674
Acyrthosiphum pisum, 154, 548
Adalia bipunctata, 186, 576, 579, 581–583
Adalia decempunctata, 583
Adelgidae, 477, 576
Adelomyrmex, 218
Adonia variegata, 583
Aedes, 636
Aedes aegypti, 39, 70, 97, 198, 636–639
Aedes albopictus, 636–637, 639
Aedes scapularis, 636
Aedes sp., 70
Aedes (Stegomyia) aegypti, 635
Aegopsis bolboceridus, 355, 358, 360, 363
Aegopsis sp., 354
Aelia, 305, 696
African horse sickness, 641
Agaonidae, 375
Ageniaspis citricola
    production system, 60
    rearing, 60–61
Ageratum houstonianum, 40
Aglossata, 273
Agria housei, nutrients in, 63
Agrotis ipsilon, 25, 38, 72, 79, 664
Agrotis orthogonia, digestibility of, 24
Agrotis subterranea, 79
    nutritional indices, 29
Ahasverus advena, 421
Ajuga remota, 126
Akabane disease, 641
Alabama argillacea, 548, 599, 614
    nutritional indices, 29, 35
Alanine, 69
Aleppo galls, 370
Aleyrodidae, 599, 614, 616
Algarobius bottimeri, 326
Algarobius prosopis, 326, 338, 344–345
Allantonematidae, 585
Allelochemicals, 163
    behavioral effects, 659–660
    bioecology and nutrition, 691–692
    physiological effects, 126, 669–670
    role in reproduction and dispersal of insects, 39–41
    seed-sucking bugs, 297
Allelopathy, 126
Allium sativum, 672
Allomerus, 225
Alnus incana ssp., 617
Alnus rubra, 200
Alphitobius sp., 217
Alternanthera ficoidea, 558

Alternaria brassica, 203
Alternaria sp., 579
Althaeus hibisci, 329, 341, 344
Alycaulini, 383
Alydidae, 295
Amaranthaceae, 698
Amaranthus caudatus, 672
Amaranthus retroflexus, 698
Amaranthus sp., 556, 558
Amaryllidaceae, 505
Amblycera, 647
Amblycerini, 327
Amblycerus, 337–338, 340
Amblycerus caryoboriformis, 340
Amblycerus dispar, 340
Amblycerus hoffmanseggi, 335–337, 344
Amblycerus robiniae, 344
Amblycerus submaculatus, 335, 337–339, 344
Amblyopone, 218
Amblyseius cucumeris, 553
Ambrosia beetles, 5
Ambrosia gall, 370
Amino acids, for insects growth, 68–69
Amorphous galls, 386
α-Amylase inhibitors, plants resistance to insects, 675–676
An. cruzii, 639
Anabrus simplex, 180–181
Anadiplosis, 375
Anadiplosis sp., 384
Anagasta kuehniella, 71, 419, 421, 423, 426–427, 431, 438, 548–550, 556, 579, 599, 614, 616
Analoma spp., 354
Anaphes iole, 75
Anaplasma marginale, 642
Anastasus spp., 553
Anastrepha, 451, 458–460
Anastrepha bistrigata, 452
Anastrepha fraterculus, 79, 457
Anastrepha ludens, 453–455, 458, 462–463
Anastrepha obliqua, 453–465
Anastrepha serpentina, 454, 458, 465
Anastrepha striata, 452, 454–455, 458, 461, 465
Anastrepha suspensa, 457, 460
Anatis ocellata, 577
Anisopteromalus calandrae, 346, 439, 440, 441
Ankylopteryx exquisita, 610
Annonaceae, 384
Anobiidae, 404, 423, 429
Anochetus, 216
Anomalochrysa, 608
Anopheles, 636, 638
Anopheles albimanus, 198, 638
Anopheles gambiae, 639
Anopheles (Kerteszia) cruzii, 636
Anopheles punctulatus, 639
Anophelinae, 635
Anoplura, 647
Ant guilds
    large-sized, 216
    medium-sized, 216–217
    neotropical, 216–226

nutritional biology, 214–216
trophic to applied myrmecology, 226–228
Antheraea pernyi, artificial diet, 55
Anthocoridae, 441, 540–541, 543–544, 554, 559–560
Anthocoris confusus, 544, 552, 556
Anthocoris nemoralis, 544, 560
Anthocoris nemorum, 544, 552, 557, 560
Anthomyiidae, 644
Anthonomus grandis, 22–23, 72, 79, 135, 168
    feeding needs, 67
    nutrients in, 63
Anthribidae, 325
Anticarsia gemmatalis, 22, 40, 75, 199, 282, 668
    nutritional indices, 29
Antiteuchus mixtus, 310
Antiteuchus tripterus, 310
Antiteuchus tripterus limbativentris, 308
Aonidiella aurantii, 134
Apanteles flavipes, nutritional indices, 29
Aphelinidae, 553
Aphelinus abdominalis, 505
Aphididae, 481, 501, 548, 554, 576, 599, 614, 616, 634
Aphidiinae, 552
Aphidinae, 477
Aphidius colemani, 552
Aphidius ervi, 523, 527
Aphidius spp., 553
Aphidoidea, 7, 374, 474, 476–477, 481, 484, 503. *See also* Sap-sucking insects
    bioecology and nutrition, 474
    feeding process of, 474
Aphidoletes aphidimyza, 553, 582
Aphids, 7
Aphis fabae, 167, 478, 481–483, 487, 489, 491, 497–499, 579, 584, 667
Aphis gossypii, 487, 488, 497, 499, 548, 557, 599, 613, 614, 616
Aphis nerii, 480, 581
Aphis pomi, 491
Aphitis melinus, 75
Aphytis chilensis, 135
Apicomplexa, 640
Apini, 6
    honey production, 253–254, 260
    resources utilization in, 249–250
Apiomerus, 551
Apion sp., 374
Apis cerana, 184
Apis mellifera, 183, 241, 244, 249, 251, 252, 253–255, 259, 262, 669
    rearing in laboratory, 52
Apis mellifera scutellata, 244
Apochrysinae, 594
Apocynaceae, 381
Apoica pallens, 134
Approximate digestibility (AD), 15, 17, 20–23
Apterostigma, 217, 223
Aquifoliaceae, 376
Arachis hypogaea, 202, 309, 311–312
Aradus cinnamomeus, 310
Araneae 9 spp., 553

Araneus diadematus, 584
Araneus ocellata, 584
Araneus quadratus, 584
Arboreal ants
    associated with carbohydrate-rich resources, 224–225
    neotropical ant guilds, 224–226
    pollen-feeding, 225–226
Arboreal predator ants, 220
Arecaceae, 326, 331
Arginine, 68
Argyrotaenia sphaleropa, 79
Argyrotaenia velutinana, 25
Artemia franciscana, 549
Artificial diet
    adult requirements, 74–76
    beginning, 81–82
    biological stimuli, 68
    chemical stimuli, 67–68
    composition, 73–76, 82
    developed and adapted, 79
    evaluation, 82–85
    examples, 79
    failures and advantages, 85
    future, 85–86
    history, 54–59
    insect rearing, 52–54
    liquid, 63–64
    nutrients, 62
    physical stimuli, 67
    as powder, 63
    preparation, 77–80
    of sap-sucking insects, 499
    semi-liquid, 63
    species reared on, 55–58
    terminology, 61–62
    types, 63–64
Artificial food, 4
Artificial media, insect rearing in, 79–81
Ascaris lumbricoides, 409
Ascia monuste, 183, 184–187, 275, 278–280
Ascia monuste orseis, 283
Asclepiadaceae, 480
Asclepias curassavica, 131, 281
Asclepias humictrata, 131
Asclepias syriaca, 131, 281
Ascorbic acid, 69, 74
Asimina spp., 286
Aspartic acid, 69
Asphinctanilloides, 224
Asphondyliini, 383
Aspidiotus nerii, 135
Aspidosperma, 381
Aspidosperma spruceanum, 383
Asteraceae, 373–374, 376–377, 381
Astragalus, 334, 339
Astragalus utahensis, 332–333
Asynarchus nigriculus, 186
Atherigona varia soccata, 665
Atlantochrysa, 608
Atriplex, 377
Atta, 223

Attagenus sp., 429
Attalea maripa, 342
Atteva fabriciella, 666
Attine, 147–148
Attini, 5
Auchenorrhyncha, 597
Aulacorthum solani, 523
Aulacorthum solanum, 505
Axenic diet, 62
Azadirachta indica, 439, 668
Azteca, 225

B

B. Homoptera, artificial diet, 57
Baccharis, 376–378
    galling species on, 378
Baccharis concinna, 380, 383
Baccharis pseudomyriocephala, 377, 380
Baccharopelma dracunculifoliae, 373, 380, 383
Bacillus, 260–261
Bacillus cereus, 198, 203, 260, 261
Bacillus circulans, 251
Bacillus licheniformis, 251
Bacillus megaterium, 198, 251
Bacillus pumilus, 251
Bacillus stearothermofilus, 259
Bacillus subtilis, 251, 259–260
Bacillus thuringiensis (Bt), 99, 136, 182, 197, 430,
    439–441, 505, 670–671, 690
Bacillus thuringiensis israelensis (Bti), 640
Bacterial agents and host plant, pre-ingestion interactions,
    197
Bacterial diseases
    gut environment interactions, 197–198
    host plant effects on, 197–199
    host plant effects on resistance to, 199
Bactrocera, 458–459
Bactrocera cucurbitae, 466
Bactrocera dorsalis, 343, 456, 459, 466
Bactrocera invadens, 459
Bactrocera neohumeralis, 458
Bactrocera oleae, 457, 459
Bactrocera tryoni, 454–456, 458
Baizongia pistaceae, 150, 154
Bandeiraea simplicifolia, 672
Bartonella quintana, 645
Basiceros, 217–218
Bathycoelia thalassina, 302, 305
Battus philenor, 283
Bauhinia, 331
Bauhinia brevipes, 383
Bauhinia purpurea, 668
Baumannia, 150
Beauveria, 585
Beauveria bassiana, 196, 201–203, 441
Bemisia spp., 548, 559
Bemisia tabaci, 201, 581, 599, 614, 616, 664, 666
Beta carotene, 72
Bethylidae, 440, 442
Bidens pilosa, 307, 550, 556, 558

Bifunctional inhibitors, plants resistance to insects,
        676–677
Bignoniaceae, 377
Bioecology and nutrition
    chemical ecology and, 163–172
    insects, 8–9
        antibiotic factors, 669–670
        antixenotic factors, 667–669
        chemical causes, 667–670
        epidermal toughness and thickness, 662–665
        integrated pest management (IPM), 687–700
        lignin, 664–665
        morphological resistance, 661–666
        pests management within context of, 692–699
        phagodeterrents, 667–669
        for phytophagous arthropods, 687–692
        pilosity of epidermis, 666
        repellents, 667
        silicon, 662–664
    and IPM tactics
        allelochemicals, 691–692
        host plant resistance, 689–690
        mixed crops cultivation, 691
        trap crops, 690–691
    phytophagous arthropods, 688–692
        insect–plant interactions, 688
        plant diversity and stability, 688–689
Biotechnology
    and resistance of plants to insects, 670–677
        α-amylase inhibitors, 675–676
        bifunctional inhibitors, 676–677
        enzyme inhibitors, 673–675
        protease inhibitors, 674–675
Biotic variation adoptions, in predatory beetles, 583–584
Biotine, 69
Blatella germanica, 69
    nutrients in, 63
Blatella germanica, 55, 184
Blattabacterium, 153
Blatta orientalis, 184
Blattella, 72
Blattidae, digestive process, 95, 107
Blissus spp., 295
Blochmannia, 5, 153, 155
    symbionts, 150, 155
Bluetongue virus, 641
Boarmia (Ascotis) selenaria, 668
Bombini, 6
    resources utilization in, 248–249
Bombus, 244–245, 248–249, 262
Bombus diversus, 248
Bombus hypnorum, 248
Bombus hypocrita, 248
Bombus ignitus, 248
Bombus perplexus, 248
Bombus rufocinctus, 249
Bombus spp., 238
Bombus terrestris, 248–249
Bombus terricola, 249
Bombycidae, 548
Bombyx mori, 35, 123, 204, 284, 548, 658

    artificial diet, 55
    dry material and energy use by, 26
    feeding needs, 67
    M. persicae, 69
    nutrition, 63, 69, 71
    rearing in laboratory, 52
Boraginaceae, 373
Borrelia recurrentis, 645
Bostrichidae, 420, 426
Bothynus spp., 354, 355
Bovine ephemeral fever, 641
Brachycera, 451, 642
Brachymyrmex, 222, 225
Bracon hebetor, 441
    artificial diet, 59
Braconidae, 440–41, 524, 552, 584
Brassica campestris, 695
Brassicaceae, 282, 355, 374, 479
Brassica juncea, 278
Brassica oleracea, 25–26, 665
Brassica rapa, 439
    ssp. *Pekinensis,* 203
Brevicoryne brassicae, 479, 503
Brevicoryne brassicae, 479, 671
Brontocoris, 7
Brontocoris tabidus, 546, 549, 555, 560–561
Bruchidius ater, 345
Bruchidius atrolineatus, 343
Bruchidius dorsalis, 343
Bruchidius sahlbergi, 333, 335, 338
Bruchidius uberatus, 333
Bruchidius villosus, 336, 340
Bruchinae, 326, 329–331, 334, 340, 344, 371, 422, 424.
        See also Seed-chewing beetles
    host plant families, 326
Bruchine, life cycle, 330
Bruchini, 327
Bruchophagus, 344
Bruchus, 327, 331
Bruchus brachialis, 345
Bruchus pisorum, 326, 340, 342–343, 345, 676
Bruchus rufimanus, 343, 345
Bubonic plague, 7
Buchnera, 5, 63, 73, 86, 110, 153, 487, 497–498
    symbionts, 150, 153–154
Buchnera aphidicola, 153–154, 497, 527
Bud galls, 385, 387
Bufo marinus, 227
Buprestidae, 404

C

C. clitelae, 388
Cadra, 421
Cadra cautella, 421, 425–426, 433
Caesalpinea peltophoroides, 574
Caesalpinioideae, 331
Caiman crocodilus, 643
Cajanus cajan, 300, 303, 309, 695, 697
Caliothrips phaseoli, 548
Calliandra brevipes, 382

Calliphora vomitoria, artificial diet, 55
Calliphoridae, 643–644
Callosobruchus, 327, 343
Callosobruchus analis, 335
Callosobruchus chinensis, 326, 333, 340, 342, 676
Callosobruchus maculatus, 326, 333, 335, 338, 341–344, 346, 671, 672, 674, 676
Callosobruchus spp., 333, 440
Callosobruchus subinnotatus, 344
Calluna vulgaris, 199
Calocalpe undulate, 281
Calochrysa, 617
Calocoris angustatus, 298
Calorimetric method, to measure food intake and utilization, 25–28
Campanulaceae, 375
Camponotus, 155, 225
Camponotus rufipes, 112
Camponotus spp., 150
Canavalia ensiformis, 668
Candida albicans, 260
Candida spp., 259, 608
Cannibalism
    conditions for, 179–184
    ecological significance, 186–187
    food
        availability and quality, 179–181
        impact, 184–187
    genetic bases, 179
    individual performance, 184–187
        benefits, 184–185
        costs and related strategies, 185–187
    in insects, 177–189
    population
        density, 182–183
        dynamics, 186–189
    in predatory beetles, 581–582
    in predatory Heteroptera, 551–552
    sexual, 184
    victim availability, 183–184
Canthon angustatus cyanellus, 406
Caprifoliaceae, 375
Carbohydrate
    arboreal ants associated with, 224–225
    for insects growth, 70–71, 75–76, 82
    rich resources, 224–225
Cardiochiles nigriceps, 132, 170
Cardiospermum corindum, 298
Cardiospermum halicacabum, 303, 304
Carebara, 222
Carnitine (vitamin Bt), 69
Carnivores insects, 7, 64–65
Carolina geranium, 199
Carpophilus, 420
Carsonella, 153
Carsonella sp., 150
Carthamus tinctorius, 558
Caryedes brasiliensis, 130, 328–329, 334, 340, 661
Caryedon, 327
Caryedon albonotatum, 338
Caryedon gonagra, 345

Caryedon gonagra, 340
Caryedon interstinctus, 340
Caryedon palaestinicus, 340
Caryedon serratus, 326
Cassia baubinioides, 338, 340
Cassia grandis, 335
Cassia leptadenia, 340
Cassia leptophylla, 327
Cassiinae, 331
Cavalerius saccharivorus, 314
Ce. cubana, 613, 619
Cecidomyiidae, 375–377, 381, 383, 553, 582
Cecropia, 225
Cedecea sp., 198
Centromyrmex, 218
Cephalonomia tarsalis, 440
Cephalonomia waterston, 440–441
Cephalotes, 225
Cephalotrigona, 250
Ceraeochrysa, 601, 619
Ceraeochrysa cincta, 619
Ceraeochrysa cubana, 597
Ceraeochrysa spp., 620
Cerambycidae, 325, 404
Cerapachys, 218
Ceratitis, 452, 464
Ceratitis capitata, 79, 184–185, 226, 453–464, 466
Ceratitis cosyra, 457, 459
Ceratopogonidae, 641–642
Cercidium floridum, 330–331, 335–336
Cerconota anonella, 79
Cercopidae, 597
Cetoniinae, 354, 360
Chaetopsila elegans, 346
Chagas disease, 7
Chalcidoidea, 375, 377
Charidryas harrissi, 283
Chauliops, 310
Cheliomyrmex, 224
Chemical ecology, and food, 5
Chenopodium album, 583
Chenopodium ambrosioides, 667
Chilocorinae, 572, 578
Chilocorini, 573, 575, 584
Chilocorus kuwanae, 579
Chilo infuscatelus, 663
Chilo partellus, 282
Chilo supressalis, 663
    artificial diet, 55
Chinavia, 305
Chlorochroa ligata, 151
Chlorochroa sayi, 151
Chlorochroa uhleri, 151
Chloropidae, 376
Chlosyne lacinia, 286
Cholesterol, 71, 73
Choline, 69, 74
Choristoneura fumiferana, 128
Chrysanthemum spp., 558
Chryseida bennetti, 346
Chrysolina quadrigemina, 128

Chrysomela knaki, 24
Chrysomelidae, 325, 422, 424. *See also* Seed-chewing
    beetles
Chrysomeloidea, 374
Chrysopa, 605, 608, 617
Chrysopa carnea, nutrients in, 63
Chrysopa coloradensis, 616
Chrysopa formosa, 610, 611
Chrysopa nigricornis, 610, 612, 617
Chrysopa oculata, 605, 611–612, 616–617
Chrysopa pallens, 610–611
Chrysopa perla, 598, 606, 608, 613, 617
Chrysopa phyllocroma, 611
Chrysopa quadripunctata, 605, 616–618
Chrysopa rufilabris, 600, 607–609, 613, 617
Chrysopa slossonae, 605, 617–618
Chrysopa sp., 610
Chrysopa spp., 598, 608
Chrysoperla, 598, 601, 604, 617, 619–620
Chrysoperla carnea, 179, 184, 186, 196, 582, 595–596,
    609
Chrysoperla carnea s. lat, 598–601, 603–608, 610–613,
    615–619, 621
Chrysoperla comanche, 608
Chrysoperla externa, 97, 594–595, 597, 599, 613, 615–616,
    618–620
    consumption of prey by, 599
    developmental time and survival, 614
    reproductive traits, 616
Chrysoperla lanata, 616
Chrysoperla nipponensis, 612
Chrysoperla plorabunda, 598
Chrysopids, 539
    adults
        feeding behavior, 609–612
        nonpredaceous, 597–598, 607–608
        nutritional requirements, 600–601
        predaceous, 608
    anatomy, 606–607
        nonpredaceous adults, 607–608
        predaceous adults, 608
    biological control, 618–619
    biosystematics, 619
    diets, 596–601
        artificial, 599–601
        cannibalism, 598
        consumption and conversion of food, 598–599
        natural, 597–598
        nonpredaceous adults, 597–598, 607
        omnivory, 598
        prey based, 607
    digestive system
        adults, 606–609
        feeding behavior and, 609–612
        glyco-pollenophagous and prey-based diets
            associated, 607–608
        internal anatomy, 606–607
        larvae, 601–605
        nonpredaceous adults, 607–608
        predaceous adults, 608
        symbiotic yeast associated, 608–609
    feeding behavior
        adults, 609–612
        habitat and food finding, 612
        long distance attraction, 611
        plant volatiles attraction, 610–611
        post-emergence movement, 610
        prey-associated volatiles, 611
        short distance attraction, 612
    food and
        artificial diets, 596–599
        development and survival, 613
        effects on performance, 612–617
        interactions with host plant of prey, 613–615
        reproduction, 615–617
    future research recommendations, 618–620
    implications for rearing, 619–620
    larvae
        cleaning and resting after feeding, 606
        digestive system, 601–605
        nutritional requirements, 600
        predatory behavior, 604–606
        prey consumption and food conversion, 598–599
    nutritional requirements
        adults, 600
        larvae, 600
    physiology, 601–609
    prey
        capture, 605
        consumption and conversion of food, 598–599, 605
        search contact and recognition, 604–605
        specificity and stability studies, 617–618
    seasonality, 620–621
    taxonomy, 595–596
    voucher specimens, 596
Chrysopinae, 594
Chrysopodes, 601, 619–620
Chrysopodes lineafrons, 620
Chrysops celatus, 643
Chrysopsinae, 642
Chrysops viduatus, 643
Chrysothamnus, 377
Chrysthamnus nauseosus consimilis, 379
Chrysthamnus nauseosus hololeucus, 379
Ci. lectularius, 645
Cicadellidae, 577, 597
Cicer arietinum, 333, 340
Cimex hemipterus, 645
Cimex lectularius, 645
Cimicidae, 645
Cimicoidea, 645
Cimicomorpha, 542, 544–546, 645
Cinara atlantica, 478, 502, 579, 583, 599
Cinara pinivora, 579, 583, 614
Cisseps fulvicollis, 167
Citrus spp., 668
Clavigralla tomentosicollis, 305
Cleptotrigona spp., 238
Cletus punctiger, 314
Coccidae, 576, 596
Coccidophilus citricola, 573, 575
Coccidulinae, 572, 578

Coccinelidae, 7, 554
Coccinella septempunctata, 186, 575, 579, 580, 581, 583–584, 585
Coccinella septempunctata bruckii, 573, 581
Coccinella serratus palaestinicus, 340
Coccinella supressalis, 70, 675
Coccinella transversoguttata, 579
Coccinella undecimpunctata, 581
Coccinellidae, 539. *See also* Predatory beetles
    food used by groups of, 578
Coccinellinae, 572, 578, 584
Coccinellini, 573, 584
Coccoidea, 374, 377, 634
Cochliomyia hominivorax
    nutrients in, 63
    rearing, 61
Colaptes campestre, 388
Coleomegilla maculata, 182, 579, 582
Coleomegilla quadrifasciata, 583
Coleoptera, 4, 6–7, 273, 325, 353, 356, 404, 418–419, 422, 429, 520, 523, 548, 554, 580. *See also* Rhizophagous beetles; Seed-chewing beetles
    artificial diet, 52–53, 56
    digestive process, 95, 111
    galls, 373–374
    mouthparts, 66
Coleoptera–Bruchidae, 325
Collabismus clitellae, 374
Collembola, mouthparts, 66
Colobopsis, 155
Colorimetric method, to measure food intake and utilization, 23–24, 28
Conotrachelus fissinguis, 341
Convolvulacea, 326, 331, 334
Copaifera, 377
Copaifera langsdorffii, 384–385
Coprophagy, role in detritus use, 406–407
Coquillettidia, 636–637
Corcyra, 421
Corcyra cephalonica, 421, 429–430, 433
Coreidae, 295
Corimelaena extensa, 302
Corimelaenidae, 295
Corymbia citriodora, 667
Corynebacterium sp., 198
Cotesia congregata, 22
Cotesia congregatus, 132
Cotesia flavipes, 527, 529
    rearing, 61, 76
Cotesia kariyai, 170
Cotesia marginiventris, 170
Covering galls, 385, 387
Cracca virginiana, 334
Creightonidris, 218
Crematogaster, 225
Crotalaria, 696
Crotalaria juncea, 359
Crotalaria lanceolata, 311–312, 695
Crotalaria spectabilis, 359
Cryptoblabes gnidiella, 79
Cryptocercidae, 405

Cryptocercus punctulatus, 107
Cryptolaemus mountrouzieri, 576
Cryptolaemus spp., 573
Cryptolestes ferrigineus, 421, 425–427, 433–434, 438, 440–442
Cryptolestes pusillus, 421
Cryptomyrmex, 218
Cryptosporidium parvum, 409
Crytochaetum iceryae, 134
Ctenocephalides felis felis, 644
Ctenocolum, 331
Ctenocolum podagricus, 344, 345
Ctenocolum tuberculatum, 328–329
Cucumis sativus, 571
Cucurbita andreana, 202
Cucurbitaceae, 7, 577
Cucurbita maxima, 202
Cucurbita pepo, 571
Culex, 636
Culex pipiens, 70
Culex pipiens molestus, 639
Culex quinquefasciatus, 636–638
Culicidae, 635–640, 644
Culicinae, 635
Culicoides, 641–642
Culicoides molestus, 642
Culicoides variipennis, 642
Culicomorpha, 635
Curculionidae, 325, 374, 389, 419–421, 424
Curculionidea, 374
Cyanocobalamin (vitamin B_{12}), 69
Cyclocephala, 361
Cyclocephala flavipennis, 354
Cyclocephala spp., 354
Cycloneda sanguinea, 575, 579, 583, 585
Cydia molesta, 166
Cydia pomonella, 135, 530
Cylindromiya euchenor, 553
Cylindromyrmex, 218
Cymbopogon spp., 667
Cynaeus angustus, 425
Cynipidae, 375, 377, 381, 384
Cyperaceae, 376
Cyphoderris sp., 184
Cyphomyrmex, 217, 223
Cysteine, 69
Cytisus scoparius, 336, 340, 345, 504

D

Dacetini predators, 218–220
Daceton armigerum, 220
Daceton boltoni, 220
Dactylurina staudingeri, 261
Daktulosphaira vitifoliae, 484
Danaus erippus, 281
Danaus plexippus, 131, 280, 582
Daphne laureola, 280
Datura wrightii, 285
De. hominis, 644
Defoliator caterpillars. *See also* Lepidoptera

acceptability, 278–280
feeding
   and digestion, 276
   on nonvegetal sources, 283–284
   periods, 283–285
food
   perception, 276–277
   utilization and selection, 282
leaf interaction, 278–285
   acceptability, 278–280
   competition and food deprivation, 281–282
   dispersal, 282–283
   feeding on nonvegetal sources, 283–284
   feeding periods, 283–285
   food utilization and selection, 282
   leaf characteristics impact, 280–281
   nutrients and allelochemicals impact, 280
   performance and preference, 278–280
   physical structure and microfauna impact, 280–281
morphology and biology, 274–277
natural enemies, 285–287
Delphastus pusillus, 581
Demodema brevitarsis, 354
Demodema spp., 354
Dendroctonus frontalis, 169
Dendroctonus micans, 665
Dendroctonus monticolae, 658
Dengue, 7
Dentroctonus frontalis, 147
Deois flavopicta, 25
Depressaria pastinacella, 128
Dermaptera, artificial diet, 56
Dermatobia hominis, 644
Dermestes, 69
Dermestes maculatus, 428
Dermestes sp., 428–429
Dermestidae, 419, 429
Desmodium tortuosum, 312
Detritivores insects, 65. *See also* Detritus
   ecological functions, 407–410
      biological control, 409–410
      leaf litter decomposition rates, 407–408
      waste removal and related functions, 408–409
   external and internal Rumen in, 407
Detritus. *See also* Detritivores insects
   adaptations as food, 403–406
      to access nutrients in low availability, 403–405
      for high availability in space and time, 405–406
   based food webs, 398
   coprophagy role in use of, 406–407
   as food resource, 398–403
      abundance, 398–400
      allocation, 402
      consequences of, 402–403
      distribution, 400–401
      population and community consequences,
         402–403
      use, 401–402
Deuteromycota, 441
Diabrotica, 669
Diabrotica longicornis, 132

Diabrotica undecimpunctata howardi, 202
Diabrotica virgifera virgifera, 201, 671
Diachasmimorpha longicaudata, 465, 530
Diactor bilineatus, 308
Diadegma, 531
Diaeretiella rapae, 132
Diaphorina citri, 167
Diaspididae, 596
Diaspis echinocacti, 573
Diatraea saccharalis, 37, 75, 599, 614
   artificial diet, 52
      preparation of, 77–79
   diet, 74
   nutritional indices, 15–16, 21, 25, 29, 37, 40
   rearing, 76
Diatraea saccharalis, 527, 529, 663
Diatrea grandiosella, 179
Dichelops melacanthus, 152, 299–300, 309–310, 312,
      698–699
Dichelops melancathus, 696
Dicrocerus furcatus, 699
Dictyoptera
   artificial diet, 56
   mouthparts, 66
Dicyphinae, 545
Dicyphus, 554
Dicyphus errans, 545
Dicyphus hesperus, 559–560
Dicyphus tamaninii, 545, 559
Dieffenbachia sequina, 672
Dietary stress and starvation, effects on entomopathogenic
      diseases, 196–197
Digestion
   of carbohydrates, 100–101
   of lipids and phosphates, 101–102
   microorganisms role, 104–105
   midgut conditions affecting enzyme activity, 105
   of proteins, 98–100
Digitonthophagus gazella, 409
Diloboderus abderus, 354, 357–359, 361–362
Diloboderus sp., 354
Dinarmus basalis, 345–346
Dinarmus colemani, 346
Dinarmus laticollis, 346
Dinarmus vagabundus, 346
Dinocampus coccinellae, 584
Dinoponera, 216
Dioclea, 334
Dioclea megacarpa, 130, 334, 340
Dione juno juno, 276
Dione moneta moneta, 274–275
Diospyros hispida, 374
Diplorhoptrum, 222
Dipsocoromorpha, 542
Diptera, 4, 6–7, 353, 404, 519–520, 523, 548, 553, 580,
      582, 633–644. *See also* Fruit flies
   artificial diet, 52–53, 56–57, 62
   digestive process, 95–96, 98, 112–113
   galls, 373, 376
   mouthparts, 66
   nutritional value, 54

Dirofilaria, 640
Dirofilaria immitis, 636
Discoid galls, 385–386
Discothyrea, 218
Disteniidae, 325
Ditrysia, 273
Diuraphis noxia, 489, 500, 502, 505
Dolichoderines, 222
Dolichoderus, 225
Dorcus rectus, 184
Dorymyrmex, 222
Drepanosiphinae, 477
Drepanosiphinae *Eucallipterus tilia,* 490
Drepanosiphum acerinum, 494
Drepanosiphum platanoides, 481, 494
Drilling insects, 371
Drosophila
    nutrients, 63
    rearing in laboratory, 52
Drosophila ampelophila, artificial diet, 55–56
Drosophila melanogaster, 71, 97, 203
Drosophila simulans, 203
Dynastinae, 354, 360–361
Dysaphis devecta, 491
Dyscinetus dubius, 355
Dyscinetus gagates, 355
Dyscinetus spp., 354
Dysdercus bimaculatus, 299
Dysdercus cingulatus, 672
Dysdercus koenigii, 302
Dysdercus maurus, 302, 308
Dysdercus peruvianus, 102, 110
Dysdercus spp., 295
Dysmicoccus brevipes, 614
Dysmicoccus cryptus, artificial diet, 62

E

Eacles imperialis magnifica, nutritional indices, 29
Eciton, 224
Ectatomma, 216, 225
Ectatomma tuberculatum, 220
Edessa meditabunda, 298, 305, 310, 696
Efficiency of convertion of digested food (ECD), 15, 17, 19–21
Efficiency of convertion of ingested food (ECI), 15, 17, 19–20
Elaeis quineensis, 168
Elasmolomus sordidus, 309
Elasmopalpus lignosellus, 74, 79
Elasmucha grisea, 308
Elasmucha putoni, 308
Eldana saccharina, 136, 663, 664
Elliptical galls, 386
Emblemasoma auditrix, 180–181
Encarsia formosa, 553
Encyrtidae, 584
Endolimax nana, 409
Endopiza viteanea, 135
Endosymbionts, 150
Engytatus nicotianae, 559

Enicocephalomorpha, 542
Entamoeba coli, 409
Enterobacter sp., 198
Enterococcus, 261
Enterococcus casseliflavus, 151
Enterococcus faecalis, 303
Enterolobium cyclocarpum, 334
Entomopathogenic diseases
    nutritional implications in insect mass rearing, 203–204
    starvation and dietary stress effects on, 196–197
    symbionts impact on, 203
Entomopathogens, host plant pathogens interactions, 203
Entomophthera spp., 553, 558
Enzyme
    inhibitors and plants resistance, 673–675
    midgut conditions and activity of, 105
Eocanthecona furcellata, 546
Ephemeroptera, mouthparts, 66
Ephestia, 421, 428
Ephestia elutella, 421
Ephestia kuehniella, 419, 549
Ephestia spp., 422
Epigaeic generalist predator ants, 216–217
Epilachna borealis, 130, 577
Epilachna cacica, 571
Epilachna paenulata, 571, 577
Epilachna spp., 577
Epilachna spreta, 571
Epilachna tredecimnotata, 39
Epilachna varivestis, 577, 669
Epilachna vigintioctopunctata, 575
Epilachninae, 572–573
Eretmocerus eremicus, 553
Ergosterol, 71
Ericaceae, 374
Erinnyis ello ello, nutritional indices, 29
Erinnyis ello, 276
Eriococcidae, 384, 579, 597
Eriophyidae, 596
Eriopis connexa, 575, 583
Erythrina, 334
Escherichia coli, 259–261, 644
Escovopsis sp., 147
Eubaptini, 327
Eucallipterus salignus, 491
Eucallipterus tiliae, 492, 494
Eucalyptus, 377
Eucalyptus cloeziana, 558
Eucalyptus spp., 439, 561
Euetheola humilis, 355, 361
Eugenia uniflora, 380
Eulophidae, 375, 584
Eunectes murinus, 643
Eupalea reinhardti, 574–575, 577
Eupelmidae, 344, 346
Eupelmus cushmani, 346
Eupelmus cyaniceps, 346
Eupelmus orientalis, 345, 346
Eupelmus vuilleti, 345
Euphaleurus ostreoides, 373

Euphorbia heterophylla, 695–696
Euphorbia spp., 126
Euplectrus sp., 525
Eurosta, 388
Eurosta solidaginis, 381, 383, 388
Eurygaster, 305
Eurytoma, 344
Eurytoma gigantea, 388
Eurytomidae, 344, 346, 375
Eurytominae, 375
Euschistus, 310
Euschistus conspersus, 302, 308, 312
Euschistus heros, 151, 152, 299–300, 303, 309, 312,
        694–699
Euschistus variolorius, artificial diet, 55
Eutheola spp., 354
Euthera tentatrix, 553
Eutrichopodopsis nitens, 307
Evonymus europaeus, 583
Exochomus quadripustulatus, 584
Exoplectra miniata, 579–580
Extrafloral nectar, and natural enemy attraction, 171

F

Fabaceae, 326, 331, 371, 373–377, 384–385, 485
Facultative symbionts, 149–150
Fagaceae, 374–375
Fanniidae, 644
Feeding
    habits
        carnivorous insects, 7
        and damage caused by stored-product pests,
            420–424
        hematophagous insects, 7–8
        insects, 6–8
        phytophagous insects, 4, 6–7
        social insects, 6
    of sap-sucking insects
        and electrical penetration graph, 499–502
        and nutrition, 498–499
        rate, 488–492
Festuca arundinacea, 281
Filz galls, 387
Flavobacterium, 497
Flavobacterium sp., 198
Fold galls, 385, 387
Folic acid, 69
Folsomia candida, 217
Food
    adaptations of detritus, 403–406
        to access nutrients in low availability, 403–405
        for unpredictable high availability, 405–406
    artificial, 4
    baits and monitoring, 436
    chemical ecology and, 5
    constituents and digestive enzymes, 484–487
    consumption, 4–5
    detritus as resource, 398–403
        abundance, 398–400
        allocation, 402
        consequences of, 402–403
        distribution, 400–401
        population and community consequences, 402–403
        use, 401–402
    digestion, 4–5
    for fruit flies, 452–453
    handling and ingestion, 102
    natural, 4
    in predatory beetles
        preferences, 580–581
        quality, 579–580
        selection, 576–581
        specificity, 577–579
        toxicity, 581
    sap-sucking insects, 484–487
        intake mechanisms and saliva composition in,
            482–484
    seed-sucking bugs, 296–298
        abundance, 298
        allelochemicals, 297
        nutritional composition, 296–297
        physical and structural aspects, 297–298
    stored-product pests
        attractants and gustatory stimuli, 427
        digestive enzymes, 430–432
        microorganisms, 434
        nutrient budget and relative growth rate, 432–433
        nutritional requirements, 427–430
        and nutrition characteristics, 425–426
        oviposition stimuli, 426–427
        physiological and behavioral adaptations,
            434–435
        search and utilization, 426–434
    utilization, 4–5
Food intake and utilization
    for growth in larval phase, 21, 30, 33–35
    meaning, 19–23
        AD, 15, 17, 20–23
        ECD, 15, 17, 19–21
        ECI, 15, 17, 19–20
        RCR, 15–16, 19–20
        RGR, 15–16, 19–20
        RMR, 15–16, 19–20
    measure, 23–29
        calorimetric method, 23–28
        direct method, 23, 28
        gravimetric method, 23, 28
        immunological method, 25
        indirect method, 23–29
        isotope method, 24, 28
        trace element method, 25
        uric acid method, 24–25
    nutritional indices, 16–23, 30–33
        experimental techniques, 17
        feces, 19
        food consumed quantity, 17–18, 36–38
        weight gains, 18–19
    for reproduction and dispersal, 36–41
Forcipomyia, 641
Forcipomyia townsvillensis, 642
Formica polyctena, 180

Formicines, 222
Frankliniella occidentalis, 201, 544, 548, 674
Frankliniella schultzei, 664
Friesella, 243
Frieseomelitta, 250, 251
Frieseomelitta varia, 261
Frugivorous insects, 7
Fruit flies
    allelochemicals and, 459
    applicability, 464–466
    behavior, 461–464
    biotic and abiotic factors, 458
    feeding, 459–461
    foodstuffs, 452–453
    nutritional needs, 453–458
        carbohydrates, 456–457
        lipids, 457
        proteins, 454–456
        symbionts, 458
        vitamins and mineral salts, 457–458
    taxonomy, 451–452
Fulgoridae, 597
Fungus grower ant, 223–224
Fungus-growing insects, 146–148
    ambrosia beetles, 147
    ant subfamily myrmicinae, 147–148
    termites subfamily, 148
Fungus-induced gall, 370

G

Galanthus nivalis, 505, 668
Galleria mellonella, 70–71, 113, 202, 548
Galls
    acari-induced, 370
    adaptive significance, 387–389
    aleppo galls, 370
    anatomy and physiology, 381–383
    classification, 385–387
    development, 383–385
    fungus-induced, 370
    herbivore insect guilds, 370–372
    host plant, 386
        location and choice, 377–379
        taxa, 377
    inducing insect, 372–376
        Coleoptera, 371–374
        Diptera, 371–373, 376
        Hemiptera, 371–374
        Hymenoptera, 371–373, 375
        Lepidoptera, 371–373, 375–376
        Orthoptera, 371–372
        Phasmatodea, 371
        Thysanoptera, 371–374
    insect-induced, 370–372
    makers, 6
    morphology, 379–381
    nematoid-induced, 370
    species of genus *Baccharis,* 378
    types, 385, 387
Gastrimargus transverses, 184

Gelechiidae, 375, 419, 549, 599, 614, 616
Genista, 581
Geocoridae, 559
Geocoris, 7, 540
Geocoris atricolor, 547
Geocoris pallens, 547, 555, 557
Geocoris punctipes, 543, 547–548, 553–557, 559–560
Geocoris spp., 547, 551, 553
Geocoris uliginosus, 557, 559
Geologic time, plants and insects across, 122–123
Geranium caroliniarum, 199
Gerridae, 551
Gerromorpha, 542
Giardia lamblia, 409
Gibbobruchus, 331
Gleditsia japonica, 343
Glossata, 273
Glossina spp., 149–150, 154
Glutamic acid, 69
Glycine max, 199, 311–312, 558
Glycine spp., 297
Gnamptogenys, 217, 218
Gnamptogenys concinna, 220
Gnamptogenys striatula, 217
Gnathocerus cornutus, 422
Gordini sp., 151
Gossypium thurberi, 127
Grapholita molesta, 135
    artificial diet, 55
Gravimetric method, to measure food intake and
    utilization, 23, 28
Green lacewings. *See* Chrysopids
Griffonia simplicifolia (GSII), 672
Grossulariaceae, 375
Guapira opposita, 385
Gut environment
    bacterial diseases interactions, 197–198
    viral diseases interactions, 199–200
Gymnandrosoma aurantianum, 74, 79

H

Habrochus sequester, 343
Haemagogus, 636
Haemagogus janthinomys, 636
Haemagogus leucocelaenus, 636
Haematobia irritans exigua, 409
Haematobia irritans, 409, 644
Haematobia thirouxi potans, 409
Haemoproteus, 641
Hairy galls, 385–386
Halyzia, 577
Hamiltonella defensa, 497, 504
Hapithus agitator, 184
Harmonia axyridis, 179–180, 186, 554, 574–575, 579,
    582–583, 585
Helianthus annuus, 558
Helicoverpa armigera, 284
Helicoverpa zea, 23–24, 40, 63, 75, 79, 128, 168, 182, 197,
    199, 282, 548, 666, 668, 670
Heliothis armigera, 199

Heliothis subflexa, 62, 132
Heliothis virescens, 170–171, 198, 282, 525–526, 528
    artificial diet, 55, 82–83
    diet, 74
    nutritional indices, 15–16, 23, 25, 29, 34, 40
Heliothis virescens, 168, 197, 556
Heliozela staneella, 376
Hematophagous insects, 7–8
Hemiptera, 6–7, 353, 418, 420, 474, 548, 554, 577, 579,
        597, 633, 645–647
    artificial diet, 52, 57
    digestive process, 96, 104, 108–111
    galls, 372–374
    rearing on natural hosts, 59
Hemyda aurata, 553
Hepatocystis, 641
Herbivore insects, 64
    guilds and galls, 370–372
Herbst, 419
Herpetogramma phaeopteralis, 664
Hesperioidea, 273
Heterobathmiina, 273
Heterocampa obliqua, 274
Heterogomphus spp., 354
Heteroneura, 273
Heteroponera, 217
Heteroponera dentinodis, 217
Heteroponera dolo, 217
Heteroptera, 633, 645–647. *See also* Predatory
        heteroptera
    artificial diet, 57
    mouthparts, 66
Heteropterans, 353. *See* also Seed-sucking bugs
    abiotic factors impact, 312–313
    adaptations and responses to changes, 313–314
    adults
        dispersal, 304–306
        food switch, 310–312
        suitable food, 309–310
    biology, 298–308
        adults dispersal, 304–306
        feeding, 298–301
        host plant choice, 304–306
        ingestion, 298–301
        mating, 302
        natural enemies and defense, 307–308
        nymph development, 303–304
        oviposition, 302–303
    biotic factors impact, 308–312
    foods
        leaves, branches and trunks, 310
        nymph-to-adult switch, 310–312
        seeds and fruits, 308–310
    managing
        crop residue management, 698–699
        feeding process interference, 694
        on host plant, 695–699
        host plant resistance, 692–693
        host plant sequences, 695
        less preferred plants as source, 697–698
        mixed crops use, 693–694
        monitoring bugs and crop colonization, 699
        in overwintering sites, 698–699
        preferred plants as traps, 696–697
        on soybean, 692–694
        specific feeding habits, 696
        trap crops use, 693, 696–697
    nymph
        adults dispersal, 304–306
        development, 303–304
        to-adult food switch, 310–312
    performance, 308–312
        abiotic factors impact, 312–313
        biotic factors impact, 308–312
        humidity, 312–313
        less suitable foods, 310
        rain and wind, 313
        suitable foods, 308–309
        temperature and light, 312
Heterorhabditis bacteriophora, 202, 585
Heterorhabditis indica, 204
Heterorhabditis megidis, 201
Heterorhabditis sp., 202
Heterospilus, 344
Heterospilus prosopidis, 344–346
Hibiscus moscheutus, 329, 341
Hierodula membranacea, 184
Hippodamia convergens, 196, 575, 579–581, 583–584
Hippodamia tredecimpuncta, 579
Hippodamiinae, 584
Hippodamini, 573, 584
Hirsutella thompsonii, 201
Histidine, 68
Hofmannophila pseudospretella, 113
Holidic diet, 62
Holometabolous insects, 520
Homalodisca coagulate, 150
Homalotylus, 584
Homoeosoma electellum, 197
Homoptera, mouthparts, 66
Homopterans, 353
Honey
    antibacterial activities, 258–259
    in Apini, 253–254, 260
    in Meliponini, 254–258, 260
    microorganisms, 259–262
    microscopy, 252–253
    physicochemical characteristics, 254–258
    production, 252–263
    stingless bee, 256–259, 261–262
Hoodia gordonii, 277
Hordeum vulgare, 500
Horismenus missouriensis, 344–345
Horismenus sp., 344, 346
Host plant
    bioecology and insect nutrition, 689–690
    compounds and disease interactions, 202
    effects on
        bacterial diseases, 197–199
        diseases caused by nematodes, 201–202
        mycoses, 200–201
        resistance to bacterial diseases, 199

viral diseases, 199–200
families of bruchinae, 326
location and choice of galls, 377–379
managing heteropterans, 695–698
crop residues management, 698–699
feeding process interference, 694
host plant sequences, 695
less preferred plant food source, 697–698
mixed crops use, 693–694
monitoring bugs and crop colonization, 699
in overwintering sites, 698–699
preferred plants traps, 696–697
soybean, 692–693
with specific feeding habits, 696
pathogens and entomopathogens interactions, 203
preingestion interactions between
bacterial agents, 197
viral agents, 199
resistance to
heteropterans on soybean, 692–693
IPM tactics, 689–690
sap-sucking insects, 480–481, 498–499
location and acceptance by, 478–479
search
olfactory stimuli role, 164–165
process in insects, 164–165
sequences in managing heteropterans, 695
taxa and galls, 377
Howardula sp., 585
Hyalophora cecropia, 72
Hydrangea hortensis, 577
Hylomyrma, 217
Hymenoptera, 6–7, 273, 344, 353, 418, 420, 440,
580. *See also* Neotropical ant guilds;
Parasitoids
artificial diet, 52, 57
digestive process, 95, 104, 111–112
galls, 373, 375
mouthparts, 66
nutritional value, 54
Hyperaspinae, 573
Hyperaspini, 575
Hyperaspis delicata, 577, 579
Hyperaspis vicinguerrae, 579
Hypericum perforatum, 128
Hypochrysa elegans, 608, 610
Hypogaeic foragers, 222
Hypogaeic generalist predator ants, 217
Hypoponera, 217
Hyposoter exigua, 132
Hypothenemus hampei, 227
Hyssopus pallidus, 530

I

Icerya purchasi, 134, 572, 581
Ichneumonidae, 440, 523–524
Idiobionts parasitoids, 516, 518–519
Ilex aquifolium, 281
Illinoia liriodendra, 491
Illinoia liriodendri, 490

Immature insects, mouthparts, 66
Immunological method, to measure food intake and
utilization, 25
Impatiens wallerana, 201
Indigofera, 696
Indigofera endecaphylla, 309
Indigofera hirsuta, 695
Indigofera suffruticosa, 695
Indigofera truxillensis, 309, 695
Inga edulis, 580
Inositol, 74
Insects
alellochemicals, 659–660
artificial diet, 4, 52–53
adult requirements, 74–76
beginning, 81–82
biological stimuli, 68
chemical stimuli, 67–68
composition, 73–76, 82
developed and adapted, 79
evaluation, 82–85
examples, 79
failures and advantages, 85
future, 85–86
history, 54–59
liquid, 63–64
nutrients, 62
physical stimuli, 67
as powder, 63
preparation, 77–79
room for preparation, 79–80
semi-liquid, 63
species reared on, 55–58
terminology, 61–62
types, 63–64
bioecology and nutrition, 8–9
antibiotic factors, 669–670
antixenotic factors, 667–669
chemical causes, 667–670
epicuticular waxes, 665–666
epidermal toughness and thickness, 662–665
integrated pest management (IPM), 687–700
lignin, 664–665
management within context of, 692–699
morphological resistance, 661–666
phagodeterrents, 667–669
for phytophagous arthropods, 687–692
pilosity of epidermis, 666
repellents, 667
silicon, 662–664
biotechnology and resistance of plants, 670–677
α-amylase inhibitors, 675–676
bifunctional inhibitors, 676–677
enzyme inhibitors, 673–675
lectins, 671–673
protease inhibitors, 674–675
cannibalism in, 177–189
carnivorous, 7
cellulose digestion, 404–405
coprophagy in detritus role in mutualisms, 406–407
defoliators. *See* Lepidoptera

development in geologic time, 122–123
diet. *See also* Artificial diet
    adult requirements, 74–76
    composition, 73–76, 82
dietetics, 8
digestion
    basic plans, 106–113
    Blattidae, 95, 107
    of carbohydrates, 100–101
    Coleoptera, 95, 111
    Diptera, 95–96, 98, 112–113
    gut morphology and function, 94–98
    Hemiptera, 96, 104, 108–111
    Hymenoptera, 95, 104, 111–112
    Isoptera, 95, 108
    Lepidoptera, 95–96, 98, 113
    of lipids and phosphates, 101–102
    microorganisms role, 104–105
    midgut conditions affecting enzyme activity,
        105
    Orthoptera, 95, 108
    overview, 102–104
    physiology of, 93–115
    of proteins, 98–100
digestive enzymes, 98–102
    midgut conditions and, 105
    secretion mechanisms, 113–114
digestive system, 106–107
drilling, 371
feeding
    habits, 6–8, 64–65
    needs, 67–68
food
    consumed quantity, 17–18, 36–38
    consumption and digestion, 4–5
    handling and ingestion, 102
    intake and utilization, 21, 30, 33–41
frugivorous, 7
fungus-growing, 146–148
growth
    ecdysis cost, 35
    food intake and utilization, 21, 30, 33–35
    instars, 21, 35–36
    larval phase, 21, 30, 33–36
    nutritional needs, 68–73
gut
    conditions affecting enzyme activity, 105
    morphology and function, 94–98
hematophagous, 7–8
host search in, 164–165
host selection stages, 658–660
induced galls, 369–389
instars for, 33–35
mining, 371
mouthparts, 64–66
    feeding habits and, 64–65
    types, 65–66
mutualisms between microorganisms and, 406–407
nutrition
    cooperating supplements principle, 63
    identity rule, 62

needs, 68–73
    nutritional proportionality principle, 62
    principles, 62–63
    symbionts and, 145–156
nutritional indices for food intake and utilization
    AD, 15, 17, 20–23
    ECD, 15, 17, 19–21
    ECI, 15, 17, 19–20
    experimental techniques, 17
    feces measure, 19
    food consumed quantity, 17–18, 36–38
    meaning, 19–23
    methods to measure, 23–29
    RCR, 15–16, 19–20
    RGR, 15–16, 19–20
    RMR, 15–16, 19–20
    value, 30–33
    weight gains, 18–19
nutritional needs for growth, 68–73, 75
    amino acids, 68–69
    carbohydrate, 70–71, 75–76, 82
    lipids, 71, 82
    mineral salts, 70, 82
    nucleic acids, 71
    nutrient storage, 72
    sterols, 71, 82
    symbionts, 72–73
    vitamins, 69–70, 82
    water, 71–72, 82
pheromone emission, 168–169
physiological effects, 659–660
phytophagous, 4, 6–7
plant interactions, 660–661, 688. *See also* Plants
plant volatiles effect on pheromone emission,
    168–169
proteins and carbohydrates proportion, 63
rearing
    artificial diet, 55–59, 76–79
    artificial media for, 79–81
    entomopathogenic diseases in, 203–204
    field collecting, 59
    in laboratory, 52–54
    mass-scale, 60–61, 203–204
    medium-sized, 60
    natural hosts, 59
    small-scale, 60
    techniques, 76–77
    ways and types, 59–61
reproduction and dispersal
    allelochemicals role, 39–41
    food intake and utilization for, 36–41
    quality of food, 36–38
    selection and acceptance of food, 38–39
social, 6
in stored grain. *See* Stored-product pests
symbionts, 5, 63
    and nutrition, 145–156
trophic interactions, 5
Integrated pest management (IPM), 8–9, 474
    insects bioecology and nutrition, 687–700
    sap-sucking insects, 502–505

biological control, 503–504
plant nutrition, 503
plant resistance, 504–505
strategies in, 436, 440, 443
tactics and bioecology and nutrition
allelochemicals, 691–692
host plant resistance, 689–690
mixed crops cultivation, 691
trap crops, 690–691
Intraguild predation
in predatory beetles, 582–583
in predatory Heteroptera, 552–554
Ipilachna tredecimnotata, 130
Ipomoea imperati, 341
Ipomoea pes-caprae, 334, 340–341
Iridomyrmex spp., 227
Ischnocera, 647
Isoleucine, 68
Isoptera, 404
artificial diet, 52, 58
digestive process, 95, 108
nutritional value, 54
Isotope method, to measure food intake and utilization, 24, 28
Italochrysa, 617
Itoplectis conquisitor
artificial diet, 55
nutrients in, 63

J

Jacaranda decurrens, 665
Jadera choprai, 299, 301, 303–304
Jadera haematoloma, 298, 302

K

Kalotermitidae, 405
Keiferia lycopersicella, 666
Kinetoplastida, 640
Klebisiella pneumoniae, 151, 198, 259, 303
Klebsiella sp., 198
Koinobionts parasitoids, 516–518, 524
Kytorhinini, 327
Kytorhinus sharpianus, 343

L

Labidus, 224
Lacanobia oleracca, 668, 672 673
Lacerate-and-flush feeding, 298
Lacewings, 7. *See also* Chrysopids
natural diet, 596–599
Lachininae, 477
Lachnomyrmex plaumanni, 221
Lachnomyrmex victori, 221
Lagerstroemia indica, 583
Lamellicornia, 354
Lamiaceae, 504, 559
Lamium purpureum, 504
Lantana camara, 379

Lariophagus distinguendus, 346
Lariophagus texan, 346
Lariophagus texanus, 346
Larrea, 377
Larval development
food intake and utilization for, 21, 30, 33–35
ecdysis cost, 35
instars, 21, 35–36
room for, 80–81
Lasiocampa quercus, 281
Lasioderma serricorne, 421, 423–424, 426, 428, 436–437
Lasiohelea, 641
Lasiopterini, 383
Lauraceae, 373
Leaf chewers, 6
Leaf cutters ant, 223
Leaf galls, 386
Lecanicillium lecanii, 201
Lectins, plants resistance to insects, 671–673
Lecythidaceae, 375
Legionary ants, 224
Leguminoseae, 580
Leishmania, 634–635
Leishmaniasis, 7
Leonurus sibiricus, 307, 695, 699
Lepidoptera, 4, 6–7, 325, 353, 418–419, 422, 424, 429, 435, 440, 519, 523, 548–549, 554, 556, 579–580, 658
artificial diet, 52–53, 57–58, 62
caterpillars
acceptability, 278–280
feeding and digestion, 276
food perception, 276–277
leaf interaction, 278–285
morphology and biology, 274–277
natural enemies, 285–287
tritrophic relationships, 285–287
digestive process, 95–96, 98, 113
feeding habits, 273–274
galls, 373, 375–376
morphology and biology, 274–277
mouthparts, 66
Tibouchina pulchra system, 383
Lepismatidae, 404
Lepitopilina heteroma, 523
Leptanilloides, 224
Leptinotarsa decemlineata, 39, 196, 666, 669–670, 675
Leptocimex boueti, 645
Leptoconops, 641
Leptogenys, 218
Leptoglossus clypealis, 302
Leptoglossus zonatus, 308
Leptopodomorpha, 542
Lestes nympha, 182
Lestrimelitta spp., 238
Leucaena leucocephala, 341
Leucine, 68
Leucoagaricus, 223
Leucochrysa, 601, 619–620
Leucochrysa spp., 620

Leucocoprinus, 223
Leucocytozoon, 640, 641
Leucoptera coffeella, 75
Leurotrigona, 250, 251
Licania cecidiophora, 370
Lignocellulosic biofuels, 442–443
Ligustrum lucidum, 303, 305, 309–310
Ligyrus ebenus, 355
Ligyrus spp., 354
Limenitis archippus, 131
Lindorus lophantae, 576
Linepithema, 222
Linepithema humile, 228
Linoleic acid, 71–74
Linolenic acid, 71, 73–74
Linum sp., 126
Liogenys fuscus, 354, 358
Liogenys spp., 354, 358
Liogenys suturalis, 354, 358
Lipaphis erysimi, 613, 672
Lipids, for insects growth, 71, 82
Liposcelididae, 423
Liposcelis, 423
Liposcelis bostrichophila, 435
Lipotrophidae, 442
Liquid food, 102
Liriomyza trifolii, 666
Litter-nesting fungus growers ant, 223–224
Locusta, 70
Locusta migratoria, 67, 132
    feeding needs, 67
Lolium multiflorum, 579
Lolium perenne, 579, 663
Lonchocarpus, 331
Lonchocarpus muehlbergianus, 344–345, 385
Longitarsus melanocephalus, 202
Lonomia circumstans, nutritional indices, 29
Lotus, 339
Loxa deducta, 309
Lucanidae, 354
Lupinus, 499
Lupinus luteus, 304, 309
Lutzomyia, 634
Lutzomyia intermedia, 634–635
Lutzomyia longipalpis, 634–635
Lutzomyia spathotrichia, 635
Lutzomyia umbratilis, 635
Lutzsimulium, 640
Lycopersicon esculentum, 201
Lycopersicon hirsutum f. *glabratum*, 201, 666, 669
Lycopersicon hirsutum f. *typicum*, 666
Lycopersicon spp., 666
Lycosidae, 553
Lygaeidae, 295, 540–543, 551, 555, 560
Lygaeus equestris, 307
Lygaeus kalmii, 302, 308
Lygus hesperus, 80, 302, 309
Lygus rugulipennis, 306
Lymantria dispar, 67, 197–198, 200, 203
Lysiloma divaricata, 341
Lysine, 68, 74

M

M. lacerata, 341
Maaladu ummarconala, 672
Macadamia integrifolia, 696
Macerate-and-flush feeding, 298
Machaerium aculeatum, 375
Machaerium uncinatum, 382
Macrolophus, 7, 540, 554
Macrolophus caliginosus, 545, 548–549, 552, 554–555,
    557, 559–560
Macrolophus pygmaeus, 540, 545, 559–560
Macrolophus spp., 549
Macrosiphoniella tanacetaria, 503
Macrosiphum aconitum, 581
Macrosiphum euphorbiae, 479, 490–491, 498–499,
    503–505, 666
Macrotermitinae, 5
Malacosoma californicum pluvial, 200
Malacosoma disstria, 197
Malaria, 7
Mallada basalis, 610
Mallophaga
    artificial diet, 58
    mouthparts, 66
Malpighiaceae, 377
Malvacea, 326, 331
Malvaceae, 377
Mammillaria *crinita*, 261
Manduca quinquemaculata, 132
Manduca sexta, 197–198, 277, 282–283, 670
    feeding behavior and nutrient selection in, 23
Manduca sexta, 131, 132, 297, 666
Manihot esculenta, 200
Mansonella, 640
Mansonella ozzardi, 640
Mansonia, 636–637
Mantis religiosa, 184
Mark galls, 385, 387
Maruca vitrata, 671
Mattese oryzaephili, 442
Mechanitis isthnia, 130
Medicago sativa, 558
Megacerini, 331
Megacerus, 334
Megacerus baeri, 335, 340–341
Megacerus discoidus, 340
Megacerus reticulatus, 340
Megalomyrmex, 217
Megalopodidae, 325
Megalotomus quinquespinosus, 304
Megoura viciae, 491, 494
Melaphis rhois, 374
Melastomataceae, 373, 376, 383
Melia azedarach, 135, 668
Melia volkensii, 669
Melipona, 238–239, 241–243, 245, 247–248, 250,
    254–255, 259, 261
Melipona asilvai, 256, 258
Melipona beecheii, 251, 254, 256
Melipona bicolor, 238–239, 262

Melipona compressipes, 256, 258
Melipona fasciata, 261
Melipona favosa, 256
Melipona favosa favosa, 256
Melipona fuscopilosa, 261
Melipona grandis, 256
Melipona mandacaia, 256, 258
Melipona marginata, 238–239
Melipona quadrifasciata, 238, 247, 251, 256, 258,
        260–261
Melipona rufiventris, 239, 261
Melipona scutellaris, 241, 256–258
Melipona seminigra, 258
Melipona solani, 259
Melipona subnitida, 257, 259–262
Melipona trinitalis, 257
Meliponini, 6
    caste determination and differentiation, 250–251
    honey production, 254–258
    larval food in, 251–252
Melittobia digitata, 523
Melolontha melolontha, 168
Melolonthidae, 354
    abiotic and biotic factors, 362–363
    exploration and performance of larvae, 361–362
    food
        environmental impact on, 361–362
        exploration, 359–361
        roots source, 355–356
    localization and selection of host plant, 359–360
    morphological and biological features, 356–359
    roots as food source, 355–356
Membracidae, 597
Menognathous insects, 66
Menorhynchous insects, 66
Mentha piperita, 439
Meridic diet, 62
Merobruchus julianus, 338
Merobruchus spp., 335
Metagnathous insects, 66
Metahycus flavus, 182
Metarhizium anisopliae, 200–203, 441
Metarhizium anisopliae var. acridum, 135
Metionine, 68
Metopolophium dirhodum, 493, 664
Metschnikowia, 608
Micrococcus, 497
Microorganisms, mutualisms between insects and,
        406–407
Microplitis croceipes, 170–171
Migdolus fryanus, 111–112
Mimosa spp., 341
Mimosa texana, 341
Mimosestes, 335
Mimosestes amicus, 330–331, 344–345
Mimosestes spp., 335
Mimosoideae, 331, 580
Mineral salts, for insects growth, 70, 82
Mining insects, 371
Miridae, 540–545, 553–555, 559–560
Monomorium, 222, 225

Mononychellus tanajoa, 200
Monophlebidae, 596
Moraceae, 373
Mormidea, 305, 696
Morphological resistance
    insects bioecology and nutrition
        epicuticular waxes, 665–666
        epidermal toughness and thickness, 662–665
        lignin, 664–665
        pilosity of epidermis, 666
        silicon, 662–664
Morus alba, 123
Morus nigra, 123
Mourella caerulea, 240
Mouthparts, 64–66
    chewing, 65
    feeding habits and, 64–65
    labial sucking, 65
    licking, 66
    piercing-sucking, 65
    shredder, 65
    sucking maxillary, 65
    types, 65–66
Murgantia histrionica, 151–152, 696
Musca autumnalis, 409
Musca domestica, 71, 103, 112, 548, 644
    midgut conditions affecting enzyme activity in, 105
    nutrients in, 63
Musca vetustissima, 409
Muscidae, 548, 643–644
Mycetagroicus, 223
Mycetosoritis, 223
Mycocepurus, 223
Myrmeleontidae, 594
Myrmica ruginodis, 584
Myrmicinae, 5
Myrmicines, 221–222
Myrmicocrypta, 223
Myrtaceae, 373–375, 377, 485
Myzus nicotiana, 499
Myzus ornatus, 481
Myzus persicae, 70, 478–479, 481–483, 485–489, 498,
        503–505, 554, 613–614, 666
    artificial diet, 55
    nutrients in, 63

N

N. circumflexus, 497
N. perilampoides, 259
Nabidae, 540–544
Nabis, 540
Nabis alternatus, 545, 553, 554
Nabis americaniformis, 545
Nabis ferus, 545
Nabis (Nabidae), 7
Nabis pseudoferus, 545–546
Nabis pseudoferus ibericus, 560
Nabis spp., 553, 557, 559–560
Nacarina, 617
Nannotrigona, 250–251

Nannotrigona testaceicornis, 247, 262
Natural diet, 62
Natural enemy
    attractions and extrafloral nectar, 171
    induced volatiles and, 169–171
    plant–enemy–herbivore-interactions, 169–171
Natural food, 4
Nauphoeta cinerea, 107
Necator americanus, 409
Neivamyrmex, 224
Neltumius arizonensis, 326
Nematocera, 580, 635
Nematodes, diseases caused by, 201–202
Nematoid-induced gall, 370
Neocapritermes opacus, 218
Neodiprion rugifrons, 130
Neodiprion sertifer, 130
Neodiprion swainei, 130
Neogregarinorida, 442
Neomegalotomus parvus, 295, 300, 303, 305, 308–310,
        697, 698
Neotropical ant guilds, 216–226
    arboreal ants, 224–226
        carbohydrate-rich resources, 224–225
        pollen-feeding, 225–226
    arboreal predator ants, 220
    with carbohydrate-rich resources, 224–225
    fungus growers, 223–224
        leaf cutters, 223
        litter-nesting fungus growers, 223–224
    generalist predators, 216–217
        epigaeic generalist predator, 216–217
        hypogaeic generalist predator, 217
    generalized
        formicines and dolichoderines, 222
        myrmicines, 221–222
        small-sized hypogaeic foragers, 222
    legionary ants, 224
    specialist, 217–220
        dacetini predators, 218–220
        predation in mass and nomadism, 218
    species with
        kinetic mandibles, 219–220
        static pressure mandibles, 218–219
    subterranean ants, 226
Neozygites tanajoae, 200
Nephus spp., 573
Nepomorpha, 542
Nerium oleander, 581
Nesidiocoris, 540
Nesidiocoris spp., 549
Nesidiocoris tenuis, 540–541, 545, 548, 560
Neuroptera, 7, 523, 594
    artificial diet, 58
    mouthparts, 66
Nezara viridula, 74, 151–152, 183, 298–305, 307–314,
        693–697, 699
Nicotiana, 499
Nicotiana tabacum, 168
Nicotinic acid, 69
Nilaparvata lugens, 664, 668, 671–673

Nineta vittata, 611
Nitidulidae, 420, 424
Nocardia, 150
Nocardia sp., 151
Noctuidae, 548, 554, 556, 597, 599, 614, 616
Nomamyrmex, 224
Nomamyrmex esenbeckii, 224
Nomuraea rileyi, 201
Nonnutritional interaction symbionts, 156
Nothochrysa, 608
Nothochrysinae, 594
Notobitus meleagris, 302
Notonectidae, 551
Noviini, 575
Novius cardinalis, 576
Nucleic acids, for insects growth, 71
Nutrient storage, for insects growth, 72
Nutrition
    and bioecology of insects, 8–9, 62–63
    and food characteristics, 425–426
    requirements of sap-sucking insects
        amino acids, 487–488
        carbohydrates, 488
        vitamins, 488
    stored-product pests requirements, 427–430
Nutritional biology and ant guilds, 214–216
Nutritional indices
    for food intake and utilization
        AD, 15, 17, 20–23
        ECD, 15, 17, 19–21
        ECI, 15, 17, 19–20
        experimental techniques, 17
        feces measure, 19
        food consumed quantity, 17–18, 36–38
        meaning, 19–23
        RCR, 15–16, 19–20
        RGR, 15–16, 19–20
        RMR, 15–16, 19–20
        weight gains, 18–19
    measure, 23–29
        calorimetric method, 25–28
        colorimetric method, 23–24, 28
        direct method, 23, 28
        gravimetric method, 23, 28
        immunological method, 25
        indirect method, 23–29
        isotope method, 24, 28
        trace element method, 25
        uric acid method, 25
    value, 30–33
Nutritional needs
    of fruit flies, 453–458
        carbohydrates, 456–457
        lipids, 457
        proteins, 454–456
        symbionts, 458
        vitamins and mineral salts, 457–458
    for insects growth, 68–73
        amino acids, 68–69
        carbohydrate, 70–71, 82
        lipids, 71, 82

mineral salts, 70, 82
nucleic acids, 71
nutrient storage, 72
sterols, 71, 82
symbionts, 72–73
vitamins, 69–70, 82
water, 71–72, 82
Nutritional physiology
  of sap-sucking insect
    feeding strategies, 480–481
    food constituents and digestive enzymes, 484–487
    food intake mechanisms and saliva composition,
      482–484
    host location and acceptance, 478–479
    host plant condition, 480–481, 498–499
    phloem feeding, 481–482
    symbionts, 496–498
  of sap-sucking insects, 477–498
Nyctaginaceae, 385
Nysius groenlandicus, 313
Nysius spp., 295
Nysius vinitor, 309
Nyssomyia whitmani, 635

O

Ochlerotatus, 636
Ochrimnus mimulus, 312
Ocimum basilicum, 439
Odontomachus, 216
Oebalus, 305, 696
Oidium sp., 577
Okanagana rimosa, 180
Oligidic diet, 62
Olivier, 419
Olla v-nigrum, 574–576, 582, 583
Omnivores insects, 65
Omphalocera munroei, 286
Onagraceae, 376
Onchocerca, 640
Onchocerca cervalis, 641
Onchocerca gibsoni, 641
Onchocerca volvulus, 640–641
Onchocercidae, 640
Onchocercosis, 640
Oncopeltus fasciatus, 131, 183, 302, 307, 309, 314
  artificial diet, 55
Ooencyrtus sp., 553
Ooencyrtus spp., 553
Oomyzus gallerucae, 170
Operophtera brumata, 128, 199
Ophyra, 644
Orchidaceae, 375
Oreina cacaliae, 168
Origanum vulgare, 669
Orius, 540, 551–552, 554–556
Orius albidipennis, 560
Orius (Anthocoridae), 7
Orius armatus, 560
Orius insidiosus, 544–545, 548–550, 552–553, 555–557,
  559–560

Orius laevigatus, 540, 544, 548–549, 552, 556, 560–561
Orius majusculus, 544, 548, 549, 553–554, 556, 560
Orius minutus, 560
Orius niger, 548
Orius perpunctatus, 558
Orius spp., 549, 557–559
Orius strigicollis, 560
Orius thyestes, 558
Orius tristicolor, 553, 555–557, 560
Orius vicinus, 548, 555
Orsodacnidae, 325
Ortaliinae, 578
Orthoptera, 6, 353
  artificial diet, 52, 58
  digestive process, 95, 108
  galls, 371
  mouthparts, 66
  nutritional value, 54
Oryctes, 405
Oryzaephilus mercator, 421, 423, 427, 435
Oryzaephilus surinamensis, 419, 421, 423, 425–430,
  433–435, 438, 440–442
Osmotic pump feeding, 298
Ostertagia ostertagi, 409
Ostrinia nubilalis, 62, 280, 668, 671, 672
  artificial diet, 55
Oxyepoecus, 217, 221
Oxyepoecus crassinodus, 221
Oxyepoecus myops, 221
Oxyepoecus plaumanni, 221
Oxyepoecus punctifrons, 217
Oxyepoecus rastratus, 221
Oxyepoecus reticulatus, 221
Oxyopes salticus, 553
Oxypeltidae, 325
Oxytrigona mellicolor, 245
Ozophora baranowskii, 302

P

Pachycondyla, 216–217, 220
Pachycondyla marginata, 218
Pachycondyla stigma, 217
Pachymerini, 327, 331
Pachymerus cardo, 328
Pandora neoaphidis, 200, 203
Panesthia cribata, 107
Pangoniinae, 642
Panstrongylus megistus, 151, 646
Pantoea agglomerans, 151, 201
Pantoea sp., 151, 198, 303
Pantothenic acid, 69
Papilionoideae, 331
Papilonoidea, 273–274
Paraponera clavata, 220
Parasitilenchus coccinellinae, 585
Parasitoids, 7
  adults, 516, 521, 530–531
  development strategies, 516–517
  first trophic level effect, 521–522
  host

interactions, 521–522
    as nutritional environment, 519–521
    restrictions, 522–529
    searching, 169–170
  immature, 517–519
  nutritional quality, 519–521
    adult, 521
    egg, 519
    larva, 519–520
    pupa, 520–521
  nutritional requirements
    adults parasitoids, 516, 530–531
    idiobionts, 516, 518–519
    immature parasitoids, 517–519
    koinobionts, 516–518
Parastrachia japonensis, 299, 308
Paratrechina, 222, 225
Paratrigona, 248
Paratrigona subnuda, 247
Parkinsonia aculeata, 326
Parlatoria oleae, 134
Paropsis atomaria, 39
Partamona, 251
Partamona helleri, 238
Parthenium hysterophorus, 558
Passalidae, 354, 401
Passiflora sp., 308
Pastinaca sativa, 128
Pectinophora gossypiella, 136
  artificial diet, 55
  nutrients in, 63
Pediculidae, 647
Pediculus humanus capitis, 644
Pediculus humanus corporis, 644
Pediculus humanus, 647
Pediobius foveolatus, 584
Pellaea stictica, 152
Peloridiomorpha, 542
Pennisetum glaucum, 558
Pentatomidae, 295, 540–544, 548, 554–555, 560, 577
Pentatomomorpha, 542, 544, 546 547
Penthobruchus germaini, 326
Perillus bioculatus, 546, 553
Periplaneta americana, 107, 184, 405
Periplaneta orientalis, artificial diet, 55
Peucetia viridans, 553
Peyerimhoffina gracilis, 611
Phaedon cochleariae, 203
Phaedon cochleariae, 201
Phalacromyrmex, 218
Phalacrotophora berolinensis, 584
Phalacrotophora fasciata, 584
Phaseolus, 338, 676
Phaseolus aureus, 335, 338
Phaseolus coccineus, 580
Phaseolus lunatus, 577, 674
Phaseolus radiatus, 341
Phaseolus spp., 297
Phaseolus vulgaris, 201, 296, 304, 309, 311–312, 333,
    342–343, 558, 577, 583, 673, 674, 676
Phasmatodea galls, 371

Phasmida, artificial diet, 58
Pheidole, 217–218, 221–222, 225
Phenylalanine, 68–69
Pheromones, 163
Phidotricha erigens, 79
Phillonorycter blancardella, 134
Philosamia cynthia ricini, artificial diet, 33
Phlaeothripinae, 374
Phlebotominae, 634–635, 647
Phlebotomus, 634–635
Phloem-sucking insects, 474
Phormia, 69
Phorodom humuli, 167
Photorhabdus, 201–202
Phrynosoma coronatum, 228
Phthiraptera, 7, 647
Phthorimaea operculella, 168
Phycitinae, 435
Phygadeuon trichops, 75
Phyllocnistis citrella, 60
Phyllomorpha laciniata, 303
Phyllophaga cuyabana, 354, 357–359, 361–364
Phyllophaga spp., 354
Phyllophaga triticophaga, 354, 357–358, 361–362
Phyllotreta cruciferae, 168
Phylloxera glabra, 580
Phylloxeridae, 477, 577
Phyrrocoris apterus, 314
Phytalus sanctipauli, 363
Phytolacca americana, 672
Phytophagous arthropods
    bioecology and nutrition of, 688–692
        insect–plant interactions, 688
        plant diversity and stability, 688–689
Phytophagous insects, 4, 6–7. *See also* Insects
    host selection by, 658–661
Phytoseiidae, 553
Picea sitchensis, 199
Picoides pubescens, 388
Picromerus bidens, 547, 560
Pieridae, 597
Pieris brassicae, 25–26, 167, 277
Pieris melete, 280
Pieris napi, 280
Pieris rapae, 23, 72, 134, 198, 277, 280, 283
Piezodorus guildinii, 299–300, 304–305, 308–309,
    694–696, 699
Piezosternum calidum, 150
Pilobolus sporangia, 409
Pinaceae, 375
Pineus boerneri, 502
Pintomyia fischeri, 635
Pinus, 577
Pinus elliotii, 502
Pinus sp., 579
Pinus species, 658
Pinus taeda, 169, 478, 502
Piper nigrum, 439
Pisum sativum, 333, 340, 342, 438, 485–486, 580
Pisum sativum, amino acids in phloem sap, 486
Pit galls, 387

Pitinus tectus, 428
Pityoborus spp., 147
Plagiodera versicolora, 187
Planipennia, 594
Plantago lanceolata, 202
Plant–herbivore interactions
    avoid plant defenses, 129–130
    constitutive volatiles and, 166–167
    generalist and specialist strategy, 131–132
    host plant manipulation, 131
    induced volatiles and, 167–168
    metabolizing and sequestering plant toxins, 130–131
    natural enemy interactions, 169–171
        extrafloral nectar and natural enemy attraction, 171
        host searching, 169–170
        induced volatiles and, 169–171
    trophic interactions, 165–169
Plant–insect interactions
    abiotic factors in tritrophic interactions, 133–135
    coevolution theory, 124–125
    defenses in plants, 128–129
        cost, 128
        factors affecting, 128
    herbivore perspective, 129–131, 165–169
        avoid plant defenses, 129–130
        generalist and specialist strategy, 131–132
        host plant manipulation, 131
        metabolizing and sequestering plant toxins, 130–131
    history, 123–124
    plant perspective, 125–129
    trophic level, 125, 132–135
Plants. *See also* Plant–insect interactions
    coevolution theory, 124–125
    defenses in, 128–130
        cost, 128
        factors affecting, 128
    development in geologic time, 122–123
    diversity and stability, 688–689
    insect interactions, 660–661, 688
    insect pheromone emission, 168–169
    manipulation, 131
    nutrition IPM perspectives, 503
    resistance to insects
        α-amylase inhibitors, 675–676
        antibiotic factors, 669–670
        antixenotic factors, 667–669
        bifunctional inhibitors, 676–677
        bioecology and nutrition, 8–9, 661–670
        and biotechnology, 670–677
        chemical causes, 667–670
        enzyme inhibitors, 673–675
        epicuticular waxes, 665–666
        epidermal toughness and thickness, 662–665
        IPM perspectives, 504–505
        lectins, 671–673
        lignin, 664–665
        morphological resistance, 661–666
        phagodeterrents, 667–669
        pilosity of epidermis, 666

        protease inhibitors, 674–675
        repellents, 667
        silicon, 662–664
    toxins, 130–131
    volatiles effect, 168–169
Plasmodiidae, 640
Plasmodium, 639
Plasmodium falciparum, 198
Platynaspis, 574
Platynota rostrana, 79
Platypodinae, 5
Plautia stali, 150, 151, 313
Plebeia, 243, 248, 250–251
Plebeia droryana, 239, 242
Plebeia emerina, 238–239
Plebeia lucii, 250
Plebeia pugnax, 238–239
Plebeia remota, 238–239, 247–248, 250–251
Plebeia saiqui, 239
Plebeia sp, 262
Plebeia spp., 238
Plebeia tobagoensis, 245
Plebeia wittimani, 257
Plecoptera, mouthparts, 66
Plectris pexa, 354
Plectris spp., 354
Plodia interpunctella, 183, 186, 196–197, 421–422, 425, 427, 429, 431, 438, 441–442, 549
Plodia interpunctella granulosis virus (*PiGV*), 197
Plunentis porosus, 303
Plutella xylostella, 40, 168, 198, 612, 665, 669
Plutellidae, 597
Poaceae, 376
Podapolipidae, 585
Podisus, 7, 307, 540, 549, 554
Podisus distinctus, 546
Podisus maculiventris, 308, 546–548, 551, 553–554, 560
Podisus nigrispinus, 546, 548–549, 553, 558, 560–561
Podisus rostralis, 548
Podisus sculptus, 546
Podisus species, 558
Polistes chinensis antennalis, 180
Polistes jadwigae, 180
Pollen, protein value, 252
Polygonaceae, 373, 375
Polyrhachis, 155
Popilia japonica, 360–361, 363
Populus tremuloides, 197
Porrycondilinae, 375
Pouch galls, 385, 387
Predator
    host searching, 169–170
    plant interactions
        phytophagy, 554–556
        in predatory Heteroptera, 554–558
    and seed-chewing beetles, 346
Predatory beetles
    abiotic and biotic adoptions, 583–584
    biology, 574–576
    cannibalism, 581–582
    defense strategies, 581

development
    adult, 576
    postembryonic, 574–576
evolution, 572–573
food
    preferences, 580–581
    quality, 579–580
    selection, 576–581
    specificity, 577–579
    toxicity, 581
intraguild competition, 582–583
morphology, 572–573
natural enemies, 584–585
taxonomy, 572–573
Predatory bugs. *See* Predatory Heteroptera
Predatory Heteroptera
    commercial biological control, 559–561
    family
        Anthocoridae, 544–545
        Lygaeidae, 547
        Miridae, 545
        Nabidae, 545–546
        Pentatomidae, 546–547
    feeding behavior
        and prey digestion, 542–544
    food influence on
        development and reproduction, 547–548
        mass rearing, 548–550
        plant for, 543
    habitat choice and distribution, 558–559
    infraoder
        Cimicomorpha, 544–546
        Pentatomomorpha, 546–547
    taxonomy, 541
    trophic relationships, 551–558
        cannibalism, 551–552
        intraguild predation, 552–554
        natural enemies, 558
        predator–plant interactions, 554–558
Prionopelta, 218
Proceratium, 218
Prociphilus tesselatus, 617
Procryptocerus, 225
Propylea quatuordecimpunctata, 581, 583
Prosopis, 344–345
Prosopis spp., 326
Prosopis velutina, 330–331
Prostephanus truncatus, 425–426
Protease inhibitors, plants resistance to insects,
    674–675
Protura, mouthparts, 66
Pselactus spadix, 665
Pseudaletia separata, 170
Pseudaletia sequax
    on artificial diet, 84
    nutritional indices, 29
Pseudaletia unipuncta, 669
Pseudoatta argentina, 216
Pseudoatta sp., 216
Pseudococcidae, 596, 614
Pseudomonas aeruginosa, 259–261

Pseudomonas ferruginea, 217
Pseudomonas fluorescens, 198
    infection, 196
Pseudomonas sp., 198
Pseudoplusia includens, 23–24, 71, 670
    nutritional indices, 26–28
Pseudotectococcus rolliniae, 374, 384
Psilids, 7
Psocidae, 597
Psocoptera, 418, 423
Psorophora, 636–637
Psudaletia sequax, 79
Psychodidae, 634–635
Psydium cattleianum, 579
Psyllidae, 474
Psyllobora gratiosa, 573, 577
Psylloborini, 572–573, 577
Psylloidea, 372–373, 384
Pteridium aquilinum, 498
Pteromalidae, 344, 346, 375, 440
Pullus spp., 573
Pulus auritus, 580
Pygiopachymerus lineola, 327, 335, 337
Pymotes, 346
Pyralidae, 419, 421, 424, 426, 429, 548–549, 579, 597, 599,
    614, 616
Pyralis farinalis, 421
Pyridoxine, 69
Pyrrhocoridae, 295
Pyrus malus, 200

Q

Quercus, 377
Quercus infectoria, 370
Quercus robur, 199
Quercus sp., 424

R

Raphanus raphanistrum, 695
Reduviidae, 541–542, 544, 645
Regiella insecticola, 203, 497, 504
Relative consumption rate (RCR), 15–16, 19–20
Relative growth rate (RGR), 15–16, 19–20
Relative metabolic rate (RMR), 15–16, 19–20
Reproduction and dispersal
    food intake and utilization, 36–41
        allelochemicals role, 39–41
        quality of food, 36–38
        selection and acceptance of food, 38–39
Reticulitermes flavipes, 201
Reticulitermes speratus, 149
Rhaebini, 327
Rhagoletis, 451–452, 458, 464
Rhagoletis basiola, 459
Rhagoletis juglandis, 456
Rhagoletis pomonella, 457
Rhamnaceae, 375
Rhinocoris tristis, 188
Rhinotermitidae, 405

Rhizophagous beetles. *See also* Melolonthidae
    food exploration, 359–361
    localization and selection of host plant by, 359–360
Rhizophagous insects, localization and selection of host
    plant by, 359–360
Rhizophagous melolonthidians, food exploration,
    359–361
Rhodillus spp., 150
Rhodius prolixus, 110
Rhodnius prolixus, 151, 646
Rhodococcus equi, 151
Rhodococcus rhodnii, 151
Rhopalidae, 295
Rhopalocera, 273
Rhopalomyia chrysothamni, 379
Rhopalosiphum maidis, 500, 504, 599, 614, 664
Rhopalosiphum padi, 489, 493–494, 502, 668
Rhus glabra, 374
Rhynchosciara americana, 103
Rhynchosia minima, 199
Rhyncophthirina, 647
Rhysobius lophanthae, 135
Rhyzobius sp., 573
Rhyzopertha dominica, 420–422, 424–427, 431–434,
    438–440, 442, 667
Riboflavin, 69
Ricinus communis, 307, 310, 695, 697
Rickettsia, 645
Rickettsia prowazekii, 644
Riptortus clavatus, 302
Riptortus linearis, 384, 696
Rodolia cardinalis, 134, 572, 581
Roll galls, 385, 387
Rollinia laurifolia, 374
Root
    feeding insects, 6
    food source for melolonthidae, 355–356
Rosa, 377
Rosaceae, 375
Rosa nutkana, 200
Rosette galls, 385, 387
Rubiaceae, 377
Rutelinac, 354, 360–361
Rynchophorus phoenicis, 168

S

Saccharomyces cerevisiae, 260, 601
Saccharomyces fragilis, 601
Saccharopolyspora spinosa, 442
Saissetia oleae, 612
Salicaceae, 373, 375
Salicicaceae, 485
Salix acutifolia, 485
Salix sp., 486
Salmonella cholerasuis, 260
Salmonella sp., 261
Santalum album, 666
Saprovore insects, 65
Sap-sucking insects
    artificial diets, 499
    digestive tract, 476–477
    energy budget, 489–492
    evolution and distribution, 474–475
    feeding
        electrical penetration graph, 499––502
        and nutrition, 498–499
        rate, 488–492
    honeydew and excretion, 496
    IPM perspectives, 502–505
        biological control, 503–504
        plant nutrition, 503
        plant resistance, 504–505
    lipids, 488
    minerals, 488
    mouthparts, 475–476
    natural increase
        changes in, 494–495
        development rate, 492–493
        intrinsic rate, 492–495
        reproductive rate, 493
        survival rate, 493–494
    nutritional budget, 488–492
    nutritional physiology, 477–498
        feeding strategies, 480–481
        food constituents and digestive enzymes,
            484–487
        food intake mechanisms and saliva composition,
            482–484
        host location and acceptance, 478–479
        host plant condition, 480–481, 498–499
        phloem feeding, 481–482
        symbionts, 496–498
    nutritional requirements
        amino acids, 487–488
        carbohydrates, 488
        vitamins, 488
    trace metals, 488
Sarcina, 497
Sarcophagidae, 643–644
Sarothamnus, 334
Sator limbatus, 331–332, 335–336, 338, 344
Sator pruininus, 335
Scapotrigona mexicana, 254
Scapotrigona nigrohirta, 254
Scapotrigona polysticta, 254
Scapotrigona postica, 254
Scaptocoris castanea, 698–699
Scaptotrigona, 240, 242–244, 248, 254–255
Scaptotrigona bipunctata, 241, 243, 247
Scaptotrigona depilis, 254, 261
Scaptotrigona pachysoma, 257
Scarabaeinae, 399, 402
Scarabaeoidea, 354, 356–357, 359
Schistocerca, 70–71
Schistocerca americana, 283
Schistocerca gregaria, 132, 135, 185, 201, 663–664
    feeding needs, 67
    nutrients in, 63
Schistocerca sp., 62, 72
Schizaphis graminum, 150, 154, 481, 485, 489, 497, 502,
    504, 599, 614, 664–665

Schwarziana, 250
Schwarziana quadripunctata, 238–239, 251
Sciobius granosus, 71
Scirpus, 376
Scolytinae, 5
Scolytus ventralis, 147
Scrophulariacea, 559
Scutelleridae, 295
Scyminae, 572, 575, 578
Scymnus apetzi, 584
Scymnus spp., 573, 577
Scymnus subvillosus, 584
Sechium edule, 571, 577
Seed chewers (borers), 6
Seed-chewing beetles, 325–346
    diapause and dispersal, 342–344
    distribution, 327–331
    enemies, 344–346
    fruits defenses, 333–334
    host plant specificity, 329–332
    intra- and interspecific competition, 340
    larval and pupal development, 330, 338–340
    morphological adaptations, 327–331
    obtaining energy, 334–335
    oviposition behavior, 335–338
    parasitoids, 344–346
    predation rate, 340–342
    predators and, 346
    reproductive performance, 342–344
    seed
        availability and, 332
        defenses, 333–334
        predation, 340–342
    taxonomy, 327–331
Seed suckers, 6
Seed-sucking bugs. *See also* Heteropterans
    biology, 298–308
        feeding, 298–301
        host plant choice, 304–306
        ingestion, 298–301
        mating, 302
        natural enemies and defense, 307–308
        nymph development, 303–304
        nymphs and adults dispersal, 304–306
        oviposition, 302–303
    digestion, 298–301
    excretion, 298–301
    food characteristics, 296–298
        abundance, 298
        allelochemicals, 297
        nutritional composition, 296–297
        physical and structural aspects, 297–298
    food utilization, 298–301
    ingestion, 298–301
Seed weevils, 6
Semiadalia undecimnotata, 576, 583, 585
Semiochemicals, trophic interactions mediated by,
        165–171
Senecio jacobea, 167
Senna alata, 338–339, 344
Senna macranthera, 339, 342

Senna multijuga, 337, 341–342, 342, 344
Sennius, 331
Sennius bondari, 335, 337, 342, 344
Sennius crudelis, 337, 344
Sennius lamnifer, 337
Sennius lateapicalis, 337
Sennius leptophyllicola, 327, 337
Sennius morosus, 338, 340
Sennius puncticollis, 337
Sennius simulans, 338, 340
Sennius sp., 328
Sennius spp., 328
Sennius subdiversicolor, 337
Sericesthis nigrolineata, 362
Sericomyrmex, 217, 223
Serine, 69
Serratia symbiotica, 497
Sesbania, 311
Sesbania aculeata, 309
Sesbania emerus, 309, 311
Sesbania rostrata, 696
Sesbania vesicaria, 298, 311–312
Sexual cannibalism, 184
Sidis spp., 573
Silene latifolia, 128
Silvanidae, 419
Simopelta, 218
Simuliidae, 640–641, 644
Simulium, 640
Simulium ochraceum, 641
Simulium pertinax, 640
Simulium vittatum, 641
Sinea, 7, 551
Sinea diadema, 551–552
Sinxenic diet, 62
Siphonaptera, 7, 633, 644–645
    artificial diet, 58
    mouthparts, 66
Sirex, 405
Sitobion avenae, 493, 494, 664
Sitophilus, 38, 155
Sitophilus granarius, 38, 419, 421–422, 425–427, 431,
        433–434, 437, 440–441
Sitophilus oryzae, 421, 426–427, 431–435, 438–441, 667
Sitophilus oryzae primary endosymbiont (SOPE),
        150, 155
Sitophilus spp., 422, 440
Sitophilus zeamais, 203, 420–421, 426–427, 434, 438,
        441–442, 667
Sitotroga cerealella, 197, 419, 421–423, 427, 438, 440,
        549, 599, 614
Social bees, 6
    foraging
        activity, 238–239
        strategies, 243–248
    honey production, 252–262
        antibacterial activities, 258–259
        in Apini, 253–254
        in Meliponini, 254–258
        physicochemical characteristics, 254–258
        in stingless bee, 256–259, 261–262

larval food, 251–252
pollen, 252
resources acquisition
  floral constancy and load capacity, 243–248
  foraging activity, 238–239
  niche width and floral resources, 239–243
  physical factors effecting, 238–239
resources utilization, 248–251
  in Apini, 249–250
  in Bombini, 248–249
  caste determination and differentiation, 248–251
  in Meliponini, 250–251
Social insects, 6. *See also* Social bees
Sodalis, 150, 155
Sogatella furcifera, 664
Solanaceae, 7, 282, 355, 374, 504, 559, 577
Solanum, 504
Solanum berthaultii, 499, 504, 666
Solanum hirsutum f. *glabratum*, 666, 669
Solanum hirsutum f. *typicum* spp., 666
Solanum lycocarpum, 374, 388
Solanum spp., 666
Solanum tuberosum, 168
Solenopsis, 217, 221–222, 225
Solenopsis invicta, 227–228, 308
Solenopsis richteri, 227
Solenopsis saevissima, 227
Solidago, 376–377
Solidago altissima, 381, 383, 388
Solidago canadensis, 558
Solid food, 102
Sorghum bicolor, 556, 696
Sorghum spp., 558
Soybean
  managing heteropterans on, 692–694
    feeding process interference, 694
    host plant resistance, 692–693
    mixed crops use, 693–694
    trap crops use, 693
Spartium, 581
Spathosternum prasiniferum, 184
Speciomerus giganteus, 332, 333, 340
Spectrophotometer-enzymatic method. *See* Uric acid method
Spermophagini, 331
Spermophagus, 328
Speyeria mormonia, 282
Sphaerotheca fuliginea, 343
Sphenophorus levis, 79, 111
Sphenostylis stenocarpa, 671
Sphinctomyrmex, 218
Spodoptera, 274
Spodoptera eridania
  nutritional indices, 29, 35
Spodoptera exempta, 662–664
Spodoptera exigua, 132, 167, 170, 197, 546, 548, 666, 670, 675
Spodoptera frugiperda, 182, 183, 185, 199–200, 280–281, 283, 614, 662, 666
  on artificial diet, 84
  diet, 74
digestion, 103
  nutritional indices, 15–16, 18, 20, 22, 24, 29, 35, 37, 40–41
Spodoptera latifascia, nutritional indices, 29
Spodoptera littoralis, 38, 179, 198, 554, 669
Spodoptera litura, 669
Staphylococcus aureus, 258–261
Staphylococcus pyogenes, 260
Starvation and dietary stress, effects on entomopathogenic diseases, 196–197
Stator beali, 331, 332
Stator generalis, 327, 334
Stator limbatus, 330
Stator sordidus, 335
Stegobium paniceum, 421, 428, 440
Stegobium sp., 429, 434
Stegomyrmex, 218
Steinernema, 201–202
Steinernema carpocapsae, 202, 585
Steinernema glaseri, 202
Steinernema riobrave, 204, 442
Steinernema riobravis, 202
Steinernematidae, 442
Stenocorse bruchivora, 346
Stenoma catenifer, 79
Stenoma scitiorella, 134
Stenorryncha, 7
Stephanofilaria stilesi, 644
Stephanoprora denticulata, 219
Sternechus subsignathus, 389
Sternorrhyncha, 372, 474, 483, 596, 597
Sterols
  for insects growth, 71
Stethorus punctillum, 580
Stethorus spp., 578
Sticholotidinae, 572, 575, 578
Sto. calcitrans, 644
Stomoxys calcitrans, 113, 644, 672
Stored grain pests. *See* Stored-product pests
Stored-product pests
  biological control, 440–442
  digestive and excretory systems, 424–425
  feeding habits and damage caused, 420–424
  food
    attractants and gustatory stimuli, 427
    digestive enzymes, 430–432
    microorganisms, 434
    nutrient budget and relative growth rate, 432–433
    nutritional requirements, 427–430
    and nutrition characteristics, 425–426
    oviposition stimuli, 426–427
    physiological and behavioral adaptations, 434–435
    search and utilization, 426–434
  grain storage and losses, 418
  growth regulators, 442
  lignocellulosic biofuels, 442–443
  management, 434–442
    bioactive compounds, 438–440
    enzyme inhibitors, 437–438
    grain composition, 437

monitoring and food baits, 436
plant resistance, 436–440
mouthparts, 424–425
nutrition
characteristics, 425–426
requirements, 427–430
physiological and behavioral adaptations
food and environmental changes, 434–435
storage ecosystem and, 418–420
Streptomyces avermectilis, 135
Stromatocyphella conglobata, 583
Strumigenys, 218–219
Strumigenys rugithorax, 219
Strumigenys schmalzi, 218
Strumigenys splendens, 219
Strumigenys subedentata, 219
Stylet-sheath feeding, 298
Subterranean ants, 226
Supputius, 7
Supputius cincticeps, 546, 548
Symbionts, 5, 63, 86
entomopathogenic diseases, 203
essential, 153–155
Blochmannia, 155
Buchnera, 153–154
SOPE, 155
Wigglesworthia, 154–155
external, 146–148
facultative, 149–150
Heteroptera in, 150–153
for insects growth, 72–73
internal, 148–53
microorganisms, 146
nonnutritional interaction, 156
and nutrition of insects, 145–156
primary, 153–155
sap-sucking insect, 496–498
secondary, 149–150
Syrphus ribessi, 72
Sytophilus oryzae, 150

T

Tabanidae, 642–644
Tabaninae, 642
Tabanoidea, 642
Tabanomorpha, 642
Tagetes erecta, 559
Talisia esculenta, 672
Tanaostigmatidae, 375
Tatuidris, 219
Tectococcus ovatus, 579
Telanepsia, 406
Telenomus calvus, 553
Telenomus mormideae, 307
Tenebrio molitor, 72, 99, 204, 217, 421, 426, 428–432, 443, 548
Tenebrionidae, 418–419, 421–422, 424–425, 548
Tenebrio spp., 428
Tenthredinidae, 376, 384
Tephritidae, 375, 381, 451–452

Tephritids, 7
Termitomyces, 148
Tetragona, 251
Tetragona clavipes, 261
Tetragonisca, 248, 254
Tetragonisca angustula, 238–239, 242–243, 247–248, 254, 257–262
Tetranychidae, 596
Tetrastichinae, 584
Tetrastichus, 584
Tetrastichus bruchophagi, 345
Thasus acutangulus, 307
Thaumatomyrmex, 218
Thea, 577
Thellungiella halophila, 665
Thellungiella parvula, 665
Themos malaisei, 112
Theocolax elegans, 440–441
Therioaphis maculata, 134, 504, 505
Thiamine, 69, 72
Thripidae, 548
Thrips palmi, 664
Thyanta calceata, 307
Thyanta pallidovirens, 151
Thyanta perditor, 307, 309
Thysanoptera, 6, 384, 548, 580
galls, 373–374
mouthparts, 66
rearing on natural hosts, 59
Thysanura, 404
mouthparts, 66
Tibraca limbativentris, 310
Tiliaceae, 485
Tineola bisselliella, 113
Tinocallis kahawaluokalani, 583
Tococa, 225
Tomoplagia rudolphi, 376
Tortricidae, 597
Torulopsis sp., 608
Torymidae, 375
Torymus atheatu, 346
Torymus sinensis, 184
Toxoneuron nigriceps, 518, 525–526, 528–529
Toxoptera aurantii, 612
Toxorhynchites, 637–638
Toxorhynchites moctezuma, 188
Toxorhynchites splendens, 637
Toxotrypana curvicauda, 456
Trace element method, to measure food intake and utilization, 25
Trachymyrmex, 217, 223
Tranopelta, 226
Trembalya sp., 150
Treonine, 68
Trialeurodes, 548
Trialeurodes vaporariorum, 666
Triaspis, 344
Triaspis thoracica, 345
Triatoma, 645
Triatoma brasiliensis, 646
Triatoma dimidiata, 646

Triatoma infestans, 151, 196, 646
Triatoma pseudomaculata, 646
Triatoma sordida, 151, 646
Triatominae, 645–648
Tribolium, 70
Tribolium, 179
Tribolium castaneum, 97, 179, 187, 419, 421–423, 425–427,
431–433, 437–438, 440–442, 443, 667
Tribolium confusum, 63, 179, 421, 423, 426–430, 432–433,
438, 441
Tribolium sp., 70
Tribolium spp., 418, 424, 440–441
Trichogramma, 523
Trichogramma atopovirilia, artificial diet, 59
Trichogramma deion, 440, 441
Trichogramma galloi, 523
artificial diet, 59
Trichogramma pretiosum, 523
artificial diet, 55, 59
Trichogramma sp., 344
Trichogramma spp., 63, 75
rearing, 61
Trichogrammatidae, 344, 440
Trichogramma turkestanica, 530
Trichoplusia ni, 23, 75, 277, 614
Trichopoda pennipes, 307
Trichoprosopon digitatum, 188
Trichoptera, 523
Trichuris trichiura, 409
Trifolium incarnatum, 199
Trifolium pretense, 203
Trigona, 244, 247, 251, 255
Trigona amalthea, 245
Trigona carbonaria, 245, 257–258
Trigona crassipes, 251
Trigona hypogea, 251, 261
Trigona necrophaga, 251
Trigona spinipes, 247
Trissolcus basalis, 307, 693
Trissolcus brochymenae, 553
Trissolcus euschisti, 553
Triticum aestivum, 307, 489, 668, 695, 697
Tritrophic interactions, abiotic factors effects, 133–135
Trogidae, 354
Trogoderma granarium, 419, 421, 427, 430, 435
Trogoderma spp., 421, 431
Trophic interactions
mediated by semiochemicals, 165–171
plant–herbivore interactions, 165–169
Trophic relationships
in predatory Heteroptera, 551–558
cannibalism, 551–552
intraguild predation, 552–554
natural enemies, 558
predator–plant interactions, 554–558
Tropiconabis capsiformis, 553
Tropidothorax leucopterus, 307
Try. equiperdum, 642
Try. vivax, 642
Trypanosoma cruzi, 646
Trypanosoma evansi, 642

Tryptophan, 68
Tuberolachnus salignus, 485, 491, 494
Tunga penetrans, 645
Tungidae, 645
Tuta absoluta, 79, 545, 560
Tynacantha, 307
Typhaea stercorea, 421
Typhlomyrmex, 218
Tyria jacobaeae, 167
Tyrosine, 69
Tytthaspis (Micraspis) sedecimpunctata, 579
Tytthaspis sedecimpunctata, 577
Tytthaspis trilineata, 579

U

Ulmus minor, 170
Umbellularia californica, 135
Uresiphita reversalis, 286
Uric acid method, to measure food intake and utilization,
24–25
Uroleucon tanaceti, 503
Uromyces azukiola, 343
Urosigalphus, 344
Urosigalphus bruchi, 345
Uscana, 344
Uscana lariophaga, 345
Uscana mukerjii, 346
Uscana semifumipennis, 344–346
Uscana senex, 345
Utetheisa ornatrix, 79, 130, 282, 283

V

Valine, 68
Vanessa cardui, 198
Vassobia breviflora, 698
Venturia canescens, 183
Verbenaceae, 374, 384
Vernonia polyanthes, 376
Vesicular galls, 385
Vesperidae, 325
Vespula sp., 669
Vibidia, 577
Vicia faba, 200, 343, 489, 498, 580, 583
Vieira, 617
Vigna angularis, 676
Vigna luteola, 199
Vigna sinensis, 277, 333
Vigna unguiculata, 305, 333, 341, 343, 696
Viral diseases
gut environment interactions, 199–200
host plant effects on, 199–200
Vitamin A, 69, 74
Vitamin C, 69, 74
Vitamin D, 69
Vitamin E, 69, 74
Vitamin K, 70
Vitamins
for insects growth, 69–70, 74, 82
Viteus vitifoliae, 484, 487

W

Wasmannia, 221, 225
Wasmannia auropunctata, 222, 228
Water, for insects growth, 71–72, 82
Whiteflies, 7, 59, 548
Wigglesworthia, 5, 153, 155
    symbionts, 150, 154–55
Wolbachia, 63, 73, 86, 155–156, 203
Wuchereria bancrofti, 636

X

Xanthogaleruca luteola, 170
Xanthomonas sp., 198
Xenic diet, 62
Xenorhabdus, 201–202
Xylocoris, 441
Xyonysius sp., 302

Y

Yponomeutidae, 597

Z

Zabrotes, 327
Zabrotes interstitialis, 335
Zabrotes spp., 335
Zabrotes subfasciatus, 326, 328, 333–334, 338, 340,
    342–343, 346, 421–423, 438, 667, 676
Zagloba beaumonti, 574–575
Zea mays, 167, 170, 556, 558
Zeller, 419
Zelus, 7, 551
Zelus cervalicus, 553
Zeugloptera, 273
Zoophagous insects, 5
Zootermopsis angusticollis, 180